Aircraft
Maintenance and Repair

Seventh Edition

Michael J. Kroes

William A. Watkins (Deceased)

Frank Delp (Deceased)

Ronald Sterkenburg

New York   Chicago   San Francisco   Lisbon   London   Madrid   Mexico City   Milan   New Delhi   San Juan   Seoul   Singapore   Sydney   Toronto

1 2 3 4 5 6 7 8 9 0   QVR/QVR   1 9 8 7 6 5 4 3

ISBN 978-0-07-180150-8
MHID 0-07-180150-2

Sponsoring Editor: Larry S. Hager
Editing Supervisor: Stephen M. Smith
Production Supervisor: Richard C. Ruzycka
Acquisitions Coordinator: Bridget L. Thoreson
Project Manager: Yashmita Hota, Cenveo® Publisher Services
Copy Editor: Upendra Prasad, Cenveo Publisher Services
Proofreader: Christine Andreasen
Indexer: Robert Swanson
Art Director, Cover: Jeff Weeks
Composition: Cenveo Publisher Services

Printed and bound by Quad/Graphics.

McGraw-Hill books are available at special quantity discounts to use as premiums and sales promotions, or for use in corporate training programs. To contact a representative, please e-mail us at bulksales@mcgraw-hill.com.

This book is printed on acid-free paper.

Aircraft
Maintenance and Repair

About the Authors

Michael J. Kroes has been an aviation practitioner and educator for over 35 years. He has held various FAA certifications, including Airframe and Powerplant Mechanic, Inspection Authorization, Designated Mechanic Examiner, and Designated Engineering Representative, and he holds a commercial pilot license. Mr. Kroes has worked for some of the leading aviation companies, including Raytheon and Allied Signal, and has also spent 25 years as a professor and department head at Purdue University. Recognized as a leading expert on FAA technician certification, he authored a comprehensive study funded by the FAA. This study was used to develop new FAA technician certification content and guidelines. Mr. Kroes resides in Pinehurst, North Carolina.

Ronald Sterkenburg is a professor in the Aviation Technology Department at Purdue University. He is a certified Airframe and Powerplant (A&P) Mechanic, holds an Inspection Authorization (IA), and is a Designated Mechanic Examiner (DME). Although his main research interests are in advanced composite materials for aerospace vehicles, he has published many articles, book chapters, and books on all types of aviation maintenance topics. Dr. Sterkenburg resides in West Lafayette, Indiana.

Contents

Preface to the Seventh Edition

In preparing this edition, the authors reviewed FAR Parts 65 and 147, FAA-H-8083-30 & 31, AC 43.13-1B & 2B, and manufacturers' and operators' manuals and data to ensure that all required areas of study were included.

This revised edition updates material from the previous edition relating to aircraft structures and systems on current operational aircraft. Figures, charts, and photos have been updated to reflect new technology. The advanced composite materials chapter has been completely rewritten to reflect state-of-the-art maintenance and repair information and techniques for advanced composite structures used in modern aircraft. Advanced automated systems such as ECAM and EICAS are discussed in most chapters to reflect their operation relative to particular aircraft systems. The structures chapter has been expanded with multiple examples for the calculation of bend allowance and the setup of shop equipment. The hydraulics chapter has been expanded to include the 5000-psi hydraulic systems used in some new aircraft models. A discussion of a centralized fault display system (CFDS) has been added to the troubleshooting chapter.

Michael J. Kroes
Ronald Sterkenburg

Preface to the Sixth Edition

Aircraft Maintenance and Repair is designed to provide aviation students with the theoretical and practical knowledge required to qualify for certification as FAA airframe technicians in accordance with Federal Aviation Regulations (FARs). This text covers the subjects categorized in the FARs under Airframe Structures and Airframe Systems and Components and may be used as a study text in connection with classroom discussions, demonstrations, and practical application in the shop and on aircraft.

Aircraft Maintenance and Repair is one of five textbooks in the McGraw-Hill Aviation Technology Series. The other books in the series are *Aircraft Powerplants, Aircraft Basic Science, Aircraft Electricity and Electronics*, and *Aircraft Gas Turbine Engine Technology*. Used together, these texts provide information dealing with all prominent phases of aircraft maintenance technology.

In preparing this edition, the authors reviewed FAR Parts 65 and 147, Advisory Circular (AC) 65-2D, AC 65-15A, and AC 43.13-1A & 2A to ensure that all required areas of study were included. Related FARs and the recommendations and suggestions of aviation maintenance instructors, aircraft manufacturers, aviation operators, and maintenance facilities were given full consideration in the revision of this text.

This revised edition retains material from the previous edition relating to structures and systems that are employed on current operational aircraft. In addition, information dealing with expanding and emerging maintenance-related technologies has been incorporated to provide a comprehensive source of information for the aviation student, technician, and instructor. Two new chapters have been added. They include information about the identification of hazardous materials, their storage, use, and disposal. Troubleshooting has been identified as its own process and is discussed generically.

Key revisions to and the expansion of the previous edition include designing repairs so that the repairs are repairable and techniques in designing repairs based upon the mechanical properties of the materials. Also, bend allowance calculations and terminology have been revised to follow the more traditional industry conventions.

Each topic covered in this series of texts is explained in a logical sequence so students may advance step by step and build a solid foundation for performing aviation maintenance activities. Students' understanding of the explanations and descriptions given in the text should be enhanced by the use of numerous photographs, line drawings, and charts. Review questions at the end of each chapter enable students to check their knowledge of the information presented.

In addition to being a classroom and shop instruction text, the book is valuable for home study and as an on-the-job reference for the technician. The materials in this text and in others in the series constitute a major source of technical knowledge for technical schools and colleges and universities.

Although this text is designed to provide information for the training of aviation personnel, the user must realize that product and aircraft manufacturers establish guidelines and procedures for the correct use and maintenance of their product or aircraft. Therefore, it is the responsibility of the user of the text to determine and follow the specific procedures recommended by the manufacturer when handling a specific product or when working on a specific aircraft or component.

Michael J. Kroes
William A. Watkins

Aircraft
Maintenance and Repair

Hazardous Materials and Safety Practices 1

AUTHORS' NOTE

Although every effort has been made to ensure that the regulations and standard practices referred to in this text are current, recommended safety practices and associated regulations are always subject to change. Since the distribution of this book is not controlled, revisions to all existing copies is impossible. As a result, the technical information, such as material safety data sheets, is included only for educational purposes and should not be used in application. In addition, there are applications that are unique in one aspect or another. In these cases the recommended practices may differ from those used as general industry standard. Before attempting any activity, the aviation maintenance technician should review the most recent regulations, recommended practices prescribed by their employer, the associated equipment manufacturer's recommendations, and the information provided by the manufacturers of any supplies being used.

INTRODUCTION

There are many specialized careers available to today's aviation maintenance technician. As with any technical career, each career path has associated with it activities that can subject the technician and others to varying degrees of harm if performed without care. This chapter is intended to help the aviation maintenance technician identify potentially hazardous materials and ways in which the potential for harm can be minimized.

Today there are tens of thousands of products used in industry, with more being developed each day. Numerous governmental agencies (and, therefore, hundreds of governmental regulations) control the development, safety requirements, and health and environmental issues related to these products. Key among these agencies are the Consumer Product Safety Commission (CPSC), the Food and Drug Administration (FDA), the Department of Transportation (DOT), the Environmental Protection Agency (EPA), and the Occupational Safety and Health Administration (OSHA). Although all these agencies have some effects that may be felt in the aviation industry, the primary impact results from the last three organizations mentioned.

Some Federal Air Regulations (FARs) refer to the DOT standards in their text and use these standards as the criteria with which the aviation industry must comply. In addition, as users of potentially dangerous chemicals, the aviation industry must comply with both the regulations of the EPA as they relate to environmental concerns and OSHA as their usage relates to the safety and health of its employees.

Since the aviation industry is by its nature predominantly interstate commerce, most businesses in the aviation industry are subject to federal regulations. In addition, most state and some local governments have also passed safety and environmental related legislation that parallels or supplements federal legislation. As a result, the regulations associated with each are quite similar. Regardless of which jurisdiction applies to the operations of the aviation business, the operation must comply with some type of hazardous-materials regulation. In some instances, more than one jurisdiction may control the operations of the business.

Because of the vastness of this subject area and the general duplication of regulations between federal, state, and local governments, discussions in this chapter are limited to federal regulations and generic handling of hazardous materials. In addition to the information found in this chapter, in later chapters the aviation maintenance technician will find more safety data related to the specific types of equipment and/or processes as they are discussed throughout the text.

HAZARDOUS MATERIALS

The aviation maintenance technician frequently must work in potentially dangerous environments. In many cases, particularly when dealing with hazardous materials, the technician may not easily recognize those hazards. Some of these dangerous environments may be caused directly by the materials with which the aviation maintenance technician must work. In addition, exposures may be caused by other activities occurring in the area that are not directly related to the technician's activities.

Hazardous materials are typically grouped into three categories: *chemical agents,* and *physical* and *biological hazards.*

Chemical Agents

Within the **chemical agents** category, four classes exist. Comprehensive Loss Management, Inc., a professional developer of and consultant for safety and health awareness systems headquartered in Minneapolis, Minnesota, has trade-marked the acronym **FACTOR**™ to help remember the classes of chemical agents. Much of the information in this chapter comes from and is included in their programs. Because each class of chemical agent requires different usage, handling, and storage techniques, it is important that the aviation maintenance technician be able to recall and identify each of these classes. FACTOR™ stands for

Flammable
And
Corrosive
Toxic
Or
Reactive

The two outside letters of the acronym FACTOR, F and R (flammable and reactive), become hazardous primarily after some outside event, condition, or substance interacts with them. For example, the necessary components for a fire to occur are fuel, oxygen, and heat. In that relationship, **flammables** are the fuel, and heat and oxygen are the outside agents.

Reactives, when combined with certain other materials, are capable of generating heat and/or gases, causing an explosion.

The inside letters of the acronym, C and T (corrosives and toxins), on the other hand, act directly on the human body when exposure occurs. Exposing the skin, eyes, and other mucous membranes (such as the nose) to these elements can cause varying degrees of harm. **Toxic agents** cause poisoning. Aviation maintenance technicians should be particularly concerned when using toxic agents, because the ultimate effects of toxic poisoning are frequently delayed. It may take weeks, months, or even years for the poisoning to become apparent; because the toxic poisons are capable of using the bloodstream to move through the body, the cause-and-effect relationship may not be easily recognized.

As a general rule, when working with flammable and reactive agents, to avoid hazardous situations the aviation maintenance technician first needs to be concerned with exposing the agents to outside materials and conditions. Personal exposure to corrosive and toxic agents is the primary concern when dealing with toxins and corrosives. Therefore, the personal safety equipment used with corrosive and toxic agents should be designed to limit contact and/or exposure. Personal safety equipment designed for use with flammable and reactive materials is designed to limit heat exposure or impact, such as flying objects in the case of an explosion. In all cases, the recommended safety equipment recommended by the agent manufacturer, the employer, or the instructor should always be used.

Table 1-1 is a partial listing of frequently used chemical agents found in the aerospace industry. The aviation maintenance technician should be aware of the labels on the materials found in the work area and read them carefully.

TABLE 1-1 Frequently Used Chemical Agents

Listing of Commonly Found Hazardous Materials in an Aviation Environment*

Aircraft Systems	Aircraft Servicing	Component Shops
System Liquids	Lubricants	Inspection
Gasolines	Dry lubricants	Liquid penetrants
Jet fuels	Spray lubricants	Dye penetrants
Hydraulic fluids	Greases	Welding
Brake fluids	Solvents and Cleaners	Argon gas
Anti-ice additives	Methyl ethyl ketone	Hydrogen gas
Gases	Toluene	Oxygen gas
Freons	Engine cleaners	Acetylene gas
Nitrogen	Carburetor cleaners	Fluxes and pastes
Oxygen	Paints and Primers	Other
Halons	Paint strippers	Compressed air
Others	Primers	Glass beads
Alcohols	Doping products	Bluing and thinner
Methanol	Lacquers	Quenching fluids
Battery acids	Enamels	Muriatic acid
Glycol	Epoxies	Locking compounds
Baking soda	Adhesives	Anti-seizing compounds
Degreasers	Fiberglass resins	Mineral spirits
Disinfectants	Gasket adhesives	Cutting fluids
Squibs	Rubber adhesives	Soldering fluxes

*This listing is a partial generic listing of potentially hazardous materials. The aviation maintenance technician should always refer to label instructions and warnings as well as MSDS data.

Flammables (and Combustibles)

Flammables are materials that may easily ignite in the presence of a catalyst such as heat, sparks, or flame. They may be in any of the three physical forms: solid, liquid, or gas. Combustible liquids are very similar to flammable liquids, but they are not as easy to ignite.

Frequently found **flammable** or **combustible** materials in the aviation industry include *fuels, paint-related products, alcohols, acetone, toluene,* and some *metal filings.*

Generally Recommended Personal Safety Equipment

- Fire-retardant clothing
- Fire extinguisher

Handling and Storage

- Limit access to open flames, sparks, hot surfaces, etc. *Note:* Static electricity may produce sparks. To avoid sparks, containers should be grounded.
- Limit quantities to the minimum needed to accomplish the desired task.
- Store the materials in approved containers only and in designated areas only.
- Store flammable toxins and corrosive toxic materials separately. The corrosive gases could attack the flammable containers, eventually leading to a leak of flammable materials.

Typical Emergency Procedures

- Turn off electrical equipment or any other potential source of sparks.
- Attempt to close shutoff valve(s).
- Remove container(s) from the area.
- For large spills, leave the area immediately and notify your supervisor.
- In case of direct contact with skin or eyes, rinse immediately with water.
- If toxic substances are inhaled, go to a fresh-air area.
- If contact is made through clothing, remove wet clothing and store it in a proper container.
- Do not attempt to remove the substance with compressed air.

Corrosives

Corrosive materials are materials that can react with metallic surfaces and/or cause burning of the skin.

Frequently found corrosives in the aviation industry include *acids* and *bases,* such as battery acids and metal-cleaning solutions. Strong acids are most normally found in a liquid form, whereas bases tend to come in powdered form.

Generally Recommended Personal Safety Equipment

- Gloves, aprons, respirator, face shield or goggles, and, sometimes, protective footwear.

Handling and Storage

- Containers must be corrosive resistant.
- Eye (goggles and/or face shields) and skin protection (such as gloves) should always be worn.
- Never add water to acid.
- Acids and bases should be stored separately.
- Eye washes and showers should be easily accessible to the work area.
- Flammable toxins and corrosive toxic materials should be stored separately. The corrosive gases could attack the flammable containers, eventually leading to a leak of flammable materials.

Typical Emergency Procedures

- Remove any corrosives that have come in contact with your skin or eyes by rinsing with fresh water (approximately 15 minutes).
- Remove any contaminated clothing.
- Go to fresh air area.
- Ventilate area.
- Check safety equipment before attempting to stop the flow of spillage by creating a dam.
- If swallowed, DO NOT INDUCE VOMITING. Drink large amounts of water. Seek medical attention immediately.

Toxins

Toxins are generally defined as any substance that can cause an illness or injury. The effects of toxins, unlike flammables and corrosives, may appear all at once (called acute effects) or may build up over time with additional exposure (chronic effects). Some toxins may dissipate over time when further exposure is eliminated, while others remain in a human's system, even after death.

Frequently found toxins in the aviation industry may be grouped into eight categories.

1. *Solvents* and *thinners* for *bluing (such as Dykem), paints, ketones, and adhesives.*
2. *Solids such as metal dust or asbestos.* Compressed air should never be used to clean metal dust from equipment or clothing. The use of compressed air may result in minute particles of material being embedded in the pores of the skin.
3. *Machine lubricants, cutting fluids, and oils.*
4. *Gases such as carbon dioxide or nitrogen.* These gases may not only possess a toxic nature but also displace the oxygen normally found in the air.
5. *Polymers, epoxies, and plastics.* Although not normally toxic in their final form, these materials possess toxic properties during the fabrication process.
6. *Sensitizers, such as epoxy systems.* Such materials react with and may destroy portions of the body's immune system. The effects of sensitizers may be cumulative, so minimal levels of exposure are recommended.
7. *Carcinogens.* Carcinogens may cause changes in the genetic makeup of a human cell, resulting in cancer.

Although the use of carcinogens is rare in the aviation industry, aviation maintenance technicians associated with cargo aircraft should pay particular attention to the cargo manifest before cleaning spillage.

8. *Reproductive hazards, such as carcinogens.* These hazards are rare in the aviation industry. Such materials may either interfere with the reproductive process (as in the cases of DBCP) or affect the developing process of the fetus (such as dimethyl acetamide).

Generally Recommended Personal Safety Equipment

- Gloves, aprons, respirator, face shield or goggles, and, sometimes, protective footwear are recommended.
- Be sure to use the environmental control systems that may already be in place, such as ventilation fans and filters.

Handling and Storage

- Minimize the release of toxic agents into the environment by capping all containers and storing them in properly ventilated areas. When toxins are used in open containers, such as dip tanks and trays, their surface areas should be kept to a minimum in order to reduce the rate of evaporation into the surrounding environment.
- Flammable toxins and corrosive toxic materials should be stored separately. The corrosive gases could attack the flammable containers, eventually leading to a leak of flammable materials.

Typical Emergency Procedures

- If there is any doubt in your mind regarding the degree of toxicity of the substance spilled, **LEAVE THE AREA IMMEDIATELY AND NOTIFY YOUR SUPERVISOR.**
- Generally speaking, if the spillage is less than 1 gal, it may be cleaned up by wiping it up with absorbent materials.

Reactives

Reactive agents are those materials that react violently with other materials (not necessarily solids). The reactions that may take place range from violent explosions to the emission of heat and/or gases.

The following reactives are frequently found in the aviation industry:

1. Oxidizers, which add oxygen to situations where high levels of heat and burning are present
   a. Peroxides
   b. Perchloric acid and chromic acid
   c. Halogens, such as bromine and iodine

2. Water-reactive materials, such as lithium, react with water and form hydrogen gases, which are very explosive.

Examples of incompatible reactive materials include

- Cyanides (frequently used in plating) and acids
- Chloride bleach and ammonia (this combination forms highly toxic chlorine gas)

Generally Recommended Personal Safety Equipment

- Gloves, aprons, respirator, and face shield or goggles are suggested.
- Be sure to use the environmental-control systems.

Handling and Storage

- Store reactive materials in a location separate from other materials. Always review the MSDS (material safety data sheet) for incompatible materials.
- Many reactives are both toxic and corrosive.

Typical Emergency Procedures

- Shut down electrical equipment whenever possible.
- If there is any doubt in your mind regarding the degree of reactivity and toxicity of the substances involved, **LEAVE THE AREA IMMEDIATELY AND NOTIFY YOUR SUPERVISOR.**

Material Compatibility with Chemical Agents

Before leaving the topic of chemical agents, it is important to realize that although some materials meet the minimum standards for protective equipment in particular applications, other materials surpass these requirements. Table 1-2 lists various types of protective equipment materials and their relative effectiveness when used with common chemical agents. Although Table 1-2 provides generally accepted data, the aviation maintenance technician should always consult the MSDS, discussed later in this chapter, for specific protective equipment requirements.

Physical Hazards

Physical hazards are those to which the aviation maintenance technician is exposed that are usually caused by the use of some type of equipment not directly controllable by the technician. Typically, this type of hazard is generated by the operation of equipment that can be detected by the human senses. However, many physical hazards that fall into this classification are not detectable by the human senses. These hazards include X rays, microwaves, beta or gamma rays, invisible laser beams, and high-frequency (ultrasonic) sound waves.

Compressed liquids and gases, such as welding oxygen and acetylene, aviator's breathing oxygen, nitrogen, and hydraulic accumulators, present another physical hazard to the aviation maintenance technician. Although some of these substances by themselves present hazards as chemical agents, placing them under pressure may create another unique hazard.

OSHA requires that areas where this exposure exists be clearly marked and that individuals exposed to these hazards

TABLE 1-2 Chemical Resistance of Protective Clothing Materials

Chemical	Neoprene	Vinyl Plastic	Rubber Latex	Nitrite	Syn Latex	Nat. Latex
			Resistance of Materials			
Alcohols	E	E	G	E	E	G
Caustics	E	E	E	E	E	E
Chlorinated solvents	G	F	NR	E	G	NR
Ketones	G	NR	G	G	G	G
Petroleum solvents	E	G	F	S	E	F
Organic acids	E	E	E	E	E	E
Inorganic acids	E	E	E	E	E	E
Nonchlorinated solvents	G	F	NR	G	G	NR
Insecticides	E	E	F	S	E	F
Inks	E	E	F	S	E	F
Formaldehyde	E	E	E	S	S	E
Acrylonitrile	E	G	E	S	E	E
Hydraulic fluid	E	E	F	S	E	F
Carbon Disulfide	NR	F	G	F	NR	G
Paint remover	F	F	NR	E	F	NR

S Superior
E Excellent
G Good
F Fair
NR Not recommended

be provided the proper safety equipment. In many cases this is easily accomplished, but in the aerospace industry particular concern should be paid to portable equipment that generates these hazards. Such equipment results in the potential for hazards to exist in areas where exposure is not usually a concern. X ray of aircraft structural parts is an example of such a situation. The aviation maintenance technician should remain conscious that potentially hazardous equipment is portable and remain vigilant for possible exposure in the work area.

Biological Hazards

Biological hazards, although not normally a major concern to the aviation maintenance technician, may occasionally exist in the work environment. Biological hazards are living organisms that may cause illness or disease. Some biological hazards also have toxic by-products. Typically, biological hazards are transmitted in the form of air droplets or spores and enter the body through contact with contaminated objects or individuals.

The practicing aviation maintenance technician in the workplace would most likely be exposed to biological hazards when working on cargo aircraft or in a cargo (baggage) compartment where breakage or leakage of biologically hazardous materials has occurred. FAA regulations require that the transportation of biologically hazardous materials be documented. When in doubt about the presence of such materials, the aviation maintenance technician should consult the aircraft's record, possibly including the cargo manifest.

OSHA'S HAZARDOUS COMMUNICATIONS STANDARDS

In 1983, the first regulation requiring employers to advise employees of potentially hazardous materials in the work place was established. This standard, the Hazardous Communications Standard (29 CFR 1910.1200), was established by OSHA and has since been expanded to include almost all employers. The law requires that all employees and their supervisors be informed about the known hazards associated with the chemicals with which they work, regardless of the quantity of the chemicals involved in the operation. These requirements are part of the various **right-to-know** regulations. As part of these right-to-know regulations, employers are required to post a notice similar to that shown in Fig. 1-1.

There are five basic requirements of a hazard-communications program:

1. *Inventory.* An inventory (list) of all hazardous materials used within the workplace must be established and maintained.

2. *Labeling.* All hazardous chemicals shall be properly labeled.

Job Safety and Health
It's the law!

OSHA®
Occupational Safety
and Health Administration
U.S. Department of Labor

EMPLOYEES:

• You have the right to notify your employer or OSHA about workplace hazards. You may ask OSHA to keep your name confidential.

• You have the right to request an OSHA inspection if you believe that there are unsafe and unhealthful conditions in your workplace. You or your representative may participate in that inspection.

• You can file a complaint with OSHA within 30 days of retaliation or discrimination by your employer for making safety and health complaints or for exercising your rights under the *OSH Act*.

• You have the right to see OSHA citations issued to your employer. Your employer must post the citations at or near the place of the alleged violations.

• Your employer must correct workplace hazards by the date indicated on the citation and must certify that these hazards have been reduced or eliminated.

• You have the right to copies of your medical records and records of your exposures to toxic and harmful substances or conditions.

• Your employer must post this notice in your workplace.

• You must comply with all occupational safety and health standards issued under the *OSH Act* that apply to your own actions and conduct on the job.

EMPLOYERS:

• You must furnish your employees a place of employment free from recognized hazards.

• You must comply with the occupational safety and health standards issued under the *OSH Act*.

This free poster available from OSHA –
The Best Resource for Safety and Health

Free assistance in identifying and correcting hazards or complying with standards is available to employers, without citation or penalty, through OSHA-supported consultation programs in each state.

1-800-321-OSHA (6742)
www.osha.gov

OSHA 3165-12-06R

FIGURE 1-1   Right-to-know poster.

3. *Material safety data sheets (MSDSs).* Material safety data sheets must be obtained for all material stored and/or used in the work area. A copy of these MSDSs must be maintained and made readily available to all employees during normal working hours. MSDSs provide detailed information concerning composition, health hazards, special handling instructions, and proper disposal practices for materials.

4. *Training.* All employees must be provided training regarding their rights under the right-to-know program, the proper handling of these materials, the labeling system used, and detection techniques.

5. *Written program.* Each employer must establish a written program that will comply with the four points just mentioned; this written program must be present at the work facility.

Material Safety Data Sheets

A material safety data sheet (MSDS) is a document provided by the material manufacturer or subsequent material processor that contains information related to the material hazard and includes safe handling and disposal procedures. The format of these sheets must be consistent with the requirements of the OSHA Hazard Communications Standard. MSDSs should be provided by the manufacturer for each hazardous material supplied by them. Normally MSDSs are provided with each shipment of a hazardous material. If one is not provided the technician should request one from the manufacturer. Most manufacturers post their MSDSs on their websites. Figure 1-2 is an example of a 16-section standardized MSDS. This format is used by most manufacturers.

The Occupational Safety and Health Administration (OSHA) specifies that certain information must be included on MSDSs, but does not require that any particular format be followed in presenting this information. OSHA recommends that MSDSs follow the 16-section format established by the American National Standards Institute (ANSI) for preparation of MSDSs.

By following this recommended format, the information of greatest concern to workers is featured at the beginning of the data sheet, including information on chemical composition and first aid measures. More technical information that addresses topics such as the physical and chemical properties of the material and toxicological data appears later in the document. While some of this information (such as ecological information) is not required by the Hazard Communication Standard (HCS), the 16-section MSDS is becoming the international norm.

As of June 1, 2015, MSDSs will be replaced by Safety Data Sheets (SDSs). The HCS will require chemical manufacturers, distributors, or importers to provide SDSs to communicate the hazards of hazardous chemical products. The new SDS will have a uniform format, and will include the section numbers, the headings, and associated information under the headings below:

Section 1: Identification includes product identifier; manufacturer or distributor name, address, phone number; emergency phone number; recommended use; restrictions on use.

Section 2: Hazard(s) Identification includes all hazards regarding the chemical; required label elements.

Section 3: Composition/Information on Ingredients includes information on chemical ingredients; trade secret claims.

Section 4: First-Aid Measures includes important symptoms/effects (acute, delayed); required treatment.

Section 5: Firefighting Measures lists suitable extinguishing techniques, equipment; chemical hazards from fire.

Section 6: Accidental Release Measures lists emergency procedures; protective equipment; proper methods of containment and cleanup.

Section 7: Handling and Storage lists precautions for safe handling and storage, including incompatibilities.

Section 8: Exposure Controls/Personal Protection lists OSHA's Permissible Exposure Limits (PELs); Threshold Limit Values (TLVs); appropriate engineering controls; personal protective equipment (PPE).

Section 9: Physical and Chemical Properties lists the chemical's characteristics.

Section 10: Stability and Reactivity lists chemical stability and possibility of hazardous reactions.

Section 11: Toxicological Information includes routes of exposure; related symptoms; acute and chronic effects; numerical measures of toxicity.

Section 12: Ecological Information*

Section 13: Disposal Considerations*

Section 14: Transport Information*

Section 15: Regulatory Information*

Section 16: Other Information includes the date of preparation or last revision.

Inventory

Even though many companies inventory the materials they use, a special inventory of hazardous materials must be maintained. Each item in the inventory should include the name of the hazardous material, location, and the approximate (or average) quantity in each area.

Materials sold in consumer form are not normally controlled. For example, if the aviation maintenance technician purchases a painted aluminum cover plate in a ready-to-install form, no MSDS would accompany the product. However, if the aviation maintenance technician purchases the aluminum sheet and paint separately, to fabricate the product, it is likely that an MSDS would accompany both the aluminum sheet and the paint.

Labeling

All hazardous materials should have identifying labels adhered to them. As a general rule, these labels should *never* be removed. In instances where materials are received in bulk form and transferred to small containers for use, two general rules apply. First, the container should be clearly labeled. Second, once a container is used for one hazardous substance, it should never be used to hold another substance.

Note: Since other agencies regulate this information, OSHA will not be enforcing Secs. 12 through 15 [29 CFR 1910.1200(g)(2)].

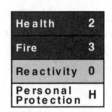

Health	2
Fire	3
Reactivity	0
Personal Protection	H

Material Safety Data Sheet
Methyl ethyl ketone MSDS

Section 1: Chemical Product and Company Identification

Product Name: Methyl ethyl ketone

Catalog Codes: SLM2626, SLM3232

CAS#: 78-93-3

RTECS: EL6475000

TSCA: TSCA 8(b) inventory: Methyl ethyl ketone

CI#: Not applicable.

Synonym: 2-Butanone

Chemical Name: Methyl Ethyl Ketone

Chemical Formula: C4H8O

Contact Information:

Sciencelab.com, Inc.
14025 Smith Rd.
Houston, Texas 77396

US Sales: **1-800-901-7247**
International Sales: **1-281-441-4400**

Order Online: ScienceLab.com

CHEMTREC (24HR Emergency Telephone), call:
1-800-424-9300

International CHEMTREC, call: 1-703-527-3887

For non-emergency assistance, call: 1-281-441-4400

Section 2: Composition and Information on Ingredients

Composition:

Name	CAS #	% by Weight
Methyl ethyl ketone	78-93-3	100

Toxicological Data on Ingredients: Methyl ethyl ketone: ORAL (LD50): Acute: 2737 mg/kg [Rat]. 4050 mg/kg [Mouse]. DERMAL (LD50): Acute: 6480 mg/kg [Rabbit]. VAPOR (LC50): Acute: 23500 mg/m 8 hours [Rat].

Section 3: Hazards Identification

Potential Acute Health Effects:
Hazardous in case of skin contact (irritant, permeator), of eye contact (irritant), of ingestion, of inhalation (lung irritant).

Potential Chronic Health Effects:
CARCINOGENIC EFFECTS: Not available. MUTAGENIC EFFECTS: Mutagenic for bacteria and/or yeast. TERATOGENIC EFFECTS: Classified POSSIBLE for human. DEVELOPMENTAL TOXICITY: Not available. The substance may be toxic to gastrointestinal tract, upper respiratory tract, skin, eyes, central nervous system (CNS). Repeated or prolonged exposure to the substance can produce target organs damage.

Section 4: First Aid Measures

Eye Contact:
Check for and remove any contact lenses. Immediately flush eyes with running water for at least 15 minutes, keeping eyelids open. Cold water may be used. Get medical attention.

Skin Contact:
In case of contact, immediately flush skin with plenty of water. Cover the irritated skin with an emollient. Remove contaminated clothing and shoes. Cold water may be used.Wash clothing before reuse. Thoroughly clean shoes before reuse. Get medical attention.

Serious Skin Contact:
Wash with a disinfectant soap and cover the contaminated skin with an anti-bacterial cream. Seek medical attention.

FIGURE 1-2   Sample MSDS. (*Sciencelab.com, Inc.*)

Inhalation:
If inhaled, remove to fresh air. If not breathing, give artificial respiration. If breathing is difficult, give oxygen. Get medical attention.

Serious Inhalation:
Evacuate the victim to a safe area as soon as possible. Loosen tight clothing such as a collar, tie, belt or waistband. If breathing is difficult, administer oxygen. If the victim is not breathing, perform mouth-to-mouth resuscitation. Seek medical attention.

Ingestion:
Do NOT induce vomiting unless directed to do so by medical personnel. Never give anything by mouth to an unconscious person. Loosen tight clothing such as a collar, tie, belt or waistband. Get medical attention if symptoms appear.

Serious Ingestion: Not available.

Section 5: Fire and Explosion Data

Flammability of the Product: Flammable.

Auto-Ignition Temperature: 404°C (759.2°F)

Flash Points: CLOSED CUP: -9°C (15.8°F). OPEN CUP: -5.5556°C (22°F) (Tag).

Flammable Limits: LOWER: 1.8% UPPER: 10%

Products of Combustion: These products are carbon oxides (CO, CO2).

Fire Hazards in Presence of Various Substances: Highly flammable in presence of open flames and sparks, of heat.

Explosion Hazards in Presence of Various Substances:
Risks of explosion of the product in presence of mechanical impact: Not available. Risks of explosion of the product in presence of static discharge: Not available. Explosive in presence of oxidizing materials, of acids.

Fire Fighting Media and Instructions:
Flammable liquid, soluble or dispersed in water. SMALL FIRE: Use DRY chemical powder. LARGE FIRE: Use alcohol foam, water spray or fog.

Special Remarks on Fire Hazards:
Ignition on contact with potassium t-butoxide. Vapor may cause a flash fire.

Special Remarks on Explosion Hazards:
Reaction with Hydrogen Peroxide + nitric acid forms heat and shock-sensitive explosive product. Mixture with 2-propanol will produce explosive peroxides during storage.

Section 6: Accidental Release Measures

Small Spill:
Dilute with water and mop up, or absorb with an inert dry material and place in an appropriate waste disposal container.

Large Spill:
Flammable liquid. Keep away from heat. Keep away from sources of ignition. Stop leak if without risk. Absorb with DRY earth, sand or other non-combustible material. Do not touch spilled material. Prevent entry into sewers, basements or confined areas; dike if needed. Be careful that the product is not present at a concentration level above TLV. Check TLV on the MSDS and with local authorities.

Section 7: Handling and Storage

Precautions:
Keep locked up. Keep away from heat. Keep away from sources of ignition. Ground all equipment containing material. Do not ingest. Do not breathe gas/fumes/ vapor/spray. Wear suitable protective clothing. In case of insufficient ventilation, wear suitable respiratory equipment. If ingested, seek medical advice immediately and show the container or the label. Avoid contact with skin and eyes. Keep away from incompatibles such as oxidizing agents, metals, acids, alkalis.

Storage:
Store in a segregated and approved area. Keep container in a cool, well-ventilated area. Keep container tightly closed and sealed until ready for use. Avoid all possible sources of ignition (spark or flame).

FIGURE 1-2 (continued)

Section 8: Exposure Controls/Personal Protection

Engineering Controls:
Provide exhaust ventilation or other engineering controls to keep the airborne concentrations of vapors below their respective threshold limit value. Ensure that eyewash stations and safety showers are proximal to the work-station location.

Personal Protection:
Splash goggles. Lab coat. Vapor respirator. Be sure to use an approved/certified respirator or equivalent. Gloves.

Personal Protection in Case of a Large Spill:
Splash goggles. Full suit. Vapor respirator. Boots. Gloves. A self-contained breathing apparatus should be used to avoid inhalation of the product. Suggested protective clothing might not be sufficient; consult a specialist BEFORE handling this product.

Exposure Limits:
TWA: 200 STEL: 300 (ppm) from ACGIH (TLV) [United States] [1999] TWA: 150 STEL: 300 (ppm) [Australia] TWA: 590 STEL: 885 (mg/m3) from NIOSH TWA: 200 STEL: 300 (ppm) from NIOSH TWA: 590 STEL: 885 (mg/m3) [Canada] TWA: 200 STEL: 300 (ppm) from OSHA (PEL) [United States] TWA: 590 STEL: 885 (mg/m3) from OSHA (PEL) [United States] Consult local authorities for acceptable exposure limits.

Section 9: Physical and Chemical Properties

Physical state and appearance: Liquid.

Odor: Acetone-like. Pleasant. Pungent. Sweetish. (Strong.)

Taste: Not available.

Molecular Weight: 72.12 g/mole

Color: Clear. Colorless.

pH (1% soln/water): Not available.

Boiling Point: 79.6 (175.3°F)

Melting Point: −86°C (−122.8°F)

Critical Temperature: 262.5°C (504.5°F)

Specific Gravity: 0.805 (Water = 1)

Vapor Pressure: 10.3 kPa (@ 20°C)

Vapor Density: 2.41 (Air = 1)

Volatility: Not available.

Odor Threshold: 0.25 ppm

Water/Oil Dist. Coeff.: The product is more soluble in oil; log(oil/water) = 0.3.

Ionicity (in Water): Not available.

Dispersion Properties: See solubility in water, diethyl ether, acetone.

Solubility: Soluble in cold water, diethyl ether, acetone.

Section 10: Stability and Reactivity Data

Stability: The product is stable.

Instability Temperature: Not available.

Conditions of Instability: Heat, ignition sources, mechanical shock, incompatible materials.

Incompatibility with Various Substances: Reactive with oxidizing agents, metals, acids, alkalis.

Corrosivity: Non-corrosive in presence of glass.

Special Remarks on Reactivity:
Incompatible with chloroform, copper, hydrogen peroxide, nitric acid, potassium t-butoxide, 2-propanol, chlorosulfonic acid, strong oxidizers, amines, ammonia, inorganic acids, isocyanates, caustics, pyrindines. Vigorous reaction with chloroform + alkali.

Special Remarks on Corrosivity: Not available.

Polymerization: Will not occur.

FIGURE 1-2 *(continued)*

Section 11: Toxicological Information

Routes of Entry: Absorbed through skin. Dermal contact. Eye contact. Inhalation.

Toxicity to Animals:
WARNING: THE LC50 VALUES HEREUNDER ARE ESTIMATED ON THE BASIS OF A 4-HOUR EXPOSURE. Acute oral toxicity (LD50): 2737 mg/kg [Rat]. Acute dermal toxicity (LD50): 6480 mg/kg [Rabbit]. Acute toxicity of the vapor (LC50): 32000 mg/m3 4 hours [Mouse].

Chronic Effects on Humans:
MUTAGENIC EFFECTS: Mutagenic for bacteria and/or yeast. TERATOGENIC EFFECTS: Classified POSSIBLE for human. May cause damage to the following organs: gastrointestinal tract, upper respiratory tract, skin, eyes, central nervous system (CNS).

Other Toxic Effects on Humans: Hazardous in case of skin contact (irritant, permeator), ingestion, or inhalation (lung irritant).

Special Remarks on Toxicity to Animals: Not available.

Special Remarks on Chronic Effects on Humans: May cause birth defects based on animal data. Embryotoxic and/or foetotoxic in animals.

Special Remarks on Other Toxic Effects on Humans:
Acute Potential Health Effects: Skin: Causes skin irritation. May be absorbed through the skin. Eyes: Causes eye irritation. Inhalation: Inhalation of high concentrations may cause central nervous effects characterized by headache, dizziness, unconsciousness, and coma. Causes respiratory tract irritation and affects the sense organs. May affect the liver and urinary system. Ingestion: Causes gastrointestinal tract irritation with nausea, vomiting and diarrhea. May affect the liver. Chronic Potential Health Effects: Chronic inhalation may cause effects similar to those of acute inhalation. Prolonged or repeated skin contact may cause defatting and dermatitis.

Section 12: Ecological Information

Ecotoxicity: Ecotoxicity in water (LC50): 3220 mg/l 96 hours [Fathead Minnow]. 1690 mg/l 96 hours [Bluegill].

BOD5 and COD: Not available.

Products of Biodegradation:
Possibly hazardous short term degradation products are not likely. However, long-term degradation products may arise.

Toxicity of the Products of Biodegradation: The product itself and its products of degradation are not toxic.

Special Remarks on the Products of Biodegradation: Not available.

Section 13: Disposal Considerations

Waste Disposal:
Waste must be disposed of in accordance with federal, state and local environmental control regulations.

Section 14: Transport Information

DOT Classification: CLASS 3: Flammable liquid.

Identification: Ethyl methyl ketone UNNA: 1193 PG: II.

Special Provisions for Transport: Not available.

Section 15: Other Regulatory Information

Federal and State Regulations:
New York release reporting list: Methyl ethyl ketone Rhode Island RTK hazardous substances: Methyl ethyl ketone Pennsylvania RTK: Methyl ethyl ketone Minnesota: Methyl ethyl ketone Massachusetts RTK: Methyl ethyl ketone New Jersey: Methyl ethyl ketone California Director's list of Hazardous Substances: Methyl ethyl ketone TSCA 8(b) inventory: Methyl ethyl ketone TSCA 8(d) H and S data reporting: Methyl ethyl ketone: Effective: 10/4/82; Sunset: 10/4/92 SARA 313 toxic chemical notification and release reporting: Methyl ethyl ketone CERCLA: Hazardous substances.: Methyl ethyl ketone: 5000 lb. (2268 kg)

FIGURE 1-2 *(continued)*

Other Regulations:
OSHA: Hazardous by definition of Hazard Communication Standard (29 CFR 1910.1200). EINECS: This product is on the European Inventory of Existing Commercial Chemical Substances.

Other Classifications:

WHMIS (Canada):
CLASS B-2: Flammable liquid with a flash point lower than 37.8°C (100°F). CLASS D-2A: Material causing other toxic effects (VERY TOXIC).

DSCL (EEC):
R11- Highly flammable. R36/37- Irritating to eyes and respiratory system. S9 - Keep container in a well-ventilated place. S16 - Keep away from sources of ignition - No smoking. S25 - Avoid contact with eyes. S33 - Take precautionary measures against static discharges.

HMIS (U.S.A.):

    Health Hazard: 2

    Fire Hazard: 3

    Reactivity: 0

    Personal Protection: h

National Fire Protection Association (U.S.A.):

    Health: 1

    Flammability: 3

    Reactivity: 0

    Specific Hazard:

Protective Equipment:
Gloves. Lab coat. Vapor respirator. Be sure to use an approved/certified respirator or equivalent. Wear appropriate respirator when ventilation is inadequate. Splash goggles.

Section 16: Other Information

References: Not available.

Other Special Considerations: Not available.

Created: 10/10/2005 08:39 PM

Last Updated: 11/01/2010 12:00 PM

The information above is believed to be accurate and represents the best information currently available to us. However, we make no warranty of merchantability or any other warranty, express or implied, with respect to such information, and we assume no liability resulting from its use. Users should make their own investigations to determine the suitability of the information for their particular purposes. In no event shall ScienceLab.com be liable for any claims, losses, or damages of any third party or for lost profits or any special, indirect, incidental, consequential or exemplary damages, howsoever arising, even if ScienceLab.com has been advised of the possibility of such damages.

FIGURE 1-2 *(continued)*

Probably the most common standardized hazardous materials identification placard used today is that of the NFPA. Although this code is intended for the use of firefighters during a fire emergency, it is another tool available to the aviation maintenance technician that may be used to avoid hazardous situations. This placarding system uses four diamonds to form another diamond (Fig. 1-3). Each diamond position identifies the degree to which a particular type of hazard is present.

The top three diamonds follow a numbering system from 1 to 4, indicating the degree of hazard.

The topmost diamond specifies the relative fire hazard. The relative fire hazard is a function of the temperature at which the material will give off flammable vapors that will ignite when they come in contact with a spark or flame. This temperature is called the **flash point.** Figure 1-3 also shows how the number code is used to express the way in which the flash point ranges are specified.

The left side of the diamond specifies the health hazard and the right side of the diamond indicates the degree of the reactivity of the material.

The bottommost diamond indicates any specific hazard and, if more than one, the major hazard that applies to this material.

This diamond coding system may also use different colors to segregate each type of hazard.

The *health hazard* diamond is *blue.*

The *flammability* diamond is *red.*

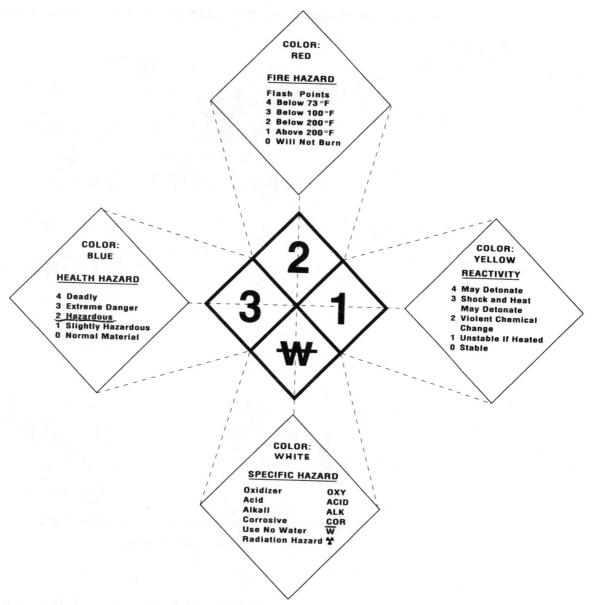

FIGURE 1-3   NFPA placard. (*Copyright © 1990, National Fire Protection Association, Quincy, MA 02269. This warning system is intended to be interpreted and applied only by properly trained individuals to identify fire, health, and reactivity hazards of chemicals. The user is referred to a certain limited number of chemicals with recommended classifications in NFPA 49 and NFPA 325M, which would be used as a guideline only. Whether the chemicals are classified by NFPA or not, anyone using the 704 system to classify chemicals does so at his/her own risk.*)

The *reactivity* diamond is *yellow*.
The *specific hazard* diamond is *white*.

The aviation maintenance technician should be conscious of the environment in which work is being accomplished. For example, aviation maintenance technicians working in confined, poorly ventilated areas should use appropriate precaution if the NFPA health hazard code is greater than 0. Aviation maintenance technicians working on hot brakes or engines should beware of the relative fire hazard code.

Many companies have their own labeling program. In cases where codes are used, the company has the obligation to identify the coding system. An example of a labeling system used by Weber Marking Systems of Arlington Heights, IL, is shown in Fig. 1-4.

In-house labeling systems are most frequently used when the operations of the company require the transferring of hazardous materials from one container to another. In most cases, the in-house labeling is used in addition to the original container labels.

Generic in-house labeling systems are also available commercially. A commercially available hazardous materials identification system produced by Labelmaster, an American Labelmark Company, Chicago, IL, is shown in Fig. 1-5. As with most marking systems, the degree of hazard severity is based upon a numbering system. American also adds

FIGURE 1-4 In-house labeling system. (*Weber Marking Systems*)

an index system using letters that correspond to the recommended personal protective equipment for use with the hazardous material.

The majority of hazardous materials with which the aviation maintenance technician will come in contact are complex mixtures of chemicals. Normally, mixtures are analyzed as a whole to determine their physical properties and health hazards, if any exist. If a mixture has not been tested as a whole, it is to be considered hazardous if it contains more than 1 percent of any hazardous material. In the case of carcinogens, the minimum component amount is 0.1 percent.

DISPOSAL AND ACCIDENTAL RELEASES OF HAZARDOUS MATERIALS

Just as the Occupational Safety and Health Act was enacted to protect the health of industrial workers, the National Environmental Policy Act of 1969 (NEPA) was enacted to protect the environment. The NEPA established the Environmental Protection Agency (EPA) to accomplish its objectives.

Businesses that use hazardous materials must be conscious of the hazards, not only while using the materials but also when disposing of them. As part of the employee's orientation to hazardous materials, proper disposal techniques should be addressed.

Creators of hazardous waste are responsible for the identification, separation, labeling, packaging, storage, shipping, and disposal of the waste produced. The EPA monitors the movement of hazardous waste from the time it is generated to the time it reaches a licensed treatment, storage, and disposal facility (TSDF).

As part of this process, generators of hazardous waste must maintain detailed records regarding hazardous-waste materials. It is therefore extremely important that the aviation maintenance technician comply with the record-keeping practices and procedures of the company.

Routine handling of hazardous materials is typically not a problem for the aviation maintenance technician because the technician either is familiar with the procedure or has time

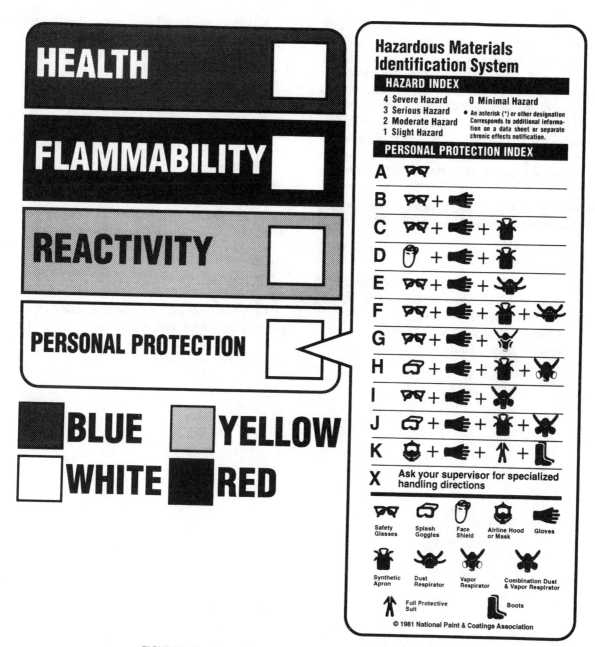

FIGURE 1-5 Generic labeling system. (*American Landmark Co.*)

during which specific instruction may be obtained. Accidental release of hazardous materials, however, is another story.

When an accidental release occurs, the typical reaction is one of panic. As part of the employer training sessions, the aviation maintenance technician will receive instructions regarding procedures to follow in case of an accidental release, but these may not be remembered, particularly when an emergency occurs.

Accidental releases most typically occur when the hazardous material is in a liquid or gaseous state. It is important that the aviation maintenance technician not equate hazardous materials with rarely used materials. Spillage of commonly used hazardous materials such as aviation fuels and lubricants is also considered an accidental release of a hazardous material.

Accidental releases, by definition, do not happen because they are planned. However, because of their potential impact, the aviation technician should plan for and anticipate their occurrence. Prior to using a hazardous material, the aviation maintenance technician should evaluate the types of accidental releases that might occur and prepare for them. A review of the MSDS prior to usage is advised.

The aviation maintenance technician should be concerned first with personal safety. If a release might have detrimental effect on other individuals, a means of notification should be established. Containment is the next priority. This may take a little imagination. For example, if the hazardous materials are in a tank, how could a leak be stopped? If a drain shutoff value was dislodged, how could the drain be quickly plugged?

Disposal and Accidental Releases of Hazardous Materials   **15**

The EPA has established reporting procedures for accidental release of hazardous materials. Whether an accidental release needs to be reported is determined by the quantity or concentration of material released. Calculations of concentration may be rather complex and typically are beyond the capability of the technician. Therefore, all accidental releases should be reported to the aviation maintenance technician's supervisor as soon as possible. Any information the aviation maintenance technician has regarding the amount released should be noted and submitted to the supervisor.

REVIEW QUESTIONS

1. What are the three general categories of hazardous materials?

2. What are the four classes of chemical agents?

3. How long does it take for a toxic agent to show its effects on the human body?

4. What types of chemical agents typically require the use of personal protective equipment designed to limit direct body contact?

5. List some of the flammable materials found in the aviation industry.

6. What type of corrosives generally come in powder form?

7. What type of corrosives generally come in liquid form?

8. How long does it take toxins to dissipate from the human body?

9. What type of toxin may cause cancer?

10. What happens when chloride bleach and ammonia are mixed?

11. To what document should the aviation maintenance technician refer regarding the potential hazards when dealing with hazardous materials?

12. Why are physical hazards not always easy to avoid?

13. In what form are biological hazards most likely to be transmitted?

14. What requires that an employee be informed about the presence of hazardous materials in the workplace?

15. What are the five basic requirements of a hazard-communications program?

Aircraft Structures 2

INTRODUCTION

A thorough understanding of the structural components of aircraft and the stresses imposed on those components is essential for the certificated aircraft technician. Such understanding ensures that the technician will design and make repairs in a manner that restores the damaged part to its original strength. It is the purpose of this chapter to familiarize the technician with the principal structural components of various aircraft and to discuss the loads that are applied to these components during operation of the aircraft.

AIRCRAFT STRUCTURAL DESIGN

Load Factors and Airplane Design

The structure of an aircraft must be strong enough to carry all the loads to which it might be subjected, including the repeated small to medium loads experienced in normal flight and the big loads experienced during extreme conditions. To fly, an airplane's exterior must have an aerodynamic shape. Into this shape must be fitted members having a high strength-to-weight ratio that are capable of sustaining the forces necessary to balance the airplane in flight. Airplanes are generally designed for a specific purpose that dictates the structural design required.

The airplane structure must be capable of withstanding much more force than that imposed by its own weight. When the purpose of a particular design is established, the designers provide structure according to strict standards established by the Federal Aviation Administration to ensure safety. In general, airplanes are designed to withstand 1.5 times the maximum expected forces. To be certified by the Federal Aviation Administration, the structural strength (load factor) on airplanes must conform with the standards set forth by Federal Aviation Regulations.

The loads imposed on the wings in flight are stated in terms of **load factor**. Load factor is the ratio of the total load supported by the airplane's wing to the actual weight of the airplane and its contents—i.e., the actual load supported by the wings divided by the total weight of the airplane. For example, if an airplane has a gross weight of 2000 lb [907 kg] and during flight is subjected to aerodynamic forces that increase the total load the wing must support to 4000 lb [1814 kg], the load factor is 2.0 (4000/2000 = 2). In this example, the airplane wing is producing lift that is equal to twice the gross weight of the airplane.

Another way of expressing load factor is the ratio of a given load to the pull of gravity, i.e., to refer to a load factor of 3 as "three g's," where g refers to the pull of gravity. In this case the weight of the airplane is equal to 1 g, and if a load of three times the actual weight of the airplane were imposed upon the wing due to curved flight, the load factor would be equal to three g's. Refer to *Aircraft Basic Science* for more information.

All airplanes are designed to meet certain strength requirements, depending upon the intended use of the airplane. Classification of airplanes as to strength and operational use is known as the category system. Aircraft may be type-certificated as *normal, utility,* or *acrobatic.*

The normal category is limited to airplanes that have a seating configuration, excluding pilot seats, of nine or less, a maximum certificated takeoff weight of 12 500 lb or less, and intended for nonacrobatic operation. Nonacrobatic operation includes: (1) any maneuver incident to normal flying; (2) stalls (except whip stalls); and (3) lazy eights, chandelles, and steep turns, in which the angle of bank is not more than 60 degrees. The normal category has a positive load factor limit of 3.8 (often referred to as the limit load factor) and a negative load factor limit of 1.52.

The utility category is limited to airplanes that have a seating configuration, excluding pilot seats, of nine or less, a maximum certificated takeoff weight of 12 500 lb or less, and intended for limited acrobatic operation. Airplanes certificated in the utility category may be used in any of the operations covered in this section and in limited acrobatic operations. Limited acrobatic operation includes: (1) spins (if approved for the particular type of airplane); and (2) lazy eights, chandelles, and steep turns, or similar maneuvers, in which the angle of bank is more than 60 degrees but not more than 90 degrees. The utlity category has a positive load factor limit of 4.4 and a negative load factor limit of 1.76.

The acrobatic category is limited to airplanes that have a seating configuration, excluding pilot seats, of nine or less, a maximum certificated takeoff weight of 12 500 lb or less, and intended for use without restrictions, other than those shown to be necessary as a result of required flight tests. The acrobatic category has a positive load factor limit of 6.0 and a negative load factor limit of 3.0.

Small airplanes may be certificated in more than one category if the requirements of each category are met.

The category in which each airplane is certificated may be readily found in the aircraft's Type Certificate Data Sheet or by checking the Airworthiness Certificate found in the cockpit.

To provide for the rare instances of flight when a load greater than the limit is required to prevent a disaster, an "ultimate factor of safety" is provided. Experience has shown that an ultimate factor of safety of 1.5 is sufficient. Thus, the aircraft must be capable of withstanding a load 1.5 times the limit load factor. The primary structure of the aircraft must withstand this "ultimate load" (1.5 × limit load factor) without failure.

Since the limit load factor is the maximum of the normally anticipated loads, the aircraft structure must withstand this load with no ill effects. Specifically, the primary structure of the aircraft should experience no permanent deformation when subjected to the limit load factor. In fact, the components must withstand this load with a positive margin. This requirement implies that the aircraft should withstand successfully the limit load factor and then return to the original unstressed shape when the load is removed. If the aircraft is subjected to a load in excess of the limit load factor, the overstress may cause a permanent distortion of the primary structure and require replacement of the damaged pans.

Aircraft Loads

Aircraft loads originate during two distinctly different operating conditions, in flight and on the ground. These distinctly different flight and ground load conditions must be considered to understand the most critical conditions for the structural components.

Flight loads are also divided into two types: **maneuvering** loads and **gust** loads. The word **maneuvering** does not necessarily imply acrobatic flight, since such routine actions as a banked turn or a stall above landing speed are considered maneuvers in the sense that the airplane is subject to loads greater than 1 g. In level, trimmed, steady cruise flight, all parts of the airplane and its contents are subject to a gravitational loading of 1 g. A passenger weighing 170 lb will exert a measured loading of 170 lb upon the airplane in level flight. This loading is carried into the airframe through the seat and floor structure.

When the airplane is being maneuvered into a 2-g banked turn, a 170-lb body will load the seat and supporting structure at 340 lb instead of the original 170 lb, since 2 × 170 = 340. In similar fashion, the turning maneuver that doubles the body load also doubles the load applied to the wings and other parts of the airplane.

Gust loads, in general, are of shorter duration than maneuver loads, but their direction change can be much faster and sometimes will appear to be almost instantaneous. It is during these times of instantaneous change that the load factors produced are the highest.

Each flight involves at least one takeoff, one landing, and usually some taxiing. Once again, the purposes of the

aircraft will determine, to a large extent, the amount of time to be spent in the air and on the ground. Usually the landing loads, rather than takeoff loads, govern the design of the gear attachment structure of an airplane, even though the allowable takeoff weight may be higher than the landing weight. Descent velocities of the particular type of airplane as well as the wing loading and the shock-absorption characteristics of the landing gear, struts, and tires will determine, in large part, the reaction at ground contact. Total reaction force divided by the weight of the aircraft is called the **landing load factor.**

An airplane is designed and certificated for a specified maximum weight during flight. This weight is referred to as the *maximum certificated gross weight.* It is important that the airplane be loaded within the specified weight limits because certain flight maneuvers will impose an extra load on the airplane structure, which, if the airplane is overloaded, may impose stresses exceeding the design capabilities of the airplane. If, during flight, severe turbulence or any other condition causes excessive loads to be imposed on the airplane, a very thorough inspection must be given to all critical structural parts before the airplane is flown again. Damage to the structure is often recognized by bulges or bends in the skin, "popped" rivets, or deformed structural members.

The *V-N* Diagram

The design of an aircraft is dictated by the anticipated use. One of the most important guidelines used by the engineer in defining that use is the diagram relating limit and ultimate load factors to forward speed, the *V-N diagram.*

A typical *V-N* diagram is shown in Fig. 2-1. The *V-N* diagram in the figure is intended to show the general features of such a diagram and does not necessarily represent the characteristics of any particular airplane. Each aircraft has its own particular *V-N* diagram with specific *V*'s and *N*'s. The flight operating strength of an airplane is presented on a graph whose horizontal scale is airspeed (*V*) and whose vertical scale is load factor (*N*). For the airplane shown, the positive limit load factor is 3.8 and the positive ultimate load factor is 5.7 (3.8 × 1.5). For negative lift flight conditions, the negative limit load factor is 1.52 and

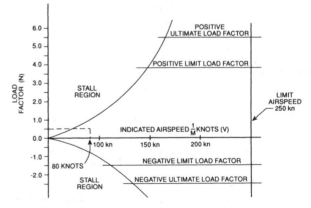

FIGURE 2-1 Typical *V-N* diagram.

the negative ultimate load factor is 2.28 (1.52 × 1.5). The never-exceed speed, which is the placard red-line speed, is 250 knots, and the wing level stall speed is 80 knots. If this airplane is flown at a positive load factor greater than the positive ultimate load factor of 5.7, structural damage will be possible. When the airplane is operated in this region, objectionable permanent deformation of the primary structure may take place and a high rate of fatigue damage is incurred.

The same situation exists in negative lift flight, with the exception that the limit and ultimate load factors are of smaller magnitude and the negative limit load factor may not be the same value at all airspeeds.

PRINCIPAL AIRCRAFT STRUCTURES

The principal aircraft load-carrying structural sections or components shown in Fig. 2-2 include the fuselage, lifting surfaces, control surfaces, stabilizers, and landing gear. The fuselage is the central aircraft component, which has a cockpit or flight deck for the crew and a section for the passengers and cargo. The lifting surfaces include the wings on airplanes and gliders and the main rotors of helicopters. Control surfaces include ailerons, rudders, elevators, flaps, spoilers, and trim tabs. Stabilizers are used to improve the pitch and yaw stability of the aircraft. The landing gear may be fixed or retractable and may use skids, wheels, floats, or skis, depending on the type of aircraft and the operating terrain.

The **stresses** (effects of applied forces) to which structural members are subjected are *compression, tension, torsion, bending,* and *shear.* The following definitions will serve to aid the student in understanding the nature of each type of stress:

Compression is the stress that tends to crush or press together. The landing gear of an aircraft is subjected to compression when the aircraft is landing.

Tension is the stress in a member when a force tends to elongate or stretch it. A bolt tightened to hold parts firmly together is subjected to tension. A cable is in tension when it is used to lift an aircraft or engine.

Torsion is the stress of twisting. Rotating shafts under load are subjected to torsion.

Bending is actually a combination of compression and tension. When a bar is bent, the portion of the bar toward the outside of the bend is subjected to tension and the portion of the bar toward the inside of the bend is subjected to compression. The wings of an aircraft are subjected to bending stresses.

Shear is the stress developed when a force tends to cause a layer of material to slide along an adjacent layer. When two strips of metal are joined by means of rivets or bolts, a tensile force applied to the opposite ends of the assembled strips in a manner tending to pull them apart will produce shear stress in the rivets or bolts.

Strain is the effect of overstressing a part or assembly to the point where a permanent deformation takes place. If an aircraft part has become strained, the part very likely will no longer be airworthy (Fig. 2-3).

When an airplane is designed, the loads that are likely to be applied to parts or assemblies of the airplane during operation are carefully computed and analyzed by engineers. This process is called **stress analysis.** The performance of the stress analysis ensures that the airplane will perform according to its approved specifications without danger of failure. In addition to stress analysis, **fatigue loading** is another area that has become an important design consideration in all classes of aircraft. The increasing performance of modern aircraft and the higher utilization rates have dictated the requirements for the primary structure to approach infinite service life. Many structural areas, especially those subjected to repeated high loads, are designed by fatigue requirements. Most modern aircraft use the damage tolerance design philosophy. The aircraft structure is designed in such a way that it has the ability to sustain defects safely until repair can be affected. The structure is considered to be damage tolerant if a maintenance program has been implemented that will result in the detection and repair of accidental damage, corrosion, and fatigue cracking before such damage reduces the residual strength of the structure below an acceptable limit.

AIRCRAFT STATION NUMBERS

In the service, maintenance, and repair of aircraft, it is necessary to establish a method of locating components and reference points on the aircraft. This has been accomplished by establishing reference lines and **station numbers** for the fuselage, wings, nacelles, empennage, and landing gear. For large aircraft, the Air Transport Association of America (ATA) has set forth **zoning** specifications in ATA-100 Specification for Manufacturers' Technical Data. Zoning is discussed later.

Fuselage Stations

Longitudinal points along the fuselage of an airplane are determined by reference to a zero **datum line** [F.S. (fuselage station) 0.00] usually at or near the forward portion of the fuselage. The position of the datum line is set forth in the Type Certificate Data Sheet or Aircraft Specification for the airplane and also in manufacturer's data. Fuselage stations for a general aviation aircraft are shown in Fig. 2-4. In this case the datum line is located before the nose of the aircraft. Some aircraft manufacturers use the engine firewall, wing, or other reference point of their aircraft as the datum line. However, it is most common to select the datum line ahead of the aircraft. Station numbers are given in inches forward or aft of the datum line. Fuselage station numbers forward of the datum line are negative (−) and station numbers aft of the datum line are positive but are not usually shown with a positive (+) sign.

1. RADOME
2. FUSELAGE NOSE LOWER STRUCTURE
3. FUSELAGE NOSE UPPER STRUCTURE
4. FORWARD SERVICE DOOR
5. FUSELAGE STA 229 TO 474 UPPER STRUCTURE
6. FUSELAGE STA 474 TO 817 UPPER STRUCTURE
7. UPPER COWL DOOR
8. PASSENGER AFT ENTRANCE STAIRWELL DOOR
9. FUSELAGE STA 817 TO 908 LOWER STRUCTURE
10. FUSELAGE STA 817 TO 908 UPPER STRUCTURE
11. DORSAL FIN
12. VERTICAL STABILIZER
13. VERTICAL STABILIZER TIP
14. REMOVABLE TIP FAIRING
15. ELEVATOR
16. ELEVATOR CONTROL TAB
17. ELEVATOR GEARED TAB

18. HORIZONTAL STABILIZER AFT SECTION
19. HORIZONTAL STABILIZER TIP ASSEMBLY
20. HORIZONTAL STABILIZER LEADING EDGE
21. RUDDER
22. RUDDER TAB
23. TAIL CONE
24. FUSELAGE TAIL STRUCTURE
25. PASSENGER AFT ENTRANCE DOOR STAIRWAY
26. PYLON AFT PANEL
27. THRUST REVERSER COWLING
28. LOWER COWL DOOR
29. PYLON CENTER PANEL
30. PYLON LEADING EDGE
31. ENGINE
32. NOSE COWL
33. FUSELAGE STA 642 TO 817 LOWER STRUCTURE
34. OVERWING EMERGENCY EXITS
35. FLAP VANE
36. SPOILER

37. WING FLAP
38. AILERON TABS
39. AILERON
40. WING TIP
41. WING MAIN STRUCTURE
42. WING SLAT
43. FLAP HINGE FAIRINGS
44. WING LEADING EDGE
45. MAIN GEAR
46. MAIN GEAR OUTBOARD DOOR
47. MAIN GEAR INBOARD DOOR
48. KEEL
49. WING-TO-FUSELAGE FILLET
50. FUSELAGE STA 229 TO 474 LOWER STRUCTURE
51. PASSENGER FORWARD ENTRANCE DOOR
52. FORWARD STAIRWELL DOOR
53. PASSENGER FORWARD ENTRANCE STAIRWAY
54. FORWARD NOSE GEAR DOORS
55. AFT NOSE GEAR DOORS
56. NOSE GEAR

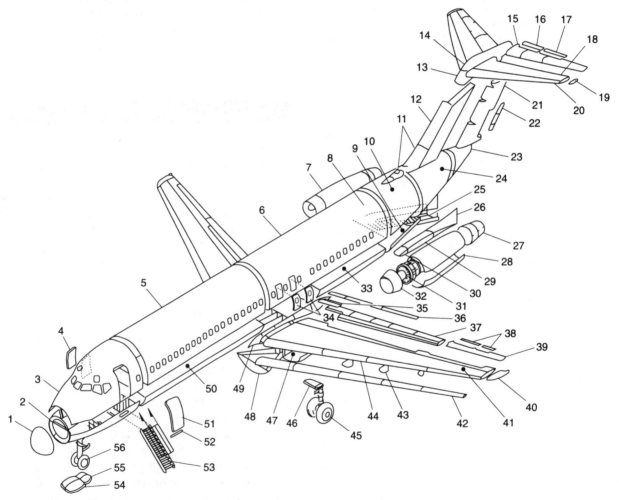

FIGURE 2-2 Airplane structural components. (*Douglas Aircraft Co.*)

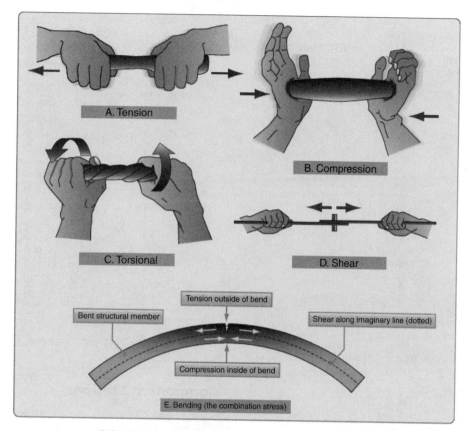

FIGURE 2-3 Stresses acting on structural members.

Wing Stations

To locate points on the wing of an airplane, the *wing station* (WS) numbers are measured from the centerline of the fuselage. This line is also called the *butt line* (BL). Wing stations are indicated in inches either right or left of the fuselage centerline. Wing stations for the wings of an airplane are shown in Fig. 2-4.

Water Line

The *water line* (WL) is a line established for locating stations on a vertical line. The term water line originated with the design and building of ship hulls and was used as a

vertical reference. Vertical measurements on an airplane may be either negative or positive, depending upon whether the point is above or below the water line. WL stations are used to locate positions on the landing gear, vertical stabilizer, and at any other point at which it is necessary to locate a vertical distance. Figure 2-5 shows how WL stations are used for location on the vertical stabilizer of an airplane.

Butt Line

As mentioned previously, the butt line (BL) is the centerline of the fuselage. Positions on the horizontal stabilizer and elevator are given butt line station numbers, as illustrated in Fig. 2-6.

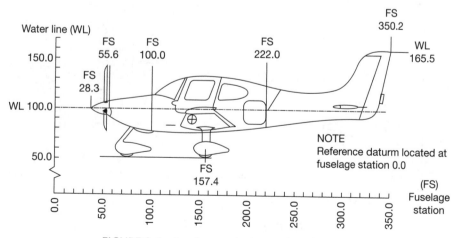

FIGURE 2-4 Fuselage and wing stations (Cirrus).

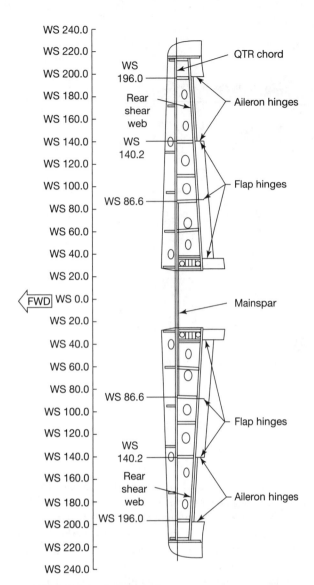

FIGURE 2-4   (continued)

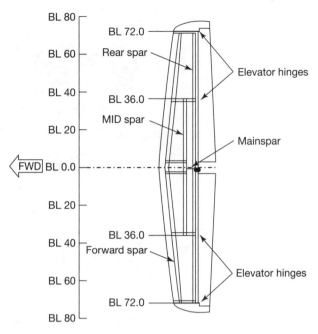

FIGURE 2-6   Butt-line stations.

Component Stations

Some aircraft components are given their own station reference lines. For example, an aileron may have aileron stations established across its span, starting with aileron station (AS) 0.00, located at the inboard edge of the aileron. Positions outboard of the inboard station are indicated in inches. Other examples of component stations are shown in Fig. 2-7 and include winglet stations, engine stations, and vertical stabilizer/rudder stations.

Summary

The two-view drawing of Fig. 2-8 shows how positions are located by means of fuselage stations, wing stations, water line, and butt line. It will be noted that a nacelle butt line

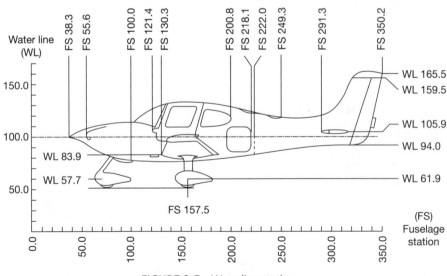

FIGURE 2-5   Waterline stations.

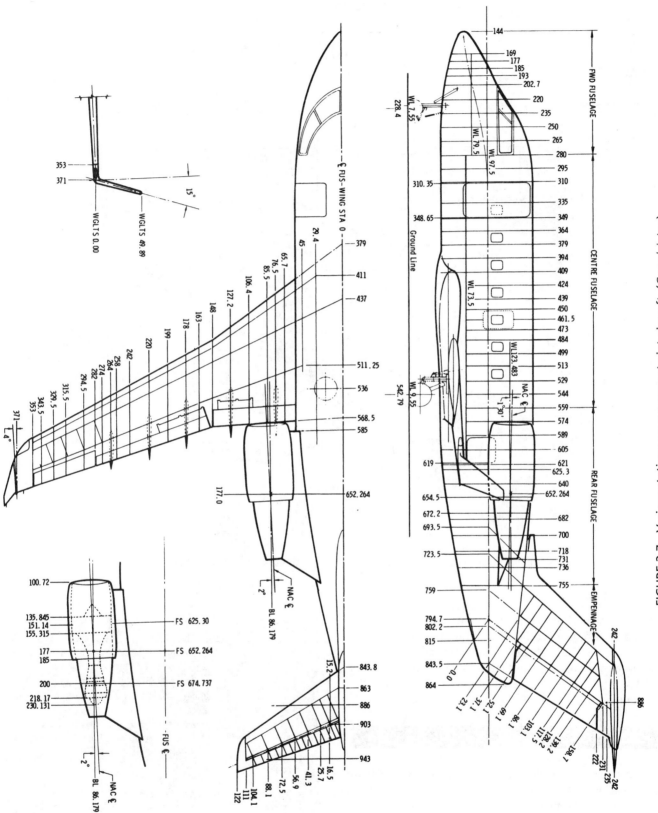

FIGURE 2-7 Various stations on a corporate jet aircraft. (*Canadair Inc.*)

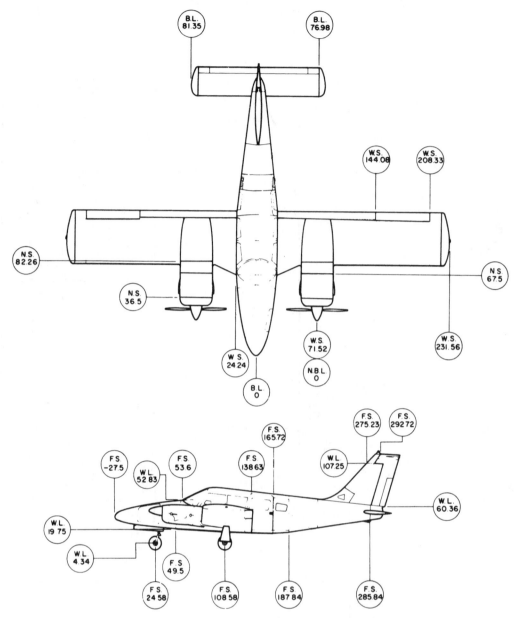

FIGURE 2-8 Use of station numbers to establish positions on an airplane. (*Piper Aircraft Co.*)

(NBL) is established at the centerline of the nacelle for locating positions in the nacelle. In the drawing of Fig. 2-8, the NBL is at WS 71.52, which is 71.52 in [181.66 cm] from the fuselage centerline. Note that the center of the nose wheel is 4.34 in [11.02 cm] above the WL, thus placing the WL at the rim of the wheel.

ZONING

As mentioned previously, zoning of large aircraft has been specified by the Air Transport Association of America in the ATA-100 Specification.

A zone is identified by one of three indicators, depending upon whether it is a major zone, major subzone, or simply a zone. Major zones are identified by three-digit numbers as follows:

Major Zone No.	Area
100	Lower half of the fuselage to the rear pressure bulkhead (below the main cabin deck)
200	Upper half of the fuselage to the rear pressure bulkhead
300	Empennage, including fuselage aft of the rear pressure bulkhead
400	Powerplants and struts or pylons
500	Left wing
600	Right wing
700	Landing gear and landing gear doors
800	Doors
900	Reserved for uncommon differences between aircraft types not covered by standard series numbers

The standard series is from 100 to 800, and the special series numbers are in the 900 bracket.

Figure 2-9 shows the major zones of a transport aircraft.

Major Subzones

Major zones are divided into major subzones by the addition of a second nonzero digit to the major zone number. For example, the major zone 300 may be subzoned as follows:

Major Subzone No.	Area
310	Fuselage aft of the pressure bulkhead
320	Vertical stabilizer and rudder
330	Left horizontal stabilizer and elevator
340	Right horizontal stabilizer and elevator

Subzones are divided by the use of a third nonzero digit in the three-digit number. The subzone 320 may, therefore, be divided into zones as follows:

Zones	
321	Vertical stabilizer leading edge
322	Vertical stabilizer auxiliary spar to front spar
323	Front spar to rear spar
324	Rear spar to trailing edge

325	Lower rudder
326	Upper rudder
327	Vertical stabilizer tip

From the foregoing, it can be seen that the entire airplane can be divided into specific zones for the identification of any area that requires inspection, maintenance, or repair. The zone numbers can be utilized in computerized maintenance record systems to simplify the processing of records and instructions.

Figure 2-10 shows the subzones of a transport aircraft wing.

NOMENCLATURE AND DEFINITIONS

The definitions given here are provided as a convenient reference for the following discussion regarding the structural components of an aircraft. A detailed discussion of aircraft nomenclature and the theory of flight can be found in the text *Aircraft Basic Science*.

Aileron Hinged sections of the trailing edge of the left and right wing, which operate in series to provide lateral control. When one aileron is raised, the opposite is lowered, producing rolling movements around the longitudinal axis of the aircraft.

Airfoil A surface such as an airplane wing, aileron, or rudder designed to obtain reaction from the air through which it moves.

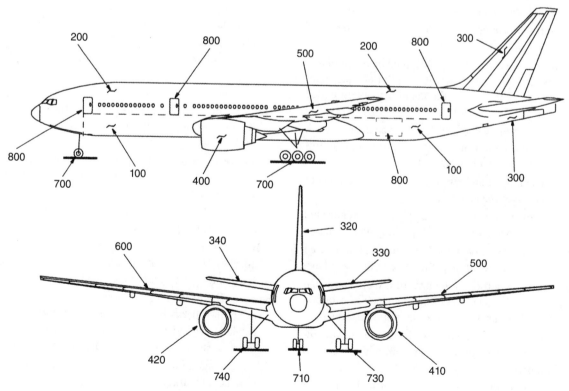

FIGURE 2-9   Major zones of a transport aircraft.

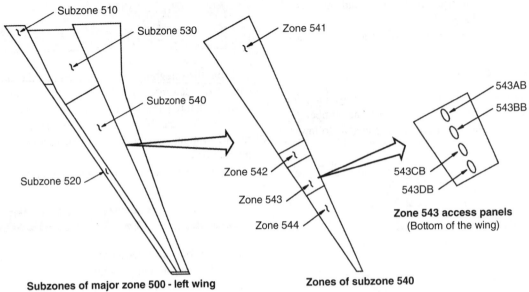

Subzones of major zone 500 - left wing

Zones of subzone 540

Zone 543 access panels
(Bottom of the wing)

FIGURE 2-10   Subzones of a tranport aircraft wing.

Bulkhead A heavy structural member in the fuselage to contain pressure or fluids or to disperse concentrated loads. A heavy circumferential frame which may or may not be entirely closed by a web.

Buttock line (butt line) A vertical reference line or plane parallel to the centerline of the airplane. Used to locate points or planes to the left or right of airplane centerline.

Cantilever A beam or member supported at or near one end only without external bracing.

Center section The middle or central section of an airplane wing to which the outer wing panels are attached.

Circumferential A frame shaped to the circumference of the fuselage diameter.

Cockpit On small aircraft, the area occupied by the pilot and passengers. On cabin airplanes, if the pilot compartment is separated from the rest of the cabin, it is often called the cockpit.

Control surface A movable airfoil or surface, such as an aileron, elevator, flap, trim tab, or rudder, used to control the attitude or motion of an aircraft in flight.

Cowl panel The hinged and removable sides of the pods or nacelles that cover the engines.

Cowling A removable cover or housing placed over or around an aircraft component or section, especially an engine.

Elevator The hinged section of the horizontal stabilizer used to increase or decrease the angle of attack of the airplane.

Empennage The aft portion of an aircraft, usually consisting of a group of stabilizing planes or fins, to which are attached certain controlling surfaces such as elevators and rudders.

Fairing A piece, part, or structure having a smooth streamlined contour; used to cover a nonstreamlined object or to smooth a junction.

Fin A term commonly applied to the vertical stabilizer (vertical fin) or any stabilizing surface parallel to the vertical centerline of the airplane. Horizontal surfaces are commonly called stabilizers.

Firewall A fireproof or fire-resistant wall or bulkhead separating an engine from the rest of the aircraft structure to prevent the spread of a fire from the engine compartment.

Flaps, leading edge Hinged section of the underside of the leading edge, which, when extended, reduces airflow separation over the top of the wing. Leading edge flaps are hinged at the leading edge of the airfoil.

Flaps, trailing edge Hinged section of the trailing edge of the wing, which can be lowered and extended. When lowered, these flaps provide the airplane with greater lift at lower speeds.

Frame A circumferential structural member in the body that supports the stringers and skin; used in semimonocoque construction.

Hat section The cross-section shape of the stringers used in the fuselage; it is a common rolled shape that looks like a top hat with the brim curled up.

Longeron A principal longitudinal member of the framing of an aircraft fuselage or nacelle. It is usually continuous across a number of points of support.

Pressure web A web that primarily seals an area in order to retain cabin pressurization.

Rib A fore and aft member of an airfoil structure (wing or aileron) of an aircraft used to give the airfoil section its form and to transmit the load from the skin to the spars.

Section Any of the larger subassemblies of the airplane that are built separately and, when joined, form the complete airplane. The airplane is broken down into smaller sections to ease production and handling problems.

Span The maximum distance, measured parallel to the lateral axis, from tip to tip of any surface, such as a wing or stabilizer.

Spar A principal spanwise beam in the structure of a wing, stabilizer, rudder, or elevator. It is usually the primary load-carrying member in the structure.

Stabilizer A fixed horizontal tail surface that serves to maintain stability around the lateral axis of an aircraft.

Station line All parts of an airplane are identified by a location or station number in inches from a beginning point. Station lines in the fuselage start forward of the nose; those for the wing usually start at the centerline of the fuselage.

Stringer Longitudinal member in the fuselage or spanwise members in the wing to transmit skin loads into the body frames or wing ribs.

Strut A supporting brace that bears compression loads, tension loads, or both, as in a fuselage between the longerons or in a landing gear to transmit the airplane loads.

Vertical fin Sometimes referred to as vertical stabilizer or fin. It is fixed to provide directional stability. The trailing edge is hinged to form the rudder.

Water line A horizontal reference line or plane parallel to the ground used to locate points vertically.

Web A thin gauge plate or sheet that, when supported by stiffening angles and framing, provides great shear strength for its weight. Used in many applications throughout an aircraft because of its strength-to-weight ratio.

FUSELAGES

The **fuselage** is the body of an aircraft to which the wings and the tail unit are attached. It provides space for the crew, passengers, cargo, controls, and other items, depending upon the size and design of the aircraft. The aircraft structure is designed to provide maximum strength with minimum weight. The fuselage should be designed to satisfy two major criteria: (1) protect the passengers in the event of a crash and (2) efficiently tie together the powerplant, wing, landing gear, and tail surface loads. This must be accomplished with interior space for passenger comfort and minimum frontal area and contour drag for maximum performance. Perhaps the most distinct feature of the fuselage is a result of its purpose: providing space for payloads. The space required creates the need for comparatively large openings within the airframe in relation to its size, whether the airplane is a light four-place craft or a large passenger transport. Around this space and function the fuselage structure is designed and built. If an airplane is of the single-engine type, the engine is usually mounted in the nose of the fuselage.

The fuselage must have points of attachment for the wing or wings, tail surfaces, and landing gear so arranged and installed that these parts can be inspected, removed, repaired, and replaced easily. The fuselage must be strong enough at the points of attachment to withstand flying and landing loads. Finally, the fuselage should be shaped to offer low resistance to the air and provide good vision for the pilot. The design of many large aircraft is such that the wing structure extends through the fuselage, thus eliminating the necessity for the fuselage to carry strictly wing-generated loads and stresses. A discussion of wing design is presented later in this chapter.

Types of Fuselages

In general, fuselages are classified into three principal types, depending upon the method by which stresses are transmitted to the structure. The three types according to this classification are *truss, semimonocoque,* and *monocoque.*

A **truss** is an assemblage of members forming a rigid framework, which may consist of bars, beams, rods, tubes, wires, etc. The truss-type fuselage may be subclassified as a *Pratt truss* or a *Warren truss.* The primary strength members of both Pratt and Warren trusses are the four longerons. As defined previously, the longeron is a principal longitudinal member of the aircraft fuselage. In the truss-type fuselage, lateral bracing is placed at intervals. The lateral structures may be classed as bulkheads, although this is not strictly true from a technical standpoint. The spaces between the bulkheads are called **bays.**

A **Pratt truss** similar to the type used in present aircraft with tubular fuselage members is shown in Fig. 2-11. In the original Pratt truss, the longerons were connected with rigid vertical and lateral members called struts, but the diagonal members were made of strong steel wire and were designed to carry tension only. In the Pratt truss shown in Fig. 2-11, the diagonal members are rigid and can carry either tension or compression.

A **Warren truss** is illustrated in Fig. 2-12. In this construction, the longerons are connected with only diagonal members.

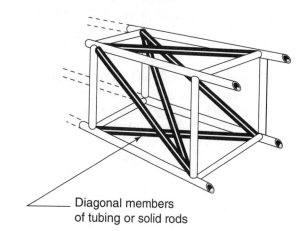

Diagonal members
of tubing or solid rods

FIGURE 2-11 Pratt truss.

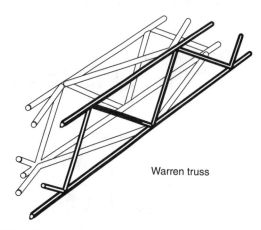

Warren truss

FIGURE 2-12 Warren truss.

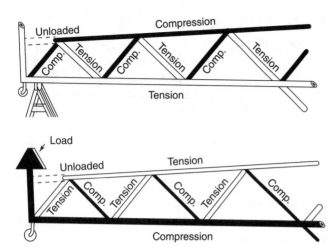

FIGURE 2-13   Reversal of loading on a truss.

Normally, all members in the truss are capable of carrying both tension and compression. When the load is acting in one direction, compression loads are carried by every other member, and the alternate members carry the tension loads. When the load is reserved, the members that previously carried tension now carry compression and those that were carrying compression now carry tension. This reversal of loading is shown in Fig. 2-13.

The determination as to the type of truss construction used in an aircraft may be academic in most cases, but if a modification to the aircraft truss structure is being considered, the type of construction can become important. In some aircraft it is very easy to identify the type of truss used, as shown in Fig. 2-14, which is a Warren truss. Because of the location of structural attachments, the Pratt truss fuselage in Fig. 2-15 is somewhat disguised, but by careful examination the characteristic vertical and diagonal members can be identified. Although these two fuselage structures are made of steel, it should be understood that a truss structure can be made of wood, aluminum, or any other structural material, according to the aircraft manufacturer's choice.

A **semimonocoque** structure consists of a framework of vertical and longitudinal members covered with a structural skin that carries a large percentage of the stresses imposed upon the structure. Figure 2-16 illustrates the construction of an all-metal semimonocoque tailboom. The vertical members

of the tailboom are called **frames,** or **bulkheads.** Between the principal vertical members are lighter **formers,** or **rings,** to maintain the uniform shape of the structure. The longitudinal members are called **stringers,** and they serve to stiffen the metal skin and prevent it from bulging or buckling under severe stress. Use of stringers has enabled aircraft designers to use aluminum skins as light as 0.020 in thickness for the primary structure on airplanes as large as light twins. Larger semimonocoque aircraft use progressively thicker skins and still maintain an equivalent stress level in the skin, along with an equally good weight-to-strength ratio. The construction of a semimonocoque fuselage is illustrated in Fig. 2-17.

A full **monocoque** fuselage, shown in Fig. 2-18, is one in which the fuselage skin carries all the structural stresses. This construction merely involves the construction of a tube or cone without internal structural members. In some cases it is necessary to have former rings to maintain the shape, but these do not carry the principal stresses imposed upon the structure. The monocoque structure is especially useful with honeycomb or foam sandwich composite structures. Due to the high stiffness of the sandwich structure, large structures can be built that do not need external reinforcements such as stringers and frames. This construction is light and stiff. Radomes used in aircraft are most often monoque structure, and the Cirrus SR20 series aircraft, shown in Fig. 2-18, uses a monoque structure. The fuselage is constructed of two large panels which are bonded to form the fuselage structure.

The monocoque structure can carry loads effectively, particularly when the fuselage is of a small diameter. As its diameter increases to form the internal cavity necessary for a fuselage, the weight-to-strength ratio becomes more inefficient, and longitudinal stiffeners or stringers are added. The result is that the most popular type of structure used in structural aircraft design today is the semimonocoque.

In many aircraft, a mixture of construction types may be found. For example, an aircraft may have a steel tube cabin structure and a semimonocoque aft fuselage structure, as in the Cessna Model 188 shown in Fig. 2-19.

General Construction of Fuselages

As explained previously, fuselages are designed with a variety of structural components. The great majority of fuselages are all metal and semimonocoque in construction. This statement

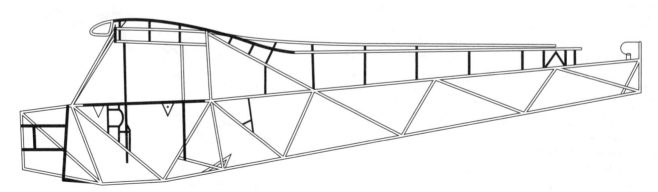

FIGURE 2-14   A steel tube fuselage using a Warren-type truss. (*Piper Aircraft Co.*)

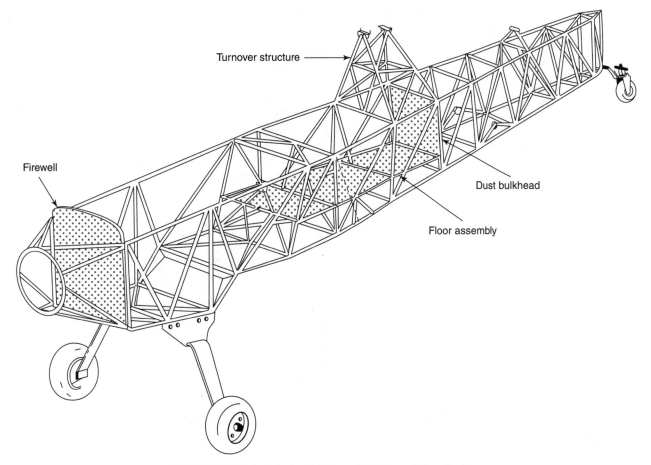

Turnover structure

Firewell

Dust bulkhead

Floor assembly

FIGURE 2-15 A typical welded-steel fuselage. (*Schweizer*)

applies to small, medium, and large aircraft. The interior structure to which the skin or plating is attached consists of longerons, frames, bulkheads, stringers, gussets, and possibly intercostal members riveted, bolted, or bonded together to form a rigid structure that shapes the fuselage. The skin or plating is riveted or bonded to the structure to form the complete unit.

The types and thicknesses of materials used for covering the fuselage of a typical light airplane are shown in the drawings of Fig. 2-20. The thickness of the skin varies according to position on the fuselage. The required thickness of material for a given section of the fuselage is determined by engineers during the design and stress analysis of the aircraft.

Fuselages for aircraft are designed with many similarities. The forward section of the fuselage usually contains the cockpit and passenger cabin. The shape of this section depends upon the passenger capacity and the performance specifications for the aircraft. The rear section, often referred to as the **tail cone,** is usually circular or rectangular in cross section and tapers toward the tail.

Transport Aircraft Fuselages

Fuselages for transport aircraft generally include a section forward of the main cabin to provide a streamlined nose, a main cabin section, which is uniformly cylindrical or oval in shape, and a tail section, which tapers to a minimal size at the

extreme rear end. These shapes are illustrated in Fig. 2-21. The materials most commonly used throughout the structure are high-strength aluminum, steel, and titanium alloys. The most extensively used materials are certain aluminum alloys, which are selected according to the particular type of load they are best suited to withstand. Fiberglass, graphite, kevlar, and honeycomb core materials are used extensively on secondary areas of the structure and on many flight control surfaces.

Transport-category aircraft fuselages are generally of semimonocoque construction, utilizing the same basic principles described for light aircraft. Due to their size, transport-size aircraft use additional components such as keel beams, shear ties, and frames. *Frames* are circumferential members generally spaced at regular intervals along the length of the fuselage. As illustrated in Fig. 2-22, frames stabilize the skin and stringers and distribute the concentrated loads. Heavy frames reinforced by beams attached to webs are usually called *bulkheads*. Stringers (also shown in Fig. 2-22) are longitudinal members spaced around the fuselage circumference; they extend the full length of the fuselage. The stringers are attached to the outboard edge of the frame and the inboard face of the skin. Floor beams, which provide the support for the cabin floor, attach to the frame and run horizontally across the fuselage. The floor-to-skin shear ties shown in Fig. 2-23 are those ties that extend longitudinally along the intersection of the floor beams and frame. A major longitudinal fuselage component in the wing center

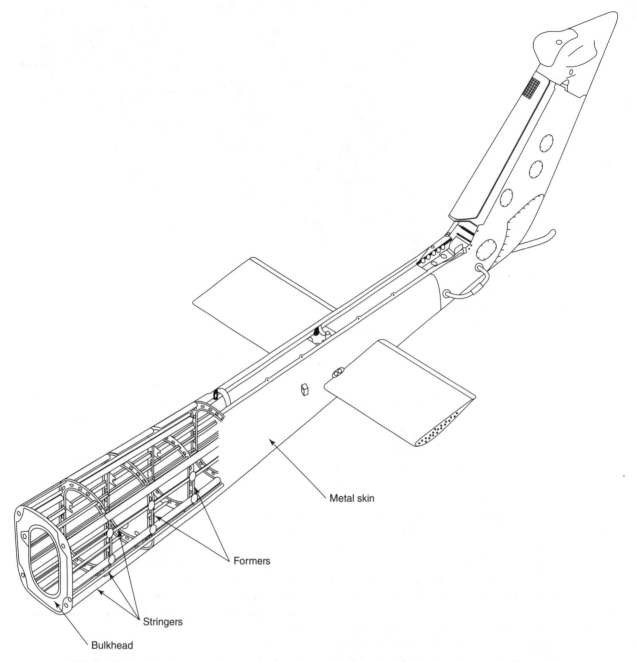

Metal skin

Formers

Stringers

Bulkhead

FIGURE 2-16 Semimonocoque construction is employed in this helicopter tailboom. (*Bell Textron*)

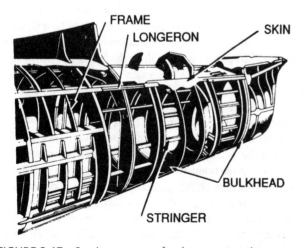

FRAME
LONGERON
SKIN

BULKHEAD

STRINGER

FIGURE 2-17 Semimonocoque fuselage construction.

FIGURE 2-18 Full monocoque composite fuselage.

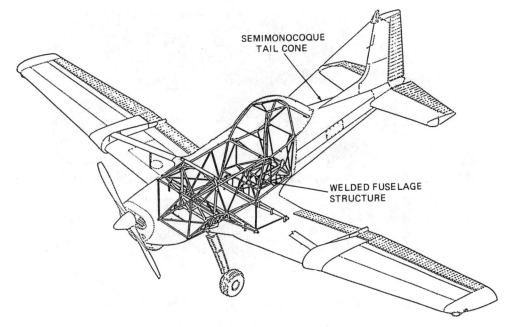

FIGURE 2-19 Aircraft with a combination of steel tube and semimnocoque construction in its fuselage structure. (*Cessna Aircraft Co.*)

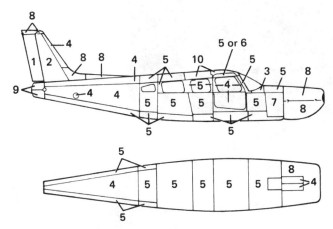

FIGURE 2-20 Materials and thickness for a light-airplane fuselage. (*Piper Aircraft Co.*)

section and wheel-well area is called a *keel beam*. The keel beam pictured in Fig. 2-24 extends along the fuselage centerline through the wheel well and under the wing center section.

Most aircraft designs are now incorporating components that are called *fail-safe* components. These major structural fittings are made in two parts and are joined together by riveting or bonding. Each half of the two structural parts is capable of carrying the full required structural load of the assembly. With this design philosophy, the failure of one of the fitting components will not result in a structural failure of the attachment fitting, which could result in separation of components in flight. This type of design is found on many corporate and transport aircraft and is illustrated in Fig. 2-25.

Because of their size, transport aircraft fuselages are commonly constructed by first building subassemblies of long panels several feet high, as shown in Fig. 2-26. These panels are then joined on a mating jig to form the circular shape of

a fuselage section, as shown in Fig. 2-27. The fuselage is then assembled by connecting the forward, mid, and aft fuselage sections.

The forward section tapers the cross section of the fuselage to the front of the aircraft. This section has provisions for the flight-crew stations.

The transport fuselage contains one or more midsection assemblies. These midsection assemblies are basically circular in shape with a constant cross-section size. The midsection contains structures to connect the fuselage to the wing and may include landing gear attachment points.

The aft section changes the cross-sectional shape of the fuselage into the size and shape necessary to join with the fuselage afterbody, or tail cone. The afterbody, or tail cone, is the point of attachment for the aft flight-control surfaces and, depending on the aircraft design, may also incorporate an engine installation area.

The fuselage sections are joined to complete the basic assembly of the fuselage. This assembly process for each of the sections and the mating of the sections is illustrated in Figs. 2-27 through 2-31.

From the illustrations of fuselage construction, it can be seen that a large number of special fixtures are used in the assembly of an aircraft fuselage. These fixtures allow for precision alignment of the structural components as each subassembly is put together. Various fixtures are also used to ensure proper alignment of the subassemblies as they are joined to form the fuselage sections, and additional fixtures are used to join the sections into the completed fuselage.

The materials and type and number of structural components used in the construction of transport aircraft fuselages vary throughout the structure according to the structural strength required and the thickness of material that is desired and acceptable. Normally, the exterior fuselage skins are of 2000 series aluminum, as shown in Fig. 2-32, to provide a

FIGURE 2-21   Fuselage shapes for large aircraft.

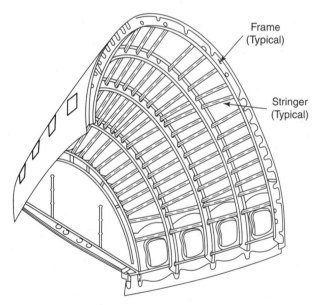

Frame (Typical)

Stringer (Typical)

FIGURE 2-22   Transport aircraft fuselage construction. (*Boeing Commercial Aircraft Co.*)

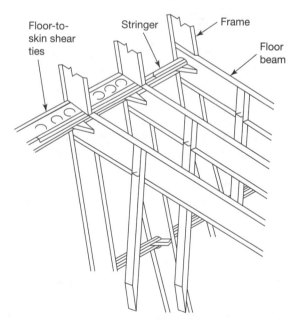

Floor-to-skin shear ties

Stringer

Frame

Floor beam

FIGURE 2-23   Structural floor supports. (*Boeing Commercial Aircraft Co.*)

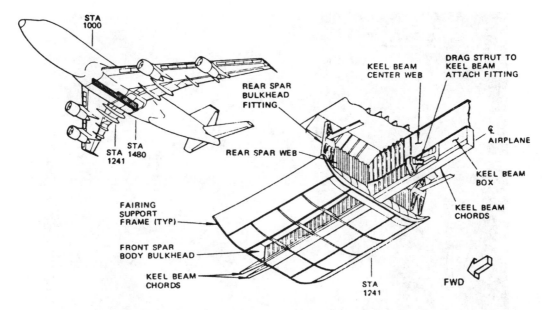

FIGURE 2-24　Fuselage keel beam. (*Boeing Commercial Aircraft Co.*)

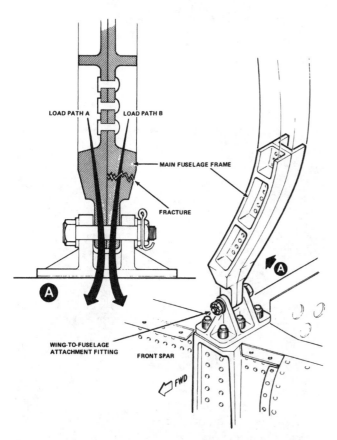

FIGURE 2-25　A fail-safe design provides two load-carrying paths for operational loads. (*Canadair Inc.*)

FIGURE 2-26　Fabrication of fuselage panels with ring riveter.

surface that is resistant to corrosion. The thickness of fuselage skins varies according to applicable loads. The heaviest skins are those in the lower fuselage surrounding the openings for the wings and landing gear doors. Most longitudinal skin splices are of the lap type. These splices are structural joints and must be pressure-sealed. Circumferential skin splices are generally of the butt-joint type. The butt joint is necessary on circumferential splices to obtain an aerodynamically smooth

surface. Prior to assembly, many of the fuselage components are treated to inhibit corrosion of the metal. These treatments may include the use of anodizing, sealants, and primer coatings. The selection of the metal alloys used in the construction is also a factor in control of corrosion. With proper selection of the metals that will be in contact with each other, dissimilar metal corrosion can be reduced or eliminated. Also, by properly designing the fuselage component assemblies, areas which might trap moisture and other corrosive substances can be eliminated.

The assembly process for a fuselage may include the use of hand-operated tools, special power equipment, mechanical fasteners, and bonding materials. Many areas of the fuselage do not adapt well to the use of heavy stationary machinery, so hand-held tools must be used to connect various portions of the fuselage structure. These tools include drills, riveting guns, metal shavers, and tools for the installation of special fasteners.

FIGURE 2-27　Internal structure of transport aircraft. (*Airbus.*)

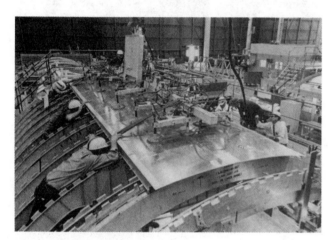

FIGURE 2-28　Handling devices for large fuselage panels. (*Northrop Corp.*)

FIGURE 2-30　Joining of fuselage sections. (*Airbus.*)

FIGURE 2-29　Fuselage join. (*Airbus.*)

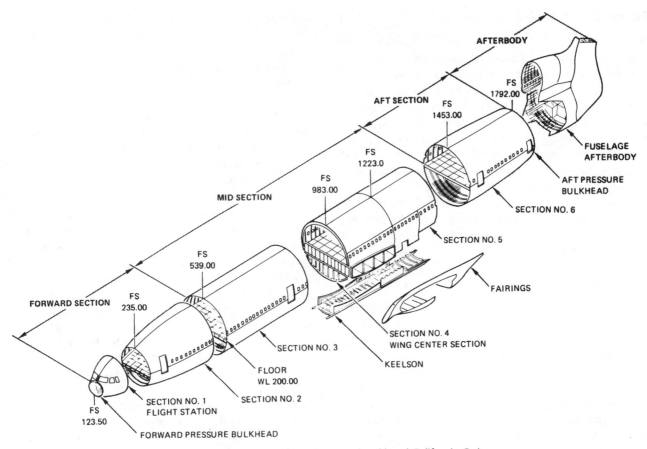

FIGURE 2-31 L-1011 fuselage sections. (*Lockheed California Co.*)

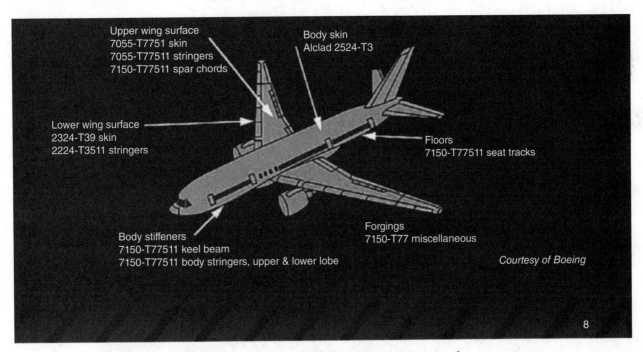

FIGURE 2-32 Aluminum alloys used for transport aircraft.

Bonded Fuselages

Bonding of structural components, which is the process of joining parts by using an adhesive rather than a mechanical fastener, has been used for several years and has some unique advantages over conventional mechanical fasteners. First, it distributes the structural loads evenly over the full bonding area and does not require that the structures be disturbed by the drilling of holes. Second, the bonding agents can incorporate corrosion-inhibiting materials in their mixture and reduce or eliminate the possibility of corrosion occurring in the bonded seams. Last, by properly designing the components, the slight amount of bonding agent that flows out from between components acts as a deflective barrier and prevents moisture or other substances from getting into the bonded seams and starting corrosion.

Composite Construction

The use of composites in aerospace has been growing for more than 30 years. Their high specific strength, excellent corrosion and fatigue resistance, and design flexibility make composites ideally suited to numerous aerospace primary and secondary structures. In their malleable state, composite materials can be formed into an infinite range of shapes and contours to support advanced aerodynamic concepts. When cured, the composite structures become exceptionally strong, stiff, and fatigue-resistant.

Fiberglass, graphite, boron, and aramid fibers, together with epoxy and other organic matrix resins, allow the designer to tailor structures to meet applied loads more effectively through deliberate fiber orientation and more efficiently by reducing fiber content where loads are light. Thus, weight may be decreased in certain structures by an average of 20 to 30 percent relative to metals, with some reductions running as high as 50 percent.

Composites have been slow to be incorporated into commercial aircraft design because of their cost and because of the industry's vulnerability to consumer liability claims. In the last generation of Boeing aircraft, the 757 and 767, composites were used only in secondary components and made up only 3 percent of the structural weight. However, in the next generation of Boeing aircraft, the 777, composites are expected to make up 10 to 20 percent of the structure and to be utilized throughout the aircraft, as illustrated in Fig. 2-33.

Composite components were used increasingly on the new aircraft introduced in the 1980s. Composite applications include flight-control surfaces, fairings, engine nacelle components, and interiors. Because of their satisfactory structural and manufacturing properties, composite materials are widely used in the interiors of commercial aircraft. In addition to meeting mechanical property and fabrication requirements, all materials used within the pressurized envelope of the aircraft must meet both the flammability resistance requirements defined by regulatory agencies and, if applicable, smoke and toxic gas–emission guidelines of the manufacturers. Additionally, visible portions of the cabin and interior components must meet stringent aesthetic requirements to satisfy the airlines and their customers. Durability and maintainability are also important considerations for interior applications.

New aircraft such as the Boeing 787 and Airbus 350, as illustrated in Fig. 2-34, are being built with all composite fuselage and wing structures. Composites make up 50 percent of the structure.

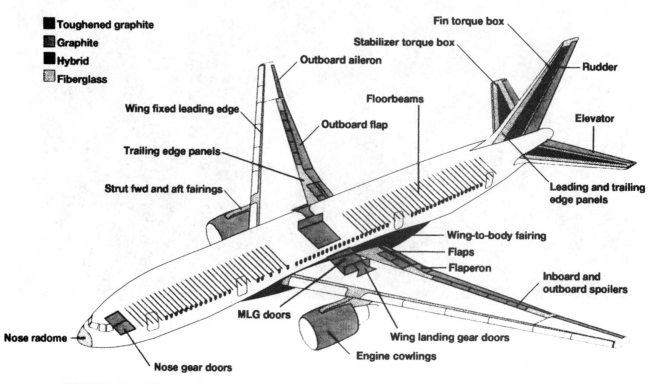

FIGURE 2-33   Utilization of composite materials on a Boeing 777. (*Boeing Commercial Aircraft Co.*)

Fuselage Components

Along with the bulkheads, formers, stringers, and skin, the aircraft fuselage incorporates many structurally reinforced areas to allow for the attachment of airframe components and to allow for openings in the fuselage for non-load-carrying items such as windows and doors. Components that are attached to the fuselage at reinforced areas include wings, landing gear, engines, stabilizers, jackpads, and antennas.

For the attachment of many components, a simple doubler or gusset plate is used. This arrangement involves cutting sheet metal to the proper size for the required flight loads, shaping it to increase its rigidity, and attaching it to the fuselage in the area beneath the component being attached. An example of this type of reinforcement is a doubler placed between stringers to support the flight loads associated with the installation of a radio antenna, as shown in Fig. 2-35.

Some attachment points require the use of special fittings such as forgings, castings, welded assemblies, or heavy sheet-metal structures to be able to withstand large loads such as occur through the wing and landing gear attachment points. An example of this type of arrangement is shown in Fig. 2-36. Note that the forgings are bolted to the surrounding bulkhead components and that the bulkhead components consist of several reinforcing layers of sheet-metal structure. The top portion

of the bulkhead assembly includes the carrythrough member for the aircraft front spar. This is a thick, formed sheet-metal member that is riveted to the bulkhead vertical members.

The structure of the aircraft vertical stabilizer may be an integral part of the aircraft fuselage or the vertical and horizontal stabilizers may be attached to castings or reinforced bulkheads by some fastening means, such as bolts, so that they can be easily removed for replacement or repair operations. Most transport aircraft have a composite material horizontal and vertical stabilizer structure.

COCKPITS, CABINS, AND COMPARTMENTS

Cockpit, or Pilot's Compartment

The **cockpit** is that portion of the airplane occupied by the flight crew. From this cockpit originate all controls used in flying the airplane. The word *controls* is a general term applied to the means provided to enable the pilot to control the speed, direction of flight, altitude, and power of an aircraft.

In designing an airplane, the engineers allow sufficient headroom, visibility, clearance for controls, and space for the movement of the hands and feet. When the space occupied by

FIGURE 2-34　Boeing 787 aircraft.

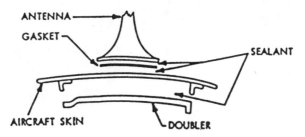

FIGURE 2-35　A doubler is used to reinforce the aircraft skin where the antenna is installed.

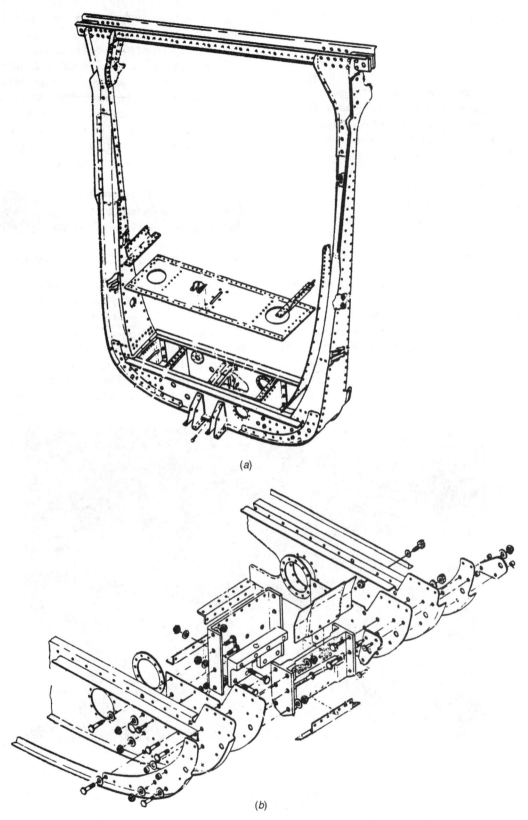

(a)

(b)

FIGURE 2-36 Attachment points for wings and landing gear on a light aircraft fuselage structure. (a) Wing-spar attachment. (b) Landing-gear attachment area. Exploded view of landing gear attachment area. (*Cessna Aircraft Co.*)

the flight crew is completely enclosed, it may be referred to as a cabin, flight deck, or crew compartment.

The following information contains some of the guidelines that the manufacturers must take into account when designing the pilot's position in an aircraft. The cockpit or flight deck must be designed so that the required minimum flight crew can reach and operate all necessary controls and switches. Some duplication of controls and instruments may be required, depending upon the type of flight operations being carried out and the regulations under which the aircraft was certificated. The structure must be watertight to the extent that no leakage of water into the crew cabin occurs when flying through rain or snow. Noise and vibration must be within acceptable limits through design and proper insulation. For commercial transport aircraft, a lockable door must be used between the crew compartment and the passenger compartment. The pilot's compartment of a turbine aircraft is shown in Fig. 2-37.

For complete information concerning the requirements for the flight deck, refer to FAR Parts 23 and 25.

Passenger Compartments

Passenger compartments must be designed and equipped to provide a maximum of comfort and safety for the passengers. This is particularly true for those aircraft certificated as air carriers for passenger service.

Passenger compartments must be adequately ventilated by means of a system that precludes the presence of fuel fumes and dangerous traces of carbon monoxide. The ventilation system is usually integrated with the heating and air-conditioning systems, and if the airplane needs to be pressurized, the pressurization system is also included in the design.

The seats in the passenger compartment of any aircraft must be securely fastened to the aircraft structure, regardless

of whether or not the safety-belt load is transmitted through the seat. Each seat must be equipped with a safety belt that has been approved by the Federal Aviation Administration (FAA).

Figures 2-38 through 2-40 show typical jet-transport passenger compartment features and dimensions.

Requirements for Doors and Exits

The doors and special exits for passenger-carrying aircraft must conform to certain regulations designed to provide for the safety and well-being of passengers. These requirements are established by the FAA in FAR Parts 23 and 25. Some of these requirements include the following.

Closed cabins on all aircraft carrying passengers must be provided with at least one adequate and easily accessible external door.

No passenger door may be located in the plane of rotation of an inboard propeller or within 5° thereof as measured from the propeller hub.

The external doors on transport aircraft must be equipped with devices for locking and for safeguarding against opening in flight either inadvertently by persons or as a result of mechanical failure. It must be possible to open external doors from either the inside or the outside even though persons may be crowding against the door from the inside. The use of inward-opening doors is not prohibited if sufficient measures are provided to prevent occupants from crowding against the door to an extent that would interfere with the opening of the door. The means of opening must be simple and obvious and must be so arranged and marked internally and externally that it can be readily located and operated even in darkness.

FIGURE 2-37 Class cockpit.

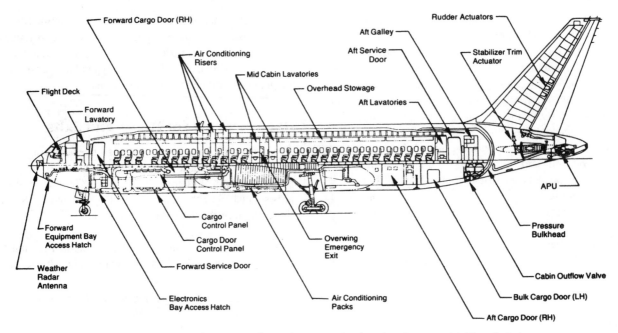

FIGURE 2-38   Fuselage arrangement of a Boeing 767-200. (*Boeing Commercial Aircraft Co.*)

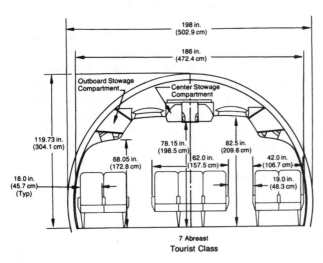

7 Abreast
Tourist Class

FIGURE 2-39   Fuselage passenger compartment cross section of a Boeing 767-200. (*Boeing Commercial Aircraft Co.*)

Reasonable provisions must be made to prevent the jamming of an external door as a result of fuselage deformation in a minor crash. External doors for a transport airplane must be so located that persons using them will not be endangered by the propellers of a propeller-operated airplane when appropriate operating procedures are employed.

Means must be provided for a direct inspection of the door-locking mechanism by crew members to ascertain whether all external doors for which the initial opening movement is outward, including passenger, crew, service, and cargo doors, are fully locked. In addition, visual means must be provided to signal to appropriate crew members that all normally used external doors are closed and in the fully locked position. This requirement is often met by placing indicator lights in the cockpit. These lights are operated by switches incorporated in the door-locking mechanisms.

All airplanes, except those with all engines mounted on the approximate centerline of the fuselage and a seating capacity of five or fewer, must have at least one emergency exit on the opposite side of the cabin from the main door. If the pilot compartment is separated from the cabin by a door that is likely to block the pilot's escape in a minor crash, there must be an exit in the pilot's compartment.

The requirements for emergency exits are determined in accordance with seating capacity of the airplane and are also set forth in FAR Parts 23 and 25.

Doors and Windows

The aircraft fuselage structure must be designed to allow for openings for doors, windshields, and windows. These openings require the use of special provisions in the area to allow the operational loads to flow around these openings. The number of doors range from only one on a light aircraft to several, such as are found on the Boeing 777 illustrated in Fig. 2-41.

The doors for light aircraft are usually constructed of the same materials used for the other major components. Typically, the main framework of a door consists of a formed sheet-metal structure to provide rigidity and strength, and to this framework is riveted the sheet-metal outer skin. The metal used is an aluminum alloy such as 2024-T4. The door frame is formed on a hydropress, stamp press, or drop hammer.

If the door is used for entrance to an upholstered cabin, the inside of the door is covered with a matching upholstered panel. Inside the door structure are located the door-latching and locking mechanisms. The upper portion of the door often contains a window made of a clear plastic similar

Passenger cabin soundproofing equivalent or
better than current industry standard.

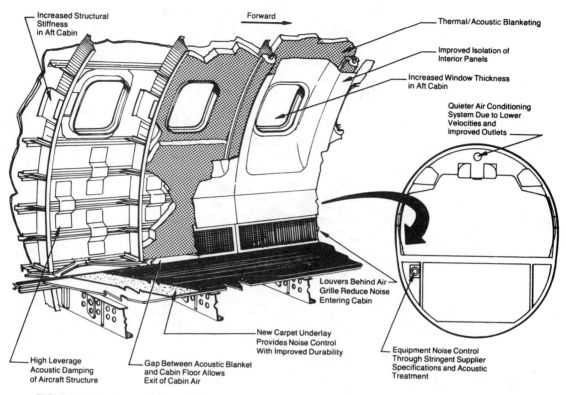

FIGURE 2-40 Acoustic control features of a Boeing 767-200. (*Boeing Commercial Aircraft Co.*)

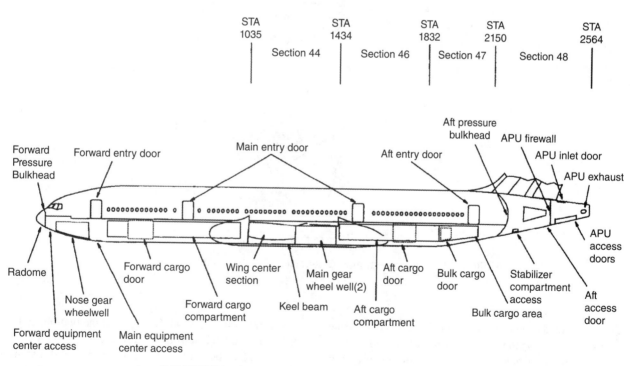

FIGURE 2-41 Access and entry doors on transport aircraft.

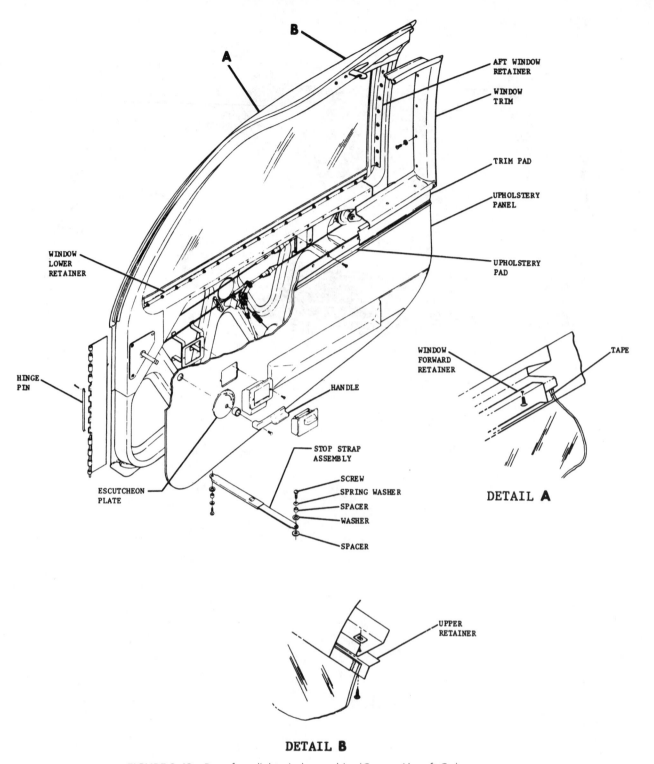

FIGURE 2-42 Door for a light-airplane cabin. (*Cessna Aircraft Co.*)

to that used for the other cabin windows. The edge of the plastic sheet used for the window is protected with waterproof material and is held in the frame by means of one or more retainers designed to form a channel. The retainers are secured by means of rivets or screws. A door for a light airplane is shown in the drawing in Fig. 2-42.

Many aircraft are equipped with stair-type doors. These doors are designed as one or two-piece doors. The one-piece door, shown in Fig. 2-43, is hinged at the bottom, and when the door is unlatched from the fuselage, it pivots outward and downward to become a stair for entering and leaving the aircraft. A two-piece version of this door has a top half, which is hinged at the top of the door frame. The top half is opened first and locked into position. The bottom half is then lowered and becomes the entrance stair. The stair-door is found on both pressurized and unpressurized aircraft.

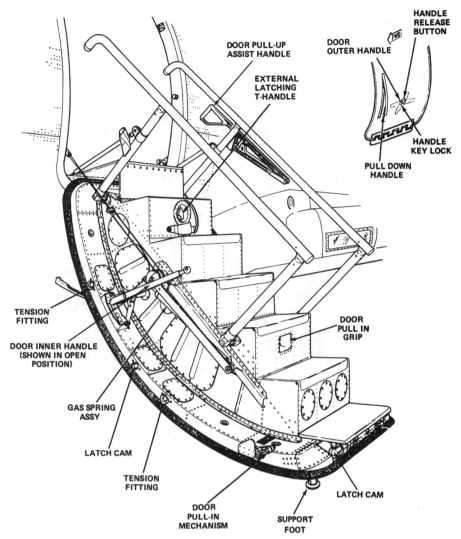

DOOR PULL-UP ASSIST HANDLE

EXTERNAL LATCHING T-HANDLE

HANDLE RELEASE BUTTON

DOOR OUTER HANDLE

FWD

HANDLE KEY LOCK

PULL DOWN HANDLE

TENSION FITTING

DOOR PULL IN GRIP

DOOR INNER HANDLE (SHOWN IN OPEN POSITION)

GAS SPRING ASSY

LATCH CAM

TENSION FITTING

DOOR PULL-IN MECHANISM

SUPPORT FOOT

LATCH CAM

FIGURE 2-43 A one-piece stair-door on a Canadair Challenger 601. (*Canadair Inc.*)

The doors for a pressurized aircraft must be much stronger and much more complex than the door for a light airplane. Typical of a door for the main cabin of a transport-category aircraft is that shown in Fig. 2-44. As shown in the drawing, the door consists of a strong framework of aluminum alloy to which is riveted a heavy outer skin formed to the contour of the fuselage. At the top and the bottom edges of the door are hinged gates that make it possible, in effect, to decrease the height of the door so it can be swung outward through the door opening.

The hinging and controlling mechanism of the door is rather complex in order to provide for the necessary maneuvering to move the door outside the airplane when loading and unloading passengers. For safety in a pressurized airplane, the door is designed to act as a plug for the door opening, and the pressure in the cabin seats the door firmly in place. To accomplish this, the door must be larger than its opening and must be inside the airplane with pressure pushing outward. This prevents the rapid decompression of the cabin that could occur if the door closed from the outside and the securing mechanism were to become unlatched.

Another type of entrance door being used in airliners is a vertical retracting door. This type of door stays inside the aircraft during operation and does not require the complex motions associated with the typical airliner hinged door. When being operated, the vertical door slides into the overhead area of the cabin, providing a clear access opening for entering and leaving the aircraft. Figure 2-45 shows this type of door.

Window openings are centered between the fuselage frames and are reinforced by aluminum doublers and a high-strength aluminum alloy window frame. Figure 2-46 illustrates where a sheet-metal doubler is used around the window of a transport aircraft. The windows for the passenger compartment of a large airplane must be designed and installed so there is no possibility that they will blow out when the compartment is pressurized. They must be able to withstand the continuous and cyclic pressurization loadings without undergoing a progressive loss of strength.

Cargo Compartments

Baggage and cargo compartments must be designed in such a manner that they will carry their approved capacity under all normal conditions of flight and landing without failure of any structural part. Each compartment must bear a placard

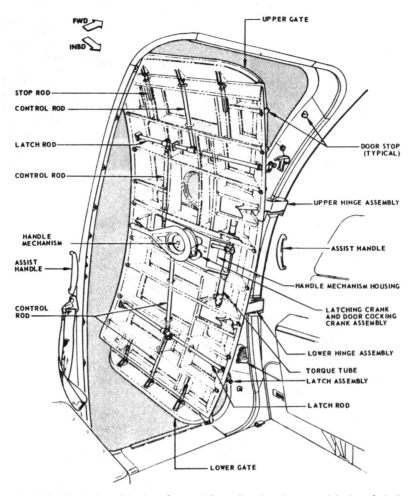

FIGURE 2-44   Main cabin door for an airliner. (*Boeing Commercial Aircraft Co.*)

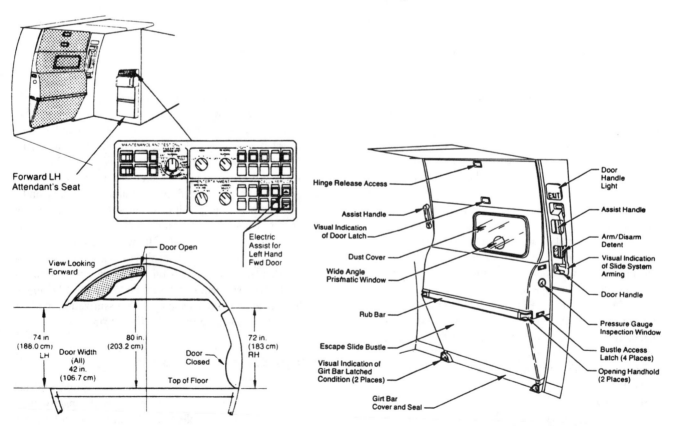

FIGURE 2-45   Main entrance door features on a Boeing 767-200. (*Boeing Commercial Aircraft Co.*)

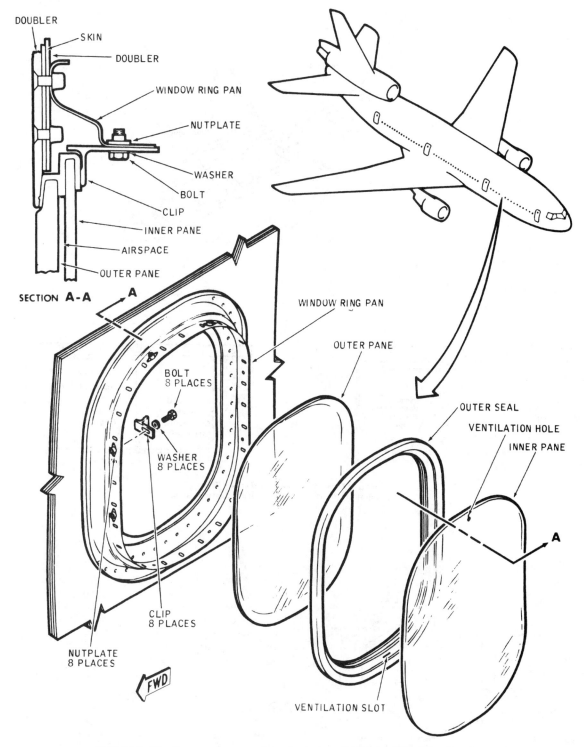

FIGURE 2-46 Passenger compartment window. (*McDonnell Douglas Corp.*)

stating the maximum allowable weight of contents as determined by the structural strength of the compartment and the flight tests conducted for certification of the aircraft. Suitable means must be provided to prevent the contents of baggage and cargo compartments from shifting.

Transport-type aircraft are provided with heavy decking and support structures, which enable them to carry heavy freight in addition to baggage for the passengers. Freight is generally "containerized"; that is, the freight is packed in containers designed

to fit the contour of the fuselage. Baggage is also containerized in many cases. Special loading equipment is employed to lift and roll cargo containers to move them into the airplane compartments. Powered rollers are built into the cargo decks to permit easy movement of the cargo containers.

When the cargo containers are in position, they are secured so they cannot shift during flight or upon landing. A typical containerized cargo arrangement is shown in Fig. 2-47. This type of aircraft can be configured for either increased passenger

Cockpits, Cabins, and Compartments   **45**

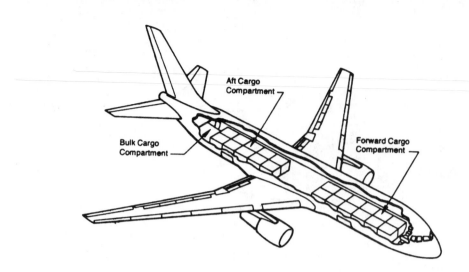

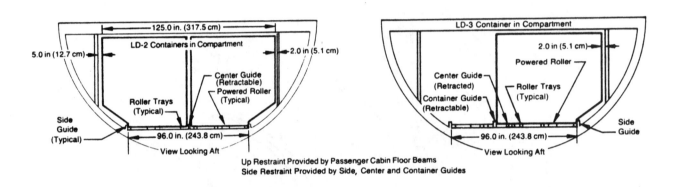

LD-2 Containers in Compartment

LD-3 Container in Compartment

Up Restraint Provided by Passenger Cabin Floor Beams
Side Restraint Provided by Side, Center and Container Guides

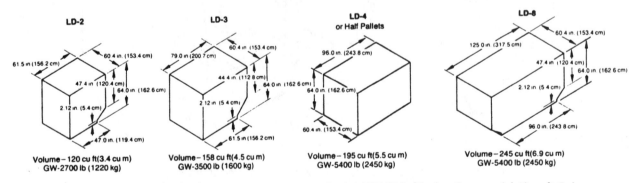

LD-2	LD-3	LD-4 or Half Pallets	LD-8
Volume—120 cu ft(3.4 cu m) GW-2700 lb (1220 kg)	Volume—158 cu ft(4.5 cu m) GW-3500 lb (1600 kg)	Volume—195 cu ft(5.5 cu m) GW-5400 lb (2450 kg)	Volume—245 cu ft(6.9 cu m) GW-5400 lb (2450 kg)

FIGURE 2-47 Containerized cargo arrangement on a Boeing 767-200. (*Boeing Commercial Aircraft Co.*)

or cargo loads, depending on the demand. Cargo compartments must meet strict requirements for fire protection, as set forth in FAR Part 25.

WINGS

The primary lifting surface of an aircraft is the wing. Although its shape may be widely varied, its function remains the same. The purpose of this section is to identify the basic types of wing design and construction that can be found in most aircraft.

Wings are attached to airplanes in a variety of locations, vertically and longitudinally. The terms *high-wing, low-wing,* and *mid-wing* all describe both airplane types and methods of wing attachment. Wing design can be divided into two types, *cantilever* and *semicantilever* (Fig. 2-48). A cantilever wing contains all its structural strength inside the

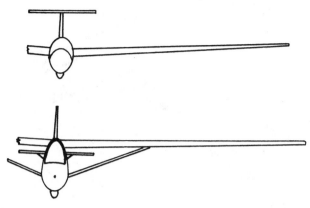

FIGURE 2-48 The top aircraft has a cantilever wing attachment, and the bottom aircraft uses a semicantilever design.

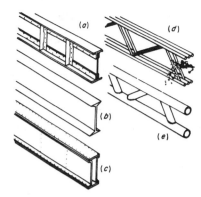

FIGURE 2-50 Metal-spar construction. (a) Built-up I-beam; (b) extruded I-beam; (c) built-up double-web spar; (d) welded-steel small-tubing structure; (e) welded-steel large-tubing structure.

wing structure and requires no external bracing. This type of wing is normally found on high-performance aircraft and on transports. The semicantilever design obtains its strength both by internal wing design and external support and by bracing from struts and wires. This type of wing is usually found on light aircraft designs and relatively slow aircraft designed to carry heavy loads. Note that all wings require spanwise members of great strength to withstand operational stresses, which are greatest during flight and upon landing. The shape and type of wing used on an aircraft are determined by considerations other than structural. The proposed usage of the aircraft will dictate the structural design of the wing. Small, low-speed aircraft have straight, nearly rectangular wings. For these wings the main load is in bending. Cost of building and maintenance also influences wing design. Airplane wing designs represent a careful balance of performance, cost, fabrication techniques, weight, and strength.

Basic Features of Wing Construction

Conventional wings are of three general types: *monospar, two-spar,* and *multispar.* True stressed-skin wings may have shear webs but no true "spars." The **monospar wing** has only one spar, the **two-spar wing** has two spars, as the name indicates, and the **multispar wing** has more than two spars. A **wing spar,** sometimes called a **wing beam,** is a principal spanwise member of the wing structure. The spars in the basic structure for the wings of a typical light aircraft are shown in Fig. 2-49. Metal spars may be made in a variety of designs. Examples of some metal-spar types are shown in Fig. 2-50. These spar shapes may be achieved through an extrusion process or may be assembled by riveting, bonding, and welding. Increasingly, all composite carbon fiber wing spars are utilized in new composite wings.

If a single spar is used, it is located near the midpoint of the airfoil chord line. If two spars are used, one is located near the leading edge and the other is located near the rear of the wing, usually just forward of the trailing-edge flight controls. Typical spar locations are shown in Fig. 2-49.

The spars include attachment points for connection to the fuselage and, if the wings are of a semicantilever design, strut

fittings located at a midpoint along the spar. If the landing gear or engine is mounted on the wing, the spars incorporate structural attachments for these components. Figure 2-51 illustrates a wing spar of transport aircraft.

A **wing rib,** sometimes called a **plain rib,** is a chordwise member of the wing structure used to give the wing section its shape and also to transmit the air loads from the covering to the spars. The ribs, which are placed at appropriate intervals along the wing span, also stabilize the spars against twisting and act as formers to hold the airfoil's shape. The rib may extend from the leading edge of the wing to the trailing edge, or it may extend only to the rear spar, as in the area ahead of a flap or aileron.

Typical metal ribs are shown in Fig. 2-51. The built-up rib is used in conjunction with metal spars and is riveted to them. The stamped rib with lightening holes and the stamped rib with a truss-type cross section are used with either metal or wood spars. The rib at the bottom of the picture is stamped in three sections and is usually riveted to metal spars. A new development in wing rib construction for transport aircraft is the machined rib. The rib structure is machined with a 5-axis CNC machining center from one large piece of aluminum. Figure 2-52 shows machined ribs inside a wing structure of transport aircraft. To assist in holding the shape of the wing, spanwise members called *stiffeners,* or *stringers,* are attached to the skin. These are usually found fairly closely spaced on the upper wing surface, which normally is in compression, stiffening the compression skin to resist the induced bending loads. Where stiffening demands become extreme, the skin may be reinforced by a corrugated panel or honeycomb sandwich instead of individual stringers. Stiffeners are illustrated in Fig. 2-53.

Typical **stressed-skin** metal construction is shown in Fig. 2-53. The skin of the wing is riveted to the ribs and stringers. It serves not only as a covering but also as a part of the basic structure of the wing. Most aircraft use aluminum as wing covering. The aluminum skin has high strength and is employed as a primary load-carrying member. The skin is quite strong in tension and shear and, if stiffened by other members, may be made to carry some compressive load. The thickness of wing skins varies widely, depending upon the local stresses

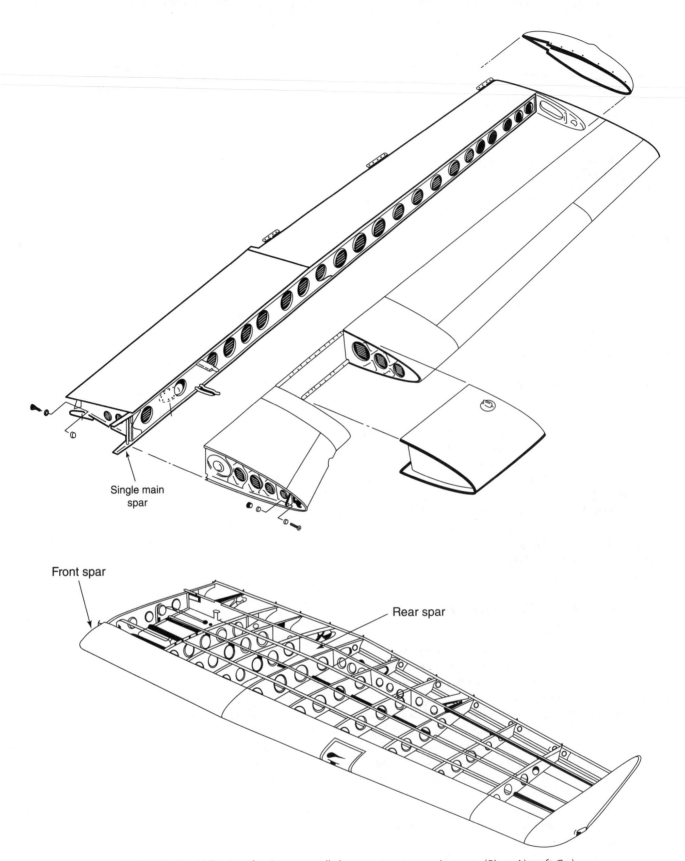

FIGURE 2-49 Light-aircraft wings normally have one or two main spars. (*Piper Aircraft Co.*)

Single main spar

Front spar

Rear spar

FIGURE 2-51   Ribs for metal wings.

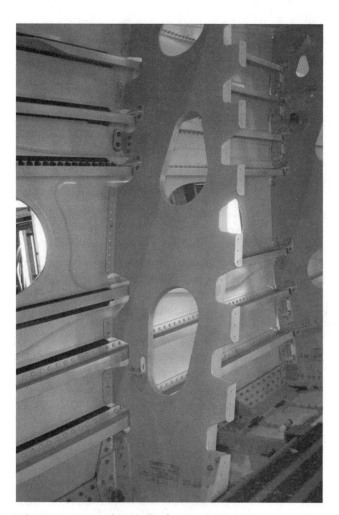

FIGURE 2-52   Machined rib of transport aircraft.

encountered in service. Thicknesses vary from as low as 0.016 in for small aircraft to as much as 0.60 in for the wings of transport aircraft. Metal skins are fairly rigid and hold their aerodynamic shape well; however, it often becomes necessary to make openings in the wing skins to allow access for service. These cutouts greatly weaken the wing, especially in its ability to resist twist, and doublers are required around the cutouts.

Fuel tanks are normally located in the inboard portion of the wing. The fuel tank may be a removable metal container, a bladder fuel cell located in a compartment, or an integral fuel tank. Figure 2-54 shows a removable tank made of metal and held in a covered bay of the wing between the spars and ribs. Bladder fuel cells are rubberlike bags, which are placed in structurally reinforced wing areas, usually between the spars. These bladder tanks, illustrated in Fig. 2-55, eliminate the need for a large access opening in the wing for installation and removal of the fuel tank. Bladder tanks are normally installed and removed by rolling them up and working them through a small opening in the wing surface. Integral fuel tanks are built into the basic wing structure and cannot be removed.

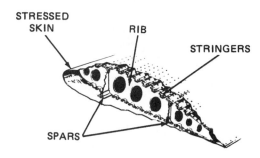

FIGURE 2-53   Reinforced stressed-skin wing construction.

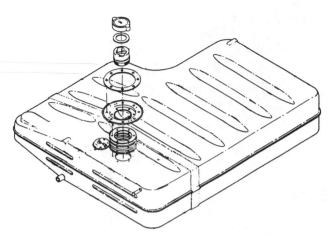

FIGURE 2-54 A typical metal fuel tank. (*Cessna Aircraft Co.*)

An exterior view of the wing for a light, high-wing monoplane is provided in Fig. 2-56. This view shows many of the details of the wing other than purely structural ones. The wing is of all-metal construction with a single main spar, formed ribs, and stringers. An inboard section of the wing, forward of the main spar, is sealed to form an integral fuel bay area. Stressed skin is riveted to the spars, ribs, and stringers to complete the structure. Note that the wing skin has many plates and doors to provide access to the interior for inspection and required service.

Since the wing shown in Fig. 2-56 is fully cantilevered, the wing-attachment fittings must be designed with adequate strength to carry the high stresses inherent in this type of construction. These attachment fittings are shown in details A and B in the illustration.

Wood Wing Construction

The wing spars for a wood wing must be made of aircraft-quality solid wood and plywood. Wood spars may be solid or may be built up, as shown in Fig. 2-57. Figure 2-58 shows

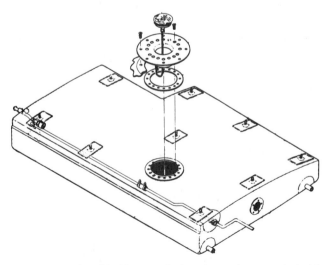

FIGURE 2-55 A bladder-type fuel tank uses fasteners to hold the cell in position inside the supporting structure. (*Cessna Aircraft Co.*)

a general design for a light-aircraft wing to be covered with fabric.

Ribs for wooden wings are shown in Fig. 2-59. The rib shown in Fig. 2-59(a) illustrates the use of compression ribs with heavy cap strips and solid plywood webs. Although the rib shown extends only from the front spar to the rear spar, a compression rib may be full size. The rib in Fig. 2-59(b) is a typical built-up or truss rib with cross bracing secured to the cap strips with plywood gussets. The cap strips are usually made of aircraft-grade spruce, which are first cut in straight strips and then bent to fit the rib configuration.

Figure 2-59(c) shows a built-up rib with a plywood web, which has been lightened by cutting large holes in the portions having the least stress. The ribs shown in Fig. 2-59(d) and (e) have continuous gussets, which eliminate the need for leading-edge strips. The only important difference between the ribs shown in Fig. 2-59(d) and (e) is the method of bracing. Observe the two names for the tie rods, or brace wires. The wires carrying drag loads are called **drag wires,** and those carrying the loads opposite drag are called **anti-drag wires,** as shown in Fig. 2-58.

Transport Aircraft Wings

The main frame of a modern transport wing consists of spars, ribs, bulkheads, and skin panels with spanwise stiffening members. Wing panel with stiffeners are shown in Fig. 2-60. The assembly of the structure may include the use of nonmetallic composite components and the bonding of metal structures as well as the use of conventional metal alloys and fasteners. The structural strength of the wing must be sufficient to carry its own weight along with the weight of the fuel in the wings, the weight of engines attached to the wings, and the forces imposed by the flight controls and landing gear. These stresses vary tremendously in direction and in magnitude during the aircraft's transitions from moving on the ground to flight operations to returning to the ground.

Figure 2-61 shows the basic structure of a modern transport wing. Transport wings consist of two or more main spars, with intermediate spars used between the main spars in some designs. These intermediate spars assist the main spars in carrying the operational loads. The front and rear spars provide the main supporting structure for fittings attaching the fuselage, engine pylons, main landing gear, and flight surfaces to the wing. Located between the spars are ribs, which, depending upon their design, may be used for purposes such as fuel bulkheads and the support of control surfaces as well as providing the airfoil shape of the wing. Figures 2-62 and 2-63 show typical locations and uses for transport wing ribs.

The auxiliary structure of the wing includes the wing tips, leading edge, and trailing edge. The leading edge of the wing incorporates leading-edge ribs, structural reinforcement members, and attachment points for components such as slats and leading-edge flaps. The trailing edge also incorporates structural members serving similar purposes to those

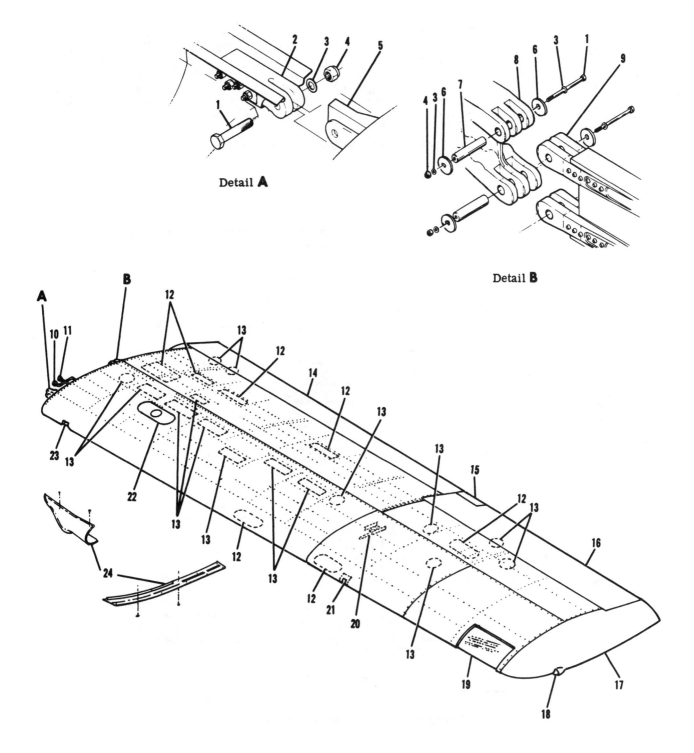

Detail **A**

Detail **B**

1. BOLT
2. FWD FUSELAGE FITTING
3. WASHER
4. NUT
5. FWD WING FITTING
6. RETAINER
7. DOWEL PIN
8. AFT FUSELAGE FITTING
9. AFT WING FITTING
10. FUEL VENT LINE
11. FUEL LINE
12. DOOR

13. PLATE
14. FLAP
15. AILERON TAB
16. AILERON
17. WING TIP
18. POSITION LIGHT
19. LANDING AND TAXI LIGHTS
20. PITOT TUBE
21. STALL WARNING UNIT
22. FUEL FILLER DOOR
23. RAM AIR INLET
24. FAIRING

FIGURE 2-56   Exterior view of an all-metal wing for a light airplane. (*Cessna Aircraft Co.*)

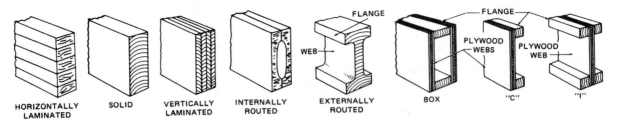

FIGURE 2-57 Wood spar sections.

HORIZONTALLY LAMINATED · SOLID · VERTICALLY LAMINATED · INTERNALLY ROUTED · EXTERNALLY ROUTED · FLANGE · WEB · BOX · FLANGE · PLYWOOD WEBS · "C" · PLYWOOD WEB · "I"

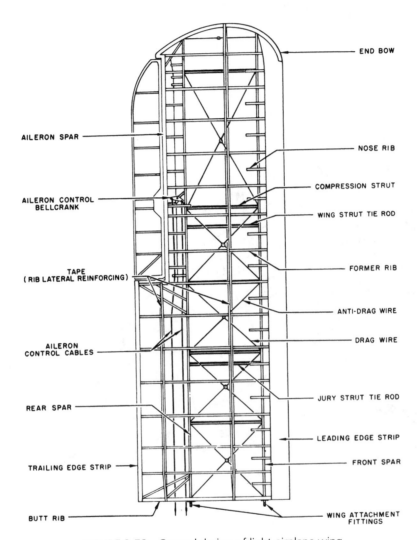

FIGURE 2-58 General design of light-airplane wing.

of the leading edge. The trailing-edge structure normally incorporates an extensive structure to carry and transmit the loads imposed by the operation of the flight controls such as flaps and ailerons.

The wing's internal structures are covered with large metal skin panels, which have spanwise stringers attached to achieve the desired structural strength of the wings. The leading edge and trailing edge are attached with permanent-type fasteners. The wing tips are removable for inspection and maintenance.

In many transports, the wing is constructed in three or more major assemblies, such as the left- and right-wing panels and a center-wing section. These sections are joined with permanent fasteners to form a one-piece wing, which is attached to the fuselage. Figure 2-64 shows wing sections being joined.

The fittings used to attach the fuselage, engine pylons, main landing gear, and flight-control surfaces to the wing are secured with interference-fit and close-tolerance fasteners and are not considered removable except for structural repair. An *interference-fit* fastener is a bolt, pin, or rivet that is slightly larger than the hole in which it is installed and must be pressed into place. The result is that there can be no "play," or clearance, between the fastener and the installation

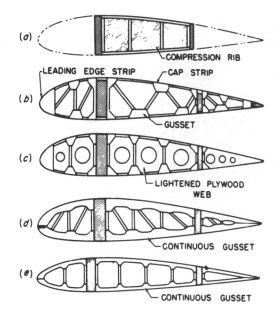

FIGURE 2-59 Ribs for wooden wings.

FIGURE 2-60 Wing panel with stiffeners.

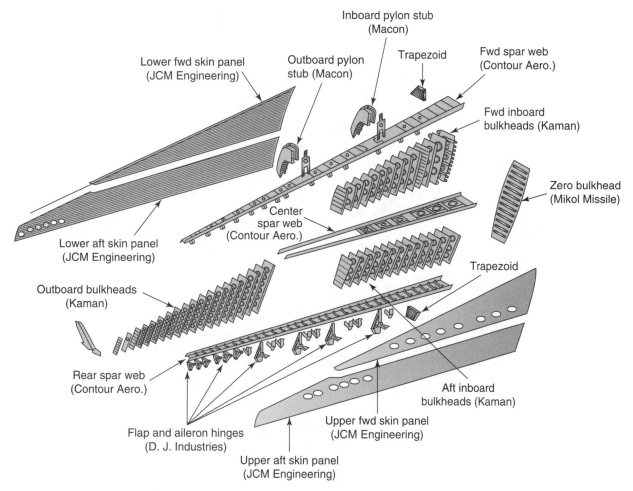

FIGURE 2-61 The internal structure of a modern transport aircraft.

regardless of expansion or contraction due to temperature changes. The fastener and the fitting essentially become one unit.

Fuel tanks are included in the basic wing structure so that the wing serves as a fuel tank. This is known as an **integral fuel tank design,** also referred to as a *wet wing.* The integral tanks are sealed with special compounds applied to fasteners and between components during assembly. The inside of an integral tank is shown in Fig. 2-65. Areas of the wings that do not contain fuel are termed *dry bays.*

Wings   **53**

FIGURE 2-62   The construction of a modern transport wing.

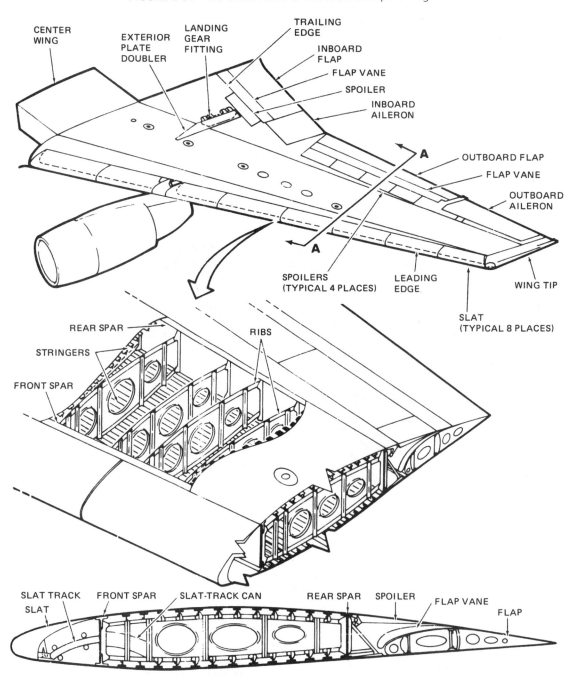

CENTER WING

EXTERIOR PLATE DOUBLER

LANDING GEAR FITTING

TRAILING EDGE

INBOARD FLAP

FLAP VANE

SPOILER

INBOARD AILERON

A

OUTBOARD FLAP

FLAP VANE

OUTBOARD AILERON

SPOILERS (TYPICAL 4 PLACES)

LEADING EDGE

WING TIP

SLAT (TYPICAL 8 PLACES)

A

REAR SPAR

RIBS

STRINGERS

FRONT SPAR

SLAT TRACK

SLAT

FRONT SPAR

SLAT-TRACK CAN

REAR SPAR

SPOILER

FLAP VANE

FLAP

FIGURE 2-63   Construction of the wing for the DC-10. (*McDonnell Douglas Corp.*)

FIGURE 2-64   Joining main wing sections. (*Airbus*)

TAIL AND CONTROL SURFACES

The stabilizers and the control surfaces of an airplane are constructed in a manner similar to the wings but on a much smaller scale. They usually include one or more primary members (spars) and ribs attached to the spars. The vertical stabilizer (fin) may be constructed as a part of the fuselage or may be a separate member that is both adjustable and removable. The tail section of an airplane, such as is shown in Fig. 2-66, includes the stabilizers, elevators, and rudder and is commonly called the *empennage*.

FIGURE 2-65   Sealing of the wing for the fuel tank. (*McDonnell Douglas Corp.*)

Horizontal Stabilizer

The horizontal stabilizer is used to provide longitudinal pitch stability to the aircraft and is usually attached to the aft portion of the fuselage. It may be located above or below the vertical stabilizer or at some midpoint on the vertical stabilizer. The stabilizer may be constructed of wood, steel tubing, sheet metal, or composite materials. The method of construction is similar to that used for wings, with spars, ribs, stringers, and a surface skin being used. Generally the rear spars are much heavier than the front spars, which is the opposite of most wing construction, because of the large loads imposed on the stabilizers from the attached rudder and elevators. Examples of different types of horizontal stabilizers are shown in Figs. 2-67 through 2-69. The horizontal stabilizer may be designed as a fixed surface attached to the tail cone or as a movable surface used to provide pitch trim.

If the stabilizer is designed to provide pitch trim, it normally is attached to the fuselage with a pivoting hinge as its rear spar. At the front spar is an attachment for a mechanical or hydraulic actuator controlled by the pilot to move the leading edge of the stabilizer up and down to change the trim of the aircraft (Fig. 2-70).

Vertical Stabilizers

The vertical stabilizer for an airplane is the airfoil section forward of the rudder; it is used to provide longitudinal (yaw) stability for the aircraft. This unit is commonly called the fin. The construction of the vertical stabilizer is very much like that of the horizontal stabilizer, and, as

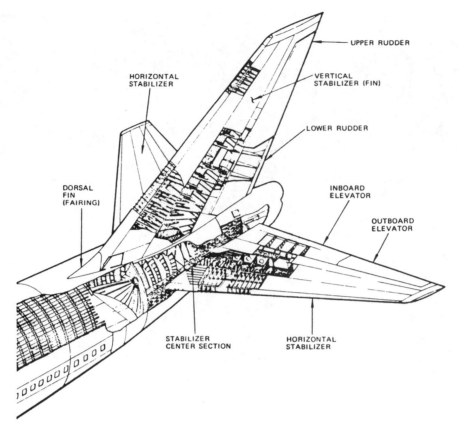

FIGURE 2-66 Aircraft tail section. (*Boeing Commercial Aircraft Co.*)

FIGURE 2-67 A welded-steel-tube horizontal stabilizer. (*Schweizer*)

mentioned previously, it may be constructed as an integral part of the fuselage. The rear structural member of the fin is provided with hinges for the support of the rudder. On many aircraft, a **dorsal fin** is installed immediately forward of the vertical stabilizer. The function of the dorsal fin is to improve the yaw stability of the aircraft and to provide a streamline fairing between the vertical stabilizer and the fuselage. A dorsal fin is shown in Fig. 2-66. Some aircraft, especially those equipped with floats or external

cargo pods, may also require the addition of a **ventral fin** on the bottom of the fuselage in the area below the vertical stabilizer, as shown in Fig. 2-71.

Control Surfaces

The **primary control surfaces** of an airplane include the ailerons, rudder, and elevator. **Secondary control surfaces** include tabs, flaps, spoilers, and slats. The construction of the control surfaces is similar to that of the stabilizers; however, the movable surfaces usually are somewhat lighter in construction. They often have a spar at the forward edge to provide rigidity, and to this spar are attached the ribs and the covering. Hinges for attachment are also secured to the spar. Where it is necessary to attach tabs to the trailing edges of control surfaces, additional structure is added to provide for transmission of the tab loads to the surface.

Control surfaces may be constructed of any combination of materials, with the more common combination being a sheet-metal structure (usually an aluminum alloy) covered with metal skin, composite structures, a steel structure covered with fabric, or a wood structure covered with plywood or fabric. Most of these types of construction are treated by some method to inhibit the deterioration of the structure and the covering and include drain holes to prevent water from becoming trapped inside the structure and causing the control surfaces to be thrown out of balance. Methods of joining the components include metal fasteners as well as adhesives and bonding agents. Most new transport aircraft are using composite and bonded structures, which include honeycomb

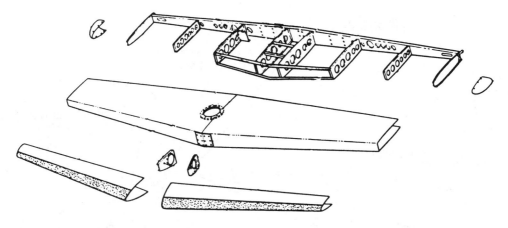

FIGURE 2-68 A typical light-aircraft aluminum-structure horizontal stabilizer. (*Cessna Aircraft Co.*)

internal components such as those shown in Fig. 2-72. These control surfaces are constructed with composite materials, including graphite, kevlar, and fiberglass. Such structures are often sealed from the atmosphere and therefore do not include drainage openings in their design.

LANDING GEAR

In any study of aircraft structures it is necessary to consider the landing gear, its construction, its arrangement, and the methods by which it is attached to the aircraft structure. The method of attachment to the aircraft structure is important because of the need for transmitting landing loads to the aircraft without overstressing portions of the aircraft structure.

Types of Landing Gear

Landing gear must be classified as either *fixed* or *retractable,* and it may also be classified according to arrangement on the aircraft.

A **fixed landing gear** is attached to the airframe so that it is held in a fixed position. Because it is always in the airstream, this type of landing gear generates a significant amount of aerodynamic drag. This arrangement is normally found on relatively low-speed aircraft and aircraft designed for simplicity of operation. A **retractable landing gear** is carried partially or completely inside the airframe structure to reduce drag. When necessary for landing, the gear is extended by some mechanism. The gear is retracted into the airframe after takeoff. An aircraft can have a combination of fixed and retractable landing gear.

The two most common configurations for landing gear are the *conventional* and *tricycle* arrangements. A conventional landing-gear configuration consists of two main landing gears located ahead of the aircraft center of gravity and a tailwheel located near the tail of the aircraft. With conventional landing gear the aircraft sits on the ground in a tail-low attitude. Tricycle landing gear consists of two main landing gears located aft of the center of gravity and a nose wheel located near the nose of the aircraft. With tricycle landing gear the aircraft sits in an approximately level flight attitude.

The vast majority of aircraft being manufactured today, including all transport-category aircraft, are of the tricycle configuration. Two views of the main landing gear for the Boeing 747 aircraft are shown in Fig. 2-73. Note that the main gear of this tricycle-gear-equipped aircraft includes two **body-gear struts** with four wheels each and two **wing-gear struts** with four wheels each. Thus there are 16 main wheels and 2 nose-gear wheels. Landing-gear units that incorporate four wheels with axles mounted on the ends of a main beam, as employed on the Boeing 747 and other large aircraft, are often referred to as "bogies." Figure 2-74 illustrates how the landing gear is attached to the aircraft structure. The attached fittings of the wing landing gear consist of the forward and aft trunnion bearing support fittings, the walking support fittings, and the side strut fittings. The trunnion fittings, which support the main landing gear strut, are installed on the rear wing spar and the landing-gear support beam. The attachment fittings are fabricated from aluminum alloy forgings.

A full discussion of landing gear is provided in Chap. 14, "Aircraft Landing-Gear Systems."

POWERPLANT STRUCTURES

Aircraft powerplants provide the source of thrust to move the aircraft through the air. Engines are also the source of vibration, heat, possible fire, and concentrated thrust and drag forces. The aircraft powerplant is usually enclosed in a housing called a *nacelle* and is attached to the aircraft by an engine mount. The engine must be isolated from the rest of the aircraft by a barrier known as a firewall.

Nacelles

An engine nacelle is used to enclose the engine in a streamlined housing to improve the aerodynamics of the aircraft, to support and protect the engine and its components, and

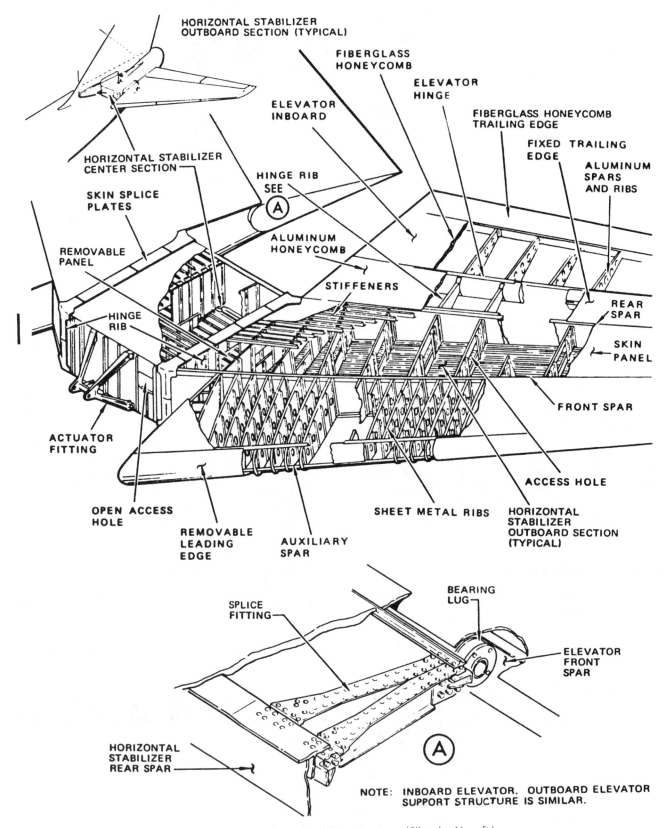

HORIZONTAL STABILIZER
OUTBOARD SECTION (TYPICAL)

FIBERGLASS
HONEYCOMB

ELEVATOR
HINGE

FIBERGLASS HONEYCOMB
TRAILING EDGE

FIXED TRAILING
EDGE

ALUMINUM
SPARS
AND RIBS

ELEVATOR
INBOARD

HORIZONTAL STABILIZER
CENTER SECTION

HINGE RIB
SEE
Ⓐ

SKIN SPLICE
PLATES

ALUMINUM
HONEYCOMB

REMOVABLE
PANEL

STIFFENERS

REAR
SPAR

HINGE
RIB

SKIN
PANEL

FRONT SPAR

ACTUATOR
FITTING

ACCESS HOLE

OPEN ACCESS
HOLE

SHEET METAL RIBS

HORIZONTAL
STABILIZER
OUTBOARD SECTION
(TYPICAL)

REMOVABLE
LEADING
EDGE

AUXILIARY
SPAR

SPLICE
FITTING

BEARING
LUG

ELEVATOR
FRONT
SPAR

HORIZONTAL
STABILIZER
REAR SPAR

Ⓐ

NOTE: INBOARD ELEVATOR. OUTBOARD ELEVATOR
SUPPORT STRUCTURE IS SIMILAR.

FIGURE 2-69   Horizontal stabilizer structure. (*Sikorsky Aircraft.*)

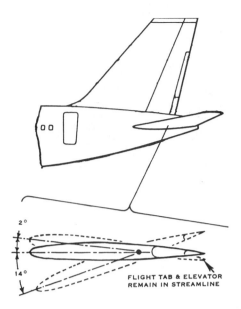

FIGURE 2-70 The horizontal stabilizer may be designed to give longitudinal trim. (*Bell-Boeing Joint Program.*)

FIGURE 2-73 Multiple landing gear for the Boeing 747.

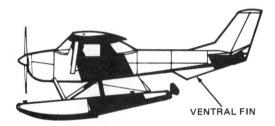

FIGURE 2-71 A ventral fin is added to the bottom of a fuselage to improve the aircraft's longitudinal stability. (*Edo.*)

to direct airflow into the engine for cooling and combustion and away from the engine for proper exhaust outflow.

The nacelle normally has removable segments, which allow for access to the engine for maintenance. The nacelle may also incorporate cowl flaps for control of engine cooling in a reciprocating engine, thrust reversers or reverser support structures for turbine engines, and auxiliary air doors for turbine engine operation at low altitudes and slow speeds. The nacelle may also incorporate components of several engine-monitoring systems such as fire detection and thrust indication.

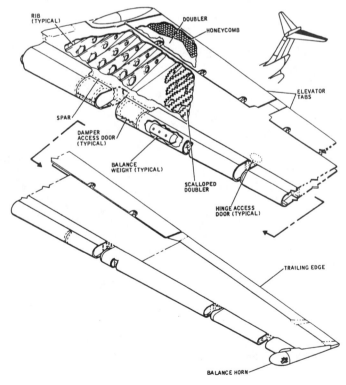

FIGURE 2-72 The elevator of a DC-9. (*McDonnell Douglas Corp.*)

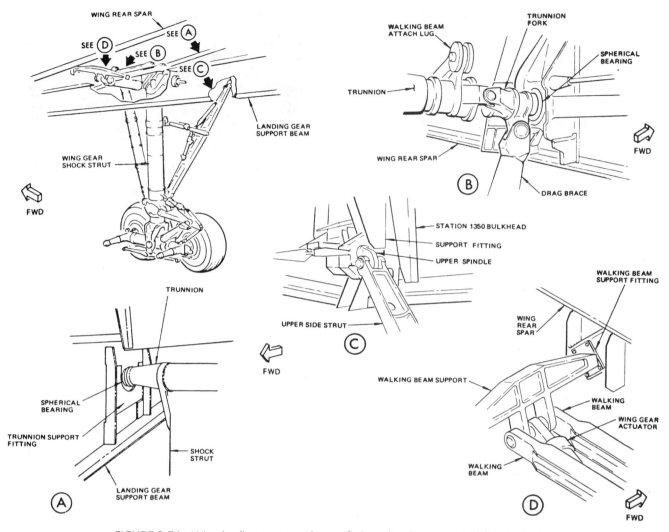

FIGURE 2-74 Wing landing-gear attachment fittings. (*Boeing Commercial Aircraft Co.*)

The nacelle may be constructed of sheet metal and/or composite components assembled through a combination of removable fasteners, rivets, and bonding.

An engine nacelle for a light airplane is shown in Fig. 2-75. Note that this nacelle is securely attached to and streamlined with the wing. Here the wing structure is reinforced to carry the extra weight and thrust of the engine.

The structure employed to attach an engine nacelle or pod to a wing or fuselage may be referred to as a **strut** or a **pylon.** Nacelle struts transmit engine loads to the wing by way of the engine mounts. The wing is reinforced at the point where the engine struts are attached. A nacelle for a Boeing 747 jet airliner is shown in Figs. 2-76 and 2-77. *Structural fuses,* illustrated in Fig. 2-77, permit the loss of an engine without the loss of a wing in cases of severe engine damage and resulting vibration and drag forces.

Firewalls

All engines, auxiliary power units, fuel-burning heaters, and other combustion equipment intended for operation in flight

FIGURE 2-75 Engine nacelle for an all-metal light twin airplane.

FIGURE 2-76 Engine nacelle for a Boeing 747.

Engine Mounts

An **engine mount** is a frame that supports the engine and attaches it to the fuselage or nacelle. Engine mounts vary widely in appearance and construction, although the basic features of construction are well standardized. Ideally, engine mounts are designed so that the engine and its accessories are accessible for inspection and maintenance.

Light-aircraft engine mounts may be of welded-steel tubing or aluminum alloy sheet metal. The construction will include some forged and metal-plate components. Two types of light-aircraft engine mounts are shown in Fig. 2-79. The vibrations that originate in reciprocating engines are transmitted through the engine mount to the airplane structure; therefore, mounts for such engines must be arranged with

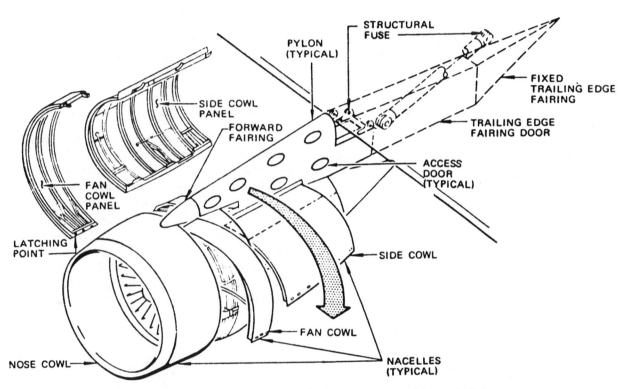

FIGURE 2-77 Turbine engine nacelle. (*Boeing Commercial Aircraft Co.*)

as well as the combustion, turbine, and tailpipe sections of turbine engines must be isolated from the remainder of the aircraft by means of firewalls, shrouds, or other equivalent means. Figure 2-78 illustrates a firewall used to protect the nacelle strut (pylon) and systems from exposure to high temperature and fire damage.

Firewalls and shrouds must be constructed in such a manner that no hazardous quantity of air, fluids, or flame can pass from the compartment to other portions of the aircraft. All openings in the firewall or shroud must be sealed with close-fitting fireproof grommets, bushings, or firewall fittings. Firewalls and shrouds are constructed of fireproof materials such as stainless steel, Inconel, or titanium. These provide protection against both heat and corrosion.

some sort of rubber or synthetic rubber bushings between the engine and mount-attaching structure for damping these vibrations. These bushings are often a part of the engine-mounting bracket and may be installed on the engine at the factory. The maximum vibration absorption is obtained when the mounting bolts are tightened so that the engine can move within reasonable limits in a torsional (rotating) direction but is restrained from any fore-and-aft movement. The torsional motion is then damped by the restraining action of the pads or cushions and the friction of the metal surfaces held by the bolts. If these bolts are too tight, the mounting structure tends to vibrate with the engine, which is undesirable. For this reason, technicians should always consult the manufacturer's service manual when tightening such bolts.

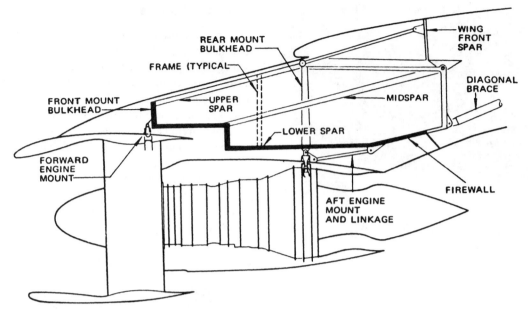

FIGURE 2-78 Engine firewall and pylon structure. (*Boeing Commercial Aircraft Co.*)

Most turbine-powered aircraft use forged metal mounts bolted to the airframe attachment points and to the engine, similar to the mount shown in Fig. 2-80.

Many transport aircraft have the engine attachments designed to allow for a quick removal and installation of a complete engine and mount assembly. This requires that fluid lines, electrical cables, control linkages, and engine-mounting attachments to the airframe be designed for easy separation at or near the firewall. This type of arrangement is referred to as a QEC (quick engine change) package.

A complete discussion of engine mounts is contained in the text *Aircraft Powerplants*.

Cowling and Fairings

The cowling and fairings are generally similar, but they differ in detail and to some extent in function. The cowling usually consists of detachable sections for covering portions of the airplane, such as engines, mounts, and other parts where ease of access is important. The engine cowling is designed to provide proper aerodynamic flow of air into the engine and to provide access to various parts of the engine. The cowling affords protection to the engine and also aids in streamlining the area covered.

Engine cowlings normally consist of formed sections, made of sheet metal, aluminum honeycomb, composite, or other fireproof material, which are attached to each other and

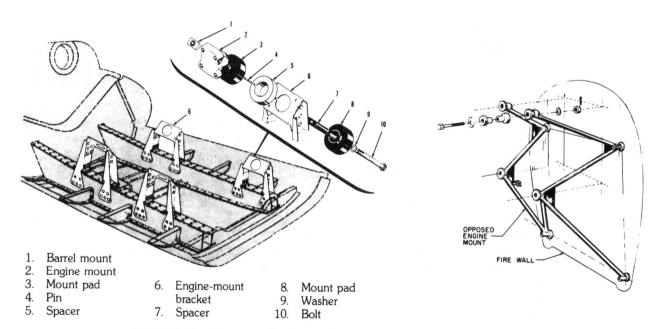

1. Barrel mount
2. Engine mount
3. Mount pad
4. Pin
5. Spacer
6. Engine-mount bracket
7. Spacer
8. Mount pad
9. Washer
10. Bolt

FIGURE 2-79 Two types of light-aircraft engine mounts. (*Cessna Aircraft Co.*)

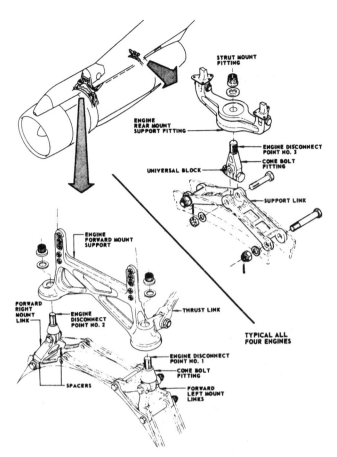

FIGURE 2-80 Engine mounts for wing-mounted turbine engine.

to the supporting structure with quick-disconnect fasteners and screws. The cowling area commonly encloses the engine completely except for the inlet and outlet areas, which provide a path for engine-cooling airflow. A turbine engine cowling is illustrated in Fig. 2-88.

Cowlings on reciprocating engines often incorporate cowl flaps so that the pilot can control the size of these cooling openings and thus the temperature of the engine. Cowlings are designed to minimize the chance of liquids collecting in the cowling area by providing paths through which the liquids can flow to drain openings or to the airflow openings. This reduces the chance of fire by removing fuel and oil and reduces any weight problem that might develop from any liquid collecting in the cowl. These fluid paths are positioned so that the liquids will not come into contact with the engine exhaust system or exhaust gases downstream of the exhaust system.

Cowling sections are normally of a size and weight convenient for one person to carry. When they are removed for inspection and maintenance work, they should be stored in numbered racks corresponding to numbers placed either temporarily or permanently on the cowling sections so that they can be replaced in the proper order.

A fairing is used principally to streamline a portion of an airplane, although it may protect some small piece of equipment or merely improve the appearance. Fairing units may be composed of several small sections or may be stamped or formed into one large section. The sections may be removable

and attached in the same manner as a cowling, or they may be permanently attached to the aircraft.

Cowlings and fairings should be handled with care so that they are not bent or broken. Long strips of large sections are sometimes not rigid enough to support their own weight. Small parts made of light-gauge material are easily damaged during maintenance work. Chafing strips, made of either fabric or fiber, may be used between pieces of installed cowling. These strips require replacement if they are worn to the point where metal parts rub together. Attachment devices must be inspected to be sure that they function easily and securely. When they become badly worn or loose, they should be replaced.

ROTORCRAFT STRUCTURES

Many of the aircraft components used on airplanes are common to the helicopter. Helicopters have a rotor system driven by the engine, and through this rotor system the helicopter generates lift and propulsive thrust. Most helicopters have only one main rotor and rely on an auxiliary rotor to overcome the torque of the main rotor and maintain directional control. A complete discussion of rotorcraft configurations and theory of flight is included in the text *Aircraft Basic Science.*

The flight and landing forces to which a helicopter are subjected are somewhat different from those of a fixed-wing aircraft. On a fixed-wing aircraft, the lift forces are concentrated on the wing and the thrust forces are centered on the powerplants. On a helicopter, both the lift and thrust forces are concentrated on the main rotor. Helicopters also encounter problems with continuously high levels of vibration due to all the rotating components, which contributes to metal fatigue.

For many years rotorcraft utilized the same methods and materials of construction as fixed-wing aircraft. However, in recent years, as the popularity of helicopters has grown, differences in the materials and construction methods utilized have become more pronounced. Problems with fatigue, vibration, and weight have led helicopter manufacturers to adopt composite-construction technology more rapidly. Early helicopter structures comprised primarily steel-tube construction. Tubular construction is very strong but has a poor strength-to-weight ratio, is very labor-intensive to build, and has poor aerodynamic qualities. Helicopter construction evolved to utilizing aluminum semimonocoque and monocoque designs. Sheet-metal construction has better aerodynamic qualities and a higher strength-to-weight ratio. A third type of construction utilizes composite materials such as graphite, kevlar, and fiberglass combined with bonded structures. Many helicopters today utilize a wide variety of materials to gain the advantages of each. Such an example is illustrated in Fig. 2-82.

Advanced composites are used more and more in commercial helicopters: The secondary structure of the Sikorsky Blackhawk is made mostly of composite materials, and the new Boeing 360 is virtually all composites. The use of composites can be attributed to better performance, lighter

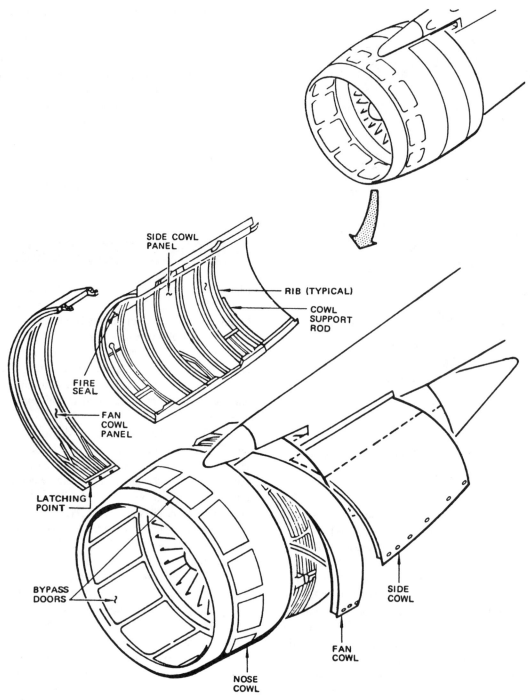

FIGURE 2-81 Engine cowling. (*Boeing Commercial Aircraft Co.*)

weight, and lower cost. Bell-Boeing's V-22 Osprey, a new type of aircraft that can take off vertically and fly forward by rotating its propeller, is constructed almost entirely of composites to enable it to meet critical design and weight criteria. The V-22 illustrated in Fig. 2-83 is the first production aircraft whose airframe is fabricated almost entirely of composite materials, primarily a graphite/epoxy solid laminate structure. Approximately 6000 lb of graphite epoxy are used in the 13,000-lb aircraft structure. Only about 1000 lb of metal are used in the entire aircraft, and that comprises mainly fasteners and copper mesh used in outer surfaces

for lightning protection. Aside from their advertised benefits, such as increased stiffness and strength with reduced weight, if composites are placed efficiently (concentrating material only where strength is needed), most structural parts can be fabricated less expensively than those made from metal.

One of the most unique features of the V-22 is the pylon support assembly. It allows the entire nacelle and rotor system to pivot at the end of the wing torque-box while supporting the rotor system, engine, and transmission. Because of their structural complexity, the pylon support and spindle

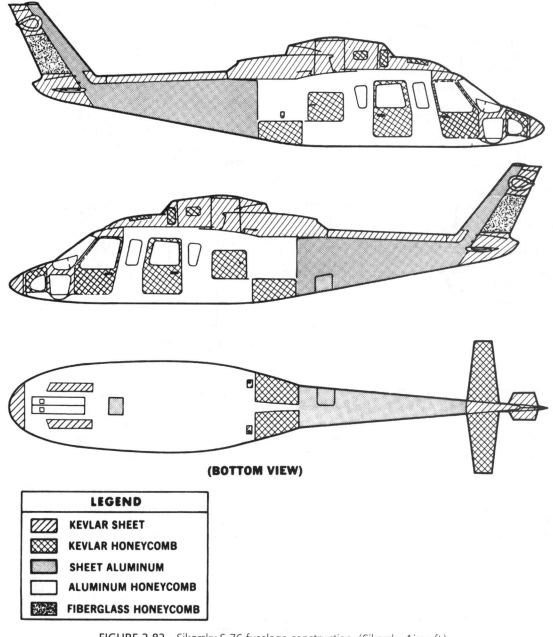

LEGEND

▨	KEVLAR SHEET
▨	KEVLAR HONEYCOMB
▢	SHEET ALUMINUM
▢	ALUMINUM HONEYCOMB
▨	FIBERGLASS HONEYCOMB

FIGURE 2-82 Sikorsky S-76 fuselage construction. (*Sikorsky Aircraft.*)

assemblies present strong challenges to state-of-the-art manufacturing technologies. The pylon support assembly comprises more than 400 plies of carbon/epoxy fabric and unidirectional tape. A tape-laying machine, capable of laying composite material on a contoured surface, has been incorporated into the system. Most of the older tape-laying machines are used only in the fabrication of flat components because of constraints imposed by their hardware and/or software. However, a combination of recent advances in computerization and innovative machine concepts has enabled the new tape layers to minimize those limitations. For example, some machines contain ten axes of control interfaced to special computer software. When the machine is programmed to lay tape over an area in a certain sequence, it does so automatically, repetitively, and flawlessly.

The process of laying composite fibers in a curved surface without disturbing the fibers is known as *natural-path tape laying*. The natural path is the path that tape tends to follow to maintain equivalent stress to all fibers. To produce composite layups that are not deformed, equivalent stress to all fibers is absolutely necessary.

Fuselage

The major assemblies of a single-rotor helicopter are the forward fuselage and the tailboom. Figure 2-83 shows the major components of a Sikorsky S-76 helicopter. The forward fuselage consists of a lower fuselage (tub) and an upper fuselage (cabin). The lower fuselage includes the integral fuel tanks and provides structural support for the flooring, retractable landing

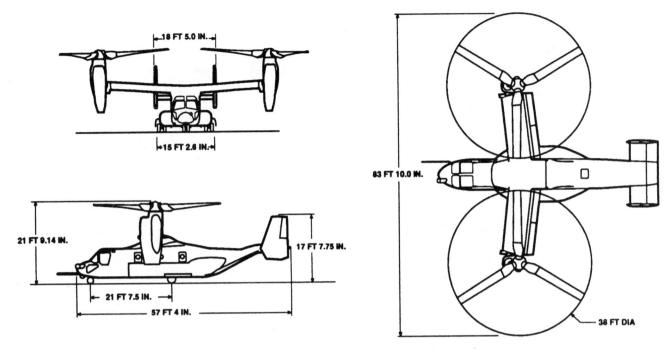

FIGURE 2-83   V-22 Osprey. (*Bell-Boeing Joint Program.*)

gear, and cargo hook (if installed). Included in the upper fuselage are the cockpit, cabin, two engine compartments enclosed by a firewall and cowling, a transmission compartment and main transmission fairings, and a baggage compartment. An electronics compartment is in the nose and an electrical compartment is aft of the baggage compartment. A flight-control enclosure is in the center aft portion of the cockpit. The cockpit, cabin, and baggage compartment are entered through doors on both sides of the fuselage. The tail cone, attached to the rear of the fuselage, supports the intermediate and tail-gear

box, drive shaft, tail rotor and its associated control system, horizontal and vertical stabilizers, and lights.

The primary purpose of the tailboom is to support the stabilizing surfaces and tail rotor of the helicopter. The tail rotor is used to control the yaw of the aircraft and to counteract the torque of the engine-transmission system. A vertical stabilizer provides stability in forward flight and relieves the tail rotor of some of its workload in forward flight. As shown in Fig. 2-84, a tail skid may project below the tail rotor and vertical stabilizer and is used to protect the tail rotor

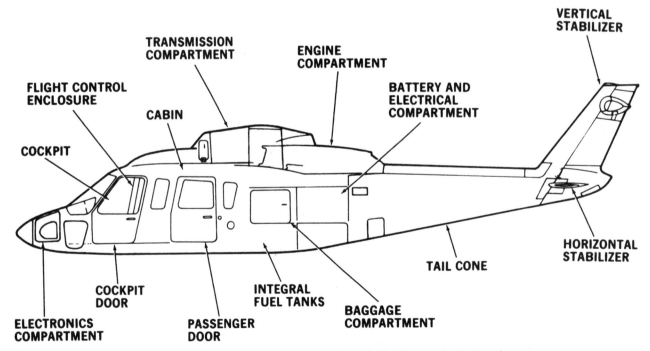

FIGURE 2-84   Major helicopter assemblies. (*United Technologies Corp.*)

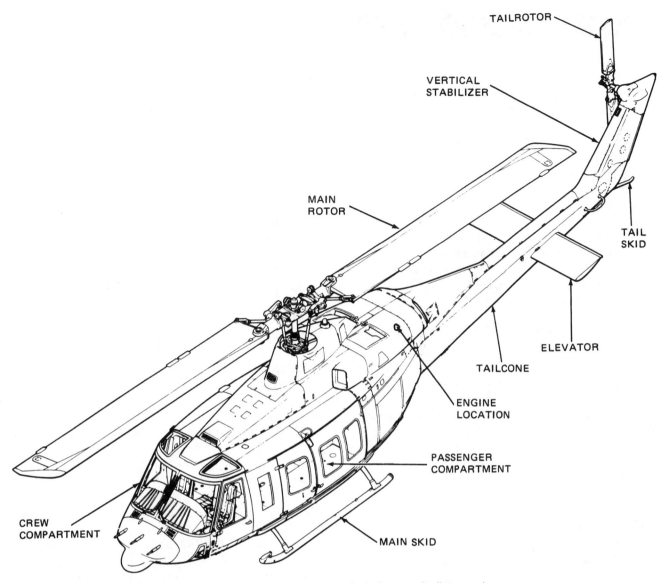

FIGURE 2-85   The components of a helicopter. (*Bell Textron.*)

from ground contact. A horizontal surface is found on most helicopters. This surface may be a fixed horizontal stabilizer, used to improve the pitch stability of the helicopter, or it may be a movable elevator, used to improve the pitch control of the craft.

Landing Gear

The function of the landing gear on a helicopter under normal conditions is to provide a support on which the helicopter can rest. The main structural requirement for the landing gear is to absorb the energy due to descent. Although landing gear is traditionally constructed of aluminum, many manufacturers are switching to crushable graphite composite material in landing-gear construction to provide additional energy absorption beyond that provided by metal. Helicopters can have landing gear that is either fixed or retractable. The fixed landing gear may consist of a pair of skids, wheels, or floats. The retractable landing gear consists of retractable wheels similar to fixed-wing aircraft.

The fixed-skid systems, shown in Fig. 2-85, may or may not incorporate a shock-absorption system. If no shock absorber is used, then the skid structure is mounted directly on the bottom of the helicopter fuselage structure. If a shock-absorber is used, it usually includes an air-oleo cylinder near each end of each skid. Both the wheel and skid landing gear have advantages and disadvantages. The skid gear shown is less complex than wheels and is very durable, but is difficult to maneuver the helicopter on the ground. For the wheel-type landing gear, an air-oleo shock absorber is normally used.

Retractable landing gear on a helicopter is similar in basic design to an airplane retraction system. Helicopters can have floats installed in place of their normal landing gear. This type of conversion is most often found on helicopters equipped with skids, due to the simplicity of the conversion. Such a conversion allows the helicopters to operate equally well from land or water.

1. Define the term *load factor*.
2. Into what categories are aircraft type-certificated?
3. What is meant by the term *ultimate load*?
4. Flight loads are typically divided into what two types?
5. What are the principal load-carrying structures of an airplane?
6. List five stresses imposed on an aircraft structure.
7. Compare *tension* and *torsion*.
8. Explain the purpose of station numbers for an airplane.
9. List four types of station numbers.
10. Where is the butt line of an aircraft?
11. How are fuselage stations indicated as being ahead of or behind the datum line?
12. Explain the purpose and principle of zoning for large aircraft.
13. How are numbers used to identify the major zones?
14. Name three types of fuselages classified according to the method by which stresses are transmitted to the structure.
15. Describe the difference between a Pratt truss and Warren truss.
16. What is the term for the longitudinal (fore-and-aft) structural members of a semimonocoque fuselage?
17. Describe the vertical structural members in a semi-monocoque fuselage.
18. What carries the principal loads in a monocoque fuselage?
19. What carries the principal loads in a semimono-coque fuselage?
20. What is a fail-safe feature in aircraft construction?
21. Discuss the advantages of bonded construction.

22. Why are composite materials currently being utilized on aircraft structures?
23. What is the primary structural design difference between a cantilever wing and a semicantilever wing?
24. Name the principal load-carrying members in a light-aircraft wing.
25. Name the principal chordwise carrying member in a light-aircraft wing.
26. What is the purpose of drag and antidrag wires?
27. How is a metal wing sealed so it may be used as a fuel tank?
28. Describe the construction of a typical stabilizer for a light airplane.
29. What is the purpose of a vertical stabilizer?
30. Name the primary and secondary control surfaces of an airplane.
31. Describe the difference between conventional landing and tricycle landing gear.
32. What type of landing gear is most commonly employed on transport category aircraft?
33. What is the purpose of an engine nacelle?
34. What is the purpose of a structural fuse on an engine pylon?
35. Of what type of materials should firewalls be constructed?
36. What is the primary purpose of engine cowling?
37. How are the forces acting on a helicopter different from those on a fixed-wing aircraft?
38. What are the advantages of composite construction on helicopters?
39. What are the functions of the tailboom on a helicopter?
40. What types of landing gear may be used on helicopters?

Fabrication and Repair of Wood Structures 3

INTRODUCTION

Wood aircraft structures combine many of the attributes associated with metal and composite structures, such as light weight, low cost, and high strength, while requiring only the minimum of special equipment for proper maintenance and repair. For this reason, many of the lighter aircraft that have been produced have made use of wood primary and secondary components, such as wing spars, ribs, and control surfaces. A great many of these aircraft are still in operation, and a few designs are still in production using wooden structural components. Figure 3-1 shows aircraft which incorporate wood in their structures.

AIRCRAFT WOODS

There are two principal types of wood, *hardwoods* and *softwoods*, and all woods may be classed as one or the other. The distinction between hardwoods and softwoods is not based on the "hardness" of the wood but rather on the cellular structure of the wood.

Softwoods

Softwoods come from trees that have needlelike or scalelike leaves and are classified as evergreens or conifers. The wood of these trees is composed primarily of fibrous cells and has a smooth, even appearance when cut in cross section. Softwood has a high strength-to-weight ratio, which makes it a very desirable structural material for use in aircraft construction. Softwood is usually used as a solid wood for spars, cap strips, and compression members and as a veneer for plywood cores.

Woods included in the softwoods used in aircraft are Sitka spruce, Douglas fir, Port Orford white cedar, and western hemlock. Sitka spruce is the wood used as a reference material to establish the suitability of other softwoods for use in aircraft construction and repair.

Hardwoods

Hardwoods come from trees that have broad leaves and are classified as deciduous because they lose their leaves each fall. The wood of these trees is composed of a mixture of large cells, causing pores in the wood, distributed among the smaller fibrous cells. These pores are often visible when the wood is cut smoothly. Hardwoods are generally heavier than softwoods and are used where their strength advantage makes the extra weight acceptable over the softwoods. Hardwoods are commonly used as solid wood for support blocks and tip bows and as veneers for the facing and core material of plywood.

Hardwoods commonly used in aircraft structures include mahogany, birch, and white ash.

Terminology for Woods

Even though the aircraft technician may not have occasion to use standard terminology for woods very often, it is considered desirable to understand the terms illustrated in Fig. 3-2.

FIGURE 3-1   Airplanes having wood structures.

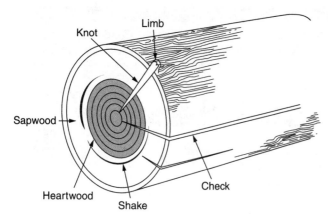

FIGURE 3-2 Nomenclature for woods.

Annual rings Concentric layers of wood that can be seen at the end of a tree trunk that has been cut perpendicular to its length. The rings are caused by the different rates of growth during each year as the seasons change.

Bark The external covering of a tree trunk or branch.

Check A radial crack that cuts across the grain lines.

Compression failures Wrinkles or streaks across the grain line caused by mechanical stress on the wood after the annual rings had grown. For detection, compression failures may require close examination with a light source aimed almost parallel to the grain structure.

Compression wood Deformed grain structure in the wood caused by mechanical stress on the tree, such as supporting the weight of a heavy limb, during growth. It is characterized by wide annual rings when compared to the normal size of the tree's growth rings.

Decay A biological growth living off of the wood and causing a breakdown in the strength of the wood. Discoloration may also be present.

Grain The lines in wood caused by the annual rings. Grain also refers to the direction of the wood fibers.

Hard knot A knot that is firmly embedded in the wood and shows no sign of coming loose.

Heartwood The center part of a tree trunk, which is dead and carries no sap. This part of the tree serves only to support the tree.

Knot The base of a limb inside the tree. A knot will cause a deviation of the grain lines as they form around the knot.

Mineral streaks Coloring in the wood caused by minerals in the soil or other naturally occurring agents during the tree's growth.

Moisture content The weight of water contained in a wood sample compared to the weight of the wood sample if all the water was removed from it.

Pin knot A knot resulting from the growth of a twig.

Pitch pocket Voids between the annual rings that contain free resin. These pockets are usually relatively small in cross section and are not to be confused with shakes, which can be extensive.

Sapwood The part of a tree that is alive or partially alive and carries sap. Sapwood begins immediately under the bark and extends to the heartwood. The sapwood is often lighter in color than the heartwood.

Shake A separation between the annual ring layers.

Spike knot A knot that was cut through parallel to the limb during the sawing operation such that the knot runs across the board.

Split A crack in the wood resulting from rough handling.

Springwood The soft, light-colored part of the annual ring. This wood ring is normally wider than the summerwood ring because of the rapid tree growth during the spring season.

Summerwood The harder and usually darker part of each annual ring. This wood is formed during the slow summer season growth.

Evaluating Wood for Aircraft Use

The primary requirement for wood that is to be used in aircraft structures is that it be sufficiently sound and of such quality that it will provide the strength required for the structure. It has been determined through research that Sitka spruce is generally the best wood for use in aircraft structures because of its combination of lightness, strength, stiffness per unit weight, and toughness when compared to other species. Because of specific requirements, other species may be used due to unique qualities within the general evaluation criteria. The following paragraphs discuss the wood characteristics that the technician must consider when selecting wood of the desired species.

There are two classifications of water in wood, *free water* and *cell water*, as shown in Fig. 3-3. **Free water** is the water that flows up and down the tree carrying nutrients. **Cell water** is water trapped within the walls of the wood cells' structure and is part of the structure of the tree. Aircraft woods are *kiln-dried* to remove all the free water and a portion of the cell water, so the resulting moisture content is between 8 and 12 percent. A moisture content above or below this range is not considered acceptable.

Kiln-dried wood is dried by placing the boards of fresh-cut wood in a precisely controlled oven and raising the

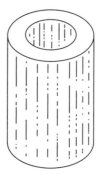 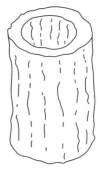

FIGURE 3-3 A wood cell before (*left*) and after (*right*) drying. When the cell water is removed, the wood shrinks and becomes stronger.

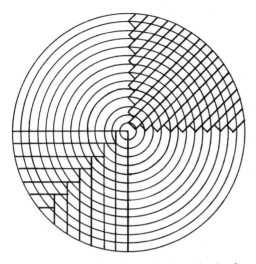

FIGURE 3-4 This illustration shows two methods of cutting a log to obtain quarter-sawed wood.

temperature to a specified level for a specified period of time. Not only does this process reduce the moisture content to the desired level, but it also kills the insects and decay-producing organisms that may have infected the wood.

The specific gravity of aircraft woods should be from 0.34 to 0.40, depending upon the type of wood. Aircraft spruce should have a specific gravity of approximately 0.36.

The grain structure of the wood must be examined to determine if the wood has been properly cut, if the grain lines are sufficiently straight, and if there is a minimum number of annual rings per inch.

The way a board is cut is important, because this affects the strength of the piece of wood and the shrinkage characteristics of the wood. Ideally, aircraft solid wood should be cut so that the annual rings are parallel to the narrow dimension of the board. This is known as **quarter-sawed**, or **edge-sawed**, wood and is illustrated in Fig. 3-4. For practical purposes, a board is considered to be quarter-sawed if the annual rings are at an angle no greater than 45° to the narrow dimension.

The slope of a grain line is determined by looking at the side of a board and noting the angle that the grain line makes with the edge of the board. Ideally, the grain lines will be parallel to the edge of the board, but a deviation or slope of 1:15 is allowed. This means that a grain line starting at the edge of the board may not move more than 1 in [2.54 cm] from the edge of the board when it is 15 in [38.1 cm] from the starting point, as shown in Fig. 3-5.

The number of annual rings per inch, or grain count, is another criterion that must be checked for aircraft-quality wood. The grain count is taken by counting the number of grain lines (annual rings) per inch on the sample. This is best done by looking at the end of a board and measuring a 1-in [2.54-cm] line perpendicular to the annual rings. The minimum grain count for most softwoods is six rings per inch [2.54 cm], with the exception of Port Orford white cedar and Douglas fir, which must have a minimum of eight rings per inch [2.54 cm].

When evaluating wood, the following defects are not acceptable: checks, shakes, splits, decay, compression wood, compression failure, and spike knots. Defects that might be acceptable, depending on their size, location, and condition, are hard knots, pin knot clusters, mineral streaks, and irregularities in grain direction. Evaluation criteria for these defects are given in Table 3-1.

Specifications for aircraft woods as given in Federal Aviation Advisory Circular (AC) 43.13-lB provide that certain minor defects, such as small solid knots and wavy grain, may be permitted if such defects do not cause any appreciable weakening of the part in which they appear. As a practical rule, aircraft technicians should not use any wood about which they have doubts. The safe policy is to use wood that is straight-grained, free from cracks, knots, or any other possible defect, and guaranteed as aircraft-quality.

Wood Substitutions

When repairing or rebuilding wood components, species substitution may be allowed if the structural strength of the component is not reduced. Table 3-2 shows types of wood that may be considered for substitution, with spruce being the reference wood. Note that the choice of a substitution may have to take into account changes in size, different bonding qualities, and different working qualities.

Plywood

Plywood is composed of an uneven number of layers (plies) of wood veneer assembled with the grain of each layer at an angle of 45° to 90° to the adjacent layers. The outside layers are called the **faces**, or the **face** and **back**, and the inner layers are called the **core** and **crossbands**. The core is the

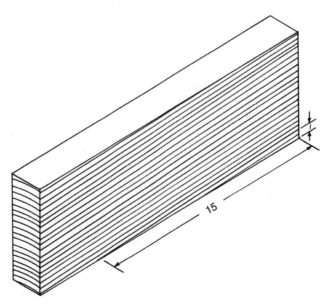

FIGURE 3-5 The maximum slope allowed in aircraft wood is 1:15.

TABLE 3-1 Wood defects

Defect	Acceptability
Checks	Not acceptable.
Compression failure	Not acceptable.
Compression wood	Not acceptable.
Cross grain	Spiral grain, diagonal grain, or a combination of the two is acceptable providing the grain does not diverge from the longitudinal axis of the material more than 1:15. A check of all four faces of the board is necessary to determine the amount of divergence. The direction of free-flowing ink will frequently assist in determining grain direction. If the deviation is greater than specified, the wood is not acceptable.
Curly grain	Acceptable if local irregularities do not exceed limitations specified for cross grain.
Decay	Not acceptable.
Hard knots	Sound hard knots up to $\frac{3}{8}$ in [9.5 mm] in maximum diameter are acceptable providing: (1) they are not in projecting portions of I-beams, along the edges of rectangular or beveled unrouted beams, or along the edges of flanges of box-beams (except in lowly stressed portions); (2) they do not cause grain divergence at the edges of the board or in the flanges of a beam more than 1:15; and (3) they are in the center third of the beam and are no closer than 20 in [50.8 cm] to another knot or other defect (pertains to $\frac{3}{8}$-in [9.5 mm] knots—smaller knots may be proportionally closer). Knots greater than $\frac{1}{4}$-in [5.7 mm] must be used with caution.
Interlocking grain	Acceptable if local irregularities do not exceed limitations specified for cross grain.
Mineral streaks	Acceptable, providing careful inspection fails to reveal any decay. Not acceptable if accompanied by decay.
Pin knot clusters	Small clusters are acceptable, providing they produce only a small effect on grain direction. Not acceptable if they produce a large effect on grain direction.
Pitch pockets	Acceptable in center portion of beam, providing they are at least 14 in [35.56 cm] apart when they lie in the same growth ring and do not exceed $1\frac{1}{2}$ in [3.81 cm] in length by $\frac{1}{8}$ in [3.18 mm] width by $\frac{1}{8}$ in [3.18 mm] in depth and providing they are not along the projecting portions of I-beams, along the edges of rectangular or beveled unrouted beams, along the edges of the flanges of box-beams. Otherwise, not acceptable.
Shakes	Not acceptable.
Spike knots	Not acceptable.
Splits	Not acceptable.
Wavy grain	Acceptable, if local irregularities do not exceed limitations specified for cross grain.

center ply, and the layers between the core and outer layers are the crossbands.

The layers of plywood are bonded with special adhesive of the synthetic resin type such as phenol formaldehyde adhesive. Flat aircraft plywood is usually assembled with a thermosetting (hardened by heat) adhesive in a large, heated hydraulic press. It must be emphasized that aircraft plywood is of much higher quality than commercial grades. Every layer of wood in a sheet of aircraft plywood must be of excellent quality to provide for uniform strength throughout.

Plywood has a number of advantages over solid wood in that it is not likely to warp, it is highly resistant to cracking, and its strength is almost equal in any direction when stresses are applied along the length or width of a panel. Its change in dimension is negligible with changes in moisture content.

The most commonly used types of plywood for aircraft manufacture are mahogany and birch. The core and crossbands may be made of basswood or a similar wood that provides adequate strength. Mahogany has a reddish-brown appearance, whereas birch is of a light yellow or cream color. Mahogany offers a better bonding surface than birch because of its porosity.

When selecting or ordering plywood for aircraft use, the technician should make sure that the wood is of aircraft quality. Some commercial plywoods appear to be as good as aircraft plywood; however, it will be found that the quality is only on the surface and the strength does not compare with the aircraft-quality product.

Laminated Wood

Laminated wood is several layers of solid wood bonded together with an adhesive. Laminated wood differs from plywood in that each layer of wood has the grain running in the same direction, whereas plywood has the grain direction of each layer at a large angle to the previous layer. Laminated wood tends to be more rigid than a piece of solid wood of the same size and is much more resistant to warpage. Laminated wood is used for components that require a curved shape, such as wing-tip bows and fuselage formers, and is used in place of solid wood, such as for solid-type wing spars.

TABLE 3-2 Aircraft woods

Species of wood	Strength properties as compared to spruce	Maximum permissible grain deviation (slope of grain)	Remarks
Spruce (*Picea*) Sitka (*P. sitchensis*), Red (*P. rubra*), White (*P. glauca*)	100%	1:15	Excellent for all uses. Considered as standard for this table.
Douglas Fir (*Pseudotsuga taxifolia*)	Exceeds spruce	1:15	May be used as substitute for spruce in same sizes or in slightly reduced sizes providing reductions are substantiated. Difficult to work with hand tools. Some tendency to split and splinter during fabrication, and considerably more care in manufacture is necessary. Large solid pieces should be avoided due to inspection difficulties. Bonding satisfactory.
Noble Fir (*Abies nobilis*)	Slightly exceeds spruce except 8% deficient in shear	1:15	Satisfactory characteristics with respect to workability, warping, and splitting. May be used as direct substitute for spruce in same sizes providing shear does not become critical. Hardness somewhat less than spruce. Bonding satisfactory.
Western Hemlock (*Tsuga heterophylla*)	Slightly exceeds spruce	1:15	Less uniform in texture than spruce. May be used as direct substitute for spruce. Upland growth superior to lowland growth. Bonding satisfactory.
Pine, Northern White (*Pinus strobus*)	Properties between 85 and 96% those of spruce	1:15	Excellent working qualities and uniform in properties but somewhat low in hardness and shock-resisting capacity. Cannot be used as substitute for spruce without increase in sizes to compensate for lesser strength. Bonding satisfactory.
White Cedar, Port Orford (*Chamaecyparis lawsoniana*)	Exceeds spruce	1:15	May be used as substitute for spruce in same sizes or in slightly reduced sizes providing reductions are substantiated. Easy to work with hand tools. Bonding difficult, but satisfactory joints can be obtained if suitable precautions are taken.
Poplar, Yellow (*Liriodendron tulipifera*)	Slightly less than spruce except in compression (crushing) and shear	1:15	Excellent working qualities. Should not be used as a direct substitute for spruce without carefully accounting for slightly reduced strength properties. Somewhat low in shock-resisting capacity. Bonding satisfactory.

ADHESIVES AND BONDING PROCEDURES

Adhesive are used almost exclusively for joining wood in aircraft construction and repair. A part is regarded as satisfactorily bonded if the strength of the joint is equal to the strength of the wood. In a strong joint, there is complete contact of adhesive and wood surfaces over the entire area of the joint and a thin, continuous film of adhesive between the wood layers unbroken by foreign particles or air bubbles.

To accomplish satisfactory bonding in aircraft wood structures, it is necessary that a number of exacting rules be observed and that all materials be of the high quality specified for aircraft woodwork. If either the adhesive or the wood is not of satisfactory quality or if the techniques employed are not correct, the bonding operations will be inferior and may result in failure.

Types of Adhesives

There are two broad categories of adhesive used in aircraft wood structure, *casein* and *synthetic resin*. The synthetic resin adhesives are commonly used in modern construction and repair operations.

Casein adhesives are manufactured from milk products, are highly water-resistant, and require the addition of sodium salts and lime to prevent attack by microorganisms. Casein adhesive should not be used anymore due to inferior performance. Synthetic adhesives should be the first option for bonding wood structure.

Synthetic adhesives are of the urea formaldehyde, resorcinol formaldehyde, phenol formaldehyde, and epoxy types. Depending on the formulation of the adhesive, it may be water resistant or waterproof and may be purchased in a liquid or powdered form. Synthetic adhesives are not attacked by microorganisms. Resorcind is the only known adhesive which is recommended and approved for use in wood aircraft structures and which fully meets necessary strength and durability requirements. Epoxy adhesives are acceptable and many new epoxy resin systems appear to have excellent working properties and are less critical of joint quality and clamping pressure.

Mixing Adhesives

The mixing of adhesives must be done in accordance with the adhesive manufacturer's instructions to ensure that the full strength of the adhesive will be available. The following discussion is meant to present guidelines for mixing adhesive so that the technician will have an idea of mixing requirements.

The container used for mixing adhesive must be of a material that will not react with the chemicals that make up the adhesive. The container and mixing tools must be clean and free of any contaminants or old adhesive.

In preparation for mixing, the ingredients are measured out in the proper proportions. These proportions may be either by weight or by volume. The ingredients may include powders and liquids, purchased as part of the adhesive, and water. The sequence of mixing may call for the powder to be added to water, water to be added to the powder, or two liquid components to be mixed in some specific sequence, such as adding a liquid catalyst to a liquid adhesive.

For mixing adhesive properly, the room temperature generally must be at or above 70°F [21°C]. The process of mixing the adhesive requires that the speed of mixing be slow enough so that air is not whipped into the mixture. Air would result in a weak adhesive joint. Once the adhesive is mixed, it may have to stand for some period of time to allow the components of the adhesive to interact before a proper adhesive joint can be formed.

Once the adhesive is ready to be used, it has a specific working life, during which it can be applied with assurance that a proper adhesive bond will form. This time is influenced by the room temperature, with higher temperatures resulting in a shorter working life. If the ambient temperature is high, the working life of the adhesive can be extended by placing the adhesive container in a water bath of cool water (no lower than 70°F [21°C]). The average working life of adhesives is 4 to 5 h at 70°F [21°C].

Surface Preparation for Adhesive Bonding

To ensure a sound adhesive bond, the wood must be properly prepared to allow full surface contact between the components being joined. The condition of the wood must be such that the adhesive bonds properly with the surface of the wood. This includes being free of any surface contaminants and having the proper moisture content.

Wood surfaces to be bonded should be smooth and true. Chapped or loosened grain, machine marks, and other surface irregularities are objectionable. Joints of maximum strength are made between two planed or smoothly sawed surfaces that are equally true.

Although the wood surface must be true prior to bonding, the method of obtaining this trueness may affect the strength of the bond. For example, softwoods should not be sanded when preparing the surface for bonding. Sanding fills the wood pores with wood dust and prevents the adhesive from properly penetrating the surface. However, hardwoods can be sanded prior to bonding without any detrimental effects on the adhesive bond. With either type of wood, filing and planing are considered proper methods to prepare the surface for bonding. There should be no more than 8 h between the time that the surface is prepared for bonding and the bonding operation takes place.

The surface to be bonded should be free of any paints, oils, waxes, marks, or particles that would interfere in any way with the proper bonding of the adhesive to the wood surface. The presence of wax on a surface can be detected by placing water drops on the surface. If they bead up, then wax is present and must be removed prior to bonding. This may be particularly useful in determining the surface condition of plywoods that may have been protected with a waxed paper.

The *moisture content* of wood when it is bonded has a great effect on the warping of bonded members, the development of checks in the wood, and the final strength of the joints. A moisture content at the time of bonding that is between 8 and 12 percent is generally regarded as satisfactory, but the higher the moisture content within this range, the better will be the joint. If the moisture content is too low, the adhesive cannot wet the surface properly, and it sometimes produces what are called **starved joints**—that is, joints not adequately bonded. Bonding increases the moisture content of the wood; therefore, the moisture added in this manner must dry out or distribute itself in the wood before the part can be machined or finished. Other factors in establishing moisture content are the density and thickness of the wood, the number of plies, the adhesive mixture, and the quantity of adhesive used.

Adhesive Bonding Procedures

A strong joint in the wood is obtained from complete contact of adhesive and wood surfaces over the entire joint area and a continuous film of good adhesive between the wood layers that is unbroken by air bubbles or by foreign particles. Under these conditions, the adhesive penetrates the pores of the wood and forms a bond that is stronger than the bond between the original wood fibers. When broken, such a joint will not separate at the adhesive bond but will fracture in the wood outside the bond.

Adhesive should be spread evenly over both surfaces forming the adhesive joint. Either a brush or a soft-edged spreader may be used to apply the adhesive. If a brush is used, careful inspection must be made after spreading the adhesive for any bristles that may have broken off.

Two different assembly methods may be employed in joining wood parts with adhesive. The **open assembly** method is often recommended because it reduces the time required for the adhesive to set up. In open assembly, the adhesive is applied to both surfaces to be joined, and the parts are not put together for a specified length of time. If pieces of wood are coated with adhesive and exposed to a free circulation of air, the adhesive thickens faster than when the pieces are laid together as soon as the adhesive is spread. This latter process is called **closed assembly**.

In bonding operations, the assembly time may be as little as 1 min or as long as 20 min, but the adhesive must remain at a satisfactory consistency throughout the period. Unless specifically stated to the contrary by the manufacturer of the adhesive, open assembly should not permit the adhesive to be exposed to the open air for more than 20 min.

Bonding Pressure

The functions of pressure on an adhesive joint are as follows: (1) to squeeze the adhesive into a thin, continuous film between the wood layers, (2) to force air from the joint, (3) to bring the wood surfaces into intimate contact with the adhesive, and (4) to hold the surfaces in intimate contact during the setting of the adhesive.

A light pressure is used with thin adhesive and a heavy pressure is used with thick adhesive. Corresponding variations in pressure are made with glues of intermediate consistencies. The pressure applied should be within the range approved for the types of wood being bonded. For example, the bonding pressure should be between 125 and 150 psi [861.75 and 1034.25 kPa] for softwoods and between 150 and 200 psi [1034.25 and 1378.8 kPa] for hardwoods.

The method of applying pressure depends on the size, shape, and contour of the surface. Pressure can be applied by the use of clamps, nails, weights, nail strips, or screws.

Bonding Temperature

Temperature of the bond line affects the cure rate of the bonded joint. Some adhesives require a minimum temperature to cure. Always follow the manufacturer's recommendations.

CONSTRUCTION AND REPAIR OF WOOD STRUCTURES

Before attempting to repair a damaged wooden aircraft structure, the technician must understand the nature of the required repair and have the correct materials and technical information required at hand to make the repair.

Nomenclature for Wooden Aircraft

The nomenclature for a wooden wing is shown in Fig. 3-6. Note that the parts are named according to standard practice for both metal and wooden wings. In the illustration, the **leading-edge strip**, the **plywood skin**, and the **corner block** are unique to wooden wing construction. Some of the

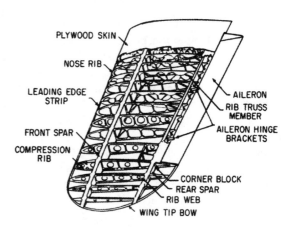

FIGURE 3-6   Nomenclature for a wooden wing.

nomenclature for a wooden fuselage is given in Fig. 3-7. Here again, the nomenclature is similar to that given for a metal fuselage of the semimonocoque type.

Bending and Forming Wood

Bending of wood is necessary to achieve the desired shape of components while maintaining the structural strength of a straight piece of wood. Any type of wood may be bent, with the degree of shaping depending upon the size of the piece, the type of wood, and the technique used in preparing the wood for bending.

Solid wood is normally bent only over a very large radius and then only when the wood is of a small cross-sectional area. Only the best, clearest, straight-grained material should be considered for bending. Woods commonly used for bent components include spruce, ash, and oak. Typical airframe components made of bent solid wood include wing-tip bows, rib cap strips, and fuselage stringers.

Laminated wood structures are commonly used to form any severely bent structure because of the ease with which the thin laminations can be formed and because of the high strength of the finished laminated structure. Laminated members, since they have a parallel grain construction, have about the same properties as solid wood, except that laminated members are usually more uniform in their strength properties and less prone to change shape with variations in

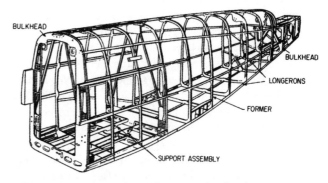

FIGURE 3-7   Nomenclature for a wooden fuselage.

moisture content. Curved laminated structures are used for items such as tip bows, formers, and bulkheads.

Plywood is formed to make leading-edge coverings and surface panels. Most curved plywood components start out as flat sheets and, through various bending operations, are formed to the desired shape. While solid and laminated structures are normally bent in only one direction, plywood is often bent in two planes by stretching it over formers, resulting in a double curvature. This double curvature is often found in areas such as fairings and wing tips.

Wood may be bent in a dry condition or after being soaked in water for some period of time. Dry bending allows the least amount of bending, whereas soaking the wood in cold water makes the wood more flexible. To increase the flexibility of the wood, it can be soaked in hot water or, for maximum flexibility, it can be heated in a steam chamber. The wood should be exposed to the steam for 1 h per inch [2.54 cm] of thickness, with a maximum of 4-h exposure. Excessive heating causes the wood to break down structurally.

Immediately after steaming, the wooden part must be bent. If the curvature is slight, the part may be bent by hand over a form of the desired shape. If the curvature is pronounced, most of the deformation (change of shape) is accomplished by compression or shortening. This is done by using a forming die and holding strap, such as the one illustrated in Fig. 3-8. The wood to be bent is fitted snugly between the bulkheads shown in the picture and then bent over the forming die. In some cases, the type of clamp shown in the lower drawing of Fig. 3-8 does not hold. It is then necessary to use a vise-type clamp with outer and inner forming dies.

The wood, having been bent, should remain in the forms until it has cooled and dried enough to keep its shape. The forms are usually made with a slightly greater curvature than that required for the finished part to provide for the tendency of the wood to straighten out somewhat after it is taken out of the forms. In addition, the forms should be designed so that they expose as much as possible of the bent piece to the drying effect of the air.

Laminated wood members that do not require severe bending, such as wing-tip bows, may be formed without steaming or any other softening preparation. If the laminations are thin enough to take the necessary bend without splitting, they are cut to size and planed on both sides. Laminations sufficient in number to make up the required thickness are coated with adhesive and clamped in a form of the necessary shape. Time is allowed for the adhesive to set and for the wood to dry thoroughly, after which the wooden part will retain the shape of the form. If there is any springback, it will be very slight. In certain cases, where it is not desirable to use very thin laminations or where the bending curvature is severe, the laminations may be steamed and bent to shape before being bonded together. A laminated component in a jig is shown in Fig. 3-9.

When curved plywood members are needed, several layers of veneer may be bent and bonded in one operation, or the prepared plywood may simply be bent.

When built-up plywood members are desired, veneer strips or sheets are bent over a form after adhesive has been applied to their surfaces. The sheets or strips are held together, often

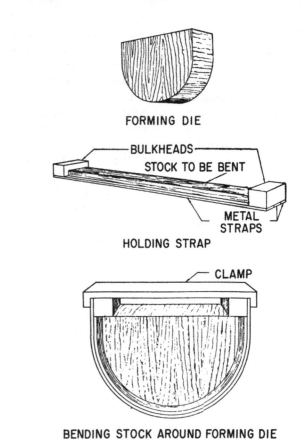

FORMING DIE

BULKHEADS
STOCK TO BE BENT
METAL STRAPS
HOLDING STRAP

CLAMP

BENDING STOCK AROUND FORMING DIE

FIGURE 3-8 Using a forming die and holding strap.

with staples, while the adhesive sets; the member then retains its shape after it has been removed from the form. The grain of each successive layer of veneer should be perpendicular to that of the adjoining layer, but in some jobs the veneer is applied on the form with the grain running at an oblique angle of about 45° from the axis of the member. If the work is done carefully, a built-up plywood member should have about the same properties as a bent laminated member.

The degree to which plywood may be bent is illustrated in Table 3-3. Note that if plywood is selected with the grain lines 45° to the face grain rather than 90° to the grain line, then a sharper bend is possible. Also note that the greater the number of veneers in a plywood of a given thickness, the sharper the bend that is possible.

Wing Spar Construction

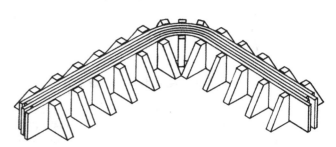

FIGURE 3-9 A laminated wood structure is normally held in a jig until the adhesive dries.

TABLE 3-3 Minimum Recommended Bend Radii for Aircraft Plywood

Plywood Thickness		10 Percent Moisture Content, Bent on Cold Mandrels		Thoroughly Soaked in Hot Water, Bent on Cold Mandrels	
		At 90° to face grain	At 0 to 45° to face grain	At 90° to face grain	At 0 to 45° to face grain
(1)	(2)	(3)	(4)	(5)	(6)
Inch	No. of plies	Inches	Inches	Inches	Inches
0.035	3	2.0	1.1	0.5	0.1
0.070	3	5.2	3.2	1.5	0.4
0.100	3	8.6	5.2	2.6	0.8
0.125	3	12	7.1	3.8	1.2
0.155	3	16	10	5.3	1.8
0.185	3	20	13	7.1	2.6
0.160	5	17	11	6	2
0.190	5	21	14	7	3
0.225	5	27	17	10	4
0.250	5	31	20	12	5
0.315	5	43	28	16	7
0.375	5	54	36	21	10

Note: Columns (1) and (3) may also be used for determining the maximum thickness of single laminations for curved members.

Wooden wing spars are constructed using several different techniques, depending on the size of spar required and the structural strength requirements. Some aircraft may include several different construction techniques along the length of one spar as the required structural strength changes.

Based on the materials used in the structure, spars can be divided into two broad categories: *solid spars* and *built-up spars.*

Solid spars use solid wood as the primary components. These spars may be made of one piece of wood that is rectangular in cross section, several pieces of solid wood laminated together, an externally routed solid piece, or an internally routed spar formed by routing out portions of two boards and then joining the routed sides of the boards together to form a spar. Examples of these different types of solid spars are shown in Fig. 3-10. Solid spars may change in their external dimensions along their length, may have areas that are not routed, such as at fittings, and may include plywood plates attached to areas requiring reinforcements, such as at fitting attachments.

Built-up spars include a combination of solid wood and plywood components. Built-up spars can be divided into three basic types—**C-beam**, **I-beam**, and **box-beam**—as illustrated in Fig. 3-11. C-beam and I-beam spars consist of a plywood web as the principal vertical member running the length of the spar. At the top and bottom of this web are located solid wood cap strips. For a C-beam, the cap strips are on only one side of the spar, whereas an I-beam has cap strips on both sides of the web. Intercostals are located vertically between the cap strips at

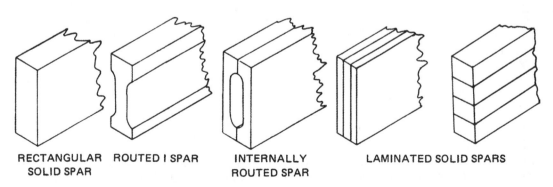

RECTANGULAR SOLID SPAR   ROUTED I SPAR   INTERNALLY ROUTED SPAR   LAMINATED SOLID SPARS

FIGURE 3-10   Solid and laminated spars.

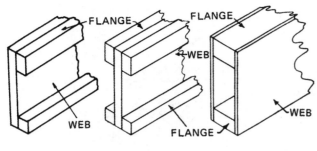

FIGURE 3-11 Built-up spars.

intervals to increase the strength and rigidity of the spar. Blocks are used between the cap strips to allow for the attachment of fittings.

A box-beam spar consists of a top and bottom solid-wood cap strip, plywood webs on the outside of the cap strips, and intercostals and blocks used for strength, stiffness, and attachment of fittings.

Spar Repairs

When a spar is damaged, the damage must be evaluated to determine if the spar can be repaired or if it must be replaced and what the economic factors are concerning the cost of a repair versus the cost of replacement. The economic factors must be decided between those doing the repair and the aircraft owner. Factors that determine the repairability of a spar include the existence of any previous repairs, the location of the damage, and the type of damage. If a spar has been repaired twice, it is generally considered to be unrepairable. If the damage is in such a location that a splice is not possible without interfering with wing fittings, as discussed later, then the spar is unrepairable. If the damage is such that the integrity of the repair will be in doubt, such as the presence of extensive decay, then the spar is unrepairable. Keep in mind that each spar must be evaluated, and no one set of rules can apply to all spars.

If a spar is determined to be repairable, the repair procedures outlined in the aircraft's maintenance manual or in AC43.13.1B should be followed. Figure 3-12 is an example of a repair of a wooden wing spar.

Rib Construction

Ribs give the wing and other airfoil sections the desired cross-sectional shape. In some wings, certain ribs take the compression load between the front and rear spars, in which case they replace the compression struts that would otherwise be used to separate members. A tapered wing may be tapered in width, tapered in thickness, or tapered in both thickness and width. Therefore, the ribs of tapered wings vary in size from wing tip to wing root, although the cross-sectional shape (airfoil section) of each rib is the same throughout in most designs.

Some ribs are built with the nose section, the center section between the spars, and the trailing-edge section as separate units, all being butted against the spar and fastened with adhesive and nails. Other ribs are constructed in one unit and slipped over the spar to their proper stations.

The rib structure may consist of the truss type with plywood-gusseted joints, the lightened and reinforced plywood type, the full plywood-web type with stiffeners, or some other special design. It is very important that no change be made in the shape of an airfoil section during construction or repair.

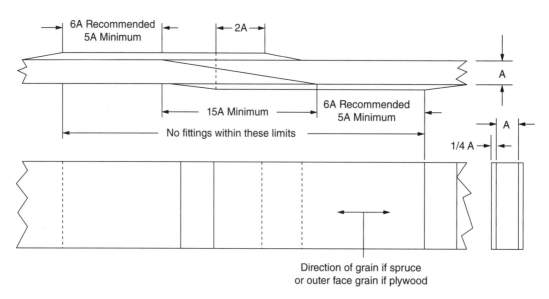

Reinforcement plates to be spruce or plywood and should be bonded only, with the ends feathered off at 5:1. Solid spars may be replaced with laminated spars or vice versa, provided the material is of the same high quality.

FIGURE 3-12 Example of repair of a solid or laminated wooden wing spar.

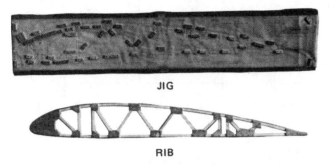

JIG

RIB

FIGURE 3-13　Rib jig and completed rib.

When making replacement ribs, it is best if they are made from a drawing furnished by the manufacturer or from a drawing made by a repair agency and certified correct by the manufacturer. However, an original rib may be used as a pattern for making a new rib if the original rib is not so badly damaged that comparison is inaccurate.

A wood rib is usually assembled in a **rib jig**. The rib jig is made by drawing a pattern of the rib on a smooth, flat plank and then nailing small blocks of wood to the plank so that they outline the rib pattern. During assembly, the cap strips are inserted between the blocks to hold them in the proper position for attachment of the vertical and diagonal members and the plywood gussets. Gussets are attached to the cap strips, verticals, and diagonals with nails and adhesive. Figure 3-13 shows a rib assembled in the jig and a completed rib.

The components of the rib are cut so that they are a "push" fit, with perfect alignment between all contact surfaces. There should not be any visible gap between components, and none of the components should require more than a gentle push to position them onto the jig board.

Once all the components are cut and their fit is checked by positioning them on the jig, adhesive is applied to all the contact surfaces and the rib is assembled in the jig. Nails are used to apply pressure to the adhesive joints covered by the gussets. Once the adhesive has set, the rib is removed from the jig, excess adhesive is removed, and gussets are added to the opposite side of the rib. The rib is finished first by filing

down any overhang of the gussets and then by varnishing the rib. Care must be exercised to prevent any varnish from getting on the wood in the areas where the rib will be bonded to the spars.

When a rib must be repaired or installed on a wing that is already assembled, it is usually necessary to build the rib on the wing. This requires great precision to make sure that the rib is the same shape and size as the original. To do this, a rib jig can be constructed, cut into sections to fit around the spars, and positioned on the wing to allow a rib to be built in place, as shown in Fig. 3-14.

Wood ribs are commonly attached to the spar by the use of adhesive and then nailing the rib to the spar through the vertical members of the ribs. Ribs should not be attached by nailing through the rib cap strips (top and bottom pieces), as this will significantly weaken the ribs. Some rib installations make use of corner blocks bonded and nailed to the spars and the ribs to hold the ribs in position. After the ribs are attached to the spars, a cap strip may be placed on the top and bottom of the spars between the ribs to further stabilize the ribs on the spars. Figure 3-15 shows an example of acceptable wing rib repairs.

Leading- and Trailing-Edge Strips

The leading edge of a wing, of an aileron, or of other airfoil surfaces may be provided with special nose ribs, stiffeners, and a covering of sheet metal or plywood to maintain its shape and carry local stresses. A metal or wooden strip is often used along the leading edge to secure the ribs against sidewise bending and to provide a surface for attaching the metal, fabric, or plywood covering. In some airplanes, the metal or plywood covering acts as the stiffener for the ribs instead of the leading-edge strip.

Figure 3-16 shows the use of a template in shaping the leading edge. The curved portion of the leading-edge strip completes the contour along the nose of the wing. It is important in forming the leading edge of the strip to use a template, corresponding to the curvature, for checking the work.

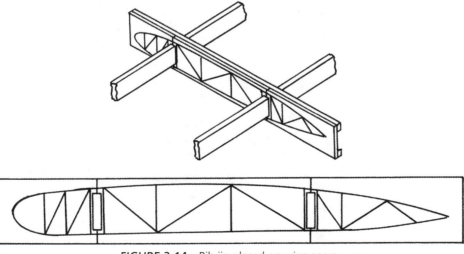

FIGURE 3-14　Rib jig placed on wing spars.

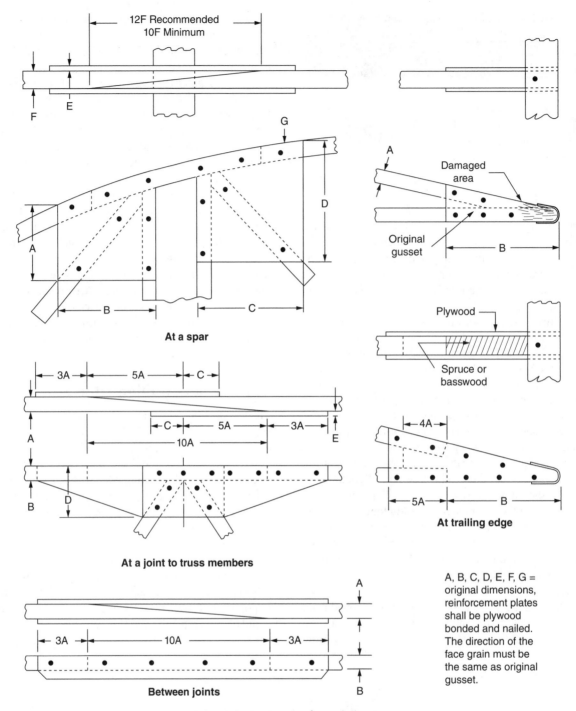

At a spar

At a joint to truss members

Between joints

At trailing edge

A, B, C, D, E, F, G = original dimensions, reinforcement plates shall be plywood bonded and nailed. The direction of the face grain must be the same as original gusset.

FIGURE 3-15 Repair of wood ribs.

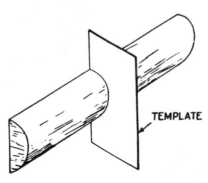

FIGURE 3-16 A template used to shape the leading edge.

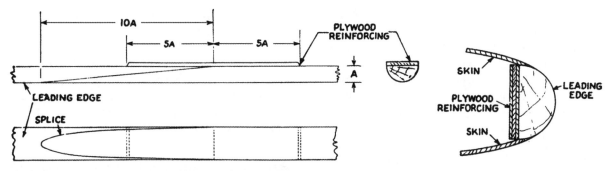

FIGURE 3-17   Repair of a damaged leading-edge section.

The trailing-edge strip of an airfoil surface may be of wood or metal. The wooden strip is made similar to the leading-edge strip. Templates may be used to indicate the camber of the rib.

The leading edges of wing and control surfaces are repaired by carefully made and reinforced splices, such as the one illustrated in Fig. 3-17, where a damaged leading-edge section of a horizontal or vertical stabilizer is repaired by splicing in a new section. To obtain additional bonding area, the skin is feathered into the strip. If this is done properly, only one reinforcing strip is required.

Wing-Tip Bow Construction and Repair

A wing tip may have any of several shapes. For example, it may be square, elliptical, or circular in plan form. If the wing tip is elliptical or circular, a wooden or metal wing-tip bow is required for attaching the plywood or fabric covering. A wooden bow for this purpose may be made of solid wood or laminations and bent to the required shape.

Figure 3-18 shows three types of wing-bow cross sections with the plywood surface and the tip bow indicated in each.

A wing-tip bow that has been badly damaged should be removed and replaced. A cracked or broken bow may be repaired by splicing in a new piece. The new piece may be spliced in at the spar. It should have the same contour as the original bow, and the splices should meet the requirements of a scarf joint.

Installation of Plywood Skin

Plywood for the skin of the airplane is cut and shaped to fit the surface to be covered, enough material being allowed to provide for scarfing (tapered-edge joints) where needed. Depending upon the design, it may be necessary to cut the skin larger than is needed and to trim it to size after it has been bonded to the frame, particularly in the case of wing tips.

The next step is to nail the plywood temporarily to the frame in one corner, make any required adjustments, and then nail in the opposite corner. These temporary nails are driven through small strips of wood so that they can be pulled out easily. The corners having been secured, the plywood is then pressed down against the framework with the hands to be sure that it is in the exact position and that all supporting members, such as ribs and spars, are properly contacted.

If possible, the plywood is then marked on the inside so that the areas of contact with the internal structure are identified. The plywood is then removed and adhesive is applied to the contact area of the structure, which is indicated by the drawn lines on the plywood. The plywood is then carefully placed back on the structure and secured with nailing strips.

The nailing strips used to hold the plywood in place until the adhesive sets are strips of wood about $\frac{3}{16}$ to $\frac{1}{4}$ in [5 to 6 mm] thick and $\frac{1}{2}$ in [12.7 mm] wide. The strips are first nailed near the center of the skin panel and then nailed outward in both directions. More nailing strips are applied, this time perpendicular to the first set, starting near the middle

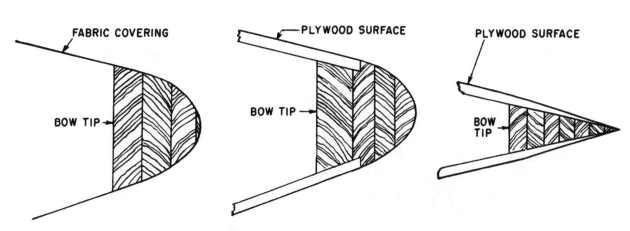

FIGURE 3-18   Wing-bow cross sections.

and working outward in whichever direction will avoid wrinkles and permit the plywood to lie smooth.

The nails are driven down tightly, and enough nailing strips are used to cover the entire surface of the supporting members. The width of the supporting member determines the number of nailing strips required. One nailing strip is used on surfaces $\frac{1}{2}$ in [12.7 mm] wide. Several nailing strips, laid side by side, are used for covering wider surfaces.

The nails are driven so that those in one strip alternate with regard to those in an adjacent strip. If the plywood skin is from $\frac{1}{16}$ to $\frac{3}{32}$ in [1.59 to 2.38 mm] thick, $\frac{5}{8}$-in [15.88-mm] nails spaced at $\frac{1}{2}$-in [38.1-mm] intervals are used for holding the skin to the ribs. The object is to space the nails as close together as possible without splitting one of the parts being bonded. Whenever nailing strips come into contact with the adhesive, waxed paper is placed under the strips to prevent them from being bonded to the member.

When a leading edge is to be covered, the plywood is first cut to the approximate size of the section to be covered. The leading edge is then bent over a form or over the wing leading edge. It is often necessary to soften the wood by soaking or steaming it prior to the bending operation. The plywood is held in place over the form or leading edge with shock cords, rubber straps, or any other apparatus that will provide an even pressure over the surface of the plywood without distorting the plywood or damaging the leading-edge structure.

When the plywood is thoroughly dry, the formed piece is fitted again and cut to the precise size. Adhesive is applied to the framework (nose ribs, spar, etc.) and the formed plywood. The plywood is placed on the frame and pressure is applied by means of nailing strips. These nailing strips may be started along one edge of the spar or along the leading-edge strip. More nailing strips are then laid over the nose ribs, starting with the center rib and working toward each end of the wing section. Soaking the nailing strips in warm water to make them bend more easily is a good treatment for those nailing strips laid over the curved portion of the leading edge.

Although nailing strips usually provide a satisfactory method for applying pressure to adhesive joints when installing plywood skins, other methods are also used, and some of these may be more effective than the nailing strips. One of these methods is to place a heavy piece of web strap around a wing immediately over a rib and the plywood after the adhesive has been applied. The strap is then tightened until the plywood skin is pressed firmly against the member to which it is being bonded.

Another method that can be used is to apply pressure to the plywood over an adhesive joint by means of shot bags or sandbags. Great care must be exercised to make sure that the plywood is pressed against the structure to which it is being attached and that there is a good adhesive film in all joints.

Repair of Damaged Plywood Structures

When the stressed plywood skin on aircraft structures requires extensive repairs, it is essential that the manufacturer's instructions be followed carefully. When the repair is made according to manufacturer's specifications, the original strength of the structure is restored.

Before proceeding with any repair operations, the type and extent of damage to a plywood skin and the internal structure should be evaluated. Information that should be taken into account in determining whether repair or replacement is necessary and the type of repairs possible include the location of the damage, the size of the damaged area, the thickness of the plywood, whether the plywood has a single or double curvature, and the location of and damage to internal structural components.

If internal components are damaged then enough of the plywood covering must be removed to allow for access to and proper repair of these internal components. The repair of internal components should follow the guidelines established for spars and ribs, replacing or splicing in new sections by the use of scarf splices and reinforcement plates.

The curvature of the plywood must be determined if the damage area is of any significant size. If a piece of paper can be laid on the surface and smoothed out without wrinkling, then the plywood has a single curvature (bent in only one direction) and preparing a replacement panel will not require any special equipment. If the paper cannot be laid on the surface without wrinkling, then the plywood has a double curvature. If this is the case, a replacement panel must be obtained from the manufacturer or manufactured by the use of a mold or form and some sort of pressure mechanism to achieve the double curvature.

TABLE 3-4 Plywood Repair Selection	
Repair	Limitation
Scarf patch	No restriction.
Surface patch	Must be entirely aft of 10% chord line or must wrap completely around leading edge and terminate aft of 10% chord line. Maximum perimeter of 50 in [127 cm] (subject to interpretation of AC43.13-1B).
Plug patch	Oval or circular in shape. Sizes permitted include 4- and 6-in [10.16- and 15.24-cm] diameter circles and 3-in by $4\frac{1}{2}$-in [7.62- by 11.43-cm] and 5- by 7-in [12.70- by 17.78-cm] ovals. Used to repair skin where no internal structure is involved.
Splayed patch	Maximum size of trimmed, round hole is $1\frac{1}{2}$ in [3.81 cm]. Maximum skin thickness is $\frac{1}{10}$ in [12.54 mm].
Fabric patch	Maximum diameter of trimmed hole is 1 in [2.54 cm]. Edge of hole may be no closer than 1 in [2.54 cm] to any structural member. Not allowed on a leading edge or frontal area of a structure.

Note: This table is intended only to aid in the selection of a repair procedure. See AC43.13-1B, Chap. 1, for specific repair instructions.

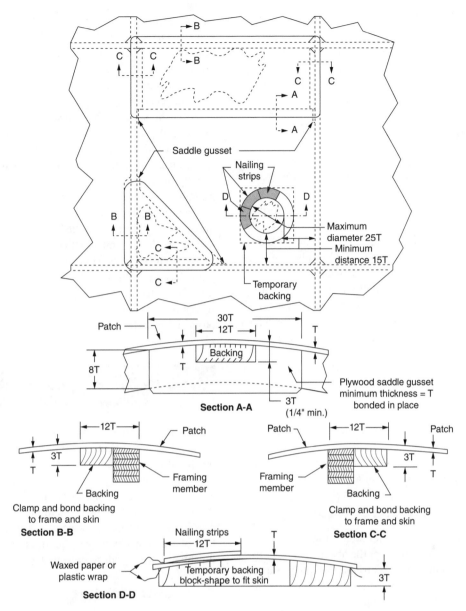

Section A-A

Section B-B

Clamp and bond backing
to frame and skin

Section C-C

Clamp and bond backing
to frame and skin

Section D-D

FIGURE 3-19 Example of scarf repair.

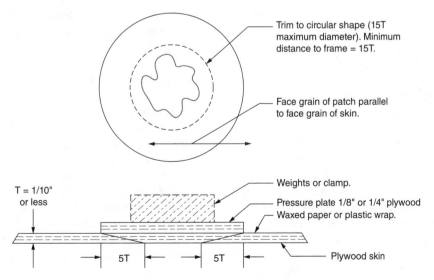

FIGURE 3-20 Example of splayed patch.

There are five repair procedures for plywood skin that can be used. They vary in complexity, and their use may be restricted based on the location of the damage, size of the damaged area, and thickness of the plywood, as shown in Table 3-4. Figure 3-19 shows an example of a scarf repair and Fig. 3-20 shows an example of a splayed patch.

Fiber Insert Nut Plates

Fiber insert nut plates are installed at various locations in the wings and fuselage for the attachment of inspection doors, fuel-tank cap covers, and other parts. If one of these nuts is damaged or if a machine screw breaks off in it, the portion of the wooden strip carrying the nut and extending about 1 in [2.5 cm] on each side should be sawed out and replaced by a new section that is bonded and nailed in place. These nut strips are $\frac{3}{8}$-in [9.53-mm] plywood strips with the nuts embedded at regular intervals. Finally, a $\frac{1}{16}$-in-[1.59-mm-] thick plywood strip is bonded and nailed over the nuts and drilled on correct centers to allow the passage of the machine screws. This is a better method of attaching nut plates than the use of wood screws.

Skis

Wooden ski runners that are fractured are usually replaced, but a split at the rear end of the runner having a length not more than 10 percent of the ski length may be repaired by attaching, with adhesive and bolts, one or more wooden cross-pieces across the top of the runner.

Finishing Repaired Wood Surfaces

Wood surfaces of an airplane structure that have been repaired must be finished to prevent the absorption of moisture, oil, or other contaminants and to prevent deterioration. In every case, the proper finish must be selected and the surface prepared in accordance with approved practice.

The interior surfaces of wooden structures such as wings and control surfaces should be thoroughly coated with spar varnish or lion oil. Two coats of either finish brushed onto the clean surface will usually be sufficient. Before the finish is applied, it is important to see that all excess adhesive, grease, oil, crayon marks, and similar contaminants are removed. Sawdust, shavings, loose wood particles, and similar material should also be removed from the inside of the structure.

When a structure is designed so that it is not possible to reach the interior after the outer skin is installed, it is necessary to finish the interior before the last pieces are fixed in place. In such cases, care must be taken to see that areas to be bonded are not coated. After assembly, it is sometimes possible to reach through inspection holes with a small "touch-up" spray gun and apply finish to small areas. Exterior surfaces, whether plywood or solid wood, are comparatively easy to finish. Such surfaces must be clean; that is, excess adhesive, grease, oil, etc., must be removed. Oil and grease spots can be removed with naphtha or a similar petroleum solvent. Small nail holes, scratches, or depressions in the wood surface should be filled with plastic wood or a similar compound, and the surface should be sanded smooth after the material has dried. Two coats of spar varnish, wood sealer, or MIL-V-6894 varnish should be applied as a base. MIL-V-6894 varnish is dopeproof and should be used when the final finish is to be lacquer or dope.

After the sealer or base varnish is applied to a wood surface and the varnish has dried, it may be necessary to sand the surface lightly to remove rough spots. The final finish can then be applied in as many coats as required. This is usually done with a spray gun. The operation of spray equipment is discussed in Chap. 5.

End-grain wood surfaces require more finish than side-grain surfaces because of the tendency of the grain to absorb the finish. It is good practice to use a wood filler before applying the varnish. The wood filler is applied after the end grain has been smoothed with sandpaper. After the wood filler has dried, a clear or pigmented sealer may be used in two or more coats. The surface can then be finished as previously described.

CARE OF AIRCRAFT WITH WOOD STRUCTURES

The earliest models of successful aircraft were constructed largely with wood structures. Fuselages were then changed to welded steel structures, but the airplane was still equipped with wood-structured wings and control surfaces. The maintenance and care of these airplanes were handled remarkably well by the technicians in both civil and military areas. The key to success in the maintenance of the wood structures was the selection of the best woods for aircraft structures and the proper finishing of the woods to prevent the absorption of moisture.

Control of Moisture

Moisture is the deadliest enemy of wood. Wood that is kept dry will seldom, if ever, deteriorate over a period of many years. Dry, in this sense, means that it holds no more moisture than its natural content under dry-air conditions. On the other hand, when unprotected wood is exposed to water for an appreciable length of time, fungus begins to grow and penetrate the wood cells. This fungus is the cause of decay, dry rot, or whatever term is used to describe deterioration due to fungus.

Moisture also has the effect of causing wood to swell. If wood is alternately wet and dry over a period of time, it will crack and warp; this will reduce its structural strength and cause stresses of various kinds.

It is apparent from the preceding discussion that one of the primary considerations in the care of airplanes with wood structures is to ensure that the wood is finished with an effective, water-resistant coating. **Spar varnishes** of the phenol formaldehyde type (MIL-V-6893) or glycerol phthalate type (MIL-V-6894) are commonly employed for the finishing of wood structures. Synthetic finishes with a polyurethane base are becoming increasingly popular for finishing purposes.

An important factor in preventing moisture from affecting wood structures is to ensure that drain holes are provided in all low points and that the holes are kept open. The drain holes will permit collected water to drain out and the area to dry. It is important to consider drain holes when repairing or recovering an airplane. Sometimes a technician may fail to take note of the position of all drain holes, with the result that some critical areas may not have drain holes.

One of the best methods for extending the service life of an airplane with wood structures is to store it in a dry, well-ventilated hangar when it is not in use. This, of course, is not always possible; however, the practice may well pay for itself in reduced maintenance costs.

Effects of Temperature

Temperature changes, although not as critical as moisture, cause stresses and dimensional changes that can lead to cracks, looseness of fittings, and deterioration of finishes. Desert conditions, with extremes of temperature and low humidity, can cause a maximum of shrinkage in wood structures. This can lead to loose fittings and separation of some bonded joints. It is incumbent upon technicians, under such circumstances, to be particularly alert to detect these conditions.

High temperatures also lead to deterioration of finishes. These temperatures lead to the evaporation of plasticizers in coatings, and this causes brittleness and cracking. In such cases, it is necessary to remove or rejuvenate finishes and restore them to optimum condition.

Low temperatures are likely to cause damage if moisture is present. Freezing of wet structures can cause rupture of fibers and cells, thus weakening the parts affected.

Operation and Handling

As with any type of airplane, one having wood structures can be damaged by improper operation and handling. Pilots must not exceed flight limits set forth for the aircraft and should use great care in landing and taxiing. Careless operation can lead to broken or cracked spars and other wood structures.

Moving an aircraft on the ground must be done with care to avoid cracking or breaking ribs and other structures in the wings and control surfaces. *Note:* Approved walkways and steps must be utilized when it is necessary to climb upon the aircraft. Lifting and pushing must be accomplished by applying pressure or force only to solid structures that can withstand the forces applied.

INSPECTION OF AIRPLANES HAVING WOOD STRUCTURES

The inspection of wood structures requires a great amount of care on the part of the technician. Because of the nature of wood, it tends to hide the beginnings of deterioration and cracks. The following discussion is designed to make the technician aware of some of the problems that may be encountered when inspecting wood structures and some methods that can be used to detect these problems.

Defects in Wood Structures

During the inspection of an airplane with wood structures, the technician must know what to look for that will indicate a defective or weak structure and the necessity for repair. The following are defects most commonly found when performing a complete inspection.

Dry Rot and Decay

Dry rot and decay are essentially the same and are caused by fungus in damp or wet wood. The wood may be black, brown, gray, or some combination of the three colors. It may be breaking down into particles, or there may be a softening of the surface. Dry rot and decay can also be detected by pressing a sharp-pointed instrument such as a scribe into the wood to determine the force necessary to penetrate the wood. If the force required is less than that required for the same depth of penetration in sound wood, it is a sign that deterioration has taken place. These conditions require replacement of the defective part.

Separated Bonded Joints

Wherever a adhesive joint is found open or separated, the structure must be rebuilt.

Deteriorated adhesive joints

Deterioration of adhesive joints is caused by aging and deterioration of the adhesive. Casein adhesive that was not treated to prevent fungus will deteriorate in the presence of moisture. Synthetic resin adhesives are not generally subject to this type of deterioration. Deteriorated adhesive joints require rebuilding of the structure affected.

Cracks

Shrinkage of the wood or stress applied to it can cause cracks. Whatever the cause, the cracked member must be replaced.

Compression Failure

Compression failure is caused by a compressive force acting essentially parallel to the grain of the wood. Compression failure is indicated by a line or lines extending across the grain where the wood fibers have been crushed. A test for a compression failure is to apply a small amount of free-running ink to the wood near the suspected break. The ink will flow along the normal grain until it reaches a compression failure. At this point it will flow cross-grain along the failure.

Surface crushing

Surface crushing is caused when the wood is struck by a hard object. This produces indentation, abrasion, and rupture of the wood fibers. Damaged parts should be replaced or repaired.

Staining

Stains that are caused by moisture indicate that an adhesive joint has failed or that the protective coating is deteriorating. This type of stain is usually dark in color and tends to expand along the grain of the wood. When water stains are found, the cause must be corrected and the affected parts replaced or repaired. Surface stains that are easily removed without removing wood do not usually require replacement of affected parts. The protective coating on such parts must be restored.

Corrosion

Corrosion of attachment bolts, screws, nails, and fittings in or on wood structures indicates the presence of moisture. Corroded parts should be replaced, and the cause of moisture intrusion should be eliminated.

Inspection Procedures

Before a major inspection is started on an aircraft with wood structures, the aircraft should be perfectly dry. In a warm, dry climate, this presents no problem; however, in other areas it is well to have the aircraft stored in a dry, well-ventilated hangar for a few days prior to the inspection. Humidity (moisture in the air) causes wood to swell, thus closing cracks and open adhesive joints so they are not easily detected.

In cases where maintenance manuals are available for an aircraft, the manufacturer's instructions should be followed. In other cases, the inspection should be carried out with a checklist and in a sequence that will ensure a thorough examination of every structural part of the aircraft.

At the start of the inspection, it is a good plan to examine the complete exterior surface of the aircraft for condition and contour. If the aircraft is covered with plywood, a noticeable defect on the surface can indicate damage inside. If the plywood is covered with fabric and a crack or split appears in the fabric, the fabric must be removed so the plywood underneath can be examined for damage.

If the surface of the plywood is warped or wavy, defects are indicated inside the structure. Access through an inspection hole or by removal of a section of plywood skin is necessary in order to examine the interior. If the plywood is not stressed—i.e., does not carry a structural load—a small amount of undulation (waviness) can be permitted.

Where openings are not already provided in the plywood cover for access to the interior or wings, it is often necessary to provide cutouts, and they should be made adjacent to members of the wing frame. The cutaway section should be made as small as possible and closed after the work is completed. Closing is accomplished by the same method as previously described for making repairs. A triangular cutout section is easiest to make, and it presents a small area of opening; however, the oval cutout has a better appearance, provided that time and space permit this type of repair.

To examine the inside of areas where direct visual inspection is not possible, it is necessary to employ aids such as a flashlight, mirror, magnifying glass, and possibly a borescope or similar instrument. The interior areas should appear clean, unstained, and solid. Adhesive seams or joints should show no cracks or separations. Adhesive joints can be checked by attempting to insert one of the thinnest leaves of a feeler gauge into the joint. If the feeler gauge can penetrate the joint, the joint must be disassembled and repaired.

The finish inside the wood structure should indicate no deterioration. If it has become opaque or has a rough milky appearance, it should be removed to reveal the condition of the wood underneath. If the wood is clean, solid, and unstained, it can be refinished with approved varnish.

Wood wings that are covered with fabric are somewhat easier to examine than those covered with plywood. In all cases they should be inspected for all the defects listed previously. If the wing is to be recovered with new fabric, a very thorough examination can be made of the adhesive joints and wood condition.

The quality of wood wing structures covered with fabric can be checked by applying moderate pressure in various areas. Grasping a rib tip at the trailing edge and attempting to move it up and down will reveal damage and failure of adhesive joints. If there is noticeable movement, the condition must be repaired after removing fabric as necessary.

Aileron attachment fittings, usually on the rear spar, are checked by the application of pressure. If they are loose or move too easily, repair or replacement is required.

The ribs on a fabric-covered wing can be checked by examination and touch. Those that can be moved or are out of shape require opening of the fabric for further examination and repair.

Elevators, rudders, stabilizers, and fittings can be inspected in the same manner as that employed for wings. Loose or corroded fittings may require replacement, and it may be necessary to repair enlarged or elongated bolt holes. Bolts must fit snugly in bolt holes.

In cases where a spar is laminated and bolts pass through the laminated section to attach a fitting, the bolts should be loosened to take the pressure off the area. This will allow any separations between the layers of wood to be revealed. Care must be taken to retighten the bolts to the correct torque after the inspection. All bolted areas should be checked for cracks, loose bolts, crushed wood, and corrosion. Defects must be repaired.

If, during inspection, adhesive deterioration is found, a major rebuilding job may be required to restore the structure to its original strength. Adhesive deterioration is indicated when a joint has separated and the adhesive surface shows only the imprint of the wood, with no wood fibers clinging to the adhesive. If the adhesive is in good condition, any separation will occur in the wood and not in the adhesive. This type of separation is usually caused by excessive stress due to a hard landing or other improper operation of the airplane.

Regardless of the type of wood structure, a thorough visual inspection should be performed with and without stress being placed on the structure. The stress inspection usually requires at least two people, one to do the inspecting and one to apply stress to the structure. The stress is applied by pushing up, pulling down, and twisting the structure by applying pressure

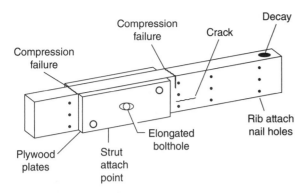

FIGURE 3-21 Likely areas to incur structural damage.

at the primary structures such as the end of the wing spars. Do not apply so much stress as to damage the structure!

While the stress is being applied and released, examine the external surfaces for any unusual wrinkles that may appear. Inside the structure, listen for unusual noises and check for unusual movement between parts and cracks that open up under stress. Also look for any powder flow from joints, indicating abrasion and breakdown of adhesive joints. If anything irregular is noted, investigate the area further until a defect is found or until the movement or sound is determined to be normal. Figure 3-21 shows the likely areas to incur structural damage.

REVIEW QUESTIONS

1. What types of trees produce *softwoods? Hardwoods?*

2. What are some uses of hardwoods in aircraft structures?

3. What is considered to be the best type of wood for aircraft structures?

4. How does *laminated wood* differ from *plywood?*

5. What are *annual rings* in wood?

6. What is the difference between *checks* and *shakes?*

7. What is meant by *compression wood?*

8. What is the maximum grain deviation or slope allowed for structural aircraft wood?

9. What is meant by the term *quarter-sawed?*

10. What is the minimum number of annual rings allowable per inch in aircraft structural woods?

11. What are the two classifications of water found in wood?

12. What should be the moisture content for aircraft woods?

13. Explain the term *kiln-dried.*

14. Describe *aircraft plywood.*

15. What are the advantages of plywood?

16. What are the most common types of plywood for aircraft use?

17. What types of adhesives are most satisfactory for aircraft wood assembly?

18. Which type of adhesive requires additives to inhibit the growth of microorganisms?

19. What is meant by the *working life* of a adhesive?

20. Describe a wood surface that is properly conditioned for bonding.

21. What is the adhesive pressure recommended for use with softwoods?

22. Describe *open assembly* of a adhesive joint.

23. Why is a nailing strip used in the installation of plywood?

24. How is plywood held tightly in place while the adhesive is drying? Give three methods.

25. What is the most important consideration in the preservation of wood aircraft structures?

26. By what methods is a wood aircraft structure protected against moisture?

27. Discuss the effects of temperatures on wood.

28. Name the principal defects that may be found in wood aircraft structures.

29. Why should an aircraft be thoroughly dry before an inspection of wood structures is started?

30. How may inspection access be provided for a closed compartment in a plywood-covered wing?

31. Discuss methods for determining whether a adhesive joint is separated.

32. What inspections should be made with respect to bolts and fittings attached to the wood spar of a wing or control surface?

33. How can it be determined whether a separation is caused by adhesive deterioration or by excessive stress?

Fabric Coverings 4

INTRODUCTION

Although the majority of modern aircraft are constructed of metal, there are still many requirements for aircraft fabrics of various types, and there are many fabric-covered aircraft certificated for operation. It is, therefore, essential that the aircraft technician be familiar with approved fabric materials and processes.

Fabric covering for aircraft has been in use for many years because of its low cost, ease of installation, ease of repair, light weight, strength, and durability. In addition, the design and construction of a fabric-covered airplane is such that it does not require the special manufacturing and repair equipment associated with metal aircraft.

Aircraft-covering fabrics are made of cotton, polyester fiber, glass fiber, and linen. Approved methods for the application of cotton are set forth in the FAA publication AC 43.13-1B and are also covered in this text. Because of the durability and strength of polyester fabrics, many aircraft owners are re-covering their aircraft with this material. Two such processes are sold under the names Stits Poly-Fiber® and Ceconite®. Covering an airplane under these processes must be accomplished according to the instructions provided by the fabric manufacturer.

To add to the variety of processes used for aircraft covering, some aircraft manufacturers have devised their own processes for covering aircraft. These often combine the cotton and polyester processes. When re-covering modern fabric-covered aircraft, consult the aircraft manufacturer's service manuals for specific information about their processes. Traditional covering materials such as cotton and linen have been replaced by polyester and fiberglass systems. These new covering systems are available for most aircraft that were designed and originally equipped with cotton grade A fabric.

FABRIC TYPES AND TERMINOLOGY

Nomenclature for Fabrics

To be able to handle, inspect, and install fabrics properly requires that the technician be familiar with some of the

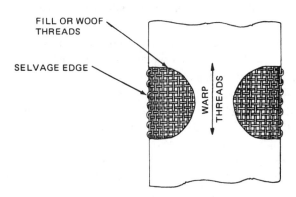

FIGURE 4-1   Warp and fill, or woof.

basic terminology used with woven fabrics. Although this list is not complete, it will serve as a basis for terms presented in later portions of this chapter. Referring to Figs. 4-1 and 4-2 will be helpful in understanding some of the following terms.

Bias A cut, fold, or seam made diagonally across the warp and fill fibers of a piece of cloth. Bias-cut fabric allows the materials to be stretched slightly for better forming to structural contours.

Bleaching A chemical process used to whiten textile materials. Grade A airplane fabric is not bleached and is usually a light

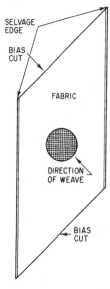

FIGURE 4-2   Bias-cut fabric.

89

cream color. Bleaching, if not properly done, can weaken a material and make it unfit for use.

Calendaring A process of ironing fabric by threading it wet between a series of heated and cold rollers to produce a smooth finish. Calendaring causes the nap to lay close to the surface. The nap is the "fuzzy" surface caused by the thousands of ends of individual fibers.

Fill The fibers of a piece of fabric that are woven into the warp fabric. These fibers run perpendicular to the length of the fabric as it comes off of a roll of material.

Mercerizing A chemical process in which cotton is exposed to the action of a strong caustic solution that tends to shrink the material and give it a silky appearance.

Selvage The naturally bound edge of a length of fabric. The selvage is where the fill fibers turn around and go back through the warp fibers during the weaving process.

Sizing A textile glue used to stiffen and protect fabrics and threads. It gives "body" to a material.

Thread count The number of threads, either warp or fill, on the edge of a piece of fabric. The thread count uses a unit of measurement of *threads per inch*.

Warp The threads in a woven fabric that run the length of the fabric.

Weight An indication of the weight of the fabric per unit of area. The weight is commonly stated as ounces per square yard.

Woof Same as *fill*.

Organic Fabrics

Organic fabrics are those made from plant materials. These include two grades of cotton fabrics and linen.

For many years, the standard approved aircraft covering has been **grade A mercerized cotton cloth.** However, cotton grade A fabric covering systems have been replaced by polyester and fiberglass systems that have superior performance and endurance. This material is identified by the SAE number AMS 3806, and the specifications are set forth in FAA Technical Standard Order (TSO) C15. Approved fabric predoped with cellulose nitrate dope is numbered MIL-C-5643 and fabric predoped with cellulose acetate butyrate dope is numbered MIL-C-5642. The minimum tensile strength of approved grade A fabric in the new, undoped condition is 80 lb/in [140 N/cm]. This means that a strip of the fabric 1 in [2.54 cm] in width must be able to support a weight of 80 lb [355.84 N] in tension without breaking.

After fabric has been used on an airplane, its minimum permissible strength is 70 percent of new strength, or 56 lb/in [98 N/cm] for grade A fabric undoped. This means that when the fabric on an airplane has deteriorated to the point where the strength is less than 56 lb/in [98 N/cm], the fabric must be replaced. Testing of fabric is discussed later in this chapter.

The military specification for grade A fabric is MIL-C-5646. A fabric having this number or the number AMS 3806 (TSO C15) has, in the past, been considered acceptable for all aircraft. However, with the advent of inorganic covering materials, grade A cotton may not be used on an aircraft originally covered with an inorganic fabric unless approval is obtained from the aircraft manufacturer or the FAA. To determine the original fabric covering material used on an aircraft, check the aircraft service manual.

Grade A fabric must have a thread count of 80 to 84 threads per inch in both length and width. The weight of the fabric must not be less than 4 oz/yd^2 [135.6 g/m^2]. The fabric is calendared after weaving to lay the nap and make the finished material smooth.

For aircraft having a wing loading of no greater than 9 lb/ft [43.94 kg/m] and a placarded never-exceed speed no greater than 160 mph [257.5 km/h], a lighter-weight fabric may be used. (Wing loading can be determined by dividing the maximum allowed gross weight of the aircraft by the wing area.) This fabric is designated as **intermediate-grade aircraft fabric** and carries the number AMS 3804 (TSO C14). The minimum tensile strength of this fabric in the new, undoped condition is 65 lb/in [113.75 N/cm]. Remember that "pounds per inch" does not have the same meaning as "pounds per square inch." The thread count for intermediate fabric is 80 minimum to 94 maximum threads per inch [32 to 37 threads per centimeter] for both warp and woof (fill).

With intermediate-grade cotton, observe the same precaution concerning aircraft originally covered with an inorganic fabric. Also, intermediate-grade fabric is no longer commonly available. Its only value is as a reference material to determine the minimum strength to which fabric can be allowed to deteriorate before the fabric must be replaced. This is discussed in detail in a later section of this chapter.

For many years early in the history of aviation, linen was commonly used for the covering of aircraft. Linen, being woven from flax fiber, is strong, light, and durable. Aircraft linen is an especially fine grade of linen cloth, and if it complies with the requirements of TSO C15, it is suitable for use on certificated aircraft originally covered with organic fabric. The British specification 7F1 meets all the requirements of TSO C15.

Inorganic Fabrics

An **inorganic fabric** is one that requires chemical processing to create the fiber. Once the fiber is created, it is woven in the same manner as used for organic material. The inorganic fabrics have two advantages over organic fabrics in that they resist deterioration by the ultraviolet rays of the sun and they resist attack by microorganisms. The only significant disadvantage associated with inorganic fabrics is the care required to ensure proper bonding of the dopes and finishing products to the fabric, as is explained later in this chapter.

There are two types of inorganic fabrics used to cover aircraft: *polyesters* (Dacron-type materials) and *fiberglass*.

Polyester fabrics are manufactured under the trade names of Stits Poly-Fiber and Ceconite. These materials come in a variety of weights, thread counts, and tensile strengths. These fabrics have become very popular as replacements for the organic materials due to their ease of installation and resistance to deterioration when compared to organic materials.

Razorback is the most widely used type of fiberglass material for covering aircraft. It has an advantage over all other types of materials in that it is impervious to deterioration, heat, and most chemicals. As a result, it is often the fabric of choice for aircraft subject to exposure to chemical environments such as agricultural operations.

Surface Tape

Surface tape, also called **finishing tape**, is usually cut from the same material that is being used to cover the airplane. Both the Stits and Ceconite covering systems use their own type of surface tape. Do not substitute cotton surface tape for aircraft covered with polyester covering systems. The edges are **pinked** (cut with a saw-toothed edge) to provide better adhesion when doped to the surface and to reduce the tendency to ravel. The edges of surface tape made from synthetic materials are often cut in a straight line, with the edges being sealed by heat to prevent raveling. Surface tape material is used to reinforce the fabric covering at openings and fittings, protect and seal rib-attachment processes, and streamline surface irregularities.

A roll of surface tape is shown in Fig. 4-3. This tape should have the same fiber, yarn size, tensile strength, and number of threads per inch as the fabric upon which it is applied. The sizing must not exceed 2.5 percent; however, approved tape will usually fulfill this requirement.

Surface tape is available as straight-cut, where the edges of the tape are parallel to the warp threads, and as bias-cut, for use on surfaces with compound curvatures, such as at wing tips and curved trailing edges. The tape is supplied in standard-length rolls (such as 100 yd [91.44 m]) and can be ordered in various widths ranging from about 1 in [2.5 cm] to 6 in [15.2 cm] or more.

FIGURE 4-3   A roll of surface tape.

FIGURE 4-4   A roll of reinforcing tape.

Reinforcing Tape

Reinforcing tape is a special product that has a much larger warp thread than fill thread. It is used over ribs between the lacing cord and fabric covering to prevent the cord from cutting or wearing through the fabric and to help distribute the air loads. Reinforcing tape is also often used for inter-rib bracing of wing structures prior to installation of the fabric covering. Reinforcing tape bearing the specification number MIL-T-5661, or equivalent, is approved for aircraft use. It is ordinarily obtainable in several widths that conform to the different widths of ribs or rib cap strips. This tape is of a material similar to the fabric covering used on the airplane. The tensile strength is at least 150 lb per ½ in [525.35 N or 53.6 kg/cm]. If a synthetic fabric or fiberglass covering is being installed, the reinforcing tape may be the standard cotton type, or it may be a special tape designed and manufactured for use with the covering being applied. In all cases, it is wise to consult the specifications applicable to the job or process concerned. A roll of reinforcing tape is shown in Fig. 4-4.

Sewing Threads and Lacing Cords

Sewing thread, for either machine sewing or hand sewing, is used to join two fabric edges together during the installation or repair of fabric covering materials.

Cords are normally heavy threads used where a significant amount of strength is required of each stitch, such as when attaching a fabric covering to wing ribs or fuselage stringers.

When an airplane is covered with a synthetic fabric under a Supplemental Type Certificate or a Parts Manufacturer's Approval, special cords and threads made of the same material as the fabric are often recommended. For example, in the Stits Poly-Fiber covering process, polyester fabric is used and polyester lacing cord and sewing threads are required.

Lacing cord, as mentioned previously, is used for lacing fabric to the structure and is often referred to as **rib-stitching cord** or **rib-lacing cord**, because it is commonly used for stitching or lacing the fabric to the wings of an airplane. Acceptable lacing cords carry the specification

FIGURE 4-5   A braided cord.

FIGURE 4-6   Braided cord with a hollow center.

numbers MIL-T-6779 or MIL-C-2520A for a linen cord and MIL-T-5660 for a cotton cord. The cord must have a minimum tensile strength of 40 lb [18.14 kg] single or 80 lb [36.29 kg] double. Both the Stits and Ceconite covering systems use their own type of lacing cord made of polyester material. Do not substitute cotton or linen materials for aircraft covered with polyester fabrics.

Lacing cord is often waxed when received. If it is not, it should be waxed lightly before use. Beeswax is suitable for this purpose. Waxing is accomplished by drawing the cord under tension across a piece of the wax.

A **braided cord**, illustrated in Fig. 4-5, is made by weaving strands of thread together to form either a solidly woven cord or one with a hollow-channel center, such as that illustrated in Fig. 4-6. Some cords have a channel made with a hollow center, which contains one or more straight, individual threads called a **core**, the purpose of which is to increase the strength of the cord and to hold the outer braided cover to a rounded contour. Figure 4-7 shows a braided cord with a core.

In the covering of aircraft, the braided cords are not commonly used; however, the technician should be aware of such materials because they are encountered occasionally. Braided cords meeting the specifications MIL-C-5649 and MIL-C-5648 are approved for use in lacing fabric to aircraft structures.

Waxed cords are used for attaching **chafing strips**. These cotton or synthetic cords may be either four-ply or five-ply, but they must be double-twist and waxed. The chafing strips are sometimes hand-sewn russet-leather reinforcing strips that are placed on movable brace wires or rods at their points of intersection and on places where chafing may occur on control cables and the tubing of the structure.

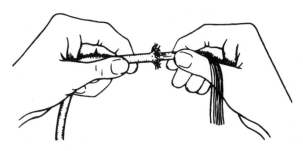

FIGURE 4-7   Braided cord with a core.

FIGURE 4-8   A metal grommet.

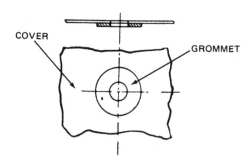

FIGURE 4-9   Installation of a plastic grommet.

In place of leather, it is common practice to use synthetic or plastic materials such as neoprene, Corprene, Teflon, and polyethylene formed as sheet or tubing to provide resistance to wear and abrasion.

Grommets

Grommets are installed where it is necessary to reinforce holes in textile materials used for drainage, lacing, or inspection. A grommet consists of one or two parts, depending upon the type. A **metal grommet**, used for lacing eyes, consists of two parts, as shown in Fig. 4-8. These parts may be made of either brass or aluminum.

Plastic grommets are used for drainage and ventilation purposes. Plain grommets are simply thin plastic washers, which are doped directly on the fabric after the first coat of dope has dried. This installation is shown in Fig. 4-9. Grommets are placed on the bottom of the trailing edges of horizontal flight surfaces, the bottom of vertical surfaces, and the low points of fuselage structures. For horizontal surfaces with a positive dihedral, the grommets are on the out-board side of the ribs. For a negative dihedral, the grommets are on the inboard side of the ribs. A neutral dihedral requires a grommet on each side of a rib. These configurations are shown in Fig. 4-10. These drain grommets allow any moisture inside the structure to drain out when the plane is on the ground or in flight.

Where exceptionally good drainage and ventilation is desired, as with seaplanes, **marine** or **seaplane grommets** are installed. A grommet of this type is shown in Fig. 4-11. This grommet is constructed with a streamlined aperture, which creates a suction and causes increased air circulation in the part to which it is applied.

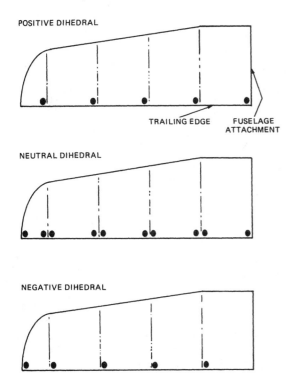

FIGURE 4-10 The bottom view of a horizontal stabilizer, showing desired drain grommet locations for different dihedral angles.

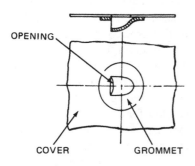

FIGURE 4-11 A seaplane grommet.

After the installation of plain plastic grommets or seaplane grommets, the fabric in the grommet opening should be cut out with a sharp knife or similar instrument. The opening should not be punched out, because this does not remove the fabric; thus the opening is likely to close and prevent proper drainage and ventilation.

Inspection Rings

Inspection rings are installed on the fabric of the fuselage or wings where it is necessary to examine fittings, internal bracing, cables, and similar items inside the covered structure. The plastic rings, about 4 in [10.2 cm] in diameter, are often doped onto the fabric in the proper location. Except when necessary for installation or adjustment, the center is not cut out until the first annual, or 100-h, inspection is made. After the inspection is made, a metal inspection cover or plate is installed in the hole. The plastic ring provides support for the inspection plate, which is held in place by

FIGURE 4-12 A plastic inspection ring.

spring clips attached to the inside surface. An inspection ring is shown in Fig. 4-12.

Special Fasteners

While the use of rib-lacing cord is the classic method of attaching fabric to wing, stabilizer, and control-surface ribs, several other methods are employed. Some of these are associated with particular aircraft designs, and some are approved through the use of Supplemental Type Certificates (STCs) or field approvals. Note that each of the methods to be discussed is used on metal ribs. Only rib lacing is used with wood ribs.

Self-tapping screws are used by some manufacturers as the standard method of attaching fabric to ribs. A plastic washer is used under the screw head to distribute the load onto the fabric surface. The screws used should be long enough to allow at least two threads of the grip (threaded part) to extend beyond the metal inside the structure. The use of screws for fabric attachment is shown in Fig. 4-13.

Pull-type rivets with large heads, shown in Fig. 4-14 are employed in some aircraft designs. These rivets require that a small hole be opened in the reinforcing tape and the fabric. The rivet is inserted through the reinforcing tape, fabric covering, and metal rib. A pulling tool is then used to expand the tail of the rivet, and the fabric attachment is complete.

Various types of metal clips have been used on aircraft as original installation methods and as approved replacement processes for original attachment methods. Fabric clips like the one shown in Fig. 4-15 have been used by aircraft manufacturers as original attachment methods. The clips are flexed and slipped through small holes in the reinforcing tape and fabric.

Martin clips, made by the Arbor Company, are often used in place of other types of attachment methods because of their simplicity and quick installation. These clips require a hole in

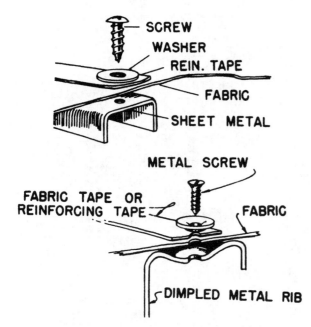

FIGURE 4-13 Use of self-tapping screws to hold fabric.

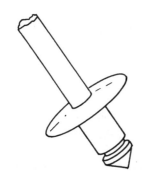

FIGURE 4-14 A fabric rivet.

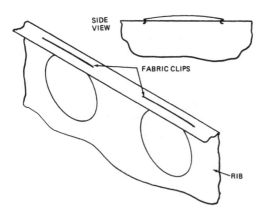

FIGURE 4-15 Fabric clips are used to hold the fabric onto the ribs in some aircraft. In this illustration, the fabric has been eliminated to show how the clips are positioned on a rib.

the fabric and the rib. The clips are flexed to reduce the size of the tooth in the wire and inserted through the opening in the structure, as shown in Fig. 4-16. The tooth then expands and locks the fabric into position on the rib. The clips are supplied in a wire roll and are cut to the lengths required for each rib.

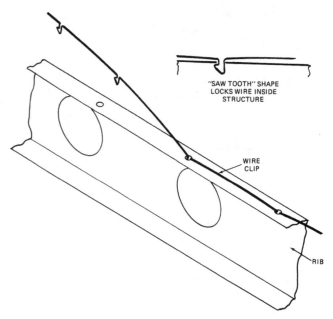

FIGURE 4-16 Martin clips are used to hold fabric to the structure in a manner similar to fabric clips. The fabric has been eliminated in this illustration for clarity.

Miscellaneous Materials

Beeswax is used to coat hand-sewing threads and rib-lacing cord prior to installation. This wax provides a coating on the thread or cord that protects it from deterioration and provides lubrication as it is pulled through the fabric.

Blued carpet tacks are sometimes used to hold the fabric covering in place on a wooden structure temporarily while the permanent hand sewing or tacking is being done. These tacks are used solely for temporary use. If driven in for permanent use, they may split, crack, and otherwise weaken the wood in which they are placed, and they eventually work loose due to the vibration of the airplane. Care must be taken that large tacks are not used with small wood structures because of the danger of splitting the wood.

Brass, tinned iron, or **Monel metal tacks** are rustproof, and they do not cause the damage that results from the use of blued carpet tacks. These tacks are sometimes used for permanently tacking the covering on wooden structures.

Pins are used for holding the edges of the covering temporarily while the cover is being shaped to the structure before the hand sewing begins. A T pin, such as the one illustrated in Fig. 4-17, is desirable because it can be inserted and removed easily while the fabric is under tension.

DOPES AND FINISHING MATERIALS

The word *dope* has many meanings, and in aircraft use it does not have a precise definition. **Aircraft dope** has been defined as a colloidal solution of cellulose acetate butyrate or cellulose nitrate; however, other solutions have been developed that serve the function of dope for sealing, tautening,

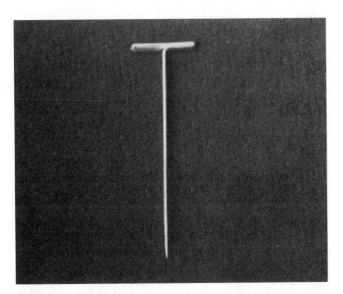

FIGURE 4-17   A T pin.

and protecting airplane fabric coverings, and these can also properly be called dopes. Because of this generalization of the term, all references to dope in this text indicate any one of several different finishing products.

Dopes are supplied as a clear coating material and as a pigmented material. Pigmented dopes have particles added that give a color to the dope. The most prominent pigment material associated with aircraft is aluminum oxide, which gives the dope a silver color and is used to reflect the sun's ultraviolet rays. Other pigmenting materials include chemical compounds of titanium, chromium, and iron.

Some pigments are prone to "bleeding" through subsequent layers of dope. That is, the pigment of the dope will move up into the next layer of paint and cause a change in the color. Red is notorious for bleeding, and if covered with a color such as white, the result will be a pink surface. Because of this problem, nonbleeding pigments have been developed and should be used when a background color such as red is to be trimmed with some other color.

Aircraft nitrate and butyrate dopes shrink as they dry. This feature is used in organic and fiberglass covering operations to aid in creating a taut fabric finish. This tautening feature is not desired when covering an aircraft with polyester material. As a result, nontautening nitrate and butyrate dopes and special dopes for polyester materials have been developed. These dopes should be used whenever the tautening action of the dope is to be kept to a minimum.

Nitrate and Butyrate Dopes

Nitrate dope is composed of nitrocellulose (similar to celluloid and gun cotton) combined with plasticizers and thinners. The plasticizers are needed to provide flexibility and resistance to cracking after the dope has cured. Nitrate dope is highly flammable in both liquid and dry states. Its principal advantages over butyrate dope are low cost, ease of application, and better adhesion.

Butyrate dope is composed of cellulose acetate butyrate with suitable plasticizers and thinners added. Butyrate dope is more fire-resistant than nitrate dope and provides greater shrinkage of fabric. Care must be taken not to apply too many coats of butyrate dope, because the fabric may become too tight and warp the structure over which the covering is applied. Some processes use a combination of a few coats of nitrate dope finished with several coats of butyrate dope.

Synthetic dopes under proprietary trade names have been developed for polyester materials and are being utilized successfully. When using such a dope, the technician must follow the manufacturer's instructions exactly.

A **quick-building (QB) dope** is designed to make an extra-heavy layer for each coat, thus making fewer coats necessary than would be the case for ordinary nitrate dope. This type of dope contains an extra-large percentage of materials that form solids upon drying.

Fungicidal Additives

Cotton and linen fabrics are subject to attacks by fungus, with the result that their strength is reduced, even though they were originally finished properly. To combat this situation, fungicides were developed for addition to the dope employed for the first coat.

A military specification, MIL-D-7850, requires that the first coat of a covering process with acetate butyrate dope be treated with the fungicide zinc dimethyldithiocarbonate. This is a powder that forms a suspension with the dope when properly mixed. The specified amount of the powder is first mixed with a small amount of dope to form a paste. Additional dope is then added and mixed with the paste. This thinned paste is then mixed with the proper amount of dope to meet the manufacturer's specifications.

Another fungicide, copper naphthonate, is also used, but it has a tendency to bleed out onto light-colored fabrics.

Fungicidal dope is applied in a very thin consistency to ensure saturation of the fabric. When this has been accomplished, the other coats can be applied in the most suitable consistency.

Aluminum Powders and Pastes

Aluminum-pigmented dope, also referred to as **silver dope**, is applied to the fabric after all the necessary components of the cover (surface tapes, inspection rings, drain grommets, etc.) have been installed. The aluminum dope contains particles of aluminum oxide, which form an aluminum layer on the surface of the fabric to reflect the ultraviolet rays of the sun. Without this aluminum pigment, the fabric would deteriorate quickly.

Aluminum-pigmented dope is made merely by adding aluminum powder or paste to the dope. For clear nitrate dope, 8 to 16 oz [227 to 454 g] of aluminum paste or 5 to 8 oz [142 to 227 g] of aluminum powder is added to 1 gal (gallon) [3.79 L] of the dope and mixed thoroughly to give the proper concentration of aluminum in the dope. To mix aluminum paste with dope, a small amount of the clear dope or dope thinner is mixed with the paste, and the mixture is worked until all lumps are completely removed. Sufficient dope is added so the mixture flows easily, and after mixing completely the mixture is added to the clear dope and the entire amount is then stirred until the aluminum is evenly distributed. To obtain the best finish, a small amount of dope thinner should be added. This helps the minute flakes of aluminum come to the surface and form a solid, light-tight layer.

Dope is available with the aluminum oxide premixed in the dope. Do not confuse this dope with pigmented dopes that may simply be giving a silver color to the dope and do not provide the ultraviolet protection for the fabric.

Rejuvenators

A **fabric rejuvenator** is a thin, dopelike finish to which powerful solvents have been added. Its purpose is to soften and penetrate old dope finishes, thus replacing some of the solvents and plasticizers that have been lost by evaporation and oxidation over a period of years. If a dope finish is not badly cracked and if the fabric under the dope is still in good condition, rejuvenation can add considerable time to the life of the covering. On the other hand, if the fabric is weak, it is likely that the rejuvenation will further weaken it and hasten the need for a re-cover job. A material specifically manufactured as a rejuvenator should be used generally for the rejuvenation of nitrate dope finishes. Thinned acetate butyrate dope can be used for rejuvenating either nitrate dope or butyrate dope finishes. This product will have more shrinking effect than nitrate rejuvenator. This makes it essential that the fabric under the dope be in very good condition.

Special Finishing Products

The use of synthetic fabrics has brought about the development of finishing products designed to be more compatible with these materials than are the conventional nitrate and butyrate products. These newer products have been developed by both the finish manufacturers and the synthetic material manufacturers as part of a covering "system."

Adhesives are often used to attach fabric to airframe structures. This has proved to be an easier method than hand-sewing operations, with no reduction in the security of attachment. Two products that are widely used for this purpose are Poly-Tak for the Stits Poly-Fiber covering process and Super Seam Cement for use with Ceconite and Razor-back glass cloth (fiberglass).

WARNING: Be aware that any product used with a covering system must be approved by the manufacturer of the covering process and the FAA. Some products are advertised as performing as well as approved products, but have not been approved for use through the FAA/PMA approval procedure. Only use quality products that carry an FAA/PMA approval on their label.

The Stits Poly-Fiber covering process makes use of several proprietary products in its covering process. These include Poly-Tak® for attaching the fabric to the structure,

Poly-Brush® for initial bonding of the finishing material to the fabric, Poly-Spray® for ultraviolet protection and buildup, and Poly-Tone® or Aero-Thane® for a durable clear or pigmented finish.

Cooper Aviation Supply Company produces several products for use with the Ceconite covering processes. These include Super Seam Cement for bonding fabric to the structure, Dac-Proofer as a first coat, and Spra-Fill as a built-up ultraviolet protective coat.

Again, care must be exercised in the selection of products for use with inorganic materials. To vary the process from what is approved by the manufacturer for that material may render the aircraft unairworthy.

Solvents and Thinners

Solvents and thinners are used in the fabric covering finishing process to clean the material prior to the application of dopes and finishing products and to reduce the viscosity of liquids so that they can be properly applied. Each type of finishing product may require the use of a specific thinner (reducer), and the product manufacturer's instructions should be consulted prior to the use of any thinners or solvents. The wrong type of thinner will usually make the coating unfit for use. When the special dopes and finishes marketed under proprietary trade names are used, it is vital that the thinner or reducer supplied or recommended by the manufacturer be employed. The following discussion deals with some of the more common solvents and thinners used with fabric covering processes.

Nitrate dope and lacquers are thinned by means of a thinner called **nitrate dope** and **lacquer thinner**. Specifications TT-T-266a or MIL-T-6094A meet the requirements for this product. Butyrate dope must be thinned with **cellulose acetate butyrate dope thinner**, MIL-T-6096A or equivalent. The technician must always be certain to use the correct thinner for the particular coating material being mixed.

Acetone is a colorless liquid that is suitable for removing grease from fabric before doping and is very useful in cleaning dope and lacquer from suction-feed cups and spray guns. It is widely used as an ingredient in paint and varnish removers but should not be used as a thinner in dope because it dries so rapidly that the doped area cools quickly and collects moisture. The absorbed moisture in the fabric then prevents uniform drying and results in **blushing**, which is moisture contamination of the dope.

Retarders

Retarder, or **retarder thinner**, is a special slow-drying thinner used to slow the drying time of dope and other finishing products. When humidity is comparatively high, rapid drying of dope causes blushing. Retarder is mixed with the dope to reduce the tendency to blush. Blushing is caused by the condensation of moisture in the surface of the dope and results in a weak and useless finish. The condition is explained more fully in a later section. Retarder for nitrate dope carries the specification MIL-T-6095A. For butyrate dope, MIL-T-6097A is used.

Consult the manufacturer's literature for approved retarders for proprietary finishing products.

FACILITIES AND EQUIPMENT FOR AIRCRAFT COVERING

The fabric covering of aircraft requires a work area specially configured for the handling of airframe structures in a controlled and fire-safe environment and a collection of quality tools and equipment that are unique to the fabric covering operations.

The Fabric Shop

The room or section of a building utilized for the preparation and installation of fabric covering is often called the **fabric shop**. This shop should be well lighted, clean, well ventilated, and of sufficient size to accommodate any size of aircraft upon which fabric is to be installed. It would be well to have the fabric shop air-conditioned or temperature-controlled for best results in covering.

The fabric shop should be capable of being sealed off from the general aircraft maintenance area, especially when the dopes and finishing products are being applied. Because of the fire hazard associated with the use of finishing products, the work area should be configured to minimize any fire hazard and adequate fire-fighting equipment should be on hand.

Tools and Equipment for Covering

The degree of success the technician will have in producing a first-class covering job often depends upon the availability of suitable tools and equipment. The tools described here are generally considered necessary in addition to the standard hand tools usually available in an aircraft repair shop.

A small **harness awl** is useful for making small holes in fabric or other similar materials.

A **magnetic tack hammer** is most useful for picking up and holding tacks that are too small to be held in the fingers while driving.

A common **pocket knife** is always useful for cutting textile materials and wood.

A variety of **needles** are required for rib lacing and hand sewing. Straight upholsterer's needles up to 16 in [40.64 cm] in length and 12-gauge diameter are needed for rib lacing thick wings. Smaller needles of the same type are used for the thinner wings. For hand sewing, both straight and curved needles are needed. Typical upholsterer's needles are shown in Fig. 4-18.

Because temporary tacks are often used to hold fabric in place on wooden structures, it is often necessary to pull such tacks. The **claw tack puller** is the best tool for this purpose.

A pair of **bent-handle trimmer's shears**, 10 or 12 in [25 or 30 cm] in length, is required for cutting fabric, tapes, etc.

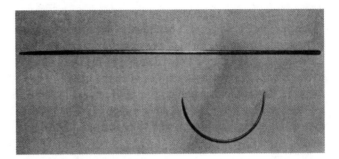

FIGURE 4-18 Upholsterer's needles.

The shears must be handled carefully to avoid damaging the cutting edges and should be used only for cutting comparatively soft materials. Trimmer's shears are shown in Fig. 4-19.

Pinking shears are needed to produce the pinked edge that is required for tape, patches, and other fabric pieces that are to be cemented or doped to another surface. The cutting edges of pinking shears must be protected when not in use and must be kept sharp. Pinking shears are shown in Fig. 4-20.

A steel **measuring tape** at least 50 ft [15.24 m] in length is necessary for measuring wings, fuselages, and lengths of fabric. The tape should be kept clean and dry.

A **sewing thimble** is useful for pushing a needle through thick seams where extra pressure is necessary.

A large **cutting table** is needed for laying out and cutting the aircraft fabric covering.

An **easel** may be used to support the airfoil in a nearly vertical position while the fabric covering is being rib-laced to the structure. In Fig. 4-21 two easels are being used to support a wing in a vertical position for rib lacing.

Trestles, or "sawhorses," are needed to support wing panels in the horizontal, or flat, position. The trestles must be adequately padded to prevent damage to the wing structure

FIGURE 4-19 Trimmer's shears.

FIGURE 4-20 Pinking shears.

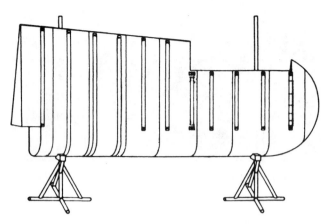

FIGURE 4-21 Two easels used to support a wing.

FIGURE 4-22 An industrial-type sewing machine. (*Singer Co.*)

and fabric. Wing panels are placed on the trestles while measuring, fitting, and installing the fabric covering. Care must be taken to see that the wing panel is supported at points where the structure is of sufficient strength to prevent damage.

A **sewing machine** used for sewing aircraft fabric is a heavy-duty, industrial-type machine larger and more durable than the family sewing machine. This type machine is shown in Fig. 4-22.

SELECTION OF FABRIC COVERING MATERIAL

The selection of the type of material to be used on an aircraft and the covering system to be used must be determined by the technician and the aircraft owner. General factors that will influence this decision include the technician's familiarity with the different systems, the operational characteristics of the aircraft, the desires of the owner, and the cost of the processes. When considering the cost of a process, take into account the initial cost of the covering operation and the life expectancy of the covering material. A system for which the price is 25 percent higher than another system but lasts twice as long may be a desirable choice.

Minimum Fabric Requirements

The minimum strength of a fabric used to re-cover an aircraft must meet the minimum strength requirements of the covering material originally used on the aircraft. When dealing with aircraft originally covered with organic materials, this evaluation can be made by determining the aircraft V_{ne} (red-line airspeed) and wing loading. The V_{ne} is found by looking at the aircraft operating limitations found in the aircraft specification. Type Certificate Data Sheet, or approved operator's handbook. If the V_{ne} is greater than 160 mph [257.5 km/h], then grade A cotton is the minimum fabric type that can be used on the aircraft.

The wing loading is found by dividing the maximum gross weight of the aircraft by the wing area. The maximum gross weight allowed can be found in the same documents used to determine the V_{ne}. The wing area might be found in the aircraft documents, or it may have to be determined by measurement of the aircraft. The wing area is the surface area of the wing as viewed from directly over the aircraft and ignores any curvature on the surface due to airfoil shape.

If the wing loading is found to be greater than 9 lb/ft² [4.39 g/cm²], then grade A cotton must be used to cover the aircraft. If the wing loading is no greater than 9 lb/ft² [4.39 g/cm²], the V_{ne} is no greater than 160 mph [257.5 k/h], and the aircraft was not originally covered with a stronger fabric, then the fabric used on the aircraft must meet the standards of intermediate-grade cotton.

It should be noted that intermediate-grade cotton is no longer commonly available, but applicants for technician's certificates are still being asked by examiners to evaluate aircraft for the minimum grade of fabric required. The minimum grade of fabric required will be useful information when evaluating fabric that is installed on an aircraft to determine when it must be replaced. This aspect of fabric evaluation is covered in a later section of this chapter.

STCs and Other Approvals

With the various inorganic processes presently available for covering an aircraft, it is often found desirable to abandon the organic fibers in favor of these more modern materials.

Stits Poly-Fiber products and Ceconite products are approved through STCs (Supplemental Type Certificates) for use on most, if not all, production aircraft having fabric coverings. If covering with these products is desired, contact the manufacturer or one of their dealers for full details on material selection. Be sure to check any material used to verify that it is the correct, approved material by examining the markings on the material. Ceconite can be identified by the word CECONITE and the type number of the fabric (101, 102, 103, etc.) stamped on the selvage of the fabric at 1-yd [0.91-m] intervals. Poly-Fiber materials can be identified by the stamping "POLY-FIBER D-101A (or D-103 or D-104), FAA PMA, STITS AIRCRAFT" on the selvage at 1-yd [0.91-m] intervals or two rows of three-line or six-line stamps spaced 24 in [0.61 m] apart in the center area, alternating each 18 in [0.46 m]. Examples of the Poly-Fiber stamps are shown in Fig. 4-23.

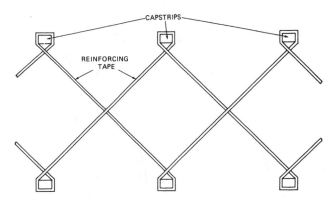

FIGURE 4-23 Examples of Stits Poly-Fiber logo used to mark Poly-Fiber fabrics. (*Stits Aircraft*)

Razorback fabric is a fiberglass material that is approved as replacement fabric for all aircraft, regardless of the type of fabric originally used. This material does not make use of an STC, but it is approved based on the requirements of FAA Advisory Circular No. 20-44. Razorback fabric can be identified by the stamping that occurs along the selvage at regular intervals.

APPLICATION OF FABRIC COVERS FOR AIRCRAFT

Fabric covers are manufactured and applied to various aircraft structural units, such as control surfaces, fuselages, and wings. Even airplanes that are generally classified as "all metal" sometimes have fabric-covered control surfaces. The instructions given here are general enough to meet the usual requirements, and at the same time specific techniques are explained in enough detail so that technicians should not encounter any serious difficulty in applying their knowledge.

Preparing the Structure

Prior to the installation of fabric on an aircraft structure, the structural integrity should be checked and all necessary repairs made. All wires and cables should be in place or "fish wires" should be in position so that after the cover is installed, the cables and wires can be pulled through the structure using the fish wires. All protrusions should be covered or cushioned so that they will not puncture the fabric covering. Any coating or finish, such as zinc chromate or spar varnish, should be covered with dope-proof paint, aluminum foil, cellophane tape, or masking tape. This prevents the cements, dopes, and other finishing products used on the fabric from attacking the protective coatings.

When covering a fuselage it is often desirable to install the headliner and all internal systems and components prior to covering it with fabric; wings should have all the ribs braced in position by installing inter-rib bracing to prevent the ribs from twisting and shifting when the fabric is installed and tightened. Reinforcing tape is often used for this purpose, as shown in Fig. 4-24.

The required type, number, and desired location of inspection openings should be determined. Factory drawings

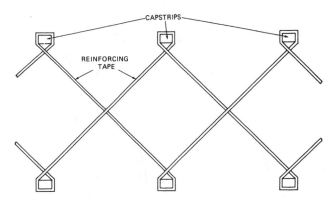

FIGURE 4-24 Application of inter-rib bracing.

can be used for this purpose, as can the old cover that was removed from the structure. If no information is available as to the proper location of these openings, make a sketch of the structure and note the locations necessary to install, adjust, and inspect the structure.

Fabric Seams

Fabric seams are used to join pieces of fabric together and to attach fabric to the aircraft structure. Machine-sewn seams are used to join large pieces of fabric together to form blankets or envelopes by using a sewing machine. There are four types of machine-sewn seams: the **plain overlap**, the **folded fell**, the **French fell**, and the **modified French fell**. The French fell and folded fell are the strongest of these seams. The modified French-fell seam is often used because of its ease of sewing. A plain overlap seam is sometimes used where selvage or pinked edges are joined. Examples of each of these seams are shown in Fig. 4-25.

Hand sewing is used to close fabric edges after the material is positioned on the structure. Before starting the hand-sewing operation, the fabric is straightened on the structure and tack stitches are placed about every 6 to 12 in [15.24 to 30.48 cm] along the opening to be sewed. Tack stitches are simply stitches through the two pieces of fabric being joined that hold the fabric in position while the hand-sewing operation is performed.

Hand sewing is performed by starting at one end of the opening, folding the fabric edges under until the edges of the

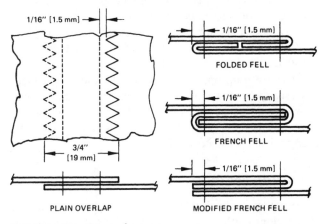

FIGURE 4-25 Types of seams.

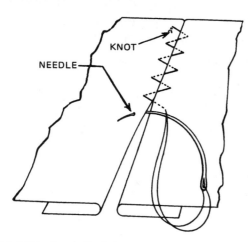

FIGURE 4-26 Baseball stitch.

fabric just touch. There should be at least ½ in [12.7 mm] of material folded under, with any excessive amount of material being cut off. The needle is pushed through the fabric no more than ¼ in [6.4 mm] from the edge of the fold, pushed through the other fold, and tied in a square knot locked with a half hitch on each side. A baseball stitch is then used, as shown in Fig. 4-26, until the opening is closed. The baseball stitches should be no more than ¼ in [6.4 mm] back from the edge of the fold and should be spaced no more than ¼ in [6.4 mm] apart along the opening. A lockstitch should be included in the hand sewing at 6-in [15.24-cm] intervals. This will prevent the whole seam from opening if a thread should break. A lockstitch is a modified seine knot, which is discussed later in this chapter. At the end of the sewing, the seam is finished with a lockstitch and a half hitch. If the original thread is not long enough to completely close the seam, the stitching should be tied off with a modified seine knot and half hitch and a new thread started at the next stitch spacing.

Doped seams are formed by using approved adhesives to attach the fabric to the structure and to attach the fabric to another piece of fabric. To attach the fabric to the structure, a strip of adhesive is brushed onto the structure and the fabric is immediately placed on the adhesive and pressed into it. Fabric is bonded to fabric in a similar manner, with a coat of adhesive being applied to the fabric and the other piece of fabric being pressed into the first coat. The amount of minimum overlap of the fabric onto the structure or another piece of fabric varies, depending upon the materials and adhesives being used, but it is in the range of 2 to 4 in [5.08 to 10.16 cm]. Consult the instructions for the adhesive and fabric being used for specific information.

Most covering operations require the use of machine-sewn seams, but machine sewing by the technician has been all but eliminated because of the availability of envelopes. In most covering processes, doped seams have eliminated the need for hand-sewing operations.

Covering Methods

There are two methods that can be used to cover a structure, the **envelope method** and the **blanket method**. Each method

has its own advantages. It is up to the technician to determine which best fits the project at hand.

The envelope method involves making or buying a sleeve that can slide over the prepared structure. The sleeve has been sewed together on a sewing machine, so only a small portion of the material must be closed by hand sewing or a doped seam.

If an envelope must be made, the component to be covered should be measured to determine the size of the envelope. When laying out the fabric, special consideration is given to the width of the fabric (selvage to selvage) and the spacing between the ribs or stringers. The machine-sewn seams should be located so that no seam is positioned where any rib lacing or other attachment mechanism will penetrate the seam.

Once the dimensions are available and the position of the seams is determined, the fabric strips are sewed together. In most instances, the seams run longitudinally on the fuselage and chordwise on wings and other surfaces. Seams perpendicular to the direction of flight are acceptable if they do not disrupt the airflow. Once the fabric strips are sewed together, the fabric is folded over and sewed to form a tube. One end of the tube is normally closed to complete the sleeve. This envelope construction process is shown in Fig. 4-27. The envelope is now ready to install on the structure.

The blanket method of covering involves the use of fabric as it comes off the roll. It is cut to size and folded over the structure to be covered, as shown in Fig. 4-28. If the structure is too large to be covered by the material as it comes off the roll, strips of the material are sewed together, as is done in the first step of making an envelope. The fabric is then wrapped around the structure and closed along the open edge by hand sewing or using a doped seam.

Installation of Fabric

Once the method of covering is determined and the structure is prepared to receive the cover, the fabric is positioned on the structure. If the covering material is organic or fiberglass, sliding an envelope onto the structure will require some care, since the envelope will be a snug fit on the structure. Polyester envelopes are loose-fitting and slide on easily. Regardless of the type of material, care will be required to prevent snagging or tearing of the fabric as it is positioned on the structure.

Once the fabric is in position, all seams are closed by hand sewing, mechanical attachments, or doped seams. The only restriction placed on the use of doped seams concerns organic fabrics. Unless the adhesive being used is approved for higher speed, doped seams may not be used with organic fabrics when the V_{ne} is greater than 150 mph [241.4 km/h] when using the blanket covering method. Also, unless otherwise approved by the manufacturer of a covering system, all doped seams should overlap the fabric by at least 4 in [10.16 cm]. Seams of this type commonly have a 4-in-[10.16-cm-] wide piece of surface tape placed over the seam if the seam is at the leading edge of a surface; a 3-in-[7.62-cm-] wide

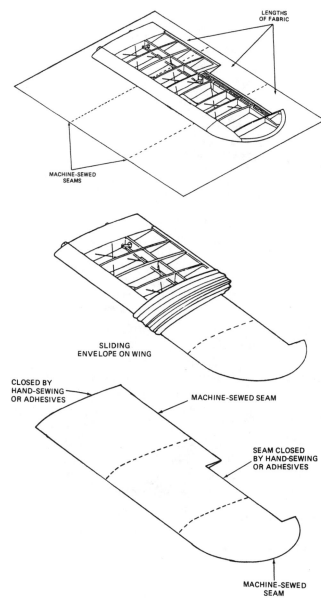

FIGURE 4-27 Prominent steps in making and installing an envelope cover. (*Christen Industries Inc.*)

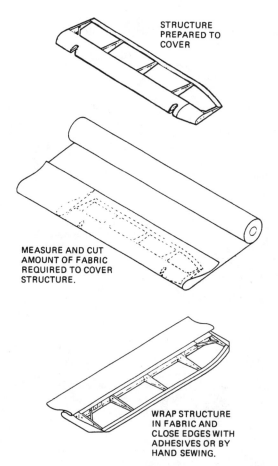

FIGURE 4-28 Covering an aileron by the blanket method.

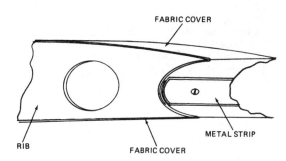

FIGURE 4-29 Metal strip used to attach fabric to a structure.

surface tape is used if the seam is at the trailing edge. A typical mechanical attachment method is shown in Fig. 4-29.

With the fabric in position, organic and polyester materials can be shrunk to some extent. Organic fabrics are preshrunk prior to the application of dopes by saturating the fabric with water. This is done by wiping down the fabric with a wet sponge and allowing the fabric to dry. Polyester fabrics are shrunk by the use of an electric iron. With the iron set for the temperature recommended by the fabric manufacturer (generally around 250°F [121.1°C]), the fabric is shrunk by moving the iron smoothly over the surface to remove all the sag and wrinkles from the fabric. Care must be taken during this process to keep the sewn seams straight and away from areas where stitching or mechanical attachment of the fabric would penetrate the seam. Complete details and steps to be followed for specific polyester products can be found in the fabric manufacturer's literature.

To give the details of each finishing process established by the inorganic fabric manufacturers at this point would be very confusing, because the processes are quite varied. Therefore, the remaining discussion deals with the organic process. A summary of current inorganic processes is given in Table 4-1. This table is for general reference only, since the processes are subject to change as the products continue their evolution.

Once the fabric is smooth and properly taut, the first coat of clear dope is applied. This should be a thin coat of nitrate dope containing a fungicide and applied with a brush to proper penetration of the fabric.

With the second coat of clear dope, install the antitear tape, reinforcing tape, drain grommets, inspection rings, and reinforcing patches.

TABLE 4-1 Summary of Fabric Covering Processes

Operation	Organic	Razorback	Poly-Fiber	Ceconite
Shrinking prior to doping.	Soak with water.	None.	Electric iron.	Electric iron.
Initial coat.	Brush on nitrate dope with fungicide added.	One coat of nitrate dope sprayed on.	Poly-Brush applied by brush or spray gun (with some limitations on the use of the gun).	Thinned nitrate dope mixed with Super Seam Cement and brushed on.
Build up coats.	Nitrate and butyrate dope. Minimum of two buildup coats, brushed or sprayed.	Spray on two to five coats of butyrate dope until fabric is tautened.	N/A.	Two brushed on coats of nitrate dope.
Installation of grommets, rib stitching, fasteners, and surface tapes.	Install with second coat of dupe.	Perform after fabric has tautened.	Attach with local coat of Poly-Brush.	Accomplish after third brush coat of nitrate dope. Use Super Seam Cement with nitrate dope for proper adhesion of tapes.
Build up coats.	One coat of clear dope, brushed or sprayed.	Apply two coats of butyrate dope by brush.	Second coat of Poly-Brush applied with spray gun.	Brush or spray on three coats of dope.
Ultraviolet protection.	Minimum of two coats of aluminum-pigmented dope.	Apply one coat of aluminum dope as a filler coat and undercoat for light-pigmented finish coats.	Minimum of three coats of Poly-Spray for buildup and ultraviolet protection, applied with a spray gun.	Apply two coats of aluminum dope.
Finishing coats (pigmented).	Minimum of three coats.	Spray on two coats of pigmented dope.	Poly-Tone or Aero-Thane applied by spray gun.	Apply three coats of pigmented dope.

Note: Poly-Fiber, Poly-Brush, Poly-Spray, Poly-Tone, and Aero-Thane are registered trademarks of Stits Aircraft.
Super Seam Cement and Ceconite are trademarks of Ceconite, Inc.
Razorback is a trademark of Razorback Fabrics, Inc.
Contact the fabric manufacturer for full details on approved covering processes.

Antitear tape is surface tape, wide enough to extend beyond the reinforcing tape, that is placed over the top of all ribs on aircraft with a V_{ne} of more than 250 mph [402.3 km/h]. For these aircraft, tapes are also placed on the bottom of all ribs within the propeller slipstream. The **slipstream** is defined as the propeller diameter plus one extra rib space.

Reinforcing tape is placed on the top and bottom of all ribs and on all stringers that are to be fastened to the fabric covering by rib lacing or mechanical fasteners. This tape should be the same width as that of the rib or stringer.

Drain grommets, inspection rings, and reinforcing patches are placed on the surface at the locations described previously in this chapter.

Rib Lacing and Other Attachment Methods

Following the application of the initial coats of dope or finishing products, the fabric covering must be attached to internal structural components by the **rib-stitching process** or mechanical processes. These processes are commonly required on wings, stabilizers, and control surfaces. Their use on fuselages is often limited to stabilizer surfaces incorporated in the fuselage basic structure.

The mechanical processes have been discussed in an early part of this chapter and include the use of screws, rivets, and metal clips. The following discussion covers the rib-stitching process, since it is the most complex of the operations. The only part of the rib-stitching process applicable to the other processes involves the determination of attachment spacing and layout procedures. For this discussion the structure considered will be a wing.

Rib lacing, or rib stitching, must be accomplished in accordance with the aircraft manufacturer's instructions. The spacing between the stitches can be determined from the covering removed from the aircraft or from the aircraft manufacturer's maintenance information. If this information is not available, the chart shown in Fig. 4-30 may be used to establish an acceptable maximum rib-stitch spacing. Note that the spacing is based on the aircraft V_{ne} and the location of the ribs in relation to the propeller slipstream. The maximum spacing allowed for aircraft with speeds just below 250 mph [402.3 km/h] and all higher airspeed values is 1 in [2.54 cm].

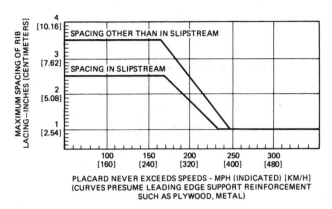

NOTES
1. IF ORIGINAL RIB STITCH SPACING CANNOT BE DETERMINED, USE SPACING INDICATED IN THESE CURVES.
2. LACING TO BE CARRIED TO LEADING EDGE WHEN VELOCITY EXCEEDS 275 MPH [440 KM/H]

FIGURE 4-30   Chart to show spacing of rib stitches.

Some manufacturers have determined that the spacing on their aircraft can be greater than that indicated on this chart. In such cases, the manufacturer's information should be followed. Be aware that these spacing values are maximums; spacing may be made closer as required, but it may not be greater.

The top of one rib should be marked with the proper stitch spacing. Start at the forward area of the rib, just aft of the leading-edge material, and mark the first stitch. The second stitch is marked at one-half of the required spacing. The rest of the stitches are marked at the maximum distance allowed or some closer spacing. The distance between the last two stitches is half-spaced. If the use of the maximum allowed spacing does not match the length of the rib, the spacing can be decreased evenly along the length of the rib, or a short-stitch spacing can be used near the front or rear area of the rib. Examples of stitch spacing are shown in Fig. 4-31.

With one rib laid out, the same pattern can be repeated on the top and bottom of all ribs. On many aircraft, the spacing between each stitch on the bottom of the rib must be reduced slightly due to the shorter rib length resulting from the difference in curvature between the top and bottom of the rib. The rib stitching should pass through the wing as perpendicular

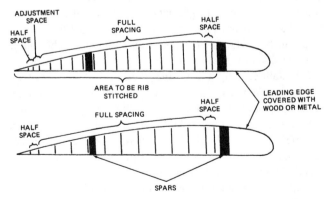

FIGURE 4-31   Rib stitching requires that the first and last stitches be half-spaced.

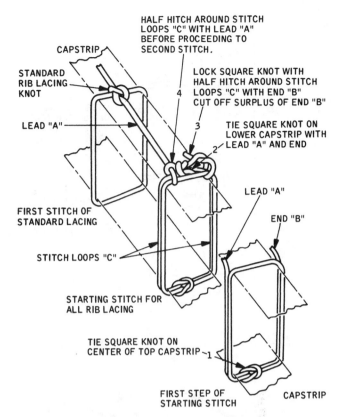

FIGURE 4-32   How a double-loop starting stitch is made.

to the wing chord line as possible. Some technicians find it desirable to prepunch the stitching holes with a needle before starting the operation.

The wing is normally placed vertically on easels with the leading edge down when performing the rib-stitching operation. Plan the operation so that the lacing knots are on top surfaces of high-wing aircraft and on the bottom of low-wing aircraft. The stitching can be started at either end of the rib but is commonly started near the leading edge. Properly waxed rib-stitching cord is used for this operation.

The starting stitch is a double loop and is performed as shown in Fig. 4-32. This stitch involves the use of square knots and half hitches. The remaining stitches are modified seine knots, which are tied as shown in Fig. 4-33. This knot is different from the seine knot. Some aircraft require that all rib stitching use a double loop for strength. Each loop involves tying the modified seine knot. The knots are to remain on the surface of the fabric and are not pulled back inside the structure unless this is approved by the aircraft manufacturer or the STC's process. The last stitch is a modified seine knot, secured with a half hitch.

If the cord should break or be too short to complete the stitching of a rib, a second piece can be attached. This is done by the use of the splice knot, as shown in Fig. 4-34. This knot should be positioned on the length of cord so that it is located inside the structure, as in Fig. 4-35. The splice knot should not be located on the surface of the fabric.

When all the rib stitching is completed, the remainder of the covering process can be performed.

Application of Fabric Covers for Aircraft   **103**

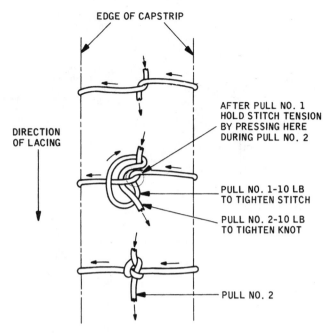

FIGURE 4-33 Making a modified seine knot.

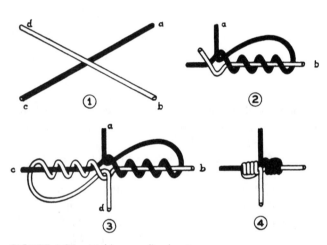

FIGURE 4-34 Making a splice knot.

Application of Surface Tape

Upon completion of rib lacing, the fabric cover should be ready for the application of **surface** (finishing) **tape**.

Surface tape should be applied over all rib lacing, seams, leading and trailing edges, and other points where reinforcement is necessary. A coat of dope is brushed on the areas where the tape is to be applied and then the tape is immediately placed on the wet dope. Another coat of dope is brushed over the tape and care is taken to see that all air bubbles are worked out from under the tape. An excess of dope will cause runs and sags and may drip through the fabric to the inside of the wing; therefore the technician must use only the amount of dope required to fill the tape and bond it securely to the surface.

If the aircraft has a V_{ne} greater than 200 mph [321.8 km/h], all surface tape placed on trailing edges should be notched at 18-in [45.72-cm] intervals. If the tape begins to separate from the trailing edge, it will tear loose at a notch

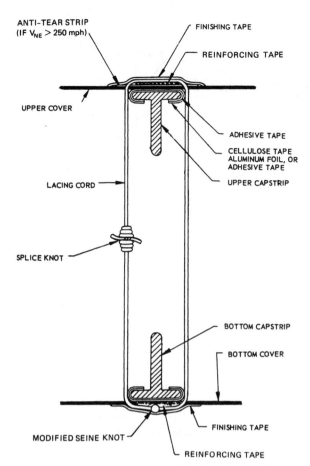

FIGURE 4-35 The splice knot should be located inside the structure, not on the surface of the fabric.

and prevent the loss of the entire strip. If the entire scrip were lost, the controllability of the aircraft could be affected.

Finishing Process

When the fabric is properly attached to the structure and all the surface tapes, grommets, and inspection rings are in place, the remaining coats of dope or finishing material can be applied. The application may be by brushing or spraying, depending on the process being used.

Using the organic process, one coat of clear dope is applied, followed by at least two coats of dope containing the aluminum pigment previously discussed. There should be enough aluminum-pigmented dope on the fabric so that light will not pass through the coating. Light blocking can be determined by placing a fire-safe light source on one side of the fabric and looking through the other side of the fabric. If light is visible, then the coating is not yet sufficient. When making this determination, all surrounding light should be blocked out on the viewing side. When the aluminum coating is sufficient, at least three coats of pigmented dope should be applied. These are normally sprayed on. Between each coat of dope, the surface can be sanded lightly to remove any irregularities. *Note:* Care must be taken when sanding around any protrusions such as rib-stitch knots so that the sandpaper does not cut through the dope and into the fabric.

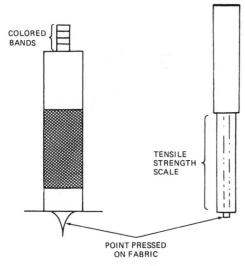

FIGURE 4-36   A Seyboth-type tester, on the left, uses colored bands to indicate the fabric strength. A Maule-type tester has a scale to indicate the fabric strength.

A thorough discussion of painting equipment, techniques, and problems can be found in Chap. 5.

FABRIC INSPECTION

The testing of fabric is used to determine if the fabric has sufficient strength to assure safe operation of the aircraft. The fabric should be tested at regular intervals, normally during each annual inspection.

Causes of Deterioration

The common aircraft fabric coverings in use today are cotton, linen, dacron, and glass cloth. Of these materials, glass is the only material that does not deteriorate under normal conditions. For this reason, glass coverings are visually inspected for damage and dope-coat deterioration only. No fabric-strength test is necessary. The rest of the covering materials deteriorate in varying amounts due to sunlight, aircraft fluids, mildew, and industrial wastes.

The portion of the fabric to be tested should be in areas where deterioration would be the most rapid. These areas include surfaces that receive the most sunlight, such as the top of the wing and the top of the fuselage. Areas painted with dark colors absorb the most ultraviolet rays from the sun. Industrialized areas are particularly detrimental to fabric coverings because of the sulfur dioxide given off to the atmosphere. Sulfur dioxide, when combined with oxygen, sunlight, and water, forms sulfuric acid, which is very caustic to fabric coverings. Hydraulic fluids, powerplant fuels and oils, and battery acids are among the aircraft fluids that take their toll on aircraft fabrics, especially the lower fuselage areas.

Testing Methods

The strength of the fabric is based on its tensile strength. Normal testing techniques involve the use of a Seyboth or Maule

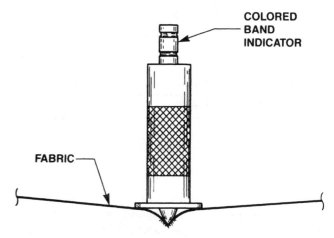

FIGURE 4-37   A punch tester should be held at right angles to the fabric being checkd.

"punch" tester. These testers are shown in Fig. 4-36. The Seyboth tester penetrates the fabric and indicates the strength of the fabric by a scale on the top of the tester. The Maule tester applies pressure to the fabric and is not normally used with enough force to penetrate airworthy fabric. A scale on the side of the Maule tester indicates the force that is being applied in pounds per inch of tensile strength.

The punch tester is a popular tool for checking fabric condition. This method has gained acceptance with maintenance technicians because the fabric is tested in the doped (as-is) condition. A sample need not be removed from the aircraft, making testing quicker and easier.

When using the punch tester, it should be kept at right angles to the fabric surface being checked, as shown in Fig. 4-37, and away from the internal aircraft structure. Care should also be exercised in selecting an area where two layers of fabric do not exist, such as seams covered with surface tapes.

The number of dope coats on the fabric's surface may affect the punch tester's indication of fabric strength. For example, the tester would be accurate on aircraft fabric with the average number of surface dope coats. Those aircraft with more than the average number of dope and/or paint coats may indicate greater fabric strength, whereas the actual strength value is less. For this reason, punch testers are considered only an approximate indication of fabric strength.

To perform an exact test of fabric, a tensile test should be made. When performing this test, a sample of fabric is taken from the weakest area of the aircraft covering. Determine the weakest fabric on the aircraft by punch testing. A strip of fabric 1 in [2.54 cm] wide and several inches long is cut from the installed material. All the dope is removed from the fabric. The fabric is then clamped at one end to a supporting fixture, and a clamp on the other end is attached to a load. The load is increased until the minimum standards are met or until the fabric breaks. Figure 4-38 shows one method of performing this test.

Required Tensile Strength

Regardless of the method used to test fabric strength, the minimum tensile strength in pounds will be the same. Referring back to the discussion of the selection of fabric for use

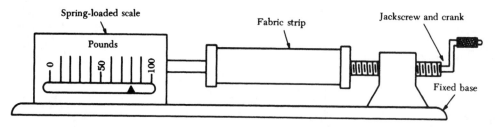

FIGURE 4-38 One method of testing the tensile strength of a fabric sample.

on an aircraft, the fabric that is installed has to meet the requirements only of the **minimum-grade** fabric that was required on that aircraft. For example, if an aircraft required only intermediate-grade cotton and was covered with grade A cotton, the grade A cotton can deteriorate to the minimum values allowed for intermediate-grade material before it must be replaced.

The minimum strength for an aircraft requiring grade A cotton is 56 lb/in [10 kg/cm]. For an aircraft requiring intermediate-grade cotton, the minimum strength is 46 lb/in [8.21 kg/cm]. Polyester fabrics should be tested to the minimum value required of the original fabric covering, which may be a higher value than for the organic materials. Razorback glass fabric does not have to be tested for tensile strength.

When the inspection is completed and if the fabric is airworthy, the holes created by the fabric testing must be repaired before returning the aircraft to service. Indentations left by the Maule tester often return to a smooth surface of their own accord.

REPAIR OF FABRIC COVERINGS

When fabric is damaged, the technician must consider several factors to determine the method of repair. First, is the damage repairable or should the entire covering be replaced? Although the damage may be repairable, if the remainder of the original fabric is only marginal in strength, it may be advisable to replace the entire covering. Another important consideration concerning the type of repair that can be performed is the V_{ne} of the aircraft. Last, where are the internal structural members in relation to the damaged area? These factors influence the selection of the type of repair to be performed.

Tears in Fabric

Tears in a fabric covering can usually be repaired by sewing and doping on a fabric patch. The objective is to restore the original strength and finish to the repaired area.

A single tear should be repaired by removing all of the pigmented and aluminized dope around the area to be covered with the patch and then sewing the tear using a baseball stitch, as shown in Fig. 4-39. The dope can be removed by softening and scraping or by sanding. The most satisfactory method is to apply a heavy coat of dope to the area and allow

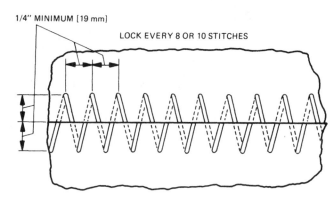

FIGURE 4-39 Sewing a tear with a baseball stitch.

it to soften the old surface dope, which can then be removed by scraping. Strong solvents such as acetone can be used to soften the old dope, but care must be taken to see that the solvent does not drip through the opening to the lower surface of the fabric, where it will cause blisters. When the cleaned surface around the tear has been sewn and the stitches locked every 8 to 10 stitches, a piece of pinked-edge surface tape or fabric is doped over the seam. The tape or fabric patch should extend at least 1½ in [3.81 cm] beyond the tear in all directions. Additional coats of dope are applied to the patch, sanding between coats to produce a smooth finish. The final costs of pigmented dope are applied and finished according to procedures explained previously. The dope used must be compatible with the original.

If a tear is of the V type, the procedure is the same as that just described; however, the sewing should start at the apex of the V in order to hold the fabric in place while the seams are completed. This is illustrated in Fig. 4-40.

Doped Repairs

Doped-on repair patches can be employed on all fabric-covered aircraft that have a never-exceed speed not greater than 150 mph [241.35 km/h]. A doped-on patch can be used for a damaged area that does not exceed 16 in [40.64 cm] in any direction. A repair of this type is made by trimming the damaged area and then removing the old dope in the area where the patch is to be applied. The patch is cut to a size that will overlap the old fabric at least 2 in [5.08 cm] for any patch not over 8 in [20.32 cm] across. For holes between 8 in [20.32 cm] and 16 in [40.64 cm], the patch should overlap the original fabric by one-quarter the distance of the major dimension of the repair.

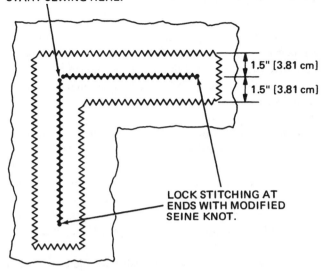

1.5" [3.81 cm]
1.5" [3.81 cm]

LOCK STITCHING AT ENDS WITH MODIFIED SEINE KNOT.

FIGURE 4-40   Method of sewing a V-type tear.

Where doped-on patches extend over a rib, the patch must be cut to extend at least 3 in [7.62 cm] beyond the rib. The patch is then laced to the rib over a new piece of reinforcing tape. The original lacing and reinforcing tape should not be removed. A piece of surface tape is placed over the new lacing on the top and bottom of the structure and the dope is built up using the sequence for a new fabric cover—clear coats, aluminum coats, and pigmented coats.

Sewn-in Patch

If the V_{ne} of the aircraft is greater than 150 mph [241.4 km/h] and the damage does not exceed 16 in [40.64 cm] in any direction, a sewn-in patch can be used. The first step is to trim out the damage in a rectangular or circular shape. The dope is then removed for at least 1½ in [3.81 cm] from the edge of the fabric in the same manner as previously discussed. A patch of material of the same type as being repaired is then cut to be sewn into the opening. This patch should be cut so that there will be at least ½ in [12.7 mm] of extra material on all sides. The patch is then marked for the size of the opening. Tack stitches are used to

hold the patch in position, with the marked edge just touching the edges of the cut opening. The patch is then sewed in using a baseball stitch, with lockstitches every 8 to 10 stitches. The ½ in [12.7 mm] of extra material is folded inside the structure during the sewing operation, as shown in Fig. 4-41, which prevents the threads of the patch from being pulled out of the fabric.

A surface patch is now cut to cover the sewn-in patch. This patch should extend beyond the cut edge of the fabric by at least 1½ in [3.81 cm]. If the cut edge is within 1 in [2.54 cm] of a structural member, the surface patch must extend at least 3 in [7.62 cm] beyond the members. The surface patch is attached by dope. Rib stitching is performed over new reinforcing tape if either the sewn-in or surface patch covers a structural member. Surface tape is placed over the rib stitching, and the patched area is finished with dope. Figure 4-42 shows a sewn-in patch repair.

Doped-on Panel Repair

When the damage to an aircraft fabric surface is greater than 16 in [40.64 cm], a panel should be doped on. In this type of repair, the old fabric is cut out along a line approximately 1 in [2.5 cm] from the ribs nearest the repair. The fabric on the leading and trailing edges is not removed unless both the top and bottom of the wing are to be repaired. The surface tape is removed from the ribs adjacent to the repair, but the lacing and reinforcing tape are left intact. The patch panel is cut to a size that will overlap the trailing edge by at least 1 in [2.54 cm], extend around the leading edge and back to the forward spar, and extend at least 3 in [7.62 cm] beyond the ribs on each side of the repair.

If the leading edge of a wing is either metal- or wood-covered, the patch may be lapped over the old fabric at least 4 in [10 cm] at the nose of the leading edge.

The area of the old fabric that is to be covered by the patch must be thoroughly cleaned, and a generous coat of new dope must be applied. The new panel is then put in place and pulled as taut as possible. A coat of dope is applied to the patch where it overlaps the old fabric. After this coat has dried, a second coat of dope is applied to the overlapped area.

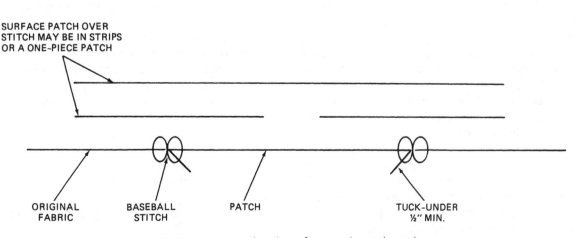

SURFACE PATCH OVER STITCH MAY BE IN STRIPS OR A ONE-PIECE PATCH

ORIGINAL FABRIC

BASEBALL STITCH

PATCH

TUCK-UNDER ½" MIN.

FIGURE 4-41   An edge view of a sewn-in patch repair.

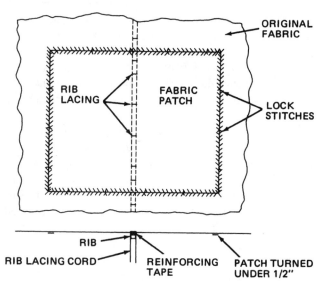

FIGURE 4-42   A sewn-in patch.

Reinforcing tape is placed over the ribs under moderate tension and is laced to the ribs in the usual manner. The rib stitches are placed between the original rib stitches. The new patch panel is then given a coat of dope and allowed to dry. Surface tape is applied with a second coat of dope over the reinforcing tape and edges of the panel. Finishing of the panel is then accomplished in the normal manner.

Sewn-in Panel Repair

If a panel repair cannot give the proper tautness by using the doped-on panel repair, a sewn-in panel repair can be performed. As with the doped-on panel, this repair is for damage exceeding 16 in [40.64 cm] in any one direction.

To perform the repair, remove the surface tape from the ribs, the leading edge, and the trailing edge adjacent to the damaged panel. Trim back the damaged fabric to within 1 in [2.54 cm] of the centerline of the adjacent ribs. Fabric should not be removed from the leading and trailing edges unless the repair involves both the top and bottom fabric surfaces. Do not remove the reinforcing tape and rib stitching at the ribs.

A patch should be cut that will extend 3 in [7.62 cm] beyond the ribs, to the trailing edge, and around the leading edge to the front spar on the opposite side of the wing.

Clean the area of the original fabric to be covered by the patch and pin or tack-stitch the patch in place. Take care to pull the patch tight and eliminate any wrinkles. The patch is now attached to the original fabric by hand sewing with the edge of the patch tucked under $\frac{1}{2}$ in [1.27 cm]. After the patch is attached, new reinforcing tape is laid over the ribs and the patch is rib-stitched to the ribs.

A coat of clear dope is now applied. Surface tapes are then laid along the sewed seam, over the rib-stitching, and at other areas appropriate for the aircraft area repaired. The surface is finished following the regular doping procedures.

This type of repair can be used to cover both the top and bottom surfaces of one or more adjacent rib bays.

Fabric Rejuvenation

In some cases it is possible to rejuvenate or restore the condition of dope coatings on an aircraft. When the dope has hardened to the extent that it is beginning to crack, it should be rejuvenated, provided that the fabric underneath is still strong enough to meet FAA specified requirements. If the fabric is at or near minimum strength, no attempt should be made to rejuvenate it.

Fabric rejuvenator is a specially prepared solution containing strong solvents and plasticizers designed to penetrate age-hardened dope and restore its flexibility. The fabric surfaces to be rejuvenated should be thoroughly cleaned to remove wax, oil, and any other contaminating material. Rejuvenator is then applied to the dry dope surface in accordance with the manufacturer's directions. The application is usually performed with a spray gun.

After the application of rejuvenator, the old dope should soften through to the fabric. Cracks then can be sealed and the surface can be allowed to set. Finishing coats of clear and pigmented dopes can then be applied in the normal manner.

There are several rejuvenator products on the market, and the manufacturer's instructions for each should be followed. Be aware that in some cases the rejuvenator will cause the fabric to sag; it will not retauten as the solvents evaporate. This may require that the surface being treated be re-covered. For this reason it is advisable to test the effect of a rejuvenator on a small surface before attempting rejuvenation of any large surfaces.

REVIEW QUESTIONS

1. Name types of fabric suitable for covering an airplane.
2. What is the difference between *warp* and *fill*?
3. What is the *selvage* of fabric?
4. What is meant by *bias-cut fabric*?
5. What is the minimum tensile strength for new grade A fabric?
6. To what extent may the tensile strength of grade A fabric decrease while it is on the airplane before it must be replaced?
7. Give the conditions that make mandatory the use of grade A fabric or the equivalent.
8. Under what conditions may intermediate-grade fabric be installed on an airplane?
9. Describe *surface*, or *finishing, tape*. For what purpose is it used?
10. What is the purpose of *reinforcing tape*?
11. Why is Dacron polyester fiber more desirable as a covering than grade A fabric?
12. What is the difference between *S-twist* and *Z-twist* threads?
13. What type of *wax* is used to coat thread when hand sewing?

14. Under what conditions should *seaplane grommets* be installed?

15. Describe the installation of a *plastic inspection ring*.

16. What methods may be used to attach fabric to wing ribs?

17. Why should the fabric in the center of a dram grommet be cut out rather than punched out?

18. What is the purpose of using aluminum-pigmented dope on aircraft fabric?

19. Compare acetate butyrate dope and cellulose nitrate dope.

20. What thinner should be used for nitrate dope?

21. What is the purpose of a *retarder*?

22. What is the cause of *blushing*?

23. At what point in a doping operation should fungicidal dope be applied?

24. List the tools and equipment needed in a fabric shop.

25. What types of needles are required for aircraft fabric work?

26. Why is it necessary to notch the edges of surface tape at the trailing edge of a wing?

27. At what intervals should hand-sewn seams be locked with a lock stitch?

28. What types of stitches are used in hand sewing?

29. Name the types of machine seams used with aircraft fabric.

30. Under what conditions can *doped seams* be used?

31. What width of lap is required for a lapped and doped seam at the trailing edge of a wing or control surface?

32. Describe the *envelope* and *blanket methods* for covering a wing.

33. Why is it a good practice to save the old fabric in large sections when re-covering an airplane?

34. Describe the purpose of *inter-rib bracing*.

35. What materials are used for dope-proofing?

36. Describe the function and installation of *antitear tape*.

37. What determines the spacing between rib stitches?

38. What type of knot is used to lock each rib stitch?

39. At what locations must *surface tape* be installed?

40. At what point in the sequence of covering operations are *inspection rings* and *drain grommets* installed?

41. What method is used to tighten polyester fabric?

42. Under what conditions is a *sewn-in patch* required for damaged fabric?

43. Under what conditions is a *doped-on patch* allowed?

44. What is the purpose of a *fabric rejuvenator*?

45. Describe various methods for testing fabric to determine the extent of deterioration.

Aircraft Painting and Markings 5

INTRODUCTION

Aircraft are normally painted to provide both protection for the aircraft surfaces and a pleasing appearance. Many different types of finishing materials, generally called *paints*, are used on aircraft. Each type serves one or more purposes, and each must be applied in some specific manner to ensure proper adhesion and an acceptable durability.

This chapter examines the equipment used to apply paints and the different paints that are commonly used for aircraft. In addition, this chapter discusses the application of paint trim and the application of aircraft registration numbers. Keep in mind that the information in this chapter is general in nature and the manufacturers of specific painting equipment and finishing products should be consulted concerning the proper use and application of their individual products.

Spray painting is the commonly used method of applying paint over large surfaces. The majority of this chapter deals with the spray-painting process.

AIRCRAFT FINISHING MATERIALS

Paint Systems

The paint on an aircraft is referred to as a paint system. A paint system is a family of products that are compatible. A typical paint system includes: cleaners, conversion coatings, primers, and top coats. Most often the individual products in a paint system are not interchangable with other products of a different paint system or a different manufacturer. Modern high-solids paint systems have replaced traditional low-solids paints due to environmental regulations. Figure 5-1 shows a popular paint system used for propeller aircraft.

The U.S. Environmental Protection Agency issued new requirements for the use of primers and topcoats with low volatile organic compound (VOC) content effective September 1998. These requirements, called the National Emissions Standards for Hazardous Air Pollutants (NESHAP), reduced the VOC content of primers to 350 g/L from 650 g/L and the VOC content of topcoats to 420 g/L from 600 g/L. The reduction in the amount and type of solvents used in high-solids primer and topcoat created a challenge for both suppliers and users in developing and qualifying acceptable paints. These new paints must meet all engineering performance and application requirements and provide a finished appearance that is acceptable to operators. VOCs are solvents that get released into the air as the paint dries and these solvents could potentially cause harm to the environment. The performance and application requirements of solvents in high-solids paint are divided into two categories: wet film properties and dry film properties. Both these properties have changed for the primer and the topcoat.

Wet Film Properties

Conventional paint (both primer and topcoat) uses large volumes of solvent to keep high-molecular-weight, long-chain hydrocarbon polymers in solution (in a liquid state). This gives the paint a longer application life and makes it easier to apply.

High-solids paint generally contains low-molecular-weight, short-chain polymers to allow for a greater solids density and lower solvent content. As a result, the paint maintains its required flow characteristics while meeting NESHAP requirements for reduced VOC content. The combination of low VOC and high-solids content enables the paint to cross-link more densely. High-solids paint forms many more chemical bonds between resin chains than does conventional paint, making it more stable. Cross-linking affects paint properties such as chemical resistance and flexibility. The flow characteristics of the paint, also called rheology, are controlled with a blend of solvents and additives. The short-chain polymers are viscous liquids that continue to flow after spraying and after all the solvent has evaporated. Controlling the rheology of the paint is critical to producing an acceptable appearance. If the paint viscosity is too low during and after application, the paint will run and sag. Conversely, if the viscosity is too high, the paint does not flow well and the result is a bumpy "orange peel" appearance.

The Eclipse paint system is an example of a high-solids polyurethane topcoat that is used by airframe manufacturers and repair stations to paint transport aircraft. This paint system provides premium gloss and high distinctiveness of image. Figure 5-2 shows the mixing instructions for this paint system.

It is the responsibility of the technician to understand the nature of the materials and procedures used in applying a paint system to a particular part of an aircraft. The technician should consult the tech sheets and MSDSs prepared by the manufacturer before the application of the paint system.

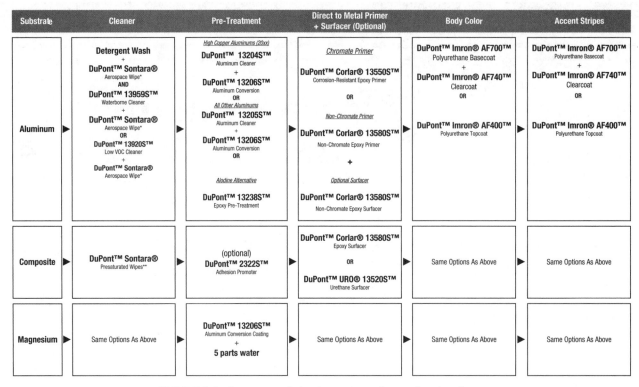

Substrate	Cleaner	Pre-Treatment	Direct to Metal Primer + Surfacer (Optional)	Body Color	Accent Stripes
Aluminum	Detergent Wash + DuPont™ Sontara® Aerospace Wipe* AND DuPont™ 13959S™ Waterborne Cleaner + DuPont™ Sontara® Aerospace Wipe* OR DuPont™ 13920S™ Low VOC Cleaner + DuPont™ Sontara® Aerospace Wipe*	*High Copper Aluminums (20xx)* DuPont™ 13204S™ Aluminum Cleaner + DuPont™ 13206S™ Aluminum Conversion OR *All Other Aluminums* DuPont™ 13205S™ Aluminum Cleaner + DuPont™ 13206S™ Aluminum Conversion OR *Alodine Alternative* DuPont™ 13238S™ Epoxy Pre-Treatment	*Chromate Primer* DuPont™ Corlar® 13550S™ Corrosion-Resistant Epoxy Primer OR *Non-Chromate Primer* DuPont™ Corlar® 13580S™ Non-Chromate Epoxy Primer + *Optional Surfacer* DuPont™ Corlar® 13580S™ Non-Chromate Epoxy Surfacer	DuPont™ Imron® AF700™ Polyurethane Basecoat + DuPont™ Imron® AF740™ Clearcoat OR DuPont™ Imron® AF400™ Polyurethane Topcoat	DuPont™ Imron® AF700™ Polyurethane Basecoat + DuPont™ Imron® AF740™ Clearcoat OR DuPont™ Imron® AF400™ Polyurethane Topcoat
Composite	DuPont™ Sontara® Presaturated Wipes**	(optional) DuPont™ 2322S™ Adhesion Promoter	DuPont™ Corlar® 13580S™ Epoxy Surfacer OR DuPont™ URO® 13520S™ Urethane Surfacer	Same Options As Above	Same Options As Above
Magnesium	Same Options As Above	DuPont™ 13206S™ Aluminum Conversion Coating + 5 parts water	Same Options As Above	Same Options As Above	Same Options As Above

FIGURE 5-1 Recommended paint system of propeller aircraft.

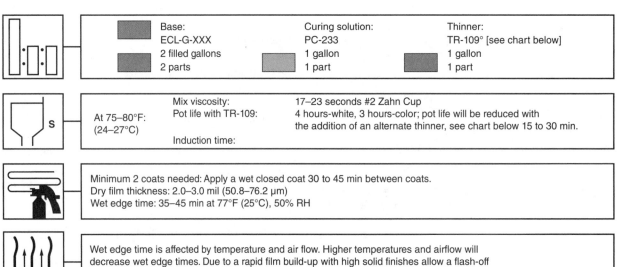

	Base: ECL-G-XXX 2 filled gallons 2 parts	Curing solution: PC-233 1 gallon 1 part	Thinner: TR-109° [see chart below] 1 gallon 1 part

At 75–80°F: (24–27°C)
Mix viscosity: 17–23 seconds #2 Zahn Cup
Pot life with TR-109: 4 hours-white, 3 hours-color; pot life will be reduced with the addition of an alternate thinner, see chart below 15 to 30 min.
Induction time:

Minimum 2 coats needed: Apply a wet closed coat 30 to 45 min between coats.
Dry film thickness: 2.0–3.0 mil (50.8–76.2 µm)
Wet edge time: 35–45 min at 77°F (25°C), 50% RH

Wet edge time is affected by temperature and air flow. Higher temperatures and airflow will decrease wet edge times. Due to a rapid film build-up with high solid finishes allow a flash-off time between the layers with a minimum of 30 min.

Drying times	77°F (25°C) Pot life	77°F (25°C) 60% RH	90°F (32°C) 40% RH	120°F (48°C) 50% RH
TR-109	3–4 h	10–12 h	8–9 h	4–5 h
TR-111	1½–2 h	7–8 h	4–5 h	3–4 h
TR-112	1–1½ h	5–6 h	2–3 h	1½–2 h
TR-113	½–1 h	2–3 h	1–2 h	< 1 h
TR-141	3 h	10–12 h	7–9 h	4–6 h

Note: TR-112 & TR-113 are recommended for touch-up areas and speed lines only and are pre-adjusted to meet specific dry times. No additional accelerator should be added. The pot life & wet edge of the top coat are dramatically reduced when mixed with TR-112 & TR-113.

Type equipment	Tip Size	Gun Air Pres.	Pot Air Pres.	Other
Conventional spray gun	.047–.055 in	45–65 psi	5–20 psi	Low transfer efficiency
HVLP	1.2–1.4 mm	8–25 psi	Varies w/pot size	Iwata, Sata, Binks Mach 1
Low Pres. Electrostatic	1.2–1.5 mm	35–45 psi	12–100 psi	PRO3500, 4500SC
Air Assisted Electrostatic	.009–.013 in	55–65 psi	1800–2500 psi	PRO4000AA Raneburg (Tips: 1.2–1.5 mm)

FIGURE 5-2 Mixing instructions for paint system.

Paint System Components

Cleaners

Surface preparation for painting of the aluminum skin begins with using an alkaline cleaner or similar product for metal degreasing, removal of paint remover residues, and final cleaning prior to paint. New regulations specified that any cleaning solvents used before painting must have a vapor pressure below 45 mm Hg. More environmental (green) products have been developed to replace traditional fast evaporating cleaners such as 100% MEK with a slower evaporating solvent blend of methyl propyl ketone and MEK and water-based cleaners. Cleaners and degreasers for composite structures could be different from cleaners used for aluminum and the technician must carefully select the correct cleaner to avoid damaging the aircraft.

After the aircraft is cleaned, water is sprayed over the airplane surface to clean it. Premature breaks in the water film are a result of contamination and this indicates that the surface is dirty and further cleaning is necessary. After the surface is free of water breaks, the airplane is ready for pretreatment.

Pretreatment

Aluminum conversion coating is applied to the aluminum skin to prevent corrosion and improve adhesion for the paint system. Alodine is a brand name that is often used for this process. This coating is applied like a wash, allowing the coating to contact the surface and keeping it wet for 2 to 5 min without letting it dry. The chromated conversion coating, alodine 1000, allows the primer to adhere to the oxide that is created on the airplane skin during conversion. Adhesion occurs when the alodine reacts with the aluminum oxide and converts it into a mixed chromate-aluminum oxide, which provides good adhesion for the primer. It then must be thoroughly rinsed with clean water to remove all chemical salts from the surface. Depending on the brand, the conversion coating may color the aluminum a light gold or green, but some brands are colorless. Traditional conversion coatings contain hexavalent chromium which is highly toxic and regulated. New non-hexavalent chromium–based processes are becoming commercially available and new water-based pretreatments are now available that give superior adhesion, improved paint flexibility, and advanced corrosion protection. After the excess alodine is rinsed off and the airplane is dried, it is ready for primer and topcoat.

Wash primers often contain phosphoric acid and are an alternative to conversion coatings and serve as a tie coat over properly prepared uncoated aluminum. The wash primer will reactivate aged or sealed anodized surfaces. Most often wash primers are used to improve the filiform corrosion resistance of an aluminum aircraft and can be used (1) as an alternative pretreatment to chemical conversion coatings, (2) for reactivation of aged anodized or chromated alloys and sealed anodized surfaces, (3) for strip ability of polyurethane systems with alkaline paint removers, and (4) to provide adhesion of subsequent polyurethane or epoxy/isocyanate primers.

Primers

Primer is applied to adhere to both the aluminum or composite surface (aircraft skin) and the topcoat. The primer protects the aluminum skin from corrosion. Zinc chromate primer, red iron oxide, and gray enamel undercoat primers were used extensively for aircraft painting operations, but due to environmental regulations and the development of newer improved epoxy primers they have largely been replaced by low-VOC high-solids nonchromate epoxy or polyurethane primers. Epoxy and polyurethane primers are two-part materials consisting of a base material and a catalyst. Epoxy and polyurethane primers are compatible with polyurthane topcoats. Most new epoxy primers contain no chromates. A example of this new type of epoxy primer is the DuPont Corlar 13580 non-chromate epoxy primer that is used with DuPont Imron topcoats. Intergard® 10206 is a waterborne two-component chromate-free epoxy primer designed to adhere to multiple substrates (e.g., steel and aluminum). This product has a very low VOC and is easily applied. It is important for the technician to follow the manufacturer's directions for the product being used in any coating systems.

Topcoats

Topcoats provide the decorative paint scheme for airplanes. Ultraviolet (UV) light degrades certain pigments and resin systems over time. In an acidic environment, such as a volcanic eruption or severe air pollution, color shift occurs (the paint changes color, usually growing darker and less brilliant), and topcoat gloss decreases. Topcoats are available in two types: one-stage and two-stage topcoats. The one-stage topcoat is applied in one operation with all the necessary chemicals in the paint. It often needs to be buffed to produce a high-gloss surface. The two-stage paint system consists of a base coat and a clear coat. The base coat (color) is sprayed over the primer first and the clear coat is applied after the base coat has dried. The two-stage topcoats provide a high-gloss finish and good UV protection.

Primers and topcoats must perform together as a paint system. These systems must be resistant to chemical attack because airplanes are often exposed to hydraulic fluid, fuel, and maintenance chemicals when in service. High-density polymer cross-linking provides the required chemical resistance in high-solids paint, but cross-linking also affects the flexibility of the paint. Flexibility and adhesion of paint systems are crucial because of the thermocycling, aerodynamic forces, and body structure stresses put on the airplane between takeoff and landing.

Content of Paint

Most paint finishes are made up of four basic components: *pigment*, *binder*, *solvent*, and *additives*.

Pigment provides the color and durability. Pigment gives the paint the ability to hide what is underneath the finish. In addition to providing durability and color, pigment may also improve the strength and adhesion of the paint.

The **binder** holds the pigment in liquid form, makes it durable, and gives it the ability to stick to the surface. The binder is the backbone of the paint.

Solvent dissolves the binder and carries the pigment and binder through the spray gun to the surface being painted. Most solvents are derived from crude oil. When used with lacquer, this solvent is called a *thinner*. When used with enamel, it is called a *reducer*.

Most aircraft finishes also include other components lumped together under the name **additives**. Additives represent only a small part of the paint, but they can have a significant effect on its physical and chemical properties. Additives are designed to speed drying, prevent blushing, improve chemical resistance, and provide a higher gloss.

Types of Finishes

The exterior of an airplane is painted with one of two basic finishes: acrylic enamels or polyurethane. The majority of all propeller, business, and transport aircraft are painted with polyurethane paint systems. Even many home-built and fabric-covered aircraft use polyurethane paint systems due to superior finish and durability. Lacquers were originally finishes developed in China to produce a high gloss on wood products. However, lacquers are no longer used to paint modern aircraft. Paint systems are specifically developed for different parts of the aircraft such as the exterior of the aircraft; the interior of the aircraft, which has strict regulations regarding smoke emissions during a fire; fuel tanks; wing surfaces; and engine compartments.

Acrylic Enamels

Acrylic enamels were developed in the 1960s and are more durable and have faster drying times than alkyd enamels. Acrylic enamel finishes provide a glossy, hard surface that has good resistance to scratching and abrasion. Their disadvantages include the longer drying time and special precautions that must be taken when repairing any damage to the surface. Various paint manufacturers supply acrylic enamels in both one- and two-part preparations with various characteristics of durability, heat resistance, and weather resistance to match the needs of different aircraft operators.

Polyurethane

Polyurethane finishing materials provide a very high gloss surface with excellent durability, weathering resistance, and abrasion resistance. Due to their resistance to chemical attack, they are often used by aircraft manufacturers in places subject to exposure to strong chemical agents, such as areas containing Skydrol-type hydraulic fluids.

These finishes normally are a two-part mixture, with a catalyst added to activate the drying, or "hardening," process. Once the catalyst is added, the material has a specific working life, after which it may not be used. These finishes are often part of a finishing "process," where the finish material provides the best life and durability when used in conjunction with a specific surface-cleaning operation, a surface priming coat, a corrosion inhibitor, and a finish topcoat, as shown in Fig. 5-3.

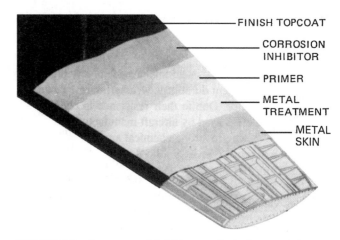

FIGURE 5-3 Sequence of finishing products for a metal aircraft. (*U.S. Paint.*)

Polyurethane finishes should not be waxed. Wax tends to yellow the finish, and it collects dirt, reducing the cleanliness of the aircraft surface. Mild soap and water are recommended for cleaning, and solvents are applied with a soft cloth to remove oil and grease.

One problem that is found with polyurethanes is the difficulty in removing them, or "stripping" large areas, with solvents and chemicals. This type of finish resists most stripping materials. Consult the finish manufacturer for specific information about removing old polyurethane coatings.

Thinners and Reducers

Thinners, also called **reducers**, are used to reduce the viscosity of finishing products so that they will flow properly through the spray gun and onto the surface being painted. Ordinarily thinners are used with enamels and reducers are utilized with lacquers.

A thinner or reducer must be properly balanced to provide proper dilution of the material, so that it will not only pass through the gun smoothly but also atomize easily as it leaves the gun. The paint material must be kept in solution long enough for it to flow out to a smooth, even surface and not allow the paint film to sag or run. The thinner or reducer must evaporate, leaving a tough, smooth, and durable paint finish.

There are two variables that affect the spraying of materials: temperature and humidity. These variables must be carefully observed and compensated for by use of the proper thinner or reducer. Of the two variables, temperature is the most crucial. Hot, dry weather produces the fastest drying time. Cold, wet, or humid weather produces the slowest drying time.

A general rule to follow in selecting the proper thinner or reducer is: the faster the shop drying conditions, the slower drying the thinner or reducer should be. In hot, dry weather, use a slower-drying thinner or reducer. In cold, wet weather, use a fast-drying thinner or reducer.

If a thinner or reducer evaporates too rapidly, the problems that can be caused are orange peel, blushing, and over-spray. If, on the other hand, a thinner or reducer evaporates too slowly, sags and runs may result.

Each finishing product may require a specific reducing liquid. Although some thinners may indicate that they are compatible with all types of finishing products or with a specific group of products, such as all enamels, it is best to only use the reducer recommended by the finish manufacturer. If the wrong thinner is used, the paint may not mix with the thinner, the thinner might prevent the paint from drying, or the thinner may attack the coats of paint beneath the coat being applied.

Epoxy and polyurenthane paint products require a catalyst, also called curing solution, to cure the paint. The catalyst is mixed per instruction with the base material. After mixing there is an induction time of 15 to 20 min before the paint can be applied to the aircraft surface. As soon as the base material is mixed with the catalyst, the curing process starts and these type of products have a limited pot life. The polyurethane paint system in Fig. 5-2 has a 4-h pot life for the color white and a 3-h pot life for colors.

Additives

With the rapid changes made in paint technology during the last decade, the utilization of paint additives has become extremely popular. Additives generally make up no more than 5 percent of the paint at most (and usually much less). Additives perform a variety of vital functions. Some additives can speed up or slow down the drying time. Other additives can raise and lower the finished gloss. Some additives perform a combination of functions, such as eliminating wrinkling, providing faster cure time, and improving chemical resistance.

Those additives that speed up the cure time and improve gloss are often referred to as *hardeners*. Those that slow drying are called *retarders*, and those that lower gloss are called *flatteners*.

Safety

The painting process generally involves the utilization of hazardous materials. It is very important that the guidelines outlined in Chap. 1 on the handling and disposal of these materials be followed. Federal, state, and local codes will dictate the type of safety equipment required for the protection of the operator and others in the area. Protection from hazards such as fire, explosion, burns, and toxic fumes must be provided through the proper utilization of safety equipment, including respirators and eye protection.

All those involved with the finishing of airplanes must keep in mind that many solvents, lacquers, and other materials used are highly flammable. Great care must be taken to ensure that ignition of flammable materials cannot take place.

Dry sanding of cellulose nitrate finishes can create static charges that can ignite nitrate fumes. The interior of a freshly painted wing or fuselage contains vaporized solvents. If a static discharge should take place in this atmosphere, it is most likely that there will be an explosion and fire. It is recommended that wet sanding be performed in every case possible because the water is effective in dissipating static charges.

Finishing materials should be stored in closed metal containers in a fireproof building. The building should be well ventilated to prevent the accumulation of flammable vapors.

The floor of a dope room or paint booth should be kept free of spray dust and sanding residue. These substances can be ignited by static discharges or by a person walking over them. It is recommended that the floors be washed with water frequently to prevent the buildup of flammable residues.

The use of a drill motor to drive a paint-mixing tool should be avoided. Arcing at the motor brushes can ignite flammable fumes.

Solvents and other solutions are often injurious if they splash on the skin or in the eyes. Technicians using these materials should be protected with clothing that covers the body, goggles for the eyes, and an air filter for the nose and mouth.

The use of respirators when spraying finishes is highly recommended, and with some finishing products it is mandatory. Some finishing materials will coat the inside of the lungs and cause health problems. Always check the product manufacturer's safety recommendations.

A forced-air breathing system must be used when spraying any type of polyurethane or any coating that contains isocyanates. It is also recommended for all spraying and stripping of any type, whether chemical or media blasting. The system provides a constant source of fresh air for breathing, which is pumped into the mask through a hose from an electric turbine pump. Protective clothing, such as Tyvek® coveralls, should be worn that not only protects personnel from the paint but also helps keep dust off the painted surfaces. Rubber gloves must be worn when any stripper, etching solution, conversion coatings, or solvent is used. When solvents are used for cleaning paint equipment and spray guns, the area must be free of any open flame or other heat source. Solvent should not be randomly sprayed into the atmosphere when cleaning the guns. Solvents should not be used to wash or clean paint and other coatings from bare hands and arms. Use protective gloves and clothing during all spraying operations.

SPRAY-PAINT EQUIPMENT

Spray-painting equipment represents a significant investment for any company. If the equipment is properly maintained and used, it will more than pay for itself. If the equipment is misused or not properly maintained, it can cost a lot for a company in terms of damage to aircraft finishes and poor customer relations. Therefore, it is important that all aircraft technicians have a basic understanding of the proper operation, use, and maintenance of aircraft spray equipment.

Spray Booth and Exhaust Systems

It is difficult to produce quality paint finishes without the use of a spray booth. Spray booths are enclosures that provide fire and health protection and are designed to provide a controlled movement of air through the spray area. This allows the incoming air to be filtered and the solvent fumes and overspray to be removed from the spray area. Spray booths come in all sizes and shapes to accommodate the types of spraying being performed. Figures 5-4a and 5-4b show a

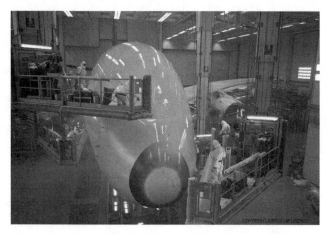

FIGURE 5-4a Paint hangar for transport aircraft. (*Airbus.*)

FIGURE 5-4b Wing painting of transport aircraft. (*Airbus.*)

paint hangar and personnel wearing respirators and protective clothing for the painting of a transport aircraft.

Compressed-Air Systems

A compressed-air system is an important component of any spray painting operation. A compressed-air system typically includes an air compressor, which produces air that is piped to an air transformer-regulator and fed to the spray gun by a flexible hose. Compressed-air systems collect water that must be drained daily. Both the regulator and compressed-air tank have drain valves for this purpose.

Air Transformer

An air transformer-regulator, shown in Fig. 5-5, is a device that removes oil, dirt, and moisture from compressed air, regulates the air pressure, indicates the regulated air pressure by means of gauges, and provides outlets for spray guns and other air tools.

The air transformer removes entrained dirt, oil, and moisture by centrifugal force, expansion chambers, impingement plates, and filters, thus allowing only clean, dry air to emerge from the outlets. The air-regulating valve provides positive control, ensuring uniform, regulated pressure. The pressure gauges indicate regulated pressure and, in some cases, main-line air

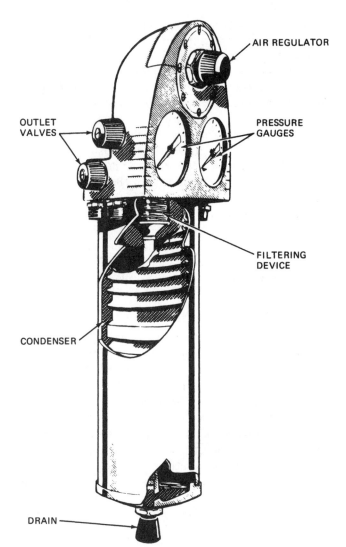

FIGURE 5-5 An air transformer. (*DeVilbiss Co.*)

pressure. Valves control air outlets for hose lines to spray guns and other air-operated equipment. The drain valve provides for elimination of accumulated sludge, which consists of oil, dirt, and moisture.

The air transformer should be installed at least 25 ft [8 m] from the compressor and should have its air takeoff from the main line rising from the top of the line. This prevents liquid water from entering the transformer air inlet. The main line should be equipped with a water trap or drains to prevent the accumulation of water. The drain at the bottom of the air transformer should be opened at least once a day to remove moisture and sludge.

Pressure Drop

One of the most important considerations that must be given in the selection and utilization of hoses and fittings is the pressure drop that will occur between the regulator and the spray gun. One of the most important factors in producing a quality paint finish is the ability to maintain the constant desired air pressure. The pressure will always drop whenever a spray gun is turned on and the air begins to flow through the hose. A pressure drop also occurs when

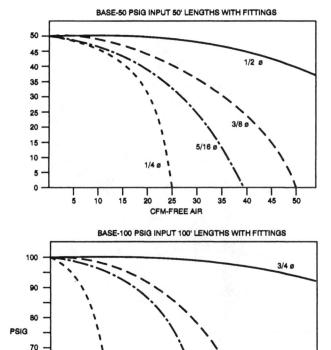

BASE-50 PSIG INPUT 50' LENGTHS WITH FITTINGS

BASE-100 PSIG INPUT 100' LENGTHS WITH FITTINGS

FIGURE 5-6 Pressure drop for various diameter hoses.

air moves through a hose. Pressure drop cannot be eliminated, but it may be reduced through the proper selection and use of hoses. The inside diameter of a hose has the greatest effect on the pressure drop, as illustrated in Fig. 5-6. As the length of the hose increases, the pressure drop will also increase, as shown in Fig. 5-6. The surface smoothness inside an air hose will greatly influence the pressure drop; a rough surface can cause as much as a 50 percent increase in the pressure drop.

Spray Equipment Classification

Spray equipment can be classified into several different types by principles of operation. Different types of equipment may be suitable to specific types of operations. An understanding of the principles of the different types of spray systems will assist the technician in matching the equipment requirements to a specific job.

Airless Spraying

Airless spraying is a method of spray application that does not directly use compressed air to atomize the paint. Hydraulic pressure is used to atomize the fluid by pumping it at relatively high pressure through a small orifice in the spray gun nozzle. The resulting high fluid velocity of the paint flowing through the orifice results in the paint material being atomized. Since air is not being used to atomize the material, the term *airless* is used to describe this method.

Airless spraying is not commonly used on aircraft. Instead, it is better suited to industrial applications where large areas are to be painted and high flow rates are required.

Air Spraying

In air spraying the paint is delivered to the spray gun by either siphon feed, gravity, or pressure feed. Compressed air is introduced at the front end of the spray gun to atomize the paint. The atomizing of the material can take place either outside the air nozzle (external mix) or inside the air nozzle (internal mix).

Environmental regulations have also affected the type of air spraying equipment. Conventional high-pressure spray guns produced a very good finish but the paint transfer to the aircraft skin was relatively low, about 35% of the paint actually transferred to the aircraft and the rest of the paint got lost in the atmosphere. Newer high-volume low-pressure (HVLP) spray equipment has largely replaced older high-pressure paint guns. These HVLP guns can transfer at least 65% of paint upward to 80% due to the lower pressure, which means less overspray and less waste. It is important for technicians to recognize the difference between the two paint gun systems because they look very similar.

Siphon feed is used only on internal-mix spray guns. In the siphon-feed gun, the siphon-type air nozzle sends a hollow column of moving air around the fluid nozzle and, in so doing, draws the paint material out of the paint gun cup, as shown in Fig. 5-7. Siphon-feed spray guns usually operate in the 30- to 60-psi range and 5- to 25-psi range for HVLP guns. Siphon spraying is best suited to smaller paint jobs with the use of light finishing materials that require good atomization. It is ideal for applications where small amounts of color or frequent color changes are required. A gravity-feed gun provides the same high-quality finish as a siphon-feed gun, but the paint supply is located in a cup on top of the gun and supplied by gravity. The operator can make fine adjustments between the atomizing pressure and fluid flow and utilize all material in the cup. Many new paint gun designs utilize the gravity-feed design. Figure 5-8 shows a gravity-feed HVLP gun.

In a pressure-feed system, the paint is delivered under air pressure to the spray gun. The pressure container may be either a cup on the gun or a tank, as shown in Fig. 5-9.

The *pressure tank*, also called a *pressure pot* or *pressure-feed tank*, is a closed metal container that provides a constant flow of paint or other material to the spray gun at a uniform pressure. These tanks range in size from 2 to 120 gal [7.57 to 454.2 L]. A pressure-feed tank is illustrated in Fig. 5-10.

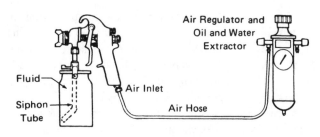

FIGURE 5-7 Siphon spray system. (*Binks Manufacturing Co.*)

FIGURE 5-8   HVLP gravity paint gun.

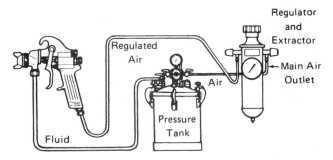

FIGURE 5-9   Pressure spray systems. (*Binks Manufacturing Co.*)

Fluid is forced out of the tank by the compressed air in the tank. To change the rate of fluid flow from the tank, the air pressure is adjusted by means of the pressure regulator. If the tank is the type without a regulator, the air pressure is controlled by an air transformer. Pressure tanks are designed for working pressures up to 110 psi [759 kPa].

As can be seen in Fig. 5-10, a typical pressure feed tank consists of a shell (1), clamp-on lid (2), fluid tube (3), fluid header (4), air-outlet valve (5), fluid-outlet valve (6), air-inlet

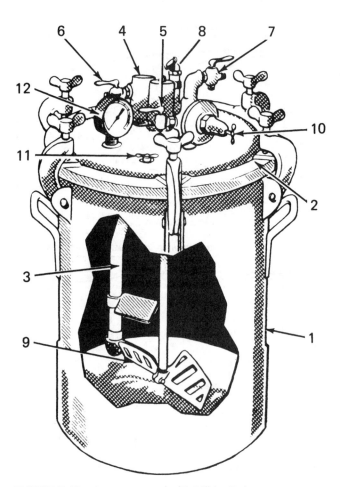

FIGURE 5-10   A pressure tank. (*DeVilbiss Co.*)

valve (7), safety relief valve (8), agitator (9), pressure regulator (10), release valve (11), and pressure gauge (12).

The agitator is an essential mechanism for a pressure tank. Its purpose is to ensure that the finishing materials being sprayed are kept in a thoroughly mixed condition. The most convenient type of agitator is driven by an air motor.

Pressure tanks are often utilized with a separate fluid container set inside the tank. This reduces the cleanup problems and makes it possible to change rapidly from one material to another.

The air pressure for which the pressure vessels are rated must not be exceeded, because its failure under pressure could be harmful to operating personnel. The delivery of compressed air supplied to the container is controlled by a pressure regulator. The air-pressure control together with the selection of the fluid-nozzle orifice size and adjustment of the fluid-flow valve provides the rate of flow of the paint from the gun. Pressure-feed painting is well suited to applications requiring medium to large quantities of paint, such as painting an entire aircraft. Many manufacturers and repair stations have switched to HVLP production spray equipment to comply with environmental regulations and to improve the transfer of paint and the reduction of waste.

Electrostatic Attraction

Electrostatic painting may have either air or airless painting systems for atomizing the paint. During the paint-atomization

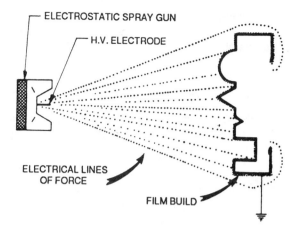

FIGURE 5-11 Electrostatic spray action. (*Binks Manufacturing Co.*)

phase, a high-voltage electrostatic charge is placed on the paint particles as they leave the nozzle. This causes the paint particles to be attracted to the nearest grounded object (the item being sprayed), as illustrated in Fig. 5-11. As the paint particles build up on the surface, they exhibit electrical resistance, preventing other particles from attaching themselves to the same place. This action forces the other particles to seek out areas that are still uncoated, resulting in a very uniform coating, which extends into areas that would not normally be coated. Electrostatic spraying generally reduces overspray and results in less paint being used to cover the surface. The limitation associated with electrostatic painting is that the parts being painted must be electrically conductive, and as such it is not generally used on items such as plastic fairings.

Spray Guns

Spray guns are manufactured in a number of different types and sizes. Sizes vary from the small airbrush guns to the heavy-duty production guns used with pressure tanks. Guns are made to spray fluids of various viscosities. Spray guns are manufactured for conventional spraying, HVLP spraying, and airless spraying. The nozzle and air cap of the conventional gun and HVLP gun are designed to mix air with the paint stream to aid in atomization of the liquid. Airstreams are also utilized to shape the pattern deposited by the gun. Airless spray guns do not mix air with the paint before it leaves the nozzle. This results in reduced overspray, less paint "fog" in the air, and a saving of paint. Regardless of the type of spray gun being used, the technician must be certain to adjust it correctly for the type of paint being applied.

One of the popular spray guns for general use in conventional spraying is the type of MBC gun manufactured by the DeVilbiss Company. A gun of this type is shown in Fig. 5-12. Figure 5-13 shows the same spray gun used with a liquid container, also called a cup. The spray gun also can be used with a pressure tank (pressure pot), in which case the liquid is fed under pressure from the tank through a hose to the gun. Spray guns manufactured by the Binks Manufacturing Company are also extensively employed for the application of all types of finishes. The choice, in any case, depends upon the preference of the operator.

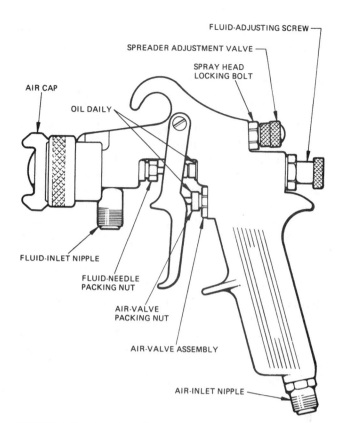

FIGURE 5-12 Drawing of a typical spray gun. (*DeVilbiss Co.*)

FIGURE 5-13 Spray gun with paint cup. (*DeVilbiss Co.*)

As shown in Fig. 5-12 a spray gun includes a **fluid adjusting screw**, which controls the **fluid valve**, a **spreader adjustment valve**, an **air-valve assembly**, an **air cap**, a **fluid inlet**, and an **air inlet**. The balance among air pressure, airflow, fluid pressure, and fluid flow is controlled by adjusting the valves concerned to provide the desired spray quality and pattern. Air flowing through the center area of the air cap and through holes near the center mixes with fluid from the fluid tip and produces an atomized spray. Air flowing from the openings in the horns of the air cap impinges on each side of the fluid spray to produce an elongated spray pattern. This pattern may be adjusted for a wide or narrow area by changing the setting of the spreader adjustment valve. The pattern can be made vertical or horizontal by turning the air cap.

If the spray gun is used with a pressure tank, the fluid is fed under pressure to the fluid inlet of the gun. The fluid pressure must be correct to obtain the proper fluid flow. The pressure is controlled by means of a regulator on the pressure tank.

The air pressure, fluid pressure, airflow, and fluid flow are adjusted to provide the best spray quality for the material being sprayed. Manufacturers often give the most satisfactory settings for the application of each product and also specify the viscosity for best results.

Using a Spray Gun

The difference between a first-class finishing job and a poor job is often determined by the use of the spray gun. An experienced technician who has mastered the techniques in handling a spray gun usually has no difficulty in producing a good finishing job. The first requirement is to have the gun properly adjusted for the type of finishing material being applied.

A good spray pattern depends on the proper mixture of air and paint. Under normal conditions, three basic adjustments will provide a good spray pattern and the desired degree of wetness. First, the size of the spray pattern should be adjusted using the spreader adjustment valve. This valve regulates the flow of air through the holes in the cap. The more the valve is open, the wider the spray pattern. Second, the fluid-adjusting screw should be adjusted to match the fluid volume to the spray-pattern size. Last, it is important to ensure that the air pressure for the gun is correctly adjusted. This pressure is normally controlled by an **air transformer**. The device acts as a pressure regulator to provide the correct pressure as adjusted by the knob on the top. This unit has fittings for the incoming air from a standard air-pressure line and one or more outlet fittings for the attachment of the air hose to the gun. Many paint guns use a regulator valve and pressure gauge to adjust the pressure at the air inlet of the spray gun (see Fig. 5-8). Conventional paint guns operate at 50 to 60 psi and HVLP paint guns operate at 10 to 25 psi. The tech sheet provided by the paint system manufacturer will prescribe the correct air pressure setting (see Fig. 5-2).

The operator may test the gun by trying it against a test surface. When the desired pattern is obtained, the operator may start applying the finish.

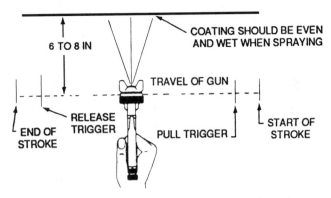

FIGURE 5-14   Correct spray-gun technique. (*Binks Co.*)

Among the conditions that the operator of a spray gun must observe are the following:

1. The gun must be held at the correct distance from the surface being sprayed (6 to 10 in [15 to 25 cm]). If the gun is held too close, it is likely to cause runs or sagging.

2. Air pressure should be correctly adjusted.

3. The trigger must be released at the end of each pass; otherwise, excessive paint buildup occurs when the direction of spray-gun movement is reversed. This procedure is shown in Fig. 5-14.

4. The gun must be moved in a straight line, parallel to the surface being sprayed. Moving the gun in an arc causes heavy paint buildup and runs in the center portion of the pass.

5. The gun must be held level and perpendicular to the surface.

6. The correct type of air cap and spray nozzle must be used. Consult the appropriate instructions for the material being sprayed.

7. The paint must be thinned to the correct consistency (viscosity). Manufacturers of aircraft finishes often recommend testing the viscosity with a Zahn viscosity cup. This test is accomplished by placing the required amount of finish in the specified cup and noting the time required in seconds for the finish to drain from the cup.

8. The surface to be sprayed must be clean. A **tack cloth** may be used to wipe away dust or other particles. A tack cloth is a specially treated soft cloth that removes all dust from a surface but leaves no wax, oil, or other material on the surface that would interfere with the adhesion of coatings.

9. The temperature and humidity must be within the proper range for satisfactory results; otherwise a variety of problems may occur.

The two most common problems in spray painting are tilting the gun and arcing the gun. If the gun is tilted, the part of the spray pattern closest to the surface will receive a heavier film and the part farthest away will get a light covering. If the gun is swung in an arc, the gun will be closer to the surface at the middle of the pass and the paint film will be heavier in that area.

A normal spray pattern for a spray gun forms an elongated oval, with a length $2\frac{1}{2}$ to 3 times the width. This is shown in Fig. 5-15(a). This pattern indicates that the spray gun is clean

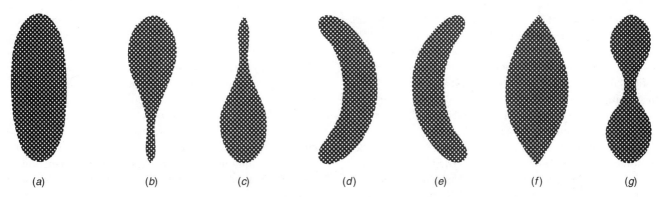

(a) (b) (c) (d) (e) (f) (g)

FIGURE 5-15   Correct and distorted spray-gun patterns.

and properly adjusted for the liquid being sprayed. The pattern at (b) is top-heavy and indicates that the horn holes are partially plugged, there is an obstruction on top of the fluid tip, or there is dirt on the air cap seat or fluid-tip seat. Causes for the bottom-heavy pattern at (c) are the same as for the top-heavy pattern, except that the obstruction is on the bottom side of the fluid tip. Heavy right or left patterns, shown in (d) and (e), indicate that one or the other of the horn holes is partially clogged or that there is dirt on either the right or left side of the fluid tip. The heavy center pattern shown in (f) indicates that the spreader adjustment valve is set for too low an airflow, or, for a pressure feed, the fluid pressure is too high for the atomization air being used, or the material flow is in excess of the cap's normal capacity. The condition can also be caused by too large or too small a nozzle. A split spray pattern (g) is due to air and fluid flows not being properly balanced. Increasing fluid pressure or decreasing spreader air should correct the split spray condition.

The best results in the use of a spray gun are obtained when the gun is properly maintained and serviced. Bearing surfaces and moving parts should be lubricated daily with a light oil. The lubrication of the gun should be done according to the manufacturer's instructions.

After each use of the gun, it is necessary that it be cleaned thoroughly to prevent the accumulation of dried paint in the nozzle and paint passages. It is not usually necessary to take the spray gun apart to clean it, because paint solvent sprayed through the gun will ordinarily clean out the paint. The outside of the gun should be thoroughly washed with solvent.

It is desirable to disassemble the gun periodically and inspect it for wear and accumulation of paint and dirt. Worn parts should be replaced.

The packing nuts around the fluid needle and air-valve stem should be kept tight, but not so tight that parts cannot move freely.

It is not wise to allow the gun to become so dirty that soaking of the complete gun becomes necessary. Also, the gun should not be left immersed in solvent or thinner. It is especially important to avoid the use of caustic alkaline solutions for cleaning, because alkalis destroy aluminum alloy components of the gun.

An unbalanced or distorted spray pattern indicates a dirty air cap. The air cap should be removed and washed thoroughly in clean thinner. If reaming of the air-cap holes is

necessary, a match stick, broom bristle, or some other soft material should be used. Do not use a hard or sharp instrument, because it may permanently damage the cap.

Spray-gun parts are not ordinarily removed, except for the air cap. When replacing the air-valve assembly, the air-valve spring should be properly seated in the recess in the body of the gun. When removing any parts that seal on a gasket, the condition of the gasket should be examined. When the fluid tip is replaced, it must be tightened carefully.

FINISHING METAL AIRCRAFT AND PARTS

Since all aircraft contain metal parts and most aircraft are of all-metal construction, it is important that the technician be familiar with certain principles involved in metal finishing. In this section, the processes of plating or the special chemical finishes are not discussed because they are not usually in the realm of general airframe maintenance. If such a process is required by the manufacturer, the work can be done by companies that specialize in this type of work.

Paint Removal

For many years, chemical stripping has been the standard procedure for removing paints, primers, and surface contaminants from the surfaces of aircraft components. Although chemical stripping is still frequently used, a new process called *dry stripping* reduces the environmental hazards associated with the use of chemicals.

Chemical Stripping

Several paint-stripping products are available, and these usually fall into two categories. **Wax-free paint remover** is a highly volatile liquid that evaporates rapidly after it is applied. For this reason, stripping large areas of an airplane surface with this material is a slow and arduous process. **Wax-type stripper** has the consistency of thick cream because of the wax, which holds the solvent in contact with the paint surface long enough to lift or dissolve the old coatings. Wax-type stripper is applied with a brush in a thick layer and is allowed to remain on the surface until the coating lifts or dissolves. Some manufacturers recommend laying a polyethylene drop cloth over the stripper-coated surface to slow the drying of the solvents. The stripper should not be removed from the surface until all the paint has lifted or softened. If an area dries before this happens, additional stripper should be applied to the dried area. When all the old finish has softened or lifted, the stripper and finish can be flushed off with water.

Acrylic lacquer will not lift and wrinkle when stripper is applied, but it will soften. When acrylic lacquer has softened, it should be scraped off with a nonmetallic scraper, and then the area should be washed with MEK, acetone, or similar solvent. A powerful spray gun should be used to spray the solvent into cracks, crevasses, and metal joints or seams. The purpose of the solvent is to remove all traces of wax that would otherwise be left by the stripper.

Polyurethane coatings are not easily removed with paint stripper; however, if the stripper is retained on the surface for sufficient time, the bond between the coating and primer is loosened and the coating can be removed. The surface should then be scrubbed with MEK, acetone, toluol (toluene), or xylol.

Paint stripper should not be allowed to come into contact with Plexiglas windows, fiberglass parts, fabric, or other plastic or porous materials. The active agent in the stripper damages some materials and is absorbed by others, neither of which is acceptable.

The principal consideration in the stripping of an airplane is to see that all wax and other residues are completely removed from the surface. This is accomplished by proper solvents, as mentioned, and by washing the aircraft with approved cleaners.

WARNING: Paint-stripping agents must not be used on metal aircraft that employ adhesive-bonded seams rather than riveted seams. Stripping material is likely to penetrate the bonded seams and render the entire structure unairworthy. Manufacturers specify methods by which seams can be masked and protected if stripping operations are necessary.

For specific chemical-stripping procedures, consult the manufacturer of the finish being removed or the stripper's manufacturer.

Dry Stripping

The *dry stripping process* utilizes a stream of nonabrasive plastic beads propelled by low-pressure (20 to 60 psi) air to remove materials from the surface that is to be cleaned. The mixture of air and plastic media is accelerated through a venturi nozzle and directed onto the work surface. When the beads strike the surface, paints and contaminants are stripped off.

The unique cleaning capability of this process lies in its ability to control the removal rate of surface buildup without scratching, etching, marring, or otherwise damaging most substrates, thus preserving surface integrity. It also leaves a good surface for painting. Depth of surface removal is controlled by adjusting the distance between the nozzle and the surface, air pressure, and media selection. Thus, only the top layer of paint may be removed, leaving a sound primer or substrate. Chemical stripping does not permit such control.

The individual plastic beads used in the new process have an irregular configuration. They are granular, with sharp, angular edges that provide an extremely effective cutting, shearing, and chipping action to remove paint, carbon, sealants, grease, oil, and dirt deposits. Several different bead sizes are available. The beads can be reused an average of 5 to 10 times, since loss due to wear and breakdown is only 10 to 20 percent per use.

Before the plastic beads were developed, there was no medium on the market that was satisfactory for stripping paint from aluminum and magnesium parts. Hard media such as silica sand or aluminum oxide damaged these metal surfaces. Although a medium made from walnut shells is softer, it disintegrates more quickly. Since plastic beads are a manufactured product, their size, hardness, and consistency can be accurately controlled. The plastic beads are softer than aluminum and magnesium aircraft parts, yet they are sharp and sufficiently hard to remove paint and even chemical-resistant polysulfide primers.

Air pressure is only one variable among many that control speed, safety, and surface finish in the dry stripping process. Other variables that may be controlled are angle of attack, distance from nozzle to workpiece, and hardness and size of the plastic media, and the nozzle size. Optimum nozzle size for airframes is generally $\frac{1}{2}$ in I.D.

There have been a few incidents of airframe damage done by inexperienced operators. The process is operator-sensitive and has its limitations. One type of potential damage is metal expansion due to friction heat buildup or outright pummeling of the delicate metal surface. Keeping a close eye on the surface is essential. Excessive dwell time is a definite factor in this type of damage.

When dry stripping thin plate, it is important to use a low angle of attack and to keep the nozzle moving quickly across the surface, thus allowing heat to dissipate. It is often necessary to go back over an area several times to do a thorough cleaning, but this is generally safer than dwelling on a given area too long. Reducing air pressure—and thus media velocity—is another way to avoid metal expansion.

A related type of damage is unwanted removal of protective coatings such as Al-Clad or anodize. Again, a lower angle of attack, a softer or finer grade of media, and lowered air pressure will prevent this.

For specific dry stripping procedures, consult the directions for the equipment being used. For composite structures, sanding is often the only option because chemical and dry stripping methods could cause loss of fiber material, delamination, or disbonding of a honeycomb sandwich structure. The technician must carefully remove the paint without removing any of the carbon fiber or fiberglass material. A combination of power sanders and hand sanding is required.

Sanding

Sanding is frequently required as a part of the surface preparation or in between the paint finishing coats. Sanding may be performed with either power sanders or by hand. Hand sanding is a simple back and forth scrubbing action with the abrasive paper flat against the surface. On flat surfaces, the abrasive paper should be backed up with a sanding block or pad to produce an even pressure on the sandpaper surface. On flat surfaces a flat, hard sanding block may be used; however, on curved surfaces and on edges, a sponge rubber pad that will follow the contours should be utilized. Two common methods of hand sanding are *wet* and *dry*. **Dry sanding** often results in the paper becoming clogged with the material being removed. In **wet sanding**, periodically dipping the paper in water will rinse away the removed material. Wet sanding generally produces a smoother finish.

Most heavy sanding is done with power sanders. There are two types of power sanders commonly used in surface finishing, the disc sander and the orbital sander. Both types may be driven by compressed air or electricity.

Corrosion Protection

A metal fuselage is usually painted for two principal reasons: corrosion prevention and appearance. **Corrosion** is an electrochemical process in which oxygen or other elements combine with metals to form various compounds such as rust, aluminum oxide, metal chlorides, and other metal salts. The first requirement for corrosion is the presence of moisture; the other is a combination of dissimilar metals or metals and chemicals. If a metal is kept dry and clean, it will not corrode; however, a little moisture in the presence of other elements can cause severe corrosion.

In the design and manufacture of aircraft structures, engineers are careful to avoid a design of any assembly of parts that requires the contact of dissimilar metals. If two dissimilar metals must be joined, the surfaces are usually insulated from each other by means of special coatings, treated tapes, or other methods.

When preparing a metal structure for refinishing, the technician must be careful to avoid any condition that can cause corrosion and must remove any corrosion-causing condition that may be found. When in continuous contact with metals, hygroscopic materials (those that absorb and hold moisture), such as leather or canvas, are very likely to cause corrosion. Such materials should be waterproofed before installation.

The improper application of aircraft finishes can lead to **filiform corrosion** (so named because it forms in fine filaments). If phosphoric acid from an etching solution or wash primer is trapped under subsequent coatings before complete conversion to the phospate film, filiform corrosion will develop under the finish coating, damaging both the metal and the finish. To avoid this condition, it is necessary that there be adequate moisture in the air or in the wash primer to complete the conversion of the phosphoric acid to phosphate. Humidity should be such that a minimum of 0.09 lb of water is present in each pound of air [90 g per kilogram of air]. This is accomplished with a relative humidity of 57 percent at 70°F [21.1°C] or a relative humidity of 49 percent at 75°F [23.9°C]. If the humidity is not great enough to provide the required moisture, a small amount of distilled water may be added to the wash primer. Since a typical wash primer thinner is alcohol, water can be mixed with it.

Another important consideration in the conversion of the phosphoric acid in the wash primer is to allow sufficient time for the conversion to take place before applying the next coating. This is normally 30 to 40 min. If filiform corrosion starts, it will continue until it is removed. If it is discovered, the finish in the affected area and possibly the entire finish of the airplane will have to be removed and all traces of the corrosion eliminated.

Surface Preparation

The preparation of a metal surface for finishing depends upon the type of paint job required. Three general situations are usually considered. The first situation is the painting of a new airplane having a bare, polished Al-Clad finish. The preparation of a new aluminum surface involves only thorough cleaning and treating with a phosphoric acid etch and/or a wash primer. The phosphoric acid etch is also referred to as a **conversion coating**.

Another condition involves refinishing an airplane over an old finish. In this case, the technician must determine what type of finish was previously applied to the airplane and then make sure that the material for the new finish is compatible with the old finish. Preparation of an old finish for refinishing usually requires sanding to remove the oxidized outer layer of finish. Any additional treatment required before application of the new finish will be described in the finish manufacturer's instructions.

The third situation that the technician may encounter is the complete stripping of the old finish and treatment of the metal surface prior to the application of the new finish.

Conversion Coatings

Conversion coatings are applied to fresh, clean metal to aid in preventing corrosion and to microscopically roughen the surface for better adhesion of additional coatings. A conversion coating for aluminum or steel can be a phosphoric acid etch, which leaves a tough, inorganic phosphate film on the metal.

The treatment of a magnesium surface requires a chromic acid etch. This solution can be prepared by mixing approximately 1 percent chromic anhydride (CrO_3), 0.78 percent calcium sulfate ($CaSO_4$), and 98.22 percent water. This solution should be brushed on the bare magnesium and allowed to remain for 1 to 3 min. It should then be washed off with water and the surface should be dried. After drying, the surface is ready for the application of a wash primer.

Wash Primers

A typical wash primer is described as a two-part butyral-phosphoric acid resin containing corrosion-inhibiting pigments. Other wash primers contain chromic acid as the etching agent. Wash primer may be applied directly to clean, bare metal, or it may be applied over the conversion coating described previously. For best results, it is recommended that the wash primer be applied over the conversion coating.

The wash primer developed by the Stits Aircraft Coatings Company utilizes as many as four types of liquids.

These are the pretreatment wash primer and metal conditioner, wash primer reducer, wash primer acid component, and wash primer retarder. To prepare the wash primer for use, the base component of the wash primer is shaken and mixed thoroughly. Then equal parts of the acid and base components are mixed and stirred thoroughly. The reducer is used to thin the mixture, if necessary, to reduce the viscosity, which should be such that the flow from a no. 2 Zahn viscosity cup requires 15 to 18 s. In warm, humid weather, when it is desired to slow the drying time, wash primer retarder is added to the mixture.

Wash primers should be applied in one very thin wash coat, not more than 0.5 mil [0.013 mm] in thickness. The wet, thin coat will penetrate the microscopic craters on the metal surface and any corrosion pits that exist.

Wash primers are manufactured in a variety of formulas by different companies, and it is essential that the technician follow carefully the directions on the container or in the manufacturer's instruction manual. An important point to observe is the pot life of the primer. After the components of the wash primer are mixed, the pot life is usually from 6 to 8 h. All material should be used during the pot-life period; unused material should be discarded.

Application of Primers

A primer serves two principal purposes: improved bonding and corrosion protection. It provides a bonding layer for the finish coatings in that the primer makes a good with the metal and the finish coating makes a good bond with the primer.

Two-part **epoxy primers** are popular for use under the urethane and polyurethane coatings. The primer is prepared for use by mixing the required quantity of catalyst with the base component. There is a waiting period (induction time) of 30 to 40 min before the mixture is ready for spraying. This time is necessary to allow the catalytic action to progress to a point where the primer is satisfactory for spraying.

Epoxy primers may be used on steel, aluminum, magnesium, or fiberglass. For the most complete protection against corrosion, the epoxy primer is applied over a wash primer.

Finish coatings should be sprayed over the epoxy primer within 24 h of primer application. Otherwise, the primer forms a hard surface to which the finish coating will not adhere properly. If more than 24 h has elapsed since the application of epoxy primer, the surface can be roughened with 400-grit wet or dry sandpaper, Scotch Brite pads, or other recommended material before the application of finish coats.

Acrylic lacquers and enamels should not be applied over an epoxy primer for at least 5 h (preferably overnight). Acrylics have a tendency to penetrate epoxy primer if it is not fully cured, and this results in a poor surface appearance. Urethane and polyurethane coatings can be sprayed on epoxy primers after 1 h curing time. These coatings soften the surface of the epoxy and form a chemical bond with the primer.

When epoxy primer materials are handled, it is important that the catalyst container be kept tightly closed except when the material is poured. The catalyst is reactive to moisture and will absorb it from the air if the container is not closed.

If the catalyst has absorbed considerable moisture and the container is resealed, a chemical action can take place that will cause the container to burst.

Typical instructions for the application of an epoxy primer are as follows:

1. See that the surface is thoroughly cleaned and dust-free.
2. Add exactly one part catalyst to two parts of the primer base component. Stir thoroughly and allow 30 min induction time before application. In high humidity, allow 1-h induction time to avoid curing agent "bloom."
3. After the two components are mixed, reduce with epoxy reducer to attain a viscosity of 19 to 21 s with a no. 2 Zahn viscosity cup. Additional thinning is required in hot weather.
4. Apply spray coats at 30-min intervals.

Instructions for the application of epoxy primer will vary for different manufacturers. For example, one instruction for a particular product requires a mixture of one-to-one catalyst and base component. This same instruction requires the application of only one coat of primer to a thickness of approximately 0.0005 in [0.013 mm]. It is most important, therefore, that the technician pay strict attention to the appropriate instructions.

Another primer specifically designed for use under urethane coatings is urethane primer. This is a two-part primer, which must be mixed a short time before use and must be used within 6 to 8 h after mixing.

Typical instructions for the application of urethane primer on Al-Clad aluminum are as follows:

1. Mix two parts urethane primer with one part urethane primer catalyst. Add urethane reducer as needed to obtain a viscosity of 18 to 22 s with a no. 2 Zahn viscosity cup.
2. Allow the catalyzed material to stand for an induction period of 30 min.
3. Spray one coat of the urethane primer on the Al-Clad aluminum surface, which has been previously coated with chromate wash primer. Allow to air-dry at 77°F [25°C] for at least 4 h before applying urethane finish coats. Urethane enamel must be applied within 48 h; otherwise the primer will need a light sanding before the enamel is applied.

Three-component polyurethane primers for external use and structural components have been developed that are chromate free and isocyanate cured. These primers are compatible with polyurethane topcoats. These primers consist of a base material, hardener, and thinner. They have a limited pot life and can be recoated in 2 h. Lightly rub the surface with Scotch brite extra-fine pads if more than 72 h expire before the topcoat is applied. Like Epoxy primers there is an induction time of 15 to 30 min after mixing to fully activate the paint system. These produts are typically applied directly over a wash primer.

In some cases, primer or enamel will **crater**, or **crawl**, as soon as it is sprayed on the surface. This means that it does not remain flat and smooth but forms small craters and ridges as if being repelled from areas of the surface. This repulsion is usually caused by oil or some other contaminant on the surface. It is usually necessary to remove the fresh coating and thoroughly clean the area with an approved solvent. In some cases, anticrater solutions are available to reduce or eliminate cratering. Such solutions must be used only as directed.

After the application of a primer, the spray equipment should be cleaned thoroughly and immediately. Some primers, such as epoxy primer, will set up hard within a short time and ruin the equipment. Spray guns, paint cups, pressure pots, and paint hoses must be disassembled to the extent possible, and every trace of the primer must be removed. Reducer or MEK is suitable for this purpose.

Topcoat Finishes

A wide variety of finish coatings are available, each with its special characteristics. A number of these finish coatings have been described in this chapter.

Lacquers, both nitrate and acrylic, are quick-drying and easily applied with a spray gun, provided that they are properly thinned with the correct reducer to meet the conditions of temperature and humidity at the time they are applied. Because of their fast-drying characteristics, they quickly dry dust-free. Care must be taken to ensure that lacquers are not sprayed over coatings that they cause to lift and wrinkle.

Enamels do not dry as fast as lacquers and should, therefore, be applied in a dust-free spray booth. Some enamels are designed for air drying, and others should be baked to cure the finish. In any event, the surface of the enamel should be dry before being exposed to conditions where dust may settle and mar the finish.

As mentioned previously, urethane and polyurethane products are being used extensively for finishing aircraft. When properly applied, they provide a high gloss ("wet look"), durability, and ease of maintenance. Typical instructions for a urethane coating are as follows:

1. Prepare the surface to be coated, as previously explained with wash primer and epoxy primer. The primer should have been applied not more than 36 h before the finishing is begun.
2. Mix catalyst in proportions given by the manufacturer. Stir the mixture thoroughly and add reducer to lower to a spray viscosity of 17 to 19 s with a no. 2 Zahn viscosity cup. Filter through a 60×48-mesh (or finer) cone filter. Allow a 20-min induction time before spraying.
3. Apply a light, wet tack coat with a spray gun and follow with two medium cross coats at 10- to 20-min intervals. The dry film thickness should be approximately 1.7 mils [0.0017 in, or 0.041 mm]. A DeVilbiss model MBC-S10 or JGA-502 spray gun with a no. 30 air cap and EX tip and needle at 45-lb [310-kPa] pressure has been tested and found satisfactory.

There are varying degrees of thickness for a sprayed top coat. Terms such as **dust**, **mist**, **single coat**, and **double coat** are commonly used to describe the spray thickness. A very light dry coat is often referred to as a dust coat. An application of slower-drying thinners or reducers over a clear coat is called a mist coat. Double coating is often used in the application of lacquers, where a second spray coating is applied immediately after the first coat is finished. Generally two or more double coats are required to apply a lacquer topcoat properly. Two or three single coats are normally required for enamel topcoats. The first coat should be allowed to become tacky before applying additional coats.

Identification of Finish Coatings

When it becomes necessary to repair or touch up an airplane finish, the nature of the previous coating must be determined in order to ensure that the new coating will adhere to the old coating without lifting or otherwise damaging it.

Various methods and techniques for identification of finishes are employed. To begin, it is a good plan to examine the airplane logbook to see if the finish is identified. If the airplane has not been refinished since new, the type of finish can be identified by the manufacturer.

Cellulose nitrate dope and lacquer dissolve when wiped with a cloth saturated with nitrate thinner or reducer. The distinct odor of cellulose nitrate also aids in making an identification. Synthetic engine oil, MIL-L-7808 or equivalent, softens cellulose nitrate finishes within a few minutes after application but has no effect on epoxy, polyurethane, or acrylic finishes.

MEK wiped onto acrylic finishes picks up pigment but does not affect epoxy or polyurethane finishes unless it is rubbed into the surface.

In all cases where a product is to be used for refinishing, the manufacturer's instructions should be consulted. Often the instructions provide information regarding the compatibility of other finishes with the finish to be applied.

Touching Up Finishes

Airplane finishes that have become scratched, scraped, cracked, or peeled should be repaired as quickly as possible in order to prevent the onset of corrosion. Any of the old finish that is loose or deteriorated should be removed. It is best to confine the operation to as small an area as possible; however, if the finish is badly deteriorated, it is best to strip an entire panel or refinish the entire airplane.

The old finish can be removed by the careful use of abrasive paper to take off the damaged coating and "feather" in the edges of the damaged area with the original coat. The surface should then be cleaned and all traces of dirt, wax, oil, and other contaminants removed. If the old finish is removed down to bare metal, the same sequence of primers and coatings should be followed as has been previously described. The same type materials as originally used on the aircraft should be used for the repair so that any interaction of products is avoided. If the original type of paint used cannot be located, select a product that will not interact with the coating being repaired.

Spray application of finishes is desirable, and the use of an air brush rather than a large spray gun may be appropriate, depending on the area being repaired. Be sure to protect the surrounding parts of the aircraft and work area from overspray.

Good-quality acrylic finishes are available in aerosol containers, and these may be used for a finish coat, provided that the undercoat material is compatible. Acrylic finish can be applied over epoxy primer if done within about 6 h of the epoxy application. If acrylic lacquer is applied over an old acrylic finish, it is considered good practice to soften the old finish with the application of acrylic thinner, either by wiping or spraying, immediately prior to applying the finish coat. The thinner must be dried before the finish is applied. Urethane enamel can be applied over an old epoxy or urethane coating but should not be applied over an old alkyd enamel surface. The surface should be thoroughly washed with an emulsion cleaner and then steam-cleaned. The surface is then sanded with 280- to 320-grit aluminum oxide or silicon carbide paper and wiped down with MEK or toluol. Urethane enamel can then be sprayed on the surface, as previously explained.

Masking and Decorative Trim

Aircraft are not painted in only one color. Most aircraft contain at least a line or two of decorative trim, with some even being painted in two-tone or multitone color combinations. Even if an aircraft does not have any decorative trim, registration numbers must be applied. The techniques for the application of trim are not complicated, but they do require some care on the part of the painter.

Templates and chalk lines are normally used in laying out the trim lines. The lines should not be laid out with a pencil or pen that can damage the surface or leave permanent marks. The surface of the finish must be hard enough (dry enough) so that this work will not damage the finish. For many finishing materials the paint must not have been on the aircraft more than a few hours or a few days when the trim color is applied. If the paint has been on too long and has cured, the trim color may not properly adhere to the surface coat.

Once the trim lines are laid out, tape is laid along the lines. The tape used should be of a type whose adhesive will not be attacked by the solvents in the trim paint. Propylene fine-line masking tape is a type that is commonly used. This tape also has a smoother edge than conventional crepe paper masking tape and gives a sharp trim line. Paper or plastic is then taped onto the trim tape so that a protective cover is placed over the surface areas next to the trim areas. This protects these surfaces from overspray.

When all the masking and protective covers are in place, the area to be painted is wiped with a clean cloth containing solvent, and after this is dry, the paint is applied.

Rubbing Compounds

Rubbing compounds contain an abrasive such as pumice, which levels the top of the finish, making it smooth. Rubbing compounds are available in various cutting strengths to be used for both hand and machine compounding.

Generally compounds with coarse particles are called **rubbing compounds,** whereas those with fine particles of pumice are called **polishing compounds**. Rubbing compounds are used to smooth out the surface being compounded and to remove fine scratches. Polishing compounds are used to smooth the finish and to bring out the gloss of topcoats, particularly with lacquers. Small areas are best done by hand; however, large areas generally are done with air-powered buffers.

Common Painting Problems

If the conditions for the application of an aircraft finish are not correct, one or more irregularities in the paint surface may show up. These irregularities are addressed here as they relate to spray painting.

Runs and **sags** are usually found on vertical or sloping surfaces and are the result of a heavy coat of paint being applied in an area. The heavy coat has such a buildup that gravity takes over and the paint runs down the surface before it can properly set. This can be caused by an improperly adjusted spray gun, by holding the gun too close, or by moving the gun too slowly.

Starved, or **thin film**, is the result of the gun being held too far away from the surface and from being moved too rapidly. An improper gun adjustment, providing too little paint or too much air, will also cause this condition.

Dry spray, or **texture**, is caused by the gun being held too far away, or by too much air pressure.

Blisters, or **bubbles**, are small raised circular areas with a layer of paint over them. These may be caused by oil or water in the air line of the spray gun. These may also be caused by applying a second coat of paint too soon over a fresh first coat that is still giving off large quantities of thinner.

Blushing means the paint appears to be cloudy or milky. It is caused by the vehicle in the paint evaporating so fast that the surface temperature of the paint is reduced to the point where the moisture in the air condenses and mixes with the paint.

Crazing is when fine splits or small cracks, often called "crow's-feet," appear in the finished surface. This is usually caused by the air temperature being too cold.

Fish eyes are small, craterlike openings in the finish. They are caused by improper surface cleaning or preparation. Many waxes and polishes contain silicone, which is the most common cause of fish eyes. Silicones adhere firmly to the paint film and require extra effort for their removal. Additives are available that will help eliminate fish eyes.

Orange peel is the term for paint that creates an irregular surface like the surface of an orange peel. The paint is of varying thickness on the surface and does not flow out properly. It may be caused by improper air pressure at the gun, the gun being held too far away from the surface, or insufficient thinning of the paint.

Peeling is the loss of adhesion between the paint and the surface or the primer coat. This may result from the surface not being properly cleaned and prepared or from the use of an incompatible primer.

Pinholes are small holes or depressions in the paint. These are caused by insufficient drying time between coats of paint or trapped solvents or moisture in the finish.

The correction for each of these problems involves eliminating the stated cause. In some cases, depending on the finishing product being used, the paint may have to be completely removed. In other cases, it may be allowed to dry for some period of time and then sanded lightly before continuing with the process. Still another solution may be to apply a corrective coat to the surface, normally consisting of a combination of paint, thinners, and retarders. In each case, follow the product manufacturer's recommendations.

REGISTRATION MARKS FOR AIRCRAFT

The finishing of certificated aircraft involves the application of required identification marks as set forth in FAR Part 45. This section gives the general requirements for such markings; however, the technician should consult the latest regulations when applying the markings to an aircraft.

All aircraft registered in the United States must be marked with nationality and registration marks for easy identification. The marks must be painted on the aircraft or otherwise affixed so they are as permanent as the finish. The marks must contrast with the background and be easily legible. The letters and numbers must be made without ornamentation and must not have other markings or insignia adjacent to them that could cause confusion in identification.

Required Markings

The registration and nationality markings for United States—registered aircraft consist of the Roman capital letter N followed by the registration number of the aircraft. The registration number is issued by the FAA when the aircraft is first registered.

When marks that include only the Roman capital letter N and the registration number are displayed on **limited, restricted, or experimental aircraft** or on aircraft that have **provisional certification,** the words *limited, restricted, experimental,* or *provisional airworthiness* must be displayed near each entrance to the cockpit or cabin. The letters must be not less than 2 in [5.08 cm] or more than 6 in [15.24 cm] in height.

Location of Marks for Fixed-Wing Aircraft

Registration marks for fixed-wing aircraft must be displayed on both sides of the vertical tail surface or on both sides of the fuselage between the trailing edge of the wing and the leading edge of the horizontal stabilizer. If the marks are on the tail of a multivertical-tail aircraft, the marks must be on the outer surfaces of the vertical sections.

If the aircraft has engine pods or components that tend to obscure the sides of the fuselage, the marks can be placed on the pods or other components.

Location of Marks for Non-Fixed-Wing Aircraft

Helicopters or other rotorcraft must have nationality and registration markings displayed on the bottom surface of the fuselage or cabin, with the top of the marks toward the left side of the fuselage and on the side surfaces of the fuselage below the window lines as near the cockpit as possible.

Airships (dirigibles and blimps) must have markings displayed on the upper surface of the right horizontal stabilizer and on the undersurface of the left horizontal stabilizer, with the top of the marks toward the leading edge, and on each side of the bottom half of the vertical stabilizer.

Spherical balloons must display markings in two places diametrically opposite each other and near the maximum horizontal circumference of the balloon.

Nonspherical balloons must have markings displayed on each side of the balloon near the maximum cross section and immediately above either the rigging band or the points of attachment of the basket or cabin-suspension cables.

Sizes of Nationality and Registration Marks

Except for the special cases discussed in FAR Part 45, the height of letters and numbers for fixed-wing aircraft must be equal and at least 12 in [30.48 cm] high.

Marks for airships and balloons must be at least 20 in [50.8 cm] in height. The marks on the bottom of the fuselage or cabin of a rotorcraft must be four-fifths as high as the fuselage is wide or 20 in [50.8 cm] in height, whichever is less. The marks on the sides of the rotorcraft cabin must be a minimum of 2 in [5.08 cm] but need not be more than 6 in [15.24 cm] in height.

The width of the characters in the markings must be two-thirds the height of the characters, except for the number 1 and the letters M and W. The number 1 must have a width one-sixth its height and the letters M and W may be as wide as they are high. The thickness of the strokes or lines in the letters and numbers must be one-sixth the height of the characters. The spaces between the characters must not be less than one-fourth the width of the characters. For 12-in-[30.48-cm] high letters, this equals a width of 8 in [20.32 cm] for all letters except W and M, which are 12 in [30.48 cm] wide, and a 1, which is 2 in [5.08 cm] wide. The space between characters is 2 in [5.08 cm]. Figure 5-16 illustrates the taping and painting of letters and trim on aircraft exterior.

FIGURE 5-16 Taping and painting of letters on aircraft exterior. (*Airbus*)

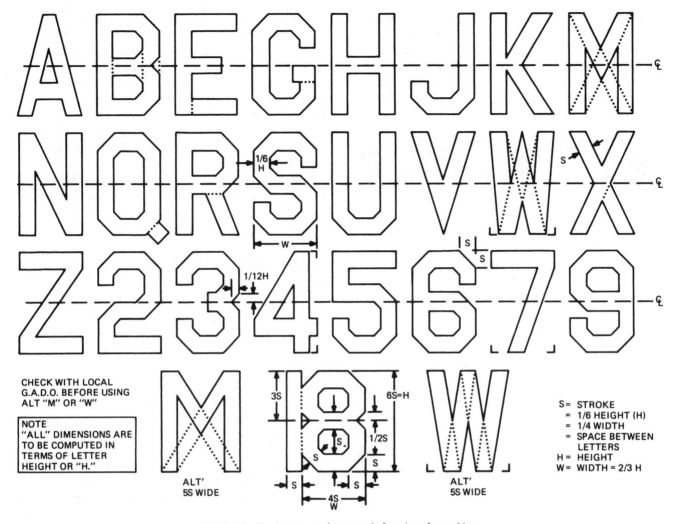

CHECK WITH LOCAL
G.A.D.O. BEFORE USING
ALT "M" OR "W"

NOTE
"ALL" DIMENSIONS ARE
TO BE COMPUTED IN
TERMS OF LETTER
HEIGHT OR "H."

S = STROKE
  = 1/6 HEIGHT (H)
  = 1/4 WIDTH
  = SPACE BETWEEN
    LETTERS
H = HEIGHT
W = WIDTH = 2/3 H

FIGURE 5-17   Letters and numerals for aircraft marking.

Laying Out Letters and Numbers

Letters and numerals for the markings on aircraft are formed as shown in Fig. 5-17. The illustration is not intended to show the correct interval or spacing between the characters but merely to indicate the type and proportions of approved characters.

To lay out numerals or letters, a master template, such as the one shown in Fig. 5-18, can be constructed. With this device it is possible to lay out any letter or numeral. For some letters or numerals it is necessary to do a small amount of additional construction, but the template will provide the principal guidelines. Some people find that an additional template measuring 2 × 12 in [5.08 × 30.48 cm] aids in making the diagonal lines of the characters. This guide can be made of Plexiglas or sheet metal.

When laying out the registration letters and numbers on the side of the aircraft, light guidelines are drawn with chalk, and after the letters are outlined properly, paint trim tape is laid along the lines and masking, and covers are placed over the rest of the aircraft to protect it from overspray. The masked-off characters are wiped with a soft cloth containing solvent to remove chalk marks and any oil or other contaminants. The surface is then sprayed with the appropriate color.

Stencils for laying out registration numbers are available, as are decals. The stencils are faster than laying out the number by hand, but additional expense is involved. The use of decals may be warranted if spraying is not practical or if the numbers are only temporary.

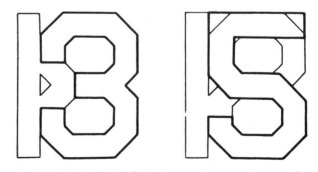

FIGURE 5-18   Template for laying out letters and numerals.

1. Describe the primary functions of a *primer.*
2. Describe a *wash primer.*
3. Why is an *epoxy-type primer* supplied in two parts?
4. What basic components make up paint?
5. What are the two types of exterior paint commonly used on aircraft?
6. What are the advantages of *polyurethane enamel*?
7. What is the function of *thinners* and *reducers*?
8. What is the purpose of using *additives* in the paint?
9. Why are respirators used in conjunction with paint spraying?
10. What are the functions of an *air transformer*?
11. How often should an air transformer be drained?
12. How do air-hose diameter and length affect pressure drop?
13. List two common types of *air-spraying equipment.*
14. Describe a *paint pressure tank.*
15. What is the advantage of *electrostatic painting*?
16. Describe the operation of a *spray gun.*
17. What are the adjustments on a spray gun?
18. How is the shape of the spray pattern adjusted?
19. What will happen if the spray gun is moved in an arc when spraying instead of parallel to the surface?
20. List two methods of stripping paint from an aircraft.
21. What precautions must be observed when chemical-stripping an aircraft?
22. What variables may be controlled when dry-stripping an aircraft?
23. Why is it necessary to avoid contact between dissimilar metals in an aircraft structure?
24. What condition can lead to *filiform corrosion*?
25. What may be done to slow drying time if relative humidity is comparatively high when wash primers are applied?
26. What is the purpose of a *conversion coating*?
27. What is the value of an etch for a metal surface before finishing?
28. What are the principal reasons for applying a primer before finishing a metal surface?
29. What precautions must be taken in the application of a lacquer finish over other finishes?
30. What causes runs or sags to appear when using a spray gun?
31. What devices may be used to assist in the laying out of aircraft markings?
32. What type of tape should be used when laying out aircraft paint trim lines?
33. What markings are required on any aircraft registered in the United States?
34. Describe the form and size of the characters used for official aircraft markings.
35. What is the total length of the area on which the N-number N1234M is installed if 12-in [30.48-cm] letters are used?

Welding Equipment and Techniques 6

INTRODUCTION

The technician may be called upon to repair important aircraft parts by welding. If the technician does not understand the process of welding or is careless, the weld may fail. It is, therefore, essential to be well acquainted with the approved welded repairs, techniques for welding, and the operation of welding equipment. The technician who is not sufficiently skilled to perform an airworthy welded repair should call upon a qualified welder to do the work. In any event, the technician must be able to inspect and evaluate the quality of any welded structure or welded repair.

FUNDAMENTALS OF WELDING

Types of Welding

Welding is a process used for joining metal parts by either fusion or forging. **Forge welding** is the process used by a blacksmith when heating the ends of wrought iron or steel parts in a forge fire until the ends are in a plastic state and then uniting them by the application of mechanical pressure. Even today this mechanical pressure is sometimes the result of blows from a heavy hammer. **Fusion welding** is the process used by welders in the aviation industry and other industries in which enough heat is applied to melt the edges or surfaces of the metal so that the molten (melted) parts flow together, leaving a single, solid piece of metal when cool. In both forge and fusion welding, the process is described as a **thermal metal-joining process** because heat is required. Only fusion-type welding is used in aircraft work.

Typically, the types of fusion welding used by the technician are **oxyacetylene** (or **oxyfuel**) **welding**, commonly called *gas welding*, **electric-arc welding**, and **inert-gas arc welding**. A variety of special welding processes and techniques have been developed, but these are generally employed only in the original manufacturing processes.

Gas welding and inert-gas arc welding are the most frequently used of all welding processes in aviation. **Gas welding** produces heat by burning a properly balanced mixture of oxygen and acetylene or other fuel as the mixture flows from the tip of a welding torch. Since the temperature of an oxyacetylene flame at the tip point of the torch may

be 5700°F [3149°C] to 6000°F [3316°C], it is apparent that it is hot enough to melt any of the common metals.

Another gaseous fuel that produces almost as much heat as acetylene is a mixture of methylacetylene and propadiene stabilized. Sold under the name of MAPP® by Airco Welding Products, this fuel is safer than acetylene because it does not become unstable at any operating pressure.

The heat required for the fusion of metal parts can be produced by an electric current. Electric welding includes *electric-arc welding*, *electric-resistance welding*, and *inert-gas-arc welding*. In **electric-arc welding**, the heat of an electric arc is used to produce fusion of the parts by melting the edges of the parts being joined and the end of the welding electrode and then allowing the molten metal to solidify in a welded joint. The arc is formed by bringing together two conductors of electricity, the edges being joined and the electrode, and then separating them slightly. **Electric-resistance welding** is a process whereby a low-voltage, high-amperage current is brought to the work through a heavy copper conductor offering very little resistance to its flow. The parts are placed in the path of the current flow, where they set up a great resistance to it. The heat generated by the current flow through this resistance is enough to fuse the parts at their point of contact. **Spot welding** and **seam welding** are common versions of electric-resistance welding.

Inert-gas-arc welding is a process in which an inert gas such as helium or argon blankets the weld area to prevent oxidation of the heated metal. This is particularly important in welding titanium, magnesium, stainless steels, and other metals that are easily oxidized when subjected to melting temperatures. Tungsten inert-gas (TIG) welding and metal inert-gas (MIG) welding are commonly used forms of inert-gas welding. The names Heliarc® and Heliweld® are trade names that have been used to designate tungsten inert-gas welding when helium is used as the inert gas.

The American Welding Society (AWS) has developed a series of codes used to standardize the identification of the various processes. The AWS process codes for welding techniques are identified in Table 6-1.

Identification of Metals

The technician must be able to identify various metals before attempting to weld them so that the proper torch tip, filler rod, and technique required can be determined. In some

TABLE 6-1 AWS Process Codes

SMAW	Shielded metal-arc welding	FRW	Friction welding
SAW	Submerged-arc welding	EBW	Electron-beam welding
GMAW	Gas metal-arc welding	LBW	Laser-beam welding
*ST	Spray transfer	B	Brazing
*B	Buried arc	TB	Torch brazing
*P	Pulsed arc	FB	Furnace brazing
*S	Short-circuiting arc	IB	Induction brazing
FCAW	Flux-cored arc welding	RB	Resistance brazing
GTAW	Gas tungsten-arc welding	DB	Dip brazing
PAW	Plasma-arc welding	IRB	Infrared brazing
RW	Resistance welding	DFB	Diffusion brazing
OFW	Oxyfuel gas welding	S	Soldering
DFW	Diffusion welding		

*Not standard AWS letter designations. Pulsed-arc and short-circuiting arc are officially designated GMAW-P and GMAW-S, respectively. But for brevity, the designations are given here as P and S. There are no official designations for spray transfer and buried arc.
Source: American Welding Society

organizations, metals may be marked with painted bands of different colors on tubes and bars or by means of numbers on sheet stock. When acquiring the material from stock, the technician should be sure that the material not used retains the material identification coding or is recoded in the same manner as the original material.

Where there are no colored bands or numbers on the metal, three types of tests are commonly used: (1) *spark test*, (2) *chemical test*, and (3) *flame test*.

In the identification of metals by means of the **spark test**, *ferrous metals* may be recognized by the characteristics of the spark stream generated by grinding the material with a high-speed grinding wheel. A ferrous metal is one that contains a high percentage of iron. In general, nonferrous metals cannot be identified by a spark test because they do not produce a large shower of sparks and, in fact, may produce almost none.

In applying the spark test the technician should obtain samples of various metals in order to compare the pattern of a known material with the pattern produced by the material being identified. When a known sample produces the same spark characteristics as the unknown piece of metal, identification is accomplished. The characteristics to be observed are: (1) the volume of the spark stream, (2) the relative length of the spark stream (in inches), (3) the color of the spark stream close to the grinding wheel, (4) the color of the spark streaks near the end of the stream, (5) the quantity of the sparks, and (6) the nature of the sparks.

In volume, the stream is described as extremely small, very small, moderate, moderately large, and large. The relative length may vary from 2 to 70 in [5 to 177 cm], depending upon the metal. For example, cemented tungsten carbide produces an extremely small volume of sparks and the stream is usually only about 2 in [5.08 cm] long. On the other hand, machine steel produces a large volume and may be about 70 in [180 cm] long. These particular figures apply when a 12-in [30.48-cm] wheel is used on a bench stand. The actual length in each case depends upon

the size and nature of the grinding wheel, the pressure applied, and other factors.

In color, the stream of sparks may be described as red, white, orange, light orange, or straw-colored. The quantity of sparks may be described as none, extremely few, very few, few, moderate, many, or very many. The nature of the sparks may be described as forked or fine and repeating (exploding sparks). In some cases, the sparks are described as curved, wavy, or blue-white, but in most instances the terms previously given apply.

Some handbooks for welders include tables showing these characteristics of the sparks, but all the terms used to describe the spark stream are only comparative. One person may describe a color as orange, whereas another person will refer to the same stream of sparks as light orange or even straw-colored. Because of this situation, the use of the known samples saves time and improves accuracy.

The **chemical test** for distinguishing between *chrome-nickel corrosion-resisting steel* (18-8 alloy) and *nickel-chromium-iron alloy* (Inconel) should be known by welders. A solution consisting of 10 g cupric chloride dissolved in 100 cm^3 hydrochloric acid is used. One drop is applied to the unknown metal sample and allowed to remain on the metal for about 2 min. At the end of this time, three or four drops of water are slowly added with a medicine dropper. The sample is then washed and dried. If the metal is stainless steel, the copper in the cupric chloride solution will be deposited on the metal, leaving a copper-colored spot. If the sample is Inconel, the spot left will be white.

A **flame test** is used to identify *magnesium alloys*. The welding flame is directed on a small sample until the metal is brought to the melting point. If the metal sample is magnesium alloy, it will ignite at once and burn with a bright glow.

Types of Weld Joints

A **joint** is that portion of a structure where separate base-metal parts are united by welding. The word *weld* is often used to refer to a *joint*. For example, a **butt weld** is a welded butt joint. The word *seam* is often used to refer to a welded joint, especially in a case of tanks and containers. Five different types of joints, illustrated in Fig. 6-1, are used to weld the various forms of metal. These are (1) butt joints, (2) tee joints, (3) lap joints, (4) corner joints, and (5) edge joints. In addition to the various types of joints, the technician should be aware that the edges of the materials to be welded may require special preparation. The type of edge preparation required depends upon the materials being welded and the techniques used in the welding process. Where appropriate, edge preparation will be discussed as a part of the description of the various welding processes.

Butt Joints

A **butt joint** is a joint made by placing two pieces of material edge to edge in the same plane so that there is no overlapping. It is called a *butt joint* because the two edges, when joined, are abutted together. There are two classifications of butt joints: **plain butt joint** and **flange butt joint**. The plain butt joint is used where the two pieces of the materials to be welded are aligned in approximately the same plane. Flange butt joints are welded using edges that are turned up 90°, producing a flange height of from one to three times the thickness of the material being welded. The flanges are fused together during the welding process. Since the flanges supply enough metal to fill the seam, a filler rod is not normally used. The result is a joint that appears similar to that of a plain butt joint.

Tee Joints

A **tee joint** is a form of joint made by placing the edge of one base part on the surface of the other base part so that the surface of the second part extends on either side of the joint in the form of a T. Filler rod is used with tee joints.

A **plain tee joint** is acceptable for most metal thicknesses in aircraft work and also may be used for heavier metals, where the weld can be located so that the load stresses will be transverse (perpendicular) to the longitudinal dimensions of the weld. The only preparation required is cleaning the surface of the horizontal member and the end of the vertical member. The weld is then made from each side with penetration into the intersection. This results in a **fillet weld**, having a general triangular cross-sectional shape. (Any weld that joins two parts that are at right angles to each other may be called a *fillet weld*. Corner joints, lap joints, and edge joints also require fillet welds.)

Lap Joints

A **lap joint** is a joint made by lapping one base over the other and is used in plate, bar, tubing, and pipe. These joints are widely used in the construction of articles fabricated from plate and sheet metal (flat, wrought metals), but a lap joint is not as efficient as a butt joint for distributing load stresses. A lap joint is commonly used where the primary load stress will be transverse (perpendicular) to the line of weld.

A **single-welded lap joint** is used for sheet, plate, and structural shapes where the loading is not severe. The same type of joint can be used for telescope splices in steel tubing, and in that application it is better than a butt joint.

A **double-welded lap joint** is used for sheet and plate where the strength required is greater than that which can be obtained when a single weld is used. This type of joint provides for great strength, when properly made, in all ordinary thicknesses of sheet and plate.

The **offset**, or **joggled**, **lap joint** is used for sheet and plate where it is necessary to have a lap joint with one side of both plates or sheets in the same plane; that is, on one side the surface is flush. This type of joint provides for a more even distribution of load stresses than either the single or double lap joint, but it is more difficult to prepare.

Edge Joints

An **edge joint** is a form of joint made by placing a surface of one base part on a surface of the other base part in such

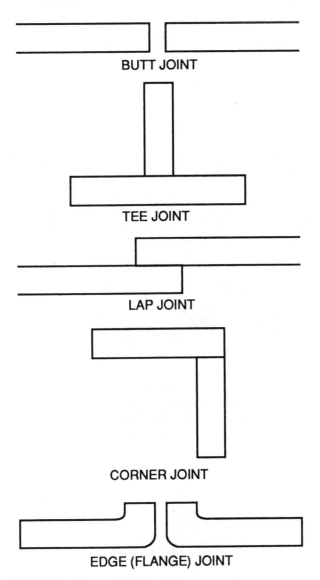

FIGURE 6-1 Types of welding joints.

BUTT JOINT

TEE JOINT

LAP JOINT

CORNER JOINT

EDGE (FLANGE) JOINT

a manner that the weld will be on the outer surface planes of both parts joined. This type of joint is not used where a high joint strength is required, but it is widely used for fittings composed of two or more pieces of sheet stock where the edges must be fastened together. This use is acceptable because the joint is not subjected to high stresses. Edge joints can be used also for tanks that are not subjected to high pressures.

Edge joints are usually made by bending the edges of one or both parts upward at a 90° angle, placing the two bent ends parallel to each other, or placing one bent end parallel to the upright unbent end and then welding along the outside of the seam formed by the two joined edges.

Corner Joints

A **corner joint** is made by placing the edge of one part at an angle on an edge or a surface of another part so that neither part extends beyond the outer surface of the other, the structure resembling the corner of a rectangle.

There are three types of corner joints for plate and sheet. The **closed type of corner joint** is used on lighter-gauge metals, where the joint is subjected only to moderate stresses. It is made without adding much, if any, filler rod because the edge of the overlapping sheet is melted and fused to form the bead.

The **open type of corner joint** is used on heavier sheet for the same purpose as a closed type of corner joint. It is made by fusing the two edges at the inside corner and adding enough welding rod to give a well-rounded bead of weld metal on the outside.

In the third case, if such an open joint is required to bear a fairly heavy load, an additional weld must be made on the inside corner to provide the necessary strength where a light concave bead has been laid on the inside.

Welding Characteristics and Nomenclature

To make a proper weld it is necessary to identify the characteristics of a correct weld. The technician should also be aware of the changes in material characteristics that may take place as a result of the welding operation.

Parts of a Weld

Figure 6-2 shows the names of the parts of a weld. The **face** is the exposed surface of the weld. The **root** is the zone at the bottom, or base, of the weld; in other words, it is the depth that fusion penetrates into the base metal at the joint. The **throat** is the distance through the center of the weld from the root to the face. The **toe** is the edge formed where the face of the weld meets the base metal; that is, it is the edge of the fusion zone in the base metal on each side of the weld. The **reinforcement** is the quantity of weld metal added above the surface of the base metal (the metal in the parts being joined) to give the weld a greater thickness in cross section. Materials welded by the electric-resistance process may not have a reinforcement, depending upon the details of the technique used.

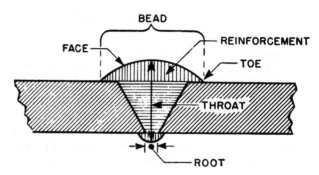

FIGURE 6-2 Nomenclature of a weld.

Other terms not illustrated in Fig. 6-2 are as follows: (1) The **leg** is the dimension of the weld metal extending on each side of the root of the joint, and (2) the **fusion zone** is the width of the weld metal, including the depth of fusion in the base metal on each side of the joint.

In Fig. 6-2, the **bead** is shown. This is the metal deposited as the weld is made. In order to have good penetration, the base metal at the joint must be melted throughout its thickness; hence a bead of weld metal should be visible on the underside of a butt joint, as shown in Fig. 6-2. A good indication of penetration in the case of a fillet weld is the presence of scale on the lower side.

Figure 6-3 consists of three drawings that illustrate the meaning of width of fusion, reinforcement, throat, root, leg, and toe.

Proportions of a Weld

The three most important proportions of a weld are (1) the **depth of penetration**, which should be at least one-fourth the thickness of the base metal; (2) the **width of the bead**, which should be between two and three times as great as the thickness of the base metal; and (3) the **height of the reinforcement**, which should be not less than one-half the

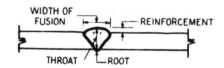

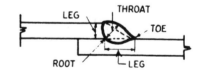

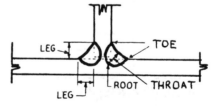

FIGURE 6-3 Details of weld nomenclature.

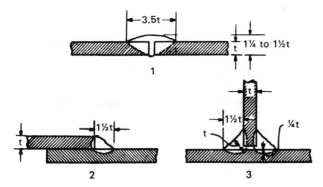

FIGURE 6-4   Preferred dimensions of a weld.

thickness of the base metal. Note that these dimensions should be considered minimums and are less than the recommended values shown in the various figures included in this chapter.

In Fig. 6-4, the butt weld shown in drawing 1 has a bead that is three to five times the thickness of the base metal. In attaching an aircraft fitting by means of a lap weld, as in drawing 2 of Fig. 6-4, the width of the fillet bead is $1\frac{1}{2}$ times the thickness of the upper sheet. In making a tee joint, as in drawing 3, the weld bead has a thickness through the throat that equals the thickness of the vertical member. The penetration of the weld into the sides of the joint is one-fourth the thickness of the base metal, and the height of the reinforcement meets the requirements previously given.

Correct Formation of a Weld

A weld must be formed correctly to provide strength and to resist fatigue in a joint. If it is not made properly, the weld may have less than 50 percent of the strength for which the joint was designed. Figure 6-5 shows correct lap joints; Fig. 6-6 shows correct tee joints; Fig. 6-7 shows good corner joints; and Fig. 6-8 illustrates properly formed butt joints.

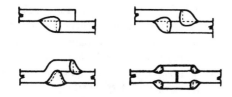

FIGURE 6-5   Properly made lap joints.

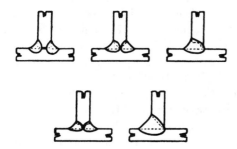

FIGURE 6-6   Correct tee joints.

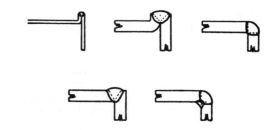

FIGURE 6-7   Good corner joints.

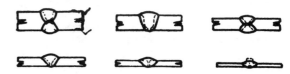

FIGURE 6-8   Properly formed butt joints.

The typical causes of improperly formed weak welds are (1) undercutting of the base metal at the toe of the weld; (2) not enough penetration; (3) poor fusion of the weld metal with the base metal; (4) trapped oxides, slag, or gas pockets in the weld; (5) overheating of the weld; and (6) overlap of the weld metal on the base metal. These incorrect conditions result from inexperience, poor technique, or carelessness. Figure 6-9 shows a large number of faults commonly found in the weld-metal formation of various joints. Any weld that has an appearance similar to one of the drawings in this illustration should be rejected.

Chemical Changes Produced by Welding

A **chemical change** occurs when a substance is added to or taken from a metal. The heat of the welding process will cause the loss of one or more of the chemical constituents of a piece of metal if the heat remains on the metal for any length of time; this loss usually will result in a reduction of such physical properties of the metal as tensile strength, ductility, and yield point. Also, if some element is added to the metal during the welding process or if there is some material change in one or more of the chemical constituents, the change will usually lower the strength of the metal.

Physical Changes Produced by Welding

A physical change is a change of any kind that takes place without affecting the chemical structure of a metal. Some of the physical changes most important in welding are changes in the melting point, heat conductivity, and rate of expansion and contraction.

The melting point is the degree of temperature at which a solid substance becomes liquid. Pure metals have a melting point, but alloys have a **melting range**. Welders should know the approximate melting points of the various metals with which they work because they must often weld together metals that have widely different melting points. If a metal includes an alloyed element, the melting point is lowered; hence the melting points given in tables for alloyed metals

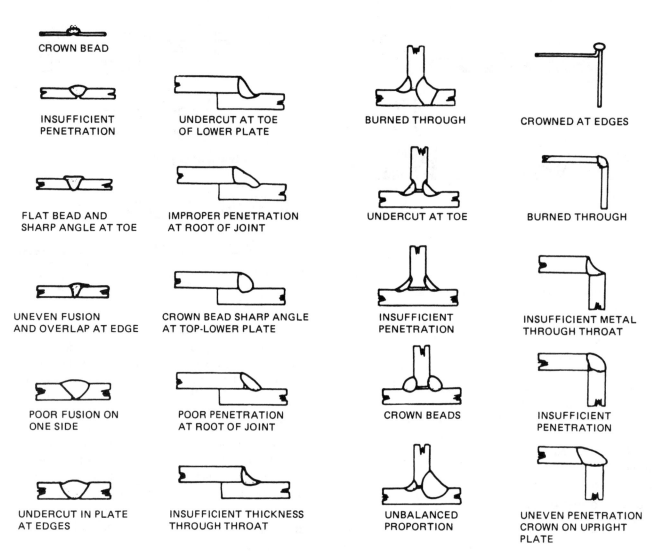

CROWN BEAD

INSUFFICIENT
PENETRATION

UNDERCUT AT TOE
OF LOWER PLATE

BURNED THROUGH

CROWNED AT EDGES

FLAT BEAD AND
SHARP ANGLE AT TOE

IMPROPER PENETRATION
AT ROOT OF JOINT

UNDERCUT AT TOE

BURNED THROUGH

UNEVEN FUSION
AND OVERLAP AT EDGE

CROWN BEAD SHARP ANGLE
AT TOP-LOWER PLATE

INSUFFICIENT
PENETRATION

INSUFFICIENT METAL
THROUGH THROAT

POOR FUSION ON
ONE SIDE

POOR PENETRATION
AT ROOT OF JOINT

CROWN BEADS

INSUFFICIENT
PENETRATION

UNDERCUT IN PLATE
AT EDGES

INSUFFICIENT THICKNESS
THROUGH THROAT

UNBALANCED
PROPORTION

UNEVEN PENETRATION
CROWN ON UPRIGHT
PLATE

FIGURE 6-9 Common faults in welding.

vary according to the proportion of alloying elements present and should be considered with this fact in mind.

Some physical changes take place in the materials being welded because the application of heat during the welding process results in expansion and contraction of the materials. The technician should keep in mind the difference between temperature and heat. Put simply, heat is the quantity of a temperature that is available. Heat is measured in British thermal units (BTUs), whereas temperature is measured in degrees.

Expansion is an increase in the dimensions (length, width, thickness) of a substance under the action of heat. If a metal structure is unevenly heated, there will be an uneven expansion, and this will produce distortion (warping) and possibly breakage. On the other hand, if the temperature is raised progressively throughout the whole mass of the object, the action is uniform and there is no distortion or breakage.

Applying this to welding, if the heat from the welding process is concentrated at one point on a metal object, the metal in the heated area tends to expand where the heat is applied, and the portion that opposes this expansion may be distorted, cracked, or severely strained.

Tables giving the properties of metals usually include the coefficients of expansion. A **coefficient of thermal expansion** of any metal is the amount that the metal will expand per inch for each degree rise in temperature. For example, aluminum has a coefficient of expansion of 0.000 012 34, whereas steel has a coefficient of expansion of 0.000 006 36. This shows that aluminum expands more than steel for each degree rise in temperature. In both these cases, the coefficient refers to a rise of 1°F [0.5556°C].

To apply this knowledge, a simple formula can be used. Let A represent the length in inches of the piece of metal, B the temperature in degrees Fahrenheit, and C the coefficient of expansion. Then, expansion in inches $= A \times B \times C$. Thus, if a piece of aluminum is 1 in [2.54 cm] long, is raised in temperature 1°F [0.5556°C], and has a coefficient of expansion of 0.000 012 34, then the expansion is $1 \times 1 \times 0.000\ 012\ 34$, or 0.000 012 34 in [0.000 031 2 cm].

Contraction is the shrinking of a substance when cooled. It is the reverse of expansion. Unless there is some restraint, materials contract as much when cooled as they expanded when they were heated, assuming that the temperature is uniform throughout.

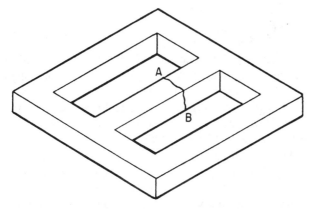

FIGURE 6-10 Welding a closed section requires heating of the entire closed area.

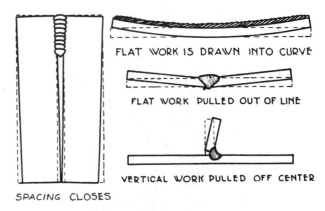

FLAT WORK IS DRAWN INTO CURVE

FLAT WORK PULLED OUT OF LINE

VERTICAL WORK PULLED OFF CENTER

SPACING CLOSES

FIGURE 6-11 Shrinkage of metal caused by welding.

In a trussed frame, whether it is in an airplane or a bridge, there is a restriction of the free movement of the metal parts. When such restrictions are present and the metal is malleable (capable of being worked into shape by hammering, rolling, or pressing), warping will take place. If the metal is brittle, it will usually crack. If the piece of metal is "open," that is, if no obstructions or restrictions hinder its free expansion and contraction, there is no danger of its being damaged from expansion and contraction. An example of open metal occurs in the case of an ordinary bar of metal, a length of unattached tubing, or some similar detached piece of metal.

If the metal is "closed," as in Fig. 6-10, there is danger from expansion and contraction. The bar that was formerly free and unattached is now the center section in Fig. 6-10, and it is fastened rigidly to a solid frame. If the break marked with the letters A and B in Fig. 6-10 is welded, provision must be made for expansion and contraction. Since the crosswise and lengthwise members of the frame are rigid, they do not permit the ends of the bar in the center to expand; hence, the only place where expansion can take place while the metal is heated during the welding process is at the point of the weld. When this portion begins to cool, the center bar contracts and shortens, but the frame in which it is placed refuses to surrender to the inward pull of the ends of the center bar. Warping occurs along the line of weld, or possibly a break occurs.

To avoid this damage, a trained welder heats the whole object before attempting to weld the break in the center piece. The whole object expands equally, pulling apart the edges of the break. The welder makes the weld and allows the object to cool. It all cools to the same extent, contracts equally, and suffers neither warpage nor breakage. Figure 6-11 shows examples of shrinkage in welded metal objects.

Conductivity

Conductivity is the physical property of a metal that permits the transmission of either an electric current, called electrical conductivity, or heat, called thermal conductivity, through its mass. The **rate of conductivity** is the speed at which a metal body will transmit either an electric current or heat through its mass. The rate of conductivity varies among metals. Radiation (heat loss) influences both the rate of heat conductivity and the area that will be affected by heat conductivity. Thus, metals that are good heat conductors may be poor radiators, and those that are good radiators may be poor conductors.

In welding, a considerable amount of heat is carried away from the point of application and is lost to the surrounding environment. For this reason, metals that have a high thermal conductivity require more heat in welding than those with a low conductivity, other things being equal.

Another thing to remember in welding is that the higher the thermal conductivity, the more extensive and the hotter will become the heated area around the weld. Therefore, more expansion can be expected with metals of high thermal conductivity, other things being equal.

Effect of High Temperatures on the Strength of Metals

Some metals have absolutely no strength or almost no strength when they are raised to extremely high temperatures. In some cases, this temperature may be far below the melting point of the metal. For example, aluminum alloys, brass, bronze, copper, cast iron, and certain alloy steels become very brittle at high temperatures near their melting points. If such metals are strained while at these high temperatures, they will break, or check, in the area that has been heated.

For example, the melting point of aluminum is 1218°F [659°C]. At 210°F [99°C], it has 90 percent of its maximum strength; at 400°F [204°C], it has 75 percent; at 750°F [399°C], it has 50 percent; at 850°F [454°C], it has 20 percent; and at 930°F [499°C], it has only 8 percent of its maximum strength. Yet at 930°F [499°C], it is still far below the melting point.

OXYACETYLENE WELDING

Oxyacetylene Welding Equipment

Oxyacetylene welding equipment may be either portable or stationary. A **portable apparatus** can be fastened on a hand truck or cart and pushed around from job to job. It consists of one cylinder containing oxygen; another containing acetylene;

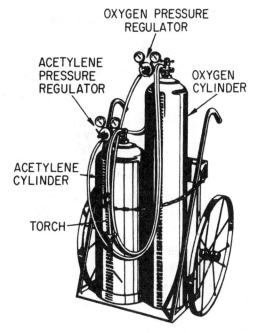

FIGURE 6-12 A portable welding outfit.

acetylene and oxygen pressure regulators, complete with pressure gauges and connections; a welding torch with a mixing head, tips, and connections; two lengths of colored hose, with adapter connections for the torch and regulators; a special wrench; a pair of welding goggles; a safety flint and file gas lighter; and a fire extinguisher. Figure 6-12 shows a portable welding outfit.

Stationary equipment is used where several welding stations are located close to each other and the stations can be supplied with gases through a manifold system. The oxygen and acetylene cylinders, located in areas separate from each other for safety, feed into the appropriate manifold through a master regulator. The master regulator sets the maximum pressure for the manifold. Each work station has a line regulator so that the welder can set the pressure as necessary for the particular task. In some shops, the acetylene does not come from cylinders. Instead, it is piped directly from an acetylene generator, an apparatus used for producing acetylene gas by the reaction of water upon calcium carbide.

Acetylene

Acetylene is a flammable, colorless gas with a distinctive odor that is easily detected, even when strongly diluted with air. It is a compound of carbon and hydrogen having the chemical symbol C_2H_2, which means that two atoms of carbon are combined with two atoms of hydrogen.

When acetylene is mixed with air or oxygen, it forms a highly combustible gas. It has a flame spread of 330 ft/s [99 m/s]. To prevent it from burning back to the source of supply during welding, the acetylene, when mixed with air or oxygen, must flow from the torch at a velocity greater than the flame spread, or the absorption of heat by the torch tip must be sufficient to prevent the flame from entering the tip.

Under low pressure at normal temperature, when free from air, acetylene is a stable compound; however, when it is compressed in an empty container to a pressure greater than 15 psi [103 kPa], it becomes unstable. At 29.4 psi [202.74 kPa) pressure it becomes self-explosive, and only a slight shock is required to cause it to explode even when it is not mixed with air or oxygen. As a general rule, the technician should never allow the acetylene pressure in the welding system to exceed 15 psi [103 kPa].

Although this gas is highly explosive, it is shipped in cylinders under high pressure with a high degree of safety. This is possible because the manufacturers place a porous substance inside the acetylene cylinder and then saturate this substance with acetone, which is a flammable liquid chemical that absorbs many times its own volume of acetylene. A cylinder containing a correct amount of acetone can be charged to a pressure of more than 250 psi [1724 kPa] with safety under normal conditions of handling and temperature.

Acetylene cylinders are available in several sizes, holding up to 300 ft³ [8.5 m³] of gas at a maximum pressure of 250 psi [1724 kPa]. The cubic feet of acetylene gas in a cylinder may be found by weighing the cylinder and subtracting the tare weight stamped on the cylinder from the gross weight; that is, the weight of an empty cylinder is subtracted from the weight of a charged cylinder. The difference is in pounds; this figure is multiplied by the weight of acetylene, 14.5 ft³/lb [411 L per 0.454 kg], to obtain the number of cubic feet in the cylinders. Figure 6-13 is an exterior view of an acetylene cylinder.

Oxygen

Oxygen is a tasteless, colorless, odorless gas that forms about 23 percent by weight and about 21 percent of volume of the atmosphere. Oxygen is an extremely active element. It combines with almost all materials under suitable

Head ring helps protect cylinder valve from damage

Sturdy steel cylinder body

Fuse plugs control the release of acetylene contents for protection should temperature outside or inside cylinder exceed 212° F (100°C) in case of a fire

Foot ring to protect cylinder from moisture and corrosion

FIGURE 6-13 An acetylene cylinder. (*Linde Div., Union Carbide Corp.*)

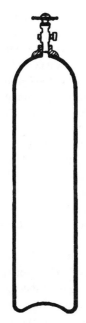

FIGURE 6-14   An oxygen cylinder.

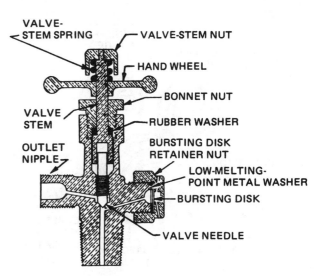

FIGURE 6-15   An oxygen cylinder valve.

conditions, sometimes with disastrous results. For example, grease and oil are highly combustible in the presence of pure oxygen; hence it is important to avoid bringing pure oxygen into contact with oil or grease. Such a mixture of oxygen and oil can produce a violent explosion. There are recorded cases of welders being killed by turning a stream of pure oxygen into a can of grease. Even grease spots on clothing may lead to explosions if they are struck by a stream of oxygen. Oxygen is necessary to make acetylene burn at a temperature high enough to melt metal in welding. In technical language, oxygen supports the combustion of the gas used in producing the welding flame.

The standard cylinder for storing and shipping oxygen gas for welding and cutting purposes is a seamless, steel, bottle-shaped container like the one shown in Fig. 6-14. It is made to withstand exceedingly high pressures. Although an acetylene cylinder is normally charged at a pressure of 250 psi [1724 kPa] at a temperature of 70°F [21.1°C], an oxygen cylinder is initially charged at the plant to a pressure of 2200 psi [15 171 kPa] at a temperature of 70°F [21.1°C].

Two sizes of oxygen tanks are generally available. The standard size is a cylinder having a capacity of 220 ft³ [6.23 m³]; the small cylinder has a capacity of 110 ft³ [3.11 m³]. Since the weight of oxygen is 0.08926 lb/ft³ [1.43 kg/m³], 11.203 ft³ [0.32 m³] equal 1 lb [0.4536 kg]. To find the quantity of oxygen in a cylinder, simply subtract the weight of an empty cylinder from the weight of a charged cylinder, and multiply the number of pounds by 11.203 to obtain the cubic feet of oxygen in the cylinder.

Figure 6-15 shows the construction of an oxygen cylinder valve assembly. A safety device (bursting disk) is contained in the nipple at the rear of the valve and consists of a thin copper-alloy diaphragm. When the cylinder is not in use, the valve is covered with a protector cap. This is an important feature in preventing the valve from being broken in handling.

Acetylene and Oxygen Regulators

Acetylene and **oxygen regulators** are mechanical instruments used to reduce the high pressure of the gases flowing from their containers and to supply the gases to the torch at a constant pressure and volume, as required by the torch tip or nozzle. Almost all regulators are available for either single-stage or two-stage pressure reduction.

In an installation where the gases are piped to the individual welding stations, only one gauge is required for each welding station, because it is necessary only to indicate the pressure of the gas flowing through the hose to the torch.

Regulators on cylinders are usually equipped with two pressure gauges. A high-pressure gauge shows the pressure of the gas in the cylinder, and a low-pressure gauge indicates the pressure of the gases flowing to the torch.

The high-pressure gauges on oxygen regulators are graduated in pounds per square inch from 0 to 3000. The low-pressure, or working, gauge for oxygen-welding regulators is usually graduated in pounds per square inch from 0 to 100, 0 to 200, or 0 to 400. Figure 6-16 shows oxygen-pressure gauges mounted on a pressure regulator.

Figure 6-17 shows acetylene-pressure gauges mounted on a pressure regulator. Acetylene regulators are designed in a manner similar to oxygen regulators, but they are not required

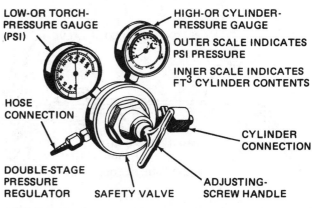

FIGURE 6-16   Oxygen-pressure gauges on the regulator.

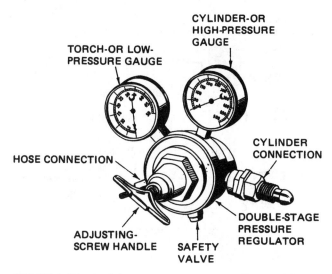

FIGURE 6-17 Acetylene-pressure gauges on the regulator.

to withstand such high pressures. The high-pressure gauge for acetylene indicates pressures up to a maximum scale value of only 400 psi. The maximum scale values on various low-pressure, or working, gauges range from 30 to 50 psi and the dial graduations have values of $\frac{1}{2}$ to 2 psi, depending upon the purpose for which the gauge is to be used.

Welding Torches

A **welding torch** is a device used to mix oxygen and acetylene together in the correct proportions and to provide a means of directing and controlling the quality and size of the flame. The welding gases flow from the inlet ports, past needle valves, through tubes in the torch to a mixing valve and then through the tip. **Needle valves** are used to regulate the volumes of acetylene and oxygen that flow into the **mixing head**. The torch head is usually located at the forward end of the handle. The mixing head is seated in the torch head and extends beyond the torch head. As its name indicates, the purpose of the mixing head is to provide for the correct mixing of the gases for the best burning conditions. The mixture of oxygen and acetylene flows from the mixing head into the tip of the torch and then emerges at the end of the tip, where it is ignited and burns to provide the welding

flame. Some welding torches use the oxygen flow to help dissipate the heat absorbed by the torch during the welding process by routing the oxygen passage through the torch; then the passageway acts as a cooling coil.

Welding torches may be divided into two principal types: (1) the **balanced-pressure type**, sometimes called the **equal-pressure type**, illustrated in Figs. 6-18 and 6-19, and (2) the **injector type**, illustrated in Figs. 6-20 and 6-21. These torches are available in different styles and sizes, and they are obtainable for use with several tip sizes, which are interchangeable. The selection of the style and size of the torch depends upon the class of work to be done. The selection of the tip size depends upon the amount of heat and the size of the flame required for the kind and thickness of the metal to be welded.

The primary difference to be considered in selecting balanced-pressure type or an injector-type welding torch is the source of the acetylene gas. If the acetylene gas is obtained from an acetylene generator, the pressure in the acetylene lines will be low. When the acetylene comes from cylinders the pressure is comparatively high. Balanced-pressure-type torches are used where the acetylene source pressure is high. The injector-type torch uses the velocity of the oxygen to cause a low-pressure effect that draws the necessary amount of acetylene into the mixing chamber, where the acetylene and oxygen are thoroughly mixed and directed to the tip.

Tip nozzles may have a one-piece hard copper tip or they may have a two-piece tip that includes an extension tube to make the connections between the mixing head and the tip. Welding tips are made in a variety of sizes and styles. Removable tips are made of either hard copper or of an alloy such as brass or bronze. The tip sizes differ in the diameter of the orifice, which provides the correct amount of gas mixture at a velocity that will produce the heat necessary to do the job.

Both the velocity and volume of gases are important. Velocity is important because it regulates the amount of heat that the technician will be able to apply to the material to be welded. The temperature is regulated by the mixing head, but the amount of that temperature or heat (measured in BTUs) is regulated by the velocity of the gases.

Too low a velocity will also allow the flame to burn back into the tip and cause a "pop" (backfire), which will blow the flame out. This will also happen if the tip gets too hot.

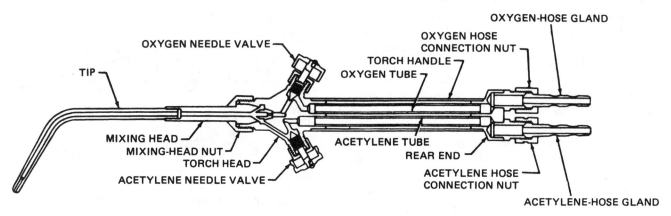

FIGURE 6-18 Cutaway drawing of a balanced-type welding torch.

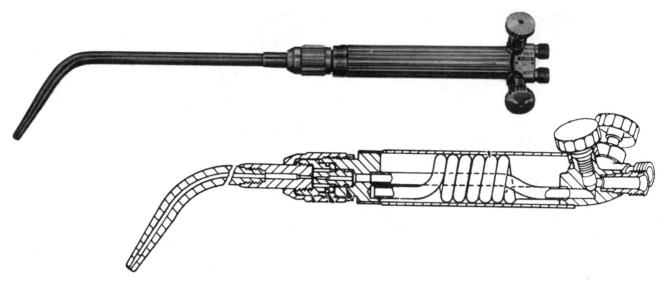

FIGURE 6-19 Photo of a balanced-type torch with a cutaway drawing to show cooling coil. (*Linde Div., Union Carbide Corp.*)

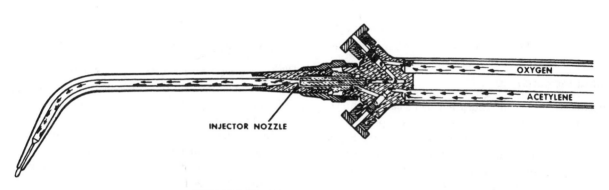

INJECTOR NOZZLE

OXYGEN

ACETYLENE

FIGURE 6-20 Injector-type welding torch.

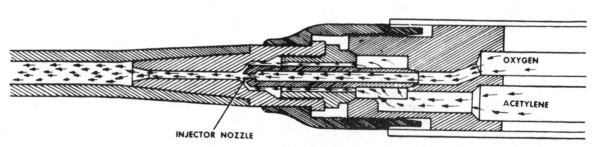

OXYGEN

ACETYLENE

INJECTOR NOZZLE

FIGURE 6-21 Mixing head for injector-type welding torch.

The manufacturer of the welding equipment supplies a table giving the approximate pressure of acetylene and oxygen for the various-size tips for an equal-pressure torch and a similar table for an injector-type torch. The torch tips should be of a proper size for the thickness of the material and the type of intersection involved. Tip sizes are designated by numbers, and each manufacturer has an individual system of numbering. Some manufacturers designate tip sizes by 0, 1, 2, 3, 4, etc., and others start with a higher number such as 20, 21, or 22. The commonly used sizes for working on steel butt welds are given in Table 6-2.

TABLE 6–2 Typical Torch-Tip Sizes

Thickness of Steel, in [mm]	Diameter of Hole in tip, in [mm]	Drill Size
0.015–0.031 [0.38–0.79]	0.026 [0.66]	71
0.031–0.065 [0.79–1.65]	0.031 [0.79]	68
0.065–0.125 [1.65–3.18]	0.037 [0.94]	63
0.125–0.188 [3.18–4.76]	0.042 [1.07]	58
0.188–0.250 [4.78–6.35]	0.055 [1.40]	54
0.250–0.375 [6.35–9.53]	0.067 [1.70]	51

FIGURE 6-22 Tip cleaner. (*Linde Div., Union Carbide Corp.*)

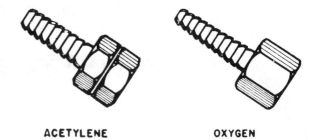

ACETYLENE   **OXYGEN**

FIGURE 6-23 Acetylene- and oxygen-hose connectors.

For any given thickness of steel or steel alloys, the heat required varies according to the angle of intersection. For example, the heat required for two pieces of metal of the same thickness intersecting at an angle of 30° is different than the heat required for welding the same two pieces intersecting at an angle of 90°.

It is important to select a small tip for light work and a larger tip for heavy work, because the size of the tip determines the amount of heat applied to the metal. If a small tip is used for heavy work, there is not enough heat to fuse the metal at the right depth. If the tip is too large, the heat is too great, and holes will be burned in the metal.

Cleaning the Torch Tip

Small particles of carbon, oxides, and metal can be removed from the tip of a welding torch with a soft copper wire, a drill of the correct size, a tip cleaner manufactured for the purpose, shown in Fig. 6-22, or any other suitably shaped device that will not damage the tip. Care must be taken to maintain a smooth, round orifice through which the gases can emerge.

If the tip becomes worn to the extent that the opening of the orifice flares out or is bell-shaped, the end of the tip should be ground square on a piece of fine emery cloth held flat against a smooth surface. The tip should be held perpendicular to the surface of the emery cloth and moved back and forth with straight strokes.

The outside surface of the tip can be cleaned with fine steel wool to remove carbon, oxides, and particles of metal. After the outside is cleaned, the orifice should be cleaned to remove any material that may have entered.

Welding Hose

A **welding hose** is a specially made rubber or synthetic rubber tube attached to the torch at one end and to a pressure regulator at the other end. It is used to carry the gases from their containers to the torch.

Several techniques are used in combination to prevent the oxygen and acetylene hoses from being connected to the wrong regulator or the wrong fitting on the torch. The acetylene hose is usually red or maroon in color, the threads on the fittings are left-handed, the fittings normally have a groove cut around the middle of the wrenching surface, and the word ACETYLENE may be found on the hose or the

letters ACE may be found on the fittings. The oxygen hose is normally green in color, the fittings are right-handed with no groove on the wrenching surface, and the word OXYGEN may be found on some hoses or the letters OXY may be on the fittings. Figure 6-23 illustrates typical acetylene- and oxygen-hose connections to show the difference.

Welding hoses and fittings should be examined regularly to see that the hoses are in good condition and that the fittings are not worn or damaged to the extent that they allow leakage of gases.

Welding, or Filler, Rod

The **welding rod**, sometimes called a **filler rod**, is filler metal, in wire or rod form, drawn or cast, used to supply the additional metal required to form a joint. During welding, the rod is melted into the joint, where it fuses with the molten base metal, the metal from the rod forming a large proportion of the actual weld metal. Welding rods are usually composed of only one metal or alloy, although rods known as **composite rods** contain more than one metal. If a rod has a very small diameter, it is usually known as a **wire**.

In selecting welding rod or wire for a particular application, the technician must be certain that the rod will be compatible with the material being welded and that the rod material will respond properly to any heat-treatment process required after the weld is completed. Additionally, some rods are available with coatings or flux to prevent contamination of welds, and other rods, such as those used for common steel applications, are coated with copper to prevent the rod from rusting.

Welding rods are 3 ft [0.91 m] long and come in several diameters, ranging from $\frac{1}{16}$ in [1.59 mm] to $\frac{1}{4}$ in [6.35 mm]. The size of the rod must be matched to the size of the material being welded. The diameter of the rod used is often the same as the thickness of the material being welded. Manufacturers of welding rods can provide specific information about their products to aid the technician in selecting the correct rod material and diameter for a specific welding application.

Safety Equipment

Welding goggles, such as those shown in Fig. 6-24, are fitted with colored lenses to keep out heat and the ultraviolet and infrared rays produced during welding. Clear lenses are provided in front of the colored lenses to protect the colored lenses from

FIGURE 6-24 Welding goggles. (*Linde Div., Union Carbide Corp.*)

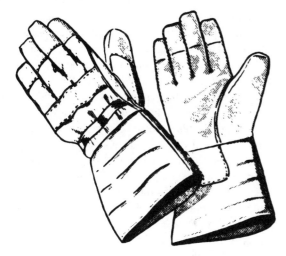

FIGURE 6-25 Welding gloves.

damage. The goggles should hug the face so closely that sparks and tiny pieces of hot metal cannot get inside.

Figure 6-25 shows a pair of **welding gloves**, commonly described as the **gauntlet style**. The material, manufacturing quality, and fit must be such that the gloves protect the hands and wrists from burns and flying sparks. They are usually made of asbestos or of chemically treated canvas.

Figure 6-26 shows a device that has several names. It may be called an **igniter**, **friction lighter**, **safety lighter**,

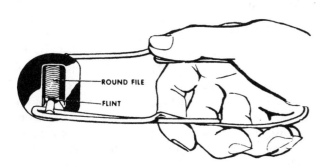

FIGURE 6-26 A friction lighter for welding torches.

or **spark lighter**. It is a hand-operated device used to light a gas torch safely. It consists of steel, a flint, a shield, and a spring. The steel, usually placed in a cup-shaped shield, resembles a file and is attached to the other end of the spring. The technician grasps the spring in one hand and compresses and releases the spring, forcing the flint to rub across the steel, thus producing sparks that light the gas coming out of the torch. This device is safe because it is composed of non-combustible material, the spark burns only for a fraction of a second, and the lighter is long enough to protect the welder's hand from the flame when properly used.

Stationary welding benches are often equipped with a pilot flame by which the welder may light the torch. If the torch should blow out while the welder is adjusting it or while welding, it is necessary merely to pass the tip of the torch over the pilot flame to reignite it.

Never use cigarette lighters or matches. Always use the spark lighter just described or a pilot flame. If a cigarette lighter or matches are used, the puff of the flame from the torch may burn the hand.

One or more portable **fire extinguishers** are kept at hand to be used if the flame from the welding torch, flying sparks, or flying pieces of hot metal set fire to anything. **Carbon dioxide** is generally the fire-extinguishing medium used because it is effective in combating gasoline or oil fires and may be used on wood and fabric fires. Carbon dioxide is often combined with a chemical powder, which aids in extinguishing fires.

Setting Up Oxyacetylene Welding Equipment

In order for efficient welding to take place, the equipment must be set up properly. This will allow a proper torch flame to be created, afford the maximum safety for the operator, and prevent unnecessary wear or damage to the equipment. In this discussion, the general procedures used for a portable system are addressed, but the basic procedures discussed are also applicable to a stationary system. When working with specific welding equipment, always follow the equipment manufacturer's procedures and recommendations.

Assembling the System

The first step in setting up portable welding apparatus is to fasten the cylinders to the cart or hand truck. The purpose of this step is to prevent the cylinders from being accidentally pulled or knocked over. The protecting cap that covers the valve on the top of the cylinder is not removed until the welder is ready to make a connection to the cylinder.

The second step in setting up the apparatus is to "crack" the cylinder valves. The welder stands beside or behind the cylinder outlet, as shown in Fig. 6-27, and opens the cylinder valve slightly for a moment and then quickly closes it. The purpose of this step is to clear the valve of dust or dirt that may have settled in the valve during shipment or storage. Dirt will cause leakage if it gets into the regulator, and it will mar the seat of the regulator inlet nipple even if it does not actually reach the regulator.

FIGURE 6-27 Correct position for the operator when opening the cylinder valve.

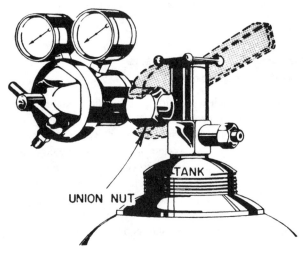

FIGURE 6-28 Wrench in position for tightening the union nut.

The third step is to connect the regulators to the cylinders. The welder uses a tight-fitting wrench to turn the union nut, as shown in Fig. 6-28, and makes certain that the nut is tight so that the gas will not leak.

The fourth step is to connect the hoses to their respective regulators. The green hose is connected to the oxygen regulator and the red hose is connected to the acetylene regulator. With the cylinder valves open, the regulator adjustment handles should be turned clockwise a sufficient amount to blow gas through the hoses and clear any dust or dirt. The handles are then turned counterclockwise until there is no pressure on the diaphragm spring and the cylinder valves are closed.

The fifth step is to connect the hoses to the torch, as shown in Fig. 6-29. The acetylene fitting, identified by the groove around the nut, has a left-hand thread and, therefore,

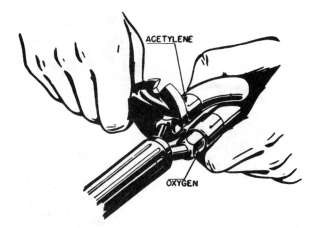

FIGURE 6-29 Connecting the gas hose to a torch.

will fit only the acetylene fitting on the torch. A tight-fitting wrench should be used to avoid damage to the nuts.

The sixth step in setting up the welding equipment is to test for leaks. This should **not** be done by using a lighted match at the joints. No open flame should be allowed in the vicinity of welding equipment except for the flame of the torch. There are several ways to test for leaks, but the best method is to apply soapy water to the joints with a brush. Before making this test, the oxygen and acetylene needle valves on the torch are closed. The cylinder valves are opened, and then the regulator adjusting screws are turned to the right (clockwise) until the working gauges show only a low pressure. The brush is dipped in the soap solution, and the solution is spread evenly over the connections. A leak is indicated by a soap bubble.

If a leak is discovered, the welder should close the cylinder valves and search for the source of trouble. It is generally sufficient to tighten the connecting unit slightly in order to stop the leak. Less common sources of trouble are dirt in the connection, which must be cleaned out, or marred seats or threads in the connection. If the seats or threads are damaged, the connections should be replaced. Having removed the trouble causing a leak, the welder must again test for leaks to be absolutely certain that none exist.

Ordinarily, the welder knows the correct tip size for the work to be done, a small opening for thin metal or a larger opening for thick metal being provided. On the assumption that the apparatus is set up, the next job is to adjust the working pressure of the gases.

Setting the Pressure

Figure 6-27 shows a welder in the correct position for opening the cylinder valve, regardless of whether the cylinder has a regulator attached or not. When a regulator is installed on the cylinder, the operator stands behind or to the side of the regulator and opens the valve slowly. If the regulator is defective, pressure may build up behind the glass and cause it to burst. This would be likely to inflict injury on anyone standing in the area immediately in front of the regulator.

When ready to open the cylinder valves, the welder should open the acetylene cylinder valve about one complete turn and open the oxygen valve all the way, slowly in both cases. If a valve wrench is used on the acetylene valve, this should be left on the valve so that the acetylene flow can be turned off quickly if a flame appears at a fitting or at a hose rupture. Then the welder sets the working pressure for the oxygen and the acetylene by turning the adjusting screw on the regulator to the right (clockwise) until the desired pressure reading is obtained on the gauge. As mentioned before, the exact pressure required for any job primarily depends upon the thickness of the metal, and this determines the size of the welding tip used. Remember, do not allow the acetylene pressure in the hose to exceed 15 psi.

Lighting the Torch

To light the torch, the welder opens the acetylene needle valve on the torch three-quarters of a turn and then uses the spark lighter to light the acetylene as it leaves the tip. The welder should do this as quickly as possible in order to prevent a large cloud of gas from developing. The flame should be large, very white, and smoky on the outer edges. If the flame produces much smoke, the welder should "crack" the oxygen needle valve very slightly and as soon as the flame appears to be under control, continue slowly to open the oxygen needle valve until a well-shaped bluish-white inner cone appears near the tip of the torch. This cone is surrounded by a second outer cone or envelope that varies in length, depending upon the size of the welding tip being used. This is known as a **neutral flame** and is represented by the upper drawing in Fig. 6-30.

Oxyacetylene Flames

A welding flame is called neutral when the gas quantities are adjusted so all the oxygen and acetylene are burned together. Theoretically, $2\frac{1}{2}$ volumes of oxygen are required to burn 1 volume of acetylene in order to produce this neutral flame, but actually it is only necessary to provide 1 volume of oxygen through the torch for 1 volume of acetylene consumed, because the remainder of the required oxygen is taken from the atmosphere. The carbon monoxide and hydrogen gas that come out of the first zone of combustion combine with oxygen from the air to complete the combustion, thus forming carbon dioxide and water vapor.

The neutral flame produced by burning approximately equal volumes of acetylene and oxygen oxidizes all particles of carbon and hydrogen in the acetylene, and it has a temperature of about 6300°F [3482°C]. This neutral flame should have a well-rounded, smooth, clearly defined, blue-white central cone. The outer cone, or envelope, flame should be blue with a purple tinge at the point and edges.

A neutral flame melts metal without changing its properties and leaves the metal clear and clean. If the mixture of acetylene and oxygen is correct, the neutral flame allows the molten metal to flow smoothly, and few sparks are produced. If there is too much acetylene, the carbon content of the metal increases, the molten metal boils and loses its clearness, and the resulting weld is hard and brittle. If too much oxygen is used, the metal is burned, there is a great deal of foaming and sparking, and the weld is porous and brittle.

A neutral flame is best for most metals. However, a slight excess of one of the gases may be better for welding certain types of metal under certain conditions. For example, an excess of acetylene is commonly used with the nickel alloys Monel and Inconel. On the other hand, an excess of oxygen is commonly used in welding brass.

A **carburizing**, or **reducing**, **flame** is represented by the middle drawing of Fig. 6-30. This occurs when there is more acetylene than oxygen feeding into the flame. Since the oxygen furnished through the torch is not sufficient to complete the combustion of the carbon, carbon escapes without being burned. There are three flame zones instead of the two found in the neutral flame. The end of the brilliant white inner core is not as well defined as it was in the neutral flame. Surrounding the inner cone is an intermediate white cone with a feathery edge, sometimes described as greenish-white and brushlike. The outer cone, or envelope, flame is bluish and similar to the outer cone, or envelope, flame of the neutral flame.

An **oxidizing flame** is represented by the lower drawing of Fig. 6-30. It is caused by an excess of oxygen flowing through the torch. There are only two cones, but the inner cone is shorter and more pointed than the inner cone of the neutral flame, and it is almost purple. The outer cone, or envelope, flame is shorter than the corresponding portion of either the neutral flame or of the reducing flame and is of a much lighter blue color than the neutral flame. In addition to the size, shape, and color, the oxidizing flame can be recognized by a harsh, hissing sound, similar to the noise of air under pressure escaping through a very small nozzle.

The **oxidizing flame** is well named. It oxidizes, or burns, most metals, and it should not be used unless its use is definitely specified for some particular purpose. Since an oxidizing flame is generally objectionable, the welder must examine the flame every few minutes to be sure of not getting an oxidizing flame. The welder does this by slowly closing the torch oxygen valve until a second cone or feathery edge appears at the end of the white central cone and then opening the oxygen valve very slightly until the second cone disappears.

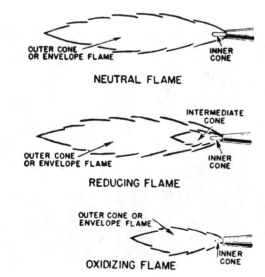

FIGURE 6-30 Neutral, reducing, and oxidizing flame.

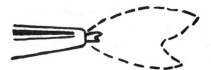

FIGURE 6-31 Flame produced by an obstructed tip.

Figure 6-31 shows a neutral or oxidizing flame with an irregular-shaped outer cone, or envelope—flame produced by an **obstructed tip**. When this flame is discovered, the welder should immediately shut down the welding apparatus and either clean the tip or replace it.

After learning to adjust the flame so that the proportions of oxygen and acetylene are correct, the welder must then learn how to obtain a **soft flame**. This flame is produced when the gases flow to the welding tip at a comparatively low speed. If the gases flow to the welding tip at a comparatively high speed, under too much pressure, they produce a **harsh flame** that is easily recognized because it is noisy. A harsh flame destroys the weld puddle and causes the metal to splash around the edges of the puddle. It is very difficult to get the metal parts to fuse properly with a flame of this kind.

If the mixture of acetylene and oxygen and the pressure are correct, the welder may still fail to obtain a soft neutral flame if the welding tip is dirty or obstructed in any manner. An obstructed welding tip does not permit the gas mixture to flow evenly, and it restricts the source of heat required to melt the metal; therefore, a good weld is very difficult to produce.

If there is any fluctuation in the flow of the gases from the regulators, the mixture will change, regardless of other conditions; hence a good welder watches the flame constantly and makes any necessary adjustments to keep it neutral and soft.

Backfire and Flashback

A **backfire** is a momentary backward flow of gases at the torch tip, causing the flame to go out and then immediately to come on again. A backfire is always accompanied by a snapping or popping noise. Sometimes the word *backfire* is used loosely to mean *flashback*, but a true backfire is not as dangerous as a flashback, because the flame does not burn back into the torch head and does not require turning off the gases.

There are five common causes of backfires: (1) there may be dirt or some other obstruction in the end of the welding tip; (2) the gas pressures may be incorrect; (3) the tip may be loose; (4) the tip may be overheated; or (5) the welder may have touched the work with the tip of the torch or allowed the inner cone of the flame to touch the molten metal (puddle).

If the tip is dirty or obstructed, it is removed and cleaned or replaced. If the gas pressures are wrong, they are adjusted. If the tip is loose, the torch is turned off and the tip tightened. If the tip is overheated, the torch is turned off and allowed to cool. If the tip touches the work or the inner cone contacts the puddle, the welder merely avoids repeating the error.

Flashback is the burning back of the flame into or behind the mixing chamber of the torch. Where flashback occurs, the flame disappears entirely from the tip of the torch and does not return. In some instances, unless either the oxygen or the acetylene or both are turned off, the flame may burn back through the hose and pressure regulator into the gas supply (the manifold or the cylinder), causing great damage. *Flashback* should not be confused with *backfire*, as explained previously. The welder must always remember that if a flashback occurs, there will be a shrill hissing or squealing, and the flame will burn back into the torch. The welder must quickly close the acetylene and oxygen needle valves to confine the flash to the torch and let the torch cool off before lighting it again. Since a flashback extending back through the hoses into the regulators is a symptom of something radically wrong, either with the torch or with the manner of its operation, the welder must find the cause of the trouble and remedy it before proceeding. Flashback arrestor valves can be installed on each hose to prevent a high-pressure flame or oxygen-fuel mixture from being pushed back into either cylinder causing an explosion. The flashback arrestors incorporate a check valve that stops the reverse flow of gas and the advancement of a flashback fire. The flashback arrestor valve is installed between the hose and torch or between the hose and the regulator. Figure 6-32 shows a flashback arrestor valve.

Shutting Down the Welding Apparatus

The procedure for shutting down the welding apparatus is as follows:

1. Close the acetylene needle valve on the torch to shut off the flame immediately.
2. Close the oxygen needle valve on the torch.
3. Close the acetylene cylinder valve.
4. Close the oxygen cylinder valve.
5. Remove the pressure on the regulators' working-pressure gauges by opening the acetylene valve on the torch to drain the acetylene hose and regulator.
6. Turn the acetylene-regulator adjusting screw counterclockwise (to the left) to relieve the pressure on the diaphragm, and then close the torch's acetylene valve.

FIGURE 6-32 Flashback arrestor valve.

7. Open the torch oxygen valve, and drain the oxygen hose and regulator.

8. Turn the oxygen-regulator adjusting screw counterclockwise to relieve the pressure on the diaphragm; then close the torch's oxygen valve.

9. Hang the torch and hose up properly to prevent any kinking of the hose or damage to the torch.

Preparation of the Metal

The elements to be welded should be properly held in place by welding fixtures that are sufficiently rigid to prevent misalignment due to expansion and contraction of the heated material. These fixtures must also positively locate the relative positions of the pieces to be welded. The parts to be welded should be cleaned before welding by sandpapering or brushing with a wire brush or by some

similar method. If the members to be welded have been metallized, the surface metal should be removed by careful sandblasting.

All mill scale, rust, oxides, and other impurities must be removed from the joint edges or surfaces to prevent them from being included in the weld metal. The edges, or ends, to be welded must be prepared so that fusion can be accomplished without the use of an excessive amount of heat.

Special Edge Preparations

In addition to cleaning the surfaces, the edges may need to be **reduced in size** with a grinding wheel or a file so that they will fuse with the smallest possible amount of heat. Whether or not a welder must reduce the size of the edges is determined by the thickness of the metal. It is apparent that the use of too much heat will burn the metal. In addition, an excessive

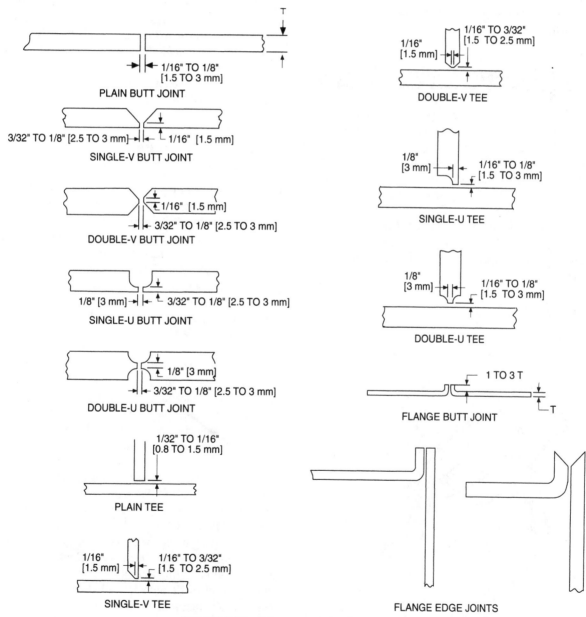

FIGURE 6-33   Edge preparation for welding.

amount of heat will radiate from the weld into the base metal and will cause it to expand at first and to contract later; this will result in warping if the metal is soft or in cracking if the metal is brittle. Reducing the volume of material at the joint reduces the amount of heat required to fuse the joint.

There are three basic types of end preparations: **flange**, **tapered**, and **U**. Tapered and U end preparation may be either single or double. Figure 6-33 shows the basic configurations of edge preparation. Note that in all cases except the tapered-edge joint, there is a portion of the edges that is not reduced. This is because during the welding process, **feathered edges** (edges brought to a point) puddle (melt) quickly, actually leaving a void. Since welding repairs usually join two pieces fixtured in order to maintain alignment, voids filled with rod are undesirable.

In the case of edge-joint edge preparation, a feathered taper is permissible because the molten metal will act as a filler during the welding process. As the edge puddles, the molten material will flow into the V created by the two opposed tapers. In cases similar to this, the use of filler rod is not normally required.

GAS WELDING TECHNIQUES

Welding may be considered both a skill and an art. Expert welders need technical understanding of the processes with which they are working and many hours of practice to develop the manual dexterity necessary to produce a quality weld. Although certificated aviation maintenance technicians are not always expected to be expert welders, they still need to know a good weld when they see one, and they should be able to perform a satisfactory welding job when it becomes necessary. Furthermore, they should know their own abilities and whether a welding specialist should be called in to do a particular repair job.

Holding the Torch

Figure 6-34 shows one method for holding the torch when welding light-gauge metal. In this method, the torch is held

FIGURE 6-34   Holding the torch for welding light metal.

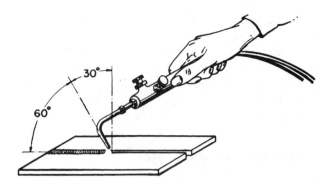

FIGURE 6-35   Holding the torch for welding heavy metal.

as one might hold a pencil. The hose drops over the outside of the wrist, and the torch is held as though the welder were trying to write on the metal.

Figure 6-35 shows how the torch can be held for welding heavier work. In this method, the torch is held as one would hold a hammer; the fingers are curled underneath and the torch balanced easily in the hand so that there is no strain on the muscles of the hand. A good way to describe the grip of the torch is to say that it should be held like a bird, tightly enough so that it cannot get away but loosely enough so it will not be crushed.

Forming the Puddle and Adding Filler Material

In the fusion welding process the base material (the materials to be joined) are brought to a molten state in the area of the joint, producing a **puddle**. The puddle is moved over the joint area, allowing the molten material from one part to intermingle with the other. To provide reinforcement to the joint, filler material is added by placing the filler rod in the puddle. The heat from the molten metal melts the filler material, and it too intermingles. Note that the filler material is melted by the puddle, not the torch flame. Melting the filler rod with the torch flame will generally result in the molten rod popping.

As the technician welds, it is important to note that the key to success in oxyacetylene welding is to control the motion of the puddle, making sure that the pattern used is consistent. The purpose of the rod is to add reinforcement material. Beginning welders have a tendency to chase the rod with the puddle instead of adding the rod to the puddle.

Another mistake frequently made by the novice welder is not adding enough rod. When all components of the welding process are properly balanced, the rod should remain in the puddle at all times. When chasing the rod, as just described, many novice welders add only one length of rod to the joint. This would be the equivalent to laying the rod along the joint and melting it as the bead is run along the joint. The welder should apply a light but constant pressure on the rod, so that more than one length of filler material is added to the joint.

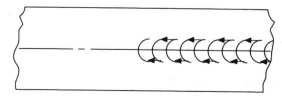

SEMICIRCULAR WELDING MOTION

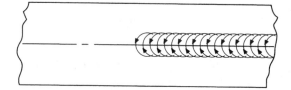

CIRCULAR WELDING MOTION

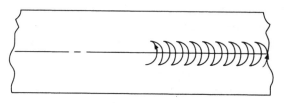

CRESCENT WELDING MOTION

FIGURE 6-36 Acceptable welding motions.

Torch Motions

The welder may use either a **semicircular**, **crescent**, or a continuous **circular motion**, shown in Fig. 6-36. Regardless of the motion, the welder keeps the motion of the torch as uniform as possible in order to make smooth, even-spaced ripples. These **ripples** are the small, wavelike marks left on the surface of the completed weld by the action of the torch and welding rod.

Novice welders frequently find it difficult to concentrate on maintaining the torch motion and adding rod at the same time. For that reason it is recommended that the novice welder use a crescent motion. Since the preferred width of bead is three to five times the thickness of the base material, the novice welder may visualize the crown and bottom of the crescent motions as surrounding the filler rod, as shown in Fig. 6-37. Using this technique allows the novice welder to concentrate on the puddle pattern and apply pressure on the filler rod to add reinforcement.

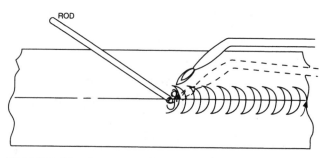

FIGURE 6-37 Crescent motion surrounds the filler rod.

Another difficulty experienced by novice welders is the control of heat to the part. The first approach to solving this difficulty is to soften the flame by reducing the pressure. Another way to solve the problem is to slightly increase the distance between the torch flame and the base material. A slight increase in distance has a significant effect upon the amount of heat absorbed by the part. Another option available to the welder, but not recommended to the novice, is to increase the rate of travel, or the speed with which the fusion process is accomplished.

Forehand Welding

Forehand welding, sometimes called **forward welding**, is a welding technique in which the torch flame is pointed forward in the direction the weld is progressing. In other words, it is pointed toward the unwelded portion of the joint, and the rod is fed in from the front of the torch or flame. Figure 6-38 shows how this is done. The welded portion of the weld joint is indicated by the crosshatched portion of the figure.

The torch head is tilted back from the flame to allow it to point in the direction that the weld is progressing. The angle at which the flame should contact the metal depends upon the type of joint, the position of the work, and the kind of metal being welded, but the usual angle is from 30° to 60°.

This forehand, or forward, technique must not be confused with the backhand, or backward, technique shown in Fig. 6-39 and explained in detail later in this chapter. The two techniques are distinctly different.

When forehand welding, filler rod is added to the pool of melting metal in front of the torch flame. The angle of the rod in relation to the torch must vary for different operations,

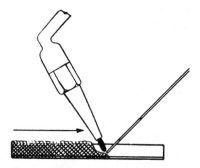

FIGURE 6-38 Forehand welding.

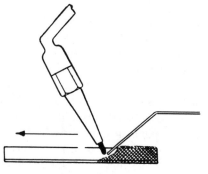

FIGURE 6-39 Backhand welding.

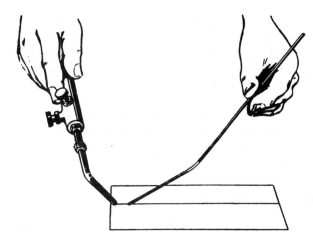

FIGURE 6-40   Welding rod bent to facilitate application.

but it is always necessary to add it to the weld by holding the end of the rod down into the molten pool of base metal formed by fusing the joint edges. If the rod is held above the pool and permitted to melt and drop into the weld, impurities floating on the surface of the molten metal will be trapped and a poor joint will be produced.

When the rod is used, it can be kept straight, or it can be bent to form a right angle near the end, as shown in Fig. 6-40.

The welding torch is brought down until the white cone of the flame is about $\frac{3}{8}$ in [3.175 mm] from the surface of the base metal. The tip of the rod is then inserted in this puddle, and as the rod melts, the molten pool is gradually worked forward.

Backhand Welding

Backhand welding, sometimes called **backward welding**, is a technique in which the flame is directed back toward the finished weld, away from the direction the weld is progressing, and the rod is fed in from the back of the torch or flame. This is the method illustrated in Fig. 6-39, which shows that the torch flame is pointed toward the finished weld (the crosshatched part of the bar) at an angle of about 60° to the surface of the work.

The welding rod is added *between* the flame and the finished weld. The flame is moved back and forth across the seam with a semicircular motion, thus breaking down the edges and the side walls of the base metal in order to fuse them to the necessary depth. In this backhand technique, the semicircular motion is directed so that the base of the arc falls toward the finished weld.

Whether the forehand or the backhand technique is used, the end of the filler rod is always held in the pool and given a slight alternating, or back-and-forth rocking, movement as metal is added from the rod to the pool. This movement of the rod must not be made too energetically. It must be controlled so that the melted metal from the pool is not shoved over onto the metal, which is not yet hot enough to receive it.

The backhand technique is preferred by most welders for metals having a heavy cross section. The metal being welded may be held in any position except for welding seams that run vertically. By using the backhand technique, the large

pool of melting metal that must be kept up at all times is more easily controlled, and the required depth of fusion in the base metal is easier to obtain.

Welding Positions

The four *welding positions* are *flat-position welding*, *vertical-position welding*, *horizontal-position welding*, and *overhead-position welding*. The welder must be able to make a good weld in any one of these four positions. A **welding position** refers to the plane (position) in which the work is placed for welding.

Two of the terms refer to welding the top or bottom surface of a work in a horizontal plane (flat work); overhead position is used when the underside of work is welded and flat position when the topside of work is welded. The other two terms refer to welding work in a vertical plane but make a distinction according to the direction of the line of weld. Thus horizontal position is used when the line runs across from side to side; vertical position when the line runs up and down. The **line of weld** is simply the path along which the weld is laid.

Figures 6-41 through 6-43 illustrate the four positions as seen from different viewpoints.

Flat-Position Welding

The **flat position** is the position used when the work is laid flat or almost flat and welded on the topside, with the welding torch pointed downward toward the work. Thus, if a weld is made with the parts to be welded laid flat on the table or inclined at an angle less than 45°, it is designated as being flat. The weld may be made in this position by the forehand

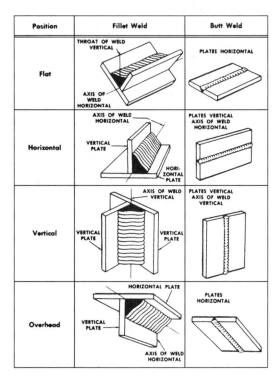

FIGURE 6-41   The four positions for welding.

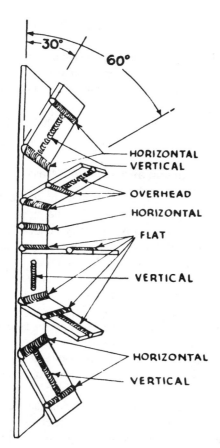

FIGURE 6-42   Welding positions as viewed from the side.

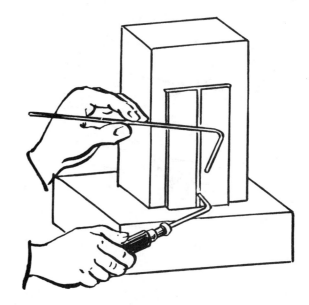

FIGURE 6-44   Adding rod when making a vertical weld.

the parts are inclined at an angle of more than 45°, with the weld running vertically, it is described as a vertical weld. The weld should be made from the bottom, with the flame pointed upward at an angle of from 45° to 60° to the seam for welding in this position. The rod is added to the weld in front of the flame, as it is in ordinary forehand welding. Figure 6-44 shows how the filler rod is added to the weld in front of the flame while making a weld in the vertical position.

Horizontal-Position Welding

The **horizontal position** is the position used when the line of weld runs across (horizontally) on a piece of work placed in a vertical or almost vertical position; the welding torch is held in a horizontal or almost horizontal position. Thus, when a weld is made with the parts in a vertical position or inclined at an angle of more than 45° with the seam running horizontally, it is called a **horizontal weld**. The seams in this horizontal position may be welded by either the forehand or the backhand technique; in either case, the flame should point slightly upward in order to aid in keeping the melting metal from running to the lower side of the seam. The welding rod should be added to the weld at the upper edge of the zone of fusion, since it dissipates some of the heat and lowers the temperature enough to help in holding the melting metal in the proper place. Figure 6-45 shows the hands of the welder holding the torch and the rod in this position.

Overhead-Position Welding

The **overhead position** is the position used when work is flat (horizontal) or almost flat and is welded on the lower side, with the welding torch pointed in an upward direction toward the work. Thus when a weld is made on the underside of the work with the seam running horizontally, or in a plane that requires a flame to point upward from below, it is described as an **overhead weld**.

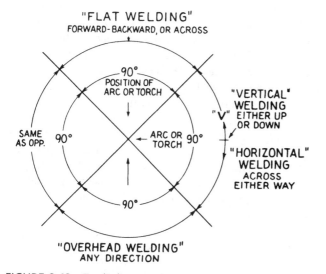

FIGURE 6-43   Torch direction for different welding positions.

or by the backhand technique, depending upon the thickness of the metal. The seam runs horizontally.

Vertical-Position Welding

The **vertical position** is the position used when the line of the weld runs up and down (vertically) on a piece of work laid in a vertical or nearly vertical position. The welding torch is held in a horizontal or almost horizontal position. Thus, when

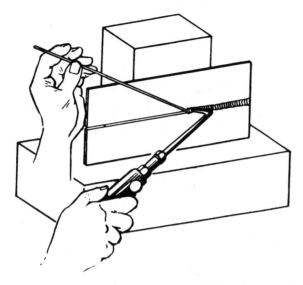

FIGURE 6-45 Positions of hand and torch for a horizontal weld.

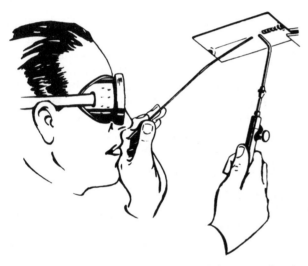

FIGURE 6-46 Positions of hand and torch for an overhead weld.

Either the forehand or the backhand technique can be used in welding seams in an overhead position. In either case, the flame must be pointed upward and held at about the same angle as it is for welding in a flat position. The volume of flame used for overhead welding should not be permitted to exceed that required to obtain a good fusion of the base metal with the filler rod. Unless the welder avoids creating a large pool of melting metal, the metal will drip or run out of the joint, thus spoiling the weld. Figure 6-46 shows a welder's hands holding the rod and torch correctly for this type of weld.

Weld Quality

The properly completed weld should have the following characteristics:

1. The seam should be smooth and of a uniform thickness.
2. The weld should be built up to provide extra thickness at the seam.

3. The weld metal should taper off smoothly into the base metal.
4. No oxide should be formed on the base metal at a distance of more than $\frac{1}{2}$ in [12.7 mm] from the weld.
5. The weld should show no signs of blowholes, porosity, or projecting globules.
6. The base metal should show no sign of pitting, burning, cracking, or distortion.

Reducing Distortion and Residual Stress

To reduce distortion and residual stress produced by welding, the expansion and contraction of the metal should be controlled. The distortion is especially noticeable in welding long sections of thin sheet metal because the thinner the metal being welded, the greater the distortion.

Four things a welder can do to control the action of those forces that adversely affect the finished weld are the following: (1) distribute the heat more evenly; (2) put a smaller amount of heat into the weld; (3) use special fixtures to hold the metal rigidly in place while it is being welded; and (4) provide a space between the edges of the joint. The nature of each welding job determines whether only one or all four of these methods should be used.

In discussing expansion and contraction, the even distribution of heat was explained. Preheating the entire metal object before welding sets free the stored-up forces and permits a more uniform contraction when the welding job is completed.

Putting a smaller amount of heat into a weld is difficult for a beginner, although an experienced welder can estimate accurately the amount of heat required. This is an example of the skill that comes from experience. One of the "tricks of the trade" is to use a method called **stagger welding**, shown in Fig. 6-47. The operator welds briefly at the beginning of the seam, skips to the center, then jumps to the end, comes back to where the first weld ended, and repeats this staggered process until the weld is finished.

Another term for stagger welding is **skip welding**, which is defined as a welding technique in which alternate intervals are skipped in the welding of a joint on the first pass and completed on the second pass or successive passes. The purpose is to prevent any one area of the metal from absorbing a great deal of heat, thus avoiding buckling and the tendency toward cracking.

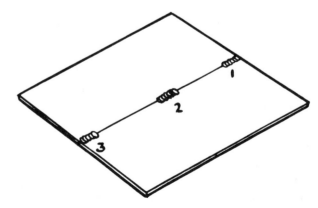

FIGURE 6-47 Stagger, or skip, welding.

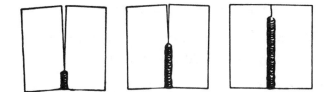

FIGURE 6-48 Tapering space to allow for metal expansion.

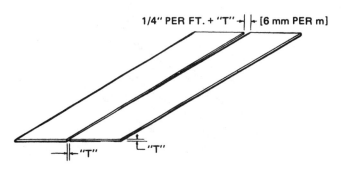

FIGURE 6-50 Spacing of metal parts for welding.

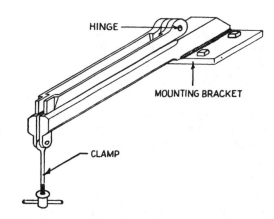

FIGURE 6-51 A welding fixture for sheet-metal butt joints.

When a welder uses special fixtures to hold the metal rigidly in place while it is being welded, excessive movements of the metal are hindered and the severe distortion resulting from expansion and contraction is prevented. A **fixture** is a rigid structure or mechanism, either wood or metal, that holds parts while they are being worked on (drilled, sawed, welded, etc.) before assembly, or that holds the component parts while they are being assembled or disassembled. Therefore, a welding fixture is simply a contrivance that holds the metal sections rigidly in place while a seam is welded. Fixtures are fastened tightly, but not so tightly that they will hinder the normal expansion and contraction of the metal at the ends of the joints. If a fixture is fastened too tightly, it causes internal stresses in the metal that lower its ability to carry heavy loads.

The fourth method for reducing distortion and residual stresses is the **careful spacing** of the pieces to be welded. Figure 6-48 shows how a **tapering space** is allowed between the pieces that are to be welded together. This tapering space is a distance equal to the thickness of the metal for every foot of seam length. For example, if a welder is to make a joint between two pieces of metal that are both $\frac{1}{2}$ in [12.7 mm] thick and 1 ft [30.45 cm] long, the space between them should be $\frac{1}{2}$ in [12.7 mm] at the ends opposite the starting point. If these same two pieces were 2 ft [60.96 cm] long, then the space at the wider ends would be twice as much, or 1 in [2.54 cm].

Figure 6-49 shows a method of providing for expansion of the base material when welding butt joints. The edges are set parallel and tack-welded. A **tack weld** is one of a series of small welds laid at intervals along a joint to hold the parts in position while they are being welded. This method is used for making short, straight seams or the curved seams on tubing, cylinders, tanks, etc.

Figure 6-50 shows how the edges for butt joints may be set and spaced when tack welding is not used; the amount of spacing depends on the kind of metal being welded and usually ranges from $\frac{1}{8}$ to $\frac{3}{8}$ in [3.18 to 9.53 mm] for each foot [30.48 cm] of the seam length. This method is better for flat sheets and longitudinal seams of cylindrical shapes.

Figure 6-51 shows a welding fixture for sheet-metal butt joints. It is useful for both flat sheets and longitudinal seams of cylindrical shapes.

Figure 6-52 shows a fixture consisting of four pieces of angle iron, which is used for welding butt joints in sheet metal. The angles for supporting the work on the lower side may be bolted or welded together. A recess of $\frac{1}{32}$ to $\frac{3}{64}$ in [0.794 to 1.191 mm] deep and from $\frac{1}{2}$ to $\frac{3}{4}$ in [12.7 to 19.05 mm] wide is machined in the center. This is labeled "milled groove" in the illustration.

The fixture shown in Fig. 6-53 is used for welding corner joints. It consists of three pieces of angle iron. The edge of the piece used to support the work on the lower side is ground or machined off to provide about $\frac{1}{16}$-in [1.59-mm] clearance for the joint. This enables the fusion to penetrate the base metal all the way through at this point. These fixtures are held in place on the metal by means of C-clamps.

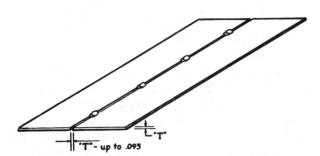

FIGURE 6-49 Tack welding to hold metal in position.

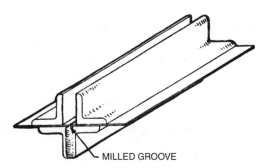

FIGURE 6-52 Use of a welding fixture.

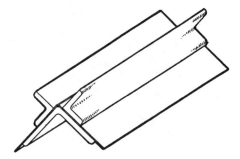

FIGURE 6-53   Welding fixture for corner joints.

In welding plate stock where the shape of the part permits, the butt joints may be set up as illustrated in Fig. 6-50, but if the shape of the work is such that the joint edges must be parallel, either the skip-welding procedure previously explained or the *step-back method* should be used for welding the joint. Either of these methods will lower the heat strains, because the heat is more evenly distributed over the whole length of the seam, thus causing the expansion and contraction to be uniform.

The **step-back method**, also called the **back-step method**, is a welding technique in which the welder welds and skips intervals between tack welds with each successive pass until the joint is completely welded. As each pass is completed, the welding is "back-stepped" or "stepped-back" to the next unwelded interval near the beginning of the weld. Figure 6-54 illustrates the step-back method. This should be studied and compared with the stagger welding shown in Fig. 6-47 and a more detailed drawing of this same "skip" method shown in Fig. 6-55.

Figure 6-56 shows that when it is not possible to design a fixture for the work and the metal is held in such a manner that normal expansion and contraction are restrained, the edges of the sheet may be bent up at the joint, as shown in the upper drawing, or a bead may be formed in each sheet parallel to the seam, $\frac{5}{8}$ to 1 in [1.59 to 2.54 cm] from the joint, as shown in the lower drawing of Fig. 6-56. Normally,

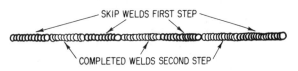

FIGURE 6-54   The step-back method of welding.

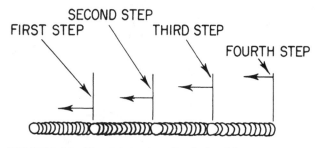

FIGURE 6-55   Details of skip welding.

FIGURE 6-56   Use of bend or bead to restrain distortion.

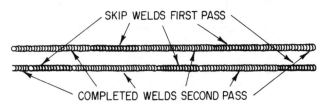

FIGURE 6-57   Reducing stresses in welding a tee joint.

either of these forms will straighten sufficiently on cooling to relieve the strain of the weld. However, if for any reason bends of this type are not practical, the welder may place chill plates or a cold pack of wet nonflammable material on the metal near the joint and parallel to the seam to reduce the flow of heat and the expansion that follows.

Figure 6-57 shows the welding procedure for reducing stresses in tee joints. The welding in this case alternates from one side of the vertical member to the other. If the plate is heavy, the tee joints are heated to a dull red on the opposite side from the side on which the weld is being made. A separate torch is used for this purpose. If done carefully, this heating causes a uniform expansion on both sides of the plate and produces an even contraction that prevents the parts from being pulled out of their correct alignment.

There are several methods for providing for the expansion in the welding of castings. The one selected in each case depends upon the type, kind, and shape of the casting and also on the nature of the break being welded. The welder who has mastered the principles and techniques already described should have no trouble in selecting the correct procedure for each job. For example, the entire casting might require preheating to a temperature that will prevent expansion strains on one job, whereas on another job it might be enough to apply local preheating—that is, heat applied only in the vicinity of the welding zone.

When preheating is applied, the parts preheated must be cooled evenly and slowly. Depending upon the design and other factors, the opposing parts or sections may be preheated to relieve the strains that come from welding. Mechanical devices such as screw jacks may be used as an additional precautionary measure. Figure 6-58 shows the application of methods of providing expansion in the welding of rectangular castings, and Fig. 6-59 shows the method of providing for expansion in the welding of circular castings.

When parts are fabricated or repaired by welding, there is usually some stress that remains. This stress should be relieved to obtain the full strength of the weld and the base metal. **Heat treatment**, which is discussed in considerable detail in *Aircraft Basic Science*, is the most reliable method of relieving stress, provided that the part can be heated in a furnace to the stress-relieving temperatures and then cooled slowly and evenly. For example, aluminum and

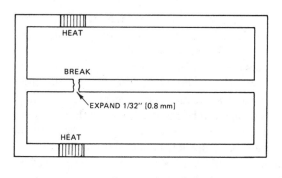

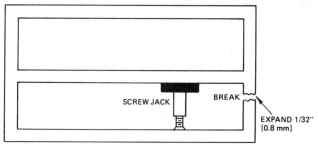

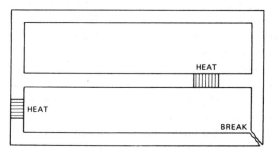

FIGURE 6-58   Providing for expansion in welding rectangular castings.

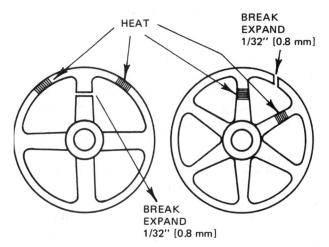

FIGURE 6-59   Providing for expansion in welding circular castings.

aluminum alloys require a temperature of from 700 to 800°F [371 to 427°C]; gray cast iron, 900 to 1000°F [482 to 538°C]; nickel-chromium-iron alloy (Inconel), 1400°F [760°C]; carbon steels up to 0.45 percent carbon, 1000 to 1200°F [538 to 649°C]; chrome-molybdenum-alloy steel, 1150 to 1200°F [621 to 649°C]; and chrome-nickel stainless steel (18-8), from 1150 to 1200°F [621 to 649°C].

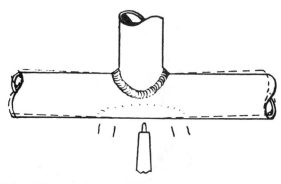

FIGURE 6-60   Application of heat to straighten a tube.

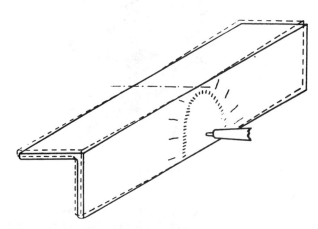

FIGURE 6-61   Use of heat to straighten an angle section.

Local heating with a welding flame may be used to relieve or eliminate distortion in structures fabricated of steel tubing, angle iron, and similar materials by bringing the metal to a red heat at the proper locations. Figure 6-60 shows a tube pulled out of alignment by weld shrinkage being given this treatment, and Fig. 6-61 shows an angle iron pulled out of alignment by weld shrinkage being given a similar application of local heat.

Practices to Avoid

- Welds should not be filed to present a smooth appearance. Filing will usually weaken the weld.
- Welds must not be filled with solder, brazing metal, or any other filler.
- When it is necessary to reweld a joint, all the old weld metal must be removed before proceeding with the new weld.
- A new weld should never be made over an old weld if it can be avoided. The heat of rewelding increases the brittleness of the weld material and causes it to lose strength.
- A weld should never be made over a joint that was previously brazed. Brazing metal (brazing filler rod) penetrates and weakens the steel.

Simple Tests for Welds

There are many reliable tests for welds, but at this stage of training, the technician should at least be able to apply one or two very simple tests to judge progress. One of these is

Gas Welding Techniques   **155**

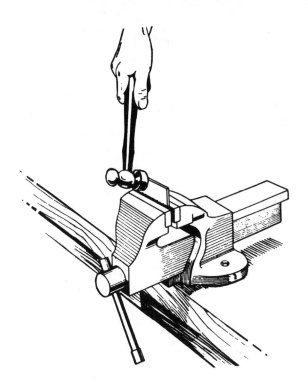

FIGURE 6-62 Bend test for a weld.

usually called the **bend test**. The welder allows the metal to cool slowly and then picks it up with a pair of pliers and clamps the metal in a vise with the weld parallel to the top of the jaws of the vise and slightly above the top of the vise, as shown in Fig. 6-62. The welder strikes the top of the metal with a hammer so that the metal is bent along the line of the weld. The weld should be bent in on itself—that is, bent so the bottom of the weld is in tension and the top is in compression. If the weld breaks off very sharply and shows a dull, dirty break and the presence of blowholes, the weld is unsatisfactory. If the weld has been made properly, the metal will not break off short. Instead, it will be distorted under the blows of the hammer until it forms an angle of at least 90° without cracking under repeated hammering.

A **visual inspection** is another simple test. The welder examines the smoothness of the bead, the amount of reinforcement (making certain that the seam is at least 25 to 50 percent thicker than the base metal), and the cleanliness of the completed weld. The contour should be even. It should extend in a straight line, and its width and height should be consistent. No pits should be present. It should be understood that a clean, smooth, fine-appearing weld is not necessarily a good weld, because it may be dangerously weak inside. However, the opposite is true; that is, if the weld is rough-pitted, uneven, and dirty looking, the weld is almost always unsatisfactory inside.

Gas Welding Aluminum, Magnesium, and Titanium

It is strongly recommended that the appropriate type of inert-gas welding be employed when aluminum, magnesium, or titanium parts must be joined by welding. These metals and their alloys oxidize very rapidly when heated, and it is difficult to protect the surfaces from oxidation in gas welding even though flux is applied liberally.

When magnesium reaches a sufficiently high temperature, it burns with a very bright flame. Care must, therefore, be taken to avoid a situation where magnesium shavings, particles, or scraps can be ignited. Titanium, in the molten state, reacts rapidly with oxides and oxygen. When being welded, it must be completely protected from oxygen.

Pure aluminum and some of its alloys can be gas welded. Designations for weldable aluminum and alloys are 1100, 3003, 4043, and 5052. Alloys 6053, 6061, and 6151 can be welded if provision is made for heat treating after welding.

Aluminum should be welded with a soft neutral flame or a slightly reducing flame when either acetylene or hydrogen is used for the welding gas. This ensures that there will be no oxygen in the flame to combine with the aluminum.

Before starting to weld aluminum, the edges to be welded must be thoroughly cleaned. Solvents can be used to remove grease, oils, or paint. The solvent must be such that it evaporates completely dry and leaves no residue. Oxides can be removed from the areas to be welded with fine emery paper. Care must be taken not to scratch the aluminum outside the weld areas because this will cause corrosion.

The preparation of the edges of aluminum sheet or plate that are to be gas welded depends upon the thickness of the material. For thicknesses of 0.060 in [1.52 mm] or less, a 90° flange can be formed on the edges. The height of the flange should be about the same as the thickness of the metal. A cross-section drawing of flange edges is shown in Fig. 6-63(a). If the metal is between 0.060 and 0.190 in [1.52 and 4.83 mm], the penetration of the weld is improved by notching the metal, as shown in Fig. 6-63(b). This practice also aids in preventing distortion of the metal due to expansion. If the metal thickness is greater than 0.190 in [4.83 mm], it should be beveled and notched, as shown in Fig. 6-63(c).

When the metal edges have been cleaned and prepared as explained, aluminum welding flux should be applied. If possible, the flux should be brushed on both the flame side and the opposite side of the metal. The welding rod should also be dipped in the flux. After completion of the weld, the flux should be removed completely by washing. If flux remains, it will cause corrosion.

As mentioned previously, the flame used for aluminum welding should be soft (mild). This type of flame is produced by adjusting the torch as low as possible without causing the tip to pop. The flame is applied to the edges of the work at an angle of approximately 45° until the aluminum begins to melt. It is difficult to tell when the aluminum has reached melting temperature because its appearance does not change appreciably. Touching the tip of the welding rod to the heated area will reveal when the aluminum is beginning to soften. If the aluminum becomes too hot, it will drop away and form a hole in the material. If the flame is too harsh, this condition will be aggravated. In any case, only an experienced welder should attempt to perform this type of work.

If possible, the material to be welded should be supported in a fixture, which will reduce distortion as much as possible.

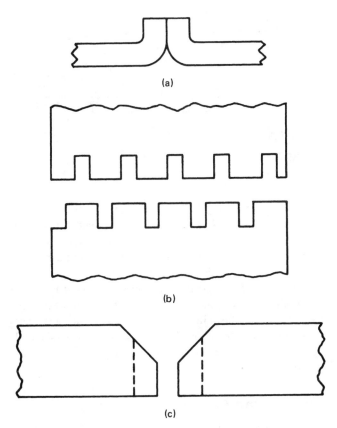

FIGURE 6-63 Edge preparation for gas welding aluminum sheet and plate.

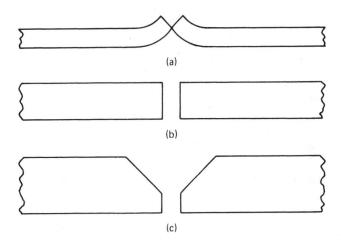

FIGURE 6-64 Edge preparation for gas welding magnesium sheet and plate.

The material should be preheated before starting the weld by passing the torch over the entire area surrounding the weld. Care must be taken not to overheat the material, because doing so will change the structure of the metal and reduce its strength.

The welding of magnesium with butt joints is accomplished in a manner similar to the welding of aluminum. (The butt weld is generally the only type of weld made with magnesium when using oxyacetylene equipment.) The edges to be welded should be cleaned and the oxides removed. Solvents are used to remove oil or grease, and fine emery cloth or paper can be used to remove the oxides.

Edge preparation of sheet magnesium differs from that of aluminum because magnesium alloys cannot be bent sharply without cracking. For sheet magnesium alloy with thickness of 0.040 in [1.016 mm] or less, the edges should be bent up in a curve, as shown in Fig. 6-64(a). For sheet thicknesses from 0.040 to 0.125 in [1.02 to 3.18 mm], the edges of the metal are left square with a space of 0.0625 in [1.59 mm] between the edges of the sheets to be welded; see Fig. 6-64(b). For magnesium plate with thicknesses between 0.125 and 0.250 in [3.18 and 6.35 mm], the edges should be beveled as shown in Fig. 6-64(c). The bevel on each edge is approximately 45°, making an included angle of 90°. A space is allowed between the sheets as shown. For magnesium plate more than 0.250 in [6.35 mm] in thickness, the edges are beveled as for the thinner plate; however, the space between the edges is increased to 0.125 in [3.18 mm].

The welding rod used with magnesium alloys must be of the same alloy as the material being welded and should be available from the manufacturer of the alloy. A protective coating is applied to the magnesium alloy welding rod to protect it from corrosion. This coating must be removed before the rod is used in welding.

The experienced welder will usually develop the most satisfactory techniques for welding magnesium based on experience. The weld may be made in much the same way as for aluminum, or the tip of the welding rod may be kept in the center of the molten pool as the weld continues. It is recommended that the weld be made in one continuous pass unless oxidation takes place. In this case, the welding must be stopped and all the oxide removed before continuing.

To prevent distortion, the edges of the joint should be tack-welded at intervals of 0.5 to 3 in [12.7 to 76.2 mm], depending upon the shape and thickness of the metal. Distortion that does occur can be straightened with a soft face mallet while the metal is still hot.

Completed welds should be allowed to cool slowly to the level where the material can be handled. Then the weld should be scrubbed clean with a stiff brush and water to remove the flux. After this the part should be soaked in hot water to dissolve and remove any traces of flux not removed by brushing. After soaking, the part should be treated by immersion in an approved acid bath to neutralize alkalinity and then rinsed in fresh water and dried as quickly as possible. Clean, bare magnesium will corrode rapidly unless given a treatment to provide a corrosion-resistant chemical coating. One such treatment is accomplished with an approved chromic-acid solution. Another treatment that is suitable for many magnesium alloys is the dichromate treatment. This involves boiling the magnesium in a solution of sodium dichromate after an acid pickle bath. This treatment is effective for corrosion prevention and provides a good base for primers and finishes. It can be used for all magnesium alloys except those containing thorium.

ELECTRIC-ARC WELDING

Electric-arc welding, also referred to as **stick-electrode welding** and technically referred to as **shielded metal arc welding** (SMAW), is used here to denote the standard arc process, which utilizes an electrode filler rod and is generally employed for welding heavy steel. This method requires a special generator to provide a low-voltage, high-amperage current for the arc. The power supply may be an electric motor-driven generator, an engine-driven generator, or a special transformer.

The electric arc is made between the tip end of a small metal wire, called the **electrode**, which is clamped in a holder held in the hand, and the metal being welded. A gap is made in the welding circuit by holding the tip of the electrode $\frac{1}{16}$ to $\frac{1}{8}$ in [1.59 to 3.18 mm] away from the work. The electric current jumps the gap and makes an arc, which is held and moved along the joint to be welded. The heat of the arc melts the metal. The arc is first caused ("struck") by touching the electrode to the metal, and then the electrode is withdrawn slightly to establish the correct gap across which the arc flows.

The several types of welding machines include motor generators, engine-driven generators, transformers, rectifiers, and combination transformers and rectifiers. Figure 6-65 illustrates a typical arc-welding power supply. Each type has its place and purpose, but the basic function of each is the same—that is, to provide a source of controlled electric power for welding. This controlled electric power has the characteristic of high amperage at low voltage. The high amperage is required to provide sufficient heat at the arc to melt the metal being welded. The voltage must be low enough to be safe for handling and yet high enough to maintain the arc. The welding machine permits the operator to control the amount of current used. This, in turn, controls the amount of heat at the arc. Some welding machines permit the operator to select either a forceful or soft arc and to control its characteristics to suit the job.

Good welders do more than simply hold the arc. They must, first of all, be able to select the correct size and type

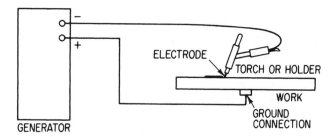

FIGURE 6-66 An arc welding circuit.

of electrode for each job. They must know which machine to use for the job and be able to set the current and voltage controls properly. They must be able to manipulate the electrode and arc to make a satisfactory weld under varying conditions. In addition, welders must have a knowledge of joint preparation, positioning the work, distortion, and many other factors that enter into the final result of a good weld. They must be skilled in both the mechanics and craft of welding. Nearly anyone can "stick two pieces of metal together," but becoming a good welder requires study, training, and practice.

The Welding Circuit

The operator's knowledge of arc welding must go beyond the arc itself to include how to control the arc, and this requires a knowledge of the welding circuit and the equipment that provides the electric current used in the arc. Figure 6-66 is a diagram of the welding circuit. The external circuit begins where the electrode cable is attached to the welding machine and ends where the ground cable is attached to the welding machine. When direct-current, straight-polarity power is used, the current flows through the electrode cable to the electrode holder, through the holder to the electrode, and across the arc. From the work side of the arc, the current flows through the base metal to the ground cable and back to the welding machine. The circuit must be complete for the current to flow, which means that it is impossible to weld if the cables are not connected to the machine or to either the electrode or the work. All connections must be firm so the current can flow easily through the entire circuit.

Welding Arc

The action that takes place in the arc during the welding process is illustrated in Fig. 6-67. The *arc stream*, seen in the middle of the picture, is the electric arc created by the current flowing through air between the end of the electrode, technically known as the **anode**, and the work, more properly referred to as the **cathode**. As the current flows through the arc stream, a column of ionized gas called the **plasma** is formed. The arc is very bright as well as hot and cannot be looked at with the naked eye without risking painful, though temporary, injury. It is, therefore, essential that the operator wear a suitable protective hood, such as the one shown in Fig. 6-68. New style helmets have an auto-darkening feature that allows the welder to see the welding torch and filler

FIGURE 6-65 Power supply unit for arc welding. (*Lincoln Electric Company.*)

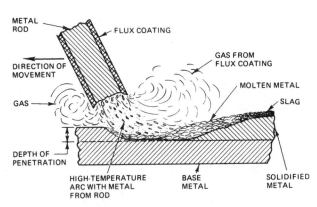

FIGURE 6-67   The action in a welding arc.

FIGURE 6-68   A welding helmet. (*Linde Div., Union Carbide Corp.*)

rod so that he/she can position them in the correct position. As soon as the intense light of the welding starts the helmet lens will automatically darken. The darkness of the lens is adjustable. Welding goggles used for gas welding will not adequately protect the eyes of the welder during TIG or MIG welding processes.

The plasma is a combination of both neutral and excited gas atoms. In its center are electrons, atoms, and ions. The heat of the arc is caused by the collision of these particles at the accelerated speeds caused by voltage drop as they move between the anode and cathode. The temperature of this arc is from 6000 to 11 000°F [3315° to 6090°C], which is more than enough to melt the metal. The temperature is most affected by three factors: the type and amount of electrical power, the distance between the anode and cathode (the gap between the electrode and the work piece), and the type of atmosphere or gases between the anode and cathode.

The arc melts the plate, or base, metal and actually digs into it, just as the water flowing through a nozzle on a garden hose digs into the earth. The molten metal forms a pool, or "crater," and tends to flow away from the arc. As it moves away from the arc, it cools and solidifies. A slag forms on top of the weld to protect it during cooling. The slag comes from the flux coating on the electrode.

Arcs for welding are produced by different kinds of power supplies and the various types of power produce arcs with different characteristics. The power supply may be designated dcsp, dcrp, or ac. Here dcsp means **direct current, straight polarity**. Straight polarity is the condition when the electrode is negative and the base metal or work is positive. Also, dcrp means **direct current, reverse polarity**, where the electrode is positive. The direction of current will have a pronounced effect on the depth and penetration of the weld.

Ac power, of course, means alternating current. With ac power, the order in which the leads from the power source are connected will make no difference. Ac power is often considered the most efficient of the power sources because it produces a higher temperature for a given amperage.

Dcsp power provides the deepest penetration for a given amperage and is used when welding rusty surfaces and tight-fit joints. The technician should be aware that this process develops the most heat.

Dcrp power produces the smoothest bead and less splatter due to the lessened penetration. Thinner materials and joints with wide gaps are most frequently welded with dcrp power.

Electrode, or Rod

In preparing to perform arc welding, the operator should seek to determine the best power source and type of welding rod for the welding being done. Very often, however, only one source of power is available, and the only choices the operator has are the direction of welding current and type of rod.

The function of the electrode is much more than simply to carry current to the arc. The electrode is composed of a core of metal rod or wire material, around which a chemical coating has been extruded and baked. The core material melts in the arc, and tiny droplets of molten metal shoot across the arc into the pool. The electrode provides filler metal for the joint to occupy the space or gap between the two pieces of the base metal. The coating also melts or burns in the arc and serves several functions. It makes the arc steadier, increases the arc force, provides a **shield** of smokelike gas around the arc to keep oxygen and nitrogen in the air away from the molten metal, and provides a flux for the molten pool, which picks up impurities and forms the protective slag. The electrodes to be used in any arc-welding operation depend upon the material being welded, the type of power being used, and the type of weld. The operator must make certain that the rod chosen will supply the strength required of the weld.

The American Welding Society (AWS) has established a standard classification system for the coding of SMAW. The coding classifies electrodes by the electrode's **tensile strength, best welding position,** and **flux-welding current** recommendations. Since most SMAW is related to the welding of mild steel, only the code as commonly associated with mild steel-core electrodes is identified here. If any other materials are to be considered, the electrode manufacturer's catalogs should be consulted for the proper electrode selection.

The AWS code for mild steel electrodes is a five- or six-position identifier, which begins with a letter. The letter designates the core material. E indicates a mild steel.

TABLE 6-3 SMAW Electrode Digit Code

AWS Numbering System		
The prefix "E" designates arc welding electrode.		
EXXXX	Electrode used for arc welding	
The first two digits of four-digit numbers and the first three digits of five-digit numbers indicate minimum tensile strength.		
E60XX	60,000-psi tensile strength	
E70XX	70,000-psi tensile strength	
E110XX	110,000-psi tensile strength	
The next-to-last digit indicates the best position.		
EXX1X	All positions	
EXX2X	Flat and horizontal positions	
The last two digits together indicate the type of coating.		
EXX10	Organic	dc only
EXX11	Organic	ac or dc
EXX12	Rutile	ac or dc
EXX13	Rutile	dc or ac
EXX14	Rutile, approx. 30% iron powder	ac or dc
EXX18	Low hydrogen and approx. 30% iron powder	dc or ac
EXX24	Rutile, approx. 50% iron powder	ac or dc
EXX27	Mineral and approx. 50% iron powder	ac or dc
EXX28	Low hydrogen and approx. 50% iron powder	dc or ac
If a suffix is used, it indicates the approximate alloy deposit.		
-A1	$\frac{1}{2}$% Mo	
-B1	$\frac{1}{2}$% Cr, $\frac{1}{2}$% Mo	
-B2	$1\frac{1}{4}$% Cr, $\frac{1}{2}$% Mo	
-B3	$2\frac{1}{4}$% Cr, 1% Mo	
-C1	$2\frac{1}{2}$% Ni	
-C2	$3\frac{1}{4}$% Ni	
-C3	1% Ni, 0.35% Mo, 0.15% Cr	
-D1 and D2	0.25–0.45% Mo, 1.25%–2.00% Mn	
-G	0.50 min Ni, 0.30 min Cr, 0.20 min Mo, 0.10 min V	
-M	1.3–1.8 Mn, 1.25–2.50 Ni, 0.40 Cr, 0.25–0.50 Mo, 0.05 max V	

Source: American Welding Society

The next two or three positions are numbers that identify the tensile strength of the core material. The fourth position is used for tensile strength identification only if the code is six positions in length.

The next-to-last position (fourth in a five-position code or fifth in a six-position code) indicates the recommended welding position. 1 indicates any position. 2 indicates the code composition is suitable for horizontal fillets and flat-position welding.

The last position in the code signifies the type of flux coating and recommended power source. Table 6-3 shows the meanings of the numbers in the last position of the code.

Table 6-4 shows a series of recommended SMAW electrodes for a variety of usages. Note also that Table 6-4 denotes the electrode's diameter and amperage range. The selection of the amperage is proportional to the diameter of the electrode. The greater the diameter of the electrode, the greater the amperage required. Although the best

amperage for any particular application will require fine tuning for a variety of variables, a good rule of thumb is to start with an amperage equal to the decimal equivalent of electrode diameter, expressed in inches, multiplied by 1000. For example, a good starting amperage for a $\frac{1}{8}$-in-diameter electrode is 125 A.

Once the arc is struck, the welder needs to maintain a gap between the material and the electrode of slightly less than the diameter of the electrode. Since the electrode is constantly shortening in length, a particular height is difficult to judge and maintain by sight. Experienced SMAW welders maintain the proper distance by sound and observations of the finished weld. A sharp crackling sound, similar to an egg frying, is an indication that the proper arc gap is being maintained. Electrode sticking to the material during the welding process is an obvious result of an arc gap that is too short. Excessive splattering and poor bead quality indicate an arc gap that is too large.

TABLE 6-4 SMAW General Electrode Selection Chart

	Pipe	Carbon Steel	Alloy Steel	Low Alloy Steel	Cast Iron	Quality	Electrode Polarity	Penetration	Best Positions	Deposit Rate	Sizes and Current Ranges (A)				AWS Spec.
											$\frac{3}{32}$ in	$\frac{1}{8}$ in	$\frac{5}{32}$ in	$\frac{3}{16}$ in	
General purpose															
E6010	Yes	Yes				X ray	dc+	Deep	All	High	40–70	75–130	90–175	140–225	A5.1
E6011	Yes	Yes				X ray	ac/dc+	Deep	All	High	50–85	70–110	80–165	110–180	A5.1
High-tensile pipe															
E7010-A1	Yes						dc+	Deep	All	High	50–90	75–130	90–175	140–225	A5.5
E7010-G	Yes						dc+	Deep	All	High	N/A	75–130	90–185	140–225	A5.5
E8010-G	Yes						dc+	Deep	All	High	N/A	75–130	90–185	140–225	A5.5
Multipass welding															
E7024		Yes		Yes			ac/dc+/dc-	Shallow	Horiz.	Highest	N/A	100–175	160–225	215–315	A5.1
E6027		Yes		Yes		X ray	ac/dc+/dc-	Shallow	Horiz.	Highest	N/A	N/A	175–240	230–300	A5.1
Sheet metal															
E6012		Yes					ac/dc-	Medium	Horiz.	Medium	N/A	80–150	110–200	155–275	A5.1
E6013		Yes					ac/dc-	Medium	Horiz.	Medium	70–105	100–135	145–200	190–260	A5.1
E7014		Yes					ac/dc-	Medium	Horiz.	Medium	75–100	100–145	135–225	185–280	A5.1
Thick section welds															
E7018		Yes	Yes			X ray	ac/dc+	Medium	All	Medium	70–100	90–170	120–225	170–300	A5.1
E7028		Yes	Yes			X ray	ac/dc+	Medium	All	Medium	N/A	N/A	170–270	210–330	A5.1
High-tensile steels															
E8018-B2		Yes		Yes		X ray	ac/dc+	Med/deep	All	High	N/A	120–170	130–225	180–290	A5.5
E8018-C3		Yes		Yes		X ray	ac/dc+	Med/deep	All	High	N/A	110–170	130–225	180–290	A5.5
E8018-C1		Yes		Yes		X ray	ac/dc+	Med/deep	All	High	N/A	90–160	120–200	180–300	A5.5
E11018-M		Yes		Yes		X ray	ac/dc+	Med/deep	All	High	70–110	90–170	120–225	160–310	A5.5
Cast irons															
Est					Nonmachinable		ac/dc+	Medium		Medium	70–110	90–150			
ENi-Cl					Machinable		ac/dc+	Medium		Medium	60–110	90–150			

Once the arc is struck and the gap is established, the welder will use bead patterns similar to those used in gas welding. Whatever the pattern used, it should be consistently maintained during the SMAW process.

Self-shielded flux-cored electrode welding is an application of shielded metal arc welding in a semiautomated application. The primary differences between self-shielded flux-cored electrode welding and SMAW is that the electrode comes in the form of a wire rather than a rod and the flux is located in the center of a tubular electrode wire rather than external to it. Since self-shielded flux-cored electrode welding does not lend itself readily to aircraft-maintenance applications, further discussion in this text is limited to similarities to other welding processes.

INERT-GAS WELDING

The term **inert-gas welding** describes an electric-arc welding process in which an inert gas is used to shield the arc and molten metal to prevent oxidation and burning. The process originally employed helium for the gas shield, and the name **Heliarc** was given the process by the Linde Division of the Union Carbide Corporation. Many improvements have been made in the original process and techniques, and new names have been assigned.

Types of Inert-Gas Welding

Tungsten Inert-Gas (TIG) Welding

Tungsten inert-gas (TIG) welding, classified as **gas tungsten-arc welding** (GTAW) by the AWS, is accomplished by means of a torch with a nonconsumable tungsten electrode. The electrode is used to sustain the arc and the molten pool of metal. Filler rod is added to the pool to develop the desired thickness of bead. Inert gas, usually argon, is fed to the weld area through the gas cup on the torch. The gas cup surrounds the electrode and directs the gas in a pattern to prevent the intrusion of oxygen and nitrogen from the air. In some cases it has been found beneficial to mix small amounts of oxygen and other gases with the inert gas to gain the best results.

Metal Inert-Gas (MIG) Welding

Another type of inert-gas welding utilizes a metal electrode, which melts and is carried into the weld pool to provide the extra thickness desired. This type of weld has been called **metal inert-gas (MIG) welding** and is classified **gas metal-arc welding (GMAW)** by the AWS. In this type of welding, the metal electrode must be of the same material as the base metal being welded. Since the electrode is consumed in this process, the electrode wire is automatically fed through the torch so that the torch is held a constant distance from the work surface. Carbon dioxide added to argon gas is commonly used for the MIG process.

Plasma-Arc Welding

Plasma-arc welding (PAW) is the third type of inert-gas welding that is discussed in any detail in this chapter. Plasma-arc welding is often used in applications where GTAW is preferred but more heat or speed is required, such as the welding of exotic metals. In plasma-arc welding the flow of the plasma is restricted but is at an increased speed through an orifice, resulting in higher temperatures and improved concentration of heat.

Advantages of Inert-Gas Welding

Inert-gas welds, because of 100 percent protection from the atmosphere, are stronger, more ductile, and more corrosion-resistant than welds made with ordinary metal-arc processes. In addition, the fact that no flux is required makes welding applicable to a wider variety of joint types. Corrosion due to flux entrapment cannot occur, and expensive post-welding cleaning operations are eliminated. The entire welding action takes place without spatter, sparks, or fumes. Fusion welds can be made in nearly all metals used industrially. These include aluminum alloys, stainless steels, magnesium alloys, titanium, and numerous other metals and alloys. The inert-gas process is also widely used for welding various combinations of dissimilar metals, and for applying hard-facing and surfacing materials to steel.

Inert-Gas Welding Joint Design

Many of the preparation and welding techniques are the same or similar for inert-gas welding as for oxyacetylene welding. This section introduces the methods and procedures that are different for inert-gas welding when compared to oxyacetylene welding operations.

Although there are innumerable welding joint designs possible, the basic types are *butt joint, lap joint, corner joint, edge joint,* and *tee joint.* Almost any TIG, MIG, or plasma-arc weld will be one or a combination of two or more of these basic types. Selection of the proper design for a particular application depends primarily on the following factors:

1. Physical properties desired in the weld
2. Cost of preparing the joint and making the weld
3. Type of metal being welded
4. Size, shape, and appearance of the part to be welded

Filler metal in the form of welding rod need not be used if proper reinforcement and complete fusion of the edges can be obtained without it. The joint designs described in this section are but a few of the many that can be successfully welded with the inert-gas method.

No matter what type of joint is used, proper cleaning of the work prior to welding is essential if welds of good appearance and physical properties are to be obtained. On small assemblies, manual cleaning with a wire brush, steel wool, or a chemical solvent is usually sufficient. For large assemblies or for cleaning on a production basis, vapor de-greasing or tank cleaning may be more economical. In any case, it is necessary to remove completely all oxide, scale, oil, grease, dirt, rust, and other foreign matter from the work surfaces.

Precautions should be taken when using certain chemical solvents such as carbon tetrachloride, trichlorethylene,

FIGURE 6-69   Square-edge butt joint.

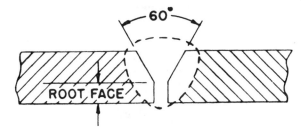

FIGURE 6-70   Single-V butt joint.

FIGURE 6-72   Flange joint.

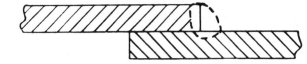

FIGURE 6-73   Lap joint.

and tetrachloroethylene, which break down in the heat of an electric arc and form a toxic gas. Welding should not be done when these gases are present, and the solvents should not be exposed to the heat of the welding torch. Inhalation of the fumes can be dangerous; hence proper ventilation equipment should be provided to remove fumes and vapors from the work area.

The square-edge **butt joint** shown in Fig. 6-69 is the easiest to prepare, and it can be welded with or without filler material, depending on the thickness of the pieces being welded. Joint fit for a square-edge butt joint should always be true enough to assure 100 percent penetration with good fusion. When welding light-gauge material without adding filler metal, extreme care should be taken to avoid low spots and burn-through. The heavier thickness generally requires filler metal to provide adequate reinforcement.

The single-V butt joint, shown in Fig. 6-70, is used where complete penetration is required on material thicknesses ranging between $\frac{3}{8}$ and 1 in [9.5 and 25.4 mm]. Filler rod must be used to fill in the V. The included angle of the V should be approximately 60°; the root face will measure from $\frac{1}{8}$ to $\frac{1}{4}$ in [3.18 to 6.35 mm], depending on the composition and thickness of the pieces being welded.

The double-V butt joint of Fig. 6-71 is generally used on stock thicker than $\frac{1}{2}$ in [12.7 mm], where the design of the assembly being welded permits access to the back of the joint for a second pass. With this type of joint, proper welding techniques assure a sound weld with 100 percent fusion.

A flange-type butt joint such as that illustrated in Fig. 6-72 should be used in place of the square-edge butt joint where some reinforcement is desired. This joint is practical only on relatively thin material, about 0.065 to 0.085 in [1.65 to 2.16 mm].

A **lap joint** has the advantage of eliminating entirely the necessity for edge preparation. Fig. 6-73 (p. 154) shows such a joint. The only requirement for making a good lap weld is that the plates be in close contact along the entire length of the joint and that the work be thoroughly cleaned, as explained previously. On material $\frac{1}{4}$ in [6.35 mm] thick or less, lap joints can be made with or without filler rod. When no filler metal is used, care must be taken to avoid low spots or burn-through. The lap-type joint is not usually recommended on material thicker than $\frac{1}{4}$ in [6.35 mm] except for rough fit-up. When so used, filler rod must always be added to ensure good fusion and buildup. The number of passes required depends on the thickness of the pieces being joined.

Corner joints are frequently used in the fabrication of pans, boxes, and all types of containers as well as for other heavier purposes. The type A corner joint, shown in Fig. 6-74, is used on material thicknesses up to $\frac{1}{8}$ in [3.18 mm]. No filler metal is required, as the amount of base metal fused is sufficient to ensure a sound, high-strength weld. Type B, as shown

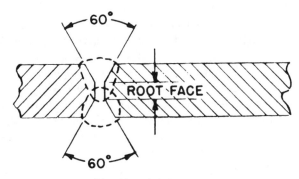

FIGURE 6-71   Double-V butt joint.

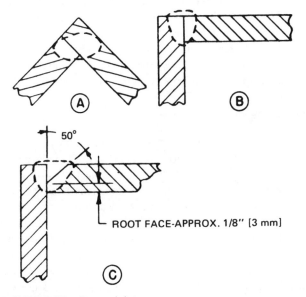

FIGURE 6-74   Corner joint.

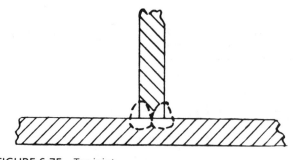

FIGURE 6-75 Tee joint.

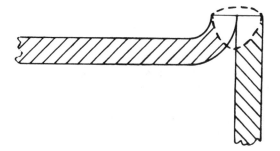

FIGURE 6-76 Edge joint at a corner.

FIGURE 6-77 Use of a backup bar with a cutout.

in the illustration, is used on heavier material that requires filler rod to provide adequate reinforcement. Type C is used on very heavy material, where 100 percent penetration is impossible without the beveled-edge preparation. The nose should be thick enough to prevent burn-through on the first pass. The number of passes required depends on the size of the V and thickness of the members being welded. On all corner joints, the pieces must be in good contact along the entire seam.

All **tee joints** require the addition of filler rod to provide the necessary buildup. Such a joint is shown in Fig. 6-75. The number of passes on each side of the joint depends upon the thickness of the material and the size of the weld desired. When 100 percent penetration is required, the welding current must be adequate for the thickness of the web material.

A typical **edge joint** is shown in Fig. 6-76. Such joints are used solely on light-gauge material and require no filler rod. Preparation is simple, and the joint is economical to weld. This type of joint should not be used, however, where direct tension or bending stresses will be applied to the finished joint because it will fail at the root under relatively low stress loads.

Weld Backup

On many TIG welding applications, the joint should be backed up. This is done for several reasons. On light-gauge material, backing is usually used to protect the underside of the weld from atmospheric contamination resulting in possible weld porosity or poor surface appearance. In addition to these functions, weld backup prevents the weld puddle from dropping through by acting as a **heat sink** and drawing away some of the heat generated by the intense arc. The backup also can physically support the weld puddle. A TIG weld can be backed up by (1) metal backup bars, (2) introducing an inert-gas atmosphere on the weld underside,

(3) a combination of the first two methods, or (4) use of flux backing painted on the weld underside.

Flat, metal backup bars are generally used on joints where the bar does not actually touch the weld zone. If the bar comes in contact with the underside of the weld, nonuniform penetration may occur and the weld underside may be rough and uneven.

A type of backup bar more commonly used is that shown in Fig. 6-77 where the surface is cut or machined out directly below the joint. On square-edge butt joints, for example, where fit-up is not too accurate and filler rod is required, a bar of this sort will protect the bottom of the weld from excessive contamination by the atmosphere and will draw the heat away from the weld zone.

Gas Tungsten-Arc Welding (GTAW)

Equipment for (GTAW)

The equipment used for **gas tungsten-arc welding** (GTAW) consists primarily of a power supply, a torch, and a gas supply, together with connecting hoses and cables. For torches that require cooling, a water system must be included.

The power supply for GTAW can be the same as that employed for standard arc (stick-electrode) welding, or it may be especially designed for inert-gas welding. A portable power supply unit for Heliarc welding is shown in Fig. 6-78. The power supply shown in Fig. 6-78 is the Linde HDA 300 welder, which can supply 300 A for a 60 percent duty cycle. The unit also includes the gas cylinder, gas-pressure and flow regulator, and tow cart. When purchased as a complete welding unit, the torch, connecting lines, and other accessories are included.

Gas tungsten-arc welding can be accomplished with direct-current, straight polarity (dcsp); direct-current, reverse polarity (dcrp); or with alternating-current, high-frequency stabilized (achf) power. Achf or dcrp power is generally recommended for magnesium and aluminum alloys and castings, whereas dcsp power is recommended for stainless steel, copper alloys, nickel alloys, titanium, low-carbon steel, and high-carbon steel. Achf or dcrp provides the best results with beryllium-copper alloys or copper alloys less than 0.040 in [0.016 mm] thick.

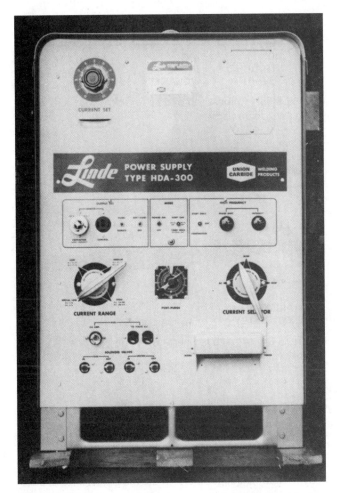

FIGURE 6-78 A portable power supply unit. (*Anderson Equipment Company*)

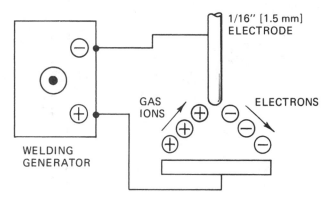

STRAIGHT POLARITY

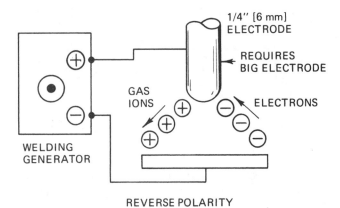

REVERSE POLARITY

FIGURE 6-79 Action of different types of welding current.

The effects of different types of current may be understood by studying the diagrams of Fig. 6-79. In direct-current welding, the welding current may be looked up as either straight polarity (SP) or reverse polarity (RP). The machine connection for dcsp welding is with the electrode negative and the work positive. The electrons will then flow from the electrode to the work, as shown in the illustration.

For dcrp welding, the connections are reversed and the electrons flow from the work to the electrode. The following explains why dcrp or achf power is required for aluminum welding. The oxide coating on aluminum has a much higher melting point than aluminum and interferes with the fusion of the metal. With dcrp power, the electrons leaving the surface of the base metal break up the oxides as they flow to the electrode. This same effect takes place during one-half of each ac cycle when achf power is used.

In straight-polarity welding, the electrons hitting the plate at a high velocity exert a considerable heating effect on the plate. In reverse-polarity welding, just the opposite occurs; the electrode acquires this extra heat, which then tends to melt the end. Thus, for any given welding current, dcrp requires a larger-diameter electrode than dcsp. For example, a $\frac{1}{16}$-in-diameter [1.59-mm] pure-tungsten electrode can handle 125 A of welding current under straight-polarity conditions. If the polarity were reversed, however, this amount of current

would melt off the electrode and contaminate the weld metal. Hence, a $\frac{1}{4}$-in-diameter [6.35-mm] pure-tungsten electrode is required to handle 125-A dcrp satisfactorily and safely.

These opposite heating effects influence not only the welding action but also the shape of the weld obtained. Dcsp welding will produce a narrow, deep weld; dcrp welding, because of the larger electrode diameter and lower currents generally employed, gives a wide, relatively shallow weld. The difference is illustrated in Fig. 6-80.

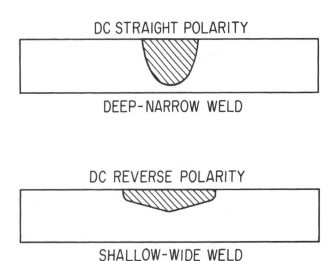

FIGURE 6-80 Effects of different polarities on the weld.

Inert-Gas Welding **165**

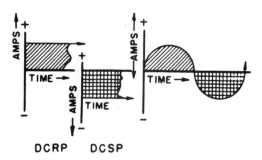

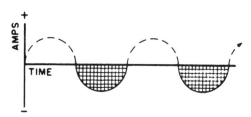

DCRP   DCSP

FIGURE 6-81 Illustration of ac wave through one cycle. (*Linde Div., Union Carbide Corp.*)

FIGURE 6-82 Rectified wave through two cycles with reverse polarity completely rectified. (*Linde Div., Union Carbide Corp.*)

FIGURE 6-83 Comparison of dcsp and dcrp with achf weld. (*Linde Div., Union Carbide Corp.*)

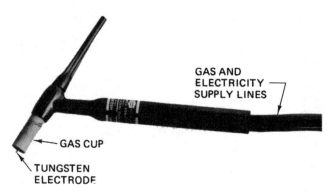

FIGURE 6-84 A torch for inert-gas welding. (*Linde Div., Union Carbide Corp.*)

One other effect of dcrp welding should be considered here, namely, the **cleaning effect**. The electrons leaving the plate or the gas ions striking the plate tend to break up the surface oxides, scale, and dirt usually present.

Theoretically, straight ac welding is a combination of dcsp and dcrp welding. This can be explained by showing the three current waves visually. As illustrated in Fig. 6-81, half of each complete ac cycle is dcsp and the other half is dcrp. Actually, however, moisture, oxides, scale, and other materials on the surface of the work tend to prevent the flow of current in the reverse-polarity direction. This effect is called **rectification**. For example, if no current at all flows in the reverse-polarity direction, the current wave will look something like the curve in Fig. 6-82.

To prevent the effects of rectification, it is common practice to introduce into the welding current a high-voltage, high-frequency, low-power additional current. This high-frequency current jumps the gap between the electrode and the workpiece and pierces the oxide film, thereby forming a path for the welding current to follow. Superimposing this high-voltage, high-frequency current on the welding current provides the following advantages:

1. The arc may be started without touching the electrode to the workpiece.
2. Better stability of the arc is obtained.
3. A longer arc is possible. This is particularly useful in surfacing and hard-facing operations.
4. Welding electrodes have a longer life.
5. The use of wider current ranges for a specific-diameter electrode is possible.

A typical weld contour produced with high-frequency stabilized alternating current is shown in Fig. 6-83, together with both dcsp and dcrp welds for comparison.

GTAW Torch

A typical torch for inert-gas welding is illustrated in Fig. 6-84. This is the Linde HW-17 torch, suitable for thin to medium metal thicknesses. The torch consists of the collet and collet body to hold the tungsten electrode, the gas lens for controlling and directing the gas flow, the gas cup, the handle through which current and gas flow, and supporting structures.

Electrode Selection

The technician must select the proper electrode diameter and shape. The chemical compositions of various types of electrodes are shown in Table 6-5. The diameter of the electrode must be large enough to carry the required amperage.

The shape of the electrode depends upon the material composition of the electrode, the amount of current being used during the process, and the characteristics of the desired joints. Figure 6-85 shows the different shapes of GTAW electrodes.

Filler Rod Selection

The filler rod must be metallurgically compatible with the metal being welded. Table 6-6 shows the recommended filler rod for various types of aluminum. The filler rod manufacturer's recommendations should be referred to when welding other or dissimilar metals.

Gas Selection

GTAW uses either argon, helium, or a mixture of the two gases. Argon provides the smoothest arc action because the arc is more stable, the arc is easier to initiate, and wider variations to the length of the arc are possible. Because argon is heavier than air, the gas envelope does not dissipate as

TABLE 6-5 The Chemical Compositions of Various Types of Electrodes

AWS Classification	Tungsten, Min, Percent	Thoria, Percent	Zirconia, Percent	Total Other Elements, Max, Percent
EWP	99.5	—	—	0.5
EWTh-1	98.5	0.8 to 1.2	—	0.5
EWTh-2	97.5	1.7 to 2.2	—	0.5
EWTh-3*	98.95	0.35 to 0.55	—	0.5
EWZr	99.2	—	0.15 to 0.40	0.5

*A tungsten electrode with an integral lateral segment throughout its length that contains 1.0% to 2.0% thoria. The average thoria content of the electrode shall be as specified in this table.
Source: American Welding Society

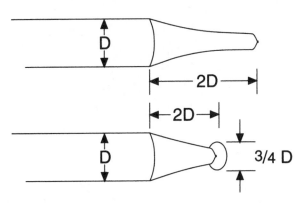

FIGURE 6-85 The different shapes of GTAW (*TIG*) electrodes. (*Lincoln Electric Co.*)

TABLE 6-6 The Recommended Filler Rod for Various Types of Aluminum

Base Metal	Recommended Filler Metal*	
	For Maximum As-Welded Strength	*For Maximum Elongation*
EC **1100**	1100 **1100, 4043**	EC,1260 **1100, 4043**
2219	2319†	
3003	5183, 5356	1100, 4043
3004	5554, 5356	5183, 4043
5005	**5183, 4043, 5356**	**5183, 4043**
5050	5356	5183, 4043
5052	5356, 5183	5183, 4043, 5356
5083	5183, 5356	5183, 5356
5086	**5183, 5356**	**5183, 5356**
5154	5356, 5183	5183, 5356, 5654
5357	5554, 5356	5356
5454	5356, 5554	5554, 5356
5456	**5556**	**5183, 5356**
6061	4043, 5183	5356‡
6063	4043, 5183	5356‡
7005	5039	5183, 5356
7039	**5039**	**5183, 5356**

Notes:
*Recommendations are for plate of "0" temper.
†Ductility of weldments of these base metals is not appreciably affected by filler metal. Elongation of these base metals is generally lower than that of other alloys listed.
‡For welded joints in 6061 and 6063 requiring maximum electrical conductivity, use 4043 filler metal.
However, if both strength and conductivity are required, use 5356 filler metal and increase the weld reinforcement to compensate for the lower conductivity of 5356.
Source: The Aluminum Assoc.

quickly as that of helium; argon is typically more economical than helium, however.

Helium, on the other hand, will raise the arc's temperature for a given arc length and amperage. These elevated temperatures are employed when GTAW is used to weld nonferrous material thicker than $\frac{1}{4}$ in [6.35 mm]. A mixture of helium and argon produces higher temperatures and arc stability.

GTAW Equipment Setup and Check

Table 6-7 shows the generic setup relationships for the welding of various thicknesses of aluminum, types of joints, and welding positions. The column "Number of Passes" refers to the multipass welding techniques discussed later in this chapter.

Before starting to weld, the entire welding setup should be thoroughly checked. It is most important to use the proper-size electrode and gas cup. All components must be functioning properly to realize the full advantage of this

type of welding. The following instructions can be used as a guide:

- Check all connections in the gas supply line for tightness. Be sure that good seals are obtained between the torch body, the cap, and the gas cap, because any air leakage into the gas stream will contaminate both the weld and the electrode. Be sure any gaskets required are in good

TABLE 6-7 The Generic Setup Relationships for the Welding of Various Thicknesses of Aluminum, Types of Joints, and Welding Positions

Plate Thickness (in)	Welding Position*	Joint Type	Alternating Current (A)	Electrode† Diameter (in)	Argon Gas Flow‡ (cfh)	Filler Rod Diameter (in)	Number of Passes
$\frac{1}{16}$	F	Square butt	70–100	$\frac{1}{16}$	20	$\frac{3}{32}$	1
	H, V	Square butt	70–100	$\frac{1}{16}$	20	$\frac{3}{32}$	1
	O	Square butt	60–90	$\frac{1}{16}$	25	$\frac{3}{32}$	1
$\frac{1}{8}$	F	Square butt	125–160	$\frac{3}{32}$	20	$\frac{1}{8}$	1
	H, V	Square butt	115–150	$\frac{3}{32}$	20	$\frac{1}{8}$	1
	O	Square butt	115–150	$\frac{3}{32}$	25	$\frac{1}{8}$	1
$\frac{1}{4}$	F	60° single bevel	225–275	$\frac{5}{32}$	30	$\frac{3}{16}$	2
	H, V	60° single bevel	200–240	$\frac{5}{32}$	30	$\frac{3}{16}$	2
	O	100° single bevel	210–260	$\frac{5}{32}$	35	$\frac{3}{16}$	2
$\frac{3}{8}$	F	60° single bevel	325–400	$\frac{1}{4}$	35	$\frac{1}{4}$	2
	H, V	60° single bevel	250–320	$\frac{3}{16}$	35	$\frac{1}{4}$	3
	O	100° single bevel	275–350	$\frac{3}{16}$	40	$\frac{1}{4}$	3
$\frac{1}{2}$	F	60° single bevel	375–450	$\frac{1}{4}$	35	$\frac{1}{4}$	3
	H, V	60° single bevel	250–320	$\frac{3}{16}$	35	$\frac{1}{4}$	3
	O	100° single bevel	275–340	$\frac{3}{16}$	40	$\frac{1}{4}$	4
1	F	60° single bevel	500–600	$\frac{5}{16} - \frac{3}{8}$	35–45	$\frac{1}{4} - \frac{3}{8}$	8–10

*F = flat; H = horizontal; V = vertical; O = overhead.
†Diameters are for standard pure or zirconium tungsten electrodes. Thoriated tungsten electrodes are not generally used for AC TIG.
‡Helium is not generally used in AC TIG welding aluminum. When helium is used, however, flow rates are about twice those used for argon.
Source: Lincoln Electric Co.

condition and are firmly in place. After welding, the electrode should have a clean, silvery appearance upon cooling. A dirty, rough electrode surface usually signifies air leakage in the torch or gas supply system.

- Check the welding current and gas-flow settings. They should be preset to the approximate values recommended for the material being welded.
- Select the proper gas cup and electrode size and electrode extension. The electrode should extend beyond the edge of the gas cup 1 to $1\frac{1}{2}$ times the diameter of the electrode. Fig. 6-86 depicts this relationship.
- Check the rate of water flow through the torch if it is the water-cooled type. Flow rates lower than those recommended decrease torch efficiency and may result in damage to the torch, particularly if the torch is being used at or near its maximum capacity.
- Check the ground connection to be sure it is securely clamped to the workpiece. The workpiece should be cleaned at the point of contact, preferably by grinding, to ensure good contact.
- Check to ensure that the postflow timer is properly set. The postflow timer allows the gas to continue flowing through the torch for a designated time after the foot pedal is released. Postflow is necessary to prevent oxidation of the electrode while the electrode is cooling. The postflow timer is typically set to 10 s.

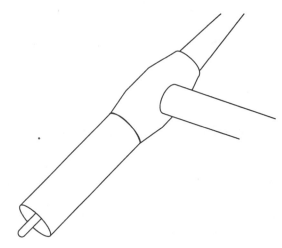

FIGURE 6-86 The electrode should extend beyond the edge of the gas cup.

Starting an Arc

There is nothing difficult about starting an arc in the proper manner. The procedure explained in this section should ensure a good start and maximum protection of the work from atmospheric contamination at the start of the welding operation.

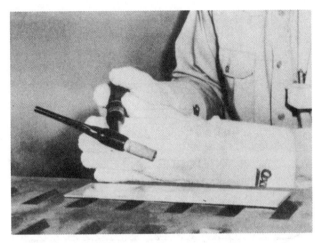

FIGURE 6-87 Torch in position ready to strike an arc.

In arc welding with a tungsten electrode (TIG), the electrode does not have to touch the work to start the arc. The superimposed high-frequency current jumps the gap between the welding electrode and the work, thus establishing a path for the welding current to follow. To strike an arc, the torch is held in a horizontal position about 2 in [5.08 cm] above the work or starting block, as shown in Fig. 6-87. The end of the torch is then swung quickly down toward the work so the end of the electrode is about $\frac{1}{8}$ in [3.18 mm] above the metal. The arc strikes at this distance. The downward motion should be made rapidly to provide the maximum amount of gas protection to the weld zone. The position of the torch when the arc strikes is shown in Fig. 6-88.

In dc welding, the same motion is used for striking the arc. In this case, however, the electrode must actually touch the work in order to start the arc. As soon as the arc is struck, the torch is withdrawn so the electrode is about $\frac{1}{8}$ in [3.18 mm] from the work. This prevents contamination of the electrode in the molten pool. High frequency is sometimes used to start a dc arc. This eliminates the need for touching the workpiece. The high-frequency current is automatically turned off by means of a current relay when the arc is started.

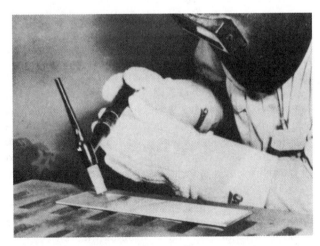

FIGURE 6-88 Torch in position when arc is struck.

The arc can be struck on the workpiece itself or on a heavy piece of copper or scrap steel and then carried to the starting point of the weld. A carbon block should not be used for striking the arc because it is likely to contaminate the electrode and cause the arc to wander when welding with a hot electrode is commenced. The action must be very rapid because the arc tends to strike before the torch is in the proper position.

To stop an arc, the torch is merely snapped quickly back to the horizontal position. This motion must be made rapidly so the arc will not mar or damage the weld surface or work.

Arc Wandering

With the torch held stationary, the points at which an arc leaves the electrode and impinges upon the work may often shift and wave without apparent reason. This is known as **arc wandering** and is generally attributed to one of the following causes: (1) low electrode-current density, (2) carbon contamination of the electrode, (3) magnetic effects, and (4) air drafts. The first two causes are indicated by a very rapid movement of the arc from side to side, generally resulting in a zigzag weld pattern. The third cause, magnetic effect, usually displaces the arc to one side or the other along the entire length of the weld. The fourth produces varying amounts of arc wandering, depending upon the amount of air draft present.

When current density of the electrode is at a sufficiently high level, the entire end of the electrode will be in a molten state and will be completely covered by the arc. When too-low current density is used, only a small area of the electrode becomes molten, resulting in an unstable arc, which has poor directional characteristics and is difficult for the operator to control. Too-high current density results in excessive melting of the electrode.

When a carbon block is used to strike the arc, electrode contamination often results. As the electrode touches the carbon, the molten tungsten on the tip of the electrode forms tungsten carbide. This has a lower melting point than pure tungsten and forms a larger molten ball on the end of the electrode. This, in effect, reduces the current density at the electrode end and arc wandering occurs. The electrode can also be contaminated by touching it to the workpiece or filler rod. When electrode contamination occurs in any form, it is best to clean the electrode by grinding, breaking off the end, or using a new electrode.

Magnetic effects are not generally encountered and are too complex to be discussed fully in this text. The most common magnetic action on the arc, however, results from the magnetic field set up by the current flowing in the work. This field may tend to attract or repel the arc from the normal path. One method for remedying this condition is to alter the position of the ground connection on the work until the effects are no longer noticed.

Making a Butt Weld

After the arc has been struck with a TIG torch, as previously explained, the torch should be held at about a 75° angle to

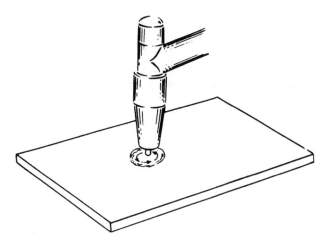

FIGURE 6-89 Forming the molten puddle.

the surface of the work. The starting point of the weld is first preheated by moving the torch in small circles, as shown in Fig. 6-89, until a small molten pool is formed. The end of the electrode should be held approximately $\frac{1}{8}$ in [3 mm] above the work. When the puddle becomes bright and fluid, the torch is moved slowly and steadily along the joint at a speed that will produce a bead of uniform width. No oscillating or other movement of the torch, except for the steady forward motion, is required.

When filler metal is required to provide adequate reinforcement, the welding rod is held at about 15° to the work and about 1 in [2.5 cm] away from the starting point. The starting point is then preheated, as explained previously, to develop the molten pool. When the puddle becomes bright and fluid, the arc is moved quickly to the rear of the puddle, and filler rod is added by quickly touching the leading edge of the puddle. The rod is removed and the arc is brought back to the leading edge of the puddle. As soon as the puddle becomes bright again, the steps are repeated. This sequence is continued for the entire length of the weld. Figure 6-90 illustrates the steps as described. The rate of forward speed and the amount of filler rod added depend on the desired width and height of the bead.

For making **butt joints** on a vertical surface, the torch is held perpendicular to the work and the weld is usually made from top to bottom. When filler rod is used, it is added from the bottom or leading edge of the puddle in the manner described previously. Figure 6-91 shows correct positioning of the rod and torch relative to the work.

Making a Lap Weld

A **lap weld**, or **joint**, is started by first developing a puddle on the bottom sheet. When the puddle becomes bright and fluid, the arc is shortened to about $\frac{1}{16}$ in [2 mm]. The torch is then oscillated directly over the joint until the sheets are firmly joined. Once the weld is started, the oscillating movement is no longer required. The torch is merely moved along

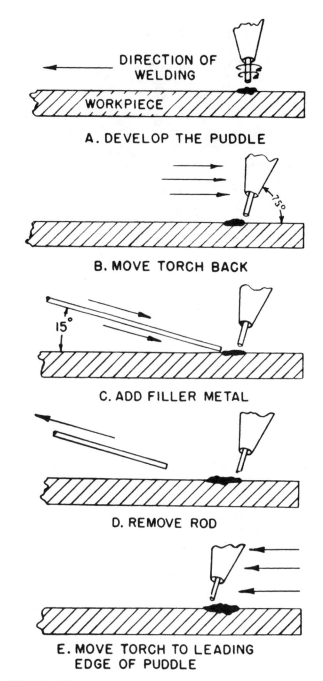

FIGURE 6-90 Steps in starting an inert-gas weld.

the seam, with the electrode held just above the edge of the top sheet.

In lap welding, the puddle developed will be boomerang or V-shaped, as shown in Fig. 6-92. The center of the puddle is called the **notch**, and the speed at which this notch travels will determine how fast the torch can be moved forward. Care must be taken to see that the notch is completely filled for the entire length of the seam. Otherwise, it is impossible to get 100 percent fusion and good penetration.

When filler metal is used, faster welding speeds are possible because the rod helps to fill the notch. Complete fusion must be obtained rather than allowing bits of filler

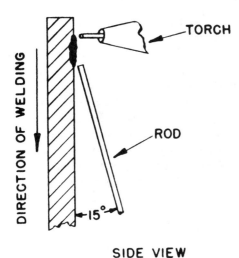

SIDE VIEW

FIGURE 6-91   Position of torch and filler rod for vertical GTAW (TIG) weld.

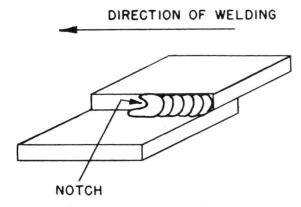

FIGURE 6-92   Shape of puddle in lap welding.

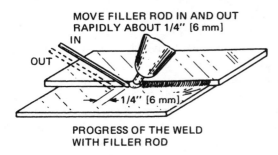

FIGURE 6-93   Procedure for lap welding with GTAW (TIG) welding.

rod to be laid into the cold, unfused base metal. The rod should be alternately dipped into the puddle and withdrawn $\frac{1}{4}$ in [6.35 mm], as illustrated in Fig. 6-93. By carefully controlling the melting rate of the top edge and by adding just enough filler metal where needed, a good uniform bead of correct proportions can be obtained.

Making a Corner, or Edge, Weld

The **corner**, or **edge**, **weld** is the easiest type to make. A puddle is developed at the starting point, and the torch is

then moved straight along the joint. Rate of travel is regulated to produce a uniform bead. A too-slow welding speed will cause molten metal to roll off the edges. Irregular or too-high speeds will produce a rough, uneven surface. No filler metal is required.

Multipass Welding

Multipass welding is generally required for welding material over $\frac{1}{4}$ in [6.35 mm] thick. The number of passes required depends upon the thickness of the material, the current-carrying capacity of the equipment involved, and the assembly being fabricated. The first pass should be a **root weld** and should provide complete fusion at the bottom of the joint. Subsequent passes can be made at higher currents owing to the backup effect of the root weld. Care should be taken to prevent inclusions between weld layers. On heavy work, it is sometimes advantageous to carry all the beads along simultaneously in a staggered arrangement to utilize the residual heat of preceding passes.

Gas Metal-Arc (MIG) Welding (GMAW)

Gas metal-arc welding (GMAW) simplistically appears to be the same process as GTAW, except that the electrode is consumable and fed through the torch automatically. However, as the technician should be aware, they are really quite different. Figure 6-94 shows the front of a GMAW power

FIGURE 6-94   Front view of a GMAW (MIG) welding power supply. (*Miller Electric Manufacturing Co.*)

FIGURE 6-95 GMAW (MIG) wire-feed system. (*Miller Electric Manufacturing Co.*)

supply and Fig. 6-95 illustrates a typical electrode wire-feeding system.

Two general techniques or processes are employed when welding with MIG (GMAW) in aviation maintenance applications. These are called *spray-arc transfer* and *short-circuit transfer.* Three other GMAW techniques exist but have either no or extremely limited aviation maintenance applications. They include *buried-arc, pulsed-arc,* and *fluxed-cored* welding. Taken as a whole, the GMAW techniques are capable of welding almost all metals in all types of configurations, in all positions.

In **spray-arc transfer** the metal from the electrode is carried to the molten pool in fine droplets by the force of the arc and by gravity. The power source for this process is constant-voltage direct current, reverse polarity. For best results with this process, the work should be level or nearly so. The shielding gas employed with spray-arc transfer is usually argon with 1 or 2 percent oxygen for stabilization of the puddle and arc. Of the five techniques mentioned here, spray-arc transfer is the most stable, offers the most directional flexibility, and is nearly splatter-free when properly accomplished.

When **short-circuit transfer** is used, the tip of the metal electrode is touched to the metal to provide a short circuit. This results in a high amperage, which melts the end of the electrode into the pool. The electrode is constantly fed into the pool to maintain the arc and transfer the metal.

When steel is welded with short-circuit transfer, the shielding gas can be pure CO_2 or a mixture of CO_2 and argon. For stainless steel, the mixture of gases is helium, argon, and CO_2.

Buried-arc transfer uses carbon dioxide as the shield for the welding area. The carbon dioxide causes the filler metal to be globular in nature. These large drops increase the frequency of short circuits during the welding process and this results in a considerable amount of splatter.

Pulsed-arc welding involves the use of high-amperage pulsations during the process in order to reduce the average current. However, highly specialized power supplies are required and extensive operator training is required.

The *fluxed-core* method mentioned here is more properly known as *gas-shielded flux-cored electrode welding.* It differs from the self-shielded flux-cored electrode welding previously mentioned in that carbon dioxide gas acts as an external shield and from the traditional MIG welding because its core is tubular (not solid). Application of the flux-cored process in aviation is extremely limited and is therefore beyond the scope of this text.

Equipment for GMAW

Typically, GMAW requires the use of DCRP with a constant voltage supply. With a constant voltage the power controller will automatically maintain the amperage at a level that will melt the electrode. These power supplies are referred to as **SVI** power supplies because they control the slope, voltage, and inductance during the welding process.

The desired voltage is a function of a variety of factors. Table 6-8 shows the typical ac voltages recommended for a variety of materials and shielding-gas combinations. In addition to these factors, variables such as the type of electrode, type of joint, thickness of the material, and position of the weld all combine to determine the proper voltage.

Slope

For a given set of operating conditions, there is a relationship between the voltage and amperage required to establish the desired arc. This relationship may be graphically displayed by a straight line drawn between two such sets of voltage-amperage relationships. The slope of this line is the **slope of the power source.** Anything that changes the resistance of the welding circuit—for example, the length of the welding cables or loose connections—changes the slope of the power source. Figure 6-96 shows the relationship between two different slopes.

In GMAW, where a constant voltage is applied, the slope of the line is used to determine the resultant current. This current controls the way in which the molten droplets leave the electrode. High currents typically result in the droplets of molten material leaving the electrode in a violent fashion. This results in excessive splattering around the weld area. Low voltages result in a calmer flow of molten material from the electrode to the base material.

Inductance

The rate at which the current changes to reflect a new combination of welding variables—for example, the initiation of an arc—is called **inductance.** If this rate of current change is rapid, excessive splattering is the result. By adding inductance to the welding circuit, the rate of current change is decreased and a smoother weld can be established and maintained.

Also associated with this phenomenon is the slope of the power source. The "flatter" the slope of the line, the less the inductance.

TABLE 6-8 Typical ac Voltages Recommended for a Variety of Materials and Shielding-Gas Combinations*

Metal	Drop transfer 1.6-mm- ($\frac{1}{16}$-in-) Diameter Electrode					Short-Circuiting Transfer 0.9-mm- (0.035-in-) Diameter Electrode			
	Argon	Helium	25% Ar-75% He	AR-O_2 (1–5% O_2)	CO_2	Argon	Ar-O_2 (1–5% O_2)	75% Ar-25% CO_2	CO_2
Aluminum	25	30	29	—	—	19	—	—	—
Magnesium	26	—	28	—	—	16	—	—	—
Carbon steel	—	—	—	28	30	17	18	19	20
Low alloy steel	—	—	—	28	30	17	18	19	20
Stainless steel	24	—	—	26	—	18	19	21	—
Nickel	26	30	28	—	—	22	—	—	—
Nickel-copper alloy	26	30	28	—	—	22	—	—	—
Nickel-chromium-iron alloy	26	30	28	—	—	22	—	—	—
Copper	30	36	33	—	—	24	22	—	—
Copper-nickel alloy	28	32	30	—	—	23	—	—	—
Silicon bronze	28	32	30	28	—	23	—	—	—
Aluminum bronze	28	32	30	—	—	23	—	—	—
Phosphor bronze	28	32	30	23	—	23	—	—	—

*Plus or minus approximately 10%. The lower voltages are normally used on light material and at low amperage; the higher voltages with high amperage on heavy material.

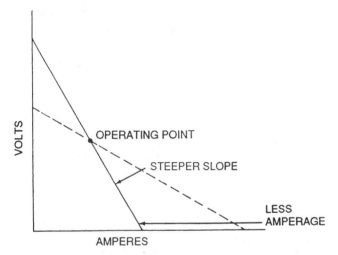

FIGURE 6-96 The relationship between two different slopes. (*American Welding Society*)

Equipment Setup

The setup used for the power supply depends upon the type of arc desired. The short arcs require less voltage (16–20 V) and result in less heat, less fluidity of the molten metal droplets, and more splatter. For short arcs the slope of the power source should be set on a "flat" setting, with the inductance between the minimum setting and half of the scale. The voltage should be set to maintain an arc that is $\frac{1}{8}$ in [3.18 mm] to $\frac{3}{32}$ in [2.38 mm] in length.

The spray arc mode requires a voltage between 20 and 30 V, needs higher heat, and results in more fluid-metal transfer and less splatter. The slope for the spray-arc mode should be "steep" and the inductance should be in the upper half of the inductance scale.

Another important factor in setting up the equipment for GMAW is the **wire-feed rate**. Since the wire is the electrode and the arc distance is controlled by the feed rate, if the wire feed rate is too great, the electrode will eventually contact the metal. If it is not fast enough, the electrode will burn back into the torch cup.

Selection of Electrode Wire

The selection of electrode wire for GMAW is similar to that of other bare filler rod. The recommendations of the equipment and wire manufacturers should always be followed. Some filler metals are very similar to the base metals, whereas others, such as the electrode wire for 6061 aluminum, are quite different.

The designation system for filler electrode wire established by the AWS is the same as that used for similar filler rod materials. Steel wire has an AWS designation of E70S-X, where the X is a sequential dash number indicating the exact wire chemistry. Aluminum wire is designated in the same way GTAW wire is designated. Stainless steel wire is again the same as the GTAW designation, except the prefix is "ER." Copper and copper alloy wires have a prefix of "ECU" and a suffix indicating the chemical composition.

Starting the Weld

Since the technician is primarily concerned with restoring the workpiece to its original condition, setting up the equipment will need to be accomplished with the use of scrap materials of the same type and thickness as those being repaired. Once preliminary setup has been established, the technician should proceed using the *scratch and retreat method*.

The **scratch and retreat method** is designed to minimize the effect of beginning and ending the weld at the weld-joint ends. As previously noted when discussing other types of fusion welding, the dissipation of heat at the ends of the weld joint is less than in the body of the joint, resulting in poorer weld joints at the ends. However, since the structural load transfer is greatest at the beginning and ends of joints, this is the area that requires the best joint. The technician should use a **scratch start technique** approximately an inch from the end of the joint and then quickly move or **retreat** to the end of the joint and then proceed, using the normal welding process. At the end of the joint the torch should be moved from the end of the joint, retreating about 1 in. over the newly welded joint before the arc is terminated.

During the welding process the arc height should be maintained at approximately the same level; however, minor variations in height should not affect the weld quality. The torch should be maintained at an angle of 70° to 85° and moved in a forehand direction. The width of the bead may be established by the welder by controlling the torch angle and the rate of travel of torch. At the ends of the joints, only the rate of travel should be used to control the bead size.

Plasma-Arc Welding (PAW)

Plasma-arc welding (PAW) is a modification to the GTAW process. It takes advantage of the plasma section of the welding arc (see "Welding Arc") by concentrating its flow through an orifice. By concentrating this area of the arc, temperatures ranging to 60 000°F have been developed, with theoretical temperatures up to 200 000°F potentially available. However, since the temperatures of GTAW exceed those normally required by the metals being welded, the main advantages of PAW are the directional stability of the plasma arc and the focusing of the arc by the orifice. Also, the temperature of the arc is not affected a great deal by minor changes in the torch's **stand-off distance**. The plasma

arc is $\frac{1}{4}$ in [6.35 mm] long, whereas the gas tungsten arc is approximately one-tenth that length.

There are two welding techniques available when plasma-arc welding, the **melt-in** technique (which is used in conventional GTAW) and the **keyhole** technique. The technician is most likely to use the low-amperage melt-in technique, since the high-amperage keyhole technique requires precise control of welding travel rates. The keyhole technique actually causes the molten puddle to extend completely through the part. The molten metal is then forced by the velocity of the arc to flow around it and to the rear of the arc stream, where the molten material solidifies. The keyhole technique is usually limited, in practice, to machine welding applications.

The power source may be the same as that used in GTAW. However, the torch must have extremely good cooling capabilities to use DCRP due to the increased temperatures.

The plasma gas, an ionized column of gas that passes from the anode to the cathode, may be supplemented by an inert shielding gas. The **orifice gas** need only be inert to tungsten (the electrode material). The supplemental **shielding gas** need not be inert, but must not adversely affect the properties of the weld joint. Typically the orifice gas is either argon (Ar), helium (He) or an argon-helium (Ar-He) mixture, depending upon the material, the material thickness, the current, and type of welding technique employed. The shielding gas is either argon, an argon-helium mixture or an argon-hydrogen mixture. Tables 6-9 and 6-10 show the recommended combinations for high- and low-current plasma-arc welding, respectively.

In plasma-arc welding the workpiece may not be required to be part of the electric circuit. The heat used in the plasma-arc process for cutting or welding is transferred to the workpiece through the plasma gas rather than the

TABLE 6-9 Recommended Combinations for High-Current Plasma-Arc Welding[*]

| Metal | | Thickness | | Welding Technique | |
		mm	in	Keyhole	Melt-in
Carbon steel	Under	3.2	$\frac{1}{8}$	Ar	Ar
(aluminum killed)	Over	3.2	$\frac{1}{8}$	Ar	75% He-25% Ar
Low alloy steel	Under	3.2	$\frac{1}{8}$	Ar	Ar
	Over	3.2	$\frac{1}{8}$	Ar	75% He-25% Ar
Stainless steel	Under	3.2	$\frac{1}{8}$	Ar, 92.5% Ar-7.5% H_2	Ar
	Over	3.2	$\frac{1}{8}$	Ar, 95% Ar-5% H_2	75% He-25% Ar
Copper	Under	2.4	$\frac{3}{32}$	Ar	75% He-25% Ar, He
	Over	2.4	$\frac{3}{32}$	Not recommended[†]	He
Nickel alloys	Under	3.2	$\frac{1}{8}$	Ar, 92.5% Ar-7.5% H_2	Ar
	Over	3.2	$\frac{1}{8}$	Ar, 95% Ar-5% H_2	75% He-25% Ar
Reactive metals	Under	6.4	$\frac{1}{4}$	Ar	Ar
	Over	6.4	$\frac{1}{4}$	Ar-He (50 to 75% He)	75% He-25% Ar

[*]Gas selections are for both orifice and shielding gases.
[†]The underbead will not form correctly. The technique can be used for copper-zinc alloys only.
Source: American Welding Society

TABLE 6-10 Recommended Combinations for Low-Current Plasma-Arc Welding*

Metal		Thickness		Welding Technique	
		mm	in	Keyhole	Melt-in
Aluminum	Under	1.6	$\frac{1}{16}$	Not recommended	Ar, He
	Over	1.6	$\frac{1}{16}$	He	He
Carbon steel	Under	1.6	$\frac{1}{16}$	Not recommended	Ar, 25% He-75% Ar
(aluminum killed)	Over	1.6	$\frac{1}{16}$	Ar, 75% He-25% Ar	Ar, 75% He-25% Ar
Low alloy steel	Under	1.6	$\frac{1}{16}$	Not recommended	Ar, He, Ar-H$_2$ (1-5% H$_2$)
	Over	1.6	$\frac{1}{16}$	75% He-25% Ar, Ar-H$_2$ (1-5% H$_2$)	Ar, He, Ar-H$_2$ (1-5% H$_2$)
Stainless steel		All		Ar, 75% He-25% Ar, Ar-H$_2$ (1-5% H$_2$)	Ar, He, Ar-H$_2$ (1-5% H$_2$)
Copper	Under	1.6	$\frac{1}{16}$	Not recommended	25% He-75% Ar, 75% He-25% Ar, He
	Over	1.6	$\frac{1}{16}$	75% He-25% Ar, He	He
Nickel alloys		All		Ar, 75% He-25% Ar, Ar-H$_2$ (1-5% H$_2$)	Ar, He, Ar-H$_2$ (1-5% H$_2$)
Reactive metals	Under	1.6	$\frac{1}{16}$	Ar, 75% He-25% Ar, He	Ar
	Over	1.6	$\frac{1}{16}$	Ar, 75% He-25% Ar, He	Ar, 75% He-25% Ar

*Gas selections are for shielding gas only. Argon is the orifice gas in all cases.
Source: American Welding Society

arc itself. This phenomenon significantly lessens the problems encountered by other inert-gas welding processes, such as arc wander, and the high velocity of the plasma gas reduces arc deflection.

Plasma-arc welders are classified as *transfer* and *nontransfer.* **Transfer** plasma-arc welders include the workpiece as part of the electric circuit. **Nontransfer** plasma-arc welders do not. Figure 6-97 illustrates the difference between the transfer and the nontransfer modes.

The function of the orifice, found in the torch, is to direct and accelerate the plasma gas toward the workpiece. In the nontransfer application the orifice material also acts as the cathode for the electric circuit.

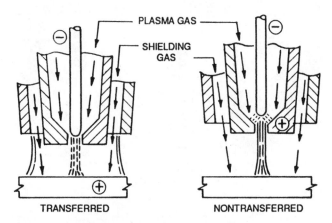

FIGURE 6-97 The difference between the transfer and the nontransfer modes. (*American Welding Society.*)

In addition to the general practical application advantages of plasma-arc welding, the primary advantage of nontransfer plasma-arc welding to the aviation maintenance technician is that the aircraft itself no longer needs be part of the electrical circuit, which is a requirement of the more traditional electric welding processes. This eliminates the concerns about magnetizing the welded material and the magnetic induction of current into electrical lines and instruments.

Electrode selection

The electrodes used in PAW are the same materials as GTAW. The electrode, however, must be round and concentric. The tip of the electrode is not extended past the orifice cup, helping to eliminate electrode contamination. The tip of the electrode is ground to an included angle between 20° and 60°, with the end being either sharp or flat.

Starting the Arc

Because the electrode does not extend beyond the orifice, the arc may not be started by using the touch-start technique. Instead a **pilot-arc system** is employed, which may use either a separate power source or the welding power source, if properly designed. Not only does the pilot arc start the plasma arc, but it is also useful in positioning the torch for the plasma-arc welding process. The pilot-arc is terminated once the plasma-arc process is initialed.

Rod Selection

The rod selection process is the same as in the GTAW welding process.

Manual Paw Welding Procedures

The technician is most likely to use the manual torch in a maintenance environment. The similarities between GTAW and PAW make this transition relatively easy. For many, plasma-arc welding is likely to be easier. The standoff distance is less critical when using PAW. PAW standoff may vary from $\frac{1}{8}$ in [3.18 mm] to $\frac{1}{4}$ in [6.35 mm] without affecting the process. The stability of the arc voltage in plasma welding provides for greater control when welding materials that require low currents. Figure 6-98 compares the volt-amperage characteristics between GTAW and PAW in argon. The relative flatness of the plasma-arc (upper) line exemplifies the stability of the relationship.

The recommended torch angle of 25° to 35° is only slightly more steep than the GTAW torch angle (15°). A forehand technique is recommended for PAW processes. The filler rod, when required, is applied in the same manner as in GTAW techniques. The torch-movement pattern is the same as GTAW as they relate to bead shape, size, and penetration. Weld joint designs are similar, as are fixturing and backup plate techniques.

Friction-Stir Welding Process

A new type of welding process used in the aerospace industry is friction-stir welding (FSW). This type of welding is a production-type solid-state (meaning the metal is not melted) welding process and is used for applications where the original metal characteristics must remain unchanged as far as possible. The welding equipment is automated to obtain a high accuracy and repeatability of the welds. This technique utilizes a nonconsumable rotating welding tool to generate

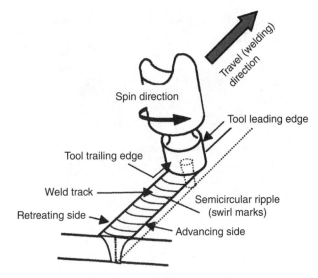

FIGURE 6-99 Friction stir welding terminology.

frictional heat and deformation at the welding location, thereby affecting the formation of a joint, while the material is in the solid state. Figure 6-99 shows the rotating tool.

The metal is fused using mechanical pressure, much like joining clay, dough, or plasticine. This process is primarily used on aluminum, and most often on large pieces which cannot be easily heat treated after the weld is completed to recover temper characteristics. The principal advantages of FSW, being a solid-state process, are low distortion, absence of melt-related defects, and high joint strength, even in those alloys that are considered non-weldable by conventional techniques (e.g., 2xxx and 7xxx series aluminum alloys). Furthermore, friction-stir (FS)–welded joints are characterized by the absence of filler-induced problems and defects, since the technique requires no filler. Finally, the hydrogen content of FSW joints tends to be low, which is important in welding steels and other alloys susceptible to hydrogen damage. The Eclipse Aerospace Company has used friction stir welding for the primary structure of the Eclipse Jet since it received FAA approval in 2002. Figure 6-100 shows the welding process.

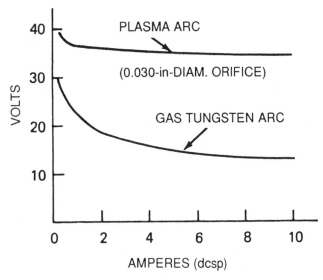

FIGURE 6-98 Comparison of the volt-amperage characteristics between GTAW (MIG) and PAW (plasma) in argon.

FIGURE 6-100 Automated friction-stir welding process.

Besides Eclipse other companies that use FSW include Boeing, Airbus, BEA, Lockheed Martin, NASA, US Navy, Mitsubishi, Kawasaki, Fokker as well as other industrial companies in the United States, Europe, and Japan.

CONCLUSION

In welding with any particular type of equipment, the technician should consult the manual supplied with the equipment to ensure that the most effective gases and techniques are employed.

In this section we have explained common types of gas and electric-arc welding. Oxyacetylene is the most common form of welding used in the aviation maintenance field. The plain electric-arc (stick-electrode) system is commonly used for heavy industrial construction in steel. Inert-gas welding, on the other hand, is used in a wide variety of precision welding on many different types of metals and alloys. For this reason it is particularly well adapted for welding structures and parts for aerospace vehicles.

There are many other types of welding, such as **submerged-arc** and **electron-beam welding**. These are specialties and are usually practiced by highly skilled welders. They are not usually done by the average airframe and powerplant technician; hence, detailed descriptions of these processes are considered beyond the scope of this text.

REVIEW QUESTIONS

1. What is meant by *fusion welding*?
2. Why is it usually necessary to add metal in the form of a welding rod while making a weld?
3. List the items necessary for a complete *portable welding outfit.*
4. Briefly describe *acetylene gas* and explain why it is used in gas welding.
5. At what pressures does acetylene gas become unstable and in danger of exploding?
6. What material is placed inside an acetylene cylinder to absorb the gas and thus ensure safe handling?
7. What is the maximum pressure of the gas in an acetylene cylinder?
8. Why is it necessary to keep oxygen equipment free from oil or grease?
9. To what pressure is an oxygen cylinder charged?
10. How can you determine the volume of gas in an oxygen cylinder?
11. Discuss the importance of a protector cap over the valve on an oxygen cylinder.
12. Describe a *gas-pressure regulator* and explain its function.
13. Why are two pressure gauges used on a cylinder gas regulator?
14. List the main differences between an *oxygen-pressure regulator* and an *acetylene-pressure regulator.*
15. How does a welder adjust the gas regulator to increase pressure?

16. Name the two principal types of *welding torches.*
17. Which type of welding torch is generally preferred for aircraft welding? Why?
18. Explain the function of the *mixing head.*
19. What determines the size of tip to be used in a welding torch?
20. How are oxygen and acetylene torch hoses identified?
21. How is the danger of incorrectly connecting the hoses avoided?
22. What are the two principal functions of *welding goggles*?
23. Why should a friction lighter be used to ignite the gas at the tip of an oxyacetylene welding torch?
24. Briefly describe the steps for setting up oxyacetylene welding apparatus.
25. What method should be used in testing for oxyacetylene leaks?
26. In lighting an oxyacetylene welding torch, what gas should be turned on first?
27. Explain how to adjust the oxyacetylene torch for a neutral flame.
28. Why is a *neutral flame* best for welding steel?
29. Describe the appearance of a neutral oxyacetylene flame.
30. What is the effect of using an oxidizing oxyacetylene flame?
31. Compare the carbonizing (carburizing) flame with the oxidizing flame.
32. What is the effect of a damaged or obstructed oxyacetylere welding torch tip orifice?
33. Explain how a gas welding torch tip can be cleaned properly.
34. Why is a soft flame better than a harsh flame for welding?
35. What are some of the causes of *popping* at the welding tip?
36. Give the procedure for shutting down the oxyacetylene welding apparatus.
37. Discuss the knowledge of welding necessary for the technician.
38. How should the welding torch be held for heavy welding?
39. Explain the difference between *forehand* and *backhand welding.*
40. Explain the purpose of the *filler rod.*
41. Approximately how far should the oxyacetylene welding flame cone be held from the base metal?
42. Describe the application of welding rod to the weld.
43. Explain circular and semicircular motions of the torch.
44. At what angle is the oxyacetylene torch flame applied to the metal?
45. Name the four common positions of a weld.
46. Describe how metal is prepared for welding.
47. Why are the edges of the metal beveled in many cases?

48. Describe a butt joint in welding.

49. Why is it not necessary to use filler rod when welding a flange butt joint?

50. Under what conditions is a double-V tee joint used?

51. Describe a *single-welded lap joint.*

52. Name the various parts of a *weld.*

53. What are the three most important proportions of a weld?

54. What should be the width of the welding bead?

55. What should be the depth of penetration in welding?

56. What are the usual causes of improperly formed weak welds?

57. Describe the chemical changes that may take place during gas welding.

58. What is the effect of expansion in welding?

59. Explain *coefficient of expansion.*

60. What precautions must be taken in welding a closed metal assembly?

61. Discuss the effect of high temperatures on the strength of metals.

62. What actions can a welder take to reduce the effects of distortion and residual stresses?

63. Explain *stagger welding* and *skip welding.*

64. What is the purpose of a *welding fixture*?

65. When welding heavy plate, how may welding stresses be reduced?

66. How is *preheating* accomplished?

67. Explain stress relieving.

68. What is meant by the spark test for the identification of metals?

69. Describe a good-quality completed weld.

70. What is meant by the bend test for a weld?

71. What type of flame should be used for aluminum and magnesium welding?

72. How are the edges of thin aluminum sheet prepared for gas welding? Aluminum plate over 0.25 in [6.35 mm] thick?

73. Describe procedures for making magnesium butt welds with a gas-welding process.

74. Describe the process of *electric-arc welding.*

75. Describe the electrode for arc welding.

76. What are two functions of the electrode?

77. What is the purpose of the coating on the electrode?

78. Describe the *welding circuit.*

79. What is the nature of the power used for electric-arc welding with respect to voltage and amperage?

80. What type of devices supply the power for arc welding?

81. Describe what takes place in the welding arc.

82. What temperature is developed in the arc?

83. Explain the three types of current that may be used for arc welding.

84. Describe *inert-gas* or *gas-arc* welding.

85. What is the nature of the electrode used with inert-gas welding?

86. Explain the difference between TIG and MIG welding.

87. Give the advantages of inert-gas welding.

88. What equipment is necessary for doing inert-gas welding?

89. What types of electrical power are used with inert-gas welding?

90. What is the difference in the effects of dcsp and dcrp current?

91. Why is dcrp or achf power required for welding aluminum and magnesium?

92. Why is a larger electrode required for dcrp welding power?

93. What are the advantages of achf power?

94. Describe two techniques that can be used with MIG welding.

95. Name five basic joints for inert-gas welding.

96. How should metal be prepared before inert-gas welding is started?

97. Why is a weld backup used with some welds?

98. List the checks that should be made before inert-gas welding is started.

99. Describe the process of striking an arc.

100. Why is it undesirable to strike the arc on a carbon block?

101. What is meant by arc *wandering*?

102. What are the principal causes of arc wandering?

103. Describe the process of starting an inert-gas weld.

104. What is meant by a *multipass weld*?

Welded Aircraft Structures and Repair 7

INTRODUCTION

As mentioned in Chap. 6, a rather wide variety of welding processes have been developed for joining various types of metals for aircraft. Oxy-fuel welding is still practiced extensively for the manufacture and repair of steel aircraft structures; however, inert-gas welding (Heliarc or TIG and MIG) is more commonly used for the welding of aluminum, stainless steels, titanium, and magnesium alloys.

This chapter deals with the conventional methods used for the repair of steel structures, which are easily performed by the technician either in the shop or in the field with portable equipment. Metal airplanes are designed so that the skin of the airplane, rather than the interior structures of steel tubing, carries a large part of the loads and stresses during flight. As a result, fewer steel structures occur in the newer airplanes, but many parts of structures are still built with steel.

CONSTRUCTION OF STEEL-TUBE ASSEMBLIES BY WELDING

In dealing with aircraft steel tubing, the technician is generally concerned with repairs rather than building a complete structure. However, extensive repairs may require that portions of the structure be removed and replaced with new structures. For this reason, some of the general considerations given to building aircraft structures are addressed in this section. For any repair, the aircraft manufacturer or the FAA should be consulted to identify the specific repair techniques acceptable for a particular aircraft.

Types of Steel Tubing

The steel tubing used most extensively for aircraft structures, engine mounts, and similar parts is **chromium-molybdenum**, also called **chrome-molybdenum** or **chrome-moly**. It is usually designated by the SAE number 4130. The principal properties of chrome-moly steels are resistance to impact, fatigue, abrasion, and high-temperature stress. They are capable of deep hardening when given a suitable heat treatment, have good machinability, and are easily welded by gas and electric-arc methods.

Aircraft steel tubing is sized according to its outside diameter and wall thickness. Tubing with an outside diameter from $\frac{3}{16}$ to $2\frac{1}{2}$ in [4.76 mm to 6.35 cm] is available, and wall thickness may vary from 0.035 in [0.89 mm] for the $\frac{3}{16}$-in tubing to 0.120 in [3.05 mm] for the $2\frac{1}{2}$-in tubing. Most diameter tubing sizes are available in various wall thicknesses. For example, $\frac{7}{8}$-in- [2.22-cm-] diameter tubing is available with several wall thicknesses, from 0.035 to 0.120 in [0.89 to 3.05 mm).

When selecting the size of tubing to use for a project, determine the type of material originally used along with the original outside diameter and wall thickness. Material that does not meet the minimum specifications of the original tubing should not be used.

Although tubing is generally thought of as being circular in cross section, steel tubing may be purchased in streamline (teardrop), oval, square, and rectangular cross section.

Joint Designs

The location where two or more pieces of tubing come together is known as a **joint**. If the pieces radiate out from the joint, the result is called a **cluster**. When reconstructing a structure, the joint design should match that of the original.

Tubing running lengthwise in a fuselage is normally of one piece or several pieces telescoped together with larger pieces at the front of the aircraft and smaller pieces toward the rear of the aircraft. The smaller pieces are slipped inside of the larger pieces and are welded in position, normally by the use of a *scarf* or *fishmouth* weld, which are discussed later. Short tubing pieces are used for the vertical, lateral, and diagonal pieces that make up the truss structure. These types of structures were discussed in Chap. 2.

Steel-tube control surfaces normally have a continuous piece of tubing making up the leading and trailing edges, with short pieces used for the ribs and diagonal members.

Before short pieces of material are joined to larger pieces to form clusters or joints, they are contoured to fit the joint as if they were going to be glued in place. At a cluster, as shown in Fig. 7-1, the perpendicular piece is cut to fit the longitudinal piece and then the diagonals are cut to fit the other pieces. Because of the feathered edges that result from such close fitting, no allowance purposely is made for expansion. When the tube is feathered to fit, extremely thin edge sections result. These thin sections absorb heat quickly and seem to disappear, leaving enough gap to allow for expansion during the welding process.

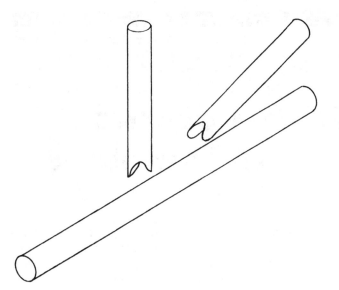

FIGURE 7-1   Tubing shapes to form a good welded joint.

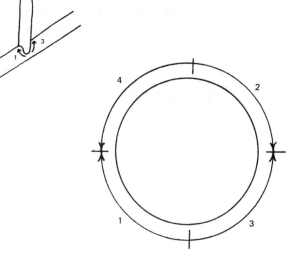

FIGURE 7-2   Sequence for a welding joint.

Alignment Fixtures

A suitable fixture must be used in the construction of aircraft steel-tube assemblies by welding. Fixtures are used to support and hold the various members of the assembly in their proper positions while they are permanently fastened. Welding fixtures should be rigid enough to withstand significant heat and related strains without losing shape as the parts are being assembled.

In some factories the fixtures for welding flat assemblies are made of boiler plate $\frac{3}{8}$ to $\frac{1}{2}$ in [9.53 to 12.7 mm] thick. These plates are fitted with suitable clamps, or blocks, to receive and hold the members of the assembly.

In a case where only one unit is to be built, the fixture for the construction of flat assemblies may be entirely made of wood, using the same concepts as a mass-production fixtures made of metal. A flat-top table can be used as the base, and wood blocks grooved to receive the members can be fastened to this base with nails, or screws. A piece of reasonably heavy iron or steel plate or thick nonflammable material is placed under the joint to protect the wood during welding.

If the tubular structure is not flat, such as a steel-tube engine mount, a more complicated fixture is required. It may have a heavy steel-base plate and a top plate of a lighter gauge, supported with angle-iron vertical members that are welded to plates to obtain the greatest amount of rigidity. Holes and clamps are provided for holding the members to be welded, and the assembly is usually left in the fixture after welding until it has cooled.

Welding Procedure

Oxyacetylene welding equipment is normally used to weld steel tubing. However, many people prefer to use inert-gas welding because of its more localized heating and the fact that there is no flame extension. Although there is no requirement for any specific process to be used, the technician should choose the process that he or she is most comfortable using and that will produce an airworthy repair. If an inert-gas process is used, check the completed structure for any residual magnetism that might affect the aircraft magnetic compass. Any residual magnetism should be removed by the use of a **degausser**.

When welding an assembly, it is desirable to tack-weld the components before completing any one weld. Tack-welding involves making three or four small equally spaced welds between each of the components. These welds help to prevent the components from shifting during the welding process. Weld each joint in turn. A joint should be welded in quarters, alternating across the weld, as shown in Fig. 7-2. This reduces the chance of the tubing warping out of alignment.

Attachment of Welded Fittings

Aircraft fittings of interest to technicians are principally small attachments or connections that are fastened to tubing members. They may be built up of one or more thicknesses of sheet metal, or they may be forged or machined from bars or billets. The method of welding fittings to tubular members depends upon the load stress they will carry in operating conditions.

Moderately stressed fittings that are not subjected to vibration are generally made of a single thickness of sheet steel and are welded to only one wall of the tube, as shown in Fig. 7-3.

Fittings or lugs for transmitting high stresses are welded to the supporting members at more than one point. High-stressed fittings attached to the main member of a structural

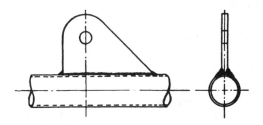

FIGURE 7-3   Simple fitting where the stress is moderate.

unit halfway between station points are welded to both walls of the tube, as shown in Fig. 7-4, where the tube is slotted and the fitting is inserted.

Fittings attached to the main members of tubular structures where brace members terminate are typically welded to the brace members. The main members and ends of the brace members may be slotted and the fitting welded, as shown in Fig. 7-5. The fitting may also be built up using two or three sections with fingers that extend to the brace members, as shown in Fig. 7-6.

Figure 7-7 shows a representative male fitting for round struts. It consists of a bearing sleeve or bushing welded to the strut end, which is reinforced with a steel plate formed around the bearing. The plate is welded to the bearing sleeve and strut with a fillet weld.

The fitting in Fig. 7-8 is a typical female strut-end connection used for round struts. This is forged and machined to fit into the strut end and is attached to the strut with a combination riveted and welded joint.

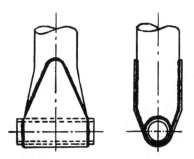

FIGURE 7-7   Fitting for the end of a strut.

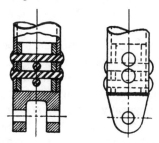

FIGURE 7-8   Female fitting for the end of a round strut.

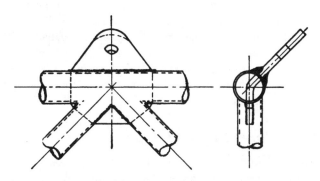

FIGURE 7-4   Inserted fitting where stresses are high.

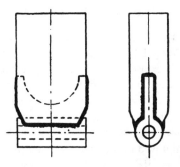

FIGURE 7-9   Forged and machined fitting for the end of a streamlined strut.

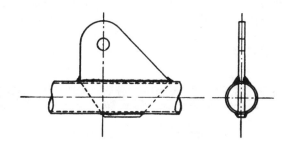

FIGURE 7-5   Inserted fitting at a station.

The fitting illustrated in Fig. 7-9 is a forged and machined male fitting for elliptical or streamlined struts. This fitting is inserted by slotting the strut end to receive its tang, and the attachment is made by means of a fillet weld.

The fitting shown in Fig. 7-10 is a typical female fitting for streamlined or elliptical struts, and it also may be used for round sections. It is built up of sheet steel and bearing sleeves. The strut end is slotted and formed to receive the fitting.

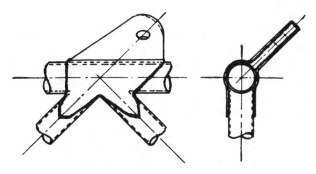

FIGURE 7-6   Fitting reinforcement with welded fingers.

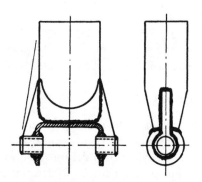

FIGURE 7-10   Female end fitting for streamlined strut.

Corrosion Protection

After welding, hollow steel structures may be filled with hot linseed oil or petroleum-base oils, under pressure, in order to coat the inside surface and discourage corrosion. This practice also helps to detect weld cracks, because the hot oil will seep through cracks that are not otherwise visible to the eye. This procedure is not applicable in all cases, but it is recommended where a large portion of the structure has been rewelded. Carefully examining all joints with a medium-power magnifying glass, at least 10 power, after first removing all scale is an acceptable method of inspection for repaired structures.

Corrosion-proofing the interior steel-tube structures can also be done satisfactorily by filling the structures with an epoxy primer or tube-sealing solution and draining before sealing. Sealing the members only, without the use of a preservative, by permanently closing all openings to prevent air from circulating through them is also considered an acceptable method for protecting the interior of repaired structures, provided that the interior of the tubing is thoroughly dried.

Finishing of Welded Structures

Once the structure has been welded, a protective coating should be applied. Prior to the application of the protective coating, all oil, grease, dirt, and welding residue should be removed. This involves the use of a wire wheel, abrasive paper, or sandblasting to remove particles and the use of a solvent to remove any grease or oil. As soon as the metal has been cleaned, it should be finished immediately and not touched by ungloved hands until at least a priming coat has been applied. If this precaution is not observed, corrosion resistance capabilities are severely reduced.

A very satisfactory protection for the exterior of tubing is obtained by spraying or brushing two coats of zinc chromate primer mixed with 4 oz [113.4 g] of aluminum-bronze powder per unthinned gallon [3.785 L] in the second coat of primer. This finish provides excellent protection against corrosion and is not affected by the application of dope for fabric-covered aircraft. Any equivalent finish using oil-base primers and enamels is acceptable. Excellent epoxy primers are now available; they should be used as directed by the manufacturer. The portions of the structure coming in contact with doped fabric should be given a protecting strip of tape or a coat of dopeproof paint if the base finish is not already dopeproof.

Visual Inspection

A visual inspection is the primary means of examining a welded steel structure. The structure is checked for proper alignment, evidence of deformation, and any indication of cracking. Alignment and deformation can be verified by a symmetry check, as discussed in Chap. 12,

and by placing a straightedge along the tubing. Cracks in the tubing and weld defects can be checked with a magnifying glass.

If bent or cracked structures are found, check around the damaged area for secondary damage. Secondary damage may be created in some areas away from the primary damage because of the transmission of loads through the structure from the primary area.

A check for internal corrosion in tubing can be made by tapping on the tubing with a small hammer. Any dull sounds or sudden changes in sound when compared to the surrounding tubing may indicate a weak structure. A center punch and hammer can be used to check if suspect tubing is corroding internally. When the center punch is held against the tubing and tapped with a hammer, badly corroded tubing will give way. In the same manner a Maule fabric tester can be used to apply pressure on the tubing. The bottom of the tubing in the lower portions of the structure are the most likely to corrode internally.

More detailed inspections of an area may be performed by *dye-penetrant inspection, fluorescent-penetration inspection, magnetic-particle inspection, X-ray inspection, ultrasonic inspection,* and *eddy-current inspection.* Chapter 9 discusses these inspection techniques in more detail.

AIRCRAFT TUBING REPAIR

Steel tubing is adaptive to aircraft construction because it is strong for its weight and can be repaired easily. In making repairs, either steel tubing or sheet stock having code numbers 1025 (carbon steel), 4130 (chrome-moly), or 4330 (nickel-chromium-molybdenum) might be used. In any case, if replacements are being made, the replacement material should be the same as the original. Repairs can be made with little equipment. All that is required is a welding apparatus, a short section of steel to replace the damaged section, and the tools to cut and prepare the metal for welding, generally a hacksaw and files.

Procedures for Weld Repairs

The parts to be welded should be cut to the proper dimensions and secured in place with clamps or fixtures to ensure correct alignment. In the case of a major rework of a fuselage, it is likely that the entire fuselage must be placed in a fixture to keep it properly aligned while the repair is being made.

The areas to be welded must be thoroughly cleaned by wire brushing, filing, or by some other method to ensure a weld free of such defects as inclusions or contamination. Only a steel wire brush should be used, because any small amount of other metals left on the surface will weaken the weld.

The torch flame should be adjusted for a reasonably soft condition; that is, the force of the flame should not be such that it will blow the molten pool of metal away from the weld. For comparatively thin tubing, the inner cone should not be more than $\frac{1}{8}$ in [3.18 mm] long.

The size of the welding torch tip is primarily governed by the thickness of the metal. Table 6-2 gives typical tip sizes.

The tip should be large enough to permit rapid welding but small enough to avoid overheating and burning the metal. Burning weakens the metal and causes it to crack when it contracts during the cooling. Rewelding is objectionable for the same reason; that is, it tends to overheat the metal and thus might lead to structural failure.

The **welding (filler) rod** usually is made of mild steel for welding chrome-moly steel because it flows smoothly and contributes to a uniform, sound weld. Flux is not required. Some welders recommend the use of a welding rod made of alloy steel, but the disadvantages of alloy steels sometimes appear to outweigh the advantages. In any event, if the repaired part requires heat treating, it is necessary that the welder use heat-treatable filler rod.

As steels increase in temperature, their strength lessens. Conversely, as they cool, they increase in strength to the temperature where brittleness begins to significantly increase. This is especially true in the case of chrome-moly steel; hence it may crack if placed under a comparatively light stress while at the elevated temperatures experienced during welding. One way to avoid this problem is to use *welding fixtures* that will not restrict the expansion and contraction of the welded members.

During welding heat is continually being dissipated into the surrounding metal. The technician compensates for this heat dissipation during the welding process. As the heat from the process approaches an edge of a joint, there is less material to absorb the heat. This results in the same amount of heat concentrating in a smaller area. The result is that burnaways occur more easily near the edges of the material. The technician may lessen this tendency by drawing the welding flame away as it approaches an edge of a joint.

The technician should never attempt to weld over an old weld. Rewelding an area only weakens the area further. Welding over a previously welded area to make it appear smoother likewise weakens the joint. Filling the weld area with any filler material, especially any applied by heat, is forbidden because not only is the area weakened, but at the same time an area of potential failure is hidden from view.

If there is a failure or damage in a welded area that requires rewelding, all the old welded area should be removed. Techniques in patching and splicing of damaged tubular structures are discussed later in this text. Regardless of the size of the failure or damage, welding may not be accomplished by rewelding over an existing weld. Rewelding an old weld weakens the material and increases its brittleness.

Welding should never be done over a brazed area or where the brazing has been removed. If a repaired area was covered with corrosion-resistant materials, such as paints, primers, and chrome, before the repair was made, this material should be replaced after the repair is completed.

For standard steel-tubing repairs, it is good practice to sandblast the welded area after welding to provide a good clean metal base for the application of primer. The weld should be primed before corrosion has had a chance to start, and the clear surface should be protected from any contamination before it is primed.

Controlling Distortion

When steel-tube structural units and steel fittings are constructed or repaired, welders can expect to experience the effects of the material's expansion or contraction during the welding process. Expansion and contraction on the base material causes strain in the material, which may ultimately result in cracking or warpage of the base material. The best way to prevent these results is to anticipate their existence and compensate for them.

Unfortunately, when a material is placed under a strain, the material's memory is affected. As the material is strained, it continually attempts to establish a new equilibrium. As the material seeks a new equilibrium, it rearranges the stresses within the part. Consequently, as the material is cooled it will not always return to its original shape or position. In addition, in order to relieve some of the stresses, the material will crack. Even if the material does not crack, it may tend to re-form itself, causing warpage.

Cracking can be prevented by reducing the strains on the weld and base metal caused by weight of the base material itself or restriction of the normal processes of expansion and contraction. For example, one end of a web member may be welded to the flange member of a truss and permitted to cool before the opposite end of the web member is welded. Also, joints of a tubular assembly in which several members terminate should be welded first and allowed to cool before the opposite ends are welded. Cooling is needed because the additional time required to weld such joints permits the heat to flow from the weld into the members, and this heat flow causes them to expand. If the members are connected to similar joints at the opposite end, it is advisable to apply heavy clamps, chill plates, or wet nonflammable material to the members, close to the weld, in order to restrict this heat flow and thus reduce the expansion.

When fittings are welded, shrinkage strains and their accompanying tendency to cause cracking can be reduced greatly by beginning the weld at the fixed end and working toward the free end of the opening or seam. Cold-rolled alloy-steel forms or heat-treated forms are annealed before welding to reduce the brittleness. Another practice is to relieve the stresses of alloy-steel parts after welding by heating the whole part uniformly to a temperature between 1150 and 1200°F [621 and 649°C] and then allowing the part to cool very slowly.

If a joint is held together by both riveting and welding, the rivet holes are lined up and the welding is completed before the rivets are driven. This procedure prevents the shearing stress on the rivets and the elongation of the holes caused by the expansion and contraction of the metal.

Warpage is controlled by using enough clamps and correctly constructed fixtures. **Progressive welding**, which is welding progressively along the joint from the beginning to the end, either continuously from start to finish or in sections, with each section tied in or joined to the next, is another

FIGURE 7-11   Jack and tube used for alignment.

method for reducing strains. A third method is to heat the member on the opposite side from which the weld is made.

Shrinkage is provided for by making the required allowance for normal shrinkage, as indicated by a trial weld under similar conditions. An allowance of $\frac{1}{32}$ in [0.794 mm] at each end of a truss or web member is often enough. This allowance is a rough, general one for both steel-tube structural units and fittings built up of sheet metal, but it is not a substitute for careful work.

Devices for Holding Tubular Structures in Alignment

Figure 7-11 is a drawing of a jack and a tube. Both the jack end and the tube end have radii designed to fit over tubular strut members. The distance between the two ends is adjusted by the position of the nut on the jack end.

Figure 7-12 is a drawing of a jack-tube fixture for holding tubular structures in alignment during repair. The jack-tube fixture is set up between struts, and a clamp is placed on the outside of the struts opposite the jack-tube fixture. This clamp is adjustable and is placed on the outside of the members directly over and parallel to the jack to prevent the assembly from spreading apart.

The original dimensions are obtained from the aircraft's drawings or by measuring the corresponding member of the airplane that is not distorted. Because of the symmetrical design of aircraft, the design of many portions of an aircraft are the mirror images of parts located on the other side of the aircraft. For example, the basic structure of the left wing is the mirror image of the right wing. The related mirror measurements may often be applied to the structure being repaired. Care should be taken by the technician that dimensions obtained in this manner are appropriate. Another source of dimensional information is other aircraft from the same manufacturer and of the same model. Extreme care should be taken by the technician when using dimensional information obtained from sources other than the aircraft's drawings. All such repairs should be approved by the FAA before the repair is undertaken.

Repairs to Fittings

When the fitting welded to a structural member is discovered to be broken or worn, it is removed and a new fitting is installed. The new fitting must have physical properties at least equal to those of the original part or unit. In removing the damaged fitting, care must be taken that the tube to which it is attached is not damaged, and the new fitting must be installed in the same manner as the original. If a damaged fitting attached to a main member, where truss members terminate, cannot be removed without weakening the structural members, the part of the tubing having the damaged fitting attached to it should be removed, and a new section and fitting should then be installed.

Repairing Bends and Dents

Most repairs on aircraft tubing require the cutting out and welding in of a partial replacement tube or the replacement of a whole section of tubing, but some repairs are relatively minor. An example is the straightening of a piece of fuselage tubing that has become slightly bent or buckled. If the bent or buckled part is not made of a heat-treated steel, it should be stronger after straightening because the cold working the metal receives in the straightening process adds to its strength.

One device useful for straightening bent or buckled tubing is a heavy-duty C-clamp, as shown in Fig. 7-13. Three blocks of hardwood are required for use with a C-clamp. These blocks are cut to fit the shape of the tubing, and the tubing grooves are lined with a soft material such as leather or heavy cloth. Another unit of required equipment is a steel bar, sufficiently strong to withstand the bending stress applied when the tubing is straightened. One of the grooved blocks is placed at each end of the bent section of tubing and the blocks are held in place by the C-clamp and bar, the latter spanning the bent area and backing up the wooden blocks, as shown in the illustration. The third block is placed under the jaw of the clamp to bear against the tubing at the point of greatest bend. Since the C-clamp jaws apply pressure to force the tubing toward the steel bar, the tubing is straightened as the clamp is tightened. The handle of the C-clamp should be turned until the tubing is bent slightly past the straight position to allow

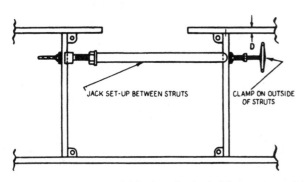

FIGURE 7-12   Fixture with jack and tube holding structure in alignment.

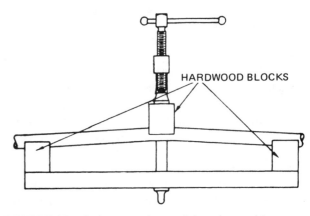

FIGURE 7-13   C-clamp used to straighten bent tubing.

for spring-back. The clamp and blocks are removed and the tubing is tested by placing a straightedge on both the side and the top of the tube. If the tube is straight, the job is completed. However, if a bend is still apparent, the operation must be repeated until the desired result is obtained.

Whenever a bent tube is straightened, the adjoining welded joints must be examined for cracks. If cracks are found, they must be repaired. They usually occur at the point where the maximum bend was corrected. Briefly, the repair is made by drilling a hole at each end of the crack and welding a split steel sleeve over the crack. The details of such a repair are given later in this chapter. Remember that the tube-straightening procedure just illustrated and described is done only if the tube is not dented, crushed, or kinked.

Another minor local defect in tubing is one in which a section of tubing has been slightly flattened; that is, it has become slightly oval-shaped or out of round. This defect is remedied by first drilling a steel block to the diameter of the tube to be re-formed. This block, which is usually about 4 in [10 cm] square, is sawed in half lengthwise through the center of the hole, and the two sections are separated. A small quantity of cup grease or oil may be applied to the blocks on the surfaces that will come into contact with the tube. The two grooved steel blocks, obtained by sawing through the original block, are clamped into position over the out-of-round portion of the tube. The clamp is gradually tightened and pressure is applied, as shown in Fig. 7-14, and the assembly is rotated around the tube until it assumes its original, normal shape. Heating the tubing to a dull red makes this procedure easier. If the out-of-round portion of the tube is longer than the length of the two steel blocks, the procedure is repeated

throughout the length of the out-of-round area until the whole length has resumed its original round shape.

If the defect is a minor, smooth dent in the tubing, not deeper than one-twentieth the tube diameter, and is clear of the middle third of the tube section, it may be disregarded. If the dent is large enough to require removal and the tubing is not out-of-round for any considerable distance, the dent may be pushed out by air pressure. An air fitting is placed in the tube at one of the self-tapping screws provided at the ends of the main steel tubes found in some aircraft so that an air pressure of about 75 psi [520 kPa] or more can be applied inside the tubing. It may be necessary to drill a hole for the installation of a temporary fitting. The hole is welded closed after the dent is removed. The dented area is heated with the welding torch until it is dull red, and the air pressure is maintained until it forces out the dent and restores the tubing to its original shape.

In some cases the combination of heat and air pressure is not sufficient to straighten the tube. Under these conditions, a welding rod is tack-welded to the center of the dented area and pulled while the area is pressurized and heated. When the dent has been pulled out in this manner, the welding rod is removed, the heated area is allowed to cool, and then the air pressure is released. The air-pressure fitting is then removed and the opening is closed.

Surface Patches

If a crack, hole, or significant dent is found in a tube, it may be repaired by one of the following methods. When a crack is located, all finish must be removed with a wire brush, steel wool, or by sandblasting. If the crack is in an original weld bead, the existing bead is carefully removed by chipping, filing, or grinding. When this has been done, the crack is welded over along the original welded line. A common error in following this procedure is to remove some of the tubing material while taking off the weld bead.

If a small crack is near a cluster joint but not on the weld bead, the finish is removed and a no. 40 (0.098-in [2.49-mm]) hole is drilled at each end of the crack to prevent the crack from spreading. A split reinforcement tube is welded in place completely around the tubing over the cracked area. The repair tubing in this case should have an inside diameter approximately the same as the outside diameter of the tubing being repaired. The repair section is cut at an angle of 30°, as shown in Fig. 7-15, before being split so it will fit over the repaired section. The repair tubing should extend a distance of at least $1\frac{1}{2}$ times the diameter of the tubing being repaired beyond each end of the crack. This is shown in the illustration. When the weld is completed and cooled, a coat of zinc chromate or other primer is applied to the area where the finish was removed. Finally, the area is given finishing coats to match the adjoining surfaces.

Under certain conditions, a dent or hole in tubing can be repaired by an external patch that does not completely surround the tubing. If a dent is not deeper than one-tenth the tube diameter and does not encompass more than one-quarter the tube circumference; if it is free from cracks, abrasions, and sharp

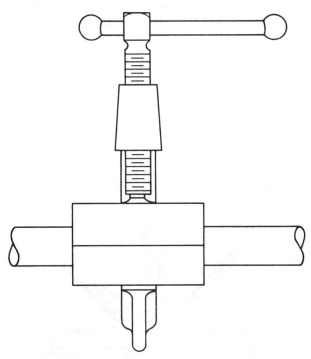

FIGURE 7-14 Use of a grooved steel block to re-form steel tubing.

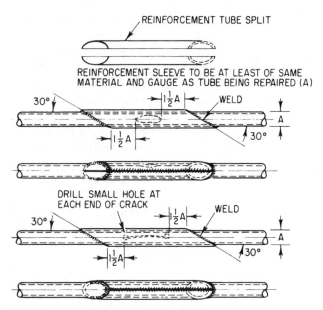

REINFORCEMENT TUBE SPLIT

REINFORCEMENT SLEEVE TO BE AT LEAST OF SAME
MATERIAL AND GAUGE AS TUBE BEING REPAIRED (A)

DRILL SMALL HOLE AT
EACH END OF CRACK

FIGURE 7-15   Split-sleeve reinforcement.

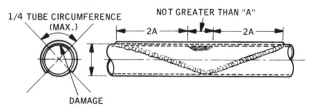

1/4 TUBE CIRCUMFERENCE
(MAX.)

NOT GREATER THAN "A"

DAMAGE

FIGURE 7-16   Welded patch for dent or hole.

corners; and if the tube can be substantially reformed without cracking before the application of the patch, then a patch such as that shown in Fig. 7-16 can be used. A hole in tubing that is not longer than the tube diameter and does not involve more than one-quarter the tube circumference can be patched in this same manner. Such a patch is not permitted in the middle third of the tube section and must not overlap a tube joint.

Heat-treated compression members can be returned to their original strength, if the dents are minor, by means of a split tube clamped in place over the damaged section, as shown in Fig. 7-17, where the split tube is called a **splint tube** because it serves as a splint for an injured section. This type of repair should be considered only for temporary purposes, and a suitable permanent repair should be made as soon as possible. For this temporary repair, the split rube should have a wall thickness equal to that of the tube being repaired, and it must be clamped tight enough to prevent both the split tube and the clamps from becoming loose in service. Such a repair should be inspected frequently until it is replaced.

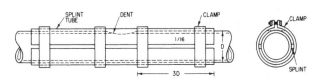

SPLINT TUBE   DENT   CLAMP   CLAMP
1/16
D
3D
SPLINT

FIGURE 7-17   Clamped-tube reinforcement for heat-treated tube.

If the dents occur in tubular members that are not heat-treated, such dents may be reinforced by welding repairs, as illustrated in Fig. 7-15. The reinforcement tube is clamped in place and welded along two sides and at the ends, as explained previously. This method of repair is satisfactory for short struts or dents in long members near the center of the span because such members are under greater bending stress near the center of the span, and their full strength must be retained.

The **split-sleeve** reinforcement illustrated in Fig. 7-15 is suitable for repairing cracks, dents, gouges, and other types of damage in structural tubing. In many cases it is necessary to straighten the member because of the bend that occurred at the time of the damage. After the member has been straightened, it is necessary that the structure, especially at the welded joints, be thoroughly inspected to detect any possible secondary damage that may have occurred. The dye-penetrant method of inspection lends itself well to this inspection. Cracks in welds must be repaired by removing the old weld and rewelding the joint.

Where damage has occurred at a cluster, a patch repair is often desirable. If tubular members, such as fuselage longerons, have sustained local damage at a cluster, they are repaired by welding on a **patch plate**, also called a *finger plate*. Such a repair is illustrated in Fig. 7-18. The patch must be of the same material and thickness as the

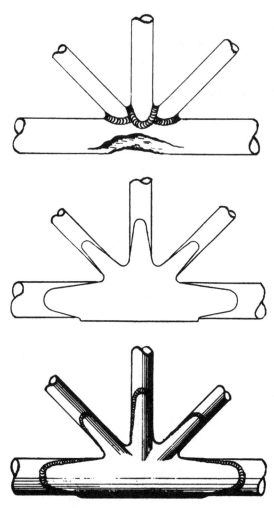

FIGURE 7-18   Patch plate repair for damage at a cluster.

injured tube and of a size sufficient to cover the damage. The fingers that extend onto the truss members should have a width equal to the diameter of the brace tube and a length equal to at least $1\frac{1}{2}$ times that of the diameter. The ends of the fingers should be rounded or pointed to prevent heating the tube to an annealing temperature in a direct cross section at these spots. Rounding the ends of the fingers also helps to distribute the load, thus reducing stress concentration. Figure 7-19 shows additional details of finger plate repairs.

To prepare a patch for the repair of damage at a cluster, it is best to make a template by cutting a piece of heavy paper in the shape of the required patch. The paper is fitted around the cluster to be repaired and marked for correct size. It is then cut out and tested by placing it in the position of the patch. When correctly shaped, a paper template is placed on a piece of sheet steel of the correct type and thickness and the steel is marked for cutting.

The patch should be trimmed so that it will extend past the dent in both directions and have fingers as wide as the brace members. All the existing finish on the part to be repaired must be removed. The patch is clamped into position and lack-welded in several places where its edges touch the tubing. It is heated, and light hammer blows with the ball end of a ball peen hammer are applied to form the patch around the repair area. The patch should not be overheated, but it must be softened enough so it can be formed to the tubing with a gap of not more than $\frac{1}{16}$ in [1.59 mm] between the tubing and the patch. The patch is then fused to all the tubes involved by welding around all the edges. When the weld is completed and cooled, the surface around the joint is refinished.

Major Welded Repairs

Unless the damage to the member of a steel-tube structure is comparatively slight, it is usually better to remove the damaged section and to weld in either a partial replacement tube or an entirely new section of tubing. Any tube cutting is done with a hacksaw and not by the oxyacetylene-flame cutting process. The manner of removing tubes and the number to be removed are determined by the location and extent of the damage. Any tubes inserted as replacements are joined at their ends by means of a splice.

Splicing in the case of partial tube replacements may be done by using an external replacement tube of the next-larger diameter, in which case the replacement tube is spliced to the stub ends of the original tubing, or it may be done by using a replacement tube of the same diameter together with either internal or external reinforcing sleeves.

If the original damaged tube includes fittings or castings that have been made especially to fit the tube, the spliced replacement tube must be of the same diameter as the original tubing, and this calls for either internal or external **reinforcing sleeves** under or over the splices. If no fittings or castings are attached to the original tubing, it is possible to use an **external replacement tube**.

The two principal types of splice welds permitted in the repair of aircraft tubing are the *diagonal (scarf) weld* and the *fishmouth weld*. A splice is never made by butt welding. The best form is the **fishmouth weld**, sometimes called a fishmouth joint. It is a tubular joint used in joining two pieces of tubing end to end, in which the edges are cut to resemble a fish's mouth. For pieces of equal diameter, a butt joint with the joining ends of both pieces cut in matching fishmouths is used in conjunction with an inner sleeve, whereas for pieces of unequal diameter, a reduction joint with only the end of the larger piece cut in a fishmouth is used. A **scarf joint** is a joint between two members in line with each other, in which the joining ends of one or both pieces are cut diagonally at an angle of about 30° from a centerline (scarf cut). In welding aircraft tubing, for example, scarf joints are used both as butt joints and reduction joints.

A **reduction joint** is the joint made between two members of unequal diameter or width, both members being on the same general plane—that is, not at an angle to each other. Reduction joints are used, for example, in the welding of aircraft tubing for joining tubes end to end for greater length, as in the construction of longerons, to repair defective sections of tubing, or to brace a section of a piece of tubing. When additional length is the main purpose, the end of the smaller tube is telescoped into the end of the larger tube far enough for adequate bracing. A welded joint of this nature is sometimes called a **telescope joint**. When repair or bracing of a central section is the main purpose, a short section of larger tubing is slipped over the smaller tube like a sleeve. Scarf joints and fishmouth joints are usually used in welding reduction joints on aircraft tubing, but occasionally a plain reduction joint with unshaped edges is used.

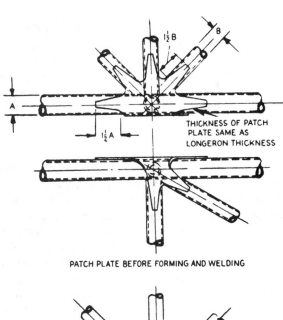

THICKNESS OF PATCH PLATE SAME AS LONGERON THICKNESS

PATCH PLATE BEFORE FORMING AND WELDING

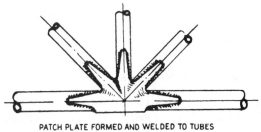

PATCH PLATE FORMED AND WELDED TO TUBES

FIGURE 7-19   Details of finger patch repair.

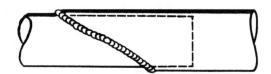

FIGURE 7-20   A scarf-butt joint.

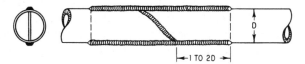

FIGURE 7-21   A scarf-butt splice with reinforcing plate.

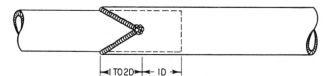

FIGURE 7-22   A fishmouth reduction splice.

Figure 7-20 shows a **scarf-butt joint**. Figure 7-21 shows a **scarf-butt splice** reinforced with a steel gusset plate. This splice is sometimes used to join the ends of the circular member of a radial engine mount. The ends of the tube are prepared with a slot to receive the gusset plate, which extends from one to two diameters on each side of the weld between the tube ends. The gusset plate should be $\frac{1}{4}$ in [6.35 mm] wider than the exterior diameter of the tube, and the cut for the scarf splice should be made at an angle of 30°.

Figure 7-22 shows a **fishmouth reduction splice** having the end of the telescoping tube cut to give a fishmouth shape. The length of the cut on the outside measurement of the tube is from one to two diameters of the smaller tube. This joint has a greater length of welded seam than a butt or scarf splice and does away with heating the tube to a welding temperature in a direct cross section. It is used for splicing continuous members of steel-tube fuselages and members of other units where tube splices of different diameters are required by the construction.

Figure 7-23 shows a **scarf reduction splice** having the end of the telescoping tube cut diagonally at an angle of 30°. This joint is used for splicing members of different diameters and resembles the fishmouth splice to the extent that the tube is not heated to a welding temperature in a direct cross section.

The fishmouth weld is stronger than the scarf (diagonal) joint because of its resistance to bending stresses. There is no single straight line of weld through the structure where

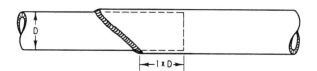

FIGURE 7-23   A scarf reduction splice.

a fishmouth weld is used: hence a straight-line break cannot occur if the part is subjected to vibration or shock. However, in some aircraft repairs it will be found that the location of the damage and its extent are such that a diagonal type of weld must be used to the exclusion of the fishmouth type.

The following precautions must be observed in splicing:

1. A cut for splicing purposes must not be made in the middle third of a section of tubing because aircraft tubing must withstand high bending stresses.

2. Only one partial replacement tube can be inserted in any one section of a structural member, because more than one would weaken the member too much. If more than one tube in a joint is damaged, the entire joint must be removed and a new, preassembled, welded joint of the correct design must be inserted.

3. If a web member is damaged at a joint so badly that it is not possible to retain, at that location, a stub long enough to permit the splicing of a replacement, an entirely new web member must be installed.

4. If a continuous longeron is damaged at a joint, the replacement-tube splices must be at locations far enough past the joint on each side to avoid the necessity of locating the splice weld too close to the joint weld. The reason is that a welded joint is weaker than the metal it joins; hence the placing of a weld close to another weld increases the already existing weakness. The correct procedure is first to cut loose the web member at the damaged joint and to remove the damaged section of longeron tubing. The replacement tubing is then spliced to the stub ends of the original longeron section. Finally, the web member is welded to the new section of longeron tubing.

5. Wooden braces are used to keep the tubes in alignment while repairs are being made. The bent or damaged tubes having been replaced by new ones, it is then important to examine the original alignment of the corresponding tubes on an undamaged airplane of the same make and model to be sure that no error exists.

Rosette Welds

A rosette weld is one of the types of welds classified as **plug welds**. A plug weld is a weld holding two lapped pieces. It is laid as a plug in a hole (slot) cut through the top piece or cut through both pieces. A rosette weld is a round plug weld. Rosette welds are generally used to fuse an inner reinforcing tube (liner) with the outer member. Where a rosette weld is used, the holes should be made in the outside tube only and be of a sufficient size to ensure fusion of the inner tube, A hole diameter of about one-quarter the tube diameter of the outer tube is adequate. If the sleeves or inner liners fit tight, the rosettes may be omitted.

Inner-Sleeve Repairs

Figure 7-24 shows a partial replacement tube spliced to the original tubing by means of **inner reinforcing sleeves**. There is a very small amount of welding to be done; hence there is little possibility of weakening or distorting the tubing.

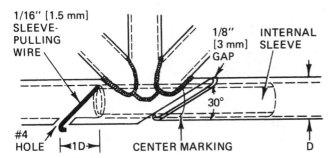

FIGURE 7-24 Splice with inner reinforcement sleeve.

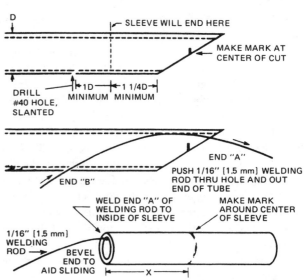

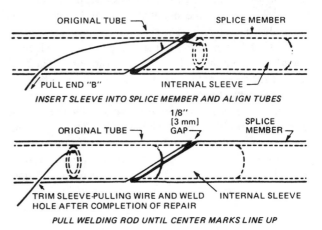

CUT AND MARK SLEEVE AND WELD ON END OF SLEEVE-PULLING WIRE

INSERT SLEEVE INTO SPLICE MEMBER AND ALIGN TUBES

PULL WELDING ROD UNTIL CENTER MARKS LINE UP

FIGURE 7-25 Details of construction for inner reinforcing sleeve repair.

In addition, a smooth outer surface is obtained for the repaired section.

Diagonal cuts are made in the damaged tubing to remove the injured portion. The cuts are located away from the middle third of the damaged section. When the part has been removed, any burr or roughness is removed from the edges of the cuts by filing. A replacement tube that matches the damaged original tubing in both wall thickness and diameter is obtained; from this tubing, a length $\frac{1}{4}$ in [6.35 mm] less than that of the removed section is cut by means of diagonal cuts on the ends. This gives a $\frac{1}{8}$-in [3.18-mm] gap between the original tubing and each end of the replacement tubing.

The next step is to cut two reinforcing sleeves. Tubing for this purpose must have the same wall thickness as the original tubing and an outside diameter equal to the inside diameter of the original tubing. The sleeves are cut across the tubing and not diagonally. Each sleeve must fit snugly inside the original tubing, leaving a clearance between sleeve and tubing of not more than $\frac{1}{16}$ in [1.59 mm]. These inner sleeves are cut long enough so that each end of a sleeve is not less than $1\frac{1}{2}$ tube diameters from the diagonal cuts in the original tubing and the replacement tubing, as shown in Fig. 7-25.

The two reinforcing sleeves having been cut, the next task is to start splicing. The sequence given next should be followed to do a good job:

1. Set up a fixture or brace arrangement to support the structure while the welding is done. Figure 7-26 shows how a brace takes the place of the damaged tubing in holding the vertical members in line during welding.

2. On the stub end of the original tube, halfway along the diagonal cut, make a small mark on the outside.

3. Measure a distance $2\frac{1}{2}$ tube diameters long, starting at the nearest end of each diagonal cut on the original tubing. Center-punch the tube at these locations, and begin drilling holes, no. 40 size, using a drill held at a 90° angle to the surface of the tubing. When the hole is started far enough to keep the drill from jumping out, slant the drill toward the cut and continue drilling at a 30° angle. Then, file off the burr from the edges of the holes, using a round needlepoint file.

4. Insert one end of a length of $\frac{1}{16}$-in [1.59-mm] welding wire through the hole just drilled and push it out through the diagonally cut-open end of the original tubing. Repeat the process at the other stub end using another wire. These wires are used to draw the sleeves into the tubing.

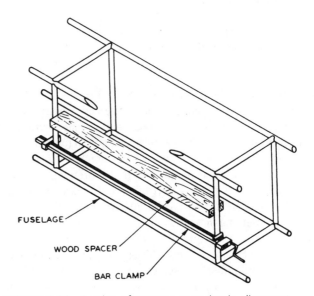

FIGURE 7-26 Bracing of structure to maintain alignment.

5. Weld the end of each wire that sticks out of the open end of the tubing to the inside of one of the inner sleeves, as shown in Fig. 7-25. The sleeve can be drawn into the tube more easily if the ends of the sleeves to which the wires are welded are first beveled.

6. Using thin paint, metal dye, or emery paper, place a narrow mark around the centers of the reinforcing sleeves.

7. Push the inner sleeve into the replacement tube so that the place where the wire is welded to the sleeve is 180° from the drilled hole. If the drilled hole is at the bottom of the tubing, the inner sleeves are located so that the place at which the wire is welded is at the top. If the inner sleeve fits too snugly in the replacement tube, cool the sleeve with dry ice or cold water. If the sleeve continues to stick, polish it with emery cloth.

8. Line up the stub ends of the original tube with the replacement tube.

9. Begin to pull the end of the wire that protrudes from the drilled hole. Pull the sleeve along until the center mark on the sleeve is directly in line with the center mark on the diagonal cut. When these two marks are in line, the sleeve is properly centered under the joint, as illustrated in Fig. 7-25. Repeat the procedure for the other sleeve at the opposite end of the replacement tube.

10. Bend the pulling wire over the edge of the hole to hold the sleeve in position. Weld the inner sleeve to the original tube stub and replacement tube at one end. This takes up the gap between the replacement tube and the original tube at one end, as shown in Fig. 7-25. A weld bead must be formed over the gap. After the joint is welded, snip the pulling wire off flush with the surface of the tube. Weld over the drilled hole.

11. When this weld over the drilled hole is cool, adjust the brace arrangement to provide for contraction and shrinkage. Having adjusted the brace, pull the sleeve into position and tack-weld the gap at the other end of the replacement tube. This holds the joint in alignment. Remove the brace to eliminate any restraint on the contraction forces at the joint. Complete the weld around the gap that has been tack-welded.

12. Weld over the drilled hole, clean the welds, and apply the appropriate protective coatings.

Apply a similar method to any splicing described elsewhere, except that sleeve-pulling wires are not needed if external reinforcing sleeves are used or where sleeves are omitted and a simple replacement tube of a larger diameter is used for splicing.

Figure 7-27 illustrates a typical longeron splice using the same diameter replacement tube. In this drawing, A is the original tube, B is the insert tube, and C is one-quarter the diameter of tube A (six rosette welds for each splice on the drilled outside tube only). It is shown that a partial replacement tube is spliced to the original tubing by means of inner reinforcing sleeves. This is a very efficient method. There is a minimum amount of welding to be done; hence there is little danger of weakening and distorting the tube from the heat of welding. In addition, this method has the advantage of presenting a smooth outside surface for the repaired section.

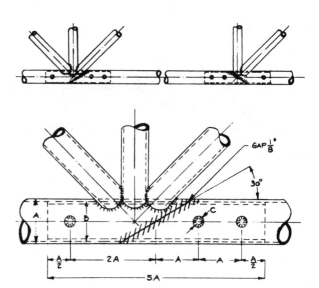

FIGURE 7-27 Longeron splice using tubing of same diameter as original.

Splicing Using Larger-Diameter Replacement Tubes

An excellent method for repairing a damaged section of tubing between stations is illustrated in Fig. 7-28. In this method a replacement tube large enough to telescope over the original tubing is used. The replacement tubing should fit snugly over the original tube. This method can be used only where there is sufficient undamaged tubing at each end of the section to allow for the necessary stub lengths. On one end a minimum of $2\frac{1}{2}$ tube diameters of length should be allowed; on the other end, a minimum of $4\frac{1}{2}$ diameters of tube length is necessary. The original tubing is cut square across at each end, as shown in the drawing.

The replacement tube must have the same wall thickness as the original tube and should have an inside diameter just large enough to slip over the stub ends of the original; in any case, it should not be more than $\frac{1}{16}$ in [1.59 mm] larger than the outside diameter of the original tube. A fishmouth end should be cut on each end of the replacement tube with an angle of 30° from the centerline. The tube should be of such a length that each end will extend a minimum distance of $1\frac{1}{2}$ tube diameters over the stub at each end.

To install the replacement tube, the long stub of the original tube is sprung sufficiently to permit the replacement tube to be slipped over the stub. The other end of the replacement tube is then aligned with the short stub and pushed over it to a distance of at least $1\frac{1}{2}$ tube diameters. When the replacement

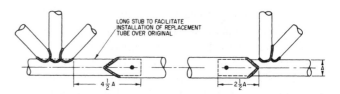

FIGURE 7-28 Replacement of tubing between stations with larger tube over stubs of original.

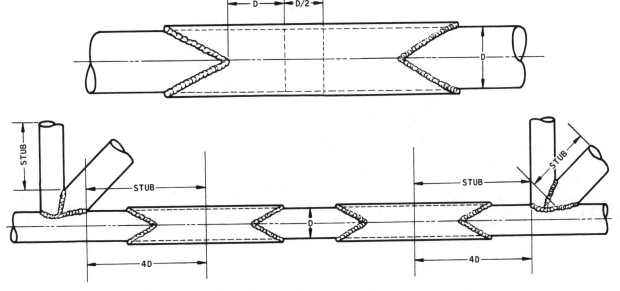

FIGURE 7-29 Fishmouth sleeves for replacement tube between stations.

tube is in the correct position, the edge of the fishmouth is welded to the original tube.

Outside-Sleeve Repairs

There are three principal methods of repairing a damaged tube, as suggested before. One is the use of an inner-sleeve splice, another is the use of a larger-diameter replacement tube, and the third is the use of a partial replacement tube of the same diameter as the original, reinforced by an **outside sleeve.** The latter method requires more welding than the other two, so there is greater danger of distortion from the heat of welding; hence it is usually the least desirable repair method. However, if neither of the other two methods can be used for any reason, this method must be employed.

Figure 7-29 illustrates a satisfactory method for repairing tubing in a bay with outside-sleeve reinforcements and an outer sleeve. The fishmouth sleeve is shown; however, a 30° scarf sleeve is also satisfactory. Remember that the fishmouth slope is cut 30° from horizontal.

To make the repair shown, the damaged portion of tubing is cut out by sawing straight across the tubing at a distance of four times the tube diameter from the adjacent welded fittings or tubes. Before the tube is removed, it is wise to brace the fuselage section to make sure that it remains in correct alignment.

A new section of tube, having the same diameter and wall thickness as the original, is cut at such a length that there is a gap of about $\frac{1}{32}$ in [0.79 mm] when it is installed to allow for expansion. The inside diameter of the outer reinforcing sleeves should be such that the sleeves will fit snugly over the other tubes, and in any case the maximum diameter difference between the inner tubing and the sleeve should be not more than $\frac{1}{16}$ in [1.59 mm].

The outer sleeves are cut with fishmouth ends or diagonally, preferably fishmouth, the length being such that the nearest portion of the sleeve end cut is at least $1\frac{1}{2}$ tube

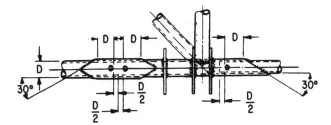

FIGURE 7-30 Fishmouth sleeve repair for splice at a station with fittings.

diameters from the ends of the original tube. The two sleeves are slipped over the replacement tube, and the tube is then aligned with the stubs of the original tubing. The sleeves are slipped out over the centers of the joints and adjusted to provide the maximum reinforcement. They are then tacked in place by welding. To reduce warpage, one sleeve is welded in place and allowed to cool before the other is welded. Rosette welds should be employed for further reinforcement unless the sleeves fit tightly.

Figure 7-30 shows the use of a tube of the same diameter as the original with fittings involved. In this method, the new section of tube is the same size as the longeron forward (left) of the fitting. The rear end (right) of the tube is cut at 30°, and it forms the outside sleeve of a scarf splice. A sleeve is centered over the forward joint as illustrated.

Replacement of a Cluster at a Station

Figure 7-31 illustrates a suitable method for replacing a damaged cluster weld by means of welded outer sleeves. The structure is braced to hold it in correct alignment, and then the original cluster is sawed out, as shown by the dotted lines. The longeron section is cut squarely across, and a new section of tubing with the same wall thickness and diameter is cut to fit the space where the old tube was removed.

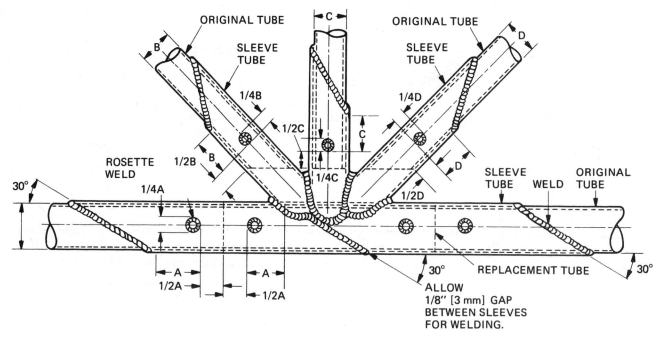

FIGURE 7-31　Replacement of a damaged cluster weld at a station.

Two sleeve sections with dimensions as shown are cut to fit over the joints and are welded together at the center and at the outer ends with 30° diagonal welds. Rosette welds are also employed to reinforce the assembly. Sleeves are also cut to fit the truss members, and these sleeves should be slipped over the ends of the truss members before the longeron weld is made. Otherwise, assembly will be impossible.

Finally the truss member sleeves are welded into place, first at the cluster and then to the original tubes. It is important to remember that one weld should be made at a time and allowed to cool before making the next in order to reduce warping.

Repair at Built-in Fuselage Fittings

When a welded structure has steel fittings built in at a cluster, the repair is somewhat more complicated than those previously described, but the same principles apply. Figure 7-32 shows a tube (sleeve) of larger diameter than the original. This makes it necessary to ream the fitting holes (at the longeron) to a larger diameter. The sleeve should extend about 6 in [15 cm] forward (left of fitting) of the joint and 8 in [20 cm] aft (right of fitting). The forward splice should be a 30° scarf splice. The rear longeron (right) should be cut off about 4 in [10 cm] from the centerline of the joint and

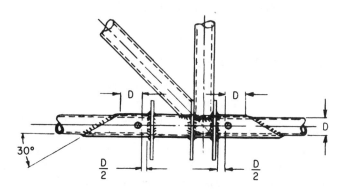

FIGURE 7-33　A simple sleeve repair at a station with fittings.

a spacer 1 in [2.54 cm] long fitted over the longeron. This spacer and the longeron should be edge-welded. A tapered-V cut about 2 in [5 cm] long is then made in the rear end of the outer sleeve. The end of the outer sleeve should be swaged to fit the longeron and then welded.

When the sections of longeron tubing on each side of the station and fittings are the same size, a simpler type of repair is possible, as shown in Fig. 7-33. The sleeve is cut and welded in place as shown, and then the fittings are welded in the correct positions.

For the condition where there is a larger difference in diameter of the longeron on each side of the fitting, the repair shown in Fig. 7-34 can be used. In this repair, a section of tubing forms an inner sleeve for the section on the left of the fitting and an outer sleeve for the section on the right of the fitting. Standard welding procedures are employed with dimensions as shown.

Repair of Streamline Tubing

Although streamline tubing can be repaired in many ways using standard welding practices, there are four methods that

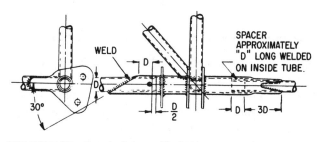

FIGURE 7-32　Repair sleeve with fittings using repair sleeve of larger diameter than original.

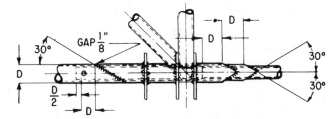

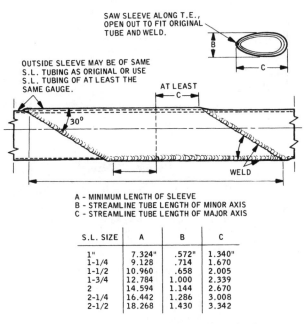

FIGURE 7-34 Repair where there is a large difference in diameter of the longeron on each side of the station.

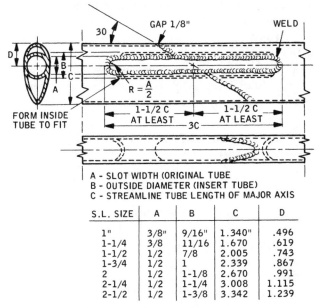

FIGURE 7-35 A streamline tube splice using an inner round tube.

A - SLOT WIDTH (ORIGINAL TUBE
B - OUTSIDE DIAMETER (INSERT TUBE)
C - STREAMLINE TUBE LENGTH OF MAJOR AXIS

S.L. SIZE	A	B	C	D
1"	3/8"	9/16"	1.340"	.496
1-1/4	3/8	11/16	1.670	.619
1-1/2	1/2	7/8	2.005	.743
1-3/4	1/2	1	2.339	.867
2	1/2	1-1/8	2.670	.991
2-1/4	1/2	1-1/4	3.008	1.115
2-1/2	1/2	1-3/8	3.342	1.239

A - MINIMUM LENGTH OF SLEEVE
B - STREAMLINE TUBE LENGTH OF MINOR AXIS
C - STREAMLINE TUBE LENGTH OF MAJOR AXIS

S.L. SIZE	A	B	C
1"	7.324"	.572"	1.340"
1-1/4	9.128	.714	1.670
1-1/2	10.960	.658	2.005
1-3/4	12.784	1.000	2.339
2	14.594	1.144	2.670
2-1/4	16.442	1.286	3.008
2-1/2	18.268	1.430	3.342

FIGURE 7-36 Streamline tube splice using a split sleeve.

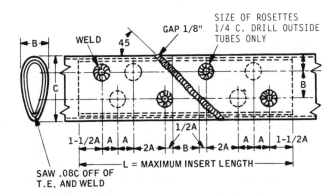

A IS $\frac{2}{3}$ B

B IS MINOR AXIS LENGTH OF ORIGINAL STREAMLINE TUBE
C IS MAJOR AXIS LENGTH OF ORIGINAL STREAMLINE TUBE

S.L. SIZE	A	B	C	L
1"	.382	.572	1.340	5.16
1-1/4	.476	.714	1.670	6.43
1-1/2	.572	.858	2.005	7.72
1-3/4	.667	1.000	2.339	9.00
2	.763	1.144	2.670	10.30
2-1/4	.858	1.286	3.008	11.58
2-1/2	.954	1.430	3.342	12.88

FIGURE 7-37 Streamline tube splice using an inside streamline tube reinforcement.

have been proven reliable and are accepted by the Federal Aviation Administration for certificated aircraft.

The method illustrated in Fig. 7-35, recommended for landing gear, utilizes a section of round tubing that is one gauge thicker than the original tube to serve as the splicing reinforcement. The streamline tubes to be spliced are cut with a 30° scarf and the ends to be joined are slotted according to the approved specifications shown in the table. The combined slots are three times the length of the major axis of the streamline tubing. A gap of $\frac{1}{8}$ in [3.18 mm] is established between the scarfed ends of the streamline tubes to provide for full penetration of the weld. The round tubing section is fitted in the slots as shown and welded in place.

Figure 7-36 illustrates the accepted splice for streamline tubing where a split sleeve is welded over the original tube to reinforce the splice. This repair is applicable to wing-and-tail surface brace struts and similar members. In this method the original tubing is cut squarely across and the split sleeve is scarfed at 30°. The outer sleeve is cut along the trailing edge and then opened up to fit over the original tubing. It is then welded at the trailing edge to the inside tubing and along the end scarfs. The dimensions are given in the table.

The splice shown in Fig. 7-37 utilizes an inside sleeve of streamline tubing as a reinforcement. This tubing is the same as the original; however, the size is reduced by cutting off the

trailing edge and rewelding. The reinforcing tube is inserted in the ends of the original tubing, which has been cut with a 45° scarf. A gap of $\frac{1}{8}$ in [3.18 mm] is allowed for welding the two ends together and to the inner sleeve. Rosette welds are used as shown to secure the inner sleeve.

Figure 7-38 illustrates the splicing method where steel plates are used to reinforce the splice. The steel plates are twice the thickness of the tubing wall. Slots are cut in the streamline tubing according to the dimensions given in the table and the plates are welded into the slots. The ends of the streamline tubing are scarfed at 30° and welded together. A gap of $\frac{1}{8}$ in [3.18 mm] is allowed to facilitate the weld.

Aircraft Tubing Repair **193**

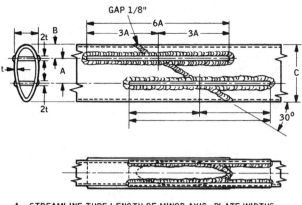

A - STREAMLINE TUBE LENGTH OF MINOR AXIS, PLATE WIDTHS.
B - DISTANCE OF FIRST PLATE FROM LEADING EDGE, 2/3A.
C - STREAMLINE TUBE LENGTH OF MAJOR AXIS.

S.L. SIZE	A	B	C	6A
1"	.572	.382	1.340	3.43
1-1/4	.714	.476	1.670	4.28
1-1/2	.858	.572	2.005	5.15
1-3/4	1.000	.667	2.339	6.00
2	1.144	.762	2.670	6.86
2-1/4	1.286	.858	3.008	7.72
2-1/2	1.430	.954	3.342	8.58

FIGURE 7-38 Streamline tube splice using steel-plate inserts.

Repair of Engine Mounts

Welded engine mounts are particularly vulnerable to damage by cracking, largely because of the vibration to which they are subjected. Landing loads on such mounts are also particularly severe because the weight effect of the engine is multiplied several times during a hard landing. For these reasons the materials used in the manufacture of engine mounts should be of high quality, and the welded joints must be uniform and strong. Engine-mount repairs are accomplished in the same manner as other tubular-steel repairs; however, replacement tubes should be large enough to slip over the original tubing. A fishmouth joint is recommended, although a 30° scarf may be used if necessary. Rosette welds are used to ensure a completely rigid joint.

Repairs to engine mounts must be governed by accurate means of checking the alignment. When new tubes are used to replace bent or damaged tubes, the original alignment of the structure must be maintained. This is done by measuring the distance between points of corresponding members that have not been distorted and by reference to the drawings furnished by the manufacturer.

If all members are out of alignment, the engine mount must be replaced by one supplied by the manufacturer; the method of checking the alignment of the fuselage or nacelle points should be requested from the manufacturer.

Minor damage, such as a crack adjacent to an engine attachment lug, may be repaired by rewelding the ring and extending a gusset or a mounting lug past the damaged area. Engine-mount rings that have been extensively damaged should not be repaired but should be replaced unless the method of repair is approved by an authorized agency.

Wing- and Tail-Surface Repairs

Built-up tubular wing- or tail-surface spars may be repaired by using any of the standard splices and methods of repair described in this chapter if the spars are not heat-treated. If they are heat-treated, the entire spar assembly must be heat-treated again to the manufacturer's specifications after the repair is completed. This is usually less practical than replacing the spar with one furnished by the manufacturer of the airplane.

Damaged wing-brace struts made from either round or streamlined tubing are usually replaced by new members bought from the original manufacturer. There is no objection from an airworthiness viewpoint to repairing such members in any correct manner. Brace struts must be spliced only adjacent to the end fittings. When making repairs to wing- and tail-surface brace members, particular attention must be paid to the proper fit and alignment.

Welded Repair of Fabric-Covered Steel Fuselage

When it is necessary to make a repair of a fabric-covered, steel-tubing fuselage without the removal of large sections of fabric, care must be taken to see that the fabric is not damaged by heat or set afire. Fabric covering is often very flammable; hence it should not be exposed to the sparks from welding, and heat conducted through the tubing should be stopped before it reaches the fabric.

To protect the fabric, it is cut and laid back away from the area to be welded. This usually requires a cut along the longeron and perpendicular to the longeron on each side. When the fabric is rolled back, it should not be creased or folded, because this will crack the dope and possibly weaken the fabric. Wet cloths are then wrapped around the tubing on all sides of the area to be repaired. These cloths prevent the heat from being conducted to the fabric. Additional wet cloth is placed over the fabric where sparks from the welding process could reach to the fabric.

After the welding is completed and the tubing is refinished, the fabric is sewed, reinforced, and doped, as has been described earlier in this text.

Certain aircraft components require special handling when being repaired by welding. This special handling may be necessary for proper restoration of component strength or for the safety of the technician when performing the repair operation.

Steel Parts That Must Not Be Welded

Airplane parts that depend for their proper functioning on strength properties developed by cold working must not be welded. These parts include (1) streamlined wires and cables; (2) brazed or soldered parts (the brazing mixture or solder will penetrate hot steel and weaken it); and (3) steel

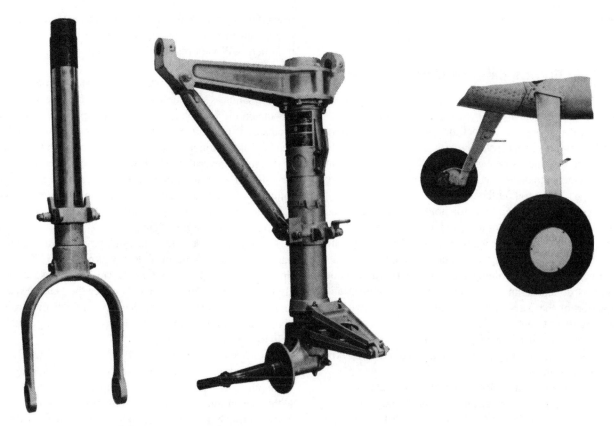

FIGURE 7-39   Landing-gear assemblies that cannot be repaired by welding.

parts, mostly of nickel alloy steels, that have been heat-treated to improve their physical properties, particularly aircraft bolts, tumbuckle ends, axles, and other heat-treated steel alloy parts.

Figure 7-39 shows landing-gear assemblies that are not generally regarded as repairable for the following reasons: (1) The axle stub is usually made from a highly heat-treated nickel-alloy steel and carefully machined to close tolerances; hence such stubs are generally replaceable and should be replaced if damaged; (2) the oleo portion of the structure is usually heat-treated after welding and is precisely machined to make certain that the shock absorber will function properly. These parts would be distorted by welding after machining.

The spring steel strut, also shown in Fig. 7-39 and installed on a number of Cessna airplanes, cannot be welded without severely reducing its strength. If a strut of this type is damaged or found to be cracked, it should be replaced with an airworthy part.

Repairable Axle Assemblies

Figure 7-40 shows three types of landing-gear and axle assemblies formed from steel tubing. These may be repaired by any standard method shown in the previous sections of this chapter. However, it is always necessary to find out whether or not the members are heat-treated. Remember that all members that depend on heat treatment for their original physical properties should be heat-treated again after the welding operation. If it happens that the heat-treat value cannot be obtained

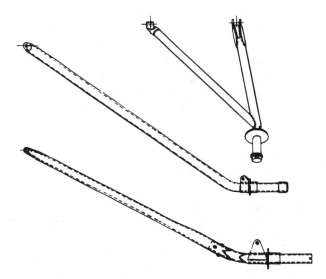

FIGURE 7-40   Three types of landing-gear and axle assemblies that can be repaired by welding.

from the manufacturer, some test must be made, such as the Brinell or Rockwell hardness test; such tests must be applied in several places of the member being tested. Welding must be done with a rod that can be heat-treated.

Damaged ski pedestals that are constructed of welded tubular steel can be repaired according to the methods described previously in this chapter. Care should be taken to note whether the pedestals are heat-treated; if they are, they should be welded with a heat-treatable rod and heat-treated again after the repair is made.

Repair of Engine Exhaust Units

Exhaust manifolds for reciprocating engines and exhaust cones and noise suppressors for turbine engines are usually constructed of stainless steel, Inconel, or other high-temperature alloy. Welding of cracks in these parts is most satisfactorily accomplished by means of inert-gas welding. It is important for the technician to determine what the alloy is and to use the correct type of filler rod for the weld.

A strong fixture is required for holding the parts of an exhaust manifold in position during construction and for maintaining the correct alignment during assembly and repair. The typical unit may have a heavy bedplate with fittings for locating and holding the parts in position while welding is done, making allowance for the expansion and contraction of the parts in a longitudinal direction without permitting them to get out of line.

The repair of exhaust manifolds usually consists of welding cracks or breaks, patching worn parts, and replacing broken or worn fittings.

Before repairing a crack or break, the metal on the inside and outside must be clean and bright, and there must be a $\frac{1}{16}$- to $\frac{3}{32}$-in [1.59- to 2.38-mm] hole drilled at each end of the crack to prevent it from spreading. If the crack is more than 2 in [5.08 cm] long, the break is tacked at 2-in intervals. When gas welding is used for the repair of exhaust units, distortion and cracking on cooling are reduced by heating the part to between 400 and 600°F [204 and 316°C]. In repairing stainless steel exhaust units, if much welding has been done, the units are preheated to between 1900 and 2000°F [1038 and 1093°C]; they are then cooled slowly and evenly to relieve the residual stresses caused by welding and to restore the metal to its original condition.

When a damaged fitting is replaced, the old fitting, including the weld-metal deposit, must be ground off and the section cleaned inside before the new fitting is made and welded in place.

It is often necessary to repair sections of the exhaust manifold that telescope together because they are subjected to considerable wear. The worn portion is removed and replaced with a new section made of sheet stock of the same gauge and kind of metal as the original portion. In gas welding, a heavy coat of flux is applied to the edges on the lower side to prevent oxidation of the weld metal and the base metal.

Repair of Fuel and Oil Tanks

When fuel and oil tanks are repaired, either by inert-gas or oxy-fuel welding, it is most important that the tank be thoroughly purged of any fuel or oil fumes before the welding is begun. Even if a fuel tank has been empty for a long time and appears to be dry inside, it is still likely that enough fumes can be released by the heat of welding to create an explosive mixture. The correct procedure in such cases is to wash the tank with hot water and a detergent and then allow live steam to flow through the tank for about 3 min. This treatment will vaporize and remove any residual fuel or oil that may be in the tank.

The welding of tanks, whether they be constructed of aluminum, stainless steel, or titanium, should be done by the inert-gas method if possible. As previously explained, this method produces a smooth weld with a minimum of distortion due to heat and also eliminates oxidation and burning of the metal.

There are several types of welded seams used in the construction of aluminum tanks for aircraft. These include butt joints, corner joints, edge joints, and lap joints. When baffle plates are riveted to the shell, the rivets are headed by welding to make them liquid tight.

Corrugated beads, formed into the sheet, adjacent and parallel to the seams, provide for expansion and contraction. These beads, acting as expansion joints, close up slightly when the metal expands under the heat of welding and straighten out on cooling enough to relieve the strains of contraction. In addition, they add stiffness to the metal and help to prevent buckling.

Figure 7-41 shows a riveted joint between the shell and baffle plate with a rivet headed to the shell by welding.

Figure 7-42 shows a joint used to secure the baffle plates to the shell in some types of tank construction. The plates either extend through a slot and are welded to the shell, or they are located at the seams in the shell and are welded to the shell when the seams are made.

Figure 7-43 shows a mechanical lock seam sealed by welding in the construction of some types of shells.

Figure 7-44 shows a joint adaptable for small seams. The turned edges provide stiffness to the sheet at the joint, tend

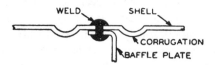

FIGURE 7-41 Welded joint with rivets to secure baffle plates.

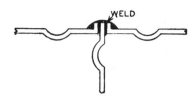

FIGURE 7-42 Welded joint to secure baffle plate in a tank.

FIGURE 7-43 A welded lock seam.

FIGURE 7-44 Welded seam with turned-up edges.

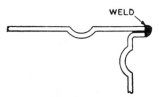

FIGURE 7-45   Corner-seam edge joint.

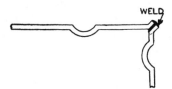

FIGURE 7-46   Flanged corner-seam joint.

FIGURE 7-47   A butt-type joint with stiffener strip.

to maintain alignment during welding, and supply additional metal for the weld.

Figure 7-45 shows how edge joints may be used for corner seams in making a tank and for welding hand-hole cover plates into the shell.

Figure 7-46 shows a corner seam widely used in tank construction. It is welded like a flanged butt joint.

Figure 7-47 shows a butt-type joint with a stiffener strip between the butt ends that is extended above the plates. Fillet welds are used to secure the whole assembly.

If a tank is repaired by gas welding, it is extremely important to remove all welding flux after a repair in order to prevent corrosion. As soon as welding is completed, a tank is washed inside and outside with great quantities of hot water and is then drained. It is then either immersed in 5 percent nitric or sulfuric acid or filled with this solution and also washed on the outside with the same solution. The acid is left in contact with the weld about 1 h and then rinsed carefully with clean, fresh water. The efficiency of the cleaning process is tested by applying some acidified 5 percent silver nitrate solution to a small quantity of rinse water that has been used for the last washing of the weld. If a white precipitate is formed, the cleaning has not been done thoroughly, and the washing must be repeated.

Cowlings

Aluminum and aluminum-alloy **cowlings** are usually repaired by patching worn spots; by welding cracks, or breaks, caused by vibration; and by replacing worn or broken fittings.

A crack, or break, is repaired by welding the fracture. Inert-gas welding should be used if possible, but a skilled operator can make suitable repairs by gas welding. All welds must have enough penetration to provide a small bead on the lower side. In repairing cracks that extend into the sheet from a hole in the cowl or from the edge, the cracks are lined

up, tacked, and welded progressively, starting from the end of the crack toward the edge or opening. A cowling having been repaired by welding must be finished on the outside to provide a smooth surface.

Worn spots may be repaired by removing the worn section and replacing it with a new piece that is cut to fit the removed portion and then formed to the required shape before installation. It is then placed in position and tack-welded at intervals of 1 to 2 in [2.5 to 5 cm] around the edges before the actual welding is performed.

SOLDERING AND BRAZING

In general, soldering may be described as either *soft* or *hard*, depending upon the type of material used for the soldering bond, which determines the temperature required in the process. The difference between soft soldering and brazing, or hard soldering, is in the temperature. By definition, if the filler metal has a melting point of more than 800°F [426.67°C], the process is called **brazing**, or **hard soldering**.

There are a variety of techniques used to apply solder to joints. These include the small-quantity and maintenance techniques of iron soldering and torch soldering, which are discussed later in this chapter, and the mass-production techniques of dip soldering, where the joint is lowered into a pot of molten solder; resistance soldering, where heat is produced by passing electric current through the joint materials; induction soldering, where the passing of a magnetic field over the joint materials produces an electric current; and furnace soldering, where the units to be soldered are passed through an electric or gas furnace. In all cases, properly designed and soldered joints use capillary action to draw the solder material into the joint.

Because the soldering processes depend upon capillary action to draw the solder into the joint, soldered joints require very small clearances. One common fault in maintenance soldering is to attempt to fill an area with solder without first minimizing the distance between the mating materials.

The solder is available in a variety of forms, such as wire, rod, bar, cakes, ribbon, foil, and powdered. For dip-soldering processes, solder is also available in pigs, ingots, and slabs.

Iron soldering requires the use of a tool with an electrically or flame-heated head or tip. The tip is usually fabricated from copper. This tool is called a soldering iron or soldering copper. The soldering copper must be clean and well tinned and the correct flux must be applied.

Soldering coppers are manufactured in many sizes and shapes. Figure 7-48 shows four common shapes used in commercial sheet-metal work.

If electric power is available, it is common practice to employ electric soldering irons because of their convenience. The electric tool contains a resistance heating element inside the hollow copper head. An electric soldering iron is shown in Fig. 7-49.

As explained previously, a soldering copper (either flame-heated or electrically heated) must be tinned before it can be used. If the copper is not tinned, the solder will not adhere to it and the heat will not be conducted in sufficient quantity to melt

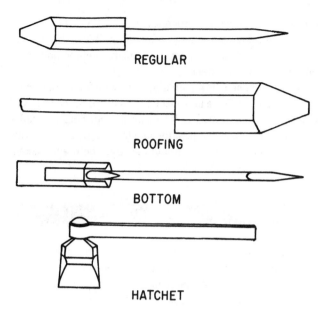

FIGURE 7-48   Common soldering coppers for commercial work.

FIGURE 7-49   An electric soldering iron.

the solder and heat the joint. To tin the copper, the tip is first filed clean and in the correct shape. After being filed to shape, the copper is heated and then filed again slightly to remove any oxide that has formed. When bright and clean, the tip is dipped into a flux before solder is applied. If rosin-core solder is used, it is not necessary to dip the tip in flux. When the entire point is covered with melted solder, the copper is ready for use.

The use of soldering irons is limited to soft soldering. Torches may be used for either soft or hard soldering. The flame of the torch should be either a neutral flame or slightly carbonizing to limit the potential for oxidation.

Sweat Soldering

Sweat soldering, also called **sweating**, is a method of soldering in which the parts to be soldered are first tinned (coated with solder) and then the melted solder is drawn between the surfaces to be soldered by capillary attraction with the application of heat. Sometimes a soldering iron is used and sometimes a neutral gas flame (torch) is used to melt and flow the solder in the joint. The torch is required in hard soldering or brazing.

Flux

Fluxes are chemical compounds that assist in cleaning the surface of the joint area of oxides, prevent the formation of

oxides during the soldering process, and lower the surface tension of the solder. Lowering the surface tension of the solder improves its wetting (or sticking) properties.

Fluxes such as borax, ammonium chloride, sodium carbonate, zinc chloride, rosin, or stearic acid unite with oxides and other impurities in molten metal and float them to the surface. In soldering, the most common fluxes are acid (hydrochloric acid in which zinc has been dissolved), rosin, and paste fluxes. The use of acid fluxes is not approved for aircraft work because of the corrosive effects of such fluxes. The fluxes recommended for aircraft work are rosin and non-acid paste flux. Excess paste flux should be wiped off after the soldering is completed.

Fluxes for hard soldering (silver soldering, brazing, and eutectic welding) are (1) powdered borax, (2) powdered borax and carbonate of soda, and (3) special patented fluxes designed for specific types of hard soldering.

Soft Soldering

In **soft soldering,** the sealing and securing of a joint between two metal pieces is accomplished with solder that consists of an alloy of tin and lead. The percentages of the two metals vary according to the particular type of solder and the strength required. Common solders are referred to as 40–60, 50–50, and 60–40, the first number being the percentage of tin (Sn) and the second number being the percentage of lead (Pb). The percentages of the two metals have a substantial effect on the melting point of the solder. Table 7-1 shows the properties of solder for various compositions of tin and lead.

Special solders may contain metals other than tin—for example, silver, antimony, or bismuth. Special solders are required for best results in certain soldering processes.

In soft soldering, the melted solder is spread over the adjoining surfaces with a soldering iron (copper) or flame. The solder does not actually fuse with the metals being joined but bonds to them on the surface; that is, the base metal does not melt. Soldering produces a relatively weak joint but is satisfactory for many purposes. It is especially useful in making airtight joints of sheet metal that do not have to withstand much pull or vibration, and it is also used to seal electrical connections. Where soldering is attempted, it is necessary to employ flux such as rosin, zinc chloride, or sal ammoniac to clean the surfaces to be soldered and prevent the formation of oxides so the solder can adhere to the metal.

The metals commonly soldered are iron, tin, copper, brass, galvanized iron, and terneplate. Aluminum, stainless steel, titanium, and other metals can be soldered under the proper conditions and with the right materials. Ordinary lead-tin solder cannot be used for all soldering. Special fluxes are available, however, that make possible a broad application of such solder.

The technique for soft soldering involves clean parts, proper fit, and correct flux. Scraping, filing, or wire brushing are the usual cleaning methods. Scale, dirt, and oxides are thoroughly removed. The cleaned surfaces are then fitted in place and coated with the proper flux, and the hot soldering copper applied with the solder to the adjoining parts. A small

ASTM Solder Classification*	Composition, wt. %		Solidus		Liquidus		Melting Range	
	Tin	Lead	°C	°F	°C	°F	°C	°F
5	5	95	300	572	314	596	14	24
10	10	90	268	514	301	573	33	59
15	15	85	225	437	290	553	65	116
20	20	80	183	361	280	535	97	174
25	25	75	183	361	267	511	84	150
30	30	70	183	361	255	491	72	130
35	35	65	183	361	247	477	64	116
40	40	60	183	361	235	455	52	94
45	45	55	183	361	228	441	45	80
50	50	50	183	361	217	421	34	60
60	60	40	183	361	190	374	7	13
70	70	30	183	361	192	378	9	17

*See ASTM Specification B32, *Solder Metal.*

additional amount of flux is often necessary to remove any oxides that may form and to help the melting of the solder that is already applied. It is essential that the soldering copper be kept in contact with the work until the solder is thoroughly sweated into the joint. Poor solder joints are often the result of insufficient heat.

When soldering a seam such as that shown in Fig. 7-50, the parts to be soldered should be tacked with drops of solder unless they are otherwise fastened together. The solder is picked up with the soldering copper and transferred to the metal, where it is deposited along the seam. Flux should be used as required to cause the solder to adhere to the metal.

To finish the seam, a hot, well-tinned soldering copper is applied with the point extending over the seam on the single thickness of metal; the heel or back of the copper is over the seam itself, at an angle of about 45°. The bar of solder or wire solder is touched to the copper while in this position. As it melts, the copper is drawn slowly along the work, keeping it at an angle and permitting it to sweat the solder into the entire width of the seam.

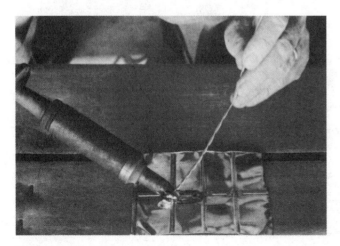

FIGURE 7-50   Soldering a seam.

Hard Soldering (Brazing)

There are seven standard classifications of commonly used filler metals for hard soldering. Each classification has a basic code followed by a dash number, which represents the AWS digit code corresponding to a particular composition. They are aluminum-silicon (BAlSi-), copper-phosphorus (BCuP-), silver (BAg-), nickel (BNi-), copper and copper-zinc (BCu- and BCuZn-), magnesium (BMg-), and precious metals (BAu- and BCo-). Table 7-2 shows the AWS chart for aluminum silicon filler materials and their uses. Note that only heat-treatable aluminum alloys may be hard soldered.

Silver soldering is the process of hard soldering most frequently used by the technician; since soft-soldered gasoline- and oil-pipe joints and similar joints fracture with repeated vibration, this type of solder is used. It is suitable for bronze parts, copper, stainless steel, and brass. Since the solder cannot be melted with a soldering copper, it is necessary to use a neutral torch flame for heating the solder and the joint. It is usually necessary to heat the parts and the solder to a red heat in order to melt the solder and cause it to flow into the joint. The correct type of flux must be used for the materials being soldered.

The process called *brazing,* in which a brass-type rod is used for "soldering" the joint, is also a form of hard soldering. Still another process is called **eutectic welding**. The term *eutectic* describes low-melting-point metals. Eutectic welding is a term of welding akin to hard soldering, in which the welding material has a lower melting point than the metal being welded; thus this type of welding can be classed as hard soldering and is similar to brazing or silver soldering.

Special Soldering Applications

The use of soft or hard soldering for the repair of aircraft parts is strictly limited. Generally these processes should not be employed for any stressed (load-bearing) part. In certain instances, silver soldering or brazing may be employed, but

TABLE 7-2 AWS Chart for Aluminum Silicon Filler Materials and Their Uses

	Al and Al Alloys	Mg and Mg Alloys	Cu and Cu Alloys	Carbon and Low-Alloy Steels	Cast Iron	Stainless Steels	Ni and Ni Alloys	Ti and Ti Alloys	Be, Zr, and Alloys (reactive metals)	W, Mo, Ta, Cb, and Alloys (refractory metals)	Tool Steels
Al and Al alloys	BAlSi										
Mg and Mg alloys	X	BMg									
Cu and Cu alloys	X	X	BAg, BAu, BCuP, RBCuZn								
Carbon and low-alloy steels	BAlSi	X	BAg, BAu, RBCuZn	BAg, BAu, BCu, RBCuZn, BNi							
Cast iron	X	X	BAg, BAu, RBCuZn	BAg, RBCuZn	BAg, RBCuZn, BNi						
Stainless steel	BAlSi	X	BAg, BAu	BAg, BAu, BCu, BNi	BAg, BAu, BCu, BNi	BAg, BAu, BCu, BNi					
Ni and Ni alloys	X	X	BAg, BAu, RBCuZn	BAg, BAu, BCu, RBCuZn, BNi	BAg, BCu, RBCuZn	BAg, BAu, BCu, BNi	BAg, BAu, BCu, BNi				
Ti and Ti alloys	BAlSi	X	BAg	BAg	BAg	BAg	BAg	Y			
Be, Zr, and alloys (reactive metals)	X BAlSi (Be)	X	BAg	BAg, BNi*	BAg, BNi*	BAg, BNi*	BAg, BNi*	Y	Y		
W, Mo, Ta, Cb and alloys (refractory metals)	X	X	BAg	BAg, BCu, BNi*	BAg, BCu, BNi*	BAg, BCu, BNi*	BAg, BCu, BNi*	Y	Y	Y	
Total steels	X	X	BAg, BAu, RBCuZn, BNi	BAg, BAu, BCu, RBCuZn, BNi	BAg, BAu, RBCuZn, BNi	BAg, BAu, BCu, BNi	BAg, BAu, BCu, RBCuZn, BNi	X	X	X	BAg, BAu, BCu, RBCuZn, BNi

Note: Refer to AWS Specification 5.8 for information on the specific compositions within each classification.

X—Not recommended; however, special techniques may be practicable for certain dissimilar metal combinations.

Y—Generalizations on these combinations cannot be made. Refer to chapters 20 and 21 for usable filler metals.

*—Special brazing filler metals are available and are used successfully for specific metal combinations.

Filler metals:

BAlSi—Aluminum silicon   BCuP—Copper phosphorus
BAg—Silver base   RBCuZn—Copper zinc
BAu—Gold base   BMg—Magnesium base
BCu—Copper   BNi—Nickel base

the repair must be approved and passed by proper government authority if it is used on a certificated aircraft.

Aluminum is difficult to solder because of its high thermal conductivity and the ease with which aluminum oxidizes. The preferred temperature range of aluminum soldering is 550°F [288°C] to 700°F [371°C].

There are two methods of limiting oxidation during the soldering process, either flux application or abrasion. The abrasion method is rather crude but effective. In the abrasive method the oxides are removed from the aluminum by scraping it, since the soldering processes is accomplished either by wire brushes or wire wheels, or scratching the surface with the copper tip. When using the abrasive technique, the solder must be applied immediately to limit oxidation of the base metals.

When fluxes are used, they must be liquid at the soldering temperature so that the flux and its impurities may be displaced by the molten filler material. Aluminum alloys used as filler material for soldering aluminum usually contain between 0.50 and 0.75 percent tin, with the remainder zinc.

Magnesium soldering is usually done only to correct surface defects and should not be used to carry any type of structural loading. Filler materials for magnesium should be selected with care, since fluxes that typically improve the wetting action during the process are not recommended when soldering magnesium. The most common filler materials for magnesium are 60 percent cadmium, 30 percent zinc and 10 percent tin, or 90 percent cadmium and 10 percent zinc. The tin lowers the melting point of the first material from 500°F [260°C] to 315°F [157°C].

To ensure that the surface is free from oxides, it should be cleaned to a luster. Preheating the base material to a temperature slightly less than the melting point of the filler material reduces stress buildup in the materials.

During the soldering of magnesium, as soon as the filler material becomes molten, it should be rubbed into the material with a wire brush or the welding tip. The rubbing improves the wetting of the magnesium, which is somewhat hampered by the lack of flux.

Stainless steel soldering presents two major difficulties during the process. The first is stainless steel's low thermal conductivity. The second is the difficulty in removing oxides from stainless steel. Stainless steel is extremely resistant to oxidation, but once it has oxidized, the oxidation is extremely hard to remove. For this reason fluxes used during stainless steel soldering are acid-based. Since acid-based solders are generally prohibited when soldering on aircraft and large quantities of heat are required, this process should be accomplished only in accordance with specific manufacturer's maintenance repair instructions and, if possible, with the part removed from the aircraft. Particular attention should be paid to the removal of all flux after the soldering process is complete.

REVIEW QUESTIONS

1. What type of steel is commonly used for tubular steel structures in aircraft?

2. If major repair work is done on a welded steel tubular fuselage, what precautions must be taken?

3. What principal factor governs the size of a *welding tip*?

4. When welding a structure that will require heat-treating, what type of filler rod must be used?

5. When welding a section of tubing into a structure, why is it a good practice to allow one end to cool after welding before welding the other end?

6. Describe a method for straightening a bent section of tubing in a fuselage.

7. Under what conditions is it permissible to ignore a dent in structural tubing?

8. Describe a method for removing a dent from steel tubing.

9. Describe a method that may be employed to detect small cracks in a welded joint.

10. Before welding a reinforcing sleeve or patch over a crack in tubing, what should be done with respect to the crack?

11. Describe the installation of a *finger plate patch*.

12. At what angle is the scarf for a repair sleeve cut?

13. Describe a *fishmouth cut* for a sleeve.

14. Why is a fishmouth joint stronger than a scarf joint?

15. When cutting a length of replacement tubing that is to be spliced with inner sleeve reinforcements, how much clearance must be allowed for each end of tubing?

16. How are inner splice sleeves placed into position for welding a replacement tube between stations?

17. What are the lengths of the original tube stubs required when splicing in a new replacement section between stations and telescoping the replacement section over the original stubs?

18. Describe a *rosette weld*.

19. What is the maximum clearance between an outer reinforcement sleeve and the tube inside it?

20. Describe the replacement of a cluster weld with the use of outside reinforcement sleeves.

21. What precautions must be taken when making a welded repair on a fabric-covered fuselage?

22. What special care must be taken with the inspection and repair of welded engine mounts?

23. Describe three methods for welding steel-attachment fittings to tubular structures.

24. Explain how a fuselage structure can be held in alignment while making welded repairs.

25. Describe three methods for splicing *streamline tubing*.

26. What type of landing-gear axle structures are not repairable by welding?

27. Why is it not permissible to weld a steel structure that has been brazed?

28. What types of nondestructive inspections can be performed to ensure that welds have been made properly?

29. How can the interior of welded steel tubing be protected against corrosion?

30. Discuss the weld repair for exhaust parts.

31. What precaution must be taken before welding fuel and oil tanks?

32. What should be done to welds that are made on the outside surface of cowlings?

33. What is meant by *soldering*?

34. What is the effect of alloying tin and lead to produce material for soft soldering?

35. What is the difference between *soft soldering* and *hard soldering*?

36. What are the three hard-soldering processes?

37. How would you tin a soldering copper?

38. What is the function of a *flux* in soldering?

39. Describe the procedure for soft soldering.

40. Why is a torch needed for hard soldering or for brazing?

41. What would be the best type of soldering to join two pieces of stainless steel?

42. Discuss the limitations on soldering as a method for aircraft repair.

Sheet-Metal Construction **8**

INTRODUCTION

In the early years of aviation, aircraft structural repair required a knowledge of woodworking, dope and fabric, and tubular and flat steel plate welding. After all-metal airplanes were developed, structural assembly and repair of aircraft involved a much greater proportion of sheet-metal work and a declining amount of work involving wood, fabric, and welding. Metal aircraft utilize aluminum alloy, magnesium, titanium, stainless steel, aluminum alloy bonded sandwich (honeycomb), glass fiber, advanced composite materials, and other exotic materials. The thicknesses are from thin sheet (up to 0.080 in [2.03 mm]) to heavy plate (0.080 in [2.03 mm] and above), sometimes an inch or more in thickness, depending on the stresses to be carried by the section of skin involved. Processing these materials has evolved from the use of simple manually controlled equipment to sophisticated processes controlled by computers.

The construction of aircraft components from sheet metal may be necessary when performing modifications to aircraft or as part of a repair operation. The basic standards and techniques presented in this chapter that deal with repair procedures are directly applicable to the assembly of new aircraft and components as well as their repair. It is important that the technician realize that the information presented here is of a general nature. For specific information about repairs to a particular model aircraft, consult the manufacturer's maintenance and structural repair manuals.

Damage to an aircraft structure may be caused by corrosion, vibration, impact of rocks or other objects (called foreign-object damage [FOD]), hard landings, excessive loads in flight, collision, and crash damage. In every case of such damage, it is the duty of the technician to make a very careful inspection of all affected parts and assess the degree of damage. The technician must then determine the type of repair required and design the repair so that the strength and other attributes of the structure are returned as closely as possible to their original values. This process requires a thorough knowledge of the structure, the materials of which it is constructed, and the processes employed in the manufacture.

Fasteners for metal structural assemblies once consisted largely of aluminum alloy rivets and plated steel bolts. As the performance of aircraft increased, strength requirements for certain structures necessitated the use of such fasteners as Hi-Shear® rivets, Huck Lockbolts®, close-tolerance bolts, and other high-strength fasteners. For assemblies where it was not possible to work on both sides to install rivets, a number of different types of blind rivets were developed. The generic use and basic installation of special fasteners are discussed in the text *Aircraft Basic Science.* Later in this chapter, factors that should be considered when using special fasteners are discussed.

DESIGN PHILOSOPHIES

When dealing with aircraft construction, there are two general groupings of sheet-metal components: *structural* and *nonstructural* components. **Structural** components are those components of the aircraft that transfer the forces exerted on the aircraft from one location to another or absorb the forces during flight. Ribs, stringers, longerons, bulkheads, and the aircraft's skin are typically structural components.

Nonstructural components are those components that do not transmit or absorb the forces of flight. **Nonstructural** components typically provide aerodynamic functions, such as a fairing, or direct air flow in a manner similar to cowlings or intake and exhaust ducts.

Although almost all aircraft sheet-metal components transmit some force, the primary consideration in aircraft design is placed upon the structural components.

As new materials and their attributes were added to sheet-metal components and more experience was obtained in aircraft design, the philosophies used in designing structural aircraft components were altered to reflect these new developments. Large transport-category aircraft design criteria are no exception.

Safe-Life Design

Originally, the recognized theory of structural design was called **safe-life**. The philosophy of safe-life design was to test the various components to failure and to use as a component's airworthy maximum life 25 percent of the average life, when tested to destructive failure.

Fail-Safe Design

The next stage in structural design philosophy was **fail-safe** construction. In fail-safe construction each component was designed to be able to accept the forces of adjacent components should their neighbor components fail. Although anticipating failure, the fail-safe design philosophy did not include a disciplined study of crack growth and residual strength characteristics of the components involved. Inspection criteria established using the fail-safe philosophy failed to specify inspection techniques and frequencies that were sufficient to detect discrepancies prior to catastrophic failure.

Another major difficulty with both safe-life and fail-safe design philosophies was that their design testing was done in a laboratory environment with near-perfect parts. In-service variables such as aircraft utilization and climate were not factored into the component's testing.

Damage-Tolerant Design

Realizing that these operating variances existed and that components are rarely in perfect or even near-perfect condition when fabricated on a production basis, the **damage-tolerant** design philosophy for sheet-metal structural components evolved. Formally acknowledged and accepted by the FAA in 1978, the damage-tolerant design philosophy accepts the existence of minor flaws in components, anticipates their growth, and establishes an inspection discipline designed to identify these flaws before they become critical to the aircraft's airworthiness. Working backward, the design engineer estimates the point at which the flaw will be identified through inspection and, based upon that criterion, designs the part (with a safety factor) to withstand the existence of that flaw. Where damage-tolerant design is not practical, safe-life philosophy designs are required.

In 1981, Advisory Circular 91.56 was issued by the FAA. The intent of this document was to provide guidelines for the issuance of **Supplemental Inspection Documents** (SIDs) for large transport-category aircraft. This advisory circular requires that engineering evaluations of large transport-category aircraft structures be made using damage-tolerant philosophies.

The basic design philosophies of AC 43.13-1A & 2A are based upon the safe-life and fail-safe design philosophies of structural design. Although the FAA, in FAR 25.571 and AC-91.56, recognizes and establishes the philosophy of damage-tolerant design, there is no requirement, at the time of this writing, to use damage-tolerant design criteria when designing a structural repair for an aircraft even if it was designed using that philosophy. However, for a variety of reasons, which are beyond the scope of this text, it is advisable to use damage-tolerant design criteria for aircraft that were designed under that criteria.

Because of the desirability of using damage-tolerant design criteria when repairing aircraft initially designed using that criteria and the fact that the FAA is investigating the advisability of establishing regulations that will require damage-tolerant repair design, some of the basics used in the damage-tolerant design philosophies are discussed in this chapter.

Application of Safety Factors

For a variety of reasons, when designing an original structure it is unwise to design a part so that the entire strength of the material is utilized under standard operating conditions. Maximum strengths are typically determined in laboratory environments, whereas the materials, when produced for production, will vary from the laboratory sample. For example, when a sheet of aluminum is heat-treated the hardness of the sheet will vary. When the sheet is quenched it is dipped into a tank. The last portion of the sheet to enter the quench will be slightly softer than the first. Engineers almost always apply a safety factor to their design to provide a margin of safety to compensate for material variances, design errors, product misuse, and other unforeseen variances from the norm.

In the case of pure tension or shear loading, the process of applying a safety factor is simple. The strength characteristic involved is simply divided by the desired margin of safety and the result is used in the appropriate calculations. The technician should be sure that the safety factor is applied to the proper strength factor.

In some instances the fastener may be subject to both shear and tension. This is more true for bolts than rivets, but the relationship should be familiar to the technician. The combination of shear and tension produces torsion. The shear-tension (torsion) relationship may be visualized using a simple graph, as shown in Fig. 8-1. By connecting the maximum shear strength on the *y*-axis and the maximum tensile strength on the *x*-axis, a line representing the maximum combinations of shear and tension is produced.

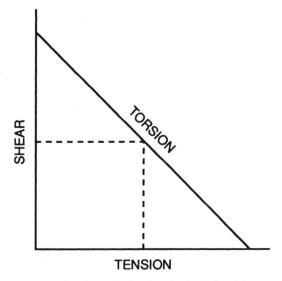

FIGURE 8-1   The shear-tension (torsion) relationship.

By adding the desired safety factor to 1 and then dividing the result into the maximum strengths, the design strength limits may be calculated.

$$P_{(\text{design})} = \frac{P_{(\text{ultimate})}}{1 + S_{\text{factor}}} \qquad (8\text{-}1)$$

$$P_{t(d)} = \frac{P_{t(u)}}{1 + S_{\text{factor}}} \qquad (8\text{-}2)$$

$$P_{s(d)} = \frac{P_{s(u)}}{1 + S_{\text{factor}}} \qquad (8\text{-}3)$$

where S_{factor} = safety factor
P = ultimate strength
d = design load; also designated by (design)
t = tensile strength
s = shear strength
u = ultimate strength; also designated by (ultimate)

The same plotting technique may be used to develop a line showing a "safe" stress-tension relationship, as shown in Fig. 8-2. Any combination of stress and tension that meet below the safe stress-tension line meets the safety-factor criteria.

Points above the line but below the maximum relationship line exceed the defined safety factor but are still functional. Points above the maximum relationship line will result in design failure.

For example, if the ultimate tensile strength (UTS) of a material is 60 000 psi and a safety factor of 25 percent is applied, the design strength is 48 000 psi.

Although, the inclusion of safety factors in repair design is usually not necessary because repairs are based upon the strength characteristics of existing materials, the technician should be aware of their use. In some specific situations the technician may want to apply an additional safety factor to one aspect of the repair or another to ensure that if failure occurs, it will take place in a specific part or location.

Margin of Safety

Another way to express the degree of safety is the **margin of safety (MS)**. The margin of safety is the ratio of the allowable (ultimate) force divided by the applied (design) force less 1.

$$MS = \frac{1}{R} - 1 = \frac{P_{t(u)}}{P_{t(d)}} - 1 \qquad (8\text{-}4)$$

where MS = margin of safety
R = ratio of design load to ultimate load
$P_{t(u)}$ = ultimate tensile strength
$P_{t(d)}$ = applied tensile force

If a material has a design load of 30 000 psi applied in tension but has an ultimate tensile strength of 40 000 psi, the margin of safety is 33 percent:

$$MS = \frac{P_{t(u)}}{P_{t(d)}} - 1 = \frac{40\,000}{30\,000} - 1$$

$$= 1.33 - 1 = 0.33, \quad \text{or} \quad 33\%$$

If the material was selected based upon other factors (such as shear) and as a result only 20 000 psi were applied in that area, the margin of safety for the tensile strength would be 100 percent. This means that the material is capable of bearing twice the load that is anticipated to be applied.

FACTORS AFFECTING SHEET-METAL PART AND JOINT DESIGN

In Chap. 2 of this book and in *Aircraft Basic Science*, some of the forces and stresses that apply to aircraft structures are discussed. All structural components of an aircraft are designed to withstand the effect of these forces during the life of the aircraft. The most basic of these forces result from the weight of the aircraft, flight loads such as lift and drag, and propulsion loads such as thrust. The magnitude of these loads varies during the life of the aircraft, depending upon the frequency and type of operation the aircraft experiences.

Landing gears experience these loads while the aircraft is static, is taxiing, and during takeoff and landing. Uplocks on retractable landing-gear aircraft experience these loads only when the landing gear is in the UP position.

The aircraft fuselage structure experiences these forces all during the life of the aircraft. Wing loads are transmitted through fuselage structure while the aircraft is static, during taxiing, takeoffs, and landings, and during flight. On the ground the fuselage must support its own weight and a portion of the weight of other aircraft components (and fuel). During taxiing, particularly during maneuvering operations, the fuselage structure is subjected to twisting forces (torsion) as the aircraft turns. In flight, the fuselage experiences a variety of forces as the wings produce lift and raise the fuselage into the air.

Stresses

The result of such forces is called **stress**. Stress is the load applied over a given area. The way in which these forces are

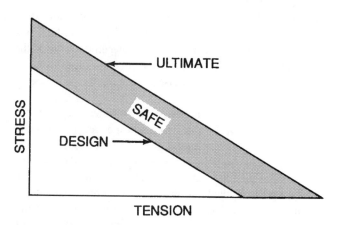

FIGURE 8-2   "Safe" stress-tension relationship.

applied determines the type of stress (compression, tension, torsion, bending, or shear).

Some stresses are most typically unidirectional, such as tension and shear. In such cases, the cross-sectional area over which the load is applied may be easily determined, and this area can be used to calculate the amount of stress. In unidirectional load application the cross-sectional area is that area of a plane that is perpendicular to the direction of the load applied.

Multidirectional loads, such as torsion and bending, are typically divided into their component parts using vector analysis (see *Aircraft Basic Science*). The individual load components may then be analyzed. The cross-sectional areas used are always those perpendicular to the direction in which the load vector is applied. Compressive loads may be either unidirectional or multidirectional.

If a piece of sheet stock 0.032 in [0.813 mm] thick by 2 in [50.8 mm] wide by 10 in [254 mm] in length has a load applied along its length, the cross-sectional area is perpendicular to the length. The cross-sectional area used to calculate the stress of this load is the thickness times the width.

Some components have **residual stresses**. In *Aircraft Basic Science* the concepts of **plasticity**, **ductility**, **malleability**, **brittleness,** and **elasticity** are reviewed. From the **law of conservation of energy** (also discussed in *Aircraft Basic Science*), it is known that energy can neither be created nor destroyed. When a component is fabricated, some of the energy required is dissipated to the environment through heat. For example, drill bits, which get hot during the drilling operations, and hot material chips carry some of this energy away from the part. The remainder of the energy is absorbed by the material being formed.

When a material is formed, the **kinetic energy** applied to the material to form it must overcome the material's resistance to being formed. This resistance to being formed may be considered a type of **potential energy**. To form material the kinetic energy applied by the process must exceed the potential energy, which exhibits itself as a resistance to forming, that exists in the material. When the forces applied during the forming process are terminated, the two energies find an equilibrium point. For example, when using a cornice brake to form aluminum, the technician will observe some **spring-back** in the material as the bending leaf is lowered. This phenomenon is known as elasticity. This process also occurs during the metal-removal process and is particularly evident if the turning or cutting tools pass over the material several times without a change in the tool's position. A lessening amount of material is removed with the passage of each tool over the workpiece.

Some of the kinetic energy of the forming process is absorbed by changes in the material's **crystalline** and **grain structures** (see *Aircraft Basic Science*). It is the continual counteracting of these two types of energy in a material that results in the residual stresses contained within a formed aircraft component. During the heat-treating processes (see *Aircraft Basic Science*), these two types of energies are redistributed to lessen their effects.

As forces are applied to a structural component, they are absorbed by the component in a manner similar to that experienced during the forming process. As the external forces are reduced, the material returns to its original shape, assuming the elastic limits are not exceeded.

When forces are applied over a given area of material, a stress exists. The unit of measure of these forces depends upon the unit of measure used in identifying the forces (pounds, grams, etc.) and the unit of measure for area (inches, feet, meters, centimeters, etc.). Since the force is applied over an area, the linear unit of measure is always squared. This results in a unit of measure typically referred to as pressure, such as pounds per square inch (psi).

Aircraft are structurally designed to withstand certain stresses. In *Aircraft Basic Science*, the ability of material to withstand certain amounts and types of stress is identified. Each material, depending upon its alloys and hardness, has empirically established stress limits. Equating the design stresses to the stress-bearing capabilities of the material (along with other considerations such as weight and a safety factor) is the function of the design engineer. One variable in the material selection is the thickness of the material. Scratches reduce the cross-sectional area of a material, thus reducing the stress-bearing capabilities of the material and resulting in either a reduction in the safety factor or, in the case of deep scratches, the ability of the material to carry the designed stresses.

In addition to the properties discussed in *Aircraft Basic Science*, the technician should be familiar with another property of materials called *bearing stress*. **Bearing stress** is a special type of compression loading, except that it is concerned with the loading of stresses only in the immediate area where the loads are exerted. In bearing stress the area of immediate concern typically is significantly smaller than the surrounding area, and these stresses result from the transmission of the applied load from one structural member to another. The transmission of the stress is normally accomplished through a mechanical fastener.

If the fastener is experiencing shear stress, the edge of the hole in the material is experiencing bearing stress. The factors that affect bearing stress include the diameter of the hole, the thickness and type of the material being joined, the distance of the hole from the edge of the material, and the magnitude of the load that is applied.

Several facts make the application of bearing loads unique. First, if the load is large enough, the hole will elongate as the load is applied. The extent of that elongation depends on the compressibility of the material, which is a function of the cross-sectional area between the edge of the hole and the edge of the part. As the edge distance increases, the cross-sectional area increases, changing the compressive stress.

Another factor that needs to be considered is that if a fastener is temporary in design, as in the case of bolts and screws, the hole is larger than the bolt. The result is that a smaller area of the hole's cross section actually carries the load. As the load increases, the material compresses (elongates), increasing the load transmission (contact) area.

Finally, bearing loads transmitted through a hole are frequently reversed, increasing wear by friction and fatigue.

A landing-gear support brace is a good example of a reversed load. On the ground the load on the brace may lock the gear in the down position, resisting loads that would tend to collapse the landing gear (compression). During retraction the same brace may force the gear to retract (tension).

Because of the mathematical complexity involved, bearing strengths are found in tables of material properties. The edge distances are noted in terms of preestablished edge distances of 1.5D and 2.0D. Table 8-1 shows some mechanical properties, including ultimate bearing strengths, for commonly used aircraft materials.

Strain

In order for strain to exist, a stress must be applied. Strain is measured by establishing a ratio of the increase in length that results from a stress divided by the original length of the material. If the unit of measure for a strain calculation is inches, the unit of measure for strain is inches per inch.

Using the example shown in Fig. 8-3, the stress area is the thickness times the width, and the strain relates to the length of the material. If the load is applied across the material's width, as shown in Fig. 8-4, the stressed cross-sectional area

TABLE 8-1 Mechanical Properties of Commonly Used Aircraft Materials (*The Aluminum Association*)

For training purposes only, DO NOT use for design.[a]

All pressures in kpi (1000 psi).

			F_{tu}^c	F_{su}	F_{bu}^b $e/D = 1.5$	$e/D = 2.0$
Sheet[d,e]						
Chrome molybdenum aircraft steel (5Cr-Mo-V)[SAE 4130, AMS 6537]						
Rc49-52			240	144	N/A	315
Rc54-56			280	168	N/A	365
Nickel-chromium steels (SAE 4718)[9Ni-4Co series]						
AMS 6424		Sheet	220	N/A	N/A	N/A
AMS 6526		Bar/Tubing	220[f]	137	346	440
Aluminum sheet						
2014-	T6		64	39	97	123
2014-	T6	Clad	61	38	93	118
2024-	T3		63	39	104	129
2024-	T3	Clad	59	37	97	121
2024-	T361		67	42	111	137
2024-	T361	Clad	61	38	101	125
6061-	T4		30	20	48	63
7075-	T6		76	46	118	152
7075-	T6	Clad	70	42	108	140
Rivets (driven)—MS20470 Type[g]						
Aluminum 2017-	T3	(D)		38		
2024-	T31	(DD)		41		
2117-	T3	(AD)		30		
5056-	H312	(B)		28		
Monel				49		
Ti-Cb				53		
A-286				90		

[a]For structural design refer to MIL-HDBK-5.
[b]Bearing values are "dry pin."
[c]Long transverse (grain direction).
[d]All sheet material 0.040 in [0.916 mm] thick, unless otherwise noted.
[e]Sheet and tubing: at least 99% of population is expected to equal or exceed this value, with a confidence of 95%, unless otherwise noted.
[f]Tension in lateral (grain direction).
[g]Rivets: at least 90% of population is expected to equal or exceed this value, with a confidence of 95%, unless otherwise noted.

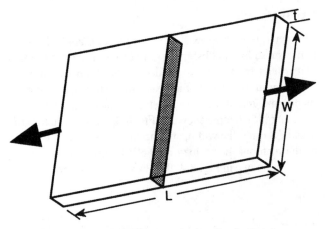

FIGURE 8-3 Strain cross section for longitudinal stress.

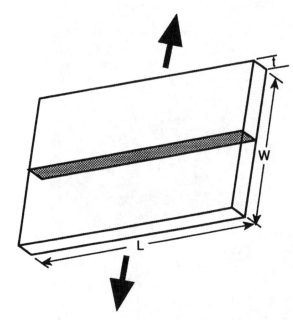

FIGURE 8-4 Strain cross section for lateral stress.

is the length times the thickness, and the strain is in relation to the width.

Fatigue

As an aircraft ages, it is subjected to a variety of stresses. These stresses come from loads applied on the ground, during taxiing, take-off and landing, aircraft pressurization, and during flight. The accumulation of these stresses over time results in a weakening of the material. This weakening is called **fatigue**.

In the discussion of elasticity in *Aircraft Basic Science*, it is stated that when the load is released from a component, the component returns to its original configuration. To the naked eye this is true, but in actuality there is a slight difference in the component's shape and grain structure each time a load is applied and then released. Because of this slight change in the material, the attributes of the material

also change slightly each time a load is applied. Unless the elastic limit of the material is exceeded, the effects of fatigue are rarely noted.

Each time a material experiences strain (growth in a linear dimension), even if the stress applied does not exceed its elastic limit, there is a minor retention of the deformation experienced. This results in the material being slightly (often minutely) different each time an additional stress is applied to a material. These differences compound until the material is no longer capable of carrying the applied stress.

A simplistic example may help explain this phenomenon. Assume that a material with a UTS of 65 000 psi is 3 in [76.2 mm] wide, 10 in [254 mm] long, and 0.032 in [0.813 mm] thick and that a tension load of 4992 lb is repeatedly applied along the length of the material that results in a 0.0001 percent increase in the length of the material. Also assume that the volume of the material always remains the same and that the increase in length results in a decrease in the thickness. In this example it is assumed that the width remains constant in an effort to simplify the mathematics. In an actual application, changes in the material's width would also have to be considered.

The load was determined by applying a safety factor (discussed earlier in this chapter) of 25 percent to the maximum load calculations.

$$\text{Design strength} = \frac{\text{max UTS}}{1.25} \tag{8-5}$$

$$\text{Design strength} = \frac{65\,000}{1.25} = 52\,000 \tag{8-6}$$

$$\begin{aligned}\text{Design load} &= \text{design tensile strength} \times \text{area}\\ &= 52\,000 \times (0.032 \times 3) = 4992 \text{ lb}\end{aligned} \tag{8-7}$$

Since the stress is applied along the length, each time the stress is applied, the thickness is decreased 0.0001 percent to compensate for the increase in length, thus maintaining the same volume.

The thickness of the material after a load is applied may be found by subtracting the amount of shrinkage caused by the application of the load from the thickness before the load was applied. The formula for the determining the thickness after one load cycle is

$$\begin{aligned}\text{Thickness}_{\text{after 1 cycle}} &= \text{thickness}_{\text{original}} -\\ &\quad (\text{thickness}_{\text{original}} \times \text{shrinkage factor})\end{aligned} \tag{8-8}$$

Factoring the equation yields

$$\begin{aligned}\text{Thickness}_{\text{after 1 cycle}} &= \text{thickness}_{\text{original}} \times\\ &\quad (1 - \text{shrinkage factor})\end{aligned} \tag{8-9}$$

For ease in reading, substitute

T_0 for thickness$_{\text{original}}$

T_1 for thickness$_{\text{after 1 cycle}}$

Sh for the shrinkage factor

The equation then becomes

$$T_1 = T_0 \times (1 - \text{Sh}) \tag{8-10}$$

The subscript indicates the number of load cycles under consideration.

To calculate the thickness after the application of the second load, the equation is

$$T_2 = T_1 \times (1 - \text{Sh}) \qquad (8\text{-}11)$$

Substituting Eq. (8-10) for the thickness after 1 cycle (T_1), the equation becomes

$$T_2 = [T_0 \times (1 - \text{Sh})] \times (1 - \text{Sh}) \qquad (8\text{-}12)$$

Rewriting Eq. (8-12) results in

$$T_2 = T_0 \times (1 - \text{Sh})^2 \qquad (8\text{-}13)$$

After calculating the thickness for additional load cycles, it would become apparent that the thickness of the material after any number of cycles (T_n) can be expressed as

$$T_n = T_0 \times (1 - \text{Sh})^n \qquad (8\text{-}14)$$

where T is thickness, n is the number of cycles, and T_0 is the original thickness.

The formula for the cross-sectional area after any number of cycles (designated by the subscript n) is

$$A_n = W \times T_n \qquad (8\text{-}15)$$

where W = width of the material (which is assumed to remain constant in this example)
T_n = the thickness after n load cycles
A_n = the cross-sectional area after n load cycles

The minimum cross-sectional area (designated $A_{\min}$) of the material that will accept a stress without failure is the ultimate tensile strength (UTS) for that the material times the area over which the load will be applied:

$$\text{UTS} \times A_{\min} = \text{actual load} \qquad (8\text{-}16)$$

Solving for $A_{\min}$ yields

$$A_{\min} = \frac{\text{actual load}}{\text{UTS}} \qquad (8\text{-}17)$$

Substituting the example assumptions yields

$$A_{\min} = \frac{\text{actual load}}{\text{UTS}} = \frac{4992 \text{ lb}}{65\,000 \text{ lb/in}^2} \qquad (8\text{-}18)$$

$$A_{\min} = 0.0768 \text{ in}^2 \qquad (8\text{-}19)$$

In determining the number of cycles required to reduce the cross-sectional area to $A_{\min}$, the area is equal to the width (W; remember for the purpose of this example the width is said to remain constant) times the thickness after n load cycles (T_n):

$$A_{\min} = W \times T_n \qquad (8\text{-}20)$$

Substituting Eq. (8-14) for T_n yields

$$A_{\min} = W \times [T_0 \times (1 - \text{Sh})^n] \qquad (8\text{-}21)$$

Dividing both sides of the equation by W, we have

$$\frac{A_{\min}}{W} = T_0 \times (1 - \text{Sh})^n \qquad (8\text{-}22)$$

Substituting the known and calculated values results in

$$\frac{0.07680}{3} = 0.032 \times (1 - 0.000\,001)^n$$

$$0.0256 = 0.032 \times (0.999999)^n$$

$$\frac{0.0256}{0.032} = (0.999999)^n$$

$$0.8 = (0.999999)^n$$

The use of logarithmic functions is discussed in *Aircraft Basic Science*. From that discussion it is known that if $a^y = x$, then $y = \log_a x$. Using that relationship and solving for n gives

$$n = 223\,143$$

After 223 143 stress cycles, the material will no longer be able to withstand the design stress of 4992 psi.

It is not important at this time that the technician be able to develop these equations. What is important is that the technician understand the concepts of fatigue used to develop this simplistic example.

Just as each material has a recognized average standard elastic limit, ultimate tensile strength, and yield point, materials also have standard fatigue lives. Typically, however, these standards are not guaranteed by the material fabricators because they are based upon averages and are provided only as guidelines for design engineers. Ultimately, each design application requires empirical testing to establish the various limits applicable to that particular design.

Cracks and Scratches

In the discussion of stress earlier in this chapter, the need to keep a part scratch- and crack-free was mentioned. In some instances these scratches and cracks are referred to as *stress risers*.

As discussed, stress is the force applied over a given area. Depending upon the way in which the force is applied, that area may be the cross-sectional area of the material. In such a case the material's thickness and width are the two linear measures used to determine the cross-sectional area.

If there are no scratches in the material (assuming a constant thickness), the stress on the material is constant throughout. However, if there is a **scratch** in the material, the thickness in the area of the scratch is reduced, lessening the cross-sectional area. Since the cross-sectional area is the denominator of the stress equation (force/area), as the area lessens the stress rises, resulting in the term **stress riser**.

Cracks in a material have much the same effect as scratches in that they reduce the stress-bearing capabilities of a material. Since a crack is, in essence, a void in the material, the area of the crack cannot carry any of the stress load. For example, assume that a crack extends one-half the distance of the material. As a result, the cross-sectional area is halved, and the stress on the remaining material is twice that for which it was designed.

The force that would normally be transmitted through the area that is now a void acts as though it wants to be transmitted as soon as possible. As a result, rather than the load being transferred equally over the remaining area, the load concentrates at the end of the crack. This concentration of force often exceeds the load-carrying capabilities of the material at the end of the crack and causes that area to crack as well, extending the length of the existing crack. This is known as **crack growth**. It is obvious that under these conditions, the longer the crack, the more the crack will tend to lengthen.

The same effect results when loads are applied at **sharp corners**. Loads concentrate at these points, causing a rapid localized increase in stress.

Stress Risers

Figure 8-5 illustrates the effect of stress concentrations created by three typical stress risers. The horizontal lines

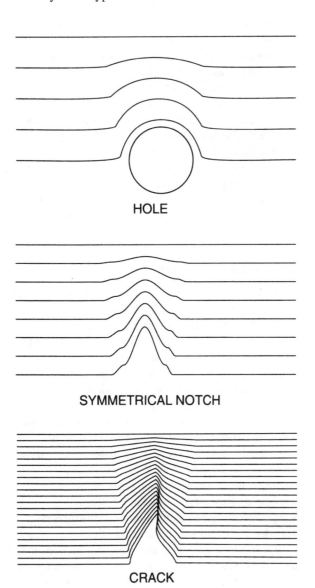

HOLE

SYMMETRICAL NOTCH

CRACK

FIGURE 8-5 Stress patterns around stress-transmission barriers.

represent the pattern of stress transmission within the part. From these illustrations it can be seen that the more gradual the material's physical change, the less likely the stresses are to concentrate. In addition, the same relationship is true for material configurations that are symmetrical about a point, as shown in the stress pattern of a "V" slot. The more erratic the disturbance to the stress pattern, as in the case of a crack, the more stresses are likely to concentrate.

Comparatively, holes concentrate stress by a factor of 3, other symmetrical patterns, by a factor of 6, and nonsymmetrical patterns, by a factor of approximately 20 times the stress applied in the immediate area of the cause. As the distance from the cause of stress increases, these factors are reduced. This reduction in force multiple continues until the distance from the disturbance cause negates the force multiple.

If two or more stress risers exist in the same general area, their net effect is greater than either of the individual stress risers. Their combined effect upon the overall stress is compounded, since the effects build upon one another. For example, a minute crack emanating from a hole may experience a stress factor of 60 times the applied load, because if the crack is very small, it is near the applied load of the hole, which is 3 times the applied load. The load applied to the crack is 3 times the applied force. The resultant force is 20 times the applied force of 3 times (the original applied load), or 60 times the original applied load. As the crack from a hole elongates, the applied load begins to decrease and continues to decrease until the multiple force caused by the hole no longer affects the crack.

Hole Generation

Although holes themselves do not create rapid localized stress risers, the process of generating a hole often does. During the drilling or punching process used for the creation of a hole, small cracks at the circumference of the hole are initiated. These small cracks introduce small stress risers into the material and begin the crack-growth process.

When drilling holes for rivets, the standard practice is to use a number size drill that is slightly larger than the diameter of the rivet. Table 8-2 shows the drill sizes used for the standard rivet diameters listed.

The number of cracks generated during the drilling process may be reduced by first drilling the hole slightly undersize (smaller than needed) and finishing the hole to the proper size with a reamer.

Relief Holes

When a piece of metal has two bends that intersect, it is necessary to provide relief holes in the metal at the intersection of the bends. If relief holes are not provided, the metal crowds together in the corners and sets up stresses that lead to cracks.

Holes have less of a tendency than cracks to result in rapid stress risers. Holes create a gradual change to the cross-sectional area of the material. The change in the cross-sectional area

TABLE 8-2 Drill-Bit Sizes Used for the Standard Rivet Diameters

Rivet size								
inches	$\frac{1}{16}$	$\frac{3}{22}$	$\frac{1}{8}$	$\frac{5}{32}$	$\frac{3}{16}$	$\frac{1}{4}$	$\frac{5}{16}$	$\frac{3}{8}$
millimeter	1.58	2.38	3.17	3.96	4.76	6.35	7.93	9.52
Drill Number	#51	#41	#30	#21	#11	F	P	W
Nominal hole diameter								
inches	0.067	0.096	0.1285	0.159	0.191	0.257	0.323	0.386
millimeter	1.70	2.43	3.26	4.03	4.85	6.52	8.20	9.80

due to a crack is more abrupt. Since the change in the cross-sectional area for a hole is more gradual than a crack, the remaining material is allowed to absorb the changes in stress more gradually.

Figure 8-6 shows how metal should be laid out and drilled to provide relief holes when the ends of the formed material are to be welded. The size of the holes is a minimum of $\frac{1}{8}$ in [3.18 mm] for metal thicknesses of 0.064 in [1.63 mm] or less and greater for thicker metals. As shown in the drawing, the holes are drilled with centers at the intersections of the inside-bend tangent lines. This position provides maximum relief, or clearance.

Figure 8-7 demonstrates the use of holes to reduce the potential of crack growth. This process, known as **stop-drilling**, changes the rate at which the stress increases from a rapid localized increase to a gradual increase.

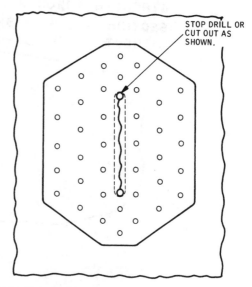

FIGURE 8-7 Stop-drilling of a crack before repair.

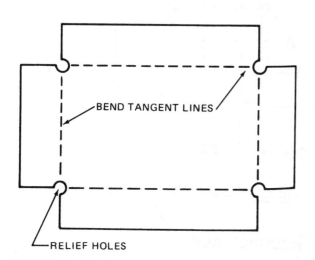

JUNCTION BOX

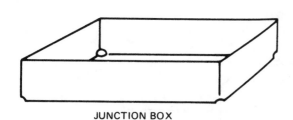

BEND TANGENT LINES

RELIEF HOLES

FIGURE 8-6 Flat layout showing location of relief holes.

Stress Corrosion

As stress is applied to a material, additional minute cracks are formed and/or existing cracks grow. As the cracks start or grow, new surface areas of the metal are exposed to the atmosphere. Because these newly exposed areas are not protected from corrosion, oxygen from the atmosphere quickly oxidizes with these areas. Since the cause of this increase in corrosion is applied stress, this phenomenon is called **stress corrosion**.

Load Transfer Through Structural Fasteners

Structural fasteners, which include various types of rivets and bolts, are used to transmit the forces applied from one piece of material to another. In the case of a repair patch, the load is transferred from the original material to the fasteners and from the fasteners to the patch material (which extends over the damaged area of the original material) to the fasteners on the opposite side of the damage, which return the load to the balance of the original material. Screws are normally not considered structural fasteners.

Holes, such as the holes required for riveting or bolts, also have an effect upon the load-carrying capabilities of material. Figure 8-8 (a duplication of Fig. 2.18 in AC 43.13-1B & 2B) demonstrates this effect. Holes reduce the cross-sectional area of the material. Examining the illustration at the top of the figure (A) shows that one-quarter ($\frac{1}{4}$) of the material's cross-sectional area is in either row of rivet holes. This results in the note that indicates that the material strength at a row is only 75 percent of a hole-free piece of material. The lower part of the figure (B) demonstrates the same

thing, except since the rivet spacing is different, the relative strength is different. The technician should take care when using this part of the figure, because it fails to note the material's strength for the middle two rows would be 67 percent of the material's strength, since the rivet pitch in those rows is 3D, compared to the 6D of the outside rows of the rivets and the 4D shown in area (A) of the same figure.

In the original design of the aircraft, the effect of holes on the load-carrying capability of material is compensated for in the design of the structural component. In repair design,

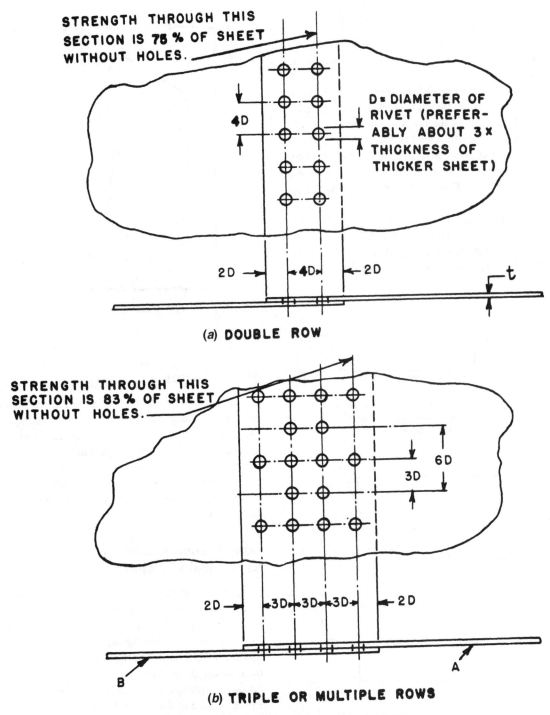

FIGURE 8-8   FAA Advisory Circular 43.13-1B & 2B.

the addition of holes to the structure will reduce the load-carrying capability of the original material.

FUNDAMENTAL CALCULATIONS FOR STRUCTURES

In the text *Aircraft Basic Science*, the properties of materials and their attributes are discussed, as well as the various empirical testing techniques used to determine their values. The results of this testing may be found in a variety of sources, ranging from manufacturer's data to technical government specifications such as MMPDS-06.

The maximum stress that any component can support is related to the UTS of the component's material. For materials in pure **tension** the maximum load is equal to the UTS of the material. Material in pure **compression** may carry a maximum load of approximately 70 percent of the UTS.

Torsion is a combination of compression and tension loads. Therefore, components subject to torsion can be assumed to be able to withstand loads no greater than 70 percent of the UTS.

Even though the UTS is of critical importance because of its direct relationship to material failure, elastic limits are also critical, particularly where aerodynamics and mating parts are concerned. Exceeding the elastic limit of aircraft skins will cause the aircraft skin to buckle, affecting aerodynamics. When mating materials with different elastic limits are stressed beyond the elastic limit of either or both of the materials, the proper functioning of the joint as originally designed may be in question.

Loads Applied in Structural Fastener Joints

When loads are applied to joints secured by fasteners, there are six types of load-carrying capabilities that must be taken into consideration. In sheet-metal applications, two of these are rarely applied (torsion and compression), so their discussion in this portion of the text is inappropriate.

The first load-application consideration is the **shear** load, which is applied to the fastener. In sheet-metal applications the loads applied to the sheet materials act in opposite directions. This action by the sheet stock results in a shearing load being applied to the fastener. For each type of fastener material, the fastener's ability to resist a shear load is directly dependent upon the diameter of the fastener.

The pulling apart of the joint places the sheet material in tension. Each material has a maximum **tensile** load-carrying capability, based upon the type, thickness, and width of the material. As holes are placed in the sheet material, the effective width of the material is reduced by the diameter of the hole. So, in determining the tensile load-carrying capability of the material, the number of fastener holes required must be considered.

At the same time a shearing load is applied to the fastener, the sheet material between the fastener's installation hole and the edge of the material must resist the tendency to rip out. This load is frequently referred to as the **tearout** load. For a given material type and thickness, the ability of the sheet material to resist the tearout load is a function of the shortest distance between the edge of the material and edge of the fastener hole.

Finally, as the tearout load is applied to the sheet material, the resistance of the fastener to shear imparts a compressive load on the sheet material. This load compresses, to some degree, the sheet material that comes in direct contact with the fastener. As a result of the increased density in the area of immediate contact, the material's resistance to tearout is increased in that localized area. This localized area of compressed material is called the **bearing** area, and the load applied in this area is called the **bearing load**. The resistance to bearing loads of a given type of material is a function of the area of contact, which by industry convention is defined as the diameter of the hole times the thickness of the material.

Determining the Cross-Sectional Area

To determine the load-carrying capability of any structure, the cross-sectional area over which the load will be applied must be calculated. In determining the cross-sectional area, the nature of the fastener must be considered. To install a standard fastener, the hole must be slightly larger than the fastener. If the fastener is a rivet, the rivet is assumed to increase the diameter of the hole during installation. Therefore, when using these calculations for riveted joints, d is equal to the diameter of the hole and not the standard diameter of the rivet.

Fastener Shear Area

$$A_s = \frac{\pi \times d^2}{4} \qquad (8\text{-}23)$$

The diameter is used here instead of a radius, since fasteners and their standard installation holes are specified by diameter. This value represents the area of the fastener that would be cut through if the fastener failed due to shear.

Tensile Area of Sheet

$$A_t = w \times T \qquad (8\text{-}24)$$

This area is the width of the sheet where the load is applied times the thickness of the sheet.

Tearout Area

$$A_e = 2 \times \left(E - \frac{d}{2} \right) \times T \qquad (8\text{-}25)$$

Tearout area is the total area of the sheet that will have to be torn if the rivet is to be separated from both splice sheets. E is the edge distance of the rivet (remember edge distance is measured from the center of the rivet shank) and $\frac{d}{2}$ represents the portion of the material included in the edge distance but is actually a void in the sheet material. Subtracting $\frac{d}{2}$ from E results in the distance from the edge of the sheet material to the edge of the rivet hole. Multiplying the actual material distance by the thickness of the material (t) results in the cross-sectional area that would be removed by putting the rivet through the material to the edge. The factor of 2 in the equation represents the two sheets (and assumes that the edge distances are equal for both sheets).

Bearing Area

$$A_b = d \times T \qquad (8\text{-}26)$$

The bearing area is the cross-sectional area of a hole where the slicing plane passes through the center of the hole and is perpendicular to the direction of the applied force.

Load-Carrying Capability Calculations

Fastener Shear Load   Fastener shear load is the cross-sectional area of the fastener [Eq. (8-23)] times the ultimate shear strength of the rivet material.

$$P_s = A_s \times F_s \qquad (8\text{-}27)$$

Substituting Eq. (8-23) for A_s gives

$$P_s = \frac{\pi \times d^2}{4} \times F_s \qquad (8\text{-}28)$$

Sheet Tensile Load   Tensile load is the ultimate tensile load of the sheet material times the tensile area [Eq. (8-24)].

$$P_t = A_t \times F_t \qquad (8\text{-}29)$$

Substituting Eq. (8-24) for A_t gives

$$P_t = w \times T \times F_t \qquad (8\text{-}30)$$

Sheet Tearout Load   Tearout load is the ultimate shear strength of the sheet times the total area that must be torn out in order to shear both of the splice sheets [Eq. (8-25)].

$$P_e = A_e \times F_s \qquad (8\text{-}31)$$

Substituting Eq. (8-25) for A_e gives

$$P_e = 2 \times \left(E - \frac{d}{2} \right) \times T \times F_s \qquad (8\text{-}32)$$

Sheet Bearing Load   Sheet bearing load is the bearing area of the sheet [Eq. (8-26)] times the ultimate bearing strength of the sheet.

$$P_b = A_b \times F_b \qquad (8\text{-}33)$$

Substituting Eq. (8-26) for A_b results in

$$P_b = d \times T \times F_b \qquad (8\text{-}34)$$

BENDING METALS

The technician frequently encounters sheet-metal fabrication requirements that require bending the metal to various shapes. When a metal is bent, it is subjected to changes in its grain structure, causing an increase in hardness.

The composition of the metal, its temper, and its thickness determine the minimum radius that may be successfully generated in the part through bending without adversely affecting the material. Table 8-3 shows the recommended radius for different types of aluminum. Note that the smaller the thickness of material, the smaller the recommended bend radius.

Alclad may be bent over radii slightly smaller than those shown. This is possible because the pure aluminum coating of aluminum is more malleable than the alloy core material and the thickness of the core material in Alclad is slightly less than that of noncoated alloy of the same sheet thickness.

Also note that as the material increases in hardness, the recommended bend radius increases. This is important to observe when working in an environment where heat-treating facilities are available.

If the material is formed immediately after the heat-treat process, the material will not have reached its full hardness because the material has not been fully aged. In *Aircraft Basic Science*, the process of aging of materials is discussed. If the metal has not reached its full hardness, it may be bent using a slightly smaller radius.

Minimum Bend Radius

Aircraft manufacturers have developed minimum bend radius charts that are included in their aircraft Structural Repair Manuals (SRMs) that the technician can use to determine a safe bend radius (Table 8-3). Keep in mind that different manufacturers use sometimes different minimum safe bend radius values. Always consult the SRM of the aircraft before bending material. Drawings used for fabrication of aircraft parts will normally call out for the radius to be used for each part. If it is not called for in the drawing, the technician should consult a minimum bend radius chart. For example, the minimum bend radius using Table 8-3 for aluminum 2024-T3 with a material thickness (Gauge) of 0.032 inches is 0.12 inches ($\approx \frac{1}{8}$ in). The same material in the annealed condition (2024-O) will have a bend radius of 0.06 inches). This example shows that it can be beneficial to anneal aluminum to make a sharper bend. The technician should always compare the bend radius in the drawing with the minimum bend radius chart to make sure that the bend can be made without cracking or weakening the material.

Characteristics of a Bend

By comparing the amount of material required to make a radius bend to the amount of material required to make an equivalent 90° bend (square corner), it can be demonstrated that radius bends require less material than do right-angle bends. Figure 8-9 illustrates the comparison.

			Minimum Inner Bend Radii		
			Aluminum		
Gauge	2024–0 5052–H34	2024–T3/T4	5052–0	7178–0 7075–0	7178–T6 7075–T6
A	0.03	0.06	0.03	0.03	0.09
0.016	0.03	0.06	0.03	0.03	0.09
0.018	0.03	0.06	0.03	0.03	0.12
0.020	0.03	0.06	0.03	0.03	0.12
0.022	0.06	0.09	0.03	0.06	0.12
0.025	0.06	0.09	0.03	0.06	0.12
0.028	0.06	0.09	0.03	0.06	0.16
0.032	0.06	0.12	0.03	0.06	0.16
0.036	0.06	0.16	0.06	0.06	0.19
0.040	0.06	0.16	0.06	0.06	0.19
0.045	0.09	0.19	0.06	0.09	0.25
0.050	0.09	0.19	0.06	0.09	0.25
0.056	0.12	0.22	0.06	0.12	0.28
0.063	0.12	0.22	0.06	0.12	0.31
0.071	0.12	0.28	0.09	0.12	0.38
0.080	0.16	0.34	0.09	0.19	0.44
0.090	0.19	0.38	0.09	0.19	0.50
0.100	0.22	0.44	0.12	0.22	0.62
0.112	0.25	0.50	0.12	0.28	0.75
0.125	0.25	0.56	0.12	0.28	0.88
0.140	0.34	0.62	0.12	0.38	1.00
0.160	0.38	0.75	0.16	0.44	1.12
0.180	0.44	0.88	0.19	0.50	1.25
0.190	0.50	0.88	0.19	0.56	1.25

TABLE 8-3 Recommmended Bending Radius for Different Types of Aluminum

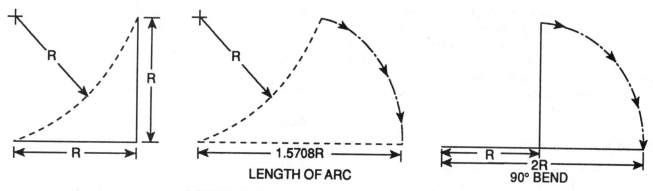

FIGURE 8-9 Comparing a square and a radiused corner.

The amount of material to make the square corner is two times the dimension R in the illustration ($2R$). The amount of material required to make an equivalent bend is (simplistically) $\frac{1}{4}$ of the circumference of a circle $\frac{90°}{360°}$ with a radius of R $\frac{\pi 2R}{4}$. If these two bends required the same amount of material, the following equation would be true:

$$2R = \frac{\pi 2R}{4}$$

Multiplying both sides by 4 yields

$$8R = \pi 2R$$

Divide both sides by R:

$$4 = \pi$$

Since π does not equal 4, this equation is not true, and one type of bend requires less material. Since π (approximately equals $\frac{27}{7}$) is less than 4, the side containing π—the radiused corner—will require less material.

When using layout techniques, the technician must be able to calculate exactly how much material will be required for the bend. If this is not done, other attributes, such as attachment holes, will almost certainly be misaligned.

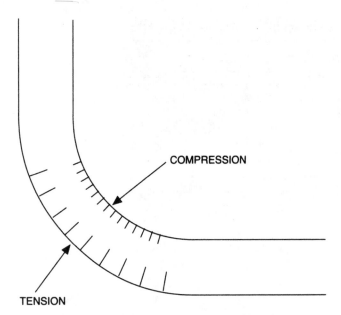

FIGURE 8-10 Forces acting on a sheet-metal bend.

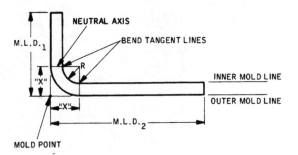

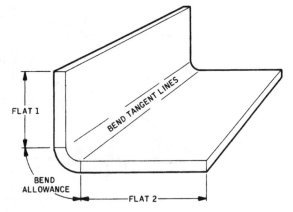

FIGURE 8-11 Nomenclature of a bend.

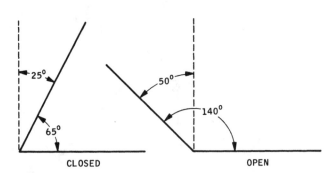

FIGURE 8-12 Closed and open angles.

When a piece of sheet metal is bent, as portrayed in Fig. 8-10, the material on the outside of the bend is in **tension** and stretches; the material on the inside of the bend is in **compression** and shrinks. Where the two forces of tension and compression meet within the metal, there is a plane that neither stretches nor compresses. This plane remains the same length after bending as it was before bending and is called the **neutral axis**.

Empirically (through experimentation) it has been shown that the actual neutral axis is approximately 44.5 percent through the thickness of the metal, measuring from the inside of the bend. The empirical neutral axis may be dimensioned from the center of the bend radius and may be calculated by adding to the bend radius 0.445 (44.5 percent) times the thickness of the metal.

Bend Nomenclature

If a material is bent around a radius, the radius will have a center. The material will have a thickness, resulting in two radii associated with each bend: an inside radius and an outside radius. When describing a bend in aviation, the term **bend radius** is used; it refers to the inside radius.

Bend Allowance

The **bend allowance**, abbreviated **BA**, is the length of sheet metal required to make a bend over a given inside radius (that is, it is the distance from the beginning to the end of the bend), measured perpendicular to the axis of the bend and along the neutral axis. The distance of the bend allowance depends upon the thickness of the metal, the radius of the bend, and the degree of the bend.

A drawing illustrating the nomenclature of a bend is shown in Fig. 8-11. As shown in the drawing, the bend begins and ends at the **bend tangent lines**, and the length of the neutral axis between the two lines is the **bend allowance**. Since the

neutral axis does not change in length, the BA can be measured on the flat part before the metal is bent. The **bend angle** is the number of degrees the material must be bent from its flat position to obtain the desired configuration. The **angle of bend** is equal to 180° less the bend angle. The distinction between the *bend angle* and the *angle of the bend* should be noted so that their proper meanings are not confused during subsequent discussions.

Figure 8-12 shows two interior angles. The interior angle on the left is 65° and is called a **closed angle** because it is less than 90°. The interior angle on the right is 140° and is called an **open angle** because it is greater than 90°. These terms are used in problems for determining setback. A 90° interior angle is neither open nor closed. In Fig. 8-12, notice that the closed-angle flange leans toward the leg. The angle is 65°, which is 25° less than 90°; hence it is said to be closed 25° or have a 25° closed bevel. The open-angle flange leans away from the leg. The angle in the illustration is 140°,

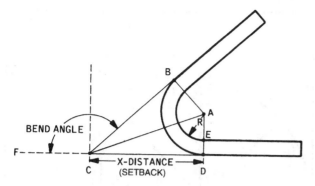

FIGURE 8-13 *X*-distance (setback).

which is 50° more than 90°; hence it is said to be open 50° or have a 50° open bevel.

The **mold-line dimensions** of a part are given from the end of the metal part to the outer mold point. The **outer mold point** is the intersection of the outer mold lines after the part is bent. This point is shown in the drawing of Fig. 8-11. When the part is bent around a given radius, the metal will not extend to the mold point, and the material required to make the part will be smaller in length than the sum of the mold-line dimensions. The **mold-line angle** is the angle formed by the mold lines. The mold-line angle is equal to 180° less the bend angle.

The **X-dimension** is the distance from the mold point to the bend tangent lines. This distance is also called **setback**. Figure 8-13 shows the *X*-distances (setbacks) for bends less than, greater than, and equal to 90°. Subtracting the setback from the mold-line dimension results in the length of the flat portion of the mold-line dimension.

Since the tangent point (by definition) is at right angles to the mold line, the simple trigonometric cotangent function (cotan(*x*) = adjacent/opposite) can be used to calculate the setback. Figure 8-13 shows the two right triangles that result when lines are drawn from the center of the radius to the bend-line tangent points and between the mold point and the center of the radius.

The distance between the center point of the radius (A) and the outer mold line (D or B) is the sum of the radius of the bend (AE) and the thickness of the sheet material (ED). This is true for both triangles. The hypotenuse for both of the right triangles is the distance from the mold point (C) to the center of the radius (R). Since both triangles are right triangles and two of their sides are identical, the third sides must be identical, as must all corresponding angles.

Since the two angles that form the mold angle are equal, they each must equal one-half the mold angle. Since the mold angle is equal to 180° less the bend angle, the two angles must also equal one-half of the quantity 180° less the bend angle. In this discussion the notation Ba is used for the bend angle. Later, when the bend allowance is discussed, the notation BA is used for the bend allowance.

Setback angle = one half of (180° less the bend angle)

$$SA = 0.5 \times (180 - Ba)$$

The cotangent of the setback angle (SA) is equal to the adjacent side divided by the opposite side. Expressed another way, the cotangent of the setback angle is equal to the setback divided by the radius plus the thickness of the material.

$$cot(SA) = \frac{setback}{radius + thickness}$$

In this case the setback is the unknown. By multiplying both sides of the equation by radius plus the thickness, the equation becomes

$$cot(SA) \times (radius + thickness) = setback$$

Development and Use of the K-Factor

Typically, however, aircraft drawings express the bend angles of a part in terms of the bend angle. Since the greater the number of calculations required, the greater the potential for miscalculation, it is often suggested that the setback be calculated in terms of the information provided on the drawing. To comply with this recommendation, the technician must use the bend angle directly in the required calculations. The technician who is interested in the logic associated with this calculation should continue reading this section. Those interested solely in the calculation's application should proceed to the next section, "K-Factor Charts."

This calculation may be accomplished by the use of some simple geometric relationships. Beginning with a simplified sketch of the part, label the mold point *A*, the radius of the bend *R*, the bend tangent point on the base line *T*, and the opposite bend tangent point *C*. Draw an arc that has a center at the mold point (*A*) so that the radius of this arc passes through the bend radius center. The point where the arc crosses the base line with the material in the flat should be labeled *B*. See Fig. 8-14.

To simplify the discussion, the basic logic of each step is identified; however, the formal geometric proof is not fully stated. Each letter identifies a conclusion, and the numbers identify the layout process required to generate the conclusion. The steps are identified first, and then the conclusion is given. For example, in order to arrive at the conclusion in step A, the constructions in step A-1 must first be completed.

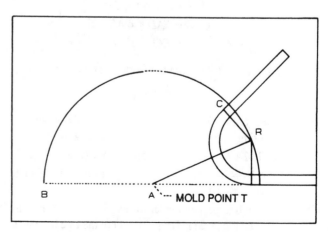

FIGURE 8-14 Geometric development of K-factors: Start.

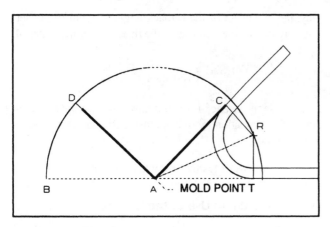

FIGURE 8-15   Geometric development of K-factors: Step A-1.

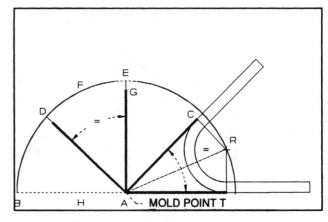

FIGURE 8-17   Geometric development of K-factors: Result A.

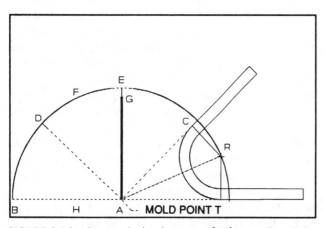

FIGURE 8-16   Geometric development of K-factors: Step A-2.

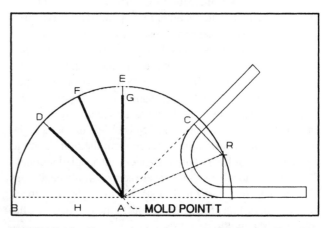

FIGURE 8-18   Geometric development of K-factors: Step B-1.

A-1.   Draw a line that is perpendicular to line *AC* (the upper mold line) passing through point *A*. Label the point of intersection with the arc *D* (Fig. 8-15).

Draw a line perpendicular to line *AB* passing through point *A*, labeling the intersection with the arc point *E* (Fig. 8-16).

A.   Angles *TAE* and *CAD* are equal (both right angles) (Fig. 8-17).

$$TAE = TAC + CAE$$
$$CAD = EAD + CAE$$
$$TAC + CAE = EAD + CAE$$

Subtracting *CAE* from both sides results in

$$TAC = EAD$$

B-1.   Bisect angle *DAE*. Use *F* to label the point where the bisecting line passes through the arc (Fig. 8-18).

B.   Angle *TAR* = angle *EAF* = angle *FAD* (Fig. 8-19).

C.   Line *FA* = line *AR*, since each is a radius of the same arc (Fig. 8-20).

D-1.   Draw a line perpendicular to line *EA* passing through point *F* (Fig. 8-21). The point where the line intersects *EA* should be labeled *G*.

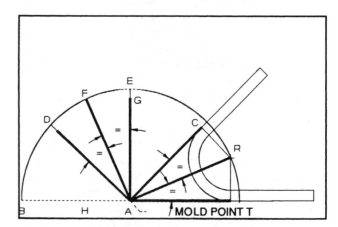

FIGURE 8-19   Geometric development of K-factors: Result B.

D.   Triangles *FGA* and *RTA* are equal. Both are right triangles with like angles and equal hypotenuses (Fig. 8-22).

E-1.   Draw a line parallel to line *GA* passing through point *F*. The intersection point with line *BT* is labeled *H* (Fig. 8-23).

E.   Triangles *FHA* and *GAF* are congruent (Fig. 8-24).

F.   Therefore, angles *GAF* and *HFA* are equal and corresponding lines *FH*, *GA*, and *AT* are equal (Fig. 8-25).

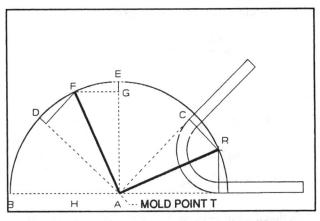

FIGURE 8-20 Geometric development of K-factors: Result C.

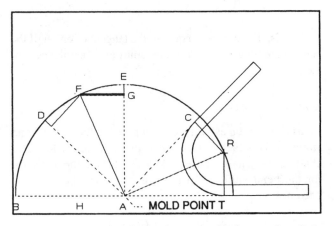

FIGURE 8-21 Geometric development of K-factors: Step D-1.

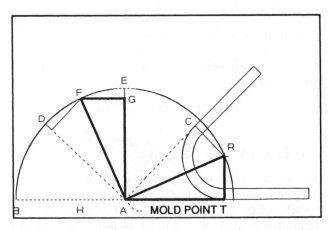

FIGURE 8-22 Geometric development of K-factors: Result D.

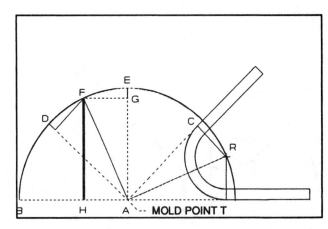

FIGURE 8-23 Geometric development of K-factors: Step E-1.

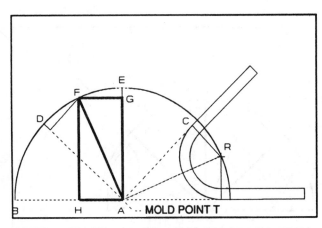

FIGURE 8-24 Geometric development of K-factors: Result E.

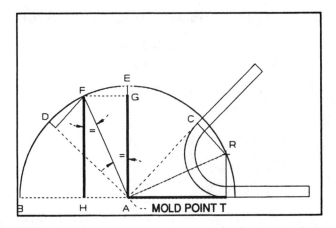

FIGURE 8-25 Geometric development of K-factors: Result F.

Lines *FA* and *RA* are equal and lines *HA*, *FG*, and *RT* are equal.

G-1. Angle *BAC* = angle *BAD* + angle *DAE* + angle *EAC* (Fig. 8-26).

G. Angle *BAE* = angle *DAC* (both are right angles) (Fig. 8-27).

$$BAE = BAD + DAE$$
$$DAC = CAE + DAE$$
$$BAD + DAE = CAE + DAE$$

Subtracting DAE from both sides of the equation gives

$$BAD = CAE$$

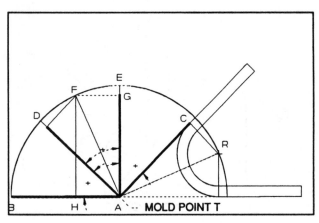

FIGURE 8-26   Geometric development of K-factors: Step G-1.

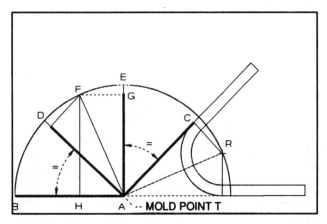

FIGURE 8-27   Geometric development of K-factors: Result G.

H. $DAF = FAE$, since triangle DAE was bisected (see B-1).

$$BAD = CAE \text{ (see F)}$$

$$BAD + DAF = CAE + FAE$$

$$BAF = FAC$$

$$BAF + FAC = BAC$$

Substituting BAF for FAC gives

$$BAF + BAF = BAC$$

$$2 \times BAF = BAC$$

$$BAF = \left(\frac{BAC}{2}\right), \quad \text{or one-half the bend angle}$$

I. $$\tan(BAF) = \frac{FH}{HA} = \tan\left(\frac{BAC}{2}\right)$$

$$FH = \tan\left(\frac{BAC}{2}\right) \times HA$$

Substituting $HA = RT$ (see step F)

$$FH = \tan\left(\frac{BAC}{2}\right) \times RT$$

Substituting $FH = AT$ (see step F)

$$AT = \tan\left(\frac{BAC}{2}\right) \times RT$$

$$RT = \text{radius} + \text{thickness}$$

$$AT = \tan\left(\frac{BAC}{2}\right) \times (\text{radius} + \text{thickness})$$

$$AT = \text{setback}$$

$$BAC = \text{bend angle}$$

$$\text{Setback} = \tan\left(\frac{\text{bend angle}}{2}\right) \times (\text{radius} + \text{thickness})$$

$$\text{Setback} = \tan\left(\frac{\text{bend angle}}{2}\right) \times (R + T)$$

That is, the setback is equal to the tangent of one-half the bend angle times the sum of the radius and the thickness of the material.

K-Factor Charts

Charts have been established that give the factor tan(bend angle/2). This factor is known as the K-factor. A K-factor chart is shown in Table 8-4. Substituting the K-factor for tan(bend angle/2) in the last equation, the equation becomes

$$\text{Setback} = K \times (R + T)$$

In using the chart it should be noted that the bend angle shown in the chart is the full bend angle, and the division of the angle is included in the calculation deriving the K-factor. Therefore, when looking up the K-factor for a bend angle of 120°, find 120° on the chart and read the K-factor in the adjacent column (1.732). Substituting this K-factor yields

$$\text{Setback} = 1.732 \times (R + T)$$

Flat Layout: Leg Length

In a flat-pattern layout, the length of any leg is the length of the mold-line dimension less its setback. If there is a radius at each end of the part, such as in a U channel, the flat-pattern length of the web is the mold length less setbacks calculated for each end.

Bend-Allowance Calculations

Bend allowance may be calculated using three different formulas, depending upon the degree of accuracy required. The most accurate formula uses the empirical location of the neutral axis. The other formulas assume that the neutral axis is halfway through the thickness of the material and differ only in the way the radius arc is defined. All three

TABLE 8-4 K-Chart

Deg.	K	Deg.	K	Deg.	K	Deg.	K	Deg.	K	Deg.	K
1	0.0087	37	0.3346	73	0.7399	109	1.401	145	3.171		
2	0.0174	38	0.3443	74	0.7535	110	1.428	146	3.270		
3	0.0261	39	0.3541	75	0.7673	111	1.455	147	3.375		
4	0.0349	40	0.3639	76	0.7812	112	1.482	148	3.487		
5	0.0436	41	0.3738	77	0.7954	113	1.510	149	3.605		
6	0.0524	42	0.3838	78	0.8097	114	1.539	150	3.732		
7	0.0611	43	0.3939	79	0.8243	115	1.569	151	3.866		
8	0.0699	44	0.4040	80	0.8391	116	1.600	152	4.010		
9	0.0787	45	0.4142	81	0.8540	117	1.631	153	4.165		
10	0.0874	46	0.4244	82	0.8692	118	1.664	154	4.331		
11	0.0963	47	0.4348	83	0.8847	119	1.697	155	4.510		
12	0.1051	48	0.4452	84	0.9004	120	1.732	156	4.704		
13	0.1139	49	0.4557	85	0.9163	121	1.767	157	4.915		
14	0.1228	50	0.4663	86	0.9324	122	1.804	158	5.144		
15	0.1316	51	0.4769	87	0.9489	123	1.841	159	5.399		
16	0.1405	52	0.4877	88	0.9656	124	1.880	160	5.671		
17	0.1494	53	0.4985	89	0.9827	125	1.921	161	5.975		
18	0.1583	54	0.5095	90	1.000	126	1.962	162	6.313		
19	0.1673	55	0.5205	91	1.017	127	2.005	163	6.691		
20	0.1763	56	0.5317	92	1.035	128	2.050	164	7.115		
21	0.1853	57	0.5429	93	1.053	129	2.096	165	7.595		
22	0.1943	58	0.5543	94	1.072	130	2.144	166	8.144		
23	0.2034	59	0.5657	95	1.091	131	2.194	167	8.776		
24	0.2125	60	0.5773	96	1.110	132	2.246	168	9.514		
25	0.2216	61	0.5890	97	1.130	133	2.299	169	10.38		
26	0.2308	62	0.6008	98	1.150	134	2.355	170	11.43		
27	0.2400	63	0.6128	99	1.170	135	2.414	171	12.70		
28	0.2493	64	0.6248	100	1.191	136	2.475	172	14.30		
29	0.2586	65	0.6370	101	1.213	137	2.538	173	16.35		
30	0.2679	66	0.6494	102	1.234	138	2.605	174	19.08		
31	0.2773	67	0.6618	103	1.257	139	2.674	175	22.90		
32	0.2867	68	0.6745	104	1.279	140	2.747	176	26.63		
33	0.2962	69	0.6872	105	1.303	141	2.823	177	38.18		
34	0.3057	70	0.7002	106	1.327	142	2.904	178	57.29		
35	0.3153	71	0.7132	107	1.351	143	2.988	179	114.59		
36	0.3249	72	0.7265	108	1.376	144	3.077	180	Inf.		

formulas calculate the circumference of the neutral-axis circle and factor the circumference by the percentage of the circumference that actually exists in the part by multiplying the circumference by the number of degrees in the bend angle and dividing by 360°, the number of degrees in the full circumference.

In flat-pattern layout, the minimum length and width of material required to produce the part are called the **developed length** and **developed width**, respectively. If either distance is affected by a radiused bend, its developed distance will be less than the sum of the related mold lines.

Geometric (0.5T) Formulas

The circumference of a circle is defined by the equation

$$C = \pi \times D = \pi \times 2R = 2\pi \times R$$

In this equation the neutral axis is considered to be halfway through the thickness of the material ($R + 0.5T$). Adjusting the basic formula to reflect the length of the neutral axis, the equation becomes

$$C = \pi \times (D + T) = 2\pi \times (R + 0.5T)$$

To calculate the bend allowance (BA), a factor for the portion of the radius arc that actually exists needs to be added, where N equals the bend angle.

$$BA = C \times \frac{N}{360}$$

$$= \pi + (D + T) \times \frac{N}{360} = 2\pi \times (R + 0.5T) \times \frac{N}{360}$$

The equations are expressed separately in terms of the diameter as

$$BA = \pi \times (D + T) \times \frac{N}{360}$$

and in terms of the radius as

$$BA = 2\pi \times (R + 0.5T) \times \frac{N}{360}$$

The benefit to the technician in the use of either of these formulas is the minimizing of the number of calculations required to obtain the desired result.

Empirical Formula Method

The **empirical formula** for determining **bend allowance** is

$$BA = (0.01743R + 0.0078T)N$$

where N is the bend angle. This formula simply eliminates all known values from the equation and substitutes the empirical radius of approximately 44.5 percent of the material's thickness in place of the factor $0.5T$. The formula can be derived as follows:

$$BA = 2\pi \times (R + 0.445T) \times \frac{N}{360}$$

$$BA = \left[\left(\frac{2 \times 3.1416}{360} \right) \times R + \left(\frac{2 \times 3.1416}{360} \right) \times 0.445T \right] \times N$$

$$BA = (0.01745R + 0.0078T) \times N$$

The difference between the calculated $0.01745R$ and the empirical $0.01743R$ is probably due in some part to experimental error and the significant digits used when using π in the calculations. In any case, the difference is insignificant in that it represents only a 0.1 percent error.

In application, there is little difference between the empirical formula and the simpler $0.5T$ method. This is because of the technician's inability to maintain close tolerances using standard manual layout techniques. The thickest material listed in Table 8-3 is 0.258 in, with a maximum radius listed of ten (10) times the thickness. Using these dimensions, the difference in the bend allowance for a 90° bend between the empirical method and the $0.5T$ method is less than $\frac{1}{32}$ in, which is typically considered the tightest tolerance attainable for a manual layout.

Flat-Pattern Layout

A flat-pattern layout is composed of all the dimensions necessary to fabricate a part from the flat. The flat-pattern dimension (FPD) adjusts for the setback where the mold-line distance includes a radius. The developed width (DW) is equal to the sum of all FPDs and bend allowances.

Each mold line has two ends; therefore, the flat-pattern dimension (FPD) for each mold line is MLD (mold-line dimension) less the setback (Sb) at end 1 less the setback of end 2. Where there is no setback, the value of the setback is zero (0) and may therefore be disregarded.

$$FPD = MLD - Sb_1 - Sb_2$$

In the case of a right-angle piece, setback applies to only one end of each leg, so the formula is simply

$$FPD = MLD - Sb_1 - 0 = MLD - Sb_1$$

As an example, consider a 90° angle fabricated from 0.040 material, having legs with a mold-line dimension of 2.000 in and a bend radius of 0.125 in. The first step is to determine the setback distance. The second step is to calculate the bend allowance. The setback and bend allowance are then used in step 3 to calculate the developed width. Step 1 uses the K-factor chart (Table 8-4). Steps 2 and 3 are calculated using both the empirical and geometric methods. The technician should note that the difference between the developed widths determined by the two different techniques is insignificant.

1. Find the setback (*X*-distance).

$$X = K \times (R + T)$$

$$= 1.000 \times (0.125 + 0.040)$$

$$= 1.000 \times 0.165$$

$$= 0.165 \text{ in } [4.191 \text{ mm}]$$

2. Find the bend allowance.

Empirical:
$$BA = [(0.01743R) + (0.0078T)] \times N$$

$$= [(0.01743 \times 0.125) + (0.0078 \times 0.040)] \times 90$$

$$= (0.00217875 + 0.0003120) \times 90$$

$$= (0.00249075) \times 90$$

$$= 0.22416750$$

$$= 0.224 \text{ in } [5.69 \text{ mm}]$$

Geometric:

$$BA = 2\pi \times \left(R + \frac{T}{2} \right) \times \frac{N}{360}$$

$$= 2\pi \times \left[0.125 + \left(\frac{0.040}{2} \right) \right] \times \frac{90}{360}$$

$$= 2\pi \times (0.125 + 0.020) \times 0.250$$

$$= 2\pi \times 0.145 \times 0.250$$

$$= 2\pi \times 0.036$$

$$= 0.072\pi$$

$$= 0.228$$

3. Find developed width.

Empirical:
$$DW = MLD_{\text{leg 1}} - SB_{\text{leg 1}} + MLD_{\text{leg 2}} - SB_{\text{leg 2}} + BA$$

$$= 2.000 - 0.165 + 2.000 - 0.165 + 0.224$$

$$= 4.000 - 0.330 + 0.224$$

$$= 4.000 - 0.106$$

$$= 3.894$$

Geometric:

$$DW = MLD_{leg\,1} - SB_{leg\,1} + MLD_{leg\,2} - SB_{leg\,2} + BA$$
$$= 2.000 - 0.165 + 2.000 - 0.165 + 0.224$$
$$= 4.000 - 0.330 + 0.228$$
$$= 4.000 - 0.102$$
$$= 3.898$$

Bend-Allowance Tables

When bend-allowance tables are available for the type of metal being used, it is possible to save calculation time. The values shown in Table 8-5 are suitable for nonferrous sheet metal such as aluminum alloy and are derived from the empirical formula mentioned earlier. Common alloys are designated 2024-T3, 2024-T4, 2017-T3, etc. The right side of the table indicates the minimum bend radius to be used for the type and thickness of material indicated. For example, bending $\frac{5}{32}$-in-thick 2017-T3 material with a radius smaller than $\frac{15}{32}$ in is not recommended.

To form a 60° bend in a sheet of aluminum alloy where the thickness of the metal is 0.040 in [0.102 mm] and the radius of the bend is $\frac{3}{16}$ in [4.76 mm], locate the thickness of the metal from the row of figures at the top of Table 8-5. Under 0.040, locate the value in the line opposite $\frac{3}{16}$, which is the radius dimension. In this case, the value is 0.003 58. To find the bend allowance, multiply 0.003 58 (the bend allowance for 1°) by 60 to determine the bend allowance for 60°. Hence, the bend allowance is $60 \times 0.003\,58$, or 0.214 80 in [5.46 mm]. This is the amount of material required to make the bend desired.

A U-channel is a very common type of sheet-metal structure that technicians will have to make or repair. The following discussion is a step-by-step example on how to calculate and lay out a flat pattern for the fabrication of the U-channel in Fig. 8-28. All the dimensions are given in the drawing and the material is 2024-T3.

Step 1. Check if the bend radius in the drawing is equal to or larger than the minimum bend radius given in Table 8-3. The minimum bend radius for 2024 T3 in Table 8-3 is 0.12 in, and the radius specified in the drawing is 0.25 in, so the bend radius is good.

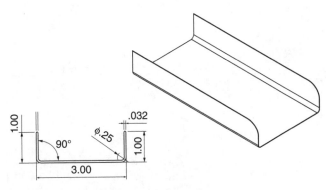

FIGURE 8-28 Sheet-metal U-channel.

Step 2. Calculate setback (SB).

Use the formula: Setback = K value × (material thickness + bend radius)

$$SB = K \times (MT + BR)$$
$$SB = 1 \times (0.032 + .25)$$
$$SB = 0.282$$

Note: The K value for a 90° angle is 1.

Step 3. Calculate the legs of the U-channel. For this example the left leg is leg 1, the middle leg is leg 2, and the right leg is leg 3.

Leg 1 = Mold line dimension (MLD) (dimension in the drawing) – SB

Leg 1 = MLD – SB

Leg 1 = 1 – .282

Leg 1 = .718 in

Note: Leg 2 has two bends and therefore it also has 2 setbacks.

$$Leg\,2 = MLD - (2 \times SB)$$
$$Leg\,2 = 3 - (2 \times .282)$$
$$Leg\,2 = 2.436\ in$$

$$Leg\,3 = MLD - SB$$
$$Leg\,3 = 1 - .282$$
$$Leg\,3 = .718\ in$$

Step 4. Calculate the bend allowance or find the number in Table 8-6.

Calculation method: $BA = (0.0173R + 0.0078T)N$

$$BA = (0.0173 \times .25 + 0.0078 \times 0.032) \times 90$$
$$BA = 0.415\ in$$

Using bend-allowance chart:

In Table 8-6 for the material thickness of 0.032 in and a bend radius of .25 in we find .004 61 per degree of bend. The bend is 90° so we multiply .004 61 × 90. The bend allowance is 0.415 in.

Note: The U-channel has two bends, so the bend allowance has to be taken into account two times for the calculation of the total developed width.

Step 5. Calculate total developed width (TDW).

$$TDW = Leg\,1 + Leg\,2 + Leg\,3 + BA\,1 + BA\,2$$
$$TDW = .718 + 2.346 + .718 + .415 + .415$$
$$TDW = 4.612\ in$$

Example of open bend

Not all bends are 90°. Some sheet-metal parts have flanges that are less than 90° and these are called open bends. The K factor needs to be used to calculate the setback and the total developed area (TDW) of these parts. Also the bend

Radius ↓ / Thickness →	.016	.020	.022	.025	.028	.032	.040	.045	.051	.064	.072	.081	.091	.128	5/32	3/16
1/32	.00067	.00070	.00072	.00074	.00077	.00079										
1/16	.00121	.00125	.00126	.00129	.00131	.00135	.00140	.00144	.00149	.00159	.00165					
3/32	.00176	.00179	.00180	.00183	.00186	.00188	.00195	.00199	.00203	.00213	.00220	.00226	.00234			
1/8	.00230	.00234	.00235	.00238	.00240	.00243	.00249	.00253	.00258	.00268	.00274	.00281	.00289	.00317		
5/32	.00285	.00288	.00290	.00292	.00295	.00297	.00304	.00308	.00312	.00322	.00328	.00335	.00343	.00372	.00394	.00473
3/16	.00339	.00342	.00344	.00347	.00349	.00352	.00358	.00362	.00367	.00377	.00383	.00390	.00398	.00426	.00449	.00528
7/32	.00394	.00397	.00398	.00401	.00403	.00406	.00412	.00417	.00421	.00431	.00437	.00444	.00452	.00481	.00503	.00582
1/4	.00448	.00451	.00454	.00456	.00458	.00461	.00467	.00471	.00476	.00486	.00492	.00499	.00507	.00535	.00558	.00636
9/32	.00503	.00506	.00507	.00510	.00512	.00515	.00521	.00526	.00530	.00540	.00546	.00553	.00561	.00590	.00612	.00691
5/16	.00557	.00560	.00562	.00564	.00567	.00570	.00576	.00580	.00584	.00595	.00601	.00608	.00616	.00644	.00667	.00745
11/32	.00612	.00615	.00616	.00619	.00621	.00624	.00630	.00634	.00639	.00649	.00655	.00662	.00670	.00699	.00721	.00800
3/8	.00666	.00669	.00671	.00673	.00676	.00679	.00685	.00689	.00693	.00704	.00710	.00717	.00725	.00753	.00776	.00854
13/32	.00721	.00724	.00725	.00728	.00730	.00733	.00739	.00743	.00748	.00758	.00764	.00771	.00779	.00808	.00830	.00909
7/16	.00775	.00778	.00780	.00782	.00785	.00787	.00794	.00798	.00802	.00812	.00819	.00826	.00834	.00862	.00884	.00963
15/32	.00829	.00833	.00834	.00837	.00839	.00842	.00848	.00852	.00857	.00867	.00873	.00880	.00888	.00917	.00939	.01018
1/2	.00884	.00887	.00889	.00891	.00894	.00896	.00903	.00907	.00911	.00921	.00928	.00935	.00943	.00971	.00993	.01072
17/32	.00938	.00942	.00943	.00946	.00948	.00951	.00957	.00961	.00966	.00976	.00982	.00989	.00997	.01025	.01048	.01127
9/16	.00993	.00996	.00998	.01000	.01002	.01005	.01012	.01016	.01020	.01030	.01037	.01043	.01051	.01080	.01102	.01181
19/32	.01047	.01051	.01051	.01055	.01057	.01058	.01065	.01070	.01073	.01083	.01091	.01098	.01105	.01133	.01157	.01236
5/8	.01102	.01105	.01107	.01109	.01112	.01114	.01121	.01125	.01129	.01139	.01146	.01152	.01160	.01189	.01211	.01290
21/32	.01156	.01160	.01161	.01164	.01166	.01170	.01175	.01179	.01183	.01193	.01200	.01207	.01214	.01245	.01266	.01345
11/16	.01211	.01214	.01216	.01218	.01220	.01223	.01230	.01234	.01238	.01248	.01254	.01261	.01269	.01298	.01320	.01399
23/32	.01265	.01268	.01269	.01273	.01275	.01276	.01283	.01288	.01291	.01301	.01309	.01316	.01322	.01351	.01374	.01454
3/4	.01320	.01323	.01324	.01327	.01329	.01332	.01338	.01343	.01347	.01357	.01363	.01370	.01378	.01407	.01429	.01508
25/32	.01374	.01378	.01378	.01381	.01384	.01386	.01392	.01397	.01401	.01411	.01418	.01425	.01432	.01461	.01484	.01562
13/16	.01429	.01432	.01433	.01436	.01438	.01441	.01447	.01451	.01456	.01466	.01472	.01479	.01487	.01516	.01538	.01617
27/32	.01483	.01486	.01487	.01490	.01493	.01494	.01501	.01506	.01509	.01519	.01527	.01534	.01540	.01569	.01593	.01671
7/8	.01538	.01541	.01542	.01545	.01547	.01550	.01556	.01560	.01565	.01575	.01581	.01588	.01596	.01625	.01647	.01726
29/32	.01592	.01595	.01596	.01599	.01602	.01604	.01616	.01615	.01619	.01629	.01636	.01643	.01650	.01679	.01701	.01780
15/16	.01646	.01650	.01651	.01654	.01656	.01659	.01665	.01669	.01674	.01684	.01690	.01697	.01705	.01734	.01756	.01835
31/32	.01701	.01704	.01705	.01708	.01711	.01712	.01718	.01724	.01727	.01737	.01745	.01752	.01758	.01787	.01810	.01889
1	.01755	.01759	.01760	.01763	.01765	.01768	.01774	.01778	.01783	.01793	.01799	.01806	.01814	.01843	.01865	

Material designations shown with brackets at right of table: 2024-O, 2017-T3, 2024-T3, 2024-T6

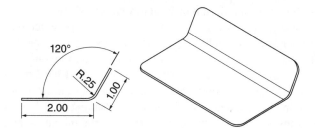

FIGURE 8-29 Open bend.

allowance for open bends is different from (less than) 90° bends. The following discussion is a step-by-step example on how to calculate the TDW of a sheet-metal angle in Fig. 8-29. All the dimensions are given in the drawing. The material is 2024 T3 and the material thickness is 0.050 in.

Step 1. Verify if the bend radius is equal to or larger than the minimum bend radius. The minimum bend radius for 0.050 2024 T3 aluminum is .19 (Table 8-3). The bend radius indicated in the drawing is .25. This means that the bend radius in the drawing is acceptable and the part won't crack during bending.

Step 2. Calculate setback (SB).

Because the bend angle is not 90° we have to look up the K value in the Table 8-6 (K-chart). The drawing indicates that the angle of bend is 120° but if you look carefully you will see that the flange is bent less than 90° from flat. We need to calculate the complement of the angle indicated in the drawing. We can calculate the actual bend angle as follows:

Actual bend angle = (180 – bend angle in drawing)

Actual bend angle = 180 – 120

Actual bend angle = 60

The K value for 60 = .5773

Use the formula: Setback = K value × (material thickness + bend radius)

SB = K × (MT + BR)

SB = .5773 × (0.050 + .25)

SB = 0.19

Step 3. Calculate the legs of the angle piece.

Leg 1 = Mold line dimension (MLD) (dimension in the drawing) – SB

Leg 1 = MLD – SB

Leg 1 = 2 – .19

Leg 1 = 1.81 in

Leg 2 = MLD – SB

Leg 2 = 1 – .19

Leg 2 = .81 in

Step 4. Calculate the bend allowance or find the number in Table 8-6.

Calculation method: BA = (0.0173R + 0.0078T)N

BA = (0.0173 × .25 + 0.0078 × 0.050) × 60

BA = 0.2829 in

Using bend-allowance chart:

In Table 8-6 for the material thickness of 0.050 in and a bend radius of .25 in we find .004 76 per degree of bend. The bend is 60° so we multiply .004 61 × 60. The bend allowance is 0.28 in.

Step 5. Calculate total developed width (TDW).

TDW = Leg 1 + Leg 2 + BA

TDW = 1.81 + .81 + .28

TDW = 2.9 in

Example of a closed bend

Some sheet-metal parts have flanges that are bend more than 90° and these are called closed bends. The K factor needs to be used to calculate the setback and the total developed area (TDW) of these parts. Also the bend allowance for closed bends is different (more) than for 90° bends. The following discussion is a step-by-step example on how to calculate the TDW of a sheet-metal angle in Fig. 8-30. All the dimensions are given in the drawing and the material is 2024 T3 and the material thickness is 0.050 in.

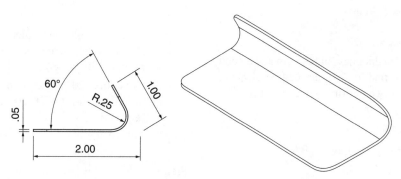

FIGURE 8-30 Closed bend.

Step 1. Verify if the bend radius is equal to or larger than the minimum bend radius. The minimum bend radius for 0.050 2024 T3 aluminum is .19 (Table 8-3). The bend radius indicated in the drawing is .25. This means that the bend radius in the drawing is acceptable and the part won't crack during bending.

Step 2. Calculate setback (SB).

Because the bend angle is not 90° we have to look up the K value in the Table 8-6 (K-chart). The drawing indicates that the angle of bend is 60°, but if you look carefully you will see that the flange is bent more than 90° from flat. We need to calculate the complement of the angle indicated in the drawing. We can calculate the actual bend angle as follows:

Actual bend angle = (180 – bend angle in drawing)

Actual bend angle = 180 – 60

Actual bend angle = 120

The K value for 120 = 1.732

Use the formula: Setback = K value × (material thickness + bend radius)

$$SB = K \times (MT + BR)$$
$$SB = 1.732 \times (0.050 + .25)$$
$$SB = 0.52$$

Step 3. Calculate the legs of the angle piece.

Leg 1 = Mold line dimension (MLD) (dimension in the drawing) – SB

Leg 1 = MLD – SB
Leg 1 = 2 – .52
Leg 1 = 1.48 in

Leg 2 = MLD – SB
Leg 2 = 1 – .52
Leg 2 = .48 in

Step 4. Calculate the bend allowance or find the number in Table 8-6.

Calculation method: BA = (0.0173R + 0.0078T)N
$$BA = (0.0173 \times .25 + 0.0078 \times 0.050) \times 120$$
$$BA = 0.56 \text{ in}$$

Using bend-allowance chart:

In Table 8-6 for the material thickness of 0.050 in and a bend radius of .25 in we find .004 76 per degree of bend. The bend is 120° so we multiply .004 61 × 120. The bend allowance is 0.55 in.

Step 5. Calculate total developed width (TDW).

$$TDW = \text{Leg 1} + \text{Leg 2} + BA$$
$$TDW = 1.48 + .48 + .56$$
$$TDW = 2.52 \text{ in}$$

J-chart for calculation of total developed width

The J-chart is often found in Manufacturers' Structural Repair Manuals, and it is a simple and quick way to find the Total Developed Width (TDW) of a flat pattern layout. The main advantage of the J-chart is that you don't have to do any calculations, remember formulas, or look up information in charts and tables. The only information that you need to know are bend radius, bend angle, material thickness, and mold line dimensions, which you can find in the drawing or you can measure them with simple measuring tools.

How to find the total developed width using a J-chart

- Place a straight edge across the chart and connect the bend radius on the top scale with the material thickness on the bottom scale.
- Locate the angle on the right-hand scale and follow this line horizontally until it meets the straight edge.
- The factor X (bend deduction) is then read on the diagonally curving line.
- Interpolate when the X factor falls between lines.
- Add up the mold line dimensions and subtract the X factor to find the TDW.

Example 1. Refer to Fig. 8-30 and determine the TDW using the J-chart.

Step 1. Place a straight edge (ruler) across the chart and connect .25 on the top scale (bend radius) with .050 on the bottom scale (thickness).

Step 2. Locate the 60° angle on the right-hand scale (angle) and follow this line horizontally until it meets the straight edge.

Step 3. Follow the curved line down and read 0.06 for the bend deduction.

Step 4. Add up the mold line dimensions and subtract the X factor (bend deduction).

$$TDW = (MLD 1 + MLD 2) - \text{X factor}$$
$$TDW = (2 + 1) - .06$$
$$TDW = 2.9$$

Example 2. Refer to Fig. 8-31 and determine the TDW using the J-chart. Note the J-chart is somewhat difficult to read for large bend angles but the accuracy obtained is sufficient for most sheet-metal projects.

Step 1. Place a straight edge (ruler) across the chart and connect .25 on the top scale (bend radius) with .050 on the bottom scale (thickness).

Step 2. Locate the 120° angle on the right-hand scale (angle) and follow this line horizontally until it meets the straight edge.

Step 3. Follow the curved line down and read .40 for the bend deduction.

Step 4. Add up the mold line dimensions and subtract the X factor (bend deduction).

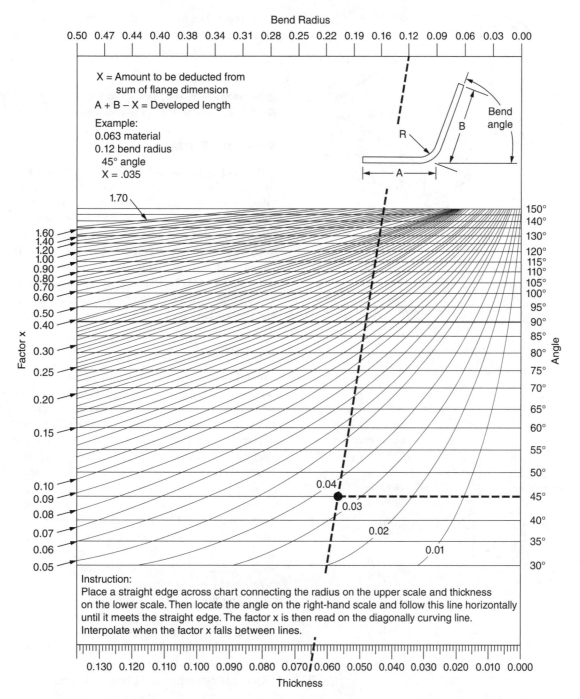

FIGURE 8-31 J-chart for calculation of total developed width (TDW).

$$TDW = (MLD\ 1 + MLD\ 2) - X\ factor$$

$$TDW = (2 + 1) - .40$$

$$TDW = 2.6$$

Bend Relief Radius

Where a sheet-metal part has intersecting bend radii, it is necessary to have a bend relief radius. This bend relief radius prevents an intense concentration of stresses that would be allowed if it were a square corner (the most abrupt cross-sectional change possible). The larger and smoother this radius the less likely a crack will form in the corner. Generally this radius is specified on the drawing (Fig. 8-32). Its positioning is important. It should be located so its outer perimeter touches the intersection of the inside bend tangent lines. This will keep any material from the bend interfering with the bend allowance area of the other bend. If these bend allowance areas intersect with each other, there would be substantial compressive stresses accumulate in that corner while bending. This could cause the part to crack while bending or at some later date because of its pre-stressed, weakened state.

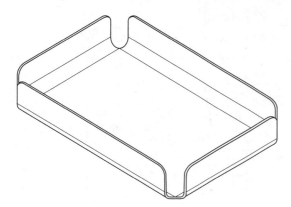

FIGURE 8-32   Box with bend radius relief holes.

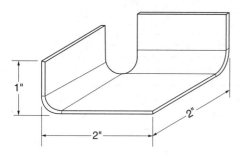

FIGURE 8-33   Bend relief radius holes.

Layout Method

Lay out the basic part using traditional layout procedures. This will give you the width of the flats and the bend allowance. It is the intersection of the inside bend tangent lines that index the bend relief radius' position. Bisect these intersected lines and move outward the distance of the radius on this line. This will be the center of the hole. Drill at this point. Finish by trimming off the remainder of the corner material. This trim out is often tangent to the radius and perpendicular to the edge (Fig. 8-33). This will leave an open corner. If the corner must be closed, or a slightly longer flange is necessary, then trim out accordingly. If the corner is to be welded it will be necessary to have touching flanges at the corners. The length of the flange should be one material thickness shorter than the finished length of the part, so only the insides of the flanges will touch.

PREPARATION FOR LAYOUT WORK

Working Surface

Before beginning aircraft layout work, the technician should find a good working surface, such as a smooth, flat bench or table. A good surface plate is ideal if a small sheet of material is to be used for the layout. In working with a large sheet of material, it is important to avoid bending it; hence it is a good practice to have a helper to assist in laying the sheet on the working surface. To protect the undersurface of the material from any possible damage, it is often advisable to place a piece of heavy paper, felt, or plywood between the material and the working surface.

Layout Process

Layout process of various kinds are applied to a metal surface for layout work so that the pattern will stand out clearly while the technician is cutting along the drawn or scribed lines. Among the fluids or coatings used are *zinc chromate, bluing fluid, flat white paint,* and *copper sulfate solution.* The coatings used should be easy to scratch away with a scriber or other marking instrument so the mark will show clearly.

Zinc chromate is a metal primer that can be sprayed on a metal surface in a thin coat. It not only serves as a good background color for the pattern used in layout work but also acts as a protection for the surface during the layout work. It tends to prevent corrosion and helps prevent scratches. Zinc chromate need not be removed from a part after the layout and forming are completed.

Bluing fluid, also called **Dykem blue,** is brushed on the metal surface. Although both scribed lines and pencil work show up clearly on a metal surface coated with zinc chromate, this is not the case with bluing fluid. The scribed lines will be clearly visible against the dark background, but pencil work is difficult to see. Chemically, bluing fluid is merely a blue or purple dye dissolved in alcohol or a similar solvent. It does not protect metal against corrosion or serve as a binder for paint; hence it must be removed from the part with alcohol or other suitable solvent.

Flat white paint, soluble in water, can be used for some types of layout work. To be sure that it will come off when water is applied, a small sample area can be painted on a piece of scrap metal. The disadvantage of using lay-out fluids and a scribe is that the material could easily be scratched by the scribe and weaken the material or cause corrosion because the protective Alclad coating has been scratched off. For these reasons many technicians use-extra-fine point permanent markers such as a Sharpie and they mark directly on the aluminum, or on the protective plastic on the aluminum or the aluminum can be taped off with masking tape and the layout can be made directly on the masking tape.

Another layout fluid is a **copper sulfate solution**. Scribed lines on iron or steel stand out clearly when this solution is brushed on the surface. Through a chemical action, a coating of copper is deposited on the iron or steel. The scribed lines show as a bright steel color through the copper-colored coating. Copper sulfate solution must not be used on aluminum or aluminum-alloy surfaces. Field use of commercial felt markers that will not damage the metal provides a "no-mess," pocketable layout dye that is removable with acetone.

Planning the Work

Having examined a blueprint from which a design is to be transferred to material, the technician should plan the job carefully. If the part to be made is small, it may be possible to make it from scrap metal that is found to be sound after having been examined carefully to make sure it has no scratches or nicks. If there is no suitable scrap available, it may be advisable to cut the part from a corner of a large sheet of metal, thereby avoiding the waste that would occur if the piece were cut from the center.

In many cases it may be advisable to do the layout on a piece of cardboard or stiff paper before marking the metal for cutting. This is particularly true where the shape is complex and there is considerable danger of mistakes. This technique is particularly adaptable to field use, where the repair is being made in or on the aircraft. It is a sign of poor quality to produce a good repair with all the layout lines and mistakes showing. The technician may attach the paper layout to the aircraft with rubber cement or other suitable temporary adhesive when checking for proper size and shape.

Reference Edges and Reference Lines

When a technician "trues up" one edge of a sheet of metal on a squaring shear or if the sheet is received with one edge already straight, that edge can be used as a **reference edge**. A reference edge provides a line from which various measurements can be made, thus increasing the uniformity and accuracy of the work. If there are two such reference edges at right angles to each other, it is even better, for then the operator can obtain a much greater degree of accuracy in layout. When possible, these reference edges should be edges of the finished part; however, if the finished part is to have an irregular outline, it may be advisable to prepare one or two reference edges even though they will disappear when that portion of the material is cut away in finishing the part.

When cutting tools are not available or when it is not practical to establish a reference edge, **reference lines** should be drawn or scribed, preferably two lines at right angles. These may be established only temporarily and have no relation to other lines except as reference lines, or they may be centerlines for holes. In a like manner, the center line for the completed part may serve as one of the reference lines. In that case, the technician merely constructs a perpendicular to the centerline of the part, thus obtaining the two desired lines.

Reference edges and reference lines may be better understood if they are regarded as **base lines** used in the construction of angles, parallel lines, and intersecting lines required in the layout. However, in the strict sense of the term, a base line is the horizontal reference line, as viewed by the observer; hence it is technically correct to speak of reference lines and edges as previously explained.

Care of Material During Layout

We have already mentioned the importance of not wasting material by cutting a part from the center of the sheet. It is likewise a matter of good judgment to avoid making cuts that extend into the center. A good technician tries to do layout work along existing edges, leaving as much material intact and usable as possible.

The technician who wears a wristwatch or rings should be careful to see that these do not scratch the material. Leaving tools and instruments on the material is another source of damage. When weights are used to hold a pattern or template on the material, they should be smooth on their lower surface or be padded with felt. They should also be free from sharp corners or projections that might accidentally injure the axis. The distance of the bend allowance depends on the material. Manufacturers of aircraft and sheet-metal parts may spray a thin coat of polyvinyl alcohol (PVA) or polyvinyl chloride (PVC) on new sheet metal to help prevent "handling rash." These green-gold coatings are water-soluble and are washed off as necessary for fabrication and painting.

HAND TOOLS FOR SHEET-METAL WORK

Most common hand tools are used at one time or another when working with sheet-metal structures. This section considers those tools that are principally associated with sheet-metal construction and repair operations not included in the discussion of general-use hand tools in the text *Aircraft Basic Science*.

Hammers

Figure 8-34 shows two types of **hammers** often used by aircraft technicians in the repair and forming of sheet-metal

FIGURE 8-34 Planishing hammers.

FIGURE 8-35   Hand nibbling tool. (*U.S. Ind. Tool & Supply.*)

FIGURE 8-37   Chassis punches.

aircraft parts. The hammers shown are used for smoothing and forming sheet metal and are commonly called **planishing hammers**. The hammer with the square face is used for working in corners. When it is necessary to work in the radius of a bend, a hammer with a cross peen is used.

Hand Nibbling Tool

A **hand nibbling tool** is used to remove metal from small areas by cutting out small pieces of metal with each "nibble." This tool is most useful in confined areas where using snips is not possible and where a minimum amount of metal distortion is desired. A hand nibbling tool is shown in Fig. 8-35.

Hole Saws

A **hole saw** is a small, circular saw, the shank of which can be inserted in the chuck of a drill motor or drill press. Several hole saws are shown in Fig. 8-36.

During use of a hole saw with a drill motor, it is sometimes necessary to drill a pilot hole and use a hole-saw pilot or a hole-saw guide with a pilot. In some instances the hole saw includes a drill, which both provides the pilot hole and then, after the hole is drilled, acts as the pilot. The former is recommended for aircraft applications because the side of the flutes on the latter's drill may rub and cut metal while acting as a pilot. The pilot is the part of the saw that keeps the saw in place while cutting the metal.

Chassis Punch

Small holes may be punched in sheet metal with a **chassis punch.** The chassis punch includes a punch-and-die assembly and a threading mechanism. The threads are in the punch and the bolt passes freely through a hole in the die. A hole is drilled in the material in the center of the desired hole location. As the bolt, which also acts as a pilot, is tightened, the

die is drawn into the punch. The punch has sharp edges that, when brought in contact with the metal, enable it to cut progressively as the screw is tightened. Smaller-size punches may be used to cut holes sufficient in size for the bolts of the larger-size chassis punches (see Fig. 8-37).

Hand Rivet Set

A **hand rivet set** is shown in Fig. 8-38. This type of set is used in commercial sheet-metal work and is also used for some hand-riveting operations in aircraft work. The holes in the hand rivet set are for the purpose of drawing rivets through the metal before riveting. A rivet is inserted through a drilled hole in the metal, and the hole in the set is placed over the protruding shank of the rivet. With the head of the rivet held against a suitable backing block, the set is struck lightly with a hammer. This draws the sheets of metal together and causes them to fit snugly against the inside of the head of the rivet. The shank of the rivet may then be headed with the hammer.

Rivet Gun and Set

The **rivet gun**, or **riveting hammer**, is the device most commonly used by the aviation maintenance technician for driving rivets. The rivet gun is equipped with a **rivet set** designed to fit the head of the rivet being driven. The set is inserted into the **set sleeve** of the gun and is held in place by means of a **retaining spring**. The retaining spring must always be in place to prevent the rivet set from leaving the rivet gun and causing injury to anyone who may be in the vicinity. During operation a piston in the gun is driven rapidly back and forth by compressed air, which is alternately directed to one side of the piston and then the other. This causes the piston to strike the set rapidly, driving the rivet against the bucking bar and forming the bucked head. The cutaway drawing in Fig. 8-39

FIGURE 8-36   Hole saws.

FIGURE 8-38   A hand rivet set.

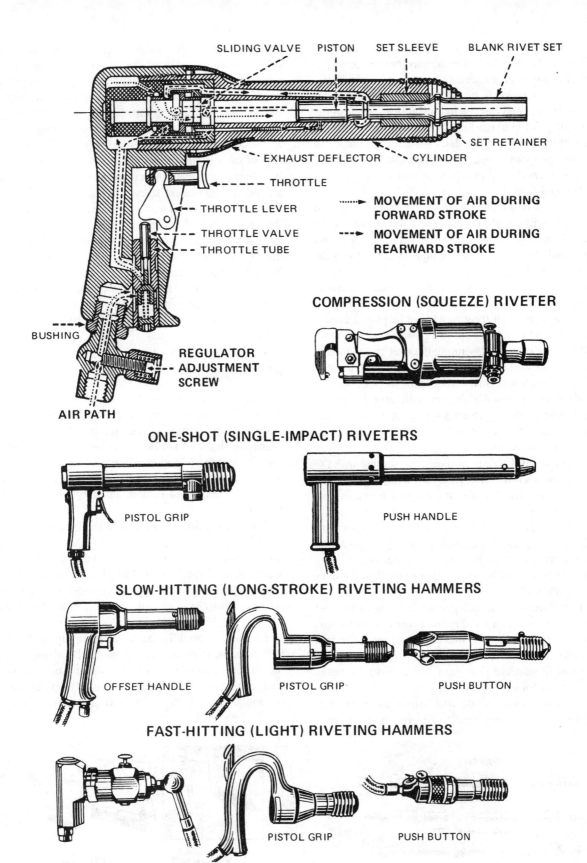

SLIDING VALVE   PISTON   SET SLEEVE   BLANK RIVET SET

SET RETAINER

EXHAUST DEFLECTOR   CYLINDER

THROTTLE

THROTTLE LEVER

THROTTLE VALVE

THROTTLE TUBE

········▶ MOVEMENT OF AIR DURING FORWARD STROKE

--·-▶ MOVEMENT OF AIR DURING REARWARD STROKE

BUSHING

REGULATOR ADJUSTMENT SCREW

AIR PATH

COMPRESSION (SQUEEZE) RIVETER

ONE-SHOT (SINGLE-IMPACT) RIVETERS

PISTOL GRIP

PUSH HANDLE

SLOW-HITTING (LONG-STROKE) RIVETING HAMMERS

OFFSET HANDLE

PISTOL GRIP

PUSH BUTTON

FAST-HITTING (LIGHT) RIVETING HAMMERS

PISTOL GRIP

PUSH BUTTON

FIGURE 8-39   Construction and types of rivet guns.

illustrates the construction and operating mechanism of a typical rivet gun and a variety of other designs for rivet guns.

Rivet guns and rivet sets perform better work and will last longer if they are properly handled and serviced, the latter meaning ordinary good care, including lubricating, cleaning, etc. A few drops of light machine oil should be placed in the air intake of the rivet gun on a daily basis. Depending upon use, the rivet gun should be disassembled and cleaned, worn parts should be replaced, and then the gun should be reassembled and lubricated.

A rivet set can be a deadly weapon. If a rivet set is placed in a rivet gun without a set retainer and the throttle of the gun is open, the rivet set may be projected like a bullet out of the gun and cause either a severe injury to a person or the destruction of equipment.

When operating the rivet gun, the set should always be placed against a pliable stationary object such as a piece of wood or aluminum. A rivet set should not be placed against steel or other hard metal when the air power is on; this practice will ruin the rivet set. Operating the rivet gun without a physical restraint needlessly extends the stroke of the piston, resulting in an abnormal extension of the set-retaining spring. Repeated operation of the gun without restraint may cause a failure of the restraining spring. A failure of the retaining spring and a lack of a physical restraint may combine to create a projectile.

Rivet guns are available in various sizes, starting with IX, the smallest, which is used for rivets of $\frac{1}{16}$ and $\frac{3}{32}$ in [1.59 and 2.38 mm] in diameter. The size of rivet guns increases progressively for rivets of larger diameter, the most commonly used size being the 3X, which is used for rivets from $\frac{3}{32}$ to $\frac{5}{32}$ in [2.38 to 3.97 mm] in diameter. For larger-size rivets, 4X and 5X rivet guns may be used. There are several larger sizes of rivet guns used for even larger rivets than the ones we have mentioned.

Rivet sets are made in many sizes and shapes to meet different requirements for riveting and to provide for the different types of rivets. The **shank** of a rivet set is the part inserted into the rivet gun. Shanks are made in uniform standard sizes to fit standard rivet guns. Figure 8-40 illustrates typical rivet sets for use with a rivet gun.

Rivet sets designed for use with universal- or brazier-head rivets have a cupped head to fit over the rivet head.

The radius of the arc used for the cupped head has a slightly larger radius than that of the rivet head to ensure that the maximum force of the gun is applied to the center of the rivet head. This causes the rivet to draw the materials together as the bucked head is being formed.

The actual force applied during the riveting operation is a function of the regulator adjustment setting and the position of the throttle (trigger). The force rate of impact may be adjusted at the gun when the gun construction includes a regulator-adjustment screw. During the riveting operation the rivet gun is subjected to considerable vibration, making the throttle difficult to control. While the rivet set is placed against a nonrigid material, the throttle is fully opened and the regulator adjustment screw is turned until the desired force is attained. Setting the regulator-adjustment screw when the throttle is fully opened limits the maximum force that may be applied during the rivet gun's operation. This limits the potential for accidental damage due to loss of throttle control during use.

Bucking Bars

A **bucking bar** is a smooth steel bar made up in a variety of special shapes and sizes and used to form a head on the shank of a rivet while it is being driven by a rivet gun. The edges are slightly rounded to prevent marring the material, and the surface is perfectly smooth. The face of the bar, placed against the shank of the rivet, is flat. Bucking bars are sometimes called **dollies**, **bucking irons**, or **bucking blocks**.

For best results, the technician should choose a bucking bar of the proper weight and shape for a particular application. A common rule of thumb is that the bucking bar should weigh approximately 1 lb [450 g] less than the number of the rivet gun with which it is being used. For example, the bucking bar to be used with a 3X gun should weigh about 2 lb [900 g]. Figure 8-41 illustrates a group of bucking bars.

Expanding bucking bars, shown in Fig. 8-42, are steel blocks whose diameters or widths can be adjusted. A bucking bar of this type is attached to the end of a hollow steel shaft containing a bar that can be twisted to expand or reduce the width of the block. It is used to buck (upset) rivets on the

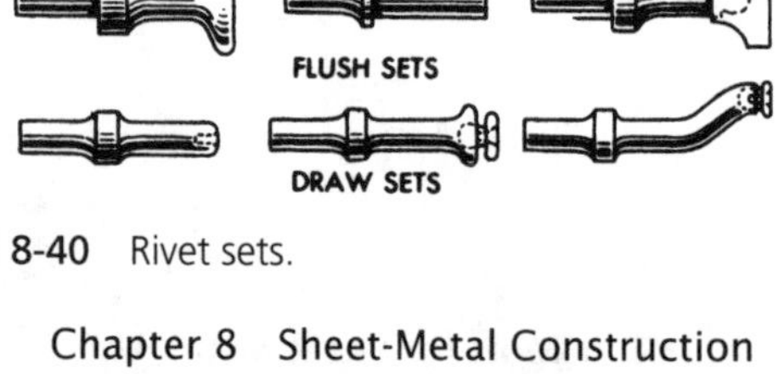

FIGURE 8-40 Rivet sets.

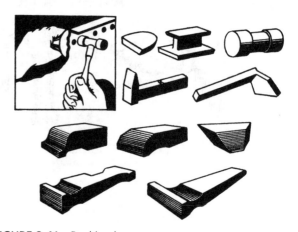

FIGURE 8-41 Bucking bars.

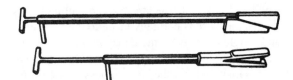

FIGURE 8-42   Expanding bucking bars.

inside of tubular structures or in similar spaces that cannot be reached by regular bucking bars. The space must be small enough for one side of the partly expanded block to press against the tip of the rivet's shank and for the other side to press against a strong supporting surface. Expanding bucking bars speed up the process of riveting in areas with limited accessibility, such as the skin on the wing section.

Sheet Fasteners

During the riveting process the sheets of metal must be held close together or the rivet will tend to expand between the sheets and leave a gap, which reduces the strength of the joint and promotes the accumulation of moisture between the sheets. This, of course, leads to corrosion. The tool designed to meet this need is called a **sheet fastener** and is quickly and easily installed. Sheet fasteners are commonly called "Clecos," a trade name applied by an early developer and manufacturer of sheet fasteners.

Sheet fasteners have been designed in many styles and shapes; however, they are presently limited to relatively few designs. One popular sheet fastener, manufactured by the Wedgelock Company, is illustrated in Fig. 8-43. Figure 8-44 shows the internal construction of the fastener.

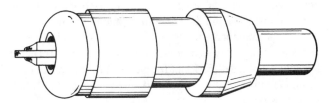

FIGURE 8-43   Sheet fastener.

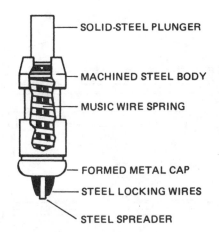

- SOLID-STEEL PLUNGER
- MACHINED STEEL BODY
- MUSIC WIRE SPRING
- FORMED METAL CAP
- STEEL LOCKING WIRES
- STEEL SPREADER

FIGURE 8-44   Construction of a sheet fastener.

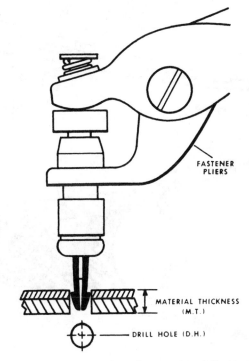

FIGURE 8-45   Inserting a sheet fastener in a drilled hole.

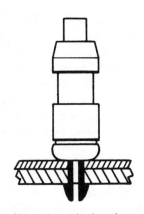

FIGURE 8-46   Locking wires gripping sheet metal.

The fastener consists of a machined steel body, in which is installed a plunger, coil spring, locking wires, and a spreader. When the plunger is depressed with the fastener pliers, the locking wires extend beyond the spreader and "toe in," reducing the diameter. The locking wires can then be inserted in a drilled hole of the proper size, as shown in Fig. 8-45. When the pliers are released, the locking wires are drawn back over the spreader. This causes the wires to separate and grip the sides of the drilled hole, as in Fig. 8-46. Removal of the fastener is accomplished by reversing the process of installation.

Hole Finder

A **hole finder** is a tool used to locate the position of holes to be placed in replacement or repair material relative to existing holes in aircraft structures or skin when the hole locations may not be more directly transferred. The tool has two leaves parallel to each other and fastened together at one end.

FIGURE 8-47   A hole finder.

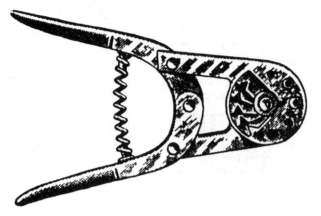

FIGURE 8-48   Rivet cutter.

The bottom leaf of the hole finder has a pin installed near the end that is aligned with a bushing on the top leaf, as shown in Fig. 8-47. The desired hole to be transferred is located by fitting the pin on the bottom leaf of the hole finder into the existing rivet hole. The hole in the new part is made by drilling through the bushing on the top leaf. If the hole finder is properly made, holes drilled in this manner will be perfectly aligned. A separate hole finder must be used for each diameter of rivet.

Rivet Cutter

A **rivet cutter**, shown in Fig. 8-48, is used to cut rivets to the proper length for a particular installation. The cutter is hand-operated, has sized holes so that the rivets are not distorted when cut, and has a gauge that can be set to the desired length of the cut rivet.

FLOOR AND BENCH MACHINERY FOR SHEET-METAL WORK

There are many sheet-metal tools that are large enough in size to be mounted on work benches or free-standing. Most aircraft-maintenance facilities have a few of the basic larger pieces of sheet-metal fabrication equipment, such as drill presses, shears, and brakes. Other large pieces of equipment used in the aviation industry are associated with particular aspects of the sheet-metal repair and fabrication processing. The purpose of discussing the various types of equipment is to make technicians aware of the large variety of tools that may be used when working with sheet-metal structures.

Short-term access to many of these tools may be gained by renting the use of the tools from companies heavily

involved in sheet-metal structures. When renting the use of these tools, the work is normally taken to the tool, since the size and weight of the equipment may prevent transport and setup of the tools at the technician's work site. In addition, technicians may find that they do not possess the skills necessary to operate many of these machines properly. In such cases, many companies will, on a contract basis, fabricate sheet-metal components to the technician's specifications. These companies are most commonly called *job shops*.

Squaring Shears

Squaring shears, shown in Fig. 8-49, are used to cut and square sheet metal. The mechanism consists of two cutting blades operated either by foot or motor power. The lower blade is stationary and is attached securely to the bed, which has a scale graduated in fractions of an inch for measuring the sheet being cut. This scale is perpendicular to the cutting blades. The other blade is mounted on the upper crosshead and is moved up and down to cut the metal stock. The upper blade is set at angle to the lower blade so the cut will take place at only one point on the metal at a time. The cut then takes place progressively across the width of the sheet.

Foot-power squaring shears usually can cut mild carbon steel up to 22 gauge. Aluminum alloys up to 0.050 in [1.27 mm] and above can be cut without difficulty. To operate, the long-bed gauge is set parallel to or at an angle with the blades according to the shape desired. The metal is placed on the bed of the machine and is trimmed off from $\frac{1}{4}$ to $\frac{1}{2}$ in [6.35 to 12.7 mm] to make a straight edge on the sheet. The trimmed edge is held against the gauge, and the sheet is sheared to size. To square the sheet, the end that has been cut last is held against the side gauge, and from $\frac{1}{4}$ to $\frac{1}{2}$ in [6.35 to 12.7 mm] is trimmed off. The sheet is turned over, with the straight edge held against the long gauge that has been set to the required distance, and then the other edge is sheared.

The parts numbered in Fig. 8-49 are as follows: (1) housing and leg, (2) lower crosshead, (3) foot treadle, (4) turnbuckle for adjusting blade, (5) T slot for gauge bolt, (6) upper crosshead, (7) upper blade, (8) bed, (9) side gauge, and (10) bolt for extension arm.

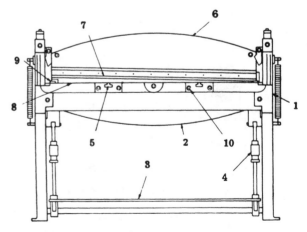

FIGURE 8-49   Squaring shear.

FIGURE 8-50   Electric power shear.

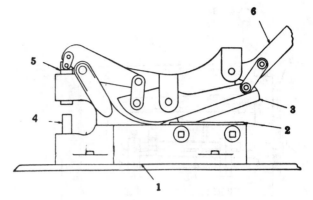

FIGURE 8-52   Slitting shears.

Squaring shears are also equipped with a "stop bar," which may be set to a desired dimension. This feature is used when a large number of pieces are to be cut to the same dimension. When using the stop, the technician should take care to protect the material as it drops after the shearing operation.

Figure 8-50 shows a power shear for cutting thicker materials.

Gap-Squaring Shears

Gap-squaring shears, shown in Fig. 8-51, resemble regular squaring shears except that the housing is constructed so that the sheet can pass completely through the machine, thus making possible the cutting of any desired length. Gap-squaring shears are used where the regular squaring shears are too narrow to split long sheets.

Slitting Shears

Slitting shears, shown in Fig. 8-52, are used to slit sheets in lengths where the squaring shears are too narrow to accommodate the work. Slitting shears of the lever type, for cutting heavier grades of sheet metal, are commonly used and are

FIGURE 8-53   Lever shears with punching attachment.

known as *lever shears*. The parts numbered in Fig. 8-52 are (1) base, (2) lower blade, (3) upper blade, (4) die, (5) punch, and (6) handle. Such shears should not be used to cut rods or bolts unless fitted with a special attachment. Some lever shears have a punching attachment on the end (opposite the blades) for punching heavy sheets, as shown in Fig. 8-53.

Throatless Shears

Throatless shears are shown in Fig. 8-54. They are usually made to cut sheet metal as thick as 10 gauge in mild carbon steel and 12 gauge in stainless steel. For aluminum-alloy

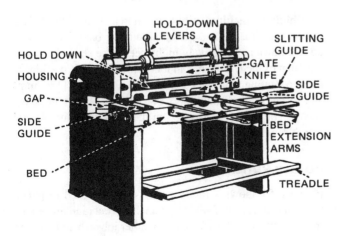

FIGURE 8-51   Gap-squaring shear.

FIGURE 8-54   Throatless shears. (*Beverly shears.*)

Floor and Bench Machinery for Sheet-Metal Work   **235**

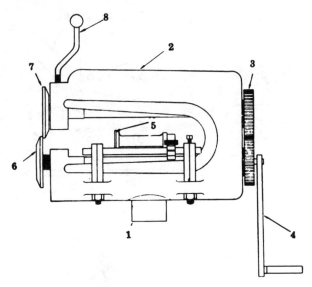

FIGURE 8-55 Rotary slitting shears.

FIGURE 8-56 Scroll shears.

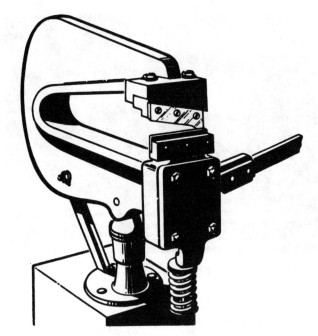

FIGURE 8-57 Portable Unishear.

stock, they can cut much heavier sheets. The frame of the throatless shear is made so that sheets of any length can be cut and the metal can be turned in any direction, allowing irregular lines to be followed or notches to be made without distorting the metal.

Rotary Slitting Shears

Rotary slitting shears are shown in Fig. 8-55. The numbered parts of the drawing are (1) shank for bench standard, (2) frame, (3) back gear, (4) hand crank, (5) adjustable gauge, (6) lower rotary cutter, (7) upper rotary cutter, and (8) cutter-adjusting screw.

Rotary slitting shears consist of a frame with a deep throat fitted with circular, disk-shaped cutters fastened to parallel shafts and connected with gears. The cutting wheels are operated by a crank or by a power-driven wheel. Such shears are used for slitting sheet metal and cutting irregular curves and circles. The edge of the sheet is held against the gauge, and the end of the sheet is pressed against the cutting wheel. The handle is turned until the full length of the strip has been cut. For irregular curves, the gauge is slid back out of the way and the handle is turned slowly; and at the same time the operator keeps the cutting wheels on the line to be cut.

Scroll Shear

A **scroll shear**, shown in Fig. 8-56, is used to cut short, straight lines on the inside of a sheet without cutting through the edge. The upper cutting blade is stationary, whereas the lower blade is movable. The tool is operated by a handle connected to the lower blade.

Unishears

A portable Unishear is shown in Fig. 8-57. Unishear is merely a trade name for a type of power shear similar to a nibbling machine. It has a high-speed, narrow, reciprocating shearing blade and adjustments of vertical and horizontal clearances, depending on the thickness of the metal to be cut. It is especially useful in cutting internal curved patterns where the radius is small. A machine for the same purpose is also made in stationary form.

Nibbling Machine (Nibbler)

A **nibbling machine (nibbler)**, shown in Fig. 8-58, is a machine with a die and a vertical cutting blade that travels up and down at a relatively high speed. The stroke is longer than that of the Unishear, and it is adjustable. This machine is used for cutting mild steel up to a thickness of $\frac{1}{2}$ in [12.7 mm] and to cut circles and curves of complex shapes in heavy sheets. The machine operates on the shearing principle and leaves rough edges.

Rotary Punch

Figure 8-59 shows a rotary punch which is used to punch holes in metal parts. The rotary punch can cut radii in corners, make washers, and perform many other jobs where holes are required. The machine is composed of two cylindrical

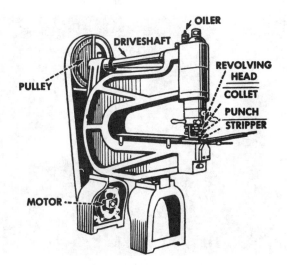

FIGURE 8-58   Nibbling machine.

FIGURE 8-59   Rotary punch.

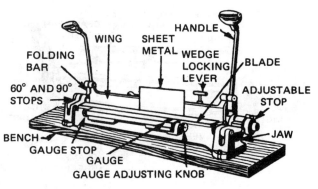

FIGURE 8-60   Bar-folding machine.

turrets, one mounted over the other and supported by the frame, with both turrets synchronized to rotate together. Index pins, which ensure correct alignment at all times, may be released from their locking position by rotating a lever on the right side of the machine. This action withdraws the index pins from the tapered holes and allows an operator to turn the turrets to any size punch desired. When rotating the turret to change punches, release the index lever when the desired die is within 1 inch of the ram, and continue to rotate the turret slowly until the top of the punch holder slides into the grooved end of the ram. The tapered index locking pins will then seat themselves in the holes provided and, at the same time, release the mechanical locking device, which prevents punching until the turrets are aligned. To operate the rotary punch, place the metal to be worked between the die and punch. Pull the lever on the top of the machine toward the operator, actuating the pinion shaft, gear segment, toggle link, and the ram, forcing the punch through the metal. When the lever is returned to its original position, the metal is removed from the punch. The diameter of the punch is

stamped on the front of each die holder. Each punch has a point in its center that is placed in the center punch mark to punch the hole in the correct location.

Folding Machines

A **bar-folding machine**, or **bar folder**, is a tool commonly used to turn (bend over) narrow edges and to turn rounded locks on flat sheets of metal to receive stiffening wires. A stop gauge is set to the width of the fold desired, the sheet metal is held against the stop, and the bending leaf of the machine is turned over by lifting the bending handle. As the bending leaf swings over, it folds the sheet. The bar-folding machine is shown in Fig. 8-60.

A **pipe-folding machine** is illustrated in Fig. 8-61. This is a machine designed to bend flanges either on flat sheets or on stock that has previously been formed into a cylindrical shape. Bar-folding and pipe-folding machines are not commonly used in aircraft sheet-metal repair.

Cornice Brake

The **cornice brake,** also called a **leaf brake,** is a machine used to make simple bends in flat sheet-metal stock. A drawing illustrating this type of brake is shown in Fig. 8-62. The cornice brake can form locks and seams, turn edges, and make bends through a wide range of angles. It operates like a bar-folding machine except that a clamping bar takes the place of the stationary jaw of the latter. The bar is raised so

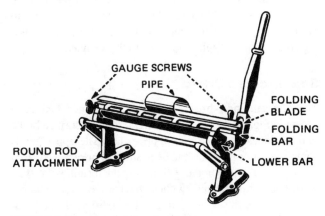

FIGURE 8-61   Pipe-folding machine.

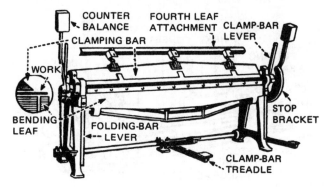

FIGURE 8-62 Cornice brake.

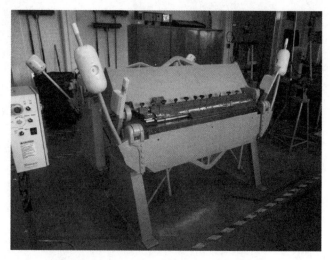

FIGURE 8-64 Box and pan brake.

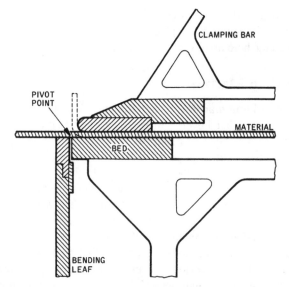

FIGURE 8-63 Positions of parts in making a bend.

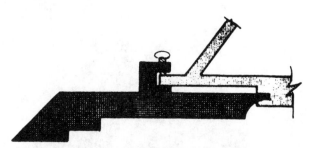

FIGURE 8-65 Upper jaw can be repositioned by loosening the thumbscrew.

that the sheet of metal can be pushed in to any desired distance, and the clamping bar is then lowered to hold the metal in place. A stop gauge is set at the angle or amount of bend to be made, and then the bending leaf is raised until it strikes the stop. When it is desired to make a bend having a given radius, a **radius bar** having the correct radius for the bend is secured to the clamping bar. Figure 8-63 shows the relative positions of the parts when making the bend.

Note that when properly set up, the distance between the radiused end of the radius bar and the pivot point of the bending leaf is equal to the thickness of the material that is to be bent. This will become an important factor when discussing an alternative "sight line" later in this chapter.

Box and Pan Brake

The **box and pan brake**, also called a **finger brake**, is shown in Fig. 8-64. This brake differs from the cornice brake in that it does not have a solid upper jaw. The upper jaw is composed of fingers that can be positioned or removed from the upper leaf of the brake. This allows the forming of boxes, pans, and similar shapes without having to distort existing sides to make the final bend. By using different combinations of fingers, the width of the upper jaw can be changed

to fit specific needs. The jaws are held on the upper leaf by a thumbscrew, as shown in Fig. 8-65.

How to Adjust a Brake?

Before a sheet-metal brake can be used to fabricate a sheet-metal part a few checks/adjustments are necessary to operate the equipment. The technician should check if the correct radius bar is installed, adjust the nose gap, adjust the clamping pressure, and check if the nose bar is parallel.

Check Nose Bar Radius

Typically the radius of the nose bar pieces is stamped on the nose pieces. If not sure, use the radius gauge to determine the correct radius. The fabrication drawing should indicate the radius size to use.

Adjust Nose Gap

The gap between the front of the nose piece and the folding bar should be 1 metal thickness. The easiest way to check is to move the folding bar up to 90° and insert a piece of sheet metal with the same thickness as the work piece in between the nose bar and the folding bar; see Fig. 8-66a. Adjust the nose gap by moving the nose gap adjusting mechanism forward or aft; see Fig. 8-66b. Some brakes have a turn knob and others like in the picture use two bolts to adjust the fore and aft movement.

FIGURE 8-66a Insert metal piece between nose gap and folding bar.

FIGURE 8-67a Insert metal in the brake.

FIGURE 8-66b Nose gap adjustment mechanism.

Adjust Clamping Pressure

Insert a piece of sheet metal of the same thickness as the metal used for the fabrication of the part in the brake (see Fig. 8-67a). Pull the handles on the side of the brake down. If the handles are too easy or too hard to pull down the clamping pressure needs to be adjusted. Adjust the nut on the clamping pressure mechanism (see Fig. 8-67b) to change the clamping pressure. The handle should have a

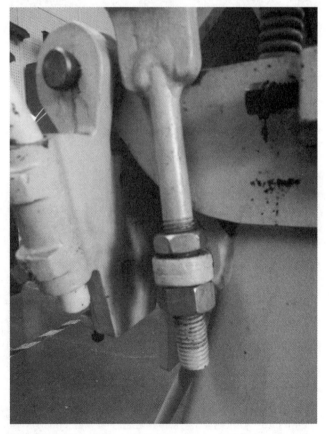

FIGURE 8-67b Adjust clamping pressure by turning the nuts.

FIGURE 8-68a Adjust for parallel alignment.

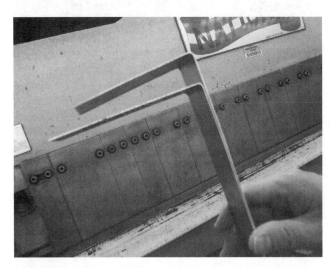

FIGURE 8-68b Check of the bend angles.

moderate resistance and the inserted sheet metal should not be able to be moved after the handles are down.

Adjusting the Nose Bar for Parallel

It is important to check that the nose gap is the same over the length of the brake. If the nose gap is different then the bend angle will be different, which will result in an uneven bend. Insert two pieces of sheet metal at the outsides of the brake (see Fig. 8-68a) and bend them to 90°. Hold the pieces on top of each other and see if they have the same bend angle (see Fig. 8-68b). If the angle is not the same, the nose gap adjusting mechanism on one side needs to be adjusted till the two pieces have the same bend angle.

Forming Roll

The **forming roll**, shown in Fig. 8-69, is used to form sheet metal into cylinders of various diameters. It consists of

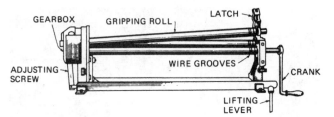

FIGURE 8-69 Forming roll.

right-end and left-end frames, between which are mounted three solid-steel rolls connected with gears and operated either by means of a power drive or a hand crank. The front rolls can be adjusted to the thickness of the metal by two thumbscrews located either on the bottom or the top of each end frame. These rolls grip the metal when it is started in the machine and carry it to the rear, or forming, roll, which is adjusted in the same manner as the front pair. The forming roll is called a **slip-roll** former. The top roll in this type is so arranged that one end can be loosened and raised, permitting the work to be slipped out at the end to prevent distortion of the metal. Small cylinders and pipes can be formed on the slip-roll former.

Turning Machine

A **turning machine** is a short-throated rotary device used on circular work to turn a narrow edge (flange) or to score a crease. The upper roll die is disk-shaped with a round edge. The lower roll is cylindrically shaped, with a semicircular groove running around its side near the outer end. This machine is sometimes referred to as a **burring**, **turning**, or **wiring machine**, since it performs all these functions, provided the correct attachments are available.

A turning machine with a variety of attachments is shown in Fig. 8-70. In this drawing the parts are (1) turning rolls, (2) elbow-edging rolls, (3) burring rolls, (4) wiring rolls, (5) elbow-edging rolls, (6) elbow-edging rolls, and (7) burring rolls. The letter A refers to a small crank screw, and the letter B refers to the crank that controls the rolls.

Beading Machines

The **beading machine**, shown in Fig. 8-71, is used to turn beads on pipes, cans, buckets, etc., both for stiffening and for ornamental purposes. It is also used on sheet-metal stock that is to be welded to prevent buckling and breaking of the metal. The numbers in the drawing refer to (1) single-beading roll, (2) triple-beading roll, (3) double-beading roll, and (4) ogee-beading roll. This machine was used extensively for aluminum oil lines on early aircraft to provide beads for holding clamped rubber tubing.

Crimping Machine

The **crimping machine**, shown in Fig. 8-72, is used to make one end of a pipe joint smaller so that several sections may

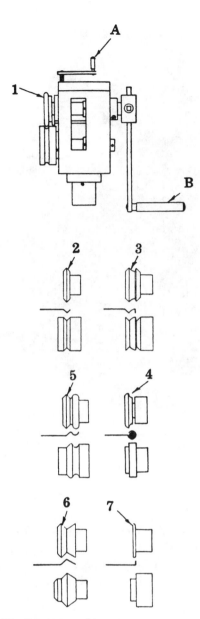

FIGURE 8-70   Turning machine.

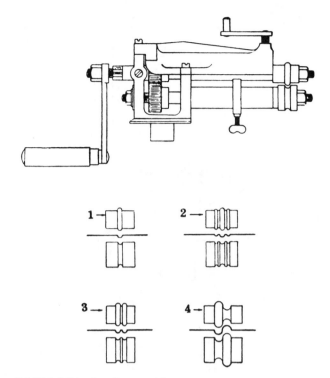

FIGURE 8-71   Beading machine.

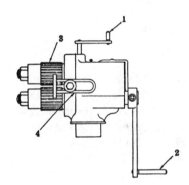

FIGURE 8-72   Crimping machine.

be slipped together. The mechanism is similar to that of the beading machine, but the rolls are corrugated longitudinally. The parts numbered in the drawing are (1) vertical adjustment handle, (2) crank, (3) crimping rolls, and (4) adjustable gauge.

Combination Rotary Machine

The **combination rotary machine** is a three-speed, motor-driven combination of the burring, turning, wiring, elbow-edging, beading, crimping, and slitting machines. This machine is illustrated in Fig. 8-73, with numbers indicating the following types of rollers: (1) thin forming rolls, (2) thick forming rolls, (3) burring rolls, (4) wiring rolls, and (5) elbow-edging rolls. This machine is supplied with a spanner wrench for quickly changing the different rolls. It is most useful where a variety of operations are to be performed.

Metal-Cutting Band Saw

A **metal-cutting band saw** is similar to a wood-cutting band saw in construction. Because of the different properties of the metals that may be cut, the saw must be adjustable in speed. The metal-cutting blades must have hard-tempered teeth, which will cut through hard metal without excessive wear or breakage. A metal-cutting band saw is shown in Fig. 8-74.

In using a band saw to cut metal, the operator should refer to a chart to determine the type of blade and the blade speed for any particular metal. Table 8-6 gives blade speeds.

Stakes

A **stake** is a type of small anvil having a variety of forms, as shown in Fig. 8-75. It can be set in a bench plate and used to bend and shape sheet metal to the desired form by hand

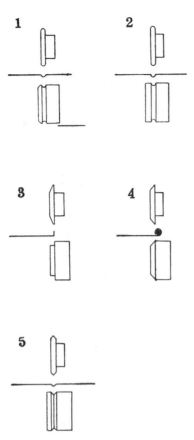

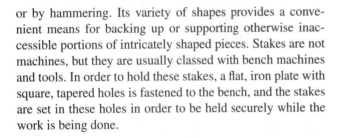

FIGURE 8-73 Combination rotary machine.

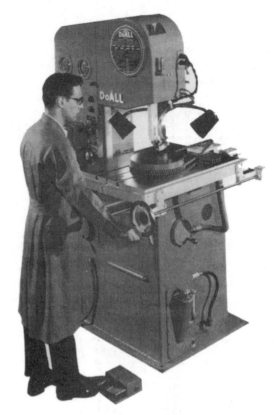

FIGURE 8-74 Metal-cutting band saw. (*DoAll Co.*)

or by hammering. Its variety of shapes provides a convenient means for backing up or supporting otherwise inaccessible portions of intricately shaped pieces. Stakes are not machines, but they are usually classed with bench machines and tools. In order to hold these stakes, a flat, iron plate with square, tapered holes is fastened to the bench, and the stakes are set in these holes in order to be held securely while the work is being done.

Drill Press

A **drill press** is a bench- or floor-mounted machine designed to rotate a drill bit and press the sharpened point of the bit against metal in order to drill a hole. The drill press is driven by an electric motor through a speed-changing mechanism, either a belt transmission or a gear transmission. A drill press is shown in Fig. 8-76.

TABLE 8-6 Pitch and Velocity Settings for Carbon-Steel Saw Blades

Material	Blade Pitch, Teeth per inch			Blade Velocity, ft/min		
	$\frac{1}{2}$ inch Thick	1 inch Thick	3 inches Thick	$\frac{1}{2}$ inch Thick	1 inch Thick	3 inches Thick
Carbon steels	10	8	4	175	175	150
Manganese steels	10	8	4	125	100	80
Nickel steels	10	10	4	100	90	80
Nickel-chrome steels	10	8	4	100	90	75
Molybdenum steels:						
4017–4042	10	4	4	135	125	110
4047–4068	10	8	4	125	100	75
Chrome-molybdenum steels	10	8	4	100	75	50
Corrosion-resistant steels, 303, 416	10	8	4	100	80	60

Note: For complete information on the operation of a power saw for metal cutting, the manufacturer's chart should be consulted.
For some types of materials a high-speed steel blade or a tungsten-carbide blade should be used.
When cutting many types of materials, cutting fluid or lubricants should be used.
To prevent tooth stripping and breaking of teeth, at least two teeth should be in contact with the work.

FIGURE 8-75   Stakes.

FIGURE 8-76   Gear-type drill press.

The belt transmission consists of two stacks of V-belt pulleys, which may vary progressively in size from 2 to 6 in [5.08 to 15.24 cm]. The pulleys are arranged so that one set decreases in size as the belt is moved up the stack and the other decreases as the belt is moved down the stack. Thus, as the belt is moved up or down the pulleys, the ratio of the motor speed to the spindle speed is changed. This is an important feature because the speed of rotation for the drill bit should vary in accordance with the type of material being drilled and the size of the hole being drilled.

The drill press spindle is either fitted with a standard chuck or provision is made for the insertion of drill bits with tapered shanks. Many drill presses are arranged so that a drill chuck with a tapered shank can be installed when the machine is driving small drills and, when large drills are used, the chuck can be removed and a drill with a tapered shank can be inserted directly into the hole in the spindle.

When used correctly, the drill press makes it possible to do precision drill work. There should be no play in the spindle, spindle bearings, or chuck, and all should be in perfect alignment. The drill point should be properly sharpened and should indicate no wobble when the machine is turned on. The work being drilled must be securely clamped to the drill press table so it cannot move during the operation. If it is not clamped to the table, it can catch in the flutes of the drill bit and spin around. Many operators have severely injured their hands and fingers with such a spinning part.

The operator of a drill press should make certain that the machine speed is adjusted correctly for the work being performed, that the drill point has the angle most suitable for the material, and that the correct drilling pressure is applied with the feed lever.

Other Power Tools

In addition to the tools already mentioned, the technician will encounter other special sheet-metal-forming equipment in major manufacturing or fabrication facilities. Some of these are the *stretcher*, *shrinker*, *metal-spinning lathe*, *drop hammer*, and *hydroform*.

A **metal stretcher** is a machine having two pairs of jaws placed close to each other. When a flange of sheet metal is placed between the jaws, the metal is first clamped tightly and then a strong force is automatically applied to increase the distance between the pairs of jaws. This stretches the metal slightly. The jaws then release, the metal is moved a short distance sideways, and the cycle is repeated. Thus, the metal is stretched in small increments along the full length to be stretched until the desired amount of stretch is attained.

A **metal shrinker** operates on the same principle as the stretcher, except that when the jaws grip the metal, they move toward each other, thus compressing and shrinking the metal.

It should be noted that the jaws of the metal stretchers and shrinkers are serrated in order to prevent the material from slipping during the process. These serrations leave indentations on the material, weakening it and producing stress risers. As a result, the use of metal stretchers and shrinkers is not recommended in the manufacture and repair of structurally critical parts. The sheet metal former in Fig. 8-77 is used for stretching, shrinking, and punching sheet metal.

A **metal-spinning lathe** is used to form sheet metal into shapes that have a circular cross section, such as spinners for propellers. The machine differs from other metal lathes in that it has no back gears, carriage, or lead screws. It is rigid, and the spindle is usually driven by a step-cone pulley. Speed is important; hence the machine is adjustable over a wide range. In general, the thicker the metal, the slower must be the speed. Forms are constructed to provide the shape required for the finished product, and the metal is shaped by spinning over these forms.

Drop hammers are used to form sheet-metal parts between a punch and die, which have previously been constructed to

FIGURE 8-77   Sheet metal former.

FIGURE 8-78   English wheel.

provide the required shape. Since expensive matching dies are required, the drop hammer is economical only for large quantities of standard production parts. The weighted hammer is generally the male portion of a forming die that is raised and then allowed to drop freely between vertical guides onto the stock, which forces it into the matching portion of the die. The hammer may be raised by a rope running over a drum, by a vertical board running between two opposed cylinders, or by air, steam, or hydraulic mechanisms.

A **hydroform** is similar to a hydraulic press. A form in the shape and contours of the sheet-metal part to be formed is made from hardwood or other durable material. The metal to be formed is placed over the form (die) and pressed into shape with a thick rubber blanket or similar device. Pressure is applied hydraulically.

English Wheel

The English wheel, a popular type of metal-forming tool used to create double curves in metal, has two steel wheels between which metal is formed (Fig. 8-78). To use the English wheel, place a piece of sheet metal between the wheels (one above and one below the metal). Then, roll the wheels against one another under a pre-adjusted pressure setting. Steel or aluminum can be shaped by pushing the metal back and forth between the wheels. Very little pressure is needed to shape the panel, which is stretched or raised to the desired shape. It is important to work slowly and gradually curve the metal into the desired shape. Monitor the curvature with frequent references to the template. The English wheel is used for shaping low crowns on large panels and polishing or planishing (to smooth the surface of a metal by rolling or hammering it) parts that have been formed with power hammers or hammer and shot bag.

FABRICATION OF SHEET-METAL PARTS

Although not often called upon to make complex parts for an aircraft, the technician should have some knowledge of the techniques employed. The making of parts from sheet metal involves cutting to correct dimensions, bending, stretching, and shrinking.

Prior to making a part, the technician must determine the dimensions of the part to be made, either from a blueprint or from the part to be duplicated. This requires the use of layout techniques described previously.

Templates

A **template** is a pattern from which the shape and dimensions of a part may be duplicated. A template may be made from galvanized iron sheet, steel sheet, aluminum sheet, or other metal. After all dimension lines are accurately marked on the template, it is cut to the required shape.

If only one part is to be made, making a template is usually not necessary. Instead, the time otherwise used to make a template is used to make the actual part. Where several parts are to be made, using a template ensures accuracy and uniformity in the parts.

Cutting Sheet Metal

A previous section of this chapter discussed the various tools that are used to cut sheet metal. The selection of the proper tool is dictated by the size of the metal to be cut, both thickness and surface dimensions, by the length and shape of the cut, and by the accessibility of the material to be cut.

If the technician is working with new metal and cutting it to be shaped and prepared for installation, large tools are normally used at least until the piece is down to a size where finishing contours are being made. If the technician is working on or in the aircraft, light, air-powered tools and hand tools are commonly used. This often involves the use of snips, nibblers, and saws to trim out damaged areas and files and drills to prepare the area for the installation of the new components.

When cutting metal, it is important for the technician to understand not only how to properly use each tool, but also the damage that can result from improper use of the tool. For example, when using some types of shears, the starting or finishing edge of the shear tears the metal before starting to cut properly. On these tools, the metal cut should be started in from the end of the cutting blade, and the cut should be stopped before running off of the other end of the shear blade. The metal can be repositioned and the cut continued, section by section, until the whole length of the piece is cut. If the technician is not sure how a tool will perform, he or she should use it on a scrap piece of metal before cutting on a piece to be used in a repair.

When using aviation snips, note that some snips are designed to cut arcs to the left, some to the right, and some straight. It is important to use the proper type of snip required for the job.

When cutting metal, the technician should wear the clothing and protective gear appropriate to the type of cutting being performed. Special consideration should be given to eye protection when cutting with power tools and when working with hand tools above eye level. Metal shavings and file particles can result in eye damage. The handling of cut sheet metal can result in cuts on the technician's skin if proper gloves and protective clothing are not worn. If possible, sharp edges should be removed from metal immediately after it is cut.

Bending the Part

In bending a sheet-metal section to produce a required bend accurately, reference lines must be used. The type and location of the reference lines depend upon the type of bending machine employed. The most common type of bending machine for aircraft sheet-metal work is the cornice brake, which was described earlier in this chapter.

Since a bend begins at the bend tangent line, this line will be positioned directly below the center of the radius bar on the brake, as shown in Fig. 8-79. In this position the bend tangent line (BTL) will be out of sight; hence another reference line must be established for visual reference. This is called the **sight line**, and its position is located the distance of the radius from the BTL. This is shown in Fig. 8-79. When the part is placed in the brake, the sight line is directly below the nose of the radius bar. This, of course, will position the BTL directly below the center of the radius bar, and the bend will be formed between the bend tangent lines.

Technicians may find a **gauge line** an easier alternative to use when dealing with larger sheets of material. The gauge line uses the same basic concept as the sight line but takes advantage of the commonality of the different types of brake presses. When a brake press is properly set up, the radius bar is one material thickness from the end of the brake's bed. By noting a position similar to the sight line but adding one

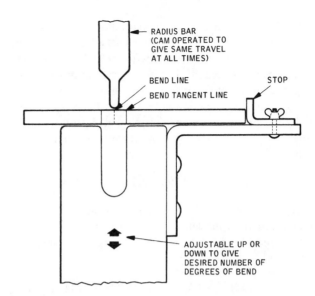

FIGURE 8-80 Bending with a press or vertical brake.

material thickness to the dimension to act as a gauge line, the technician may line up the gauge line with the edge of the bed of the brake.

If a production machine such as a press brake is used, the reference line used to position the part is called a **bend line** and is located between the two bend tangent lines. The positioning of metal on a press brake is shown in Fig. 8-80.

It must be emphasized that great care is needed to produce an accurate bend. The placing of the metal, the setting of the brake, and the actual bending operation all require precision to ensure that the bend will meet the specifications required.

Once the metal is in the correct position in the brake, the bending leaf is raised until the desired angle is formed. The metal will have to be bent slightly more than the desired angle, since the metal will spring back slightly when the bending leaf is lowered. The exact amount of overbending required depends on the type of metal being worked and its thickness.

When holes are to be drilled in any of the legs or flanges of a part before the part is bent, the hole centerline dimensions are treated the same as mold-line dimensions. This is illustrated in Fig. 8-81. To find the exact location of a hole

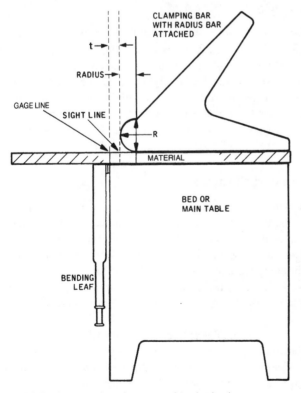

FIGURE 8-79 Locating sheet metal in the brake.

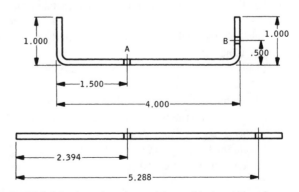

FIGURE 8-81 Locating the positions of holes with reference to bends.

center from the left edge of the part, the mold-line dimensions and the hole-center dimensions are added and then the setback of all bends between the left edge and the hole center are subtracted. In Fig. 8-81, it is necessary to determine the developed width of the metal to form the part illustrated.

To locate the center of hole A from the left edge, the mold-line dimension of the left leg is added to the mold-line dimension from the left leg to the center of the hole and the setback of the left bend is then subtracted.

$$1.000 + 1.500 - 0.106 = 2.394 \text{ in } [60.8 \text{ mm}]$$

Next, to locate the center of hole B, all the mold-line dimensions from the left edge of the part to the center of hole B are added and the setback distances of both bends are subtracted.

$$1.000 + 4.000 + 0.500 - 0.212 = 5.288 \text{ in } [134.32 \text{ mm}]$$

The setback distance 0.212 is the sum of the setback distances for both bends.

Bending Sheet Metal

Whenever possible, metal should be annealed before bending to reduce any internal stresses in the material. In addition, it is preferable that bends be made at a 90° angle to the grain of the material. If the material is annealed prior to forming, it must be returned to the appropriate hardness by heat-treating.

Bending of sheet metal has been discussed before; however, certain practices should be reemphasized. If possible, a bend in sheet metal should be made across the grain. This will reduce the tendency of the metal to crack. On bare aluminum sheet, the grain can be seen; however, on clad sheet aluminum, the grain is not readily apparent. Normally, the grain of clad aluminum sheet runs lengthwise with a full sheet. Furthermore, the manufacturer's identification letters and numbers are aligned with the grain.

The procedure in making a simple bend is as follows:

1. Cut the required metal sheet to the dimensions required for the part; deburr and smooth the edges.
2. Determine the correct radius for the bend.
3. See that the correct radius bar is installed on the brake.
4. See that the brake is clean and free of metal particles and scraps.
5. See that the sight line is correctly marked on the metal to be bent.
6. Install the metal in the brake with the sight line correctly positioned under the radius bar.
7. Clamp the metal in place with the clamping bar.
8. Set the stop on the brake for the required bend angle, making allowance for spring-back.
9. Rotate the bending leaf to the stop.
10. Remove the bend part and check the bend for accuracy.

If the bend radius is too large to be made on a standard leaf break, it may be possible to accomplish it on a vertical

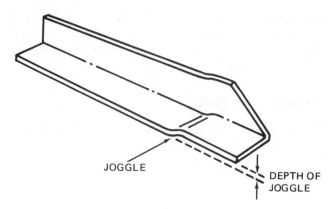

FIGURE 8-82   A joggled part.

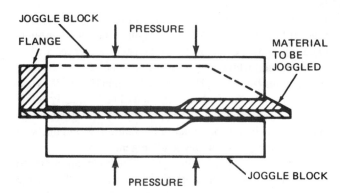

FIGURE 8-83   How a joggle is made.

brake equipped with the correct radius bar. If a still-larger-bend radius is required, a metal roller can be used.

Joggle

When a reinforcing angle is attached to metal sheet across a splice, it is usually necessary to **joggle** the angle. This is accomplished with joggle bars of the correct dimension or with a shop-made device. Joggle bars or dies can be made of either metal or hardwood. If a large number of parts are to be joggled, metal should be used to make the joggling tool.

A joggled part is shown in Fig. 8-82. The principle involved in the making of a joggle is illustrated in Fig. 8-83.

Curved Flanges

Making **curved flanges** involves the stretching or shrinking of the flange. Fig. 8-84 shows examples of each. To make a curved inside flange, it is necessary to shrink the flange metal. This can be accomplished in several ways. One method for shrinking the flange is to use a V-block made either of hardwood or metal. If the block is made of metal, it is necessary that the edges of the V be rounded off to avoid damaging the part.

To shrink the flange, the angle is placed on the V-block, as shown in Fig. 8-85, and the flange is struck on the edge with moderate blows from a soft hammer. The blows are directed slightly toward the inside of the angle to prevent the flange from bending outward. The part is moved continuously back

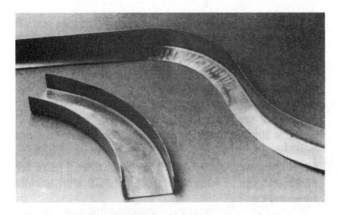

FIGURE 8-84   Different types of curved flanges.

hammer or a metal stretcher. The flange is placed flat against the bed of the hammer and moved continually back and forth under the hammer blows until the desired curve is obtained. If the metal work hardens too much, annealing it is necessary before continuing the stretching process. If a power hammer is not available, the flange can be hammered by hand to stretch the metal. Shrinking and stretching of metal flanges can be done most effectively and easily by means of shrinking or stretching machines, as described earlier.

To form a curved part with a flange, as shown in Fig. 8-86, a flat layout is made and the metal is cut to the required dimensions, as shown in Fig. 8-87. Form-blocks are made of hardwood or metal with a size and shape to match the part. The metal is clamped between the form-blocks in a vise as shown, and the flange is formed with a shrinking mallet. A hardwood wedge block is held against the metal on the side opposite the mallet strokes to prevent the metal from buckling. Mallet strokes are applied evenly along the flange curve, gradually shrinking the metal until the flange is formed.

Metal Bumping

The fabrication of parts involving compound curves, when done by hand, is accomplished by **bumping**. This process requires the use of a rounded wooden mallet and either a sandbag or a die. Parts such as fairings are usually formed on the sandbag.

A typical sandbag for metal bumping is a leather pouch about 18 in [45 cm] square filled with sand and securely

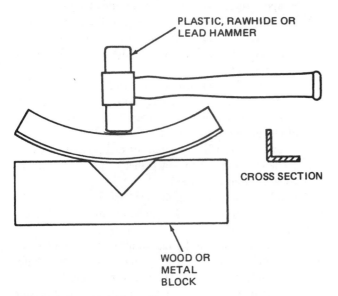

FIGURE 8-85   Shrinking a flange.

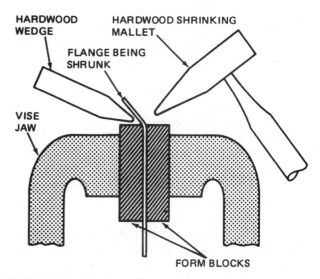

FIGURE 8-86   Forming a flange on a curved part.

and forth across the V as the blows are struck to provide a uniform bend. If bends occur in the flange, they are removed by lightly planishing against a flat wooden block. If a **metal shrinker** is available, the curved flange can be made more easily than by hand.

To make a part with an external curved flange, the flange must be stretched. Usually this is done by using a power-driven

FIGURE 8-87   Flat layout for curved part to be flanged.

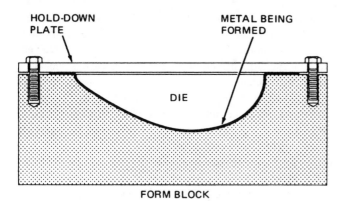

HOLD-DOWN PLATE   METAL BEING FORMED

DIE

FORM BLOCK

FIGURE 8-88 Use of a form block and hold-down plate in forming a small part with compound curves.

sewed to prevent leakage of the sand. Before being filled with sand, the bag is coated on the inside with a plastic material to prevent the sand from working into the pores of the leather.

To form a part, the technician selects a piece of soft aluminum alloy of the correct size and thickness for the part. After the metal is trimmed, the edges should be smoothed. With the metal against the sandbag, the technician carefully applies blows with the bumping mallet until the desired shape is attained. Templates are used to check the progress of the forming and the final shape. Upon completion of the forming, the part is trimmed to final dimensions and the edges are smoothed with a file. Irregularities caused by bumping the former surface may be removed with **hand metal rollers**. A set of these metal rollers consists of two oval-shaped rollers held against each other by means of a spring steel frame. An adjustment is provided to increase or decrease the pressure of the rollers against the surface of the formed metal. The rollers may be made of steel or hardwood.

Small parts with compound curves are usually made with a die, or form-block. The die can be made of hardwood or metal, depending upon the number of parts to be made. It consists of a block into which a cavity of the correct shape has been cut. A hold-down plate of a size and shape to fit the die is required to hold the metal while it is being shaped. The shape of the cavity in the die is checked with templates to ensure that it is correct.

To form a part, the metal is clamped between the hold-down plate and die, as shown in Fig. 8-88. The metal is then slowly forced into the cavity with a correctly shaped hammer or other forming tool.

RIVETING

A **rivet** is a metal pin or bar with a cylindrical shank, used for fastening two or more pieces of metal together. The metal pieces to be joined have holes of the proper size drilled through them. The shank of the rivet is inserted through one of these holes. One end of the rivet has a head formed previously by the manufacturer. The size and shape of the head are chosen to fit the requirements of the application.

After the rivet is inserted through the holes in the metal, a **bucked head** is formed on the end opposite the manufactured head. This bucked head is formed by any of the various methods described in this chapter. Figure 8-89 illustrates popular head styles and standard head markings for aircraft rivets.

AC 43.13-1B & 2B contains a description of basic solid rivets. *Aircraft Basic Science* also discusses standard aircraft rivets and their codes, as well as a variety of special fasteners. Technicians who use aircraft rivets should memorize this code, and they should check the symbol on each rivet that they use. By so doing, they will avoid the possibility of rivet failure in the aircraft structure.

In addition to the MS and AN codes discussed in *Aircraft Basic Science,* technicians working on large aircraft will encounter National Aerospace Standard (NAS) codes on aircraft blueprints and assembly drawings. The NAS-523 rivet code is illustrated in Fig. 8-90 and is used to describe rivet and installation specifications for a particular assembly. Technicians will find instructions for the rivets to be used in a particular repair setup with coding in four quadrants designated NW (upper left), NE (upper right), SW (lower left), and SE (lower right).

The upper left (NW) quadrant designates the rivet part number, either AN or MS, and the material of which it is made. In the NW section of the chart, note that the code letters BJ identify an MS20470AD (AN470AD) rivet, which is a universal-head rivet made of aluminum alloy 2117-T3. Only a few codes are shown in this chart, since a total review is beyond the scope of this text. For a complete listing, the technician should consult NAS523.

The upper right (NE) quadrant specifies the diameter of the rivet and the required positioning of the manufactured head of the rivet. The letters *N* and *F* are used to indicate that the manufactured head be placed on the near or far side of the repair, respectively.

The lower left (SW) quadrant provides dimple and countersink information. Letters and numbers as shown are placed in the SW quadrant to indicate to the technician exactly what type of installation is to be made.

The lower right (SE) quadrant of the symbol gives the fastener length and indicates whether a spot weld may be used as an alternative method.

The column on the right of the quadrant shown in Fig. 8-90 is included here to provide additional information and examples for the technicians.

In addition to standard code numbers for fasteners and fittings, manufacturers often design their own items of hardware and apply their own part numbers. The technician must use the manufacturer's structural repair manual and be sure to employ the parts specified.

Drilling Holes for Fasteners

To make a good fastener joint, it is essential that the fastener hole be prepared properly. There are three steps to preparing the fastener hole: drilling, reaming, and deburring. Drilling provides the initial hole, reaming enlarges the hole to the proper size and finish condition, and deburring

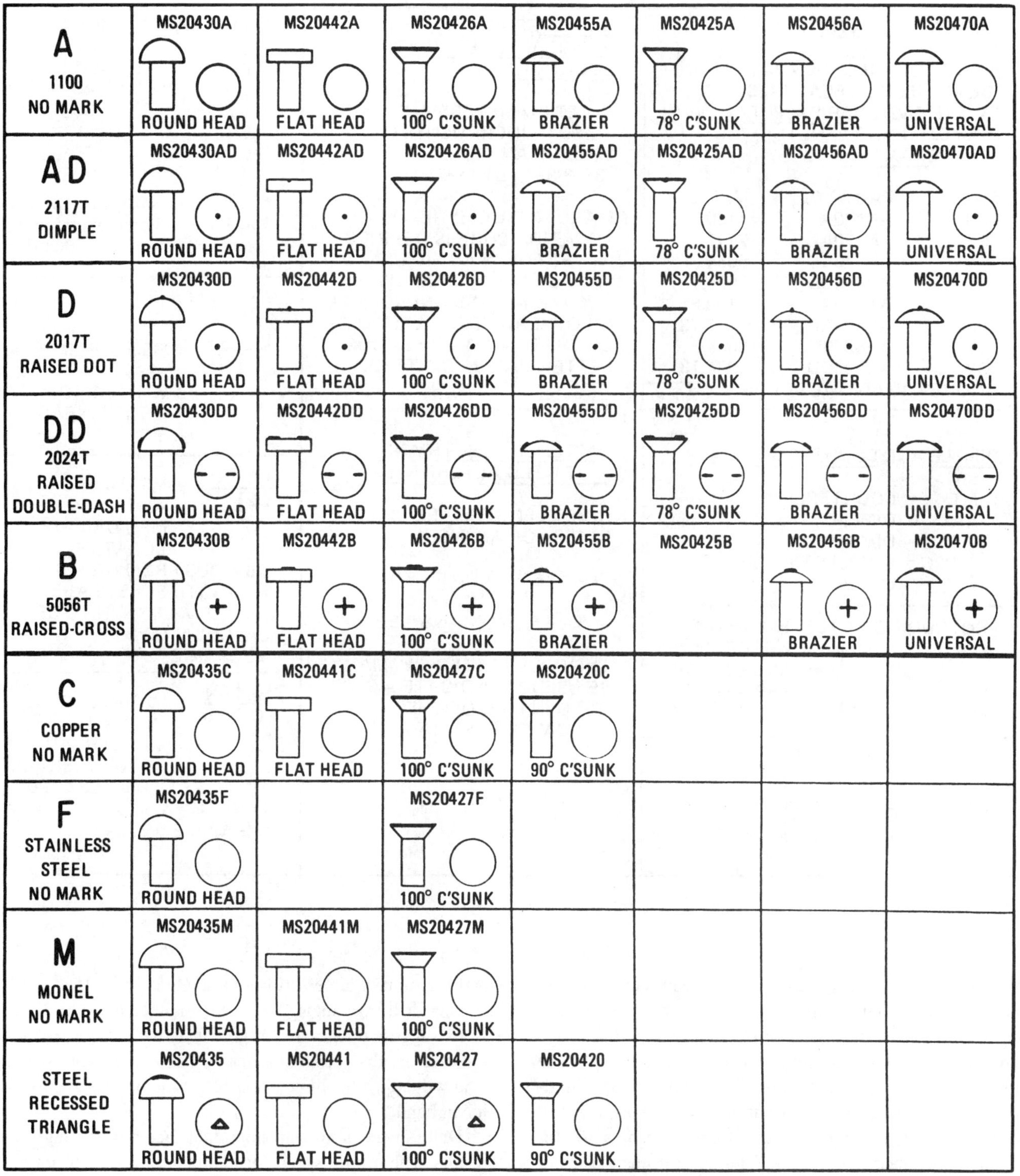

A 1100 NO MARK	MS20430A ROUND HEAD	MS20442A FLAT HEAD	MS20426A 100° C'SUNK	MS20455A BRAZIER	MS20425A 78° C'SUNK	MS20456A BRAZIER	MS20470A UNIVERSAL
A D 2117T DIMPLE	MS20430AD ROUND HEAD	MS20442AD FLAT HEAD	MS20426AD 100° C'SUNK	MS20455AD BRAZIER	MS20425AD 78° C'SUNK	MS20456AD BRAZIER	MS20470AD UNIVERSAL
D 2017T RAISED DOT	MS20430D ROUND HEAD	MS20442D FLAT HEAD	MS20426D 100° C'SUNK	MS20455D BRAZIER	MS20425D 78° C'SUNK	MS20456D BRAZIER	MS20470D UNIVERSAL
D D 2024T RAISED DOUBLE-DASH	MS20430DD ROUND HEAD	MS20442DD FLAT HEAD	MS20426DD 100° C'SUNK	MS20455DD BRAZIER	MS20425DD 78° C'SUNK	MS20456DD BRAZIER	MS20470DD UNIVERSAL
B 5056T RAISED-CROSS	MS20430B ROUND HEAD	MS20442B FLAT HEAD	MS20426B 100° C'SUNK	MS20455B BRAZIER	MS20425B	MS20456B BRAZIER	MS20470B UNIVERSAL
C COPPER NO MARK	MS20435C ROUND HEAD	MS20441C FLAT HEAD	MS20427C 100° C'SUNK	MS20420C 90° C'SUNK			
F STAINLESS STEEL NO MARK	MS20435F ROUND HEAD		MS20427F 100° C'SUNK				
M MONEL NO MARK	MS20435M ROUND HEAD	MS20441M FLAT HEAD	MS20427M 100° C'SUNK				
STEEL RECESSED TRIANGLE	MS20435 ROUND HEAD	MS20441 FLAT HEAD	MS20427 100° C'SUNK	MS20420 90° C'SUNK			

FIGURE 8-89 Head styles and markings for aircraft rivets.

the hole finishes the process by removing any rough edges that remain.

The Process of Drilling

During the drilling process, the material is first sheared and then routed away from the hole through the flutes of the drill. As the drill begins to penetrate the material, the process gradually changes from a shearing action to a ripping action.

The results of this ripping action are seen as burrs on the edge of the hole. What the technician does not see are the minute cracks that are generated during the process.

Determining Fastener Hole Size and Drill Size

After selecting the fastener size, the first requirement for a perfectly drilled hole is the use of a drill that is ground accurately.

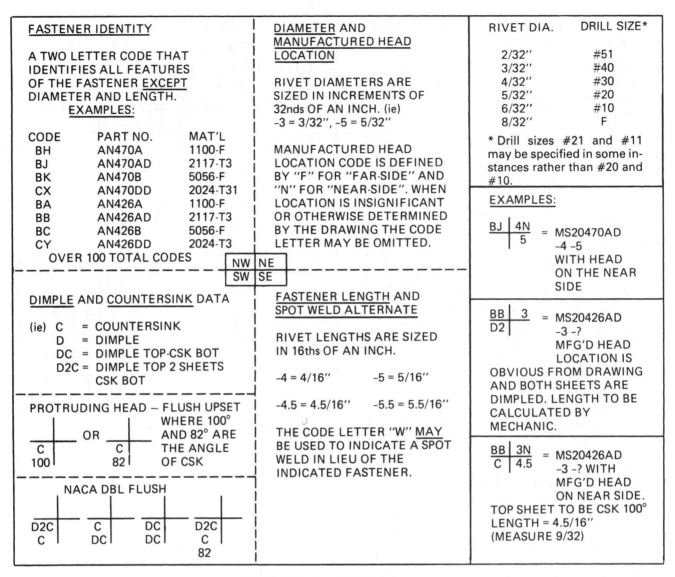

FIGURE 8-90   The NAS-523 rivet code.

New drills usually have a satisfactory point, but after they are worn, they should be sharpened or discarded. The dimensions for a correctly ground point are shown in Fig. 8-91. Observe the **drill-point angle** (118°) and the **drill-rake angle** (12°). For soft materials—for example, soft aluminum, lead, wood, and plastics—it is better to have the drill sharpened with a smaller drill-point angle, such as 90° for medium-soft materials and 45° for very soft materials. For very hard and tough materials such as steel, stainless steel, and titanium, a larger drill-point angle (125 to 150°) and a smaller drill-rake angle (10°) are recommended. The dimensions suggested here are for guidance and are not meant to indicate an absolute requirement. The experienced technician will adjust the dimensions to get the best results for the job at hand.

Drill speed is also an important factor in getting good results. The proper speed for aluminum alloy will not produce the best results with stainless steel or titanium. Drill speed determines the rate at which the outer cutting edge of the drill is moving across the material being cut. For example, a $\frac{1}{8}$-in [3.18-mm] drill having a circumference of 0.3927 in [9.97 mm] turning at the rate of 1222 rpm will have a cutting speed of 40 ft/min [1219.2 cm/min]. When harder materials are drilled, slower speeds are required. In addition, a cutting and cooling lubricant is needed. Lubricating oil, lard oil, water-soluble oil, and others are used. Table 8-7 shows drill and cutting speeds for various-size drills. This table provides information for commonly used drill sizes at various recommended cutting speeds. For values not shown on the chart,

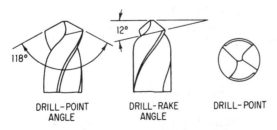

FIGURE 8-91   Dimensions for a correctly ground drill point.

250   Chapter 8   Sheet-Metal Construction

TABLE 8-7	Drill and Cutting Speeds							
Cutting Speed, ft/min [cm/min]	30 [914]	40 [1219]	50 [1524]	60 [1829]	70 [2134]	80 [2438]	90 [2743]	100 [3048]
Diameter, in [mm]					rpm			
$\frac{1}{16}$ [1.59]	1833	2445	3056	3667	4278	4889	5500	6111
$\frac{3}{32}$ [2.38]	1222	1630	2038	2445	2853	3260	3667	4074
$\frac{1}{8}$ [3.18]	917	1222	1528	1833	2139	2445	2750	3056
$\frac{3}{16}$ [4.76]	611	815	1019	1222	1426	1630	1833	2037
$\frac{1}{4}$ [6.35]	458	611	764	917	1070	1222	1375	1528
$\frac{5}{16}$ [7.95]	367	489	611	733	856	978	1100	1222
$\frac{3}{8}$ [9.33]	306	407	509	611	713	815	917	1019
$\frac{7}{16}$ [11.13]	262	349	437	524	611	698	786	873
$\frac{1}{2}$ [12.7]	229	306	383	458	535	611	688	764

it is necessary simply to compute or extrapolate from the ones shown.

The values given in Table 8-7 are not required but are recommended for optimum results. A cutting speed of 100 ft/min [3048 cm/min] is recommended for aluminum alloys; however, lower speeds can be used very satisfactorily. For stainless steel and titanium, a cutting speed of 30 ft/min [914.4 cm/min] is recommended, but a lower or higher speed can be used. Care must be taken with the harder and tougher materials to avoid too much speed and pressure, which will result in overheating the drill and rendering it useless.

In drilling larger holes of $\frac{3}{16}$ in [4.76 mm] or more, it is wise to drill a pilot hole first. The pilot drill should not be more than one-half the diameter of the final hole. This is particularly true when drilling harder materials. Before using a drill bit, the technician should examine it to see that it is straight, that the point conforms to required standards, and that the shank is not scored or otherwise damaged.

The location of a hole to be drilled may be indicated by marking with a pencil or, in the case of heavy sheet stock, by making a slight indentation with a center punch. For holes that must be held within extremely close tolerances, a **drill fixture** is normally used. This device holds the drill accurately in position while the hole is being drilled.

When beginning to drill a hole, the technician must be very careful to hold the drill perpendicular to the material being drilled and must also steady the drill and motor so that the drill will not move away from the correct position and damage the adjacent material. It is common practice to start the drill by placing it in position and turning it by hand before turning on the electric or air power to operate the motor. By this method the hole will be started, and the drill will usually remain in the proper position. Figure 8-92 shows a technician holding the drill properly for starting to drill a hole.

Figure 8-93 illustrates properly and improperly drilled holes. The left and middle drawings show holes that are clean and in good alignment. The right drawing shows two holes that were drilled at an angle and would not be suitable for riveting.

FIGURE 8-92 Proper method for starting a drill.

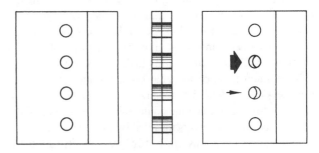

FIGURE 8-93 Properly and improperly drilled holes.

A hole is not complete until it is both drilled and burred. **Burring** is the process of removing rough edges and chips from a newly drilled hole. It is usually done by hand with a drill larger than the hole, or it can be done with a special burring tool, which is merely a piece of metal with sharp edges. When two or more sheets are drilled at the same time, it is necessary to remove chips and burrs from between the sheets. Figure 8-94 illustrates the results of leaving material between drilled sheets. Removal of burrs from drilled holes (deburring) may be accomplished with a manufactured **deburring tool**, a countersink using a very light cut, or other

FIGURE 8-94 Material between sheets of metal.

tool that will clear the edges of a drill or punched hole. Care must be taken to remove only the rough edges and chips from the hole.

Installing Rivets

The installation of common rivets consists of drilling holes slightly larger (0.001 to 0.003 in [0.025 to 0.076 mm]) than the rivet shank in the parts to be joined, removing the burrs from the edges of the holes, inserting the rivet, and driving the rivet. A no. 40 drill is used for a $\frac{3}{32}$-in [2.382-mm] rivet, a no. 30 drill is used for a $\frac{1}{8}$-in [3.175-mm] rivet, and a no. 21 drill is used for a $\frac{5}{32}$-in [3.97-mm] rivet. Note that the first dash number of a rivet (dia.), when added to the first number of the drill size, will equal 7. A -2 rivet requires a no. 51 drill; a -3 rivet, a 40 drill; a -4 rivet, a 30 drill; a -5 rivet, a 21 drill; and a -6 rivet, an 11 drill. This is referred to as the 7-rule and is coincidental. It gives the technician a method to help remember drill sizes, but it applies only to rivet sizes -2 through -6. The rivet is usually driven by means of a pneumatic hammer and a bucking bar to "back up" the rivet.

Countersinking

To install countersunk rivets, it is necessary to provide a conical depression in the surface of the skin so that the head of the rivet is flush with the surface. This depression is made by means of a **countersink** when the skin is sufficiently thick and by dimpling when the skin is thin. The use of a machine countersink is limited by the size of the rivet and the thickness of the skin. Generally, sheet metal should not be machine-countersunk entirely through the sheet. For sheet metal of 0.040 to 0.051 in [1.02 to 1.28 mm] thickness, it is common practice to countersink not more than $\frac{2}{3}$ the thickness of the sheet. For repairs on an airplane, the specifications for use of machine countersinking may usually be determined from the rivets installed by the manufacturer.

A countersink for use in a drill press or drill motor is shown in Fig. 8-95. The pilot of the countersink ensures that the countersunk portion of the hole will be properly centered.

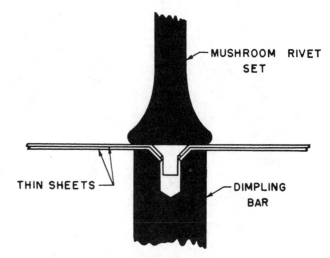

FIGURE 8-96 Use of a dimpling bar for dimpling.

It is good practice to use an adjustable stop on the drill motor or drill press to ensure that the depth of the countersink will be accurate. A sheet of metal should never be countersunk through more than 99 percent of its thickness.

Dimpling

Dimpling for countersunk rivets is a common practice when using a relatively thin skin such as 0.016 to 0.025 in [0.41 to 0.64 mm] in thickness. **Dimpling** can be accomplished with a dimpling bar and flush set, as shown in Fig. 8-96. The rivet head is the die that forms the dimple. When thin skin is attached to a heavier structural member, the heavy member is subcountersunk and the skin is dimpled into the countersunk depression, as illustrated in Fig. 8-97. For production work in a factory, dimpling often is accomplished with dimpling dies used in a pneumatic squeeze riveter.

It is sometimes necessary to dimple heavy sheet in a highly stressed part of the airplane in order to retain the maximum strength of the sheet. A process called **hot**

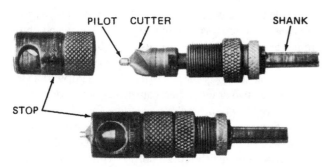

FIGURE 8-95 Countersink.

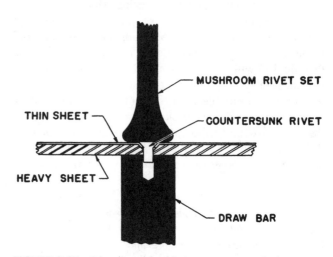

FIGURE 8-97 Dimpling thin skin into countersunk sheet.

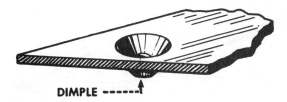

FIGURE 8-98   Dimple formed with heat.

FIGURE 8-99   Automatic hot-dimpling machine.

dimpling has been developed for this purpose. Hot dimpling is performed with a special hot-dimpling machine, consisting of heated dies that can be pressed together pneumatically to form a dimple, as shown in Fig. 8-98. A process wherein the sheet metal is caused to *flow* to the shape of dies is called **coin dimpling**.

Figure 8-99 shows an automatic hot-dimpling machine manufactured by Aircraft Tools, Inc. The operator sets the controls of the machine according to charts supplied by the manufacturer, which give the temperatures and pressures required for various types of thicknesses of materials.

The material, having been previously drilled, is placed over the stationary die, with the pilot of the die projecting through the hole in the material. The operator then presses on the foot control of the machine. This brings upper and lower dies toward each other; thus they press on the material and their heat is transferred to the material. As the material becomes heated sufficiently, the pressure of the dies causes

it to be formed. This pressure comes from a compressed-air system. The initial pressure on the dies is limited to prevent the material from being deformed before it has been heated sufficiently. After the material reaches the forming temperature, additional pressure is applied automatically to the dies to complete the forming operation. This pressure is maintained for a predetermined number of seconds, and then it is automatically released.

Another method for hot dimpling employs a resistance-heating machine. The dies of the machine are electrodes, which pass a current through the metal to be riveted and cause it to heat. When the metal has been heated sufficiently, full pressure is applied to the dies to form the dimple.

The rivets are installed by the use of a rivet gun and a bucking bar. The sizes of the rivet gun and bucking bar are selected to match the size of the rivet.

The various rivet guns can be adjusted to deliver the required blow for each size of rivet. The most desirable practice is to adjust the gun so that the bucked head of the rivet will be properly shaped, using as few blows of the rivet gun as possible. When the rivet gun is adjusted with too light a blow, the rivet may be work-hardened to such a degree that the head will not be formed properly without cracking the rivet.

A bucking bar is used as shown in Fig. 8-100. The bar is held firmly against the shank of the rivet while the rivet gun with the correct set is applied to the manufactured head. It is essential that the bucking bar be placed against the shank of the rivet before the rivet is driven. If the operator of the rivet gun starts to drive the rivet before the bucking bar is in place, the sheet in which the rivet is being installed will be damaged.

The correct installation of a rivet is dependent upon the proper use of the bucking bar as well as the rivet gun. The face of the bucking bar must be held square with the rivet, or the rivet may "clinch"; that is, the bucked head will be driven off center. Sometimes the operator can control the formation of the bucked head by carefully tilting the bucking bar. Both the rivet gun and the bucking bar must be firmly in place against the rivet before the throttle of the gun is opened to drive the rivet. Figure 8-101 illustrates improperly installed rivets.

There are several names used to identify the head formed on the shank of the rivet during the bucking operation. These names include bucked head, formed head, shop head, and buck tail.

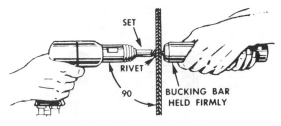

FIGURE 8-100   Bucking a rivet.

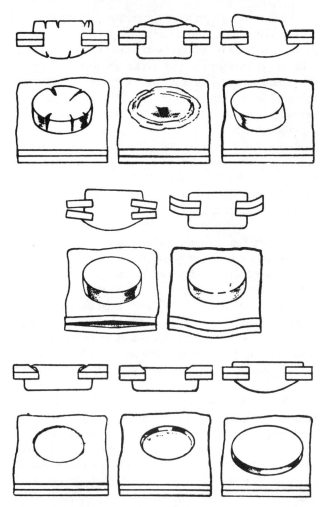

FIGURE 8-101   Improperly installed rivets.

Dimensions of Installed Rivets

When rivets are installed in a standard repair, it is necessary that certain minimum dimensions be observed. Figure 8-102 shows the desired dimensions for bucked rivet heads and length of rivet. The drawing shows 0.65*D* as the height of the bucked head; however, a minimum height of 0.50*D* is acceptable.

With experience, a visual inspection will tell the technician if the rivet has been upset properly. A gauge, such as that shown in Fig. 8-103, can also be used. One gauge like this is needed for each diameter rivet. The hole in the gauge is the diameter of the bucked head, and the thickness of the gauge corresponds to proper head height. The gauge is placed over the bucked head, and irregularities are immediately apparent.

FIGURE 8-102   Desired dimensions for a rivet.

FIGURE 8-103   Rivet gauge.

Shaved Rivets

On modern, high-speed aircraft, it is necessary to remove every possible cause of drag from the outer surface of the airplane skin. For years flush (countersunk-type) rivets were installed in skin and other structural sections exposed to airflow. To obtain the most nearly perfect surface, shaved-riveting techniques were developed.

In preparation for shaved riveting, standard rivet holes are drilled in the metal to be riveted. This may be done by the manual method or by automatically programmed machines. On the outer surface of the metal, the holes are countersunk with a 60° tool instead of the conventional 100° countersink.

Standard rivets are installed with the rivet head inside the metal skin, and the shank of the rivet is driven to form a head in the conical depression on the outer surface. The forming of the rivet shank to fill the depression can be done with a standard rivet gun and smooth-faced bucking bar, but, during production, it is often done with automatic machines.

After the rivet is driven sufficiently to fill the countersunk hole completely, the excess rivet material projecting above the surface of the skin is shaved with a small rotary mill called a **rivet shaver**, shown in Fig. 8-104. With this tool, the surface of the skin and rivet is made extremely smooth so drag will be reduced to the minimum. For a manual process, the rivet shaver is held in the hand as one would hold a drill motor. It is prevented from cutting too deeply by means of a carefully adjusted stop.

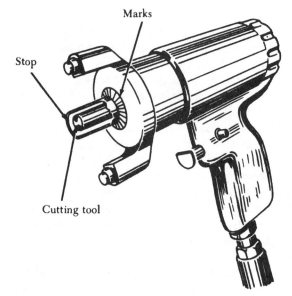

Marks

Stop

Cutting tool

FIGURE 8-104   A rivet shaver.

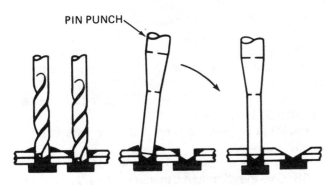

PIN PUNCH

FIGURE 8-105 Removing a rivet.

Removing Rivets

In the repair of sheet-metal aircraft it is often necessary to remove rivets. However, great care must be used, or damage may be done to the metal from which the rivets are removed. Rivets are removed by drilling through the manufactured head with a drill one size smaller than the shank of the rivet. The technician must make sure that the drill is started and held in the exact center of the rivet head. The drill should penetrate no further than the base of the rivet head, or the rivet hole may be enlarged by the drill. Usually the rivet head will come off as soon as the drill has penetrated the proper distance. If the rivet head does not come off, a pin punch the same diameter as the hole may be used to snap the head off, as shown in Fig. 8-105. After the head of the rivet is removed, the shank may be pushed or driven out with the pin punch.

When replacing rivets in a hole where a rivet previously has been installed, it is necessary to ascertain that the hole has not been enlarged beyond the correct tolerance for the rivet being installed. If the hole is too large, it should be drilled to the correct size for the next larger rivet to be used.

REVIEW QUESTIONS

1. Explain the difference between *safe-life* and *fail-safe* design.
2. What is a *safety factor*?
3. How is *stress* measured?
4. Describe *strain*.
5. What is *fatigue*?
6. What is the effect of a hole in a load-carrying structural member?
7. What is the purpose of a *relief hole*?
8. What is the purpose of *stop-drilling cracks*?
9. Why are relief holes required at the corners in a box structure?
10. Explain some of the causes of cracks in sheet-metal structures.
11. If cuts are made that form a sharp angle between two edges of sheet metal, what should be done to prevent stress concentration?
12. Describe how a load is transmitted through a *rivet joint*.

13. Describe *rivet shear* and how it is calculated.
14. What is *tensile load*?
15. Describe *tearout*.
16. What is a *bearing load*?
17. Define *bend allowance*.
18. With reference to a bend, define *mold line, set-back,* and *bend tangent line.*
19. Explain *closed* and *open* angles.
20. What is meant by *X-distance*?
21. How would you find *X*-distance (setback) using the K-chart?
22. Give the empirical formula for bend allowance.
23. Compute the bend allowance for a 60° bend with a $\frac{1}{4}$-in [6.35-mm] radius when the thickness is 0.050 in [1.27 mm].
24. Compute the developed width of a piece of aluminum-alloy sheet that is to be formed into a Z section with two bends of 110° each around a radius of $\frac{3}{8}$ in [9.33 mm], when the thickness of the metal is 0.060 in [1.52 mm], the mold-line distance for each flange is 2 in [50.8 mm], and the mold-line distance for the inside flat section is 3 in [76.2 mm].
25. What precautions must be used in handling sheet metal during layout?
26. What precautions should be observed when marking sheet metal with a black lead pencil?
27. What precaution must be observed with reference to scribed lines on sheet-metal parts?
28. Explain how a *hand planishing hammer* is used.
29. For what purposes are *soft mallets* employed?
30. What is the purpose of a *hand nibbler*?
31. What is a *chassis punch*?
32. What determines the size of rivet gun and bucking bar used?
33. Explain how the *rivet gun, rivet set,* and *bucking bar* are used in installing rivets.
34. How is the size of a rivet gun related to rivet size?
35. Describe the care and service of a rivet gun.
36. Describe a bucking bar and what determines the weight of the bucking bar to be used for a particular job.
37. Describe the purpose and construction of a *sheet fastener.*
38. Explain how a *hole-finder* is constructed and how it works.
39. What is the purpose of a *rivet cutter*?
40. Describe squaring shears and explain their use.
41. For what purpose is the *gap-squaring shear* used?
42. Explain the use of *slitting shears.*
43. What is the advantage of *throatless shears*?
44. For what type of work would *rotary slitting shears* be used?
45. What is a *Unishear*? A *nibbling machine*?
46. Compare a *bar-folding machine* and a *cornice brake.*
47. What is the purpose of the *radius bar* on a cornice brake?
48. How does a *box* and *pan brake* differ from a *cornice brake*?

49. For what special purpose is the *forming roll* or *slip-roll former* designed?

50. In operating a metal-cutting band saw, what determines blade pitch and speed to be used?

51. Describe a *drill press* and its use.

52. Explain the use of a *drop hammer.*

53. Explain the construction of a *template.*

54. In bending a part on a cornice brake, discuss the importance of *bend tangent line, sight line,* and *gauge line.*

55. Discuss the bending of sheet metal.

56. How can very large-radius bends be made?

57. What is the purpose of a *joggle*?

58. How is a joggle made?

59. Describe how a curved inside flange is made.

60. How is an external, or outside, curved flange made?

61. How is a flange made on a curved edge?

62. Describe two metal-humping processes.

63. Discuss the NAS codes found on aircraft blueprints.

64. Explain the importance of a properly sharpened drill point.

65. What is the importance of drill speed with respect to materials being drilled?

66. What preparation is necessary for installing *flush rivets*?

67. Describe the process known as *hot dimpling.*

68. Give the minimum dimensions for a driven rivet.

69. What is meant by a *shaved rivet,* and how is rivet shaving accomplished?

70. When is a *rivet shaver* used?

71. Explain how a rivet should be removed from an assembly.

Sheet-Metal Inspection and Repair 9

INTRODUCTION

Proper maintenance of an aircraft's structural integrity can be a demanding task. Aircraft are subjected to a variety of influences, which can affect their ability to withstand the rigors of flight. Failure or potential failure of sheet-metal structural components is not always easy to identify. Once potential failure is identified, the technician must be able to evaluate the extent of the discrepancy, determine its cause, determine the proper corrective action or repair, and then implement that action.

In this chapter, the processes used to identify discrepancy through aircraft inspection are briefly reviewed. The importance of determining the cause of the discrepancy is discussed, and the techniques for repair of basic structural components are presented. In developing techniques for sheet-metal repair, both the guidelines for repair found in AC 43.13-1B & 2B and the relationship of the mechanical properties of the materials involved in the repair are discussed. In this chapter the technician will develop an understanding of how the FAA guidelines were derived and how to implement them in practical application.

SHEET-METAL INSPECTION

The process of aircraft inspection is discussed in *Aircraft Basic Science*. In Chap. 8 of this text, when discussing aircraft-structure design philosophies, the process of inspection as an integral part of the damage-tolerant design philosophy was identified. In 1981, the FAA issued Advisory Circular 91.56, which established guidelines for supplemental inspections for large transport-category aircraft. Whether the technician is involved with 100-h inspections, annual inspections, progressive inspections, or supplemental inspections, a knowledge of the various inspection techniques and processes is an integral part of a successful inspection.

Aircraft sheet-metal and structural components are subject to a variety of forces during their uses, as well as the weakening that comes from exposure to the environment. Simple aging also plays a significant part in the deterioration of an aircraft during its operational life. Aircraft structural inspectors must be able to identify potential problems ranging from intergranular corrosion to missing fasteners, using processes ranging from a simple visual inspection to sophisticated testing equipment.

An inspection's scope may be general, as in a preflight inspection, or very specific, as called for by an Airworthiness Directive. The parameters used by the inspector to judge "airworthiness" may be simple, as in "cracked or not cracked" observations, or the inspector may be looking for a condition that falls within a specific range, such as clearances between certain parts. When a range is specified, the range is called a **tolerance**.

A tolerance may be specified in a variety of fashions, as discussed in *Aircraft Basic Science*. A tolerance indicates a range within which a particular attribute is considered acceptable. It is important to note that in establishing tolerance, considerations regarding the aircraft's usage and inspection frequency are included. As a result, if the technician's observation or measurement of a given attribute falls within the tolerance, the observation or measurement is in compliance with the tolerance. When determining compliance within a tolerance, any one observation or measurement within a tolerance is just as much in compliance with the tolerance as any other observation or measurement within the tolerance range.

Traditionally, inspections are classified as destructive or nondestructive. Maintenance by its very nature is concerned only with nondestructive testing, since destructive testing eliminates the serviceability of the part or material being tested. Although destructive testing plays a significant part in aircraft design analysis, it is not used by the technician as part of any airworthiness inspection process. Therefore, the discussion that follows is limited to nondestructive inspection techniques.

Nondestructive Testing

Nondestructive inspection (NDI) is also referred to as nondestructive testing (NDT). The philosophy of NDI is to verify the presence of certain attributes without causing the material to fail. In effect, this is an effort to validate the existence of other attributes that have been determined to identify the existence of yet others. For example, if a metal has a certain composition, hardness, and thickness (three attributes), it has been determined to have a specific ultimate tensile strength (another attribute).

Visual Inspection

The most obvious form of NDI is the visual check. This check may be performed with the naked eye or assisted by magnification. Magnification is specified in terms of power. The most frequently used magnification level employed in aviation is 10 power, designated as "10X."

Many inspection documents begin with the instruction to clean the aircraft or the area of the aircraft to be inspected. It is, however, advisable that technicians involved in general visual-type inspections observe the aircraft before the cleaning process is begun. Clues to discrepancies may often be removed during the cleaning process, making discrepancy identification more difficult. For example, loose or improperly installed countersunk rivets disrupt the airflow around the rivet. As the air flows around these rivets, dirt in the air accumulates, leaving what appears to be a dirt trail around the rivet. It is much easier to identify these dirt trails prior to cleaning the aircraft than after the cleaning process is completed. It should be noted that a pre-cleaning inspection, although often beneficial, does not satisfy the requirements of a traditional complete inspection.

The technician can also use the sense of touch to help identify discrepant items. Running the hand or fingernails over a surface can assist the technician in finding cracks.

More detailed inspections of an area may be performed by *dye-penetrant inspection, magnetic-particle inspection, X-ray inspection, fluorescent-penetrant inspection, ultrasonic inspection,* and *eddy-current inspection.* These inspection processes are used to detect discrepancies that are not detectable using only the human senses.

Dye-Penetrant Inspection

Inspection of a metal structure is easily accomplished by means of **dye-penetrant inspection**. In this process, the dye penetrates any small cracks or fissures and then seeps out when a developer is applied to the joint. Thus the crack is revealed as a bright red line.

Fluorescent-Penetrant Inspection

Fluorescent-penetrant inspection can be used for detecting cracks or other flaws in a welded structure. A liquid containing a fluorescent material is applied to the part to be inspected and is allowed to penetrate cracks, laps, and other discontinuities. The part is then washed with a suitable solvent and dried, after which a developing powder is applied to draw the penetrant to the surface. Excess powder is brushed off, and the part is examined under ultraviolet light (black light). Cracks and other flaws are revealed as fluorescent markings.

Magnetic-Particle Inspection

Magnetic-particle inspection (Magnaflux) by means of magnetic powder applied to a magnetized part is an efficient, practical, and nondestructive method that will reveal the presence of tiny cracks and other flaws in a part. The surface

to be examined should be reasonably smooth and free from scale, because it is difficult to find cracks in the irregular surface of the weld metal. Sandblasting is a suitable method for cleaning the surface of metal parts in preparation for the magnetic particle inspection.

Magnetization of tubular clusters and other welded joints in tubular structures is usually accomplished by means of cables wrapped in coils around the area to be inspected. The technician must follow the appropriate instructions to ensure that the magnetization is produced in the correct direction. After the inspection the magnetization must be neutralized with an alternating-current field coil.

Radiological Inspection

X-ray inspection was limited in value in the past because of the inaccessibility of many joints and the necessity of taking exposures from several angles to make certain that all defects were found. However, the results are very satisfactory and the recent developments in this field have reduced the cost and time. The use of radioactive cobalt "bombs" has made it possible to X-ray joints at almost any location.

Ultrasonic Inspection

Ultrasonic inspection techniques apply high-frequency sound waves to the part being inspected. These sound waves are reflected from the opposite side of the material or from any flaw that they encounter. Wave signals from the flaw are compared with the normal wave to determine the location and size of the flaw.

Eddy-Current Inspection

In an **eddy-current inspection**, electrical currents are generated in the part by means of electromagnetic waves. The electrical current flows in the part in a circular fashion, similar to the eddies observed when draining a bath tub. If a flaw exists, the indicator will show a value different from the normal response. A well-qualified operator can diagnose the response to determine the nature of the flaw.

These descriptions are condensed but provide an overview of the inspection techniques used or observed by the technician. For a more complete description of these techniques, the reader is referred to the text *Aircraft Powerplants.*

SHEET-METAL REPAIR

In Chap. 8, the elements of structural design for fastener joint design were discussed. AC 43.13-1B & 2B provides a series of charts derived from these design concepts that are available to the technician for use in the application of the principles discussed in this chapter.

These charts are designed to meet the "as strong as the original" criteria of the Advisory Circular, but by their nature must be very general. The FAA states in the Advisory Circular and throughout its other related publications that the technician must first comply with the provisions of the **aircraft manufacturer's Structural Repair Manual** if applicable and, failing that, refer to **AC 43.13-1B & 2B** and **MMPDS-06**, *Metallic Materials and Elements for Flight Vehicle Structure*, as approved data for the repair design. If the repair is a **major repair**, the repair must be approved by the FAA, through the use of the FAA's Form 337. Because the charts used as approved data are general in nature, it is always a good idea to have the Form 337 approved prior to beginning any repair work.

Before proceeding with any repair not covered by the aircraft manufacturer's Structural Repair Manual, the aviation maintenance technician should have the related Form 337 approved. See *Aircraft Basic Science* for a discussion regarding administration and completion of FAA Form 337.

Later in this chapter, recommended repair practices are discussed. The discussion of each repair consists of three parts: (1) the techniques using AC 43.13-1B & 2B minimum criteria; (2) the application techniques recommended for use by the technician, and (3) the calculations used in developing the associated charts.

Fundamentals of Rivet Repair Design

Purpose of the Repair

The primary purpose of a repair to sheet-metal parts is to return the damaged area to its original strength. Fasteners join the patch material to the original material in such a manner that the loads applied to the original material are transferred through the fasteners to the patch material. These loads are then transmitted through the patch material across the damaged area, and then the loads are returned to the original material through fasteners. The transmission of these loads for a fastener repair is accomplished at a lap joint. A patch typically consists of two lap joints, one to pick up the load and another to return the load. Figure 9-1 illustrates the basic load-transfer concept. For most applications a single-layer patch is sufficient.

Reinforcement Plates and Plugs

The technician will frequently be required to fabricate repairs that do not alter the basic aerodynamics of the original material. In such instances a plug and reinforcement plate (sometimes called a doubler) may be used. The plug is used to maintain the surface plane and does not carry any loads. The reinforcement plate is the medium by which the loads are transferred. The actual thickness of the reinforcement plate will depend upon a relationship between the load carried and the type and number of rivets used. (Figure 9-12 shows a simple plug and reinforcement-type repair.)

Reinforcement plates are also used to provide additional strength to specific areas of a large section of an aircraft's skin. Multi-engine propeller aircraft frequently have doublers on the aircraft's fuselage in the plane of rotation of the propellers. These doublers provide extra strength in the areas where chunks of ice might strike the fuselage as they break loose from the propeller.

Classifications of Sheet-Metal Damage

Several aircraft manufacturers have developed classifications of damage that are used to determine the corrective action necessary to return an aircraft to an airworthy condition. Each manufacturer establishes the damage level based on the aircraft and the area of the aircraft involved. The classifications of damage are *negligible*, *repairable*, and *replacement*.

Negligible damage is damage that does not affect the airworthiness of the aircraft. This level of damage may be allowed to exist or can be repaired by minor patching operations. Typical of this type of damage are dents in the skin that are not cracking, cracks in areas of low stress that can be covered by a 2-in [5.08-cm] circle, and surface scratches in low-stress areas.

Repairable damage is damage that might affect the airworthiness of the aircraft and could result in a loss of function of a component or system if not repaired. Repairable

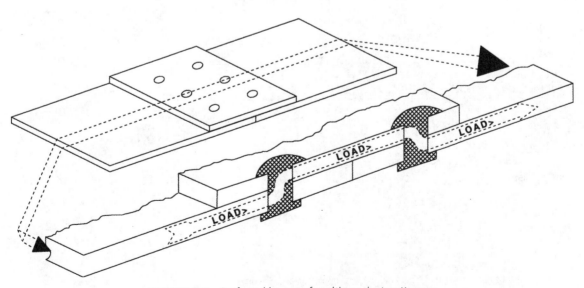

FIGURE 9-1 Preferred layout of multirow rivet patterns.

damage can be repaired by the use of a patch or by the insertion of a replacement component. This might include holes in the skin and cracked or broken formers and stringers that are not significantly deformed.

Replacement damage is damage that cannot be practically repaired and where repairing is specifically prohibited. This type of damage includes extensive corrosion, parts that are twisted or warped beyond usable limits, and components requiring alignments fixtures for proper repair.

Types of Damage

Before any attempt to make a repair is undertaken, the technician should first attempt to identify the cause and nature of the damage. If the cause was the result of an external or extraordinary condition, the technician should proceed with the repair in an effort to restore the original attributes of the structure without increasing the weight of the vehicle any more than necessary.

If, however, the cause of the damage is due to normal operations, consideration needs to be given to increasing the appropriate attribute or attributes of the structure. Although strength is an important attribute, the technician must be careful not to equate rigidity with strength. In certain situations the flexibility of the structure may play an important role in determining the strength of an aircraft component. In such cases the technician should report the damage to the appropriate individuals. If the cause of the damage cannot be reasonably assumed, the technician should obtain assistance from the manufacturer in developing a repair philosophy. Any repairs not approved by the manufacturer must be approved by the FAA. AC 43.13-1B & 2B consists of approved data that may be used by the technician to develop repair and should be referred to in the document requesting

FAA approval. The presence of a repair design or philosophy in AC 43.13-1B & 2B does not imply FAA approval in any specific application.

Repair-Material Selection and Interchangeability

Whenever possible the technician should use material of the same type and thickness as the original material. If this is not practical, FAA-approved aircraft manufacturer's Structural Repair Manuals may list approved substitution materials. Table 9-1 lists generally acceptable material substitutions. In the left-hand column the table lists materials that may be substituted for and across the top lists the materials that may be used as a substitute. Each cell, the intersection of each material row and the substitute material column, has a thickness compensation factor. By multiplying the original material's thickness by the compensation factor, the recommended thickness of the substitute material is attained. Note also that when bare material is substituted for clad material, additional corrosion protection is recommended.

General Design Guidelines

The FAA has established general rivet repair design guidelines, published in AC 43.13-1B & 2B. The guidelines include:

- **Rivet replacement.** In replacing rivets, the original size should be used if this size will properly fit and fill the holes. If not, the holes should be drilled or reamed for the next-larger rivet. Care should be taken when enlarging a rivet hole that the minimum criteria for rivet spacing and edge distances (discussed later in this chapter) are maintained.

TABLE 9-1 Metal Substitution Chart of Sheet Materials with Thickness Adjustment Factors 0.016 in [0.40 mm] to 0.125 in [3.18 mm]

Original Sheet Material	Substitute Materials									
	7075–T6	7075–T6 Clad	2024–T3		2024–T3 Clad		2024–T4 2024–T42		2024–T4 Clad[5] 2024–T42 Clad	
General Footnotes	1,2,6	1,2	3,6	4,6	3	4	3,5,6	4,5,6	3	4
2024–T3	1.00	1.00	XXXX	XXXX	1.09	1.10	1.00	1.10	1.03	1.14
2024–T3 Clad	1.00	1.00	1.00	1.00	XXXX	XXXX	1.00	1.00	1.03	1.10
2024–T4	1.00	1.00	1.00	1.00	1.00	1.00	XXXX	XXXX	1.00	1.14
2024–T4 Clad	1.00	1.00	1.00	1.00	1.00	1.00	1.00	1.00	XXXX	XXXX
7076–T6	XXXX	1.10	1.20	1.78	1.30	1.83	1.20	1.78	1.24	1.84
7076–T6 Clad	1.00	XXXX	1.13	1.70	1.22	1.76	1.13	1.71	1.16	1.76

Notes:
[1]Substitutions are not valid for pressurized areas, wing interspars, or center section structural members.
[2]All thicknesses.
[3]Up to and including 0.063 in [1.60 mm].
[4]Greater than 0.71 in [1.80 mm].
[5]2024-T4 and 2024-T42 are equivalents.
[6]Bare materials substituted for clad materials may require supplemental corrosion resistance.

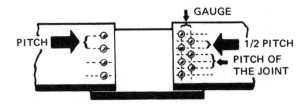

FIGURE 9-2   Rivet pitch and gauge.

- **Rivet diameter.** The rivet diameter for a sheet-metal joint should be approximately three times the thickness of the heavier sheet or somewhat larger for thin sheets.
- **Rivet spacing.** The space between rivets in a single row is called **pitch**, and the distance between rows of rivets is called **gauge**. These terms are illustrated in Fig. 9-2 The spacing between any two rivets is measured from the center of the shank of one to the center of the shank of the other. **Edge distance** is the distance between the center of the rivet shank and the nearest edge of the material. Edge distance is illustrated in Fig. 9-2. Note that the rivets at the ends of the rivet rows must meet edge-distance requirements in two directions.

The minimum spacing for aircraft rivets specified by the FAA in AC 43.13-1B & 2B is three times the diameter of the rivet shank, except for two row applications, where the rivet spacing should be 4D. The minimum edge distance is two times the diameter of the rivet shank, as shown in Fig. 9-3. Although the minimum edge distance for rivets is given as two times the diameter of the rivet shank, it is recommended that the edge distance be not less than $2\frac{1}{2}$ times the rivet shank diameter when the rivet is of the countersunk type. This will ensure adequate strength of material along the edge of the sheet.

It is general practice to limit the maximum pitch (space between rivets in a single row) to 24 times the thickness of the sheet metal. For example, if the thickness of the sheet metal is 0.083 in [2.11 mm], 24×0.083 in $= 1.992$ in, or 2 in [50.8 mm] for practical purposes.

- **Repair width.** The width of the repair should be twice that of the damaged area.

General Design Assumptions

There are several basic processes that should be included in the design and application of a fastener-secured repair. After the abbreviated listing of the processes necessary to

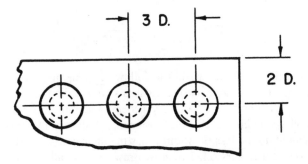

FIGURE 9-3   Minimum rivet spacing.

develop a design philosophy and the development of the philosophy, the concepts leading to these assumptions are discussed in greater detail.

- **Stop-drill all cracks.**
- **Round all corners.**
- **Be sure the thickness of the patch material used in a repair is at least equal to that of the original material.**
- **Design rivet patterns so that the rivet rows are parallel with the crack(s) and perpendicular to the relative load vector.**

The structural loads that relate to aircraft rivet repairs may be expressed as a single load (using vector analysis; see *Basic Aircraft Science*) even when there is more than one load. The general direction of the crack may be used to estimate the directional vector that represents the effective load.

The effective load vector may be assumed to be directly perpendicular to a line drawn from the beginning to the end of the crack for a single crack. If the crack is multidirectional, more than one load vector may be used. As a general rule, in establishing a rivet pattern each force vector must be compensated for on an individual basis.

- **Rivet spacing should be equal in all directions (symmetrical) so that concentrations of the load are not permitted.**
- **The width of the patch material** (W_p) **should exceed the width of the damage** (W_d) so that no significant loads are applied to the ends of the crack where stress risers will be created. At a minimum, the rivet pattern should allow for one load-bearing fastener at a distance equal to the pitch of the rivets on each side of the damaged area. Later in this chapter ways to safely minimize the width of the patch material below the general rule of twice the width of the damage are discussed.
- **Repairs should be repairable.** AC 43.13-1B & 2B refer to many minimum and/or maximum dimensional requirements. These requirements are frequently misunderstood as being the "nominal" dimension. When the technician improperly uses these dimensions as "nominal" instead of their proper interpretation (as a minimum or maximum), the potential for actually fabricating a repair that does not comply with the acceptable data of AC 43.13-1B & 2B is extremely high.

For example, assume that a technician uses the minimum edge distance of 2D for a repair using a single row of four MS20470AD3 rivets. The edge distance would be $2 \times \frac{3}{32} = \frac{3}{16}$ in. The conscientious technician locates the centers of these holes as close to $\frac{3}{16}$ in from the edge as possible. Unless the aviation technician makes a conscious decision to error in one direction only, chances are about one-half that the center punch to locate the rivet hole will actually be slightly less than the minimum acceptable edge distance. In addition, if rivets have to be removed after the installation is complete, the potential for damaging the original hole exists. If this situation exists, which requires drilling the hole to the next larger size,

TABLE 9-2 Rivet Layout Diameters for Repairable Repairs

Dia. Designator Number	Rivet Diameter, in	Layout Diameter, in
2	$\frac{2}{32}$	$\frac{4}{32}$
3	$\frac{3}{32}$	$\frac{5}{32}$
4	$\frac{4}{32}$	$\frac{6}{32}$
5	$\frac{5}{32}$	$\frac{8}{32}$
6	$\frac{6}{32}$	$\frac{8}{32}$
8	$\frac{8}{32}$	$\frac{10}{32}$

a simple rivet replacement is no longer acceptable per AC 43.13-1B & 2B criteria.

It is, therefore, recommended that the technician, after the minimal rivet diameter is established, lay out the rivet repair as though a rivet larger than the minimal rivet diameter were actually going to be installed. Table 9-2 shows the recommended diameters for rivet layout. The actual installation, however, would be made with the originally calculated minimal rivet diameter. Following this practice whenever possible has one major advantage. The replacement of rivets to the next-larger-size rivet provides a repair that, even if the hole is slightly mislocated originally, will comply with typically calculated or AC 43.13-1B & 2B acceptable practices.

In the examples of rivet repairs used later in this chapter, the references to *"Applied"* refer to the concept of designing repairs so that they are repairable.

Repair Layout Techniques

Once the basic design parameters for a repair have been established, the technician will need to develop a layout for the installation of the fasteners. Regardless of the technique used in the development of these parameters, the following layout criteria must be established before the layout process can begin: the width of the repair (W) (discussed previously), the size of the fastener hole to be used in designing the repair (D), the minimum spacing between the fasteners (S) and the edge of the repair (E), and the total number of fasteners required by the repair (T_r). Within these parameters the technician needs to determine the number of rows of fasteners and the actual fastener spacing to be used.

Note that if the repair is to be designed so that it is repairable, the diameter (D) refers to a driven rivet diameter larger than the rivet that will actually be installed. Whenever lowercase d is used, it refers to the **driven diameter** of the rivet that will actually be installed.

Number of Rows of Fasteners (N_r)

The number of rows (R) used in a joint is determined by the width of the joint repair (W), the diameter of the layout rivets (D), the minimum rivet spacing (S), the edge distance (E) [assumed to be $2D$], and the number of fasteners (T_r) required for the repair. To determine the length of a rivet row

(L), the assumed edge distance ($2D$) is twice subtracted from the width of the joint.

$$L = W - 2E \qquad (9\text{-}1)$$

Substituting $2D$ for E yields

$$L = W - 4D \qquad (9\text{-}2)$$

The edge distance is subtracted twice because each end of a row requires an edge distance.

Dividing the length of the rivet row (L) by the rivet spacing (S) yields the number of spaces (N_s) available within the row.

$$N_s = \frac{L}{S} \qquad (9\text{-}3)$$

However, because the row begins and ends with a hole, there is one more hole required than spaces. Adding 1 to the result of the equation and then taking the integer part has the effect of algebraically rounding to the next largest number of fasteners and eliminating fractional fasteners. The result is the maximum number of fasteners in a row (F_r):

$$F_r = \text{INT}\,(N_s + 1) \qquad (9\text{-}4)$$

Substituting Eq. (9-3) for N_s yields

$$F_r = \text{INT}\left(\frac{L}{S} + 1\right) \qquad (9\text{-}5)$$

Substituting Eq. (9-2) for L gives

$$F_r = \text{INT}\left(\frac{W - 4D}{S} + 1\right) \qquad (9\text{-}6)$$

The number of rows (N_r) is then the total number of fasteners required (T_r) divided by the maximum number of fasteners in a row (F_r) rounded to the next whole number.

$$N_r = \text{INT}\left(\frac{T_r}{F_r} + 1\right) \qquad (9\text{-}7)$$

Substituting Eq. (9-6) for F_r gives

$$N_r = \text{INT}\left(\frac{T_r}{\text{INT}((W - 4D)/S + 1)} + 1\right) \qquad (9\text{-}8)$$

If the technician wishes to follow the general guideline for pitch and gauge, two calculations are required to determine the number of rows. The general guidelines state that both the pitch and gauge of fasteners installed in two rows should not be less than four times the diameter of the rivet ($4D$). All repairs other than two-row repairs call for pitch and gauge to be not less than three times the diameter of the fastener (3D). Unless the choice is obvious, the most efficient way to determine the desired number of rows is to perform the calculations related to the two-row spacing (4D) first. This is accomplished by substituting 4D for S in Eq. (9-8).

$$N_r = \text{INT}\left(\frac{T_r}{\text{INT}((W-4D)/4D+1)}+1\right) \quad (9\text{-}9)$$

If two-row spacing (4D) will not accommodate the total number of fasteners required, then 3D spacing must be used and substituted for S, whether one or three or more rows are required.

$$N_r = \text{INT}\left(\frac{T_r}{\text{INT}((W-4D)/3D+1)}+1\right) \quad (9\text{-}10)$$

In order to use the minimum number of fasteners, it may be necessary to stagger the rows so that the center of a fastener in an adjacent row is in a position perpendicular to the center of the rivet spacing between the adjacent rows (see Fig. 9-1).

Actual Fastener Spacing (S_a)

Once the number of fasteners in a row is established, it is a relatively simple matter to determine a symmetrical pattern for the rivets within a row. To develop equal spacing between fasteners in a row, divide the length of the row (L) by the number of fasteners in the row (F_r). The result is a symmetrical spacing between fasteners within the row. This process will have to be repeated for each row that varies in either length or quantity of fasteners in a row.

$$S_a = \frac{L}{F_r} \quad (9\text{-}11)$$

REPAIR PRACTICES

The use of the proper size and spacing of rivets and their proper installation determines whether a riveted assembly will withstand operational stresses. If any of these items is insufficient, the structure may fail, resulting in damage to the structure and injury to the aircraft occupants.

The use of specific rivet sizes and placements is dictated by the type and amount of stress placed on a structure and the size of the material being used. The following information is based on standard industry practices. For specific rivet-selection and repair procedures, consult the aircraft manufacturer's manuals.

When a fastener joint needs to be laid out, there are a number of variables which need to be considered. These include the following:

Size of the fastener (D)

Spacing of the fasteners (S)

Total number of fasteners required (T_r)

Number of fasteners in a row (F_r)

Number of rows (N_r)

Tensile strength of the sheet with holes (T_s)

In combining all these factors into a single repair based upon the mechanical properties of the sheet and fastener materials, a number of interdependent relationships are involved. For example, the number of fasteners in a row is dependent upon the tensile strength of the sheet in the weakest row. The tensile strength of the weakest row depends upon the number of rivets in the row.

To simplify the use of these relationships in the process of designing a repair, tables have been developed for use by the technician. Simplifying the design process, however, is not without penalty. To minimize the number of tables needed to develop a repair design, as many as 10 different materials in up to 6 different forms for use in 4 different types of structural repairs are all combined into one table. To ensure that all possible combinations of these factors may successfully be used in a repair requires that the weakest of all possible combinations be used as the minimum requirement. The result is that repairs designed using these tables often require more fasteners than necessary, needlessly adding weight to the aircraft. The installation of these excess fasteners also results in increased material and labor costs.

Another difficulty with using these tables is that they were developed during the period when the majority of aircraft were designed using safe-life structural criteria. Under the safe-life philosophy, stronger—which is typically equated to more rigid—is generally considered as good or better than the original. However, modem aircraft, which require a certain degree of flexibility to accommodate the pressurized cabin, are designed using the damage-tolerant design philosophy, under which more rigid is not necessarily as good or better.

It is important for the technician to note that major repairs designed using these data must be approved via FAA Form 337 before the repair design may be considered airworthy. Although these tables are considered approved data in support of a repair design, they are not approved for all conditions. Unfortunately, the conditions under which these tables are and are not approved data are not specified in their accompanying text.

The following section of this chapter discusses the development of the repair designs using fasteners based upon the mechanical properties published in MMPDS-06. The technician, working in an environment where the general-purpose tables just described are typically considered appropriate, may wish to proceed to the section entitled "Using Repair Design Tables."

Repair Design Using Mechanical Properties

In the discussion that immediately follows, references are made to principles established in Chap. 8. When the equations developed in Chap. 8 expressing these principles are referred to, their numbers are given for review, if necessary.

Fastener Size

The maximum load that the sheet can withstand at the rivet joint is equal to the bearing load of the sheet. If the rivet's shear load-carrying capability (P_s) is less than that of the sheet-bearing load (P_b), the rivet will shear before the sheet. If the rivet's shear load-carrying capability (P_s) is greater than that of the sheet-bearing load (P_b), the sheet will fail before the rivet.

Occasionally in repair design, the user must determine which is most desirable, whether the sheet or rivet should fail first. By calculating the optimum diameter of the rivet (where both P_b and P_s are equal), the technician may select the maximum strength available for each option performing only one calculation. The probability of making the calculation that results in a whole number is relatively small, so using the next-size-diameter rivet will make P_s greater than P_b, so the sheet will fail by bearing failure first. The lower rivet size will make P_b greater than P_s, so the rivet will shear first.

$$P_b = P_s \qquad (9\text{-}12)$$

Substituting Eqs. (8-34) for P_b and (8-28) for P_s gives

$$d \times t \times F_b = \frac{\pi \times d^2}{4} \times F_s \qquad (9\text{-}13)$$

Multiply both sides of the equation by 4:

$$4 \times d \times t \times F_b = \pi \times d^2 \times F_s \qquad (9\text{-}14)$$

Divide both sides of the equation by d:

$$4 \times t \times F_b = \pi \times d \times F_s \qquad (9\text{-}15)$$

Divide both sides of the equation by ($\pi \times F_s$)

$$d = \frac{4 \times t \times F_b}{\pi \times F_s} \qquad (9\text{-}16)$$

Simplify:

$$d = \frac{4}{\pi} \times t \times \frac{F_b}{F_s} = 1.27 \times t \times \frac{F_b}{F_s} \qquad (9\text{-}17)$$

$$d = 1.27t \left(\frac{F_b}{F_s} \right) \qquad (9\text{-}18)$$

EXAMPLE Use 2024-T3 clad sheet ($F_b = 97\,000$ psi) and 2017-T4 (AD) rivet ($F_s = 30\,000$ psi).

$$d = 1.27 \times t \times \left(\frac{97\,000}{30\,000} \right) = 3.93t$$

*WARNING: The example equations are valid only for the assumed parameters and should not be used in other applications.

For the sheet to fail first, use $4t$ for the diameter of the rivet. For the rivet to shear first, use $3t$ for the diameter of the rivet. However, in this case extreme care should be used when selecting the $3t$ rivet size to ensure that the diameter of the rivet will support the flight loads.

In determining the size of rivets to be used in any aircraft repair, the technician must comply with the provisions of FAA publications. These set forth the policies and regulations of the FAA relative to the repair, maintenance, and overhaul of aircraft and engines. In the repair of military aircraft, the technician should follow military standards, as set forth in technical orders and handbooks.

Rivet Spacing

The objective of establishing a minimum edge distance and rivet spacing is to ensure that the sheet material does not tear out before bearing failure; to do this, the tearout load of the sheet must be equal to or greater than the bearing load:

$$P_b = P_e \qquad (9\text{-}19)$$

Substituting Eq. (8-34) for P_b and formula (8-32) for P_e gives Eq. (9-20).

Note that D is substituted for d, since the repair is intended to be repairable.

$$D \times t \times F_b = 2 \times \left(\frac{E - D}{2} \right) \times t \times F_s \qquad (9\text{-}20)$$

Divide both sides of the equation by t:

$$D \times F_b = 2 \times \left(\frac{E - D}{2} \right) \times F_s \qquad (9\text{-}21)$$

Simplify:

$$D \times F_b = 2EF_s - \frac{2D}{2} \times F_s \qquad (9\text{-}22)$$

$$DF_b = 2EF_s - DF_s \qquad (9\text{-}23)$$

Add DF_s to both sides of the equation:

$$DF_b + DF_s = 2EF_s \qquad (9\text{-}24)$$

Simplify:

$$D(F_b + F_s) = 2EF_s \qquad (9\text{-}25)$$

Divide both sides of the equation by $2F_s$:

$$E = \frac{D(F_b + F_s)}{2F_s} \qquad (9\text{-}26)$$

EXAMPLE Use 2024-T3 clad sheet, for which $F_s = 37\,000$ psi, $F_b = 97\,000$ psi.

$$E = \frac{D \times (97\,000 + 37\,000)}{2 \times 37\,000} = 1.81D$$

*WARNING: The example equations are valid only for the assumed parameters and should not be used in other applications.

Increasing the edge distance increases the tearout capabilities of the joint.

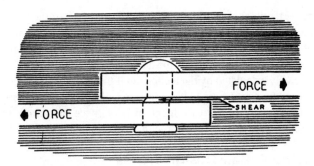

FIGURE 9-4 Shear on a rivet.

Determining the Number of Rivets Required per Lap Joint for a Repair (T_r)

The number of rivets required for each lap joint in any repair is determined by the strength necessary for the riveted joints. This strength is based upon two considerations. First, the **shear strength** of the rivets to be used must be determined. The shear on a rivet is the load that tends to cut the rivet in two parts, as shown in Fig. 9-4. Second, the **tensile strength** of the sheet metal must be determined. These two forms of strength, considered together, constitute the basis for determining how many rivets are needed.

Tensile Strength of Drilled Sheet.

The tensile strength of a sheet of a particular piece of material depends upon the lowest cross-sectional area of the sheet that is perpendicular to the plane in which the load is applied. The amount of load that may be applied to any solid sheet is relatively easy to determine. It is the shear strength of the material times the thickness of the material times the width of the material.

Calculating the tensile strength of a sheet of material with holes is only slightly more complex. The cross-sectional area of a sheet with holes is the width of the sheet less the diameter of each of the rivet holes affecting the cross section. The adjusted width is then multiplied by the ultimate tensile strength of the material.

$$P_t = (W - F_r D) t F_t \qquad (9\text{-}27)$$

where W_d = width of the damaged area
D = diameter of the rivet holes sized for replacement rivets
F_r = number of rivets in the densest row [the row that has the greatest number of rivets; see Eq. (9-6)]

In situations where more than one pattern of rivet holes exists (i.e., multiple rows of rivets with a different number or rivets in one or more rows) the row with the least cross-sectional area (i.e., the most rivets) is used to determine the tensile strength of the rivet joint sheet.

Note that the actual width of the patch would exceed that of the damaged area. However, by using the damage width at this point, the minimum tensile strength of the sheet for both a splice and a patch may be calculated by a single formula.

Number of Fasteners Required per Lap Joint (T_r).

The number of fasteners required to transfer the load from one sheet to another is equal to the tensile load of the sheet (P_t) at the joint divided by the shear load for each fastener (P_s) or the bearing load of the sheet (P_b), whichever denominator is least. This results in the highest number of rivets required to sustain the load.

$$T_r = \frac{P_t}{P_s} \qquad (9\text{-}28)$$

or

$$T_r = \frac{P_t}{P_b} \qquad (9\text{-}29)$$

Since the technician, when calculating the number of rivets required, is concerned with repair, as opposed to original design or replacement, the tensile load of the sheet (P_t) is calculated based upon the condition of the original material. The original material is used because the load that is required to be transferred is equal to the existing load applied during the aircraft's operations.

In the following equations, the original material is assumed to be solid (i.e., without existing holes). Because the presence of existing holes would lessen the load-carrying capability of the original sheet, calculating P_t based upon a solid sheet errs on the conservative side when calculating the number of rivets required. As a result, for repair Eq. (8-30) is used. If the nature of the repair requires the technician to make more exacting calculations, the technician should substitute Eq. (9-27) for P_t.

To minimize the number and complexity of the calculations required of the technician, it is suggested that the values of P_s and P_b be calculated first. Then the equation with the lower of the two values should be used. Because they are in the denominator of the equation, the higher number of rivets (T_r) will be the result of using the lower value.

For completeness of this discussion, both calculations are derived next. For P_t/P_s, the equation is

$$T_r = \frac{P_t}{P_s} = P_t \times \frac{1}{P_s} \qquad (9\text{-}30)$$

Substituting Eq. (8-30) for P_t and Eq. (8-28) for P_s gives

$$T_r = w \times t \times F_t \times \frac{1}{\dfrac{\pi D^2 F_s}{4}} \qquad (9\text{-}31)$$

Note that D is substituted for d in Eq. (8-28) because the repair is assumed to be designed to be repairable.

Equation (9-31) simplifies as follows:

$$T_r = w \times t \times F_t \times \frac{4}{\pi D^2 F_s} \qquad (9\text{-}32)$$

or

$$T_r = \frac{4 w t F_t}{\pi D^2 F_s} \qquad (9\text{-}33)$$

Dividing 4 by π yields

$$T_r = \frac{1.27wtF_t}{D^2 F_s} \qquad (9\text{-}34)$$

Substituting Eq. (8-30) for P_t and Eq. (8-34) for P_b yields

$$T_r = \frac{wtF_t}{DtF_b} \qquad (9\text{-}35)$$

Simplifying Eq. (9-35) yields

$$T_r = \frac{wF_t}{DF_b} \qquad (9\text{-}36)$$

Remember that this is the number of rivets required to transfer the load from one piece of material to another at a lap joint. So, the actual number of rivets required for a typical patch with two lap joints would be twice the number calculated here.

Width of Patch Repairs

The width of a patch must be wider than the width of the damaged area if material of the same type and thickness is used for the patch. This is because the tensile-strength capability of the damaged area is the width of the area times the thickness of the material. The tensile-strength capability of the patch is the width of the patch less the diameters of the fastener holes

times the thickness of the material. So the width of the patch must exceed the width of the damaged area by at least the sum of the diameters of the fastener holes plus the rivet spacing for those fastener holes outside the damaged area. This value may be easily calculated by the technician if the additional weights caused by a repair are to be minimized. However, the general rule of thumb previously mentioned of twice the damaged width is sufficient to meet structural requirements.

Using Repair Design Tables

Tables have been prepared to designate the number or rivets necessary to restore the strength to a given section of sheet aluminum alloy when using 2117-T3 rivets. Table 9-3 is an example for rivets from $\frac{3}{32}$ to $\frac{1}{4}$ in [2.38 to 6.35 mm] in diameter and aluminum-alloy sheet thicknesses from 0.016 to 0.128 in [0.406 to 3.25 mm]. When such tables are available, it is a simple matter to determine the number of rivets necessary for any particular repair.

If it is desired to repair a 2-in [5.08-cm] break in a sheet of 0.025-in [0.635-mm] aluminum-alloy skin on an airplane, the number of rivets would be determined as follows:

1. Select the size of rivet. Since the riveted sheet is 0.025 in [0.635 mm] thick, the rivet diameter must be at least three times this amount. This requires a rivet of at least 0.075 in [1.91 mm] in diameter. The next larger standard rivet is $\frac{3}{32}$ in [2.38 mm]; hence this is the size to be used.

TABLE 9-3   FAA AC 43.13-1B & 2B, Fig. 2.28; Number of Rivets Required

Thickness t in Inches	No. of 2117–AD Protruding Head Rivets Required per Inch of Width W					No. of Bolts
	$\frac{3}{32}$	$\frac{1}{8}$	$\frac{5}{32}$	$\frac{3}{16}$	$\frac{1}{4}$	AN–3
0.016	6.5	4.9	-----------	-----------	-----------	-----------
0.020	6.9	4.9	3.9	-----------	-----------	-----------
0.025	8.6	4.9	3.9	-----------	-----------	-----------
0.032	11.1	6.2	3.9	3.3	-----------	-----------
0.036	12.5	7.0	4.5	3.3	2.4	-----------
0.040	13.8	7.7	5.0	3.5	2.4	3.3
0.051	-----------	9.8	6.4	4.5	2.5	3.3
0.064	-----------	12.3	8.1	5.6	3.1	3.3
0.081	-----------	-----------	10.2	7.1	3.9	3.3
0.091	-----------	-----------	11.4	7.9	4.4	3.3
0.102	-----------	-----------	12.8	8.9	4.9	3.4
0.128	-----------	-----------	-----------	11.2	6.2	3.2

Notes:
[a] For stringers in the upper surface of a wing, or in a fuselage, 80 percent of the number of rivets shown in the table may be used.
[b] For intermediate frames, 60 percent of the number shown may be used.
[c] For single lap sheet joints, 75 percent of the number shown may be used.
Engineering Notes: The above table was computed as follows:
[1] The load per inch of width of material was calculated by assuming a strip 1 in wide in tension.
[2] Number of rivets required was calculated for 2117-AD rivets, based on a rivet allowable shear stress equal to 40 percent of the sheet allowable tensile stress, and a sheet allowable bearing stress equal to 160 percent of the sheet allowable tensile stress, using nominal hole diameters for rivets.
[3] Combinations of sheet thickness and rivet size above the heavy line are critical in (i.e., will fail by) bearing on the sheet; those below are critical in shearing of the rivets.
[4] The number of AN–3 bolts required below the heavy line was calculated based on a sheet allowable tensile stress of 70 000 psi and a bolt allowable single shear load of 2126 lb.

2. Refer to Table 9-3 and note that when the thickness t of the sheet is 0.025 in [0.635 mm], the number of $\frac{3}{32}$-in [2.38-mm] rivets should be at least 8.6 per inch of width W of the repair. The break to be repaired is 2 in [50.8 mm] long; hence 17.2 rivets are required. Therefore, 18 rivets are used on each side of the repair to restore the required strength.

Remember that this is the number of rivets required to transfer the load from one piece of material to another at a lap joint. So, the actual number of rivets required for a typical patch with two lap joints would be twice the number calculated here.

Although the minimum edge distance for rivets is given as two times the diameter of the rivet shank, it is recommended that the edge distance be not less than $2\frac{1}{2}$ times the rivet shank diameter when the rivet is of the countersunk type. This will ensure adequate strength of material along the edge of the sheet.

The layout for the repair discussed here could look like the one shown in Fig. 9-5. There can be variations in the design of a layout, provided the basic requirements of edge distance, rivet size, and rivet spacing are met. Observe in the illustration that rivets are spaced at a greater distance than the minimum.

Load Transmission: Multirow Rivet Patterns

So far the discussion regarding rivets and load transfer has assumed that each rivet in the repair transfers an equal load.

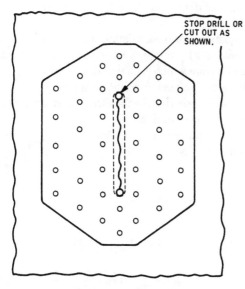

FIGURE 9-5 Layout for a typical rivet repair.

However, in application, the rivets in each row of rivets actually transfer a different load. The first row of rivets transfers the most load, with decreasing loads being transferred by each subsequent row of rivets.

From previous discussions, it is known that the rivets of an appropriately designed joint will not shear under the design load. Therefore, each rivet is capable of and actually transfers its maximum capabilities until the load remaining to be transferred is less than the capabilities of the rivet. Figure 9-6 indicates this load-carrying sequence.

The first load transfers its maximum transfer capabilities, which is the cross-sectional area of the rivet times the load applied ($L_{tr1} = ND \times L_{r1}$). The remaining load is the original load less the load transferred by the first row of rivets ($L_{r2} = L_{r1} - L_{tr1}$). The load applied to the second row of rivets is therefore less than the load applied to the first row. The load actually applied to the second row of rivets is the cross-sectional area of the rivets times the number of rivets times the actual load applied. The remaining load is the load applied to the second row of rivets less the actual load transferred. This continues until the entire load is transferred.

Stacked Doublers

The actual load-transfer process, when row transfer is considered, allows the technician to do some interesting things in an effort to save weight. Typically, the technician uses patch material of equal or greater thickness than the original. In repair applications where the repair is required because of the structural failure of a component, the repair material always requires a strength equal to or greater than the original. The weight-conscious technician may use his or her knowledge of load transfer to save weight in the design phase of the rivet repair.

The patch material needs to carry only the actual load that is transferred. Therefore, the thickness of the patch material required at the first row needs to be only thick enough to support the load L_{tr1}.

For the second-row transfer, the patch material needs to be capable of carrying the load previously transferred (L_{tr1}) plus the load transferred by the second row of rivets. The same is true for all the rows involved in the rivet repair.

The weight-conscious aviation technician has basically two options available to take advantage of this load-by-rivet-row transmission. The steps (increases in material thickness) may be fabricated by various metal-removal techniques or by "stacking" pieces of different-thickness materials upon each other. The combination of material pieces indicated in the last option

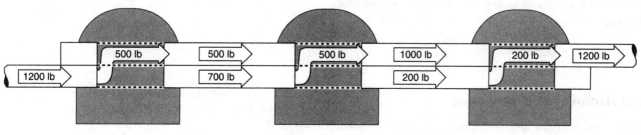

FIGURE 9-6 Transfer of load in multiple-row applications.

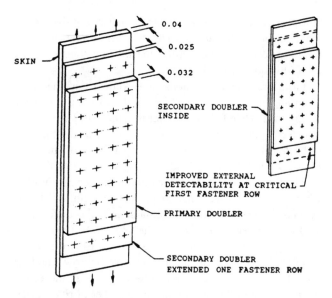

0.04

0.025

SKIN

0.032

SECONDARY DOUBLER
INSIDE

IMPROVED EXTERNAL
DETECTABILITY AT CRITICAL
FIRST FASTENER ROW

PRIMARY DOUBLER

SECONDARY DOUBLER
EXTENDED ONE FASTENER ROW

FIGURE 9-7 Stacked doubler.

is frequently referred to as **stacked doublers**, which are shown in Fig. 9-7.

Loads Applied to Multiple Planes in Shear

The earlier discussion regarding load calculations and their application needs to be expanded in order to include the use of multiple sheets of material to carry applied design loads. Loads applied to a joint may be applied in a single plane or in multiple planes. When more than two sheets make up a fastener joint, there are multiple shear planes. Figure 9-4 shows the double shear experienced by a rivet joint that has two shear planes.

Multiple planes result in a sharing of the applied loads by the materials involved. The benefit in multiple-plane load applications is that the total load is divided between the multiple planes and, as a result, the mechanical properties of the materials associated with each load plane do not need to be as large as would be required in a single-load-plane application. The sum of the load-carrying capability of each load plane is the total load-carrying capability of the joint.

There is, however, a negative impact of multiple-plane load application. Since the load planes are adjacent, separated only by one material thickness, a portion of the load applied to one plane carries over into the next plane. The thicker the material, the less this effect. In most materials handbooks, including MMDDS-06, the shear strengths for two planes are established. These two-plane shear strengths are significantly lower than single-plane shear strengths; therefore, the technician should consult these handbooks for the proper values. The double shear strengths are, however, typically much more than half the single-shear strengths, making the advantages of multiple-plane loads worthy of consideration in some applications.

Determining Design Loads

As discussions regarding rivet-repair design are put into practice, the technician may find it useful to determine the

aircraft's design loads. This may be accomplished by analyzing the aircraft's original rivet diameter in relation to the type and thickness of the material used. When applying this technique in rivet-repair design, however, the technician should be assured that the original design loads are sufficient. If the damage causing the need for a repair is nonload related, such as holes caused by a foreign object, this process will provide the needed data. However, if the cause of the repair was insufficient load transmission under normal flight operations, the technician should consult the aircraft manufacturer.

To determine the design load, begin by finding the rivet's diameter and material type. The rivet's material may be determined by examining the rivet head. The thickness must also to be determined. If the material type cannot be determined, the aircraft manufacturer must be consulted. The maximum shear strength of the rivet needs to be calculated [refer to Eq. (8-28)]. The maximum bearing load and tensile loads of the sheet material are calculated [refer to Eqs. (8-34) and (8-30), respectively]. Determine the number of rivets per inch. This is most accurately accomplished by counting the number of rivets over a lengthy span and dividing the number of rivets by the number of inches in the span.

The three loads that determine a rivet pattern are the tensile load carried by the material, the bearing load applied to the material, and the shear load placed on the rivet. By revising some of the equations previously developed and assuming the other loads are maximized, the technician can determine the maximum of any of the individual loads.

The equations for these relationships are shown next. For the technician who might wish to verify these equations, the equation from which each was derived is noted. In these equations w is assumed to be the width of the joint, $t =$ the thickness of the material, $d =$ the diameter of the driven rivet, and $T_r =$ the number of rivet.

To solve for the maximum tension load $F_{t(max)}$ with $F_{b(max)}$ at maximum, we use Eq. (9-36):

$$T_r = \frac{wF_{t(max)}}{DF_{b(max)}}$$

Multiplying both sides by $DF_{b(max)}$ yields

$$T_r DF_{b(max)} = wF_{t(max)} \tag{9-37}$$

Dividing both sides by w gives

$$\frac{T_r DF_{b(max)}}{w} = F_{t(max)} \tag{9-38}$$

To solve for the maximum sheet-bearing load $F_{b(max)}$ with $F_{t(max)}$ at maximum, we begin with Eq. (9-37):

$$T_r DF_{b(max)} = wF_{t(max)}$$

Dividing both sides by $T_r D$ gives

$$F_{b(\max)} = \frac{wF_{t(\max)}}{T_r D} \qquad (9\text{-}39)$$

To solve for the maximum tension load $F_{t(\max)}$ with $F_{s(\max)}$ at maximum, we begin with Eq. (9-34):

$$T_r = \frac{1.27wtF_{t(\max)}}{D^2 F_{s(\max)}}$$

Multiplying both sides of the equation by $D^2 F_{s(\max)}$ yields

$$T_r D^2 F_{s(\max)} = 1.27wtF_{t(\max)} \qquad (9\text{-}40)$$

Dividing both sides by $1.27wt$ gives

$$\frac{T_r D^2 F_{s(\max)}}{1.27wt} = F_{t(\max)} \qquad (9\text{-}41)$$

To solve for the maximum shear load $F_{s(\max)}$ with $F_{t(\max)}$ at maximum, we use Eq. (9-40):

$$T_r D^2 F_{s(\max)} = 1.27wtF_{t(\max)}$$

Dividing both sides by $T_r D^2$ gives

$$F_{s(\max)} = \frac{1.27wtF_{t(\max)}}{T_r D^2} \qquad (9\text{-}42)$$

To solve for the maximum shear load $F_{s(\max)}$ with $F_{b(\max)}$ at maximum, we use Eq. (9-18):

$$d = 1.27t\left(\frac{F_{b(\max)}}{F_{s(\max)}}\right)$$

Multiplying both sides by $F_{s(\max)}$ gives

$$dF_{s(\max)} = 1.27tF_{b(\max)} \qquad (9\text{-}43)$$

Dividing both sides of the equation by d simplifies the equation to

$$F_{s(\max)} = \frac{1.27tF_{b(\max)}}{d} \qquad (9\text{-}44)$$

To solve for the maximum sheet-bearing load $F_{b(\max)}$ with $F_{s(\max)}$ at maximum, we use Eq. (9-43):

$$dF_{s(\max)} = 1.27tF_{b(\max)}$$

Dividing both sides of the equation by $1.27t$ yields

$$\frac{dF_{s(\max)}}{1.27t} = F_{b(\max)} \qquad (9\text{-}45)$$

In some applications, the calculated load may exceed the ultimate strength of the material as determined by the mechanical properties. Since this is never the case in a properly designed aircraft, the maximum of any of the mechanical properties is that property's ultimate strength. Therefore, if the mechanical property being calculated exceeds its mechanical properties, the ultimate strength should be used in all subsequent equations.

Using the calculated maximums or the materials' related ultimate strength, whichever is least, and repeating the calculations allows the technician to approximate the design load factors. The more often these calculations are made the closer the approximation will become.

RIVET-REPAIR DESIGN

For each step in the design process for rivet repair, three procedures are described. The first, indicated as "*General*" (see below), follows the general guidelines established by the FAA and uses minimums as the design criteria. The second procedure, marked "*Applied*," uses the FAA minimums but is designed so that the rivet layout is repairable. The third technique for rivet layout, marked "*Calculated*," shows the results if the repair was designed using the calculations previously discussed.

Before proceeding with an explanation regarding how to use this information, the information provided in Fig. 9-8 should be reviewed. The strength data provided in Fig. 9-8 are accurate but not complete. In area (B) of Fig. 9-8 the strength

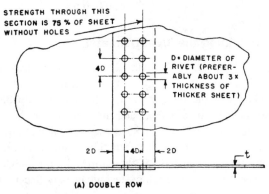

(A) DOUBLE ROW

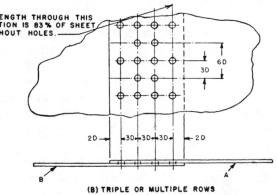

(B) TRIPLE OR MULTIPLE ROWS

FIGURE 9-8 Riveted sheet-metal splices.

comment in the upper left corner is limited only to the second row from the left and the row of rivets to which the arrow is pointing. The strength of the sheet is determined by the strength at its weakest point, which in this case is not the rows to which the arrow is referring. The relative sheet strength of the (B) joint is 66.7 percent, which results from the $3D$ spaced rivets.

Splicing of Sheets

Figure 9-8, which is a duplication of Fig. 2.18 in AC 43.13-1B & 2B, is an example of how sheet-metal materials might be spliced. The technician should note the difference between a "splice" and a "patch." A splice is the mating of two or more pieces of material so that the load is transferred from one material to another. A patch consists of two or more lap joints, where the loads applied to original material are first transferred from the original material to a doubler (which may be the patch material) and then from the doubler back to the original material. In aircraft design, a transition doubler may be a stinger, rib, or flange.

When splicing sheets, the splice should be designed as illustrated in the following example. The example makes the following assumptions:

Width of sheet (length of splice): $W = 12$ in [30.48 cm]

Sheet material:	2024-T3 Alclad sheet
	0.032 in [0.813 mm] thickness
	UTS = 59 000 psi
	UBS @ 1.5D = 97 000 psi
	USS = 37 000 psi
Rivet material:	AN470AD rivets
	2117 = T4
	USS = 30 000 psi

Determining the Rivet Diameter

In AC 43.13-1B & 2B, para. 2.99h, the discussion regarding rivet size uses the phrase "approximately three times the thickness of the thicker sheet." This relationship between the thickness of the sheet and the diameter of the rivet is not a minimum. If a technician uses the calculations previously discussed and these calculations result in a rivet diameter slightly smaller than "three times the thickness of the thicker sheet" of material, the technician may elect to use the smaller-diameter rivet.

General. Select the rivet diameter approximately three times the sheet thickness. Thus, $3 \times 0.32 = 0.096$ in [2.44 mm]. Since the next standard size greater than 0.096 in is $\frac{1}{8}$ in, use $\frac{1}{8}$-in [3.175-mm] 2117-T4 (AD) rivets. Unless design criteria dictates otherwise, when the calculated rivet diameter exceeds that of a standard-size rivet, the next largest standard-size rivet is used.

Applied. The minimum diameter of the rivets to be installed is $\frac{1}{8}$ in [3.18 mm], but for the layout work that follows the criteria for a rivet two standard sizes over (or $\frac{3}{16}$ in [4.76 mm]) will be used for the rivet layout.

Calculated. By applying Eq. (9-18), which is $d = 1.27t(F_b/F_s)$, the technician may calculate the diameter of the rivet required.

$$d = 1.27 \times 0.032 \times \left(\frac{97\,000}{30\,000} \right) = 0.131$$

Although, when using the AC 43.13 charts, the rivet-diameter selection was $\frac{1}{8}$ in [3.18 mm], using a $\frac{5}{32}$-in [3.97-mm] rivet would better meet the criteria when considering the mechanical properties of the materials involved. However, the calculated rivet size is only 0.006 in [0.15 mm] over the diameter of the $\frac{1}{8}$-in [3.18-mm] "charted" rivet diameter, and if the driven diameter of the charted rivet is considered, a difference of 0.002 in [0.05 mm] between the calculated and charted rivet exists.

Since the mechanical properties of the materials used for the repair do not reflect a safety factor, it is fairly safe to assume that the $\frac{1}{8}$-in [3.18-mm] rivet is capable of handling the loads involved. The technician could, however, calculate the *design* load applied to the repair area by examining the existing rivet diameters and their spacing, as indicated in Eqs. (9-37) through (9-45).

In continuing this example, the $\frac{1}{8}$-in [3.18-mm] rivet will be assumed to meet the design load criteria.

Determining the Number of Rivets

When determining the number of rivets, the ability of the rivets to withstand a shear load is equated to the ability of the rivets to withstand an equal bearing or tensile load, whichever is least.

General. Determine the number of rivets required per inch of width W from Table 9-3. The number of rivets per inch equals 6.2; hence the total number of rivets required is $12 \times 6.2 = 74.4$, or 75 rivets. Unless design criteria dictate otherwise, when the number of rivets exceeds a whole number, the number of rivets is always rounded up. The technician should note here that this example deals with a splice. A splice differs from a patch repair, which must have twice the number of rivets because it must first transmit the load to the patch and then return it to the skin on the other side of the damaged (or removed) area.

The technician may apply footnote (c) of Table 9-3, if appropriate. For single-lap joints, 75 percent of 74.4 rivets, or 56 (when rounded), may also be successfully used.

Applied. Since the diameter of the layout -6 rivets is larger than the -4 rivets, the -6 rivets are capable of withstanding greater shear forces. Because -4 rivets will be initially installed and their load-bearing capability is less than that of an equal number of -6 rivets, the number of rivets used in the repair should be that calculated in the general procedure. A review of Table 9-3 verifies this fact. Note that for a given thickness of material, as the diameter of the rivet increases, the number of rivets decreases.

Calculated. The load-carrying capability of the sheet material is generally assumed to be the load applied to the repair area. However, the technician might wish to develop a maximum-load applied criteria by analyzing the aircraft's existing rivet pattern and using these concepts to calculate the applied load. If this process and the general procedure are followed (i.e., additional spacing is not added to the rivet pattern in order to accommodate repairs to the repair), this will usually result in a duplication of the existing rivet pattern. For the purposes of this example, the load-carrying capabilities of the original material will be assumed to be the applied load.

When using this technique, the technician should take care in determining the cause for the repair. If rivet shear (as a result of flight loads) is the cause, the existing rivet pattern is insufficient and should not be used to determine applied loads. This also applies to the general guidelines established by the FAA in AC 43.13-1B & 2B.

By using Eq. (9-34), if the shear strength of the rivet [see Eq. (8-29)] is greater than the bearing strength of the sheet [see Eq. (8-35) or Eq. (9-36) if the opposite is true], the number of rivets required may be calculated. In this example the rivet shear strength is 389.1 lb and the bearing strength of the sheet at a $\frac{1}{8}$-in [3.18 mm] hole is 398.8 lb. When doing these calculations, the drilled-hole diameter rather than the rivet's predriven diameter must be used. Because the shear strength of the rivet is less than the bearing strength of the sheet, Eq. (9-34) is used.

Using Eq. (9-34) to find T_r, the number of rivets required is 58. If this was a patch-type repair, twice as many rivets per side of the joint, or 116 rivets, would be required. Remember, when doing these calculations, the drilled-hole diameter and not its predriven diameter must be used.

Determining Rivet Spacing and Layout

In this step, the layout is accomplished based upon an equalization of shear strength to the bearing or tensile capabilities of the sheet.

General. Using the minimum-diameter rivets of $\frac{1}{8}$ in [3.18 mm] to establish a rivet pattern, to find the number of rows required, the edge distance, which is a constant, $2D$, must be subtracted twice from the length of the splice. This results in an available rivet-row width of 11.500 in [2.92 cm]. The number of rivets that may be placed in a row is one more than the row length divided by the rivet spacing used. This is because the row length both starts and ends with a rivet. In the case of $3D$ rivet spacing, that means 31 rivets per row maximum. $4D$ spacing would accommodate 24 rivets per row maximum. Since $4D$ spacing is used only in two-row applications and 75 rivets are needed, the splice may not be accomlished in two rows. This is also the case even if footnote (c) (Table 9-3) is applied and 56 rivets were used to accomplish the repair. Therefore, $3D$ spacing must be used.

Using a $3D$ pitch, based upon 75 rivets, there would be 25 rivets per row with a pitch of 0.46 in [11.68 mm], edge

distance of 0.250 in [6.35 mm], and a gauge of 0.375 in [9.53 mm]. Figure 9-9 shows this layout.

Based upon a $3D$ pitch with 56 rivets, there would be three rows of rivets, two each with 19 rivets and a third with 18 rivets. Figure 9-10 shows this layout. Whenever multiple rows are used in a rivet pattern and the rows contain different numbers of rivets, one of the rows with the lesser pitch, or the most number of rivets, should be nearest the edge of the material.

Applied. This step of the rivet-layout procedure is the step that makes the applied process different from the general process. In this section the edge distance and rivet spacing are calculated to comply with the minimum diameter of a rivet two sizes larger than was previously determined. In applying the "repairable repair" concept, the layout is based upon the edge distance and rivet spacing required by a rivet two standard sizes larger, or $\frac{3}{8}$ in [9.53 mm].

The edge distance for a $\frac{3}{8}$-in [9.53-mm] (-6) rivet is $\frac{3}{4}$ in [19.05 mm]. The rivet spacing may be either $3D$ or $4D$, so it needs to be calculated twice. The $3D$ spacing is 0.281 in [7.14 mm]. The $4D$ spacing is 0.375 in [9.53 mm].

To find the number of rows required, the edge distance, which is a constant $2D$, needs to be subtracted twice from the length of the splice. This results in an available rivet row width of 11.625 in [2.95 cm]. The number of rivets that

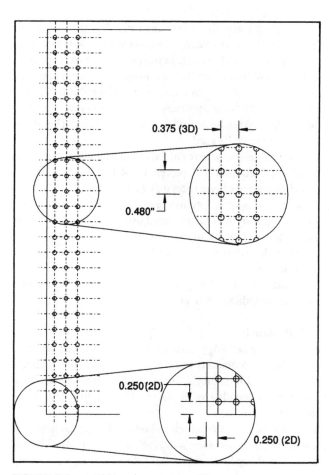

FIGURE 9-9 A 75-rivet layout using $3D$ spacing.

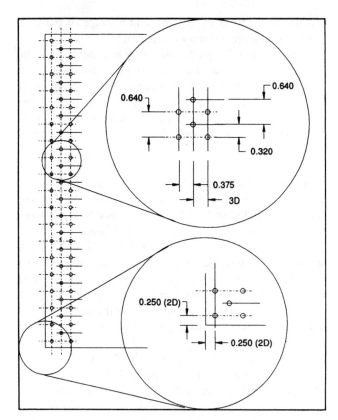

FIGURE 9-10　A 56-rivet layout using 3D spacing.

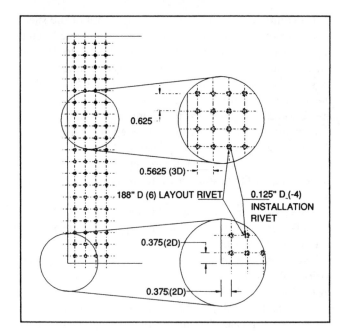

FIGURE 9-11　An acceptable layout for -4 rivets, using -6 spacing.

may be placed in a row is then one more than the row length divided by the rivet spacing used, since the row length both starts and ends with a rivet. In the case of 3D rivet spacing, that means 42 rivets per row maximum. 4D spacing accommodates 32 rivets per row maximum. Since 4D spacing is used only in two-row applications and 75 rivets are needed, the splice may not be accomplished in two rows. Therefore, 3D spacing must be used.

However, if footnote (c) of Table 9-3, is applied, the number of rivets required is 75 percent of 74.4, or 56 (55.8) rivets, so a rivet pattern of two rows may be used.

Since the applied section uses larger rivets for layout, the rivet spacing will be increased using AC 43.13-1B & 2B guidelines, so the spacing will exceed the minimums established by the Advisory Circular when $-4\frac{1}{8}$-in-diameter rivets are installed.

Figure 9-11 shows a layout that is acceptable for -6 rivets and accommodates 75 rivets.

Calculated.　Using Eq. (9-26), the technician can calculate the proper edge distance for this particular application. These calculations result in a suggested edge distance of 0.217 in [5.51 mm]. Since this distance is less than the 2D recommended by AC 43.13-1B & 2B, the 2D distance should be used.

At this point the technician should take special note that the terms *edge distance* and *rivet spacing* as used in the *general* and *applied* processes refer to a distance measured

from the center of the rivet hole. In calculating rivet spacing and edge distances, the definition of these terms must be altered.

In the splice example given previously, if flush rivets are used, it is recommended that an edge distance of $2\frac{1}{2} D$ (rivet diameter) be used. For universal-head rivets, an edge distance of not less than 2D is satisfactory.

In splice applications the width for the rivet layout is equal to W. However, in the case of a patch, the width of the rows should be twice the width of the damaged area. The technician should also keep in mind that when adding holes to a sheet of material, such as an aircraft skin, the tensile load-carrying capability of the material is reduced [see Eq. (9-27)].

Repairs for Small Holes

Small holes in sheet-metal skin may be repaired by means of a patch plate or a flush patch if the damage does not affect ribs or other structural members. The rough edges of the hole may be smoothed with a file, cut away with a **hole saw**, or punched with a **chassis punch**.

The patch for a small hole can be riveted to the outer surface of the skin, or it may be made flush, as shown in Fig. 9-12. In either type of patch the number of rivets should conform to the patterns shown in Fig. 9-12, which illustrates patches for 1-, 2-, and 3-in [2.54-, 5.08-, and 7.62-cm] round holes.

Where a flush patch is used, the patch is placed on the inner side of the skin. A plug is cut to fit the hole and is riveted to the patch. The rivets should be of the flush type, as previously described in this chapter.

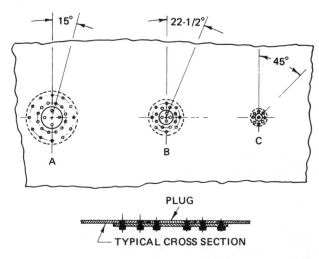

FIGURE 9-12 Flush patch for a small hole.

DIMENSIONS OF CIR. PATCH PLATES

DESIGNATION	A	B	C
DIAMETER (OUTSIDE)	7 1/2" [19 cm]	5" [13 cm]	2 1/2" [6 cm]
DIA. INNER RIVET CIRCLE	4" [10 cm]	3" [8 cm]	1 3/4" 4.5 cm]
DIA. OUTER RIVET CIRCLE	6 1/2" [16.5 cm]	4" [10 cm]	–
DIAMETER (INSIDE)	3" [8 cm]	2" [5 cm]	1" [2.5 cm]
NO. OF RIVETS (A-17ST-5/32)	24	16	8
PLUG RIVETS TO BE USED AS REQUIRED			

Repair design of small holes using a patch is similar to the splice except for three considerations. First, *W* is the diameter of the hole in the original material. Second, because this is a patch, the number of rivets required for the entire repair is at least twice the number required for the splice. Note that the outside diameters of the circular patch doublers are all greater than two times the diameter of the hole. The rivets should be equally spaced around the hole.

Third, no rivet should be placed in a line perpendicular with the vector that represents the application of the load. Rivets on a plane perpendicular to the application of the load will neither pick up nor deposit a load.

Replacement of Skin Panels

In cases where damage to stressed skin has occurred over an extensive area, it is often necessary to replace an entire panel. The original panel is removed by carefully drilling out the rivets at the seams. A new panel of the same material and thickness is cut to the same size as the original. The rivet pattern at the seams must conform to the original pattern. In cases where a portion of a panel is replaced and different rivet patterns are used on the opposite edges of panels, it is best to copy the pattern of the stronger seam. Before a damaged panel is replaced, the interior of the structure must be inspected carefully. All damaged ribs, bulkhead, or other structures must be repaired before replacing the skin panel.

Repairs of Sheet-Metal Ribs

Typical repairs for formed sheet-metal and built-up ribs are shown in Fig. 9-13. In making repairs of the type shown,

the technician must use the correct number of rivets of the proper size and material. The replacement material and material used in making reinforcements must be of the same type as that used in the original structure. Furthermore, the material must have the same heat treatment as the original. The thickness of the repair material must be the same or greater than that of the original.

Repairs for formed sheet-metal rib-cap strips are illustrated in Fig. 9-14. The repairs shown are indicative of the types of repairs required; however, many different types of repairs can be made as long as the strength and durability are adequately restored.

Stringer and Flange Splices

Splices for stringers and flanges are shown in Fig. 9-15. The original material is shown unshaded; the reinforcing material is shaded. Remember that **stringers** are the longitudinal supporting members to which the skin of the fuselage or wing is attached. The stringers are attached to the bulkheads or beltframes (formers), which are principal structural members of the assembly and are designed to take both compression and tension loads. Therefore, these riveting principles must be followed:

1. To avoid eccentric loading and buckling in compression, splicing or reinforcing parts are placed as symmetrically as possible about the centerline of the member. Attachment is made to as many elements as necessary to prevent bending in any direction.

2. So that reduction of strength under tension of the original member is avoided, the rivet holes at the end of the splice are made small—that is, not larger than the original skin-attaching rivets—and the second row of rivets is staggered back from the ends.

3. To prevent concentrating the loads on the end rivet and the consequent tendency toward progressive rivet failure, the splice member is tapered at the ends. This also has the effect of reducing the stress concentration at the ends of the splice.

4. When several adjacent stringers are spliced, the splices should be staggered if possible.

5. The diameter of rivets in stringers should be between two and three times the thickness of the leg but should not be more than one-quarter its width.

Repairing Cracked Structures

Methods for repairing cracked structures are shown in Fig. 9-16. This illustration shows repairs at the intersection of ribs and spars at both the leading edge and the trailing edge of a wing or other airfoil. Reinforcing plates must be of the same alloy and approximately $1\frac{1}{2}$ times the thickness of the original material. In every case where cracks are repaired, the cracks should be stop-drilled before installing any reinforcements. **Stop-drilling** can be defined as the process of drilling a small hole at the extreme end of a crack to prevent the crack from progressing

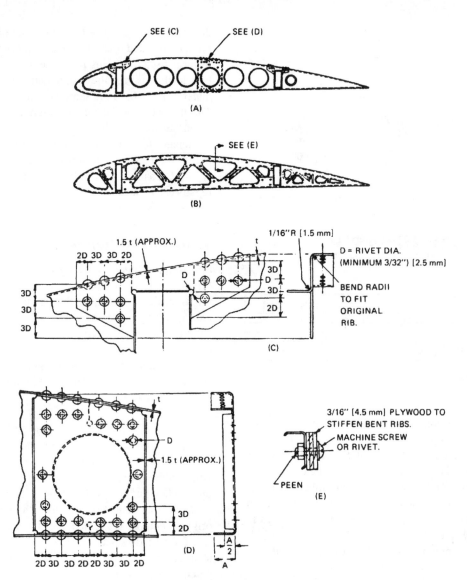

FIGURE 9-13 Repairs for sheet-metal ribs.

farther into the material. The hole at the end of the crack removes the sharp stress-concentration area.

The condition causing cracks to develop at a particular point is stress concentration at the point combined with the repetition of the stress, as would occur with vibration. Stress concentrations are caused by nicks, scratches, or incorrect design factors. Complete failure of wing structures has been caused by stress concentrations where material has been cut to form a notch. In all repairs, the technician must make sure that material is not cut to form a sharp angle between two edges and that where two edges come together to form an angle, the material is rounded ("radiused") to a radius sufficient to prevent stress concentrations. The radius should be made as smooth as possible.

Members of aircraft structures that have developed cracks at fittings can be repaired as shown in Fig. 9-17. The treatment of cracks in these repairs is the same as described previously.

Structural Repair Manual Repairs

All certificated aircraft will have a structural repair manual (SRM) that is prepared by the manufacturer and approved by the FAA. The SRM will contain repair procedures that the technician can perform without contacting the original equipment manufacturer (OEM). If the damage exceeds the damage limitations in the SRM the technician should contact the engineering department of the company or the OEM. Figure 9-18 shows a complex repair that uses multiple doublers and a filler to repair damage to the aircraft skin. Figure 9-19 shows a flush skin repair that consists of multiple repair parts. Figure 9-20 illustrates a SRM repair of a cracked beam, and Figure 9-21 shows a SRM fuselage skin repair for a transport aircraft. Note that the technician does not have to calculate rivet spacing, number of rivets, material thickness, and rows of rivets. This information is all contained in Table 9-4. The technician determines the thickness

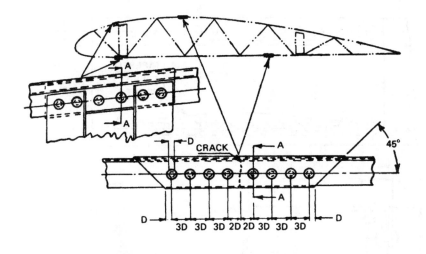

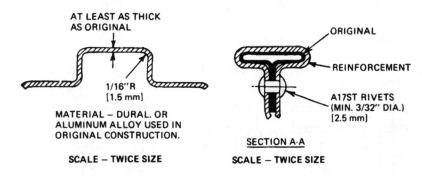

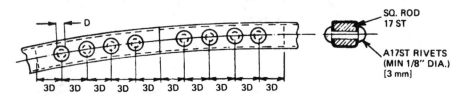

FIGURE 9-14 Repairs for metal rib-cap strips.

of the original fuselage skin and based on this information he/she determines the thickness of the repair doublers, rivet spacing, number of rows, type of rivets, and rivet size. The technician will use the information in Figure 9-21 to repair the damage to the aircraft.

Chem-Milled Skin Repair

A chem-milled structural member varies in thickness from end to end or side to side. Therefore, repair requires a procedure slightly different from standard procedures. The repair material must be as thick as the thickest part of the chem-milled structure. The repair material is applied to the thickest part of the damaged member, using normal riveting procedures. Shimming is used to fill the gap between the repair material and the thin part of the member. The shim material is secured with rivets that pass through the damaged part, the

shim material, and the repair material. Figure 9-22 shows a typical chem-milled skin repair.

Special Repairs

Where specific instructions for sheet-metal structural repairs are not available in the manufacturer's manuals, which should always be the first choice of the technician in determining the type of repair to use, publications such as AC 43.13-1B & 2B published by the FAA and MIL-HDBK-5E may be used for the development of a repair design for unique applications. Whenever specific instructions for a major sheet-metal structural repair are not available in the manufacturer's manuals, the technician must obtain FAA approval for the repair by submitting FAA Form 337. The prudent technician will make it a habit to obtain approval prior to beginning the installation of such repairs.

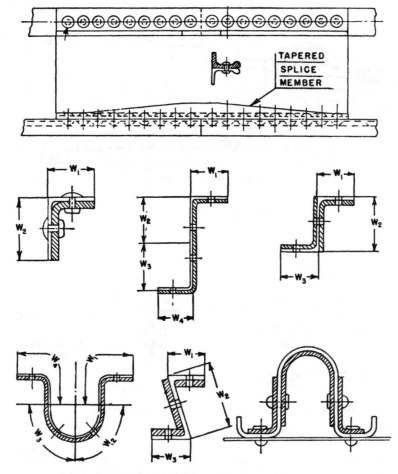

FIGURE 9-15   Splices for stringers and flanges.

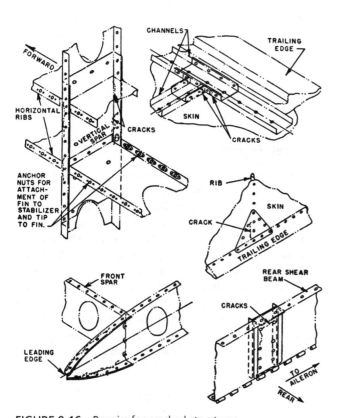

FIGURE 9-16   Repairs for cracked structures.

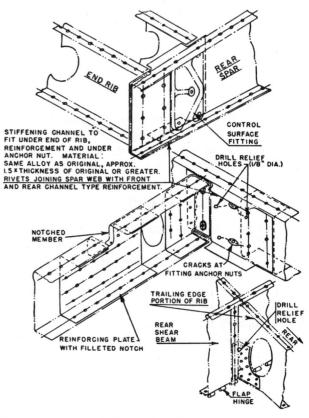

FIGURE 9-17   Repairs for cracks at fittings.

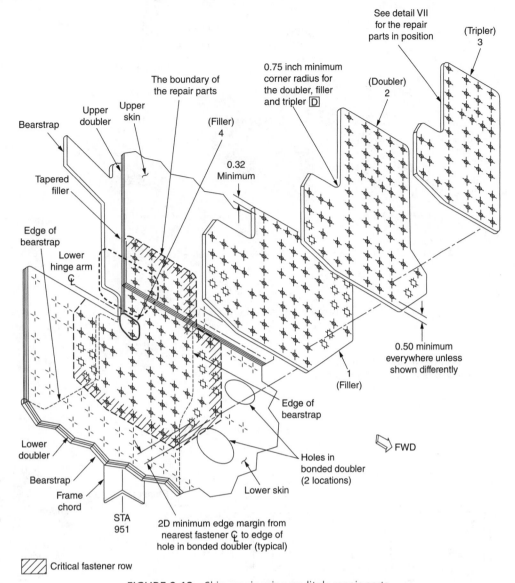

FIGURE 9-18 Skin repair using mulitple repair parts.

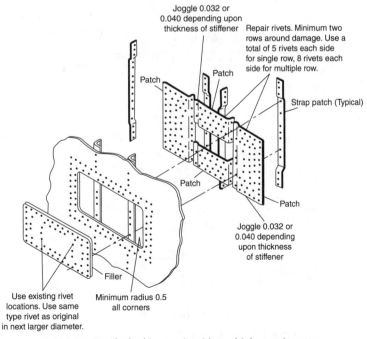

FIGURE 9-19 Flush skin repair with multiple repair parts.

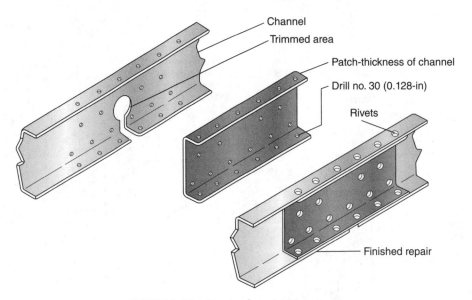

FIGURE 9-20 Repair of cracked beam.

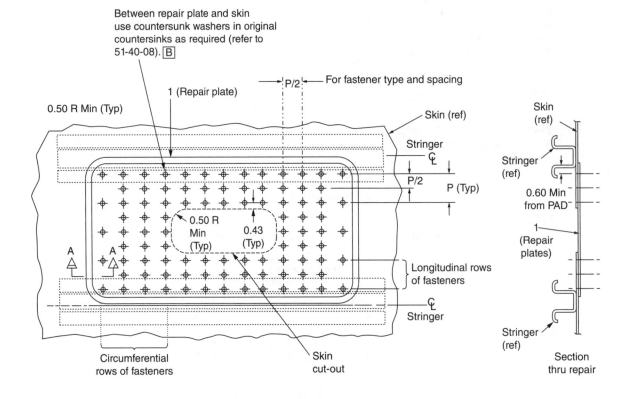

Between repair plate and skin
use countersunk washers in original
countersinks as required (refer to
51-40-08). B

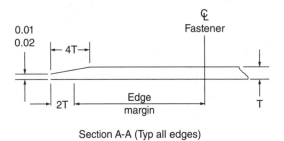

Section A-A (Typ all edges)

FIGURE 9-21 External repair of fuselage.

TABLE 9-4 Structural Repair Design Data

Repair Material		
Part	*Qty*	*Material*
1 Repair Plate	1	Same material and heat treat as original skin; use two gauges greater than original skim – 0.050 Gauge min

Fastener Requirements for Repairs Using Solid Rivets

Original Minimum Skin Pocket Gauge in Repair Area	Fasteners	Number of Rows		Spacing (P)
		Longitudinal	Circumferential	
0.036	BACR15CE5D	3	$3\frac{1}{2}$	1.56 to 1.61
0.040	BACR15CE5D	3	$3\frac{1}{2}$	1.56 to 1.61
0.045	BACR15CE5D	3	$3\frac{1}{2}$	1.56 to 1.61
0.050	BACR15CE6D	3	$3\frac{1}{2}$	1.87 to 1.94
0.056	BACR15CE6D	3	$3\frac{1}{2}$	1.87 to 1.94
0.063	BACR15CE6D	4	$4\frac{1}{2}$	1.87 to 1.94
0.071	BACR15CE8D	4	$4\frac{1}{2}$	2.38 to 2.56
0.080	BACR15CE8D	4	$4\frac{1}{2}$	2.38 to 2.56

Fastener Requirements for Repairs Using Blind Rivets

Original Minimum Skin Pocket Gauge in Repair Area	Fasteners	Number of Rows		Spacing (P)
		Longitudinal	Circumferential	
0.036	MAS1738E4	3	$3\frac{1}{2}$	1.56 to 1.61
0.040	MAS1738E4	3	$3\frac{1}{2}$	1.56 to 1.61
0.045	MAS1738E4	3	$3\frac{1}{2}$	1.56 to 1.61
0.050	MAS1738E5	3	$3\frac{1}{2}$	1.87 to 1.94
0.056	MAS1738E5	3	$3\frac{1}{2}$	1.87 to 1.94
0.063	MAS1738E5	4	$4\frac{1}{2}$	1.87 to 1.94
0.071	MAS1738E6	4	$4\frac{1}{2}$	2.38 to 2.56
0.080	MAS1738E6	4	$4\frac{1}{2}$	2.38 to 2.56

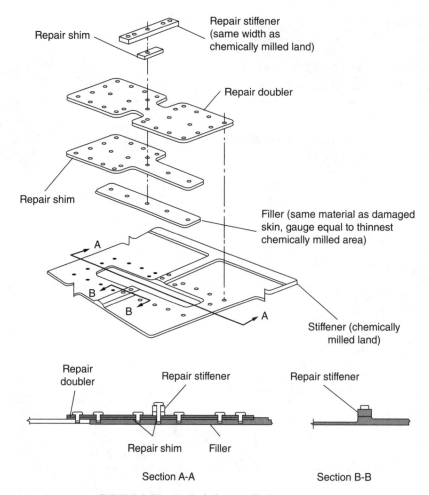

Repair shim

Repair stiffener (same width as chemically milled land)

Repair doubler

Repair shim

Filler (same material as damaged skin, gauge equal to thinnest chemically milled area)

Stiffener (chemically milled land)

Repair doubler

Repair stiffener

Repair stiffener

Repair shim   Filler

Section A-A

Section B-B

FIGURE 9-22 Typical chem-milled skin repair.

REVIEW QUESTIONS

1. List the activities of a technician in developing a repair.

2. What is a *tolerance*?

3. When is a material considered acceptable for service?

4. What are the two basic classifications of *testing*?

5. Why is destructive testing not used in aviation maintenance?

6. What is the philosophy of nondestructive testing?

7. What power of magnification is most frequently used by the technician?

8. Why is it a good idea to give an aircraft entering an inspection an abbreviated inspection before it is cleaned?

9. Describe the *dye-penetrant* inspection process.

10. Describe the basics of *fluorescent-penetrant inspection*.

11. Describe how *magnetic-particle inspection* works.

12. What common name is applied to *radiological inspection*?

13. How are sound waves used in *ultrasonic inspection*?

14. What indicates a possible crack when using *eddy-current inspection* techniques?

15. What is the primary document that the aviation technician is to follow when making a sheet-metal repair?

16. If the document referred to in Question 15 does not supply the necessary information, the technician may refer to what other documents for assistance?

17. Major repairs not found in the aircraft manufacturer's manuals must be submitted on what form to the FAA for approval?

18. When should the document referred to in Question 17 be submitted?

19. What is the primary purpose of a sheet-metal repair?

20. How does a fastener repair work?

21. At what part of a patch are the loads transferred?

22. How many lap joints are there in a *patch repair* and what is their purpose?

23. When might a plug and reinforcement plate repair be used?

24. What are the classifications of aircraft damage?

25. Define *negligible damage*.

26. Define *repairable damage*.

27. Define *replacement damage*.

28. Why is it important for the technician to identify the cause of any damage?

29. Whenever possible the technician should use what types of materials?

30. When identical replacement materials are not available, the technician should first refer to what document(s)?

31. What is the procedure for using the material-substitution table (Table 9-1)?

32. What are the general design assumptions of an approved sheet-metal repair design?

33. Describe how a repairable repair may designed.

34. After a repair has been laid out using repairable repair design techniques, what size rivets are actually installed?

35. What determines the number of rows in a rivet pattern?

36. When calculating the length of the longest possible rivet row, why must the technician subtract the edge distance twice?

37. How is the maximum number of rivets in a row determined?

38. How is the number of rows for a repair determined?

39. If the technician follows the guidelines established by the FAA in AC 43.13-1B & 2B, how many calculations are required to determine the number of rows and why?

40. If the adjacent rows of rivets have different numbers of rivets, how are they best laid out?

41. Why does the rivet spacing, regardless of technique used to determine the rivets required, vary from the calculated or general 3D and 4D guidelines?

42. What is *rivet pitch*?

43. What is *rivet gauge*?

44. How are rivet pitch and gauge measured?

45. What is *edge distance*?

46. What adjustment to the rivet-edge distance is recommended for countersunk rivets?

47. What is the generally accepted maximum rivet pitch?

48. How is a load transferred on a multirow rivet repair?

49. What type of repair utilizes load transfer knowledge to minimize the weight of a patch repair?

50. What is the benefit of using more than one sheet of material to carry a load?

The following questions refer to the tables and charts used for rivet-repair design, as found in AC 43.13-1B & 2B.

51. How is the recommended rivet diameter determined?

52. What is the edge distance recommended by AC 43.13-1B & 2B?

53. What is the rivet pitch recommended by AC 43.13-1B & 2B?

54. What is the rivet gauge recommended by AC 43.13-1B & 2B?

55. To what type rivets is Table 9-3 limited?

56. Describe the step in determining the number of 2117-T3 rivets required to repair a 3-in [7.62-cm] crack in a sheet of 0.032-in [0.813-mm] 2024-T3 clad aluminum using Table 9-3.

The following questions refer to designing rivet repairs using the mechanical properties of the materials involved.

57. When calculating the mechanical properties for joints, what diameter is used and why?

58. From what document may the mechanical properties be obtained?

59. What mechanical properties determine the rivet diameter and how are they related to each other?

60. Why is edge distance and rivet spacing important?

61. What mechanical-property relationships determine edge distance and rivet spacing?

62. Increasing the edge distance increases what other attribute of a riveted joint?

63. How is the tensile strength of the sheet in a lap joint determined?

64. What mechanical-property relationships determine the number of rivets in a lap joint?

65. When attempting to determine the maximum possible load that an aircraft was designed to encounter in a particular area, what must the technician consider?

Plastics 10

INTRODUCTION

Historically, plastics have been used for windshields, windows, landing-light covers, and interior furnishings of aircraft for many years. Because of their low cost and ease of maintenance, plastic materials are very likely to continue to be used in the foreseeable future.

FUNDAMENTALS OF PLASTIC MATERIALS

Characteristics of Plastics

Plastics can be classified by many methods. One generally accepted method of classification is by their response to heat.

Thermosetting plastics are plastic compounds or solutions that require the application of heat to set up properly, or "harden." Once these materials have set, further application of heat does not allow them to be formed in a controllable manner. Any further addition of heat normally results in deformation or structural weakening. Thermosetting materials include many bonding materials that fall under the broad plastic classifications and many materials that can also be called resinous.

Thermoplastic materials are those that soften with the application of heat. When softened, these materials can be shaped as required and they retain their structural strength when cooled. This characteristic allows this type of material to be used to form items such as windshields and landing-light covers. This type of material includes transparent plastics such as Plexiglas®. (Plexiglas is a registered trademark of the Rohm and Haas Company.)

Thermoplastic materials are those that the technician is most likely to be called upon to handle and repair; therefore, the remainder of this section deals with these types of materials.

The basic characteristics of clear plastic materials such as Plexiglas must be understood by the mechanic so that the material is handled properly.

Plastics are used in place of glass for windows because they are lower in weight and there is no reduction in clarity. Plastic is much more resistant to breaking than glass, and when it does break the edges are dull, reducing the chance of injury to personnel. Plastic is a poor conductor of heat and thus provides some level of thermal insulation. Plastic has the disadvantage of readily accepting a static charge and thus attracting dust and dirt particles. Also, being softer than glass, plastic is more easily damaged by surface abrasion than glass. This sensitivity is substantially reduced in some types of hardened plastics.

Types of Clear Plastics

There are two general classifications of clear plastics used in aircraft: **acrylics** and **cellulose acetates**. The technician needs to be able to distinguish between these two materials in order to install and repair them properly. Acrylic materials, such as Plexiglas, are the more modern materials and have replaced acetate materials in most usages. Acrylic plastics used in aviation applications must comply with the appropriate military specification: MIL-P-6886 for standard acrylics; MIL-P-8184 for craze-resistant acrylics; and MIL-P-5425 for heat-resistant acrylics.

Although these materials have been described here as clear, they are available in tints to reduce light penetration. This is most evident when a green- or gray-tinted plastic is used for aircraft windshields. This tinting serves the same purpose as the pilot wearing sunglasses; that is, it reduces the level of light intensity entering the cockpit.

The two materials can be identified by their visual and burning characteristics and by their reaction to certain chemicals and solvents.

When viewed edge-on, acrylic material appears clear or the color of the tinting if the material is tinted. Some clear pieces of acrylic material appear slightly green or blue-green. Acetate material appears yellow when viewed edge-on. When burned, acrylic burns with a clear flame and gives off an aroma described as fairly pleasant or fruitlike. Acetate burns with a heavy black smoke and a strong, pungent odor. If a sample of acrylic material is rubbed with a cloth moistened in acetone, the plastic turns white. Acetone softens acetate but does not affect its clarity. If zinc chloride is placed on acetate, it turns the material milky, but it has no effect on acrylic.

Storage and Protection

Plastics are normally supplied in sheets measuring up to 4 by 8 ft [1.22 by 2.44 m]. To protect the surface of the sheet, it is covered with a masking paper. The masking paper can be peeled back to inspect the plastic as necessary and then replaced without difficulty. If the masking is to be reused, the adhesive side of the paper should not be allowed to touch another area of the paper where the adhesive is present. These two adhesive sides are very difficult to separate. If the masking paper adhesive has dried out and will not peel off of the surface easily, aliphatic naphtha can be used to moisten the paper. This will soften the adhesive so it can be removed. Once the masking is off, the area can be washed with clean water to remove any residue of the naphtha.

Parts that have been formed are often coated with a masking material that is sprayed on to the surface. When dry, this material has a feel and elasticity similar to rubber cement. This material can be removed by peeling up a corner and then using an air gun to lift the masking.

When working with plastic material that is protected by masking, it is advisable to remove only as much material as necessary to prepare and mount the component. When the component is installed and all handling is completed, the remainder of the masking material should be removed. This provides the maximum of surface protection.

Plastic sheets should be stored in racks or on flats. When stored in racks, the sheets should be stored on edge at a 10° angle from the vertical, as shown in Fig. 10-1. The bottom edges should rest on blocks about 3 in [7.62 cm] wide, no more than 42 in [1.07 m] apart. If stored flat, plastics should be supported underneath, thinner and smaller sized pieces placed on top of thicker and larger-sized pieces; see Fig. 10-2.

Formed pieces should be stored in individual racks. If formed pieces are to be stacked, be sure that there is no possibility of damage resulting from the stacking process. Use padding as necessary to prevent damage.

Any plastic that is stored should have a spray-type or paper masking installed. Care should be taken to remove rough edges from plastics that have been cut before placing them in a storage rack or stack. Do not allow any dirt, plastic, metal, or other particles to get between pieces of plastic, as this can damage the plastic surface even with a masking installed.

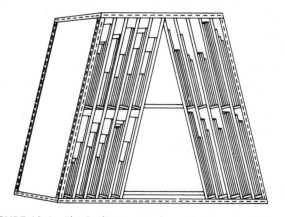

FIGURE 10-1   Plastic sheets stored on edge in racks. (*Rohm and Haas Company.*)

STACKING PLEXIGLAS SHEETS

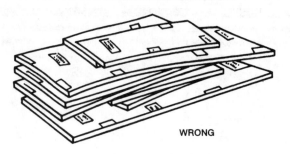

WRONG

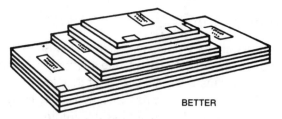

BETTER

FIGURE 10-2   The correct and incorrect way to store plastic sheets on the flat. (*Rohm and Haas Company.*)

WORKING WITH PLASTIC MATERIALS

Cutting Plastics

When cutting plastics, the technician must keep in mind that excessive stress on an area will cause the plastic to crack. Also, when using some types of power tools for cutting, chipping will occur along the edge of the cut. For these instances, the guidelines drawn on the plastic should be far enough from the finish edge so that the chipping will not be on the finished edge. Chipping can be avoided somewhat by feeding the plastic slowly through the cutting blade.

One of the characteristics of plastic is that it is an insulator. Therefore, any heat buildup in the cutting area does not dissipate into the surrounding material but is held in the cutting area. This causes the plastic to melt and flow onto the cutting blade, preventing further cutting of the material. To counteract this quality, some sort of cooling may have to be used when cutting plastics, especially when cutting thick pieces. Cooling can be achieved by the use of an air stream aimed at the cutting area or by a cutting fluid if the operation allows the use of a fluid without electrical or other dangers.

Any of the following power tools can be used to cut plastic if proper care is taken: radial arm saws, table saws, band saws, jig saws, and scroll saws. Obviously, the smoother the cutting operation, the less chance there is of chipping or cracking the plastic. For this reason, circular saws are preferred for straight cuts and band saws are preferred for curved cuts.

It is important that all cutting surfaces run true and that the workpiece be held securely. All saw teeth should have a uniform rake and height. The cutting teeth should either be hollow ground to allow for clearance between the side of the cutting tooth and the workpiece, or the teeth should be set to provide sufficient side clearance. Skip tooth saws, which are relatively soft and have extra gullet capacity, are also recommended. The enlarged gullet provides additional room to compensate for large chip formations.

The feed and speed relationship for the type of plastic and cutting teeth is important to maintain in order to prevent melting the plastic and compacting of the cutting chips.

Short, straight cuts can be made manually in plastics such as Plexiglas by using a straight edge and a scribe. Scribe a deep line across the material to be cut. Place a rod or some type of edge directly under the scribe line. Press evenly and quickly down on both sides of the scribe line and the plastic will break along the scribe line.

When cutting plastics, especially with power tools, wear eye protection. When working around power tools, avoid loose clothing that could be entangled in the tools.

Drilling Plastics

When drilling plastics, a modified high-speed drill bit is recommended. For drilling completely through plastic materials, the bit should be reprofiled to have an included angle of 60°, and the rake angle should be cut to zero. This allows the bit to scrape the plastic rather than cut the plastic. If the bit cuts into the plastic, it can grab and break the surface being drilled. For drilling partway through plastic, shallow holes with depth-to-diameter ratios of 3:1 or less should be drilled with a bit having a tip angle of 90°. If the ratio is 3:1 or greater, the tip angle should be 118°.

Most drilling operations do not require the use of a coolant. If deep holes are being drilled, a coolant may be required. The plastic can be cooled by the use of an air jet, water, or a water-soluble oil mixed with water. If a pilot hole is used to guide the full-size bit, the pilot hole can be filled with a coolant. The pilot hole helps guide the full-size bit along with providing the coolant supply. Masking paper around a hole drilled using a liquid coolant should be removed immediately after drilling the hole to prevent a residue from being left on the surface.

When thin plastic is drilled, it should be backed up with a wood block to prevent chipping or breaking of the plastic as the drill bit exits the plastic. Also, when plastic is drilled, the drill motor should rotate at as high a speed as possible using only light to moderate pressure on the bit to maintain its position and a slow rate of feed. When appropriate, the plastic being drilled should be clamped in a fixture for safety. Always wear proper eye protection when drilling and observe safety rules concerning the use of liquid coolants around electrically operated drills and equipment.

Forming Plastics

Thermoplastic materials can be shaped with or without the use of heat, depending on the severity of the forming required. When cold-forming material, the plastic can be bent if the radius of the bend is at least 180 times the thickness of the material. This allows only a slight shaping to the plastic. For most applications heat is used to soften the material and form it to the desired shape. Plastic may be formed by the use of a hot oil bath, a heater strip, a hot air chamber, or a heat gun. The particular device used is dictated by the type and amount of forming necessary.

An **oil bath** is convenient to use when a small piece of plastic must be formed, such as when making a surface or plug patch for a contoured surface. The oil bath is heated in an appropriate-sized container to the proper forming temperature for the plastic being used. The oil should be in the range of 250 to 350°F [121 to 177°C]. Water cannot be used as it boils away at 212°F [100°C]. The plastic piece is placed in the bath, and after it softens it is placed on a form to achieve the desired shape.

A **heater strip** is used to heat a straight line on the plastic so that it can be bent along that line. These heaters work well for thinner plastics, since the width of the heated area is fairly small.

For forming large pieces, a **heated air chamber** can be used. This chamber must be large enough to accommodate the plastic piece being formed and the forming structure. The chamber is heated, and the plastic is formed on the forming structure. The structure may be in the form of a mold or the components may be free-blown by the use of air pressure. Once the plastic is shaped, it is allowed to cool gradually and then is trimmed to the proper size.

It is important that the technician be aware of the minimum forming temperature for the material being formed. If, during the forming process, any of the material involved in the form drops below this temperature, cold-forming results. If the material is formed outside the cold-forming parameters, stress will result in the part. If a hot-forming process is used, it is very likely that the part requires forming beyond the cold working parameters.

Although annealing, which is discussed later in this chapter, is available to reduce these stresses, typically this is not a viable solution to the problem because to remove the stresses, the material will have to be elevated to a temperature high enough to allow the material to deform.

This hot-air chamber method is used in shops for one-time forming operations. The mechanisms used for production forming of plastics are beyond the scope of this text.

A **heat gun** is used to soften small areas of a plastic piece so that it can be shaped to fit a particular installation. For example, when installing a plastic windshield, the shapes of the corners do not always align properly with the aircraft structure. A heat gun can be used to soften the corners and allow the windshield to be fitted to the airframe.

Acrylic plastic comes in either a **preshrunk** or **unshrunk** condition. Unshrunk plastics are subject to shrinkage during the hot-forming process. Plexiglas G is an unshrunk acrylic plastic. When sufficient heat is applied to the material, it will shrink 2.2 percent in length and width and increase approximately 4 percent in thickness.

Since most acrylic plastic formed parts require an excess length and width far greater than 2.2 percent of the material's length and width, shrinkage is a design consideration only when the part must be formed to its finished dimensions.

The form block used to shape the part during the hot-forming process must include adjustments to this expansion and contraction. This is so that when the plastic material cools to room temperature, its dimensions will be in accordance with the finished part's required dimensions.

When forming any plastic, keep in mind that the material must be heated to approximately 300°F [149°C]. If this material is touched when heated, bare skin will be burned. This is a problem especially when working with an oil bath, as the hot oil will stay on the skin. The appropriate protective clothing and eye protection must be worn when forming plastics.

Many aircraft canopies are formed by using a vacuum-forming method. This method can be used with a mold or without a mold. During the vacuum-forming process without using forms a panel, which has cut into it the outline of the desired shape, is attached to the top of a vacuum box. The heated and softened sheet of plastic is then clamped on top of the panel. When the air in the box is evacuated, the outside air pressure forces the hot plastic through the opening and forms the concave canopy. It is the surface tension of the plastic that shapes the canopy. For more complex shapes a female mold is used. The female mold is placed below the plastic sheet and the vacuum pump is connected. When air from the form is evacuated, the outside air pressure forces the hot plastic sheet into the mold and fills it.

Bonding of Plastics

Bonding of aircraft plastics—in particular, aircraft windshields and windows—is required for repairs and for the attachment of certain components such as vent windows. If a plastic joint is properly shaped and the adhesive is properly applied, a strong joint is formed.

When preparing the surfaces to be joined, they must be cut to the proper shape, formed as necessary, and checked to be sure that all surfaces and edges to be joined are in full contact. The edges do not need to be roughed, since the bonding is caused by actually melting the pieces together chemically.

The adhesive commonly used for plastic windows and windshields is methylene chloride alone or in combination with other chemicals. The specific cement used with a plastic varies, depending on the specific type of plastic. A cement syrup can be made by mixing plastic shavings with the cement. The syrup is sometimes desirable, since it is easier to handle than the liquid cement, especially when working on vertical surfaces and on the bottom of surfaces.

Adhesive is applied on the full mating surface, and the two parts are then held in contact until the adhesive has set. The pressure used should be just enough to ensure full contact, about 1 psi [6.9 kPa]. Excessive pressure could result in distortion, since the cement works by softening the materials and allowing the surfaces to intermingle.

Once the cement has dried, the strength of the joint can be increased by applying heat to the joint area. The bonding of the joint introduces the solvent cement into the material and leaves a softened area at the joint. By heating the joint, the solvents in the plastic move outward into the surrounding

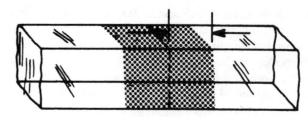

(a) ROOM TEMPERATURE EQUILIBRIUM

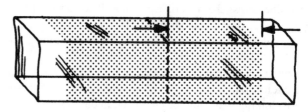

(b) EQUILIBRIUM AFTER HEAT TREATMENT

FIGURE 10-3   Effect of heat treatment on a plastic joint.

material, causing the joint to harden. This action is shown in Fig. 10-3.

Annealing Plastics

To properly complete almost any work accomplished with acrylic plastics, the piece needs to be annealed. Annealing relieves internal stresses, provides greater dimensional stability, and improves the plastic's resistance to crazing. In addition, annealing should be accomplished prior to bonding, particularly if a machining-type operation was performed in preparation for bonding. Annealing should also be the final finishing operation performed before the part is ready to install. All polishing and other final finishing processes should be completed before annealing.

The annealing of acrylic plastics consists of soaking (heating) the material at an elevated temperature and then cooling it slowly. See Table 10-1. The soak time (see Table 10-2) depends on the type of plastic and its thickness. Even though the annealing temperature is less than the forming temperature, annealed formed parts should be supported so that the weight of the material itself and any clamping do not place the material under stress during the annealing process.

As with the heat-forming process, the technician should be aware that unshrunk acrylic plastics will be subject to dimensional changes as a result of the annealing process.

INSTALLATION, MAINTENANCE, AND REPAIR OF PLASTIC MATERIALS

Cleaning Plastics

The best way to clean plastic surfaces is by flowing fresh water across the surface and then using the hand to gently remove any particles adhering to the surface. A mild soap and water solution can be used if required. The use of a cloth when cleaning plastics should be done with care to

TABLE 10-1 The Recommended Cooling Rates for Some Materials (*Rohm and Haas Company*)

Thickness (Inches)	Maximum Cooling Rate (°F/H)	Plexiglas G, Plexiglas II, and Plexiglas 55				Plexiglas I-A			
		230°F	210°F	195°F	175°F	195°F	175°F	160°F	140°F
0.060 to 0.150	120	$\frac{3}{4}$	$\frac{1}{2}$	$\frac{1}{2}$	$\frac{1}{4}$	$\frac{3}{4}$	$\frac{1}{2}$	$\frac{1}{2}$	$\frac{1}{4}$
0.187 to 0.375	49	$1\frac{1}{2}$	$1\frac{1}{4}$	$\frac{3}{4}$	$\frac{1}{2}$	$1\frac{1}{2}$	$1\frac{1}{4}$	$\frac{3}{4}$	$\frac{1}{2}$
0.500 to 0.750	25	$3\frac{1}{4}$	$2\frac{1}{4}$	$1\frac{1}{2}$	$\frac{3}{4}$	3	$2\frac{1}{4}$	$1\frac{1}{2}$	$\frac{3}{4}$
0.875 to 1.125	18	$4\frac{1}{4}$	3	2	1	4	3	$2\frac{1}{4}$	1
1.250 to 1.500	13	$5\frac{3}{4}$	$4\frac{1}{2}$	3	$1\frac{1}{2}$	$5\frac{3}{4}$	$4\frac{1}{2}$	3	$1\frac{1}{2}$
1.750	11	$7\frac{1}{4}$	$5\frac{1}{2}$	4	2	$7\frac{1}{4}$	$5\frac{1}{2}$	4	2
2.000	10	8	6	$4\frac{1}{2}$	$2\frac{1}{2}$	8	6	$4\frac{1}{2}$	$2\frac{1}{2}$
2.250	9	$8\frac{3}{4}$	$6\frac{1}{2}$	5	3	$8\frac{3}{4}$	$6\frac{1}{2}$	5	3
2.500	8	$9\frac{3}{4}$	$7\frac{1}{4}$	$5\frac{1}{2}$	$3\frac{1}{4}$	$9\frac{3}{4}$	$7\frac{1}{4}$	$5\frac{1}{2}$	$3\frac{1}{4}$
3.000	7	$10\frac{1}{2}$	8	6	$3\frac{3}{4}$	$10\frac{1}{2}$	8	6	$3\frac{3}{4}$
3.250	6	$12\frac{1}{4}$	$9\frac{1}{4}$	7	4	$12\frac{1}{4}$	$9\frac{1}{4}$	7	4
3.500	6	$12\frac{1}{4}$	$9\frac{1}{4}$	7	4	$12\frac{1}{4}$	$9\frac{1}{4}$	7	4
3.750	6	$12\frac{1}{4}$	$9\frac{1}{4}$	7	4	$12\frac{1}{4}$	$9\frac{1}{4}$	7	4
4.000	5	15	11	8	5	15	11	8	5

Time in Hours to Cool Plexiglas from the Indicated Annealing Temperature at the Maximum Permissible Rate to the Removal Temperature of 120°F

prevent trapping particles in the cloth that can damage the surface as the cloth is moved over the surface. A clean soft cloth, not a shop rag (even if it is clean), should be used. Shop rags often contain small metal particles even after they have been cleaned. These metal particles will damage the surface.

There are several commercial cleaners available that can be used to clean plastics. These commercial materials often include a wax and an antistatic component that reduces the static buildup on the plastic. Commercial waxes that do not contain cleaning grit can be used to polish clean plastic surfaces. These waxes will fill in some of the fine scratches

TABLE 10-2 The Recommended Soaking Cycle and Temperature for Several Types of Plexiglas (*Rohm and Haas Company*)

Thickness (Inches)	Plexiglas G†, Plexiglas II, and Plexiglas 55					Plexiglas I-A‡				
	230°F§	210°F§	195°F§	175°F	160°F**	195°F§	175°F§	160°F§	140°F	120°F
0.060 to 0.150	2	3	5	10	24	2	3	5	10	24
0.187 to 0.375	$2\frac{1}{2}$	$3\frac{1}{2}$	$5\frac{1}{2}$	$10\frac{1}{2}$	24	$2\frac{1}{2}$	$3\frac{1}{2}$	$5\frac{1}{2}$	$10\frac{1}{2}$	24
0.500 to 0.750	3	4	6	11	24	3	4	6	11	24
0.875 to 1.125	$3\frac{1}{2}$	$4\frac{1}{2}$	$6\frac{1}{2}$	$11\frac{1}{2}$	24	$3\frac{1}{2}$	$4\frac{1}{2}$	$6\frac{1}{2}$	$11\frac{1}{2}$	24
1.250 to 1.500	4	5	7	12	24	4	5	7	12	24
1.750	5	5	7	12	24	5	5	7	12	24
2.000	6	6	8	13	24	6	6	8	13	24
2.250	7	7	9	14	25	7	7	9	14	25
2.500	8	9	11	15	27	8	9	11	15	27
3.000	10	11	12	17	28	10	11	12	17	28
3.250	12	13	14	17	29	12	13	14	17	29
3.500	12	13	14	19	30	12	13	14	19	30
3.750	14	14	16	20	31	14	14	16	20	31
4.000	16	17	18	22	33	16	17	18	22	33

Heating Time in Hours for Plexiglas Placed in a Forced-Circulation Air Oven Maintained at the Indicated Temperature*

* Includes period of time required to bring part up to annealing temperature, but not cooling time; see Table 10-1.
† Unshrunk Plexiglas G should not be heated to annealing temperatures above 200°F, since dimensional changes larger than 1% may occur.
‡ Unshrunk Plexiglas I-A should not be heated to annealing temperatures above 160°F, since dimensional changes larger than 1% may occur.
§ Formed parts may show objectionable deformation when annealed at these temperatures.
** For Plexiglas G and Plexiglas II only. Minimum annealing temperature for Plexiglas 55 is 175°F.

that appear on plastic and improve cue visual clarity to some degree. Machines should not be used when polishing as their high speed of operation may heat the plastic and cause distortion.

Plastic should not be rubbed with a dry cloth, since this is likely to scratch the surface and can cause a buildup of static electricity.

Solvents such as lacquer thinner, benzine, carbon tetra-chloride, acetone, and other similar materials should never be used on plastic sheet because they will penetrate the surface and cause **crazing**. Crazing is the formation of a network of fine cracks in the surface of the material. The effect of crazing is to destroy the clarity of vision through the material and to weaken the structural strength.

Manufacturer's manuals usually specify the types of cleaning agents suitable for the plastic windshields and windows for specific aircraft. Certain neutral petroleum solvents such as kerosene or aliphatic naphtha cleaning fluid are sometimes specified for the removal of oil or grease. If these are used, the surface should be dried as quickly as possible and then washed with water and a mild detergent. A detergent conforming to MIL-C-18687 is recommended. Aromatic naphtha should never be used.

Occasionally, paint may need to be removed from the clear plastic parts, either from paint splatters or overspray. The technique used to remove the paint depends upon the type of paint to be removed. There are three techniques that may be used to remove paint: solvents, sandblasting, and trialene soap.

When using **solvents,** the technician should take great care in selecting the solvent to be used. Some solvents may cause crazing. A commonly used generic solvent for paint removal is cyclohexanone. A number of brand-name paint removers are also available. When the paint removal is complete, care must be taken to remove all the paint remover from the plastic. Residual amounts of paint remover could cause crazing when other surface treatments are used.

The use of **sandblasting** to remove paint is typically not applicable to aviation uses of clear plastics because it weakens the plastic and clouds the vision through the plastic.

Trialene soap may be used to remove acrylic-based paints from Plexiglas as well as paint that has been in place for an extended period.

When dealing with chemicals to remove paint from plastics, the technician should follow the instructions of the plastic's manufacturer and the paint-remover manufacturer as well as all the safety precautions listed in the product-usage instructions and material safety data sheets.

Installation of Plastic Windows and Windshields

The first step when the replacement of a windshield or window is necessary is the removal of the mechanism holding in the old component. This structure may be a panel or retainer strip that is riveted or held in by bolts or screws. The old component should be removed as carefully as possible so that it is not damaged and can be used for a reference

as to the location of holes and contour when installing the replacement piece.

The replacement piece should have the masking peeled back from the edges about 1 in [2.5 cm]; then the piece is placed in the installation to determine if trimming is necessary. If the component fits into the structure, then the mounting can be completed. If the component does not fit properly into the structure, the edges that need to be trimmed can be marked. The edge should then be trimmed with a band saw, hand file, or belt sander to within about $\frac{1}{4}$ in [6.35 mm] of the trim line and then the fit should be rechecked. After the fit check, continue the trimming until the new piece fits properly into the structure channels. It may be necessary to fit and trim the new windshield or window several times in order to achieve a proper fit. It is better to do this in small steps rather than rush the job and remove too much material. After the material is fitted, a heat gun may have to be used to heat local areas and adjust the contour slightly to reduce any installation stresses caused by the flexing required to fit properly.

The fitted piece should fit a minimum of $1\frac{1}{8}$ in [2.86 cm] into the channel, but it must be at least $\frac{1}{8}$ in [3.18 mm] from the bottom of the channel to allow for expansion and contraction of the plastic material. The exact dimensions vary from one aircraft to another. When fitting the piece, do not force the panel into position, as this may cause the plastic to crack or craze. Once the new windshield or window is fitted, it should be removed and a padding of rubber or felt should be attached to the edges that will be in the channel.

Holes for the installation of bolts, screws, and fittings should be drilled through the plastic. These holes should be $\frac{1}{8}$ in [3.18 mm] larger in diameter than the bolt or screw to be installed to allow for expansion and contraction of the plastic panel. The panel is then positioned in the structure, and the bolts and screws are installed. The bolts and nuts are tightened until just snug and then backed off one full turn to allow for expansion and contraction.

Once the panel is in position, the required retainer strips are installed using rivets, screws, and/or bolts. The remainder of the masking is removed and the panel is cleaned.

Each aircraft service manual should address the removal and installation of plastic windshields and windows. The specific instructions for a particular aircraft should be followed.

Some composite aircraft do not use fasteners to install the windshields. The windshields are adhesive bonded to the airframe structure. Figure 10-4 shows the removal instructions for an acrylic windshield that is adhesively bonded to the structure.

Inspection of Plastic Components

The inspection of windshields and windows is visual. The surfaces should be checked for cracks and surface damage. Surface abrasion should be checked, and, if it is not deep, this abrasion can be removed by an abrasion-removal process such as the Micro-Mesh process. Any crazing (small fissures in the plastic) should be evaluated as to structural weakening of the plastic and interference with pilot vision. If excessive crazing is found, the

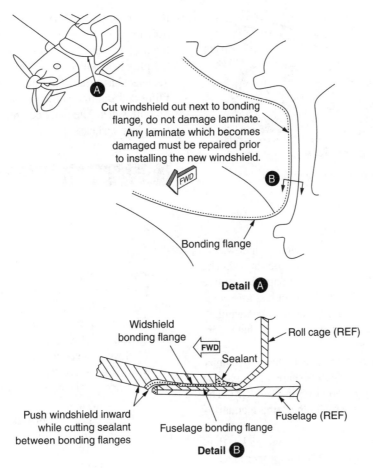

Cut windshield out next to bonding
flange, do not damage laminate.
Any laminate which becomes
damaged must be repaired prior
to installing the new windshield.

Bonding flange

Detail A

Widshield
bonding flange

Roll cage (REF)

Sealant

Fuselage (REF)

Push windshield inward
while cutting sealant
between bonding flanges

Fuselage bonding flange

Detail B

FIGURE 10-4 Adhesively bonded acrylic windshield.

windshield or window should be replaced. Any damage that is in the pilot's normal area of vision and that cannot be restored to a clear vision area is cause for replacement of the plastic component.

Repair of Cracks

Cracks commonly appear on the edges of a piece of plastic radiating out from a pressure point or a mounting hole. If the crack is in the pilot's normal vision area, the panel should be replaced. If the crack is not in the normal vision area, the crack should be stop-drilled, as shown in Fig. 10-5, and a patch installed, as in Fig. 10-6.

The surface patch of the same thickness as the original material should be cut to extend beyond the edges of the

FIGURE 10-5 Stop-drill cracks in plastic.

crack at least $\frac{3}{4}$ in [1.91 cm] and formed to lay fully on the surface contour. The edges of the patch should be tapered as shown and the patch spread with adhesive and positioned on the surface of the panel being repaired. Pressure is then applied to the patch for several hours. After 24 h, the patch can be polished to achieve clear edges.

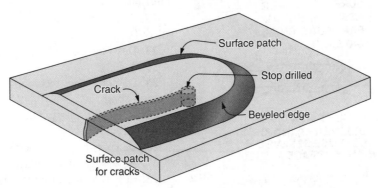

Surface patch

Stop drilled

Crack

Beveled edge

Surface patch
for cracks

FIGURE 10-6 A surface patch for a crack in plastic.

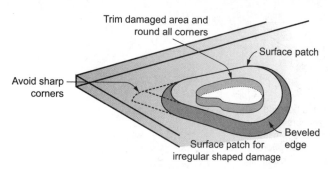

FIGURE 10-7   A surface patch for a hole in plastic.

Repair of Holes

When a hole is found in a plastic windshield or window, any cracks radiating from the holes should be stop-drilled. The hole is then dressed with a file and abrasive paper to remove rough edges. If a surface patch is to be installed, this completes the hole preparation. If a plug patch is to be installed, the hole should be trimmed out to a circle or oval that tapers outward, as shown in Fig. 10-7.

For a surface patch, a patch is prepared in the same manner as for a patch over a crack: The patch is shaped and extends beyond the edge of the damage at least $\frac{3}{4}$ in [1.91 cm]. The patch is then attached with adhesive in the same manner as used for the patch over a crack.

For a plug patch, a plug is cut from material thicker than the original material. The plug is cut to the proper size, as shown in Fig. 10-8, and the taper is more severe than that of

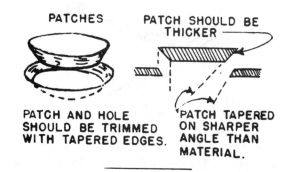

FIGURE 10-8   The plug-patch repair process.

the prepared hole. The plug is heated and pressed into the hole to allow the plug to match the edges of the hole. After the plug has cooled, it is removed, adhesive is applied, and the plug is installed in the hole. Pressure is applied until the adhesive has dried. The plug can then be trimmed down by filing and sanding so that it is flush with the original panel on both sides. The patched area is then buffed and polished to a clear surface.

REVIEW QUESTIONS

1. What is a *thermosetting plastic*?
2. What is a *thermoplastic material*?
3. List the advantages of plastics over glass.
4. What are some of the disadvantages of plastic as compared to glass?
5. What are the two general classifications of clear plastics used in aircraft, and which is more common?
6. How can the technician tell the difference between acrylic and cellulose plastics without damaging them?
7. How are plastic sheets protected while in storage?
8. If the masking paper adhesive has dried out and will not peel off of the surface easily, what can be done?
9. Describe the thermal conductivity of plastics in general and its effect upon working with plastics.
10. How is the low thermal conductivity of plastics counteracted?
11. Describe the process for drilling completely through plastics.
12. What adjustments should be made to standard drilling practices when drilling partially through plastics?
13. How can heat be applied to plastics for forming?
14. Under what conditions does the technician need to be concerned with the preshrinkage of plastics?
15. What is a cement syrup and what are its components?
16. What is the effect of annealing a plastic?
17. When should annealing be accomplished?
18. Describe the technique for cleaning a plastic surface.
19. Why should machines not be used to polish plastic surfaces?
20. What is crazing?
21. What are the three techniques that can be used to remove paint?
22. Describe the technique for removing paint with solvents.
23. Describe the technique for removing paint by sandblasting.
24. Describe the technique for removing paint with trialene soap.
25. When fitting a window or windshield, how are the edges finished?
26. Describe the preparation of mounting holes for plastics.
27. How is a hole in a plastic windshield repaired?
28. How is a plug fitted to a hole?

Advanced Composite Materials 11

INTRODUCTION

The use of composite materials in aircraft has increased greatly in the last few years. The use of composite materials has increased from a few percent in early military aircraft to over 50 percent per weight for newer civilian aircraft models such as the Boeing 787 and Airbus 350 which use carbon fiber as the primary material for fuselage and wing structures. Figure 11-1 shows the material composition of a modern transport aircraft.

LAMINATED STRUCTURES

Composite materials consist of a combination of materials that are mixed together to achieve specific structural properties. The individual materials, such as **resin and fibers**, do not dissolve or merge completely in the composite material. The properties of the composite material are superior to the properties of the individual materials it consists of. Normally, the components can be physically identified and exhibit an interface with one another. Advanced composites used in aircraft construction contain strong, stiff, engineered fibers embedded in a high-performance **matrix**. Composite materials are often selected to manufacture aircraft parts and structures

due to favorable mechanical properties. Common carbon fiber components on today's production aircraft are: fairings, flight control surfaces, landing gear doors, leading and trailing edge panels on the wing and stabilizer, interior components, and vertical and horizontal stabilizer primary structure. On transport aircraft the primary wing and fuselage structure are made of carbon fiber materials. Figure 11-2 shows a solid laminate using reinforced fibers and a matrix material.

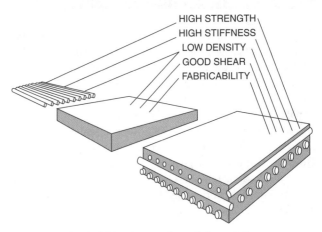

FIGURE 11-2 Solid laminate using reinforced fibers.

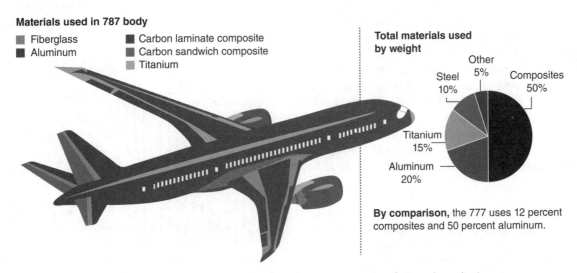

Materials used in 787 body

- Fiberglass
- Aluminum
- Carbon laminate composite
- Carbon sandwich composite
- Titanium

Total materials used by weight

Other 5%
Steel 10%
Composites 50%
Titanium 15%
Aluminum 20%

By comparison, the 777 uses 12 percent composites and 50 percent aluminum.

FIGURE 11-1 Material composition of modern transport aircraft. See also color insert.

MAJOR COMPONENTS OF A LAMINATE

Fiber: The fibers in a composite are the primary load-carrying element of the composite material. The composite material will be strong and stiff in the direction of the fibers. As a result unidirectional composites have predominant mechanical properties in one direction and are said to be anisotropic. Components made from fiber-reinforced composites can be designed so that the fiber orientation produces optimum mechanical properties, but they can only approach the true isotropic nature of metals.

Matrix: The matrix supports the fibers and bonds them together in the composite material. The matrix transfers any applied loads to the fibers, keeping the fibers in their position and chosen **orientation**. The matrix also gives the composite environmental resistance and determines the maximum service temperature of a composite structure.

STRENGTH CHARACTERISTICS

Composite materials have different strength characteristics from metals because the strength characteristics depend to a large extent on the fiber orientation and the designer could design a material that has different strength characteristics relative to natural reference axes inherent in the material. Metals such as aluminum or steel have the same strength characteristics in all directions, which is called **isotropic**. Many composite structures are designed to be quasi-isotropic. **Quasi-isotropic** material has approximately the same properties as an isotropic material. An example is a composite laminate with the fibers orientated in the 0°, 90°, + 45°, and –45° direction to simulate isotropic properties.

Fiber Orientation

The strength and stiffness of a composite laminate depend on the orientation sequence of the plies. The practical range of strength and stiffness of carbon fiber extends from values as low as those provided by fiberglass to as high as those provided by titanium. This range of values is determined by the orientation of the plies to the applied load. Proper selection of ply orientation in advanced composite materials is necessary to provide a structurally efficient design. The part might require 0° plies to react the axial loads, +/– 45° plies to react shear loads, and 90° plies to react side loads. Because the strength design requirements are a function of the applied load direction, ply orientation and ply sequence have to be correct.

- **Unidirectional**. The fibers run in one direction and the strength and stiffness are in the fiber direction only. Prepreg tape is an example of a unidirectional ply orientation.
- **Bidirectional**. The fibers run in two directions, typically 90° apart. A plain weave fabric is an example of a bidirectional

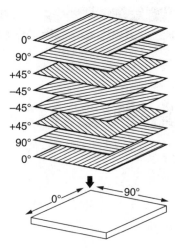

FIGURE 11-3   Quasi-isotropic lay-up.

ply orientation. These ply orientations have strength in both directions but not necessarily the same strength.

- *Quasi-isotropic*. The plies of a quasi-isotropic lay-up are stacked in a 0°, –45°, 45°, and 90° sequence or in a 0°, –60°, and +60° sequence; see Fig. 11-3. These types of ply orientations simulate the properties of an isotropic material such as aluminum or titanium. Figure 11-3 shows a quasi-isotropic lay-up.

Warp Clock

Positive angles can be measured in either clockwise or counterclockwise direction. The ASTM and MIL-HDBK-17 standard is to measure positive angles in the clockwise direction when looking at the layup surface. Figure 11-4 shows a warp clock which indicates the ply direction. **Warp** is the longitudinal fibers of a fabric. The warp is the high-strength direction, due to the straightness of the fibers. A warp clock is used to describe direction of fibers on a diagram, spec sheet, or manufacturer's sheets. If the warp clock is not available on the fabric, the orientation will be defaulted to zero as the fabric comes off the roll. Therefore, 90° to zero will be across the width of the fabric. 90° to zero is also called the **fill direction**.

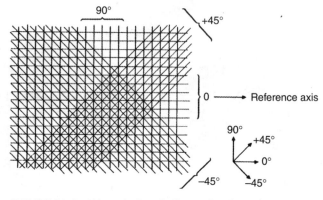

FIGURE 11-4   Warp clock to indicate ply orientation.

Laminate	Code
90	$[90/0_2/45]_S$
0	
0	
45	
45	
0	
0	
90	

FIGURE 11-5 Symmetrical lay-up.

Symmetry and Balance

Composite laminate structures must be **symmetrical** and **balanced** to reduce stresses and to avoid distortion of the laminate. A lay-up needs to be symmetrical or the lay-up will either curve or twist during the curing process. Symmetry of lay-up means that for each layer above the midplane of the laminate there must exist an identical layer (same thickness, material properties, and angular orientation) below the midplane. To achieve balance, for every layer centered at some positive angle there must exist an identical layer oriented at a negative angle with the same thickness and material properties. If the laminate contains only 0° and/or 90° layers it satisfies the requirements for balance. Figure 11-5 shows a symmetrical laminate.

Types of Fiber

The most common used dry fibers for the aerospace industry are: **fiberglass**, **carbon fiber**, and **aramid (Kevlar)**. All materials should be stored on a roll off the ground in a dry environment at room temperature. Figure 11-6 shows a storage rack with dry reinforcements. The dry fabric will be impregnated with a resin just before the repair work starts. This process is often called wet lay-up. The main advantage of using the wet lay-up process is that the fiber and resin can be stored for a long time at room temperature. The composite can be cured at room temperature or an elevated temperature

FIGURE 11-6 Dry fiber materials. See also color insert.

FIGURE 11-7 Dry fiberglass fabric material. See also color insert.

cure can be used to speed up the curing process and increase the strength. Disadvantages: process is messy and properties are less than pre-preg material properties.

Fiberglass

Fiberglass is often used for secondary structures on aircraft such as fairings, radomes, wing tips, helicopter rotor blades, and interior components. The properties of fiberglass are less than carbon fiber but the cost of fiberglass is substantially lower than carbon fiber and therefore fiberglass is an economical substitution if high strength and stiffness are not required. Fiber glass is available as continuous rovings, yarn for fabrics or braiding, mats, and chopped strand. The most common form of fiberglass is **"E" glass**. Electrical glass or E-glass is identified as such for electrical applications, but is used for many other applications as well. **S-glass and S2-glass** identify structural fiberglass that have a higher strength than E-glass. Advantages of fiberglass are: lower cost than other composite materials, chemical or galvanic corrosion resistance, and electrical properties (fiberglass does not conduct electricity). Fiberglass is distinguished by its white color as seen in Fig. 11-7, but it is also available in different colors.

Kevlar®

Kevlar® is DuPont's trademark name for aramid fibers. Aramid fiber is used as a fabric material for aerospace applications. Aramid is an organic fiber that is 43 percent lighter than and twice as strong in tension as fiberglass. Aramid fibers are generally weak in compression and are difficult to cut due to fuzzing. The main disadvantage is that the aramid fibers are hygroscopic and can absorb 8 percent of their weight of water. Aramid is recognized by its yellow color as shown in Fig. 11-8. Kevlar® 49 is used in reinforced plastics for secondary structures on some aircraft.

Carbon/Graphite

Carbon fibers, commonly called graphite, are made from organic materials. Graphite fibers have a high potential for

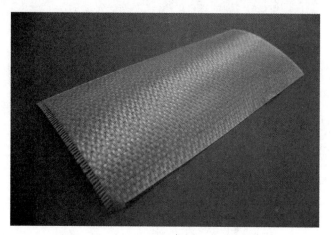

FIGURE 11-8 Aramid (Kevlar) fabric material. See also color insert.

causing galvanic corrosion when used with metallic fasteners and structures. Composites made from carbon fiber are five times stronger than grade 1020 steel for structural parts, yet are still five times lighter. In comparison to 6061 aluminum, carbon fiber composites are seven times stronger and two times stiffer, yet 1.5 times lighter. Carbon fiber composites have fatigue properties superior to all known metals, and, when coupled with the proper resins, carbon fiber composites are one of the most corrosion-resistant materials available. Carbon fiber is used for structural aircraft applications such as floor beams, stabilizers, flight controls, and primary fuselage and wing structure. Advantages are high strength, low fatigue, and corrosion resistance. Disadvantages are lower conductivity than aluminum. Carbon fibers are available in continuous filament-spooled fiber, milled fiber, chopped fiber, woven fabrics, felts, veils, and chopped fiber mattes. The size of the carbon fiber tow bundle can range from 1000 filaments (1K) to more than 200K. Generally, aerospace carbon fibers are available in bundles of 3K, 6K, 12K, and 24K filaments. Carbon fiber is recognized by its black color as shown in Fig. 11-9.

Boron Fibers

Boron fibers are very stiff and have a high tensile and compressive strength. The fibers have a relative large diameter and do not flex well and are therefore only available as a pre-preg tape product. An epoxy matrix is often used with the boron fiber. Boron fibers are often used to repair cracked aluminum aircraft skins because the thermal expansion of boron is close to aluminum and there is no galvanic corrosion potential. The boron fiber is difficult to use if the parent material surface has a contoured shape. The boron fibers are very expensive and can be hazardous for personnel. Boron fibers are used primarily in military aviation applications.

Ceramic Fibers

Ceramic fibers are used for high-temperature applications such as turbine blades in the turbine section of a gas turbine engine. The ceramic fibers can be used to temperatures up to 2200°F.

Lightning Protection Fibers

High-energy lightning strikes can cause substantial damage to composite aircraft structures. An aircraft made from aluminum alloy is quite conductive and is able to dissipate the high currents resulting from a lightning strike. However, aircraft made from composite materials need additional protection because, for instance, carbon fibers are 1000 times more resistive than aluminum to current flow, and epoxy resin is 1 million times more resistive. Many different types of conductive materials are used to protect composite aircraft ranging from nickel-coated graphite cloth to metal meshes, as shown in Fig. 11-10, to aluminized fiberglass to conductive paints. Aluminum mesh is often used with fiberglass and Kevlar materials but copper mesh is required for carbon fiber material because aluminum will corrode quickly if used on top of carbon fiber. If a composite aircraft gets damaged, in addition to a normal structural repair, the technician must also recreate the electrical conductivity designed into such a part.

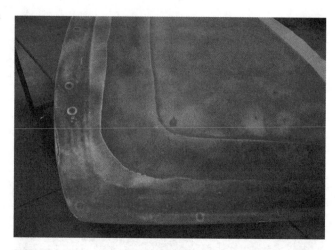

FIGURE 11-9 Carbon fiber material. See also color insert.

FIGURE 11-10 Aluminum mesh for lightning strike protection. See also color insert.

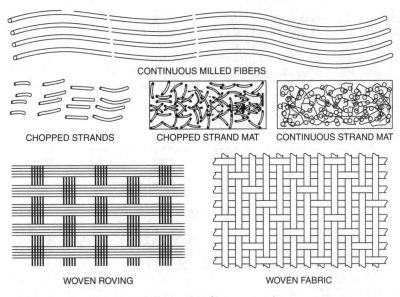

CONTINUOUS MILLED FIBERS

CHOPPED STRANDS   CHOPPED STRAND MAT   CONTINUOUS STRAND MAT

WOVEN ROVING   WOVEN FABRIC

FIGURE 11-11   Dry fiber materials.

Fiber Forms

All product forms generally begin with spooled unidirectional raw fibers packaged as continuous strands. An individual fiber is called a **filament**. The word **strand** is also used to identify an individual glass fiber. Bundles of filaments are identified as tows, yarns or rovings. Roving and tows are used to identify groups of long continuous fibers. A roving or tow is a single grouping of filament or fiber ends, such as 20-end or 60-end glass rovings. All filaments are in the same direction and they are not twisted. Carbon tows are usually identified as 3K, 6K, or 12K tows. K means 1000 filaments. When individual fiberglass filaments are twisted they are called yarns. Carbon fiber and Kevlar fibers are not twisted. Most fibers are available as dry fiber that needs to be impregnated with a resin before use or pre-preg materials where the resin is already applied to the fiber. Figure 11-11 shows the different forms of fiber material available. The most common forms of fiber products for the aerospace industry are unidirectional products (also called tape) and bidirectional woven products (also called fabric).

Unidirectional (Tape)

High-strength structural applications are often manufactured from unidirectional material. In unidirectional material, all the fibers are running in the same direction and the material has only strength properties in the fiber direction. Several layers of unidirectional material are required to create a laminate that has uniform strength in all directions. Unidirectional materials are typically available as pre-preg materials. Figure 11-12 shows a unidirectional carbon fiber material. Figure 11-13 shows unidirectional and bidirectional fibers.

FIGURE 11-12   Unidirectional material (carbon fiber). See also color insert.

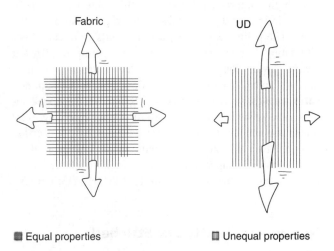

Fabric   UD

▦ Equal properties   ▥ Unequal properties

FIGURE 11-13   Bidirectional (fabric) and unidirectional (tape) properties.

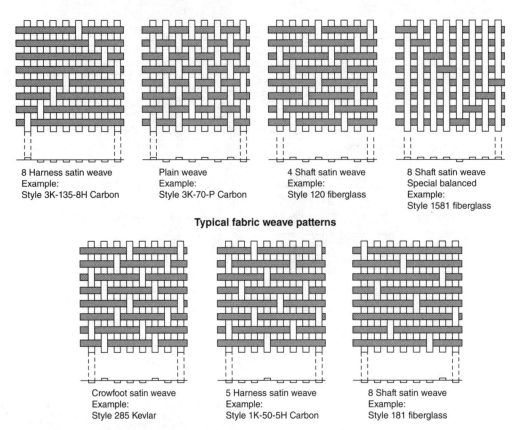

Typical fabric weave patterns

8 Harness satin weave
Example:
Style 3K-135-8H Carbon

Plain weave
Example:
Style 3K-70-P Carbon

4 Shaft satin weave
Example:
Style 120 fiberglass

8 Shaft satin weave
Special balanced
Example:
Style 1581 fiberglass

Crowfoot satin weave
Example:
Style 285 Kevlar

5 Harness satin weave
Example:
Style 1K-50-5H Carbon

8 Shaft satin weave
Example:
Style 181 fiberglass

FIGURE 11-14 Basic weave styles for composite bidirectional products.

Bidirectional (Fabric)

Bidirectional materials, often called **fabric**, are woven materials that use tows of composite material. A weaving process similar to what is used to manufacture textile products is used to make the fabric material. Fabric constructions offer more flexibility for lay-up of complex shapes than straight unidirectional tapes offer. For aerospace structures, tightly woven fabrics are usually the choice to save weight, minimizing resin void size, and maintaining fiber orientation during the fabrication process. The more common fabrics are plain, basket, or satin weaves. The **plain weave**, which is most highly interlaced, is therefore the tightest of the basic fabric designs and most resistant to in-plane shear movement. The plain weave construction results from each fiber alternating over and then under each intersecting tow. **Basket weave**, a variation of plain weave, has warp and fill yarns that are paired: two up and two down. With the common **satin weaves**, such as 5 harness or 8 harness, the fiber bundles traverse both in warp and fill directions changing over/ under position less frequently. Satin weaves have less crimp and are easier to distort than a plain weave and are used for more complex shapes. Fabrics are widely available as dry fiber or pre-preg materials. Figure 11-14 shows fabric weave styles.

Nonwoven (Knitted or Stitched)

Multiaxial fabrics are used more frequently for the manufacture of composite components. These fabrics consist of one or more layers of continuous fibers held in place by a secondary, nonstructural stitching thread. The stitching process allows a variety of fiber orientations, beyond the simple 0/90° of woven fabrics, to be combined into one fabric. Figure 11-15 shows a stitched nonwoven fabric.

Matrix Materials

The resin is often referred to as the matrix of a composite. The role of the matrix is to support the fibers and bond them together in the composite material. It transfers any applied loads to the fibers and keeps the fibers in their position and chosen orientation. The matrix also gives the composite environmental resistance and determines the maximum service temperature of the finished component. The two main classes of resins are: **thermoset** and **thermoplastic**.

Thermosetting Resins

Thermoset materials take a permanent set or shape when cured. Cure temperatures range from room temperature to 350°F for polyesters, vinyl esters, epoxies, and cyanate esters, around 450°F for bismaleimide (BMI) materials, and close to 700°F for polyimides. Cure pressures are usually less than 100 pounds per square inch (psi). The most common type of thermoset resin used for aircraft construction is **epoxy** because of its excellent mechanical properties, extended service temperature range, and ease of part manufacture. The advantages of epoxies are high strength and modulus, low levels of volatiles, excellent adhesion, low shrinkage,

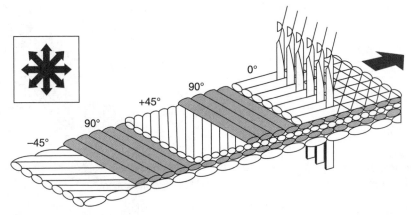

FIGURE 11-15 Stitched multiaxial fabric.

good chemical resistance, and ease of processing. Their major disadvantages are brittleness and the reduction of properties in the presence of moisture. **Bismaleimide (BMI)** resins are used when a higher service temperature is required. BMIs have epoxy-like processing characteristics yet have a higher temperature use limit. BMIs require higher cure temperatures than used for epoxies, typically 375 to 450°F. BMIs are used for aero engines and high-temperature components. **Phenol-formaldehyde** resins were first produced commercially in the early 1900s for use in the commercial market. Urea-formaldehyde and melamine-formaldehyde appeared in the 1920s to 1930s as a less expensive alternative for lower temperature use. Phenolic resins are used for interior components because of their low smoke and flammability characteristics. **Polyimide** resins excel in high temperature environments where their thermal resistance, oxidative stability, low coefficient of thermal expansion, and solvent resistance benefit the design. Their primary uses are circuit boards and hot engine and aerospace structures. A polyimide may be either a thermoset resin or a thermoplastic. Polyimides require high cure temperatures, usually in excess of 550°F [~290°C]. Consequently, normal epoxy composite bagging materials are not usable, and steel tooling becomes a necessity. Polyimide bagging and release films, such as Kapton and Upilex, replace the lower-cost nylon bagging and polytetrafluoroethylene (PTFE) release films common to epoxy composite processing. Fiberglass fabrics must be used for bleeder and breather materials instead of polyester mat materials.

Curing Stages of Thermoset Resins. Thermosetting resins use a chemical reaction to cure. There are three types of curing stages which are called A, B, and C stage.

A stage. The components of the resin (base material and hardener) have been mixed but the chemical reaction has not started. The resin is in the A stage during a wet lay-up procedure.

B stage. The components of the resin have been mixed and the chemical reaction has started. The material has thickened and is tacky. Pre-preg materials are in the B stage. To prevent further curing the resin is placed in a freezer at 0°F.

In the frozen state the resin of the pre-preg material will stay in the B stage. The curing will start when the material is removed from the freezer and heated up again.

C stage. The resin is fully cured. Some resin systems cure at room temperature and others need an elevated temperature cure cycle to fully cure.

Thermoplastic Resins

Thermoplastic materials are non-curing systems that can be reshaped or reformed after the part has been processed or consolidated. Reshaping is accomplished by reheating the part until the material softens and applying pressure to produce a newly molded shape. Thermoplastics are consolidated at temperatures up to 800°F and pressures up to 200 psi. Polyetheretherketone (PEEK) is an often used resin system for the aerospace industry to make structural parts. Semicrystalline thermoplastics possess properties of inherent flame resistance, superior toughness, good mechanical properties at elevated temperatures and after impact, and low moisture absorption. They are used in secondary and primary aircraft structures. Combined with reinforcing fibers, they are available in injection molding compounds, compression-moldable random sheets, unidirectional tapes, towpregs, and woven pre-pregs. Fibers impregnated include carbon, nickel-coated carbon, aramid, glass, quartz, and others. Amorphous thermoplastics are available in several physical forms, including films, filaments, and powders. Combined with reinforcing fibers, they are also available in injection molding compounds, compressive moldable random sheets, unidirectional tapes, woven pre-pregs, etc. The fibers used are primarily carbon, aramid, and glass. The specific advantages of amorphous thermoplastics depend upon the polymer. Typically, the resins are noted for their processing ease and speed, high temperature capability, good mechanical properties, excellent toughness and impact strength, and chemical stability. The stability results in unlimited shelf life, eliminating the cold storage requirements of thermoset pre-pregs.

Preimpregnated Products (Pre-pregs). Pre-preg material consists of a combination of a matrix and fiber reinforcement.

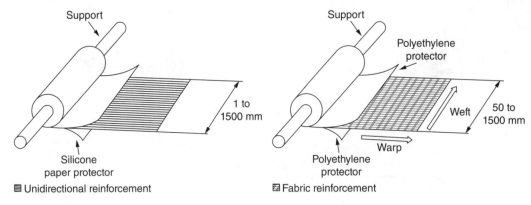

FIGURE 11-16 Pre-preg materials.

It is available in unidirectional form (one direction of reinforcement) and fabric form (several directions of reinforcement). All the major families of matrix resins can be used to impregnate various fiber forms. The resin is then no longer in a low-viscosity stage, but has been advanced to a B-stage level of cure for better handling characteristics. The following products are available in pre-preg form: unidirectional tapes, woven fabrics, continuous strand rovings, and chopped mat. Pre-preg materials must be stored in a freezer at a temperature below 0°F to retard the curing process. Pre-preg materials are cured with an elevated temperature. Many pre-preg materials used in aerospace are impregnated with an epoxy resin and they are cured at either 250 or 350°F. Figure 11-16 shows pre-preg materials.

Adhesives

Epoxy-based adhesives are the most used materials for bonding or repair of composite structures. Epoxy adhesives impart high-strength bonds and long-term durability over a wide range of temperatures and environments. Some epoxy adhesive systems cure at room temperature, while others require elevated temperatures to achieve optimum bonding characteristics. Adhesives are available as film adhesives, paste adhesives, and foaming adhesives.

Film Adhesives

Structural adhesives for aerospace applications are generally supplied as thin films supported on a release paper and stored under refrigerated conditions (−18°C, or 0°F). Film materials are frequently supported by fibers that serve to improve handling of the films prior to cure, control adhesive flow during bonding, and assist in bondline thickness control. **Adhesive films** have a separator film or backing paper which is applied to keep the material from sticking to itself. Film adhesives are tacky at room temperature and above. When applying film adhesives, it is important to prevent or eliminate entrapped air pockets between the aircraft surface and adhesive film by pricking bubbles or "porcupine" rolling over the adhesive prior to application. Figure 11-17 shows the application of film adhesives.

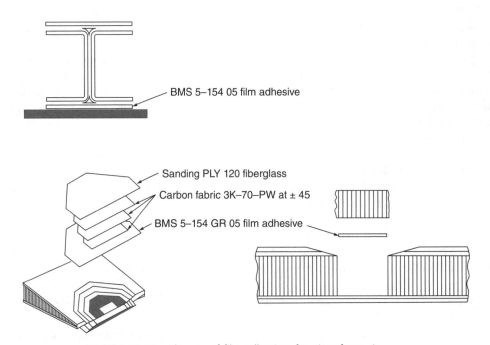

FIGURE 11-17 The use of film adhesives for aircraft repair.

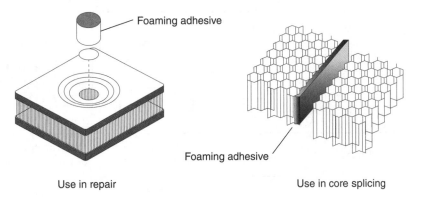

Use in repair   Use in core splicing

FIGURE 11-18 The application of foaming adhesives.

Paste Adhesives

Paste adhesives are used in the repair of composite parts as filler materials, for bonding repair core sections in place, and for bonding repair patches. Paste adhesives are two-component systems requiring careful weighing and thorough mixing to ensure strength is not compromised. Compared to film adhesives, paste adhesives have the advantage of lower-temperature cure cycles and are easier to store and ship. However, they have the disadvantage of lower strength, poor bond-line thickness control, and higher overall repair weight when compared to film adhesives. Application of paste adhesives can be accomplished by brush, by spreading with a grooved tool, or by extrusion from cartridges or sealed containers using compressed air.

Foaming Adhesives

Foaming adhesives consist of a thin unsupported epoxy film containing a blowing agent. During the rise to cure temperature, an inert gas is liberated causing an expansion or foaming action in the film. The expansion must be performed under positive pressure to prevent over-expansion and reduced strength. Following expansion, the adhesive is cured into strong highly structured foam. It is used as a lightweight splice material for splicing honeycomb core repair sections. Like film adhesives, foaming adhesives have the disadvantage of requiring a high-temperature cure cycle and must be shipped and stored at or below 0°F. They are sensitive to moisture, temperature, and contamination in the uncured state and must be handled and stored properly to prevent degradation. Figure 11-18 shows how foaming adhesives are used for the splicing and repair of honeycomb structures.

DESCRIPTION OF SANDWICH STRUCTURES

A sandwich construction is a structural panel concept that consists in its simplest form of two relatively thin, parallel face sheets bonded to and separated by a relatively thick,

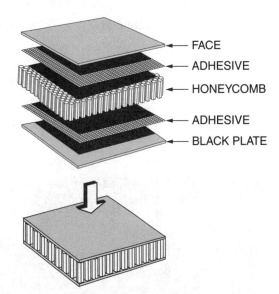

FIGURE 11-19 Honeycomb sandwich construction.

lightweight core. The **core** supports the **face sheets** against buckling and resists out-of-plane shear loads. The core must have high shear strength and compression stiffness. Honeycomb's beneficial strength-to-weight and stiffness-to-weight ratios compared to other materials and configurations are unmatched. Figure 11-19 shows a basic honeycomb sandwich construction.

Properties

Sandwich construction has high bending stiffness at minimal weight in comparison to aluminum and composite laminate construction. Most honeycomb sandwich structures are anisotropic; that is, properties are directional. As shown in Fig. 11-20, sandwich construction, especially honeycomb core construction, is structurally efficient, particularly in stiffness critical applications. Doubling the thickness of the core increases the stiffness over 7 times with only a 3% weight gain, while quadrupling the core thickness increases stiffness over 37 times with only a 6% weight gain.

	Solid Metal Sheet	Sandwich Construction	Thicker Sandwich
Relative Stiffness	100	700 7 times more rigid	3700 37 times more rigid
Relative Strength	100	350 3.5 times as strong	925 9.25 times as strong
Relative Weight	100	103 3% increase in weight	106 6% increase in weight

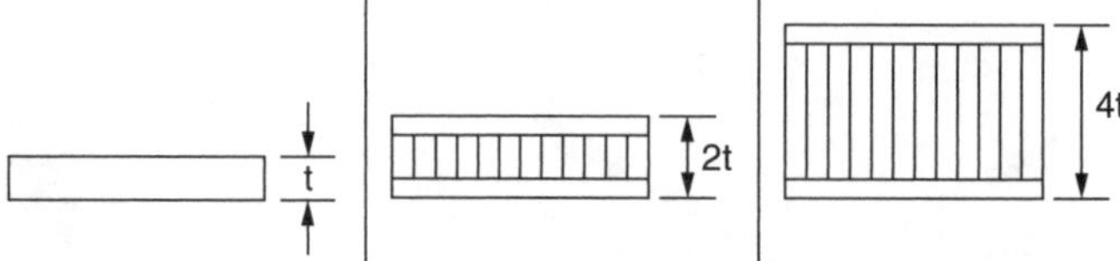

A striking example of how honeycomb stiffens a structure without materially increasing its weight.

FIGURE 11-20 The stiffening effect of honeycomb sandwich structure.

FIGURE 11-21 Aluminum honeycomb sandwich structure. See also color insert.

FIGURE 11-22 Honeycomb sandwich core materials. See also color insert.

Facing Materials

Most honeycomb structures used in aircraft construction use aluminum, fiberglass, Kevlar, or carbon fiber face sheets. Carbon fiber face sheets cannot be used with aluminum honeycomb core material because they will cause the aluminum to corrode. Titanium and steel are used for specialty applications in high-temperature constructions. The face sheets of many components such as spoilers and flight controls are very thin, sometimes only 3 or 4 plies. Figure 11-21 shows a honeycomb structure with aluminum face sheets and aluminum honeycomb core.

Core Materials

The most common core materials used in aerospace are aluminum and Nomex (Fig. 11-22). Nomex is paper that is impregnated with a phenolic resin. Aluminum core cannot

be used with carbon fiber face sheets because the aluminum will corrode.

Honeycomb

Honeycomb core cells for aerospace applications are usually hexagonal. The cells are made by bonding stacked sheets at special locations. The stacked sheets are expanded to form hexagons. The direction parallel to the sheets is called ribbon direction. Honeycomb material is available in several different kinds of materials. Metallic and nonmetallic materials are used in combination with carbon fiber, fiber glass, and Kevlar face sheets. Nonmetallic materials used are: fiberglass-reinforced honeycombs, Nomex, and Korex.

Honeycomb cell structures are available in several different styles depending on the application. The **hexagonal structure** is the most common type and used when the part is relatively flat. **Flexicore** is used when the honeycomb core

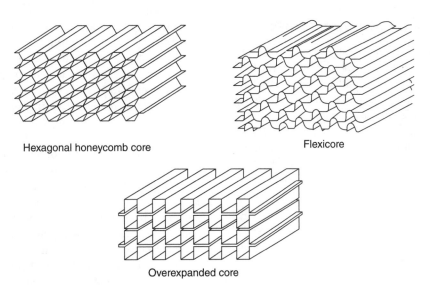

Hexagonal honeycomb core

Flexicore

Overexpanded core

FIGURE 11-23 Honeycomb core cell configurations.

material needs to be formed in a tighter contour than is possible with hexagonal honeycomb. The flexible core is used for parts that have a compound contour. Figure 11-23 shows the different styles of honeycomb core cells. Honeycomb core is available with different cell sizes. Small sizes provide better support for sandwich face sheets. Honeycomb is also available in different densities. Higher density core is stronger and stiffer than lower density core.

Foam

Foam cores are used on homebuilts and general aviation aircraft to give strength and shape to wing tips, flight controls, fuselage sections, wings, and wing ribs. Foam cores are not commonly used on commercial-type aircraft. Foam cores include a wide variety of liquid plastic materials that are filled with chemically released or mechanically mixed gas bubbles to produce rigid forms. Several types of foam core are available including polyvinyl chloride (cross-linked and uncross-linked), polyurethane, ebonite, polyimide, polymethacrylimide, and styrofoam. Carbon foam, a notable core material used for medium- to high-temperature applications, is made from coal. The mechanical properties of foam cores vary in an approximately linear fashion with material density. Foams tend to have lower mechanical properties than honeycomb, for a given weight.

Balsa Wood

Balsa is a natural wood product with elongated closed cells; it is available in a variety of grades that correlate to the structural, cosmetic, and physical characteristics. The density of balsa is less than one-half of the density of conventional wood products. However, balsa has a considerably higher density than the other types of structural cores. Modern aircraft floor panels, galley structures, and wardrobes may be fabricated from sandwich panels of balsa core faced with fiberglass or aluminum.

FIGURE 11-24 Composite laminate structure with stringers.

Description of Laminated Structures

Honeycomb sandwich might be the most efficient structural design for composite aircraft structures, but it also has some disadvantages: low impact damage resistance, water intrusion, and difficult to repair especially with fasteners. For these reasons new aircraft designs such as the Boeing 787 and Airbus 350 use a **laminate** type of composite structure for the wing and fuselage primary structure instead of a honeycomb structure. The reasons for this are that laminate structures are more **damage tolerant**, easier to repair, and easier to manufacture with automated manufacturing techniques. A laminate design consists of a laminate face sheet reinforced by external **stringers** and reinforcements as shown in Fig. 11-24. This design is similar to a semi-monoque design used for aluminum aircraft and sometimes called black aluminum. However, often the reinforcements and stringers are co-cured with the face sheet which eliminates secondary bonding processes or the use of fasteners.

Manufacturing and In-Service Damage

Damage from a great many causes, such as **scratching, gouging, impact, abrasion, erosion, local chemical attack or overheating**, may affect only a few surface plies over a large area or a depth of many plies in a smaller area. Failure to use protective devices, such as padded fixtures and sacrificial plies, may cause additional damage. Inadvertent or accidental impact may occur during the handling of parts during maintenance. The types of damage range from small surface scratches to more severe defects that would include punctures. Once damage is identified, the extent of damage must be determined and then it can be classified to determine the disposition of the defective part. Improper damage removal and repair techniques can impart greater damage than originally sustained to the composite material.

Manufacturing Defects

Manufacturing damage includes anomalies such as **porosity, microcracking, and delaminations** resulting from processing discrepancies and also such items as inadvertent edge cuts, surface gouges and scratches, damaged fastener holes, and impact damage. Examples of flaws occurring in manufacturing include a contaminated bondline surface or inclusions such as pre-preg backing paper or separation film that is inadvertently left between plies during lay-up. The inadvertent (non-process) damage can occur in detail parts or components during assembly or transport or during operation. A part is called resin rich if too much resin is used; for non-structural applications this is not necessarily bad but it will add weight. A part is called resin starved if too much resin is bled off during the curing process or if not enough resin is applied during the wet lay-up process. Resin-starved areas are indicated by fibers that show to the surface. The 60:40 fiber-to-resin ratio is considered optimum. Damage can occur at several scales within the composite material and structural configuration. This ranges from damage in the matrix and fiber to broken elements and failure of bonded or bolted attachments. The extent of damage controls repeated load life and residual strength and is, therefore, critical to damage tolerance.

Fiber Breakage. This defect can be critical because structures are typically designed to be fiber dominant (i.e., fibers carry most of the loads). Fortunately, fiber failure is typically limited to a zone near the point of impact, and is constrained by the impact object size and energy. Only a few of the service-related events listed in the previous section could lead to large areas of fiber damage.

Matrix Imperfections. These usually occur on the matrix-fiber interface, or in the matrix parallel to the fibers. These imperfections can slightly reduce some of the material properties but will seldom be critical to the structure, unless the matrix degradation is widespread. Accumulation of matrix cracks can cause the degradation of matrix-dominated properties. For laminates designed to transmit loads with their fibers (fiber dominant), only a slight reduction of properties is observed when the matrix is severely damaged. Matrix cracks, a.k.a. micro-cracks, can significantly reduce properties dependent on the resin or the fiber/resin interface, such as interlaminar shear and compression strength. For high-temperature resins, micro-cracking can have a very negative effect on properties. Matrix imperfections may develop into delaminations, which are a more critical type of damage**.**

Delamination and Debonds. **Delaminations** form on the interface between the layers in the laminate. Delaminations may form from matrix cracks that grow into the interlaminar layer or from low-energy impact. **Debonds** can also form from production non-adhesion along the bondline between two elements and initiate delamination in adjacent laminate layers. Under certain conditions, delaminations or debonds can grow when subjected to repeated loading and can cause catastrophic failure when the laminate is loaded in compression.

Combinations of Damages. In general, impact events cause combinations of damages. High-energy impacts by large objects (i.e., turbine blades) may lead to broken elements and failed attachments. The resulting damage may include significant fiber failure, matrix cracking, delamination, broken fasteners, and debonded elements. Damage caused by low-energy impact is more contained, but may also include a combination of broken fibers, matrix cracks, and multiple delaminations.

Flawed Fastener Holes. Improper hole drilling, poor fastener installation, and missing fasteners may occur in manufacturing. Hole elongation can occur due to repeated load cycling in service. Due to the low bearing loads of composite structures flawed fastener holes need to be repaired to design loads.

In-Service Defects

Many honeycomb structures such as wing spoiler, fairings, flight controls, and landing gear doors have thin face sheets which have experienced durability problems that could be grouped into three categories: low resistance to impact, **liquid ingression**, and **erosion**. These structures have adequate stiffness and strength, but low resistance to a service environment where parts are crawled over, tools dropped, and where technicians are often unaware of the fragility of thin-skinned sandwich parts. Damages to these components, such as core crush, impact damages, and disbonds, are quite often easily detected with a visual inspection due to their thin face sheets. Damages are sometimes allowed to go unchecked, often resulting in growth of the damage due to liquid ingression into the core.

The repair of parts due to liquid ingression can vary depending upon the liquid, of which water and **Skydrol** (hydraulic fluid) are the two most common. Water tends to create additional damage in repaired parts when cured unless all moisture is removed from the part. Most repair material systems cure at temperatures above the boiling point of water, which can cause a disbond at the skin-to-core interface wherever trapped water resides. For this reason,

core drying cycles are typically included prior to performing any repair. Some operators will take the extra step of placing a damaged but unrepaired part in the autoclave to dry so as to preclude any additional damage from occurring during the cure of the repair. Skydrol presents a different problem. Once the core of a sandwich part is saturated, complete removal of Skydrol is almost impossible. The part continues to weep the liquid even in cure such that bondlines can become contaminated and full bonding does not occur. Removal of contaminated core and adhesive as part of the repair is highly recommended.

Erosion capabilities of composite materials have been known to be less than that of aluminum and, as a result, their application in leading edge surfaces has been generally avoided. However, composites have been used in areas of highly complex geometry, but generally with an erosion coating. The durability and maintainability of some erosion coatings are less than ideal. Another problem, not as obvious as the first, is that edges of doors or panels can erode if they are exposed to the air stream. This erosion can be attributed to improper design or installation/fit-up. On the other hand, metal structures in contact or in the vicinity of these composite parts may show corrosion damage due to: inappropriate choice of aluminum alloy, damaged corrosion sealant of metal parts during assembly or at splices or insufficient sealant, and/or lack of glass fabric isolation plies at the interfaces of spars, ribs, and fittings.

There has been a concern with **UV radiation** affecting composite structures. Composite structures need to be protected by a top coating to prevent the effects of UV light. Special UV primers and paints have been developed to protect composite materials. Many fiberglass and Kevlar parts have a fine aluminum mesh for lightning protection. This aluminum mesh often corrodes around the bolt or screw holes. The corrosion affects the electrical bonding of the panel, and the aluminum mesh will need to be removed and new mesh installed to restore the electrical bonding of the panel. Figure 11-25 shows a panel where the aluminum mesh around the fastener holes is severely corroded.

Nondestructive Inspection of Composites

Damage to composite components is not always visible to the naked eye and the extent of damage is best determined for structural components by suitable **nondestructive test (NDT)** methods also called nondestructive inspection (NDI). NDI methods range from simple visual inspections to very sophisticated automated systems with extensive data handling capabilities. It should be noted that composites are generally more difficult to inspect than metals because laminated structures contain multiple ply orientations with numerous ply drop-offs. In addition, it is important that technicians conducting NDI be trained and certified in the method they are using. Table 11-1 shows what type of defects can be detected with what type of NDI methods.

Visual Inspection

Visual inspections are often used to inspect metallic and composite aircraft. Large areas can be inspected quickly and if damage is suspected more detailed inspection methods can be used to verify the damage. Flashlights, magnifying glasses, mirrors, and borescopes are employed as aids in the visual inspection of composites. These aids are used to magnify defects which otherwise might not be seen easily and to allow visual inspection of areas that are not readily accessible. Shining a flash light under an angle is a great aid to find damage. Resin starvation, resin richness, wrinkles, ply bridging, discoloration (due to overheating, lightning strike, etc.), impact damage by any cause, foreign matter, and blisters are some of the discrepancies readily discernible by a visual inspection. Visual inspection cannot find internal flaws in the composite, such as delaminations, disbonds, and matrix crazing. More sophisticated NDT is needed to detect these types of flaws.

Audible Sonic Testing (Coin Tapping)

Coin tapping, also called **tap testing**, with a coin or small hammer is often used to find subsurface damage to thin honeycomb or laminate structures. The method consists of lightly tapping the surface of the part with a coin or other suitable object. The acoustic response of the damaged area is compared with that of a known good area. An acoustically flat or dull response is considered unacceptable. A dull sound is a good indication that some delamination or disbond exists, although a clear, sharp tapping sound does not necessarily ensure the absence of damage, especially in thick panels. The acoustic response of a good part can also vary dramatically with changes in geometry; therefore, the geometry and interior construction of the part must be known before performing the tap test. If results are questionable, the technician can compare the results to a tap test of other like parts which are known to be good or use another inspection method. The **automated tap test** uses a solenoid instead of a hammer and this type of device will give a more

FIGURE 11-25 Corrosion of aluminum lightning strike mesh on composite panel. See also color insert.

07/24/2008 10:36 AM

TABLE 11-1   Comparison of NDI Methods for Fault Detection

Method of Inspection	Type of Defect							
	Disbond	Delamination	Dent	Crack	Hole	Water ingestion	Overheat and burns	Lightning strike
Visual	X*	X*	X	X	X		X	X
X-ray	X	X		X		X		
Ultrasonic TTU	X	X						
Ultrasonic pulse ECHO		X				X		
Ultrasonic bondtester	X	X						
Tap test	X†	X†						
Infrared thermography	X‡	X‡				X		
Dye penetrant								
Eddy current				X§				
Shearography	X‡	X‡		X§				

Note:
*For defects that open to the surface.
†For thin structure (3 plies or less).
‡The procedures for this type of inspection are being developed.
§This procedure is not recommeded.

consistent reading than the manual tap test. Figure 11-26 shows a technician performing a tap test of a nose radome.

Ultrasonic Inspection

Ultrasonic inspection has proven to be a very useful tool for the detection of internal delaminations, voids, or inconsistencies in composite components not otherwise discernible using visual or tap methodology. There are many ultrasonic techniques;

FIGURE 11-26   Coin (tap) testing of nose radome.

however, each technique uses <u>sound wave energy</u> with a frequency above the audible range. A high-frequency (usually several megahertz) sound wave is introduced into the part and may be directed to travel normal to the part surface or along the surface of the part, or at some predefined angle to the part surface. Different directions are used as the flow may not be visible only from one direction. The introduced sound is then monitored as it travels its assigned route through the part for any significant change. Ultrasonic sound waves have properties similar to light waves. When an ultrasonic wave strikes an interrupting object, the wave or energy is either absorbed or reflected back to the surface. The disrupted or diminished sonic energy is then picked up by a receiving transducer and converted into a display on the test equipment. The display allows the operator to comparatively evaluate the discrepant indications against those areas known to be good. To facilitate the comparison, reference standards are established and utilized to calibrate the ultrasonic equipment. The most common types of ultrasonic test equipment are through **transmission, pulse echo, phase array, and bond tester.** Figure 11-27 shows basic through transmission and pulse echo inspection techniques, and Fig. 11-28 shows equipment that could be used for phase array and pulse echo ultrasonic inspections. Figure 11-29 shows a large through transmission set-up to test large surfaces.

Radiography

Radiography, or X-ray, as it is often referred to, is a very useful NDI method in that it essentially allows a view into the interior of the part. This inspection method is accomplished by passing X-rays through the part or assembly being tested while recording the absorption of the rays onto a film sensitive to X-rays. Low-energy X-ray, gamma ray, or neutron emissions penetrate the test article. Variations in material thickness or density and the presence of flaws affect the intensity of the transmitted radiation. Voids and other discontinuities attenuate the radiation to different

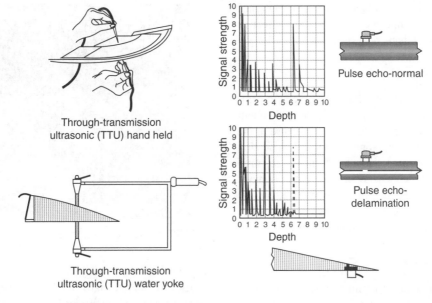

Through-transmission
ultrasonic (TTU) hand held

Through-transmission
ultrasonic (TTU) water yoke

Pulse echo-normal

Pulse echo-
delamination

FIGURE 11-27   Through transmission and pulse echo ultrasonics.

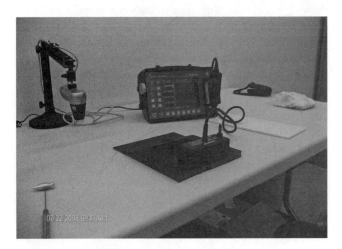

FIGURE 11-28   Phase array and pulse echo inspection equipment. See also color insert.

degrees and are revealed as contrasting light and dark areas (shadows) on the recorded image. Defects less dense than surrounding material (such as voids in a composite laminate) absorb less radiation and are shown on X-ray film as darker areas as compared to nearby images. Defects more dense than surrounding areas absorb more radiation (such as water in honeycomb sandwich assemblies) and are shown as lighter areas on X-ray film when compared to nearby images.

Thermography

Thermography comprises all methods in which heat-sensing devices are used to measure temperature variations for parts under inspection. Most common sources of heat are heat lamps or heater blankets. The basic principle of thermal inspection consists of measuring or mapping of surface temperatures when heat flows from, to, or through a test object. All thermographic techniques rely on differentials in thermal conductivity between normal, defect-free areas and those

FIGURE 11-29   Through transmission ultrasonic equipment.

having a defect. Normally, a heat source is used to elevate the temperature of the article being examined while observing the surface heating effects. Because defect-free areas conduct heat more efficiently than areas with defects, the amount of heat that is either absorbed or reflected indicates the quality of the bond. The type of defects that affect the thermal properties include disbonds, cracks, impact damage, panel thinning, and water ingress into composite materials and honeycomb core. Thermal methods are most effective for thin laminates or for defects near the surface.

Damage Assessment

After the damage to the aircraft has been inspected the damage needs to be assessed and a decision must be made about if and how to repair the damaged structure, the nature of the repair (permanent or temporary), and the needed inspection after the repair and during the residual life of the repaired structure. This decision depends upon where the damage is detected, the accuracy of damage characterization, the means available for determining the severity of the damage, and designing and performing an adequate repair.

Damage Categories

Permanent repair. A permanent repair restores the stiffness of the original structure, withstands static strength at the expected environments up to ultimate load, ensures durability for the remaining life of the component, satisfies original part damage tolerance requirements, and restores functionality of aircraft systems. Other criteria applicable in repair situations are to minimize aerodynamic contour changes, minimize weight penalty, minimize load path changes, and be compatible with aircraft operations schedule.

Interim repairs. Repair design criteria for temporary or interim repairs can be less demanding, but may approach permanent repairs if the temporary repair is to be on the airplane for a considerable time. Most users of aircraft and original equipment manufacturer (OEMs) prefer permanent repairs, if at all possible, as the temporary repairs may damage parent structure, necessitating a more extensive permanent repair or part scrapping. All temporary repairs have to be approved before the aircraft can be restored to operational status.

Cosmetic repair. Cosmetic repairs do not affect the structural integrity of the component or aircraft. A cosmetic repair is carried out to protect and decorate the surface. This will not involve the use of reinforcing materials.

Composite Repairs

The damage to an aircraft structure needs to be repaired or the part must be replaced after the damage has been inspected and evaluated. Primary structure on large transport aircraft cannot be replaced easily or not at all and the damage must be repaired on the aircraft using portable curing equipment. Smaller secondary structure parts such as flight controls, fairings, and interior parts can be removed from the aircraft and be repaired in the composite shop where ovens and autoclaves are available to restore the structural integrity of the part.

Hand Tools

The technician involved in composite repair work will have to use hand tools such as utility knives and pizza cutters to cut

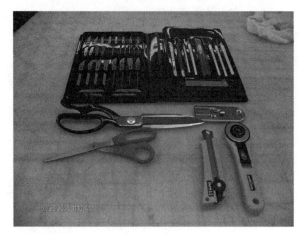

FIGURE 11-30 Cutting tools.

dry fabrics, vacuum bag materials, and pre-pregs. Knives with quick-change, retractable blades are preferred because blades must be changed often and they enhance safety. Scissors are used to cut dry fabrics. Materials made from Kevlar are more difficult to cut than fiber glass or carbon and tools will wear more quickly. Special scissors for Kevlar are available. Figure 11-30 shows a series of hand tools. Electric-driven cutters are also available for thicker fabrics and higher production.

Air Tools

Air-driven power tools such as drill motors, routers, grinders, and orbital sanders are used for composite repair. Electric motors are not recommended because carbon is a conductive material that can cause an electrical short circuit. If electric tools are used they need to be of the totally enclosed type. The technician should wear a respirator, protective clothing, and gloves when drilling, grinding, or sanding composite materials. Figure 11-31 shows an orbital sander.

Caul Plate

Caul plates are mostly made from aluminum or steel and are used to support the part during the cure cycle in an oven or autoclave. The composite shop will have a large assortment of caul plates for various jobs. A thin aluminum caul plate is

FIGURE 11-31 The use of an orbital sander for repair work. See also color insert.

often used when a repair is made on the aircraft using a heat bonder and vacuum pressure. The caul plate is placed in the vacuum bag on top of the bleeder ply separated by a solid parting film (release material). The heater blanket will be placed on top of the caul plate. The caul plate will provide a more uniform heated area and it will leave a smoother finish of the composite laminate. Either release film or mold release is used to prevent the caul plate from adhering to the part.

Vacuum Bag Materials

The vacuum bag process is the most common way to apply pressure to a composite repair that is performed on the aircraft. This process requires several types of **processing materials** to be used. These materials do not become part of the repair and will be discarded after the repair process. Release films, release agents, fiberglass bleeders, random matt polyester breathers, sealant tape, and vacuum bag film are commonly used. Vacuum bag materials should be stored in a dry and clean room on a rack as shown in Fig. 11-32.

Release Agents

Release agents, also called **mold release**, are used on caul plates, molds, and tools to ensure that the part will release from the caul plate, mold, or tool after the curing process. Due to environmental regulation water-based agents have been introduced that are easier to use. Release agents can contaminate the repair process and could cause poor bonding, and for this reason the application of release agents should be carried out away from the repair area.

Bleeder Ply

Bleeder plies are used in a composite repair with a **vacuum bag** to bleed off excess resin and remove air and volatiles that are trapped in the repair area. Bleeders are always used with wet lay-up repairs and sometimes with pre-preg repairs depending on the resin content of the pre-preg.

FIGURE 11-32 Vacuum bag processing materials. See also color insert.

Net resin pre-pregs do not require a bleeder and a bleeder should not be used because this will cause a resin-starved (dry) part or repair. The number of bleeder plies required for the repair will determine how much resin is bled off. Typical materials for a bleeder are either fiberglass or polyester.

Peel Ply

Peel plies are made from polyester or nylon and are used to create a ready surface for bonding purposes. A peel ply will be placed on the area of the lay-up that needs to be secondary bonded to another part. Just before the part will be bonded to the structure the peel ply will be removed. The removal of the peel ply creates a rough surface that is desirable for the bonding process. Uncoated and coated peel plies are available and for bonding purposes the uncoated peel plies are preferred. Some silicon-coated peel plies can leave an undesirable contamination on the surface which will actually decrease the bonding strength. The preferred peel ply material is polyester which has been heat-set to eliminate shrinkage. The uncoated peel plies can be difficult to remove after the curing process.

Sealant Tape

Heat-resistant sealant tape, also called sticky tape, is used to seal the vacuum bag to the part, or tool. The type of sealant tape to be used depends on the cure temperature of the part. Some sealant tapes can only be used up to 180°F while others could be used up to 650°F. Always check the temperature rating of the tape before use.

Perforated Release Film

Perforated release film is used to allow air and volatiles to escape out of the repair area, and it prevents the bleeder ply from sticking to the part or repair. Perforated release films are available with different size holes and hole spacing depending on the amount of bleeding required. An alternative for perforated release film is porous Teflon-coated fiberglass. This material has no holes in it but it allows air and volatiles to escape through the porous surface.

Solid Release Film

Solid **release films** are used so that the part or repair patch will not stick to the working surface of the tool or caul plate. Solid release film is also used to prevent resins from bleeding through and damaging the heat blanket or caul plate if they are used. Most release films are made from polyester, nylon, or Teflon-coated fiberglass.

Breather Material

The **breather material** creates a path for air to escape out of the vacuum bag. It is placed on top of the repair or on top of the heater blanket if used. It is important that if a bleeder is used for the repair that the bleeder contacts the breather. Failure to do this results in insufficient bleeding of the repair area. For temperatures below 350°F random mat polyester is used. The 4-oz type is used for vacuum pressure only or autoclave pressure below 50 psi. For pressures above 50 psi in the autoclave a 10-oz type must be used. For temperatures above 350°F the polyester material needs to be replaced with fiberglass material.

Vacuum Bag Film

The vacuum bag film seals off the repair from the atmosphere. Vacuum bag film is available in different materials and temperature ratings and it is important to choose the correct temperature rating for the repair. Some bagging films melt already at 250°F. Some vacuum bag materials do stretch more than others and if the vacuum bag does not stretch enough and the bag will "bridge" in the corners then the technician needs to make **pleats** in the bag to follow the contour of the part as shown in Fig. 11-33.

Vacuum Equipment

A vacuum pump is used to evacuate the air and volatiles from the vacuum bag so that atmospheric pressure will consolidate the plies. Most composite shops use a large central vacuum pumps that supplies vacuum to several repair stations. For repairs on the aircraft a portable electric-driven vacuum pump or the built-in vacuum pumps in some heat bonders can be used. Small shop air venturi pumps are also available. A vacuum port is installed in the vacuum bag; the bottom part is placed in the vacuum bag on top of the breather and the bagging film is cut above the bottom part and the top part of the vacuum port is installed with a quarter turn. Special vacuum hoses are used to connect the vacuum port to the pump. If these hoses are used in the autoclave or oven they need to be able to handle the temperature and pressure inside the oven or autoclave. A vacuum pressure regulator is sometimes used to lower the vacuum pressure during the bagging process to enable correct fitting of the bag on the repair.

Vacuum Bagging Techniques

The most common way to apply pressure to a composite repair is to use a **vacuum bag technique**. The repair patch

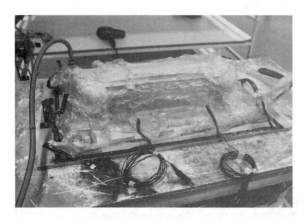

FIGURE 11-33 Pleats in vacuum bag. See also color insert.

is cured under pressure generated by drawing a vacuum in the space between the lay-up and a flexible film placed over it. The vacuum bag process consists of materials that are not part of the final repair: release films, bleeder, breather, vacuum bag film, and sealant tape. After the repair is completed the ancillary materials are discarded. Both single-side vacuum bag and envelope bag techniques are used depending on the application. After the vacuum bag is installed, a leak check must be performed to check if the vacuum bag holds vacuum. The repair manual will specify the minimum required vacuum and the maximum allowable leakage per minute.

Bleeding Techniques

When the **wet lay-up** process is used for a composite repair procedure the air and excess resin that are in the lay-up need to be removed. Wet lay-ups tend to be resin rich and it is easy to introduce air during the mixing of the resin and impregnating of the repair plies. The amount of bleeding can be controlled through the use of bleeding cloth in the vacuum bag. The repair manual will specify how many bleeder plies to use. Some pre-pregs are made with a higher percentage of resin than is required and these pre-pregs also need to be bled during the curing process. Most pre-pregs, however, are net resin pre-pregs and just enough resin is added to the fabric and no further bleeding is required. These net resin pre-pregs need to have a no-bleed vacuum bag set up.

Vertical (or Top) Bleed Out

The traditional way of bleed out using a vacuum bag technique is to use a perforated release film and a breather/bleeder ply on top of the repair. The holes in the release film allow air to breathe and resin to bleed off over the entire repair area. The amount of resin bled off depends on the size and number of holes in the perforated release film, the thickness of the bleeder/breather cloth, the resin viscosity and temperature, and the vacuum pressure. This technique works well for lay-ups with a low number of plies. If the lay-up is thick the top plies tend to be resin starved and the bottom plies tend to be resin rich.

No Bleed Out

Pre-preg systems with 32 to 35 percent resin content are typically **no-bleed systems**. These pre-pregs contain exactly the amount of resin needed in the cured laminate; for this reason resin bleed-off is not desired. Bleed out of these pre-pregs will result in a resin-starved repair or part. Many high-strength pre-pregs in use today are no-bleed systems. No bleeder is used and the resin is trapped/sealed so that none bleeds away. A sheet of solid release film (no holes) will be placed on top of the pre-preg and taped off at the edges with flash tape. A yarn of fiberglass is placed at a few places under the flash tape to allow air to get out from the lay-up. A breather and vacuum bag are installed to compact the pre-preg plies.

Single-Side Vacuum Bagging

The easiest way to apply a vacuum bag is to use a **single-side bag technique**. The vacuum bag is applied with sealant tape (tacky tape) to one side of the damaged aircraft part and a vacuum is applied. Figure 11-34 shows the single-side vacuum bag technique for a repair.

Envelope Bagging

Envelope bagging is a process in which the part to be repaired is completely enclosed in a vacuum bag or the bag is wrapped around the end of the component to obtain an adequate seal. It is frequently used for removable aircraft parts such as flight controls, access panels, etc., and when a part's geometry and/or the repair location makes it very difficult to properly seal the vacuum bag on the surface. When the original part is not holding vacuum an envelope bag could be used to eliminate leaks. In some cases, a part may be too small to allow for the installation of a single-side bag vacuum, whereas other times the repair is located on the end of a large component that must have a vacuum bag wrapped around the ends and sealed all the way around as shown in Fig. 11-35.

Alternate Pressure Application

In some applications the vacuum bag cannot be used and alternative ways to apply pressure are utilized. **Shrink tape**,

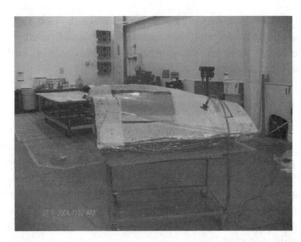

FIGURE 11-35 Envelope bag application for composite repair. See also color insert.

C-clamps, shot bags, and weights are sometimes used for curing and bonding operations.

Shrink Tape

Shrink tape is commonly used with parts that have been filament wound such as tubes and cylinders. The tape is wrapped around the completed lay-up, usually with only a layer of release material between the tape and the lay-up. Heat is

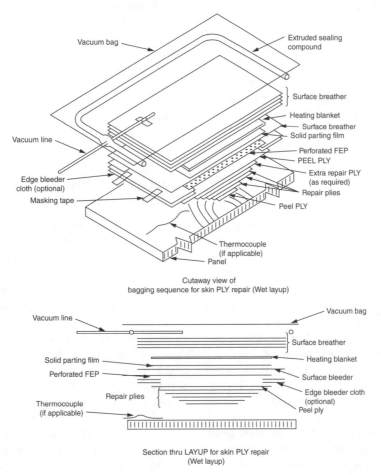

FIGURE 11-34 Single-side vacuum bag process.

applied to the tape with a heat gun to make the tape shrink. This process can apply a tremendous amount of pressure to the lay-up. After the shrinking step the part is placed in the oven for cure.

C-Clamps

When there is no room for a vacuum bag or the vacuum bag will crush the part, C-clamps can be used to apply pressure. C-clamps are especially useful for secondary bonding operations. Clamps should always be used with pressure distribution pads as damage to the part may occur if the clamping force is too high. Spring clamps can be used in applications wherein resin squeeze-out during cure would require C-clamps to be retightened periodically.

Shot Bags and Weights

At times shot bags or weights could be the only option to apply pressure to a lay-up, but the effect is limited due to the low pressure that can be applied and the fact that the part or repair needs to be horizontal.

Hand Lay-up

Most aircraft composite repair work consists of **hand lay-up** techniques. The technician uses knives and scissors to cut the material to size and stacks the plies in the repair area. The wet lay-up process and the pre-preg process are the two processes most often used to repair composite parts and aircraft structures. Wet lay-up processes are mostly used for the repair of damaged nonstructural composite structures. The mechanical properties of wet lay-up processes are less than pre-preg hand lay-up processes. Hand lay-up greatly depends on the skill of the technician who performs the repair. Common mistakes are inclusion of backing materials and incorrect ply orientation.

Wet Lay-ups

During the wet lay-up process a dry fabric will be impregnated with a resin. Typically a two-part resin system is used which is mixed just before the repair is going to be made. The fabric is placed on a piece of release film and impregnated with a resin system using a squeegee or brush. After the fabric is impregnated another piece of release film is placed on top of the fabric and a squeegee is used to remove air bubbles. The technician can lay out the repair plies and ply orientation on the release film. After this the repair plies are cut, stacked in the correct ply orientation, and vacuum bagged. Wet lay-up repairs are often used with fiberglass for nonstructural applications. Carbon and Kevlar dry fabric could also be used with a wet lay-up resin system. Many resin systems used with wet lay-up cure at room temperature are easy to accomplish and the materials can be stored at room temperature for long times. The disadvantage of room temperature wet lay-up is that it will not restore the strength and durability of the original structure and parts that were cured at 250 or 350°F during manufacturing. Some wet lay-up resins use an elevated temperature cure and have improved properties.

Pre-preg

Most primary structures of transport composite aircraft are made from pre-preg materials that are cured at elevated temperatures. Pre-preg repair materials with the same properties as the original structure are required to repair these primary structures. Pre-preg material systems consist of a dry fiber material that is impregnated with a resin during the pre-preg manufacturing process. The resin system is already in the B stage of cure and for this reason the material needs to be stored in a freezer below 0°F to prevent further curing of the resin. The material is typically placed on a roll and a backing material is placed on one side or both sides of the material so that the pre-preg will not stick together. The pre-preg material is tacky at room temperature and will adhere to other plies easily during the stack-up process. The technician will have to remove the pre-preg material from the freezer and let the material thaw which might take 8 h for a full roll or less time if the roll was divided in several kits. After the material is thawed and removed from the bag, the material can be cut in repair plies, stacked in the correct ply orientation, and vacuum bagged. Technicians should remember to remove the backing material when stacking the plies because it is easy to forget, and this introduces a serious defect to the part or repair. Many pre-preg materials use epoxy or BMI resins that need to be cured at an elevated cure cycle. Autoclave, curing ovens, and heat bonders are required to cure pre-preg materials.

Repairs consisting of four or less plies can be made in one stack-up but if more plies are required for thicker laminate structures consolidation is necessary because large quantities of air can be trapped between each pre-preg layer. This trapped air can be removed by covering the lay-up with a perforated release film, a breather ply, and applying a vacuum bag. Apply the vacuum for 10 to 15 min at room temperature. Some manufacturers use the double vacuum debulk concept to create thick ply stack-ups. The double vacuum debulk (DVD) process is discussed later in this chapter.

Storage and Handling of Pre-pregs, Film Adhesives, and Foam Adhesives

Reinforcements, preimpregnated with an epoxy resin system or other type of thermoset matrix material, will generally require **freezer storage**. **Storage life** and maximum time out of the freezer, prior to use, are specified by material manufacturers. These materials must be stored in sealed, moisture-proof bags in accordance with the applicable technical data. A record of out-time shall be maintained for all pre-preg materials. The **shelf life** of material can be improved by cutting large batches into smaller sections (kitting) the first time it is thawed. This method will permit the bulk of the material to be kept frozen while some pieces are thawed for use, thereby reducing waste caused by exceeding out-time requirements. When kitting, lay the material flat or wrap it around a tube which is at least 3 inches in diameter and 2 inches longer than the material; do not fold the material as this will distort it. All refrigerated and frozen item packages/containers must be sealed before storing and warmed to room temperature before opening; failure to warm to room temperature will result in moisture absorption which will degrade the resin. Backing or separator film must not be removed from the repair material until it is ready for use. If pre-preg materials were received without dry ice in the shipping container, determine how much time it was out of the freezer then add that time to the out-time log. Wear clean, impermeable (latex, nitrile, etc.) gloves when handling pre-pregs. If powdered gloves are all that are available, ensure exterior of gloves are cleaned before handling repair materials and hands are washed thoroughly immediately after removal of gloves to prevent contaminating the area and materials.

Uncured pre-preg materials have time limits for storage and use. The maximum time allowed for storing pre-preg material at low temperature is called the storage life which is typically 6 months to a year. The material can be tested and the storage life could be extended by the material manufacturer. The maximum time allowed for material at room temperature before the material cures is called the mechanical life. The recommended time at room temperature to complete lay-up and compaction is called the handling life. The handling life is shorter than the mechanical life. The mechanical life is measured from the time the material is removed from the freezer until the time the material is returned to the freezer.

Bonding Processes

Bonding is the most efficient way to join composite parts together and several different bonding processes are utilized. Each bonding process has pros and cons and careful selection of the correct process for the application is required.

Co-curing

Co-curing is a process wherein two parts are simultaneously cured. The interface between the two parts may or may not have an adhesive layer. Co-curing of honeycomb panels often result in poor panel surface quality which could be prevented by using a secondary surfacing material co-cured in the standard cure cycle or a subsequent "fill-and-fair" operation. A typical co-curing application is the simultaneous cure of a stiffener and a skin as used for transport aircraft. Adhesive film is frequently placed into the interface between the stiffener and the skin to increase fatigue and peel resistance. Principal advantages derived from the co-curing process are excellent fit between bonded components and guaranteed surface cleanliness.

Secondary Bonding

Secondary bonding utilizes pre-cured composite detail parts, and uses a layer of adhesive to bond two pre-cured composite parts. Pre-cured laminates undergoing secondary bonding usually have a thin nylon or fiberglass peel ply cured onto the bonding surfaces. Both film adhesives and paste adhesives are used in the secondary bonding process.

Co-bonding

In the **co-bonding process**, one of the detail parts is pre-cured, with the mating part being cured simultaneously with

the adhesive. Film adhesive is often used to improve peel strength.

Mixing Resins

Resin systems used for aircraft repair are usually two-part epoxy systems. The A and B components need to be mixed carefully. Resin systems are measured by weight and a scale is used to add the correct weights of part A and part B. Failure to use the correct mixing ratio will most likely result in incorrect curing. The resin must be mixed slowly and fully for at least 3 min. Air will enter into the mixture, if the resin is mixed too fast. If the resin system is not fully mixed, the resin may not cure properly. Make sure to scrape the edges and bottom of the mixing cup to ensure that all resin is mixed correctly. Do not mix large quantities of quick-curing resin. These types of resin systems produce heat after they are mixed. Mix only the amount of material that is required. Mix more than one batch, if more material is needed than the maximum batch side. Always use the product technical data sheet for correct usage. Figure 11-36 shows the mixing of a two-part resin system.

Impregnation Techniques

Dry fibers are impregnated with a resin and the technician can use a brush, squeegee, or roller. It is important not to push too hard on the brush or squeegee during the impregnation process because this will distort the fiber direction and affect the strength of the repair. It is important to put the right amount of resin on the fabric. Too much or too little resin will affect the strength of the repair. Air that is put into the resin during mixing or not removed from the fabric will also reduce the repair strength.

FIGURE 11-36 Weighing resin parts A and B for correct ratio. See also color insert.

Fabric Impregnation with a Brush or Squeegee

A brush or squeegee can be used to impregnate dry fabric with a resin system. The technician will first cut the fabric to the correct size and place the dry fabric on a sheet of release film. Resin is applied in the middle of the fabric and a brush or squeegee is used to carefully spread out the resin to cover all of the dry fabric. All of the fabric should be wetted. A second sheet of release film is placed on top of the wetted fabric and a clean squeegee is used to carefully remove air bubbles. The technician can make the lay-out for the repair on top of the release film with a marker and cut the plies. Most wet lay-up processes have a room temperature cure but extra heat will speed up the curing process and increase the properties. Temperatures up to 150°F are used to speed up the curing process. Figures 11-37a through c shows the fabric impregnation process.

Fabric Impregnation Using a Vacuum Bag

The vacuum-assisted impregnation method is used to impregnate repair fabric with a two-part resin while enclosed inside a vacuum bag. This method is preferred for tight-knit weaves and when near optimum resin-to-fiber ratio is required. Compared to squeegee impregnation, this process

FIGURE 11-37a   Apply resin on top of the fabric. See also color insert.

FIGURE 11-37b   Use squeegee to impregnate the fabric. See also color insert.

FIGURE 11-37c   Place release film on top of wet fabric and use squeegee to remove air from the lay-up. See also color insert.

reduces the level of entrapped air within the fabric and offers a more controlled and contained configuration for completing the impregnation process.

Procedure

- Place vacuum bag sealing tape on the table surface around the area that will be used to impregnate the material. The area should be at least 4 in larger than the material to be impregnated.
- Place an edge breather cloth next to the vacuum bag sealing tape. The edge breather should be 1 to 2 in wide.
- Place a piece of solid release film on the table. The sheet should be 2 in larger than the material to be impregnated.
- Lay the fabric on the parting film.
- Weigh the fabric to find the amount of resin that is necessary to impregnate the material.
- Pour the resin onto the fabric. The resin should be a continuous pool in the center area of the fabric. See Fig. 11-38a.
- Place a second piece of solid parting film over the fabric. This film should be the same size or larger than the first piece.
- Place and seal the vacuum bag, install vacuum port, and apply vacuum to the bag. Allow 2 min for the air to be removed from the fabric. See Fig. 11-38b.

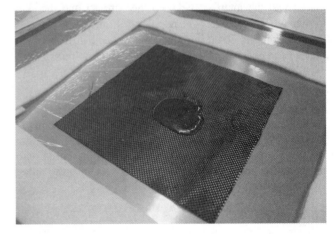

FIGURE 11-38a   Pour resin in the middle of the fabric. See also color insert.

FIGURE 11-38b Apply release film and vacuum bag. See also color insert.

FIGURE 11-38c Use squeegee to impregnate the fabric. See also color insert.

- Sweep the resin into the fabric with a squeegee. Start from the center and slowly sweep the resin to the edge of the fabric. The resin should be uniformly distributed over all of the fabric. See Fig. 11-38c.
- Open the vacuum bag and remove the fabric and cut the repair plies.

Curing of Composite Materials

The **curing** of a repair is as important as the curing of the original part material. Unlike metal repairs, where the materials are premanufactured, the technician is responsible for the manufacture of composite repair material. This includes all storage, processing, and quality control functions. For aircraft repairs, the cure cycle starts with material storage. Materials which are stored incorrectly can begin to cure before they are used for a repair. All time and temperature requirements must be met and documented. The aircraft structural repair manual should be consulted to determine the correct cure cycle for the part that needs to be repaired.

Room Temperature Curing

Room temperature curing is the easiest curing method. No curing equipment is necessary. However, the disadvantages of a room temperature cure cycle are: it takes a long time for epoxy resins to cure at room temperature and it will not restore either the strength or the durability of the original 250 or 350°F cure components. Room temperature cure resin systems are often used for wet lay-up fiber glass repairs for noncritical components. Room temperature cure time can be accelerated by the application of heat.

Elevated Temperature Curing

Composite repairs that use epoxy or BMI resins typically need to be cured at an **elevated temperature** to achieve their optimum mechanical properties. Autoclaves and curing ovens are used in the composite shop to cure repairs and heat bonders are used for on-aircraft repairs. These types of equipment use thermocouples and have a programmable computer to control the cure temperature. The technician could select a preprogrammed cure cycle recipe on the curing device or write a new program if required. Typical curing temperatures for composite materials are 250 or 350°F. The curing cycle consists of at least three segments. The first segment is the ramp up. The ramp up is often between 1 to 5°/min. The second segment is the dwell or also called soak time. This is the time the part needs to be at the maximum required temperature. The third segment is the cool down and 1° to 5° ramp down is common. Sometimes additional ramp ups and dwell time periods are added to the cure recipe. Always consult the repair manual or technical data sheet for the correct cure recipe. Figure 11-39 shows the cure recipe programmed on a heat bonder. Figure 11-40 shows the completed autoclave cure recipe.

Curing Equipment

Autoclaves and curing ovens are the types of equipment used to cure composite repairs in the composite shop. Heat bonders are used if repairs have to be made on the aircraft and an elevated cure temperature is required. Sometimes heat lamps and hot air modules are used to cure repairs on the aircraft.

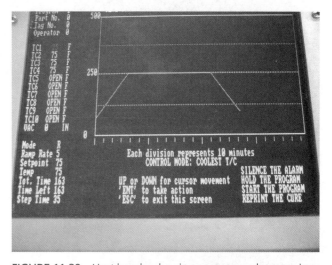

FIGURE 11-39 Heat bonder showing programmed cure recipe.

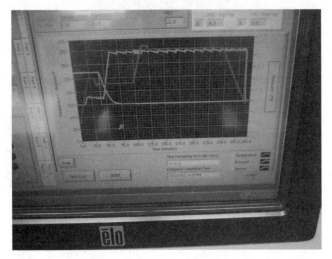

FIGURE 11-40 Completed autoclave curing recipe. See also color insert.

FIGURE 11-42 Autoclave. See also color insert.

Curing Oven

Composite repair shops use programmable **curing ovens** to cure composite parts and repairs. The curing oven is equipped with a programmable controller to accurate control a preset curing recipe. Curing ovens can control oven temperature and vacuum. The part is vacuum bagged before it is put into the oven and the oven vacuum system will apply full vacuum during the cure. Figure 11-41 shows a curing oven.

Autoclave

Autoclaves are used to manufacture most structural aircraft parts and are used by all the original equipment manufacturers (OEM) and maintenance repair stations (MRO). Autoclaves can control temperature, vacuum, pressure, and cooling. Depending on the sophistication of the autoclave these variables can be accurately controlled and adjusted to achieve an optimum cure recipe. Typical autoclave temperature goes up to 650°F and the maximum pressure could be as high as 250 psi. Most epoxy resins are cured at a maximum temperature of 350°F and a pressure of 50 to 100 psi. BMIs

use higher temperatures. Autoclaves are often equipped with a nitrogen system instead of using compressed air if the autoclaves are used for temperatures above 250°F. Nitrogen is an inert gas and is used to prevent fires in the autoclave during high-temperature and high-pressure runs. Figure 11-42 shows an autoclave at a repair station.

Heat Bonder

Heat bonders are used to cure composite and metal-bonded repairs on the aircraft. The heat bonder is a programmable temperature control unit that can control heat blankets, heat lamps, or hot air modules. The technician can write a cure recipe or select a program that is stored in memory. Most heat bonders also can supply and monitor vacuum. The heat bonder has alarms that go off to warn the technician that something is wrong with either the temperature or vacuum. During the cure cycle the technician should be within hearing distance of the alarm to take action immediately if the alarm goes off. The heat bonder will save the cure data and these data can also be printed out to go in the aircraft maintenance records. Figure 11-43 shows a heat bonder controlling a heat blanket.

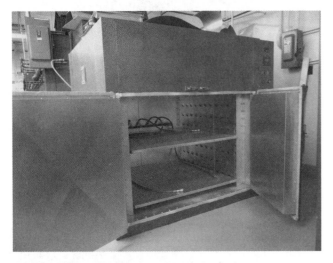

FIGURE 11-41 Curing oven. See also color insert.

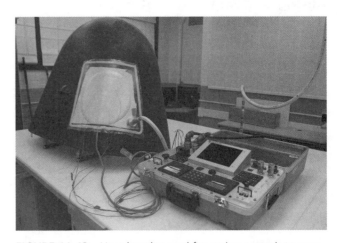

FIGURE 11-43 Heat bonder used for curing a repair to nose radome. See also color insert.

Heat Lamps

Infrared heat lamps are used for curing composites repairs when the curing temperature is below 150°F. They are ineffective for higher temperatures because it is difficult to control the heat applied with a lamp. Heat bonders and thermocouples are often used to control the heat lamps. Lamps tend to generate high-surface temperatures quickly and their temperature profile must be studied by performing a heat survey of the repair area before the repair is accomplished. Figure 11-44 shows the use of heat lamps for curing of composites.

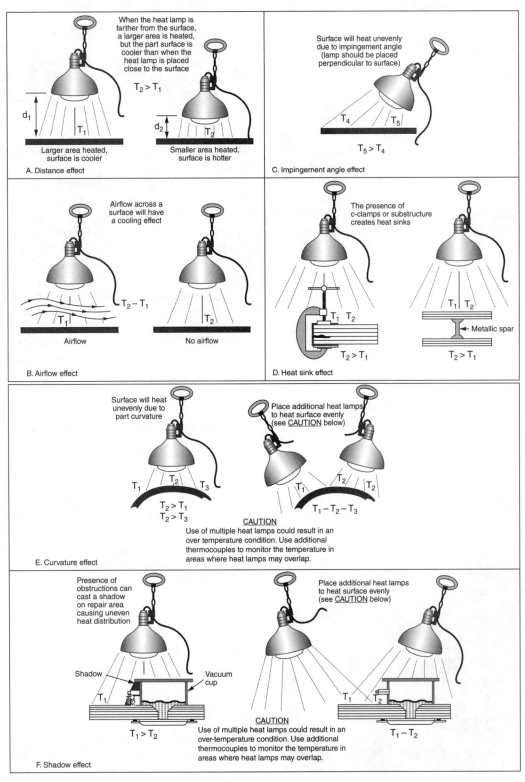

FIGURE 11-44   Heat lamp instructions for composite repair.

Hot Air Modules

Hot air module (HAM) units are generally controlled either manually or in conjunction with a heat bonder controller. They can be used to heat large areas of any contour or material. HAMs will supply controlled heat to over 900°F. HAMs are often used to heat the back side of large repair areas to reduce heat sinks. HAMs lend themselves very well to heating and curing of materials that require noncontact such as paints and coatings. When using hot air for curing of composite laminates, care should be taken not to insulate the repair by using excessive breather materials.

Heater Blanket

Heat blankets are available in many different shapes, contours, and styles. They are used to cure repairs on the aircraft. These electrical blankets are controlled by a heat bonder and thermocouples. Heat is transferred from the blanket to the aircraft via conduction. Consequently, the heat blanket must conform to and be in 100 percent contact with the part. This is usually accomplished using vacuum bag pressure. Lack of contact between the heat blanket and the area to be cured will result in inadequate heating of the repair area and an overheating of the heat blanket (Fig. 11-45).

Thermocouples

Thermocouples are used with autoclaves, curing ovens, and heat bonders to measure the temperature of the equipment of part. A thermocouple is a thermoelectric device used to accurately measure temperatures. Thermocouples consist of a wire with two leads of dissimilar metals; the metals are joined at one end. Heating the joint produces an electric current; this current is sent to the heat device controller.

Composite Repair

The repair of composite structures can be divided into two groups: bonded repairs and bolted repairs. Bonded repairs do not introduce stress concentrations by drilling fastener holes

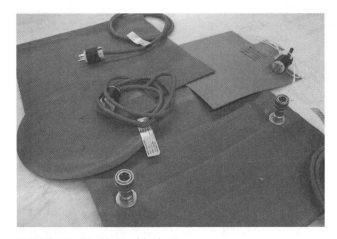

FIGURE 11-45   Heater blankets.

TABLE 11-2   Bonded versus Bolted Repair Matrix

Bonded versus Bolted Repair	Bolted	Bonded
Lightly loaded structures: laminate thickness less than 0.1 in		x
Highly loaded structures: laminate thickness between $\frac{1}{8}$ and 0.5 in	x	x
Highly loaded structures: laminate thickness larger than 0.5 in	x	
High peeling stresses	x	
Honeycomb structure		x
Dry surfaces	x	x
Wet and/or contaminated surfaces	x	
Disassembly required	x	
Restore unnotced strength		x

for patch installation and can be stronger than the original structure. The disadvantage of bonded repairs is that most repair materials require special storage, handling, and curing procedures. The preferred bonded repair method is the scarf repair. In this repair method the parent material will be tapered to create a flush repair. Bolted repairs are quicker and easier to fabricate than bonded repairs. They are normally used on composite skins thicker than 0.125 in to ensure sufficient fastener bearing area is available for load transfer. They are seldom used in honeycomb sandwich assemblies due to the potential for moisture intrusion from the fastener holes and the resulting core degradation, and low bearing area of thin face sheets. Bolted repairs are heavier than comparable bonded repairs, limiting their use on weight-sensitive flight control surfaces. Table 11-2 shows the criteria for bonded or bolted repair.

Composite Honeycomb Sandwich Repairs

Honeycomb sandwich structures are used for many aircraft components such as fairings, flight controls, radomes, and engine cowlings. These lightweight sandwich structures are susceptible to damage. Because sandwich structure is a bonded construction and the face sheets are thin, damage to sandwich structure is usually repaired by bonding. Repairs to sandwich honeycomb structure use similar techniques for the most common types of face sheet materials such as fiberglass, carbon, and Kevlar. Kevlar is often repaired with fiberglass.

Minor Core Damage (Filler and Potting Repairs)

Minor damage to a honeycomb sandwich structure can be repaired with a **potted repair**. The maximum size for a potted repair is typically smaller than 0.5 in. The honeycomb material could be left in place or could be removed, and is filled up with a potting compound to restore some strength. Potted repairs do not restore the full strength of the part. Epoxy resins filled with hollow glass, phenolic or plastic microballoons, cotton, flox, or other materials are often used as potting compounds. The potting compound can also be used as filler for cosmetic repairs to edges and skin panels. Potting compounds are also used in sandwich honeycomb panels as hard points for bolts and screws. The potting compound is heavier than the original core and this could affect flight control balance. The weight of the repair must be calculated and compared with flight control weights and balance limits set out in the structural repair manual.

Bonded Scarf Repair Example of Honeycomb Sandwich Panel

Honeycomb sandwich structures such as radomes, fairings, spoilers, and flight controls are often damaged and need to be repaired. The common causes of damage are: impact damage, moisture intrusion, lightning strikes, and erosion. A detailed step-by-step repair procedure is discussed below.

Step 1. Evaluate the damage through visual and NDI inspection and determine the proper repair procedure. Minor damage could be repaired using a potting repair and more substantial damage to the honeycomb sandwich structure must be repaired by replacing the core and repairing the face sheet. If the damage is so extensive that repair is not possible or economically feasible then it should be considered to remove the part from the aircraft (if it is removable) and replace with new part. The tap test or ultrasonic tests are often used to determine the extent of the damage. The damaged area can be much larger than what can be found with a visual inspection.

Step 2. The structure must be dried before the repair procedure can start. Sandwich or honeycomb panels are particularly prone to water absorption and retention due to the increased air volume in these panels. Cracked coatings, impact damage, and loose fasteners all provide paths for moisture intrusion. It is extremely important to ensure water and moisture are removed prior to performing any bonding, especially one incorporating a heated curing process. Standing water, water vapor, and absorbed moisture in laminates and honeycomb can produce the following damaging effects: blistering, delamination, microcrack damage, porosity, and weak bondline strength in the repair adhesive. Water intrusion can corrode the aluminum core. Water can travel from one core cell to the next and can travel far inside the structure over time. Sometimes small holes need to be drilled in the face sheet to drain the water before the drying operation starts. Remaining water can cause damage to the composite structure when it freezes at high altitude. Another problem with moisture is that if the repair is cured above 212°F the water can boil and the boiling pressure can disbond the face sheets from the core resulting in more damage.

Drying of the repair area can be accomplished in an autoclave or oven if the part can be taken off the aircraft. For parts of the aircraft that are not removable the water can be removed with a heater blanket and a vacuum bag as shown in Fig. 11-46. It is important to slowly turn up

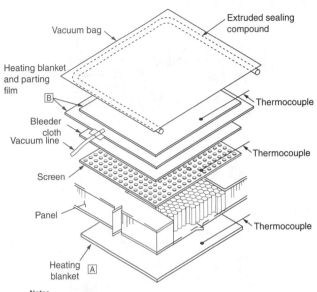

Notes

A Preferred location of heating blanket when opposite face is accessible.

B Alternate location of parting film and heating blanket when opposite side is inaccessible. This location may be used for an additional heating blanket to accelerate watch removal.

FIGURE 11-46 Water removal using vacuum bag procedure.

the heat during the water removal procedure. Consult the maintenance manual for drying times and temperatures.

Step 3. Remove the damaged core and face sheet using grinders, core knives, and routers; see Fig. 11-47. Be careful not to unnecessarily enlarge the repair area during the process. Use a vacuum cleaner to remove all debris and dust.

Step 4. Honeycomb sandwich structures with thin face sheets (four plies or less) are most efficiently repaired by using a scarf-type repair. A flexible disk sander with sandpaper pad is used to taper sand a uniform taper around the cleaned-up damage. Sometimes the repair instructions give a taper ratio and other times a ply overlap is given. After the taper sanding is complete the area needs to be cleaned with a vacuum and approved solvent. Figure 11-48 shows a single-sided taper-sanded repair.

Step 5. Use a knife to cut the replacement core. The core plug must be of the same type, class, and grade of the original core. The direction of the core cells should line up with the honeycomb of the surrounding material. The plug must be trimmed to the right length and be solvent washed with an approved cleaner before installation. A potting compound or a foaming adhesive can be used to install the replacement core plug. Usually the core replacement will be cured with a separate curing cycle and not co-cured with the patch. The plug must be sanded flush with the surrounding area after the cure. See Fig. 11-49.

Step 6. Read the aircraft maintenance manual to determine the correct repair material, ply orientation, and the number of plies required for the repair. Typically, one more ply than the original number of plies is installed. The repair plies must be installed with the same orientation as

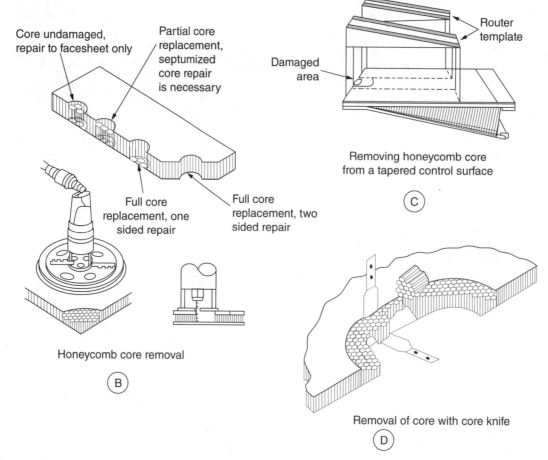

Core undamaged, repair to facesheet only

Partial core replacement, septumized core repair is necessary

Damaged area

Router template

Full core replacement, one sided repair

Full core replacement, two sided repair

Removing honeycomb core from a tapered control surface

C

Honeycomb core removal

B

Removal of core with core knife

D

Instructions

1. Remove the face sheet and/or core with a power router, using a router template to protect the undamaged part of the face sheet. Refer to SRM 51–10–02 for instructions on the use of a router and template.

   Note: The router can be adjusted to remove: One of the face sheets only, a face sheet and part of the core, a face sheet and all of the core, or both the face sheets and core. See detail B.

2. If you are routing tapered part, you can use wedge shaped router templates. This will permit the router to cut the core material parallel with the lower surface. See detail C.

3. It is permitted to remove honeycomb core with a care knife. See detail D.

FIGURE 11-47 Removal of damaged laminate and core.

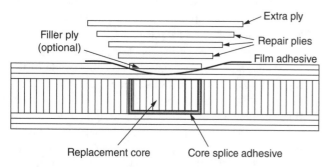

Extra ply

Filler ply (optional)

Repair plies

Film adhesive

Replacement core

Core splice adhesive

FIGURE 11-48 Single-sided taper-sanded repair.

the plies of the original structure. Repairs use either wet lay-up or pre-preg material. Impregnate the plies with a resin system for the wet lay-up repair or remove the backing material from the pre-preg material. Often extra plies of fiberglass are added on top of the repair. These plies are not structural but are there to be sanded down after the

repair is cured to achieve a smooth surface. Figure 11-50 shows the placement of repair plies.

Step 7. A vacuum bag (see Fig. 11-51) will be used to apply pressure to the repair to consolidate the repair plies. A heater blanket will be placed in the vacuum bag if an elevated temperature needs to be used to cure the repair. Most structural repairs use either 250 or 350° cure cycles.

Step 8. Most aircraft structural repairs use an elevated heat cure and autoclaves or curing ovens are preferable if the part can be removed from the aircraft. A heater blanket and a heat bonder are used if the repair has to be made on the aircraft as shown in Fig. 11-43.

Step 9. Remove the vacuum bag and inspect the repair visually and with NDI equipment. A tap test might be sufficient for thin face sheets or ultrasonic or thermography inspection methods could be used. After inspection the repair patch will be sanded and a paint system will be applied. Figure 11-52 shows a completed repair on a nose radome.

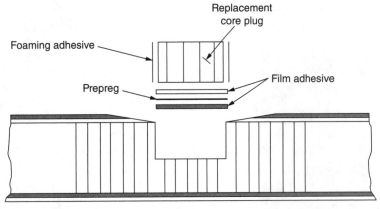

Replacement
core plug

Foaming adhesive

Prepreg

Film adhesive

Section Thru Repair area-partial depth core replacement
section A–A

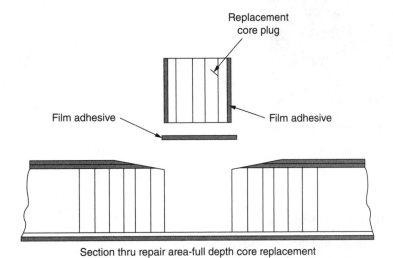

Replacement
core plug

Film adhesive

Film adhesive

Section thru repair area-full depth core replacement
section B–B

FIGURE 11-49   Replacement of damaged core.

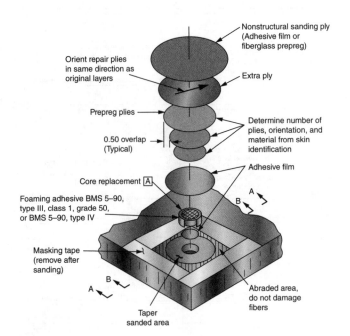

Nonstructural sanding ply
(Adhesive film or
fiberglass prepreg)

Orient repair plies
in same direction as
original layers

Extra ply

Prepreg plies

Determine number of
plies, orientation, and
material from skin
identification

0.50 overlap
(Typical)

Adhesive film

Core replacement Ⓐ

Foaming adhesive BMS 5–90,
type III, class 1, grade 50,
or BMS 5–90, type IV

Masking tape
(remove after
sanding)

Abraded area,
do not damage
fibers

Taper
sanded area

FIGURE 11-50   Repair ply placement.

FIGURE 11-51   Vacuum bag procedure. See also color insert.

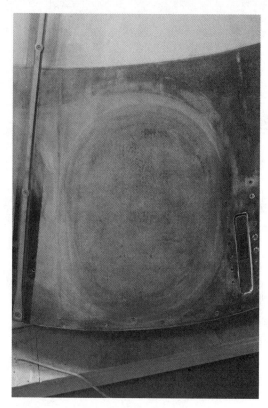

FIGURE 11-52 Repair of a nose radome. See also color insert.

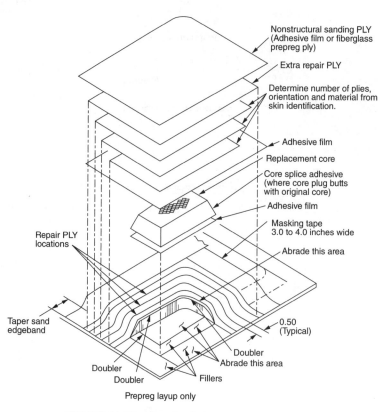

FIGURE 11-53 Repair of corner ramp.

Repair of Honeycomb Ramp Corner

Honeycomb panels are often damaged at the corners due to erosion, impact damage, and rough handling during maintenance. Honeycomb structures need to treated carefully and always put on racks and never on the ground during maintenance inspections. Figure 11-53 shows an example of a repair of a corner ramp.

Repair of Sandwich Structures with Foam Core

New generations of general aviation aircraft such as the Cirrus SR 20 are made from a composite structure that consists of a fiberglass sandwich structure with a foam core. The repair procedures for these aircraft are similar to those for honeycomb sandwich structures. Figure 11-54 shows the repair scenario for these types of structures. Figure 11-55 shows the repair technique on the aircraft.

Solid Laminates

Solid laminate structures are used extensively for new transport aircraft. Solid laminate structures ranging from 4 plies to over 100 plies are used. Thinner laminate structures can effectively be repaired with a bonded repair. Typically the repair area is scarf sanded and a flush patch is used to repair the structure. Thicker laminate structures are often repaired with a bolted repair.

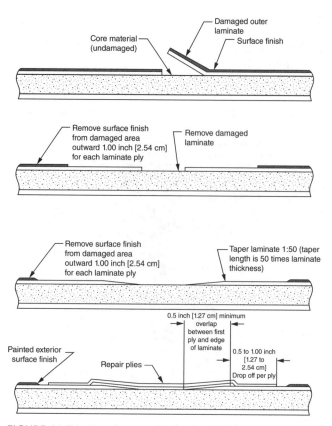

FIGURE 11-54 Repair scenarios for bonded foam cores.

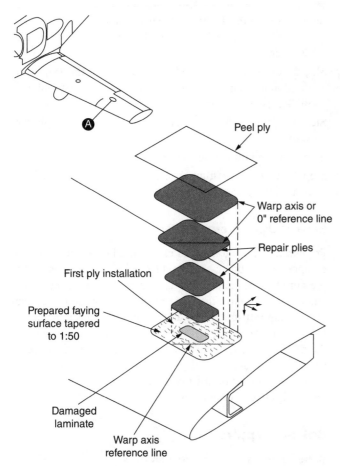

FIGURE 11-55   Placement of repair plies for bonded foam core repair.

Bonded Patch Repairs

Bonded repairs of carbon fiber laminate structures are often made with pre-preg materials but lower cure temperature alternative repair methods using wet lay-up are often available. The original structure is often scarved or stepped for a flush repair but external patch repairs could also be used. The advantage of the **stepped repair** is that the adhesive does not run down the bondline but a good step repair is hard to accomplish with manual tooling. Figure 11-56 shows an external patch repair and Fig. 11-57 shows a scarf-type repair. Figure 11-58 shows a stepped repair.

Resin Injection Repairs

Resin injection repairs are sometimes used to repair minor damage to aircraft structures. Resin injection repairs are used on lightly loaded structures for small damages to a solid

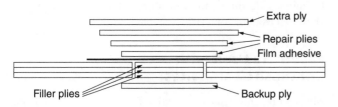

FIGURE 11-56   External patch repair for laminate structure.

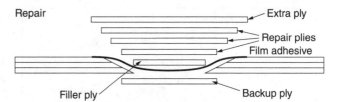

FIGURE 11-57   Scarf repair for laminate structure.

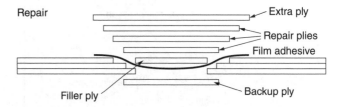

FIGURE 11-58   Stepped repair for laminate structure.

laminate due to delamination. Two holes are drilled on the outside of the delamination area and a low-viscosity resin is injected in one hole until it flows out the other hole. Resin injection repairs are sometimes used on sandwich honeycomb structure to repair a face sheet disbond. Disadvantages of the resin injection method are that the fibers are cut as a result of drilling holes, and it is difficult to remove moisture from the damaged area and to achieve complete infusion of resin. Field reports have indicated that resin injection repairs are often unsuccessful. Figure 11-59 shows the basic principle of a resin injection repair.

Composite Patch Bonded to Metal Structures

Composite materials can be used to structurally repair, restore, or enhance aluminum, steel, and titanium components. Bonded composite doublers have the ability to slow or stop fatigue crack growth, replace lost structural area due to corrosion grindouts, and structurally enhance areas. Boron epoxy, GLARE®, and graphite epoxy materials have been used as composite patches to restore damaged metallic wing skins, fuselage sections, floor beams, and bulkheads. Secondarily bonded pre-cured doublers and in situ cured doublers have been used on a variety of structural geometries ranging from fuselage frames to door cutouts to blade stiffeners. Vacuum bags are used to apply the bonding and curing pressure between the doubler and metallic surface. Film adhesives using a 250°F [121°C] cure are used routinely to bond the doublers to the metallic structure. Metal treatment for bonding is critical and care must be taken to create an active surface for bonding in addition to removing contaminates.

Repairs to Bonded Metal Structures

Some aircraft use a bonded metal construction. Adhesives are used to attach metal parts together instead of rivets and bolts. These metal structures could be repaired with techniques

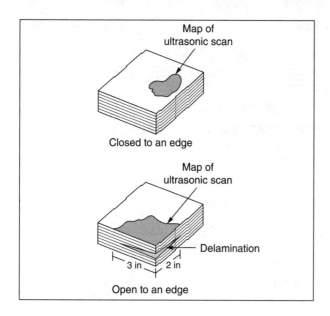

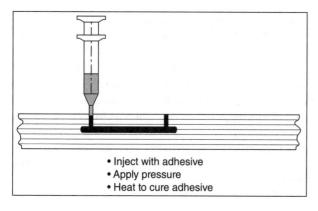

- Inject with adhesive
- Apply pressure
- Heat to cure adhesive

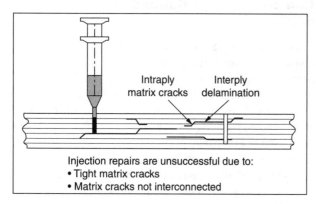

Injection repairs are unsuccessful due to:
- Tight matrix cracks
- Matrix cracks not interconnected

FIGURE 11-59 Resin injection repair for laminate structures.

similar to damage to riveted metal structures. A typical repair for the leading edge of a wing is shown in Fig. 11-60.

Step 1. Trim out the damaged area in a rectangular pattern and deburr.

Step 2. Place the repair doubler beneath the wing skin, as shown in Fig. 11-60.

Step 3. Holding the repair doubler in place with clamps, drill $\frac{1}{8}$-in holes through the wing skin, spacing the holes $\frac{5}{8}$ in apart, center to center.

Step 4. Secure the doubler to the wing leading edge with clecos and dimple the holes. Install blind rivets to secure the doubler.

Step 5. Place the preformed filler flush with the skin over the doubler. The filler must be of the same material and thickness as the skin.

Step 6. Hold the filler in place, drill holes, and dimple the skin.

Step 7. Install blind rivets to secure the filler.

Step 8. Use epoxy filler to create a smooth surface and restore the paint system.

Repairs Using Pre-cured Patches

Pre-cured composite patches can be used to repair damaged composite structures; see Fig. 11-61. The advantages of the pre-cured patch are that they can be made in advance in the autoclave, they have the same mechanical properties as the original aircraft structure, and it is relatively easy to bond the patch to the aircraft. The difficulty in this type of repair is the fit of the patch to the aircraft structure. The patch needs to follow the contour of the aircraft perfectly to accomplish a good bondline. For this reason these patches work best on flat or lightly contoured structures.

Bolted Repairs

Bolted repairs are often used on thick composite laminate structures such as primary wing, stabilizer, and fuselage structures on transport aircraft. The repair scenario for the bolted repair is similar to repairing an aluminum aircraft. Both composite patches and metallic patches are used. Titanium patches are often used with aircraft with a carbon fiber structure. The technician will have to prepare the external patch, layout the rivet pattern, drill the fastener holes, and install the fasteners. Advantages of the bolted repair are that technicians are already familiar with the repair technique, it is quicker to accomplish without the need for a temperature controlled environment, and heat cure is required. The main disadvantage of bolted repairs is that the new holes created in the parent structure weaken the structure by creating stress concentrations that become damage initiation sites. Figure 11-62 shows an example of a bolted repair. Conventional rivets installed with rivet guns, such as those used in a metal repair, cannot be used because the impact of the rivet gun will damage the composite structure. Laminate structure is often reinforced with stringers and stiffeners, and these are often damaged at the same time when the laminate skin is damaged. These stringers are often co-cured with the skin and cannot easily be replaced and need to be repaired. Repair techniques are similar to metal stringer repairs. Figure 11-63 shows an example of a stringer repair.

Fasteners Used with Composite Laminates

Many companies make specialty **fasteners** for composite structures and several types of fasteners are commonly used;

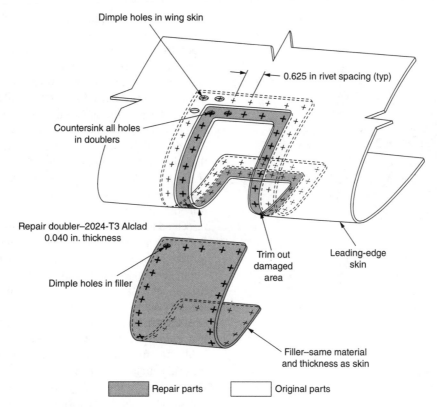

Dimple holes in wing skin

0.625 in rivet spacing (typ)

Countersink all holes
in doublers

Repair doubler–2024-T3 Alclad
0.040 in. thickness

Dimple holes in filler

Trim out
damaged
area

Leading-edge
skin

Filler–same material
and thickness as skin

Repair parts   Original parts

FIGURE 11-60   Leading edge repair of bonded metal structure.

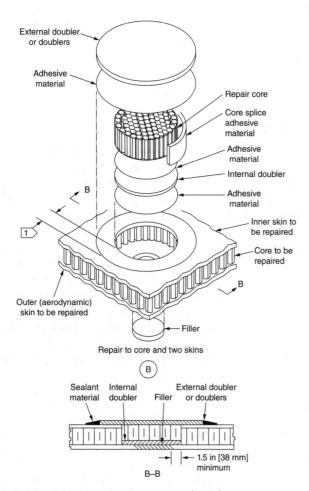

External doubler
or doublers

Adhesive
material

Repair core

Core splice
adhesive
material

Adhesive
material

Internal doubler

Adhesive
material

Inner skin to
be repaired

Core to be
repaired

Outer (aerodynamic)
skin to be repaired

Filler

Repair to core and two skins

Ⓑ

Sealant
material

Internal
doubler

Filler

External doubler
or doublers

1.5 in [38 mm]
minimum

B–B

FIGURE 11-61   Repair using procured patches.

threaded fasteners, lock bolts, blind bolts, blind rivets, and specialty fasteners for soft structures such as honeycomb panels. The main differences between fasteners for metal and composite structures are the materials and the footprint diameter of nuts and collars. Neither fiberglass-reinforced nor Kevlar fiber–reinforced composites cause corrosion problems when used with most fastener materials. Composites reinforced with carbon fibers, however, are quite cathodic when used with materials such as aluminum or cadmium, the latter of which is a common plating used on fasteners for corrosion protection.

Fastener Materials

For repair of composite parts the choice of fasteners is limited to titanium, Monel, or stainless steel. Titanium alloy Ti-6Al-4V is the most common alloy for fasteners used with carbon fiber–reinforced composite structures. Ti-3Al-2.5V and commercially pure titanium are used for some components of fastening systems. When higher strength is required, materials such as A286 or alloy 718 that have been strengthened by cold working can be used.

Drilling

Drilling holes in composite materials is different from drilling holes in metal aircraft structures. Different types of drill bits, higher speeds, and lower feeds are required to drill precision holes. Composite materials are drilled with drill motors operating between 2000 and 20 000 RPM, and a low feed rate is required. Drill motors with a hydraulic dash pod or other type

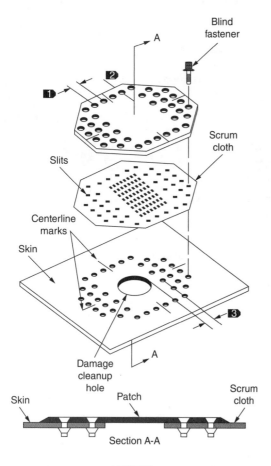

LEGEND

1 Fasteners must have minimum spacing of 4D and maximum of 6D

2 Fasteners must have minimum edge distance of 3D

3 Fasteners to edge of damage cleanup hole must be a minimum of 3D

Note:
Drill pilot holes (0.128-in diameter) in patch first. Enlarge holes to final size after transferring pilot holes to composite skin.

FIGURE 11-62 Example of a bolted repair.

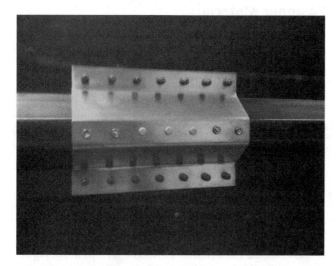

FIGURE 11-63 Stringer repair with titanium formed channel. See also color insert.

of feed control are preferred because they restrict the surging of the drill as it exits the composite materials. Structures made from carbon fiber and epoxy resin are very hard and abrasive and special flat flute drills or similar four-flute drills are required. Aramid fiber (Kevlar)/epoxy composites are not as hard as carbon but are difficult to drill unless special cutters are used because the fibers tend to fray or shred unless they are cut clean while embedded in the epoxy. Special drill bits with "clothes pin" points and "fish tail" points have been developed that slice the fibers prior to pulling them out of the drilled hole. If the Kevlar/epoxy part is sandwiched between two metal parts, standard twist drills can be used. The quality of the drilled hole is critical; therefore, each hole should be visually inspected to ensure correct hole size and location and proper normality, as well as the absence of unacceptable splintering or fiber breakout, fiber pullout, delaminations, back counterboring, evidence of heat damage, and microcracking. Delamination, separation of bonded composite plies, may occur at the drill entrance due to too fast of a feed rate for the speed, or at the drill exit due to too much feed force on unsupported materials. Splintering or fiber breakout conditions, fibers that break free from the surrounding resin matrix, may occur at the drill entrance or exit, especially in unidirectional material. Figures 11-64 show drill types used for composite materials. Larger holes can be cut with diamond-coated hole saws or fly cutters, but only use fly cutters in a drill press and not in a drill motor. A hole saw can easily remove damage from 1 to 3.5 inches in diameter in a very neat, uniform manner. The material being cut should be well supported and firmly backed up with wood or an aluminum plate to prevent backside breakout. Hole saws may also be used to cut through the core for damage removal, but only when the damage removal cut is to be made perpendicular with the skin.

Drill Types

Drilling and trimming of composite materials is difficult because standard tool steels will dull rapidly in the process. Problems can occur because of the way drills are shaped and sharpened. As a drill dulls, it tends to push against the material rather than cut, causing layer delimination. The drill bit should be shaped in a spade form (Fig. 11-64a) or a long tapered form sometimes referred to as a dagger drill (Fig. 11-64b). Diamond-tipped equipment will allow more cuts to be taken per tool. When drilling carbon fiber material, it is best to use a high-speed, automatic feed drill motor. Kevlar is difficult to drill and a special ground drill bit is used (Fig. 11-64c). These sickle-shaped Klenk drill bits will first pull on the fibers and then shear them, which results in a better quality hole. This type of drill bit is best when drilling aramid materials as it minimizes fiber fuzz.

Countersinking

Countersinking a composite structure is required when flush head fasteners are to be installed in the assembly. For metallic structures, a 100° included angle shear or tension head fastener has been the typical approach. In composite structures, two types of fasteners are commonly used: a 100° included

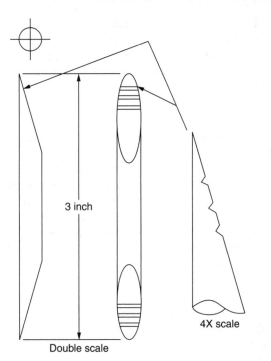

3 inch

Double scale

4X scale

FIGURE 11-64a Dagger bit.

FIGURE 11-64b Spade drill.

FIGURE 11-64c Klenk point bit for aramid (Kevlar) fibers (brad point).

angle tension head fastener or a 130° included angle head fastener. Carbide cutters are used for producing a countersink in carbon/epoxy structure. For carbon/epoxy structure, the countersink cutters usually have straight flutes similar to those used on metals. For Kevlar fiber/epoxy composites, S-shaped positive rake cutting flutes are used. A microstop countersink gauge should be used to produce consistent countersink wells.

Repair Safety

Composite materials can often be **toxic**. The typical health areas at risk are vision, dermal (skin), and respiratory. Resins, solvents, initiators (i.e., MEKP), dust, and particles commonly cause visual risks. Visual problems are accompanied by symptoms such as watering, redness, swelling, itching, and burning of the eyes. Chemical goggles and glasses with side shields are effective in protecting the eyes from direct contact with the risks mentioned above. However, vapors coming from these materials can still be harmful. Dermal risks are commonly caused by fibers, coatings, chemicals (resin, solvents, and mold release), foams, and dust. Dermal problems are accompanied by symptoms such as redness, rash, itching, burning, and dryness or cracking of the skin. Protection from these risks are

done by covering exposed skin, gloves (latex and other types), and barrier cream. Respiratory risks are commonly caused by (1) fumes and vapors (chemical and generated by grinding) and (2) dust and particles (grinding and glass bubbles). Respiratory problems are accompanied by symptoms such as headache, nausea, tightness in the chest, troubled breathing, shortness of breath, and dizziness. Avoiding prolonged exposure and selecting products with the least amount of toxicity can help reduce these risks. Other protective equipment such as a dust mask (dust and particle protection only), charcoal respirator (vapor protection for a limited period only), and supplied air mask can help in reducing respiratory risks.

Personal protection equipment (PPE) (eye protection, gloves, aprons, respirators, etc.) should be worn during handling and repair of advanced composite materials. The PPE required to safely handle a specific material is provided in the supplier's material safety data sheet (MSDS) and should be consulted to determine the PPE to use. During composite repair work the skin needs to be protected from hazardous materials. Always wear gloves and clothing that offer protection against toxic materials. Use only approved gloves that protect your skin and do not contaminate the composite material. Always wash your hands prior to using the toilet or eating. Chemicals could remain on your hands that will burn sensitive skin. Make sure that the gloves do not contaminate your composite materials. Damaged composite components should be handled with care. Single fibers can easily penetrate the skin, break off, and become splinters lodged in the skin. Sharp tweezers and a magnification lens are generally required to remove the splinter. If the splinter is hard to remove, seek medical attention immediately. Accidental chemical spills, runaway exothermic reactions, and heat blanket fires are some of the potential emergencies that can be encountered during the manufacturing and repair process.

Vacuum downdraft tables, power tools with vacuum attachments, and portable vacuum systems should be used when drilling and trimming operations are performed on cured composites. Dust is created during composite processing that could cause respiratory problems that may not manifest themselves for years. Using compressed air to blow composite dust should be avoided. Blowing the dust results in the dust mixing with the air being breathed, resulting in ingestion of the materials through the lungs. Do not use compressed air to clean dust from the body or clothing. This may force dust particles into the pores of the skin.

Fire Protection

Most solvents used for composite repair are flammable. Close all containers with solvents and store in a fireproof cabinet when not used. Make sure that solvents are kept away from areas where static electricity can occur. Static electricity can occur during sanding operations or when bagging material is unrolled. It is preferable to use air-driven tools; if electric tools are used make sure that they are of the enclosed type. Do not mix too much resin. The resin could overheat and start smoking caused by an exothermal process. Always make sure that a fire extinguisher is nearby to fight a fire.

1. Describe a composite material.
2. List some examples of aircraft components that are made of composite materials.
3. What is a *quasi-isotropic lay-up*?
4. What is the typical fiber orientation of a quasi-isotropic lay?
5. Explain the difference between *laminate symmetry* and *laminate balance*.
6. What are the three most common fiber materials used for aerospace structures?
7. Explain why lightning protection fibers and materials are necessary for composite aircraft.
8. Explain the differences between *tows*, *yarns*, and *rovings*.
9. What is the difference between tape and fabric products?
10. List the most common types of weave styles.
11. Describe the role of the *matrix*.
12. What are the two main classes of resin systems?
13. Name the most common types of *thermosetting* resins.
14. Explain the three curing stages of thermosetting resin systems.
15. Explain briefly what *thermoplastic* resin systems are and how they differ from thermosetting resin systems.
16. What is a *prepreg material*?
17. Name three types of adhesive systems used in the aerospace industry.
18. Explain the difference between a *honeycomb sandwich* and a *laminate structure*.
19. When is flexcore used as a core for a part?
20. Name the typical types of manufacturing damage.
21. Name the typical types of in-service damage.
22. Explain the difference between a delamination and a debond.
23. Why are core drying cycles necessary before a part can be repaired?
24. What types of tools are used when performing a visual inspection?
25. Explain the *coin tap inspection* method.
26. Explain the *ultra sonic inspection* method.
27. What is the difference between a pulse echo, through transmission, and phase array ultrasonic inspection?
28. What type of damage can be found with an X-ray inspection?
29. Explain the basic principle of thermography.
30. What type of defects can be found with thermography?
31. Name the three *damage* categories.
32. Explain why air tools should be used instead of electric tools.

33. What kind of processing materials are used to vacuum bag a composite repair?
34. In what situation are *peel plies* used?
35. Explain what needs to be done if the vacuum bag is "bridging."
36. Explain the *vacuum bag technique* used to repair aircraft structures.
37. How can the bleeding of resin during a cure of a wet lay-up be controlled?
38. Explain what a *no-bleed system* is.
39. Explain the *envelope vacuum bag technique*.
40. Explain the *heat shrink method* to apply pressure to a lay-up.
41. Briefly explain the *wet lay-up* process.
42. What type of materials must be stored in a freezer to retard curing?
43. What is the typical storage life of prepreg?
44. Explain the differences between *co-curing*, *co-bonding*, and *secondary bonding*.
45. What are the disadvantages of room temperature curing?
46. What type of resin systems need to be cured at an elevated temperature?
47. Name four types of curing equipment to cure composite materials.
48. Describe what an *autoclave* can control during the curing process.
49. Why are *heat lamps* not efficient above 150°F?
50. *Heat blankets* are used with what type of curing equipment?
51. Composite repairs can be divided into what two groups?
52. What are the advantages of a *bonded repair*?
53. What are the advantages of a *bolted repair*?
54. What type of repair is typically used for *honeycomb sandwich structure*?
55. Explain the *potted repair process*.
56. Describe the repair steps used for a *bonded repair*.
57. What are the advantages of a *stepped repair*?
58. What are the disadvantages of a *resin injection repair*?
59. Describe the repair steps for a bolted repair of a composite structure.
60. What is the difference between *fasteners* used for metal and composite structures?
61. What type of fastener material is used for aircraft using a carbon fiber structure?
62. What type of drill bit is used to drill Kevlar?
63. Describe what type of personal protective equipment must be worn by the technician when handling composite materials.

Assembly and Rigging 12

INTRODUCTION

The **assembly** of an aircraft refers to the joining of parts or subassemblies by various means until the entire aircraft is in condition for operation. **Rigging** is the alignment of aircraft parts or sections to obtain proper flight characteristics. A certain amount of rigging is necessary during the assembly of an aircraft, but even after final assembly, certain rigging adjustments must be made. Thus, there is some overlap between the assembly operation and the rigging operation.

AIRCRAFT ASSEMBLY

Assembly Procedures

Major assemblies, such as the fuselage, wings, engine nacelles, and landing gear, are normally constructed as complete subunits. These subunits are inspected for airworthiness and for conformance to applicable specifications. During the manufacture of an airplane, the inspected and approved subunits are brought into the final assembly line at specified stations, where they are joined to the main structure.

The actual assembly of the aircraft varies greatly according to the type, make, and model, but a few general principles apply. In a factory, all operations can be planned for the greatest efficiency, but in a repair shop, where the jobs vary to a great extent, this type of planning cannot be done as easily.

Although the assembly of different aircraft requires many varying procedures, some basic guidelines should be followed:

1. Always use the correct type and size of hardware. For example, standard bolts should not be used where high-strength bolts are required. A low-strength bolt could result in component separation in flight. Do not use a stack of washers to make up for a bolt being too long. Either use a bolt of proper length or determine what other fitting should be installed on that bolt.

2. Insert bolts and clevis pins in the proper direction, especially in the area of moving structures. If a bolt is installed backward, the threaded end and the nut may interfere with the operation of a control surface.

3. Use only the correct type of hardware safety mechanism. Unless specified by the manufacturer, fiber locknuts should not be used where the bolt will be subject to rotation, such as when used as a control surface hinge. This can cause the nut to back off the bolt. Use the recommended safety mechanism, usually a cotter pin and, occasionally, safety wire.

4. If no direction for bolts to be inserted is specified or indicated by component configuration, they should be installed with the bolt head up or forward. This will reduce the chance of the bolt falling out of a fitting if the nut should come off.

5. Never force components together without checking to determine why the force is necessary. Some components are a push-fit, but exercise caution when performing this type of assembly to avoid damage to components.

6. Check the location and routing of all cables, hoses, tubing, etc., before assembling components. This will help prevent crossed control cables and the mixing of various fluid lines.

7. Follow all recommended safety precautions when working with aircraft on jacks, hoists, and assembly supports. Do not allow anyone to be in a position where he or she could be injured if a support or lifting mechanism should fail.

If the aircraft has been disassembled for repair or shipment, it should be cleaned after disassembly and all the attaching hardware checked for condition and replaced as necessary. When the structure is disassembled, all the attaching hardware in airworthy condition should be installed in the fittings on one of the separated structures. This avoids the problem of trying to locate the correct bolt, washer, or screw from a pile of hardware removed from the aircraft.

An assortment of tapered drifts is useful for lining up bolt holes. If a drift of the correct size is not available, an undersized bolt can be used for this purpose. During assembly operations, the technician must not distort or overstress any part in order to make the bolt holes line up. If parts will not fit together properly, the cause of the misfit must be determined and corrected.

The disassembly and assembly of complex parts and subunits of an aircraft are usually clearly described in the manufacturer's maintenance manual. In every case where manufacturer's instructions are available, the technician should follow them in detail.

FIGURE 12-1 Fuselage cradle. (*Piper Aircraft Co.*)

The final assembly of an aircraft usually starts with the fuselage subassembly and progresses through various subassemblies until the entire aircraft is assembled. After the fuselage is assembled, it is usually placed in a jig or on a fixture or jig for support, as shown in Fig. 12-1, while the other subassemblies are attached to it. Depending upon the particular design and type of aircraft, assemblies such as the tail section, wings, landing gear, and other parts are joined to the fuselage.

A fuselage is provided with fittings or attachments for the purpose of hoisting or jacking. These fittings may be a permanent part of the aircraft or they may be separate attachments that can be installed when it is necessary to lift the aircraft. When it is necessary to hoist or jack the aircraft, the technician must make sure that the correct procedure is used and that the proper fittings are available. The manufacturer's manual gives detailed instructions for handling the aircraft.

In some cases, the aircraft fuselage may be hoisted by means of a **sling**. In fastening a lifting sling to a fuselage, the attachments must be correctly located. The "station points" on either the fuselage or the engine mount are often used as lifting points; however, the lifting points designed by the manufacturer should be used when available. The sling may attach to fittings on the upper portion of the fuselage, or it may be constructed of webbing that passes under the fuselage at specified points and is connected to the hoisting fittings at the top as shown in Fig. 12-2. If the aircraft is to be lifted by a lifting strap placed around the fuselage, the strap should be well padded to avoid surface damage. The strap should be placed at a reinforced position such as at a bulkhead or former so that undue stress will not be placed on unreinforced skin areas, causing structural damage.

Following the assembly of the fuselage, the next component normally assembled is the landing gear. The landing gear may attach directly to the fuselage or it may attach to the wing assembly or engine mount. The landing gear of an aircraft includes the wheels, together with their shock absorbers and devices necessary for the attachment of the gear to the wings or fuselage. To attach the main landing gear to a fuselage, the fuselage is raised high enough so that the gear can be attached. The best way to do this for a particular aircraft is given in the manufacturer's instructions. In some cases a hoist is used; in others the fuselage is jacked up or placed in a suitable jig or cradle.

After the landing-gear struts are properly attached to the aircraft, all fittings are tightened and safetied. If bonding connections are necessary, these are secured. Hydraulic lines for the brakes and electrical wiring for safety switches are connected and secured. Other attachments are installed as required. When the gear is completely installed with wheels in place and tires properly inflated, the airplane weight can be supported on the gear.

Following assembly of the landing gear to the aircraft, the subsequent assembly operations can vary considerably. The manufacturer's manual may specify the sequence of assembly events, or the choice may be up to the technician. In some cases it may be preferable to install the empennage before installing the wings; for other aircraft, the reverse may be better.

The methods of attaching the wings will vary. In most cases the wings will attach to a wing center section or fuselage structure assembly. The method of attaching the wings will vary from a simple four-bolt attachment, as shown in Fig. 12-3, to a multiple-point attachment, as is illustrated in Fig. 12-4. An example of how the various empennage components are installed is shown in Fig. 12-5.

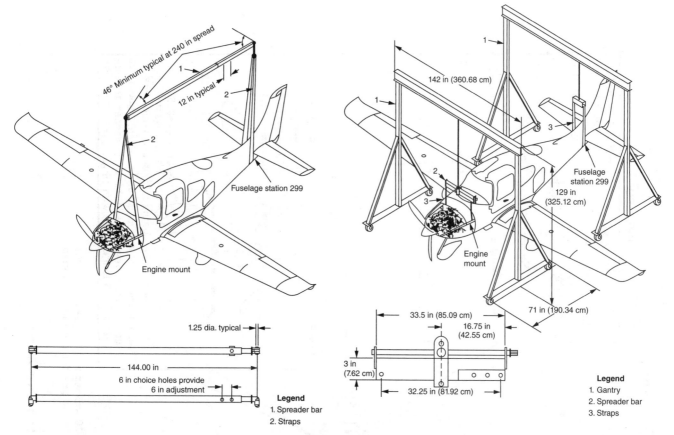

FIGURE 12-2 Aircraft hoisting. (*Cirrus Aircraft Corporation.*)

AIRCRAFT RIGGING

Aircraft rigging involves two principal types of operations. First, the aircraft structure must be rigged for correct alignment of all fixed components. The fuselage is aligned at the time of manufacture in the assembly jigs. All parts are correctly positioned in the assembly jig and then they are riveted, bolted, or welded into a complete assembly. Some types of fuselages require realignment at major overhaul periods or after damage. Wings and other large structures are aligned and assembled in jigs and fixtures to ensure correct shape and positioning of attachment fittings. When the major components are assembled, they are aligned with each other.

The second type of rigging is the alignment of control surfaces and the controls that move the surfaces. These operations require the adjustment of cable length, cable tension, push-pull rods, bellcranks, cable drums, and various other parts. Angular deflection of control surfaces must be measured with protractors or other measuring devices to ensure that the movements comply with the appropriate specifications.

Leveling the Aircraft

When rigging an aircraft, it may be necessary to establish the aircraft in a level attitude prior to checking and adjusting wings and control surfaces. The various aircraft manufacturers have devised several methods by which the technician can establish the level attitude of the aircraft. Keep in mind that the level attitude of the aircraft includes both a longitudinal level and a lateral level position. Once the aircraft is level longitudinally and laterally, the components can be rigged.

One method used on many light aircraft is to set a spirit level on a longitudinal structural member to establish the longitudinal level position and another level across specific structural members to establish the lateral level position, as shown in Fig. 12-6. This same basic procedure is accomplished in some aircraft by the installation of two nut plates on the side of the fuselage. Screws can be placed in these nut plates, and longitudinal level is determined when a spirit level placed on the extended screws is level. This arrangement is shown in Fig. 12-7.

Some aircraft make use of a plumb bob and a target to establish the aircraft level on both axes. This is done, as shown in Fig. 12-8, by suspending a plumb bob from a specified structural member and adjusting the aircraft until the plumb bob is centered on the target.

Another method used is to attach a permanent spirit level to the aircraft for each of the two axes. These levels are normally located in an equipment compartment or a wheel well and may have an accuracy as great as $\frac{1}{8}°$.

If an aircraft is not level, the aircraft may be leveled by the use of supports under the aircraft, such as jacks or tail stands. The inflation of tires and struts can be adjusted, as can the fuel loads in wings and fuselage tanks.

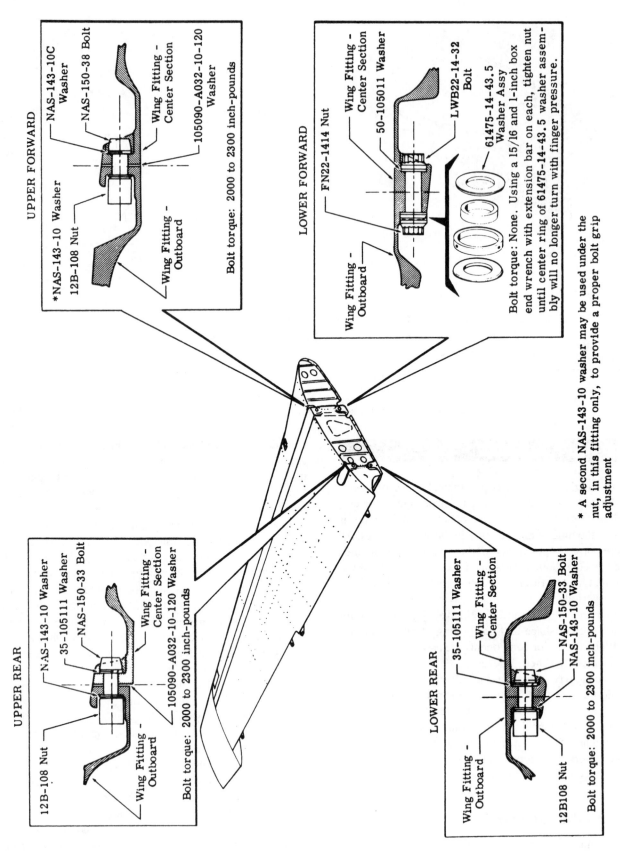

UPPER FORWARD

*NAS-143-10 Washer
NAS-143-10C Washer
12B-108 Nut
NAS-150-38 Bolt
Wing Fitting - Center Section
Wing Fitting - Outboard
105090-A032-10-120 Washer

Bolt torque: 2000 to 2300 inch-pounds

LOWER FORWARD

FN22-1414 Nut
Wing Fitting - Center Section
50-105011 Washer
LWB22-14-32 Bolt
Wing Fitting - Outboard
61475-14-43.5 Washer Assy

Bolt torque: None. Using a 15/16 and 1-inch box end wrench with extension bar on each, tighten nut until center ring of 61475-14-43.5 washer assembly will no longer turn with finger pressure.

UPPER REAR

12B-108 Nut
NAS-143-10 Washer
35-105111 Washer
NAS-150-33 Bolt
Wing Fitting - Center Section
Wing Fitting - Outboard
105090-A032-10-120 Washer

Bolt torque: 2000 to 2300 inch-pounds

LOWER REAR

35-105111 Washer
Wing Fitting - Center Section
NAS-150-33 Bolt
NAS-143-10 Washer
Wing Fitting - Outboard
12B108 Nut

Bolt torque: 2000 to 2300 inch-pounds

* A second NAS-143-10 washer may be used under the nut, in this fitting only, to provide a proper bolt grip adjustment

FIGURE 12-3 Wing attachment fittings. (*Beech Aircraft Co.*)

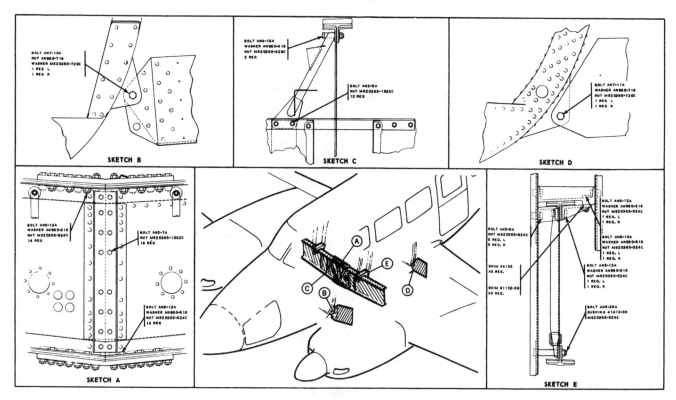

FIGURE 12-4 Wing installation. (*Piper Aircraft Co.*)

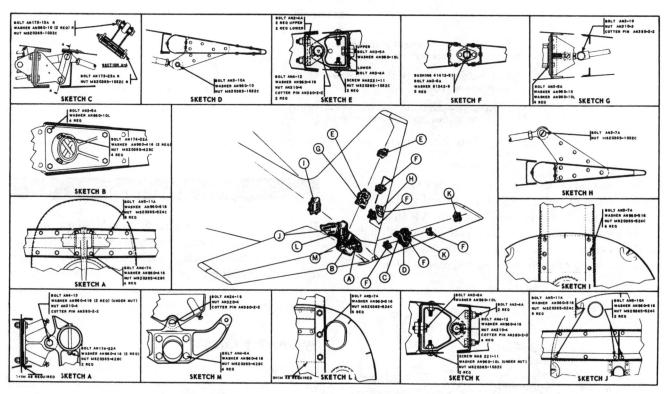

FIGURE 12-5 Empennage component installation. (*Piper Aircraft Co.*)

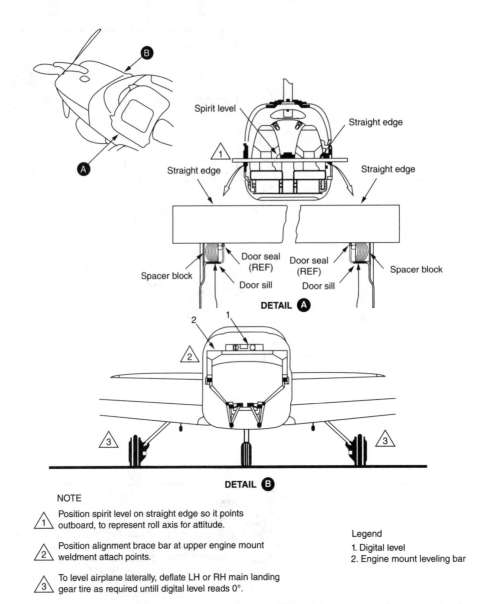

Spirit level

Straight edge

Straight edge

Straight edge

Door seal (REF)

Door seal (REF)

Spacer block

Door sill

Door sill

Spacer block

DETAIL A

DETAIL B

NOTE

⚠1 Position spirit level on straight edge so it points outboard, to represent roll axis for attitude.

⚠2 Position alignment brace bar at upper engine mount weldment attach points.

⚠3 To level airplane laterally, deflate LH or RH main landing gear tire as required untill digital level reads 0°.

Legend
1. Digital level
2. Engine mount leveling bar

FIGURE 12-6 Use of spirit levels to establish the aircraft level. (*Cirrus Aircraft Corporation.*)

Regardless of the leveling means used, care must be taken not to place a support or jack at a structurally weak area, such as on a skin panel between bulkheads. Additionally, if the air pressure is reduced in tires or struts to establish the level attitude or if fuel is transferred for this purpose, the aircraft must be properly serviced before it is returned to operation.

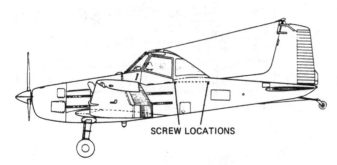

SCREW LOCATIONS

FIGURE 12-7 Special nut plates and screw locations may be used to establish the aircraft level. (*Cessna Aircraft Co.*)

FIXED-SURFACE ALIGNMENT

The fixed surfaces of an aircraft include wings, stabilizers, and tailbooms. These components must be checked for alignment with the fuselage and with each other to determine if the aircraft is properly rigged. This process involves a symmetry check longitudinally and laterally and angular adjustments such as incidence and dihedral.

Symmetry Check

From time to time, it is necessary or advisable to check the alignment of a fully assembled aircraft. A misaligned condition can seriously affect the flying qualities of the aircraft. Misalignment may not be apparent by casual observation and may be found only by alignment checks using instruments. Alignment checks will normally also be made after any major structural repairs or after the airplane has been subjected to severe conditions, such as

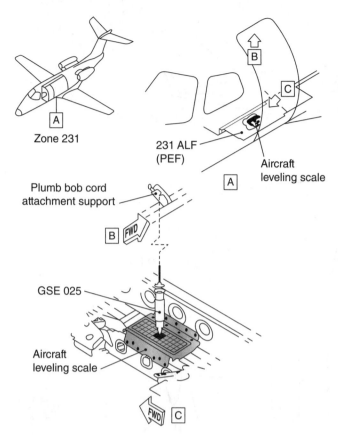

FIGURE 12-8 Leveling the aircraft with plumb bob. (*Embraer SA.*)

heavy landings, extreme turbulence, overspeeding, or violent maneuvers.

Alignment should also be checked if any of the following conditions are observed: unusual flight characteristics, wrinkled or buckled skin, areas of loose or sheared fasteners, or areas of badly fitting panels or inspection plates. If a pilot reports poor flight characteristics that cannot be corrected by rerigging of the controls, it is possible that the fuselage and wings are out of alignment or that a fixed surface is out of alignment with the aircraft. A **symmetry check** is used to verify this alignment.

Aircraft symmetry is determined by first leveling the aircraft and then measuring the distances from reference points on the aircraft central axis to reference points on the adjustable components. Figure 12-9 shows the symmetry lines used for one specific airplane. Alignment points are numbered 1 to 6, left and right. As can be seen from the drawing, measurements are made from points 1, left and right, to points 3, left and right; from points 2, left and right, to points 5, left and right, and point A. Other measurements are made as shown. If all measurements are within the tolerances given in the maintenance manual, the aircraft is in correct alignment. If any of the measurements are not within tolerance, the alignment must be adjusted as specified.

The vertical symmetry of an aircraft can be checked in a like manner by using a measuring tape to check the distance between the tip of the vertical stabilizer and the horizontal stabilizer, as shown in Fig. 12-9.

When checking an aircraft's symmetry, a drawing of the aircraft is helpful so that the values measured can be recorded and the lines of measurement indicated. Some technicians prefer to lay out the reference points on the floor of the hangar, as shown in Fig. 12-10. A plumb bob is suspended from the reference point and the position is marked with chalk. The measurements are then taken from these floor markings. This method is especially useful for large aircraft. As adjustments are made, a plumb bob suspended from the reference point immediately indicates the amount of adjustment.

When an aircraft has suffered major structural damage to the fuselage, wings, or control surfaces, it is necessary to partially disassemble the structure in order to make repairs. In such cases, it is common practice to mount the structure in a rigid steel frame (jig) to keep it in alignment while reassembling it.

One of the first steps in preparing to mount a structure in a jig is to establish an undamaged mounting point or points on the structure that can be used as a reference for locating the structure accurately and in proper alignment. Such mounting points could be the landing gear attachment points, the wing attachment points, or some other points strong enough to support the structure. When the primary reference point has been established, other points are established from reference to blueprints or the manufacturer's dimensional specifications.

When a fuselage or any other structure is secured in the jig with all attachment points accurately located, assembly can begin. During the process of replacing damaged parts such as skins, formers, and stringers, the structure should be checked from time to time for alignment. This will ensure that when the assembly is completed, the structure will be in alignment.

Effects of Rigging on Flight

The purpose of rigging an aircraft correctly is to attain the most efficient flight characteristics possible. A properly rigged and trimmed airplane will fly straight and level, "hands off," at its normal cruising speed. If air currents should disturb the stable flight attitude, it will correct itself and resume straight and level flight.

If the aircraft is out of rig, meaning that components are not properly aligned, then the total drag of the aircraft will be increased or the amount of control movement available will not provide the correct response. For example, if the wing chord line is not properly aligned in relation to the fuselage longitudinal axis, the drag of the aircraft will be increased, resulting in a higher power setting being required to overcome the drag, or if the power setting is not increased, the airspeed will be lower. If flight controls do not have enough range of motion, the pilot may not be able to control the motion of the aircraft properly. If the range of control surface travel is too great, the pilot may be able to deflect the control surface enough to place excessive stresses on the structure, resulting in failure of the structure.

The correct rigging of the wings and control surfaces on an aircraft is essential to the stability of the aircraft in flight. Stability around the longitudinal axis is provided by rigging the

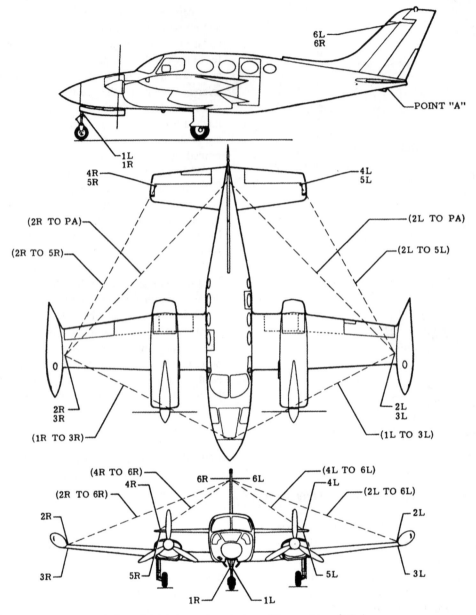

FIGURE 12-9 Alignment measurements for a light-twin-engine airplane. (*Cessna Aircraft Co.*)

wings and ailerons correctly. The wings on a monoplane are rigged for *dihedral*, *angle of incidence*, and *washin* or *washout*. **Dihedral** is the angle between the lateral plane of a wing and the horizontal plane. It is adjusted by raising or lowering the outer end of the wing. On an aircraft with cantilevered wings, the dihedral is built into the wing and is not usually adjustable. Dihedral produces stability around the longitudinal axis because of the difference in vertical wing lift when the aircraft rolls slightly off of level flight. When a wing with dihedral lowers in flight, the wing angle with the horizontal plane decreases and the vertical wing lift increases. At the same time, the wing on the opposite side of the aircraft is increasing its angle with the horizontal, and its vertical lift decreases. Thus the low wing will rise and the high wing will lower.

The **angle of incidence** of a wing is the angle formed by the intersection of the wing chord line and the horizontal axis of the aircraft. The angle of incidence must be correct if

the aircraft is to fly most efficiently. An aircraft is designed with an angle of incidence that will produce maximum lift and minimum drag at normal cruising speed. If the angle of incidence is not correct, the drag increases. The fuselage will not be in perfect alignment with the flight path but will be slightly nose up or nose down, depending upon whether the angle of incidence is too small or too great.

A complete discussion of the effects of dihedral and the angle of incidence can be found in the text, *Aircraft Basic Science*.

Checking and Adjusting Dihedral

To check the dihedral angle of a wing, the fuselage should be leveled. A dihedral board can then be used at a specified location to check the amount of dihedral in a wing. The use of a dihedral board is shown in Fig. 12-11.

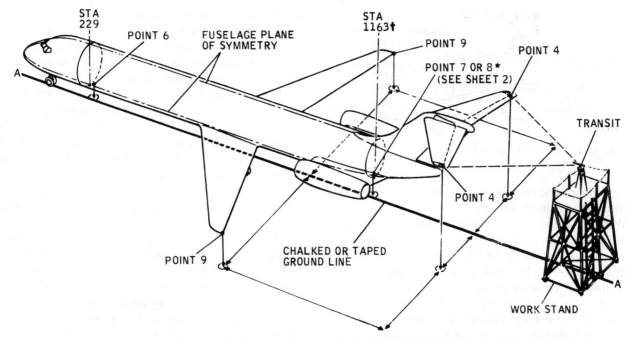

FIGURE 12-10 Wing and empennage alignment. (*Douglas Aircraft Co., Inc.*)

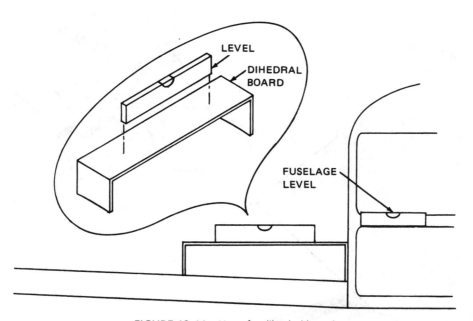

FIGURE 12-11 Use of a dihedral board.

Some light aircraft use a vertical measurement to determine dihedral. In one type of high-wing aircraft, a string is stretched from one wing tip to the other wing tip, centered on the front spar. When the string is drawn tight, the distance from the string to the top of the fuselage is measured to determine if the dihedral is correct.

Some aircraft with cantilever wings have the wings manufactured without any provision for adjustment of the dihedral or angle of incidence. This is accomplished by the use of manufacturing jigs, which ensure that all wing and fuselage joints will be identical, thereby ensuring the correct angles will be set. Other wing and fuselage fittings allow for some change in the angle of incidence of the wing during installation.

Semicantilever wings have external braces, either wires or struts. Because struts are the most common arrangement, they are addressed here, and the adjustment of wires is addressed in the section dealing with biplanes.

Many struts have fittings threaded into the end of the strut that allow the length of the strut to be changed. Thus the dihedral can be adjusted by increasing or decreasing the length of the strut.

Checking and Adjusting Incidence

To determine if the angle of incidence for the wing or horizontal stabilizer is correct, the fuselage is leveled and an incidence board, as shown in Fig. 12-12, is used, or a universal

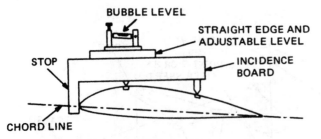

FIGURE 12-12 An incidence board.

protractor is placed on a designated surface of the wing or stabilizer and the angle of incidence is measured.

Washin is an increase in the angle of incidence of the wing from the root to the tip. If a wing has an angle of incidence of 2° at the root and an angle of incidence of 3° at the outer end, it has a washin of 1°. **Washout** is a decreasing angle of incidence from the root to the tip of a wing. Washin and washout are employed to give the wings on each side of an aircraft a slightly different amount of lift in order to aid in counteracting the effect of engine and propeller torque. Washout is also used to provide improved stall characteristics. If the angle

of incidence is greatest at the wing roots, the root sections of the wings will stall before the tips. This characteristic gives the pilot warning of a stall before the complete stall occurs; the pilot will still have control of the aircraft with the ailerons and be able to make any desired corrections.

Washing a wing "in" or "out" is the process of changing the angle of incidence of the wing. Washing out a wing decreases the lift a wing produces. Washing in a wing increases the lift the wing produces. If an airplane flies, hands off, with the right wing low, the right wing can be washed in and the left wing can be washed out to correct this condition.

On a semicantilever wing the washin and washout of the wing can be adjusted by changing the length of a strut if a strut is used for the rear spar and a strut is used for the front spar. By changing the length of only one of these struts through the adjustment fitting, the twist in the wing will change.

Some cantilevered wings also provide a means for changing the angle of incidence, as is shown in Fig. 12-13. This adjustment uses an eccentric bushing fitted into the rear spar fuselage attachment fitting. By rotating the bushing, the rear spar is raised or lowered slightly, changing the angle of incidence of the wing. Once the wing is adjusted,

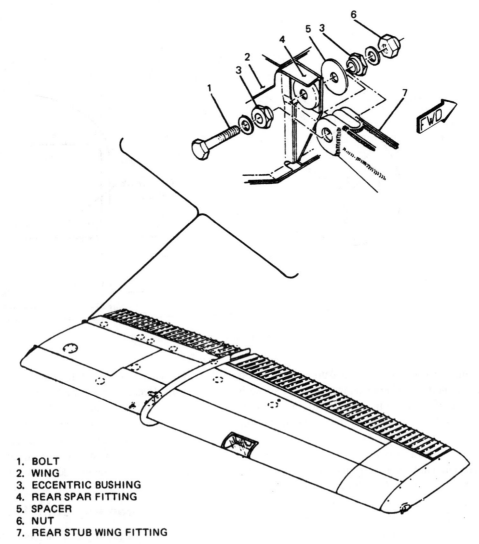

1. BOLT
2. WING
3. ECCENTRIC BUSHING
4. REAR SPAR FITTING
5. SPACER
6. NUT
7. REAR STUB WING FITTING

FIGURE 12-13 An eccentric bushing is used to adjust the wing angle of incidence. (*Cessna Aircraft Co.*)

the attachment bolt is fitted through the bushing and the bushing no longer rotates.

Biplane Wing and Fuselage Alignment

Although the biplane is considered by many to be an obsolete type of aircraft, there are hundreds of these airplanes still in operation for pleasure, sport, and agricultural work. A typical biplane is shown in Fig. 12-14.

The information a technician must have for rigging a biplane includes *stagger, angle of incidence, dihedral angle,* and *decalage.*

FIGURE 12-14   A typical biplane.

Stagger is the longitudinal difference in the positions of the leading edges of the wings of a biplane. If the leading edge of the upper wing is ahead of the leading edge of the lower wing, the stagger is *positive.*

Decalage is the difference between the angles of incidence of the upper and lower wings. If the upper wing has a greater angle of incidence than the lower wing, the decalage is said to be *positive.*

The first step in rigging a biplane is to level the fuselage, both laterally and longitudinally, in a location free from wind. The front of the fuselage should be supported in a cradle or by means of jacks. If the fuselage is supported by means of the landing gear, the flexibility of the shock struts and tires will permit movement, which is not desirable.

The second step in rigging the biplane is to install the **center section** of the wing on the fuselage. The center section is lifted into position above the fuselage, either by means of a suitable hoist or by hand. Depending upon the size of the center section, either two or four persons are required to lift the center section into position safely. While the center section is being held in position, the **cabane struts** and **stagger struts** are attached, either with temporary bolts or the regular assembly bolts. The **cross-brace** wires should also be installed at this time in order to support the center section laterally.

The next step is to adjust and secure the center section. It must be adjusted for lateral position, stagger, and angle of incidence. These adjustments must be made in an orderly sequence, and secondary adjustments may have to be made because of the effect of one adjustment upon another. The

first operation is to adjust the cross-brace wires until the center section is aligned with the center line of the fuselage. The symmetry of the center section with the fuselage is checked with a plumb line, a straightedge, a spirit level, and a steel measuring tape or scale. When the spirit level is placed laterally across the top of the center section, a level reading should be obtained. Plumb bobs dropped from identical points on each end of the center section should indicate that the center section is centered laterally. Measuring from the plumb line to reference points on the fuselage will show whether the center section is centered.

The stagger of the center section may or may not be adjustable. If fixed-length stagger struts are used on the airplane, no adjustment can be made. However, if the stagger struts are adjustable or if the airplane has **stagger** and **drift wires** between the cabane struts, adjustment must be made. Adjustment of the stagger struts or wires will move the center section forward or aft. The position is checked by dropping a plumb line from the leading edge of the center section on each side of the fuselage and measuring the distance from the plumb line to a fixed reference point, such as the front fitting for the lower wing.

The final check on the rigging of the center section is to see that all wires have the correct tension, all lock nuts are properly tightened, and the correct bolts are installed in all fittings. Safetying of nuts and bolts must be checked.

The next step in the rigging of the biplane is the attachment of the lower wings. A padded support is placed where it can support the outer end of the lower wing, and then the wing is lifted into place and attached at the fuselage fittings. The wing attachment bolts are inserted with the heads forward as a standard practice.

The **interplane struts** are attached to fittings on the top of each lower wing near the outer end. The struts must be held in an almost vertical position until the upper-wing panels are attached to the fittings on the ends of the center section. The upper ends of the struts are then attached to the fittings on the lower side of each upper wing. When this is done, both the upper and lower wings are supported by the stand or other support under the lower wing. The upper wing is supported through the interplane struts.

After the wing panels are in place, the **landing wires**, sometimes called ground wires, are installed between the fittings at the tops of the cabane struts and the fittings at the lower ends of the interplane struts. These wires are tightened to support the weight of both wing panels on each side of the fuselage. After the landing wires are tightened, the supports can be removed from under the lower wing.

The **flying wires**, which carry the wing load in flight, are installed between the fittings at the butt end of the lower wing and the fittings at the upper ends of the interplane struts. These wires are tightened just enough to take up the slack.

The dihedral angle of the wings is established by adjusting the landing wires. A bubble protractor or a dihedral board with the correct angle is used for checking the dihedral angle. When the dihedral angle is correct, the flying wires are tightened. The wires are of the streamline design and are manufactured with right-hand threads at one end and left-hand threads at the other end. Thus, by turning the wire in one direction, it is tightened; by turning it in the opposite direction, it is loosened. Tightening the landing wires increases the dihedral angle. After the wires are all tightened to the correct tension, the dihedral angle should be rechecked to see that it is correct.

When adjusting flying and landing wires, a nonmetallic wire wrench should be used. Do not use pliers or any metal tool to adjust the wires, since these will damage the wire surface and can cause wire failure.

The stagger at the outer ends of the wings should be checked to determine that it is correct. Some biplanes use a fixed N-strut arrangement such that no adjustment of stagger is required. If the airplane has incidence wires between the interplane struts, it is necessary to adjust these wires by means of turnbuckles or other means to set the correct stagger at these points.

After the wings and empennage are adjusted and set according to specifications, the airplane must be flown to determine whether the rigging is exactly as it should be or whether additional adjustments are required.

AIRCRAFT FLIGHT CONTROLS

An aircraft is equipped with movable surfaces, or airfoils, which provide stability and control during flight. These controls are normally divided into two categories, the *primary flight controls* and the *secondary flight controls*. The three primary controls for an airplane are the ailerons, elevators or stabilators, and rudder or rudders. The primary flight controls are responsible for maneuvering the aircraft about its three axes as illustrated in Fig. 12-15.

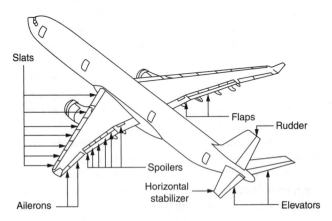

FIGURE 12-15 Aircraft flight controls. (*Airbus.*)

Because aircraft are often capable of operating over a wide speed range and with different weight distributions, secondary flight controls, also called auxiliary flight controls, have been developed. Some of these surfaces, called **tabs**, allow the flight crew to reduce or eliminate the pressure that they must apply to the flight controls. Other surfaces fall in a group termed **high-lift devices**, which include flaps, slats, and slots. These features allow the lift and drag characteristics of the aircraft wing to be changed to allow

slow-speed flight for takeoff and landing and high-speed flight for cruising. Still a third group of surfaces is used to reduce lift and generate drag. This group includes spoilers and speed brakes.

Ailerons

Ailerons, shown in Fig. 12-15, are primary flight-control surfaces utilized to provide lateral (roll) control of aircraft; that is, they control aircraft movement about the longitudinal axis. They are usually mounted on the trailing edge of the wing near the wing tip. Large jet aircraft often employ two sets of ailerons, one set being approximately midwing or immediately outboard of the inboard flaps, and the other set being in the conventional location near the wing tips. Figure 12-16 shows a transport aircraft wing with this aileron configuration. The outboard ailerons become active whenever the flaps are extended beyond a fixed setting. As the flaps are retracted, the outboard aileron control system is "locked out" and fairs with the basic wing shape. Thus, during cruising operations at comparatively high speeds, only the inboard ailerons are used for control. The outboard ailerons are active during landing or other slow-flight operations. Many transport aircraft use "drooping ailerons." During extension of the flaps, the ailerons droop to increase the lift of the wing. When drooped, the ailerons are fully operational for roll control.

Aileron control systems operated by the pilot through mechanical connections require the use of balancing mechanisms so that the pilot can overcome the air loads imposed on the ailerons during flight. Balancing of the ailerons can be achieved by extending part of the aileron structure ahead of the hinge line and shaping this area so that the airstream strikes the extension and helps to move the surface. This is

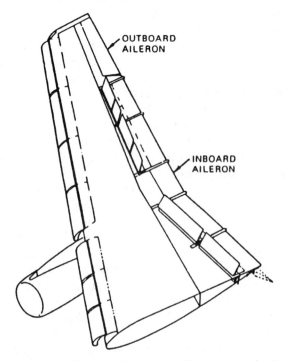

FIGURE 12-16 The L-1011 uses two ailerons on each wing. (*Lockheed California Co.*)

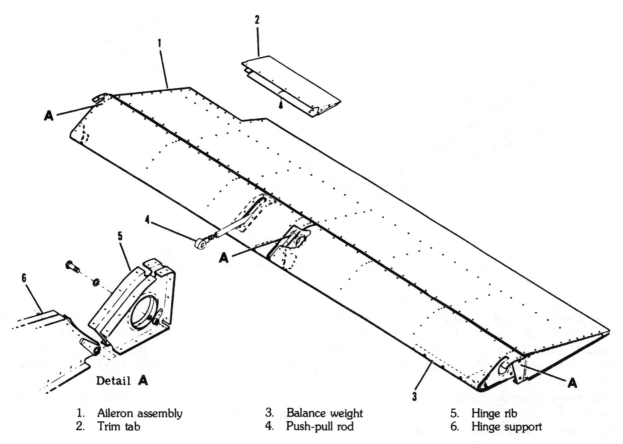

1. Aileron assembly
2. Trim tab
3. Balance weight
4. Push-pull rod
5. Hinge rib
6. Hinge support

FIGURE 12-17 Aileron showing set-back hinge line.

known as *aerodynamic balancing*. Another method that may be used is to place a weight ahead of the hinge line to counteract the flight loads. This is known as *static balancing*. Some aircraft may use a combination of these techniques. Figure 12-17 shows an aileron that uses the aerodynamic balance method, and Fig. 12-18 shows ailerons using the mass-weight balance method. Transport category aircraft use hydraulically operated ailerons and may not employ these forms of balancing. If the transport control system is designed to allow the pilot to operate the ailerons without hydraulic assistance, then some method of balancing or control by control tabs is used.

In many aircraft, the operation of ailerons causes the aircraft to yaw against the direction of the control movement; that is, a movement of the control for a left roll causes the airplane to yaw to the right. This is because the aileron that moves downward creates lift and drag, whereas the aileron that moves upward reduces lift and creates much less drag. This drag causes a condition called **adverse yaw**. To overcome adverse yaw, a number of modifications have been made in the design and operation of some aileron systems.

One method of combating adverse yaw is to design the aileron so that a substantial section extends forward of the hinge line. This forward section moves down into the airstream

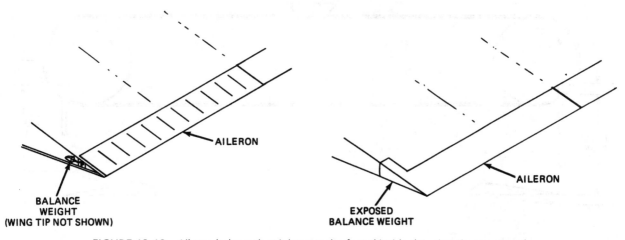

FIGURE 12-18 Aileron balanced weights may be found inside the wing tip or exposed.

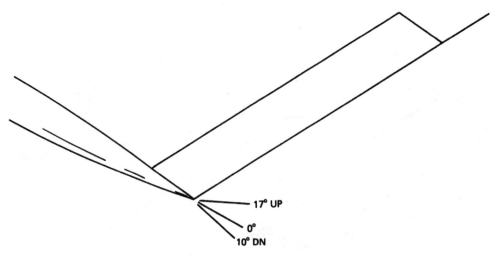

FIGURE 12-19 Differential aileron movement allows each aileron to move up a greater amount than it can move down.

when the aileron is moved up and creates drag to balance the increased drag of the down aileron. The top portion of the forward section of the aileron is rounded off in order that it will not extend upward on the down aileron. This type of aileron is called a *Friese aileron.*

Another, more common, method for controlling adverse yaw through the operation of the ailerons is to design the control systems so there is differential movement between the ailerons. The aileron moving upward moves approximately twice as far as the aileron moving downward, as shown in Fig. 12-19. In this way the drag on the up aileron tends to balance the drag on the down aileron. The difference in the amount of aileron travel between the upward and downward movement is caused by a differential control. This can be accomplished by several methods, one being

the placement of the control-rod connections on the drive wheels. An example of differential control movement is illustrated in Fig. 12-20. In this illustration, it can be seen that the movement of the control stick in one direction will cause an aileron to move up a greater distance than the other aileron moves down. When the control stick is moved in the opposite direction, the opposite effect must occur.

Rudder

The **rudder** is the flight-control surface that controls aircraft movement about its vertical axis. The rudder is constructed very much like other flight-control surfaces, with spars, ribs, and skin, and is mounted on the vertical fin, as illustrated in Fig. 12-15.

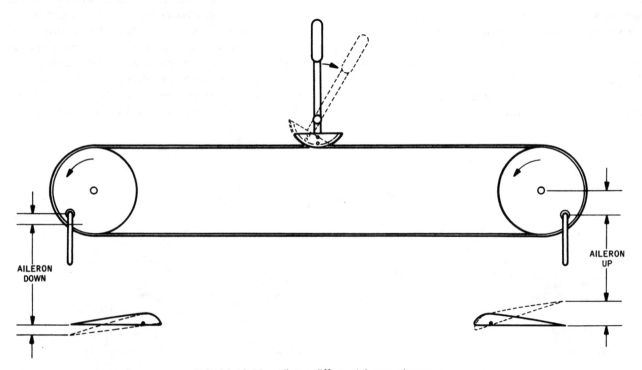

FIGURE 12-20 Aileron differential control system.

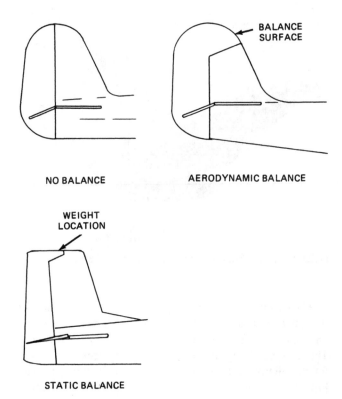

NO BALANCE   **AERODYNAMIC BALANCE**

WEIGHT LOCATION

STATIC BALANCE

FIGURE 12-21   Different rudder configurations.

Rudders are usually balanced both statically and aerodynamically to provide for greater ease of operation and to eliminate the possibility of flutter. Note that some light-aircraft rudders may not use any balancing method. Different rudders for light aircraft are shown in Fig. 12-21.

Rudders for transport aircraft vary in basic structural and operational design. Some are single structural units operated by one or more control systems. Others are designed with two operational segments that are controlled by different operating systems and provide a desired level of redundancy.

The rudder shown in Fig. 12-22 consists of one segment and a rudder trim tab. The rudder is attached to hinge brackets mounted on the rear spar of the vertical stabilizer.

Elevators

Elevators are the control surfaces that govern the movement (pitch) of the aircraft around the lateral axis. They are normally attached to hinges on the rear spar of the horizontal stabilizer. The construction of an elevator is similar to that of other control surfaces, and the design of the elevator may be unbalanced or balanced aerodynamically and/or statically. Typical elevator installations for light aircraft and transports are shown in Figs. 12-23 and 12-15.

Combination Control Surfaces

Some aircraft use combination control surfaces that combine the operation of at least two control and/or stabilizer surfaces into one component. By the use of combination surfaces, the construction of the aircraft can be simplified and the desired control response can be achieved. Examples of these types of control surfaces include *stabilators*, *ruddervators*, and *flaperons*.

A **stabilator** combines the function of a horizontal stabilizer and an elevator. This type of surface is used primarily

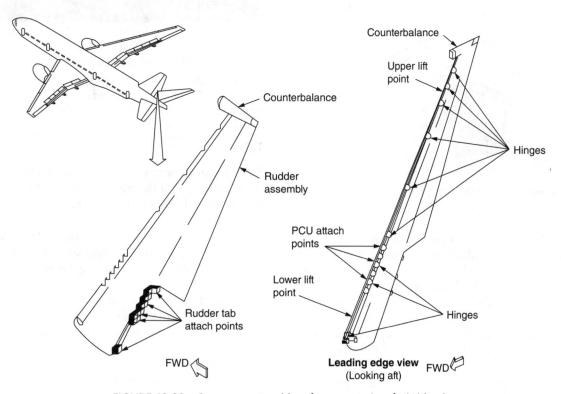

FIGURE 12-22   One segment rudder of transport aircraft. (*Airbus.*)

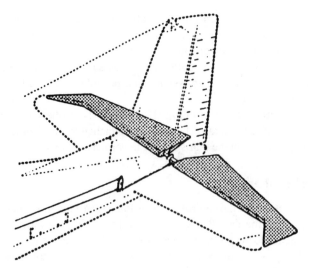

FIGURE 12-23 The elevator of a light aircraft. (*Cessna Aircraft Co.*)

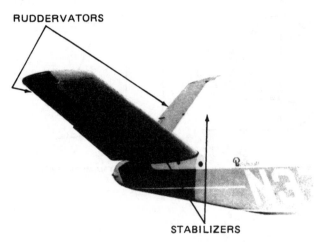

FIGURE 12-25 Ruddervators on a Beechcraft Bonanza.

on light-aircraft designs and on some high-performance military aircraft. The stabilator usually incorporates a static balance weight on an arm ahead of the main spar. This weight can project into the aircraft structure or be carried on the forward portion of the tips of the stabilator. A stabilator is normally equipped with an antiservo tab, which doubles as a trim tab. The antiservo tab moves in the same direction that the control surface is moved to aid the pilot in returning the stabilator to the trimmed neutral position. A typical stabilator for a light aircraft is shown in Fig. 12-24.

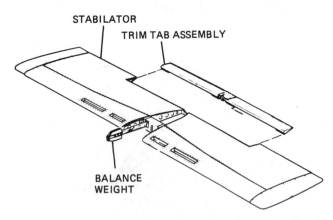

FIGURE 12-24 Stabilator for a light airplane. (*Piper Aircraft Co.*)

Ruddervators are flight-control surfaces that serve the functions of the rudder and elevators. The surfaces are mounted at an angle above horizontal, as shown in the photograph of Fig. 12-25. When serving as elevators, the surfaces on each side of the tail move in the same direction, either up or down. When serving as a rudder, the surfaces move in opposite directions, one up and one down. When combined rudder and elevator control movements are made by the pilot, a control-mixing mechanism moves each surface the appropriate amount to get the desired elevator and rudder effect.

Flaperons are surfaces that combine the operation of flaps and ailerons. These types of control surfaces are found on some aircraft designed to operate from short runways. The flaperon allows the area of the wing normally reserved for the aileron to be lowered and creates a full-span flap. From the lowered position the flaperon can move up or down to provide the desired amount of roll control while still contributing to the overall lift of the wing.

The Boeing 777 uses inboard flaperons that provide roll control at cruise speed when the outboard ailerons are locked out. During takeoff, approach, and landing the flaperons droop to provide more lift. When drooped the flaperons still provide full roll control capability.

SECONDARY FLIGHT-CONTROL SURFACES

The number and complexity of the secondary control surfaces on a particular aircraft depend on the type of operation and flight speeds for which the aircraft is designed. Figure 12-26 shows the secondary flight control surfaces found on a typical jet transport aircraft.

Tabs

Tabs are small secondary flight-control surfaces set into the trailing edges of the primary surfaces. These are used to reduce the work load required of the pilot to hold the aircraft in some constant attitude by "loading" the control surface in a position to maintain the desired attitude. They may also be used to aid the pilot in returning a control surface to a neutral or trimmed-center position. Figure 12-27 demonstrates the tab action.

Tabs can be fixed or variable, and the variable tabs can be designed to operate in several different manners. There are many different types of tabs and tab-operating systems. Figure 12-28 shows different types of control-tab configurations.

A fixed trim tab, shown in Fig. 12-29, is normally a piece of sheet metal attached to the trailing edge of a control surface.

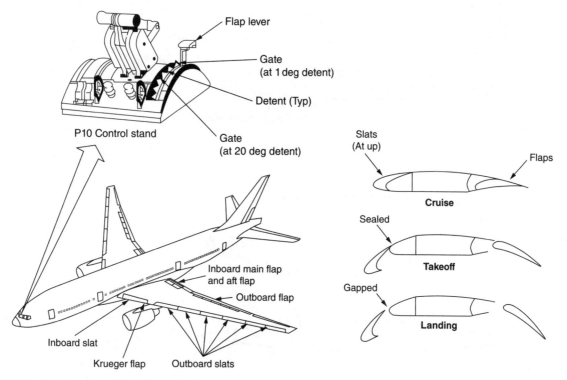

FIGURE 12-26   The location of high lift control on transport aircraft.

This fixed tab is adjusted on the ground by bending it in the appropriate direction to eliminate cabin flight-control forces for a specific flight condition. The fixed tab is normally adjusted for zero-control forces while in cruising flight. Adjustment of the tab is a trial-and-error process, in which the aircraft must be flown and the trim tab adjusted based on the pilot's report. The aircraft must then be flown again to see if further adjustment is necessary. Fixed tabs are normally found on light aircraft and are used to adjust rudders and ailerons.

Controllable trim tabs are found on most aircraft with at least the elevator tab being controlled. These tabs are normally operated mechanically by a cable and chain system, electrically by a screwjack mechanism or a motor to drive the cable and chain system, or hydraulically through actuators. When the pilot wishes to change the attitude of the aircraft with the trim system, he or she activates the trim-control system and causes the trim tab to be deflected in the direction opposite to the desired movement of the

control surface. When the trim tab is deflected into the airstream, the air tries to push the tab back flush with the control surface. Since the control mechanism prevents the tab from being pushed back flush, the whole control surface is moved.

Controllable trim tabs are adjusted by means of control wheels or cranks in the cockpit, and an indicator is supplied to denote the position of the trim tab. If the tabs can be operated electrically or hydraulically, they will incorporate some instrumentation to indicate tab position. The mechanisms employed for the operation of the aileron trim control on a typical airplane are shown in Fig. 12-30. Note that the trim is adjusted by a control wheel located in the cockpit. The wheel is turned to the right to lower the right wing and vice versa. The control wheel turns a sprocket and drives a sprocket chain. The chain turns another sprocket wheel connected through a shaft to a miter gear. Through the pair of gears, the axis of rotation is changed 90° and the motion is delivered to another sprocket. This sprocket wheel drives a chain connected to cable fittings. The cable is connected by means of conventional turnbuckles and routed through pulleys to the chain that drives the aileron trim-tab actuator. A similar mechanism can be used to trim the elevator and rudder.

The foregoing description provides a sample of a typical trim-tab system, although many systems on large aircraft are driven by means of electric actuators.

Servo tabs are used to aid the pilot in the operation of flight controls. When the pilot moves a primary flight control, a servo tab deflects in the proper direction to aid the pilot in moving the control surface. This reduces the force that the pilot must supply to the control system to maneuver the aircraft.

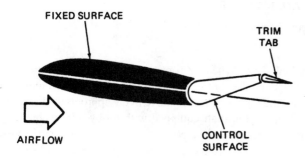

FIGURE 12-27   Trim tabs must be adjusted opposite to the desired movement of the surface being controlled.

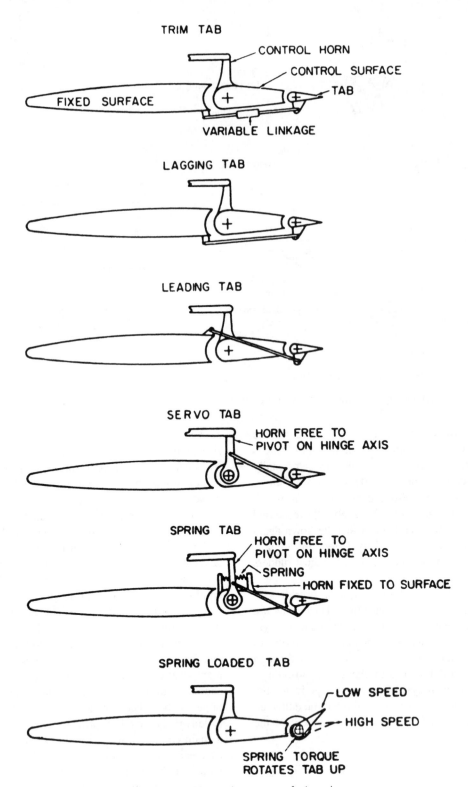

FIGURE 12-28 Various types of trim tabs.

An antiservo tab is used to aid the pilot in returning a surface such as a stabilator to the neutral position and prevent it from moving to a full deflection position due to aerodynamic forces. This type of tab has the opposite effect of a servo tab. The antiservo tab also often serves as the pitch trim tab by allowing the pilot to adjust the neutral trim position.

A control tab is used on some transport aircraft as a manual backup to flight controls that are normally operated hydraulically. When in a manual reversion mode, the pilot can operate the control tabs, and by their tab action they will cause the flight controls to move in the appropriate direction.

Transport aircraft equipped with fly-by-wire technology often do not have trim tabs. The trim function is performed by the servo valves that position the primary control surface in the desired position based on pilot and computer inputs. The trim function on these aircraft is often fully automated.

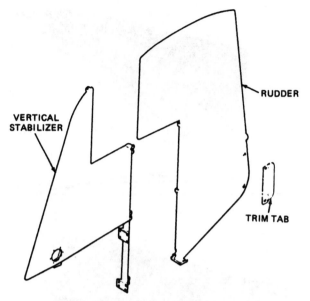

FIGURE 12-29 A fixed trim tab is adjusted on the ground for the average flight conditions. (*Ayres Corp.*)

Flaps

A **wing flap** is defined as a hinged, pivoted, or sliding airfoil, usually attached near the trailing edge of the wing. The purpose of wing flaps is to change the camber of the wing and in some cases to increase the area of the wing, thus permitting the aircraft to operate at lower flight speeds for landing and takeoff. The flaps effectively increase the lift of the wings and, in some cases, greatly increase the drag, particularly when fully extended.

Various configurations for wing flaps are shown in Fig. 12-31. The **plain flap**, in effect, acts as if the trailing edge of the wing were deflected downward to change the camber of the wing, thus increasing both lift and drag. If the flap is moved downward sufficiently, it becomes an effective air brake. The plain flap may be hinged to the wing at the lower side, or it may have the hinge line midway between the lower and upper surfaces.

The **split flap**, when retracted, forms the lower surface of the wing's trailing edge. When extended, the flap moves

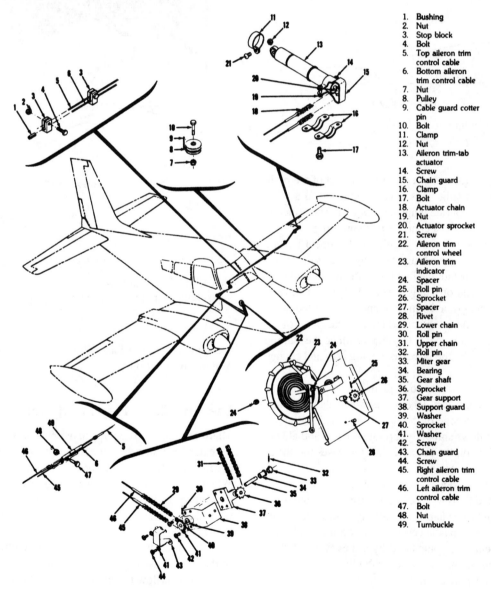

1. Bushing
2. Nut
3. Stop block
4. Bolt
5. Top aileron trim control cable
6. Bottom aileron trim control cable
7. Nut
8. Pulley
9. Cable guard cotter pin
10. Bolt
11. Clamp
12. Nut
13. Aileron trim-tab actuator
14. Screw
15. Chain guard
16. Clamp
17. Bolt
18. Actuator chain
19. Nut
20. Actuator sprocket
21. Screw
22. Aileron trim control wheel
23. Aileron trim indicator
24. Spacer
25. Roll pin
26. Sprocket
27. Spacer
28. Rivet
29. Lower chain
30. Roll pin
31. Upper chain
32. Roll pin
33. Miter gear
34. Bearing
35. Gear shaft
36. Sprocket
37. Gear support
38. Support guard
39. Washer
40. Sprocket
41. Washer
42. Screw
43. Chain guard
44. Screw
45. Right aileron trim control cable
46. Left aileron trim control cable
47. Bolt
48. Nut
49. Turnbuckle

FIGURE 12-30 Aileron trim-tab control system. (*Cessna Aircraft Co.*)

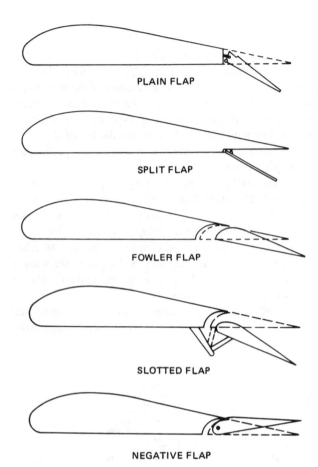

FIGURE 12-31 Configurations for wing flaps.

PLAIN FLAP

SPLIT FLAP

FOWLER FLAP

SLOTTED FLAP

NEGATIVE FLAP

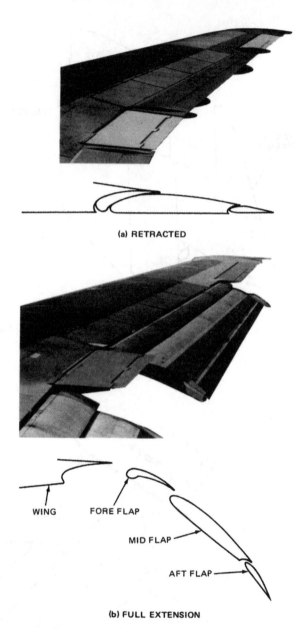

(a) RETRACTED

WING   FORE FLAP

MID FLAP

AFT FLAP

(b) FULL EXTENSION

FIGURE 12-32 The retracted and extended position of the flap segments in a typical multiple-flap system.

downward and provides an effect similar to that of the plain flap. Plain flaps and split flaps may be attached to a wing with three or more separate hinges, or they may be attached at the lower surface with a continuous piano hinge.

The **Fowler flap** and others with similar operation are designed to increase substantially the wing area as the flap is extended. When retracted, the flap forms the trailing edge of the wing. As this type of flap is extended, it is moved rearward, often by means of a worm gear, and is supported in the correct position by means of curved tracks. The effect of the Fowler flap, when extended, is to greatly reduce the stalling speed of the aircraft by the increase in wing area and change in wing camber.

A **slotted flap** is similar to a plain flap except that as the flap is extended, a gap develops between the wing and the flap. The leading edge of the flap is designed so that air entering this gap flows smoothly through the gap and aids in holding the airflow on the surface. This increases the lift of the wing with the flap extended.

Some aircraft designs incorporate combinations of Fowler and slotted flaps to greatly increase the lift and drag of the wing. When the flap is initially extended, it moves aft on its track. Once past a certain point on the track, further aft movement is accompanied by a downward deflection, which opens up the slot between the flap and the wing. Many turbine transport aircraft use this basic design, with several slot openings being used to improve the airflow over

the wing and flap surfaces. Figure 12-32 illustrates this type of flap combination.

Leading-Edge Flaps and Slats

Many airplanes are equipped with **leading-edge flaps** that are extended when the wing flaps are employed. The leading-edge flap, when retracted, forms the leading edge of the wing. When extended, the flap moves forward and down to increase the camber of the wing and provide greater lift at low flight speeds. The arrangement of a leading-edge flap is shown in Fig. 12-33.

Some aircraft have fixed slots built into the leading edge of the wings, usually only in the area of the ailerons. These slots allow airflow to be directed over the top of the wing at high angles of attack. This reduces the stalling speed of that

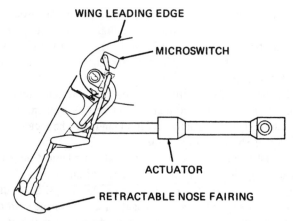

FIGURE 12-33   Leading-edge flap. (*Boeing Commercial Aircraft Co.*)

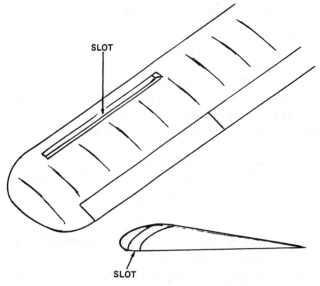

FIGURE 12-34   Slots are fixed openings in the leading edge of the wing that improve aileron control at high angles of attack.

portion of the wing and improves aileron control when flying at high angles of attack. When at normal flight attitudes, the slots have no significant effect on the flight characteristics of the aircraft. This type of design eliminates the mechanism required for the slats and eliminates the possibility of asymmetrical extension of the slats. A wing with a slot is shown in Fig. 12-34.

The use of **slats** on the leading edge of high-performance wings is a common method of reducing stalling speed and increasing lift at comparatively slow speeds. The slat forms the leading edge of the wing when not extended and creates

a *slot* at the leading edge when extended. The slot permits air from the high-pressure area under the leading edge to flow up through the leading edge and to be directed along the top of the wing. This effectively reduces the possibility of stall at lower speeds. A drawing to illustrate the operation of slats is shown in Fig. 12-35. Slats that extend to form slots may be actuated aerodynamically or by mechanical controls.

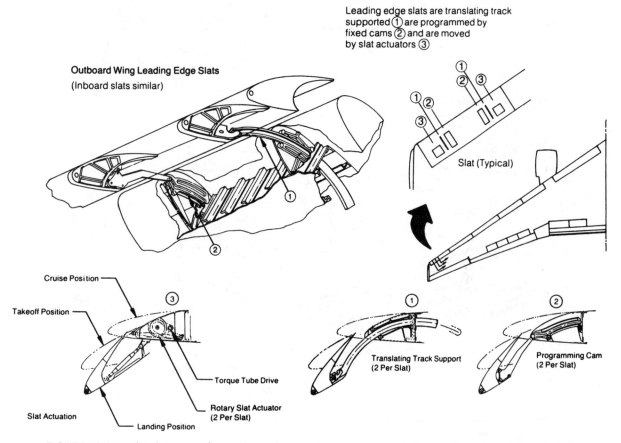

FIGURE 12-35   The slats move forward and down as they are extended. (*Boeing Commercial Aircraft Co.*)

Spoilers and Speed Brakes

Spoilers, also called *lift dumpers*, are control surfaces used to reduce, or "spoil," the lift on a wing. Spoilers are located on the upper surface of wings and are one of two basic configurations. The more common configuration on large turbine aircraft, shown in Fig. 12-36, is to have a flat-panel spoiler laying flush with the surface of the wing and hinged at the forward edge. When the spoilers are deployed, the surface rises up and reduces the lift. The other configuration, shown in Fig. 12-37, is common among sailplanes and has the spoiler located inside the wing structure. When the spoilers are deployed, they rise vertically from the wing and spoil the lift.

Flight spoilers are used in flight to reduce the amount of lift that the wing is generating, allowing controlled descents without gaining excessive air speed. Depending on the aircraft design, the spoilers may also be operated by the pilot's control wheel or stick. When the pilot moves the control left or right for a roll movement, the spoilers on the wing toward the center of the turn (upward-moving aileron) move upward and aid in rolling the aircraft into the turn. In some aircraft designs, the spoilers are the primary flight control for roll.

Ground spoilers are used only when the aircraft is on the ground and are used along with the flight spoilers to greatly reduce the wing's lift upon landing. They also increase the aerodynamic drag of the aircraft after landing to aid in slowing the aircraft.

Spoilers can be controlled by the pilot through a manual control lever, by an automatic flight-control system, or by an automatic system activated upon landing. The location of spoilers is shown in Fig. 12-38.

Speed brakes, also called divebreaks, are large drag panels used to aid in control of the speed of an aircraft. They may be located on the fuselage or on the wings. If on the fuselage, the speed brake is located on the top or the bottom of the structure. If speed brakes are deployed as a pair, one is on each side of the fuselage. If located on the wings, speed brakes are deployed symmetrically from the top and the bottom of the wing surface to control the speed of the aircraft as well as to act as spoilers to decrease the lift of the wings. See Fig. 12-39.

On some aircraft designs, particularly gliders and sailplanes, there may not be any clear distinction between a spoiler and a divebreak because one control surface may serve the purpose of both actions—i.e., to decrease lift and increase drag.

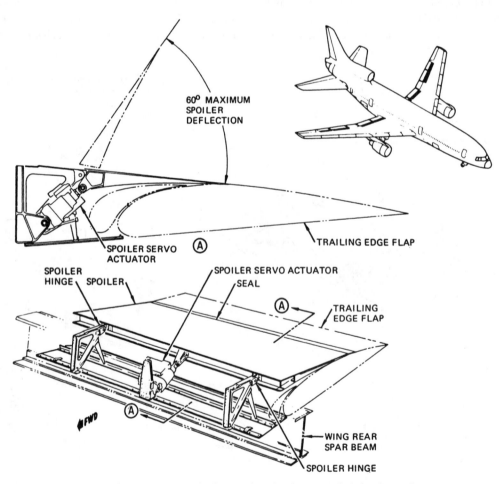

FIGURE 12-36 Spoilers are commonly designed with a hinge at their leading edge. (*Lockheed California Co.*)

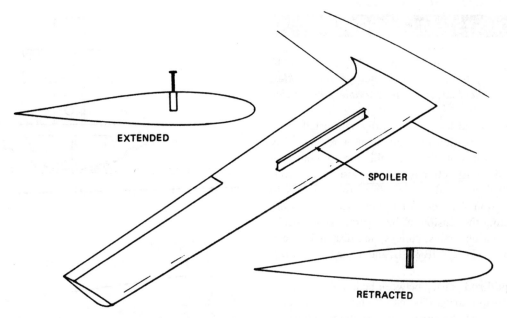

FIGURE 12-37 Some aircraft, such as sailplanes, have the spoilers arranged so that they rise vertically out of the wing.

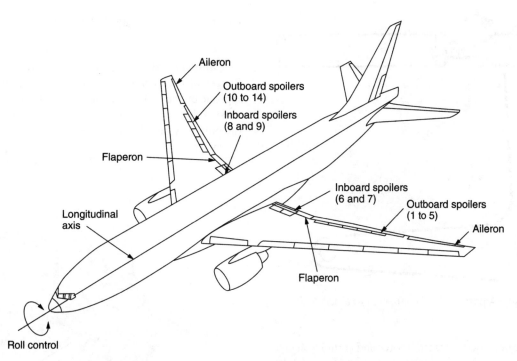

FIGURE 12-38 The location of ground and flight spoilers on transport aircraft.

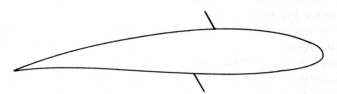

FIGURE 12-39 Speed brakes on a wing open on the top and bottom of the wing.

CONTROL-SYSTEM COMPONENTS

Several components are utilized in assembling a simple control system. Among these are *cables, pulleys, turnbuckles, push-pull rods, bellcranks, quadrants, torque tubes, cable guards,* and *fairleads.*

Aircraft-control **cables** are generally fabricated from carbon steel or corrosion-resistant steel wire and may consist of either a flexible or nonflexible type of construction. The flexible construction is generally used for aircraft controls. **Pulleys** are used in aircraft-control systems to change the direction of a cable. **Turnbuckles** are commonly used for adjusting the tension of the control cables. Additional information on cables, pulleys, and turnbuckles can be found in FAA AC 43.13-1B & 2B, and in the text *Aircraft Basic Science.*

The **push-pull rod** is used between bellcranks and from bellcranks to **torque arms** ("horns") to transmit the force and motion from one to the other. A push-pull rod connected to a bellcrank is shown in Fig. 12-40. Push-pull rods are also called **control rods** because they are often used in control systems.

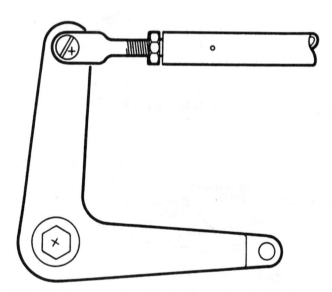

FIGURE 12-40   A push-pull rod connected to a bellcrank.

A **bellcrank** is used to transmit force and permit a change in the direction of the force. In the illustration of Fig. 12-40, the effect of the bellcrank operation can be seen.

A **quadrant** serves the same purpose as a wheel; however, the quadrant moves through a relatively small arc, perhaps as much as 100°. A quadrant, shown in Fig. 12-41, is often employed at the base of a control column or control stick to impart force and motion to a cable system.

A **torque tube** is a hollow shaft by which the linear motion of a cable or push-pull tube is changed to rotary motion. A torque arm, or horn, is attached to the tube by welding or bolting and imparts a twisting motion to the tube as the arm is moved back and forth. This is illustrated in Fig. 12-42.

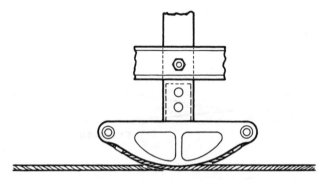

FIGURE 12-41   A quadrant used for movement of control cables.

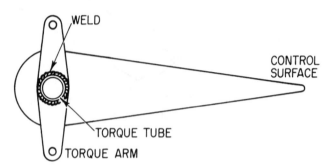

FIGURE 12-42   A torque arm, or horn.

Cable guards, or **guard pins**, are installed in the flanges of pulley brackets to prevent the cable from jumping out of the pulley, as shown in Fig. 12-43. The guard must be located so it does not interfere with the rotation of the pulley. A guard pin can be either a bolt, a cotter pin, or a clevis pin.

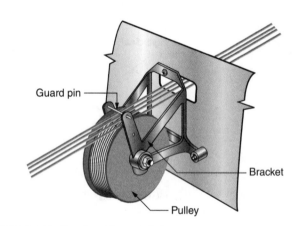

FIGURE 12-43   A cable guard pin.

A **fairlead** serves as a guide to prevent wear and vibration of a cable. The fairlead is made of phenolic material, plastic, or soft aluminum and is either split or slotted to permit the installation of the cable. These units must be installed in such a manner that there is no contact between the cable and the aircraft structure. The principal functions of fairleads are to dampen vibration, maintain cable alignment, and to seal openings in bulkheads. In no case should the fairlead be permitted to deflect a cable more than 3°, and it is good practice to install fairleads so cable deflection is as small as possible. Since cables are tightened to the extent that from 25 lb [111.25 N] to more than 160 lb [712 N] is exerted on the cable, any appreciable deflection at a fairlead will cause excessive wear of the cable and the fairlead. A fairlead is shown in Fig. 12-44.

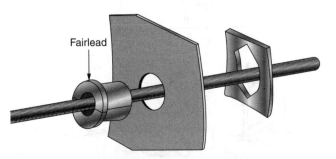

FIGURE 12-44   A fairlead.

In a pressurized turbine aircraft, cables leading from a pressurized section of the airplane to a nonpressurized section must have **air-pressure seals** installed where the cable passes through the bulkhead. A seal of this type is shown in Fig. 12-45.

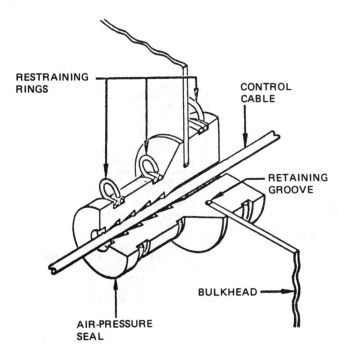

FIGURE 12-45   An air-pressure seal for cable installations in pressurized airplanes. (*Boeing Commercial Aircraft Co.*)

Typical Control System for a Light Aircraft

Most light aircraft are controlled in flight by the use of the three standard primary control surfaces, consisting of the ailerons, stabilator, and rudder. Operation of these controls is through the movement of the control column-tee bar assembly and rudder pedals shown in Figs. 12-46 and 12-47. On the forward end of each control column is a sprocket assembly. A chain is wrapped around the sprockets to connect the right and left controls and then back to idler sprockets on the column's tee bar, which in turn connect to the aileron primary control cables, as shown in Fig. 12-48. The cables operate the aileron bellcrank and push-pull rods. Rotation of the control wheels is changed to angular movement of the ailerons through the cables, bellcranks, and aileron control-tube assembly.

An elevator control system that utilizes push-pull control rods (control tubes) is illustrated in Fig. 12-49. When the control wheels are moved forward and rearward, the control arm at the lower end of the control column assembly is moved. This motion is transmitted through the push-pull rods to the bellcrank at the rear of the airplane, which, in turn, transmits the motion to the rear elevator-control tube assembly, thus moving the elevator up and down. Adjustments of the elevator movement and travel are made by means of adjustable rod ends on the control tubes.

The rudder-control system utilizes a rudder pedal assembly, illustrated in Fig. 12-47. Force applied to the rudder cables is transmitted through cables to a torque tube assembly at the base of the rudder, as shown in Fig. 12-50.

A simple mechanically controlled flap system is illustrated in Fig. 12-51. Force applied to the flap handle is transmitted through cables to a torque tube, which applies the angular flap motion.

Typical Control System for a Large Aircraft

The primary flight control system (PFCS) for the Boeing 777 uses a fly-by-wire control system and there is no mechanical connection between the flight surfaces and the cockpit controls. The PFCS receives commands from the flight crew and the autopilot and the electric/hydraulic servo valves will cause the control surfaces to move. The PFCS controls ailerons, flaperons, spoilers, horizontal stabilizer, elevators, and a rudder with trimtab, as illustrated in Fig. 12-52. The flight controls are controlled by flight computers and powered by the three airplane hydraulic systems. There is no manual backup system when the hydraulic power is lost. Flaperons and spoilers assist the ailerons in providing roll control and the spoilers also operate as speedbrakes. A variable-pitch horizontal stabilizer assists the elevators for long-term correction of the pitch attitude of the airplane. High lift for takeoff, approach, and landing is provided by trailing edge flaps, leading edge slats, and Krueger flaps.

The pilots or the autopilot commands control the PFCS. The pilots can override the autopilot. In the manual operation the position transducers change the pilots' manual commands of the control wheel, the control columns, the rudder pedals, and the speedbrake lever to analog electrical signals.

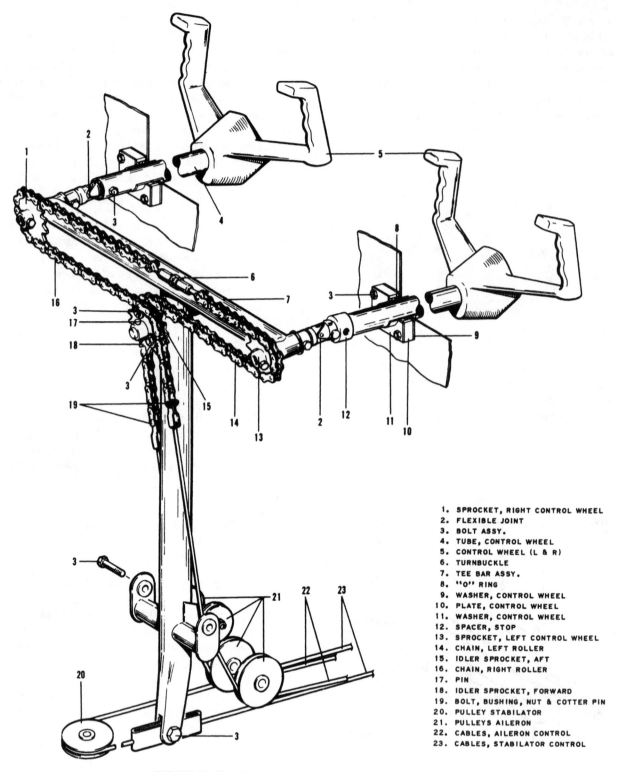

FIGURE 12-46 Control column assembly. (*Piper Aircraft Co.*)

1. SPROCKET, RIGHT CONTROL WHEEL
2. FLEXIBLE JOINT
3. BOLT ASSY.
4. TUBE, CONTROL WHEEL
5. CONTROL WHEEL (L & R)
6. TURNBUCKLE
7. TEE BAR ASSY.
8. "O" RING
9. WASHER, CONTROL WHEEL
10. PLATE, CONTROL WHEEL
11. WASHER, CONTROL WHEEL
12. SPACER, STOP
13. SPROCKET, LEFT CONTROL WHEEL
14. CHAIN, LEFT ROLLER
15. IDLER SPROCKET, AFT
16. CHAIN, RIGHT ROLLER
17. PIN
18. IDLER SPROCKET, FORWARD
19. BOLT, BUSHING, NUT & COTTER PIN
20. PULLEY STABILATOR
21. PULLEYS AILERON
22. CABLES, AILERON CONTROL
23. CABLES, STABILATOR CONTROL

These signals go to the four actuator control electronics (ACEs). The ACEs change the signals to digital format and send them to the three primary flight computers (PFCs). The PFCs have interfaces with the airplane systems through the three flight controls ARINC 629 buses. In addition to command signals from the ACEs, the PFCs also receive data from the airplane information management system (AIMS), the air data inertial reference unit (ADIRU), and the secondary attitude air data reference unit (SAARU). Figure 12-53 shows the PFC control laws.

The PFCs calculate the flight control commands based on control laws and flight envelope protection functions. The control laws supply stability augmentation in the pitch and yaw axes and flight envelope protections in all three axes.

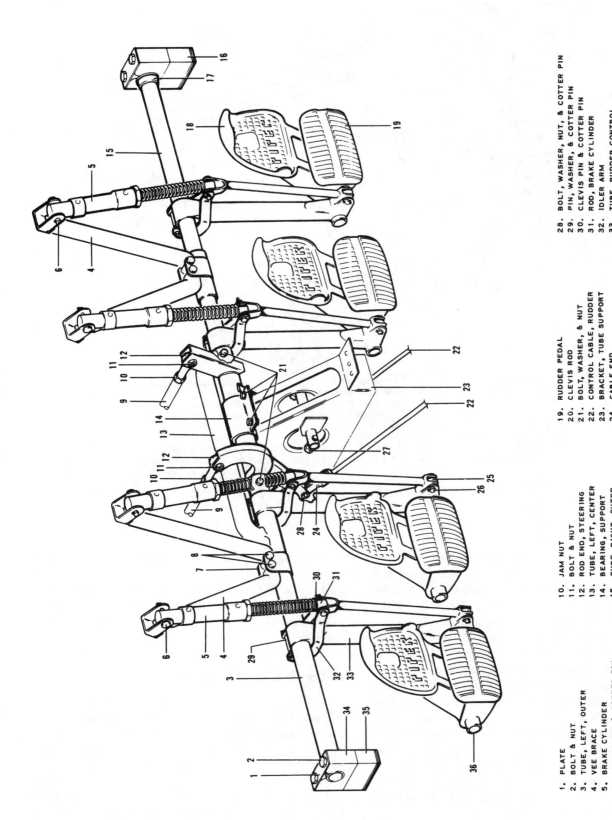

1. PLATE
2. BOLT & NUT
3. TUBE, LEFT, OUTER
4. VEE BRACE
5. BRAKE CYLINDER
6. CLEVIS PIN & COTTER PIN
7. BRACKET
8. BOLT, WASHER, & NUT
9. ROD, NOSE WHEEL STEERING

10. JAM NUT
11. BOLT & NUT
12. ROD END, STEERING
13. TUBE, LEFT, CENTER
14. BEARING, SUPPORT
15. TUBE, RIGHT, OUTER
16. SUPPORT BLOCK, LOWER
17. WASHER, SPACER
18. BRAKE PEDAL

19. RUDDER PEDAL
20. CLEVIS ROD
21. BOLT, WASHER, & NUT
22. CONTROL CABLE, RUDDER
23. BRACKET, TUBE SUPPORT
24. CABLE END
25. ROD END
26. CLEVIS PIN & COTTER PIN
27. STOPS, RUDDER PEDAL

28. BOLT, WASHER, NUT, & COTTER PIN
29. PIN, WASHER, & COTTER PIN
30. CLEVIS PIN & COTTER PIN
31. ROD, BRAKE CYLINDER
32. IDLER ARM
33. TUBE, RUDDER CONTROL
34. SUPPORT BLOCK, UPPER
35. SUPPORT BLOCK, LOWER
36. TUBE, RUDDER CONTROL

FIGURE 12-47 Rudder pedal assembly. (*Piper Aircraft Co.*)

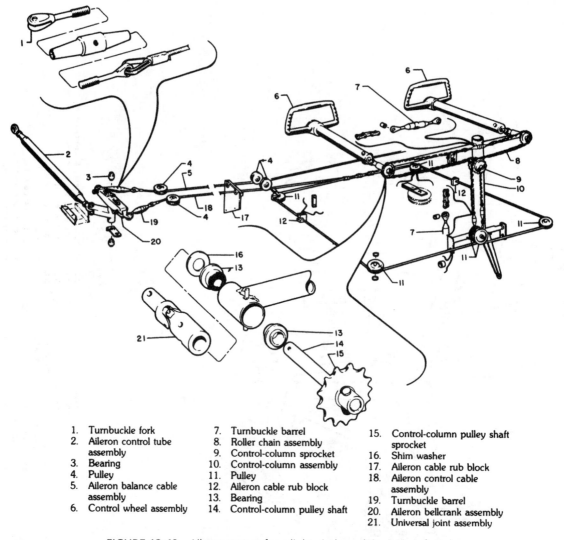

1.	Turnbuckle fork	7.	Turnbuckle barrel	15.	Control-column pulley shaft	
2.	Aileron control tube	8.	Roller chain assembly		sprocket	
	assembly	9.	Control-column sprocket	16.	Shim washer	
3.	Bearing	10.	Control-column assembly	17.	Aileron cable rub block	
4.	Pulley	11.	Pulley	18.	Aileron control cable	
5.	Aileron balance cable	12.	Aileron cable rub block		assembly	
	assembly	13.	Bearing	19.	Turnbuckle barrel	
6.	Control wheel assembly	14.	Control-column pulley shaft	20.	Aileron bellcrank assembly	
				21.	Universal joint assembly	

FIGURE 12-48   Aileron system for a light airplane. (*Piper Aircraft Co.*)

The digital command signals from the PFCS go to the ACEs. The ACEs change these command signals to analog format and send them to the power control units (PCUs) and the stabilizer trim control modules (STCMs). The ACEs and the PCUs form control loops which control the surfaces based on the PFCs commands. One, two, or three PCUs operate each control surface. One PCU controls each spoiler, two PCUs control each aileron, flaperon, and elevator, and three PCUs control the rudder. The PCUs contain a hydraulic actuator, an electrohydraulic servo valve, and a position feedback transducer. When commanded, the servo valve causes the hydraulic actuator to move the control surface. The position transducer sends a position feedback signal to the ACEs. The ACEs then stop the PCU command when the position feedback signal equals the commanded position. Two STCMs control hydraulic power to the motors and brakes of the horizontal stabilizer.

In the autopilot operation the PFCs receive autopilot commands from all three autopilot flight director computers (AFDCs). The PFCs use the autopilot commands in the same manner as the pilots' manual commands. In addition,

the PFCs supply the backdrive signals to the backdrive actuators through the AFDCs. The backdrive actuators move the control wheels, control columns, and rudder pedals in synchronization with the autopilot commands. The movement of the flight deck controls supplies visual indications to the flight crew.

PFCS Modes of Operation

The PFCS has three modes of operation: normal, secondary, and direct. Normal mode operates when the necessary data are available for the PFCs and the ACEs. All the control laws, protection functions, and the AFDCs operate. When the PFCS detects the loss of important air and attitude data, the PFCS operation changes to secondary mode. The PFCS and the ACEs operate but the PFC control laws and protection functions downgrade. The autopilot cannot operate in secondary mode. In direct mode, the PFCs are not used. The ACEs set the position of the control surfaces in direct response to analog pilot control inputs. Figure 12-54 illustrates the PFCS system.

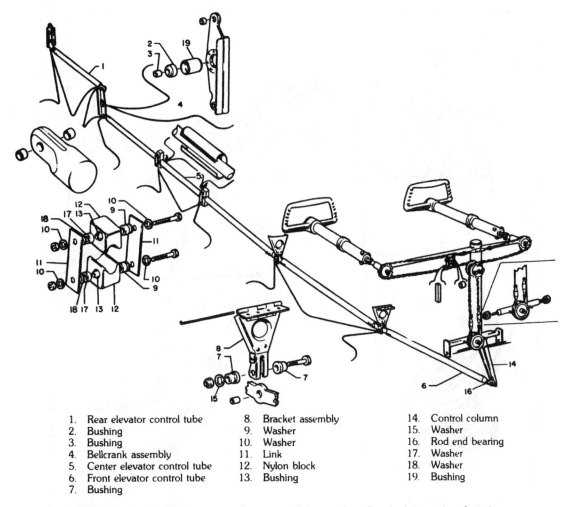

1.	Rear elevator control tube	8.	Bracket assembly	14.	Control column
2.	Bushing	9.	Washer	15.	Washer
3.	Bushing	10.	Washer	16.	Rod end bearing
4.	Bellcrank assembly	11.	Link	17.	Washer
5.	Center elevator control tube	12.	Nylon block	18.	Washer
6.	Front elevator control tube	13.	Bushing	19.	Bushing
7.	Bushing				

FIGURE 12-49   Elevator control system utilizing push-pull rods. (*Piper Aircraft Co.*)

Rudder Control System

The rudder is manually commanded by the pilot and automatically by the autompilot. The flight crew uses two conventional rudder pedals to control yaw. Signals from pedal position transducers go to the flight computers (PFCs) and the flight computers will send a signal to the servo valves (PCU) to position the rudder. The PFCs use position data from the rudder pedals, data from the air data inertial reference unit (ADIRU) and the airplane information management system (AIMS) to calculate control surface commands. The position transducers on the power control unit actuator pistons supply position feedback so that the flight control surface stops at the desired position. Two gust suppression pressure transducers measure wind gust on the vertical stabilizer. This information is used to move the rudder to reduce the effect of the side gust. Figure 12-55 illustrates the rudder control system. A rudder feel and centering mechanism is installed in the rudder control system because the rudder system is a fly-by-wire system and there is no mechanical and aerodynamic feedback. The feel and centering mechanism supplies feel and centering forces to the yaw control system.

The rudder trim control actuates the rudder trim actuator to let the flight crew trim out unwanted rudder pedal forces.

When the pilots move the rudder trim control, the rudder trim actuator moves the rudder pedals and the position transducers through the feel and centering mechanism, and the rudder pedal position transducers send a signal to move the rudder. Figure 12-56 shows the rudder and rudder trim components.

Displacement of either set of rudder pedals sends a signal to the three rudder servo valves and actuators (PCU) which move the rudder. Each rudder PCU operates with hydraulic power from a different hydraulic system. The rudder PCUs are located on the rear spar of the vertical stabilizer.

The control signals from the rudder pedals and trim control to the rudder actuators are modified by a rudder ratio changer. As the airspeed increases, the ratio changer reduces the rudder deflection that results from the pilot's rudder input. The ratio changer receives air data computer airspeed information and provides control signals through the flight computer to the power control unit (PCU). The thrust asymmetry compensation (TAC) function controls the rudder and the rudder trim if there is a thrust asymmetry from the engines. The primary flight computer uses engine thrust data from the engine data interface units (EDIUs) to calculate the yaw correction. The TAC authority limit is 60 percent of

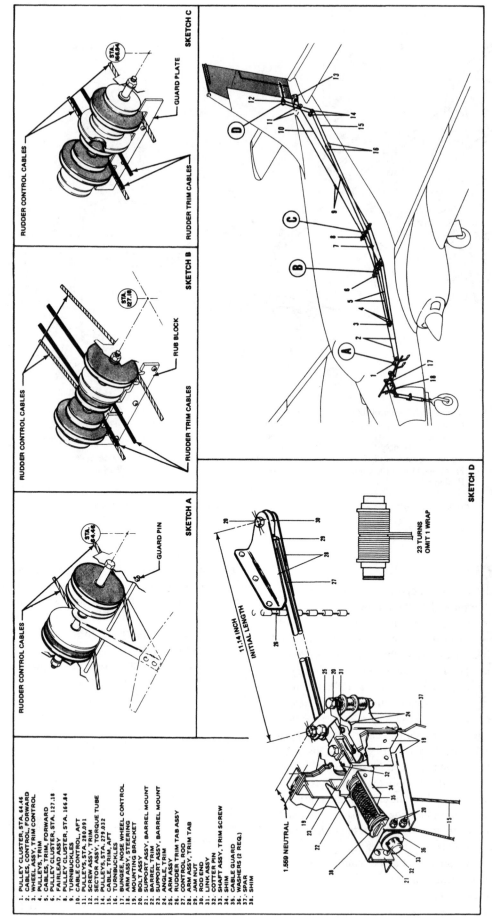

FIGURE 12-50 Rudder-control system. (*Piper Aircraft Co.*)

1. PULLEY CLUSTER, STA. 64.46
2. CABLES, CONTROL, FORWARD
3. WHEEL ASSY, TRIM CONTROL
4. PULLEYS, TRIM
5. CABLES, TRIM, FORWARD
6. PULLEY CLUSTER, STA. 127.18
7. FAIRLEAD ASSY
8. PULLEY CLUSTER, STA. 166.84
9. TURNBUCKLES
10. CABLE CONTROL, AFT
11. PULLEYS, STA. 210.091
12. SCREW ASSY., TRIM
13. SECTOR ASSY., TORQUE TUBE
14. PULLEYS, STA. 279.032
15. CABLE, TRIM, AFT
16. TURNBUCKLES
17. BUNGEE, NOSE WHEEL CONTROL
18. ARM ASSY., STEERING
19. MOUNTING BRACKET
20. BOLT ASSY
21. SUPPORT ASSY., BARREL MOUNT
22. BARREL TRIM
23. SUPPORT ASSY., BARREL MOUNT
24. ANGLE, TRIM
25. ARM ASSY
26. RUDDER TRIM TAB ASSY
27. CONTROL ROD
28. ARM ASSY., TRIM TAB
29. JAM NUT
30. ROD END
31. LINK ASSY
32. COTTER PIN
33. SHAFT ASSY., TRIM SCREW
34. SHIM
35. CABLE GUARD
36. WASHERS (2 REQ.)
37. SPAR
38. SHIM

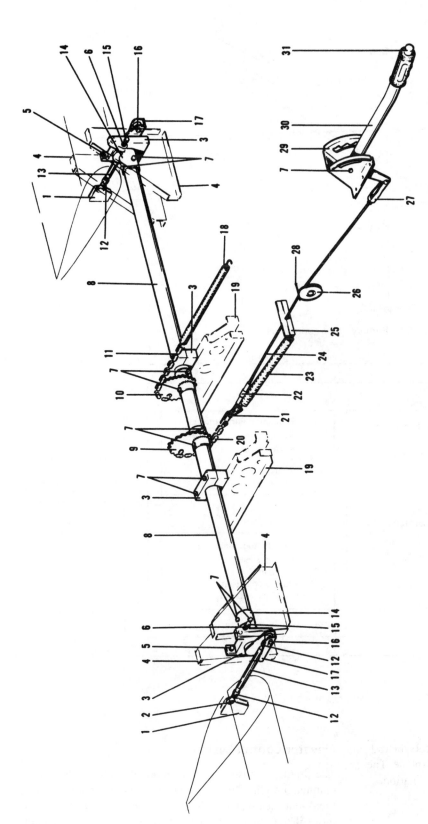

1. BRACKET, ROD ATTACHMENT
2. BOLT, WASHER, NUT & COTTER PIN
3. BLOCK, BEARING
4. BRACKET, BEARING BLOCK
5. BOLT, BEARING BLOCK ATTACHMENT
6. LOCK NUT
7. BOLT ASSY.
8. TORQUE TUBE
9. SPROCKET, TENSION SPRING
10. SPROCKET, RETURN SPRING
11. CHAIN, RETURN SPRING

12. JAM NUT
13. ROD, FLAP CONTROL
14. FITTING, TORQUE TUBE STOP
15. SCREW, FLAP ADJUSTMENT
16. BOLT, WASHER & BUSHING
17. CRANK, ARM, TORQUE TUBE
18. SPRING, RETURN
19. BRACKET, BEARING BLOCK
20. CHAIN, TENSION SPRING
21. CLEVIS BOLT, BUSHING, NUT &
    COTTER PIN

22. TURNBUCKLE
23. SPRING, TENSION
24. CABLE, FLAP CONTROL
25. RUB BLOCK
26. PULLEY
27. CLEVIS BOLT, NUT, WASHER, &
    COTTER PIN
28. SAFETY COTTER PIN
29. BRACKET, FLAP HANDLE
30. HANDLE, FLAP
31. BUTTON, FLAP RELEASE

FIGURE 12-51   Mechanical flap control system. (*Piper Aircraft Co.*)

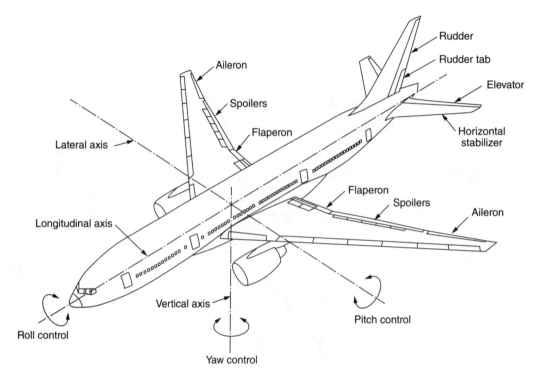

FIGURE 12-52   Primary flight control system.

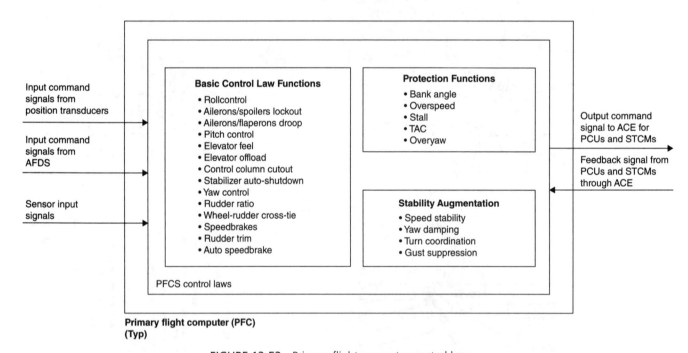

Basic Control Law Functions

- Rollcontrol
- Ailerons/spoilers lockout
- Ailerons/flaperons droop
- Pitch control
- Elevator feel
- Elevator offload
- Control column cutout
- Stabilizer auto-shutdown
- Yaw control
- Rudder ratio
- Wheel-rudder cross-tie
- Speedbrakes
- Rudder trim
- Auto speedbrake

Protection Functions

- Bank angle
- Overspeed
- Stall
- TAC
- Overyaw

Stability Augmentation

- Speed stability
- Yaw damping
- Turn coordination
- Gust suppression

Input command signals from position transducers

Input command signals from AFDS

Sensor input signals

Output command signal to ACE for PCUs and STCMs

Feedback signal from PCUs and STCMs through ACE

PFCS control laws

**Primary flight computer (PFC)
(Typ)**

FIGURE 12-53   Primary flight computer control laws.

available rudder. The yaw damper function controls the rudder to dampen the Dutch roll sensitivity of the airplane. The yaw damper function also controls the turn coordination.

Pitch Control

The pitch control consists of an elevator system for short-term correction of the pitch attitude of the aircraft and the stabilizer system that controls the long-term pitch attitude of the aircraft.

Elevator Control System

The flight crew uses two conventional control columns to command pitch. They connect to torque tubes joined by a break-out mechanism. This allows one control column to move if the other jams. The flight crew can make manual commands to the elevator in the normal mode with the pitch trim switches. This is possible only if the airplane is in the air. Six position transducers send signals to primary flight computers (PFCs). The PFCs use data from the

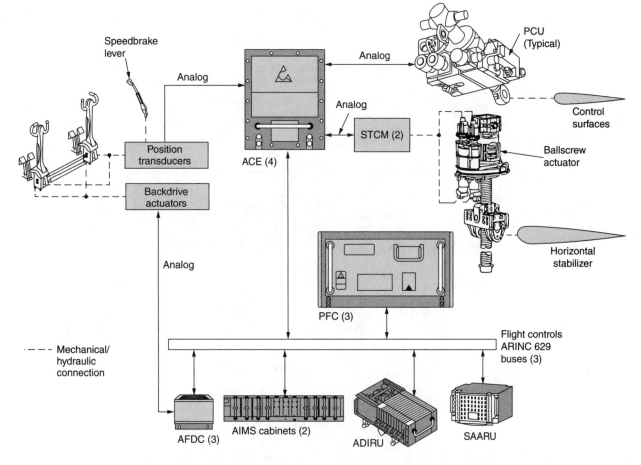

FIGURE 12-54 Primary flight control system.

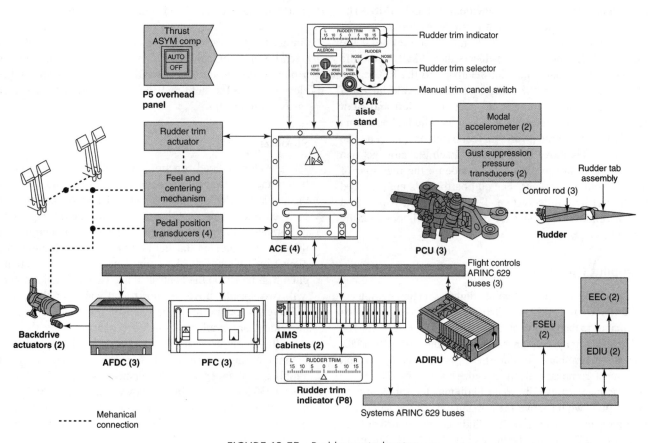

FIGURE 12-55 Rudder control system.

Control-System Components   **359**

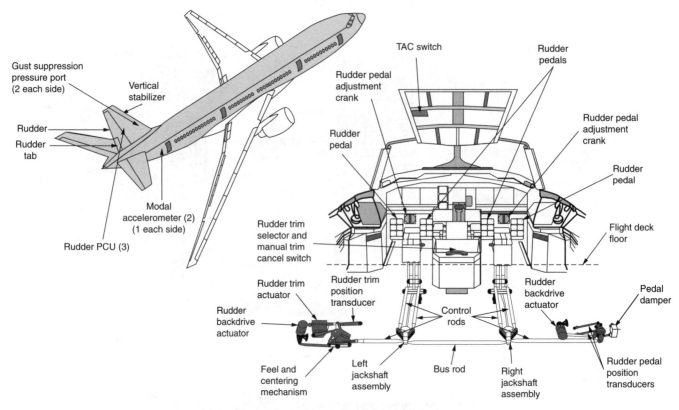

FIGURE 12-56 Rudder and rudder trim components.

following sources to calculate control surface commands: column position, air data inertial reference unit (ADIRU), and airplane information management system (AIMS). The PFCs send command signals through the ACE to the power control units (PCUs). The PCUs move the elevators; see Fig. 12-57. The position transducers on the PCU hydraulic pistons send actuator position feedback to the PFCs. The PFCs use data from the ADIRU to calculate elevator feel commands. The feel command goes to the left and right feel actuators on the elevator feel units. The feel actuators supply to the column a variable feel based on airspeed. The elevator PCUs move the elevators. Each elevator has two PCUs, which are located side by side on the rear spar of the horizontal stabilizer. The outboard PCUs are powered by the left hydraulic system. The left inboard is powered by the center system and the right inboard is powered by the right system.

Autopilot Operation

When engaged, the autopilot flight director computers (AFDCs) control airplane pitch. The AFDCs supply pitch commands to the PFCs to command the elevator surfaces. The PFCs also supply column backdrive commands. The PFCs send these commands to the AFDCs, which supply column movement commands to the backdrive actuators. The backdrive actuators move the control columns the same amount as the pilot would move them manually for the same surface movement. This gives the flight crew visual feedback when the AFDC moves the flight control surfaces. The flight

crew can override the autopilot with sufficient force on the control column to overcome the backdrive. Protection functions such as overspeed and stall protection functions in the PFCs supply flight envelope protection in both manual and autopilot operation. The overspeed protection supplies a pitch-up elevator command. The stall protection supplies a pitch-down elevator command. Figure 12-57 shows the elevator control system and Fig. 12-58 shows the elevator system components.

Stabilizer

The elevator off-load function automatically sets the position of the stabilizer during manual flight or in autoflight. In this mode, the PFCs monitor elevator deflection and transfer pitch changes to the stabilizer. Once the stabilizer goes to its commanded position, the elevator moves to neutral. The flight crew command the stabilizer to move with either the pitch trim switches on the control wheels or the alternate pitch trim levers on the control stand. Two hydraulic motors with two speeds move the horizontal stabilizer through a jackscrew mechanism. Four rates of horizontal stabilizer movement are available which are controlled by the primary flight computer based on air/ground and airspeed data. In flight, during airplane acceleration, the stabilizer moves at a high rate of speed when the airspeed is below 230 knots. Above 230 knots, the stabilizer moves at a low rate. Stabilizer position is indicated on the Engine Indicating and Crew Alerting System (EICAS). Figure 12-59 shows the stabilizer control system.

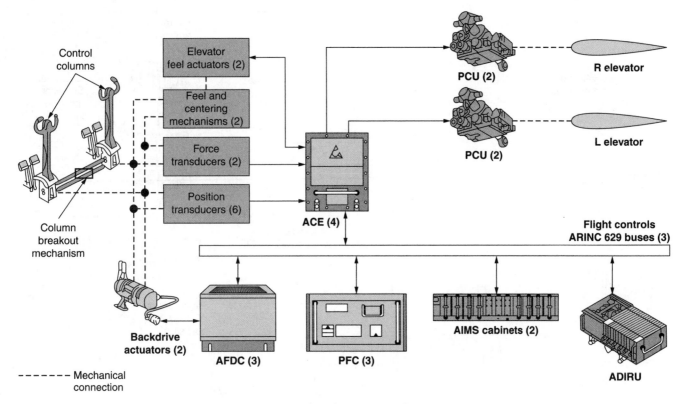

FIGURE 12-57 Elevator control system.

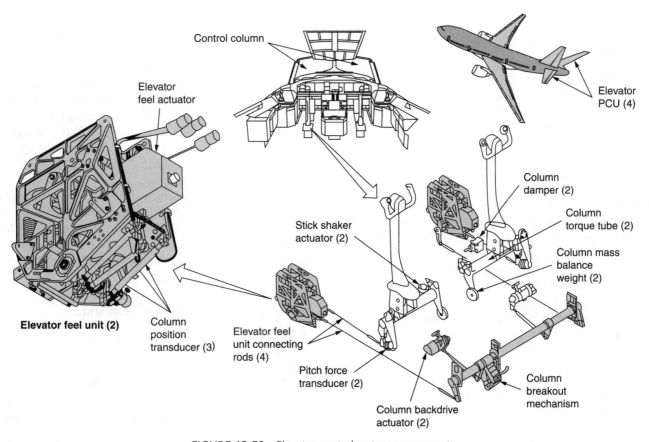

FIGURE 12-58 Elevator control system components.

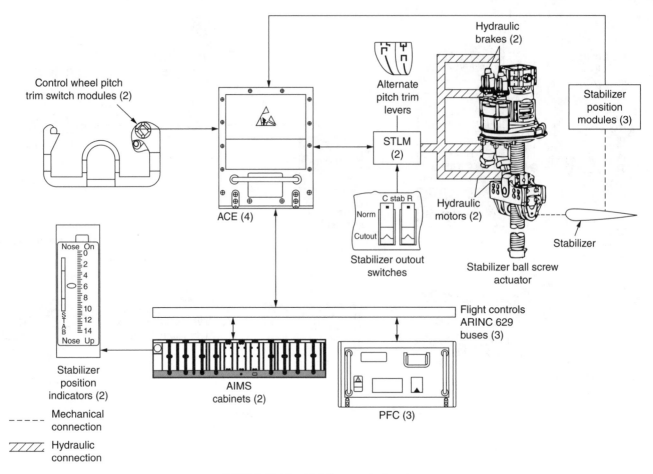

FIGURE 12-59 Stabilizer control system.

Roll Control

The ailerons and flaperons control the flight attitude of the airplane about the longitudinal axis. They also increase the wing lift with the high lift system during takeoff, approach, and landing. The ailerons are on the wing outboard trailing edge. The flaperons are standard inboard ailerons which also operate as flaps. They are between the inboard and the outboard flaps on the wing trailing edge. During roll control, the aileron and flaperon on one wing move up, and the aileron and flaperon of the other wing move down. The pilot manually commands a roll correction with the rotation of the control wheels. When engaged, the autopilot automatically commands the ailerons and the flaperons. During autopilot operation, actuators backdrive the control wheels. Figure 12-60 illustrates the roll control system. The spoilers also supply roll control. On the ground and during flight below cruise speed, the ailerons and flaperons are fully operational. At cruise speed, the ailerons fair to the wing surface and lock out. The flaperons supply roll control. During extension of the flaps, the ailerons and flaperons droop to increase the lift of the wing. The ailerons and flaperons of both wings move down. When drooped, the ailerons and flaperons are fully operational for roll control.

The flight crew uses two conventional control wheels to control roll. A wheel jam breakout mechanism supplies a mechanical link between the control wheels. If a control wheel jams, the other continues to control. Six position transducers change the flight crew commands of the control wheel to analog electrical signals. Figure 12-61 shows the flight deck components of the roll control system. These signals go to the four actuator control electronics (ACEs). The ACEs change the signals to digital format and send them to the three primary flight computers (PFCs) through the flight controls ARINC 629 buses. The PFCs use the control wheel position data, along with data from the ADIRU and AIMS, to calculate control surface commands. The PFCs send the digital commands to the ACEs, which change them to analog signals. The ACEs send the analog position commands to the power control units (PCUs), which move the control surfaces. The position transducers on the actuator pistons supply position feedback to the ACEs.

Manual Operation of the Aileron Trim

The aileron trim switches and the trim actuator permit the flight crew to trim out unwanted control wheel forces. When the pilots move the aileron trim switches on the aisle stand, the switches send a signal to the aileron trim actuator. The trim actuator moves the control wheels and the position transducers through the feel and centering mechanism. This changes the neutral position of the control wheels. The transducers send a signal to the ACEs and PFCs to move the ailerons and flaperons.

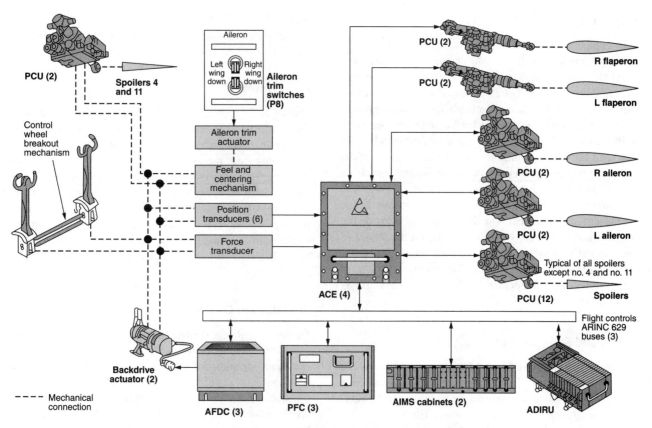

FIGURE 12-60 Roll control system.

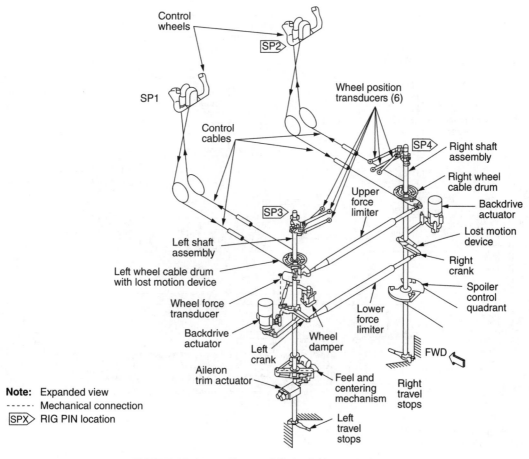

FIGURE 12-61 Roll control flight deck components.

Autopilot Operation

When engaged, the autopilot flight director computers (AFDCs) control the roll rate and attitude of the airplane. The AFDCs supply roll commands to the PFCs through the flight controls ARINC 629 buses. The PFCs use these inputs to calculate position commands for the ailerons and flaperons in a similar manner as for the manual operation. The PFCs also calculate wheel backdrive commands. The PFCs send these commands to the AFDCs, which calculate control wheel movement commands for the backdrive actuator. The backdrive actuator moves the wheels the same amount as the pilot would move them manually for the same surface movement. The flight crew can override the autopilot through the use of enough force on the wheel to overcome the backdrive actuator. A bank angle protection (BAP) function in the PFCs supplies protection in both manual and autopilot operation when the bank angle is more than 35°.

Spoilers and Speedbrakes

The spoilers help the ailerons and flaperons control airplane roll about the longitudinal axis. They also supply speedbrake control to reduce lift and increase drag for descent and landing. There are seven spoilers on each wing. The five outboard spoilers are forward of the outboard flap. The two inboard spoilers are forward of the inboard flap. Each spoiler has an assigned number, from left to right, of 1 through 14. Each spoiler is part of a symmetrical pair, for example, spoiler pair 4 and 11. One ACE and one hydraulic source control a pair.

Roll Control

During roll control, the spoilers on one wing move up and the spoilers on the other wing stay down. The pilots manually control roll with the rotation of the control wheels. When engaged, the autopilot automatically commands the spoilers. During autopilot operation, the backdrive actuators backdrive the control wheels. In roll control, all spoilers except for 4 and 11 are fly-by-wire. Spoilers 4 and 11 receive mechanical signals for roll control. Symmetrical spoiler pairs fair and lock out if one actuator in the pair fails. Spoilers 5 and 10 fair and lock out as a function of altitude and airspeed.

Speedbrake Control

During speedbrake control, the spoilers on both wings move symmetrically, as shown in Fig. 12-62. The pilots manually command speedbrake control with a conventional speedbrake lever on the aisle stand. The autopilot does not control the speedbrake function. In speedbrake control, all spoilers are fly-by-wire, including spoilers 4 and 11. The auto speedbrake system supplies automatic extension or retraction of the speedbrakes during landings and refused takeoffs.

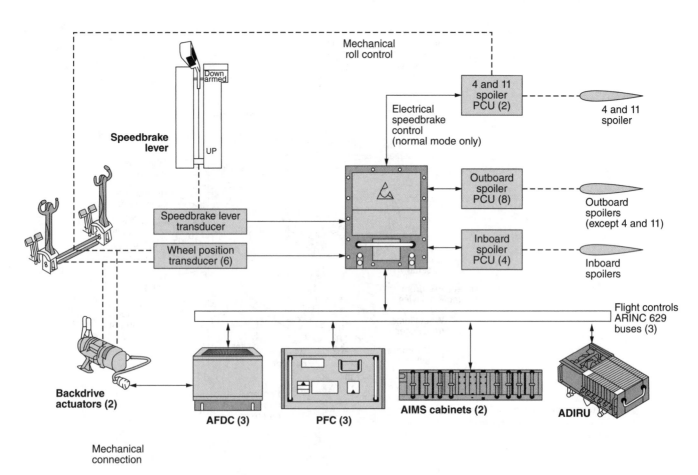

FIGURE 12-62 Spoiler and speedbrake control.

Flap/Slat Control System

The leading edge (LE) slats improve the takeoff and landing performance of the airplane. When the slats extend, they increase the lift of the wing. They also increase the angle of attack (ADA) at which the wing stalls. Each wing has an inboard slat, a Krueger flap, and six outboard slats. The slats are part of the high lift control system (HLCS). The HLCS electrically controls the slats with a fly-by-wire system. The slats operate in sequence with the flaps. Figure 12-63 shows the slats control system. At cruise speed and altitude, slat extension is inhibited. The autoslat function extends the slats to improve the airplane stall performance near stall conditions. The slats are monitored for skew or asymmetry. When there is a skew or asymmetry, the slat drive shuts down. Figure 12-64 shows the slat drive system.

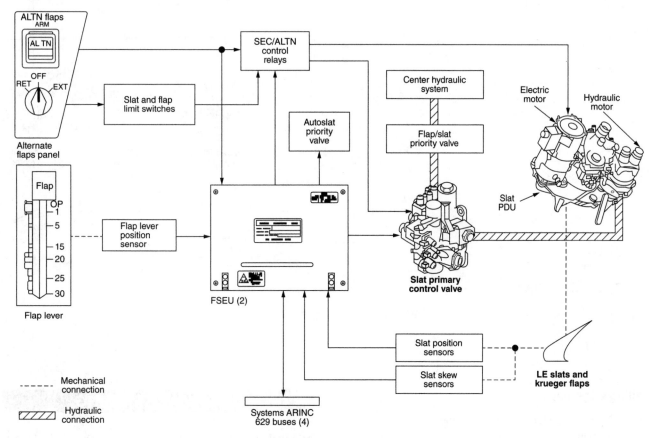

FIGURE 12-63 Slat control system.

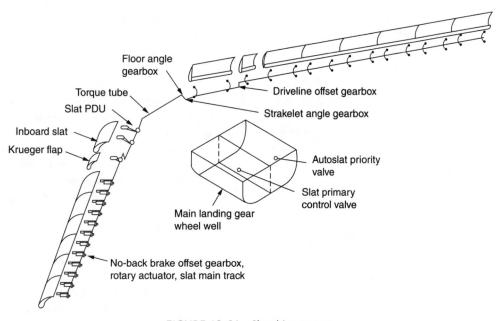

FIGURE 12-64 Slat drive system.

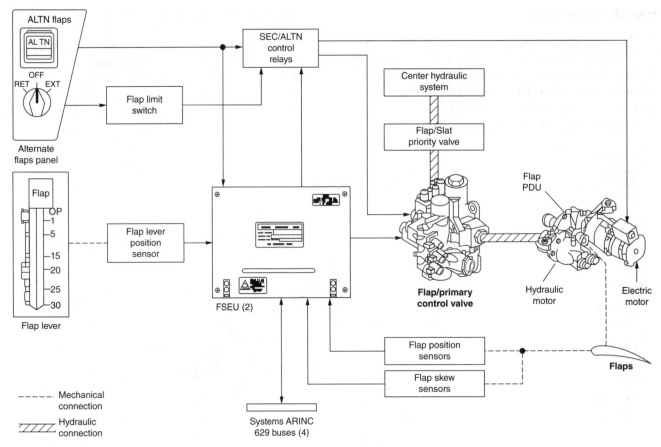

FIGURE 12-65 Trailing edge flap system.

Trailing edge flaps

The trailing edge (TE) flaps improve the takeoff and landing performance of the airplane. When the flaps extend, they increase the lift of the wing. They also help decrease the airspeed during landings and approaches. The flaps are part of the high lift control system (HLCS). The HLCS electrically controls the flaps with a fly-by-wire system. Figure 12-65 shows the trailing edge flap control system. During takeoff and landing, the pilots use the flap lever to command the flaps to move to an extended position. During cruise, the pilots command the flaps to retract to the up position. Each wing has an inboard flap and an outboard flap. The inboard flap is double-slotted and has a main flap and an aft flap. The outboard flap is single-slotted and is one piece (see Fig. 12-65).

Normal operation

The flaps operate in sequence with the slats. At cruise speed and altitude, flap extension is inhibited. The load relief function retracts the flaps to prevent structural damage to the flaps at high airspeeds. The flaps are monitored for skew or asymmetry. When the system detects a skew or asymmetry, the flap drive shuts down. Figure 12-66 shows the trailing edge flap drive system.

CONTROL SURFACE RIGGING

Although the actual rigging of controls for any particular airplane should be accomplished according to the instructions of the manufacturer, there are certain operations common to almost all systems and by which rigging of the majority of small airplanes can be accomplished. The following guidelines may be helpful in the installation and rigging of control surfaces.

1. It is recommended, though not always necessary, to level and place the airplane on jacks during the rigging and adjustment of controls.
2. Remove turnbuckle barrels from cable ends before withdrawing cables through the structure.
3. Tie a cord to the cable end before drawing the cable through the structures to facilitate the reinstallation of the cables.
4. Label all cable ends, etc., before disconnecting the cables.
5. When turnbuckles have been set to correct cable tension, no more than three threads should be exposed from either end of the turnbuckle barrel.
6. Cable tension should be taken with the appropriate surface control in its neutral position.

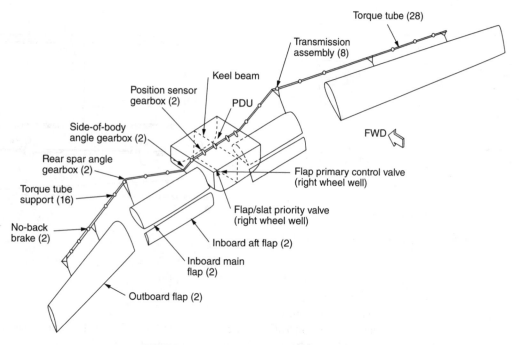

FIGURE 12-66 Trailing edge flap drive system.

7. The manufacturers of some aircraft provide rigging holes for the rigging of flight controls, as shown in Fig. 12-67. The hole in the bellcrank is lined up with the hole in the rigging fitting. A pin is placed through the two holes to prevent the bellcrank from moving. The mechanism is then rigged to achieve a neutral control surface setting and a neutral cabin control position. Another way of achieving a neutral control surface position is to clamp the surface in a neutral position, as shown in Fig. 12-68, or to use a straight-edge rigging tool, as shown in Fig. 12-69.

Among the objectives to be accomplished during the rigging procedure are (1) correct cable tensions, (2) balance or synchronization between dual controls, (3) synchronization of the cockpit control with the control surfaces to which it is linked, and (4) setting the range of control surface movement.

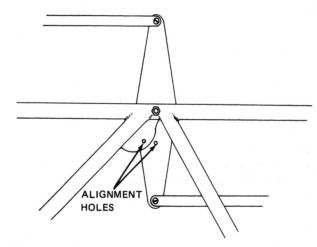

FIGURE 12-67 An alignment hole used for control surface rigging.

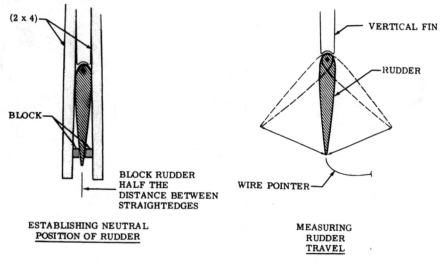

FIGURE 12-68 Centering and measuring rudder travel. (*Cessna Aircraft Co.*)

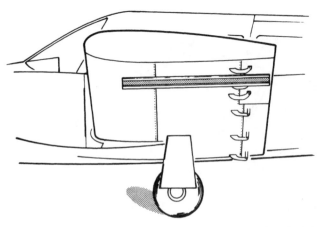

FIGURE 12-69 Straight-edge rigging tool. (*Piper Aircraft Co.*)

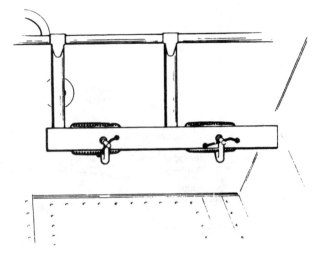

FIGURE 12-70 Cockpit controls locked in neutral position. (*Piper Aircraft Co.*)

The objective in rigging a control system is to have the cockpit control in neutral at the same time that the control surface involved is in neutral. Usually the cockpit control is locked in the neutral position by means of the **control lock** in the airplane, if this lock position is neutral, or by an installed locking arrangement such as a bar, block, or rod with clamps, as shown in Fig. 12-70, that will hold the control in neutral.

The cables are rigged by means of the turnbuckles so that the bellcrank assemblies are in the neutral position. A special rigging tool such as the one shown in Fig. 12-71 may be needed to determine the bellcrank's neutral position.

Cable tensions must be set at the same time the control units are positioned because any change in control cable tension will likely produce a change in control surface position. Cable tension is measured by means of a **tensiometer**, or cable-tension indicator, as shown in Fig. 12-72. The correct riser for the cable size is installed on the indicator, and then

the indicator is hooked over the cable. The control lever is moved up against the case, and the reading on the dial is noted. This reading is located on the conversion chart supplied with the instrument, and the cable tension is shown opposite the indicator reading in the column for the size of cable being checked. The cable-tension reading obtained may have to be corrected for ambient temperature through the use of a temperature-correction chart as shown in Fig. 12-73. It may be necessary to check and adjust cable tensions with seasonal changes during the year as the ambient air temperatures change. This change in temperature can cause the airframe structure to expand or contract an amount different from that of the control cables, due to the difference in coefficients of expansion of the airframe and cable materials. These changes may cause the cable tension to be

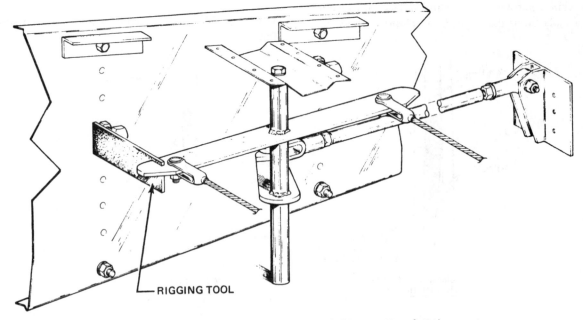

RIGGING TOOL

FIGURE 12-71 Bellcrank rigging tool. (*Cessna Aircraft Co.*)

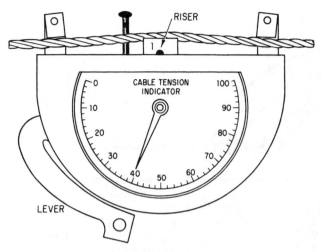

FIGURE 12-72 Use of a tensiometer.

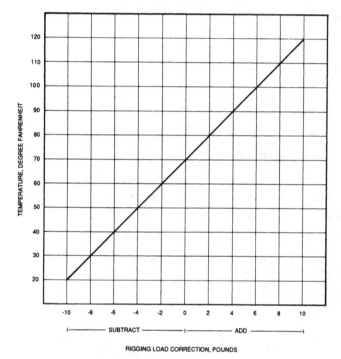

FIGURE 12-73 Cable-tension temperature-correction chart. (*Piper Aircraft Co.*)

too great or too small, resulting in excessive system wear or loss of control effectiveness. Some large aircraft incorporate cable-tension regulators in their design to correct for these variations automatically.

The tension of control cables must be within the range specified by the manufacturer. Excessive tension places undue stress on the cables, pulleys, pulley brackets, and all other parts associated with the support of the control system. During operation, these stresses can lead to failure of the system. Excessive tension also increases the wear on cables, pulleys, bearings, and other parts. In addition, excessive tension increases the difficulty of moving the controls, thus reducing the pilot's ability to control the aircraft smoothly and accurately.

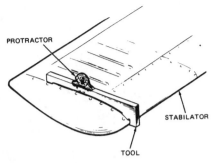

FIGURE 12-74 Use of rigging tool.

To measure control surface travel, a special rigging tool that lays on the control surface, as illustrated in Fig. 12-74, may be required. The travel is measured by means of a universal protractor, as shown in Fig. 12-75, or a digital protractor, as illustrated in Fig. 12-76.

The range of control-surface movement is given in the manufacturer's instructions or the Type Certificate Data Sheet. Control surface travel is limited by adjustable stops. Various types of control surface stops for mechanical flight-control systems may be used. Three methods are shown in Figs. 12-77, 12-78, and 12-79. In Fig. 12-77, travel stop bolts are installed in nut plates that are mounted on a bulkhead. The stop is adjusted in or out as necessary to obtain the proper control travel. The flat plates on the bellcrank allow the control to move until the plate contacts the stop bolt. In Fig. 12-78, a stop bolt is mounted in the control stick. The bolt length protruding from the fork is adjusted to achieve the proper control movement. In Fig. 12-79, one of four positions can be selected on the bellcrank stops. The hole in the bellcrank is off center, which gives a different stop value for each side of the stop.

The control cables are then adjusted and tensioned so the control surface affected is in neutral. All adjusting elements, such as turnbuckles and rod ends, must be checked after rigging to see that they are within limits. Rod ends must be screwed into or onto the rod at a distance sufficient to prevent the insertion of safety wire through the inspection hole. Turnbuckles must not have more than three threads showing outside the barrel.

Another point to remember in the final inspection of a control rigging job is to see that no cable splice or fitting can come within 2 in [5.08 cm] of a fairlead, pulley, guide, cable guard, or other unit that could cause the control to jam. The control should be moved to its extreme position both ways in making this inspection.

After all adjustments are completed, appropriate safeties (safety wire, cotter pins, jam nuts, etc.) should be installed.

BALANCING CONTROL SURFACES

The control surfaces for new airplanes are properly balanced, both statically and aerodynamically, at the factory. After the airplane undergoes overhaul, painting, or repair of the control surfaces, the static balance may be altered to

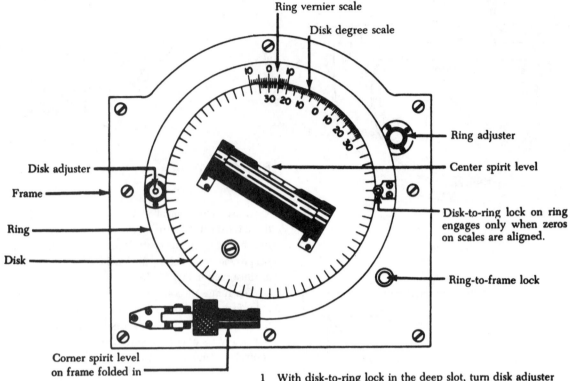

1   With disk-to-ring lock in the deep slot, turn disk adjuster to lock disk to ring.

2   Move control surface to neutral. Place protractor on control surface and turn ring adjuster to center bubble in center spirit level (ring must be unlocked from frame).

3   Lock ring to frame with ring-to-frame lock.

4   Move control surface to extreme limit of movement

5   Unlock disk from ring with disk-to-ring lock.

6   Turn disk adjuster to center bubble in center spirit level.

7   Read surface travel in degrees on disk and tenths of a degree on vernier scale.

FIGURE 12-75   A universal protractor may be used to measure alignment and movement angles.

FIGURE 12-76   Digital electronic protractor. (*Kell-Strom Tool Co.*)

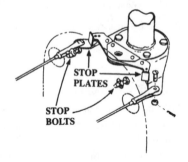

FIGURE 12-77   Bolt stops located on a bulkhead. (*Cessna Aircraft Co.*)

the extent that **flutter** will occur in flight. Flutter can lead to excessive stresses in flight, with the result that the control surface, its attachments, and the structure near the attachments can suffer damage such as cracks or complete failure. It is, therefore, necessary that the control surface balance be

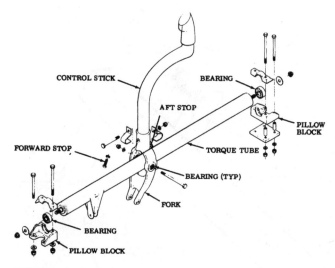

FIGURE 12-78 Bolt stops located in a control stick. (*Ayres Corp.*)

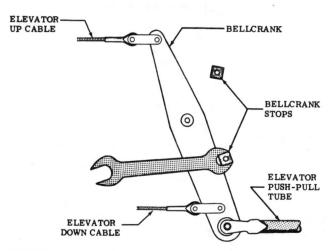

FIGURE 12-79 The use of square bellcrank stops. (*Cessna Aircraft Co.*)

checked whenever any operation is performed on a control surface that can change the static balance.

A control surface balance is checked by placing the surface on suitable bearings at the hinge line and measuring the moment forward or aft of the hinge line. At this time the

control surface should be complete with all parts, as it would be in flight. A new, unpainted surface should be painted before balancing. Trim tabs, pushrods, and all other parts should be installed.

A diagram to illustrate the balance-checking procedure for a light airplane is shown in Fig. 12-80. In this drawing, the control surface is supported at the hinge line on a knife-edge support. A scale and screw jack are placed under the trailing edge of the surface. The screw jack is adjusted up or down to provide the correct angular reading on the bubble protractor, which is placed on top of the surface. After the correct angle is established, the protractor is removed and the reading of the scale is taken. The weight of the screw jack is subtracted from the scale reading to give the weight of the surface. The weight of the control surface at the scale support point is multiplied by the distance D to give the value of the moment. If the moment is too great, additional weight will have to be installed on the surface forward of the hinge line. Another method used to balance control surfaces is the utilization of a balancing tool, as shown in Fig. 12-81. This balancing tool is adapted to most control surfaces. As illustrated in Fig. 12-82, it will provide information on the amount of weight that must be added or removed from the control surface. Balance weights for a stabilator and an aileron are shown in the drawings of Fig. 12-83.

In some cases, balance weights consist of lead plates attached to the leading edge of the control surface. This type of installation is shown in Fig. 12-84.

INSPECTION AND MAINTENANCE

The following inspections are typical of those that should be made for control systems; however, the inspections for a particular type of aircraft should follow the instructions given in the manufacturer's service or maintenance manual.

1. Examine all cables for wear or corrosion. Wear will be most apparent at or near pulleys, the ends of cable fittings, fairleads, and other points where the cable may come into contact with another part of the system. Broken strands of cable can be detected by wrapping a rag around the cable and moving the

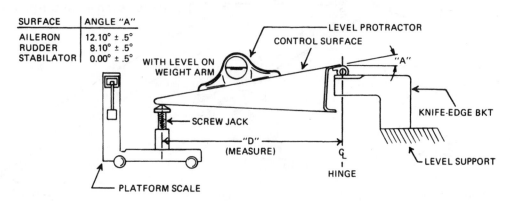

SURFACE	ANGLE "A"
AILERON	12.10° ± .5°
RUDDER	8.10° ± .5°
STABILATOR	0.00° ± .5°

FIGURE 12-80 Balance-checking procedure.

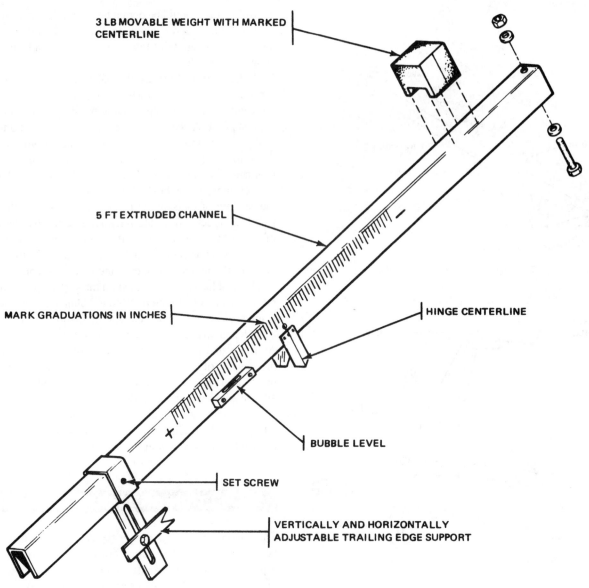

3 LB MOVABLE WEIGHT WITH MARKED CENTERLINE

5 FT EXTRUDED CHANNEL

MARK GRADUATIONS IN INCHES

HINGE CENTERLINE

BUBBLE LEVEL

SET SCREW

VERTICALLY AND HORIZONTALLY ADJUSTABLE TRAILING EDGE SUPPORT

FIGURE 12-81   Control surface balance tool. (*Piper Aircraft Co.*)

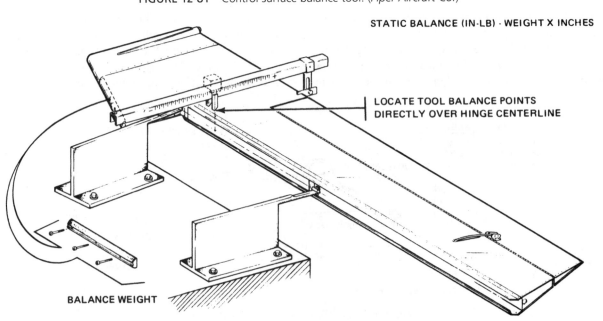

STATIC BALANCE (IN-LB) - WEIGHT X INCHES

LOCATE TOOL BALANCE POINTS DIRECTLY OVER HINGE CENTERLINE

BALANCE WEIGHT

FIGURE 12-82   Using control surface balancing tool on rudder. (*Piper Aircraft Co.*)

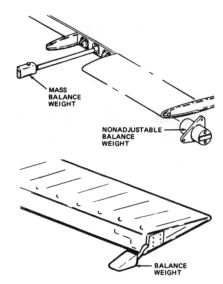

FIGURE 12-83 Balance weights for stabilator and aileron. (*Piper Aircraft Co.*)

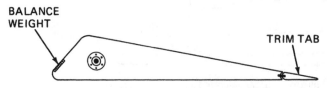

FIGURE 12-84 Balance weight on an aileron.

rag back and forth. Cables that are badly worn, have broken strands, or are appreciably corroded should be replaced.

2. Examine all pulleys for wear, cracks, and alignment. If a pulley is worn to an appreciable extent or cracked, it should be replaced. The pulleys should turn freely when the control cables are moved. If a pulley is out of line, it will cause wear to both the pulley and the cable. The mountings for such pulleys should be corrected and the cable carefully examined for wear.

3. Where cables pass through fairleads or guides, the deflection of the cable should be noted. If it is more than 3°, a correction must be made. The wear of the cable and the fairlead should be checked.

4. Wear of pulley bearings, bearing bolts, bushings, clevis pins, and all other moving parts should be checked. Replacement must be made of all parts worn beyond specified limits.

5. Cable tension should be checked by means of a tensiometer. The cable tension is adjusted by means of turnbuckles.

6. The system should be checked to see that no cable fitting comes within 2 in [5.08 cm] of a pulley, fairlead, or guide when the control is moved to its limits.

7. Control surface travel should be checked to verify that it corresponds to the Type Certificate Data Sheet. Travel can be adjusted by means of the stop bolts and/or rod ends. The control surface stops should be set to make contact before the cockpit control has reached the end of its travel.

8. After all adjustments are made, all safetying of turnbuckles, clevis pins, nuts, etc., must be examined for correct application and effectiveness. Defective safetying must be corrected. The

9. Upon completion of inspection, adjustment, and service, the control system should be given an operational check. The controls should move smoothly and easily through their full range of travel and should not exhibit any looseness or play. The systems must be checked for direction of control movement because it is often possible to cross cables and cause reverse movement. This is particularly true of aileron systems. When the right rudder pedal is pressed, the rudder should move to the right; when the control is pulled back, the elevator should move upward; and when the wheel is moved to the right, the left aileron should move down and the right aileron should move up.

HELICOPTER FLIGHT CONTROLS

Helicopter flight controls allow control of (1) movement about the three axes of the aircraft, (2) the engine power, and (3) the rotor system lift. The controls consist of the cyclic control, antitorque control, and throttle and collective control.

The *cyclic control* looks like a conventional stick control for an airplane. It allows the pilot to operate the main rotor system swash plate, which tilts and causes the rotor blades to increase or decrease their blade angle at appropriate positions in the rotor plane of rotation. This causes the rotor to tilt and increase the rotor lift on one portion of the rotor disk and decrease the lift on the opposite portion of the disk. With this difference in lift across the disk, the direction of lift of the rotor disk is changed. The rotor tilts and causes the helicopter to move in the appropriate direction, left or right laterally or forward or back longitudinally.

The tail rotor thrust is controlled by the pilot's foot pedal, called the **antitorque pedal** or **rudder pedal**. As the pilot changes the power being delivered to the main rotor system or the amount of pitch in the main rotor blades, he or she also alters the angle of the tail rotor blades to maintain the heading of the aircraft. If no correction is made, the nose of the aircraft will swing to one side or the other, the direction being determined by the pilot's control inputs and the rotor system direction of rotation.

The engine throttle control is usually incorporated in the twist grip located in the grip of the *collective control*. When the powerplant is under manual control of the pilot, the grip must be twisted to increase or decrease engine power, as the situation requires. Most helicopters have a governor or coordinator mechanism that maintains an appropriate level of power output based on the rotor system load and the position of the collective control.

The collective control lever is located on the left side of the pilot's seat and moves up and down to control the amount of lift being generated by the main rotor system. This control causes all the blades to increase their blade angle and lift when the control is pulled up and decrease their blade angle and lift when it is moved down. Pilot cockpit controls are shown in Fig. 12-85.

A few helicopter designs use a dual rotor system, where two main rotors are attached at opposite ends of the aircraft and synchronized so that the rotor blades intermesh. The

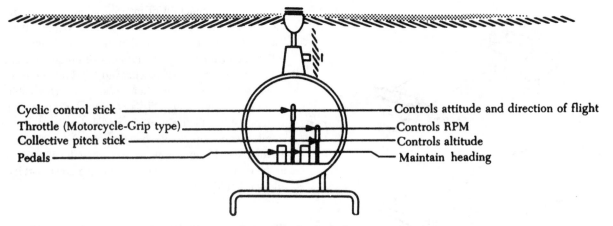

FIGURE 12-85 The flight controls of a helicopter and their purpose.

Cyclic control stick — Controls attitude and direction of flight
Throttle (Motorcycle-Grip type) — Controls RPM
Collective pitch stick — Controls altitude
Pedals — Maintain heading

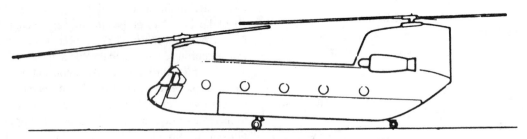

FIGURE 12-86 A dual-rotor helicopter. (*Boeing Vertol.*)

two rotors rotate in opposite directions, counteracting each other's torque, and the helicopter does not require a tail rotor. The cockpit controls are the same as for a single-rotor helicopter, but the operation of the rotor head–control mechanisms is much more intricate than for a single-rotor system. A dual-rotor helicopter is shown in Fig. 12-86. A complete discussion of helicopter configurations and aerodynamics can be found in the text *Aircraft Basic Science*.

Rotor Systems

Helicopter rotor systems are major aircraft assemblies used to convert the engine power into lift, propulsive force, and directional control. Helicopters can have one or two main rotors. If the aircraft has one main rotor, it will have an auxiliary rotor system to counter the main rotor system torque.

The rotor blades may be constructed of wood, metal, composite materials, or some combination of these materials. Most modern rotor blades incorporate bonded metal components in combination with composite components for strength and fatigue resistance.

Rotor systems are driven by the aircraft powerplant through a transmission. The transmission is used to reduce the high engine rpm to a usable rotor rpm. The transmission also provides power to aircraft system components such as fluid pumps and electrical generators. Belts, hydraulic couplings, free-turbine engines, and mechanical clutches are used on helicopters to connect the powerplant to the transmission.

A freewheeling unit is used to disconnect the engine from the rotor system any time the engine power output is reduced to a point where it is no longer driving the rotor system. When

the freewheeling unit disconnects the engine from the rotor system, the helicopter enters a type of flight known as *autorotation*, where the air moving upward through the rotor system causes the rotors to turn. The aircraft can thereby descend to a landing with the rotor system providing lift, propulsion, and directional control, even though the engine may have failed.

Main Rotor Systems

There are three basic designs for the main rotor system. These are the *fully articulated*, *semirigid*, and *rigid* rotor systems.

The **fully articulated** rotor system normally has three or more rotor blades, and each blade can move by three different motions, independent of the other blades in the system. One motion, called **flapping**, is allowed through the rotor blade flap hinge, located near the rotor hub. This allows the blade to rise and fall as it rotates around the hub. The **lead-lag** motion is through the drag hinge. The drag hinge allows the blade to move ahead of (lead) or fall behind (lag) the normal axis of the hub extension. Each blade is also free to rotate about its central axis. This is called **feathering**. So, each blade can feather, lead or lag, and flap independently of the other blades. All this motion is required so that the blades can change their angle of attack and speed through the air as the aircraft speed changes and different control inputs are fed to the rotor system from the pilot's controls. A fully articulated rotor head is shown in Fig. 12-87.

A **semirigid** rotor system, shown in Fig. 12-88, uses only two rotor blades. The blades are rigidly attached to the central hub. A teetering hinge is used to connect the hub to the rotor shaft. Through this teetering hinge the blades can

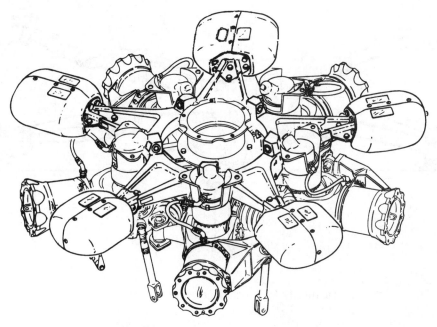

FIGURE 12-87　A fully articulated rotor head. (*Sikorsky.*)

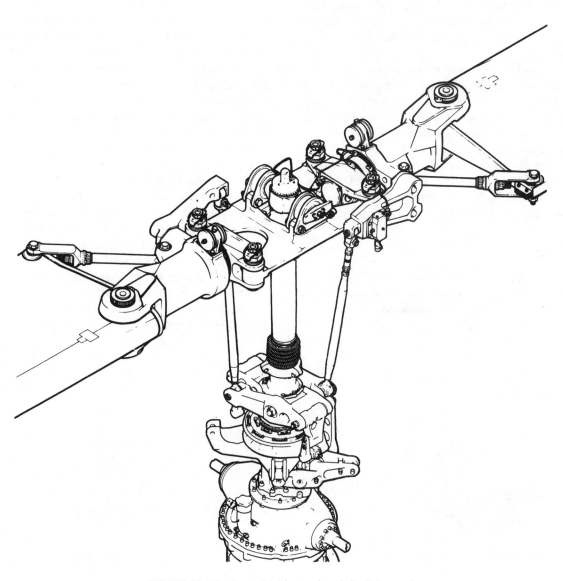

FIGURE 12-88　A semirigid rotor head. (*Bell Textron.*)

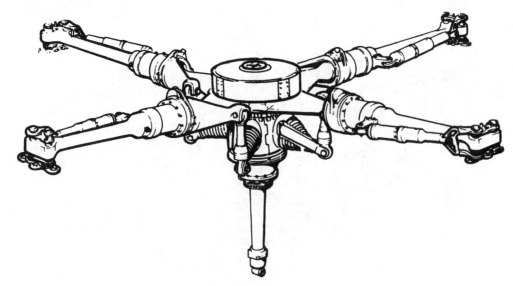

FIGURE 12-89   A rigid rotor head. (*Westland.*)

flap, one up and one down, and feather, one increasing pitch and one decreasing pitch, as a unit. If the hub has the blades underslung—that is, the blades are on a lower rotational plane than the pivot point of the hub—a drag hinge may not be required.

Figure 12-89 illustrates a **rigid** rotor system. This type of rotor system does not allow the blades to flap or drag. They can only feather. This design can use any number of blades and has become very popular for use in many modern helicopter designs when combined with composite blades.

The rotor system contains mechanical adjustments that the technician can use to correct the track of the tip paths of the rotor blades. If the blades are not in track—that is, if each blade does not follow the same path as the other blades—then a low-frequency vibration will exist during flight.

Tail Rotor

The conventional **tail rotor** consists of a rotor of two or more blades located at the end of the tailboom. The tail rotor is used with single-rotor helicopters to counteract yawing movement resulting from the torque effect of the engine-transmission–main rotor system. It does this by generating a side thrust at the end of the tailboom that tries to rotate the helicopter about its vertical axis. This yawing thrust is varied to counter the yawing movement imposed by the main rotor torque. During cruising flight and hovering changes in altitude, the tail rotor corrects for torque. The tail rotor can also be used to change the heading of the helicopter when it is in a hover.

The Fenestron tail rotor system uses what might be called a ducted fan antitorque system. This system uses a multibladed fan mounted in the vertical fin on the end of the tailboom. This arrangement reduces aerodynamic losses due to blade tip vortices, reduces the blanking of the antitorque control system by the vertical stabilizer, and reduces the chance of people walking into the tail rotor.

The **ring guard** is an adaptation of the conventional tail rotor system, which has a ring built around the tail rotor. This ring acts as a duct for the tail rotor, increases the safety of the tail rotor for ground personnel, and eliminates the vertical stabilizer blanking effect. These three types of antitorque controls are shown in Fig. 12-90.

The NOTAR system (NO TAil Rotor), illustrated in Fig. 12-91, was developed by McDonnell Douglas Helicopter

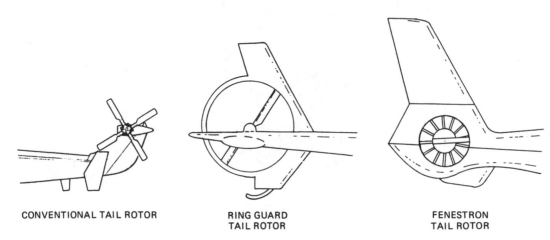

CONVENTIONAL TAIL ROTOR   RING GUARD TAIL ROTOR   FENESTRON TAIL ROTOR

FIGURE 12-90   Various types of tail rotors.

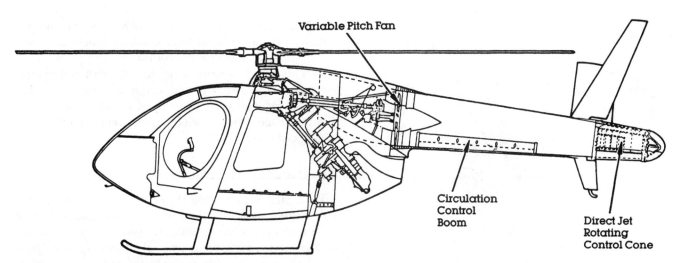

FIGURE 12-91   A NOTAR helicopter uses a flow of air instead of a tail rotor to control the yawing motion. (*McDonnell Douglas Helicopter Co.*)

Company to eliminate the hazards, maintenance, and noise of the conventional tail rotor. This system uses a ducted airflow in a tail cone to generate a predetermined amount of lift on one side of the tail cone and a controllable rotating cold-air exhaust duct to counteract variations in the main rotor system torque.

Rigging of Helicopters

Controls for helicopters in some characteristics are similar to those for airplanes; however, the differences are greater than the similarities. Controls for a helicopter cause the aircraft to rise, move forward, sideways, or backward, and change heading or turn. The main rotor provides lift and thrust, whereas the tail rotor counteracts the torque of the main rotor and provides directional control.

The main rotor is controlled by two principal systems, the *collective-pitch control* and the *cyclic-pitch control*. The **collective-pitch control** changes pitch on all blades of the main rotor simultaneously. Collective pitch is adjusted by raising or lowering the collective-pitch lever (stick).

Collective pitch and engine power control the vertical flight of a helicopter. The collective-pitch control mechanism for the Bell Model 206L helicopter is shown in Fig. 12-92. The collective control utilizes the center servo and control tubes to raise and lower the swash plate to which the pitch-change link rods are attached. Adjustments are made on the control tube and link, as shown, to provide the correct pitch range for the rotor blades.

The **cyclic-pitch control** is employed to change the pitch or angle of the plane or disk through which the main rotor blades rotate. The flying helicopter will move in the direction in which the rotor disk is tilted. The disk is caused to tilt because the cyclic control changes the blades' pitch angles to different values as the blades rotate through their paths of travel around the disk. An upward force caused by increase of pitch on a blade at a point 90° after the blade has passed the forward point of the helicopter will cause the rotor disk to tilt forward due to gyroscopic precession. Thus, when the

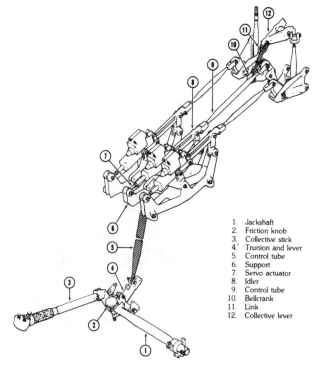

1. Jackshaft
2. Friction knob
3. Collective stick
4. Trunion and lever
5. Control tube
6. Support
7. Servo actuator
8. Idler
9. Control tube
10. Bellcrank
11. Link
12. Collective lever

FIGURE 12-92   Collective-pitch control for a helicopter. (*Bell Helicopter-Textron.*)

cyclic control stick is moved forward, the up force on the rotor is applied at one side, and the rear of the disk rises.

A cyclic-pitch control system is shown in Fig. 12-93. Adjustments for the proper range of cyclic control are made by means of the control tubes indicated in the drawing.

The direction in which a helicopter is pointed is controlled by the **antitorque rotor** (tail rotor). The control system for the tail rotor is shown in the drawing of Fig. 12-94. The purpose of the system is to change the pitch of the rotor blades, thus changing the sideward thrust exerted by the rotor. The rotor speed remains essentially constant when the helicopter is in flight.

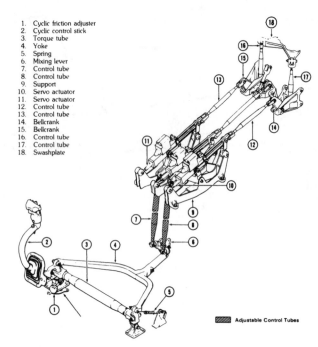

1. Cyclic friction adjuster
2. Cyclic control stick
3. Torque tube
4. Yoke
5. Spring
6. Mixing lever
7. Control tube
8. Control tube
9. Support
10. Servo actuator
11. Servo actuator
12. Control tube
13. Control tube
14. Bellcrank
15. Bellcrank
16. Control tube
17. Control tube
18. Swashplate

▨▨▨ Adjustable Control Tubes

FIGURE 12-93 Cyclic-pitch control system. (*Bell Helicopter-Textron.*)

Note from the drawing that the tail rotor pitch is changed through the operation of control rods, walking beams, and bellcranks. To establish the correct range of pitch for the rotor blades, adjustments are made in the push-pull control-rod ends. The system should be rigged so the foot pedals in the cockpit are in the neutral position when the tail rotor thrust balances the torque of the main rotor in normal operating conditions.

Helicopter Vibrations

The design and operating characteristics of helicopters are such that a variety of vibrations occur as a natural result of the forces generated by engines, rotors, transmissions, and other moving parts. Some types of vibrations below specified levels of intensity are expected and are acceptable. Vibrations that are abnormal or above a certain intensity cause discomfort to the crew and passengers and are likely to be damaging to the structure of the aircraft. These vibrations must be controlled or eliminated by adjusting, balancing, or repairing the areas affected.

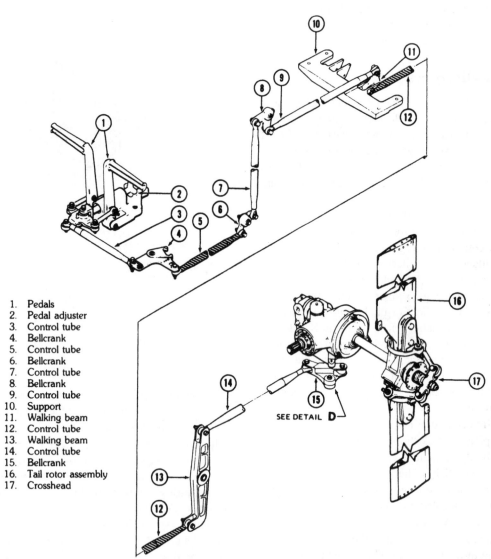

1. Pedals
2. Pedal adjuster
3. Control tube
4. Bellcrank
5. Control tube
6. Bellcrank
7. Control tube
8. Bellcrank
9. Control tube
10. Support
11. Walking beam
12. Control tube
13. Walking beam
14. Control tube
15. Bellcrank
16. Tail rotor assembly
17. Crosshead

SEE DETAIL **D**

FIGURE 12-94 Tail rotor–control system. (*Bell Helicopter-Textron.*)

Vibration frequencies in a helicopter are designated as *extremely low*, *low*, *medium*, and *high*. Extremely low frequency vibrations are less than one vibration cycle per revolution of the main rotor. One vibration per revolution is abbreviated 1/rev or 1:1. The extremely low vibration is limited to pylon rock and is controlled, primarily, by the design and shock mounting of the transmission.

Low-frequency vibrations, 1/rev or 2/rev, are associated with the main rotor. The 1/rev vibrations are of two basic types, vertical or lateral. In a vertical vibration, the entire helicopter tends to move up and down, and in a lateral vibration one side of the helicopter moves up while the opposite side moves down, the movements occurring alternately in time with the vibration frequency. A 1/rev vertical vibration is caused by one blade developing more lift at a given point than the other blade develops at the same point. This is due to incorrect pitch and track.

Lateral vibrations due to an unbalance in the rotor are of two types: spanwise and chordwise. Spanwise unbalance is caused by one blade and hub being heavier than the other. A chordwise unbalance occurs when there is more weight toward the trailing edge of one blade than the other.

Rigidly controlled manufacturing processes and techniques eliminate all but minor differences between blades, resulting in blades that are virtually identical. The minor differences that remain will affect flight but are compensated for by adjustments of trim tabs and pitch settings.

Medium frequencies of 4/rev or 6/rev are inherent vibrations associated with most rotors. An increase in the level of these vibrations is caused by a change in the capability of the fuselage to absorb vibration or a loose component such as the skids or landing gear vibrating at that frequency. Changes in the fuselage vibration absorption can be caused by such conditions as fuel level, external stores, structural damage, structural repairs, internal loading, or gross weight. Abnormal vibrations in the medium-frequency range are nearly always caused by something loose.

High-frequency vibrations can be caused by anything in the helicopter that vibrates or rotates at a speed equal to or greater than that of the tail rotor. Generally, determining the cause of a high-frequency vibration should begin with an examination of the tail rotor and a check of the rotor track. If these are satisfactory, the next step is to remove the rotor and balance it on a suitable stand. If tail rotor balance is satisfactory, then an inspection of the complete drive system should be made. By observing the drive shafting with the cover removed during operation, the technician can detect a bent drive shaft, misalignment of the drive shafting, and other possible defects.

The vibrations associated with a helicopter will be affected by the number of blades on the rotor. It is, therefore, necessary that technicians examine the maintenance manual and vibration analysis for the type of helicopter they are troubleshooting. The general considerations discussed in this section apply primarily to a two-blade rotor system.

Tracking and Balancing the Main Rotor

Tracking of a helicopter rotor simply means determining if one blade follows the path or track of the other blade or

blades as they rotate during operation. The two principal methods for rotor tracking are *stroboscopic light tracking* and *flag tracking*.

Electronic equipment has been developed that greatly aids in the accurate tracking and balancing of helicopter rotors. Equipment of this type is manufactured by the Chadwick-Helmuth Company and is recommended by many manufacturers and operators. This electronic equipment is designated by the trade name Vibrex Track and Balance System and is illustrated in Fig. 12-95.

The Vibrex Track and Balance System is used to correct track and balance by developing data in flight through the use of accelerometers and stroboscopic lights. The signals from these devices are referenced to rotor position by means of a magnetic pickup and interrupter system. The data obtained are applied to circular charts which serve as computer devices through which the adjustments required and the amount of balance correction needed are determined.

The items required for analyzing the operation of a main rotor are a **magnetic pickup** on the fixed swashplate, **interrupters** on the rotating swashplate, **retroreflective targets** on the rotor blade tips, **accelerometers** on the airframe, a Strobex light, and an **electronic balancer**, which includes the **Phazor**.

The magnetic pickup consists of a coil of wire wound on a permanent magnet. The center post of the magnet, on which the coil is wound, is one pole of the magnet. The cylindrical shell of the pickup is the other pole. The magnetic pickup is secured to the fixed swashplate of the main rotor and generates an electrical pulse when its magnetic flux lines are interrupted by the interrupter, which is secured to the rotating swashplate and passes in proximity to the pickup gap. Interrupters must be magnetic material but must not be magnets.

Electrical pulses from the magnetic pickup are used to trigger the Strobex light each time an interrupter passes to provide light for viewing tip targets, as shown in Fig. 12-96, and observing the track and lead or lag of the blades. The pulses also provide an azimuth reference for the Phazor section of the balancer against which the accelerometer signal is measured to determine **clock angle**.

An accelerometer is a piezoelectric crystal device. A piezoelectric crystal produces electrical current when subjected to stress. When it is attached to a vibrating part, it generates an electrical signal representative of the physical motion of the point to which it is attached. The voltage varies from plus to minus as the point vibrates back and forth, and the amplitude of the signal is proportional to the amplitude of the vibration.

Retroreflective target material is self-adhesive tape with a coating that reflects light back to its source. The balancer is the electronic circuitry that receives the signals from the accelerometers and magnetic pickup and then converts these inputs to readings that are transferred to the charts from which corrective information regarding the location and amount of weight required is obtained. Some helicopter blades have a provision for adding or removing balance weight near the tip of

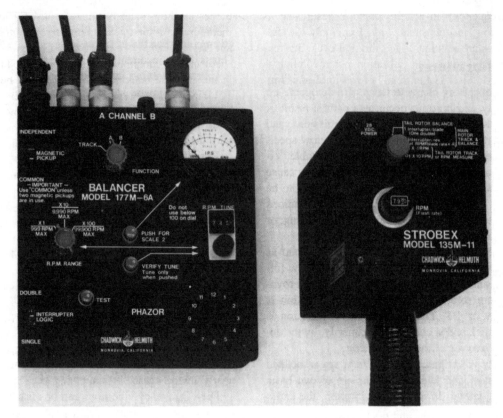

FIGURE 12-95 Vibrex Track and Balance System. (*Chadwick-Helmuth Co.*)

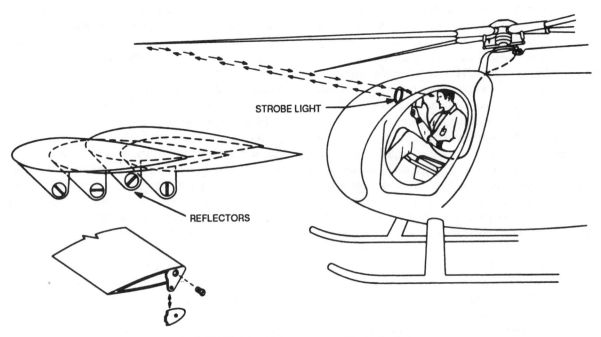

FIGURE 12-96 Blade tracking with strobe light.

the blade; other rotors have balance weights installed near the hub. The maintenance manual for the helicopter specifies the type of balances used and how they are to be installed. The technician must be certain to follow the instructions for the make and model of helicopter being serviced.

Another method for tracking rotor blades is the flag-tracking method illustrated in Fig. 12-97. In this method a tracking flag is constructed from aluminum or steel tubing. The flag portion should be made of strong, lightweight fabric tape. The reinforcing tape used in aircraft fabric work is a suitable material. The main rotor blade tips are colored with grease pencils, using a different color on each tip.

The technician holding the tracking flag should stand with the flag in front and the advancing main rotor blades

FIGURE 12-97 Checking main rotor track with a flag. (*Bell Helicopter-Textron.*)

behind. The technician should be able to see the pilot in order to receive preplanned signals to obtain a track. The technician, upon receiving a nod from the pilot to track the main rotor blades, will slowly rotate the tracking flag into the plane of the tip path. When the main rotor blades tips have touched the flag, the flag is immediately tipped away from the rotor plane. The relative vertical position of the main rotor blade tips will be indicated by transfer of the colored marks from the blade tips to the flag. There should be only one mark for each main rotor blade. It is recommended that two tracks be taken prior to making any adjustments. A gust of wind or a slight movement of the controls or helicopter may cause erroneous indications. Colored tracking marks will be left on the flag, thus indicating an out-of-track condition, which may be corrected by adjusting the pitch-link assembly.

REVIEW QUESTIONS

1. What is the difference between aircraft assembly and rigging?
2. What precautions must be taken in the hoisting of an aircraft?
3. Give a brief description of the procedure for aligning an aircraft structure.
4. What is the purpose of a jig?
5. Briefly describe the conditions that would require an alignment check for a complete aircraft.
6. What methods may be used to determine when an aircraft is level?
7. What device may be used to check *aircraft dihedral*?
8. What tools might be used to check an aircraft *angle of incidence*?
9. How may a "wing heavy" condition be corrected?
10. Define *stagger* and *decalage* in a biplane.
11. What is the effect of shortening the landing wires on a biplane?
12. What loads do flying wires carry on a biplane?
13. What flight controls are considered the *primary* flight controls?
14. What design features of an aileron system are employed to overcome adverse yaw?
15. Describe the function of *ruddervators*.
16. List some examples of *secondary flight controls*.

17. Discuss the use of *fixed trim tabs*.
18. Explain the operation of *controllable trim tabs*.
19. How does a controllable trim tab differ from a fixed tab?
20. If the trim tab on a rudder is moved to the right, what effect does it have on the flight of the aircraft?
21. Explain the operation of a servo *tab*.
22. What is the purpose of *wing flaps*?
23. What are the uses of *flight spoilers*?
24. What is the primary purpose of *ground spoilers*?
25. Describe a *fairlead* and give the limit of cable deflection permitted at a fairlead.
26. What devices are used to prevent escape of air around cables passing through a bulkhead in a pressurized aircraft?
27. Give the basic requirements for turnbuckles after completing *control system rigging*.
28. How is the cockpit control for a control system held in the neutral position while rigging?
29. How are *cable tensions* measured?
30. How is the travel of a control surface measured?
31. What is the effect of temperature on cable tension?
32. What are the effects of overtightening control cables?
33. Explain the purpose of *control surface stops*.
34. How are cable rod ends checked for proper thread engagement?
35. Why is it necessary to balance control surfaces?
36. Describe how the balance of a control surface is accomplished.
37. How may broken strands in a control cable be detected?
38. Describe the inspection of pulleys in a control system.
39. How close to a pulley, fairlead, or guide in a control system may a cable fitting or splice be allowed to approach?
40. What conditions are checked during the inspection of a control system?
41. What is the final inspection required for control systems?
42. What would be the effect of *crossed control cables* in an aileron system?
43. Where are the functions of the *rotors* on a helicopter?
44. What type of control movement induces *forward flight* in a helicopter?
45. How is *vertical flight* controlled in a helicopter?
46. How is *heading control* maintained in a single-rotor helicopter?
47. What part of the helicopter is associated with *low-frequency vibrations*?
48. What are the likely causes of *high-frequency vibrations* in a helicopter?
49. List the different methods for tracking main rotor blades.

Aircraft Fluid Power Systems 13

INTRODUCTION

Hydraulic and pneumatic systems in aircraft provide a means for the operation of large aircraft components. The operation of landing gear, flaps, control-boost systems, and other components is largely accomplished with hydraulic power systems. Pneumatic systems are used in some aircraft designs to perform the same type of operations performed by hydraulic systems. However, the majority of aircraft that have pneumatic systems use them only as backup systems for the operation of hydraulic components when the hydraulic system has failed.

PRINCIPLES OF HYDRAULICS

Hydraulics is a division of the science of fluid mechanics that includes the study of liquids and their physical characteristics, both at rest and in motion. The type of hydraulics applied to aircraft and other aerospace-vehicle systems is called **power hydraulics** because it involves the application of power through the medium of hydraulics. Among the uses of hydraulic systems in aerospace-vehicle systems are the operation of landing gear and gear doors, flight controls, brakes, and a wide variety of other devices requiring high power, quick action, and/or accurate control.

Hydraulic Terms

It is necessary to understand the exact meaning of hydraulic terms in order to understand hydraulic principles and their application to hydraulic systems. These terms are defined as follows.

Area is a measurement of a surface. In aircraft hydraulics the technician is concerned with the areas of piston heads. Knowing this area, the amount of force required to actuate a mechanism can be determined. Area is generally measured in square inches in the English system and square centimeters in the metric system.

Force is the amount of push, pull, or twist on an object. The force in a hydraulic system is derived from the pressure acting on the area of a piston head. In the English system, force is measured in pounds; in the metric system, it is measured in grams, kilograms, or newtons (N). To measure the force of hydraulics, we must be able to determine **force per unit area**. This is called **pressure** and is measured in pounds per square inch (psi) or kilopascals (kPa).

Stroke (length) is a measurement of distance expressed in inches or centimeters, and it represents the distance a piston moves in a cylinder.

Volume (displacement) is a measure of quantity, expressed in cubic inches or liters, which represents the amount of fluid contained in a reservoir or displaced by a pump or actuating cylinder.

Fluid is any substance that is liquid or gaseous in form. A liquid is a fluid whose particles form a definite volume. The term *hydraulic fluid* is used in this text as the common name for the fluid used in aircraft hydraulic systems and devices.

In general, fluids *expand* when they are heated and *contract* when they are cooled. If a fluid is confined so that it cannot escape when it is heated, pressure on the walls of the confining vessel will increase. Cooling of a confined fluid under pressure will cause a decrease in pressure.

Relationship of Terms

The terms *area, pressure, force, stroke,* and *volume* are mathematically related. This relationship establishes the foundation upon which hydraulic systems are based. It permits the technician to determine the operating pressures required for certain units in a system, the size of pump required, the requirements for strength of the material in system units, the size of tubing required, and the area and length of stroke of the actuating cylinders. Consider the relationship of force, pressure, and area. If any two of these factors are known, it is possible to calculate the third. Force equals pressure times area ($F = P \times A$), pressure equals force divided by area ($P = F/A$), and area equals force divided by pressure ($A = F/A$). A simple aid for solution of problems involving these factors is the diagram shown in Fig. 13-1. For example, suppose a force of 100 lb is exerted on a piston whose area is 4 in^2. That pressure is the amount of force per unit of area expressed in psi; therefore, on each square inch of the piston there are 25 lb of force, or 25 lb/in^2 (psi). The indicated mathematics is: Divide force (100 lb) by area (4 in^2), and the answer obtained will be 25 psi [172.4 kPa], which represents pressure (P). If force is unknown, cover F and multiply A by P. If area is unknown, cover A and divide F by P.

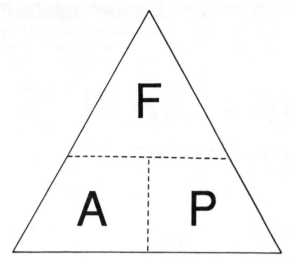

FIGURE 13-1   Relationship of force, area, and pressure.

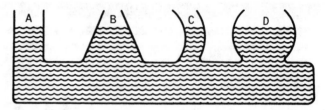

FIGURE 13-2   Liquid seeks its own level.

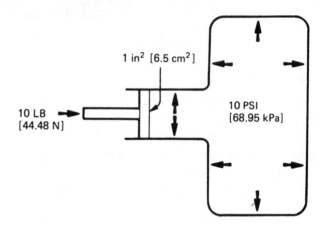

FIGURE 13-3   Fluid pressure is the same in all directions.

In general, and for practical purposes, liquids are regarded as being incompressible. This means that the volume of a given quantity of a liquid will remain constant even though it is subjected to high pressure. Because of this characteristic, it is easy to determine the volume of hydraulic fluid required to move a piston through its operating range. For example, if a piston is 4 in [10.16 cm] in diameter and its stroke is 10 in [25.4 cm], the volume of liquid necessary to move the piston through its full stroke is 125.67 in³ [2.06 L]. This is determined as follows.

The volume of the cylinder through which the piston moves is equal to the area of the piston head multiplied by the length of the cylinder. The area of the piston head is determined by the formula $A = \pi r^2$; therefore, for the piston in the example, $A = 3.1416 \times 2^2 = 12.567$. Multiplying this value by 10, we obtain the volume 125.67 in³ [2.06 L]. We know then that it will require 125.67 in³ of hydraulic fluid to move the piston through its 10-in [25.4-cm] stroke.

Because of the relative incompressibility of a liquid, we know that a given output volume from a hydraulic pump will provide an equal volume of fluid at the operating unit. For example, if a hydraulic pump discharges 100 in³ [1.639 L] of fluid through a filled connecting line between the pump and an actuating cylinder, the piston in the cylinder will have to move through sufficient distance to provide a volume of 100 in³ [1.639 L] to accommodate the fluid.

Hydraulic fluids and other liquids expand as temperature increases; therefore, safeguards must be provided in hydraulic systems to allow for the expansion and contraction of fluid as temperature changes.

A confined liquid will seek its own level, as shown in Fig. 13-2. Here the liquid is in a container open at the top and is subjected only to the force of gravity. Assuming that the container shown in the illustration is level, the pressure is the same at all points on the bottom of the container. The liquid surfaces at A, B, C, and D are all equidistant from the bottom of the container.

A basic principle of hydraulics is expressed in Pascal's law, formulated by Blaise Pascal in the seventeenth century.

This law states that a confined hydraulic fluid exerts equal pressure at every point and in every direction in the fluid. The law holds under static conditions and when the force of gravity is not taken into consideration. In Fig. 13-3 if the piston has a face area of 1 in² [6.45 cm²] and a force of 10 lb [44.48 N] is applied to it, the fluid in the container will exert 10 psi [68.95 kPa] in all directions on all surfaces within the container. In actual practice the weight of the fluid would cause a small increase in the pressure on the bottom and lower sides of the container.

When liquids are in motion, certain dynamic characteristics must be taken into consideration. One of the principal factors in liquid motion is **friction**. Friction exists between the molecules of the liquid and between the liquid and the pipe through which it is flowing. The effects of friction increase as the velocity of liquid flow increases. The result of friction can be seen in the simple experiment illustrated in Fig. 13-4.

As the liquid flows from the container through a pipe that is open at the end, the pressure of the liquid decreases

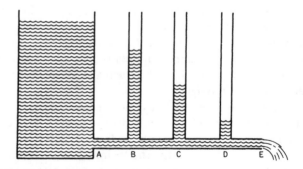

FIGURE 13-4   Reduction of pressure because of friction.

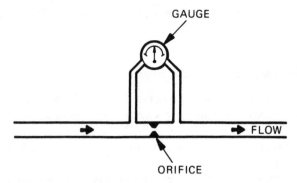

FIGURE 13-5 Differential pressure across an orifice.

progressively until it becomes 0 psi at the open end of the pipe. If the pressure at A is 4 psi [27.58 kPa], then it will be 3 psi [20.69 kPa], 2 psi [13.79 kPa], 1 psi [6.895 kPa], and 0 psi at B, C, D, and E, respectively. As liquid velocity increases through a given length of pipe, the pressure differential between the ends of the pipe will increase. The same is true when a liquid is flowing through a restriction in a pipe. For this reason, the rate of fluid flow can be determined by measuring the pressure differential on opposite sides of a given restrictor. This is illustrated in Fig. 13-5. The differential pressure reading on the gauge will increase as liquid velocity through the restrictor increases, and the reading of the gauge can be converted to gallons per minute or some other rate measurement.

Friction in a moving liquid produces heat, and this heat represents a loss of energy in a hydraulic system. According to the **law of conservation of energy,** which states that energy can neither be created nor destroyed, energy converted to heat must be subtracted from the total energy of the moving liquid. Therefore, if a hydraulic pump is discharging hydraulic fluid at a rate and pressure equivalent to 3 hp [2.24 kW] and in the system the equivalent of 0.2 hp [0.149 kW] is converted to heat, the power available for useful work is reduced to 2.8 hp [2.09 kW].

Performing Work with a Liquid

One of the principal advantages of hydraulics is the fact that force can be multiplied to almost any degree by the proper application of hydraulic pressure. In the diagram of Fig. 13-6, a piston and cylinder with a diameter of 2 in [5.08 cm] is used to develop a force of 2500 lb [11 125 N] by acting through a cylinder with a diameter of 10 in [25.4 cm]. The area of the 2-in [5.08-cm] piston is 3.1416 in² [20.27 cm²].

The area of the 10-in piston is then 5×3.1416, or 78.54 in² [506.71 cm²]. When a force of 100 lb [444.8 N] is applied to the small piston, a pressure of 100/3.1416, or 31.83 psi [219.47 kPa] is developed in the system. The force exerted by the large piston is then 31.83×78.54, or approximately 2500 lb [11 120 N]. Note that the areas of circles are proportional to the squares of the diameters; therefore, the force developed by one piston driving another is also proportional to the squares of the diameters. In the foregoing problem, the square of the diameter of the smaller piston is 2×2, or 4, and the square of the larger piston is 10×10, or 100. The ratio is then 4: 100, or 1.25. Since the force applied to the small piston is 100 psi [689.5 kPa], the force delivered by the large piston is 100×25, or 2500 lb [11 125 N].

Since energy cannot be created or destroyed, the multiplication of force is accomplished at the expense of distance. In the foregoing problem, the ratio of force multiplication is 25:1. The distance through which the pistons move must then be in an inverse ratio, or 1:25. That is, the large piston will move one twenty-fifth the distance the small piston moves. If the small piston is connected to a fluid input with check valves so it can act as a pump, it can be moved back and forth, and during each forward stroke it will move the large piston a short distance. This action can be observed in the hydraulic jacks used to raise airplanes.

In some aircraft hydraulic systems, fluid pressures of as much as 5000 psi [34 475 kPa] are employed. With a pressure of this level, a very small actuating cylinder can exert tremendous force. For example, a cylinder having a cross-sectional area of 2 in² [12.9 cm²] can exert a force of 10 000 lb [44 480 N].

A Simple Hydraulic System

Basically, a hydraulic system requires a source of hydraulic power (the pump); pipes or hoses to carry the hydraulic fluid from one point to another; a valve mechanism to control the flow and direction of the hydraulic fluid; a device for converting the fluid power to movement (actuating cylinder or hydraulic motor); and a reservoir to store the hydraulic fluid.

A **simple hydraulic system** is shown in the diagram of Fig. 13-7. The pump (P) draws hydraulic fluid from the

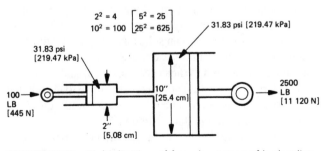

FIGURE 13-6 Multiplication of force by means of hydraulics.

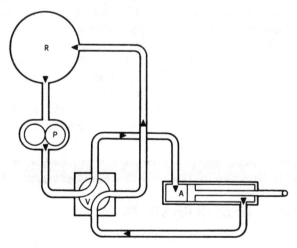

FIGURE 13-7 A simple hydraulic system.

Principles of Hydraulics **385**

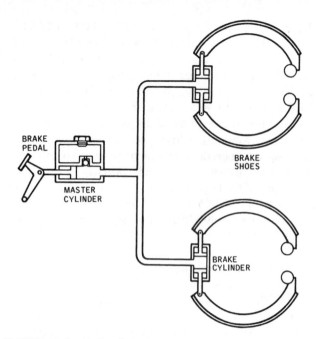

FIGURE 13-8 Hydraulic brake system.

reservoir (R) and directs it under pressure to the four-way selector valve (V). When the selector valve is in the position shown, the fluid will flow into the left end of the actuating cylinder (A) and force the piston to the right. This moves the piston rod and any device to which it is connected.

As the piston moves to the right, the fluid to the right of the piston is displaced and flows out the port at the right end of the cylinder, through the tubing to the valve, and from the valve to the reservoir. When the valve is rotated one-quarter turn, the reverse action will take place.

It must be emphasized that the diagram shown is not intended to illustrate an actual system but only to illustrate the principle of a hydraulic system. In an actual system, a pressure regulator or relief valve is necessary between the pump and the valve in order to relieve the pressure when the cylinder reaches the end of its travel. Otherwise, the pump would be damaged or the tubing would burst.

Another simple system is illustrated in Fig. 13-8. This system is similar to a hydraulic brake system. Hydraulic fluid is stored in the cylinder and is directed through a valve into the master cylinder as it is needed. When the brake pedal is depressed, the fluid is directed to the brake cylinders; these cylinders push the shoes apart, thus causing them to bear against the brake drum and provide brake action. When the pedal is released, springs attached to the brake shoes cause the shoes to contract and push inward on the brake cylinders, thus causing some of the fluid to return to the master cylinder.

HYDRAULIC FLUIDS

Purposes of Hydraulic Fluids

Hydraulic fluids make possible the transmission of pressure and energy. They also act as a lubricating medium, thereby reducing the friction between moving parts and carrying away some of the heat.

Types of Hydraulic Fluid

There are three principal types of hydraulic fluids: *vegetable-base fluids, mineral-base fluids,* and *phosphate ester–base fluids.*

Vegetable-base fluids are usually mixtures containing castor oil and alcohol and are colored blue or blue-green or are almost clear. They are considered obsolete and are not generally found in any hydraulic-power systems but may still be found in some older brake systems.

Mineral-base fluids consist of a high-quality petroleum oil and are usually colored red. They are used in many systems, especially where the fire hazard is comparatively low. Small aircraft that have hydraulic power systems for operating wheel brakes, flaps, and landing gear usually use mineral-base fluids conforming to MIL-0-5606. Mineral-base fluids are less corrosive and less damaging to certain parts than other types of fluid.

A mineral-base, synthetic hydrocarbon fluid called Braco 882 has been developed by the Bray Oil Company and is used extensively by the military services in place of MIL-0-5606. The fluid is red in color and meets the specifications of MIL-0-83282. This fluid has the advantage of increased fire resistance, and it can be used in systems having the same types of seals, gaskets, hoses, etc. that are used with petroleum base MIL-0-5606. The seals required for mineral-base fluid may be synthetic rubber, leather, or metal composition.

Phosphate ester–base fluids utilized in most transport category aircraft are very fire resistant. Although phosphate ester fluids are extremely fire resistant, they are not fireproof. Under certain conditions phosphate ester fluids will burn.

The continual development of more advanced aircraft has resulted in modification to the formulation of phosphate ester–base fluids. The continual modification of the fluid specifications has resulted in the utilization of Type I, II, III, and now Type IV fluids. Typical examples of current Type IV fluids are Skydrol LD-4 and Skydrol 500B-4. These fluids are colored purple.

Two distinct classes of Type IV hydraulic fluid exist. The class definition is according to the airframe manufacturer's hydraulic fluid specification. The classes are: Class 1, low density (Skydrol LD-4) and Class 2, high density (Skydrol 500B-4). Class 1 fluids are less dense and offer a weight savings, whereas Class 2 fluids possess handling characteristics that are beneficial in some aircraft hydraulic systems.

Seals, gaskets, and hoses used with the phosphate ester–base fluids are made of butyl synthetic rubber or Teflon fluorocarbon resin. Great care must be taken to see that the units installed in the hydraulic system are of the type designed for fire-resistant fluid. When gaskets, seals, and hoses are replaced, positive identification must be made to ensure that they are made of butyl rubber or an approved equivalent material, such as Teflon fluorocarbon resin.

Fire-resistant hydraulic fluid will soften or dissolve many types of paints, lacquers, and enamels. For this reason, areas that may be contaminated with this type of fluid must be finished with special coatings. When any of the fluid is spilled, it should immediately be removed and the area washed.

Handling Hydraulic Fluids

In addition to any other instructions given in the airplane manufacturer's manual, the following precautions should be observed in the use of hydraulic fluids.

1. Mark each airplane hydraulic system to show the type of fluid to be used in the system. The filler cap or filler valve should be marked so that it is immediately apparent to a technician what type of fluid should be added to the system.

2. Never, under any circumstances, service an airplane system with a type of fluid different from that shown on the instruction plate.

3. Make certain that hydraulic fluids and fluid containers are protected from contamination of any kind. Dirt particles quickly cause many hydraulic units to become inoperative and may cause severe damage. If there is any question regarding the cleanliness of the fluid, do not use it. Containers for hydraulic fluid should never be left open to the air longer than necessary.

4. Never allow hydraulic fluids of different types to become mixed. Mixed fluid will render a hydraulic system useless.

5. Do not expose fluids to high heat or open flames. Vegetable-base and mineral-base fluids are highly flammable.

6. Avoid contact with the fluids. Although the vegetable- and mineral-based fluids do not cause any irritation for most people when in contact with skin, the phosphate ester fluids can cause severe skin irritation. If skin contact occurs, wash the fluid off with soap and water. Consult a physician if irritation persists. Take precautions to prevent any hydraulic fluid from getting in the eyes or from being inhaled as a mist or vapor. Phosphate ester fluids cause the greatest amount of irritation in these areas. For eye contact, flush well with water and, if a phosphate ester fluid is involved, apply an anesthetic eye solution. Reaction to inhaled vapors (coughing and sneezing) stops after the vapor or mist is eliminated. For all cases of eye contact and inhalation of hydraulic fluids, consult a physician.

7. Wear protective gloves and a face shield whenever handling phosphate ester fluids and whenever working around any hydraulic lines that are under pressure.

HYDRAULIC RESERVOIRS

A hydraulic reservoir is a tank or container designed to store sufficient hydraulic fluid for all conditions of operation. Usually the hydraulic reservoir must have the capability of containing extra fluid not being circulated in the system during certain modes of operation. When accumulators, actuating cylinders, and other units do not contain their maximum quantities of fluid, the unused fluid must be stored in the reservoir. On the other hand, when a maximum amount of fluid is being used in the system, the reservoir must still have a reserve adequate to meet all requirements. Reservoirs in hydraulic systems that require a reserve of fluid for the emergency operation of landing gear, flaps, etc., are equipped with **standpipes**. During normal operation, fluid is drawn through the standpipe. When system fluid is lost, emergency fluid is drawn from the bottom of the tank.

Hydraulic reservoirs vary in complexity from nothing but a can with a vent hose on top of the cap and an outlet at the bottom to sophisticated designs incorporating filters, quantity indicators, and pressure relief systems. Reservoirs can be broken down into two basic types, *in-line* and *integral*, and these can be further classified as *pressurized* and *unpressurized.*

Hydraulic fluid reservoirs provide a compartment to store hydraulic fluid when it is not in use in a system, and they also provide sufficient fluid to make up for normal losses of fluid by seepage past seals. Reservoirs are not designed to be completely filled; they must allow for an air space above the fluid level to allow for expansion of the fluid when it is heated during system operation.

A reservoir will provide some means of checking the fluid level and of being replenished. The quantity-indicating method may be nothing more than a dipstick on the filler cap, or it may consist of a remote indicating system that displays the quantity on the aircraft flight deck. Replenishment is normally accomplished by adding fluid directly to the reservoir through a filler opening. When multiple reservoirs are used in an aircraft for independent or redundant hydraulic systems, the reservoirs may be filled from a common manifold.

In-Line Reservoirs

In-line reservoirs are those that are separate components in the hydraulic system. This is the most common type of reservoir. These can be pressurized or unpressurized.

Unpressurized reservoirs are normally used in aircraft flying at lower altitudes, such as below 15 000 ft [4583 m], or in aircraft whose hydraulic systems are limited to those associated with ground operations, such as brakes.

Pressurized reservoirs are commonly found in aircraft designed for high-altitude flight where atmospheric pressure is low. The most basic rule of hydraulics states that fluid cannot be pulled; it can only be pushed. At sea level the 14.7 psi of atmosphere provides the force to push the fluid from the reservoir to the pump. As altitude increases, atmospheric pressure decreases. With little or no pressure on the fluid, it tends to foam, causing air bubbles to form in the low part of the system. When an aircraft is operating at high altitudes, the pump will be starved for fluid unless some means of pressurization is used. Therefore, to provide a continuous supply of fluid to the pumps, the reservoir is pressurized. Along with providing a positive feed to the hydraulic pumps, a pressurized reservoir reduces or eliminates the foaming of the fluid when it returns to the reservoir. The reservoir may be pressurized by spring pressure, air pressure, or hydraulic pressure. The desired pressure to be maintained ranges from approximately 10 psi to 90 psi.

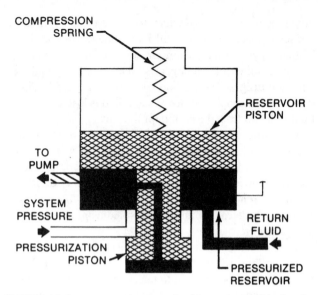

FIGURE 13-9 A reservoir pressurized by spring and hydraulic pressure. (*Bell Helicopter Textron.*)

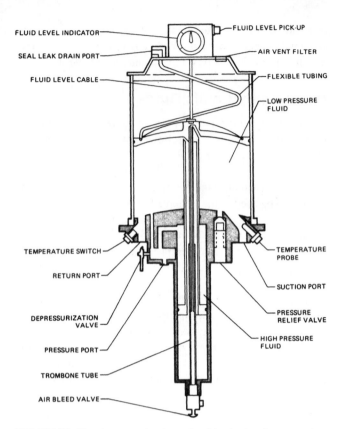

FIGURE 13-10 A reservoir pressurized by hydraulic pressure. (*Lockheed-California Co.*)

A simplified drawing of a spring-pressurized reservoir is shown in Fig. 13-9. The compression spring on top of the reservoir piston pressurizes the fluid in the reservoir and provides a positive pump inlet pressure for initial start-up of the pump. When the pump is running, system pressure (3000 psi [20 685 kPa] in this case) applied to the small area on top of the pressurization piston increases the pressure inside the reservoir to 75 psi [517.13 kPa]. This assists the spring in providing a positive pressure to the pump inlet and prevents cavitation of the pump.

Turbine engine bleed air or a venturi-type aspirator can be used to pressurize a reservoir with air. Bleed air can be fed through a pressure regulator to establish the proper pressure and then into the top of the reservoir. When using a verturi-type aspirator, the low-pressure section of the venturi draws air into the reservoir and increases the pressure. The use of air pressure acting directly on the fluid eliminates the need for any elaborate chambering of the reservoir, as is required by the other methods; the reservoir is simply pressurized, with the air settling out to the top of the airtight reservoir.

Hydraulic pressure can be used to pressurize the reservoir in a manner similar to that used to assist the spring in pressurizing the reservoir. In Fig. 13-10 hydraulic pressure entering the pressure port flows through the depressurization valve and into the area just above the pressurization piston. The force on this piston causes the reservoir piston to try to move downward, which pressurizes the reservoir. The high pressure used is 3000 psi [20 685 kPa], which gives a reservoir pressure of 85 psi [586 kPa]. The depressurization valve is used to equalize the pressure in the two piston areas during servicing. An electric pump is used to pressurize the system before engine start so that the inlet lines of engine-driven pumps are under positive pressure, which provides a positive supply of fluid to the pump and reduces pump wear.

Another example of a hydraulic reservoir for a transport-category airplane is illustrated in Fig. 13-11. The reservoir

shown is cylindrical in shape with a corrugated, perforated shield around the shell and has a fluid capacity of 2.5 U.S. gal [9.48 L]. A **manifold**, equipped with four main ports, a small pressure-line port, and a thermoswitch probe port, is welded to the bottom of the reservoir shell. The four main ports of the manifold are internally connected in tandem pairs, with both pairs opening into the supply fluid section of the reservoir. This configuration of ports makes possible either left or right installation of the reservoir by reversing the location of port connector unions and reducers. When connected in the system, one pair of ports delivers fluid to the suction ports of the engine-driven pump, the electrically driven auxiliary pump, and the ground service hand pump. One of the other pair of ports receives system return fluid, and the fourth port is equipped with a valve to drain the reservoir. A **sight glass**, located below the **relief-and-bleed valve** in the upper portion of the **diaphragm-guide cylinder** above the main portion of the reservoir, provides an indication of excessive accumulation of air in the reservoir. An instruction plate, mounted adjacent to the sight glass, and a pointer attached to the top of the relief-and-bleed valve above the sight glass, provide fluid-level instructions and direct fluid-level indications for system pressurized and depressurized conditions. A fluid-quantity transmitter, bracket, and actuating linkage, not shown in the illustration, is mounted on the inboard side of the reservoir cover. The lower end of the linkage is attached to the transmitter rotor, while the upper end is attached to the **fluid-level pointer** at the top of the relief-and-bleed valve. Fluid-level changes in the reservoir

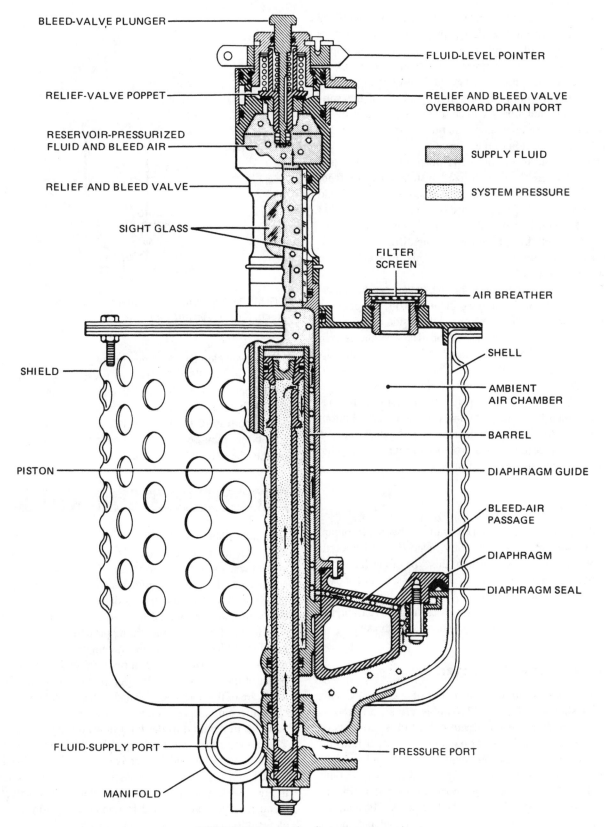

BLEED-VALVE PLUNGER

RELIEF-VALVE POPPET

RESERVOIR-PRESSURIZED FLUID AND BLEED AIR

RELIEF AND BLEED VALVE

SIGHT GLASS

SHIELD

PISTON

FLUID-SUPPLY PORT

MANIFOLD

FLUID-LEVEL POINTER

RELIEF AND BLEED VALVE OVERBOARD DRAIN PORT

SUPPLY FLUID

SYSTEM PRESSURE

FILTER SCREEN

AIR BREATHER

SHELL

AMBIENT AIR CHAMBER

BARREL

DIAPHRAGM GUIDE

BLEED-AIR PASSAGE

DIAPHRAGM

DIAPHRAGM SEAL

PRESSURE PORT

FIGURE 13-11   Hydraulic system reservoir.

raise or lower the relief-and-bleed valve and pointer, thus extending or retracting the linkage and changing the position of the rotor in the fluid-quantity transmitter. The transmitter delivers a fluid-quantity signal to the fluid-quantity indicator in the flight compartment.

The relief-and-bleed valve, located at the top of the diaphragm guide and above the sight glass, is provided to relieve excessive reservoir pressure and to bleed off accumulated system air. The relief valve is set to relieve pressure above 47 psi [324.07 kPa]. A drain line is attached to the relief-and-bleed valve to conduct excess pressure and bleed air overboard. An **air breather** is provided in the reservoir cover forward of the diaphragm guide. It allows the upper, ambient-air section of the reservoir to breathe air in and out as the reservoir diaphragm lowers and raises with pressurization of the hydraulic system and operation of the various subsystem actuators. A filter screen is provided in the breather to remove atmospheric impurities from the air that enter the upper portion of the reservoir.

Internally, the reservoir is equipped with a piston and diaphragm assembly that utilizes system pressure from the small pressure-line port to maintain a pressure head on the supply fluid. This pressure is from 28 to 30 psi [193 to 207 kPa] and is reduced from the system pressure of 3000 psi [20 685 kPa] by means of the difference in area between the piston and diaphragm. This difference ratio is approximately 100:1.

The application of force to the diaphragm can be understood by a study of the drawing of Fig. 13-11. Observe the arrows that indicate fluid pressure to the inside of the piston shaft and out through holes at the top. The pressure is then exerted downward between the barrel and the piston to the surface at the lower end of the barrel. The 3000-psi [20685-kPa] pressure on this small area is balanced by the 28- to 30-psi [193- to 207-kPa] pressure against the bottom of the diaphragm, which has about 100 times the area at the bottom of the barrel. As fluid enters the reservoir, the diaphragm is forced upward inside the reservoir shell, thus carrying the diaphragm guide and barrel upward.

Integral Reservoirs

Integral reservoirs are combined with the hydraulic pump. These types of reservoirs are often found in small aircraft, where the compact arrangement of this type of mechanism is desirable. An example of this is the brake master cylinder used with many light-aircraft systems. As shown in Fig. 13-12, the upper portion of the assembly serves as the reservoir and the lower portion serves as the pump to operate the brake.

Reservoir for a Helicopter

A hydraulic reservoir for one model of a Sikorsky helicopter is shown in Fig. 13-13. This reservoir, referred to by the manufacturer as a **hydraulic fluid tank**, consists of an upper housing and a lower housing. The upper housing consists of a filler neck, an adapter, and a window sight gauge. The filler neck consists of a cap, which screws onto the neck, a scupper with an overboard drain, and a strainer, which is secured to the inside of the filler neck. The adapter, on top of the upper housing, consists of an adapter housing with a micronic filter element and a vent line that runs from the adapter fitting to the scupper. The lower housing incorporates a baffle plate, micronic filter element, drain plug, and relief valve with a differential cracking pressure of 6 to 8 psi [41.37 to 55.16 kPa]. Fluid returning to the tank circulates around the baffle plate and passes through the micronic filter element to the supply portion of the tank.

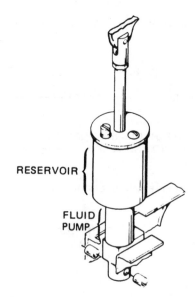

FIGURE 13-12   A reservoir in the same assembly as the fluid pump. (*Cessna Aircraft Co.*)

If the filter element becomes clogged to the extent that a 6- to 8-psi [41.37 to 55.16-kPa] pressure builds up in the return portion of the tank, the relief valve opens to allow fluid to bypass the filter element and flow directly from the return portion of the tank to the supply portion. The tank (reservoir) may be drained by connecting a coupling and hose to the external supply coupling.

HYDRAULIC FILTERS

Hydraulic filters are required to filter out any particles that may enter the hydraulic fluid. These particles may enter the system when it is being serviced or during wear of operating components. If these contaminants were allowed to remain in the circulating fluid, they could damage the seals and cylinder walls, causing internal leakage and prevent components such as check valves from seating properly. The number and location of filters in a hydraulic system depend on the specific model aircraft, but they are normally found at the inlet and outlet of the reservoir and the pump outlet. Commonly used filters are of the micronic type and porous metal type.

A **micronic filter** contains a treated paper element to trap particles in the fluid as the fluid flows through the element. The micron filters can be designed to filter out particles as small as 3 μm (A micrometer is equal to 0.0000394 in, or 0.0001 mm.) A micron filter assembly is shown in cutaway in Fig. 13-14. Also, a magnetic-type element that will attract metal particles can be used in conjunction with the paper element to make a dual-element type filter.

Porous metal filters are composed of metal particles joined together by a sintering process. These filters can trap particles as small as 5 μm in size.

Most hydraulic filter assemblies are located in the pressure and return lines. In-line filters are generally constructed much like the one illustrated in Fig. 13-15. The filter illustrated consists of a head assembly, a bowl assembly, and a 15-μm

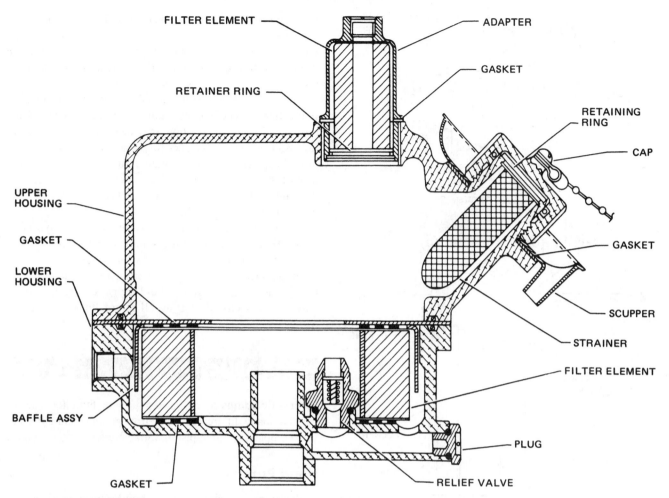

FIGURE 13-13 Hydraulic reservoir for a helicopter. (*Sikorsky Aircraft Corp., Division of United Technologies Corp.*)

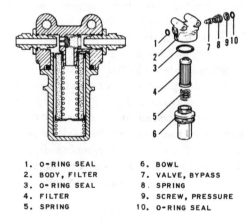

1. O-RING SEAL	6. BOWL
2. BODY, FILTER	7. VALVE, BYPASS
3. O-RING SEAL	8. SPRING
4. FILTER	9. SCREW, PRESSURE
5. SPRING	10. O-RING SEAL

FIGURE 13-14 Hydraulic filter.

paper filter element. The head assembly contains the fluid line connections and a shutoff valve to facilitate replacement of the filter element. A bypass valve in the filter prevents the system from becoming inoperative should the filter become clogged. Many filters incorporate a "pop-out" differential pressure indicator, such as the one shown in Fig. 13-15, to allow ready identification of a clogged filter. This pop-out indicator may

also be electrically connected to the cockpit to provide notification of a clogged filter. The bowl assembly is mounted on the bottom of the head assembly and is sealed with an *O ring*. This is the housing that contains the filter element.

When filters are serviced, all pressure should be removed from the hydraulic system. When removing the bowl and element, care should be taken to prevent prolonged contact of the fluid with the aircraft, clothing, or skin—especially if a phosphate ester fluid is being used. If a micronic element is used, this is replaced with a new element; the old element can be opened to check for contamination. If the contamination indicator pin has popped out, then the fluid and filters downstream from the filter must be checked for contamination and the system flushed if required. If a porous metal element is used, this should be cleaned or replaced in accordance with the appropriate service manual.

Heat Exchangers

Because of the high pressures involved in many hydraulic systems and the high rates of fluid flow, the hydraulic fluid becomes heated as the subsystems are operated. For this reason it is often necessary to provide cooling for the fluid. The **heat exchanger**, shown in Fig. 13-16, is a heat

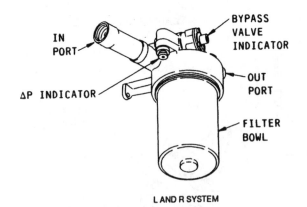

L AND R SYSTEM

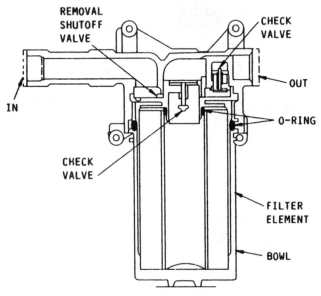

FIGURE 13-15   Hydraulic filter assembly.

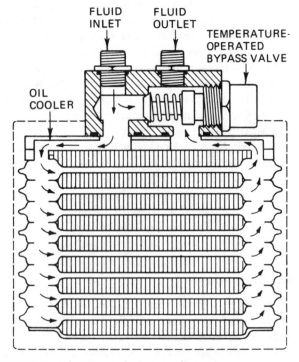

FIGURE 13-16   Heat-exchanger cooling unit.

radiator similar in design and construction to an oil cooler for an engine. Note that the heat exchanger is equipped with a temperature-operated bypass valve to increase the fluid flow through the cooling element as temperature rises.

Heat exchangers are often installed in the pumpcase drain return lines to cool the hydraulic fluid before it enters the reservoir. Cooling of the fluid may be provided by different means. The heat exchangers for some aircraft are installed in fuel cells and cool the hydraulic fluid by transferring the heat of the fluid to the fuel. Other aircraft utilize air to cool the fluid, with ram air used in flight and engine bleed air used when the airplane is on the ground.

The temperature-operated bypass valve in the hydraulic cooler fluid inlet controls the volume of return fluid circulating through the fluid cooler. As fluid temperatures rise, the bypass valve starts to close, porting return fluid through the hydraulic cooler. At high fluid temperatures, the bypass valve is fully closed, porting all return fluid through the cooler.

HYDRAULIC PUMPS

Hydraulic pumps are designed to provide fluid flow and are made in many different designs, from simple hand pumps to very complex, multiple-piston, variable-displacement pumps.

Hand Pumps

A diagram of a **single-acting hand pump** is shown in Fig. 13-17. This diagram illustrates the basic principle of a piston pump. When the handle is moved toward the left, the piston movement creates a low-pressure condition and draws fluid from the reservoir through the check valve and into the cylinder. Then when the handle is moved toward the right, the piston forces the fluid out through the discharge check valve. The check valves allow the fluid to flow only in one direction, as shown by the arrows.

A double-acting piston-displacement type of hand pump is shown in the drawing of Fig. 13-18. The IN port from the reservoir is connected to the center of the cylinder, where there is a space between the two pistons and surrounding the shaft connecting the pistons. In each piston is a check valve (nos. 1 and 2) and a passage that allows fluid to flow from the center chamber to the spaces at each

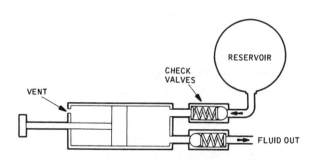

FIGURE 13-17   Single-acting hand pump.

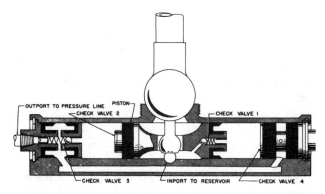

FIGURE 13-18   A double-acting, piston-displacement pump.

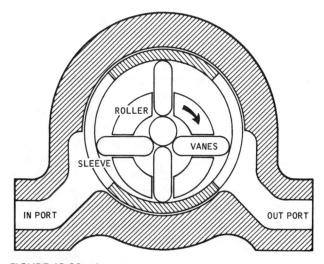

FIGURE 13-20   A vane-type pump.

end of the dual-piston assembly. When the pump handle is moved to the right, the piston assembly moves to the left, forcing fluid out though check valve 3 into the system. The check valve in the left-hand piston is held closed by fluid and spring pressures. As the piston assembly moves to the left, a low-pressure area is created in the chamber in the right end of the cylinder, and this causes fluid to flow through check valve 1 into the chamber. Check valve 4 is held in the closed position by spring and fluid pressure. When the pump handle is moved to the left, the piston assembly moves to the right and the fluid is forced out of the right-hand chamber through check valve 4 into the system. Note that a fluid passage connects the outlet chambers at each end of the cylinder.

Gear-Type Pump

A **gear-type pump** is shown in the drawing of Fig. 13-19. This pump is classed as a **positive-displacement pump** because each revolution of the pump will deliver a given volume of fluid, provided the pump is not worn and no leakage occurs. One of the two gears is driven by the power source, which could be an engine drive or an electric-motor drive. The other gear is meshed with and driven by the first gear. As the gears rotate in the direction shown, fluid enters the IN port to the gears, where it is trapped between the gear teeth and carried around the pump case to the OUT port. The fluid cannot flow between the gears because of their closely meshed design; therefore, it is forced out through the OUT port.

Vane-Type Pump

The **vane-type pump** is also classed as a positive-displacement pump because of its positive action in moving fluid. This pump is illustrated in the drawing of Fig. 13-20 and consists of a slotted rotor located off-center within the cylinder of the pump body with rectangular vanes free to move radially in each slot. As the rotor turns, the vanes are caused to move outward by centrifugal force and contact the smooth inner surface of the casing. Since the rotor is eccentric with respect to the casing, the vanes form chambers that increase and decrease in volume as the rotor turns. The inlet side of the pump is integral with the side of the casing in which the chambers are increasing in volume. Thus the fluid is caused to enter the chambers because of the low-pressure area created by the expanding chambers. The fluid is carried around the casing to the point where the chambers begin to contract, and this section of the casing is connected to the output port of the pump. The contraction of the chambers forces the fluid into the outlet port and system.

Gerotor Pump

A **gerotor pump**, shown in Fig. 13-21, consists of a housing containing an eccentric-shaped stationary liner, an internal

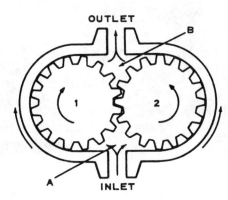

FIGURE 13-19   A gear-type pump.

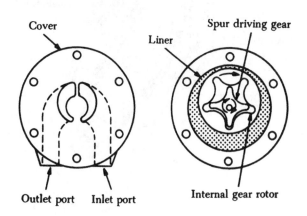

FIGURE 13-21   Gerotor-type power pump.

gear rotor having five wide teeth of short height, a spur driving gear having four narrow teeth, and a pump cover, which contains two crescent-shaped openings. One opening extends into an inlet port and the other connects to an outlet port. In Fig. 13-21 the pump cover is shown with the inner surface visible to display the inlet and outlet openings inside the pump. When the cover is turned over, the inlet is on the left and the outlet is on the right.

During operation, the gears turn clockwise. As the cavities move from the bottom of the gear to the top of the gear, they increase in volume, resulting in a vacuum being created in this area. As this cavity passes the inlet port, fluid is drawn into the cavity. As this cavity moves to the right side of the pump, the size of the cavity is decreased and fluid is forced out of the pump via the outlet port.

Multiple-Piston Pump

One of the most widely used hydraulic pumps for modern aircraft is the **axial multiple-piston pump.** This type of pump is shown in Fig. 13-22. The pump consists of a drive shaft to which the pistons are attached by means of ball sockets, a cylinder block into which the pistons are inserted, and a stationary valving surface, which fits closely against one end of the cylinder block. The drive shaft is connected to the cylinder block by means of a universal link to rotate the cylinder block with the drive shaft. The axis of rotation for the cylinder block is at an angle to the axis of rotation of the drive shaft; therefore, the pistons are caused to move in and out of the cylinders as rotation occurs. The pistons on one side of the cylinder block are moving outward, thus increasing the volume of the cylinder spaces; on the other side of the cylinder block the pistons are moving inward. The valve plate is made with two slots such that one slot

is bearing against the side of the cylinder block on which the pistons are moving away from the valving surface; the other slot is bearing against the side on which the pistons are moving toward the valving surface. The slots in the valving surface are connected to inlet and outlet chambers to provide fluid feed to the pistons on the inlet side and an outlet for the pressure fluid on the other side. As the drive shaft rotates, the pistons on one side of the cylinder block draw fluid through the valving surface slot, and the pistons on the other side force the fluid through the outlet slot.

A cutaway illustration of a **fixed-delivery** pump manufactured by Sperry Vickers, Division of Sperry Rand Corp., is shown in Fig. 13-23. **Fixed delivery** means that the pump will normally deliver a fixed amount of fluid at a given number of revolutions per minute. A **variable-delivery** pump is shown in Fig. 13-24. This pump is designed so the alignment

FIGURE 13-23 A fixed-delivery piston pump. (*Sperry Vickers Div., Sperry Rand Corp.*)

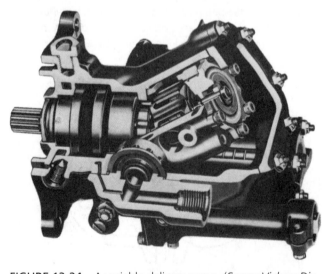

FIGURE 13-24 A variable-delivery pump. (*Sperry Vickers Div., Sperry Rand Corp.*)

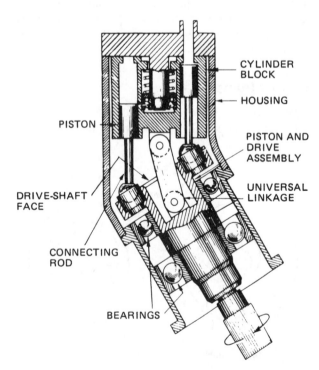

FIGURE 13-22 Cutaway drawing of an axial piston pump.

of the rotational axis of the cylinder block can be changed as desired to vary the volume of fluid being delivered at a given rpm. By changing the angle of the rotational axis, the stroke of the pistons is decreased or increased; therefore, the volume of fluid pumped during each stroke of the pistons is reduced or increased. If the axis of the cylinder block is parallel to the axis of the drive shaft, no fluid will be delivered, because there will be no movement of the pistons within their respective cylinders.

In-Line Variable Delivery Pump

Transport aircraft use in-line variable delivery pumps for their hydraulic systems. The components of a variable delivery piston pump are illustrated in Fig. 13-25. These pumps could be engine driven, electrical driven, or air driven. A variable-delivery pump adjusts fluid output depending on system demands. The pump output is changed automatically by a pump compensator within the pump. The first stage of the pump consists of a centrifugal pump which boosts the pressure before the fluid will enter the second stage piston pump. Figure 13-26 is a schematic of a typical pump used on a large transport aircraft and Fig. 13-27 shows the control function of this pump including the depressurization solenoid and blocking valve. The control function is the bottom-left portion of Fig. 13-26.

An in-line variable delivery piston pump operates on the same principle as the multiple-piston pumps described in the foregoing section: however, the drive shaft is parallel to the pistons. The movement of the pistons necessary to create a pumping action is caused by a shoe-bearing plate in the yoke assembly. As the cylinder block rotates the piston shoe continues following the yoke-bearing surface and it pushes the pistons into the cylinders, causing them to eject the fluid into the **out** port. The pistons on the descending portion of the yoke can draw fluid from the **in** port of the pump. The angle of the yoke as established by the compensator valve determines the amount of fluid expelled during each stroke of the pistons, thus providing the variable delivery required. Each piston completes this cycle on each revolution of the drive shaft, providing a continuous, non-pulsating flow of fluid to the system. Check valves are located at the outlet of the each piston to prevent pressurized fluid from flowing back into the

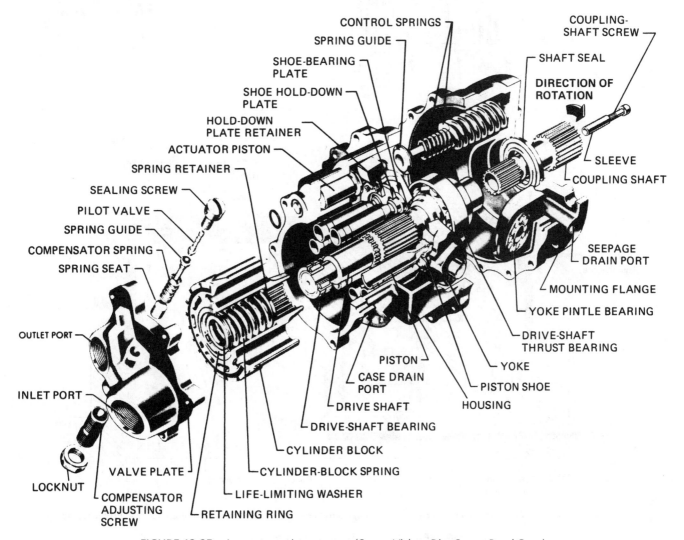

FIGURE 13-25 A cam-type piston pump. (*Sperry Vickers Div., Sperry Rand Corp.*)

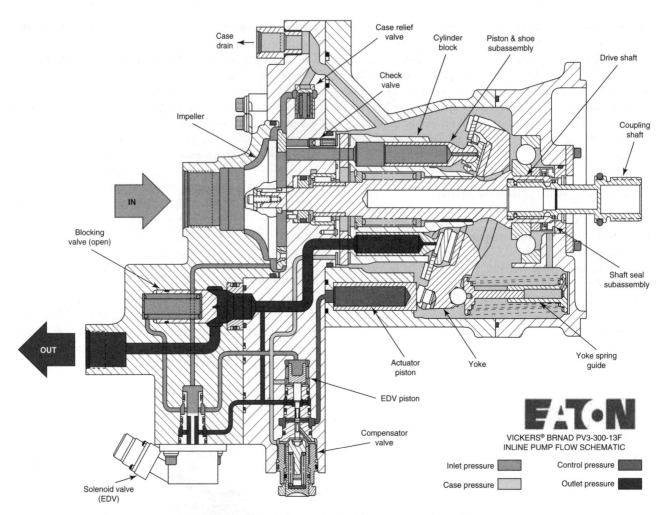

FIGURE 13-26   In-line variable delivery pump schematic.

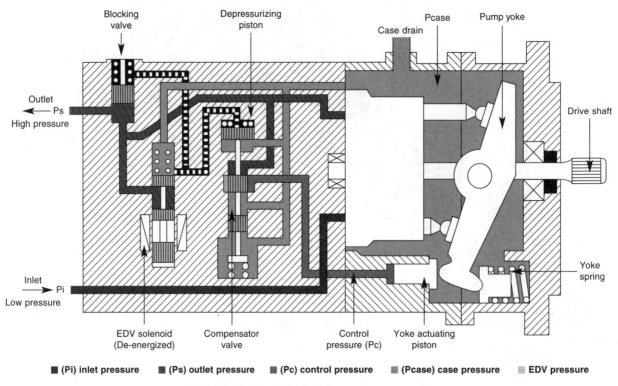

■ (Pi) inlet pressure   ■ (Ps) outlet pressure   ■ (Pc) control pressure   ■ (Pcase) case pressure   ▨ EDV pressure

FIGURE 13-27   Variable delivery pump control.

cylinders as the pistons move back in the cylinders. Internal leakage keeps the pump housing filled with fluid for lubrication of rotating parts and cooling. The leakage is returned to the return side of the hydraulic system through a case drain port. The case valve relief valve protects the pump against excessive case pressure, relieving it to the pump inlet.

The control mechanism for the variable-delivery pump is shown in Fig. 13-27. Remember that the output of this type of pump is determined by the angle of the shoe-bearing plate in the yoke assembly which produces a reciprocating action of the piston pumps. The yoke angle is changed by pressure on the yoke-actuating piston. Below system pressure of 2850 psi the compensator valve is in the up position and no high pressure can flow to the yoke actuator piston. In this position the pump delivers maximum flow. When the pressure increases above 2850 psi the compensator valve cracks open (valve moves down) and high pressure can act on the yoke actuator piston which reduces the yoke angle and as a result the output of the pump is decreased. When system pressure reaches 3025 psi the compensator valve is all the way open and high pressure acting on the yoke actuating piston will move the yoke to the zero position. The output of the pump decreases to zero and the unit pumps only its own internal leakage. The pump will provide a variable flow depending on the system pressure between 2850 and 3025 psi. The higher the pressure, the lower the flow.

Depressurized Mode

The pump in Fig. 13-26 is outfitted with a depressurization and blocking feature. This depressurization and blocking feature can be used to reduce the load on the engine during start-up and, in a multiple pump system, to isolate one pump for check-out purposes. When the solenoid valve is energized, the EDV solenoid valve will move up against the spring force and the outlet fluid is ported to the piston on the top of the compensator valve (depressurizing piston). The high-pressure fluid pushes the compensator valve spool down beyond its normal metering position. High pressure can flow directly to the yoke actuator valve which places the yoke in its zero position and reduces the output of the pump to zero flow. At the same time outlet fluid is also ported to the blocking valve spring chamber, which equalizes pressure on both sides of its plunger. The blocking valve closes due to the force of the blocking valve spring and isolates the pump from the external hydraulic system.

Ram Air Turbines

A method used on many turbine transport-category aircraft to power a pump in the event of engine and electrical system failure is the utilization of a ram air turbine (RAT), such as the one shown in Fig. 13-28. The RAT installation typically consists of an air turbine, a speed-governing device, and a variable-volume pump. The RAT can usually be deployed manually or may automatically deploy in the event of engine failure.

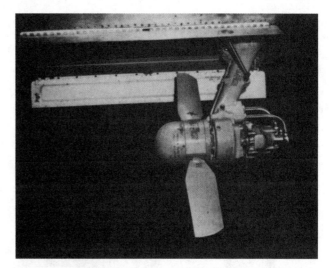

FIGURE 13-28   A ram air turbine in the extended position. (*Lockhead-California Co.*)

PRESSURE-CONTROL DEVICES

Numerous devices have been designed to control pressure in hydraulic systems; among these are *pressure switches, pressure regulators, relief valves,* and *pressure-reducing valves.*

Pressure Switches

Electrically operated **pressure switches** are used in hydraulic systems with electrically driven pumps to maintain system pressure within set limits. The pressure switch is set to open the electrical circuit to the pump motor when system pressure builds up to correct values, causing the pump to stop. As pressure drops to a lower value, the pressure switch closes the circuit to start the pump operating again. Pressure switches are also used in hydraulic systems to control the operation of warning and protective devices. The switch may turn on a light to warn the pilot of insufficient pressure, or it may turn off a pump to avoid exhausting reservoir fluid through a broken line. Pressure switches come in various types. For example, there are the Bourdon-tube type, the piston type, and the diaphragm type.

The *Bourdgn-tube type pressure switch* illustrated in Fig. 13-29 is frequently used to control pressure within a hydraulic system. The flexible steel finger attached to the small end of the Bourdon tube moves outward as the tube (a) begins to uncoil. This finger presses against the toggle plate (b) until the desired system pressure is reached, at which time it will cause the toggle to pivot rapidly, thereby opening the contact points and breaking the electrical circuit.

Pressure Regulators

A **pressure regulator** is designed to maintain a certain range of pressures within a hydraulic system. Usually the pressure regulator is designed to relieve the pressure on the

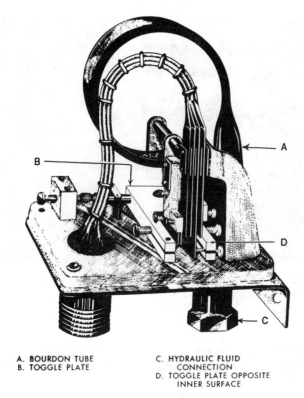

A. BOURDON TUBE
B. TOGGLE PLATE
C. HYDRAULIC FLUID
   CONNECTION
D. TOGGLE PLATE OPPOSITE
   INNER SURFACE

FIGURE 13-29   Bourdon tube pressure switch.

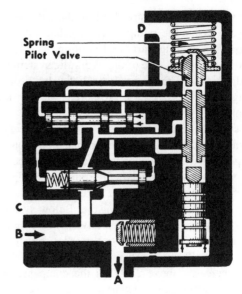

FIGURE 13-30   An unloading-type pressure regulator.

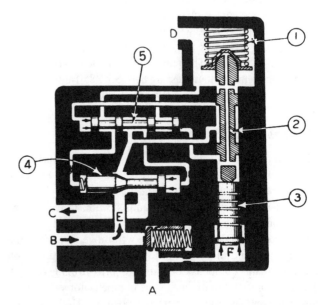

FIGURE 13-31   Unloading valve in the "kicked-out" position.

pressure pump when it is not needed for operating a unit in the system. Some pressure regulators are also called **unloading valves** because they unload the pump when hydraulic pressure is not required for operation of landing gear, flaps, or other subsystems. Continuous pressure on the pump increases wear and the possibility of failure.

A pressure regulator that is also an unloading valve is illustrated in Fig. 13-30. In this view, the unit is operating to supply fluid for charging an accumulator (hydraulic-pressure storage chamber) and to supply fluid pressure for operating units in the system. Fluid flows into port B and out of port A to the system. The check valve is off its seat because of the fluid pressure being exerted by the pressure pump. When the pressure in the accumulator builds up to the maximum level for the system, the same pressure is exerted in chamber F, Fig. 13-31. This pressure moves the plunger (piston) (3) upward to raise the **pilot valve** (2) against the pressure of the spring (1). In Fig. 13-30, observe that fluid pressure is applied to one of the pilot-valve chambers through a passage from the inlet line, around an annular groove surrounding the unloading valve [(4) in Fig. 13-31], and to the pilot valve. As the pilot valve is raised, this pressure is ported and directed through a passage to the left end of the **directional spool** (5). This spool moves to the right and causes hydraulic pressure to be directed against the right end of the unloading spool, thus moving it to the left. This permits the main flow of fluid to go from port B, through passage E, and out the return port C. Under this condition the power pump is unloaded because the fluid has free flow back to the reservoir. The regulator is said to be "kicked-out" in this position. The check valve has seated because of spring pressure and holds

the pressure locked in the system. In the drawings, port D is the "bleed-off" port, which permits fluid from the chambers at the ends of the directional spool and unloading spool to escape and allow movement of the spools as required.

When a subsystem is operated and fluid pressure in the pressurized part of the system drops to a pre-determined level, the pilot valve and plunger will move back to the lower position, directing pressure against the right end of the directional spool, as shown in Fig. 13-32. The directional spool moves to the left and causes fluid pressure to be directed to the left end of the unloading spool, moving it to the right and blocking the return line. Pump pressure then builds up and opens the check valve (bullet valve) to allow fluid flow to the operating system through port A. This position is often referred to as "kicked-in."

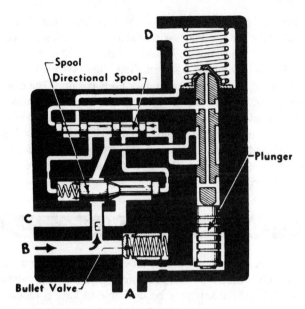

FIGURE 13-32 Moving to the "kicked-in" position.

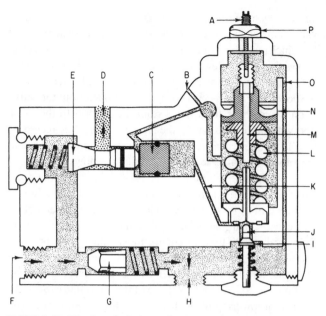

FIGURE 13-33 Bendix balanced-type regulator.

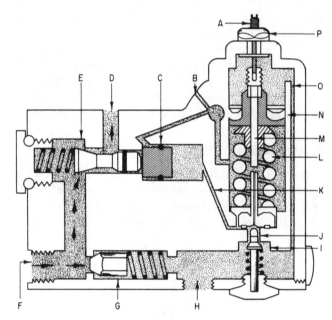

FIGURE 13-34 Bendix pressure regulator in the "kicked-out" position.

Another type of pressure regulator that serves a purpose similar to the one just described is the balanced type. A drawing of this regulator in the kicked-in position is shown in Fig. 13-33. In this drawing, the bypass valve E is held closed by spring pressure and system pressure. Fluid from the power pump enters at port F, and the pressure forces the check valve G off its seat and allows the fluid to flow out port H to the accumulator and the system. When the system-operating requirements are met, the pressure continues to build up from the pump and throughout the area in operation. This pressure bears against the poppet valve I and also increases above the piston N because of the sensing line O. The piston N has greater area than the poppet valve I; therefore, the force downward increases faster than the force upward. At a certain point the force downward becomes equal to the force upward and the valve is then said to be in the **balanced** condition. As pressure continues to increase, the downward force becomes greater than the upward force and the rod M moves downward against the force of the spring L. The hollow piston rod M moves downward and contacts the poppet valve, thus forcing it off its seat. Pressure fluid then can flow through the passage K to act against the directional valve C, which pushes the bypass valve E off its seat. When this occurs, pressure fluid entering port F can flow out through port D to the reservoir, and the pressure entering port F drops to the free-flow level. Check valve G is then immediately seated by spring pressure to trap the high pressure in the system. This is the kicked-out position, as illustrated in Fig. 13-34.

For proper operation, the regulator should remain in the kicked-out position until the pressure in the system has dropped to the lower operating level. This it will do, because the pressure for the initial kick-out was great enough to overcome the force of the spring L and the pressure against the poppet I. Since the poppet has been lifted from its seat, the only force necessary to keep the valve in the kicked-out position is that on the piston N acting against the spring L.

Therefore, the pressure in the system will drop substantially before the valve kicks in again.

The kick-out pressure of the valve is adjusted by turning the adjusting screw A. This changes the effect length of the rod below the piston and changes the amount of force necessary to bring about the kicked-out condition.

Relief Valve

A **relief valve** is comparatively simple in construction, and its function is to limit the maximum pressure that can be developed in a hydraulic system. Thus it acts as a safety valve similar in function to one that would be found in an air- or steam-pressure system. During operation, the relief valve remains closed unless the system pressure exceeds that for which the valve is adjusted. At this time the valve opens and allows the fluid to flow through a return line to the reservoir.

A drawing to illustrate the construction of a relief valve is shown in Fig. 13-35. During normal operation, the valve is on its seat and fluid flows from the IN port to the OUT port without restriction. As the pressure on the line increases to a level above that for which the valve spring is adjusted, the valve lifts off its seat and the fluid then flows through the valve and out the return line. The pressure at which the relief valve lifts is called the *cracking pressure.* The design of the valve must be such that it will not rapidly open and close and cause chattering, since this would damage the system. Relief valves are used to control maximum system pressure and to control pressure in various parts of the subsystems. For example, a relief valve, called a **wing-flap overload valve**, is often placed in the DOWN line of the wing-flap subsystem to prevent lowering of the flaps at too high an airspeed. The pressure in the DOWN line will rise above a specified level because of air pressure against the wing flaps if the airspeed is too great. The wing-flap overload valve will then open and allow excess pressure to be relieved, thus causing the down movement of the flaps to stop.

When several relief valves are incorporated in a hydraulic system, they should be adjusted in a sequence that will permit each valve to reach its operating pressure. Thus, the highest-pressure valves should be adjusted first; the others are then adjusted in the order of descending pressure values.

Thermal Relief Valves

A **thermal relief valve** is similar to a regular system relief valve; however, such valves are installed in parts of the hydraulic system where fluid pressure is trapped and may need to be relieved because of the increase caused by higher temperatures.

During the flight of an airplane, it is quite likely that fluid in many of the hydraulic lines will be at a low temperature. When the airplane lands, this cold fluid will be trapped in the landing-gear system, the flap system, and other systems because selector valves are in the neutral, or OFF, position. The fluid-temperature increase due to warm air on the ground results in fluid expansion and could cause damage unless thermal relief valves are incorporated in the systems. Thermal relief valves are adjusted to pressures that are above those required for the operation of the systems; therefore, they do not interfere with normal operation.

PRESSURE-REDUCING VALVES

Requirements of some parts of the system may demand that the designer utilize a lower pressure than the normal system operating pressure. It may be desirable to have a reduced operating pressure to prevent overloading some structures. A pressure-reducing valve will fill this need. The proper valve will reduce system pressure to the desired level; it will also relieve thermal expansion in the section of the system that it isolates. Figure 13-36 shows how a pressure-reducing valve is positioned to bring about a lower pressure for operation of the actuating cylinder.

The pressure-reducing valve illustrated in Fig. 13-37 has three ports. One is connected to system pressure, one is connected to the system return line, and the third is the reduced-pressure port. The pressure-reducing spring (A) is holding the reservoir return valve (B) and the poppet (D) to the left. Fluid under system pressure enters the system pressure port (C), where it passes around the unseated poppet and out the reduced-pressure port (E). As the fluid going out port (E) builds pressure, hydraulic force is transmitted back through the hollow poppet (D) and exerts a force on the reservoir return valve (B). When this force overcomes spring force, the reservoir return valve is pushed to the right, allowing the poppet spring (F) to seat the poppet. This prevents system fluid from going out the reduced-pressure port (E) to build up a further pressure. The inlet pressure (C) has no effect on the poppet itself because the areas on each end of the poppet exposed to the pressure

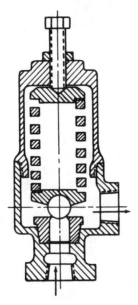

FIGURE 13-35 Construction of a relief valve.

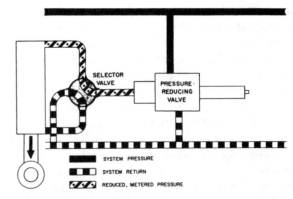

FIGURE 13-36 Typical pressure-reducing valve installation.

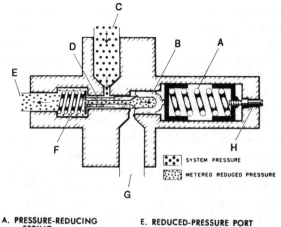

FIGURE 13-37 Pressure-reducing valve.

A. PRESSURE-REDUCING SPRING
B. RESERVOIR RETURN VALVE
C. SYSTEM PRESSURE PORT
D. POPPET
E. REDUCED-PRESSURE PORT
F. POPPET SPRING
G. RESERVOIR RETURN PORT
H. PRESSURE-ADJUSTING SCREW

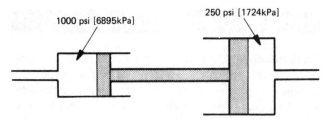

FIGURE 13-38 Principle of the debooster.

are the same; therefore, the forces exerted on the poppet are balanced. The pressure exerts an unbalanced force only on the area of the reservoir return valve (B).

In actual operation, this pressure-reducing valve will close when the desired pressure is reached. When the actuator is in operation under reduced pressure, the valve will vary its opening to meter the fluid at the speed required to maintain the desired pressure.

Another type of pressure-reducing valve is a debooster valve used in an aircraft brake system to reduce system pressure; in addition to reducing pressure it will provide for a higher volume of fluid flow to the brakes for rapid application of braking forces.

A review of two basic formulas, $F = P \times A$ and $V = A \times L$, will assist in providing a better understanding of deboosters. To determine the value of a force produced by pressure that acts on a given area, the formula $F = P \times A$ is used, where F is the total force, P is the value of the pressure expressed in pounds per square inch (psi), and A is the area of the surface exposed to the pressure. To find the volume of fluid required to move a piston within a cylinder or the amount exhausted, the formula $V = A \times L$ is used, where V equals the total volume delivered per stroke expressed in cubic units (usually cubic inches), A is equal to the surface area of the piston, expressed in square inches, and L is equal to the length of the stroke or the distance through which the piston moves as it discharges fluid.

A debooster valve operates by the differential area of two pistons. If a small-area piston is connected by a rod to a large-area piston, the two pistons will be capable of developing pressure in inverse proportion to their areas. Figure 13-38 illustrates the debooster principle. If the area of the small piston is 1 in² [6.45 cm²] and the area of the large piston is 4 in² [25.8 cm²], the large piston can transmit a pressure of only one-quarter that of the small piston. When 1000 psi [6895 kPa] is applied to the small piston, a force of 1000 lb [4448 N] will be exerted through the rod to the large piston. Since the large piston has a 4-in² [25.8-cm²] area, a force

of 1000 lb [4448 N] will develop a pressure of only 250 psi [1723.8 kPa] in the large cylinder.

A debooster consists of two pistons of different sizes fastened together in a cylinder housing machined for each piston, as illustrated in Fig. 13-39. Movement of one piston causes the other also to move. The main feature of the debooster is the piston, which has two different areas. The upper area (N) and the lower area (H) will have a given ratio. The extra fluid that is discharged will cause rapid application of brakes. The outlet port (D) and line must be larger than the inlet port to accommodate the extra flow.

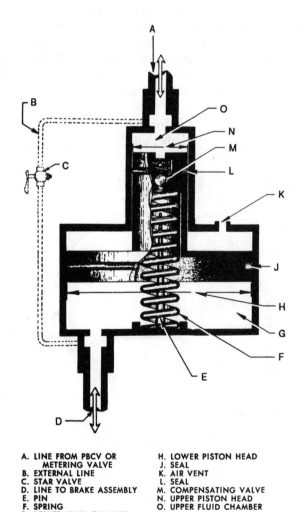

A. LINE FROM PBCV OR METERING VALVE
B. EXTERNAL LINE
C. STAR VALVE
D. LINE TO BRAKE ASSEMBLY
E. PIN
F. SPRING
G. LOWER FLUID CHAMBER
H. LOWER PISTON HEAD
J. SEAL
K. AIR VENT
L. SEAL
M. COMPENSATING VALVE
N. UPPER PISTON HEAD
O. UPPER FLUID CHAMBER

FIGURE 13-39 Typical debooster.

ACCUMULATORS

An **accumulator** is basically a chamber for storing hydraulic fluid under pressure. It can serve one or more purposes. It dampens pressure surges caused by the operation of an actuator. It can aid or supplement the system pump when several units are operating at the same time and the demand is beyond the pump's capacity. An accumulator can also store power for limited operation of a component if the pump is not operating. Finally, it can supply fluid under pressure to make up for small system leaks that would cause the system to cycle continuously between high and low pressure.

Accumulators are divided into three types according to the means used to separate the air and fluid chambers. The three types are *diaphragm, bladder,* and *piston.*

Diaphragm-Type Accumulator

Since hydraulic fluid is incompressible, practically speaking, some means is necessary to provide sustained pressure on the fluid if effective energy storage is to be maintained. For this purpose compressed air or an inert gas is used. The usual construction of the accumulator is such that a volume of compressed air is applied to a volume of fluid so the fluid will continue to be under pressure. The fluid and air are separated by a diaphragm so air cannot enter the hydraulic system.

The **diaphragm-type accumulator** consists of a metal sphere separated by a synthetic-rubber diaphragm, as shown in Fig. 13-40. The sphere is constructed in two parts, which

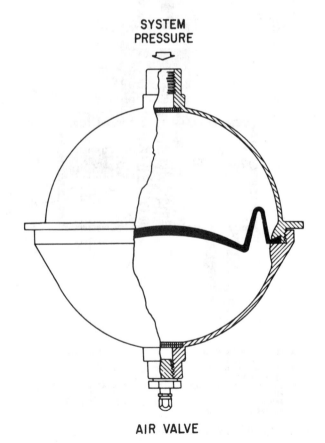

FIGURE 13-40   A diaphragm-type accumulator.

are joined by means of screw threads. At the bottom of the sphere is an air valve, such as a Schrader valve, and at the top is a fitting for the hydraulic line. A screen is placed at the fluid outlet inside the sphere to prevent the diaphragm from being pressed into the fluid outlet.

During operation of the accumulator, the air chamber is **preloaded,** or **charged,** with air pressure (approximately one-third maximum system pressure). As soon as a very small amount of fluid is forced into the fluid side of the accumulator, the system pressure gauge will show the pressure in the air chamber. This provides a means for checking the **air charge** (pressure) in the accumulator. If the system is inactive and the main pressure gauge shows zero pressure, a few strokes of the hand pump will cause the pressure gauge to rise suddenly to the charge pressure in the accumulator. The accumulator charge can also be checked by the reverse method. For example, if the system gauge shows a pressure of 1500 psi [10 342.5 kPa] when the system pump is not operating and the brakes are depressed and released a number of times, the pressure will decrease to the accumulator charge pressure and then will suddenly fall to zero.

It should be noted that some aircraft hydraulic systems monitor the hydraulic pressure by indicating the pressure on the air side of the accumulator. When the system has no hydraulic pressure, the gauge for the system indicates the accumulator air pressure. As soon as the hydraulic pressure is greater than the air charge, the air is compressed to the value of the hydraulic system pressure.

Bladder-Type Accumulator

The **bladder-type accumulator** usually consists of a metal sphere in which a bladder is installed to separate the air and the hydraulic fluid. The bladder serves as the air chamber, and the space outside the bladder contains the hydraulic fluid. The construction of a bladder-type accumulator is shown in Fig. 13-41. The air valve is at the bottom of the sphere; the fluid port is at the top. Initially, the bladder is charged with air pressure, as specified in the aircraft manual. When fluid is forced into the accumulator, the bladder collapses to the extent necessary to make space for the fluid, depending upon the fluid pressure.

Piston-Type Accumulator

Many modern hydraulic systems employ **piston-type accumulators** because they require less space than an equivalent spherical accumulator. A piston-type accumulator is shown in Fig. 13-42. Note that this unit consists of a cylinder with a free piston inside to separate the air from the hydraulic fluid. The piston is equipped with seals that effectively prevent the air from leaking into the fluid chamber and vice versa.

Servicing Accumulators

Accumulators should be checked for proper air charge at regular intervals. With no hydraulic pressure on the system,

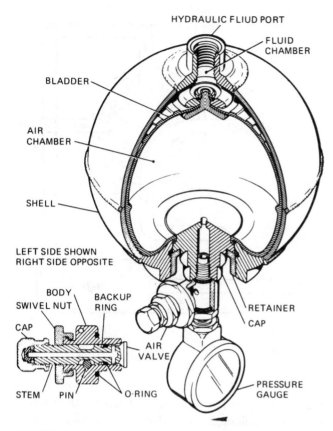

FIGURE 13-41   System accumulator.

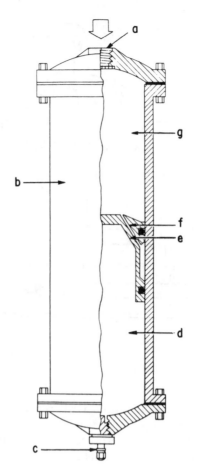

a. Fluid port
b. Cylinder
c. High-pressure air valve
d. Air chamber
e. Piston assembly
f. Drilled passage
g. Fluid chamber

FIGURE 13-42   A piston-type accumulator.

the accumulator should have a charge of between one-third and one-half of the system's operating hydraulic pressure. Specific values are given by the aircraft manufacturer. Many accumulators are equipped with permanent pressure gauges, but some require the use of a hand-held gauge.

If the accumulator must be charged, hydraulic pressure is removed from the system before the accumulator is charged. Nitrogen is the gas commonly used, but dry air may be used in some cases.

Removal and Installation of Accumulators

Great care must be exercised in the servicing and repair of hydraulic systems. This is particularly true of the high-pressure systems, which operate at pressures in excess of 3000 psi [20 685 kPa]. Before an accumulator or any other unit is removed, the technician must make certain that all the pressure in the system has been relieved. This is accomplished by operating one of the subsystems until all pressure is gone and the main pressure gauge reads zero pressure. The air pressure in the accumulator is reduced to zero by opening the air valve in accordance with the manufacturer's instructions. The accumulator can then be disconnected; however, provision should be made for fluid drainage. If the system contains a synthetic, fire-resistant fluid, such as Skydrol, the fluid must not be permitted to drain onto painted areas or other parts where the fluid can cause damage.

The installation of an accumulator is usually the reverse of removal. The air chamber of the accumulator is downward when the accumulator is installed. Sometimes the air charge

is placed in the accumulator before installation. In any event, the manufacturer's instructions should be followed. Particular care must be taken to see that all seals, valves, and fittings are of the proper type for the fluid being used in the system.

If hydraulic fluid is found in the air chamber of an accumulator, there is a leak between the two chambers. In such cases, the accumulator must be removed and repaired.

SELECTOR VALVES

Selector valves are used to direct the flow of hydraulic fluid to or from a component and achieve the desired operation. These valves fall into one of four general types: *rotary, poppet, spool* or *piston,* and *open-center-system selector valves.* The valves may be positioned by the pilot directly, by an electrical or electronic control, by hydraulic pressure, or by pneumatic pressure.

Rotary Valve

When a **rotary four-way valve** is in the position shown in Fig. 13-43(a), the fluid will flow from the valve at the top

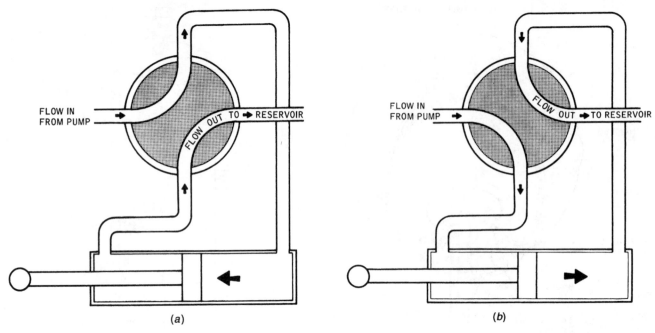

FIGURE 13-43   A rotary four-way valve.

port and will cause the actuating cylinder to be extended. When the valve is rotated 90°, as in Fig. 13-43(b), the fluid to and from the actuating cylinder will be in the opposite direction, and the cylinder will retract.

Poppet Valve

Another type of flow-control valve is shown in Fig. 13-44. The valve is shown operating a landing-gear system. This type of valve can be used to operate an actuator for any aircraft system. In this valve assembly, individual **poppet valves** are used to open and close the ports to change the direction of fluid flow. The valves are operated by cam lobes on cam rod C. Fluid enters the valve through line P from the

pressure pump, and with the gear control in the DOWN position, it passes through open valve 4 and on to the actuating cylinder. As the piston moves to the left, fluid from the left end of the actuating cylinder flows through passage B and open valve 1 to the return chamber and back to the reservoir through the return line R.

If the gear-control handle is placed in the neutral position, cam lobe 8 will open poppet valve 3, and cam lobes 6 and 9 will close poppet valves 1 and 4, assuming that the poppets are equipped with springs to keep them closed unless they are lifted by a cam lobe. When poppet valve 3 is open, fluid flow will pass from the pressure chamber PC to the neutral chamber NC and from there to the return manifold, thus permitting the fluid to flow freely and thereby reducing the load

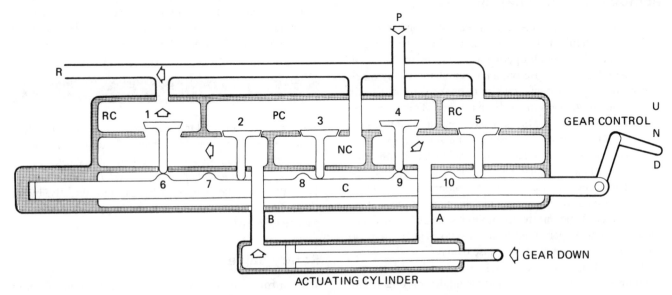

FIGURE 13-44   Poppet-type four-way valve.

on the pressure pump. In the neutral position, valves 2 and 5 are closed because cam lobes 7 and 10 are not yet in position to open them.

When the gear-control handle is placed in the UP position, valves 2 and 5 will be open and the others will be closed. Fluid will therefore flow through poppet valve 2 and out the passage B to the left end of the actuating cylinder. The piston will move to the right and force fluid through passage A and poppet valve 5 to the return chamber and return line to the reservoir.

Some poppet valve assemblies are arranged with valves in a radial position, and they are opened and closed by means of a rotary cam unit. The results are the same in any case. Poppet valves are also manufactured with electric controls, and the individual valves are opened and closed by means of solenoids.

Spool or Piston Selector Valves

A schematic drawing of a typical **spool-type selector valve** is shown in Fig. 13-45. The three positions of the valve are shown to illustrate the passages for fluid in the OFF, DOWN, and UP positions. Note that there is no fluid flow when the valve is in the OFF position; therefore, the valve must be used in a system where a pressure regulator or variable delivery pump is employed. Otherwise a high pressure would build up and cause excessive wear or other damage to the pressure pump. The use of this valve is not restricted to landing-gear systems.

A simple **piston-type selector valve** is illustrated in the drawing of Fig. 13-46. In the first drawing, the valve is in the OFF position, and fluid flow is blocked because both port A and port B are blocked by the piston. In the second view, port A is open to allow fluid to flow out to an actuating cylinder; the return flow from the cylinder enters port B and flows out port B to the reservoir. The third view shows the reverse position where fluid flows out port B and back through port A. The center of the piston rod is provided with a drilled passage, which allows the return fluid to flow to the right and out through the return port R.

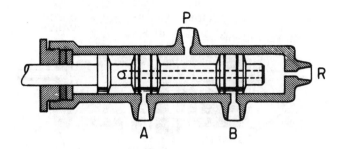

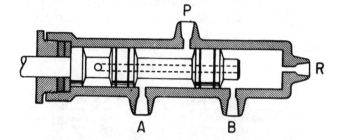

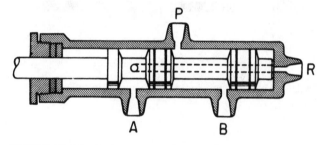

FIGURE 13-46   A piston-type selector valve.

Open-Center-System Selector Valve

The details of an open-center hydraulic system are given later in this chapter; however, it is important to understand that in this type of system, all selector valves are connected in series to a common fluid supply and return line. The valve shown in Fig. 13-44 is suitable for an open-center system.

Like other forms of selector valves, the **open-center selector valve** provides a means of directing hydraulic fluid under pressure to one end of an actuating cylinder and of simultaneously directing fluid from the opposite end of the actuating cylinder to the return line. The advantage is that the valve automatically returns to neutral when the actuating cylinder reaches the end of its stroke. The fluid output of the power pump is directed through this valve to the reservoir when the valve is in neutral position.

The valve illustrated in Fig. 13-47 consists of a housing that has four ports, a piston, two metering pins, two check valves, and a spring-loaded roller and cam-arrangement that is attached to the end of the piston. The roller and cam arrangement is designed to hold the piston in either the operating position after it has been engaged or the neutral position.

In the illustration, the sliding piston is in one of the two operating positions. The sliding piston has been moved manually to the right and is held in position by the spring-loaded cam mechanism, Fluid under pressure flows from the inlet port A through port D to one side of the hydraulic actuating piston,

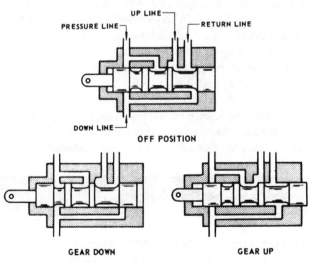

FIGURE 13-45   A spool-type selector valve.

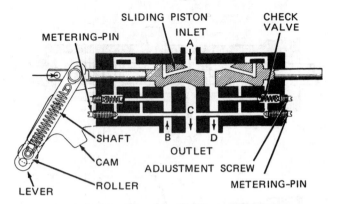

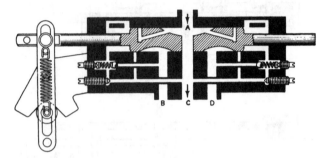

FIGURE 13-47   An open-center system selector value.

FIGURE 13-48   Open-center selector valve in the neutral position.

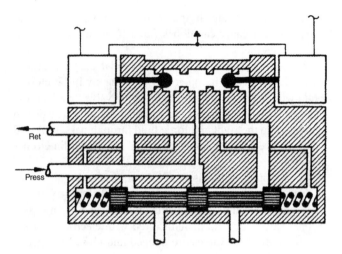

FIGURE 13-49   Solenoid-operated servo valve.

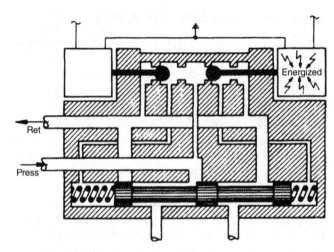

FIGURE 13-50   Servo control valve with one solenoid energized.

moving the actuating piston to its fully extended or retracted position. Fluid returning from the opposite side of the piston enters the selector valve at B and discharges through the return port C to the reservoir.

Figure 13-48 shows the same valve in the neutral position, where it is held by the lever and cam arrangement. The position of the lever, which rotates about the shaft, is determined by the position of the roller, which rolls on the cam. In the neutral position, the inlet port A is connected directly to the return port C, thus allowing the fluid to flow freely through the valve.

The valve is automatically returned to the neutral position by action of fluid pressure, which opens the relief valve and admits fluid pressure to the end of the piston. This pressure forces the valve piston back to the neutral position. The action takes place for either position of the valve.

Electrohydraulic Servo Control Valve

Electrohydraulic servo control (EHSC) valves are used extensively in transport aircraft that use a fly-by-wire flight control system. Hydraulic valves are operated without a mechanical connection. These types of valves use electrical solenoids to operate the control valve. In Fig. 13-49 both solenoids are not energized and hydraulic fluid pressure can flow past the pilot valves and apply pressure on both sides of the main spool valve. Because the areas of the spool valve ends and the springs are identical the main spool valve remains in the neutral position and the cylinder ports at the bottom of the valve are connected with the return line and

the pressure line is blocked. This type of valve is called a closed center valve.

In Fig. 13-50 the right solenoid is energized and the right pilot valve will move to the left and will block fluid pressure from going to the right side of the main spool valve and will connect the right side of the main spool valve with the return line. The main spool valve will move to the right because the pressure on the left side of the main spool valve is now higher than on the right side. When the main spool valve moves to the right the pressure line will be connected with the left cylinder port. When the right solenoid is de-energized the pressure will again be allowed to go to the right side of the main spool valve and the hydraulic pressure on both sides will be equal and the main spool valve moves to the neutral position because of spring pressure.

AUTOMATIC-OPERATING CONTROL VALVES

An **automatic-operating control valve** is one that is designed to operate without being positioned or activated by any force outside of the hydraulic fluid pressure or flow.

These valves are located in line with the system flow and function to perform operations such as prevent or restrict flow in a line, allow flow at the proper time, and change control of components between independent pressure systems.

Orifice or Restrictor Valves

An **orifice** is merely an opening, passage, or hole. A **restrictor** can be described as an orifice or similar to an orifice. A **variable restrictor** is an orifice that can be changed in size so its effect can be altered. The size of a fixed orifice must remain constant, whereas a variable restrictor permits adjustment to meet changing requirements.

The purpose of an orifice, or a variable restrictor, is to limit the rate of flow of the fluid in a hydraulic line. In limiting the rate of flow, the orifice causes the mechanism being operated by the system to move more slowly.

Figure 13-51 is a drawing of an orifice. This form of the device is merely a fitting that contains a small passage and has threaded ends. When fluid flows through the central passage, it will limit the rate of flow. An orifice of this construction may be placed in a hydraulic line between a selector valve and an actuating cylinder to slow the rate of movement of the actuating cylinder.

Figure 13-52 is a drawing of a variable restrictor. In this drawing, there are two horizontal ports and a vertical, adjustable needle valve. The size of the passage through which the hydraulic fluid must flow may be adjusted by screwing the needle valve in or out. The fact that the passage can be varied in size is the feature that distinguishes the variable restrictor from the simple fixed orifice.

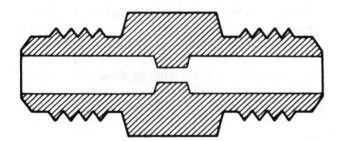

FIGURE 13-51 Drawing of an orifice.

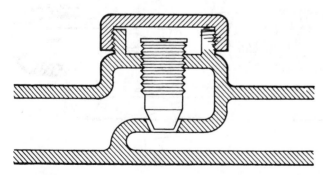

FIGURE 13-52 A variable restrictor.

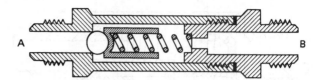

FIGURE 13-53 A ball check valve.

Check Valves

It is often necessary to prevent hydraulic fluid flow in one direction while permitting free flow in the opposite direction. The **check valve** is designed to accomplish this purpose. The construction of a simple ball check valve is shown in Fig. 13-53. Fluid pressure at port A will tend to push the ball off its seat and allow the fluid to flow through the valve. When the pressure is applied at port B, the ball will hold firmly on its seat and prevent the flow of fluid. Check valves are used as individual units in hydraulic systems, and they are also used as components of more complex valves and devices to control the flow of fluid in a given direction.

During the installation of a check valve, the technician must observe the direction of flow indicated on the body of the valve. Usually there is an arrow on the body or case of the valve to show the direction of the free fluid flow.

Orifice Check Valve

An **orifice check valve** is designed to provide free flow of hydraulic fluid in one direction and restricted flow in the opposite direction. One of the most common applications of this device is in the UP line of a landing-gear system. Since landing gear is usually quite heavy, it tends to fall too rapidly upon lowering, unless some means of restricting its movement is utilized. Since the UP line of a landing-gear actuating cylinder is the return line for hydraulic fluid from the actuating cylinder to the reservoir, any restriction in this line will limit the movement of the gear. That is, the gear movement must await the flow of the return fluid as it moves toward the down position.

An orifice check valve is also used for certain flap-control systems. Because of the air pressure on the flaps during flight, there is a continuous force tending to raise the flaps to a streamlined position. It is therefore advisable in some systems to restrict the UP movement by placing an orifice check valve in the DOWN line of the flap system.

The construction of an orifice check valve is illustrated in Fig. 13-54. When the valve is on its seat, fluid flow can occur only through the center orifice, but when the fluid flow is in the opposite direction, the valve moves off its seat and there is free flow of the hydraulic fluid through multiple openings.

The improper installation of an orifice check valve in a landing-gear system can cause serious problems. If the valve is installed in the reverse position, the movement of the landing gear is restricted as the gear is raised, but there is free flow as it is lowered. The hydraulic pressure in the system plus the force of gravity causes the gear to lower with excessive speed; when it reaches the end of its travel, the inertial

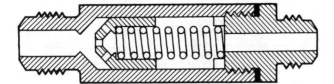

FIGURE 13-54 Construction of an orifice check valve.

forces are likely to damage the aircraft structure because of the sudden stop.

Metering Check Valve

A **metering check valve**, sometimes called a **one-way restrictor**, serves the same purpose as an orifice check valve. However, the metering check valve is adjustable, whereas an orifice check valve is not.

A drawing of a metering check valve is shown in Fig. 13-55. This unit has a housing, a metering pin, and a check-valve assembly. The pin is adjusted to hold the ball slightly off its seat. When fluid enters port B, it forces the ball away from its seat and then flows out through port A to the actuating cylinder. When the flow of fluid is reversed, the fluid entering from the actuating cylinder flows through the tiny opening between the ball and its seat, thus restricting the flow. By adjusting the metering pin in or out with a screwdriver, the rate at which the fluid can return from the actuating cylinder is controlled, because the position of the metering pin changes the width of the opening between the ball and its seat.

Hydraulic Fuse

The **hydraulic fuse** shown in Fig. 13-56 is a device designed to seal off a broken hydraulic line and prevent excessive loss of fluid. It permits normal flow in a line; but if the flow increases above an established level, the valve in the fuse closes the line and prevents further flow.

In reference to the figure, fluid enters the fuse through the passage at the right and flows between the outer case and the inner cylinder. It then passes through cutouts in the left

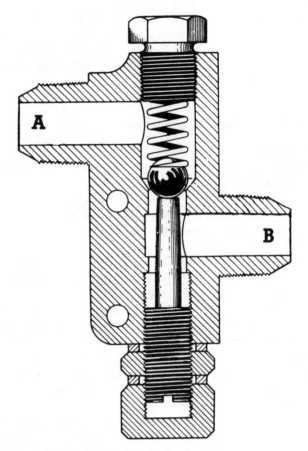

FIGURE 13-55 A metering check valve.

end of the inner cylinder and out though the center opening. At the right end of the inner cylinder is a metering orifice, which permits a small amount of fluid to enter the cylinder behind the poppet valve.

During normal operation the pressure of this fluid is approximately the same as the pressure on the opposite side of this poppet; therefore, there is no movement of the poppet. As fluid flow increases, the pressure differential across the poppet also increases. If this differential becomes excessive, the poppet moves to the left and closes the exit port of the fuse, thus stopping the fluid flow. The fuse remains closed

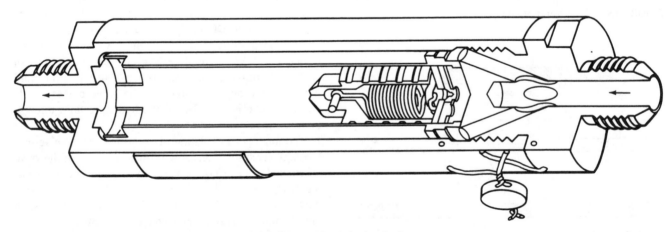

FIGURE 13-56 A hydraulic fuse.

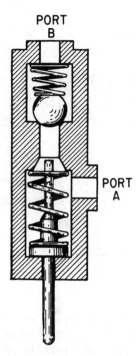

FIGURE 13-57 A sequence valve.

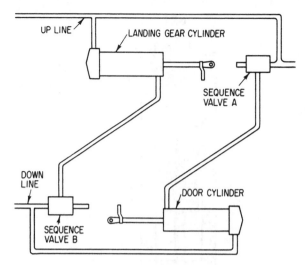

FIGURE 13-58 Landing-gear system with sequence valves.

only as long as a substantial pressure differential exists. If the pressure differential decreases to a certain predetermined level, the spring unseats the poppet and permits normal flow to resume.

Sequence Valve

A **sequence valve** is sometimes called a **timing valve** because it times certain hydraulic operations in proper sequence. A common example of the use of this valve is in a landing-gear system, where the landing-gear doors must be opened before the gear is extended and the gear must be retracted before the doors are closed.

Figure 13-57 is a drawing of a sequence valve. It is essentially a bypass check valve that is automatically operated. There is a free flow of hydraulic fluid from port A to port B, but the flow from B to A is prevented unless the ball is unseated by depressing the plunger.

Figure 13-58 is a schematic diagram of a landing-gear system with sequence valves. During the retraction of the landing gear, the fluid flows under pump pressure from the selector valve to the landing-gear cylinder and to sequence valve A. In this position, sequence valve A is closed, thus preventing the fluid from entering the door cylinder. As the landing-gear actuating-cylinder piston approaches the end of its travel, either the piston rod or some other part of the landing-gear mechanism depresses the plunger of the sequence valve, thereby permitting the fluid to flow to the door-actuating cylinder and to close the doors. The fluid displaced at the other end of the landing-gear actuating cylinder by the motion of the piston passes to the DOWN line through sequence valve B. The flow is not restricted, because it is flowing directly from port A to port B in the sequence valve illustrated in Fig. 13-57.

During the extension of the landing gear, the fluid under system pressure flows through the DOWN line and enters the door-actuating cylinder, but it is prevented from moving the landing-gear actuating-cylinder piston because sequence valve B is closed. However, when the door actuating-cylinder piston reaches the limit of its travel, the plunger on sequence valve B is depressed, and a passage is opened for the fluid to enter the landing-gear actuating cylinder and extend the gear. The fluid displaced by the motion of the door's actuating-cylinder piston flows to the UP line through sequence valve A without restriction.

Sequence valves are sometimes used in conjunction with hydraulically operated landing-gear UP locks and DOWN locks. In such instances, the sequence valve either blocks the fluid from the landing-gear actuating cylinder until the unlocking cylinder has released the lock or prevents the fluid from entering the locking cylinder until the landing gear has reached its fully retracted or extended position. Sometimes the sequence valve and the unlocking cylinder are manufactured as one unit.

To illustrate the operation of a sequence valve in a landing-gear system, the drawing of Fig. 13-59 is provided. This drawing shows a view of the sequence valve in the open position. If the valve is installed with port B connected to the landing gear UP line and port A connected to the gear door-actuating cylinder, fluid entering at port B cannot pass through to the actuating cylinder until the landing gear has been retracted and the valve plunger pressed in, as shown in the drawing. Therefore, the gear doors cannot be closed until the gear is retracted.

For the part of the system extending the landing gear, the sequence valve is installed with port B connected to the DOWN line of the landing-gear actuating cylinder. The plunger of the valve is not depressed until the gear doors are open; therefore, the gear cannot be extended until the door operation is nearly complete.

If, in the installation and adjustment of the landing gear system, the sequence valves are not set correctly, it is possible for the landing gear to strike the gear doors.

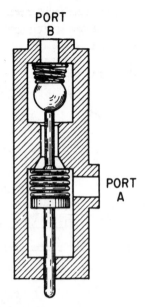

FIGURE 13-59   Sequence valve in open position.

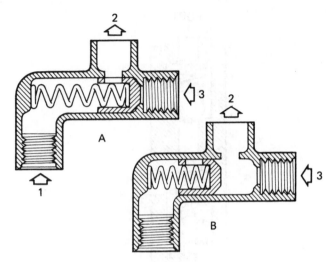

FIGURE 13-60   Operation of a shuttle valve.

Shuttle Valve

Quite frequently in hydraulic systems it is necessary to provide alternative or emergency sources of power with which to operate critical parts of the system. This is particularly true of landing-gear systems in the case of hydraulic-pump failure. Sometimes the landing gear is operated by an emergency hand pump and sometimes by a volume of compressed air or gas stored in a high-pressure air bottle. In either case it is necessary to have a means of disconnecting the normal source of hydraulic power and connecting the emergency source of power. This is the function of the **shuttle valve**.

The operation of a shuttle valve is shown in the drawings of Fig. 13-60. Port 1 of the valve is the normal entrance for hydraulic fluid from the pressure system. Port 2 is the outlet leading to the DOWN line of the landing-gear actuating cylinder. In view A the valve is in the normal position with free passage of fluid from port 1 to port 2. If main system pressure fails and it is desired to lower the landing gear, the landing-gear selector valve is placed in the DOWN position

and emergency pressure is applied to port 3. As explained previously, this pressure can come from an emergency pump or from a high-pressure air bottle. The emergency pressure forces the shuttle valve to the left, as in view B, and opens port 2 to port 3; thus fluid or air is permitted to flow through the valve to the DOWN line of the actuating cylinder. Return fluid from the cylinder will flow through the normal UP line, back through the main selector valve, and to the reservoir.

The installation of a shuttle valve in a landing-gear system is shown in the schematic drawing of Fig. 13-61. The DOWN line is blocked when emergency pressure is applied, and emergency fluid or air enters the cylinder, flowing from port 3 to port 2 and into the actuating cylinder.

Priority Valve

A **priority valve** is a sequence valve that is operated by hydraulic pressure rather than by a mechanical means. This valve is used to allow one actuator to operate and complete its operation before allowing a second component to operate. This gives the first component a priority over the second, resulting in the name "priority valve." A priority valve is represented in Fig. 13-62, and a simplified system using a

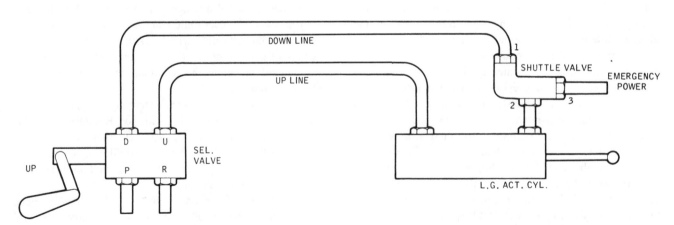

FIGURE 13-61   Location of a shuttle valve in a landing-gear system.

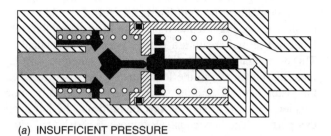

(a) INSUFFICIENT PRESSURE

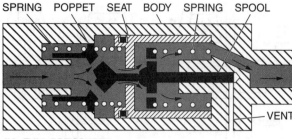

(b) FULL PRESSURE

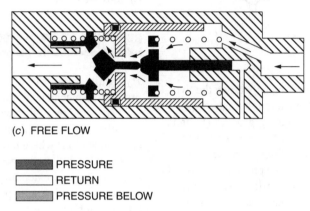

(c) FREE FLOW

▮ PRESSURE
▯ RETURN
▨ PRESSURE BELOW

FIGURE 13-62 A priority valve.

priority valve is shown in Fig. 13-63. When actuators A and B are both selected at the same time, pressure is available at the intersection of the lines. The priority valve in the line for actuator B requires that full-system pressure be applied to it to overcome the spring that is holding the spool closed. Since fluid is flowing to actuator A, the pressure is below the operating value of the valve. Once actuator A has stopped moving and the hydraulic pressure is up to near full value, the poppet has enough pressure applied to it to off-seat the spool valve and allow fluid to flow to actuator B. When fluid flows in the opposite direction, the seat shifts to the left and a free flow of fluid occurs.

The valve shown is a priority valve used for sequencing, but these valves are also used to give one component priority over another component in unrelated operations. For example, in some aircraft, a priority valve is used to give the flight-control actuators priority to system pressure over the landing gear and flap systems.

Flow Equalizers

The hydraulic **flow equalizer** is a unit that hydraulically synchronizes the movement of two actuating cylinders. To do this, it divides a single stream of fluid from the selector valve into two equal streams, causing each cylinder to receive the same rate of flow and both to move in unison. The flow equalizer also combines two streams of fluid at an equal rate; therefore, it synchronizes the actuating cylinders in both directions. Since this unit equally divides the combined flow, it is said to be *dual acting*. This unit is actually two constant-flow valves joined together with check-valve features to provide for reverse flow.

Figure 13-64 illustrates how the flow equalizer synchronizes the movement of the actuating cylinders in both directions. In the left-hand illustration, the equalizer is *splitting* the flow to extend to the pistons. The right-hand illustration shows the valve *combining* the two streams coming from the cylinders as the pistons retract. The two cylinders will move at the same rate and will always be synchronized even though unequal loads and pressure exist.

The type of flow equalizer shown in Fig. 13-65 is used to divide the pump's fluid output in the two different systems. Fluid enters the inlet port (C), where it pushes down the plug (D) and uncovers the two fixed orifices in the sleeve (B).

FIGURE 13-63 Location of priority valve in a system.

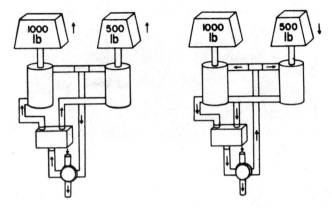

FIGURE 13-64 Flow equalizer installation.

The fluid then splits and tends to flow equally down the two side passages (A and E). It then flows through the two splitting check valves (G and P) and metering grooves (K and O) in the valve body. It then flows out the exit ports (J and N) to the actuating cylinders. Any difference in the rate of flow between the two passages results in a pressure differential between these passages. Then the free-floating metering piston (L) shifts to equalize the internal pressures, and the flow equalizes. If the pressures are different in the exit ports (J and N), there will also be a difference in the pressure drops across the metering orifices (K and O), with the greatest drop in the orifice leading to the line with least pressure in it.

To illustrate this equalizing action, assume that the actuating cylinder attached to the left-hand exit port (J) moves easier than the one attached to the right-hand exit port (N), The fluid would then try to flow faster out the left-hand exit port, but in doing so it must also flow faster through the left-hand fixed orifice, creating a greater pressure drop.

Fluid leaving the exit port (N) meets with more resistance, so the flow down the right-hand passageway would tend to be slower, causing a reduced pressure drop across the right fixed orifice and creating a greater pressure in passage A. This momentary pressure differential forces the free-floating metering piston (L) to the left, causing the space between the piston land (H) and the metering groove (K) to become smaller, which restricts the flow of fluid out the left-hand exit port (J). The flow equalizer imposes a restriction of the fluid that has a tendency toward faster flow, and the two streams are equalized.

HYDRAULIC ACTUATORS

Hydraulic actuators are devices for converting hydraulic pressure to mechanical motion (work). The most commonly utilized actuator is the actuating cylinder; however, servo actuators and hydraulic motors are also employed for special applications where modified motion is required. Actuating cylinders are used for direct and positive movement such as retracting and extending landing gear and the extension and retraction of wing flaps, spoilers, and slats. Servo actuators are employed in situations where accurately controlled intermediate positions of units are required. The servo unit feeds back position information to the pilot's control, thus making it possible for the pilot to select any control position required. The servo actuator is used to move large control surfaces such as the rudder, elevator, and ailerons. Servo units are also used to aid the pilot in the operation of cyclic and collective pitch controls in a helicopter.

Actuating Cylinders

The design of **actuating cylinders** is determined by the functions that they are to perform. Different types of actuating cylinders are shown in Fig. 13-66. The cylinder shown in Fig. 13-66(a) is a **single-acting** cylinder. Hydraulic pressure

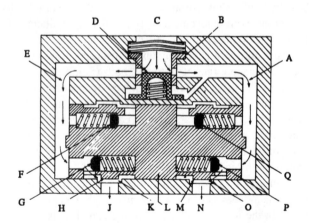

A. SIDE PASSAGE
B. SLEEVE
C. INLET PORT
E. SIDE PASSAGE
F. COMBINING CHECK
   VALVE
G. SPLITTING CHECK
   VALVE
H. PISTON LAND
J. EXIT PORT

K. METERING ORIFICE
L. FREE-FLOATING
   METERING PISTON
M. PISTON LAND
N. EXIT PORT
O. METERING ORIFICE
P. SPLITTING CHECK
   VALVE
Q. COMBINING CHECK
   VALVE

FIGURE 13-65 Flow equalizer—splitting position.

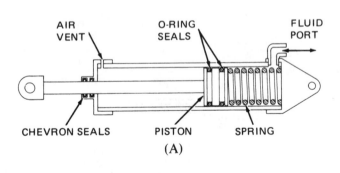

(A)

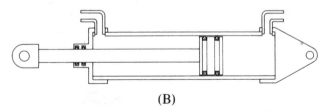

(B)

FIGURE 13-66 Actuating cylinders.

is applied to one side of the piston to provide force in one direction only. When hydraulic pressure is removed from the piston, a return spring moves the piston to its start position.

A **double-acting** actuating cylinder is designed so hydraulic pressure can be applied to both sides of the piston. Thus, the cylinder can provide force in either direction. A double-acting cylinder is shown in Fig. 13-66(b). This type of cylinder is used for the operation of retractable landing gear, wing flaps, spoilers, etc.

Many variations of the basic actuator cylinder have been designed, usually for special applications. The technician will find information on these in the applicable manufacturer's maintenance manuals.

Internal-Lock-Type Actuator

The **internal-lock actuator** allows the hydraulically operated mechanism to be locked in one of the extreme positions without the use of an external locking device. Figure 13-67 illustrates a typical actuating cylinder (B) incorporating internal locks and shows the actuator in both the locked (upper drawing) and unlocked (lower drawing) positions. In the locked position (upper), the locking segment (M) is positioned behind the lip of the concave head of the piston (C) when fully retracted. The lock spring (G) positions the lock rod piston (J)

to the right; the lock rod piston, in turn, extends the toggle (L) and positions the locking segment (M).

To unlock and extend the piston (C), fluid under pressure enters the extend port (F) and is directed through passageways to the head of both the piston (C) and the lock rod piston (J). Sine the piston (C) cannot move until the actuator is unlocked, the lock-rod piston (J) must first be moved to the left. Although the pressure is equal on both sides of the head of the lock-rod piston (J), the difference in working areas allows it to be moved to the left. This action compresses the lock spring (G) and allows the toggle (L) to collapse, so that the main actuating piston (C) can force the locking segment (M) inward as the piston begins to extend to the right. The piston (C) movement then allows the retainer spring (K) (which was compressed by the piston on the retract cycle) to move the segment retainer (E) to the right, as shown in the bottom portion of Fig. 13-67. This position holds the locking segment (M) in the retainer (D) while the piston (C) is in the extended position. This prevents the locking segment (M) from interfering with the piston (C) on the retract cycle.

To retract and lock the piston, fluid under pressure enters the retract port (A) and is directed to the piston head. When the piston (C) has retracted a specified distance, the piston head contacts the segment retainer (E) to move it to the left. As the segment retainer moves to the left, it compresses the

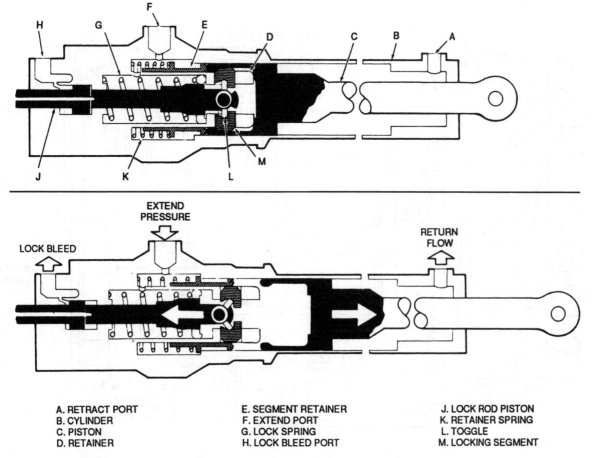

A. RETRACT PORT	E. SEGMENT RETAINER	J. LOCK ROD PISTON
B. CYLINDER	F. EXTEND PORT	K. RETAINER SPRING
C. PISTON	G. LOCK SPRING	L. TOGGLE
D. RETAINER	H. LOCK BLEED PORT	M. LOCKING SEGMENT

FIGURE 13-67 Internal-lock-type actuating cylinder.

retainer spring (K). After the piston (C) moves sufficiently to the left (fully retracted position), the lock spring (G) positions the lock rod piston (J) to the right, thus extending the toggle (L). This, in turn, positions the locking segment (M) behind the lip of the concave piston head to lock the actuator in the locked position.

The lock bleed port (H), which is connected to the return line, ensures the positive movement of the lock rod piston (J) by preventing the small amount of operational leakage around the shaft from building up pressure and creating a fluid lock. The hollow portion of the lock and piston shaft can be used for the attachment of a cable operated control so that the actuator can be unlocked manually during an emergency or ground operation.

This type of actuator can be used in many hydraulically operated subsystems and is ideal for hydraulically operated doors.

Servo Actuators

Aircraft equipped with fly-by-wire systems often use servo actuators to position the flight control surfaces. These are called power control units (PMUs) on Boeing aircraft or hydraulic jacks on Airbus aircraft. The PMU consists of an electrohydraulic control valve (EHCV) and an integrated hydraulic actuator. Motion transducers on the control column send a signal through the flight computer to the PMU

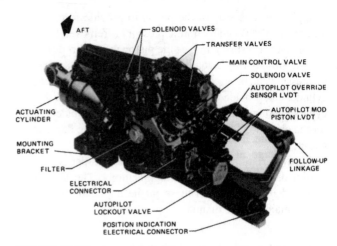

FIGURE 13-68 A servo actuator.

which in turn moves the flight control. An actuator position feedback signal will be sent back to the flight computer.

A **servo actuator** is designed to provide hydraulic power to aid the pilot in the movement of various aircraft controls. Such actuators usually include an actuating cylinder, a multiport flow-control valve, check valves, and relief valves together with connecting linkages. Figures 13-68 and 13-69 show the components of a servo actuator. When the pilot moves a cockpit control in a particular direction, force is applied to the flow control valve in the actuator, causing it

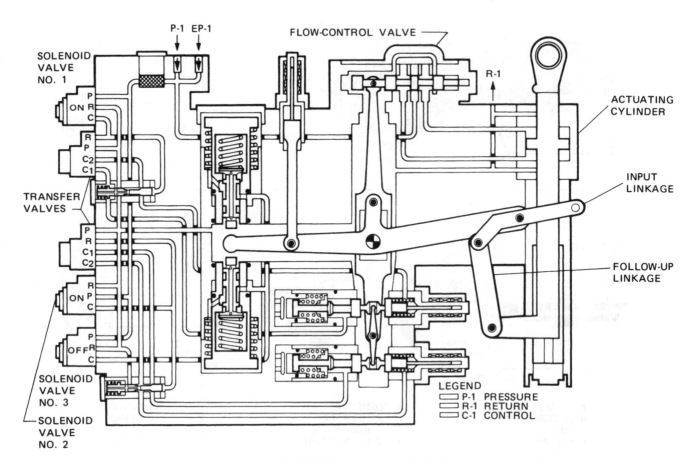

FIGURE 13-69 A servo actuator.

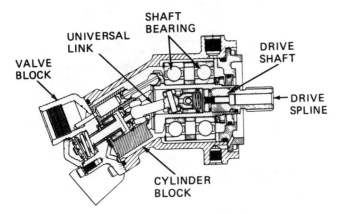

FIGURE 13-70 A hydraulic motor.

to direct fluid pressure to the actuating cylinder. The piston in the actuating cylinder moves in a direction to assist the pilot in moving the flight control. As the piston moves to the position called for by the pilot's control, the linkage between the piston and the flow-control valve moves the flow-control valve back to a neutral position and stops the flow of hydraulic fluid and, therefore, the movement of the piston. The reverse action takes place when the pilot moves the cockpit control in the opposite direction.

Hydraulic Motors

A **hydraulic motor** is illustrated in Fig. 13-70. Note that the motor is almost identical to the multiple-piston hydraulic pump described previously. When hydraulic fluid pressure is applied to the input port, it forces the pistons downward in the cylinders. Since the cylinder block axis is at an angle to the axis of the output shaft, the extension of the piston rods causes both the cylinder block and the output shaft to rotate. The two assemblies are connected by a universal linkage at the center. As each piston reaches its fullest extension, the cylinder is rotated past the input slot in the valve plate and on to the return slot. The piston is then moved inward in the cylinder, and the hydraulic fluid is discharged to the return line. The rpm of the hydraulic motor depends upon hydraulic fluid pressure and the load on the motor.

A hydraulic motor has the advantage over an electric motor of being able to operate through a wide range of speeds from 0 rpm to the maximum for the particular motor. Variable-speed electric motors can provide some flexibility in the rate of actuation; however, they lose efficiency as speed decreases.

HYDRAULIC PLUMBING COMPONENTS

As indicated previously, the various operating devices of a hydraulic system require tubing, fittings, seals, couplings, and hoses to transmit fluid pressure from unit to unit. To prevent leakage and ensure effective operation, various types of seals are employed. These are described in this section; however, there are so many different types that this section

is limited to those that are likely to be encountered by the aircraft technician. Information regarding hydraulic lines and fittings can be found in the text, *Aircraft Basic Science*.

Seals and Packings

The purpose of hydraulic packing rings and seals is to prevent leakage of the hydraulic fluid and thereby preserve the pressure in the aircraft hydraulic system. Seals are installed in most of the units that contain moving parts, such as actuating cylinders, valves, and pumps. Packing rings, seals, and gaskets are made in a variety of shapes and sizes.

Seals used in hydraulic and pneumatic systems and components are of two general classes, *packings* and *gaskets*. **Packing rings** are used to provide a seal between two parts of a unit that move in relation to each other. The **gasket** is used as a seal between two stationary parts. This classification will apply in most cases; however, deviations will be found in some instances.

The handling, installation, and inspection of rings, seals, and gaskets in an aircraft hydraulic system are extremely important because a scratched or nicked packing ring can cause the failure of a unit in the system at a critical time.

Typical shapes for hydraulic packing rings, seals, and gaskets are shown in Fig. 13-71. These drawings show the standard O-ring seal, chevron seal, universal gasket, and crush washer.

The **O-ring seal** is probably the most common type used for sealing pistons and rods because it is effective in both directions. A color code on the O-ring seal identifies the type of system in which it is used and, generally, the manufacturer. O-rings have either a series of colored dots arranged clockwise or a colored stripe around the outer circumference. The first dot, reading in a clockwise direction or the stripe indicates the system in which the O-ring is used. The following color codes are used: Red indicates use in the fuel systems only; blue indicates use in hydraulic and pneumatic systems; and yellow indicates use in oil systems only. Note: A white dot appearing first indicates a nonstandard seal. The second dot indicates the system. A white dash with a yellow dot 90° from the dash indicates a system using Skydrol.

Generally, the O-ring seal requires no adjustment when it is installed. Figure 13-72 shows a typical O-ring installation. However, a few precautions must be observed at installation, or early failure will result. Always install new O-rings with each disassembly. Do not use old rings that have been in service. The seal must be the correct size and installed without defects such as cuts, nicks, or other flaws. The seal and O-ring groove must be lubricated freely with the same type

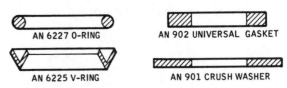

FIGURE 13-71 Shapes for packing rings, seals, and gaskets.

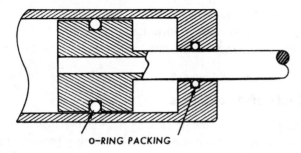

FIGURE 13-72 Typical O-ring installation.

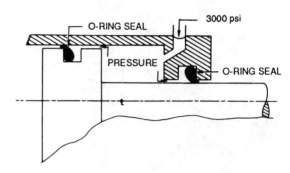

FIGURE 13-73 O-ring seal, pinched condition.

of hydraulic fluid that is used in the system unless otherwise specified in the repair manual. Care must be exercised when installing O-rings to prevent scratching or cutting the seal on threads or sharp corners. Also, make certain that the O-ring is not installed in a twisted condition; otherwise it will not function correctly.

The O-ring seal should not be used alone in a system where the pressure is greater than 1500 psi [10342.5 kPa]. If the hydraulic pressure is too great, the ring can become pinched between the moving part and stationary part of the unit, as is shown in Fig. 13-73, thus damaging the ring and destroying the seal. When the O-ring is used in higher-pressure systems, **backup rings** are installed. Backup rings are used in conjunction with seals subjected to high pressure. Their purpose is to prevent the high pressures from pinching the seal between the moving and stationary parts, such as between a piston

and cylinder wall. They give additional strength to the seal. Backup rings are usually made of Teflon.

If pressure is exerted on the seal in one direction only, the backup ring should be placed on the side away from the pressure. Where the pressure is exerted alternately in each direction, these backup rings are installed on both sides of the O-ring. Proper installation of backup rings is shown in Fig. 13-74.

The V-shaped (chevron) seal and the U-shaped seal must be installed with the same care as that used for other types of seals. The size of the seal must be correct and there must be no nicks, scratches, or other damage on the seals. When high pressure is exerted in only one direction, one set of seals is sufficient, but when the pressure is alternately applied, first in one direction and then in the other, two sets of seals are required because this type of seal works effectively in only

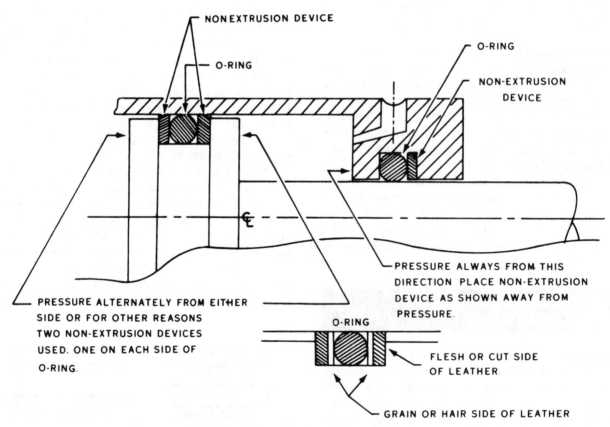

FIGURE 13-74 O-ring seal with backup ring.

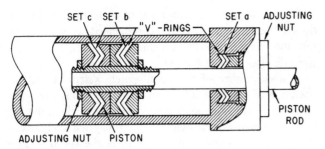

FIGURE 13-75 Installation of a dual set of chevron seals.

one direction. The installation of a dual set of chevron seals is shown in Fig. 13-75.

The general procedure to follow in installing V-shaped or U-shaped seals is as follows:

1. Install one ring at a time, being sure that it is seated properly. Never install such rings in sets.

2. Use shim stock (very thin metal sheet) to protect the packing rings if the packing crosses sharp edges or threads. The shim stock should be from 0.003 to 0.010 in [0.076 to 0.254 mm] thick. It is rolled and then placed over the threads or sharp edges. After the packing is installed, the shim stock is removed.

3. If the unit in which the packing is being installed does not have an adjustable packing gland nut, insert metal shims of graduated thickness behind the adapters to hold the packing securely in place.

4. If the unit in which the packing is being installed has an adjustable packing gland, adjust the gland nut until the V-ring stack is held together firmly but not squeezed. The gland nut is then loosened to the first lock point.

5. Whenever possible, the technician should check the unit by hand for free operation after installation before the hydraulic pressure is applied.

6. In all cases, the technician should consult the manufacturer's instructions regarding the installation of seals.

HYDRAULIC SYSTEMS FOR AIRCRAFT

Every aircraft hydraulic system has two major parts or sections, the power section and the actuating section (or sections). The power section provides for fluid flow, regulates and limits pressure, and carries fluid to the various selector valves in the system. The actuating section or subsystems are the sections containing the various operating units, such as the wing flaps, landing gear, brakes, boost systems, and steering mechanisms.

The power section may either be an *open* or a *closed* system using an engine-driven pump or a pump driven by an electric motor. Pressure developed by the pump in an open system is controlled by one of the following three valves: an open-center valve, a power-control valve, or a pump-control valve. An open system means there is a free path for fluid flow back to the reservoir until one of these units is actuated and directs fluid flow from the selector valve. Fluid flow is

then directed to the actuators (cylinders, motors, etc.), and pressures will build sufficiently to move the actuating units.

All three types of open system valves are automatically shifted by hydraulic pressure to the open position after the units have been actuated. All three type valves employ time-lag or resistance features to prevent them from shifting to "bypass" too soon, which would result in a loss of fluid flow and pressure to the units before the hydraulic cycle has been completed. When the power valve, pump-control valve, or the open-center valve is in the bypass position, the pump is said to be *idling* under no load and with little power-input requirement.

In a closed system the pressure developed by the engine-driven or electric motor-driven power pump may be regulated by a pressure regulator (unloading valve), by the power pump (which would then have an integral control valve), or by a pressure switch (which would shut off the pump driven by an electric motor when the desired system pressure is reached). The closed system will always have fluid stored under pressure whenever the power pump is operating. However, after system pressure is built up to a predetermined value, the load is automatically removed from the pump by an unloading valve called the pressure regulator or the integral control valve of the pump (called the compensator valve, which controls the pump output). In this way the pump is allowed to idle when no units are in operation and until there is further demand upon the system.

Open Systems

An **open system** is one having fluid flow but no appreciable pressure in the system whenever the actuating mechanisms are idle. Fluid circulates from the reservoir, through the pump, through the open valves, and back to the reservoir, as illustrated in the upper drawing of Fig. 13-76. This system is also called an open-center system. Selector valves in an open system (using an open-center valve) are always connected in series with each other, an arrangement whereby the pressure line goes through each selector valve. Fluid is allowed free passage through each selector valve and back to the reservoir until one of the selector valves is positioned to operate a mechanism, as illustrated in the lower drawing of Fig. 13-76. Fluid is then directed to the actuator, and pressure is allowed to build up in the systems. A pump-control valve or a power-control valve is placed in a line that goes directly from the pump to the reservoir. When closed, these valves block the flow to the reservoir and force the flow to a two-position valve and then to one of the actuating units.

A typical open system is shown in Fig. 13-77. This system is a combination closed and open system, with both parts of the system receiving their fluid supply from the same engine-driven constant-volume power pump. The pump supply is divided equally to the two systems by a flow equalizer. Other units included in the closed section of the system are discussed at a later time.

Open systems develop no pressure except when a mechanism is being operated; the pressure is then metered by the selector valve and limited by a relief valve. Each subsystem has an individual relief valve that limits the maximum pressure of the system.

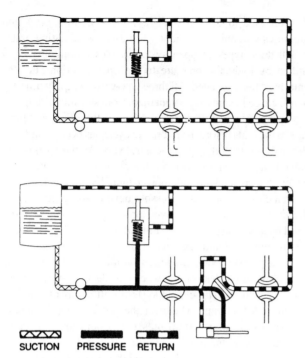

SUCTION   PRESSURE   RETURN

FIGURE 13-76   Basic open-center system.

One of the advantages of the open-center system is that it does not require expensive and complicated pressure regulators; the power pump can be a simple gear pump, although a fixed-displacement piston pump may be used.

A disadvantage of the open-center system is that the operation of only one subsystem at a time is possible without interference from other systems. For example, if the flap subsystem precedes the landing-gear subsystem and the two systems are operated at the same time, the speed of gear operation will be limited by the amount of fluid returning from the flap-actuating cylinder. As soon as the flap operation is complete and the flap-selector valve kicks out, the landing-gear operation proceeds at its normal rate.

Open systems are generally found only on light, general aviation aircraft. Transport-category aircraft require more complex systems, which may have several units operating at the same time.

Closed Systems

A **closed system** is one that directs fluid flow to the main system manifold and builds up pressure (stores fluid under pressure) in that portion of the system that leads to all the selector valves. There are two basic types of closed systems.

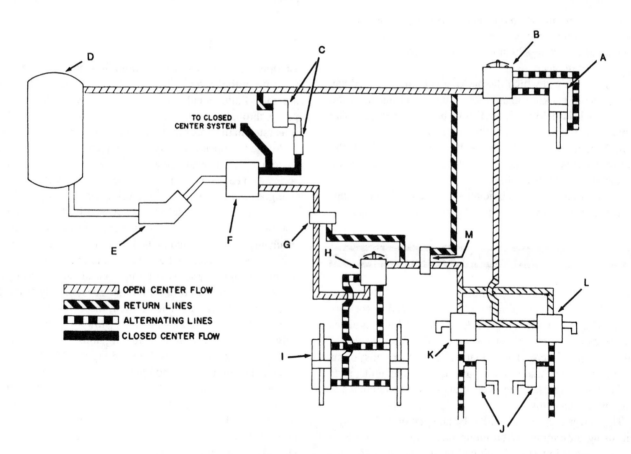

OPEN CENTER FLOW
RETURN LINES
ALTERNATING LINES
CLOSED CENTER FLOW

A. ELEVATOR ACTUATING CYLINDER
B. ELEVATOR BOOST CONTROL VALVE
C. RELIEF VALVE AND CHECK VALVE
D. RESERVOIR
E. POWER PUMP

F. FLOW EQUALIZER
G. RELIEF VALVE
H. AILERON BOOST CONTROL VALVE
I. AILERON ACTUATING CYLINDERS

J. RELIEF VALVES
K. BRAKE BOOST CONTROL VALVE
L. BRAKE BOOST CONTROL VALVE
M. RELIEF VALVE

FIGURE 13-77   Power section with open-center flow.

One has a constant-volume pump and a pressure regulator to control the pressure at a working range and to "unload" the pump when there is no flow requirement and pressure builds up in the system manifold; the second closed system utilizes a variable-volume pump and directs the flow to the system manifold, similar to the constant-volume system. The output of the variable-volume pump is controlled by an integral control valve. This valve reduces the pump flow to zero when no units are operating in the system and pressure is built up in the storage chambers, called accumulators. The pump may be driven by either an engine or by an electric motor. The means of pressure control and the means of driving the pump may vary, but all closed systems are basically the same. Any number of subsystems may be incorporated in a closed system, but subsystems differ from the open system in that the selector valves are arranged in parallel rather than in series. The pressure regulator (as used in constant-volume systems) directs the flow of the pump (or pumps) either to the system or to the reservoir, depending upon the existing system pressure. System pressure is always maintained between the kick-out and kick-in settings of the regulator when the actuating mechanisms are not in operation.

When a selector valve is positioned to operate a mechanism, the system pressure drops until it reaches the kick-in settings of the regulator, at which time the pump again directs the flow into the system to increase the system pressure until the maximum setting of the regulator is reached again. The accumulator used in the system stores fluid under pressure, stabilizes system pressure, and ensures smooth operation of the regulator. A system relief valve safeguards the system if the regulator fails.

Multiple power pumps are used in many hydraulic closed systems. Normally they are used in multiengine aircraft, where they can be driven by separate engines. This ensures hydraulic system operation in the event of an engine failure or the failure of one of the pumps. Figure 13-78 shows how dual power pumps are connected into the system. Notice that both pumps supply the same system by combining their volume output into a common pressure manifold. The pumps may be either the constant-volume type or the variable-volume type.

Figure 13-79 is a schematic drawing of a direct pressure system that has its pressure regulated by an unloading valve (pressure regulator) operating in conjunction with an accumulator. This system is also termed a **closed-center system**.

When the actuating equipment is not in use, pressure is relieved from the continuously operating engine-driven pump by means of the unloading valve after the accumulator is charged to the correct level. As soon as a subsystem is operated, pressure first comes from the accumulator for operation; when accumulator pressure has dropped to a predetermined level, the unloading valve kicks in and directs pump flow to the operating system. A study of the diagram shows that the pump output flows directly through the unloading valve and back to the reservoir when no pressure is required. When a subsystem (brakes, landing gear, or flaps) is operated, pump pressure flows through the subsystem and back to the reservoir through the subsystem return line.

The main relief valve in the system is located between the unloading valve and the return line. If the unloading valve should become stuck in the kicked-in position, the excess pressure would be bypassed through the relief valve to the reservoir. The pressure gauge would show higher than normal pressure, and the system would probably be making noise because of the operation of the relief valve. The relief valve is usually set at least 100 psi [689.5 kPa] above the normal operating pressure.

Light-Aircraft Hydraulic Systems

A schematic diagram of a hydraulic system for a light, multiengine airplane is shown in Fig. 13-80. The diagram shows that the engine-driven hydraulic pump (10) draws hydraulic fluid from the reservoir (15) and pumps it through the pressure port of the "Powerpak" assembly into the landing-gear-selector pressure chamber. When the two selector valves are in the neutral position, the fluid travels from the landing-gear selector valve (1) through the flap selector valve (2) and back to the reservoir.

The Powerpak assembly is a modular unit that includes the reservoir, relief valve, hand pump, landing-gear selector valve, wing-flap selector valve, filters, and numerous other small parts essential to the operation. When both selector valves are in the neutral position, the system acts as an open-center system, in that the fluid flows first through one selector valve and then through the other before returning to the reservoir. During this time the fluid flows freely at a reduced pressure. Since the fluid supply line runs first through the landing-gear selector valve, the flaps cannot be operated while the landing-gear subsystem is in operation. Each selector valve has a separate return line to allow fluid from the actuating cylinder or cylinders to flow back to the reservoir.

The diagram of Fig. 13-81 shows the fluid flow when the landing-gear selector is placed in the UP position. When the selector valve is placed in either operating position, it is held in the position by a detent consisting of a ball, O-ring, plunger, and spring. The ball snaps into a groove in the spool, which operates the poppet valves and prevents the spool

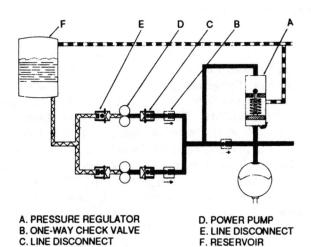

A. PRESSURE REGULATOR
B. ONE-WAY CHECK VALVE
C. LINE DISCONNECT
D. POWER PUMP
E. LINE DISCONNECT
F. RESERVOIR

FIGURE 13-78 Closed system using dual pumps.

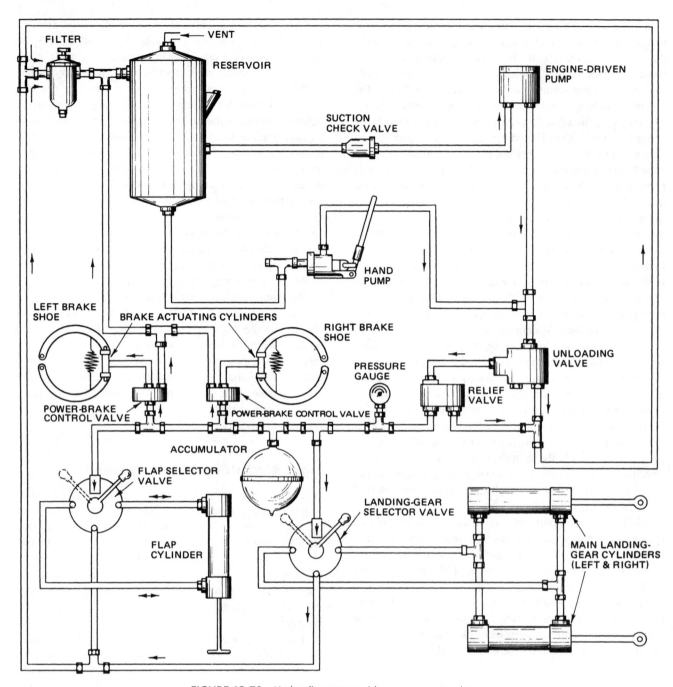

FIGURE 13-79 Hydraulic system with a pressure regulator.

from moving linearly until the operation is complete. When the actuating cylinder reaches the end of its movement, pressure builds up to approximately 1150 psi [7829 kPa]. This pressure acts against the plunger of the detent mechanism and relieves the pressure of the spring, thus allowing the ball to pop out of the groove. A spring then causes the spool and selector lever to return to the neutral position.

Figure 13-82 shows the fluid flow in the system when the wing-flap selector valve is placed in the DOWN position. The action is generally the same as that described in the foregoing paragraph. Note that the fluid to the flap-selector valve has passed through the landing-gear selector valve first and that the landing-gear valve is in the neutral position.

The fluid used in the just-described system is petroleum type, MIL-0-5606. The operating pressure is 1150 psi [7829 kPa], and the main relief valve "cracking" pressure is 1250 to 1300 psi [8618.8 to 8963.5 kPa]. Thermal relief valves are set to open at 2000 to 2050 psi [13790 to 14134.8 kPa] or 1850 to 1900 psi [12755.8 to 13100.5 kPa], depending upon the particular model of Powerpak. The main reservoir capacity is approximately 4.5 p [2.13 L], with an emergency supply of 0.95 p [0.45 L].

Another example of a comparatively simple hydraulic system is illustrated in Fig. 13-83. Since this system serves one function only, raising and lowering the landing gear, it requires no selector valves. Direction of fluid flow is controlled by the reversible, electrically driven pump.

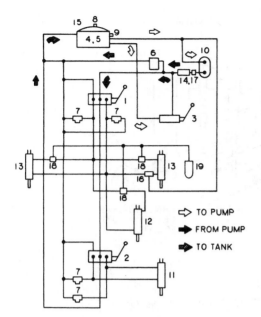

1. Landing-gear selector
2. Flap selector
3. Hand pump
4. Emergency fluid-trap stand pipe
5. Emergency hand-pump check filter
6. Main relief valve
7. Thermal relief valve
8. Atmospheric vent
9. Filter and filtering port
10. Pump
11. Flap-actuating cylinder
12. Nose-gear actuating cylinder
13. Main-gear actuating cylinder
14. System check
15. Reservoir
16. Antiretraction valve
17. Hydraulic filter
18. Shuttle valves
19. Nitrogen bottle

⇨ TO PUMP
➡ FROM PUMP
⬗ TO TANK

FIGURE 13-80 Schematic diagram of the hydraulic system for a light twin-engined airplane. (*Piper Aircraft Co.*)

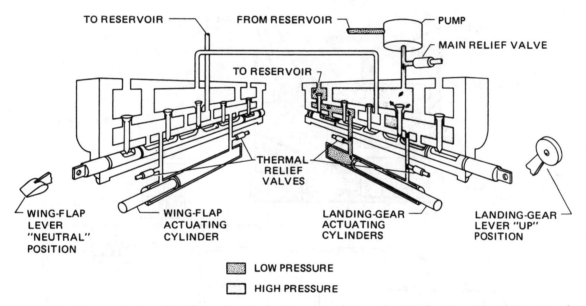

FIGURE 13-81 Diagram to show fluid flow when the landing-gear selector valve is in the UP position. (*Piper Aircraft Co.*)

The reservoir, gear pump, relief valves, reversible motor, filter, and shuttle valve are all integrated into one assembly. A pressure switch is installed on a cross fitting connected to the pump-mount assembly. This switch opens the electrical circuit to the pump solenoid when the gear fully retracts and the pressure in the system increases to approximately 1800 psi [12 400 kPa]. The switch continues to hold the circuit open until the system pressure drops to approximately 1500 psi [10 340 kPa], when it again operates to build up pressure as long as the gear-selector handle is in the UP position. The DOWN position of the selector does not affect the pressure switch.

The pump is a gear-type unit driven by a 14-V reversible motor designed to operate in a pressure range of 2000 to 2500 psi [13 790 to 17 237 kPa]. A **thermal relief valve** is connected to the gear-up pressure line to prevent excessive pressure due to thermal expansion of the fluid. This valve maintains pressure in the system up to 2350 + 50 psi [16 230 + 345 kPa]. An additional relief valve is incorporated in the system to return hydraulic fluid to the reservoir if pressure should exceed 4000 psi [27 436 kPa]. The shuttle valve, located in the base of the pump, allows fluid displaced by the actuating cylinder pistons to return to the reservoir without back pressure.

The system includes a free-fall valve that allows the landing gear to drop in the event of a system malfunction. The valve is controlled manually or by a gear backup extension device that is operated by a pressure-sensing chamber.

Hydraulic Systems for Aircraft   **421**

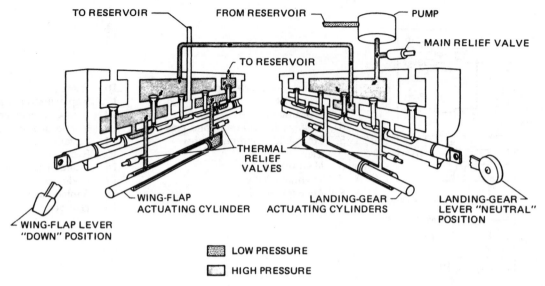

FIGURE 13-82 Fluid flow when the wing-flap selector valve is in the DOWN position. (*Piper Aircraft Co.*)

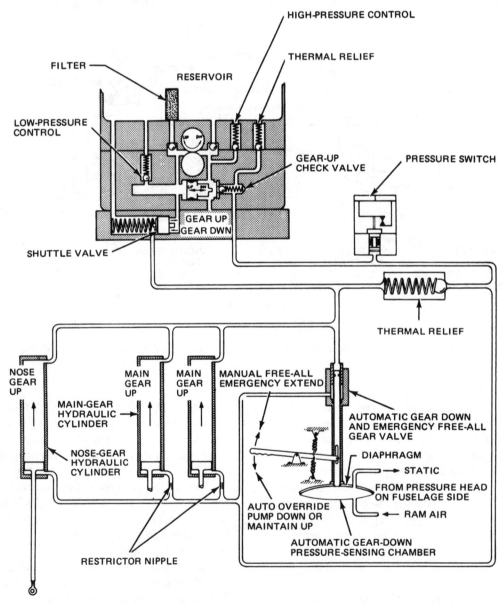

FIGURE 13-83 Hydraulic system for landing-gear operation. (*Piper Aircraft Co.*)

This chamber moves the free-fall valve to lower the gear, regardless of gear-selector handle position, depending upon engine power (propeller slipstream) and airspeed. Landing-gear extension occurs even if the selector is in the UP position at airspeeds below approximately 118 mph [190 km/h] with engine power off. The device also prevents the gear from retracting at airspeeds below approximately 93 mph [150 km/h] at sea level with full power, even though the selector switch may be in the UP position. The sensing device operation is controlled by a differential air pressure across a flexible diaphragm, which is mechanically linked to the hydraulic free-fall valve and an electric switch that control the pump motor. A high-pressure and static air source for actuating the diaphragm is provided in a mast mounted on the left side of the fuselage above the wing. Manual override of the device is provided by an emergency gear lever. The emergency free-fall lever must be held in the UP position to retract the gear and in the DOWN position to extend the gear.

HYDRAULIC SYSTEM FOR THE BOEING 777 AIRLINER

The hydraulic system described here represents the installation of one particular model of the Boeing 777 airplane. The Boeing 777 incorporates three separate and independent hydraulic power systems. These are designated: left system, center system, and right system. Each hydraulic system has a pressurized reservoir. The left system receives fluid under pressure from the left engine–driven pump (EDP) and an AC-motor pump (ACMP). The center system receives fluid under pressure from two AC-driven pumps, two air-driven pumps (ADP), and the ram air turbine (RAT). The right system receives fluid under pressure from the right engine–driven pump and an AC motor pump. Figure 13-84 is a block diagram that shows the interconnections between units in each system and shows what aircraft subsystems are powered by what system. Note that there is no interconnection between the three systems. Two supply shutoff valves controlled by the engine fire switches are installed downstream of the reservoir to stop the flow of hydraulic fluid to the engine area in case of an engine failure or fire.

Modular filter units are used in each system to combine a number of the smaller system components in one case to simplify maintenance. Each pump has its own filter module. For instance, the EDPs have a filter module that contains a temperature transducer, pressure transducer, ground servicing connection, filters, pressure relieve valve, and check valves for the direction of flow. Both the pump pressure output and the pump case drain line are routed through the filter module. The case drain line fluid is returned to the reservoir after being cooled in the heat exchanger. The ACMP has a filter module that contains filters, check valves, temperature transducer, and pressure transducer. The hydraulic fluid used for the Boeing 777 is Skydrol 5 (type V), a fire-resistant, phosphate ester-base, synthetic fluid. Servicing of any airplane's hydraulic system must be accomplished according to

the specific instructions issued by the manufacturer for the particular airplane. On the Boeing 777, the servicing station is located in the aft right-wing fairing area and is accessible through an access door. The filing equipment consists of a manually operated hand pump, hydraulic reservoir (system) selector valve, reservoir fill filter, and reservoir fill quantity gauge. The arrangement of the equipment in the filling area is shown in Fig. 13-85.

Left and Right Hydraulic Power System

The left and right hydraulic power supply systems are illustrated in the schematic drawing of Fig. 13-86. The left and right systems are very similar and therefore only the left system is discussed in this section. The left system is powered by an engine-driven pump mounted on engine 1 and an AC motor pump. The left system includes the equipment necessary to store, pressurize, deliver, control, monitor, and filter the hydraulic fluid to operate the systems previously noted. Hydraulic fluid for the system is stored in a reservoir that is pressurized by bleed air routed through a reservoir pressure module consisting of filters, check valves, restrictor valve, and pressure regulator to ensure a positive supply of hydraulic fluid to the pumps. The EDP and ACMP are variable delivery pumps that supply fluid to the various systems upon demand. The ACMP normally operates only when there is high hydraulic system demand. The EDP is equipped with an electrically controlled depressurizing valve to stop the flow of hydraulic fluid. Each pump has a filter module that contains various smaller power supply components such as filters, check valves, and temperature and pressure transducers. Hydraulic pump pressure and temperature information; reservoir pressure, temperature, and quantity information; and system pressure information is available on the airplane information management system (AIMS). Figure 13-87 shows the system information on AIMS. The EDP filter module also contains the system pressure relief valve. The pressure relief valve protects the system against damage in case a malfunction permits the pressure to rise to an abnormally high level. A hydraulic heat exchanger is installed in the pump case drain line to cool the hydraulic fluid that comes out of the pump case before it is returned to the reservoir.

Hydraulic Reservoir

The hydraulic system reservoirs (Fig. 13-88) of the left and right system contain the hydraulic fluid supply for the hydraulic pumps. The reservoir is pressurized by bleed air through a reservoir pressurization module; see Fig. 13-89.

The engine-driven pump (EDP) draws fluid through a standpipe. The electrical pump (ACMP) draws fluid from the bottom of the reservoir. If the fluid level in the reservoir gets below the standpipe, the EDP cannot draw any fluid any longer and the ACMP is the only source of hydraulic power. The reservoir can be serviced through a center servicing point in the fuselage of the aircraft. The reservoir has a sample valve for contamination testing purposes,

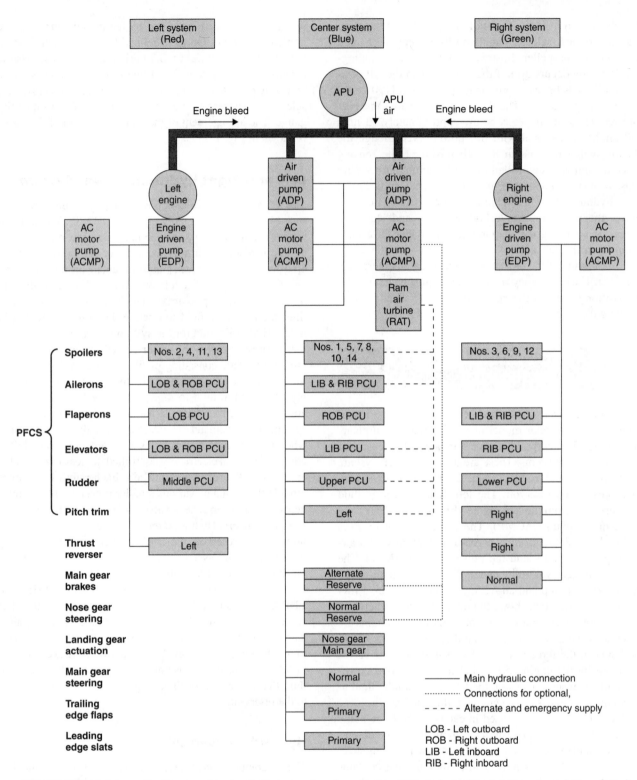

FIGURE 13-84 Schematic diagram of hydraulic power system for a Boeing 777 airplane. (*Boeing Commercial Aircraft Co.*)

a temperature transmitter for temperature indication on the flight deck, a pressure transducer for reservoir pressure, and a drain valve for reservoir draining.

Modular Unit

There are two EDP modular filter units, four ACMP modular filter units, two ADP modular filter units, one RAT modular filter unit, and three return filter modular units. The EDP modular filter unit for the left system is shown in Fig. 13-90. As shown in this drawing this unit contains filters, check valves, a relief valve, ground service disconnect, pressure and temperature transducers, and differential pressure indicators. By bringing these units together in one unit, the maintenance of the system is simplified. A schematic flow diagram of the modular unit is shown in Fig. 13-91.

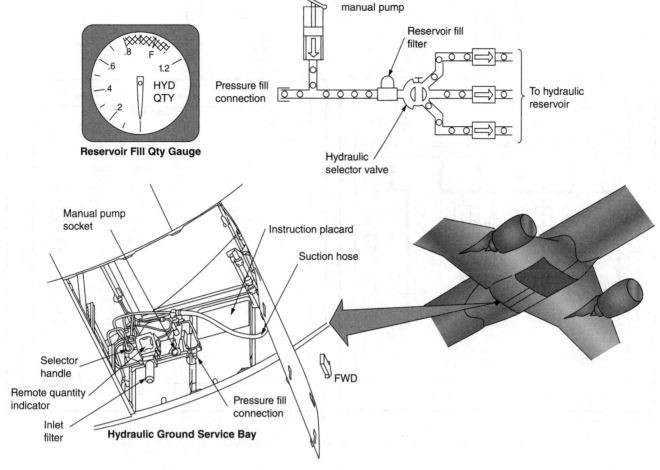

FIGURE 13-85   Hydraulic equipment in the filling area. (*Boeing Commercial Aircraft Co.*)

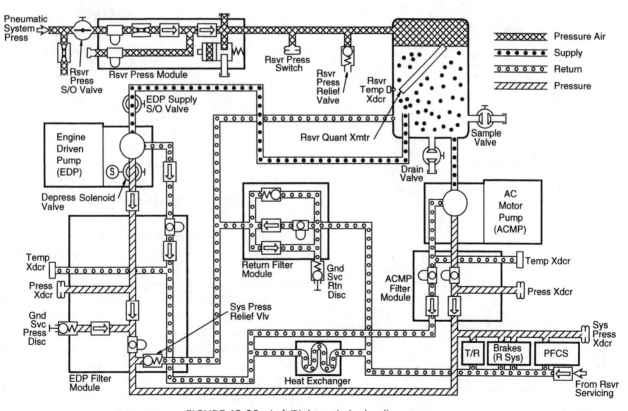

FIGURE 13-86   Left/Right main hydraulic systems.

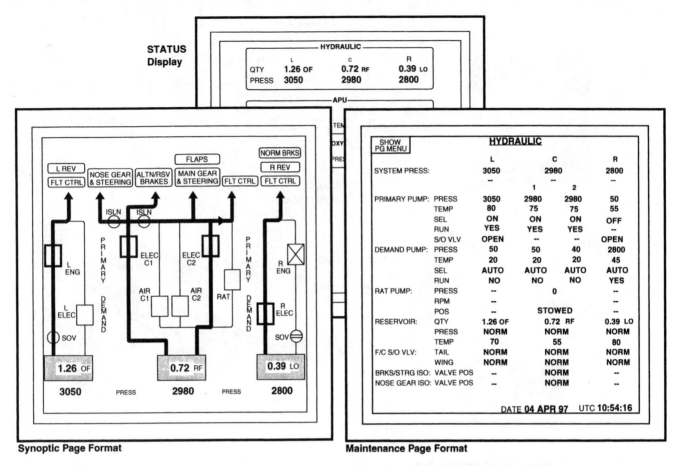

STATUS Display

HYDRAULIC			
	L	C	R
QTY	1.26 OF	0.72 RF	0.39 LO
PRESS	3050	2980	2800

APU

Synoptic Page Format

Maintenance Page Format

HYDRAULIC					
		L	C		R
SYSTEM PRESS:		3050	2980		2800
		--	--		--
			1	2	
PRIMARY PUMP:	PRESS	3050	2980	2980	50
	TEMP	80	75	75	55
	SEL	ON	ON	ON	OFF
	RUN	YES	YES	YES	
	S/O VLV	OPEN	--	--	OPEN
DEMAND PUMP:	PRESS	50	50	40	2800
	TEMP	20	20	20	45
	SEL	AUTO	AUTO	AUTO	AUTO
	RUN	NO	NO	NO	YES
RAT PUMP:	PRESS	--	0		--
	RPM	--			--
	POS	--	STOWED		--
RESERVOIR:	QTY	1.26 OF	0.72 RF		0.39 LO
	PRESS	NORM	NORM		NORM
	TEMP	70	55		80
F/C S/O VLV:	TAIL	NORM	NORM		NORM
	WING	NORM	NORM		NORM
BRKS/STRG ISO:	VALVE POS	--	NORM		--
NOSE GEAR ISO:	VALVE POS	--	NORM		--

DATE **04 APR 97**   UTC **10:54:16**

FIGURE 13-87 Hydraulic system indications shown on Aircraft Information Management System (AIMS).

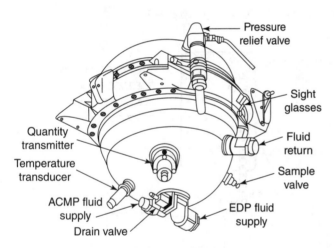

FIGURE 13-88 Hydraulic reservoir of left and right system.

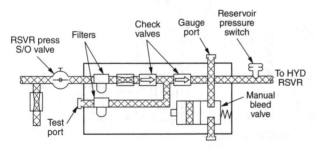

FIGURE 13-89 Reservoir pressurization module.

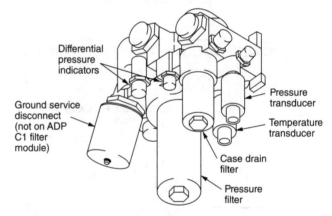

FIGURE 13-90 EDP modular filter module.

The left system return filter module removes particles from the hydraulic fluid before the fluid returns to the reservoirs. Each module has these components: replaceable filter, automatic filter bowl removal shutoff valve, differential pressure indicator (red pop-up indicator), two check valves, bypass relief valve with indicator, and ground service disconnect. Return fluid from airplane hydraulic systems and the heat exchangers goes through the return filter modules before it goes into the hydraulic reservoirs again. The left and right return filter modules have a negative pressure loop. The negative pressure loop and the check valves permit backward flow of fluid from the reservoir to the system

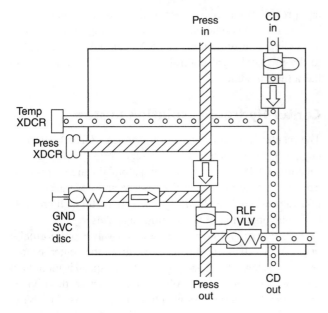

FIGURE 13-91 EDP modular filter module schematic.

without back-flush of the filter. Bypass relief valves open to permit continued hydraulic flow if the filter clogs. The relief valve starts to open at approximately 165 psi.

Heat Exchanger

A heat exchanger in each hydraulic system cools the case drain hydraulic fluid from the EDPs and the ACMPs. This extends the service life of the fluid and the hydraulic pumps. The heat

exchangers use aluminum-finned tubes to transfer heat from the fluid to the fuel. Fuel in the left and right main tanks cools the hydraulic fluid. The heat exchanger for the left hydraulic system is in the left main fuel tank between ribs 10 and 11. The heat exchanger for the right hydraulic system is in the right main fuel tank between ribs 10 and 11.

Accumulator

Four hydraulic accumulators absorb pressure changes caused by tail flight control power control unit (PCU) operation. This increases the life of the hydraulic system components. The right and center hydraulic systems each have a single hydraulic accumulator in the pressure lines to the tail flight controls. The left hydraulic system uses two accumulators that connect in the tail flight control pressure lines. All accumulators are the same and have a volume of 50 cubic inches. The accumulators have a hydraulic fluid side and a gas side. A fluid port connects the fluid side to the hydraulic line. Nitrogen is used to service the gas side of the accumulator. The hydraulic accumulators are on the forward side of the aft bulkhead of the stabilizer compartment. The hydraulic accumulator service panel is on the aft side of the stabilizer compartment aft bulkhead. Figure 13-92 shows the location of the accumulators in all three hydraulic systems.

Valves

The valves incorporated in the left system are the relief valve, hydraulic supply shutoff valve, reservoir pressure shutoff valve, reservoir drain valve, and reservoir sample valve.

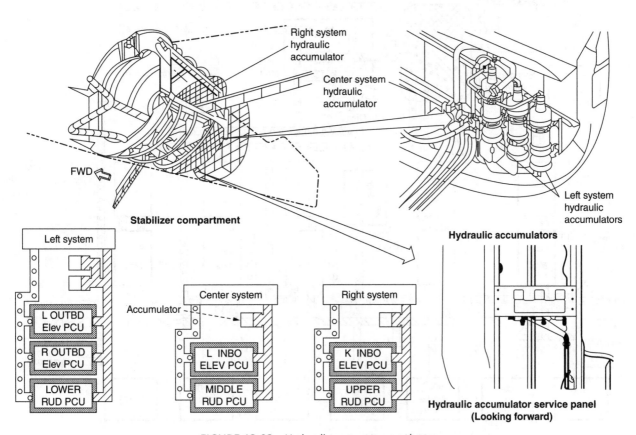

FIGURE 13-92 Hydraulic system accumulators.

The purpose of the relief valve is to protect the system against excessive pressure.

The hydraulic supply shutoff valves are provided to stop the flow of hydraulic fluid to the engine area. These valves are electrically operated and are automatically shut off when the engine fire switch is operated.

The reservoir pressure shutoff valve is a manual shutoff valve placed in-line before the pressurization module.

The reservoir drain valve is a manual valve that is used to drain the fluid out of the reservoir.

The reservoir sample valve is a manual valve that is used to take a sample of hydraulic fluid for fluid analysis. Never use the drain valve for collecting a sample because the bottom of the reservoir could contain more contaminants.

Center Hydraulic System

Many of the components in the center system are identical to the components in the left and right system. Only the different system components are discussed. Figure 13-93 shows the center system schematic.

Pumps

The center system has two AC motor pumps (ACMPs), two air-driven pumps (ADPs), and the ram air turbine (RAT) that can provide hydraulic power to the hydraulic subsystems. The AMPs only come on when there is a high demand such as landing gear retraction. The RAT can only supply power to the primary flight control surfaces (PFCS). Each pump has a filter module.

Center Hydraulic Isolation System

The center hydraulic isolation system (CHIS) supplies engine burst protection and a reserve brakes and steering function. CHIS operation is fully automatic and relays control the electric motors in the reserve and nose gear isolation valves. When the CHIS system is operational it prevents hydraulic operation of the leading edge slats.

The reserve and nose gear isolation valves are normally open. Both valves close if the quantity in the center system reservoir is low (less than 0.40) and the airspeed is more than 60 knots for more than 1 s. When CHIS is active, this divides the center hydraulic system into different parts. The NLG actuation and steering and the leading edge slat hydraulic lines are isolated from center system pressure. The output of ACMP C1 goes only to the alternate brake system.

The output of the other center hydraulic system pumps goes to the trailing edge flaps, the MLG actuation and steering, and the flight controls. If there is a leak in the NLG actuation and steering or LE slat lines, there is no further

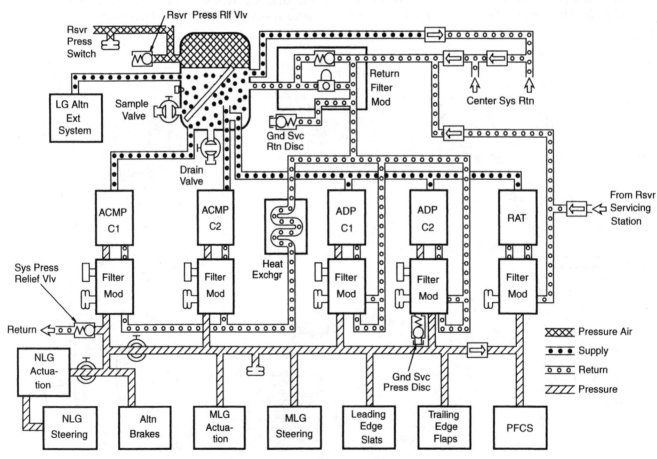

FIGURE 13-93 Center main hydraulic system.

loss of hydraulic fluid. The alternate brakes, the trailing edge flaps, the MLG actuation and steering, and the PFCS continue to operate normally.

If there is a leak in the trailing edge flaps, the MLG actuation and steering, or the flight control lines, the reservoir loses fluid down to the standpipe level (0.00 indication). This causes a loss of these systems. But, the alternate brake system continues to get hydraulic power from ACMP C1. If there is a leak in the lines between ACMP C1 and the alternate brake system, all center hydraulic system fluid is lost.

Nose Gear Isolation Valve

The nose gear isolation valve will open for any of these conditions:

- Airspeed is less than 60 knots.
- Pump pressures for ACMP C2, ADP C1, ADP C2, and the RAT is less than 1200 psi for 30 s.
- Left and right engine rpm is above idle for 30 s.
- Left and right EDP pressure is more than 2400 psi for 30 s.
- The nose landing gear (NLG) is not up, the NLG doors are not closed, and the landing gear lever is not up for 30 s.

The first condition permits the flight crew to operate the NLG steering when airspeed is less than 60 knots (decreased rudder control authority during taxi). The second condition permits operation of the NLG actuation and steering if the hydraulic leak is in the part of the center hydraulic system isolated by the reserve isolation valve. The third condition permits operation of the NLG actuation and steering if there has not been an engine burst and the other hydraulic systems are pressurized. The nose gear isolation valve opens when pressure is necessary at the NLG. If the NLG is not fully retracted or the NLG doors are not closed, the nose gear isolation valve opens to let the NLG complete the retraction. When the landing gear lever is moved to the down position, the nose gear isolation valve opens to let the NLG extend with center system pressure.

The reserve and nose gear isolation valves open again automatically when the center system quantity is more than 0.70 and airspeed is less than 60 knots for 5 s. Both valves also reset when the center system quantity is more than 0.70 and both engines and both engine-driven pumps operate normally for 30 s. Figure 13-94 shows the center hydraulic isolation system.

Ram Air Turbine

The ram air turbine (RAT) supplies an emergency source of hydraulic power to operate the flight controls. The RAT also is an emergency source of electrical power. The RAT extends

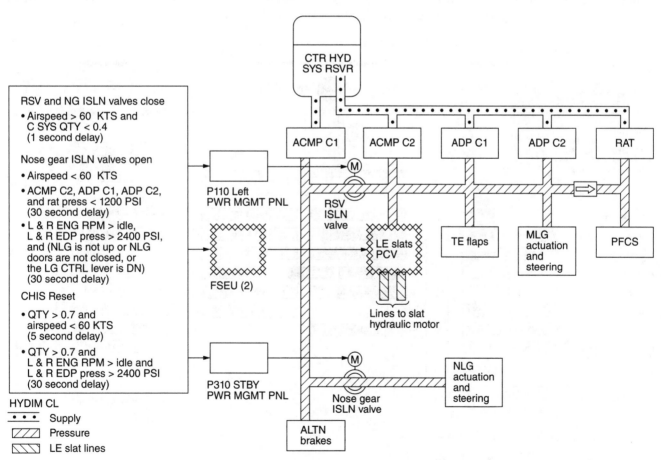

Main hydraulic systems-center hydraulic isolation system-functional description

FIGURE 13-94 Center hydraulic isolation system.

automatically in flight with a loss of hydraulic pressure in the three hydraulic systems. The RAT can also be manually extended from the flight deck.

HYDRAULIC SYSTEM FOR THE AIRBUS 380

The Airbus A380 hydraulic system consists of two hydraulic systems (green and yellow). A third hydraulic system used on previous Airbus aircraft is replaced by an electrohydraulic system made of a combination of electrohydraulic actuators (EHAs) and electrical backup hydraulic actuators (EBHAs). For the first time in civil aircraft history the A380 will use a high-pressure 5000-psi hydraulic system. The fluid remains a standard phosphate-ester type as used on most civil aircraft. The A380 hydraulic system integrates a total of eight Vickers PV3-300-31 (EDPs). The pumps are pressure-compensated, variable displacement type, and deliver 42 gpm at 3775 rpm. Their displacement is 47 ml/rev. The EDPs have a disengagement clutch, a feature not currently found on any commercial or military aircraft pump. In cases where a pump is malfunctioning while the plane is on the ground, the clutch can segregate that pump from the system. This allows the plane to be dispatched using the remaining seven pumps. The pump can only be reconnected on the ground and this is a maintenance function.

In this aircraft the electro motor pumps (EMP) will be used for ground purpose only and are located by couple in the inner pylon. Electrical systems can power the local electrohydraulic power packs. The flight control systems can be operated by electro-hydrostatic actuators (EHAs) and electrical backup hydraulic actuators (EBHAs). The brake and nose wheel steering system can be operated by local electrohydraulic generation systems (LEHGS). Figure 13-95 shows a hydraulic block diagram of the A380 hydraulic system.

The green and yellow hydraulic circuits are fully independent. Four engine-driven pumps (EDPs) per circuit pressurize the hydraulic fluid at 5000 psi in normal operation. Each circuit has a single hydraulic reservoir (RSVR). There are two EDPs per engine and each engine has one fire shut-off valve (FSOV) to supply hydraulic fluid to both EDPs on that engine. The green and yellow circuits generate hydraulic power to operate: the primary flight controls (F/CTL), the secondary flight controls, the landing gear, and the braking and steering system.

Backup Hydraulic Power

As a backup to the normal hydraulic circuits, hydraulic pressure can be locally produced for the flight controls and landing gear using local electrohydraulic power generators, supplied by the electrical network. Note that the RAT supplies only electrical power. The local electrohydraulic generation system comprises EHA and EBHA to operate primary flight controls and three LEHGS to operate braking and steering systems. Figure 13-96 shows the location of the EHAs and EBHAs.

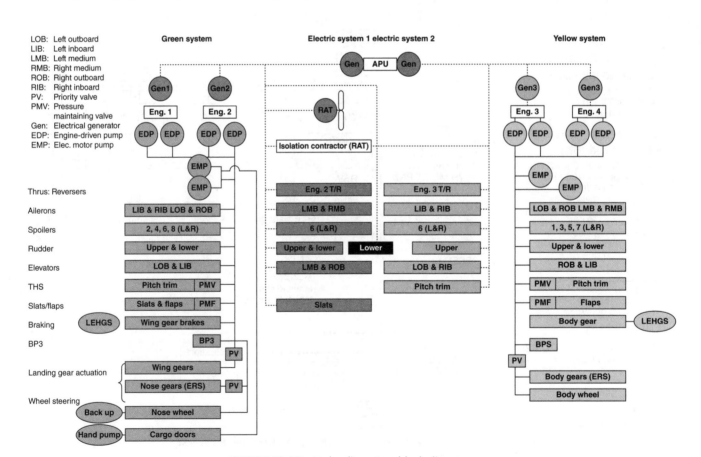

FIGURE 13-95 Hydraulic system block diagram.

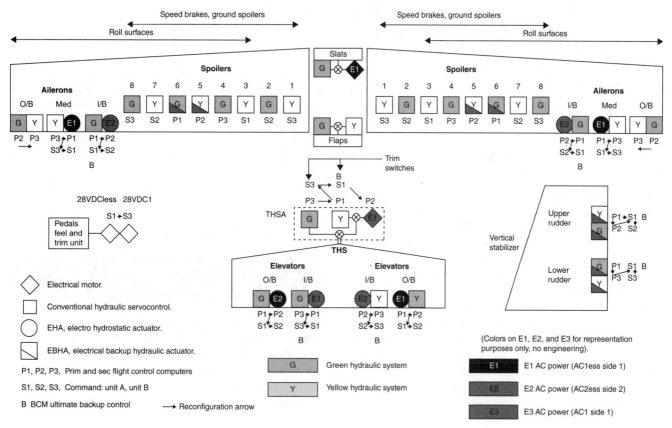

FIGURE 13-96 Location of EHAs and EBHAs.

Electro-Hydrostatic Actuator

The EHA shown in Fig. 13-97 is a self-contained unit consisting of a servo valve, electromotor and hydraulic pump, accumulator/reservoir, and a hydraulic actuator (ram). The EHAs are a backup power source for the inboard and medium ailerons and elevators flight surfaces. During normal flight with hydraulic systems operating the EHA only have a dampening function. When there is a hydraulic failure the EHA will automatically control the flight control surfaces.

Electric Backup Hydraulic Actuator

EBHAs are hydraulically powered in the normal mode by the aircraft hydraulic systems and electrically powered in

backup mode. Four EBHAs are used to power the two rudder surfaces. The EBHAs are normally powered by the green or yellow hydraulic systems and the electric motor and pump are inoperative. In case of a failure of the hydraulic system the EBHAs automatically switch to backup mode using the integrated electric motor and pump to operate the actuator piston (ram). Figure 13-98 shows a schematic diagram of an EBHA.

Local Electrohydraulic Generation System

The LEHGS is a decentralized fluid power generation system on the A380 and is used as a backup power source for braking and steering. Signals from an electronic control unit (ECU)

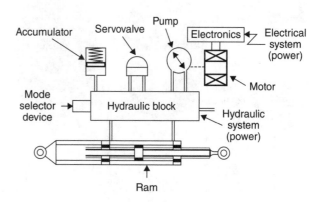

FIGURE 13-97 Electro-hydrostatic actuator.

FIGURE 13-98 Electrical backup hydraulic actuator.

activate a small electrically driven micro-pump, which is located close to the system to be controlled. The micro-pump provides 5000 psi (350 bar) of local hydraulic pressure over short runs of small-diameter lightweight tubing for braking and steering. The LEHGS consists of a reservoir, electric motor, pump, valves, filter, and controls. The LEHGS supplies hydraulic fluid power to the brakes and steering actuators in case the normal main hydraulic systems fail.

Auxiliary Hydraulic Power Generation

EMPs pressurize each hydraulic circuit, provided the aircraft is on the ground with all engines shut down. For the cargo doors operation, one of the EMPs in the green system operates to generate hydraulic power. An isolation valve closes to isolate the main hydraulic users from the cargo doors hydraulic circuit. A similar isolation valve is fitted to the yellow circuit for the body gear steering system, which can be pressurized by the yellow EMPs. A manually operated backup auxiliary hydraulic pump can be used to operate the cargo doors. Ground connections located in the inner pylons can also supply the green and yellow circuits with hydraulic pressure. Figure 13-99 shows the hydraulic power schematic of the A380 including the EMPs and the isolation valves.

Ground Service Panel

The service panel is situated in the hydraulic bay of the A380 aircraft and integrates 11 different equipment packages, which are used to monitor and maintain the two hydraulic systems. The main components of the system include the pressure and level gauges used to monitor the reservoir air-pressurization and fluid levels, a hand pump used to refill the hydraulic reservoir, and some valves used to control the fluid and the air pressurization during the refilling operation of the hydraulic reservoirs.

Two hydraulic system monitoring units (HSMUs), one for each hydraulic circuit, control and monitor the green and yellow hydraulic circuits, respectively. They also give indication on the control and display system (CDS).

HYDRAULIC SYSTEM FOR THE BELL 214ST HELICOPTER

The hydraulic power systems on the Bell 214ST helicopter consist of three separate and independent systems: two flight-control hydraulic systems, called PC1 and PC2, and a utility hydraulic system. The two flight-control systems operate in parallel, so that if either system should fail, the remaining system will provide full control capabilities. The utility system is used

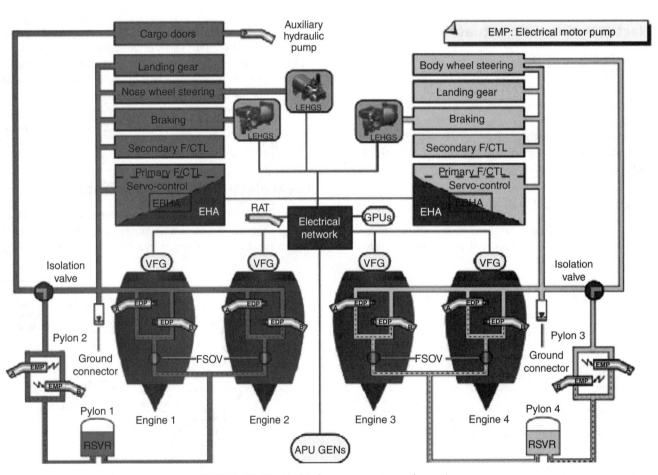

FIGURE 13-99 Hydraulic power system schematic.

only to power the cooling fan for the gearbox and transmission oil cooler. Each system is equipped with a pump, closed-circuit reservoir, hydraulic module, distribution circuit, and ground test circuit.

The hydraulic power for each system is generated by its own variable-displacement hydraulic pump. The PC1 pump is powered by the helicopter transmission, and the PC2 and utility pumps are powered by the combining gearbox. The pumps pressurize their system to 3000 psi [20 685 kPa], and they also assist the reservoir spring in pressurizing the reservoir. Each system is equipped with temperature and pressure sensors which warn the pilot of malfunctions. Protection from excessive pressure is provided by a pressure-relief valve in each system. The utility system is not discussed further, since it is very similar to the flight-control systems.

PC1 and PC2 Hydraulic Systems

Except for the power source, the PC1 and PC2 hydraulic systems are identical. The somewhat higher revolutions per minute

of the combining gearbox causes the pump for the PC2 to have a slightly greater capacity for work than the PC1 pump. The PC1 and PC2 systems are shown in Fig. 13-100.

The hydraulic pumps are located side by side on the cabin roof in the area shown in Fig. 13-101. A schematic diagram of the PC1 system is shown in Fig. 13-102. Fluid from the reservoirs is pressurized by the pumps and passes through a check valve and a pulsation dampener, which removes surges from the pump output. The fluid next flows to the hydraulic module, first passing through a filter assembly. If the filter becomes clogged so that a pressure differential of approximately 70 psi [483 kPa] exists across the filter, a red button extends from the assembly. The filter is not designed to bypass when it becomes clogged.

From the filter, fluid flows to a shutoff valve and a solenoid pilot valve. The pilot can shut down either system by opening the solenoid valve with a cockpit toggle switch. When the solenoid is energized, pressure is directed to the shutoff valve and system pressure is routed to the return line, depressurizing the system downstream of these control valves. If power

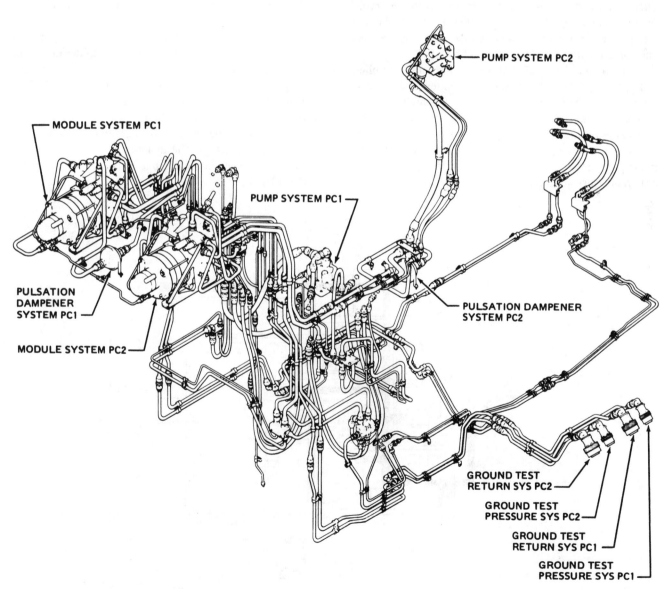

FIGURE 13-100   The hydraulic system for a helicopter—actuators not shown. (*Bell Helicopter Textron.*)

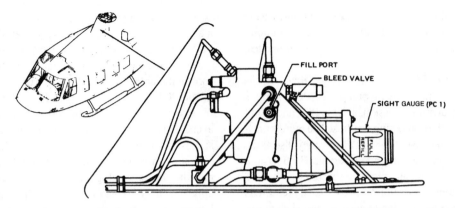

FIGURE 13-101 Hydraulic reservoir and filter assembly for a helicopter. (*Bell Helicopter Textron.*)

to the solenoid is interrupted, the shutoff valve is closed and full system pressure is restored. Electric interlocks prevent both PC1 and PC2 from being shut down in this manner at the same time.

Excess pressure in the system is released by the system relief valve, and the relieved fluid returns to the reservoir. The relief valve begins to open at 3500 psi [24 132 kPa] and is fully open at 3850 psi [26 345.8 kPa]. The valve reseats when pressure drops to 3250 psi [22 408.8 kPa].

Downstream of the relief valve, the pressure is applied to a pressure switch, a pressure transmitter, and the reservoir pressurization line. The pressure switch operates a cockpit annunciator when pressure is lost and disables the solenoid switch for the opposite system. The pressure transmitter drives a pressure gauge in the cockpit. The reservoir pressure is maintained at 75 psi [517 kPa] by the system pressurization piston.

After passing the reservoir pressurization line, the fluid exits the module and flows to the flight-control actuators. Return fluid from the actuators is routed back to the module, through the return filter in the module, and into the reservoir. The return filter has a red button to indicate when the filter is clogged, but unlike the pressure filter, the return filter can be bypassed.

From the reservoirs, the fluid is fed back to the pumps. A temperature switch and a temperature transmitter in the line between the reservoirs and the pumps send signals to cockpit displays. The transmitter allows the pilot to monitor the fluid temperature, and the switch activates an annunciator light to warn of excessive heat in the fluid.

The system can be operated by a ground cart through the ground test connection. The quantity of fluid in the reservoirs is indicated at the sight gauge on the module, as shown in Fig. 13-101.

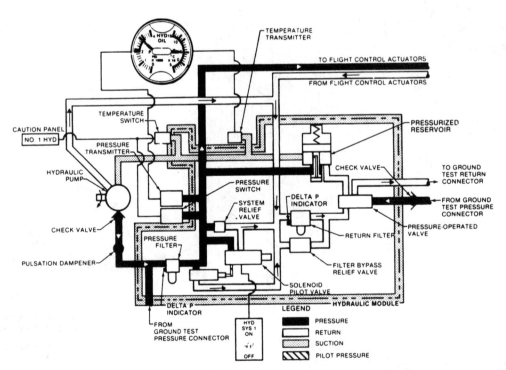

FIGURE 13-102 Schematic of a helicopter hydraulic system. (*Bell Helicopter Textron.*)

PNEUMATIC SYSTEMS FOR AIRCRAFT

Aircraft pneumatic systems are used primarily as emergency sources of pressure for many of the hydraulically actuated subsystems. The principle of operation for a pneumatic power system is the same, with one important exception, as that of a hydraulic power system. The air in a pneumatic system is compressible; therefore, the pressure in the system can reduce gradually from the maximum system pressure to zero pressure. In the hydraulic system, as soon as the accumulator fluid has been used and the pump is not operating, the fluid pressure immediately drops from accumulator pressure to zero pressure. The entire pneumatic system, including the air-storage bottles, can act to store air pressure. In the hydraulic system, the only pressure-fluid storage is in the accumulators, and the pressure is supplied by compressed air or gas in the air chamber of the accumulator.

The air in a pneumatic system must be kept clean by means of filters and also be kept free from moisture and oil droplets or vapor. For this reason, liquid separators and chemical air driers are incorporated in the systems. Moisture in a pneumatic system may freeze in the low temperatures encountered at high altitudes, resulting in serious system malfunctions.

Another important feature of a pneumatic system is that there is no need for return lines. After the compressed air has served its purpose, it can be dumped overboard, which saves tubing, fittings, and valves.

Pneumatic-System Description

The pneumatic system described in this section is utilized in the Fairchild F-27 aircraft. It provides power for operation of the landing-gear retraction and extension, nose-wheel centering, propeller brakes, main-wheel brakes, and passenger entrance-door retraction. The system described includes only the development and delivery of compressed air to each component or subsystem, not the actual operation of the component or subsystem. The pneumatic power in the airplane is delivered by one of two systems; the primary system or the emergency system.

The power section of the primary pneumatic system is that portion located in each engine nacelle shown in Fig. 13-103. It consists of a gearbox-driven compressor, bleed valve, unloading valve, moisture separator, chemical drier, back-pressure valve (right nacelle only), and a filter. In addition, each nacelle contains a shuttle valve, disk-type relief valve, and a ground charging connection to aid in ground maintenance or initial filling. Each power section independently supplies compressed air to the primary and emergency systems. The air in the primary system is stored in two storage bottles, and the system delivers the air for normal operation components, as required by directional valves. A schematic diagram of the pneumatic power system is shown in Fig. 13-104.

A 100-in³ [1.64-L] bottle is used for the main-wheel brakes, and a 750-in³ [12.3-L] bottle is used for gear operation, nose-wheel centering, propeller brakes, and passenger-door retraction. A pressure-relief valve is installed to

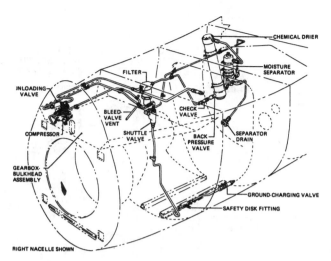

FIGURE 13-103 Pneumatic system in engine nacelle. (*Fairchild Industries, Inc.*)

protect the system from excessive pressure buildup. In the air-supply tube from the large primary bottle, an air filter is installed to filter the air for the primary system. An isolation valve, which is in the tube between the large primary bottle and the smaller brake bottle, permits maintenance to be performed downstream of the valve without discharging the large bottle. On the downstream side of the isolation valve, a pressure-reducing valve is used to reduce system pressure from 3300 to 1000 psi [22 753 to 6895 kPa]. All components except the pressure gauges and the power-section components are located in the fuselage pneumatic compartment.

Air-Pressure Sources

Two views of the compressor for the pneumatic system are shown in Fig. 13-105. One compressor is located in each engine nacelle. The compressors are of four-stage, radial design, providing a delivery pressure of 3300 psi [22 753 kPa] and 2 ft³/min [5.66 m³/min] at sea-level intake pressure on a standard day. As shown in the drawings, the cylinders and pistons of the compressor diminish in size from no. 1 to no. 4. The cylinders are caused to reciprocate in the proper sequence by a cam-assembly mechanism, which rotates with the crankshaft.

The arrangement of the crankshaft and cam in the four-stage compressor is such that when the no. 1 cylinder is on the compression stroke, the no. 2 cylinder is on the intake stroke and the compressed air enters the no. 2 cylinder. When the no. 2 cylinder compresses, the air enters the no. 3 cylinder, which is on the intake stroke. When the no. 3 cylinder compresses, the air is delivered to the no. 4 cylinder, which is on the intake stroke. The no. 4 cylinder then compresses, and the air is delivered to the system. Thus the air has been moved from the no. 1 cylinder through the four stages to be further compressed in each stage. The interstage lines are finned to provide cooling and reduce the heat of compression.

Ducted ram air is provided for cooling of the finned cylinders and the finned interstage lines. Oil pressure from the gearbox provides lubrication for the compressor through a drilled

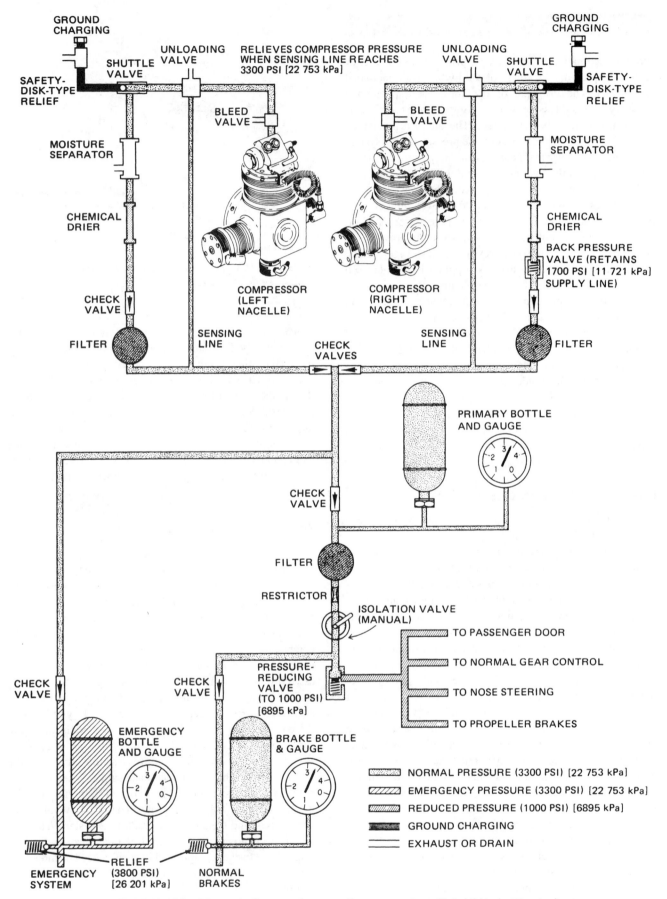

FIGURE 13-104 Schematic diagram of pneumatic power system. (*Fairchild Industries, Inc.*)

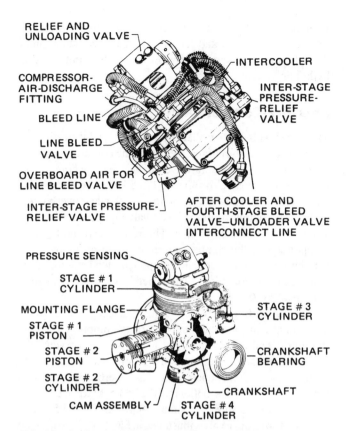

RELIEF AND
UNLOADING VALVE

COMPRESSOR-
AIR-DISCHARGE
FITTING

BLEED LINE

LINE BLEED
VALVE

OVERBOARD AIR FOR
LINE BLEED VALVE

INTER-STAGE PRESSURE-
RELIEF VALVE

INTERCOOLER

INTER-STAGE
PRESSURE-
RELIEF
VALVE

AFTER COOLER AND
FOURTH-STAGE BLEED
VALVE—UNLOADER VALVE
INTERCONNECT LINE

PRESSURE SENSING

STAGE #1
CYLINDER

MOUNTING FLANGE

STAGE #1
PISTON

STAGE #2
PISTON

STAGE #2
CYLINDER

CAM ASSEMBLY

STAGE #3
CYLINDER

CRANKSHAFT
BEARING

CRANKSHAFT

STAGE #4
CYLINDER

FIGURE 13-105   Four-stage pneumatic compressor.

passage in the mounting flange. Air is compressed by stages 1 through 4 of the compressor, whereas overpressure protection is provided by means of relief valves between stages 1 and 2 and 2 and 3. Compressed air from stage 4 is then routed by an intercooler line to the bleed valve mounted on the compressor.

Although the version of the F-27 being discussed makes use of an engine-driven air compressor to power the pneumatic system, other aircraft commonly make use of other power sources. Air for the pneumatic operation of controls can be provided by the engine compressor bleed air. Some aircraft use this medium-pressure air (100 to 150 psi [689 to 1034 kPa]) for operation after passing it through a pressure regulator and conditioners such as filters and driers. Other aircraft use this air to operate air turbine motors, which allows an increase in operation pressure on the pump side of the air turbine motor. Still other aircraft make use of a hydraulic motor or an electric motor to operate a high-pressure pneumatic pump. This high-pressure air is then regulated, conditioned, and allowed to flow on to the systems to be operated.

Air-storage bottles are used for emergency operations when hydraulic or pneumatic pressure sources have been lost. The bottles are normally pressurized with nitrogen, but some aircraft require the use of carbon dioxide. These bottles will power a system, normally a landing-gear extension system or a brake system, for only a brief period of time. The bottles must be recharged on the ground.

Storage Bottle

The primary air-storage bottle for the F-27 is constructed of steel with a plastic coating on the inner surface to provide a longer service life. The bottle is mounted above the fuselage pneumatic panel and is fitted with three ports. It has a volume of 750 in^3 [12 290.3 cm^3] and stores air at 3300 psi [22 753.5 kPa] for operating landing-gear, propeller-brake, nose-steering, and rear-door-retracting subsystems. The air pressure is reduced to 1000 psi [6895 kPa] before being routed to subsystems other than brakes. The bottle is provided with a standpipe, which permits the withdrawal of air without the danger of allowing any accumulated moisture to enter the operating systems. Accumulated moisture can be drained from the bottle by means of a drain valve located on the bottom.

Aircraft that use the pneumatic system only for emergency purposes have the storage bottles equipped with a charging valve for ground servicing and a control valve to release pressure into the system being controlled. The control valve may be an on-off type valve, where all of the bottle's contents are released at one time, or it may be such that the pilot can control the amount of pressure being applied to a system. The on-off control is found on landing "blow-down" systems, whereas the controllable valve is typically used for brake systems.

Moisture Control

In a pneumatic system it is of the utmost importance that the air in the system be completely dry. Moisture in the system can cause freezing of operating units, interfere with the normal operation of valves, pumps, etc., and cause corrosion. It is for this reason that moisture separators and chemical driers are used in pneumatic systems.

In each compressor pressure line, a mechanical moisture separator, shown in the drawing of Fig. 13-106 is installed to remove approximately 98 percent of any moisture and/or oil that may pass from the compressor. The separator is an aluminum tubular chamber, mounted vertically on the outboard side of the right nacelle and on the inboard side of the left nacelle.

Two valves are installed in the bottom of the separator. An inlet air pressure of 750 psi [5171 kPa] maximum closes the drain valve, whereas the inlet valve remains closed, thus preventing air from entering the separator. When pressure reaches 900 + 150 psi [6205.5 + 1034 kPa], the inlet valve opens and allows air to flow. The inlet valve stays open as long as inlet pressure is above 1050 psi [7240 kPa]. The air entering the inlet port flows through the inlet valve seat, guide, and tube assembly and through the orifices in the retainer assembly onto the baffle. The moisture droplets are deflected by the baffle, separated from the air stream, and caused to settle at the bottom of the air chamber against the closed drain valve.

As long as the inlet air pressure remains above 400 + 150 psi [2758 + 1034 kPa], the flow of air is through the inlet port, guide, tube assembly, and outlet port. When inlet air is reduced to 0 psi, the residual air flows through the ball check of the retainer assembly and out the open drain valve, pushing the collected moisture ahead of it. The separator is equipped with a safety disk that will burst at 5800 to 6200 psi [39 991 to 42 749 kPa] as a safety measure against over-pressurization.

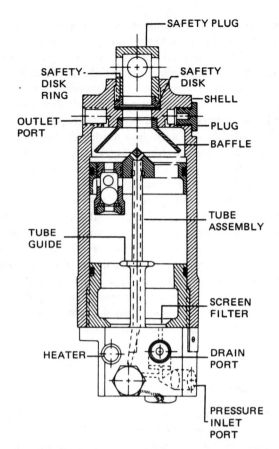

FIGURE 13-106 Moisture separator.

In the unit base is a small thermostatically controlled heater that prevents freezing of any moisture in the bottom of the air chamber. The heater is automatically operated by a hermetically sealed thermostat, located in the moisture-separator body. The thermostat closes the electrical circuit at 35°F [1.67°C] to provide heating and opens the contacts at 85°F [29.44°C] maximum. To function properly, the moisture separator must be mounted vertically with the overboard drain at the bottom.

A tubular, chemical-drier housing is installed in each nacelle downstream of the mechanical moisture separator. Each drier has an inlet and outlet port and contains a desiccant cartridge. Air is directed through this replaceable cartridge, and any moisture that the mechanical moisture separator has failed to remove will be absorbed by the dehydrating agent in the cartridge. The cartridge with dehydrating agent specification MIL-D-37 16 incorporates two bronze filters, one at each end, which allow a 0.001-in [0.03-mm] maximum-size particle to pass through.

Filters

Three sintered pneumatic filters are used, one in each compressor circuit and one in the primary circuit. The **filter** is a vertically mounted unit, containing a replaceable filter element of stainless steel that removes foreign matter of 10 μm or larger from the compressor output air.

Other types of filters commonly found in pneumatic systems include micronic and wire-screen filters. The micronic filter has a replaceable cartridge, whereas the wire-screen filter can be cleaned and reused. The basic construction of these filters is the same as the micronic filter units used in hydraulic systems.

Pressure-Control Valves

The **bleed valve**, controlled by compressor lubricating oil pressure, directs the compressed air to the compressor circuit relief valve and unloading valve. In the event the compressor oil pressure drops below 40 psi [275.8 kPa], the bleed valve will direct compressed air from the fourth stage overboard.

Pressure output from the bleed valve is routed to a **relief valve** in the unloading valve. The relief valve protects all components of the compressor circuit from excessive pressure buildup in the event any component downstream of the compressor malfunctions. The relief valve is set to open at 3800 psi [26 201 kPa].

The **unloading valve** contains a sensing valve and a directional control valve, and it directs compressor output through dehydration equipment to the system or vents the output overboard. The directional control valve, controlled by the sensing valve, opens and vents overboard when system pressure reaches 3300 psi [22 753.5 kPa] and closes when system pressure is less than 2900 psi [19 955.5 kPa].

A **back-pressure valve**, illustrated in Fig. 13-107, is installed in the pressure tube of the right nacelle only, to ensure that the engine-driven air-compressor output is kept at a predetermined value of 1700 + 150 psi [11 721.5 + 1034 kPa]. This is to provide and maintain a fixed back pressure of approximately 1700 psi upstream of the air-drying equipment. The back-pressure valve is similar in construction and operation to a check valve and consists of a valve seat, piston, and spring. It is installed in the outlet port of the chemical drier and is directly connected to the inlet port of a check valve. The right-engine compressor or a ground charging unit must be used for normal charging of the pneumatic system.

Poppet-type, spring-loaded, line-check valves are installed in the primary system, and each is identified by part number according to tube size. One valve is located in the inlet tube just before the point where the tube enters the primary bottle inlet port.

Another check valve is installed in the supply tube leading to the brake storage and is located on the left-center edge of the fuselage pneumatic panel. Two spring-loaded, poppet-type, line-check valves are installed in each pneumatic power system in each nacelle. One valve is installed in the outlet end of the back pressure valve; the other valve is installed in the upper outlet port of the "T" fitting.

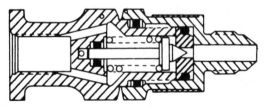

FIGURE 13-107 Back-pressure valve.

A spring-loaded, piston **relief valve** is installed in the primary system. The valve is preset, safetied, and tagged by the manufacturer for a cracking pressure of 3800 + 50 psi [26 201 + 345 kPa]. It protects the primary system from excessive pressure buildup in the event of thermal expansion or compressor power system malfunction.

This relief valve is designed to relieve pressure as a metering device instead of an instantaneous on-off valve. Its rate of relief automatically increases in direct proportion to the amount of overpressure. It reseats when the pressure decreases to approximately 3400 psi [23 400 kPa].

As mentioned previously, the air pressure employed for the operation of the landing-gear, propeller-braking, nose-wheel-steering, and passenger-door-retraction subsystems is reduced from 3300 [22 753.5] to 1000 psi [6895 kPa]. This is accomplished by means of the **pressure-reducing valve** installed in the primary pressure supply line. This valve is mounted on the pneumatic power panel just below the isolation valve, as shown in the drawing of Fig. 13-108. The pressure-reducing valve contains dual springs and poppets, which not only reduce pressure but also provide for pressure relief of the reduced pressure should the valve fail. A cross section of the valve is shown in Fig. 13-109.

Additional protection is afforded the reducer relief valve by a restrictor elbow upstream of the isolation valve. The restrictor prevents excessive flow to the reducer and reducer relief valve.

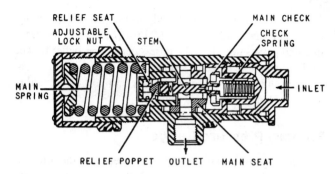

FIGURE 13-109   Pressure-reducing valve.

Flow-Control Valves

Installed in both the right and left nacelles is a two-position, three-port **shuttle valve**. The valve functions to direct air to the primary system from the compressor or the ground charging valve while preventing airflow from escaping through the lines not being used for supplying air.

The pneumatic power system is provided with an **isolation valve** so that the air-pressure storage section of the system can be isolated from the subsystems; thus work on the subsystems can be performed without discharging the entire system. The isolation valve is a two-position (open-close) type installed on the top right side of the pneumatic panel. The valve contains a spring-loaded, poppet-type plunger, which is cam- and lever-operated. The location of the valve in the system is shown in Fig. 13-104. Connected to the valve lever is a push-pull rod-and-latch assembly, which prevents closing the pneumatic panel door while the valve is in the closed position. This arrangement eliminates the possibility of operating the airplane while the air pressure is shut off.

Selector valves for pneumatic systems are similar in construction and operation to those used for hydraulic systems.

Miscellaneous System Components

The tubing used for the pneumatic system follows the pattern described for hydraulic systems. For the high-pressure sections of the system, stainless-steel tubing is employed. This type of tubing is used where the pressures reach or exceed 3000 psi [20 685 kPa]. The steel tubing conforms to specification MIL-T-8504 and varies in size from $\frac{3}{16}$ to $\frac{3}{8}$ in [4.76 to 9.33 mm] outside diameter (OD). Steel tubing is used to carry pressurized air from the compressor to the three storage bottles, pressure gauges, emergency brakes, and brake valve and to the emergency landing-gear-selector valve. Each end of the tubing is connected by the use of MS flareless fittings. Aluminum tubing is used to carry reduced pressurized air throughout the remainder of the system. Bonding strips are used as necessary to reduce static buildup and interference.

Pneumatic Manifold

An aluminum manifold having seven ports is mounted just below and to the left of the pressure-reducing valve on the pneumatic panel. As shown in the drawing of Fig. 13-108,

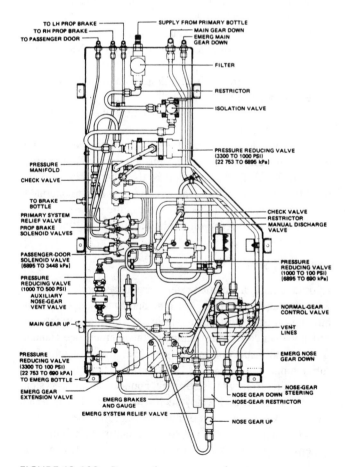

FIGURE 13-108   Pneumatic power panel.

the manifold is connected directly to the pressure-reducing valve. The manifold receives 1000 psi [6895 kPa] air pressure and passes this air to the nose-steering, propeller-brake, landing-gear, and passenger-door subsystems through a network of aluminum tubing. The lower end port of the manifold is sealed by a threaded plug.

Primary Pressure Gauge

A direct-reading pressure gauge is connected to the primary air-storage bottle and is mounted in the pneumatic instrument panel on the right side of the crew compartment to provide a visual indication of the pressure in the primary bottle.

Ground Charging Valve

A **ground charging valve** is installed in both the left and right nacelles to provide a means of initial filling or ground charging the entire pneumatic system during the period when the engines are not operating. Normal ground charging is accomplished by using the right nacelle charging valve. If the system pressure is below 1000 psi [6895 kPa], the isolation valve should be closed before charging. The left nacelle charging valve should be used only during maintenance pressure checking of nacelle components or servicing the system with a dry-air source of supply.

The charging valve also acts as a protective device against an excessive rate of inlet airflow. The valve serves as a restrictor by providing a fixed orifice through which all ground-supplied compressed air must pass.

Safety-Disk Fitting

Just upstream of the ground charging valve in each nacelle is a two-port safety-disk fitting incorporating a disk-type pressure-relief assembly. The disk will rupture at approximately 4300 to 4950 psi [29 650 to 34 000 kPa], allowing air to escape through drilled passages in the retainer.

SUMMARY OF HYDRAULIC SYSTEM MAINTENANCE PRACTICES

The maintenance of hydraulic and pneumatic systems should be performed in accordance with the aircraft manufacturer's instructions. The following is a summary of general practices followed when dealing with hydraulic and pneumatic systems.

Service

The servicing of hydraulic and pneumatic systems should be performed at the intervals specified by the manufacturer. Some components, such as hydraulic reservoirs, have servicing information adjacent to the component.

When servicing a hydraulic reservoir, the technician must make certain to use the correct type of fluid. Hydraulic fluid type can be identified by color and smell; however, it is good practice to take fluid from an original marked

container and then to check the fluid by color and smell for verification. Fluid containers should always be closed except when fluid is being removed. This practice reduces the possibility of contamination. Funnels and containers employed in filling the hydraulic reservoir must be clean and free of dust and lint.

When a system has lost a substantial amount of fluid and air has entered the system, it is necessary to fill the reservoir, purge the system of air, then add fluid to the FULL mark on the reservoir. Air is purged from the system by operating all subsystems through several cycles until all sounds of air in the system are eliminated. Air in a system causes a variety of sounds, including banging, squealing, and chattering. The reservoir should be filled before the air is purged.

When cleaning filters or replacing filter elements, be sure there is no pressure on the filter. When changing hydraulic filters, wear the appropriate protective clothing and use a face shield as necessary to prevent fluid from contacting the eye. After filters have been cleaned or replaced, the system may have to be pressure-checked to locate any leaks in the filter assembly. Hydraulic fluid leaks are readily apparent because of the appearance of fluid. Large pneumatic leaks are indicated by the sound of escaping air and the feel of an airflow. Small pneumatic leaks should be located by the use of a soap-and-water solution.

Flushing a Hydraulic System

When inspection of hydraulic filters indicates that the fluid is contaminated, flushing the system is necessary. This should be done according to the manufacturer's instructions; however, a typical procedure for flushing the system is as follows:

1. Connect a ground hydraulic test stand to the inlet and outlet test posts of the system.
2. See that the fluid in the test unit is the same type as that used in the aircraft.
3. Pump clean, filtered fluid through the system and operate all subsystems until no further signs of contamination are found upon inspection of the filters. The airplane must be raised on jacks so the landing gear can be operated.
4. Disconnect the test stand and cap the ports.
5. See that the reservoir is filled to the FULL line.

Inspections

Hydraulic and pneumatic systems are inspected for leakage, worn or damaged tubing, worn or damaged hoses, wear of moving parts, security of mounting for all units, safetying, and any other condition specified by the maintenance manual. A complete inspection includes an operational check of all subsystems.

Leakage from any stationary connection in a system is not permitted, and if any such leak is found, it must be repaired. A small amount of fluid seepage may be permitted on actuator piston rods and rotating shafts. In a hydraulic system a thin film of fluid in these areas indicates that the seals are being properly lubricated. When a limited amount of leakage is allowed at any point, it is usually specified in the appropriate manual.

Tubing must not be nicked, cut, dented, collapsed, or twisted beyond approved limits as explained previously. The identification markings or lines on flexible hose will show whether the hose has been twisted.

All connections and fittings associated with moving units should be examined for play evidencing wear. Such units should be in an unloaded condition when they are checked for wear.

Accumulators should be checked for leakage, air or gas preload, and position. If the accumulator is equipped with a pressure gauge, the preload can be read directly. Otherwise, the system pressure gauge can be used as explained previously. The accumulators should be mounted with the air chamber upward.

An operational check of the system can be performed using the engine-driven pump, an electrically operated auxiliary pump if such a pump is included in the system, or a ground test unit. The entire system and each subsystem should be checked for smooth operation, unusual noises, and speed of operation for each unit. The pressure section of the system should be checked with no subsystems to see that pressure holds for the required time without the pump supplying the system. System pressure should be observed during operation of each subsystem to see that the engine-driven pump will maintain required pressure.

Troubleshooting and Maintenance Practices

The troubleshooting of hydraulic and pneumatic systems varies according to the complexity of the system concerned and the components in the system. It is, therefore, important that the technician refer to the troubleshooting information furnished by the manufacturer.

Lack of pressure in a system can be caused by a sheared pump shaft, defective relief valve, pressure regulator or unloading valve stuck in the "kicked-out" position, lack of fluid in a hydraulic system, check valve installed backward, or any condition that permits free flow back to the reservoir or overboard. If a system operates satisfactorily with a ground test unit but not with the system pump, the pump should be examined.

If a system fails to hold pressure in the pressure section, the likely cause is the pressure regulator or unloading valve, a leaking relief valve, or a leaking check valve.

If the pump fails to keep pressure up during operation of the subsystem, the pump may be worn or one of the pressure-control units may be leaking.

High pressure in a system may be caused by a defective or improperly adjusted pressure regulator or unloading valve or by an obstruction in a line or control unit.

Unusual noise in a hydraulic system, such as banging and chattering, may be caused by air or contamination in the system. Such noises can also be caused by a faulty pressure

regulator, another pressure-control unit, or a lack of proper accumulator action.

Maintenance of hydraulic and pneumatic systems involves a number of standard practices together with specialized procedures set forth by manufacturers. The technician may be called upon to replace valves, actuators, and other units, including tubing and hoses. In every case, the technician must exercise care to prevent system contamination; avoid damage to seals, packings, and other parts; and apply proper torque in connecting fittings.

Flexible hose, particularly the type with a Teflon fluorocarbon resin tubing, tends to take a set after a period of time. During the process of removing and reinstalling such hose sections, the technician should make no effort to straighten or change the shape of the hose. It is recommended that the hose be tied with a piece of cord or safety wire to hold its shape. When flexible hose is installed, the metal sleeve of the hose fitting should be held with a wrench while the nut is tightened to the fitting. Otherwise, the hose may be twisted and damaged. Twisted hose can be detected by observing the woven wire braiding or the painted lines on the outside of the hose.

Overhaul of hydraulic and pneumatic units is usually accomplished in approved repair facilities; however, replacement of seals and packings may be done from time to time by technicians in the field. When a unit is disassembled, all O-ring and chevron seals should be removed and replaced with new seals. The new seals must be of the same material as the original and should carry the correct manufacturer's part number. No seal should be installed unless it is positively identified as the correct part and its shelf life has not expired.

When installing seals, the technician must exercise care to ensure that the seal is not scratched, cut, or otherwise damaged. When it is necessary to install a seal over sharp edges, the edges should be covered with shim stock, plastic sheet, or masking tape. Sharp-edged metal tools should not be used to stretch the seals and force them into place.

When it is necessary to replace metal tubing, the old section of tubing can be used as a template or pattern for forming the new part. In many cases, the tubing section can be procured from stock or from a parts dealer by specifying the correct part number.

The replacement of hydraulic units and tubing always involves the spillage of some hydraulic fluid. The technician should take care to see that the spillage of fluid is kept to a minimum by closing valves, if available, and by plugging lines immediately after they are disconnected. All openings in hydraulic and pneumatic systems should be capped or plugged to prevent contamination of the system. It is particularly important to see that fire-resistant hydraulic fluid is not allowed to get on painted surfaces or other material that would be damaged. Fire-resistant fluids (phosphate ester type) have a powerful solvent characteristic and should be immediately removed from any surface that may be damaged, including the skin.

The importance of the proper torque applied to all nuts and fittings in a system cannot be overemphasized. Too much torque will damage metal and seals, and too little torque will result in leaks and loose parts. In every case, the technician should procure torque wrenches with the proper range before assembling units of a system.

Additional information regarding aircraft plumbing, including plumbing for hydraulic and pneumatic systems, is provided in the text *Aircraft Basic Science*.

REVIEW QUESTIONS

1. What is the principal difference between a *hydraulic power system* and a *pneumatic power system*?
2. Explain how force is developed through the use of hydraulic fluids.
3. Explain the multiplication of force by means of hydraulics.
4. Explain the relationship between *volume, area,* and the *length of stroke*.
5. Name three types of *hydraulic fluids* and describe their basic characteristics.
6. What precaution must be taken in regard to spills on a painted surface with a phosphate ester fluid such as Skydrol?
7. Describe a *hydraulic reservoir*.
8. What is the purpose of a *standpipe* in a reservoir?
9. Why is it necessary to pressurize some reservoirs?
10. Give three methods by which a hydraulic reservoir can be pressurized.
11. What is the purpose of a hydraulic filter?
12. What are different *filter elements* that may be found in hydraulic systems?
13. What is the purpose of the "pop-out" indicator pin found on many filters?
14. Describe the purpose of a *heat exchanger* and how it functions.
15. Why is a gear-type pump described as a positive displacement pump?
16. In an axial multiple-piston pump, what causes the reciprocating motion of the pistons?
17. Describe two types of *variable-delivery pumps* and explain the operation of each.
18. Explain the purpose of a *ram air turbine*.
19. Describe the purpose of a *pressure regulator*.
20. How does a pressure regulator or unloading valve serve to prolong the life of the system pump?
21. List the different types of selector valves.
22. Explain how *poppet valves* control fluid flow.
23. Explain the operation of *spool-type valves*.
24. How are the selector valves in an open-center system connected with respect to one another?
25. Describe an *orifice-check valve*. For what purpose would such a valve be used?
26. Explain sequence valve operation and purpose.
27. What can occur if a sequence valve is not properly adjusted?
28. Under what conditions is a *shuttle valve* required?
29. Explain the operation and use of a *hydraulic fuse*.

30. Explain the operation and use of a *priority valve*.
31. Describe the operation of a *relief valve*.
32. What determines the cracking pressure of a relief valve?
33. Why is a wing-flap overload valve installed in the down line of a flap-actuating system?
34. Why are thermal-relief valves needed in certain sections of a hydraulic system?
35. Explain the principle of a *debooster valve*.
36. Explain the purpose of a *flow equalizer*.
37. What are the two functions of an *accumulator*?
38. Describe a piston-type accumulator.
39. What precautions should be taken before removing an accumulator from a hydraulic system?
40. What is a *hydraulic actuator*?
41. Describe a double-acting hydraulic actuating cylinder.
42. How does a *servo actuator* differ from an *actuating cylinder*?
43. Compare a hydraulic motor with a piston-type pump.
44. Discuss the purpose of colored markings on O-ring seals.
45. When O-ring seals are used in high-pressure systems, what devices are often necessary?
46. Compare a closed-pressure hydraulic system with an open-center system.
47. List the principal hydraulic systems in a Boeing 777 aircraft.
48. What is the function of a power-transfer unit on the 777 aircraft?
49. List the principal hydraulic systems on an A380 aircraft.
50. Name the three principal hydraulic systems for the Boeing 777 airplane.
51. What types of pumps are employed in the hydraulic power systems for the Boeing 777 airplane?
52. What systems are served by system A in the Boeing 777?
53. What is the purpose of the ground interconnect valve?
54. List the advantages of a pneumatic power system.
55. What is the difference in the principles of operation between a pneumatic system and a hydraulic system?
56. What are uses for pneumatic bottles in a hydraulic system?
57. For what functions is the pneumatic power system on the Fairchild F-27 airplane used?
58. How is the high-pressure air stored in the Fairchild F-27 pneumatic system?
59. Describe the operation of the four-stage compressor.
60. What is the purpose of the isolation valve in the F-27 pneumatic system?
61. Explain the need for dry air in a pneumatic system.
62. Describe the operation of the *moisture separator*.
63. What is a *desiccant*?

64. What is the function of the *back-pressure valve*?

65. How is lowered pressure obtained from the high-pressure air source?

66. Explain the use of the ground-charging valve.

67. List the principal consideration necessary when servicing a hydraulic reservoir.

68. What conditions can cause banging or chattering in a hydraulic system?

69. What conditions may cause lack of pressure in a hydraulic system?

70. What may cause the pressurized section of a hydraulic system to lose pressure?

Aircraft Landing-Gear Systems 14

INTRODUCTION

Landing-gear designs for aircraft vary from simple, fixed arrangements to very complex retractable systems involving many hundreds of parts. This chapter examines typical examples of landing gear and the systems by which they are operated.

LANDING-GEAR CONFIGURATIONS

The majority of aircraft are equipped with landing gear that can be classified as either *tricycle* or *conventional*.

Tricycle landing gear is characterized by having a nose wheel assembly and two main gear assemblies, one on each side of the aircraft, as shown in Fig. 14-1. This arrangement places the aircraft fuselage in a level attitude when the aircraft is on the ground. In this attitude the pilot has good forward visibility and the cabin area is level, making it easier for passengers to move inside the aircraft on the ground. This configuration also makes the aircraft stable during ground operations and easy to control. This is especially important during takeoff and landing.

Conventional-geared aircraft, illustrated in Fig. 14-2, have two main wheel assemblies, one on each side of the aircraft, and a tail wheel. This arrangement is normally associated with older aircraft and those designed for rough field operations. This arrangement has the advantage of reduced drag in the air and reduced landing-gear weight. There is some loss of forward visibility for the pilot when maneuvering on the ground due to the aircraft nose-high attitude. This configuration is less stable on the ground and requires more skill when taxiing and during takeoff and landing when compared to a tricycle-geared aircraft.

CLASSIFICATION OF LANDING GEAR

The **landing gear** of an airplane serves a number of very important functions and is classified by a number of different characteristics. It supports the airplane during ground operations, dampens vibrations when the airplane is being taxied or towed, and cushions the landing impact. The landing of an airplane often involves stresses far in excess of what may be considered normal; therefore, the landing gear must be constructed and maintained in a manner that provides the

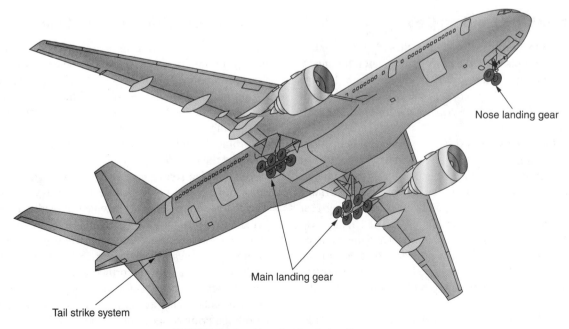

Nose landing gear

Main landing gear

Tail strike system

FIGURE 14-1   Aircraft tricycle landing gear.

FIGURE 14-2 Aircraft with conventional landing gear.

strength and reliability needed to meet all probable landing conditions.

The landing gear of an airplane consists of main and auxiliary units, either of which may be fixed (nonretractable) or retractable. The **main landing gear** provides the main support of the airplane on land or water. It may include a combination of wheels, floats, skis, shock-absorbing equipment, brakes, retracting mechanism, controls, warning devices, cowling, fairing, and structural members needed for attachment to the primary structure of the airplane.

The **auxiliary landing gear** consists of tail or nose landing-wheel installations, skids, outboard pontoons, etc., with the necessary cowling and reinforcements.

Nonabsorbing Landing Gear

Nonabsorbing landing gear includes those types of landing gear that do not dissipate the energy of the aircraft contacting the ground during landing. They only temporarily store the energy and quickly return it to the aircraft. These types of gear include *rigid landing gear*, *shock-cord landing gear*, and *spring-type gear*.

A **rigid landing gear** is commonly found on helicopters and sailplanes. This gear is rigidly mounted to the aircraft with no specific component to cushion the ground contact other than through the flexing of the landing gear or airframe structure. An example of this type of gear is the landing-gear skid of the helicopter shown in Fig. 14-3. The landing gear of a sailplane is often no more than a wheel and tire mounted directly on the airframe. The tire provides some cushioning effect, but no mechanism is designed specifically to soften the ground contact. Both these types of aircraft rely on pilot skill for a soft ground contact.

When rubber **shock cord** is used, the landing-gear struts are usually made of steel tubing mounted in such a manner

that a stretching action is applied to tightly wound rubber cord. When landing shock occurs, the cord is stretched, thus storing the impact energy of landing. The stored energy is gradually returned to the aircraft during the landing roll.

The landing-gear struts for some aircraft consist of single, tapered strips or tubes of strong spring steel or composite material. In the **spring-type gear**, one end of the strut is bolted to the heavy structure of the aircraft under the cabin area, and the axles are bolted to the opposite ends. A landing gear of this type is shown in Fig. 14-4.

On ground contact during landings, the gear flexes and stores the impact energy. As with the shock-cord gear, the stored energy is returned to the aircraft during the landing roll.

Shock-Absorbing Landing Gear

Shock-absorbing landing gear dissipates the impact energy of landing through some means. Most of these types of landing gear do this by forcing a fluid through a restriction. The movement of this fluid generates heat, and the heat is radiated into the surrounding atmosphere, thus dissipating the landing energy. The two types of shock-absorbing landing gear commonly used are the *spring-oleo* and the *air-oleo* types.

Spring-oleo struts, not usually found in modern aircraft, consist of a piston-type structure and a heavy, coiled spring. The piston-and-cylinder arrangement provides an oil chamber and an orifice through which oil is forced during landing. When the airplane is airborne, the strut is extended, and the oil flows by gravity to the lower chamber. When the plane lands, the piston with the orifice is forced downward into the cylinder and the oil is forced through the orifice into the upper chamber. This action provides a cushioning effect to absorb the primary shock of landing. As the strut collapses, the coil spring is compressed, thus providing additional

FIGURE 14-3 A skid-equipped helicopter. (*Susan Gorden & Bill Schroeder.*)

cushioning. Thus the spring supports the aircraft weight on the ground and during taxiing, and the oleo strut absorbs the shock of landing.

The principle of operation of an **air-oleo** strut is similar for all struts, regardless of the type of fluid metering employed. During compression of the strut at landing, an orifice provides a restriction of fluid flow, and this reduces the rate at which the piston (inner cylinder) can move into the outer cylinder. This provides a cushioning effect to reduce the shock of landing. As the fluid flows through the orifice into the upper chamber, the air in the upper chamber is compressed to the point that the entire weight of the aircraft is supported by the air in the landing-gear struts. The compressed air then acts as a shock absorber during the time that the aircraft is taxiing.

The **oleo shock strut** can be used for either nonretractable or retractable landing gear. This type of gear is generally constructed as shown in Fig. 14-5, although there are many variations of the design.

Fixed Gear

Nonretractable (fixed) landing gear is generally attached to structural members of the airplane with bolts, but it is not actually "fixed," because it must absorb stresses; therefore, the wheels must move up and down while landing or taxiing in order to absorb shocks. The landing gear is often equipped with a fairing where it joins the fuselage or wing to reduce the drag (air resistance). Chafing strips are used to prevent excessive wear between the sections of the fairing, because there is usually some motion between these sections. **Wheel pants** are often used to cover the wheel and tire to reduce their drag. Nonretractable landing gear may have bracing, or it may be of the cantilever type without any additional bracing.

Fixed landing gear is usually found on small aircraft and aircraft where aerodynamic cleanliness for an efficient cruise configuration is not a major factor.

Retractable Gear

Retractable landing gear was developed to eliminate, as much as possible, the drag caused by the exposure of the landing gear to airflow during flight. Usually the landing gear is completely retractable (that is, it can be drawn entirely into the wing or fuselage); however, there are aircraft in which a portion of the gear wheels is still exposed after the gear is retracted. The direction of retraction varies. On some airplanes, the retraction is toward the rear; on others the landing gear folds inward toward the fuselage; and on still others it folds outward toward the wing tips. The method of retraction also varies, although modern aircraft usually have a gear that is power operated. The retraction is normally accomplished with hydraulic or electric power. In addition to the normal operating system, emergency systems are usually provided to ensure that the landing gear can be lowered in case of main-system failure. Emergency systems consist of backup hydraulic systems, stored air or gas that can be directed into actuating cylinders or mechanical systems that can be operated manually, or free-fall gravity systems.

It must be emphasized that the landing gear of an airplane is of primary importance in the safe operation of the aircraft; because of this, the technician must be especially careful in the inspection and maintenance of landing-gear systems. Since retractable-type landing gear is much more complex than fixed gear, it is essential that each operating component of the gear be carefully examined at required intervals to reduce the possibility of failure.

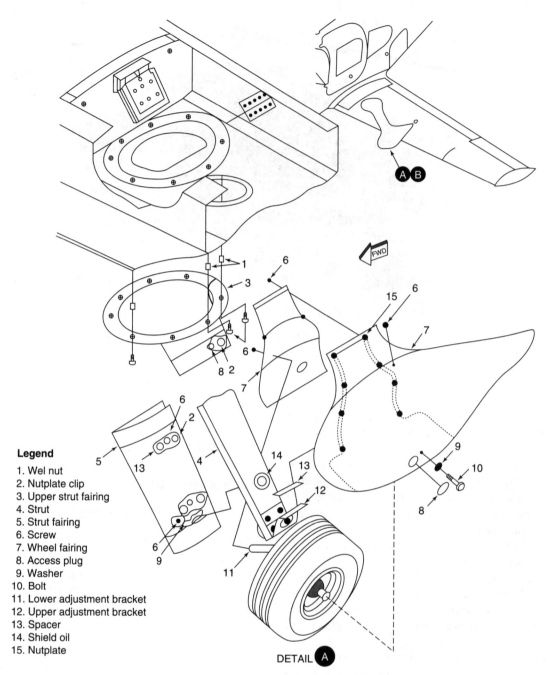

Legend

1. Wel nut
2. Nutplate clip
3. Upper strut fairing
4. Strut
5. Strut fairing
6. Screw
7. Wheel fairing
8. Access plug
9. Washer
10. Bolt
11. Lower adjustment bracket
12. Upper adjustment bracket
13. Spacer
14. Shield oil
15. Nutplate

DETAIL **A**

FIGURE 14-4   Spring-type landing gear.

Hulls and Floats

Airplanes operated from water may be provided with either a single float or a double float, depending upon the design and construction; however, if an airplane is actually a flying boat, it has a **hull** for flotation and then may need only wing-tip floats. **Amphibious** airplanes have floats or a hull for operating on water and retractable wheels for land operation.

Skis are used for operating on snow and ice. The skis may be made of wood, metal, or composite materials. There are three basic styles of skis. A conventional ski, shown in Fig. 14-6, replaces the wheel on the axle. The shock cord is used to hold the toe of the ski up when landing. The safety

cable and check cable prevent the ski from pivoting through too great an angle during flight.

The wheel ski is designed to mount on the aircraft along with the tire. The ski has a portion cut out that allows the tire to extend slightly below the ski so that the aircraft can be operated from conventional runways with the wheels or from snow or ice surfaces using the ski. This arrangement has a small wheel mounted on the heel of the ski so that it does not drag on conventional runways.

In retractable wheel-ski arrangements, the ski is mounted on a common axle with the wheel. In this arrangement the ski can be extended below the level of the wheel for landing on snow or ice. The ski can be retracted above the bottom of the wheel for operations from conventional runways.

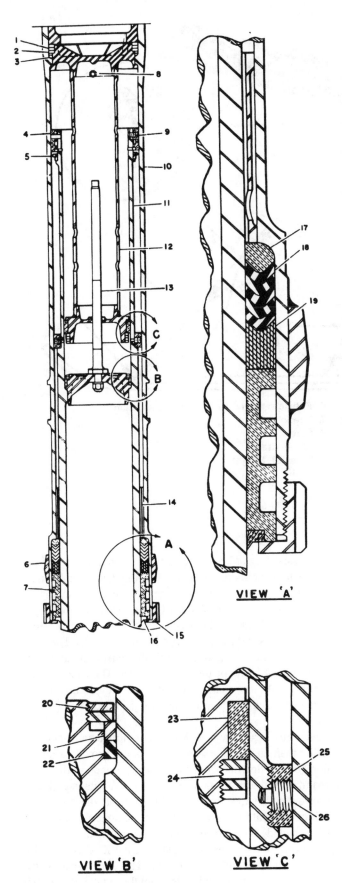

FIGURE 14-5   Air-oil shock strut.

A hydraulic system is commonly used for the retraction-system operation.

LANDING-GEAR COMPONENTS

Landing-gear assemblies are made up of various components designed to support and stabilize the assembly. The following terms identify many of these components. The terms are presented here as they relate to retractable landing-gear systems. When the terms are used with fixed landing gear, the exact use of the components may vary. Be aware that different manufacturers will occasionally use different terms for the same basic components.

Trunnion

The **trunnion**, shown in Figs. 14-7 and 14-8, is the portion of the landing-gear assembly attached to the airframe. The trunnion is supported at its ends by bearing assemblies, which allow the gear to pivot during retraction and extension. The landing-gear strut extends down from the approximate center of the trunnion.

Struts

The **strut** is the vertical member of the landing-gear assembly that contains the shock-absorbing mechanism. The top of the strut is attached to, or is an integral part of, the trunnion. The strut forms the cylinder for the air-oleo shock absorber. The strut is also called the **outer cylinder**.

The **piston**, shown in Fig. 14-9, is the moving portion of the air-oleo shock absorber. This unit fits inside the strut, and the bottom of the piston is attached to the axle or other component on which the axle is mounted. Other terms used for the piston are **piston rod**, **piston tube**, and **inner cylinder**.

In the strut shown in Fig. 14-5, the ring-seal nut (1), the compressing-ring seal (2), and the ring seal (3) form the group that seals the air pressure in the upper part of the strut. The upper bearing (4) keeps the inner cylinder (11) aligned with the outer cylinder (10). The snubber valve (5) releases when the weight is off the landing-gear strut to allow the strut to extend. The outer torque collar is (6). The lower bearing (7) helps keep the inner cylinder aligned inside the outer cylinder. The filler plug (8) is used to plug the hole through which the cylinder is filled with hydraulic fluid. The bearing lock screw (9) is used to maintain the position of the upper bearing on the inner cylinder.

The metering pin (13), extending into the piston tube (12), restricts the flow of fluid from the lower part of the cylinder to the upper part when the cylinder is being compressed during landing or taxiing. The metering pin varies in cross section to provide different rates of fluid flow during the compression stroke of the strut. Thus the movement of the strut in compressing is restricted to the rate determined to be most effective throughout the compression stroke.

The bearing spacer ring (14) is located between the lower part of the outer cylinder and the inner cylinder. The bearing

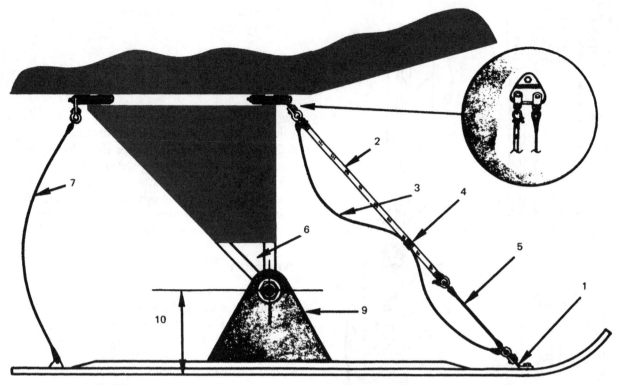

1. Fitting	6. Fabric removed to facilitate inspection
2. Shock Cord	7. Check Cable
3. Safety Cable	8. Clevis
4. Tape	9. Ski Pedestal
5. Crust-cutter Cable	10. Pedestal Height

FIGURE 14-6 A typical ski installation.

(packing) nut (15), at the bottom of the cylinder, holds the lower bearing and the packing ring seals in position. The bearing nut may also be referred to as a gland nut. The wiper ring (16) prevents dirt or other foreign material from being drawn into the cylinder as the piston moves into the cylinder during compression. Some struts include a scraper ring below the wiper ring to provide greater protection.

View A of Fig. 14-5 is an enlarged drawing of the seals and bearing at the lower end of the shock strut, marked A in the first drawing. The upper packing spacer (17), the neoprene packing (18), and the lower packing spacer (19) are compressed together to form a tight seal between the inner and outer cylinders.

View B is an enlarged drawing of the seal nut (20), compressing ring seal (21), and the ring seal (22), which are shown within the circular arrow marked B in the first drawing.

View C is an enlarged drawing of the piston bearing (23), piston-bearing lock nuts (24), the stop nut (25), and the nut lock screw (26), shown within the circular arrow marked C in the upper left drawing.

As mentioned previously, the shock struts absorb the shocks of landing and taxiing. The initial shock of landing is cushioned by the hydraulic fluid being forced through the small aperture at the metering pin, as illustrated in Fig. 14-10.

As the shock strut is compressed, the tapered orifice rod permits a diminishing rate of hydraulic-fluid flow from the inner-cylinder chamber to the upper side of the piston. Landing and taxi shocks are also cushioned by the increasing volume of hydraulic fluid above the piston, which further compresses a volume of gas in the upper end of the outer cylinder. The shock strut is serviced with hydraulic fluid through an air valve in the upper inboard side of the outer cylinder. The shock strut is serviced with dry, clean air or nitrogen through the air valve to the specified shock-strut extension provided on a servicing chart.

Air-oleo struts are also designed to utilize an orifice tube rather than an orifice and metering pin to provide a restriction for the flowing from the lower chamber to the upper chamber of the strut upon compression. A cutaway view of such a strut is shown in Fig. 14-11. Note that the strut has an **orifice plate** at the bottom of the orifice tube. Some air-oleo struts include both a metering pin and an orifice tube.

During the strut-extension stroke, the fluid is forced to return through the metering orifice. This is designed to prevent the strut from extending too rapidly on takeoff or during a bad landing and is often referred to as a **snubbing** action. If the rate of strut extension is too great, the shock that occurs due to impact at the end of the stroke can cause damage to the landing gear and the supporting structure in the aircraft.

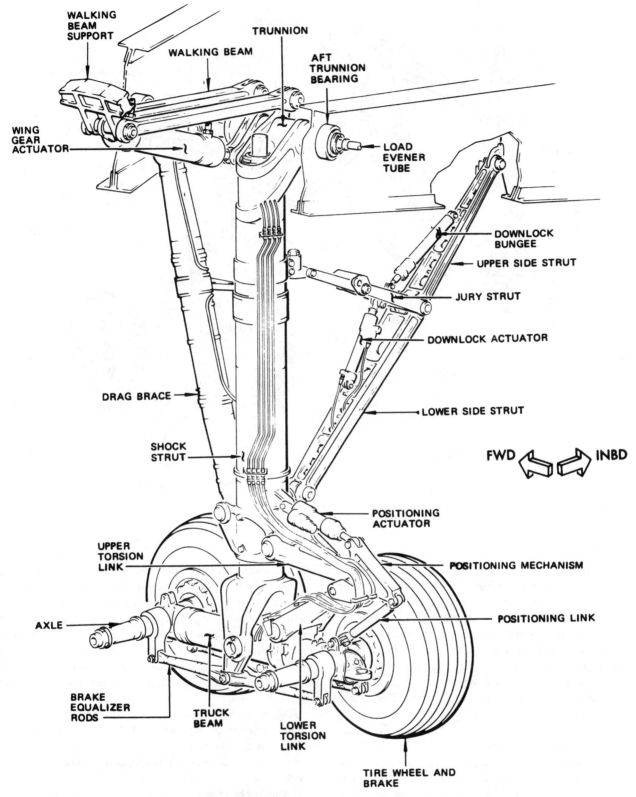

FIGURE 14-7 Landing-gear trunnion assembly.

The main-gear shock strut for a transport aircraft, illustrated in Fig. 14-12, is made of steel and consists of an **outer** and **inner cylinder**, a **piston** attached to the upper inner side of the outer cylinder, and an **orifice rod** attached to the inner cylinder. This construction varies slightly from others; however, the functional principle is the same as that for other air-oleo-type shock struts.

Upper and **lower bearings** in the shock strut provide sliding surfaces and an air-oil seal between the inner and outer cylinders. Between the lower bearing and the spacer,

Landing-Gear Components **451**

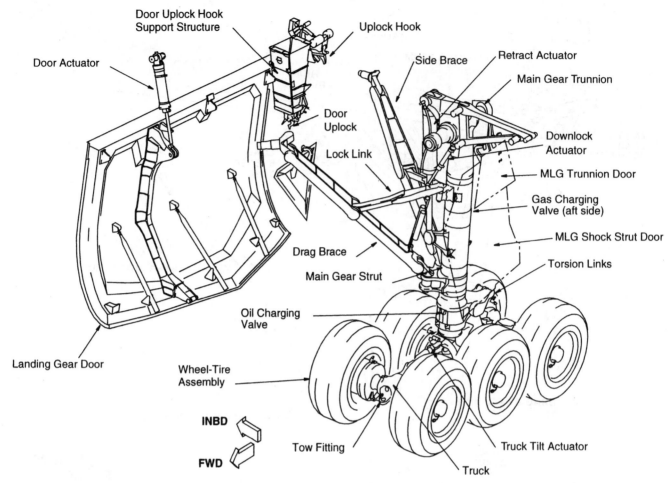

FIGURE 14-8 The main landing gear for a Boeing 777. (*Boeing Aerospace Co.*)

a **seal adapter** with annular grooves is installed. A D-ring and an O-ring seal with backup rings are inserted in the grooves to provide the air-oil seal between the cylinders.

The oleo struts for the main gear and the nose gear are usually similar in function. However, since the nose wheel must be able to swivel as the airplane turns on the ground, the mounting of the nose gear differs from that of the main gear. To provide for this requirement, the nose-gear strut is mounted in an outside tube or other structure. The strut is supported by bearings mounted in the top and bottom of the structure. The bearings and seals are held in place by means of gland nuts and/or snap rings.

Since the nose gear can swivel, provision must be made to establish the wheel alignment straight ahead when the gear is retracted and when landing. This is accomplished by means of external, mechanical centering devices such as springs or by **centering cams**, illustrated in Fig. 14-9, inside the strut. When cams are employed, a cam ring is mounted around the top of the piston to form the upper cam and another cam ring is mounted inside the bottom of the outer cylinder. When the strut is fully extended, the cam faces mate and hold the nose wheel in the centered position.

Oleo struts are provided with high-pressure **air valves** mounted in the top of the strut. The valve may contain a high-pressure core similar in appearance to the core found in a tire valve but specially constructed for high pressure, or it may be the type without a core. Both of these types of valves are illustrated in Fig. 14-13. In either case, the instructions on the use of the valve in servicing the strut are set forth in the manufacturer's instructions. **A tire valve core must never be used**.

Air-oleo struts always have an instruction plate permanently attached to the outside of the strut or nearby. This plate specifies the type of hydraulic fluid to be used in the strut and gives instructions for inflation, deflation, and filling with fluid.

Another type of strut arrangement, called an **oil snubber gear**, is shown in Fig. 14-14. This design consists of an oleo strut with an air chamber that is not pressurized above atmospheric pressure. The oleo mechanism, called an **oil snubber shock strut**, absorbs the landing energy. A large tubular drag link spring provides cushioning during taxi operations and fore-and-aft stability for the nose gear.

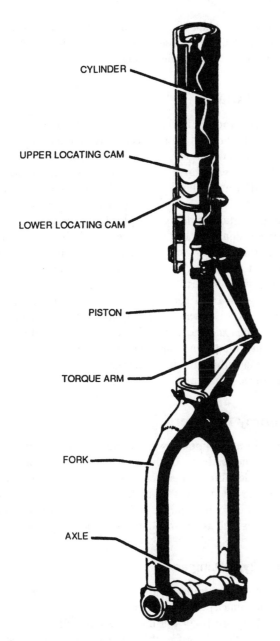

FIGURE 14-9   Nose-gear shock strut.

Torque Links

Torque links, often referred to as a *scissors assembly*, are two A-frame-type members, as shown in Fig. 14-9, used to connect the strut cylinder to the piston and axle. The torque links restrict the extension of the piston during gear retraction and hold the wheels and axle in a correctly aligned position in relation to the strut.

The upper torque link is connected to a clevis fitting on the lower forward side of the shock strut. The lower torque link is connected to a clevis fitting on the axle. The upper and lower torque links are joined together, as shown in the drawing, by a bolt and nut spaced with washers. Each link is fitted with flanged bushings. The gap between the flanged ends of the bushings is taken up by a spacer washer. This spacer establishes the alignment of the wheel. When main wheel

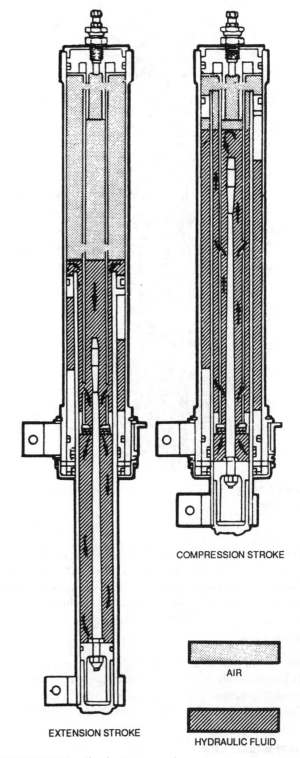

COMPRESSION STROKE

EXTENSION STROKE

AIR

HYDRAULIC FLUID

FIGURE 14-10   Shock-strut operation.

alignment is checked and is found to be incorrect, the correction is made by installing a spacer of different thickness.

Truck

The **truck** is located on the bottom of the strut piston and has the axles attached to it. A truck is used when wheels are to be placed in tandem (one behind the other) or in a dual-tandem

Labels on Figure 14-9: CYLINDER, UPPER LOCATING CAM, LOWER LOCATING CAM, PISTON, TORQUE ARM, FORK, AXLE

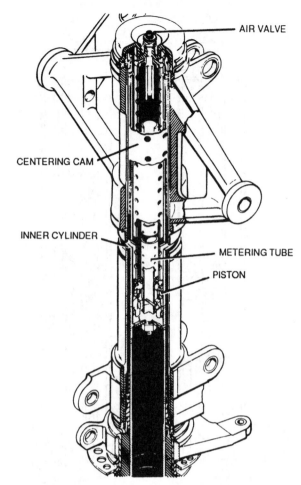

FIGURE 14-11 Landing-gear shock strut of the metering-tube type.

AIR VALVE

CENTERING CAM

INNER CYLINDER

METERING TUBE

PISTON

arrangement. The truck can tilt fore and aft at the piston connection to allow for changes in aircraft attitude during takeoff and landing and during taxiing. The truck shown in Figs. 14-15 and 14-8 is often referred to as a **bogie**.

Drag Link

A **drag link** is designed to stabilize the landing-gear assembly longitudinally. If the gear retracts forward or aft, the drag link will be hinged in the middle to allow the gear to retract. This is also called a **drag strut** and is shown in Fig. 14-15.

Side Brace Link

A **side brace link**, shown in Fig. 14-8, is designed to stabilize the landing-gear assembly laterally. If the gear retracts sideways, the side brace link is hinged in the middle to allow the gear to retract. This is also called a **side strut**.

Overcenter Link

An **overcenter link** is used to apply pressure to the center pivot joint in a drag or side brace link. This prevents the link from pivoting at this joint except when the gear is retracted, thus preventing collapse of the gear during ground operation.

The overcenter link is hydraulically retracted to allow gear retraction. This component is also called a **downlock** and a **jury strut** and is illustrated in Fig. 14-8.

Swivel Gland

A **swivel gland** is a flexible joint with internal passages that route hydraulic fluid to the wheel brakes and the bungee cylinder of a landing gear. Swivel glands are used where the bend radius is too small or space limitations prevent the use of coiled hydraulic lines. A swivel gland, which is illustrated in Fig. 14-16, may be mounted on a bracket secured to the main-gear trunnion fitting. The gland remains stationary and is the terminus of the stationary hydraulic lines. The movable portion of a swivel gland is connected to the hydraulic lines that are routed down the strut of the bungee cylinder and the wheel brakes.

In Fig. 14-16, it can be seen that the gland consists of annular grooves separated from one another by means of slipper rings and packings to isolate the pressure fluid from the return fluid. Thus the gear can be raised and lowered without disturbing the fluid passage to and from the brakes and bungee cylinder.

Shimmy Dampers

The **shimmy damper** is a hydraulic snubbing unit that reduces the tendency of the nose wheel to oscillate from side to side. Shimmy dampers (dampeners) are usually constructed in one of two general designs, *piston type* and *vane type*, both of which might be modified to provide power steering as well as shimmy damper action.

Piston-Type Dampers

A **piston-type shimmy damper** is simply a hydraulic cylinder containing a piston rod and piston and filled with hydraulic fluid. Figure 14-17 illustrates the principle of operation. The piston has an orifice that restricts the speed at which the piston can be moved in the cylinder. When the damper piston rod is connected to a stationary structure of the nose gear (air-oleo cylinder) and the damper cylinder is attached to the rotating part (air-oleo piston) of the gear, as shown in Fig. 14-20, any movement of the nose wheel alignment to the right or left causes the piston in the shimmy dampener to be moved inside the cylinder. If the movement is relatively slow, there will be little resistance from the shimmy damper; however, if the movement is rapid, there is strong resistance because of the time it takes to cause the fluid in the cylinder to flow through the orifice from one side of the piston to the other. Shimmy dampers are usually provided with an overload feature, such as some type of relief valve, to release pressure if a turning load becomes great enough to cause structural damage.

Large shimmy dampers may incorporate a temperature-compensating system, which allows the fluid in the damper to expand and contract with changes in temperature without overpressurizing the system. The hollow piston rod is vented

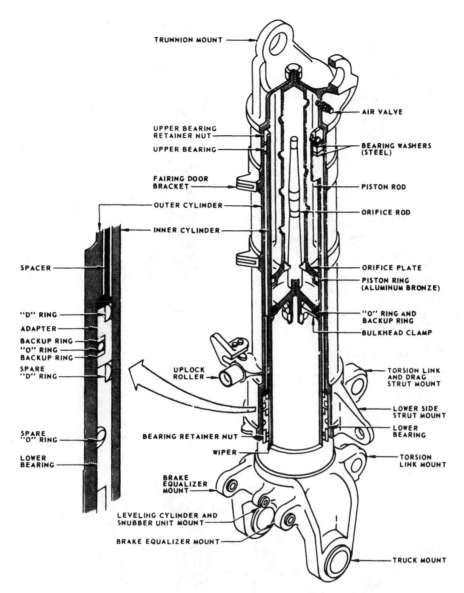

FIGURE 14-12 Cutaway drawing of main landing-gear shock strut.

to the oil chamber, with one end of the rod sealed and a compensating piston, spring, and plug being installed in the other end. When servicing the damper, the piston is inserted a specified distance into the rod. With the unit full of fluid, the spring and plug are then installed. This places a slight pressure load on the system. If the ambient temperature increases, the fluid expands and pushes the piston against the spring. If the ambient temperature decreases, the fluid reduces in volume and the spring pushes the piston further into the rod to prevent the formation of a vacuum or an air pocket in the damper.

Vane-Type Dampers

Vane-type dampers are designed with a set of moving vanes and a set of stationary vanes, as shown in the drawing of Fig. 14-18. The moving vanes are mounted on a shaft, which extends outside the housing. When the shaft is turned, the chambers between the vanes change in size, thus forcing

hydraulic fluid from one to the other. The fluid must flow through restricting orifices, providing a dampening effect to any rapid movement of the vanes in the housing. The body or housing of the vane-type damper is usually mounted on a stationary part of the nose landing gear, and the shaft level is connected to the turning part. Thus any movement of the wheel alignment to the right or left causes a movement of the vanes in the shimmy damper.

Damper Inspections

Shimmy dampers do not require extensive maintenance; however, the technician should check them for leakage and effectiveness of operation. If the damper has a fluid-replenishment reservoir, the fluid quantity should be checked periodically and fluid of the specified type added if necessary.

When inspecting shimmy dampers, the mount bolts and fittings should be checked closely for any evidence of wear. Many aircraft use bushings in the fittings so that the fit of the

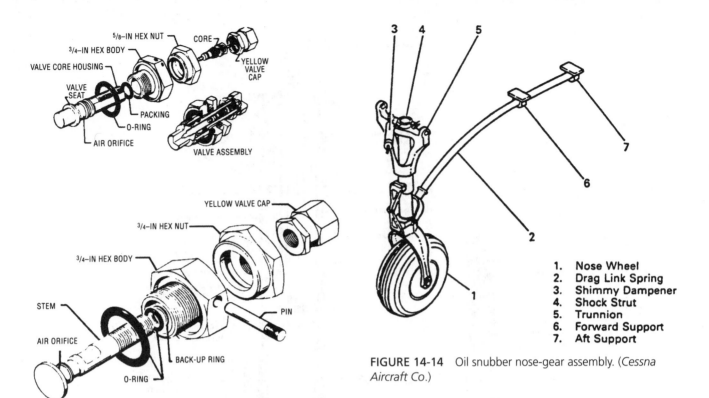

FIGURE 14-13   High-pressure air valves.

FIGURE 14-14   Oil snubber nose-gear assembly. (*Cessna Aircraft Co.*)

1. Nose Wheel
2. Drag Link Spring
3. Shimmy Dampener
4. Shock Strut
5. Trunnion
6. Forward Support
7. Aft Support

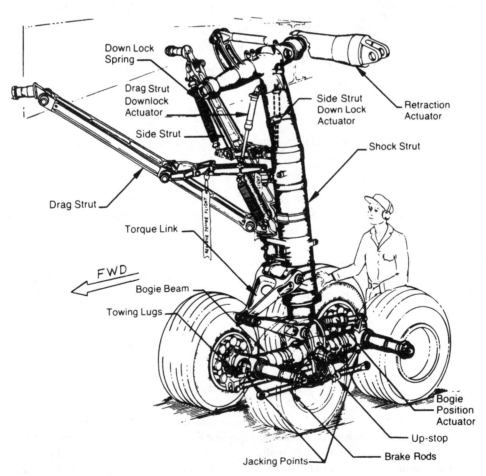

FIGURE 14-15   The main landing gear for a Boeing 767. (*Boeing Commercial Aircraft Co.*)

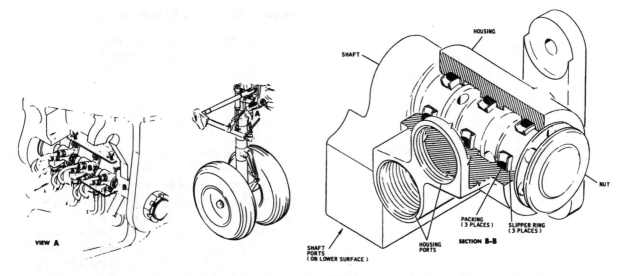

FIGURE 14-16 Swivel glands for the transfer of fluid pressure. *(McDonnell Douglas Corp.)*

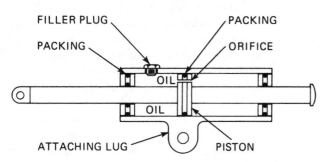

FIGURE 14-17 Drawing of a piston-type shimmy damper.

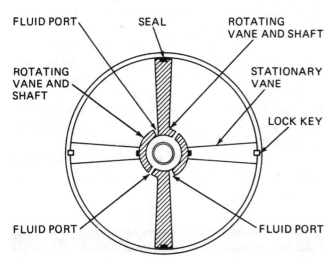

FIGURE 14-18 Principle of a vane-type shimmy damper.

bolts in the fittings can be renewed by replacing the bolts and bushings. If these mountings are allowed to become worn, the damper will be loose on the nose wheel and will allow shimmying to occur.

STEERING SYSTEMS

Steering systems are used to control the direction of movement of an aircraft while taxiing. Although a few aircraft have free-castering nose wheels, the vast majority of aircraft are equipped with some sort of steering system.

Mechanical Steering Systems

Mechanical steering systems are found on small aircraft where the pilot can press on the rudder pedal and cause the nose wheel or tail wheel to turn without any form of powered assistance. Some aircraft have the rudder pedals directly linked to the nose wheel steering arm, whereas others use a spring interconnect. An example of a system using a spring-operated bungee rod is shown in Fig. 14-19. When the pilot pushes on the rudder pedal, the spring inside the rod is compressed and applies pressure to the gears on the steering collar. As the aircraft rolls forward, the spring pressure causes the nose gear to turn.

A nose-gear steering system for a single-engine light airplane with retractable gear is shown in Fig. 14-20. This drawing shows one method by which the steering mechanism is engaged with the nose-wheel strut when the gear is extended and how it will disengage when the gear is retracted. The steering bellcrank is connected to the steering rods and through the rods to the rudder pedals. When the gear is extended, the steering bellcrank engages the steering arm, which is attached to the upper part of the strut. Moving the steering arm (through movement of the rudder pedals) will turn the nose-wheel strut and the nose wheel for steering.

Tail Wheel

Many older aircraft and some special-use aircraft have a conventional landing-gear configuration. This configuration requires the use of a **tail wheel**. The tail wheel is mounted

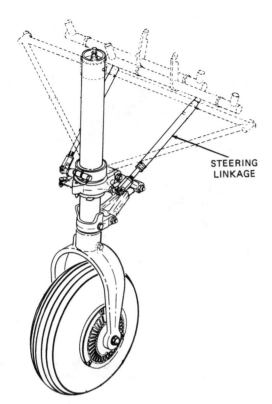

FIGURE 14-19 Steering links for a light-aircraft nosewheel. (*Cessna Aircraft Co.*)

Power Steering Systems

Power steering systems are used for aircraft that require large amounts of force to be applied to the nose wheel to achieve efficient steering control. This includes all large aircraft.

Power steering systems can be controlled by the pilot's rudder pedals, by a steering wheel in the cockpit, or by a combination system allowing full-system travel with a steering wheel and a small degree of directional control with the pedals. Operation of either of the controls causes an actuator on the nose wheel to turn the nose wheel and change the direction of movement. A follow-up system is used to provide only as much nose wheel deflection as the pilot requires based on the amount the pedal or steering wheel is deflected. Referring to Fig. 14-23, notice that when the wheel or pedal is operated, the balance of the follow-up differential mechanism is upset, and this unit pivots and actuates the steering-control valve. The steering-control valve directs hydraulic pressure to the appropriate steering cylinder and releases pressure from the other cylinder. When the nose wheel starts to rotate, the cable around the steering collar on top of the nose-wheel strut moves and returns the follow-up differential mechanism to a neutral position. This centers the steering-control valve and stops rotation of the nose wheel. If the cockpit control is moved a large amount, the follow-up differential mechanism is deflected an equal amount, requiring that the cylinders turn the nose wheel a large amount to return the follow-up unit and steering control valve to a neutral position. A similar operation would occur if the control were moved a small amount. The nose wheel would move only a little before the follow-up differential mechanism returned to the neutral position.

In the system shown in Fig. 14-23, the towing bypass valve is used by maintenance personnel to bypass the steering commands so that the aircraft can be towed without damaging the steering operating mechanism.

Very large transport aircraft such as the Boeing B777 and Airbus A380 also use power steering systems for the main landing-gear system to decrease tire scrub and the U-turn radius. The body gears on the A380 are steerable and two of the six wheels on the main gear of a B777 are steerable. Figure 14-23 shows the steering system for the main gear of the B777. While the nose gear is actuated by a cable system, the main gears use electrical signals from the nose wheel tiller transducers to control an electrohydraulic valve and actuator.

on a short spring, oleo, or other assembly on the bottom of the fuselage near the rudder. The tail wheel may be fixed in alignment with the fuselage longitudinal axis, or it may be designed to rotate, allowing the aircraft to turn easily.

Fixed-alignment tail wheels are found only on aircraft such as gliders, which are not normally taxied. A tail wheel that can rotate may be steerable, full-swiveling, and lockable. A steerable tail wheel responds to cabin rudder controls to aid in controlling the direction of movement of an aircraft on the ground. A full-swiveling/nonsteerable tail wheel is not controllable and pivots freely on its mounting. Most tail wheels are equipped with a full-swivel capability that disengages the steering control when the wheel pivots through more than about 45° from the aircraft centerline. This allows tight turns to be completed. The tail wheel is reengaged by returning the aircraft to a straight line of taxi with differential braking. A lockable mechanism is used with some tail wheels to aid in directional control during takeoff and landing. This mechanism locks the tail wheel in alignment with the aircraft longitudinal axis. When the lock is disengaged, the tail wheel returns to its full-swivel or steerable operation. A steerable tail wheel with a full-swivel capability is shown in Fig. 14-21.

For tail-wheel aircraft, the rudder-control cables are connected to the tail-wheel steering arms through springs. When the rudder is deflected, one spring is stretched and the aircraft begins to turn as it rolls forward. A tail-wheel system is shown in Fig. 14-22.

RETRACTION SYSTEMS

The purpose of retractable landing gear is to reduce the drag of the aircraft or to adapt the aircraft for landing on different surfaces, as with retractable wheels used with floats. Various systems have been designed to retract the gear. These include mechanical systems, hydraulic systems, and electric systems.

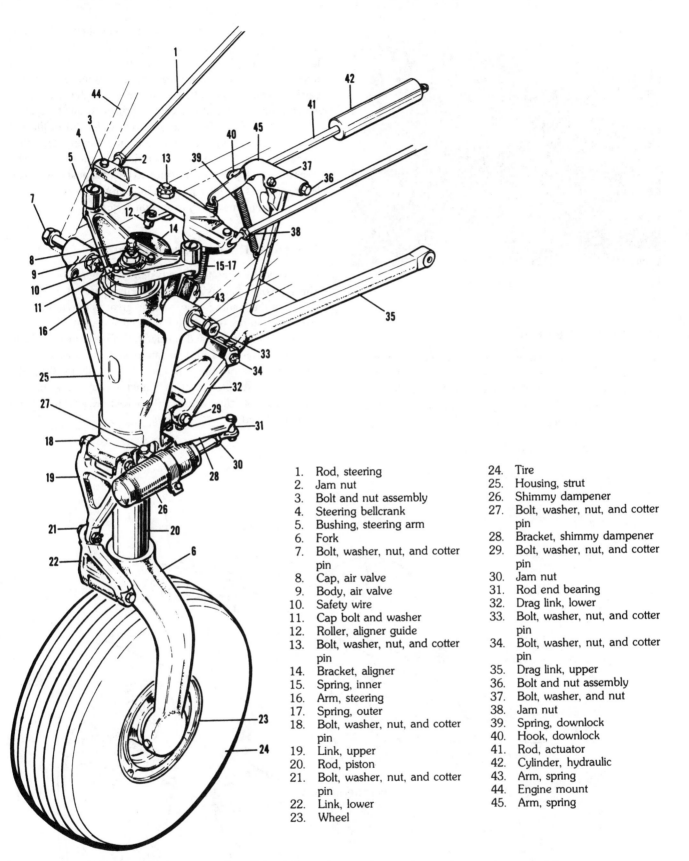

1. Rod, steering
2. Jam nut
3. Bolt and nut assembly
4. Steering bellcrank
5. Bushing, steering arm
6. Fork
7. Bolt, washer, nut, and cotter pin
8. Cap, air valve
9. Body, air valve
10. Safety wire
11. Cap bolt and washer
12. Roller, aligner guide
13. Bolt, washer, nut, and cotter pin
14. Bracket, aligner
15. Spring, inner
16. Arm, steering
17. Spring, outer
18. Bolt, washer, nut, and cotter pin
19. Link, upper
20. Rod, piston
21. Bolt, washer, nut, and cotter pin
22. Link, lower
23. Wheel

24. Tire
25. Housing, strut
26. Shimmy dampener
27. Bolt, washer, nut, and cotter pin
28. Bracket, shimmy dampener
29. Bolt, washer, nut, and cotter pin
30. Jam nut
31. Rod end bearing
32. Drag link, lower
33. Bolt, washer, nut, and cotter pin
34. Bolt, washer, nut, and cotter pin
35. Drag link, upper
36. Bolt and nut assembly
37. Bolt, washer, and nut
38. Jam nut
39. Spring, downlock
40. Hook, downlock
41. Rod, actuator
42. Cylinder, hydraulic
43. Arm, spring
44. Engine mount
45. Arm, spring

FIGURE 14-20   Nose gear for a light airplane, showing the steering mechanism. (*Piper Aircraft Co.*)

FIGURE 14-21 A steerable tail wheel.

Mechanical System

Some older aircraft use a **mechanical retractable landing-gear system** and many current-production light aircraft make use of a mechanical emergency extension system. A mechanical system is powered by the pilot moving a lever or operating a crank mechanism.

A lever arrangement is found on many older aircraft and involves the use of a lever, approximately 2 ft [0.61 m] long, located in the cabin, which moves through an arc of 90° to retract or extend the gear. When the lever is moved from the vertical to the horizontal position, the lever unlocks the gear, and, through the use of overcenter springs, torque tubes, and bellcranks, retracts the gear. The gear is locked in the UP position by securing the cabin lever in a locking device on the cabin floor. The opposite operation extends the gear. The lever locks in the vertical position to lock the gear down. The lever serves as the gear-position indicator. This system does not require an emergency extension system.

Electrical Retraction System

An electrically operated gear system is shown in Fig. 14-24. This type of system is often used on light aircraft, where the weight of the landing gear is not so great as to require large operating motors or complex hydraulic systems. When the gear selector is positioned, an electric motor is energized and operates a gear that rotates a cam plate or spider that opens the doors, positions the gear, and closes the doors. Once the gear is in the selected position, a microswitch breaks the circuit to the motor and causes the appropriate gear indication to be displayed.

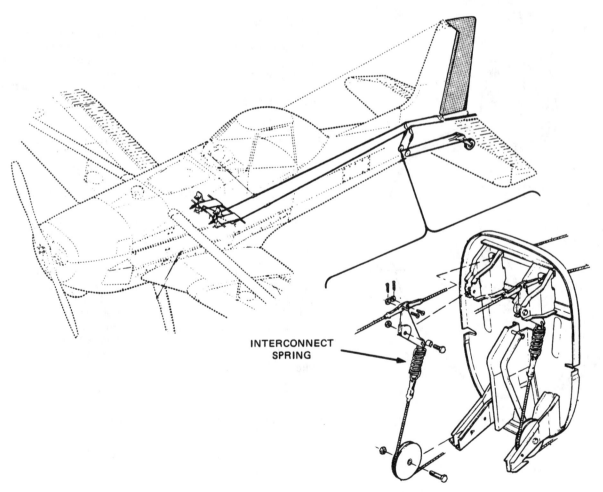

INTERCONNECT SPRING

FIGURE 14-22 Tail-wheel steering arrangement. (*Cessna Aircraft Co.*)

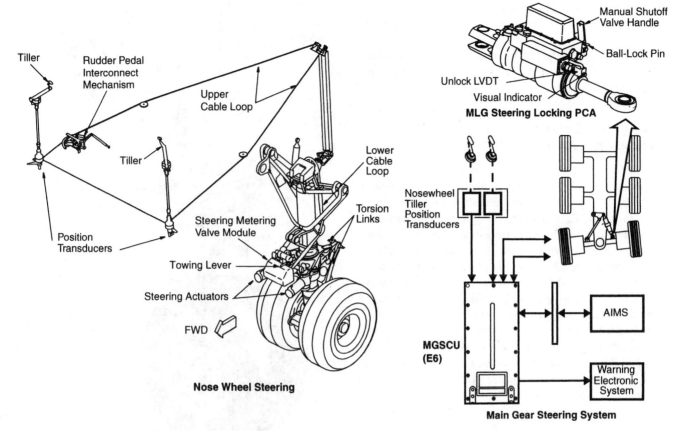

FIGURE 14-23 Nose wheel and main gear steering. (*Boeing Aerospace Co.*)

Hydraulic Retraction System

A **hydraulic retractable landing gear** makes use of hydraulic pressure to move the gear between the retracted and extended positions. Although this system is commonly used for all sizes of aircraft, it is used exclusively where the landing gear is large and could not economically be operated by any other method.

The power for the operation of the system may be generated by engine-driven pumps, electrically operated pumps, or, for emergency operation, hand- or wind-driven turbine pumps.

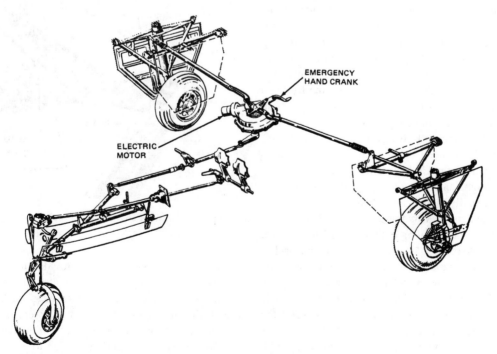

FIGURE 14-24 Electrically operated landing-gear system. (*Beech Aircraft Co.*)

Detailed information concerning the operation or the power section of hydraulic landing gear is given in Chap. 13 of this text.

Electrically operated pumps are often found on light aircraft, whereas transport aircraft rely primarily on engine-driven pumps.

Emergency Landing Gear Extension Systems

One of four methods is used to extend the landing gear if hydraulic power is lost, preventing hydraulic extension of the landing gear. (1) Some aircraft make use of an air bottle to "blow" the gear down, substituting air pressure for hydraulic pressure. This pneumatic extension has the disadvantage that the system must be bled of all air before being returned to service. (2) Some aircraft make use of a mechanical system, where the operation of a hand crank or ratchet performs the extension operation. (3) Other aircraft have a separate hydraulic system, powered by various methods, including a hand pump, to extend the gear. (4) The fourth method of emergency extension that appears to be becoming very popular for aircraft is the use of a mechanical system to release the UP locks, allowing the gear to free-fall into the down-and-locked position.

Landing-Gear Operation

One type of retractable main landing gear is illustrated in Fig. 14-25. The assembly consists principally of the shock strut; the wheel; the brake assembly, the trunnion, and side brace; the torque link, or "scissors"; the actuating cylinder, the down-and-up locks; and the bungee system.

To retract the gear, the actuating cylinder is extended by hydraulic pressure. Since the actuating cylinder can provide greater force during extension of the cylinder than it can during retraction because of the greater piston area exposed to fluid pressure, the extension movement of the actuating cylinder is used to retract the gear. Retraction of the gear requires greater force because of gravity. Extension of the actuating cylinder causes the gear to rotate on the trunnion pin until the gear is approximately in a horizontal position. When the gear reaches the full UP position, a pin on the strut engages the UP latch and locks the gear in the UP position.

When the gear is extended, the first movement of the actuating cylinder releases the UP lock. This permits the gear to fall of its own weight, and the actuating cylinder acts to snub the rate of fall. Usually there is an orifice check valve in the UP line of the landing-gear hydraulic system; this restricts the fluid flow from the actuating cylinder to the return line, thus slowing the rate of gear descent. As the gear approaches the DOWN position, the actuating cylinder moves it to the full DOWN position. In the DOWN position, a blade engages the DOWN-LOCK track and slides into the DOWN-LOCK latch, as shown in Fig. 14-26. The DOWN lock prevents the gear from retracting after it has been lowered.

The landing gear shown in Fig. 14-25 is equipped with a pneumatic bungee system for emergency operation. The

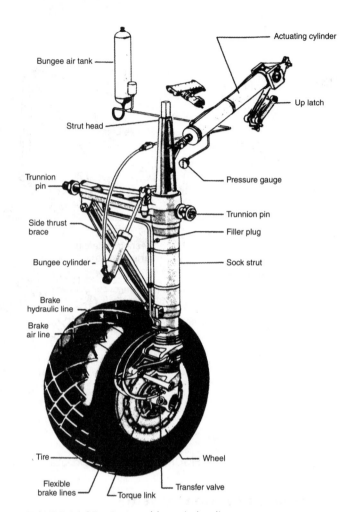

FIGURE 14-25 Retractable main landing gear.

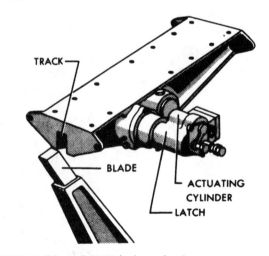

FIGURE 14-26 A DOWN-lock mechanism.

purpose of the system is to provide air or gas pressure to lower the gear in the event of hydraulic power failure. The bungee tank is charged with air or gas at a high pressure, and when it becomes necessary to lower the gear in an emergency, the air or gas is released from the tank by means of a valve and is carried through tubing to a special bungee cylinder. This cylinder provides enough force to lock the gear in the DOWN position. In many systems, the air or gas is directed to a shuttle valve, which blocks off the hydraulic system and opens the DOWN line to the main-landing-gear actuating cylinder, which then lowers and locks the gear. In any case, the landing-gear control handle must be placed in the DOWN position.

When the operation of a retractable gear system includes the opening and closing of gear doors, an associated system controlled by sequence valves is often used to operate the doors. The sequence of operation is (1) opening of doors and (2) lowering of gear. During retraction, the gear retracts and then the doors close. The doors can be operated through the hydraulic system or by a mechanical linkage in connection with the movement of the landing gear mechanism.

In many designs, the landing-gear doors are closed when the gear is extended or retracted. In such cases, the doors must operate twice for either retraction or extension of the gear.

Some aircraft do not use main inboard doors but retract the main gear so that it is flush with the underside of the wing. A brush seal or an inflatable pneumatic seal is used to seal the edges of the wheel flush with the bottom of the wheel well.

Landing-Gear-Position Indicator Systems

Another important feature of retractable landing gear is the safety mechanism that prevents gear retraction while the airplane is on the ground. This safety system often consists of an electric circuit, which includes switches, sometimes called "squat switches," operated by the extension and compression of the landing-gear struts. As long as the gear struts are compressed, the switches are open, and an electrically operated lock prevents the raising of the gear. When the airplane leaves the ground, the gear struts extend and the switches are closed. This permits operation of the landing-gear control level to raise the gear.

Another safety feature included with retractable landing-gear systems is a **warning horn**. If the gear is in the retracted position and the throttle is retarded to a below-cruise-power setting, as is done when landing, the warning horn sounds and warns the pilot that the landing gear is not in the down position. There are various types of landing-gear **position indicators**. These show whether the landing gear is up or down and whether it is locked. One type of indicator has a needle or a miniature wheel for each unit. The indicator is built into a single instrument on the instrument panel and arranged so that the needle or wheel moves along with the actual movement of the landing gear from the DOWN to the UP position (and in the same direction), thus simplifying its reading by the pilot. Many airplanes have indicator lights, such as a green light that glows when the gear is down and locked, a red light to indicate that the gear is not fully up or down, and an amber light to indicate that the gear is fully retracted and locked. If lights are used, a green light is always included, but only one light (either the red or the amber) is used with it.

Many new aircraft are equipped with a glass cockpit and traditional analog instruments for landing-gear indication and warning systems have been replaced with a software-driven electronic display. The engine information and crew alerting system (EICAS) is used to display engine and other aircraft system information. Rather than signaling a system failure by turning on a light behind a translucent button, failures are shown as a list of messages in a small window near the other EICAS indications. Figure 14-27 shows an EICAS display with gear indicators and messages. The EICAS system also has

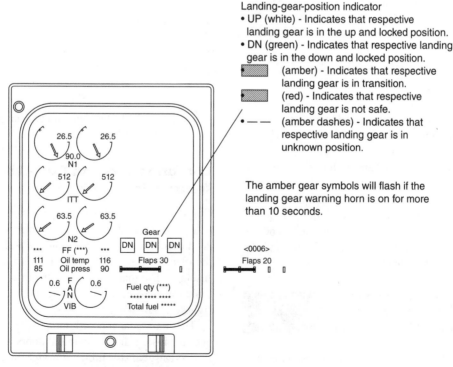

FIGURE 14-27   EICAS landing-gear indications and messages.

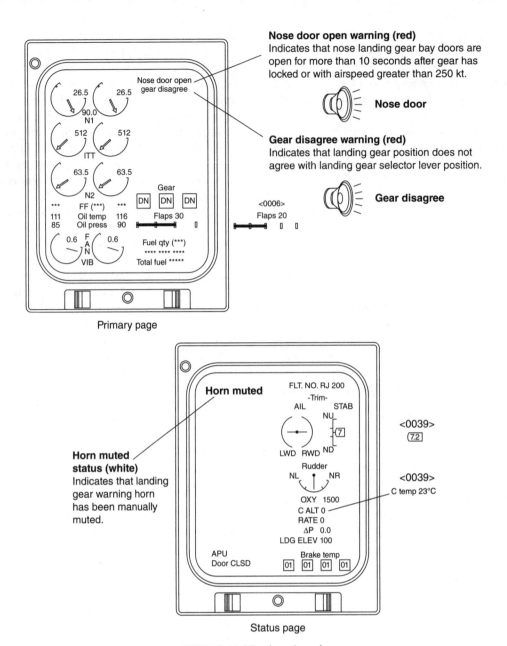

Nose door open warning (red)
Indicates that nose landing gear bay doors are open for more than 10 seconds after gear has locked or with airspeed greater than 250 kt.

Nose door

Gear disagree warning (red)
Indicates that landing gear position does not agree with landing gear selector lever position.

Gear disagree

Primary page

Horn muted

Horn muted status (white)
Indicates that landing gear warning horn has been manually muted.

Status page

FIGURE 14-27   (continued)

a maintenance page which can be accessed by maintenance to analyze landing-gear information. Figure 14-28 shows a brake, steering and gear maintenance page.

During the operation of an airplane with retractable landing gear, the system that raises and lowers the gear must operate without fail, and the gear must remain in the down-and-locked position after it is lowered. During required inspections, the airplane should be placed on jacks and the gear operated to ensure that the operating system and the DOWN and UP locks function effectively.

TRANSPORT AIRCRAFT LANDING-GEAR SYSTEMS

Transport aircraft, both corporate jets and airliners, have similar hydraulic landing-gear systems. The following discussions

will acquaint the technician with typical landing-gear systems for this category of aircraft.

Corporate Jets and Dual-Wheeled Transports

Corporate jet aircraft and many airliners have a fully retractable tricycle landing-gear mounting, with two wheels on each gear. The nose gear is a dual-wheel, steerable type with the oleo strut mounted in the forward, lower section of the fuselage. The main gear consists of two pairs of wheels and brakes on oleo struts mounted in the wing-root area aft of the right- and left-wing rear spars.

The main-gear inboard doors, when used on the aircraft, and the nose-gear compartment doors are connected to their respective gear assemblies through linkages that sequence the doors to the open position during gear travel and to

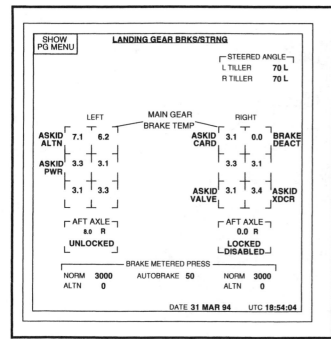

Brake and Steering Maintenance Page Format

Gear Maintenance Page

FIGURE 14-28 Landing gear indications.

the closed position at the end of each cycle. The nose-gear attached doors and the main-gear outboard doors move with the gear. Each wheel well is completely enclosed by doors when the gear is retracted, and the major portion of each wheel well is enclosed after the gear is extended.

Gear retraction and extension and the mechanical release of the door latches are controlled by the landing-gear control lever. The hydraulic power system actuates the gear, door latches, bungee cylinders, brakes, and the nose-wheel steering system.

The nose-wheel steering system is hydraulically controlled through its full range of travel by a steering wheel located on the captain's left console. The rudder pedals can be used for limited steering control to either side of the neutral position for directional control during takeoff and landing. When the steering cylinders are in the neutral position, they act as shimmy dampers.

Each main-gear wheel is fitted with a hydraulic power disk brake. The brakes are controlled by the brake-control valves, which are operated by a cable system connected to the brake pedals. An electrically controlled **antiskid system** provides a locked-wheel protection feature and affords maximum efficiency to the brake system. The antiskid-system dual-servo valve meter applies pressure to the brakes as required to provide maximum braking effect without skidding.

An electrically monitored indicating and warning system provides the flight crew with all the necessary landing-gear and gear-door-position indications. These indications are visible in the cockpit.

Each main gear consists of two wheels and brakes attached to an oleo (air-oil) strut, which is mounted on a support fitting on the rear spar in each wing-root area. Typical main-gear assemblies are shown in Figs. 14-29 and 14-30.

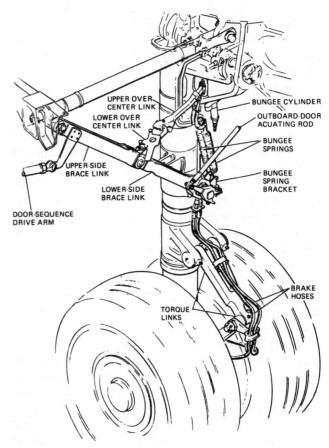

FIGURE 14-29 Main landing gear for a DC-9 airplane. (*McDonnell Douglas Corp.*)

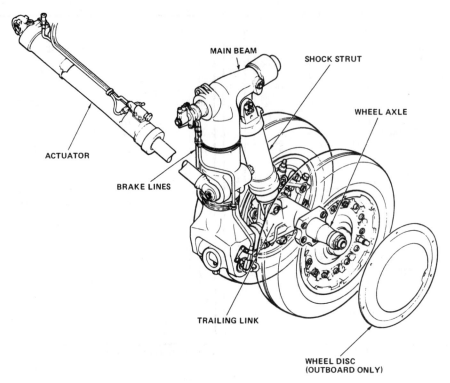

FIGURE 14-30 Main landing gear for a Canadair Challenger 601. (*Canadair Inc.*)

The main gear in Fig. 14-29 is supported laterally by the **side brace links**. These are locked in the DOWN position by **overcenter links**, which are driven hydraulically by a **bungee cylinder** and mechanically by the **bungee springs**. The side brace links and the overcenter links fold up along the rear strut during gear retraction. The main gear is hydraulically raised to the UP position by a main-gear actuating cylinder and is held in the UP position by the main-gear doors and latches during flight.

The main gear in Fig. 14-31 uses the actuator to extend and retract the gear. An internal lock mechanism in the actuator locks the gear in the extended position. An external mechanical UP lock is used to hold the gear in the retracted position.

A typical **main-gear** door arrangement and the doors' positions in relation to the landing gear are shown in Fig. 14-31. The doors enclose the main-gear wheel wells and a section of the wing-root area when the gear is retracted. There are two doors for each gear, designated as the **inboard door** and the **outboard door**. The inboard door, the larger of the two, is square, with the outboard portion curved upward to conform to the shape of the adjoining fuselage structure. The core of the door is honeycomb construction strengthened by heavy frames. The door is exceptionally sturdy because it must support the weight of the main-gear assembly in the retracted position. A single hook-and-roller-type latch assembly holds the door in the CLOSED position. The door is actuated by a single hydraulic cylinder that attaches to a clevis on the forward frame and to the bracket attached to the shear web.

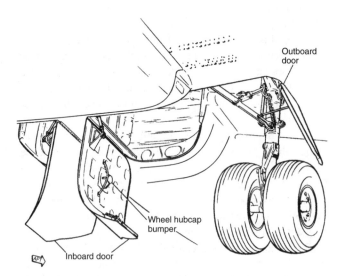

FIGURE 14-31 Transport aircraft main-gear doors. (*McDonnell Douglas Corp.*)

The outboard door is a smaller, irregular-shaped, honeycomb-core-type assembly attached to the lower wing section by a standard piano-type hinge. The door is linked directly to the gear strut by a pushrod and follows the gear strut during the gear extension and retraction operations.

The **nose gear** is typically a dual-wheel, steerable, oleo shock-strut unit that retracts forward and up into the fuselage during flight and is completely enclosed by doors when retracted. An example of the unit and its associated parts is shown in Fig. 14-32.

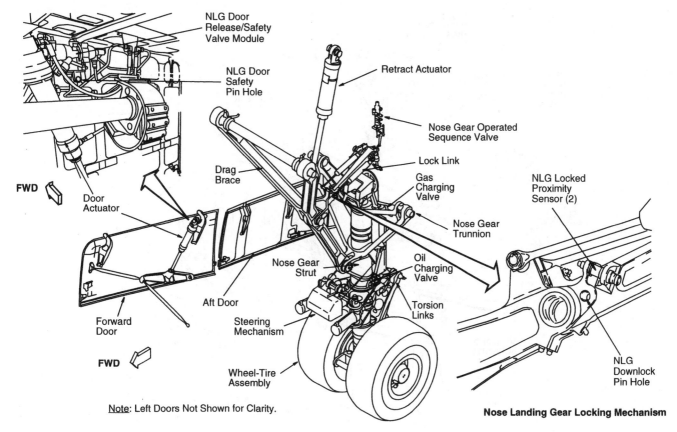

FIGURE 14-32 Nose landing gear.

The nose-gear strut is supported by the side braces and the cross arm, which form an integral part of the strut casting, and by the drag links. The wheels are mounted on an axle that is attached to the lower end of the strut piston. Overcenter linkage, which is attached between the drag links and the fuselage structure, locks the drag links and strut in the DOWN position. The nose-gear actuating cylinder is attached to a cross-arm crank and to the fuselage structure on the left side of the assembly. The cylinder is compressed when the gear is extended.

The nose-wheel steering cylinders are mounted on the upper forward side of the strut cylinder body, and the pistons are attached to the piston and axle by the torque links.

The nose-gear doors may be mechanically operated by the nose-wheel extension and retraction, or, as shown in Fig. 14-33, some doors may be operated hydraulically. In this example, the forward doors are opened hydraulically. The gear then extends or retracts and moves the mechanically linked aft door with it. The forward doors then close once the nose gear is fully extended or retracted. The sequence of operation of the forward doors and the landing-gear movement is controlled by the sequencing operation of the electrically operated selector valve. The selector valve receives door- and gear-position signals from the microswitches and proximity switches and, according to these signals, directs hydraulic pressure to the actuators in the proper sequence.

The nose-gear strut supports the forward portion of the airplane, absorbs landing shock, and steers the airplane through ground maneuvers. The strut is supported by the side braces and cross arm, which are part of the strut casting, and by the drag links. The major components of the strut are the **cylinder body**, the **steering cylinder**, and the **piston**. The cylinder body is the major support for the strut, with the side braces and cross arm as an integral part of the body. The steering cylinder is installed inside the body between the body and the piston and mounts on the steering bosses on the top side. The piston is inside the steering cylinder and is the shock absorber and support for the axle and wheels. Torque links are attached to the steering cylinder and to the axle boss. They transfer steering motion from the steering cylinder to the wheels and limit the extension length of the piston during gear retraction. The main steering cylinder described here must not be confused with the two cylinders called the **right steering cylinder** and the **left steering cylinder** shown in Fig. 14-32. These two units are **actuating cylinders**.

The side braces and cross arm support the nose-gear strut vertically and laterally. A torque-arm extension on the left side brace is the attach point for the nose-gear actuating cylinder. The drag link is attached to the forward side of the strut body and supports the strut longitudinally. The gear is locked in the DOWN or UP position by the overcenter linkage. Internally, the nose-gear strut contains a metering pin and strap assembly, a metering plate and plate support, a piston seal adapter with static and dynamic O-rings and D-rings, and a piston guide.

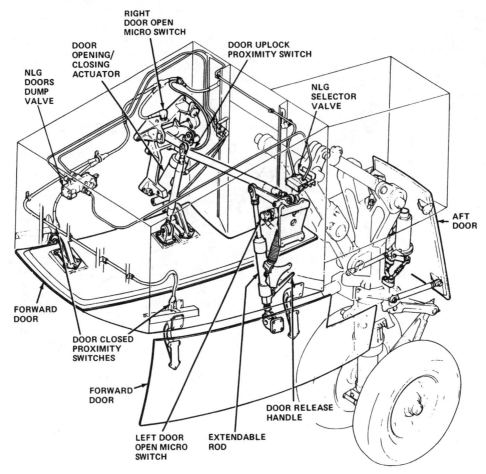

NLG
DOORS
DUMP
VALVE

DOOR
OPENING/
CLOSING
ACTUATOR

RIGHT
DOOR OPEN
MICRO SWITCH

DOOR UPLOCK
PROXIMITY SWITCH

NLG
SELECTOR
VALVE

AFT
DOOR

FORWARD
DOOR

DOOR CLOSED
PROXIMITY
SWITCHES

FORWARD
DOOR

DOOR RELEASE
HANDLE

LEFT DOOR
OPEN MICRO
SWITCH

EXTENDABLE
ROD

FIGURE 14-33 Nose-gear doors and operating mechanism for a Canadair Challenger 601. (*Canadair Inc.*)

The **nose-gear linkage** consists of the **drag links, overcenter linkage**, and **torque-tube assembly**. These are illustrated in Fig. 14-34. The drag links are the longitudinal-axis support members for the nose-gear strut in the DOWN position. They consist of three tubular links. The upper two links are connected to the nose-gear support structure by a torque tube that acts as a pivot point. The lower drag link attaches to the upper two links at the midway point and connects to the nose-gear-strut clevis on the lower forward side of the cylinder body. The junction point of the three drag-link members is also the connecting point for the lower **overcenter link**. The overcenter links hold the drag links and the nose-gear strut in the downlocked position. The links lock over center when the gear is fully extended and when the gear is fully retracted. When it is desired to retract gear that is in the down-and-locked position, the overcenter links are released by the hydraulically actuated **bungee cylinder** to permit gear retraction. The bungee cylinder, which is attached to the upper overcenter link, can be seen in Fig. 14-32.

When the gear is in the UP position, the overcenter links are normally released by the same bungee cylinder, but, in the absence of hydraulic power, they can be released by a manually operated trip lever and roller that is connected to the alternative extension control lever mechanism. The two

bungee springs are a mechanical backup for the hydraulically operated bungee cylinder.

The upper two nose-gear drag links are fitted with cranks at the pivot axis points. These cranks are the connecting points for the nose-gear forward-door drive linkage. The forward doors are cycled to the open position during gear travel and are closed at the end of each cycle of gear travel.

A service and auxiliary power unit (APU) shutdown panel is attached to the NLG gear in Fig. 14-35. This panel can be used by maintenance personnel to close the NLG doors, shut down the APU in case of an APU fire, and communicate with flight crew.

The **ground-sensing-control mechanism** provides a mechanical means of establishing a ground or flight mode, with the functions of various systems differing as the mode changes. The mechanism, shown in the drawings of Fig. 14-36, consists of a few simple linkages and a two-way, closed-circuit cable actuated by the nose-gear links during gear-strut extension and compression. When the airplane takes off, the strut extends and establishes the flight mode. When the airplane is on the ground, the strut is compressed and the ground-operation mode is established.

The linkage for the ground-sensing-control mechanism on the nose gear consists of a lever, a spring-loaded rod, and

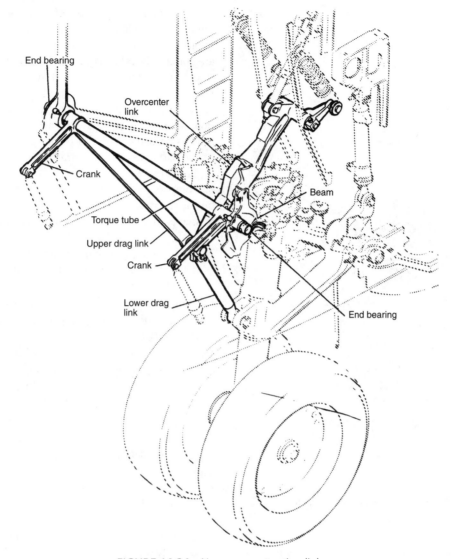

FIGURE 14-34   Nose-gear actuating linkage.

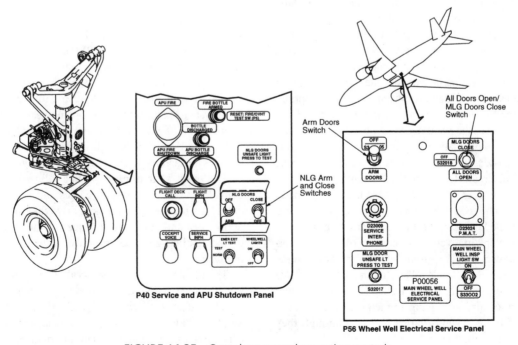

P40 Service and APU Shutdown Panel

P56 Wheel Well Electrical Service Panel

FIGURE 14-35   Gear door ground operation controls.

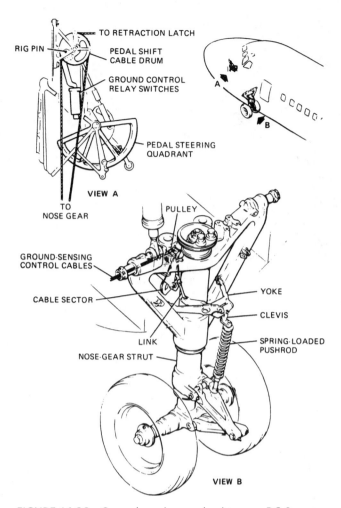

FIGURE 14-36 Ground-sensing mechanism on a DC-9.

a cable sector. The linkages in the flight compartment consist of a cable drum and shaft that drives the pedal steering-override mechanism and a striker that actuates the ground control relays. The two-way, closed-circuit cable system runs from the sector on the nose gear to the cable drum on the rudder pedal-override mechanism. The **antiretraction latch cable** takes off from the same cable drum.

When the nose-gear strut extends, the nose-gear upper torque link draws the linkage downward, pulling the spring-loaded rod assembly down, and rotates the cable sector clockwise. The cable drives the pedal-override cable drum and shaft, which disengages the pedal steering mechanism. At the same time, the ground-sensing control relays are actuated, causing various electrical circuits to assume the flight mode.

When the nose-gear strut is compressed, a ground mode is established. The pedal steering system is actuated to permit rudder-pedal steering, the antiretraction latch is engaged to prevent inadvertent operation of the landing-gear control lever, and the ground-control relays assume the ground-operations mode. The feature that prevents retraction of the gear while the airplane is on the ground is required on all aircraft having retractable landing gear. The design of these systems varies widely, although they all serve the same function.

Some aircraft rely on electrical position-sensing systems rather than mechanical systems to determine if the aircraft is on the ground or in flight. The sensors in these systems are commonly called **squat** or **WOW switches** (WOW representing weight on wheels). If the aircraft weight is on the landing gear, the squat switch keeps the gear from being retracted by maintaining an open electrical path to the retraction system. The squat-switch system may use one or more sensors on one or more of the landing-gear legs, depending on the particular model aircraft.

The proximity-sensing systems on transport aircraft can be complex systems that interact with many other systems on the aircraft, for example, the proximity-sensing system on one particular aircraft has the following functions:

Normal landing-gear-positioning control. The PSS provides the signals that command the landing gear to extend and retract and the nose gear doors to open and close.

Landing-gear-position indication. The PSS monitors landing-gear-position and provides indication and position status to the EICAS.

Weight-on-wheels indication. The PSS monitors landing-gear strut compression and provides indication of air or ground status to the aircraft systems that require the information.

Fuselage door indication. The PSS monitors the fuselage door positions and provides inputs to the EICAS for indication on the primary page, status page, and door synoptic page.

Thrust reverser indication. The PSS monitors and reports to EICAS the (stowed/unstowed) status of the left and right thrust reversers.

The landing-gear extension and retraction system is composed of two major systems, designated as the **mechanical control system** and the **hydraulic control system**. These are illustrated in Fig. 14-37.

In more complex systems the mechanical control system is divided into four sections. These are (1) the landing-gear control-valve cable system; (2) the main-gear inboard door-sequence follow-up valve system; (3) the alternative, or bypass, valve system; and (4) the subsystems.

The **landing-gear control-valve cable system** is a dual-cable system that runs from the landing-gear control-lever cable drums in the flight compartment to the landing-gear control-valve cable drums in the right wheel well. This cable system operates the landing-gear control valve during normal operation.

The **main-gear door-sequence follow-up valve system** is made up of two push-pull cables and associated linkages on the main-gear side braces to drive the main-gear inboard doors open and closed. The sequence valves are also linked to the landing-gear control-valve cable drums for initial motivation.

The **alternative**, or **bypass-valve**, **system** is a dual-cable system. One cable drives the bypass-valve cable drum, which actuates the bypass valve, the door-latch releases, and the door skids. The other cable actuates the nose-gear over-center release mechanism. The alternative system (bypass

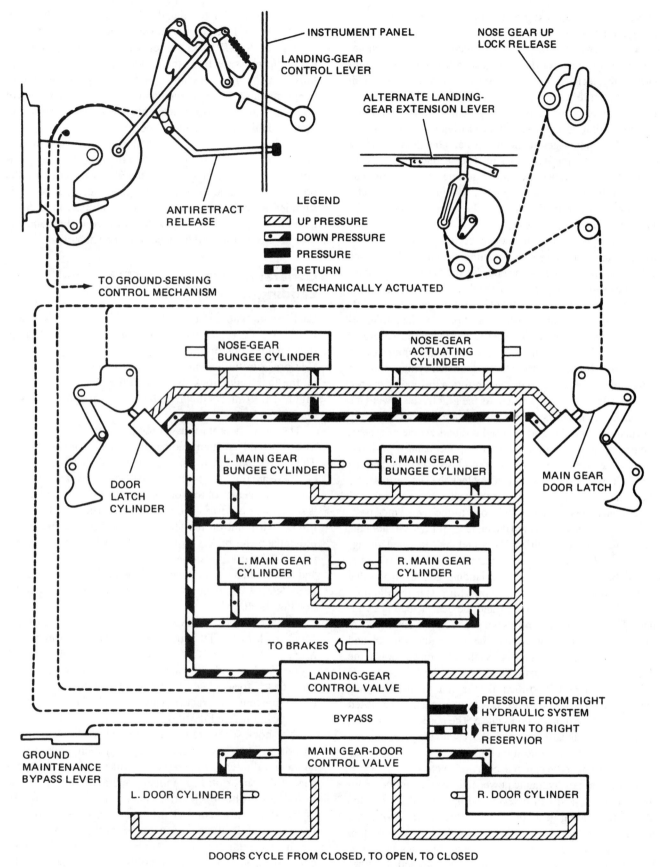

FIGURE 14-37 Landing-gear mechanical and hydraulic control systems.

valve) is used as a backup system to free-fall the gear in case of hydraulic power failure.

Some newer aircraft have replaced mechanical cable-operated systems for landing-gear retraction and extension with electrical controlled systems that use electrical/hydraulic servo valves. Onboard computers are used to control and monitor the entire landing-gear system operation.

The hydraulic control system is divided into two sections. The landing-gear-control valve and the door-sequence valves compose the basic system. The second section is the alternative, or bypass, system mentioned previously. The bypass valve is used on the ground to relieve door-cylinder pressure and permit the doors to be opened manually for ground-maintenance access.

The landing-gear-control valve ports hydraulic pressure to the main- and nose-gear actuating cylinders, the bungee cylinders, and the main-gear door-latch cylinders. The door-sequence valves port pressure to the door-actuating cylinders.

When the landing-gear-control lever is placed in the GEAR-UP position, the cable system positions the landing-gear-control valve to port up pressure to the main and nose-gear actuating cylinders, UNLOCK pressure to the bungee cylinders, and UNLOCK pressure to the doorlatch cylinders. At the same time the cable-drum linkages position the door-sequence valves in the door-open position. As the gear travels toward the UP position, the side braces fold up, driving the door follow-up mechanism toward the door-closed position. As the gear nears the end of upward travel, the follow-up mechanism positions the door-sequence valves to the door-closed position, and the doors close. The door-latch cylinders lock the latches to secure the doors, which, in turn, support the gear.

The nose-gear linkage will move to the overcenter position when the gear is up, and the bungee springs will hold the overcenter locked condition. To extend the gear, the landing-gear-control lever is placed in the GEAR-DOWN position. The cable system positions the landing-gear-control valve and the door-sequence valves to port (direct) hydraulic fluid to the door-actuating cylinders and the gear-actuating cylinders. The fluid pressure moves the gear-actuating cylinders in the DOWN direction and causes the door-latch cylinders and the bungee cylinders to unlock. As the gear moves down, the door follow-up mechanism begins to drive the door-sequence valves toward the UP position. By the time the gear is down and locked, the door-sequence valves have reached the closed position and the inboard doors close. The nose-gear doors are actuated mechanically through linkage with the gear; they are sequenced to the open position during gear travel and to the closed position at the end of each travel cycle.

In the event of hydraulic power failure or a jammed control valve, the gear can be extended by free-fall. When the alternative landing-gear control lever (located in a floor well to the right of the center pedestal) is pulled up, two backup systems are actuated. One cable system operates the nose-gear mechanical release to free the nose gear, and the other system operates the bypass valve, the door latches, and the door skids. The gear will then free-fall and lock in the DOWN position. The bungee springs ensure the locked condition of the gear in the DOWN position.

A schematic for a manual release system is shown in Fig. 14-38. In the example shown, when the release handle is operated, the UP locks are released, dump valves remove hydraulic pressure from the actuators, and the gear free-falls by gravity into the down-and-locked position. Before operating this type of extension mechanism, system control must be properly positioned to ensure correct operation and prevent damage to system components.

Smaller and less complicated aircraft may use a system that relies on electrical control rather than mechanical operation of the control valves. When the gear-selector control is positioned to retract or extend the gear, electrical power is supplied to hydraulic control–selector valves, which direct hydraulic pressure in the proper sequence to the landing-gear door actuators and the landing-gear actuators.

General Description of B777 Landing-Gear System

Each main landing gear of the B777 has a six-wheel truck. A drag brace and a side brace hold each gear in the extended position. Lock links hold the drag brace and side brace in the extended position. Over-center toggles, on the lock links, lock the gear in the extended position. Figure 14-39 shows the main landing-gear components of the B777.

The MLG side brace assembly and the MLG drag brace assembly hold the main landing gear in the extended position. They also supply lateral support to the main landing gear.

There are three axles on each MLG truck. A jacking point is under the center of the forward and aft axles. A tow fitting is on the forward part of the MLG truck. There is also a tow fitting jack point on the aft part of the MLG truck. Brake rods attach the brakes to the MLG shock strut to prevent rotation. The truck positioner actuator tilts the MLG truck up and down when the airplane is in the air. The main landing-gear-steering components steer the aft axle 8° left or right. See Fig. 14-40.

The MLG truck positioner actuator retracts to move the MLG truck to the TILT position and extends to move the MLG truck to the STOW position. When the landing gear is down and locked, the truck positioner actuator moves the truck to 13° forward wheels up (airplane in the air). During gear retraction, the truck moves to the 5° forward wheels down position. See Fig. 14-41.

The shock struts are standard forged steel air-oil shock absorbers. They have an inner cylinder which moves inside an outer cylinder. Compressed dry nitrogen is in the upper part of the shock strut. A mineral hydraulic fluid is in the lower part of the shock strut. Torsion links on the aft part of the shock strut connect the inner and outer cylinders. Figure 14-42 illustrates the main part of the shock strut.

The nose landing gear is a conventional two-wheel gear. A drag strut assembly holds the nose gear in the extended or retracted position. A lock link assembly moves to the over-center position to lock the drag strut in either position. Figure 14-43 shows the main components of the NLG.

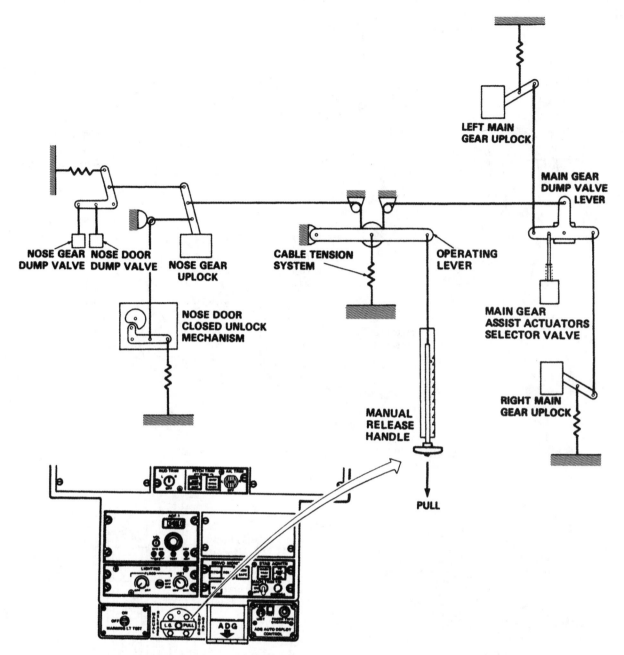

FIGURE 14-38 Schematic of an emergency gear–extension system. (*Canadair Inc.*)

Main Landing-Gear Retraction

When the landing-gear lever is selected to the UP position, the MLG selector/bypass valve moves to the UP position (Fig. 14-44). Hydraulic pressure then goes to the retract lines. The pressure goes to the MLG drag brace and MLG brace down lock actuators. These actuators start retract to unlock the gear down locks. Pressure also goes through the MLG drag brace-operated sequence valve and the MLG up lock-operated sequence valve. This pressure extends the MLG truck positioner actuator to the STOW position.

Pressure through the sequence valves also retracts the MLG door lock actuator to unlock the main gear door. The MLG door actuator gets extend pressure through the MLG door priority/relief valve. The MLG door starts to open. When the MLG door is almost all the way open, the MLG door-operated sequence valve moves to OPEN. Pressure then goes to the MLG retract actuator to retract the gear. When the gear starts to retract, the MLG drag race-operated sequence valve moves to the NOT DOWN position. As the MLG goes into the wheel well, a roller on the landing-gear strut moves the MLG up lock mechanism to the LOCKED position. This moves the MLG up lock-operated sequence valve to the locked position. This removes pressure from the MLG truck positioner actuator. Pressure trapped in the MLG truck positioner actuator keeps the MLG truck in the STOW position. Pressure then goes to the MLG up lock actuator to make sure the MLG up lock mechanism locks.

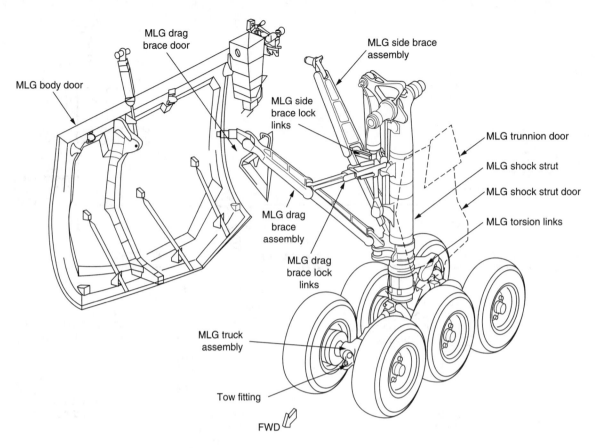

FIGURE 14-39   Main landing gear components of the B777.

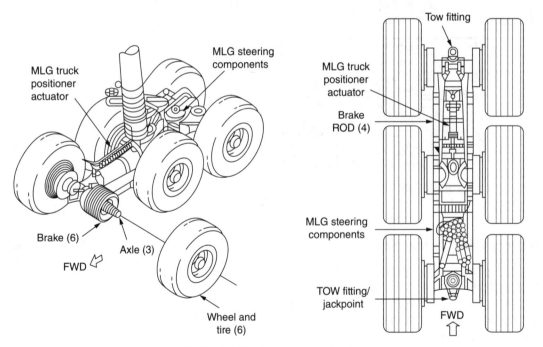

FIGURE 14-40   Landing-gear truck.

Pressure through the MLG door-operated sequence valve also goes to brake system components for the gear retract braking function.

Pressure also goes to the close side of the MLG door actuator and the lock side of the MLG door lock actuator.

The door starts to close. When the door is almost closed, a roller on the door starts to move the MLG door uplock mechanism to the locked position. Pressure in the MLG door lock actuator moves the uplock mechanism over-center to the locked position.

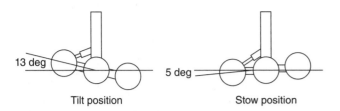

FIGURE 14-41 MLG tilt and stow position.

Main Landing-Gear Retraction

When the landing-gear lever is selected in the down position the MLG selector/bypass valve moves to the down position. This permits hydraulic pressure to go to the extend lines (Fig 14-45).

Pressure goes through the MLG drag brace-operated sequence valve and the MLG up lock-operated sequence valve. This pressure goes to the MLG truck positioner actuator to hold the MLG truck in the STOW position. This pressure also goes to the MLG door lock actuator to unlock the MLG door. The MLG door actuator then gets extend pressure through the MLG door priority/relief valve. The MLG door starts to open.

When the MLG door is almost all the way open, the MLG door-operated sequence valve moves to OPEN. Pressure then goes to the MLG uplock actuator to unlock the MLG uplock. The MLG extends by its own weight and by airloads. It is not pressurized during extension. When the main landing gear is 20° from fully extended, the MLG drag brace-operated sequence valve moves to DOWN. Hydraulic pressure then retracts the MLG truck positioner actuator which moves the

MLG truck to the TILT position. Hydraulic pressure also goes to the MLG drag brace and MLG side brace down lock actuators to lock the gear down.

Pressure from the MLG drag brace-operated sequence valve goes through the MLG up lock-operated sequence valve and the MLG door release/safety valve module. This pressure goes to the MLG door actuator and the MLG door lock actuator. The MLG main gear door closes. When the door is almost closed, a roller on the door starts to move the MLG door up lock mechanism to the locked position. Pressure in the MLG door lock actuator moves the uplock mechanism over-center to the locked position.

Alternate Landing-Gear Extension

The alternate extension system is used to extend the landing gear when there is no center hydraulic system pressure or if normal extension fails. The alternate extension system operates independently of the normal extension and retraction system.

The hot battery bus supplies power to control and operate the alternate extension system. A hydraulic pressure pack operated by an electric motor pressurizes fluid from the center hydraulic system. This fluid then goes to door release/safety valve modules for each landing gear and unlocks the landing gear doors. Fluid then goes to alternate up lock release actuators for each landing gear and unlocks the up locks for the landing gear. The landing-gear doors open and the landing-gear extend by airloads and their own weight. The landing-gear tires may contact the landing-gear doors during an alternate extension. The alternate extend pressure switch stops operation of the power pack

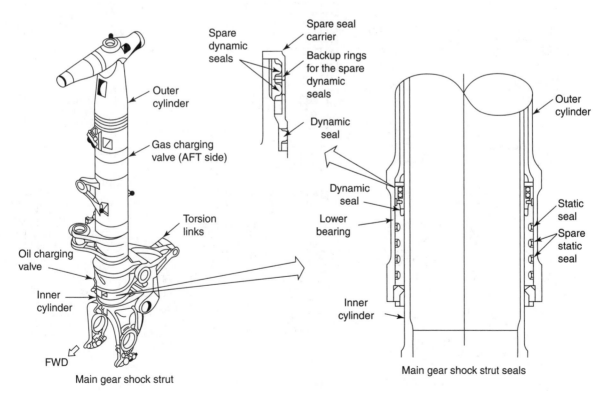

FIGURE 14–42 MLG shock strut.

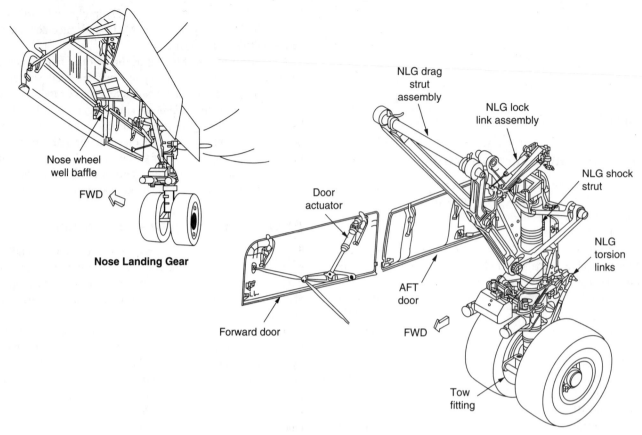

FIGURE 14-43 NLG components.

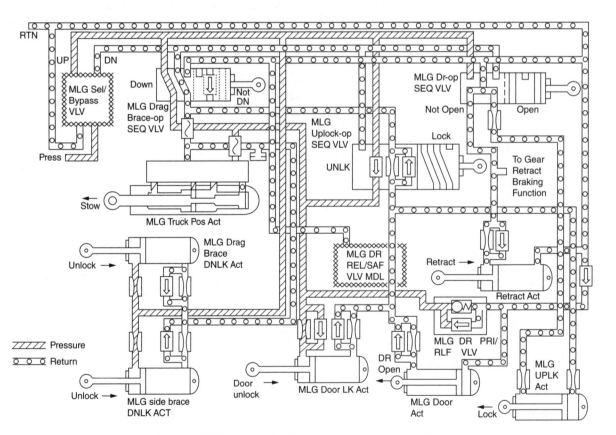

FIGURE 14-44 Main landing-gear retraction schematic.

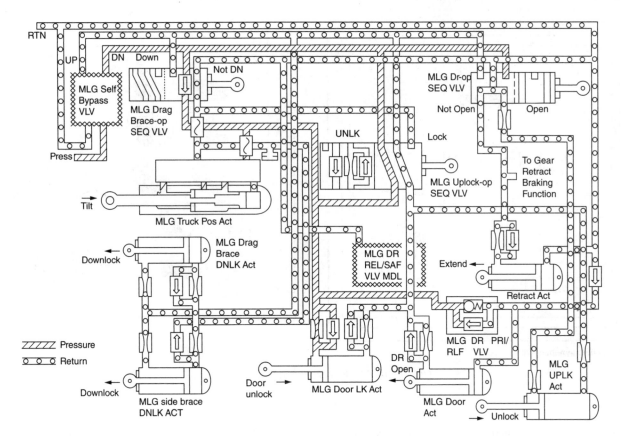

FIGURE 14-45 MLG extension schematic.

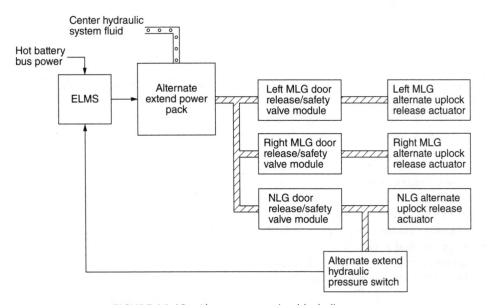

FIGURE 14-46 Alternate extension block diagram.

after the extend cycle is complete. Figure 14-46 shows a block diagram of the alternate extension system.

Landing-Gear-Position Indication and Warning System

The landing-gear-position indication and warning system shows landing-gear position on the flight deck displays. The landing-gear-position indication and warning system uses proximity sensors that send information to a proximity sensor electronics unit (PSEU). The PSEUs send position data to the AIMS and EICAS system. Figure 14-47 shows landing-gear positions on the EICAS display.

Helicopter Landing Gear

Since most helicopters do not have a substantial need to be streamlined for speed enhancement, the basic skid gear is the most common type of landing in use on rotorcraft. Figure 14-48 shows the basic components of the skid type gear.

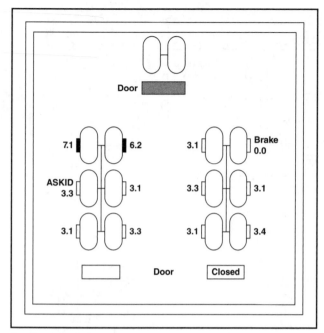

Landing gear synoptic format

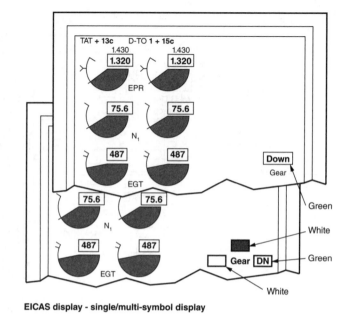

EICAS display - single/multi-symbol display

FIGURE 14-47   EICAS landing gear indications.

While skid gear construction appears very simple, there are unique airworthiness problems associated with it. There is the possibility the gear may impart a low-frequency vibration into the airframe, causing an unacceptable ride for the passengers and detrimental conditions to the airframe components.

Another situation associated with a helicopter and not with fixed-wing aircraft is a particular condition known as **ground resonance**. This condition is caused when the helicopter is on the ground and the vibrations (frequencies) combine in an additive nature to a destructive level that can cause the aircraft to bounce or roll over and be destroyed.

Some helicopters are equipped with tuning forks on the rear ends of the skid gear to dampen vibrations of the gear. Along with the shock absorbers attached to the skid gear, the tuning forks reduce the tendency toward ground resonance.

Due to the vertical landing nature of helicopters there are special inspections required on skid gear. The landing gear must be inspected for **deflection** if the aircraft has descended vertically too rapidly or has been subjected to a side load when landing.

The skid protectors (shoes), illustrated in Fig. 14-48, need to be inspected regularly during maintenance. Normal landings will wear the skid protectors. If the wear continues into the skid tubes, it may render the skid gear unairworthy and require replacement.

The helicopters that have skid gear may need to be moved on the ground to position the aircraft for maintenance, parking, or other reasons. To accomplish this task, **ground handling wheels** are used. The ground handling wheels are attached to the horizontal skid, as shown in Fig. 14-48. The aircraft then can be moved by the technicians on the ground.

Although most helicopters are furnished with skid gear, some newer helicopters are equipped with retractable landing gear. The components and operation of this gear are the same as for fixed wing aircraft.

INSPECTION AND MAINTENANCE OF LANDING GEAR

A thorough inspection of landing gear involves the careful examination of the entire structure of the gear, including the attachments to the fuselage or wings, struts, wheels, brakes, actuating mechanisms for retractable gear, gear hydraulic system and valves, gear doors, and all associated parts. It is recommended that the technician follows the instructions given in the manufacturer's manual for the aircraft being inspected. This is particularly important for the more complex types installed on aircraft with retractable gear.

Fixed-Gear Inspection

Fixed landing gear should be examined regularly for wear, deterioration, corrosion, alignment, and other factors that may cause failure or unsatisfactory operation. During a 100-h or annual inspection of fixed gear, the airplane should be jacked up so the gear does not bear the weight of the aircraft. The technician should then attempt to move the gear struts and wheels to test for play in the mounting. If any looseness is found, the cause should be determined and corrected.

When landing gear that employs rubber shock (bungee) cord for shock absorption is inspected, the shock cord should be inspected for age, fraying of the braided sheath,

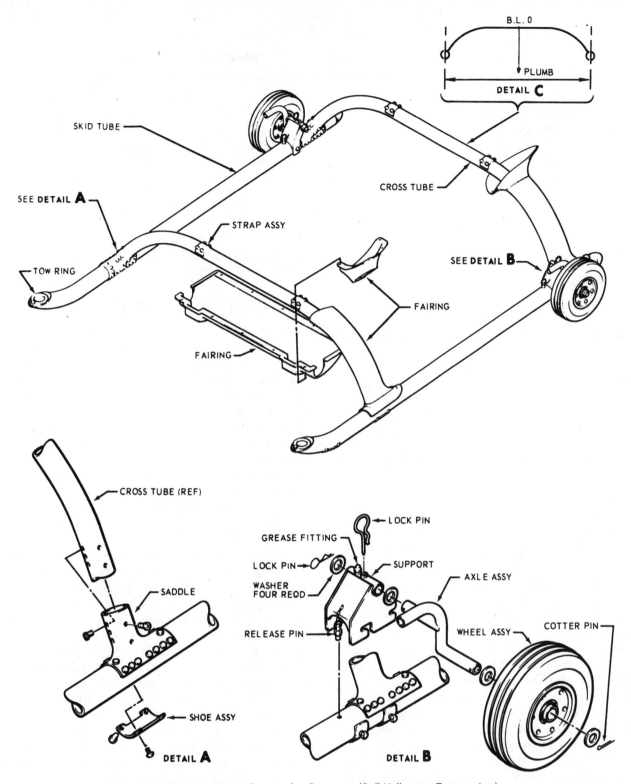

FIGURE 14-48 Helicopter landing gear. (*Bell Helicopter-Textron, Inc.*)

narrowing (necking) of the cord, and wear at points of contact with the structure. If the age of the shock cord is near 5 years or more, it is advisable to replace it with new cord, regardless of other factors. Cord that shows other defects should be replaced, regardless of age.

The cord is color-coded to indicate when it was manufactured, thus giving the technician the information needed to determine the life of the cord. According to MIL-C-5651A, the color code for the year of manufacture is repeated in cycles of 5 years. Figure 14-49 shows the colors of the code threads for each year and quarter of year.

The color coding is composed of threads interwoven in the cotton sheath that binds the strands of rubber cord together. Two spiral threads are used for the year coding and one thread is used for the quarter of the year. If a technician inspected a shock-cord installation in 1991 and found that the cord had two yellow threads and one blue thread spiraling around the sheath, this would indicate that

YEAR	COLOR	QUARTER	COLOR
1985 and 1990	Black	1st	Red
1986 and 1991	Green	2nd	Blue
1987 and 1992	Red	3rd	Green
1988 and 1993	Blue	4th	Yellow
1989 and 1994	Yellow		
1990 and 1995	Black		

FIGURE 14-49 Bungee cord color codes.

the cord was manufactured in 1989 during April, May, or June.

Shock struts of the spring-oleo type should be examined for leakage, smoothness of operation, looseness between the moving parts, and play at the attaching points. The extension of the struts should be checked to make sure that the springs are not worn or broken. The piston section of the strut must be free of nicks, cuts, and corrosion.

Air-oil struts should undergo an inspection similar to that recommended for spring-oleo struts. In addition, the extension of the strut should be checked to see that it conforms to the distance specified by the manufacturer. If an air-oil strut "bottoms"—that is, it is collapsed—the gas charge has been lost from the air chamber. This is probably due to a loose or defective air valve or defective O-ring seals.

CAUTION: **Before an air-oil strut is removed or disassembled, the air valve should be opened to make sure that all air pressure is removed. Severe injury and/or damage can occur as the result of disassembling a strut when even a small amount of air pressure is still in the air chamber.**

The method for checking the fluid level of an air-oil strut is given in the manufacturer's maintenance manual. Usually the fluid level is checked with the strut collapsed and all air removed from the strut. With the strut collapsed, the fluid level should be even with the filler opening unless other indications are given by the manufacturer. In all cases, the technician must use the correct fluid when refilling or replenishing the fluid in a strut. Most airplanes with oleo struts use MIL-05606 fluid.

The entire structure of the landing gear should be closely examined for cracks, nicks, cuts, corrosion damage, or any other condition that can cause stress concentrations and eventual failure. The exposed lower end of the air-oleo piston is especially susceptible to damage and corrosion, which can lead to seal damage, because the strut is compressed and the piston moves past the strut lower seal, causing the seal to leak fluid and air. Small nicks or cuts can be filed and burnished to a smooth contour, eliminating the point of stress concentration. If a crack is found in a landing-gear member, the part should be replaced.

All bolts and fittings should be checked for security and condition. Bolts in the torque links and shimmy damper tend to wear and become loose due to the operational loads placed on them. The nose-wheel shimmy damper should be checked for proper operation and any evidence of leaking.

All required lubrication should be performed in accordance with the aircraft service manual.

Inspection of Retractable Landing Gear

Inspection of retractable landing gear should include all applicable items mentioned in the foregoing discussion of inspection for fixed gear. In addition, the actuating mechanisms must be inspected for wear looseness in any joint, trunnion, or bearing, leakage of fluid from any hydraulic line or unit, and smoothness of operation. The operational check is performed by jacking the airplane according to manufacturer's instructions and then operating the gear retracting and extending system. Particular attention must be given to the location of the approved **jacking points**. The jacking points (pads) are placed in locations where the strength of the structure is adequate to withstand the concentrated stress applied by the jack. Additional information on jacking aircraft can be found in the text *Aircraft Basic Science.*

During the operational test, the smoothness of operation, effectiveness of up-and-down locks, operation of the warning horn, operation of indicating systems, clearance of tires in wheel wells, and operation of landing-gear doors should be checked. Improper adjustment of sequence valves may cause doors to rub against gear structures or wheels. The manufacturer's checklist should be followed to ensure that critical items are checked. While the airplane is still jacked up, the gear can be tested for looseness of mounting points, play in torque links, condition of the inner strut cylinder, play in wheel bearings, and play in actuating linkages.

The proper operation of the antiretraction system should be checked in accordance with manufacturer's instructions. Where safety switches are actuated by the torque links, the actual time of switch closing or opening can be checked by removing all air from the strut and then collapsing the strut. In every case, the adjustment should be such that the gear control cannot be placed in the UP position or that the system cannot operate until the shock strut is very near the fully extended position.

Alignment of Main-Gear Wheels

The alignment of main-gear wheels should be checked periodically to ensure proper handling characteristics during landing, taxiing, and takeoff and to reduce tire wear. Wheel alignment is generally checked for camber and toe.

Camber is the amount wheels are tilted from the vertical. If the top of the wheel tilts outward, the camber is positive, and if it tilts inward, the camber is negative. Toe is the amount wheels are angled from the horizontal axis. The wheels of an aircraft are toed in if lines drawn through the center of the wheels, perpendicular to the axles, cross ahead of the wheels. As the airplane moves forward, toe-in causes the wheels to try to move closer together. The opposite is true for toe-out.

Wheel alignment of oleo-equipped landing gear is adjusted by means of shim washers installed between the torque links at the joint between the upper and lower links. For spring steel landing gear, alignment is adjusted by the use of shim plates between the gear leg and the axle mount.

The following is a typical alignment-check procedure for a light airplane:

Place a straightedge of sufficient length to reach across the front of both main landing-gear wheels. Butt the straightedge against the tire at the hub level of the wheels. Jack the airplane up just high enough to obtain a dimension of 6.5 in [16.51 cm] between the centerline of the strut piston and the centerline of the center-pivot bolt of the gear torque links. A support is utilized to hold a straightedge in the desired location, as shown in Fig. 14-50.

Place a square against the straightedge land check whether its outstanding leg bears on the front and rear sides of the brake disk. If it touches both the forward and rear flange, the landing gear is correctly aligned with 0° toe-in. The toe-in for this particular gear is $0° \pm \frac{1}{2}°$.

Wheel toe-in and camber check
Measure dimension "X" and "Y". Toe-in is difference between X and Y. (i.e. Y − X)
Measure camber by reading protractor level held vertically against outboard flanges of wheel.

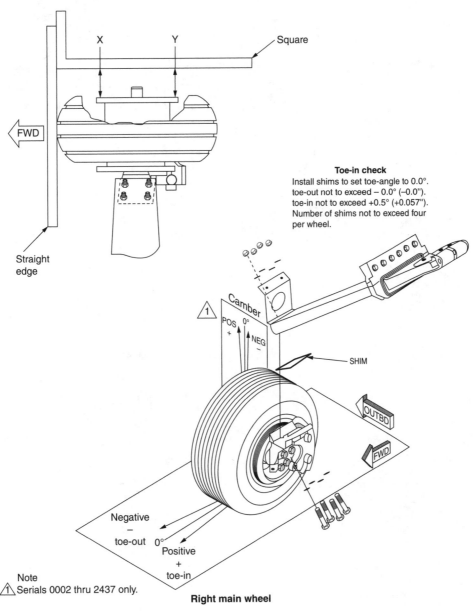

Toe-in check
Install shims to set toe-angle to 0.0°.
toe-out not to exceed − 0.0° (−0.0").
toe-in not to exceed +0.5° (+0.057").
Number of shims not to exceed four per wheel.

Note
⚠ Serials 0002 thru 2437 only.

Right main wheel

FIGURE 14-50 Wheel toe-in and chamber check.

If the square contacts the rear side of the disk, leaving a gap between it and the front flange, the wheel is toed out. If a gap appears at the rear flange, the wheel is toed in.

To rectify the toe-in and toe-out conditions, remove the bolt connecting the upper and lower torque links and remove or add spacer washers to move the wheel in the desired direction. Should a condition exist where all spacer washers have been removed and moving the wheel in or out is still necessary, then the torque link assembly must be turned over. This will put the link connection on the opposite side, allowing the use of spacers to go in the same direction.

Some aircraft landing gear is checked for alignment with the full weight of the aircraft on the gear. The gear is placed on greased plates, as shown in Fig. 14-51, to eliminate any ground friction affecting the alignment check. A straight-edge and square are used as described previously. In this instance the gear is checked for both toe-in and toe-out as well as camber, as shown in Fig. 14-50. Some aircraft use tapered shim plates between the landing gear leg and the axle to adjust toe-in, toe-out, and camber.

Inspection and Repair of Floats and Skis

Inspection of floats and skis involves examination for damage due to corrosion, collision with other objects, hard landings, and other conditions that may lead to failure. Tubular structures

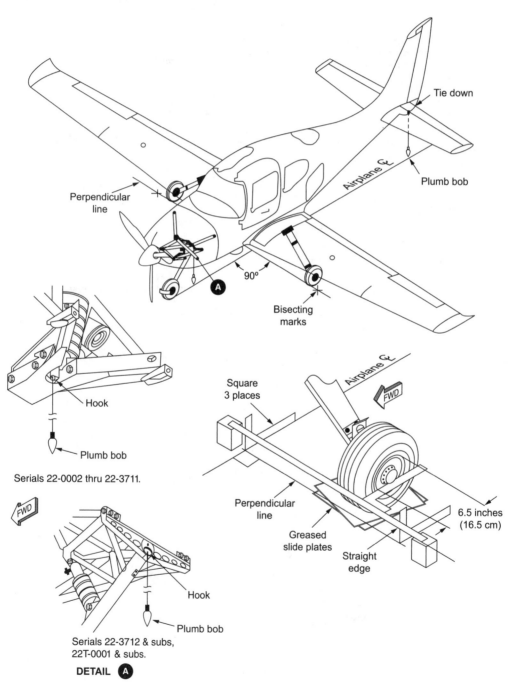

FIGURE 14-51 Landing gear alignment check.

for such gear may be repaired as described in the section "Aircraft Tubing Repair."

Floats should be carefully inspected for corrosion damage at periodic intervals, especially if the airplane is flown from saltwater. If small blisters are noticed on the paint, either inside or outside the float, the paint should be removed and the area examined. If corrosion is found to exist, the area should be cleaned thoroughly, and a coat of corrosion-inhibiting material should be applied. If the corrosion penetrates the metal to an appreciable depth, it is advisable that a patch be applied in accordance with approved practice. Special attention should be given to brace wire fittings and water rudder–control systems.

If the hull or floats have retractable landing gear, a retraction check should be performed along with the other recommendations mentioned for retractable landing-gear systems. Sheet-metal floats should be repaired using approved practices described elsewhere in this text; however, the seams between sections of sheet metal should be waterproofed with suitable fabric and sealing compound. A float that has undergone repairs should be tested by filling it with water and allowing it to stand for at least 24 h to see if any leaks develop.

Skis should be inspected for general condition of the skis, cables, bungees, and fuselage attachments. If retractable skis are used, checks in accordance with the general practices for retractable gear should be followed. Manufacturers furnish approved repair instructions for repair of skis.

TIRES AND WHEELS

The tires and wheels of an airplane are subject to severe stresses during landing and in taxiing over rough ground. Failure of a tire or wheel can lead to extremely serious accidents, often resulting in the complete destruction of the aircraft and injury or death to the crew and passengers. For these reasons, the technician must make certain the wheels and tires of the aircraft are in good condition for all environments under which the aircraft may operate.

Aircraft Tire Operation Characteristics

Heavy loads combined with high speeds and a high percent of tire deflection on landings make the operating conditions of aircraft tires extremely severe.

Figure 14-52 shows various types of tires and their speed versus load operation ranges. The operation range for aircraft tires covers the upper right-hand corner, meaning that maintenance practices and operation techniques that work well for passenger tires are not acceptable for aircraft tires.

Heavy loads and high speeds cause the heat generation in aircraft tires to exceed that of all other tires and can have a very detrimental effect. Rubber, the major material used in a tire, is a good insulator and, therefore, dissipates heat slowly. The internal heat generation is significantly affected by taxi speed, inflation pressure, and taxi distances. High taxi speeds and improper inflation pressures will reduce the tire life substantially.

An aircraft tire is designed so that the internal tensile forces on each layer of fabric are uniform in an unloaded condition. When a tire is deflected (due to weight or impact), the tensile forces on the outer plies will be higher than those on the inner plies. Due to this force difference from the outer to inner plies, shear forces are developed between the various layers of fabric. Underinflating or overloading a tire will increase these shear forces and cause separations in the shoulders and lower sidewalls of the tires, rapidly decreasing the life of the tire.

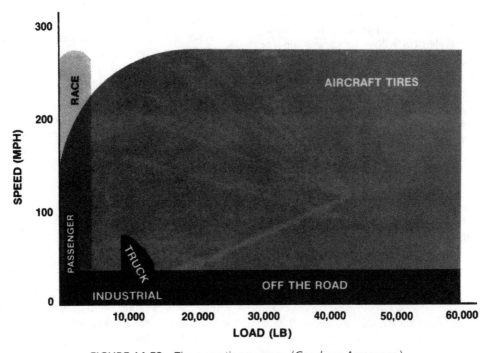

FIGURE 14-52   Tire operating rangers. (*Goodyear Aerospace.*)

Aircraft Tire Nomenclature

Aircraft tires are manufactured in a variety of sizes and strengths, and the correct types are specified by the manufacturers according to the size and landing speed of the aircraft involved. Construction is similar to that of automobile and truck tires. The number of fabric plies in the tires varies considerably. **Ply rating** is an index of tire strength. The term is used to identify a given tire with its maximum recommended load. The higher the rating, the greater the load a tire will carry. The aircraft tire's ply rating does not necessarily represent the actual number of nylon fabric plies in the tire. Rather, it varies according to the tire's strength of index for a specific type of airplane service. The majority of tires in use today on light aircraft are of the tube type whereas most transport aircraft tires are of the tubeless type. Tube-type tires make use of inner tubes to hold the air charge. **Balance marks** are placed on tires in the form of a red dot on the side of the tire at the lightest point.

In the past, aircraft tires were classified by type numbers according to performance, as shown in the following table:

Type	Design and Rating
I	Smooth contour
II	High pressure
III	Low pressure
IV	Extra low pressure
V	Not applicable
VI	Low profile
VII	Extra high pressure, low speed
	Extra high pressure, high speed
VIII	Extra high pressure, low profile, low speed
	Extra high pressure, low profile, high speed

Tires classified as Types I, II, IV, and VI are being phased out because these classifications are inactive for new designs. Tires classified as Types III, VII, and VIII are manufactured under the provisions of FAR 37.167 and are approved under Technical Standard Order (TSO) No. C62b. Such tires are required to be permanently marked, as shown in Fig. 14-53, with the brand name or name of the manufacturer responsible for compliance and the country of manufacture if manufactured outside the United States; the size, ply rating, and serial number; the qualification test speed and skid depth when the test speed is greater than 160 mph [257.6 km/h] and the word "reinforced" if applicable; and the applicable TSO number. Type III tires and those specified as low-speed tires are approved for ground speeds of less than 160 mph [257.6 km/h].

The system of classifying tires by type is being phased out. A new system is being used for all new sizes, in which tires are designated by a three-part size designation and a prefix:

(Prefix)(nominal outside dia.) × (nominal section width) − (bead dia.)

"Metric" tires use millimeters for the OD and width. Radial tires have an "R" in place of the "−" between the width and the bead. The prefixes are determined as shown below:

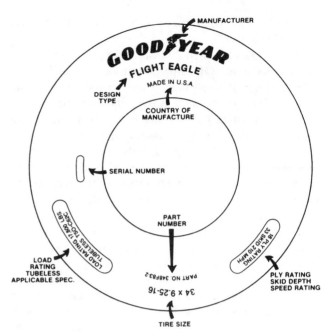

FIGURE 14-53 Tire markings. (*Goodyear Aerospace.*)

Width Ratio	Bead Ledge	Prefix
50% to 60%	15°	C
60% to 70%	15°	B
60% to 70%	5°	H
70% and higher	5°	No prefix

Width ratio = section width/rim width, expressed as a %

Bead ledge = angle at the base of the bead

In all cases of tire replacement, the technician must determine that the correct size and type of tire specified for the aircraft is installed.

Aircraft tires may be **recapped** (have new tread added) by FAA-approved repair stations. When a tire is recapped, the tire structure is checked for soundness and then a new layer of tread is attached to the original carcass. The use of recapped tires is not an acceptable practice on all aircraft. This is not as a result of any decrease in tire strength, but rather the fact that recapped tires may be slightly larger than new tires. If recapped tires are used on retractable-gear aircraft, the slightly oversize tire may jam in the wheel well. This could result in the aircraft having to make a gear-up landing. The technician must check the aircraft manufacturer's manual when determining if recapped tires may be used on a specific aircraft.

Tire Structure

A **tire** is a layered structure designed to withstand the shock and sudden acceleration of landing, the heat of brake applications, and the abrasion of landing and turning. As shown in Fig. 14-54 there are many different components of a tire.

The **steel wire beads** serve as the foundation for the attachment of the fabric plies and provide firm mounting

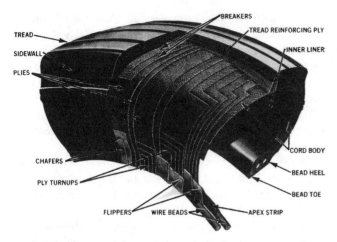

FIGURE 14-54   The parts of a tire. (*Goodyear Aerospace.*)

the tire and wheel. An inner layer of rubber is used to form an air seal on tubeless tires and to prevent tube chafing on tube-type tires. **Breakers** and the tread-reinforcing plies are used to increase the structural strength on some tires. All the parts of the tire except the tread make up the tire carcass. On tubeless tires the **inner liner** is a layer of less permeable rubber that acts as a built-in tube and prevents air from seeping through the casing plies. For tube-type tires, a thinner rubber liner is used to prevent the tube from chafing against the inside ply. The **beads**, which are made of steel wires embedded in rubber, anchor the plies and provide firm mounting surfaces on the wheel. The **bead heel** is the outer bead edge, which fits against the wheel flange. The **bead toe** is the inner bead edge closest to the tire center line.

The **tread** is the rubber compound on the surface of the tire, which is designed for runway contact. The pattern, or grooves, in the tread are designed for maximum traction in widely varying runway conditions. Tread design is also based on operational requirements of an aircraft. The design may be smooth, ribbed, or a special pattern to meet specific operational conditions.

Today, most tread designs have a continuous groove pattern with three to six ribs, based on tire size and type of service. This circumferential rib design, which is illustrated in Fig. 14-55,

surfaces on the wheel. The **plies** are diagonal layers of rubber-coated nylon cord fabric, laid in alternating directions, that provide the strength of the tire. The ply material wraps around the wire beads. The apex strips and flippers prevent the bead from abrading the plies. **Chafers** are used to protect the tire during mounting and demounting, to insulate the carcass from brake heat, and to provide a good seal between

FIGURE 14-54   Rib-tread design. (*Goodyear Aerospace.*)

FIGURE 14-56 Deflector tire. (*Goodyear Aerospace.*)

provides a good combination of long tread wear, traction, and directional stability, especially on hard-surface runways. Placement of the grooves is designed to allow water and slush to pass through the tire "footprint," minimizing skidding or hydroplaning on wet surfaces.

The tread may also be reinforced with fabric plies in the tread rubber or undertread. In addition, a chine or deflector, as shown in Fig. 14-56, may be molded as an integral part of the sidewall. The chine deflects runway water away from rear-mounted engines. For a single nose-wheel installation, a tire with a deflector on each side may be used to direct water away from the fuselage.

Tire Storage

Tires should be stored in areas that are cool, dry, and dark. Moisture causes the rubber and ply material to deteriorate. Exposure to ozone, such as from electric motors, battery chargers, and other electrical equipment, can cause the rubber to age rapidly. Petroleum products, such as oil, grease, hydraulic fluid, gasoline, and solvents, attack the rubber.

To prevent distortion, tires should be stored vertically in racks. If they must be stacked on their sides, the stacks should be no more than four high.

Tubes

Tire tubes are made of rubber sections that are vulcanized together. Ribs on the tube are designed to provide traction between the tire liner and the tube so that the tire will not rotate in relation to the tube, causing abrasion damage. An **air valve** is vulcanized to the tube to allow for inflation and deflation.

Aircraft Wheel Construction

Airplane wheels are constructed by a number of methods. Among these are (1) forming of heavy-sheet aluminum alloy,

(2) casting of aluminum or magnesium alloy, and (3) forging of aluminum or magnesium alloy. One type of wheel assembly is shown in the exploded drawing of Fig. 14-57. This is a two-piece, cast-alloy wheel designed for a light airplane. The two wheel halves are joined together with the brake disk by means of the wheel through-bolts. Bearing cups are pressed into recesses in each wheel half. Grease seals and grease-seal retainers are installed outside each bearing to prevent the escape of grease.

Another type of wheel for a light aircraft is shown in Fig. 14-58. This wheel is composed of a hub with a flange on each side. The bearings and grease-seal arrangement are similar to those in Fig. 14-57.

A main-wheel assembly for a corporate turbine aircraft is shown in the drawing of Fig. 14-59. This assembly is designed for use with tubeless tires. Note that the wheel is composed of a flange that bolts to the hub. The hub and flange are held together with 15 high-strength bolts and self-locking nuts. This wheel is designed for a tubeless tire, so a packing is installed between the two wheel halves to provide an air seal. The wheel hub contains an inflation valve for inflating and deflating the tire.

The hub also has five keys that engage and drive the tangs of the rotating brake disks. The tires are protected against blowouts resulting from excessive pressure created by heat by three fusible plugs equally spaced around the wheel. The plugs are installed in an accessible location to allow replacement without removing the tire. Wheel design features may include a deflector over the fusible plug to prevent it from causing damage to the aircraft when it is forced out of the wheel assembly.

The wheel rotates on two tapered roller bearings, the cups of which are shrink-fitted into the hub of each wheel half. The bearing cones are retained in the wheel halves and protected from dirt and moisture by bearing seals.

As discussed in previous wheel descriptions, aircraft wheels are commonly made in two pieces, inboard and outboard, to facilitate the changing of tires. Typical of such a wheel for a large turbine aircraft is that shown in the drawing of Fig. 14-60. This is a cutaway drawing of the main landing-gear wheel and is described as a Goodyear 40 × 14, Type VII, forged-aluminum, split-type wheel. The wheel halves are joined by 12 high-tensile steel bolts with self-locking nuts. An air seal is provided by an O-ring installed in a groove in the outer wheel half to prevent loss of air from the tubeless tire. Standard, tapered roller bearings are installed in each wheel half. The inner bearings are protected against loss of lubricant and the entrance of dirt by moisture-resistant seals. On the inside half of the wheel are installed the steel keys that hold the brake rotor disks. These are secured to the wheel forging by means of screws. The wheel is equipped with three fusible plugs equally spaced around the wheel.

Nose wheels for aircraft are constructed and serviced in a manner similar to main wheels. Nose wheels are often smaller in diameter and width than main wheels, and only a few aircraft use brakes on the nose wheel. If a nose wheel is not designed to accept a brake assembly, a fusible plug is not used. A pressure-release device may still be used.

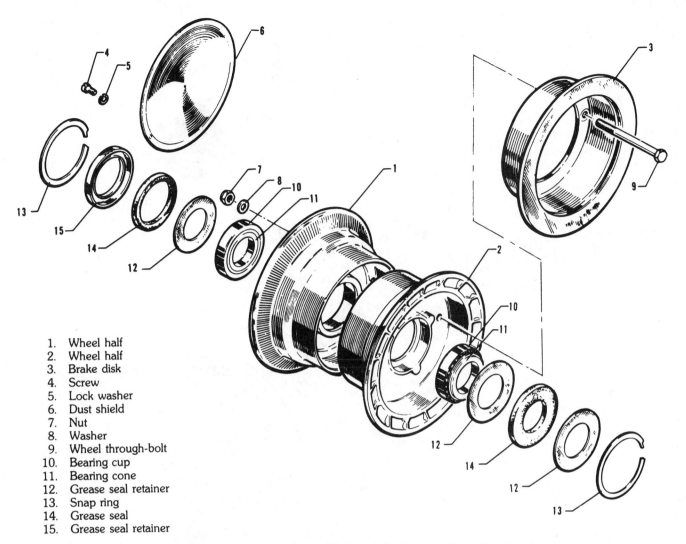

1. Wheel half
2. Wheel half
3. Brake disk
4. Screw
5. Lock washer
6. Dust shield
7. Nut
8. Washer
9. Wheel through-bolt
10. Bearing cup
11. Bearing cone
12. Grease seal retainer
13. Snap ring
14. Grease seal
15. Grease seal retainer

FIGURE 14-57   Main-wheel assembly for a light airplane. (*Piper Aircraft Co.*)

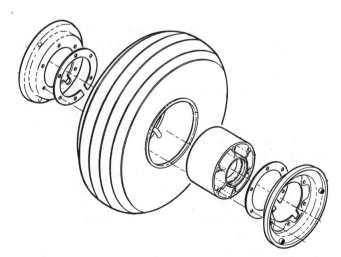

FIGURE 14-58   A hub-and-flange wheel assembly. (*Cessna Aircraft Co.*)

Demounting and Mounting Tires

Regardless of the type or size of tire involved, the same basic procedures and the exercise of certain precautions will assist the tire to function properly. The general procedures are not difficult, and it is necessary only that the technician use good judgment and follow the instructions provided by both the manufacturer of the airplane and the manufacturer of the tire.

Before a wheel is removed from the airplane, the tire should be deflated. This is an important safety procedure. Deflation is accomplished either by removing the core from the air valve or as specified in the service instructions. The wheel is then removed according to instructions, the technician being careful to see that the bearings are protected from damage and dirt.

With the wheel lying flat, the tire beads are broken away from the rims by applying pressure in even increments around the entire sidewall as close to the tire beads as possible. Sharp tools should not be used to pry the tire loose because the tire air seal will likely be damaged and the wheel can be nicked or scratched. It is recommended that

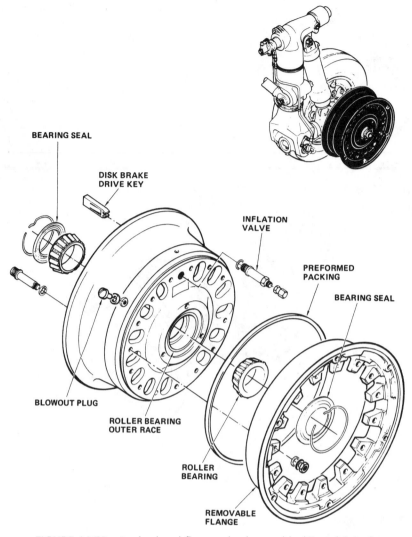

FIGURE 14-59 A wheel-and-flange wheel assembly. (*Canadair Inc.*)

Labels in figure:
BEARING SEAL
DISK BRAKE DRIVE KEY
INFLATION VALVE
PREFORMED PACKING
BEARING SEAL
BLOWOUT PLUG
ROLLER BEARING OUTER RACE
ROLLER BEARING
REMOVABLE FLANGE

an approved demounting machine be used to free tire beads from wheel rims.

The nuts and bolts holding the wheel halves together should then be removed and the wheel halves should be separated from the tire. The O-ring air seal or packing found in tubeless-tire installations should be carefully removed and placed in a protected area where it will not be damaged or contaminated with dirt.

The wheel, tire, and tube (if used) should be inspected in accordance with the manufacturer's guidelines. After the components are determined to be airworthy, the tire can be mounted on the wheel.

If a tube-type wheel and tire are being assembled, the following general procedure is used. Before placing the tube in the tire, the tube should be lightly dusted with tire talc. The talc prevents the tube from sticking to the inside of the tire. To obtain the proper balance of the tire and tube, the inner tube should be placed in the tire so that the yellow strip (the heavy spot) on the tube is located adjacent to the red dot (the light spot) on the tire. Mating of the tire and tube in this manner should bring the unit into the balance tolerance required. Before the tube is placed in the tire, the tube should

be inflated sufficiently to give it shape but not so much that the rubber is stretched.

If a tubeless wheel and tire are being assembled, the following general procedure is used. The tire should be examined to make sure the word "tubeless" is molded on the sidewall. The beads of the tire must be inspected to see that there are no nicks, cuts, or other conditions that can cause air leaks between the tire and the wheel rim. The air-seal mating surface of the wheel should be cleaned with denatured alcohol. Small particles of sand or dirt in this area can cause leakage of air. The air valve in the rim should be examined to make sure that the seal between the valve stem and the wheel rim is intact.

The wheel air seal (O-ring) should be lubricated with a light coat of grease. This permits the seal to slide easily into the seal seat when the two halves of the wheel are joined. The seal should be spread evenly around the wheel and not twisted. If the seal is reused, it should be installed as nearly as possible in the original position.

With only a few exceptions, the remainder of the installation is the same for both types of assemblies. The wheel hub and flanges are now assembled on the tire. On the

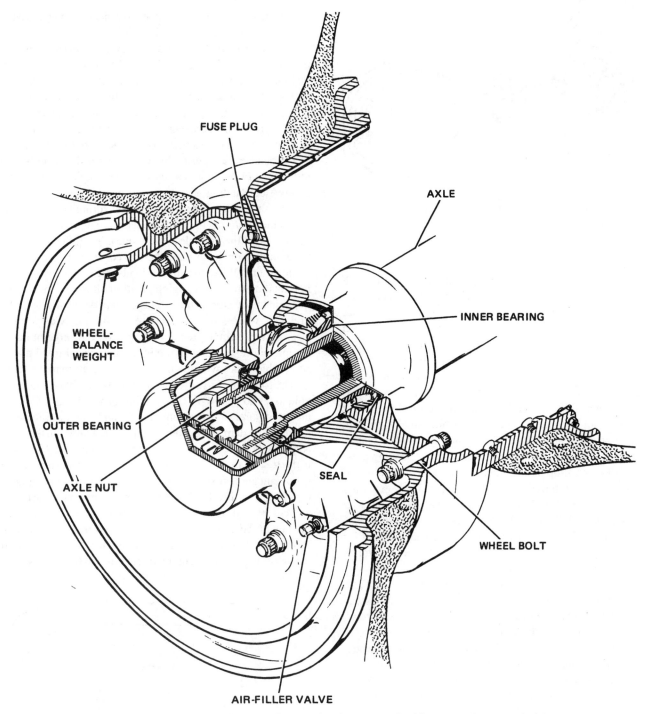

FUSE PLUG

AXLE

INNER BEARING

WHEEL-BALANCE WEIGHT

OUTER BEARING

AXLE NUT

SEAL

WHEEL BOLT

AIR-FILLER VALVE

FIGURE 14-60   Cutaway drawing of a main landing-gear wheel for a Douglas DC-9 airplane.

tube-type tire, the side of the hub with the air valve is placed in the tire first, the valve is pulled into position on the wheel half, and the second half of the wheel is inserted into the tire. For a tubeless tire, the wheel halves are placed on the tire, taking care to align the tire balance mark (red dot) with the air valve in the wheel. In both cases the inside of the center of the wheel (where the axle goes through) should be checked to be sure that the tube or the packing seal is not pinched between the wheel halves.

Bolts are then installed, taking care that the bolts and nuts are on the appropriate side. In some installations the use of an antiseize compound may be required. The technician should check the appropriate aircraft maintenance manual. If countersunk washers are used under the bolt heads, they are usually positioned with the countersink toward the bolt head. The bolts are then brought down flush with the wheel using an alternating tightening sequence. Once all the bolts are flush, an alternating torque sequence is used to tighten the bolts. The bolts are first torqued to about 25 percent of final torque, and then torque is brought up to the final installation torque in increments of 20 to 25 percent, using an alternating torque sequence.

Tires and Wheels   **489**

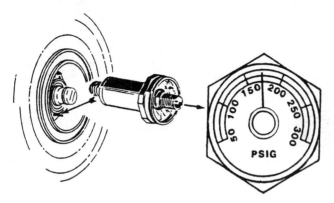

FIGURE 14-61 Integral tire inflation gauge. (*Boeing Commercial Aircraft Co.*)

The tire should be placed in an inflation cage before being inflated. This prevents damage and injury if the wheel or tire is defective and ruptures during inflation. With the valve core removed, a tube-type tire is inflated slightly and then allowed to deflate. This allows any wrinkles in the tube to straighten out. The valve core is then installed and the tire is inflated to operational pressure. A tubeless tire is first inflated to partial pressure to seat the tire bead to the wheel. It may be necessary to use a strap around the surface of the tread to prevent the tire from expanding radially. This aids smaller tires in seating properly against the wheel. Once the bead is seated, the strap is removed and the tire is inflated to operating pressure. Although shop air is commonly used on small aircraft, nitrogen is the preferred gas for tire inflation and is generally used for large aircraft. Many large aircraft are equipped with an integral tire inflation valve and pressure gauge such as is shown in Fig. 14-61. This built-in pressure gauge allows for monitoring of the pressure while servicing the tire or a quick check of the tire pressure at any time.

Tires should be allowed to set, fully inflated, for 12 to 24 h before being installed on an aircraft. This allows any small leaks to become evident and allows the tire to "grow." Because of this growth the larger size causes the pressure to drop slightly. The tire is brought back up to pressure, removed from the cage, and installed on the aircraft.

If the tires are to be installed on an aircraft with retractable gear, a retraction check should be performed if there is any doubt as to proper clearance between the tire and the wheel well structure when the wheel is retracted.

During the installation of tires on automobile wheels, a soap solution is sometimes used to make the bead of the tire slip more easily over the rim. This should not be done with airplane tires because it may cause slippage of the tire on the rim, particularly upon landing. If the tire slips on the rim, the valve stem will be pulled out of the tube and the tire will deflate. If tubeless tires slip on the rim, the air seal will be broken and the tire will deflate.

Tire and Wheel Inspection

The inspection of tires and wheels should be performed periodically to ensure that they are not damaged or worn excessively.

The frequency and extent of inspections depends on the type of aircraft and flying activity involved.

The surface condition of a tire can be inspected with the tire on the aircraft. The tread should be checked for abnormal wear. If the tread is worn in the center of the tire but not on the edges, this indicates that the tire is overinflated and the operational air pressure should be reduced. On the other hand, a tire worn on the edges, but not in the center, indicates underinflation. These situations are shown in Fig. 14-62.

Certain aircraft normally wear the tread unevenly due to the type of landing gear used. For example, aircraft with spring steel landing gear have the main wheels suspended out from the side of the fuselage on a spring steel gear leg. When the aircraft is in flight, the wheel hangs down with the outer edge of the tire below the inside edge of the tire. When the aircraft is on the ground, the two edges are perpendicular to the ground. When the aircraft lands, the tires slide outward in transition from the hanging position, thus wearing off the tread on the outside of the tire faster than on the inside of the tire. This wear is normal. The tread is normally allowed to wear down until the groove pattern is on the verge of being completely removed. The minimum amount of tread that is acceptable varies with the tire and the aircraft manufacturer.

The following are different types of tire damage and some guidelines to follow in determining airworthiness. Consult the tire or aircraft manufacturer's manuals for specific information. Any damage that does not penetrate through the tread and into the carcass plies is generally acceptable. However, if the tread damage involves peeling tread, thrown tread, or deep damage across the tread that could lead to peeling or throwing, the tire should be removed from service. Blisters and bulges indicate a separation within the tread or between ply layers. These tires should be removed from service. Flat spots indicate excessive use of brakes for deceleration and when turning. If sufficient tread remains and the tire does not

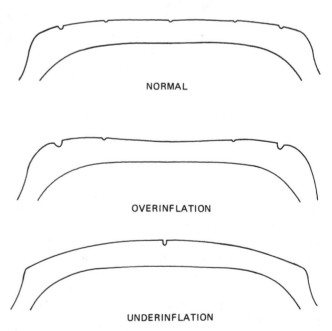

FIGURE 14-62 Examples of tread wear indicating overinflation and underinflation.

cause an out-of-balance condition, this is generally acceptable. If plies are showing, insufficient tread remains, or the tire is out of balance, then the tire should be replaced.

Damage to the sidewalls that is acceptable includes cuts, small cracks, and weather checking (a random pattern of shallow cracks), provided that the cord is not exposed or damaged.

Damage that is visible when the tire is removed from the wheel includes bead damage and inner-liner damage. Damage in these areas is more critical for tubeless tires than for tube-type tires because of the air-sealing requirements. Check with the manufacturer for specific information as to airworthiness standards.

The recapping of tires is a standard industry practice. To be recapped, the tire carcass must be in airworthy condition. Some ply damage may be acceptable, depending on the type of tire and the number of plies involved. Tires can be recapped as often as five times, but most are not recapped more than three times because of carcass deterioration.

Operators such as airlines, through experience with the operating characteristics of tires, establish criteria for the recapping of tires. For example, an airline may find that a particular tire can make approximately 200 landings for each time it is recapped and that it can be recapped at least 10 times before the carcass becomes unairworthy.

Tires that are recapped must be permanently marked to provide required information. Whenever it is necessary to replace the area containing the original markings, such markings must be replaced. In addition, each retreaded tire must display the letter R followed by a number such as 1, 2, or 3 to signify the sequential number of recaps that have been applied. A marking must be applied to display the speed-category increase if the tire is qualified for increased speed in accordance with the requirements of FAR 37.167 (TSO-C62). The month and year of recapping must be shown, together with the name of the person or agency that applied the recap. Each repaired or recapped tire should not exceed the static unbalance limits as set forth in TSO-C62.

Tubes should be inspected for chafing, thinning, and elasticity. The valve stem should be checked for flexibility where it attaches to the tube. The tube can be checked for leaks by inflating slightly and then placing it in a water tank.

Inner tubes can usually be repaired in the same manner as those used for automobiles. It is recommended that the repair patches be vulcanized and that an excessive number of patches be avoided. With too many patches, the weight of the patching material can affect the balance of the wheel-and-tire assembly.

Airplane wheels receive very severe stresses at times; therefore, it is important that they be kept in airworthy condition. Unsatisfactory conditions for aircraft wheels include cracks, dents, abrasion, corrosion, distortion, looseness of assembly bolts, and worn or loose bearings. The degree of any such condition should be checked to determine whether the wheel should be repaired or replaced.

Cracks in wheels usually require replacement of the wheel. Small dents in formed sheet-metal wheels do not necessarily impair the strength of the wheel. However, if the wheel is dented or distorted to the extent that it wobbles, it should be repaired or replaced.

To thoroughly inspect a wheel, the wheel must first be disassembled and then cleaned of all paint, grease, dirt, and corrosion.

Inspection of large cast or forged wheel parts is accomplished with fluorescent penetrant, dye penetrant, eddy-current, and ultrasonic methods. All parts are examined for cracks, nicks, cuts, corrosion, damaged threads, and any other damage, including excessive wear. Damage that exceeds the manufacturer's limits requires replacement of the part involved. Cracked wheel forgings or castings must be replaced. Particular attention is given to tire bead seats to ensure that they are smooth and will not allow leakage of air.

In the inspection and repair of any aircraft wheel, the technician must determine whether the wheel is made from aluminum, magnesium, or some other metal, because the treatment varies in some respects with different metals. The manufacturer's maintenance manual sets forth procedures and processes that are correct for a particular model of wheel.

Split-type wheels must be examined to determine the amount of wear in the assembly bolt holes. If the bolts have loosened during operation, they can very quickly enlarge the holes in the wheels. Sometimes enlarged or elongated holes can be repaired with inserts; however, the repair must have the approval of the manufacturer or the FAA.

Wheel bearings are inspected by removing the wheels and examining the bearings and bearing cups (races). The bearings should be washed in a suitable petroleum solvent and then dried. The rollers should show no signs of wear, discoloration, or roughness and should rotate freely in their cages. The bearing cups should be smooth with no signs of ripple, cracks, scoring, galling, corrosion, discoloration, or any other damage. If the bearings are in good condition, they can be reinstalled in accordance with the manufacturer's instructions.

Bearing cups are pressed into the wheel assembly. If they are damaged, they must be pressed out and replaced. If the bearing cup is loose in the wheel, the bearing recess in the wheel is oversize and the wheel must be replaced.

During inspection of the wheel bearings, the felt grease retainers should be examined for effectiveness. If they are hardened or glazed, they should be reconditioned or replaced with new ones.

Wheel Installation

There are various procedures used for the installation of wheel assemblies on an aircraft. The following provides some general guidelines.

The axle should first be cleaned and inspected. Surface damage and any damage to the axle threads are standard inspection items, along with the condition and security of bolts holding the axle onto the landing-gear leg. The wheel bearings are cleaned and packed with approved grease. The wheel and tire are inspected and assembled. The wheel assembly and bearings are carefully slid onto the axle and backed against a rear collar or stop on the axle. The retaining nut is next installed and safetied. Many aircraft have specific torque requirements for the wheel-retaining nuts. These torque requirements may have two values specified. The nut is first tightened to the higher value to seat the bearing; then the retaining nut is backed off and tightened to the lower value specified. While tightening wheel retaining nuts, the wheel should be rotated.

Great care must be exercised to see that the wheel-retaining nuts are not overtightened. In the absence of specific instructions, the wheel-retaining nut is tightened until bearing drag is felt. The nut is then backed off about one serration (castellation) or one-sixth turn before bending up the tab on the tab-lock washer or installing the cotter pin.

The grease cover or wheel cover, if used, is then installed. During this installation any required brake, air-pressure sensors, and speed-sensor components are installed and connected as appropriate for the specific aircraft.

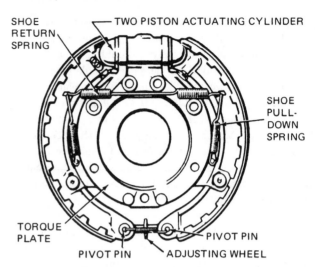

FIGURE 14-63   A dual-servo-type brake assembly.

DESIGN AND OPERATION OF BRAKE ASSEMBLIES

The various brake designs discussed in this section reflect the variety of braking capabilities required for different-size aircraft. Light aircraft can rely on a single disk brake or a simple shoe brake because the landing speeds are slow and the aircraft is light in weight. Large aircraft, such as transports, land at high speeds and weigh several tons. These aircraft require very powerful multisurfaced brakes in order for the brakes to be effective at slowing the aircraft.

Internal Expanding-Shoe Brakes

Shoe brakes will still be encountered on older aircraft and some current models of homebuilt aircraft. The types of internal expanding-shoe brakes are (1) the *one-way*, or *single-servo*, type and (2) the *two-way*, or *dual-servo*, type. **Servo action** in a brake of this type means that the rotation of the brake drum adds braking energy to the brake shoes and makes them operate more effectively and with less effort by the pilot.

In **single-servo** brakes, the servo action is effective for one direction of the wheel only, as contrasted with a **dual-servo**, or **reversible, type**, which operates and may be adjusted to give servo action in either direction. Both types are supplied with either single-shoe or two-shoe construction. Brake-shoe assemblies are attached to the landing-gear strut flange by means of bolts through the torque plate on the axle, which has as many as 12 equally spaced holes, with bolts in only one-half of the holes. The alternate holes are used to permit a variation of the position of the brake assembly to ensure that the brake cylinder is at the top or highest position on the assembly. The brake drum is attached to the wheel and rotates with it.

A dual-servo brake assembly is shown in the drawing of Fig. 14-63. As explained previously, dual-servo brakes are effective for either direction of wheel rotation; therefore they can be interchanged between the left and right wheels

of the airplane and are effective for both forward and backward motion of the airplane. These brakes may be operated hydraulically, mechanically, or pneumatically.

Expander-Tube Brakes

Another type of brake that may still be found on older aircraft is the **expander-tube-type brake**. Side and end views of an expander-tube-type brake are shown in Fig. 14-64. Figure 14-65 illustrates the principle of operation. Each expander-tube brake consists of four main parts: (1) **brake frame**, (2) **expander tube**, (3) **return springs**, and (4) the **brake blocks**. The single-type brake has one row of blocks around the circumference and is used on small aircraft. The duplex-type, expander-tube brake has two rows of brake blocks and is designed for larger aircraft. An inner fairing, or shield, fits between the **torque flange** on the axle and the brake frame to protect the frame against water.

The brake expander tube is a flat tube made of synthetic-rubber compound and fabric. It is stretched over the circular brake frame between the side flanges, and it has a nozzle that is connected with the hydraulic-fluid line by means of suitable fittings.

The brake blocks are made of a material similar to that used for molded brake linings. The blocks have notches at each corner to engage with lugs on the brake frame and to prevent movement with the brake drum as it rotates. There are grooves across the ends of each block, and the flat return springs are inserted in these grooves. The ends of the springs fit into slots in the side flanges of the brake frame, holding the blocks firmly against the expander tube and keeping them from dragging when the brake is released. The expander-tube brake is hydraulically operated and can be used with any conventional hydraulic brake system. When the brake pedal is pressed, the fluid is forced into the expander tube. The frame prevents any expansion either inward or to the sides. The pressure of the fluid in the tube forces the blocks radially outward against the brake drum. When the pressure is released, the springs in the ends of the blocks tend to force the fluid out of the expander tube and to pull the blocks away

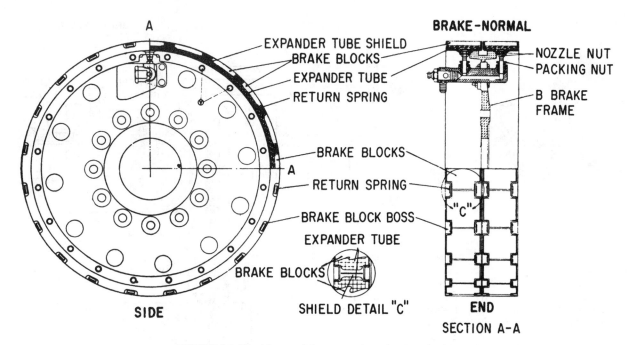

FIGURE 14-64 Views of the expander-tube-type brake.

FIGURE 14-65 Principle of operation of the expander-tube brake.

from contact with the brake drum. This action is increased by the tube itself, since it is molded slightly smaller in diameter than the brake frame and tends to contract without the help of the springs. Each block is independent in its action; therefore, there is no buildup of servo action and no tendency to grab.

During the inspection and servicing of expander-tube brakes, the technician must make sure that no hydraulic pressure is applied to the brakes when a brake is not enclosed in its drum. If the brake blocks are not restrained by the drum when hydraulic power is applied, the retaining grooves at the ends of the blocks will be broken and the blocks will pop out.

When the brake blocks are worn to their allowable limits, they are easily replaced. The return springs that retain the blocks are removed by pressing down on one end clip with a screwdriver or other tool and sliding the springs out of the rectangular holes in which they are held. When all the blocks have been removed, the entire assembly is cleaned and inspected before installing new blocks. The new blocks are installed one at a time with the return springs to hold them in place.

If the expander tube is found to be damaged upon removal of the brake blocks, the tube must be replaced.

Single-Disk Brakes

One of the most popular types of brakes, especially for smaller aircraft, is the **single-disk brake**. An exploded view of such a brake is shown in Fig. 14-66a and a cutaway is shown in 14-66b. This brake is manufactured by the Goodyear Aerospace Company.

The main disk (1) of the brake shown in Fig. 14-66a is held in the wheel by means of teeth or keys around the outer rim of the disk, causing it to turn with the wheel but allowing it limited in-and-out movement on the keys. On each side of the disk are located the linings (2), which bear against the disk when the brakes are applied, causing the wheel to slow down or stop.

One lining of the brake is mounted in a recess in the plate attached to the main axle structure. The other lining (2) is mounted against the piston (11) and moves according to the amount of hydraulic pressure applied to the piston. In Fig. 14-66a, three pistons are incorporated in the brake housing (25); therefore, three linings must be mounted on the opposite side of the disk to back up the movable linings. Single-disk brakes may be constructed with as many separate pistons and linings as deemed advisable for the airplane for which they are designed. Each piston is equipped with separate sets of linings, which bear against the brake disk (1) when the brakes are applied.

Another type of single-disk brake, manufactured by the Cleveland Wheels and Brakes Co., has achieved widespread use on single and light twin aircraft and is illustrated in the exploded view of Fig. 14-67. This disk, which is rigidly bolted to the wheel, is illustrated in Fig. 14-58 and is located between

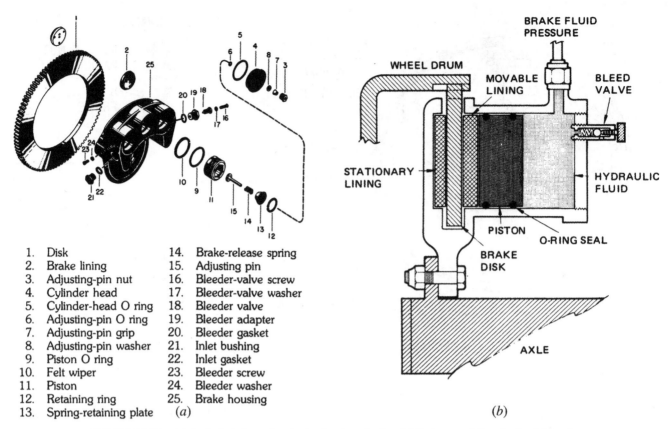

1. Disk
2. Brake lining
3. Adjusting-pin nut
4. Cylinder head
5. Cylinder-head O ring
6. Adjusting-pin O ring
7. Adjusting-pin grip
8. Adjusting-pin washer
9. Piston O ring
10. Felt wiper
11. Piston
12. Retaining ring
13. Spring-retaining plate

14. Brake-release spring
15. Adjusting pin
16. Bleeder-valve screw
17. Bleeder-valve washer
18. Bleeder valve
19. Bleeder adapter
20. Bleeder gasket
21. Inlet bushing
22. Inlet gasket
23. Bleeder screw
24. Bleeder washer
25. Brake housing

(a)

(b)

FIGURE 14-66   (a) Exploded view of a single-disk-type brake. (b) Cutaway of the single-disk brake.

the two linings. This brake utilizes contoured brake blocks as linings for both the pressure plate and back plate, sometimes called the **anvil**. Pressure is applied to the pressure plate by means of two round pistons mounted in the brake cylinder. When the brake is assembled, the linings are riveted to the pressure plate and back plate. They are replaced when one segment becomes worn beyond limits or if signs of uneven wear are evident.

Most hydraulic brake systems require a method for the removal of air from the system. In Figs. 14-66a and 14-67, items 16 through 20 and 15 through 17, respectively, make up the **brake bleeder valve** assembly. In order to bleed the air from the brakes, the valve is opened slightly and hydraulic fluid under pressure is applied to the piston. A complete discussion of brake bleeding is found at the end of this chapter.

Although modern aircraft operate the single-disk brake with hydraulic power, some older aircraft are designed with mechanically operated single-disk brakes.

Multiple-Disk Brakes

The brakes in Fig. 14-68 are used on a B777 aircraft. Each MLG truck has six brake units to stop the aircraft. The brake assembly is a rotor-stator unit that operates using hydraulic pressure. The assembly uses carbon disks as rotors and stators. The rotors and stators are compressed between the pressure plate and the end plate assembly to slow or stop the airplane. Self-adjusting pistons apply brake system hydraulic pressure to the pressure plate. The pistons automatically

adjust for brake wear. The brake units mount on bushings which ride on replaceable landing-gear axle sleeves. The center brake attachment points connect directly to the lower strut at the same place the brake rods attach. The forward and aft brake attachment points connect to the brake rods. The brake rods transmit brake torque to the strut. They also allow brake rotation as the truck position changes. Two indicator pins on the inboard side of the brake housing show brake wear.

Segmented Rotor-Disk Brakes

Segmented rotor-disk brakes are heavy-duty brakes designed for use with high-pressure hydraulic systems using power brake–control valves or power boost master cylinders. Braking action results from several sets of stationary, high-friction-type brake linings making contact with rotating (rotor) segments. This action is the same as occurs with multiple-disk brakes. An exploded view of this type brake is shown in Fig. 14-69, and a cross section of the brake is shown in Fig. 14-70.

A carrier assembly is the brake component that is attached to the landing-gear shock strut flange and on which all of the other components are mounted. The piston cups and pistons are placed in two grooves, which act as cylinders, in the carrier assembly. The automatic adjusters, which compensate for lining wear, are threaded into holes equally spaced around the face of the carrier. Each adjuster is composed of an adjuster pin, adjuster clamp, return spring, sleeve, nut, and clamp hold-down assembly.

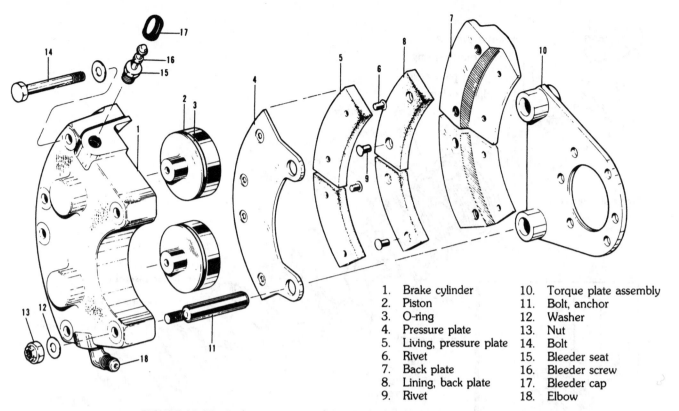

1. Brake cylinder	10. Torque plate assembly
2. Piston	11. Bolt, anchor
3. O-ring	12. Washer
4. Pressure plate	13. Nut
5. Living, pressure plate	14. Bolt
6. Rivet	15. Bleeder seat
7. Back plate	16. Bleeder screw
8. Lining, back plate	17. Bleeder cap
9. Rivet	18. Elbow

FIGURE 14-67   Brake components for a single-disk brake. (*Piper Aircraft Co.*)

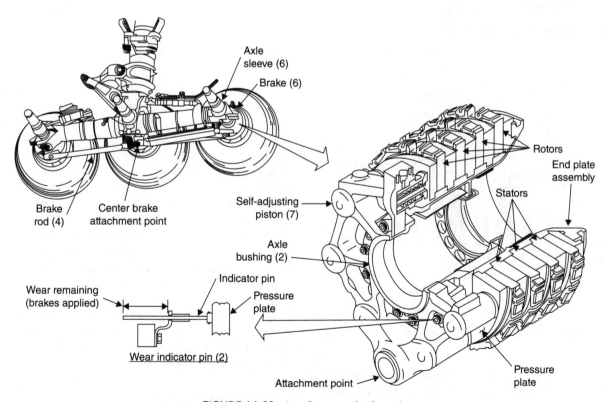

FIGURE 14-68   Landing-gear brake unit.

Design and Operation of Brake Assemblies   **495**

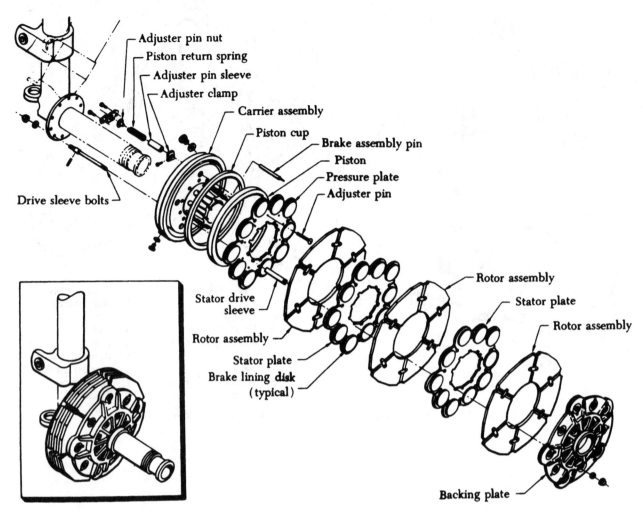

FIGURE 14-69   Exploded view of a segmented rotor brake.

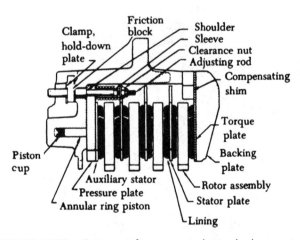

FIGURE 14-70   Cutaway of a segmented rotor brake.

The pressure plate is notched to fit over the stator drive sleeve. This component is stationary on the sleeve. An auxiliary stator plate fits next to the pressure plate and has brake-lining material attached to the side away from the pressure plate.

The rotor segment plate is installed next. This part is notched on the outside to mate with the wheel and rotate with it. The plate is made up of several segments, as shown. Alternating stator plates, with brake-lining material on both sides, and rotor assemblies are installed until the proper number of each is in place. After the last rotor segment plate is in position, a compensating shim is installed to space the back plate out from the carrier, and then the back plate is installed. The back plate contains brake lining on the side toward the rotors. The compensating shims allow the brake linings to wear down until the piston is out of travel. The shims are then removed, causing the pistons to be moved back into the cylinder, and more of the available brake lining can now be used.

Because of the gap between the rotor segments and the space between the lining sections, more brake cooling can be achieved than is possible with the multiple-disk brake, allowing more braking action to be achieved before a limiting temperature is reached.

Carbon Composite Brakes

The introduction of carbon brakes represents a significant advance in brake technology. New carbon composite brakes weight 40 percent less, yet can function in even higher temperatures with more reliability than conventional steel segmented rotor brakes. Carbon composite brakes have become a long-life, lightweight alternative to steel brakes since their introduction to commercial aviation as standard equipment on the Concorde.

Three types of carbon fiber reinforcements can be used in carbon composite brakes in a process that can take up to 5 months to make each carbon disk. Carbon fibers are molded into a precise shape and the resulting disks then are baked in special ovens, which introduce natural gas. The brake material absorbs the carbon given off by the natural gas, making the brakes even more dense with carbon. A carbon fiber rotor is illustrated in Fig. 14-71.

Carbon composites possess several unique properties that permit the combining of the overall brake disk functions of friction surface, heat sink, and structural member into a single unit. The material has the unique property that its strength does not decrease at elevated temperatures. This property, coupled with low thermal expansion, yields a brake heat sink whose operating temperature is limited effectively only by the temperature limits of the surrounding structure. When rubbed against itself, a carbon composite can perform excellently as a friction material. It provides a high heat-storage capability for each pound. This is important because in a rejected takeoff, steel brakes are designed to reach a temperature of about 2000°F. Carbon brakes have the capability of exceeding 3000°F. The maximum temperature for carbon brakes ranges from 3200 to 3400°F. In addition, good thermal conductivity characteristics serve to dissipate the heat rapidly.

Another major factor is weight. Carbon brakes weigh 40 percent less than aircraft brakes with conventional steel rotors and linings, meaning greater fuel efficiency and the ability to carry a heavier payload in commercial aircraft. The carbon brakes that can be selected for the Boeing 767-300 save nearly 300 lb per aircraft. At the present time carbon brakes are more expensive to produce and are generally found only on long-range aircraft, where weight savings can have a significant impact on cruise performance.

Carbon also offers reduced maintenance costs. An average commercial airliner will make between 700 and 3000 landings a year. Steel brakes normally last about 1000 landings before they must be replaced. Carbon brakes will last longer (1200–1500 landings) and, due to fewer brake changes, significantly reduce maintenance requirements.

Braking Heat Energy

Stopping a high-speed aircraft, either upon landing or as required for an aborted takeoff, involves the conversion of a great amount of kinetic energy to heat at the brakes and main wheels. This energy may be specified in foot-pounds or joules.

Brake limitation charts have been prepared for some airplanes to give crew members and maintenance personnel a means of determining how to deal with hot brake situations safely and effectively. One such chart for a comparatively small jet airplane is shown in Fig. 14-72. The specified purpose of the chart is to avoid in-flight fires and to ensure adequate brake capacity at all times for a rejected takeoff.

Note in the chart of Fig. 14-72 that the factors used in determining the amount of energy absorbed by the brakes in a given situation are (1) the indicated airspeed in knots at the time the brakes are applied, (2) the gross weight of the airplane, and (3) the density altitude at the airport where the braking occurs. The proper use of the chart will establish a condition zone for any particular braking event. For each zone, a particular set of requirements is set forth. These are as follows:

- Zone I, normal zone: Below 1.0 million ft-lb [138 000 kg·m]

  1. No special requirement under normal operations.

- Zone II, normal zone: 1.0 to 2.05 million ft-lb [138 000 to 282 900 kg·m]

  1. Delay subsequent takeoff as indicated by chart.

- Zone III, caution zone: 2.05 to 4.0 million ft-lb [282 900 to 552 000 kg·m]

  1. Move the airplane to clear the active runway, because uneven braking could cause one or more tires to deflate if energy is in the upper range.
  2. Use brakes sparingly to maneuver.
  3. Do not set parking brake.
  4. Allow brakes to cool for the time indicated by chart.
  5. After brakes have cooled, make a visual check of brakes.

- Zone IV, danger zone: Over 4.0 million ft-lb [552 000 kg·m]

  1. Clear the runway immediately as fusible plugs will blow 2 to 30 min after stop.
  2. Do not apply dry chemical or quench until fusible plugs have released tire pressure.
  3. Do not approach for $\frac{1}{2}$ h or until fusible plugs have blown.
  4. When artificial cooling is not used, 2 to 3 h are required for brakes to cool enough for safe removal.
  5. Tires, wheels, and brakes must be replaced.

The dotted lines in the chart of Fig. 14-72 are to show how a particular braking event can be evaluated and the appropriate zone located. The steps are as follows:

1. Locate aircraft gross weight on the chart.
2. Project vertically to the indicated airspeed line.
3. Project from the airspeed line horizontally to intersect the zero-altitude line.
4. Project parallel to the nearest density line to intersect the correct altitude vertical line.
5. Project horizontally to give correct zone, kinetic energy, and ground cooling time.

The example demonstrated on the chart is for an airplane with a gross weight of 15 000 lb [6804 kg]. The indicated airspeed at the time the brakes were applied was 110 kn. The density altitude was 5000 ft [1524 m]. When these values are applied to the chart, it is revealed that the kinetic energy is 2 350 000 ft-lb [3 186 172 J] in the caution zone, and the cooling time is 28 min.

If an airplane is equipped with thrust reversers, the energy absorbed by the brakes is reduced from the values

FIGURE 14-71 Carbon composite brake disks. (*Bendix, Wheels and Brakes Div.*)

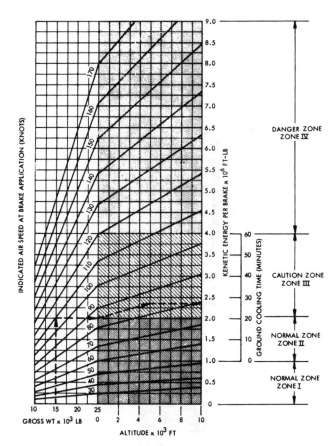

FIGURE 14-72 Brake limitation chart. (*Goodyear Aerospace.*)

systems are arranged with a pneumatic backup system for operation in case of hydraulic-fluid loss or failure of hydraulic pressure.

Independent Brake Systems

An independent brake system, such as is shown in Fig. 14-73, is usually found on small aircraft. This system is self-contained and independent of the aircraft's main hydraulic system. The basic components of this type of system are a reservoir, a master cylinder operated by the brake-control pedal or handle, a brake assembly on the wheel, and necessary lines, hoses, and fittings. Expander-tube, shoe, or disk-brake assemblies may be used with this type of system.

The **reservoir** is a storage tank that supplies the fluid to compensate for small leaks in the connecting lines or cylinders. The reservoir may be a part of the master cylinder or it may be a separate unit, as shown in the drawing. It is vented to the atmosphere to provide for feeding the fluid to the master cylinders under the force of gravity; therefore, the fluid must be kept at the correct level, or air will enter the system and reduce its effectiveness.

The **master cylinder** is the energizing unit. There is one for each main landing-gear wheel. The master cylinder is actually a foot-operated, single-action reciprocating pump, the purpose of which is to build up hydraulic fluid pressure in the brake system.

One type of master cylinder for light aircraft is illustrated in Fig. 14-74. It is a simple but effective unit, normally connected by a linkage to the brake pedal mounted on the rudder pedal. The hydraulic fluid enters the master cylinder through

shown here. The amount of reduction depends upon the time that the thrust reversers are operated.

AIRCRAFT BRAKE SYSTEMS

Brake-actuating systems for aircraft can be classified as mechanically operated, hydraulically operated, pneumatically operated, or electrically operated. Electrically operated brake systems are the newest application and are used for the B787 aircraft. These brake-by-wire systems use electrical signals from the brake pedal transducers and the onboard computer to actuate electrical brake actuators. These systems do not use hydraulic fluid at all. All brake-actuating systems provide for applying brakes either on one side of the aircraft or all the aircraft brakes by operating foot pedals or hand levers.

Mechanical brakes are found on only a few of the older, small airplanes. A mechanical brake-actuating system includes pulleys, cables, and bell cranks for connecting the foot pedals to the brake-shoe operating mechanism.

In some airplanes, the hydraulic brake system is a subsystem of the main hydraulic system. In other airplanes, there is an entirely independent brake system. Many of the large airplanes have a power brake system that is a subsystem of the main hydraulic system. The smaller airplanes usually have an independent, master brake-cylinder system.

Pneumatic brake systems utilize air pressure instead of fluid pressure to operate the brakes. Some hydraulic brake

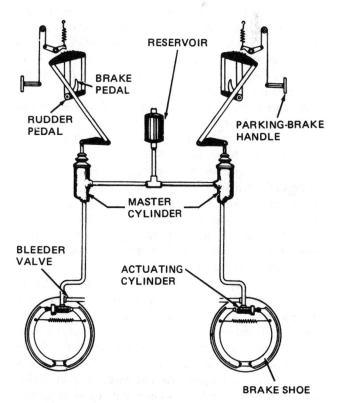

FIGURE 14-73 A shoe-type brake system.

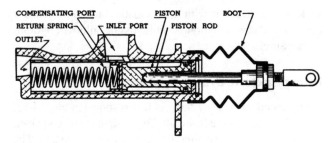

FIGURE 14-74 A master cylinder.

the **inlet port** and **compensating port** from the external reservoir, which supplies the master cylinders for both the right and left brake systems. The application of the brake forces the piston into the cylinder and causes hydraulic fluid to flow toward the brake-actuating cylinder in the wheel. The illustration shows the cylinder in the horizontal position, but when it is installed in the aircraft, it is in a vertical position with the eye of the piston rod downward. When the piston moves against the return spring, the compensating port is closed and the fluid in the cylinder is trapped under pressure. Continued pressure applied through the brake pedal forces the fluid pressure to the brake-actuating cylinder and applies the brake. When the force is removed from the brake pedal, the piston is returned to the OFF position by means of the return spring, and the compensating port is again open to the reservoir. The compensating port permits the fluid to flow toward or away from the reservoir as temperature changes, thus preventing a buildup of pressure when the brake is off.

With this type of master cylinder, the brakes are locked in the ON position for parking by means of a ratchet-type lock that is constructed as part of the mechanical linkage between the foot pedal and master cylinder. If an increase of temperature occurs, expansion increases the volume of fluid. This is compensated for by means of a spring built into the linkage. To unlock the brakes, the pilot applies enough force to the brake pedals to unload the ratchet-type lock.

Another type of master cylinder is used on a number of light aircraft. This cylinder is illustrated in Fig. 14-75. These cylinders are mounted on the rudder pedals, as shown in the drawing of Fig. 14-68. In the illustration of Fig. 14-75, this type of master cylinder incorporates a fluid reservoir (8) on the top of the cylinder (11) within the same body (7). A plastic filler plug (18) is used to close the opening in the cover (4), which is threaded into the body. The filler plug is not vented because sufficient ventilation is provided by clearance between the piston rod (3) and the piston-rod opening through the cover boss (6).

With the exception of the piston return spring (12), all internal operating parts are assembled onto the piston rod. These parts are the piston (15), piston spring (14), "lock-o-seal" (16), and compensating sleeve (17). A seal between the piston and the cylinder walls is provided by the O-ring (9) installed in a groove around the piston. As pressure is applied to advance the piston rod into the cylinder, the piston remains stationary until the lock-o-seal is seated on the piston, which requires a 0.040-in [1.016-mm] movement of the

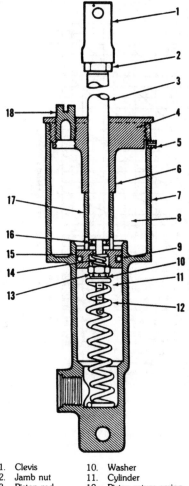

1. Clevis	10. Washer
2. Jamb nut	11. Cylinder
3. Piston rod	12. Piston return spring
4. Cover	13. Nut
5. Setscrew	14. Piston spring
6. Cover boss	15. Piston
7. Body	16. Lock-o-seal
8. Reservoir	17. Compensating sleeve
9. O ring	18. Filler plug

FIGURE 14-75 A Goodyear master cylinder.

piston rod. Proper operation of the master cylinder depends upon this seating action. When the lock-o-seal is seated, fluid cannot pass the piston, and with continued movement of the piston rod forcing the piston into the cylinder, pressure in the cylinder is increased. At any time during the stroke that force on the piston is eased, the piston return spring will tend to keep the piston seated against the lock-o-seal, maintaining pressure in the cylinder. As the force is further eased, allowing the piston-return spring to force the piston to retreat, the upper end of the compensating sleeve will contact the cover boss; thus the piston is forced to unseat itself from the lock-o-seal. This allows additional fluid from the reservoir to enter the cylinder. This positive unseating also allows unrestricted passage of fluid from the cylinder to the reservoir while the piston is in the static position. This action is to compensate for any excess fluid that may be present in the system due to pumping or from thermal expansion.

Mechanical linkages are required to transmit the energy of the foot to the master cylinder. Most airplanes have the master cylinders mounted on the rudder pedals, although a

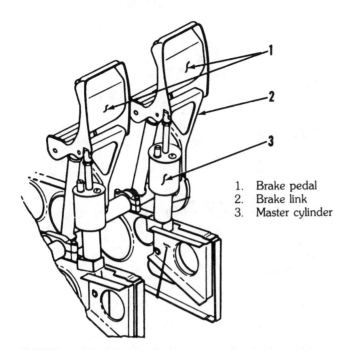

FIGURE 14-76   Master cylinders mounted on brake pedals.

1. Brake pedal
2. Brake link
3. Master cylinder

few aircraft have the master cylinders mounted at a distance from the pedals. A system of rods, levers, bell cranks, and cables is often employed to carry the mechanical energy to the master cylinders. In Fig. 14-76 the **brake pedals** are toe brakes mounted on the rudder pedals. When the brake pedals are pressed, the linkage causes the master cylinder piston to move into the cylinder and force fluid into the brake lines. When brakes of this type are pressed, it is necessary for the pilot to balance the force on one pedal with equal force on the other pedal unless he or she wishes to turn the airplane. The brakes and rudder control are operated independently.

Other control arrangements make use of **heel brakes**, operated by the pilot pressing his or her heel on the brake pedal, and a **central hand-brake lever**, which operates all brakes at the same time.

The **fluid lines** may consist of flexible or rigid tubing or a combination of both. Usually flexible tubing is employed with retractable gear systems and between the movable parts of the shock strut, as shown in Fig. 14-77.

The **brake-actuating** mechanism of the brake assembly causes braking action to occur when pressure from the master cylinders is transmitted to them. The **parking-brake mechanism** is a subassembly of the usual hydraulic brake system. The control for the mechanism is in the pilot's compartment and usually consists of a pull handle or lever. When the brake pedals are depressed and the parking-brake lever is pulled back by hand, the brakes are locked in the ON position. Depressing the brake pedals or releasing a parking-brake handle, depending on the system, again releases the brakes. Depending upon the type of master cylinder used, depressing the pedals will either build up enough pressure to unseat the parking valve, or it will unload a ratchet-type parking lock.

With respect to parking brakes, the setting of these brakes when the main brakes are hot may cause serious damage. Hot brakes should be allowed to cool before the parking brakes are applied.

Power Boost Systems

Power boost systems are used on aircraft that have high landing speeds or are too heavy for an independent brake system to operate efficiently but that do not require the power of a power brake system. This system, shown in schematic form in Fig. 14-77, uses hydraulic system pressure to operate the brakes.

When the pilot depresses the brake pedal, the power boost master cylinder opens a metered line to allow hydraulic system pressure to flow to the brakes. The metering mechanism is either a tapered pin or variable size orifice. The further the pedal is depressed, the more fluid flows through the metering device, with a resulting increase in braking action. When the pedal is released, the pressure inlet line is blocked and the master cylinder ports the pressure in the brake line to a hydraulic system return line.

This system receives pressure from the main hydraulic system through a check valve, which prevents loss of fluid and pressure if the main hydraulic system should fail. Between the check valve and the master cylinder is an accumulator that stores fluid under pressure to operate the brakes for a few braking cycles if hydraulic system pressure is lost. The accumulator also ensures adequate brake response if, at the time of application, the main hydraulic system has other demands placed upon it.

A shuttle valve may be installed in the brake line with a pneumatic bottle available for emergency operation of the brakes if all hydraulic power is lost.

Power Brake System

A **power brake system** is used to operate the brakes of large aircraft where the independent and power boost systems are not adequate. The pilot operates the system by depressing the brake pedal. This causes a power brake–control valve to direct hydraulic system pressure to the brakes and operate the brake assembly. The brake pedal is connected to the power brake control valve through an arrangement of cables, pulleys, bell cranks, and linkages.

Modern transport aircraft often use an electrohydraulic brake control valve. Electrical signals will be sent to the brake system computer which will generate an electrical signal based on brake pedal position, wheel speed, and autobrake setting.

The power brake control valve for a transport aircraft is illustrated in Fig. 14-78. These are also called **brake-metering valves.** One metering-valve assembly is used for each main landing-gear brake.

In a typical system, four hydraulic lines are attached to each valve. These lines are for pressure, return, brakes, and automatic braking. Valve ports are opened or closed by operating a circular grooved, sliding valve rod (spool).

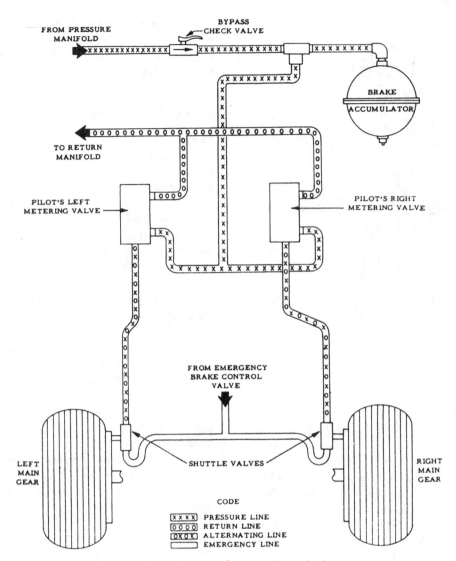

FROM PRESSURE MANIFOLD

BYPASS CHECK VALVE

BRAKE ACCUMULATOR

TO RETURN MANIFOLD

PILOT'S LEFT METERING VALVE

PILOT'S RIGHT METERING VALVE

FROM EMERGENCY BRAKE CONTROL VALVE

LEFT MAIN GEAR

RIGHT MAIN GEAR

SHUTTLE VALVES

CODE
XXXX PRESSURE LINE
OOOO RETURN LINE
OXOX ALTERNATING LINE
EMERGENCY LINE

FIGURE 14-77 Schematic of a power boost brake system.

The linkage end of the valve rod projects beyond the valve body, whereas the opposite end is supported in a sealed **compensating chamber**.

When the brake pedals are depressed, an inward movement is imparted to the metering valve rod through the mechanical linkage and cables. As the rod moves in, the return port is closed, and the pressure port is opened to direct hydraulic fluid pressure to the brakes. A passage through the valve rod permits the hydraulic fluid under pressure to enter a compensating chamber enclosing the inner end of the valve rod. Pressure acting on the end of the rod creates a return force tending to close the valve. This return force varies with the intensity of braking force and provides "feel" at the pedals. The desired braking effort is obtained by depressing the pedals a greater or lesser distance. Cable stretch and adjustment of pedal position permits the valve rod to move back until both pressure and return ports are closed. At this point the braking effort remains constant. This condition is shown schematically in Fig. 14-78. Releasing the brake pedals allows the pressure in the compensating chamber to move the valve rod out and open the brake line to the return line.

As pressure in the brake line falls, the brakes are released, and return force on the valve rod is relieved.

Automatic braking to stop the rotation of the wheels during retraction is provided by a small-diameter piston-actuating cylinder attached to the metering valve. The cylinder is connected to the landing-gear-retract hydraulic line. When the landing-gear control is placed to UP, hydraulic pressure is directed to the automatic cylinder and the piston extends. One end of the piston rod rests on the valve rod; therefore, extension of the piston opens the metering valve and applies the brakes.

Debooster Valve

A **brake debooster valve** is installed in systems where the high pressure of the hydraulic system is used to operate brakes that are designed to work with lower pressure. This valve is positioned in the hydraulic line between the power brake–control valve and the brake assembly. High pressure from the power brake–control valve is applied to the small piston of the debooster valve, as shown in the schematic drawing in Fig. 14-79. This exerts a force on the large piston in the

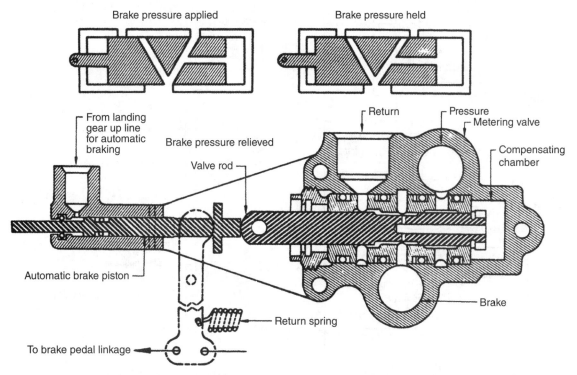

FIGURE 14-78 A power brake–control valve for a transport aircraft.

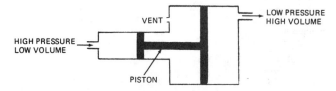

FIGURE 14-79 Debooster valve.

low-pressure chamber. The pressure in the large chamber is reduced in proportion to the difference in sizes of the small and large pistons. If the hydraulic system operates at 3000 psi [20 685 kPa] and the small piston has an area of 1 in² [6.45 cm²], 3000 lb [1360.8 kg] of force are applied to the small piston and, through the solid linkage, to the large piston. If the large piston has an area of 3 in² [19.35 cm²], the 3000 lb [1360.8 kg] of force on 3 in² [19.35 cm²] results in a fluid pressure of 1000 psi [6895 kPa].

In reducing the pressure in the system, the debooster valve also increases the volume of flow in the low-pressure chamber. In the preceding example, if the piston assembly moves 1 in [2.54 cm], the high-pressure system changes in volume by 1 in³ [16.39 cm³], whereas the 1-in [2.54-cm] movement of the low-pressure piston causes a change in volume of 3 in³ [49.16 cm³]. This increase in volume is used to obtain the desired amount of brake cylinder travel without exerting a high pressure on the system.

Multiple Power Brake-Actuating System

The brake system for a transport aircraft involves many components and a number of subsystems and is described briefly

here to provide the technician with a general understanding of how such a system operates. The schematic diagram of Fig. 14-80 shows the operating components and subsystems of an Airbus A330 brake system. As can be seen in the drawing the brake and steering control unit (BSCU) is the heart of the system. The BSCU receives signals from the brake pedals, autobrake system, antiskid system, hydraulic system, landing-gear lever, landing-gear position (down and locked), reference speed, ground spoiler, brake pressure, brake temperature and wheel speed. Based on these parameters it will generate an electrical signal that is sent to the selector valve and servo valve. The main wheels are equipped with carbon multidisk brakes which can be actuated by either of two independent brake systems. The normal system uses green hydraulic pressure while the alternate system uses the blue hydraulic system backed up by a hydraulic accumulator. An antiskid and autobrake system is also installed. Braking commands come either from the brake pedals (pilot action) or the autobrake system (deceleration rate selected by the crew).

Antiskid System

The antiskid system provides maximum braking efficiency by maintaining the wheels at the limit of an impending skid. At skid onset brake release orders are sent to the normal and to the alternate servo valves as well as to the electronic centralized aircraft monitor (ECAM) system which displays the released brakes. Full braking performance is achieved only with brake pedals at full deflection. The antiskid system is deactivated below 10 Kt (ground speed). An on/off switch in the flight station activates or deactivates the antiskid system.

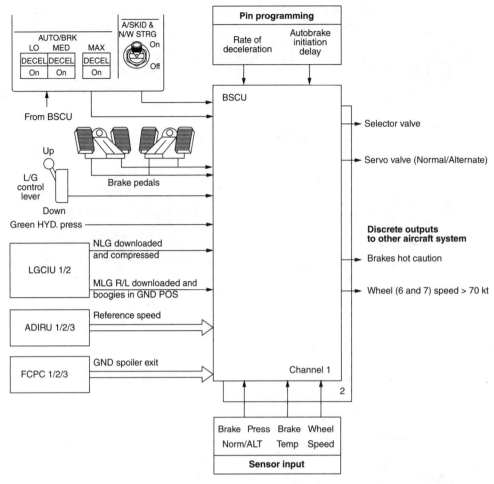

FIGURE 14-80 Power brake system of transport aircraft.

Principle of Operation

The speed of each main gear wheel (given by a tachometer) is compared with the aircraft speed (reference speed). When the speed of a wheel decreases below .88 time reference speed, brake release orders are given to maintain the wheel slip at that value (best braking efficiency). In normal operation, the reference speed is determined by the BSCU from the horizontal acceleration from the air data inertial reference units (ADIRU) ADIRU 1 or ADIRU 2, or ADIRU 3. In cases all ADIRUs have failed, reference speed equals the maximum of either main landing-gear wheel speeds. Deceleration is limited to a default value of 2.5 m/s^2 (8.2 ft/s^2). Figure 14-81 shows a schematic diagram of an antiskid system.

Autobrake System

The purpose of the autobrake system is to reduce the braking delay in the event of an accelerate-stop to improve performance and to establish and maintain a selected deceleration rate during landing, to improve comfort and reduce crew workload. The system is armed by the crew. They can select low (LO), medium (MED), or maximum (MAX) by pressing on a push button provided all the following arming conditions are met: green hydraulic system pressure available, antiskid electrically powered, no failure in the brake system,

at least two primary computers (PRIM) available, and at least one ADIRU available. Automatic braking is activated by the ground spoiler extension command. In addition, for MAX mode the nose landing-gear compressed signal is required.

Brake System Operation

There are four modes of brake operation: normal brakes, alternative brakes with antiskid, alternative brakes without antiskid, and parking brakes (Fig. 14-82).

Normal Brake Operations

During normal braking the antiskid system is operative and auto brake is available. The brake system is in the normal mode when the green system hydraulic pressure is available, MLG is in ground condition, and antiskid switch is on. The braking is controlled through the BSCU via the pedals or autobrake system. When the landing-gear lever is selected up the normal brake system stops the rotation of the wheels before they go into the wheel wells.

Alternate Brakes with Antiskid Protection

The conditions for alternate brakes are blue hydraulic system pressure is available and green hydraulic system

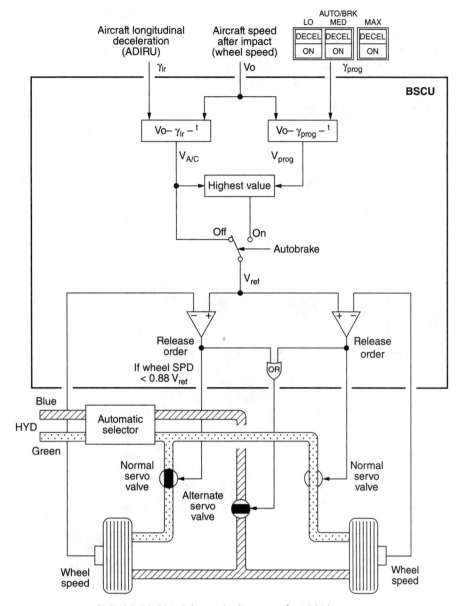

FIGURE 14-81 Schematic diagram of antiskid system.

pressure is insufficient, antiskid switch is on, and parking brake is not on. Autobrake is not available with alternate brakes. The automatic switching between the green and blue system is achieved by an automatic hydraulic selector (see Fig. 14-82). Control is achieved by the pedals through the auxiliary low hydraulic pressure distribution line acting on the dual valves. The BSCU controls the antiskid system via the alternate servo valves.

Alternate Brakes without Antiskid

Alternate brakes without antiskid is available when the auto brake and antiskid system are inoperative. The reasons that the antiskid system is deactivated are: antiskid system switch off, power supply failure, or BSCU failure. The problem can also be caused by low pressure in the blue and green hydraulic systems in which case the brake pressure is supplied by the accumulator only. The flight crew controls

the brakes with the foot pedals acting on the dual valves. Alternate servo valves are fully open and the pilot must limit the brake pressure by referring to the triple indicator on the center instrument panel to avoid wheel locking. The accumulators are dimensioned to supply at least seven full brake applications.

Parking Brake

Pressure for the parking brake system is supplied by the blue hydraulic system or accumulator via the dual shuttle valve. Alternate servo valves are open allowing full-pressure application. The accumulator maintains the parking brake pressure for at least 12 h. If the parking brake is set and no blue hydraulic or accumulator pressure is available, then the normal braking system can be applied via the brake pedals. Blue accumulators can be pressurized by pressing the blue electric pump switch.

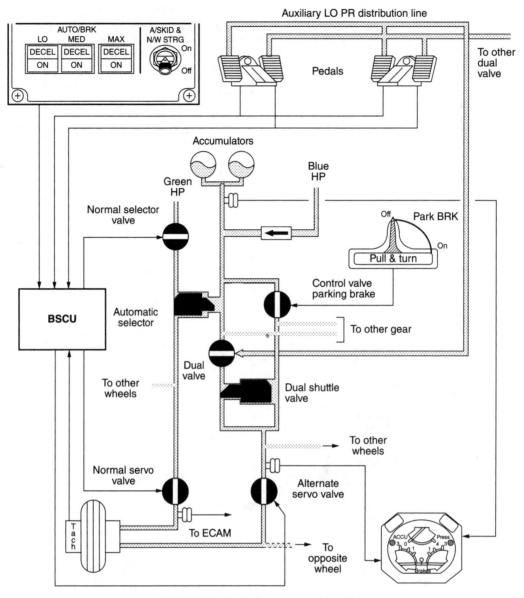

FIGURE 14-82 Brake system schematic.

ECAM Indications

The ECAM display shows brake system data for the flight crew. Figure 14-83 shows the ECAM wheel page with messages.

1. *Normal brake indication.* The normal brake indication appears amber in case normal braking has failed, antiskid switch is in the off position (associated with ECAM caution), or both BSCU channels have failed (associated with ECAM caution).

2. *Antiskid indication.* The antiskid indication appears amber and is associated with an ECAM caution in case of total BSCU failure or when the antiskid switch is off or in case of antiskid failure detected by the BSCU or in flight with at least one engine running when green and blue systems have failed.

3. *Autobrake indication.* The autobrake indication will display green when autobrake is armed and amber in case of autobrake system failure or failure of both BSCU channels.

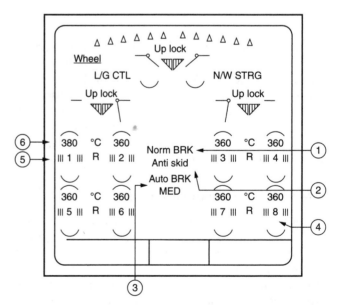

FIGURE 14-83 ECAM wheel page

MAX, MED, or LO indicates the selected rate (green). Not displayed when autobrake is faulty.

4. *Wheel number identification.* The wheel number identification indicates the position of the wheel. It appears in white on the display.

5. *III indications.* The III indications appear in green when: in flight, the landing gear is extended and the antiskid is valid, or on the ground, when antiskid is activated and the brakes are released. The message appears in amber when there is residual pressure or a brake release fault. The R (release) indication is always in white.

6. *Brake temperature indications.* The brake temperature indications are normally green. A green arc appears on the hottest wheel, when the brake temperature exceeds 100°C. The text turns amber when the corresponding brake temperature exceeds 300°C. The arc also turns amber on the hottest wheel.

BRAKE MAINTENANCE

The proper inspection and maintenance of the brake system is vital to the safe operation of the aircraft. Without proper brake operation, the aircraft may not be able to stop in the runway length available. On the other hand, if the brakes lock up, a tire may blow out and cause loss of control of the aircraft, with resulting damage and injury.

The technician should always refer to the aircraft manufacturer's instructions when inspecting and maintaining brake systems. The information presented here is meant as an introduction to brake maintenance tasks.

WARNING: **Brake systems may include compounds using asbestos in the construction of the brake-lining material. Use appropriate respiratory and other health-protective measures when working with these systems.**

Inspection of Brakes

Brake inspection is accomplished at the same time that wheels are removed for other inspections. The linings for single-disk-type brakes consist of flat cylinders or thick disks made of a tough abrasion-resistant material similar to that used for the linings of brake shoes. The manufacturer specifies the amount of wear that may be permitted for these linings before it is necessary to replace them. It is usually possible to inspect these linings without removing the wheel.

Some brake linings cannot be fully cured during manufacture, and the curing operation must be completed after the linings are installed on the aircraft. The curing process involves accelerating the aircraft to a moderate speed and then applying moderate-to-heavy braking force to heat up the linings. This situation is commonly found on certain models of single-disk brakes. If this curing operation is not performed, the linings will not provide maximum performance. The brake manufacturer's information should be consulted to determine if this process is required for specific brake models and for specific curing procedures.

The brake disks should be inspected for pitting and grooving. If these surface defects exceed the manufacturer's limits, the disk must be resurfaced or replaced.

During inspection of single-disk-type brakes, the technician should check the retaining clips that hold the brake disk in the wheel on some types of disks. If the disk is distorted or the clips are broken, the brake disk will not stay in the correct position and brake failure will result.

Multiple-disk brakes are manufactured in a number of configurations, generally for use on large aircraft. It is, therefore, necessary that the technician follow the specifications given in the manufacturer's maintenance manual when inspecting and servicing such brakes. Wear of multiple-disk brakes is determined by noting the extension of the **wear indicator pin**. When the pin has reached the minimum allowable dimension, maintenance action is required. During major inspections brakes are disassembled and disks examined for wear and for damage resulting from heat. The lugs or keys holding the rotor disks in the wheels are examined for wear and security. The actuating parts, such as the pistons and pressure plates, must also be examined for condition. Leaking pistons must be repaired as directed.

The hydraulic systems that actuate brakes should be inspected for the same defects that may be found in any other hydraulic system. Aluminum-alloy and steel tubing should be examined for wear at contact points and for security of the mounting attachments. Fittings should be checked for leakage of fluid. If it is necessary to replace a fitting or a section of tubing, the technician must make certain to install the correct part and should check the part number and then compare the part with the one removed. It must be remembered that fittings are made of steel or aluminum alloy, that they may be flare-type or flareless fittings, and that the threads may be for pipe fittings, flare-type fittings, or flareless fittings. Proper identification is, therefore, extremely important.

Brake systems may be serviced with different types of fluid, and the technician must be sure to use the correct type. Large aircraft often have power brakes supplied with the same fluid in the main hydraulic system. It is, therefore, likely that the brake hydraulic system will employ Skydrol or an equivalent fire-resistant fluid.

Flexible hoses used in the brake system must be examined for swelling, sponginess, leakage, and wear of the outer covering. A brake hose that has become soft or swollen is likely to cause "spongy" brakes, because the hose will expand and take up some of the fluid volume. The effect is similar to the presence of air in the system.

During the inspection of brakes and brake systems, the technician should follow up on any discrepancies, defects, and malfunctions reported by the pilot or other crew member. Among items that may be reported are dragging brakes, grabbing brakes, fading brakes, excessive pedal travel, and the brake pedal slowly moving down when brakes are applied. The causes of these malfunctions vary to some extent, depending upon whether the brakes are of the drum, expander-tube, single-disk, or multiple-disk type. The causes also vary according to the method by which hydraulic pressure is applied to the brakes.

Dragging brakes can be caused by air in the brake hydraulic system, broken-down or weak return springs, sticking return pins, or defective valves. When brakes are released, all hydraulic pressure should be released from the brake cylinders; however, if a valve sticks closed or is plugged, the pressure may not be released and the brakes will drag.

Grabbing brakes are usually caused by oil or some other foreign matter on the disks and linings. In addition, worn disks and drums can cause grabbing.

Fading brakes are usually caused by the condition of the lining. If the brakes have been overheated and the linings burned, glazing is likely to take place on the surface of the linings, and this condition can result in brake fade.

Excessive brake pedal travel can be caused by worn brakes, lack of fluid in the brake system, air in the system, or improperly adjusted mechanical linkages. In some very large aircraft, the brake-control cylinders are located in the wheel wells in the wings. The control cylinders are connected to the brake pedals through a system of levers, cables, and pulleys. If this system is worn or not adjusted properly, excessive pedal travel will result.

A leaking piston seal in a brake master cylinder will cause the pedal to slowly creep down while pedal pressure is applied. In such a case, the master cylinder should be overhauled.

Bleeding of Brakes

When a brake line has been disconnected for any reason, or after overhaul or replacement of any part of the brake hydraulic system, it is usually necessary to bleed the system to remove air from lines, valves, and cylinders. This is accomplished by flowing brake fluid through the system. Air in the brake hydraulic system causes spongy brakes and may cause the brakes to drag. This problem occurs because the air contracts and expands as pressure is applied or released, thus creating the spongy feeling and reducing the positive pressure that should be available for brake operation.

Aircraft brake systems usually incorporate bleeder fittings for the attachment of bleeder hoses. The fittings are installed at the brake assembly.

One method used to bleed brakes is the **gravity method**. In gravity brake bleeding, a clear plastic tube is connected to the brake bleeder fitting, and the free end of the tube is submerged in a container of clean brake fluid. The reservoir for the system is filled with the correct type of fluid, and the brakes are operated to pump fluid through the lines, valves, and cylinders. Air bubbles are seen through the clear plastic tube as the fluid passes out and into the container. When the fluid flows with no air bubbles passing through the line, the bleeder valve is closed and the bleeder tube is disconnected. The reservoir is then refilled to the proper level with the correct type of fluid. This is known as gravity bleed and is illustrated in Fig. 14-84. For any type of brake system, the manufacturer's instructions should be followed to ensure that the bleeding procedure is correct.

For some aircraft, a **pressure-bleeding system** is employed. For this type of bleeding, brake fluid is forced

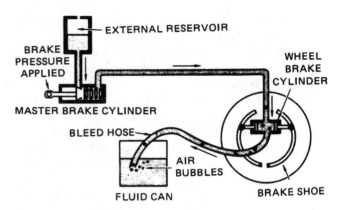

FIGURE 14-84 Gravity method of bleeding brakes.

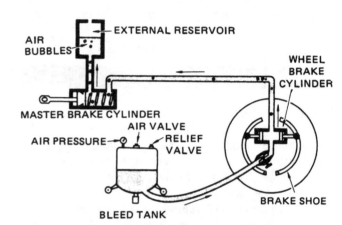

FIGURE 14-85 Pressure method of bleeding brakes.

through the system by means of a pressurized reservoir connected to the system brake-bleed fitting. Fluid is forced through the system and back to the reservoir until no bubbles appear in the fluid outflow. This is illustrated in Fig. 14-85.

REVIEW QUESTIONS

1. What are two different configurations of *landing gear*?

2. By what different methods may landing gear be classified?

3. Describe a *spring steel landing gear*.

4. What is the difference between *spring-oleo struts* and *air-oleo shock struts*?

5. What is the purpose of the *metering pin* in an air-oleo shock strut?

6. What provision is made in the design of an air-oleo strut to prevent sand and dirt on the piston from being drawn into the cylinder?

7. What cushions the shock of landing when an air-oil strut is installed in the landing gear?

8. Does fluid or air support the weight of the aircraft with air-oleo struts during taxiing?

9. What is done to prevent an oleo strut from extending too rapidly and causing impact damage?

10. What features must a nose gear strut assembly have that are not required for main gear?

11. What is the function of *torque links*?

12. What is the function of a *shimmy damper* installed with a nose gear?

13. Describe the construction of a piston-type shimmy damper.

14. What is the function of the *main-gear swivel glands*?

15. Explain the purpose of the *overcenter linkage* in the nose-gear mechanism.

16. Describe a landing-gear *truck* assembly.

17. How are power steering systems typically controlled by the pilot?

18. What is the purpose of the follow-up differential mechanism in a nose-wheel steering system?

19. What different operational capabilities may be incorporated in a tail-wheel steering installation?

20. What power sources are employed for the operation of retractable landing gear?

21. What is a squat *switch*?

22. What device is used to warn a pilot that the landing gear has not been extended prior to landing?

23. What type of warning signal is often used in an airplane in addition to position indicators?

24. What are the four methods that may be used for the emergency extension of the landing gear?

25. With what type of landing gear are most helicopters equipped?

26. How can the age of rubber shock cord be determined?

27. Under what conditions should rubber shock cord be replaced?

28. Describe an operational test of landing gear.

29. What are the main parts of a tire's structure?

30. What markings are required on aircraft tires?

31. What markings are required on a retreaded tire?

32. What are the important points in the mounting of tires with inner tubes?

33. How should tires be stored?

34. What is the purpose of an *inflation cage*?

35. Why should newly mounted tires be allowed to set for several hours before being installed on an aircraft?

36. How are overinflation and underinflation indicated in the tread-wear pattern?

37. Of what materials are airplane wheels usually constructed?

38. Describe the construction of a wheel for a light airplane.

39. What is the advantage of having wheels constructed in two parts?

40. What means are employed to prevent air from leaking through the joint between tubeless wheel halves?

41. What is the purpose of *fusible plugs* in aircraft wheels?

42. What conditions should be checked in the inspection of airplane wheels?

43. Name four types of brakes used on aircraft.

44. Describe the principle of operation for the *expander-tube brake*.

45. How is braking action developed with the *single-disk-type brake*?

46. Describe how braking action is developed in a *multiple-disk brake*.

47. What are the advantages of *carbon composite brakes*?

48. Describe a *brake master cylinder*.

49. Describe the operation of a debooster valve.

50. Where in a brake system is flexible tubing used?

51. When should a parking brake not be applied?

52. Briefly describe the operation of an antiskid system.

53. Why is it not possible to land an airplane with wheels locked if the antiskid system is armed?

54. Describe a wheel-speed transducer.

55. What may be the cause of dragging brakes, grabbing brakes, and fading brakes?

56. What would be the effect of a leaking seal in a master cylinder?

57. Why is it necessary to bleed brakes?

Aircraft Fuel Systems 15

INTRODUCTION

The aircraft fuel system is used to deliver fuel to the engines safely under a wide range of operational conditions. The system must be able to safely hold the fuel, allow filling and draining of the tanks, prevent unwanted pressure buildups in the system, protect the system from contamination, and ensure a steady supply of fuel to the engine.

The system must also provide a means of monitoring the quantity of fuel on the aircraft during flight and, in some aircraft, a means of checking fuel pressures, temperatures, and flow rates.

All these capabilities must be carried out without compromising the safety of the aircraft or its occupants. This chapter discusses methods used to meet these requirements.

For a full discussion of the characteristics of fuels and fuel additives, refer to the text *Aircraft Powerplants.*

REQUIREMENTS FOR FUEL SYSTEMS

The purpose of a fuel system is to deliver a uniform flow of clean fuel under constant pressure to the carburetor or other fuel-control unit. This supply must be adequate to meet all engine demands at various altitudes and attitudes of flight. Recommended installations employ gravity-feed or mechanical pumping systems.

The location of the various units in the fuel system must be such that the entire fuel supply, except that designated as **unusable fuel,** is available for use when the airplane is in the steepest climb, in the best angle of glide, or in any reasonable maneuver.

Reliability

Each fuel system must be constructed and arranged to ensure a flow of fuel at a rate and pressure established for proper engine and auxiliary power-unit functioning under each likely operating condition. Auxiliary power units are usually installed only in large transport-type aircraft. In case the certification of the aircraft involves unusual maneuver approval, the fuel system must perform satisfactorily during these maneuvers.

Each fuel system must be arranged so that any air that may be introduced into the system as a result of fuel depletion in a tank will not result in power interruption for more than 20 s for a reciprocating engine and will not cause the flameout of a turbine engine.

To ensure that fuel systems and components meet the requirements of applicable Federal Aviation Regulations, systems are tested under actual or simulated operating conditions. These tests involve the individual components as well as the complete system.

Fuel-System Independence

Each fuel system must meet engine-operation requirements by allowing the supply of fuel to each engine through a system that is independent of parts of another system supplying fuel to any other engine. That is, the fuel system must be designed so that any one engine can operate without being affected by fuel-system problems for another engine.

Fuel-system independence for a multiengined airplane is further ensured by the requirement that each fuel system be arranged so that the failure of any one component (other than a fuel tank) will not result in the loss of power of more than one engine or require immediate action by the pilot to prevent the loss of power of more than one engine. If a single fuel tank or a series of tanks interconnected to function as a single tank is used on a multiengined airplane, there may be a separate tank outlet for each engine, and each outlet may be equipped with a **shutoff valve.** The shutoff valve may also serve as the firewall shutoff valve if the line between the valve and the engine compartment does not contain more than 1 qt [0.94 L] of fuel, or any greater amount shown to be safe, that can drain into the engine compartment. The tank or series of tanks shall have at least two vents arranged to minimize the probability of both vents becoming obstructed simultaneously.

Filler Caps

Filler caps must be designed to minimize the probability of incorrect installation or loss of the caps in flight. Some fuel caps incorporate vents that keep the fuel tank at atmospheric pressure. The filler cap or the area immediately next to the cap should be placarded with the word FUEL and the proper type and the minimum grade of fuel approved for use in the aircraft.

Lightning Protection

The fuel system must be designed and arranged to prevent the ignition of fuel vapor within the system by direct or swept lightning strikes to areas where these are likely to occur. In addition, the design must be such that fuel vapor cannot be ignited at fuel vent outlets.

Fuel Flow

The fuel flow for a gravity-fed system must be at least 150 percent of the takeoff fuel consumption of the engine. For pressure pump systems, the fuel flow for each reciprocating engine must be at least 125 percent of takeoff fuel flow. These quantities are established by appropriate tests.

For transport-category airplanes, each fuel system must provide at least 100 percent of the fuel flow required under each intended operating condition and maneuver. The systems are tested under established conditions to ensure that the flow rate is adequate. If a flowmeter is in the system, it is blocked off to see if the full flow requirements are met when fuel is bypassing the flowmeter.

If an engine can be supplied with fuel from more than one tank, the fuel system for each reciprocating engine must supply the full fuel pressure to that engine not more than 20 s after switching to any other tank containing usable fuel. This would occur when an engine malfunction becomes apparent due to the depletion of the fuel supply in any tank from which the engine fuel can be fed. In the case of a turbine engine that can be supplied from more than one tank, in addition to having appropriate manual switching capability, the system must provide an automatic switching capability. This must be designed to prevent interruption of fuel flow to an engine without attention from the flight crew when any tank supplying fuel to that engine is depleted of usable fuel during normal operation. This requires, of course, that another tank contain usable fuel for the engine in question.

If fuel can be pumped from one fuel tank to another in flight, the fuel tank vents and the fuel-transfer system must be designed so that no structural damage to the tanks can occur because of overfilling. The unusable fuel quantity for each fuel tank and its fuel-system components must be established at not less than the quantity at which the first evidence of engine malfunction occurs under the most adverse fuel-feed condition for all intended operations and flight maneuvers involving fuel feed from that tank.

It must be impossible, in a gravity-feed system with interconnected tank outlets, for enough fuel to flow between the tanks to cause an overflow from any tank vent under any condition of intended operation or flight maneuver. The system is tested with full tanks to ensure that this requirement is met.

Hot-Weather Performance

Fuel systems must perform satisfactorily in hot-weather operation. Systems must be free of **vapor lock** when operating with fuel at 110°F [43°C] under critical conditions. Vapor lock is a condition of fuel starvation that can occur in a reciprocating engine fuel system in which the fuel in the fuel line is heated enough to cause it to vaporize, forming a bubble of fuel vapor in the line blocking fuel from flowing to the engine. The tanks of transport-type aircraft should be pressurized to prevent vapor lock under all operating conditions. Tests are made under takeoff and climb conditions to ensure that the system meets all requirements. Fuel-boost pumps installed in the bottom of a fuel tank are an effective method for reducing the possibility of vapor lock.

Requirements for Fuel-System Indicators, Warning Lights, and Controls

As explained in other sections of this chapter, fuel systems for aircraft require a number of indicators, depending upon the type and design of the system involved. Indicators that may be required are fuel-quantity indicators, fuel-pressure indicators, fuel-temperature indicators, fuel-flow indicators, and others. The indicators must be located at points where they are readily available to the appropriate member or members of the flight crew.

If the unusable fuel supply for any fuel tank exceeds 1 gal [3.79 L] or 5 percent of the tank capacity, whichever is greater, a red arc must be marked on the fuel quantity indicator for that tank extending from the calibrated zero reading to the lowest reading obtainable in level flight.

Indicators, either gauges or warning lights, must be provided at pressure fueling stations to indicate failure of the automatic shutoff means to stop fuel flow at the desired level.

Fuel tanks with interconnected outlets and airspaces may be treated as one tank and need not have separate fuel-quantity indicators. Each exposed sight gauge used as a fuel-quantity indicator must be protected against damage.

If a fuel-flowmeter system is installed in a fuel system, each metering component (fuel-flow transmitter) must have a means of bypassing the fuel flow if malfunction of the metering component severely restricts fuel flow.

There must be a means to measure fuel pressure in each system supplying reciprocating engines at a point downstream of any fuel pump except fuel-injection pumps. If necessary for the proper fuel-delivery pressure, there must be a connection to transmit the carburetor-air-intake static pressure to the proper fuel-pump relief-valve connection. If this connection is required, the gauge balance lines must be independently connected to the carburetor inlet pressure to avoid erroneous readings.

Fuel-pressure warning lights are installed in appropriate locations to inform members of the flight crew that fuel pressure has dropped below a safe level as the result of depletion, fuel-pump failure, ice in the system, or a clogged fuel filter. In some cases an aural warning system is provided in connection with the warning light.

Each fuel-tank selector control must be marked to indicate the position corresponding to each tank and to each cross-feed position. If safe operation requires a specific sequence for the use of any tanks, that sequence must be marked on or adjacent to the selector for those tanks. Each valve control for each engine must be marked to indicate the position corresponding to each engine controlled.

Each **emergency control,** including each fuel jettisoning and fluid-shutoff control, must be colored red.

FUEL TANKS

Fuel tanks are used to store the fuel for the aircraft until it is used by the engines. The following section discusses the requirements and components associated with fuel tanks to ensure a proper supply of fuel for the engines.

Fuel-Tank Requirements

Fuel tanks for aircraft may be constructed of aluminum alloy, fuel-resistant synthetic rubber, composite materials, or stainless steel. The material selected for the construction of a particular fuel tank depends upon the type of airplane in which the tank will be installed and the service for which the airplane is designed. Fuel tanks and the fuel system, in general, must be made of materials that will not react chemically with any fuels that may be used. Aluminum alloy, because of its light weight, strength, and ease with which it can be shaped and welded, is widely used in fuel-tank construction. Many aircraft now use synthetic-rubber bladders for fuel cells. These cells are light in weight and give excellent service if maintained according to the manufacturer's instructions.

Metal fuel tanks generally are required to withstand an internal test pressure of $3\frac{1}{2}$ psi [24.13 kPa] without failure or leakage and at least 125 percent of pressure developed in the tank from any ram air effect. Furthermore, they must withstand without failure any vibration, inertia loads, and fluid and structural loads to which they may be subjected during operation of the aircraft. Fuel-tank strength and durability are proven by means of extensive pressure and vibration tests prescribed in FAR Parts 23 and 25.

Fuel tanks located within the fuselage of a transport aircraft must be capable of withstanding rupture and the inertia forces that may be encountered in an emergency landing. These forces are specified as 4.5 g (gravity) downward, 2.0 g upward, 9.0 g forward, and 1.5 g sideward. Such tanks must be located in a protected position so that exposure to scraping action with the ground will be unlikely. For pressurized fuel tanks, a means with fail-safe features must be provided to prevent buildup of an excessive pressure differential between the inside and outside of the tank.

Fuel tanks must be equipped with sumps to collect sediment and water. On transport-type aircraft the sump capacity must be at least 0.10 percent of the total tank capacity or $\frac{1}{16}$ gal [0.24 L], whichever is greater; and on other types of aircraft the sump capacity must be 0.25 percent of the tank capacity or $\frac{1}{16}$ gal [0.24 L], whichever is greater. The construction of the tank must be such that any hazardous quantity of water in the tank will drain to the sump when the airplane is in the ground attitude. If the fuel system for small aircraft is supplied with a sediment bowl that permits ground inspection, the sump in the tank is not required. Fuel sumps and sediment bowls must be equipped with an accessible drain to permit complete drainage of all the sump or sediment bowl. The drain must discharge clear of all portions of the airplane and must be provided with means for positive locking of the drain in the closed position, either manually or automatically.

Fuel tanks are required to have an expansion space of not less than 2 percent of the tank capacity. In the case of nontransport aircraft, if the tank vent discharges clear of the aircraft, no expansion space is required. The construction of the tank must be such that filling the tank expansion space is not possible when the airplane is on the ground. Fuel tanks that may be filled through a pressure fitting under the wing are equipped with automatic shutoff devices to prevent overfilling the tank.

The fuel-tank filler connection must be designed in such a manner that any spilled fuel will drain clear of the airplane and will not enter the wing or any portion of the fuselage. This is usually accomplished by means of a fuel-tight scupper and overboard drain.

The filler connection must be marked with the word FUEL, the minimum grade of fuel, and the capacity of the tank. Each fuel-filling point, except pressure-fueling connection points, must have a provision for electrically bonding the airplane to ground fueling equipment.

The total usable capacity of the fuel tanks for non-transport-category aircraft must be sufficient for not less than $\frac{1}{2}$ h of operation at rated maximum continuous power. Fuel tanks having a capacity of 0.15 gal [0.569 L] for each maximum (except takeoff) horsepower are considered suitable. Fuel-quantity indicators must be calibrated to show 0 gal when the usable fuel supply is exhausted.

The fuel capacity for transport-type aircraft is determined by the manufacturer in accordance with the intended use of the aircraft. Provision is made for the maximum range of the aircraft plus more fuel for conditions where the aircraft is diverted to an alternate field.

A number of special requirements are provided for the installation of fuel tanks in aircraft. These requirements are established principally to provide for safety, reliability, and durability of the fuel tank.

The method for supporting tanks must be such that the fuel loads are not concentrated at any particular point of the tank. Nonabsorbent padding must be provided to prevent chafing and wear between the tank and its supports. Synthetic-rubber bladders or flexible fuel-tank liners must be supported by the structure so that they are not required to withstand the fuel load. The interior surface of the tank compartment must be smooth and free of any projections that might damage the liner unless the liner is provided with suitable protection at possible points of wear. A **positive pressure** must be maintained within the vapor space of all bladder cells under all conditions of operation, including the critical condition of low airspeed and rate of descent likely to be encountered in normal operation. Pressure is maintained by means of a tank vent tube, the open end of which is faced into the wind to provide continuous ram air pressure.

Fuel-tank compartments must be ventilated and drained to prevent the accumulation of flammable fluids or vapors. The compartments adjacent to tanks that are an integral part of the airplane structure must also be ventilated and drained.

Fuel tanks must not be installed on the engine side of the fire wall or in any fire zone unless the construction and arrangement are such that the zone can be proved to be as safe as it would be if the tank were not in the zone. Not less than $\frac{1}{2}$ in [12.7 mm] of clear air space must be provided between the fuel tank and the fire wall and between the tank and any fire zone. No portion of the engine nacelle skin that lies immediately behind a major air-egress opening from the engine compartment may act as the wall of an integral tank. This requirement is to prevent the spread of fire from the engine compartment to the fuel tanks.

Fuel tanks must not be installed in compartments that are occupied by passengers or crew, except in the case of single-engine airplanes. In this case, a fuel tank that does not have a capacity of more than 25 gal [94.75 L] may be located in personnel compartments if adequate drainage and ventilation are provided. In all other cases, the fuel tank must be isolated from personnel compartments by means of fume-proof and fuel-proof enclosures.

Spaces adjacent to tank surfaces must be ventilated to avoid fume accumulation due to minor leakage. If the tank is in a sealed compartment, ventilation may be limited to drain holes large enough to prevent excessive pressure resulting from altitude changes.

Each fuel tank must be vented from the top part of the expansion space. Each vent outlet must be located and constructed to minimize the possibility of its being obstructed by ice or other foreign matter and to prevent the siphoning of fuel during normal operation.

The venting capacity must allow the rapid relief of excess pressure differences between the interior and exterior of the tank. Air spaces of tanks with interconnected outlets must be interconnected to maintain equal pressure in all interconnected tanks.

Vent lines must be installed in such a manner that there are no undrainable points in any line where moisture can accumulate with the airplane in either the ground or level-flight attitude. No vent may terminate at a point where the discharge of fuel from the vent outlet will constitute a fire hazard or from which fumes may enter personnel compartments. The vents must be arranged to prevent the loss of fuel, except fuel discharged because of thermal expansion, when the airplane is parked on a ramp.

Carburetors with **vapor-elimination** connections must have vent lines to lead vapors back to one of the fuel tanks. If there is more than one tank and if it is necessary to use these tanks in a definite sequence for any reason, the vapor vent return line must lead back to the fuel tank to be used first unless the relative capacities of the tanks are such that return to another tank is preferable.

For acrobatic-category airplanes, excessive loss of fuel during acrobatics, including short periods of inverted flight, must be prevented. Fuel must not siphon from the vent when normal flight has been resumed after any acrobatic maneuver for which the airplane is approved.

Fuel-tank outlets must be provided with **fuel screens.** If a fuel booster pump is installed in the tank, the pump inlet must be provided with a screen. The screen for reciprocating-engine-powered aircraft must have 8 to 16 meshes per inch [2.54 cm]. Turbine-engine-powered airplanes require fuel tank screens adequate to prevent the passage of any object that could restrict fuel flow or damage any fuel system component. The clear area of each fuel tank screen or strainer must be at least five times the cross-sectional area of the outlet line, and the diameter of the strainer must be at least that of the tank outlet. Each finger strainer must be accessible for inspection and cleaning.

Types of Fuel Tanks

There are various basic types of fuel tanks designed for use in aircraft. The specific type chosen when designing the aircraft is a result of the available technology at the time the aircraft was designed, the size and shape of the tank area, and the types of operations for which the aircraft is designed.

Fuel tank construction can be divided into three basic types: *integral, rigid removable,* and *bladder.*

Integral Fuel Tanks

An **integral fuel tank,** shown in Fig. 15-1, is a tank that is part of the basic structure of the aircraft. Integral fuel tanks have commonly been located in the wing or fuselage, as illustrated in Fig. 15-2, but may be located in other locations such as the horizontal stabilizer. Most of these tanks cannot be removed because they are an integral and permanent part of the aircraft structure. Integral fuel tanks are formed by using structural members of the wing to form a fuel-tight tank. This is accomplished through the use of the wing skin, spars, ribs, stiffeners and stringers to form the tank and by applying sealing materials to areas where these members are joined. Sealing materials are generally synthetic rubber, either in a molded form or in an uncured state. The use of sealant is held to a minimum to avoid excess weight.

Integral fuel tanks are normally constructed of the same material as the surrounding aircraft structure and sealed with a fuel-proof sealing compound. Integral fuel tanks must be provided with access panels in the skin that may be removed in order to inspect the condition of the fuel tank. A typical access-door installation is shown in Fig. 15-3. The tank-access doors pictured fit inside the tank to seal the oval cutouts in the lower wing surface. A molded rubber seal fits between the inner skin and the door. A clamp ring mates into recesses on the door and the skin to secure the door. The assembly is flush to the lower skin when installed. A knitted aluminum gasket provides electrostatic bonding of the access doors.

Note: The aircraft must be defueled prior to removal of the tank-access doors.

Baffles are frequently installed inside fuel tanks. The primary purpose of a baffle is to reduce sloshing of the fuel. A baffle is attached to the tank structure and has several holes cut in it to allow free flow of fluid; sufficient structure

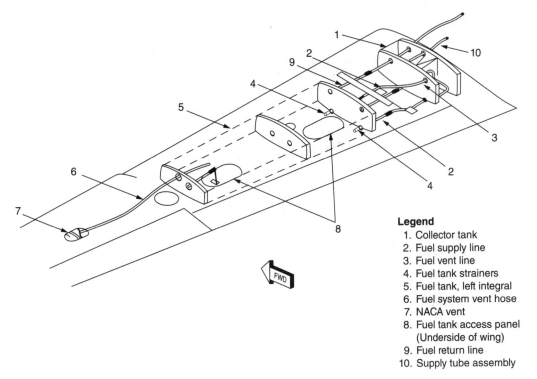

Legend
1. Collector tank
2. Fuel supply line
3. Fuel vent line
4. Fuel tank strainers
5. Fuel tank, left integral
6. Fuel system vent hose
7. NACA vent
8. Fuel tank access panel
   (Underside of wing)
9. Fuel return line
10. Supply tube assembly

FIGURE 15-1   An integral fuel tank.

must remain to prevent the rapid movement characteristic of sloshing, which is particularly noticeable when the aircraft encounters turbulent air or makes rapid changes in attitude. The baffles may have check valves, as illustrated in Fig. 15-4, installed in various members to reduce the rate of fuel flow toward the wing tips during airplane attitude changes and to maintain fuel in the boost-pump area of the fuel tank. The check valve shown is hinged at the top; it opens in the inboard direction only.

When an integral type of tank is used in a wing, the aircraft is said to have a "wet wing." This type of tank is found on some high-performance light aircraft and on most turbine-powered transport-category aircraft.

Rigid Removable Tanks

A **rigid removable fuel tank** is one that is installed in a compartment designed to hold the tank. The tank must be fuel-tight, but the compartment in which it fits is not fuel-tight. The tank is commonly made of aluminum components welded together.

The compartment in which a rigid removable fuel tank is installed is structurally complete and does not rely on the fuel tank for structural integrity. The tanks are normally held in the compartment by several padded straps, as shown in Fig. 15-5. If the compartment is in a wing or fuselage area separate from the cockpit-cabin area, then an access panel is used to cover the fuel-tank compartment.

As shown in Fig. 15-5, the tank includes in its structure one or more fuel-feed lines near the bottom on the inboard side or on the bottom, a fuel-drain fitting at the tank low point when the aircraft is in a ground attitude, a fuel-vent line near the top of the tank, a fuel-filler cap or tube, and provision for a quantity indicator.

This type of fuel tank is found on the more inexpensive light aircraft.

Bladder Fuel Cells

A **bladder fuel cell,** or **tank,** is essentially a reinforced rubberized bag placed in a non-fuel-tight compartment designed to structurally carry the weight of the fuel. The bladder incorporates all the components required for removable rigid fuel tanks, such as a vent, drain, quantity indicator, etc.

The airframe compartment for the bladder is structural, and a fairly small opening is provided for the insertion of the bladder. The bladder must be rolled up, inserted into the compartment, and then unrolled. The bladder is held in place by several "buttons" or "snaps," which attach the bladder to the top, bottom, and sides of the compartment. A typical bladder and its components are shown in Fig. 15-6. These types of fuel tanks are found on many medium- to high-performance light aircraft and some turboprop and turbine-powered aircraft.

Surge Tanks

Surge fuel tanks are normally located on transport-category aircraft and are constructed the same as integral fuel tanks. Surge tanks, which are normally empty, are designed to contain fuel overflow and prevent fuel spillage, particularly when fueling the aircraft. Surge tanks are also an integral part of the fuel-tank venting system discussed later in this chapter. The typical location of surge tanks is shown in Fig. 15-2.

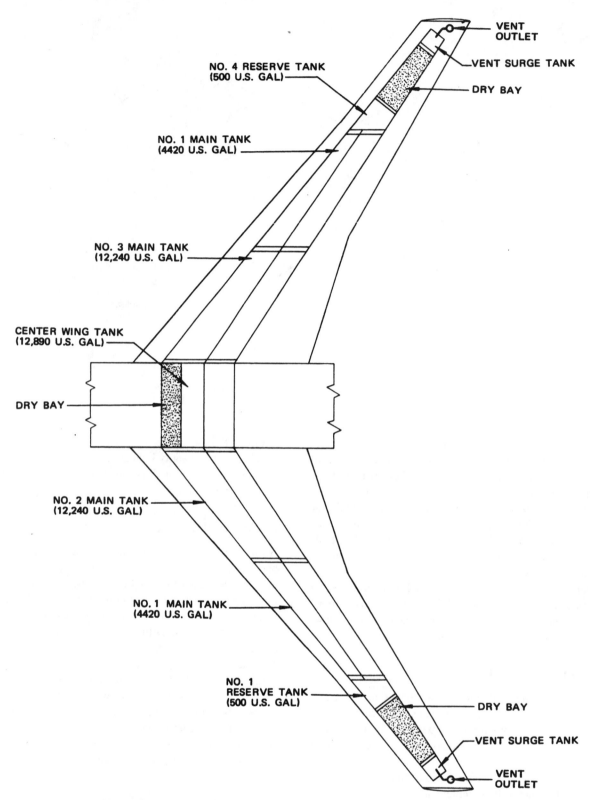

FIGURE 15-2   Integral fuel-tank arrangement. (*Boeing Commercial Aircraft Co.*)

Fuel-Tank Dripsticks

In addition to the fuel-quantity-indicating system, it is desirable to have a provision to visually determine the fuel quantity. On light aircraft this may be accomplished by viewing the tank through the fuel-filler-cap opening, but on large aircraft this would be extremely difficult. For this reason calibrated **dripsticks** are usually located on the lower surface of the wing.

A dripstick for measuring fuel is shown in Fig. 15-7. This dripstick is a nonelectric, hand-operated indicator that may be used to visually gauge the amount of fuel in the tanks

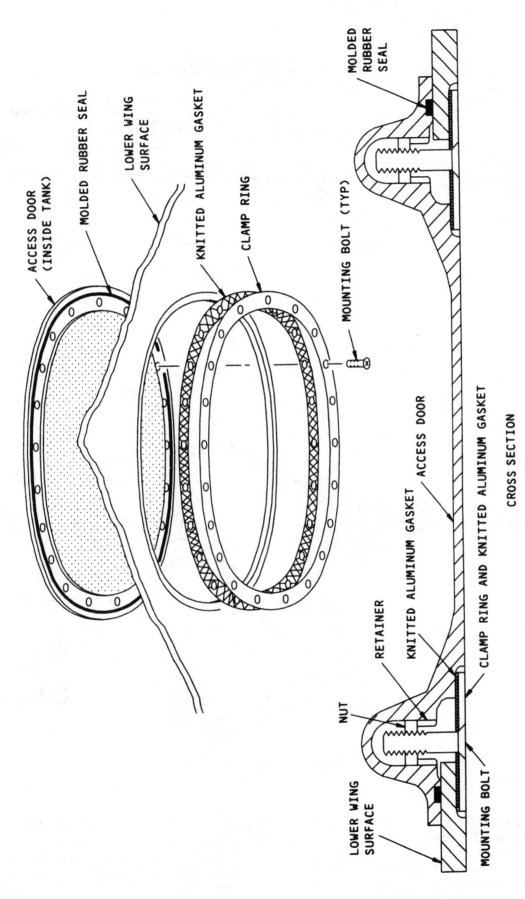

ACCESS DOOR (INSIDE TANK)

MOLDED RUBBER SEAL

LOWER WING SURFACE

KNITTED ALUMINUM GASKET

CLAMP RING

MOUNTING BOLT (TYP)

MOLDED RUBBER SEAL

ACCESS DOOR

CLAMP RING AND KNITTED ALUMINUM GASKET

NUT

RETAINER

KNITTED ALUMINUM GASKET

LOWER WING SURFACE

MOUNTING BOLT

CROSS SECTION

FIGURE 15-3   Integral fuel-tank access door. (*Boeing Commercial Aircraft Co.*)

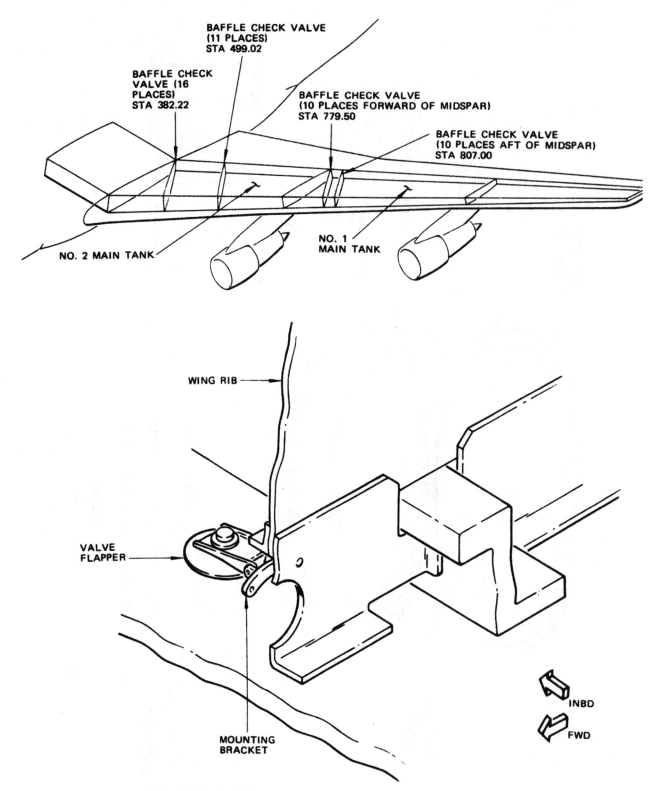

BAFFLE CHECK VALVE
(11 PLACES)
STA 499.02

BAFFLE CHECK
VALVE (16
PLACES)
STA 382.22

BAFFLE CHECK VALVE
(10 PLACES FORWARD OF MIDSPAR)
STA 779.50

BAFFLE CHECK VALVE
(10 PLACES AFT OF MIDSPAR)
STA 807.00

NO. 2 MAIN TANK

NO. 1
MAIN TANK

WING RIB

VALVE
FLAPPER

MOUNTING
BRACKET

INBD

FWD

FIGURE 15-4   Baffle check valve. (*Boeing Commercial Aircraft Co.*)

without risk of spillage. The dripstick consists of a hollow fiberglass tube graduated in inches, gallons, kilograms, or liters. To determine the fuel quantity in the tank, the measuring stick is unlocked and slowly lowered until the top of the stick is level with the fuel quantity, at which point fuel flows into the upper end of the stick and drips out a small hole in the lower end. The amount that the dripstick has been lowered is then compared to a fuel-capacity table supplied with the aircraft to determine the fuel quantity in the tank. The dripstick can be removed without draining the fuel tank,

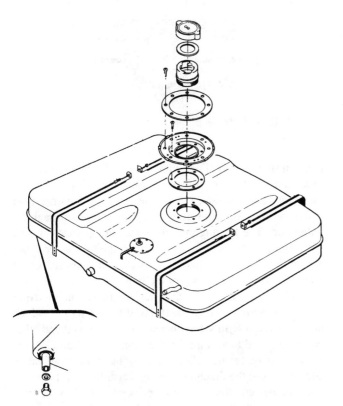

FIGURE 15-5 A rigid removable fuel tank. (*Cessna Aircraft Co.*)

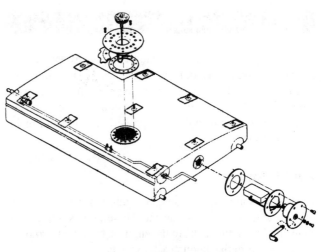

FIGURE 15-6 A bladder-type fuel cell. (*Cessna Aircraft Co.*)

since the fitting inside the tank incorporates a removal check valve. The dripsticks are located beneath the wing, as shown in Fig. 15-7.

In general, fuel-measuring dripsticks are less accurate than the fuel-quantity-indication system. The measuring sticks are generally used only as an alternative means of measuring fuel quantity. The airplane should be in a nearly level attitude when the sticks are used.

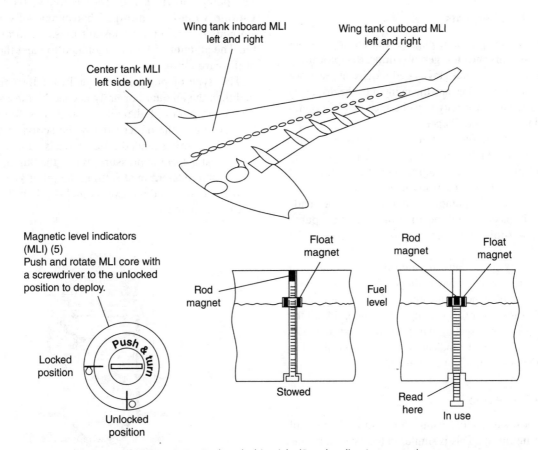

FIGURE 15-7 Fuel-tank dripstick. (*Bombardier Aerospace.*)

FUEL-SYSTEM COMPONENTS

The fuel system consists of many components, which are used to pump the fuel and control the direction of flow after it leaves the tank. The number and types of these units vary among aircraft, but the basic purpose of these items must be understood by the technician to properly inspect and maintain a fuel system.

Fuel Pumps

Fuel pumps are used to move fuel through the fuel system when gravity flow is insufficient. These pumps are used to move fuel from the tanks to the engines, from tanks to other tanks, and from the engine back to the tanks.

The principle of operation of fuel pumps is the same as for hydraulic and other types of pumps. However, because of the flammability of the fluid being pumped, the materials and design of the pumps are such as to virtually eliminate the danger of fuel ignition.

Fuel pumps are often classified according to the purpose of the pump such as **boost pump, scavenge pump,** or **cross-feed pump.** The technician should realize that any type of pump operation can be adapted to serve any of the pump purposes. In the following discussion we look at pumps and classify them according to their method of operation.

Fuel-Pump Requirements

Fuel systems for reciprocating engines and turbine engines require **main pumps** and **emergency pumps.** Reciprocating-engine systems that are not gravity-fed require at least one main pump for each engine, and the pump must be driven by the engine. The pump capacity must be such that it supplies the required fuel flow for all operations.

For turbine-engine fuel systems, there must be at least one main pump for each turbine engine. The power supply for the main pump for each engine must be independent of the power supply for each main pump for any other engine. For each main pump, a provision must be made to allow the bypass of each positive-displacement pump other than a fuel-injection pump approved as a part of the engine.

An emergency pump must be immediately available to supply fuel to the engine if any main pump fails. The power supply for each emergency pump must be independent of the power supply for each corresponding main pump. If both the normal pump and emergency pump operate continuously, there must be a means to indicate a malfunction of either pump to the appropriate flight crew members.

Vane-Type Fuel Pump

One of the most satisfactory pumps for positive delivery of fuel is the **vane pump.** This is similar to the hydraulic vane pump described in Chap. 13. A schematic diagram of such a pump is shown in Fig. 15-8. The rotor holds the sliding vanes

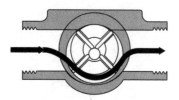

FIGURE 15-8 Vane-type fuel pump.

and is installed in the liner with its axis of rotation eccentric to the axis of the liner. When the rotor is turning, the vanes maintain a constant contact with the surface of the liner. Fuel enters the inlet port and is forced by sliding vanes through the outlet port. The floating pin aligns the sliding vanes against the surface of the liner. In one position the two lower vanes extend from the rotor, whereas the two upper vanes are forced into the rotor by the surface of the liner.

Fuel pumps vary in general design. There may be from two to six sliding vanes; sometimes the aligning pin, which holds the vanes against the liner surface, is omitted and other means are used to accomplish the same result. Some pumps use the syphon (metal bellows) type of balanced relief valve, and others use the diaphragm type. Regardless of the variations in design, the operating principle of all vane-type engine-driven fuel pumps remains the same.

Variable-Volume Pumps

The pump shown in Fig. 15-9 is known as a **variable-volume vane-type pump.** This pump delivers varying amounts of fuel under constant pressure to the carburetor. The amount of fuel is regulated to meet the demands of the carburetor.

This type of pump is designed to deliver much more fuel than the amount normally needed by the engine. Fuel enters the pump at the inlet side and is forced around the housing to the outlet side by the action of the sliding vanes. The spring-loaded relief valve is adjusted so that it releases at a specific pressure. When the pump pressure is above the predetermined setting, the relief valve is forced up from its seat and the excess fuel is relieved to the inlet side of the pump.

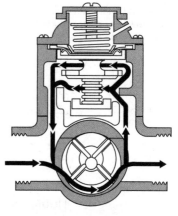

FIGURE 15-9 Variable-volume vane-type pump.

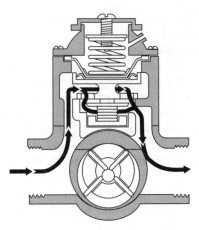

FIGURE 15-10   Fuel bypassing the fuel pump.

This relief valve has a diaphragm, which has two functions: (1) It provides venting to the atmosphere or to supercharger pressure, and, by means of a balancing action, (2) it helps maintain a constant discharge pressure no matter how much the pressure on the suction side of the pump may vary. When the diaphragm-type relief valve is used in a pressure-type fuel system, the pilot may complain that the fuel pressure is normal on the ground but increases with altitude. This may occur when the air vent on the fuel pump is partly clogged.

The **bypass valve,** shown in the pump diagram of Fig. 15-10, provides a means for the boost pump to force fuel around the vanes and rotor of the main pump for starting the engine or for emergency operation if the main pump fails. The bypass valve is held in the closed position by a spring when the pump rotor is turning.

The relief-valve spring in some pumps is located within a metal bellows to compensate for variations in atmospheric pressure. Other types use a fuel-resistant synthetic-rubber diaphragm instead of bellows. When relief valves are designed to compensate for atmospheric-pressure variations, they are known as *balanced relief valves.*

The chamber housing the diaphragm of the fuel pump incorporates a tapped hole to which a vent line is attached. As the airplane climbs to higher altitudes, the atmospheric pressure on the fuel in the tank decreases, resulting in a lowered fuel-pump-inlet pressure. To compensate for this lower pressure, the relief-valve diaphragm is vented to the atmosphere or to the fuel tanks. Thus, the decrease in atmospheric pressure in the tanks is compensated for by the pressure-relief valve. Sometimes the diaphragm is vented to the carburetor deck or to the carburetor air scoop. This subjects the diaphragm to atmospheric pressure plus ram pressure. Ram pressure is pressure that is developed in the carburetor air scoop by the forward speed of the airplane. The vent has a restrictor fitting in the line. If a diaphragm is ruptured, the restrictor limits the loss of fuel through the vent when the pump-inlet pressure is greater than atmospheric pressure or restricts the amount of air entering the pump inlet when the pump-inlet pressure is less than that of the atmosphere.

To adjust the pressure setting on a fuel pump, the adjusting-screw locknut is loosened and the adjusting screw is turned

FIGURE 15-11   An electrically operated centrifugal fuel pump.

clockwise to increase pressure or counterclockwise to decrease pressure. Usually, markings on the pump body show the correct direction to turn the adjusting screw to increase or decrease pressure. If no markings are visible, turn the screw very slightly in one direction while noting the effect on the pressure gauge. When the pressure has been set, the locknut should be tightened and safetied with the proper gauge safety wire.

Centrifugal Pumps

Centrifugal pumps are used to move fuel from fuel tanks to engines and other tanks. These pumps are electrically operated and may be designed to operate at one speed or may be designed for the crew member to select one of several speeds, depending on the operational situation. A centrifugal pump is shown in Fig. 15-11.

The centrifugal pump pressurizes the fuel by drawing fuel into the center inlet of a centrifugal impeller and expelling it at the outer edge of the impeller, as shown in Fig. 15-12. This type pump may be designed to be mounted with the electric motor on the outside of the fuel tank or for the entire pump and motor to be located inside the fuel tank, in which case it is referred to as a **submerged pump.** Because a centrifugal impeller is used, fuel can flow through the pump when the pump is not in operation. This eliminates the need for any bypass mechanism.

For ease of pump maintenance, most pumps are installed in or on the fuel tank with a mechanism that allows the pump to be removed and installed without draining the fuel tank. The mechanisms that close off the fuel lines and allow pump removal may be automatically operated as the pump is removed or may have to be activated by the technician prior to pump removal.

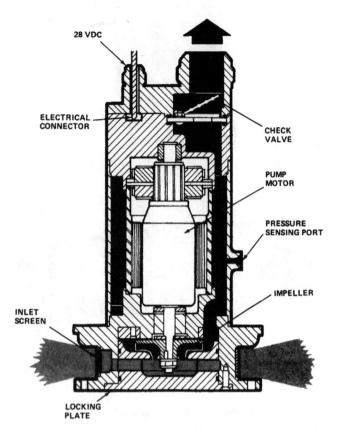

FIGURE 15-12 Cutaway of a centrifugal fuel pump.

Ejector Pump

An **ejector pump** is normally used to scavenge fuel from remote areas of fuel tanks and to provide fuel under pressure to an operating engine fuel-control unit. This type of pump has no moving parts but relies on the flow of returned fuel from the engine-driven pump to pump fuel. Figure 15-13 is an illustration of an ejector pump.

The ejector pump works on the venturi principle. Engine-driven fuel pumps supply the engine fuel-control unit with more volume than is necessary for operation in order to ensure that the engine is never starved for fuel. Excess fuel from this pump is routed back to the motive flow inlet of the ejector pump. This returned fuel is at a high pressure, on the order of 300 psi [2068.5 kPa], but at a low volume. As the motive flow fuel exits the ejector nozzle in the venturi area, a pressure drop is created and fuel from the inlet screen is drawn into the low-pressure area. The motive flow continues on through the venturi and draws the fuel from the tank with it. This induced flow of fuel is at a pressure of about 30 psi [206.9 kPa] but is of great enough volume to supply the engine-driven fuel pump.

If the ejector pump is used as a fuel-scavenge pump, the induced flow is delivered to an area of the fuel tank next to the tank outlet.

Since ejector pumps require a motive flow for operation, they do not begin operating until the engines are running. Centrifugal pumps are used in conjunction with ejector pumps to start the engine. After the engines are started, the centrifugal pumps are turned off and the ejectors maintain the required flow.

Fuel Strainers and Filters

Because of the ever-present possibility of fuel contamination by various types of foreign matter, aircraft fuel systems are required to include **fuel strainers** and **filters.** The fuel is usually strained at three points in the system: first, through a finger strainer or boostpump strainer in the bottom of the fuel tank; second, through a master strainer, which is usually located at the lowest point in the fuel system; and third, through a strainer in the carburetor or near the fuel-control unit.

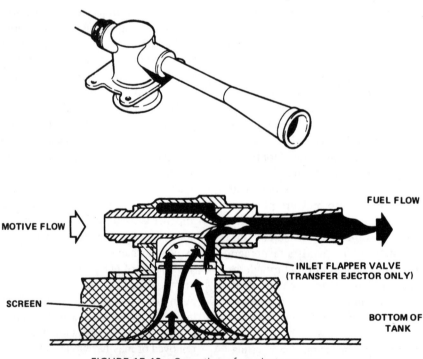

FIGURE 15-13 Operation of an ejector pump.

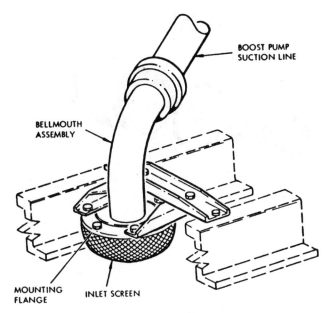

FIGURE 15-14 Boost-pump inlet screen.

A fuel strainer or filter is required between the fuel-tank outlet and the inlet of either the fuel-metering device (carburetor or other fuel-control unit) or an engine-driven positive-displacement pump, whichever is nearer the fuel-tank outlet. An example of an inlet screen designed to keep foreign material from reaching the fuel boost pump is shown in Fig. 15-14. This screen consists of a large bellmouth assembly mounted on the inlet end of the boost-pump suction line. The strainer or filter must be accessible for draining and cleaning and must incorporate a screen or an element that is easily removable.

The strainer or filter must have a sediment trap and adequate provision for easy drainage. It must be mounted in such a manner that its weight is not supported by the connecting lines or by the inlet or outlet connections of the strainer or filter. The fuel-flow capacity of the strainer or filter must be such that the flow of fuel will remain adequate even with the fuel contaminated to the level that has been established for that particular engine. The mesh size and the flow ports must be adequate to filter the fuel properly and also allow adequate fuel flow for the engine.

The tank strainer keeps larger particles of foreign matter in the tank from entering the system lines. The main strainer, located at the lowest point in the system, collects foreign matter from the line between the tank and the strainer and also serves as a water trap. The filter, or screen, in the carburetor, fuel-injection unit, or near the fuel control in a turbine-engine system removes the extremely fine particles that may interfere with the operation of the sensitive valves and other operating mechanisms. Some filters are so fine that they remove all particles larger than 40 μm in diameter (1 μm is one-thousandth of a millimeter).

A main fuel screen for a light airplane is shown in Fig. 15-15. This unit includes a sediment bowl and a screen and is called a gascolator. It consists of a cast-metal top, or cover, the screen, and the bowl, together with the necessary assembling parts. The bowl is attached to the cover by a clamp and thumbnut. Fuel enters the unit through the inlet on the top,

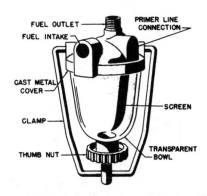

FIGURE 15-15 Main fuel screen for a light aircraft.

filters through the screen, and exits through a connection to the carburetor inlet. At designated periods the bowl is drained and the screen is removed for inspection and cleaning. The tank and carburetor outlet (finger) screens are also cleaned periodically, depending upon the installation, type of strainer, and inspection procedure established for the system.

An exploded view of one of the fuel-strainer assemblies for a light twin-engine airplane is shown in Fig. 15-16.

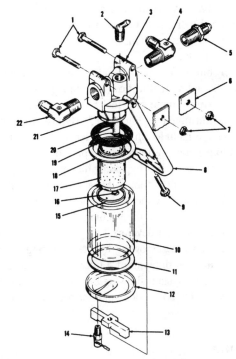

1. Bolt	11. Lower gasket
2. Elbow	12. Cap
3. Body assembly	13. Stiffener
4. Fitting for left engine installation	14. Drain valve
5. Fitting for right engine installation	15. Lid assembly
6. Mounting bracket	16. Retainer spring
7. Nut	17. Filter
8. Arm assembly	18. Upper gasket
9. Bolt	19. Flat screen
10. Glass bowl	20. Standpi...
	21. Filt...
	22. Elb...

FIGURE 15-16 Fuel-strainer assembly for a lig... airplane. (*Cessna Aircraft Co.*)

Fuel-System Compone...

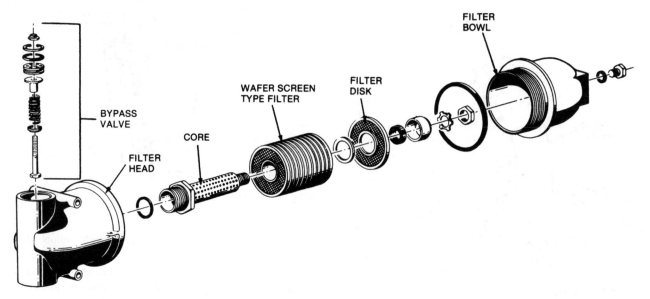

FIGURE 15-17 Fuel-filter bypass valve.

This fuel strainer is located in the nacelle of each engine and is attached to the lower center part of the fire wall. A quick-drain valve is incorporated at the bottom of the unit to permit the drainage of a small amount of fuel during the preflight inspection. This is to ensure that no water will reach the carburetor.

Fuel filters for turbine-engine systems include bypass valves that allow fuel to flow even though the filter becomes clogged. An example of a fuel filter equipped with a bypass valve is illustrated in Fig. 15-17. **Bypass warning system** switches are included in many filter assemblies to provide a signal to the cockpit notifying the pilot or other crew member that a particular filter is bypassing fuel or is nearing a condition where it will bypass fuel. A bypass warning switch in some filters is operated by the movement of the bypass valve as it opens. In other filters, the warning switch is operated by the pressure differential developed as pressure on the input side of the filter increases owing to filter clogging. Figure 15-18 is an installation drawing of a micronic filter assembly for a turbine-engine system showing the filter bowl, filter element, filter drain tube, bowl packing, filter head, and electrical connections for the bypass warning switch.

Drains

Aircraft fuel systems must be provided with **drains** such that the entire system can be drained with the airplane in its normal ground attitude. Drains are available at fuel strainers, as shown in Fig. 15-16, and at sumps as well as at other locations.

Fuel sump drain valves are used for draining accumulated moisture from fuel tanks and for draining trapped fuel remaining after defueling. Each drain valve must be located so that it will discharge clear of all parts of the airplane. The drains must have manual or automatic means for positive locking in the closed position. Each drain must have a drain valve that is readily accessible and can be opened and closed easily.

The drains must be located or protected to prevent fuel spillage in the event of a landing with landing gear retracted.

A typical example of a drain valve utilized on transport-category aircraft is shown in Fig. 15-19. This valve consists of a poppet valve, flapper valve, and screen protecting the unit from contaminants that may block the drain passages. The valve is spring-loaded closed. To drain the fuel sump, the poppet is pushed up to open the valve and allow fuel to drain through the drain hole in the center of the valve. The valve incorporates a flapper valve that allows the poppet assembly to be removed without defueling the fuel tank.

Fuel-Selector and Shutoff Valves

Fuel-selector valves provide a means of shutting off the fuel flow, selecting the tank from which to draw fuel in a multiple-tank installation, transferring fuel from one tank to another, and directing the fuel to one or more engines in a multiengined airplane. One or more of these valves are used to shut off all fuel flow to each engine. Such a valve must be positive and quick-acting, and the valve controls should be located within easy reach of the pilot (or pilots) or the flight engineer. In multiple-tank installation, the valve arrangement should be so arranged that each tank can be used separately. No shutoff valve may be on the engine side of any firewall. There must be means to guard against inadvertent operation of each shutoff valve and to allow flight crew members to reopen each valve rapidly after it has been closed.

Each shutoff valve and fuel-system control must be supported so that loads resulting from its operation or from accelerated flight conditions are not transmitted to the lines connected to the valve. Each valve and fuel-system control must be installed so that gravity and vibration will not affect the selected position.

Each fuel valve handle and its connections to the valve mechanism must have design features that minimize the possibility of incorrect installation.

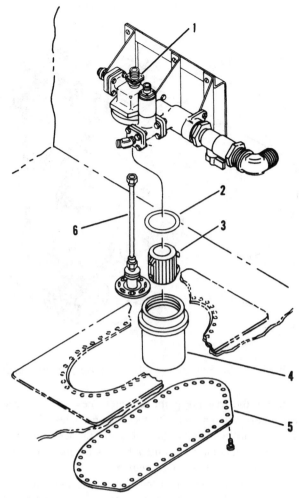

1. Electrical connection for fuel pressure indicator on filter head
2. Packing
3. Filter element
4. Filter bowl
5. Access cover
6. Filter drain hose

FIGURE 15-18   Fuel-filter assembly for a turbine-engine aircraft. (*Sikorsky Aircraft Corp., Division of United Technologies Corp.*)

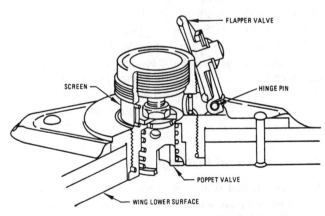

FIGURE 15-19   Fuel-tank drain valve. (*Boeing Commercial Aircraft Co.*)

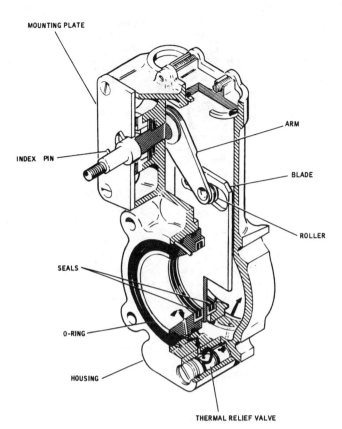

FIGURE 15-20   Fuel shutoff valve. (*Douglas Aircraft Co.*)

Fuel-shutoff valves for large aircraft, shown in Fig. 15-20, are interconnected with fire protection systems so fuel may be shut off automatically if there is an overheat or fire situation in any engine nacelle. The valve shown is a manually operated gate valve that is located in each engine fuel-supply line. The valves are operated by the fire-control handles in the cockpit and are designed to totally shut off all fuel flow to the engine. The valve consists of a valve body, valve arm, and slotted blade. These arrangements must comply with the regulations covering fire-suppression systems.

Fuel-selector valves for light airplanes are usually simple in design, construction, and operation. The selector valve shown in Fig. 15-21 consists of a cylindrical aluminum housing with several fuel-inlet ports around its circumference. The outlet port is located on the end. A center shaft, or plug, directs fuel from one of the inlet ports to the outlet connection. This center shaft is connected by linkage to the cockpit control. When the control is in the OFF position, the center shaft in the valve does not line up with any fuel connections around the circumference of the selector casing.

An exploded view of a cam-operated ball-type valve for light aircraft is shown in Fig. 15-22. As the cam is rotated, it is possible to turn on one or both of the tanks to which the valve is connected. This type of valve is usually employed where a main tank and an auxiliary tank are connected to the system.

Fuel-control valves in large transport aircraft are usually operated by electric motors. An electrically operated valve assembly of this type includes a reversible electric motor

Fuel-System Components   **525**

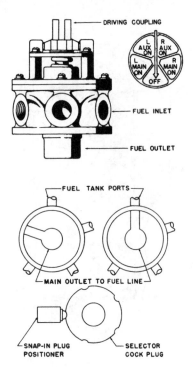

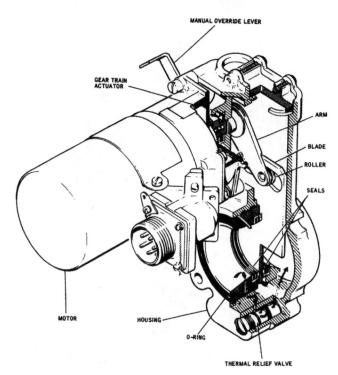

FIGURE 15-21   Fuel-selector valve for a light airplane.

FIGURE 15-23   Fuel-control valve. (*Douglas Aircraft Co.*)

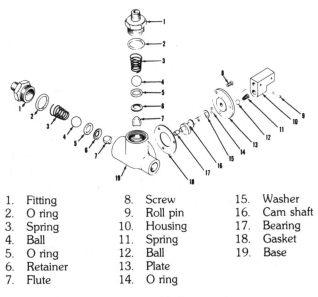

1.	Fitting	8.	Screw	15.	Washer
2.	O ring	9.	Roll pin	16.	Cam shaft
3.	Spring	10.	Housing	17.	Bearing
4.	Ball	11.	Spring	18.	Gasket
5.	O ring	12.	Ball	19.	Base
6.	Retainer	13.	Plate		
7.	Flute	14.	O ring		

FIGURE 15-22   Cam-operated ball-type valve.
(*Cessna Aircraft Co.*)

linked to a sliding valve assembly, as shown in Fig. 15-23. The motor moves the valve **gate** in and out of the passage through which the fuel flows, thus shutting the fuel off or turning it on. The valves are usually placed near the tanks with which they are associated.

Fuel Heaters

Fuel heaters are used with fuel systems for turbine engines to prevent ice crystals in the fuel from clogging system filters. Water in turbine fuel does not settle out quickly, as it does in gasoline. If the temperature of the fuel in the tanks

is below the freezing point of water, these water particles will freeze. As the fuel with these ice particles tries to pass through the fuel filters, they filter out the ice and become clogged. This causes the filter bypass to open and allows unfiltered fuel to flow to the engine.

To prevent these filters from being clogged, the fuel is passed through a fuel heater prior to entering the filter. The fuel heater is a heat exchanger that uses engine compressor bleed air, engine oil, or, if the hydraulic system generates sufficient heat, hydraulic fluid to heat the fuel above freezing and melt the ice crystals. The fuel and water mixture can then flow to the engine and be burned. The ratio of water to fuel is very slight and does not affect the operation of the engine.

Filler Caps

The filler cap must provide a tight seal and be designed so that it cannot come off in flight. Fuel-tank venting may be provided through the fuel cap. A typical fuel cap is illustrated in Fig. 15-24. The fuel cap that is illustrated is removed by lifting and rotating the handle located in the center of the cap. A chain prevents the cap from being lost when removed. When installed, the fuel cap is flush with the surface of the wing and a tight seal is provided by the O-ring, as shown in Fig. 15-24.

Fuel Lines and Fittings

In an aircraft fuel system, the various components are connected by means of aluminum alloy, copper, or other types of tubing and flexible hose assemblies with approved connecting fittings. A typical fuel-hose assembly is shown in Fig. 15-25. This hose is made of synthetic rubber and is

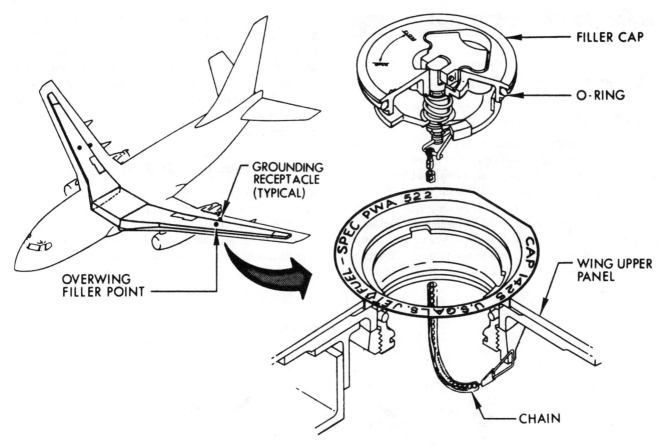

FILLER CAP

O-RING

GROUNDING
RECEPTACLE
(TYPICAL)

OVERWING
FILLER POINT

WING UPPER
PANEL

CHAIN

FIGURE 15-24   Fuel filler cap. (*Boeing Commercial Aircraft Co.*)

SYNTHETIC
COVER

FIBER
BRAID

SYNTHETIC
TUBE

FIGURE 15-25   Cutaway of a fuel hose. (*The Weatherhead Company.*)

reinforced with fiber braid embedded in bonding material. The hose conforms to MIL-H-5593A and is connected with an AN-773 hose end assembly. Another hose assembly suitable for fuel systems is shown in Fig. 15-26. This is the Weather head 3H-241 hose assembly equipped with reusable-type end fittings. This hose has a working temperature range of −40 to +300°F [−40 to +149°C] when used with fuel. The hose is covered with stainless-steel braid. Fuel lines must be of a size that will provide the required fuel-flow rate under all conditions of operation.

The installation of the fuel lines must be such that there are no sharp bends—that is, bends with a radius of less than

three fuel-line diameters. Fuel lines must be routed so that there are no vertical bends where water or vapor can collect. A bend downward and then upward will permit water to collect, and a bend upward and then downward will permit vapor to collect.

If a copper line is used in a fuel system, it should be annealed after bending and at overhaul periods; however, aluminum lines should not be annealed. Parts of the fuel system attached to the engine and to the primary structure of the airplane must be connected by means of flexible hose assemblies. Flexible hose assemblies should also be used to connect a stationary unit with a unit subject to vibration.

Fuel lines must be securely anchored to prevent vibration during operation of the engine. Usually the mountings and clamps installed by the manufacturer will be adequate. Metal lines should be bonded to prevent radio interference. All fuel lines and fittings must be of an approved type.

Additional information on fuel-line selection, fabrication, routing, and inspection may be found in the text *Aircraft Basic Science*.

Fuel-System Instruments

A discussion of fuel-quantity-indicating systems, along with pressure, flow rate, and temperature monitoring instruments, is given in Chap. 17.

FIGURE 15-26   Hose assembly with reusable fittings.

Fuel-System Components   **527**

All fuel systems can be classified into one of two broad categories: **gravity-feed** and **pressure-feed.**

Gravity-Feed Fuel Systems

A **gravity-feed** fuel system uses the force of gravity to cause fuel to flow to the engine fuel-control mechanism. For this to occur, the bottom of the fuel tank must be high enough to assure a proper fuel-pressure head at the inlet to the fuel-control component (i.e., carburetor) on the engine. In high-wing aircraft this is accomplished by placing the fuel tanks in the wings. An example of this type of system is shown in Fig. 15-27. In this example fuel flows by gravity from the wing tanks through the feed lines to the fuel-selector valve.

After passing through the selector valve, the fuel flows through the fuel strainer and then continues on to the carburetor. Fuel for the primer is taken from the main fuel strainer. Since both tanks may feed fuel to the engine simultaneously, the space above the fuel must be interconnected and vented outside the aircraft. The vent line normally is routed to the underside of the wing, where the possibility of fuel siphoning is minimized.

Pressure-Feed Fuel Systems

A **pressure-feed** fuel system, a simple version of which is shown in Fig. 15-28, uses a pump to move fuel from the fuel tank to the engine fuel-control component. This arrangement is required because the fuel tanks are located too low for sufficient head pressure to be generated or because the tanks are some distance from the engine. The system in Fig. 15-28

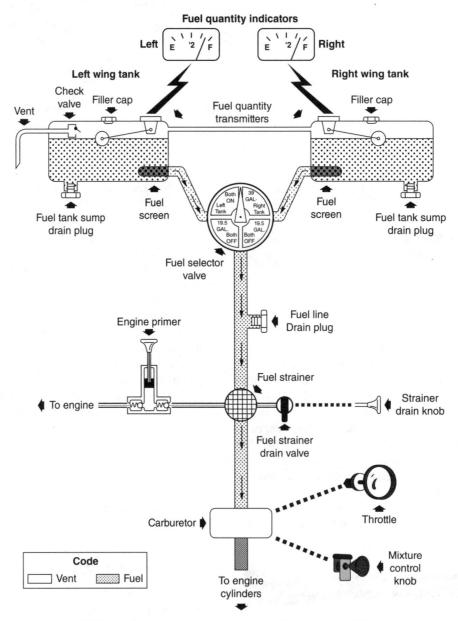

FIGURE 15-27 Gravity-feed fuel system. (*Cessna Aircraft Co.*)

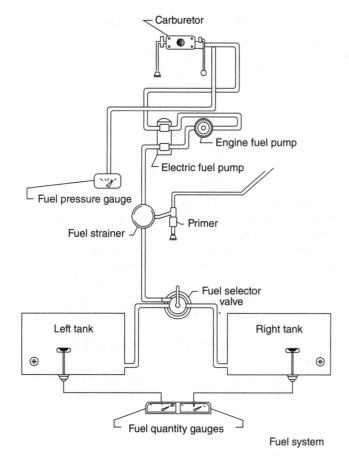

FIGURE 15-28   A light aircraft pressure-feed fuel system. (*Piper Aircraft Co.*)

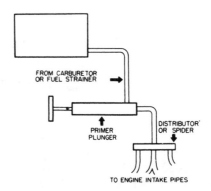

FIGURE 15-29   Priming system for a light aircraft engine.

is for a low-wing aircraft, where the wing tanks are on the same approximate level as the carburetor. In Fig. 15-28 fuel flows from the tanks through separate fuel lines to the fuel-selector valve. After leaving the selector valve, the fuel flows through the fuel strainer and into the electric fuel pump. Note that the engine-driven fuel pump is in parallel with the electric pump so that the fuel can be moved by either pump without the need for a bypass valve. The fuel boost pump supplies fuel for starting the engine, and the engine-driven pump supplies the fuel pressure necessary for normal operation. During high-altitude operation, takeoff, and landing, the boost pump is operated to ensure adequate fuel pressure.

Most large aircraft and aircraft with medium- to high-powered engines require a pressure-feed system, regardless of fuel-tank location, because of the large volume of fuel that must be delivered to the engines at a high pressure.

When reference is made to high pressure in the fuel-feed system, the value is on the order of 20, 30, or 40 psi [137.9, 206.9, or 275.8 kPa], not in the thousands of pounds per square inch as in hydraulic systems.

FUEL SUBSYSTEMS

Fuel subsystems are not involved in the direct flow of fuel from the tank to the engine but are used to prevent damage to the system, improve the operational capability of the aircraft, and eliminate hazardous conditions. All the systems discussed here are not used on all aircraft, but some, such as vent systems, are required on all aircraft fuel systems.

Primers and Priming Systems

Unlike an automobile engine, non-fuel-injected, reciprocating aircraft engines must often be primed before starting because the carburetor does not function properly until the engine is running. For this reason it is necessary to have a separate system to charge, or prime, the cylinders with raw fuel for starting. This is accomplished by the **priming system.** The usual arrangement is to have the primer draw fuel from the carburetor inlet bowl or fuel strainer and direct it to a distributor valve, which, in turn, distributes the fuel to the various cylinders. Figure 15-29 illustrates the action of a typical priming system used on a small aircraft engine. Other priming systems, used on internally geared supercharged engines, discharge the fuel into the supercharger diffuser section through a jet located on the carburetor or in the diffuser section.

The priming system shown in Fig. 15-29 has an undesirable feature in that the pump-body assembly is located in the cockpit. This causes the addition of fuel lines in the pilot's compartment and presents a fire hazard. For this reason, remote-control electric-solenoid primer valves have been developed. The solenoid valve is usually located on or near the carburetor and is wired to a control switch in the cockpit. When the switch is engaged, the solenoid-actuated valve opens and allows fuel, under booster-pump pressure, to enter the priming system. The valve is returned to the closed position by means of a spring. Engines equipped with fuel-injection systems do not require separate priming systems because the injection system pumps fuel directly into the intake ports of the engine.

Vent Systems

Fuel vent systems are designed to prevent the buildup of pressure in the fuel tanks and allow proper flow of fuel from the tanks. The fuel vent system maintains the fuel tanks at near-ambient atmospheric pressure under all operating conditions. Overpressurizing the tanks can cause structural damage. Underpressurizing can cause component malfunction and engine fuel starvation.

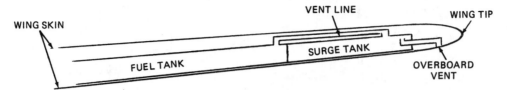

FIGURE 15-30 Typical arrangement of a surge tank.

The vent system for a light aircraft may be as simple as a hole drilled in the fuel cap or a tube running from the top portion of the fuel tank to the outside of the aircraft, as shown in Fig. 15-27. Some vent systems may have their openings pointing in the direction of flight to supply a slight pressure to the tank and aid the fuel flow to the engine.

Large aircraft have a vent system that allows each tank to vent to a high point in the system once tank pressure has exceeded some specified valve [normally in the 2- to 6-psi (13.8- to 41.4-kPa) range]. If tank pressure exceeds the predetermined value, a tank vent valve opens and vents the tank to a special vent compartment called a **surge tank, vent compartment,** or **vent box.** This compartment allows fuel to vent from the tanks, which prevents pressure buildup as well as fuel spilling overboard. The compartment is vented through a standpipe, which extends to the top of the

tank, as shown in Fig. 15-30. The compartment cannot fill under any normal conditions.

Many fuel-tank vent lines contain a float valve. The fuel vent float valve prevents fuel from entering the vent system during aircraft attitude changes. The valve illustrated in Fig. 15-31 consists of a float mounted on an arm, hinged at the mounting flange. The mounting base is installed over the inlet to the vent line. A valve is attached to the float arm contact; it closes the opening to the vent line when the float is contacted by rising fuel. The valve is normally held in the open position by the weight of the float and arm.

Vent lines are often discharged overboard the aircraft through a vent scoop, as shown in Fig. 15-32. A vent scoop is an aerodynamically designed duct that will maintain ambient pressure within the tank under all operating conditions.

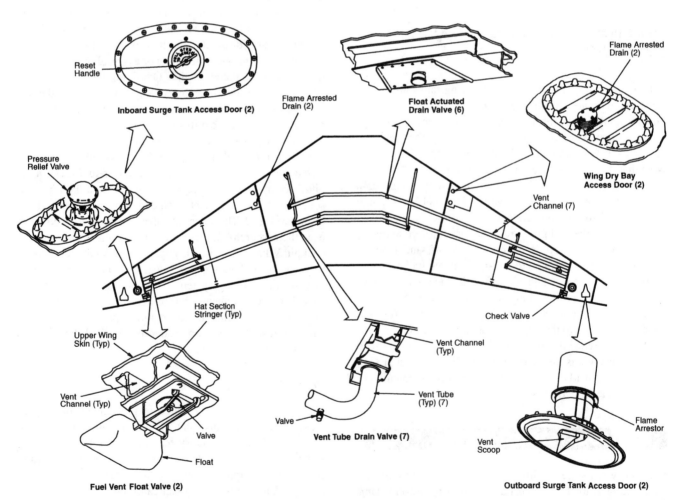

FIGURE 15-31 Fuel tank vent system.

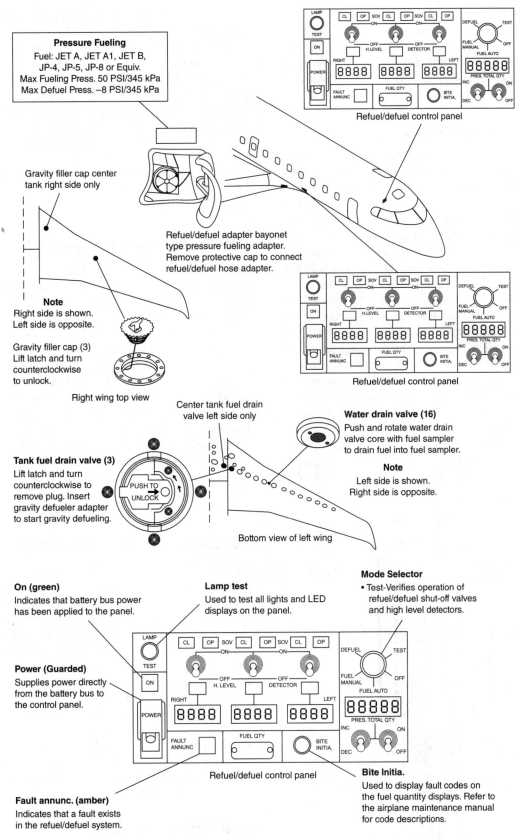

Pressure Fueling

Fuel: JET A, JET A1, JET B, JP-4, JP-5, JP-8 or Equiv.
Max Fueling Press. 50 PSI/345 kPa
Max Defuel Press. −8 PSI/345 kPa

Refuel/defuel control panel

Gravity filler cap center tank right side only

Refuel/defuel adapter bayonet type pressure fueling adapter. Remove protective cap to connect refuel/defuel hose adapter.

Note
Right side is shown. Left side is opposite.

Gravity filler cap (3) Lift latch and turn counterclockwise to unlock.

Right wing top view

Refuel/defuel control panel

Center tank fuel drain valve left side only

Water drain valve (16)
Push and rotate water drain valve core with fuel sampler to drain fuel into fuel sampler.

Tank fuel drain valve (3)
Lift latch and turn counterclockwise to remove plug. Insert gravity defueler adapter to start gravity defueling.

PUSH TO UNLOCK

Note
Left side is shown. Right side is opposite.

Bottom view of left wing

On (green)
Indicates that battery bus power has been applied to the panel.

Lamp test
Used to test all lights and LED displays on the panel.

Mode Selector
• Test-Verifies operation of refuel/defuel shut-off valves and high level detectors.

Power (Guarded)
Supplies power directly from the battery bus to the control panel.

Refuel/defuel control panel

Bite Initia.
Used to display fault codes on the fuel quantity displays. Refer to the airplane maintenance manual for code descriptions.

Fault annunc. (amber)
Indicates that a fault exists in the refuel/defuel system.

FIGURE 15-32 Single-point pressure fueling station. (*Bombardier Aerospace.*)

A flame arrester is usually mounted in the vent line between the surge tank and the vent scoop, as shown in Fig. 15-32. The arrester is a passive device, consisting of a stainless steel honeycomb core that acts as a heat sink to cool a flame below the ignition point. This device prevents an external flame from spreading into the surge tank.

Pressure Fueling Systems

A **pressure fueling system** is also referred to as a **single-point fueling system.** Technically there is a difference between these terms in that a single-point fueling system for a large aircraft uses only one fueling connection to fill all the fuel tanks. A pressure fueling system may use the same equipment to fuel the aircraft, but the aircraft may have separate fueling connections for each fuel tank. For the purpose of further discussion here, a pressure fueling system will be considered the same as a single-point fueling system.

A pressure fueling system uses pressure from the fueling station or truck to force fuel into the aircraft tanks. This is done through a fueling fitting located on the side of the fuselage or under a wing. A fueling-control panel may be located next to the fitting or be located inside the aircraft. The control panel allows the operator to monitor the fuel quantity in each tank and control the fuel valves for each tank. Through this panel the operator can fill each tank to the desired quantity. A fueling panel is shown in Fig. 15-33.

When the fueling connection is made at the fueling panel and the valve on the fuel hose is opened, the fueling manifold is pressurized with fuel. The operator can then select which tanks are to be filled by opening electrically operated fill valves.

When a tank becomes full, a float-operated shutoff valve such as the one shown in Fig. 15-33 is closed by the rising fuel level in the tank and prevents any more fuel from entering a tank. The operation of these valves not only prevents overfilling any tank, but it also allows all tanks to be filled from one position with a minimum of operator activity. The person doing the fueling can close the fill valve for any tank when the desired quantity is indicated or can allow all the tanks to be filled and terminate the fueling operation only when all the tanks indicate full and no more fuel is flowing into the aircraft.

Fuel Cross-Feed and Transfer System

On most multiengine aircraft, the fuel manifolds are connected in such a manner that any fuel tank may supply fuel to any engine. The main advantage of this type of system is its flexibility. Should an engine fail, its fuel is immediately available to supply the demand of the other engines. Also, if a fuel tank becomes damaged and loses fuel, the corresponding engine can be supplied with fuel from other tanks through the cross-feed manifold. It also makes it possible to transfer fuel from any tank to any other tank for the purpose of balancing the weight distribution and to maintain an acceptable position for the airplane's center of gravity.

A simple cross-feed system is illustrated in Fig. 15-34a. The cross-feed valve is in its normally closed position, with the fuel being supplied from the main tanks. However, the cross-feed valve may be opened at any time that it becomes necessary to feed an engine from an opposite fuel tank.

CG Control—Trim Tank Transfer

Some large transport aircraft use a trim tank transfer system to control the aircraft's center of gravity (CG). A trim tank is utilized in the horizontal stabilizer and fuel can be transferred to and from this tank (aft and forward transfer). The movement of fuel changes the aircraft's CG. When the aircraft is at cruise attitude the system optimizes the CG position to increase fuel economy by reducing drag. The normal operation is automatic, but the flight crew can manually select a forward transfer. The fuel control and management computer (FCMC) calculates the aircraft's CG and compares it to a target value. This target depends on the aircraft's actual weight. Based on this calculation, the FCMC determines the quantity of fuel to be moved aft or forward in flight. Figure 15-34b shows the trim tank schematic for aft transfer and Fig. 15-34c shows the forward transfer.

Fuel Jettison System

The fuel jettison system comprises a combination of fuel lines, valves, and pumps provided to dump fuel overboard during an in-flight emergency to reduce the weight of the airplane to allowable landing weight or to dump all the fuel except the reserve quantity required for landing. For transport-type aircraft, if the maximum takeoff weight for which the

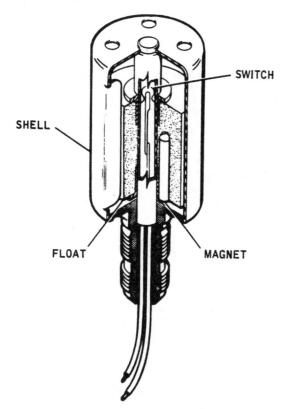

FIGURE 15-33 Float-operated shutoff valve. (*McDonnell Douglas Corp.*)

SWITCH

SHELL

FLOAT

MAGNET

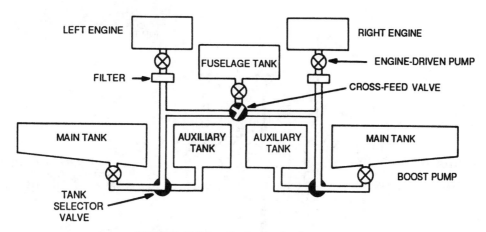

FIGURE 15-34a   Fuel cross-feed system.

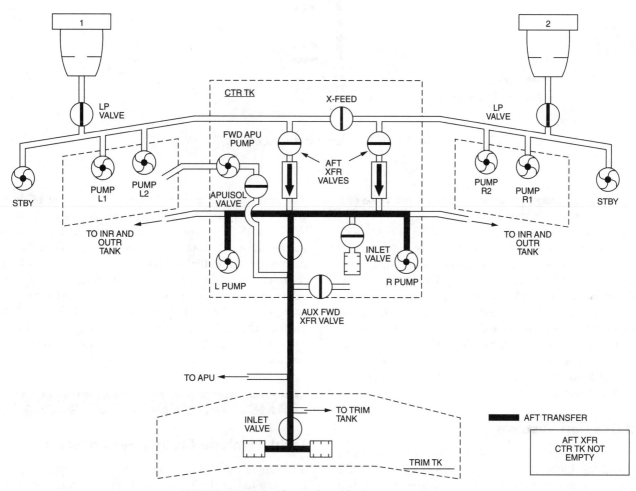

FIGURE 15-34b   Aft transfer of fuel to trim tank.

aircraft is certified is greater than 105 percent of the certified landing weight, provision must be made to jettison enough fuel to bring the weight of the airplane down to the certified landing weight. The average rate of fuel jettisoning must be 1 percent of the maximum takeoff weight per minute, except that the time required to jettison the fuel need not be less than 10 min.

The fuel jettison system and its operation must be free of fire hazards, and the fuel must discharge clear of any part of the airplane. Fuel fumes must not enter the airplane during operation of the jettison system. During operation of the system, the controllability of the airplane must not be adversely affected. Additional detailed requirements of the jettison system are specified in Federal Aviation Regulations.

Depending on the aircraft design, the force for fuel jettison is by gravity or by centrifugal pumps in the fuel tanks. Control valves have to be positioned to configure the fuel manifolds for dumping. Most aircraft dump the fuel from

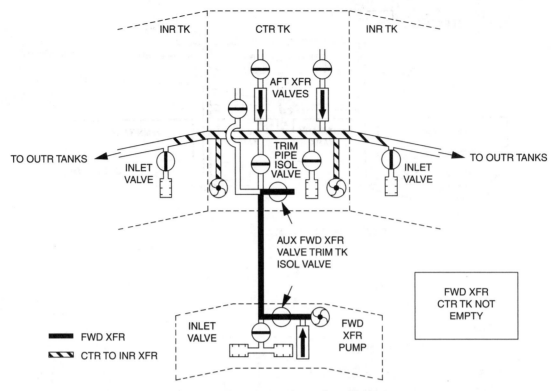

FIGURE 15-34c Forward transfer of fuel.

openings at the wing tips, but some aircraft make use of a dump chute that extends down from the wing.

A typical fuel jettison system is illustrated in Fig. 15-35. In this system fuel from the tanks is pumped into the fuel jettison manifold, which is continuous from wing tip to wing tip, with each end terminating at a fixed fuel jettison nozzle. The continuous manifold permits dumping of fuel from the opposite wing should the nozzle valve on one side fail to operate. Fuel flow through the manifold is controlled by the fuel jettison nozzle. The nozzle valve is an electrically operated valve controlled from the cockpit. After fuel passes through the nozzle valve, it flows into the jettison nozzle to be discharged into the air.

Oil-Dilution Systems

It is a well-known fact that oil congeals in cold weather, and such oil in the working parts of an engine makes the engine difficult to turn and, therefore, hard to start. A method known as **oil dilution** is employed to facilitate starting aircraft engines when the temperature is very low. With this method, a length of tubing connects the carburetor or other fuel-pressure source through a solenoid valve to the oil "Y" drain or elsewhere to the oil system.

The oil is diluted after the engine has been operated and is fully warmed up—for example, at the end of a flight before shutting down the engine. With the engine running at a fast idle, the oil-dilution-valve switch is turned on and held in the ON position for the time specified in the operator's manual. After dilution the engine is immediately shut down with the dilution valve open so that the fuel will remain in the oil.

The next time the engine is started, the diluted oil flows freely to all bearing surfaces, and the engine turns easily.

Figure 15-36 is a simplified schematic diagram showing the arrangement of an oil-dilution system. When the oil-dilution switch in the cockpit is closed, the valve is lifted from its seat by the solenoid. Fuel then flows through the inlet port and valve. It exits through the outlet port and is directed to the oil-system connection. The heat of the engine, after a few minutes of operation, vaporizes the fuel, and the fuel vapors pass out through the oil-system breathers.

TYPICAL AIRCRAFT FUEL SYSTEMS

Light Airplane Gravity-Feed System

The simple fuel system for a light single-engine airplane is shown in Fig. 15-37. The fuel tanks for this system are mounted in the wing roots and are constructed of welded aluminum alloy. The fuel tanks are held securely in place by means of padded steel straps attached to the wing spars. The padding is treated to make it nonabsorbent. The fuel filler caps are on the top of the wing, and the fuel outlets for the tanks are at the lowest point in the inboard end of the tanks. Fuel flow is by gravity to the fuel valve located forward of the fire wall. From the fuel valve the fuel flows to the "gascolator" (sump and screen assembly) in the lowest point of the system. From the sump, the fuel lines lead to the carburetor and to the primer pump, which is located on the instrument panel.

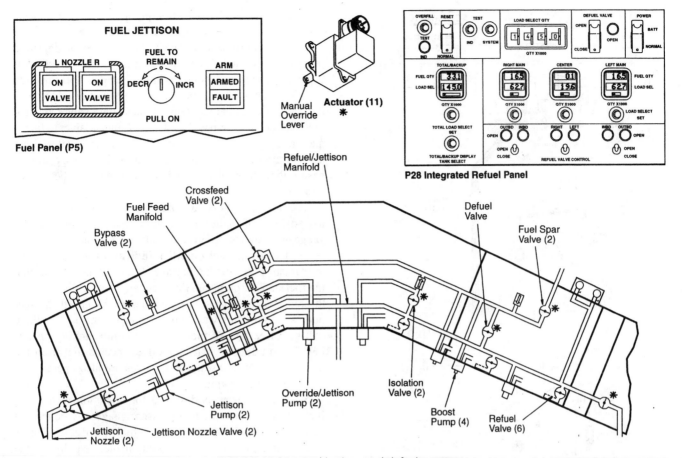

FIGURE 15-35　Fuel jettison and defuel systems.

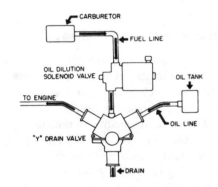

FIGURE 15-36　Arrangement for an oil-dilution system.

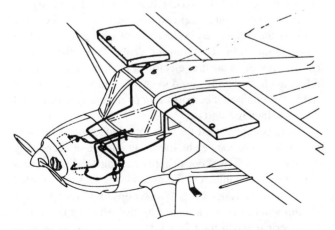

FIGURE 15-37　Fuel system for a light single-engine airplane.

Light Airplane Pressure-Feed System

The pressure-feed fuel system for a typical light airplane is shown in Fig. 15-38. This fuel system has two fuel cells made of fuel-resistant synthetic rubber. The cells are held in place in the wing compartments by means of snap fasteners. The fuel cells are vented to the atmosphere through vents in the fuel caps and vent lines. Metal fuel lines run from the fuel cells to the reservoir tanks, which are mounted in the side walls of the fuselage just behind the firewall.

From the reservoir tanks, fuel flows to the selector valve. The selector valve has four positions: RIGHT, LEFT, BOTH, and OFF. These four positions allow the pilot to select the fuel tanks used to supply the engine. OFF is used to shut off fuel flow to the area forward of the firewall when the aircraft is parked and when the system is being serviced.

From the fuel selector the fuel flows to the electric fuel pump. When the electric fuel pump is operated, the fuel system downstream of the pump is pressurized. This pump is used as a backup for the engine pump if it should fail and for starting the engine. The pump does not restrict flow when it is not running, which allows the engine-driven pump to draw fuel from the tanks.

The fuel next flows through a strainer to remove any contamination. From there it moves through flexible lines to the engine-driven fuel pump, which boosts the pressure up to the value needed by the fuel-control unit. Excess fuel volume generated by the pump is returned through a check valve to

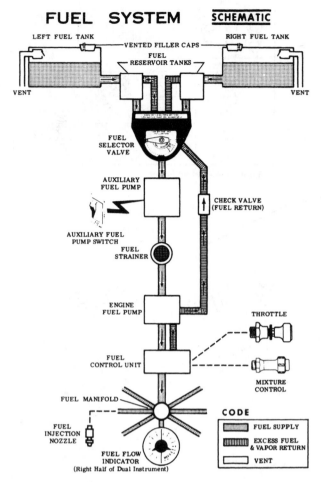

FUEL SYSTEM SCHEMATIC

LEFT FUEL TANK   RIGHT FUEL TANK
VENTED FILLER CAPS
FUEL RESERVOIR TANKS
VENT   VENT
FUEL SELECTOR VALVE
AUXILIARY FUEL PUMP
CHECK VALVE (FUEL RETURN)
AUXILIARY FUEL PUMP SWITCH
FUEL STRAINER
ENGINE FUEL PUMP
THROTTLE
FUEL CONTROL UNIT
MIXTURE CONTROL
FUEL MANIFOLD
FUEL INJECTION NOZZLE
FUEL FLOW INDICATOR
(Right Half of Dual Instrument)

CODE
FUEL SUPPLY
EXCESS FUEL & VAPOR RETURN
VENT

FIGURE 15-38 Fuel system for a high-performance single-engine aircraft. (*Cessna Aircraft Co.*)

the fuel selector. The fuel-control unit is controlled by the throttle and mixture controls in the cabin.

Fuel from the fuel-control unit is delivered to the engine cylinders and to a pressure-operated fuel flow indicator in the cabin area.

Fuel System for a Light Twin Airplane

A schematic diagram of the fuel system for a light twin airplane is shown in Fig. 15-39. The fuel system shown is employed with models of the airplane that utilize direct fuel injection for the engines.

The fuel supply is contained in two fuel bladders, one located in each wing-tip tank. The total fuel capacity of each bladder is 51 gal [193.3 L], of which 50.5 gal [191.4 L] is usable. The wing-tip tanks are streamlined aluminum alloy shells designed to give adequate structural support to the fuel bladders. The tanks are easily removed from the wing tips to which they are secured by means of bolts through the attaching fittings. An electrically operated boost pump is incorporated in each fuel tank to furnish fuel pressure for priming, starting, and emergency use in case the engine-driven pump should fail. From the boost pumps, fuel is fed to the selector valves located in the wings and then through the fuel strainers and engine-driven pumps to the fuel-injection unit.

The engine-driven fuel-injection-pump assembly includes a vapor separator and a pressure-regulating valve.

A **vapor-return line** is provided between the fuel-injection pumps and the fuel tanks. This eliminates the collection of vapor and resulting vapor lock in the pump or fuel-injection unit. The vapor-return lines carry some fuel in addition to the vapor back to the fuel tanks. Quick-drain valves are provided in each wing-tip tank and in each fuel strainer to facilitate draining of water or sediment. These are used to drain a small amount of fuel during each preflight inspection to ensure that all water and sediment are removed before the airplane is flown.

The fuel-system diagram in Fig. 15-39 illustrates the relative positions of all the units of the system. The fuel-quantity-gauge units are located in each tank, and they are electrically connected to the **fuel-quantity indicators** located on the instrument panel of the airplane. It can be noted also that the fuel-booster pumps in the tanks are electrically operated through switches that are located in the cockpit. The booster pumps are provided with bypass arrangements so that there will be a free flow of fuel when the pumps are not operating. The oil-dilution solenoid valves receive fuel from the main fuel-strainer cases. These valves are also operated by switches in the cockpit.

Since this system is designed for operation with direct fuel-injection units, the fuel-pressure gauges are connected between the fuel controls and the fuel manifolds. This is to give an indication of the pressure actually applied to the injection nozzles, thus providing the pilot with an accurate measure of engine power.

Fuel System for Regional Jet

The fuel system shown in Fig. 15-40a is used in a regional jet aircraft. The aircraft has two integral wing tanks and one integral center tank as shown in Fig. 15-40b. In flight, as the wing tank fuel quantity decreases, the fuel system computer (FSC) will automatically transfer fuel from the center tank to the wing tanks to maintain lateral balance. Two collector tanks are located in the forward section of the center wing tank. Fuel from each wing tank is fed under pressure to its respective collector tank by scavenge ejectors. The collector tank capacity is 10 gal (38 L) and when the tank is full, excess fuel is vented back to the respective wing tank. Fuel can also be fed from the wing tanks to the associated collector tanks by gravity. There is no migration of fuel from the center tank into the collector tanks. A main fuel ejector in each collector tank is immersed in fuel and is used to ensure a positive supply of fuel to the engines. The boost pumps normally supply fuel to the engines for start.

The tanks are vented through interconnecting vent lines to NACA scoops located on the lower surface of each wing. In flight, the NACA scoops supply ram air to slightly pressurize the wing tanks. On the ground, the tanks are vented to atmosphere through the NACA scoops to prevent pressure buildup within the tanks caused by the refueling process or from thermal expansion of the fuel.

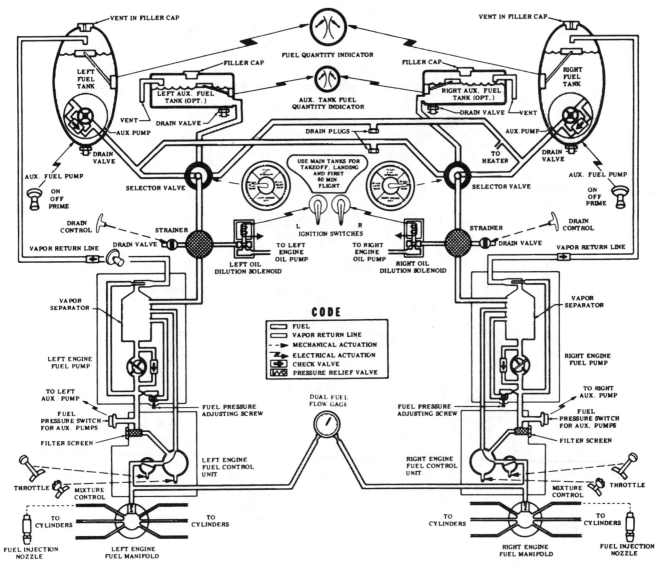

FIGURE 15-39 Fuel system for a light twin airplane. (*Cessna Aircraft Co.*)

Fuel transfer from the center tank to the wing tanks is provided by transfer ejector pumps to maintain the wing tanks at full capacity as long as possible. This is an automatic function with no manual control. The ejectors are powered by fuel pressure tapped from the engine supply lines via the fuel transfer shutoff valves which are automatically controlled by the fuel system computer (FSC). The FSC commands the respective transfer shutoff valve to open when the associated wing tank fuel quantity falls below 94% of full, and commands it to close when the tank quantity reaches 97%. The FSC will cycle the transfer system on and off until the center tank is empty. If the fuel imbalance between the wing tanks exceeds 400 lb [181 kg], a FUEL IMBALANCE caution message is displayed on the EICAS primary page. If the total fuel quantity is less than 900 lb [408 kg] the fuel quantity indication on the primary page turns amber. If the fuel imbalance between the wing tanks exceeds 800 lb [360 kg], a FUEL IMBALANCE caution message is

displayed on the EICAS (engine indicating and crew alert system) primary page and both wing tank quantity indicators turn amber. If one wing's fuel quantity is less than 450 lb [204 kg], then that wing's fuel quantity indicator will turn amber. In the event of wing tank gauging failure, the FSC will use the high-level sensors, located at the top of each tank, to control the fuel transfer operations.

To correct fuel imbalance and to maintain aircraft lateral stability, the FSC automatically initiates fuel cross flow upon detecting a fuel imbalance between wing tanks. The cross flow/APU (auxiliary power unit) pump located within the center tank provides powered cross flow in either automatic or manual mode. In automatic mode, the FSC controls the cross flow operation. If the computer detects a fuel imbalance between the wing tanks of 200 lb [90 kg], the cross flow/APU pump is activated automatically and the required cross flow shutoff valve is opened to correct the fuel imbalance. Cross flow operations continue until 50 lb [23 kg] imbalance is reached.

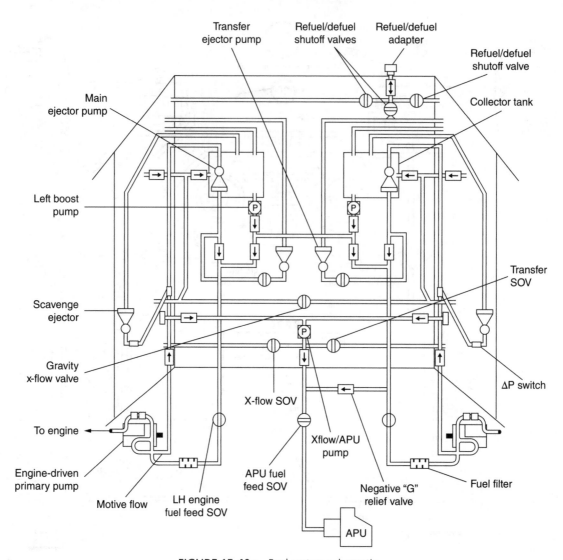

FIGURE 15-40a Fuel system schematic.

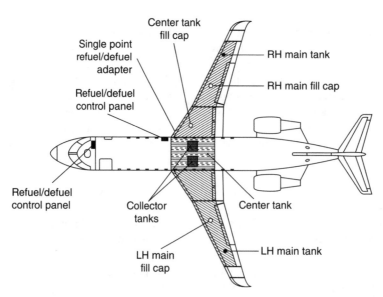

FIGURE 15-40b Integral fuel tank layout.

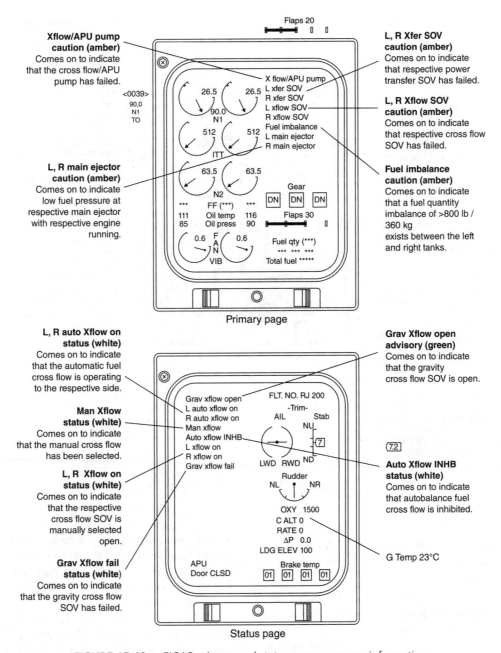

FIGURE 15-40c   EICAS primary and status page messages information.

Engine Indicating and Crew Alert System

The aircraft uses an engine indicating and crew alert system (EICAS) to display fuel system messages and cautions. Figure 15-40c shows some of the fuel system messages displayed on the EICAS primary and status page and Fig. 15-40d shows the EICAS fuel page, which displays the fuel system. The flight crew can check the fuel system operation on this page.

Fuel Systems for Large Turbine Aircraft

Although it is not possible in a standard textbook to provide a detailed description of the fuel system for a large turbine aircraft, the following description will provide an understanding of its size and complexity.

Figure 15-41 is a schematic diagram of the fuel system in one wing of a turbine transport aircraft. The fuel system is an arrangement of fuel tanks and component systems, which ensure that the engines are supplied with fuel in the proper amount at all times.

The following components of the fuel system are illustrated in Fig. 15-41.

Fueling receptacle The attachment for the pressure fueling hose. The receptacle is fitted with a valve that is automatically closed when the fueling hose is disconnected.

Restricting orifice A flow-limiting device to prevent excessive fuel flow during pressure fueling.

Manually operated shutoff valves Valves that are provided at the pressure-fueling station to permit a positive closing of fuel lines.

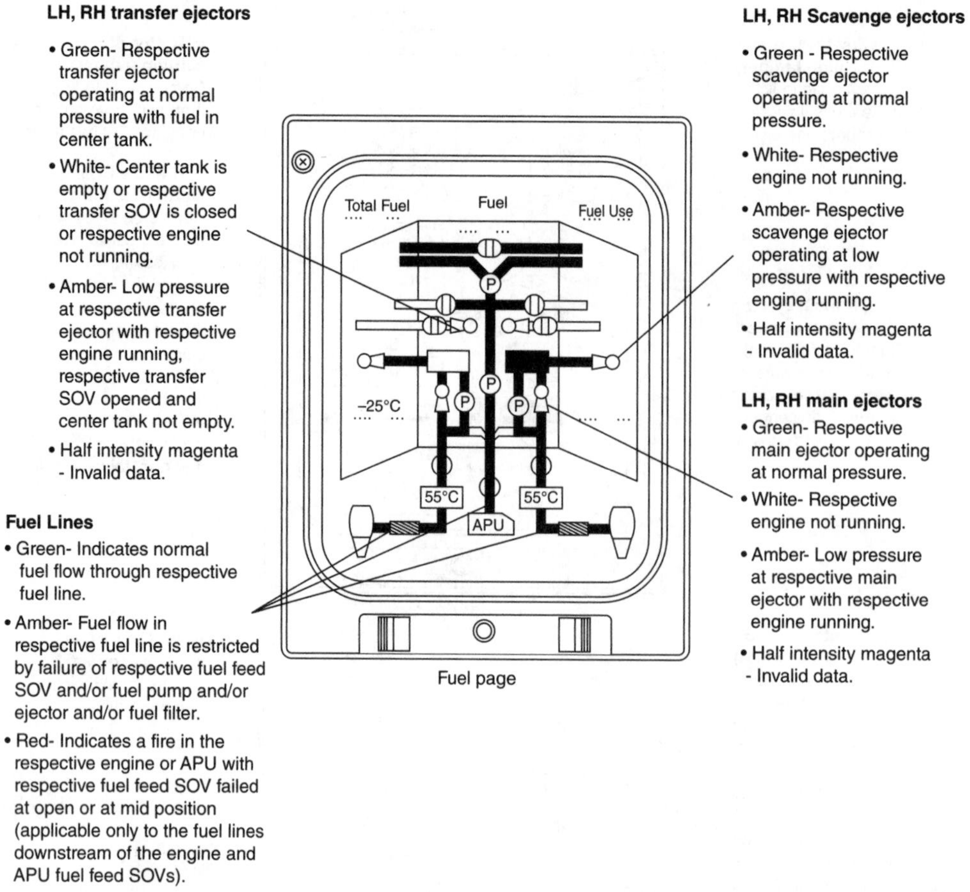

LH, RH transfer ejectors

- Green- Respective transfer ejector operating at normal pressure with fuel in center tank.
- White- Center tank is empty or respective transfer SOV is closed or respective engine not running.
- Amber- Low pressure at respective transfer ejector with respective engine running, respective transfer SOV opened and center tank not empty.
- Half intensity magenta - Invalid data.

Fuel Lines

- Green- Indicates normal fuel flow through respective fuel line.
- Amber- Fuel flow in respective fuel line is restricted by failure of respective fuel feed SOV and/or fuel pump and/or ejector and/or fuel filter.
- Red- Indicates a fire in the respective engine or APU with respective fuel feed SOV failed at open or at mid position (applicable only to the fuel lines downstream of the engine and APU fuel feed SOVs).

LH, RH Scavenge ejectors

- Green - Respective scavenge ejector operating at normal pressure.
- White- Respective engine not running.
- Amber- Respective scavenge ejector operating at low pressure with respective engine running.
- Half intensity magenta - Invalid data.

LH, RH main ejectors

- Green- Respective main ejector operating at normal pressure.
- White- Respective engine not running.
- Amber- Low pressure at respective main ejector with respective engine running.
- Half intensity magenta - Invalid data.

FIGURE 15-40d EICAS fuel page.

Fueling-level-control shutoff valve A valve that automatically closes the fueling line to a tank when that tank is filled to its maximum level.

Fueling-level-control pilot valve A valve that closes during pressure fueling when the full fuel level is reached. The closing of this valve causes pressure to be applied to the fueling-level-control shutoff valve, thus causing it to close (see Fig. 15-43).

Motor-driven valve Slide valves operated by electric motors. They are used for fuel control throughout the system.

Fuel-flow transmitter The electrically operated unit that senses the rate of fuel flow to each engine.

The fuel is distributed in four main tanks, two outboard reserve tanks, and a center-wing tank. The fuel tanks and all system components are suitable for use with any acceptable fuel that conforms to the engine manufacturer's specifications. This fuel-tank arrangement is shown in Fig. 15-42.

All the fuel tanks are located in the interspar area of the wing structure between the front and rear spars. The wing skin and ribs form the walls of the tanks, and the intermediate ribs serve as baffle plates to prevent excessive sloshing and rapid weight shifting of the fuel. The four main tanks and two reserve tanks are completely integral

with the wing structure. The center-wing tank consists of two integral sections at the wing roots and a center-wing cavity containing bladder-type fuel cells. All fuel tanks are made fuel-tight by the use of sealing compound and sealed fasteners.

A **fuel vent system** provides positive venting to the atmosphere of all fuel tanks, fuel cells, and cavities, thereby preventing excessive internal or external pressures across the tank walls during all flight maneuvers. Venting of the four main tanks, two reserve tanks, and integral sections of the center-wing tank is accomplished with sealed spanwise upper-wing skin-stiffener ducts, interconnected with appropriate drains and vent tubing that connect to vent ports located in critical areas of the tanks. The vent system is connected to a surge tank at the wing tip that is vented overboard through a vent scoop. The center-wing cavity is vented overboard through a separate vent and drain system. The cavity vent system maintains atmospheric pressure on the outside of the bladder-type fuel cells. The fuel cells are vented into the integral sections of the center-wing fuel tank.

The **engine fuel-feed system** consists of fuel lines, pumps, and valves, which distribute the fuel to the engines. This system includes four tank-to-engine fuel-feed systems that are interconnected by a fuel manifold such that fuel may be delivered from any main tank or the center-wing tank to any or all engines. The two reserve tanks store fuel and

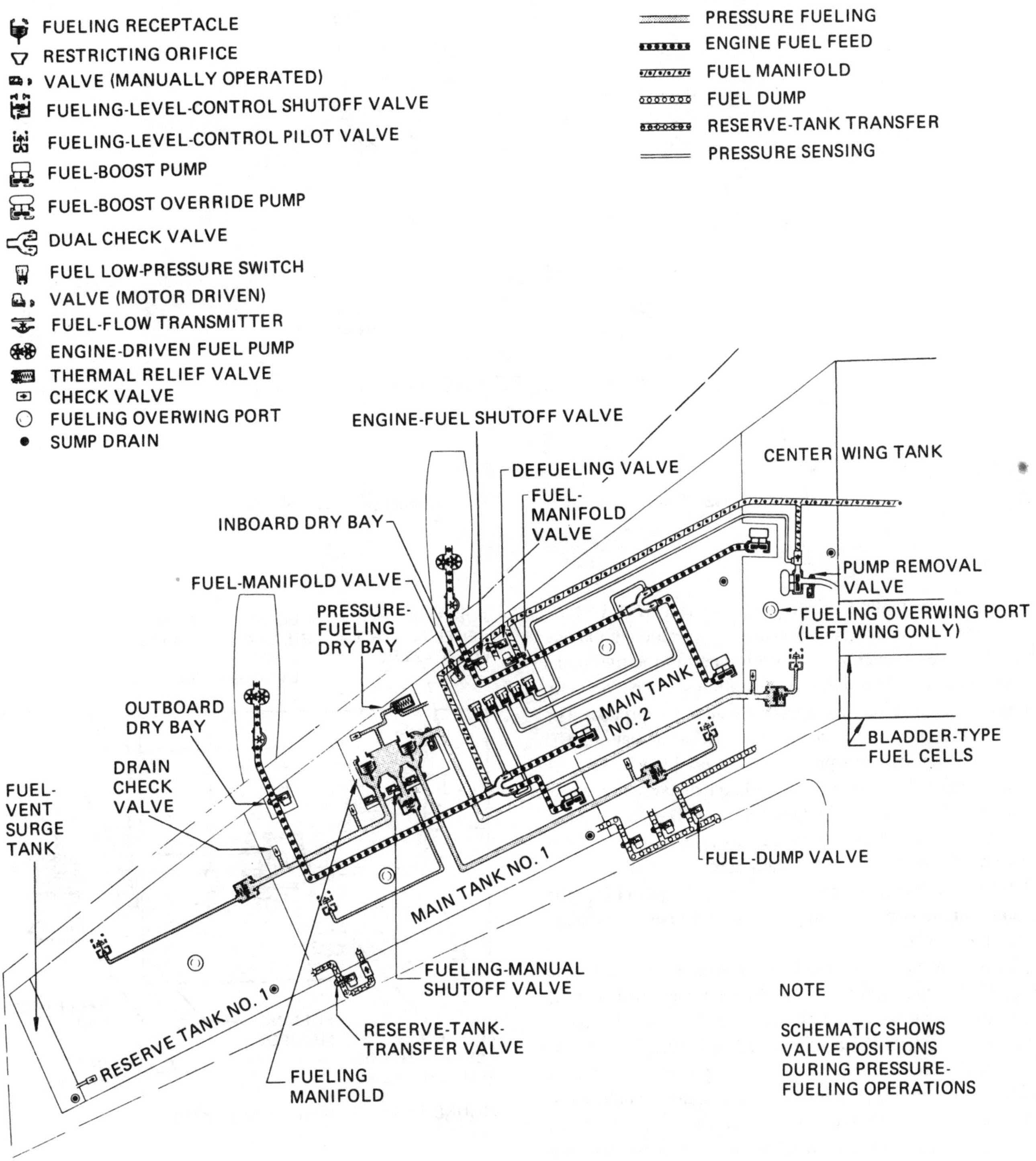
FUELING RECEPTACLE
RESTRICTING ORIFICE
VALVE (MANUALLY OPERATED)
FUELING-LEVEL-CONTROL SHUTOFF VALVE
FUELING-LEVEL-CONTROL PILOT VALVE
FUEL-BOOST PUMP
FUEL-BOOST OVERRIDE PUMP
DUAL CHECK VALVE
FUEL LOW-PRESSURE SWITCH
VALVE (MOTOR DRIVEN)
FUEL-FLOW TRANSMITTER
ENGINE-DRIVEN FUEL PUMP
THERMAL RELIEF VALVE
CHECK VALVE
FUELING OVERWING PORT
SUMP DRAIN
PRESSURE FUELING
ENGINE FUEL FEED
FUEL MANIFOLD
FUEL DUMP
RESERVE-TANK TRANSFER
PRESSURE SENSING
ENGINE-FUEL SHUTOFF VALVE
DEFUELING VALVE
FUEL-MANIFOLD VALVE
CENTER WING TANK
INBOARD DRY BAY
FUEL-MANIFOLD VALVE
PUMP REMOVAL VALVE
PRESSURE-FUELING DRY BAY
FUELING OVERWING PORT (LEFT WING ONLY)
MAIN TANK NO. 2
OUTBOARD DRY BAY
BLADDER-TYPE FUEL CELLS
DRAIN CHECK VALVE
FUEL-VENT SURGE TANK
FUEL-DUMP VALVE
MAIN TANK NO. 1
FUELING-MANUAL SHUTOFF VALVE
RESERVE TANK NO. 1
RESERVE-TANK-TRANSFER VALVE
FUELING MANIFOLD
NOTE
SCHEMATIC SHOWS VALVE POSITIONS DURING PRESSURE-FUELING OPERATIONS

FIGURE 15-41 Fuel system for a turbine transport-category aircraft. (*Boeing Commercial Aircraft Company.*)

supply the fuel by gravity flow to main tanks 1 and 4 through electrically operated transfer valves. The fuel-feed line from each main tank is pressurized by two boost pumps that are controlled by separate switches and independent circuits so that engine operation will not be affected by power failure to any single boost pump. The center-wing-tank boost pumps, known as the **fuel-boost-override** pumps, will override the main-tank boost pumps to supply fuel through the manifold to the engines. The distribution of fuel to the engines is controlled by electric-motor-driven slide valves in the fuel lines.

The valves are classified into three groups: (1) **engine fuel-shutoff valves,** which shut off fuel to the engines; (2) **fuel-manifold valves,** which control manifold distribution; and (3) **reserve-tank-transfer** valves, which control fuel from the reserve tanks to main tanks 1 and 4. All these valves are controlled by manually operated switches located on the flight engineer's panel in the cockpit.

Pressure fueling is accomplished from a station located on the lower surface of each wing between the inboard and outboard engine nacelles. Each pressure-fueling station

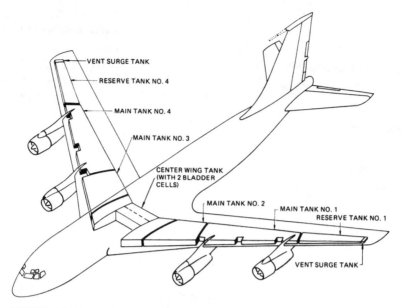

FIGURE 15-42   Fuel-tank arrangement for a four-engine turbine aircraft.

consists of two fueling-nozzle-ground connectors, two fuel-ing receptacles, and a fueling manifold. Four fueling manual shutoff valves at each fueling station permit the fuel tanks to be serviced individually. Each station services the main and reserve tanks on its respective side of the airplane. The center-wing tank is serviced from either wing station. The maximum fuel-delivery pressure is 50 psi [344.75 kPa]. A **restricting orifice plate** at each fueling manual shutoff valve limits the flow rate for the tank serviced. The approximate flow rates for the various tanks are as follows—main tanks 1 and 4: 234 gal/h [885 L/h]; main tanks 2 and 3: 227 gal/h [859 L/h]; and reserve tanks 1 and 4: 65 gal/h [246 L/h]. The center-wing-tank rate is 250 gal/h [946 L/h] per side, except on certain airplanes.

Each tank has a **fueling-level-control pilot valve** (see Fig. 15-43) that actuates a **fueling-level-control** shutoff valve to shut off the fuel flow automatically when the maximum tank capacity has been reached. A **fueling preset system** permits the tanks to be fueled to a predetermined level without constant observation of the fuel-quantity indicators. The fueling-preset-system controls are located on the flight engineer's panel. Preset switches and potentiometers are actuated in conjunction with fuel-quantity indicators. The preset system incorporates signal lights at the fueling station to inform the refueling operator that the desired quantity of fuel has been pumped into the various tanks.

The **fuel-dump system**, consisting of lines, valves, dump chutes, and chute-operating mechanism, provides the means whereby fuel can be dumped in flight. Each wing contains an extendable dump chute, which must be fully extended before fuel can be dumped. Each chute is fed fuel by dump lines from the main and center-wing tanks on their respective sides of the airplane. The reserve tank fuel is transferred to and dumped with the main outboard tank fuel. When all dump valves are open, the system will dump all fuel in excess

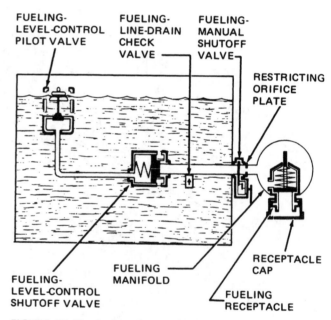

FIGURE 15-43   Fuel-level-control system.

of 10 848 lb [4921 kg] at an average rate of approximately 2500 lb [1134 kg] per min. The 10 848 lb of fuel remaining in the fuel tanks is required as a safety measure to allow the airplane adequate flight time to reach a suitable airport and make a safe landing. The principal reason for dumping fuel is to reduce the landing weight of the airplane prior to making an emergency landing.

The **fuel-quantity-indicating system** incorporates electric capacitance tank probes mounted internally in each fuel tank. The probes have a compensator for fuel-density variations. The probes feed signals to the fuel-quantity indicators on the flight engineer's lower panel. These indicators are calibrated to show the weight of fuel remaining in each tank.

The **fuel-indicating system** utilizes a **fuel-flow transmitter** to send electrical signals to the fuel-flow indicators at the pilot's station. The fuel-flow indication makes it possible for the pilot to determine individual engine-power output and also the flight time remaining before the airplane must land and refuel.

The **fuel-temperature-indicating system** indicates when there may be danger of ice crystals forming in the fuel. It must be remembered that turbine aircraft fly regularly at altitudes of over 30 000 ft [9144 m] and so the fuel reaches temperatures well below the freezing point of water. The fuel-temperature-indicating system provides a means for checking the temperature of fuel in main tank 1 and in the fuel line at each engine. The system consists of an indicator and a **five-position selector switch** on the flight engineer's lower panel. A temperature bulb is located in main tank 1 and in the outlet side of the fuel filter at each engine.

Each engine has a manually controlled engine fuel–deicing system, which removes any ice formed in the fuel before it reaches the fuel-control unit. The principal components of each deicing system are a **fuel filter,** an **air-control valve,** and **fuel-deicing heater.** The control switches are located on the flight engineer's lower panel.

The **fuel filter** is fitted with a pressure warning switch that detects filter icing and then turns on the fuel-icing-indicator light on the flight engineer's lower panel. The **air-control valve** controls the flow of high-pressure discharge air from the engine compressor to the heater through which all the fuel passes. A fuel-temperature bulb is located on the outlet side of the filter to monitor the exit temperature.

All fuel tanks have an exterior filler cap for individual tank servicing. The caps are located on the top of the wings, as shown in Fig. 15-41, thus making it possible to fill the tanks when pressure fueling is not available.

The main tanks have baffle check valves installed in the bottom of the inboard ribs to allow fuel to flow inboard toward the boost pumps during normal flight attitude and to prevent fuel flow away from the pump area during turbulence or airplane maneuvers. This provision ensures that engine stoppage will not be caused by a momentary cutoff of fuel flow.

Pressure defueling of the main tanks or center-wing tank is accomplished through a defueling valve located in the inboard dry bay of each wing. The fuel boost pumps in the main tanks or the override pumps in the center-wing tank deliver fuel to the defueling valves through their respective fuel-manifold valves. The defueling rate is approximately 50 gal/min [189.5 L/min] for each tank. Alternatively, fuel can be drawn from the tanks through the defueling valve by means of the truck defueling pump. The reserve tanks are defueled through the reserve-tank-transfer valve to the adjacent main tanks. Residual fuel in each tank can be drained through fuel-sump drain cocks located in the bottom of each tank.

Another fuel system layout for a large transport aircraft is shown in Fig. 15-44. This fuel system has the following subsystems: fuel storage, fuel indicating, pressure refuel, engine fuel feed, APU fuel feed, jettison, and defuel.

Fuel Storage

The fuel storage system has three fuel tanks: left and right main tanks (Fig. 15-45), and a center tank. Surge tanks collect fuel overflow. The fuel overflow drains into the main tanks. Each main tank has a dry bay that does not hold fuel. Wing ribs divide the fuel tanks into bays, and reduce the movement of fuel during airplane maneuvers. Access doors and cutouts are entrances into the airplane fuel tanks for inspection or component repair. Sump drain valves are at the low point of each fuel tank. The drain valves are used to get fuel samples, remove water and contaminants from the fuel tanks, drain all the fuel that remains after defueling, and to check for fuel in a surge tank before an access door is opened.

Fuel Vent System

The fuel vent system keeps the pressure of the fuel tanks near the pressure of the outside atmosphere. A large pressure difference can damage the wing structure. Drains let fuel in the vent system return to the fuel tanks so the engines or APU can use it. Flame arrestors ensure that a flame cannot enter the fuel tanks through the vent system. If a flame arrestor becomes blocked, pressure relief valves in the surge tank open to create a vent.

Pressure Refueling

The pressure refueling system transfers fuel from the refuel adapters to the main and center tanks. The pressure refuel system is operated from the integrated refuel panel (IRP) on the left wing. There are refuel adapters on both wing but only the left wing station has the IRP. There are six refuel valves, two in each tank. The valve bodies are on the rear spar, inside the tanks. The valve actuators are on the rear spar, outside the fuel tanks. The refuel/jettison manifold connects the two refuel stations to all the refuel valves. It supplies fuel from the refuel adapters to each tank. There are two refuel/jettison manifold drain valves. One is in each half of the center tank. Access to the valve is through the first fuel tank access door. There are two refuel/jettison manifold vacuum relief valves. They are on the refuel/jettison manifold, near the outboard refuel valve in each main fuel tank.

Engine Feed System

The engine fuel feed system supplies fuel to the engines from the main and center tanks. The engine fuel feed system uses these components to supply fuel to the engines: fuel pumps, fuel pump pressure switches, cross feed valves, and spar valves. There is one fuel suction bypass valve in each main tank. The forward and aft cross feed valves isolate the left and right sides of the engine feed manifold. Both valve bodies are in the left side of the center tank. The engine feed manifold connects the output of the fuel pumps to the engine fuel main supply line. There are two override/jettison pumps that attach to the rear spar of the center tank. There are four water scavenge jet pumps and two center tank fuel scavenge

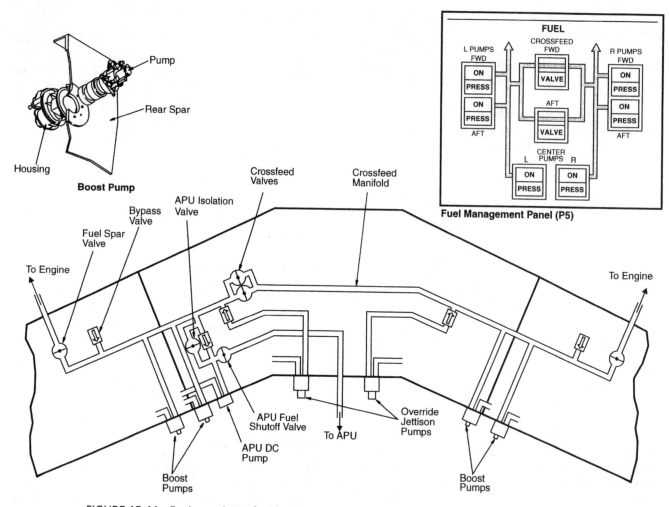

FIGURE 15-44 Engine and APU fuel feed system of transport aircraft. (*Boeing Commercial Aircraft.*)

jet pumps. Each main tank has a forward and an aft boost pump. The boost pumps are on the rear spar. Figure 15-46 shows the location of the center fuel tank components.

APU Feed System

The APU fuel feed system normally supplies fuel to the APU. It can also supply fuel to the left engine when there is no pressure in the left engine feed manifold and the left engine is not operating. The APU normally receives fuel from the left engine feed manifold. The engine feed manifold receives fuel from any of the main or center tank fuel pumps. The fuel flows through a check valve and the APU fuel shutoff valve to the APU fuel line. During APU starts and operation, if there is no fuel pressure in the left engine feed manifold, the APU dc fuel pump automatically turns on. This supplies fuel through the APU fuel valve to the APU fuel line. Figure 15-47 shows the APU fuel feed components.

Defueling

The defuel system moves fuel from the airplane tanks to the refuel station. It also moves fuel from one airplane tank into another (tank-to-tank transfer). There are two ways to

get fuel out of the tanks: the airplane fuel pumps (pressure-defuel) or ground pumps (suction-defuel).

Fuel Jettison

The fuel jettison system dumps fuel overboard to reduce the landing weight. There are two override/jettison pumps in the center tank and one fuel jettison pump in each main tank. Jettison can occur from the center tank or the main tanks. Fuel from the tank flows through the refuel/jettison manifold. The center tank uses the override/jettison pumps as engine feed pumps and jettison pumps. When the jettison system operates, the pumps put fuel through the fuel jettison isolation valves into the refuel/jettison manifold. The fuel goes out of the manifold through the fuel jettison nozzle valves and jettison nozzles. The main tanks have pumps that operate only during fuel jettison. These pumps deliver fuel directly into the refuel/jettison manifold. This fuel also leaves the manifold through the fuel jettison nozzle valves and jettison nozzles.

Fuel Indicating System

The fuel indicating system has these subsystems: fuel quantity indicating system, fuel measuring sticks, fuel temperature

Fuel Capacity*	Pounds**	Gallons	Kilograms	Liters
Main Tank (each)	62,300	9,300	28,200	35,200
Center Tank	83,100	12,400	37,600	47,000
Total	209,560	31,000	94,000	117,400

* 777-200 Airplane
** Density = 6.76 LB/GAL

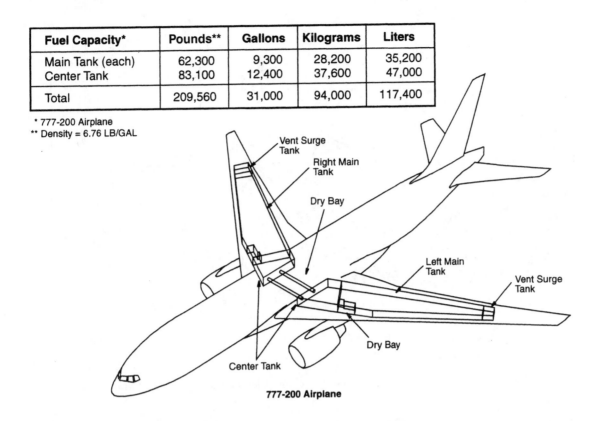

777-200 Airplane

Fuel Capacity*	Pounds**	Gallons	Kilograms	Liters
Main Tank (each)	64,000	9,560	29,000	36,200
Center Tank	174,900	26,100	79,000	98,800
Total	305,687	45,220	137,000	171,200

* 777-200B (IGW) Airplane
** Density = 6.76 LB/GAL

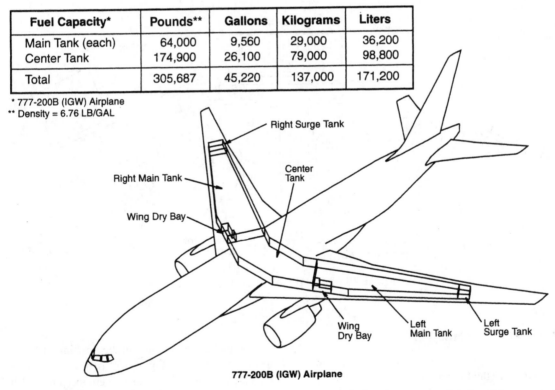

777-200B (IGW) Airplane

FIGURE 15-45   Fuel tank arrangement on transport aircraft. (*Boeing commercial aircraft.*)

indicating, and fuel pressure indicating. The fuel quantity indicating system (FQIS) has the following functions: measures the fuel volume, calculates the fuel quantity, controls refuel operations, and shows when there is water in the tanks. Eight measuring stick assemblies install through the bottom of each wing. Six are in each main tank and two are in each section of the center tank. A fuel quantity system is discussed in detail in Chap. 17.

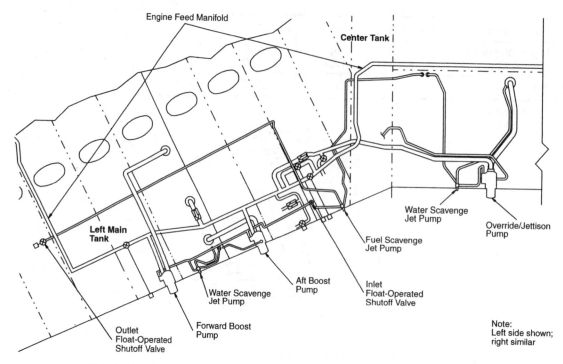

FIGURE 15-46 Center tank fuel and water scavenge jet pumps of transport aircraft.

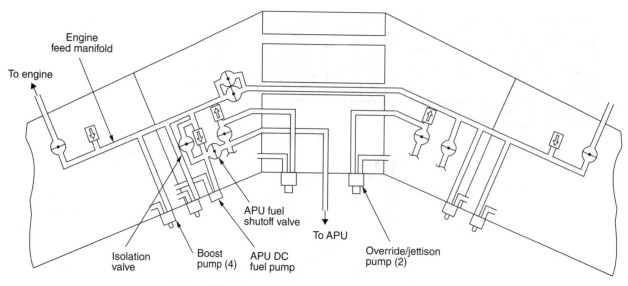

FIGURE 15-47 APU fuel feed components.

Helicopter Fuel Systems

Helicopter fuel systems include both gravity- and pressure-feed systems and cover the full spectrum of complexity.

The small gravity-feed system shown in Fig. 15-48 consists of a fuel tank mounted in the area above the engine and behind the cabin. The tank is baffled to reduce sloshing and has all the connections and components necessary for a basic fuel tank.

Fuel flows through the coarse finger screen mounted at the tank outlet and then through an on-off valve. From the on-off valve, the fuel passes through a filter (gascolator) and then flows on to the engine carburetor.

The fuel system shown in the drawing of Fig. 15-49 and in the schematic of Fig. 15-50 is described as an example of a fuel system for a turbine-engine-powered helicopter. Observe that the system includes many of the features of a turbine transport airplane system, as described previously.

The fuel system shown in these two figures is utilized on a model of the Sikorsky S-61N helicopter with a two-tank system. This system is an open-vent type consisting of two fuel tanks and associated pumps, filters, fuel lines, valves, and indicating components. The tanks are rubberized (synthetic) fabric fuel cells installed in tank compartments, where they are adequately supported by the structure below

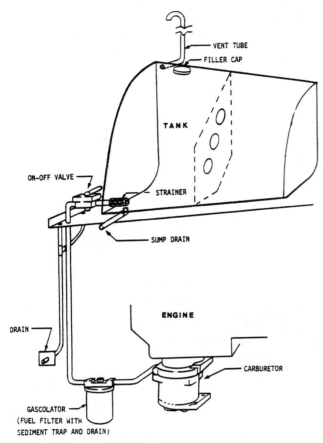

VENT TUBE

FILLER CAP

TANK

ON-OFF VALVE

STRAINER

SUMP DRAIN

ENGINE

DRAIN

CARBURETOR

GASCOLATOR
(FUEL FILTER WITH
SEDIMENT TRAP AND DRAIN)

FIGURE 15-48 Fuel system for a light helicopter. (*Robinson Helicopter Company, Inc.*)

the cabin floor panels. The forward tank supplies fuel to the no. 1 engine and the heating system, and the aft tank supplies fuel to the no. 2 engine. Separate filler caps on the left side of the helicopter permit gravity filling of the individual tanks if pressure fueling is not available. Each tank has three vents which terminate to the atmosphere of the left side of the fuselage.

A pressure fueling system allows the fuel tanks to be filled automatically by means of a pressure-fueling adapter on the left side of the helicopter. In conjunction with the pressure-fueling system, a fuel-level selector system, controlled from the cockpit, actuates the fueling shutoff valve in the pressure-fueling system to shut off fuel at a preselected level. Pressure fueling is controlled from the cockpit, with the exception of the additional controls outside the fuselage that are used to precheck the pressure-fueling high-level shutoff valves at the start of the pressure-fueling operations.

The fuel-distribution system is such that each engine is provided with a specific fuel-supply source, as required by the FARs. However, flexibility of operation is accomplished by means of the **cross-feed system** and valve. The cross-feed valve is controlled from the cockpit and offers the pilot the option of operating both engines from tank if necessary. Fuel is pumped from each tank to its respective engine by two submerged motor-driven booster pumps. Each pair of pumps is installed in a **fuel-collector can** installed in the forward part of each tank. During operation, a portion of the

fuel from the booster pumps is routed back to a **fuel ejector** in the tank. The fuel ejector consists of a venturi through which fuel is pumped. The low pressure caused by the flow through the venturi draws additional fuel from the ejector location in the tank, and this fuel is delivered to the collector cans in the tank. This ensures that there will always be an adequate amount of fuel in the boost pump area, regardless of the aircraft attitude, to avoid fuel starvation of the engines.

Fuel flow from the tanks is to a lower fuselage fuel control, which consists of a manual shutoff valve and a fuel filter. Fuel, piped from the filter to the engine through the upper fuselage, passes through a motor-operated valve that is an integral part of the emergency fuel shutoff system.

The fuel system is provided with adequate drains to ensure disposal of drained fuel away from the aircraft. Drain lines provide an overboard outlet for fuel leakage at engine accessories, for condensation, for a failed fuel-pressure transmitter, and for any fuel that remains in the engine after shutdown. The drain lines from each engine are grouped and routed directly downward within the fuselage to ports below the hull. Fuel-cell draining is accomplished through individual defueling valves provided in the bottom of each tank. Any fuel that remains in the fuel tanks after draining and any accumulated moisture within each fuel tank may be removed by opening the individual sump drain cocks. The engine centrifugal fuel pump and fuel filters are provided with drain cocks under the hull to drain accumulated moisture.

The fuel system for the helicopter is provided with indicating systems for all needed information regarding operation of the system and failures within the system. The indications provided are **fuel quantity,** warning when fuel level is at the point where 20 to 30 min of operation is possible, **boost-pump pressure, clogged fuel-filter warning,** warning of **overheat** in the engine compartment, warning of **low pressure** in the line between the emergency shutoff valve and the engine-driven pump, and **fuel pressure** to the engine fuel nozzles.

Two fuel-quantity indicators are mounted in the upper left side of the instrument panel. They are calibrated in pounds × 100 and provide for a maximum of 1400 lb [635 kg] of fuel each. The transmitters (probes) are of the capacitance type as described previously and are located in the fuel cells. The transmitting probe installed in the rear tank is provided with a thermistor sensor, which provides the signal for **low-level fuel,** indicated at the **master warning panel.**

Warning of insufficient pressure from the booster pumps is provided by pressure switches that activate warning lights on the **fuel-management panel** located in the cockpit. The **low-pressure warning lights,** which indicate low fuel pressure to the engine-driven pumps, are also located on the fuel-management panel.

Warning of clogged fuel filters is indicated by **fuel bypass** lights on the **master warning panel.** These lights are activated through pressure switches incorporated in the filter bypasses.

Warning of an overheat condition in the engine compartment is indicated by the illumination of lights in the **fire emergency shutoff selector handles** and by illumination

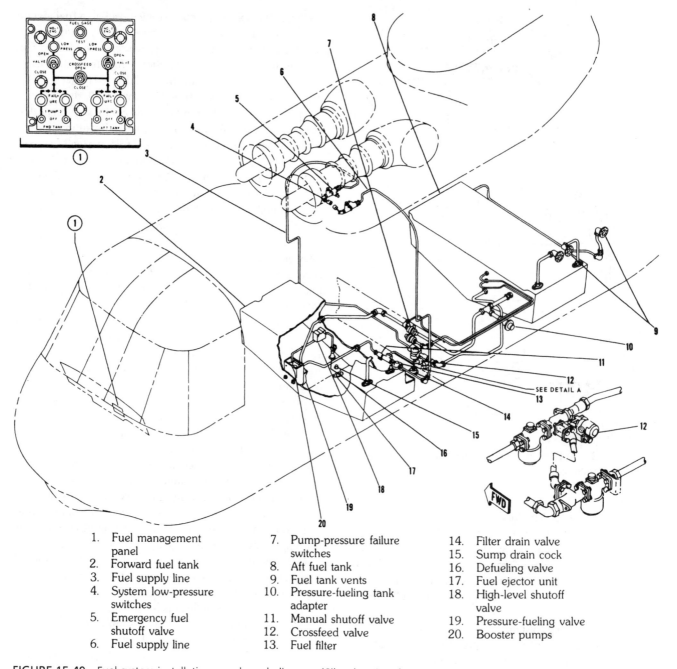

1. Fuel management panel
2. Forward fuel tank
3. Fuel supply line
4. System low-pressure switches
5. Emergency fuel shutoff valve
6. Fuel supply line
7. Pump-pressure failure switches
8. Aft fuel tank
9. Fuel tank vents
10. Pressure-fueling tank adapter
11. Manual shutoff valve
12. Crossfeed valve
13. Fuel filter
14. Filter drain valve
15. Sump drain cock
16. Defueling valve
17. Fuel ejector unit
18. High-level shutoff valve
19. Pressure-fueling valve
20. Booster pumps

FIGURE 15-49 Fuel-system installation on a large helicopter. (*Sikorsky Aircraft Corp., Division of United Technologies Corp.*)

of lights on the **fire warning panel.** Fuel piped through the upper fuselage en route to the engine passes through a motor-operated valve **(emergency shutoff valve),** which is an integral part of the emergency fuel-shutoff system. These valves are shown in the schematic drawing of Fig. 15-50.

INSPECTION, MAINTENANCE, AND REPAIR OF FUEL SYSTEMS

The proper and regular inspection of aircraft fuel systems is critical to the safe operation of the aircraft. The failure of a component may result in an engine failure due to insufficient fuel being delivered or in a fire on the aircraft. Although the

following text provides general information concerning fuel systems, aircraft manufacturers provide specific guidelines as to the methods and frequency of maintenance actions for their respective aircraft.

Fuel-System Inspections

The preflight inspection of a fuel system includes checking the fuel tanks visually for quantity of fuel. This requires the removal of tank caps and looking into fuel tanks. Sometimes a dripstick must be used in order to determine that the proper quantity of fuel is in the tanks. Another most important preflight inspection of the fuel system is the opening of fuel drains. All fuel drains should be opened

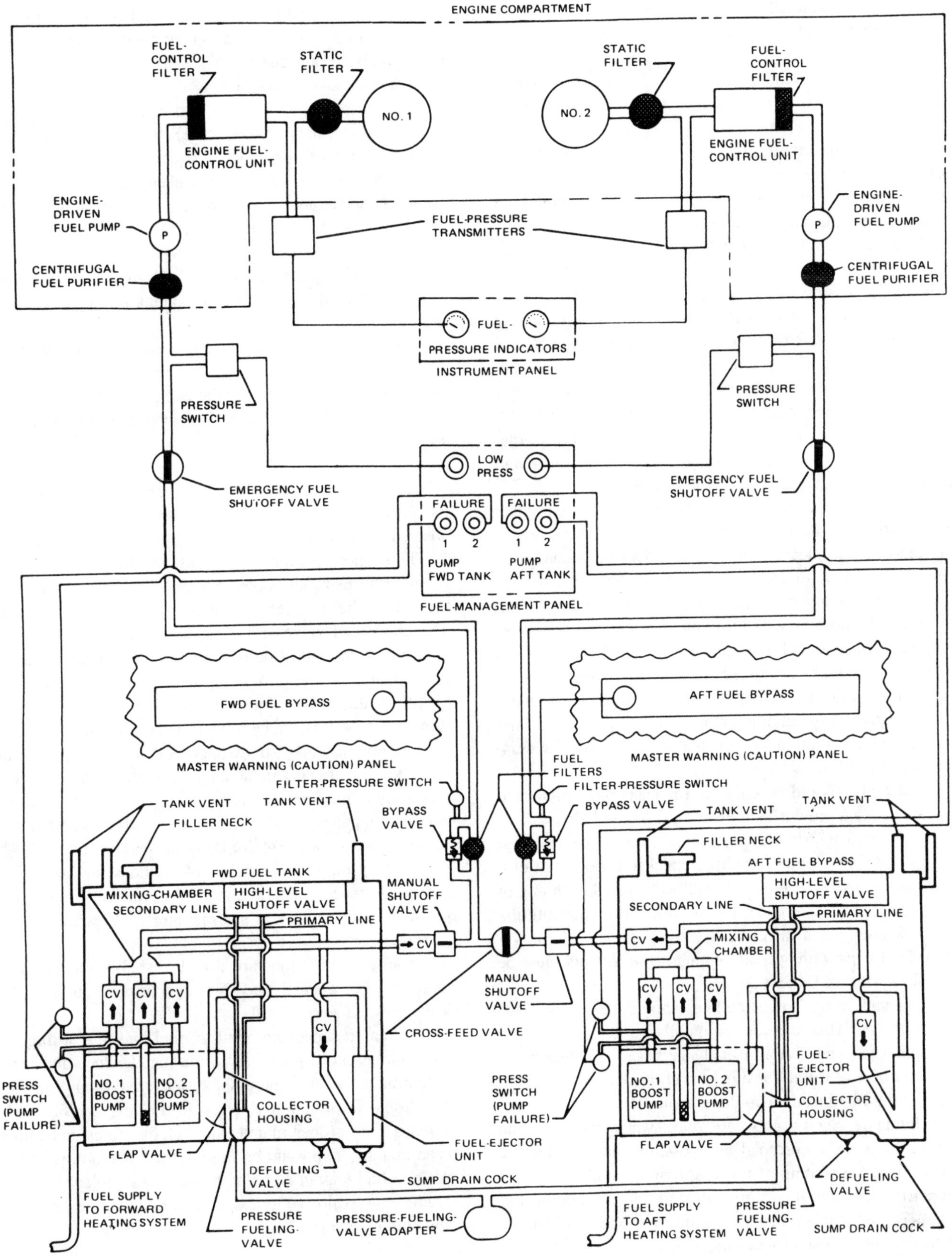

FIGURE 15-50 Schematic diagram of fuel system for the S-61N helicopter. (*Sikorsky Aircraft Corp., Division of United Technologies Corp.*)

for a few seconds to allow any accumulated water or sediment to drain out of the system. Other preflight inspections may be established by the owner or operator of the aircraft under particular circumstances; in each case, the required inspection should be carried out before the airplane is flown.

During engine run-up, the fuel pressure and the proper operation of booster pumps should be checked. If any malfunction is found in the fuel system, the airplane should not be flown until the malfunction is corrected.

A 50-h routine inspection of the fuel system probably will not require more than a visual inspection for signs of fuel leakage and draining of the main fuel strainer. During the visual inspection, the security of all lines and fittings should be checked. Special inspections may be required after 50 h of operation for a new engine or a new airplane. These inspections are recommended because metal particles and other foreign matter may have accumulated during the first 50 h of flight. In such instances it is advisable to clean all fuel strainers, drains, and vent lines. All fittings, mountings, and attachments should be checked for security.

The 100-h and annual inspections require a complete examination of the fuel system. All fuel strainers should be removed and cleaned, all drains should be opened, and sumps should be drained to make sure that all sediment and water is removed from the system. The carburetor float chamber should be drained to remove water and sediment, and the carburetor fuel filter should be cleaned. All fuel lines and hoses should be examined carefully for condition, security, and wear due to rubbing on any portion of the structure. Any unsatisfactory condition must be corrected. The primer system and oil-dilution system should be checked for satisfactory operation.

Fuel tanks should be checked for leaks, corrosion, and microorganism growth, as previously discussed.

Fuel tank caps should be examined for proper venting; that is, the vent opening should be checked for obstructions. If a cap is unserviceable, it should be replaced with one of the correct make and model for the installation. Substitution of the wrong type of cap can cause loss of fuel or fuel starvation. Fuel caps with special venting arrangements must be replaced with the same type.

Fuel tank vents and vent lines should be checked for obstructions. This can be accomplished by passing air through such vents. It is usually necessary to disconnect one end of a vent line for this inspection. Fuel-system vent lines should be installed so there are no low points that are not drainable. Such points can collect water and cause the tubing to become corroded and possibly plugged. During inspection, the routing of the vent tubing should be examined to ensure that it conforms to the requirements for fuel lines. Tubing that is plugged should be removed, thoroughly cleaned, and checked for corrosion damage.

Fuel-shutoff valves and tank-selector valves should be checked for effectiveness and accuracy of valve handle position. Leakage can be checked by turning a valve off, disconnecting the line downstream of the valve, and turning on the booster pump briefly. If fuel flows with the valve turned off, the valve should be repaired or replaced. Fuel-selector valves should be rotated through their tank positions to see if fuel flows freely from each tank when the valve is positioned at the detent indicating that tank.

Fuel-pressure, fuel-quantity, and fuel-temperature gauges should be given an operational check. If there is any indication of malfunction or inaccuracy, the gauge should be removed for bench checking. Fuel-quantity gauges are checked on the aircraft in accordance with manufacturer's instructions. Typical instructions for such a check are as follows:

1. Completely drain the fuel tank, including unusable fuel.
2. Put the amount of fuel in the tank that is specified as unusable.
3. Check the quantity gauge to see that it reads zero.
4. Add fuel to the tank in increments of 10 gal [37.9 L] and check the quantity gauge at each increment to see that it accurately indicates the correct amount of fuel in the tank within acceptable limits.

During a test of the fuel-quantity gauge as just described, the airplane should be in a normal level-flight attitude. Otherwise a false reading will be given by the quantity gauges.

The proper operation of a multitank fuel system can be checked by using the boost pumps to move fuel from one tank to another through the various combinations available. The cross-feed valve or valves are set as desired, and the appropriate fuel boost pumps are turned on. The increase in fuel quantity in the tank being filled is noted on the fuel-quantity gauge. If the line to the tank being filled includes a flow indicator, this can be used to determine fuel flow.

Fuel-pressure and fuel-temperature warning systems are checked by artificially inducing the conditions in the systems that should produce the appropriate alarm. Fuel-pressure warning lights are easily checked by noting whether they come on as fuel pressure drops during shutdown.

The foregoing are examples of methods for making the indicated inspections and checks. In practice, the methods given in the manufacturer's or operator's instructions should be followed.

Required inspections are usually listed in the manufacturer's service manual, and all special inspections should be accomplished as specified.

The maintenance and repair of fuel lines and fittings are the same as those practiced in the maintenance and repair of hydraulic systems and other systems requiring plumbing. Additional information is given in the text *Aircraft Basic Science*. The proper maintenance of fuel lines and fittings is particularly important because of the flammability of fuel. A small fuel leak in a confined area of an airplane can soon produce an explosive atmosphere, which can be ignited by any kind of spark.

Fuel-Tank Inspections and Repairs

Fuel-tank inspections involve visual inspection for any evidence of leaks. Gasoline has an identification dye added, and this dye concentrates around fuel leaks, making them easy to spot. Turbine fuel leaks are harder to spot when they are fresh, but the collection of dirt on the leak and discoloration make the leaks evident in a short period of time.

Fuel leaks are classified as a **stain**, a **seep**, a **heavy seep**, and a **running leak**, as shown in Fig. 15-51. The general rule for these leaks is that any leak in an enclosed area is considered a fire hazard, and the aircraft should not be flown. Stains, seeps, and heavy seeps are not flight hazards when on the outside of the aircraft and away from ignition sources. A running leak on the outside of the aircraft is considered a fire hazard. All leaks that are evident should be investigated to determine their cause, and if they are not a fire or flight hazard, they should be repaired during the next regular maintenance activity. All leaks should be considered a fire hazard until evaluated, and each situation should be evaluated based on the aircraft manufacturer's recommendations.

When inspecting rigid removable and bladder-type fuel-tank installations, fuel leaks appearing on the skin of the aircraft are cause to inspect the fuel tank. First check all fittings for security and evidence of leaking. Also check the gasket where the fuel-quantity sensor is mounted in the tank. Check the fuel-filler neck and fuel cap for any evidence of leakage. If no cause can be found, it may be necessary to remove the fuel tank from the aircraft and check it for leaks.

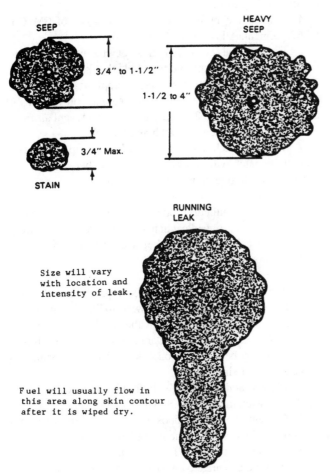

FIGURE 15-51 Classification of fuel leaks. (*Cessna Aircraft Co.*)

Leaks should show up with a visual inspection. On occasion, the tank or bladder may have to be pressurized with air and brushed with soapy water to locate leaks. Perform this inspection in accordance with the manufacturer's instructions.

Inspections of integral fuel tanks are similar to those for other types of fuel tanks. These tanks often have access panels on the surfaces, which can leak. The aircraft manufacturer will recommend a sealing compound to be used when installing these panels. If the seal is broken, the compound must be replaced with new material.

The interior surfaces and especially the bottoms of fuel tanks should be examined for the growth of microorganisms—that is, bacteria and fungi. If dark-colored, slimy growths are found, the tank should be cleaned thoroughly. Some manufacturers recommend the use of denatured ethyl alcohol as a cleaning agent. After the microbial growths are removed, the metal surfaces of the tank should be checked for corrosion. If corrosion has penetrated the metal surfaces beyond manufacturer-specified limits, the tank should be repaired.

Before actual repair procedures are described, it is important to consider certain safety precautions. The fuel vapor present in all empty fuel tanks is explosive; therefore, caution should be observed in their repair. The following steps are recommended to eliminate fire and explosion hazards. The tank drain plug should be removed, and live steam should be circulated through the tank by placing the steam line in the filler neck opening. This steaming should continue for at least $\frac{1}{2}$ h.

In another common procedure, hot water [150°F (65.56°C)] is allowed to flow through the tank from the bottom and out the top for 1 h. **Never apply heat to a tank when it is installed in an airplane.** Use of hot water applies only to welded or soldered metal tanks, not to riveted ones. The use of hot water in a riveted tank may loosen the sealing compound between joints and cause leaks. Although heat is not required for rivet repairs, the tank should be purged with an inert gas, as described next, to prevent ignition of fuel vapors by any sparks such as from the use of tools.

The **purging of fuel tanks** with an inert gas is an acceptable procedure. This is accomplished by first filling the tank with water and then introducing CO_2, nitrogen, or some other inert gas into the tank as it is drained. An adapter plug is prepared to fit the filler opening of the tank, and a hose from the gas supply is connected to the adapter plug. The gas under controlled pressure flows into the tank as the fuel drains. When the tank is completely drained, it is filled with the gas, thus preventing any chance of combustion.

Metal fuel tanks are repaired by welding, soldering, or riveting. The repair method used, of course, depends on the construction of the tank. Tanks constructed of commercially pure aluminum 3003S, 5052SO, or similar metals may be repaired by welding. Heat-treated aluminum alloy tanks are generally put together by riveting. When it is necessary to repair a riveted tank with a riveted patch, a sealing compound that is insoluble in gasoline should be used to seal the patch.

Leaks in tanks may be located by plugging all the outlets except one and admitting air pressure of about $2\frac{1}{2}$ psi [17.24 kPa]. The application of soapy water on the seams of the tank will cause bubbles to form wherever a leak exists. The areas of leakage should then be marked and the repair made by using the proper method. One type of solder repair used on stainless steel tanks is accomplished by applying a cover patch of terneplate over the crack and soldering it in place with soft solder. The patch is cut and formed so that it overlaps the damaged area by approximately $\frac{1}{2}$ in [1.27 cm]. Figure 15-52 illustrates this method.

After a soldering repair is made, the tank must be flushed with warm water to remove all traces of the soldering flux and any solder beads that may have fallen into it. Similarly, after welding a fuel tank, it is necessary to remove all the welding flux to prevent corrosion. This is accomplished by immediately washing the inside and outside surfaces of the tank with hot water. After the water bath, the tank is drained and then filled with a 5 percent nitric or sulfuric acid solution. This solution should be allowed to remain in the tank for 1 h, after which the tank should be rinsed with clean fresh water. The outside weld also should be washed with the acid solution. To ensure that the water has removed all the corrosive elements, a small quantity of 5 percent silver nitrate should be placed in a sample of the rinse water. If a heavy white precipitate results, the rinsing operation should be repeated.

Minor seepage in a riveted tank can be repaired with a sealing compound applied externally.

Another often-used method for correcting minor seepage in a metal tank is **sloshing** the inside of the tank with an approved sloshing compound. The tank is thoroughly cleaned inside and then a specified amount of the compound is poured into the tank and sloshed until the interior surface is completely coated. The remaining compound is poured from the tank, and the tank is allowed to cure for the time specified by the manufacturer of the compound. If an integral tank is to have a sloshing compound applied, the compound will have to be applied by brushing or spraying.

Dripping or running leaks in a riveted tank must be repaired by tightening the metal seam. This is done by drilling out all worn and damaged rivets near the leak and replacing them with new rivets. The new rivets should be squeezed using a pneumatic riveter if such a tool is available and can be applied correctly to the rivet. It is good practice to dip the new rivets in a sealing compound before installation. This compound should also be applied to the seam in the same manner as for stopping seepage. If it is necessary to rivet a patch in place, a sealing compound that is insoluble in gasoline should be used.

The repair of synthetic-rubber fuel bladders or cells is usually described in the manufacturer's overhaul manual. Typical of such instructions for an outside patch are the following:

1. Use a piece of synthetic-rubber-coated outside repair material (U.S. Rubber Company 5136) large enough to cover the damage at least 2 in [5.08 cm] from the cut in any direction. Buff this material lightly and thoroughly with fine sandpaper, and wash with methyl ethyl ketone (U.S. Rubber Company 3339 solution) to remove buffing dust.
2. Cement the buffed side of the patch with two coats of black rubber cement (3M Co. EC678). Allow each coat to dry 10 to 15 min.
3. Buff the cell area to be patched lightly and thoroughly with fine sandpaper, and wash with methyl ethyl ketone to remove buffing dust.
4. Cement the buffed area with two coats of 3M Co. EC678 cement. Allow to dry 10 to 15 min.
5. Freshen the cemented area of the patch and the cemented area of the cell with methyl ethyl ketone.
6. While the cement is still tacky, apply the edge of the patch to the edge of the cemented area and roll or press it down $\frac{1}{2}$ to 1 in [1.27 to 2.54 cm] at a time to prevent air from being trapped between the patch and the cell.
7. Seal the patch and a $\frac{1}{2}$-in [1.27-cm] strip of the cell around the patch with one coat of black rubber cement, and allow the patch to remain undisturbed for 6 h.

A damaged synthetic-rubber fuel cell must be patched inside as well as outside when the damage goes entirely through the cell wall. Typical instructions for the inside patching process are as follows:

1. After the damaged area has been patched on the outside of the cell and the repair allowed to stand a minimum of 6 h, the cell is ready to have the patch applied on the inside of the cell. The damaged area to which this patch is to be applied may be pulled through the filler-neck opening to make the repair simpler.
2. Lightly and thoroughly buff with a piece of Buna nylon sandwich material (U.S. Rubber Company 5063).
3. Cement the patch opposite the red-fabric side with two coats of the same cement used for the outside patch, allowing each coat to dry 10 to 15 min.
4. Buff the cell area to be patched lightly and thoroughly with fine sandpaper; then wash off the dust with methyl ethyl ketone.

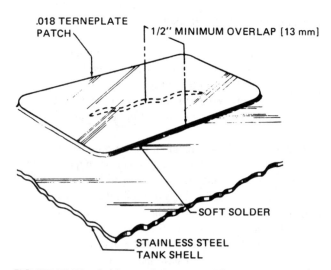

FIGURE 15-52 Solder patch for a metal fuel tank.

5. Coat the buffed area of the cell with two coats of cement, allowing each coat to dry 10 to 15 min.

6. Freshen the cemented areas of the patch and cell with methyl ethyl ketone.

7. While the cemented areas are still tacky, apply the edge of the patch to the edge of the cemented area, centering the patch over the cut in the fuel cell. Hold part of the patch off the cemented area, and roll or press it down $\frac{1}{2}$ in [1.27 cm] across at a time to avoid trapping air between the patch and the cell.

8. Remove the red fabric from the patch by moistening it with methyl ethyl ketone.

9. Seal-coat the patch and a $\frac{1}{2}$-in [1.27-cm] strip of the cell around the patch with two coats of the black cement. Allow the first coat to dry 15 min and allow the second coat to dry 12 h or more, so that when the cell is in its original position, the patching area will not stick to other areas of the cell.

Thick-walled fuel cells such as those installed in transport aircraft can be repaired using methods and materials similar to those just described. In every case, however, the repair should be accomplished in accordance with manufacturer's instructions and by a person trained in the procedures.

In general, the reinstallation of fuel tanks or cells is accomplished by reversing the process of removal. When metal tanks are reinstalled, the felt padding should be examined for condition. If new felt is installed, it should be treated to make it nonabsorbent. Rawhide leather, which contains free alkalies, is unsuitable for padding an aluminum tank because excessive corrosion will take place where the pad comes in contact with the metal. The tanks must be anchored securely and vibration held to a minimum, especially where aluminum alloy tanks are used, since such tanks are subject to cracking.

Metal tanks are subject to both internal and external corrosion, and steps should be taken to prevent this. Zinc chromate or epoxy primers are satisfactory.

Synthetic-rubber bladders and cells are held in place by lacing cord, snaps, or other fasteners. The reinstallation procedure must follow the manufacturer's instructions to ensure that the cells are properly supported in the tank compartment. The area in which the cell is installed must be inspected to make sure that there are no sharp edges or protrusions to cause wear or other damage to the cells. Any portion of the structure that may cause damage must be covered with nonabsorbent padding.

Fuel tanks must be capable of withstanding certain pressure tests without failure or leakage. These pressures may be applied in a manner simulating the actual pressure distribution in service.

Conventional metal tanks with walls that are not supported by the aircraft structure must withstand a pressure of 3.5 psi [24.13 kPa]. If there is any doubt regarding the airworthiness of a particular fuel tank, it may be tested by applying air pressure through a pressure regulator set to 3.5 psi [24.13 kPa]. If this is done, great care must be taken to avoid excessive pressure.

TROUBLESHOOTING

Accurate and rapid troubleshooting on a fuel system necessitates a complete understanding of the operation and purpose of the entire system and its various units. Most aircraft technicians are able to make quick and satisfactory repairs after a problem has been located but may have difficulties when trying to trace the source of the problem. As in all troubleshooting, if the technician follows a definite plan or procedure in locating a defective unit, time is saved, and the long process of haphazardly removing unit after unit with the hope that eventually the trouble will be found is eliminated. Additional information on the theory and practice of troubleshooting is given in Chap. 19.

Troubleshooting Examples

Several typical examples of fuel-system troubleshooting procedures are as follows.

The daily line inspection shows an adequate fuel supply. Check of the strainer indicates a normal flow of fuel through the strainer drain, but when the boost pump is turned on, the fuel-pressure-indicator needle fluctuates slightly and does not register the correct pressure.

Since the flow of fuel to the strainer is normal, the trouble could not be in the tank screen, the selector valve, or the lines to the strainer. The trouble must exist between the strainer and the carburetor. Because the fuel-pressure connection is taken from the float-chamber side of the carburetor screen, improper fuel pressure at the gauge would indicate that the fuel is restricted at some point between the carburetor bowl and the strainer. This would call for the removal of the carburetor screen. If it is clear, hose fittings between the carburetor and the strainer are loosened and the hoses are disconnected in succession. After each hose is disconnected, the boost pump is operated to check for a free flow of fuel at this point. The restriction in the line is thus located systematically.

Another example of a troubleshooting procedure is as follows: An engine is started on the primer charge but does not continue to run unless the boost pump is left on. The fuel-pressure indicator shows normal pressure as long as the pump is operated, but it becomes erratic and the reading drops when the boost pump is turned off.

The trouble should be approached logically. In the discussion earlier in the text, it was shown that the boost pump uses the same line as the engine-driven pump to force fuel to the carburetor. It was also shown that a bypass valve is incorporated in the engine-driven fuel pump to allow fuel under boost-pump pressure to flow around the engine-driven pump rotor and vanes. The lines and units of the system are not blocked, since the engine will operate under boost-pump pressure. Thus the trouble is in the engine-driven pump.

In general, when a normally operating fuel pump suddenly ceases to function, it should be removed for dismantling and inspection. However, before doing this, the pressure-relief-valve setting should be readjusted while the boost pump is being operated. If this does not correct the trouble, the procedure for removing the pump should be followed.

Another troubleshooting procedure involves an airplane that is equipped with a gravity-feed fuel system and a float-type carburetor. The engine is started, runs normally for a few minutes, and then stops. After a short period of time, the engine is restarted; it runs for a few minutes and again stops.

Judging from the symptoms, it is apparent that for one reason or another the supply of fuel is restricted. When the engine is not running, the carburetor float chamber fills with fuel, but when the engine is started, the supply drops off sharply. The first step in this particular case would be to isolate the part of the fuel system causing the trouble. The gas-colator, or strainer drain, is opened and shows a restricted fuel supply. Thus it is determined that the fault is located somewhere in the system between the tank and the strainer. The next step is to isolate the trouble still further by disconnecting a line on the engine side of the shutoff cock. If this is done and it is seen that the flow of fuel is still restricted, the elimination procedure is continued; but before the tank screen is removed, the tank vent is checked. In this case it was found to be clogged. Other units, such as the carburetor, can cause the same symptoms to appear. Therefore, it is better to follow a definite procedure that isolates the trouble rather than to start checking the first unit that comes to mind.

REVIEW QUESTIONS

1. What requirements have been established to ensure reliability of fuel systems for aircraft?
2. What is meant by *fuel-system independence*?
3. What are the requirements for lightning protection in fuel systems for aircraft?
4. Define *unusable fuel supply*.
5. Describe an *integral fuel-tank design*.
6. How is a *bladder fuel cell* attached to the airframe?
7. What internal pressures must a metal fuel tank be able to withstand?
8. Describe a rigid removable fuel tank.
9. What is the required capacity of a *fuel-tank sump*?
10. What markings are required at the fuel filler cap location?
11. What is the general requirement for fuel-tank capacity?
12. Why is a positive pressure required within the vapor space of bladder fuel cells, and how is this pressure maintained?
13. Describe the arrangement required for the venting of fuel tanks.
14. Give the requirements for fuel-tank outlets.
15. How does an *ejector fuel pump* work?
16. Give the fuel-pump requirements for a fuel system.
17. Describe a *vane-type fuel pump*.
18. Describe a *variable-volume fuel pump*.
19. Explain the need for a *bypass valve* in a fuel pump.
20. What provision is made in a fuel-pump relief valve to compensate for changes in atmospheric pressure?
21. Describe a centrifugal fuel pump.
22. Where are centrifugal fuel-boost pumps for aircraft usually located?
23. Describe the requirements for fuel strainers and filters.
24. Explain the operation of a filter bypass warning system.
25. Why are fuel shutoff valves for large aircraft interconnected with the fire protection systems?
26. Why are *fuel heaters necessary*?
27. What are the purposes of a *fuel vent system*?
28. Explain the need for an *oil-dilution system* and describe the operation of such a system.
29. What becomes of the fuel mixed with the oil when oil dilution is employed?
30. What provision is made to prevent overfilling of fuel tanks when a pressure fuel system is employed?
31. What is the principal purpose of a *fuel jettison system*?
32. What are the purposes of fuel-tank selector controls?
33. What is the purpose of a *vapor-return line* in a fuel system?
34. What is the function of a *fuel-priming system*?
35. Explain the purpose of a *cross-feed system*.
36. What is the purpose of the *fuel-collector cans* in the fuel tanks of the Sikorsky S-61N helicopter?
37. What fuel-warning indications are provided in the fuel system for the S-61N helicopter?
38. What inspections should be made of a fuel system immediately before each flight?
39. What must be done when growth of microorganisms is found in a fuel tank?
40. How are *fuel-selector valves* checked for effectiveness?
41. List the four classifications of *fuel leaks*.
42. What precaution must be observed before repairing a metal fuel tank by soldering or welding?
43. Describe the purging of a fuel tank with inert gas.
44. How can leaks in a fuel tank be discovered?
45. How can a seam in a riveted metal fuel tank be sealed?

Environmental Systems 16

INTRODUCTION

Environmental systems are those aircraft systems used to make the interior environment of the aircraft comfortable and/or habitable for human beings. Depending on the type of aircraft and altitude of operation, this may involve only supplying a flow of fresh air through the cabin by using air vents and scoops. If the temperature must be adjusted for crew and passenger comfort, some method of heating or cooling the cabin interior is required. If the aircraft is to be operated at high altitude, pressurization is necessary to make the environment acceptable to the occupants of the aircraft. Emergency oxygen systems are required on pressurized aircraft. This emergency oxygen supply is necessary to prevent injury to passengers and crew if the cabin should lose pressurization and, as a result, go to a high-altitude environment.

HEATING SYSTEMS

Aircraft **heating systems** range in size and complexity from simple heat exchangers for small, single-engine aircraft through combustion heaters used with larger aircraft to compressor bleed-air systems for turbine-engine-powered aircraft. The system being used may be independent of other systems or integrated into a complete aircraft environmental-control package.

Exhaust Heating Systems

The simplest type of heating system, often employed on light aircraft, consists of a heater muff around the engine exhaust stacks, an air scoop to draw ram air into the heater muff, ducting to carry the heated air into the cabin, and a valve to control the flow of heated air. Such a system is shown schematically in Fig. 16-1. This system is also the most basic type of heat exchanger.

A **heat exchanger** is any device by which heat is transferred from one independent system to another independent system; see Fig. 16-2. This is usually accomplished by allowing the heat in one system to pass into a portion of its containment material and then passing air over (or through)

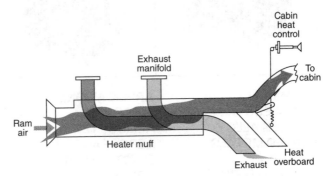

FIGURE 16-1 Exhaust heating system for a light airplane.

that material, warming the air. One system is typically closed; the other is open. A closed system is one in which the primary heat-transfer medium is in a sealed system. The open system is not as tightly sealed and is often ultimately open to the atmosphere. A heat exchanger may be used to reduce or control the level of heat in a system or to transfer heat from one system where it is a by-product into another system that requires the additional heat to function.

The radiator used by older liquid-cooled aircraft engines (much like an automobile) is a type of heat exchanger (see Fig. 16-3). The heat of the engine is absorbed by the liquid that runs through the engine in a closed system. The liquid, the primary heat-transfer medium, then passes through the radiator, where the heat in the liquid is transferred to the metal fins of the radiator. As air passes over the radiator's fins, the heat is transferred to the atmosphere. At high altitudes the same process is used to control the temperature inside the aircraft. In addition to the radiator, a heating coil is installed in the cabin. When additional heat is desired inside the aircraft, air is allowed to pass over the heating coil and enter the cabin. Both the radiator and heating coils are types of heat exchangers.

The exhaust heating system for a light airplane, the Piper Cherokee Six, is shown in Fig. 16-4. This drawing also shows the ventilating system for the airplane.

The heating system consists of the muffler and heat shroud, ducting to the air box and windshield defroster outlets, and ducting to the heat outlets in the cabin. The amount of heat delivered to the cabin is controlled from the cockpit.

ALUMINUM	**APPLICATIONS**	**FUNCTIONS**	**COMBUSTION HEATERS**
STAINLESS STEEL	AIRCRAFT	ENGINE AND	MILITARY
COPPER ALLOY	HELICOPTERS	TRANSMISSION	AIRCRAFT
PLATED METALS	AEROSPACE	OIL COOLERS	COMMERCIAL TRUCK
	ELECTRONICS	FUEL HEATERS	
PLATE AND FIN	TRUCK	AIR-TO-AIR	
SHELL AND TUBE	AUTOMOTIVE	COLD PLATES	
PLATE TYPES	MILITARY	EVAPORATORS	
SPECIAL	COMMERCIAL	CONDENSERS	
CONFIGURATIONS		SURFACE COOLERS	

FIGURE 16-2 Industrial heat exchanger. (*Southwind Corporation.*)

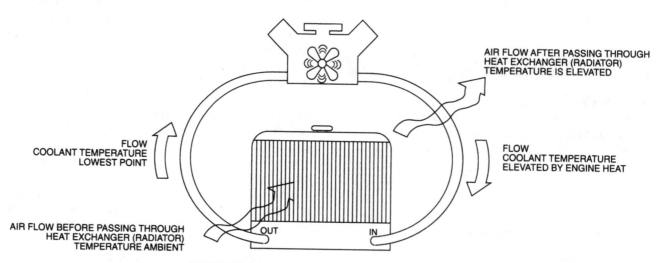

FIGURE 16-3 Automotive radiator as a heat exchanger.

Exhaust heating systems must be given regular inspections to ensure that exhaust fumes cannot enter the cabin of the airplane. This requires that the shrouds or muffs around the exhaust pipes or mufflers be removed to inspect for cracks through which exhaust fumes can enter the heater ducts. Some manufacturers recommend that this be done every 100 h of operation. One method of checking for cracks is to pressurize the exhaust pipes with compressed air and apply soapy water to all areas where cracks may possibly occur. If there is a crack, the air will cause soap bubbles to form at the crack. When performing this pressure check of the exhaust system, be sure to follow the aircraft manufacturer's instructions so that the exhaust system is not damaged by the application of excessive pressure. Additional information on exhaust systems and exhaust heaters is given in the text *Aircraft Powerplants*.

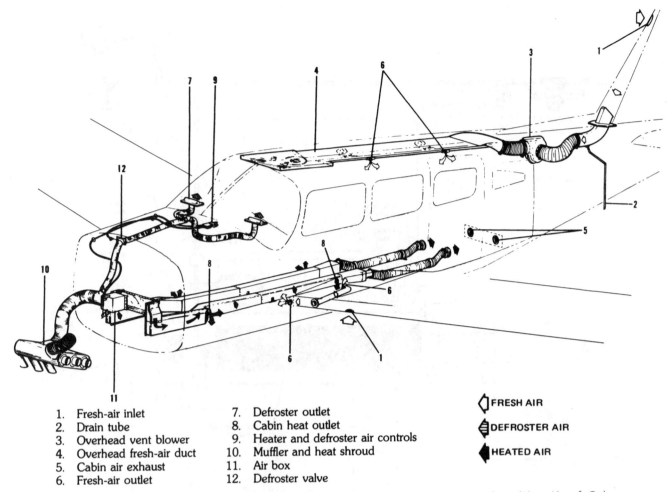

1. Fresh-air inlet
2. Drain tube
3. Overhead vent blower
4. Overhead fresh-air duct
5. Cabin air exhaust
6. Fresh-air outlet
7. Defroster outlet
8. Cabin heat outlet
9. Heater and defroster air controls
10. Muffler and heat shroud
11. Air box
12. Defroster valve

◁ FRESH AIR
◀ DEFROSTER AIR
◀ HEATED AIR

FIGURE 16-4 Arrangement of an exhaust heating system in a six-passenger, light airplane. (*Piper Aircraft Co.*)

Combustion Heater Systems

In larger and more expensive aircraft, **combustion heaters**, also referred to as **surface combustion heaters**, are often employed to supply the heat needed for the cabin. This type of heater burns airplane fuel in a combustion chamber or tube to develop the required heat. Air flowing around the tube is heated and carried through ducting to the cabin. A typical South Wind combustion heater system installed in a light twin-engine airplane is shown in Fig. 16-5. The heater is mounted in the right side of the nose section. Fuel is routed from a tee in the fuel cross-feed line through a filter and a solenoid supply valve to the diaphragm-type heater fuel pump. The fuel pump is operated by the combustion air-blower motor, which is mounted above the heater assembly. This pump provides the heater with sufficient fuel pressure, and no auxiliary boost pump assistance is necessary for proper operation of the heater.

The heater fuel pump and all external fittings on the heater are enclosed in metal housings that are vented and drained as a precaution against fire in the event of leaky fittings. Fuel passes from the heater fuel pump through a solenoid valve to the combustion chamber spray nozzle. When the cabin heater switch is placed in the HEAT position, current is supplied to the combustion air blower and to the ventilating fan.

The fan actuates the cam-operated breaker points, which start the spark plug firing. As the combustion air-blower air increases, the vane-type valve at the inlet of the combustion chamber opens. This actuates a microswitch, which, in turn, operates the solenoid valve, thus allowing fuel to spray into the heater, where the spark plug ignites the fuel. When the heated air flowing from the heater to the cabin exceeds the temperature for which the thermostat is set, the thermostat closes the solenoid valve and stops fuel flow to the heater. The heater thermostat cools, and the solenoid valve opens again to allow fuel to flow to the heater. Heated air flows from the heater, and the thermostat again causes the solenoid valve to close. This cycling on and off continues, and the heater thereby maintains an even temperature in the cabin.

The heater combustion chamber is completely separate from the ventilating system to prevent any exhaust gases from contaminating the cabin air. All exhaust gases are vented overboard through an exhaust tube directly beneath the heater.

Another combustion-type heater is shown in Fig. 16-6. This is the Janitrol heater, which is also used in twin-engine airplanes. A cutaway view, showing the combustion action of the heater, is provided in Fig. 16-7. The heater incorporates a spray nozzle and a spark plug to initiate combustion. Aviation gasoline is injected into the combustion chamber through

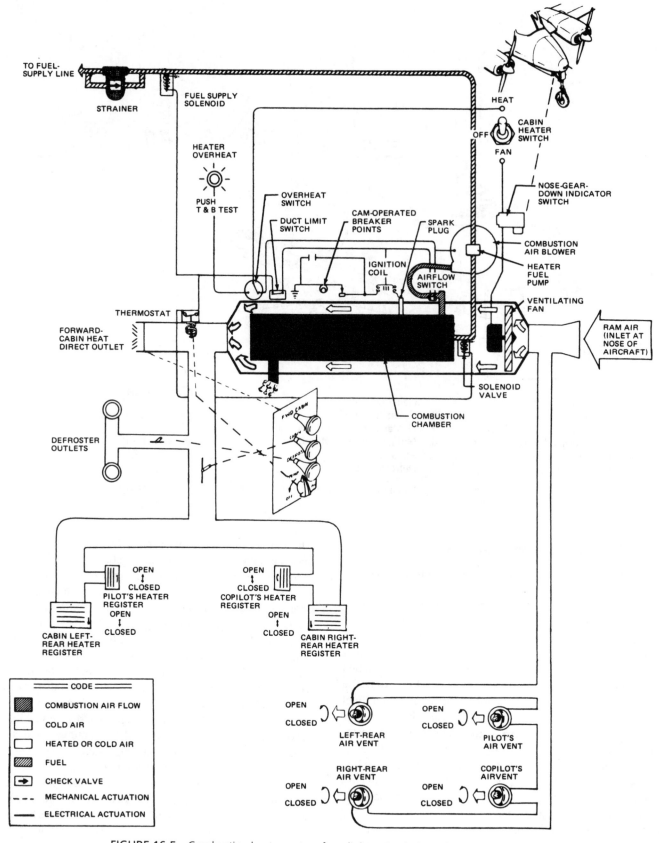

FIGURE 16-5 Combustion heater system for a light twin airplane. (*Cessna Aircraft Co.*)

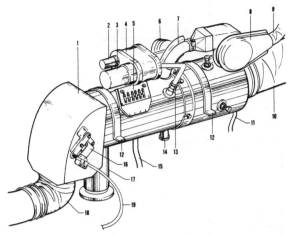

1. Heater output duct
2. Ignition vibrator
3. Ignition coil
4. Ignition unit
5. Terminal strip
6. Pressure switch
7. Elbow adapter
8. Combustion air blower and motor
9. Air inlet tube
10. Air duct hose
11. Water drain tube
12. Heater clamp
13. Spark plug
14. Heater fuel drain tube (hot fuel)
15. Heater fuel drain tube (cold fuel)
16. Adjustable duct switch
17. Exhaust shroud
18. Heat outlet duct
19. Duct switch control cable

FIGURE 16-6 A combustion heater assembly for a light airplane. (*Piper Aircraft Co.*)

the spray nozzle, and this results in a cone-shaped fuel spray, which mixes with combustion air and is ignited by the spark plug. Electrical energy for the spark plug is produced by a high-tension ignition coil, which is supplied from the airplane dc power system. Combustion air enters the combustion chamber tangent to the inner surface. This angle imparts a whirling action to the air, in turn producing a whirling flame that is stable and sustains combustion under the most adverse conditions. The burning gases travel the length of the combustion tube, flow around the inside of the inner tube, pass through crossover passages into an outer radiating area, and then travel the length of this surface and out the exhaust. Fresh air passes through spaces between the walls of the heated tubes to pick up heat, which is delivered to the cabin as required.

The electrical control circuit for the heater is shown in Fig. 16-8. A study of this circuit reveals that when the heater switch is turned to HEAT, electrical current flows through the switch to the air-valve microswitches, overheat switch, combustion air-pressure switch, duct switch, cycling switch, and fuel-cycle valve. At the same time, current flows to the ignition unit. If any of the switches is open, the unit cannot operate.

The duct switch can be adjusted for the desired temperature. When the heater temperature reaches a predetermined level, the duct switch opens, closing the fuel-cycle valve and turning off the ignition. When the temperature drops to a given level, the duct switch closes, thus turning the heater on again.

When the heater switch is turned to the FAN position, only the fan operates. This position is used on the ground when a flow of air without heat is desired.

Typical inspections of a combustion heater system include a visual inspection of the ducting and chamber for obstructions and conditions; inspection of the igniter plug for condition, gap, and operation; a check for any leakage in the airflow or fuel system; an operational check of individual system components (i.e., blower, fuel solenoids, temperature sensors); and an operational check of the complete system. Refer to the aircraft maintenance manual for specific inspection requirements for a particular model aircraft and heater.

Electric Heating Systems

Electrically operated heaters are used on some aircraft to provide heat in the cabin area when the aircraft is on the ground with the engines not running. Aircraft that typically incorporate this type of system include the smaller turbo-prop-powered aircraft.

The auxiliary electric heat system draws air from the inside of the aircraft cabin by the use of a recirculation fan. The air then passes over electrically heated coils and flows back into the cabin through the aircraft heat supply ducting.

The system incorporates safety features into the system design. For example, an airflow switch must sense that the recirculation fan is moving air through the heater before the heater can operate. Also, overheating of the system turns off the heater and illuminates an annunciator light. After an overheat, the heater does not return to operation until the pilot has cycled the control switch.

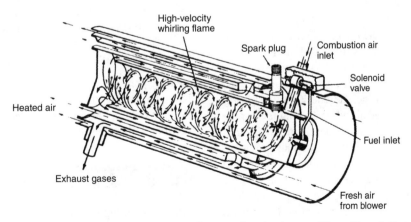

FIGURE 16-7 Burning action in a surface combustion heater. (*Piper Aircraft Co.*)

Heating Systems **559**

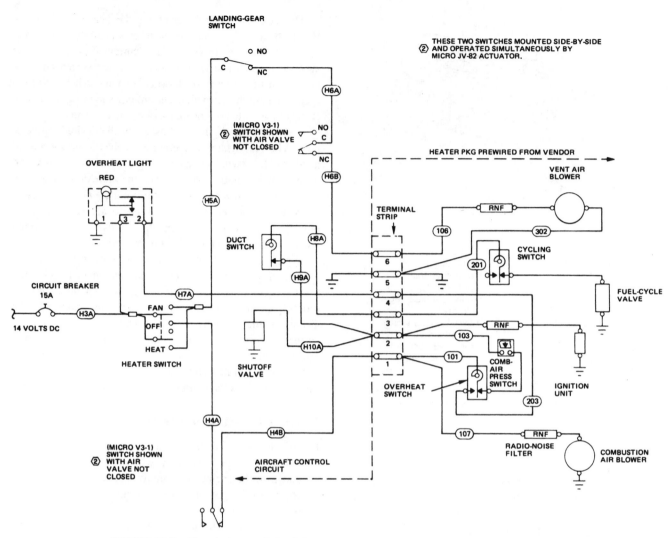

FIGURE 16-8 Electrical control circuit for a combination heater system. (*Piper Aircraft Co.*)

Heating with Bleed Air

Unpressurized turbine aircraft normally make use of compressed air from the turbine engine to provide the hot air for cabin heating. When a turbine engine compresses air prior to directing it to the engine combustion chamber, the air temperature of this air is increased by several hundred degrees Fahrenheit by the compressing action. Some of this hot compressed air, called *bleed air*, can be diverted to a cabin-heating system.

The cabin-heating system consists of ducting to contain the flow of air, a chamber where the bleed air is mixed with ambient or recirculated air, valves to control the flow of air in the system, and temperature sensors to prevent excessive heat from entering the cabin. Additional components used in a bleed-air heating system include check valves to prevent a loss of compressor bleed air when starting the engine and when full power is required of the engine, recirculation fans to move the air through the ducts and provide a flow of ambient or cabin air to the mixing chamber, and engine sensors to eliminate the bleed system if one engine of a multiengine aircraft becomes inoperative. The bleed-air heating system of a single-engine turbine aircraft is shown in Fig. 16-9.

The basic flow of air from the system is as follows: bleed air from the engine compressor flows through check valves to the mixing chamber through a temperature control valve; recirculation or ambient air is directed to the mixing chamber through a control valve; the mixed air is directed past a temperature sensor to cabin flow control valves; and the heated air flows into the cabin air from the distribution ductwork.

If any of the temperature sensors in the system sense overheating, the system is shut down, and crew action is required to restore operation to the system.

Heating Pressurized Aircraft

Heating pressurized aircraft is commonly achieved by regulating the temperature of the air used to pressurize the cabin. For this reason, heating pressurized aircraft is covered in the section "Cabin Pressurization Systems."

CABIN-COOLING SYSTEMS

Aircraft-cooling systems, also called **air-conditioning systems**, are used to reduce the temperature inside an aircraft

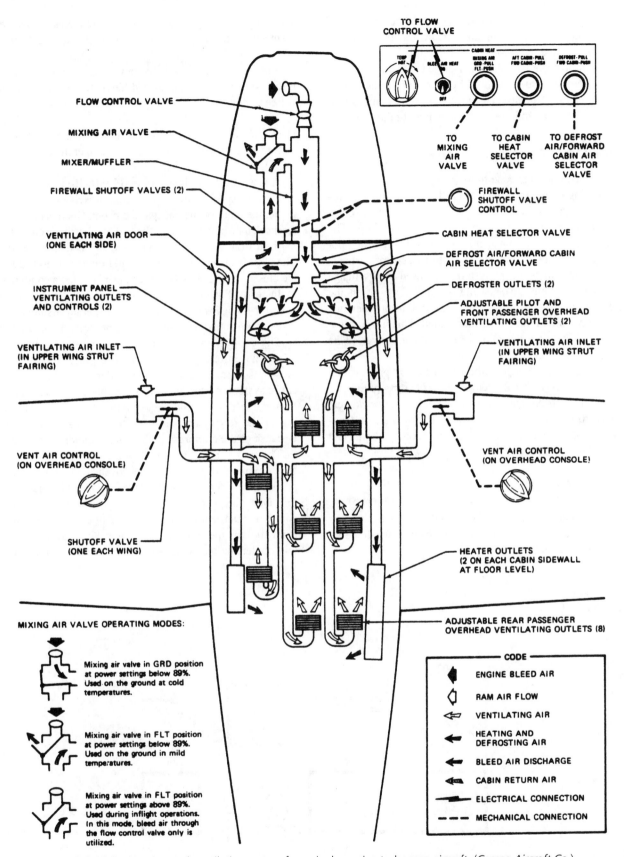

FIGURE 16-9 Heating and ventilation system for a single-engine turboprop aircraft. (*Cessna Aircraft Co.*)

for crew and passenger comfort. The two basic methods of reducing the temperature of aircraft are the freon *vapor-cycle* machine and the *air-cycle* machine.

The **vapor-cycle** machine is a closed system using the evaporation and condensation of freon to remove heat from the cabin interior. The **air-cycle** machine uses the compression and expansion of air to lower the temperature of the cabin air.

Vapor-Cycle Cooling Systems

The vapor-cycle air-conditioning system is used in reciprocating-engine-powered aircraft and in smaller turboprop light jet aircraft that do not make use of air-cycle machines to reduce the cabin interior temperature. The operation of vapor-cycle machines is controlled by the pilot and may incorporate automatic cutout or interrupt systems. These cutout and interrupt systems are used to disengage the refrigerant compressor during demand for high engine power output, such as during takeoff operations or when one engine of a multiengine aircraft has failed. This interruption in compressor operation allows all available power to be used to maintain controllable flight.

In *Aircraft Basic Science*, where the nature of gases is discussed, it is shown that the temperature of a gas is directly proportional to its pressure and/or volume. In addition, as heat is added to a solid, it becomes liquid and then a gas. Conversely, as heat is extracted from a gas, it becomes a liquid and then a solid. This concept leads to the following principles:

As liquids change to a gas (or are vaporized), they absorb heat. This heat is called the **latent heat of vaporization**.

As a given quantity of gas is condensed to a liquid, it emits heat in the same amount that it absorbs when being changed from a liquid to a gas.

When a gas is compressed, its temperature increases, and when the pressure on a gas is decreased, its temperature decreases.

Another law of science used in vapor-cycle cooling systems is that when two materials have different temperatures and heat is free to flow between them, they will attempt to equalize:

Heat transfers only from a material having a given temperature to a material having a lower temperature.

A vapor-cycle cooling system (Fig. 16-10) takes advantage of these laws of nature using two heat-exchangers to control the temperature of the cockpit and cabin. One heat exchanger works in a fashion similar to the one previously discussed, taking heat from the closed system, and is called an **evaporator**. The other heat exchanger draws heat from the air and adds it to the closed system. This heat exchanger is called a **condenser**.

Instead of water in the heat exchanger, a special fluid called a *refrigerant* is used. The refrigerant, usually freon, takes two forms during the cooling process, liquid and gas.

The cooling process starts at the compressor, where the refrigerant is in a gaseous form. The function of the compressor is to push the refrigerant, under pressure, through the entire system. As the gas enters the condenser, heat is drawn from the refrigerant and passed to the atmosphere. The cooling of the refrigerant causes it to condense into a liquid. Because of the compressor, the liquid is under pressure.

The pressurized liquid is then metered into tiny droplets by an **expansion valve**. Because of the change in form, the pressure past the expansion valve is lowered. The droplets then enter the evaporator, where they draw heat from the air and then change into a gas. As a result of heat being drawn from the air, the temperature of the air is decreased. It is this cooler air that is introduced to the cabin for cooling.

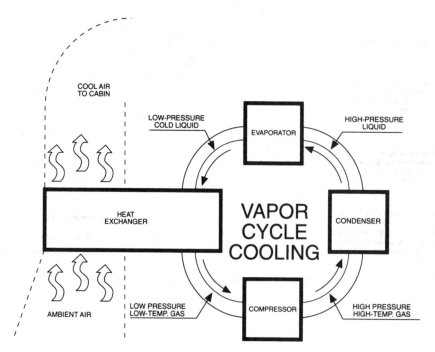

FIGURE 16-10 Basic vapor-cycle cooling.

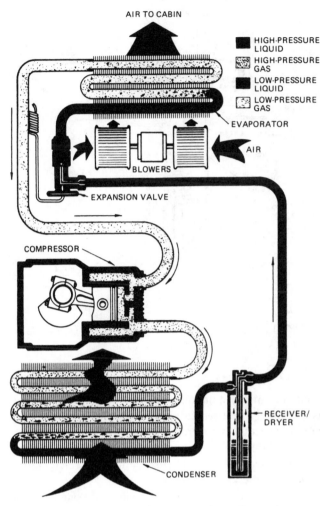

FIGURE 16-11 Diagram of a vapor-cycle cooling system. (*A R A Manufacturing Co.*)

heat from the vapor. On an airplane in flight the airflow for the condenser coils of the vapor-cycle system may rely on airflow entering through inlets in the wings or fuselage, or the coils may be mounted on a structure that is lowered into the airstream during flight. On the ground, the air must be caused to flow through the condenser by means of a blower or some type of an ejector.

As the refrigerant vapor is cooled in the condenser, it becomes a liquid and flows to the **receiver-dryer-filter**. The receiver-dryer-filter is essentially a reservoir containing a filter and a desiccant. A **sight glass** is usually located on top of the receiver to allow observation of the fluid flow through the unit. If bubbles are seen in the fluid, the system refrigerant is known to be low and requires replenishment. A cutaway drawing of a typical receiver-dryer-filter is shown in Fig. 16-12. In some systems, the dryer-filter is separate from the receiver.

Refrigerant liquid flowing from the receiver is under high pressure and has been cooled by the condenser. In some

A more detailed description of a vapor-cycle machine and its components follows.

An actual system includes controlling devices to provide for changes in cooling demand and changes in operating conditions. A schematic diagram of a typical system is shown in Fig. 16-11. A system of this type usually employs Refrigerant 134A, also known as Freon R134A, as the refrigerant. It is a gas at standard pressure and temperature but becomes a liquid when its pressure is increased sufficiently or when the temperature is reduced to the required level. The range of pressures and temperatures at which Refrigerant 134A liquefies or becomes a gas makes it an ideal medium for producing the temperatures desirable for an automobile or aircraft cabin.

The refrigeration cycle, beginning with the **compressor**, involves the compression of the refrigerant gas, which is comparatively cold and at a low pressure as it leaves the evaporator and flows to the compressor. The compressor is the source of energy required for the operation of the system.

The gas leaving the compressor is at a high temperature and pressure. In a typical range of operation, the pressure of the gas leaving the compressor may be as high as 180 psig [1241 kPa] or more. From the compressor, the hot, high-pressure gas (vapor) flows into the **condenser**, which is a heat radiator through which cool air is passed to remove

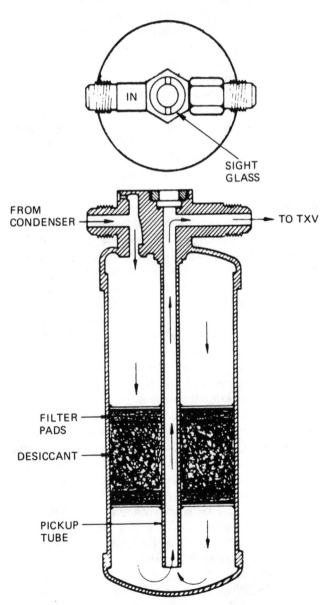

FIGURE 16-12 Cutaway drawing of a receiver-filter. (*A R A Manufacturing Co.*)

systems, the liquid refrigerant at this point is passed through a **subcooler**. The subcooler is a heat exchanger through which the refrigerant liquid flowing to the **expansion valve** and the cold vapor flowing from the **evaporator** are passed through separate coils in the subcooler. Since the fluid is warmer than the vapor, heat from the fluid passes to the vapor. The effect is to subcool the fluid and superheat the vapor.

Subcooling is the process of reducing the temperature of a liquid below the temperature at which it was condensed, the pressure being held constant. If the liquid is subcooled 20 to 30°F [11 to 17°C] below the condensing temperature, this condition ensures that there will be no premature vaporization ("flash-off") before the liquid reaches the expansion valve.

The subcooler, in adding heat to the vapor from the vaporizer to the compressor, further superheats the vapor, thus reducing the possibility of condensation taking place in the vapor and prevents liquid from reaching the compressor. Liquid entering the compressor would interfere with its performance and would probably cause damage to the compressor. When liquid refrigerant appears between the vaporizer (evaporator) and the compressor, the condition is called "slugging." The presence of slugging indicates that the controls for the system are not functioning correctly and the system will not perform efficiently.

The high-pressure liquid refrigerant, after leaving the receiver-dryer, passes through the **thermal expansion valve**. This valve consists of a variable orifice through which the high-pressure liquid is forced. Low pressure exits at the outlet side of the expansion valve, through the evaporator, and to the inlet of the compressor. As the liquid refrigerant passes through the expansion valve into the low-pressure area, it begins to break up into droplets, and by the time it leaves the evaporator, it is a gas. When the liquid refrigerant becomes a gas, it absorbs heat, thus producing the cooling effect desired. The size of the droplet affects the rate at which the heat may be transferred. The smaller the droplets for a given quantity of refrigerant, the more surface area is exposed for heat transfer. The cabin air to be cooled is carried by ducting into one side of the evaporator and out the other side to the cabin.

The orifice for the thermal expansion valve is adjusted automatically by the pressure from a thermal sensor, which senses the temperature of the gas leaving the evaporator. If the gas is warmer than it should be, thus not providing sufficient cooling to the cabin, the expansion valve provides greater restriction, producing smaller droplets and, hence, greater cooling. If the gas is too cool, the expansion valve provides less restriction of the liquid, producing larger droplets and, therefore, less cooling capability. A drawing of an expansion valve is shown in Fig. 16-13.

After the gas leaves the evaporator, it flows to the compressor and the cycle begins again. The evaporator provides the cooling for cabin air, and the condenser dissipates the heat developed when the gas is compressed.

As can be noted from this discussion, the vapor-cycle refrigeration system has a high-pressure section from the compressor to the thermal expansion valve and a low-pressure section from the thermal expansion valve to the

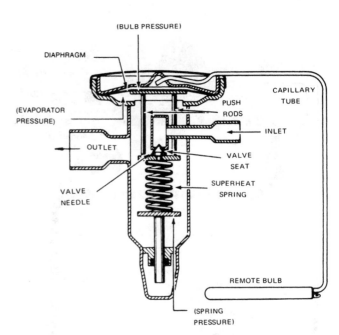

FIGURE 16-13 Drawing of an expansion valve. (*A R A Manufacturing Co.*)

inlet side of the compressor. The high-pressure section is commonly referred to as the **high side**, and the low-pressure section is referred to as the **low side**. It is important for the technician to remember these terms because they are used in the service instructions for refrigeration systems of the vapor-cycle type.

An operational check involves running the engine of the aircraft, turning on the refrigeration system, and checking the temperature at the evaporator. The sight glass on the top of the evaporator-dryer should be checked for bubbles or foam, which would indicate lack of refrigerant. If a refrigeration system operates satisfactorily for a few minutes and then stops cooling, it is an indication of water in the system. The water has frozen at the expansion valve and stopped the flow of refrigerant. Systems are provided with pressure sensors, which control an electric circuit that disengages the compressor clutch when the pressures in the high side become too great.

Servicing of a vapor-cycle refrigeration system requires the use of a **service manifold**, which makes it possible to "plug in" to the system as required to test pressures, add refrigerant, and purge, evacuate, and recharge the system. The service manifold usually has a high-pressure gauge, which should have a capacity of as high as 500 psig [3447.5 kPa], and a compound low-pressure gauge, which can provide readings ranging from −30 in Hg (inches of mercury) [−101.59 kPa], which is equivalent to 30 in Hg (Fig. 16-14) below ambient pressure (0 in Hg absolute) and up to 60 in Hg or more [203.3 kPa or more] above ambient pressure. The manifold also includes three fittings, one for the low side, one for the high side, and one by which refrigerant may be added or removed or through which the system can be purged and evacuated. The service manifold is illustrated in Fig. 16-14. Note that the manual valves in

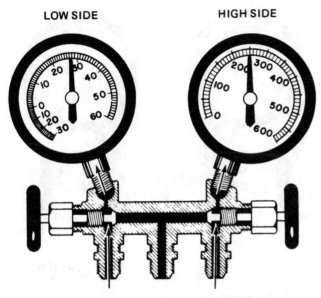

LOW SIDE   HIGH SIDE

**BOTH VALVES CLOSED—GAUGES SHOW
HIGH-SIDE AND LOW-SIDE PRESSURE**

FIGURE 16-14 Service manifold for a vapor-cycle cooling system. (*A R A Manufacturing Co.*)

the manifold are arranged in such a manner that the gauges will give indications of pressures in the system even though the valves are closed. The purpose of the valves is to allow connection of either the high, the low, or both sides of the system with the center service connection.

Vapor-cycle systems are provided with **service valves** to which the service manifold can be connected. These valves may be of the Schrader type (similar to a shock-strut or hydraulic accumulator valve) or some other type. In any case, the technician must know what type of valve is employed in the system and must be sure to have the correct type of fitting on the service manifold hose so it will properly connect to the service valve. The service valves in the system provide for connection to the low side and to the high side. The valves are often on top of the compressor, but they may be at other points in the system. The technician must be sure to connect the low-pressure side of the service manifold to the service valve for the low side of the system and the high-pressure side of service manifold to the high-side service valve.

When it becomes necessary to change a unit in a vapor-cycle refrigeration system, the refrigerant must be released. This is called **purging** the system.

The recommended procedure for purging the refrigeration system is first to connect the service manifold to the low-side and high-side service valves. The manifold port should be connected to a vacuum pump connected to a closed container. As mentioned previously, the technician must ensure that the valve fittings are correct for the type of valves involved. The service instructions provide this information. The manual valves on the service manifold must be *closed* before connecting the manifold to the system. After the service manifold is connected to the system, either one or both of the service valves are "cracked" (opened slightly),

with the vacuum pump on, to allow a slow escape of the refrigerant into a closed container. The escaping gas should be monitored for escaping compressor oil, which indicates that the system is being purged too quickly. If oil is discharging with the gas, the discharge is too rapid and the manual valves should be closed slightly. When both pressure gauges on the service manifold indicate zero pressure, the system is purged and can be disassembled.

When one or more units are removed from the system, all fittings should be capped immediately to prevent the entrance of any detrimental material. Anything except pure, dry refrigerant and refrigerating system oil can cause malfunction of the system. Hoses that may become contaminated should be replaced with new hoses.

After a system has been reassembled and is once again a closed, sealed system, it must be evacuated (pumped down) with a suitable vacuum pump. First, however, the drying agent in the receiver-dryer must be replaced. As explained previously, this drying agent will absorb any traces of moisture remaining in the system.

The vacuum pump is connected to the center fitting of the service manifold, as shown in Fig. 16-15. The manual valves on the manifold are opened and the pump is started. The reason for evacuation of the system is that all air and

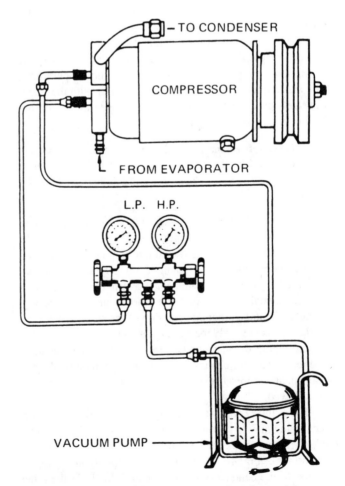

FIGURE 16-15 Vacuum pump connected to the service manifold. (*A R A Manufacturing Co.*)

attendant moisture must be removed from the system. As pressure decreases below atmospheric, the boiling point of water decreases. It is recommended that the pressure of the system be reduced to −29.0 in Hg [−98.2 kPa], that is, below standard sea-level pressure. This applies when the service is being performed at sea-level pressure. At this pressure all water in the air is completely vaporized, and an extremely small amount of air or water exists in the system. Usually about 30 min with a good vacuum pump is required to pump down the system to the desired level.

After the system is evacuated, as described previously, the manual valves on the service manifold are closed. The low-pressure gauge on the manifold will register the vacuum pressure in the system.

Following the evacuation (pump-down) of the system, the system must be recharged with the required amount of R-134A. Refrigerant is sold in 15-oz [425-g] sealed containers with fittings to which appropriate valves can be attached. It is also sold in larger containers and cylinders. For a small-capacity system, the 15-oz can (also known as a 1-lb can) is used because it is convenient and lets one determine the amount of refrigerant placed in the system.

The R-134A refrigerant is odorless and nontoxic at normal temperatures. It does, however, displace air and, therefore, has adverse environmental effects. Also, R-134A refrigerant converts to a toxic gas at high temperatures. For these reasons, purging of the refrigeration system should always be done in a closed system. In addition, in case of leakage in the evacuation system, purging the system should be accomplished in an open area where there is adequate ventilation and no chance of exposure to high temperatures. The prudent technician will become familiar with the material safety data sheet for the particular refrigerant used before beginning any service of the system.

CAUTION: Refrigerant in a liquid state can cause eye and skin damage due to the low temperature created as the refrigerant evaporates. Use a face shield and appropriate clothing to prevent injury.

Recharging the system may be accomplished using either of two techniques. The first and most desirable technique is to use a charging system that allows the technician to measure precisely the amount of refrigerant being added. The aircraft's maintenance manual or system manufacturer's service manual should specify the amount and type of refrigerant for the most efficient operation of the system. After the refrigerant is precisely measured, it may then be added to the system, under the appropriate pressures. Proper procedures and techniques for changing refrigerant specified by the aircraft manufacturer and/or the equipment manufacturer must be followed.

If the more sophisticated equipment is not available, recharging may be accomplished by connecting the container of refrigerant to the center fitting of the service manifold, as shown in Fig. 16-16. First, the appropriate hose is connected to the center fitting of the manifold. Then the other end of the hose is connected to the valve that is connected to the container of refrigerant. Remember, at this time the

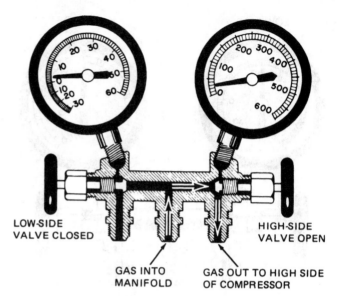

FIGURE 16-16 Arrangement of the service manifold for recharging the system. (*A R A Manufacturing Co.*)

system is in the evacuated condition and both manual valves on the service manifold are closed. After all connections are secure, the valve on the refrigerant container is turned in the assigned direction. This pierces the container and permits the gaseous refrigerant from the top of the container to flow into the service hose. The refrigerant container should be held in an upright position to ensure that only gas flows into the system. If the container is inverted, liquid refrigerant will flow, and this must not be permitted in the low side of the system.

After the container valve is opened, the fitting of the hose from the container should be slightly loosened at the service manifold. Air and gas should be permitted to escape for a few seconds to drive the air out of the service hose. Remember that air and moisture, which air contains, are both detrimental to the proper operation of the system. The service connection at the manifold is retightened, and the manual valve on the high side of the system is opened. The container can now be inverted to allow liquid refrigerant to flow into the high side of the system. Very quickly the gauge on the low side of the system should move out of the vacuum indication, showing that gaseous refrigerant is flowing through the evaporator to the low side of the compressor. If this does not occur, the system is blocked and must be unclogged before the procedure is continued.

After the prescribed amount of refrigerant has been placed in the system, the service manifold should be removed and the service fittings should be capped. The system may then be given an operational test.

It is emphasized here that the procedures described in the foregoing are general in nature and are provided so the technician will have an understanding of the principal considerations in the service and operation of vapor-cycle refrigeration systems. In servicing a particular system, the appropriate service instructions should be followed.

Environmental Control System of Light Jet

Small jet aircraft are often equipped with a vapor cycle cooling system due to the limited bleed air capability of the engines. Figure 16-17 gives an overview of the components. The heat exchanger/condenser pack and the compressor drive module are installed in the rear of the aircraft. One evaporator is installed in the cabin and one in the cockpit.

Vapor Cycle System

Figure 16-18 shows the individual components of the vapor cycle system (VCS). During operation of the vapor cycle system Freon R134A refrigerant is compressed in the compressor drive module. The Freon is a high-pressure high-temperature gas at this stage. When the Freon enters the condenser it is cooled by ram air in flight and a fan on the ground and it changes to a high-pressure liquid. The Freon flows to the expansion valve where the Freon is turned into fine droplets and the Freon will change into a low pressure gas in the evaporator. The low-pressure gas will return to the compressor and the vapor cycle will begin again.

Aircraft Heating and Cooling

During the heating cycle when no cold air is required the vapor cycle cooling compressor is automatically turned off by the environmental control system (ECS). Hot bleed air will flow from the engine compressor section when the pressure regulator and shutoff valve is opened. This hot high-pressure air will pass through the ram air cooled air to air heat exchanger and the air will be cooled. Modulating valves can mix the hot bleed air and the cooled air coming from the heat exchanger to adjust the air temperature. If cooling of the aircraft is required, the air in the cockpit and the cabin will flow through the vapor cycle cooling evaporators and the air will be cooled. Figure 16-19 shows the ECS schematic.

Temperature Control

Figure 16-20 shows the digital control function of the environmental control system. The heart of the system is the ECS controller which receives inputs from cockpit temperature and fan controls, duct temperature sensors, flow control shutoff valve (FCSOV), and monitoring data from the VCS. Based on these inputs the ECS controller will send signals to the vapor cycle system compressor and ground cooling fan to turn on or off. The ECS controller also send signals to the temperature-modulating valves (TMVs) to control the bleed air temperature.

The ECS temperature controller is a digital electronic assembly that consists of two independent circuit boards, one for the cabin and one for the cockpit. The ECS temperature controller provides the electrical power for duct temperature sensor/switch (TSS), cabin/cockpit temperature sensor, and for both TMVs. The controller also has a software-independent manual-control circuit for controlling the TMV, if the manual mode is selected from the cockpit control panel. It monitors the bleed air duct temperature as well as the zone temperature, utilizing software to perform temperature control. The actual zone temperature is compared to the pilot-selected zone temperature. The controller then modulates the TMV in order to drive the actual duct temperature to the desired duct temperature required by the respective aircraft zone. In the cooling mode the controller utilizes

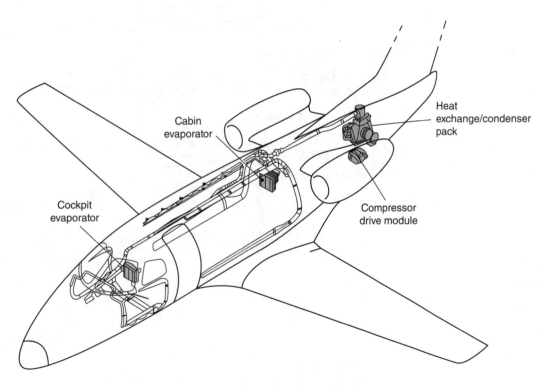

FIGURE 16-17 Vapor cycle cooling air-conditioning components.

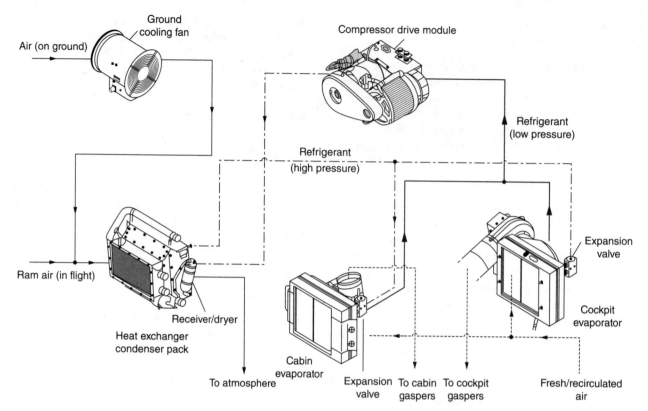

FIGURE 16-18 Vapor cycle system components.

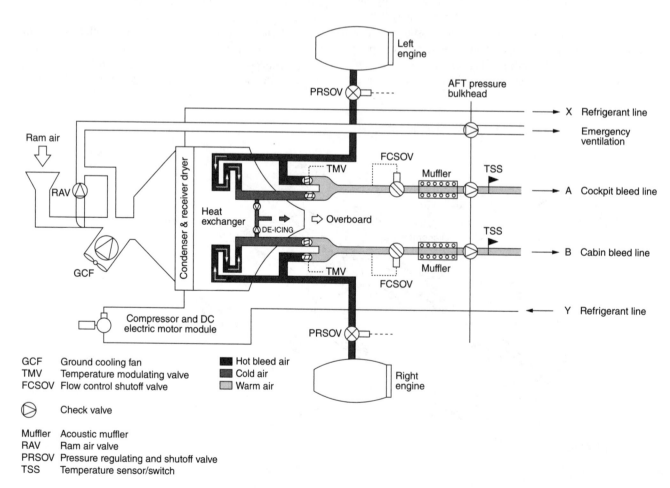

GCF — Ground cooling fan
TMV — Temperature modulating valve
FCSOV — Flow control shutoff valve

⊘ — Check valve

Muffler — Acoustic muffler
RAV — Ram air valve
PRSOV — Pressure regulating and shutoff valve
TSS — Temperature sensor/switch

■ Hot bleed air
■ Cold air
▨ Warm air

FIGURE 16-19 Environmental control system schematic.

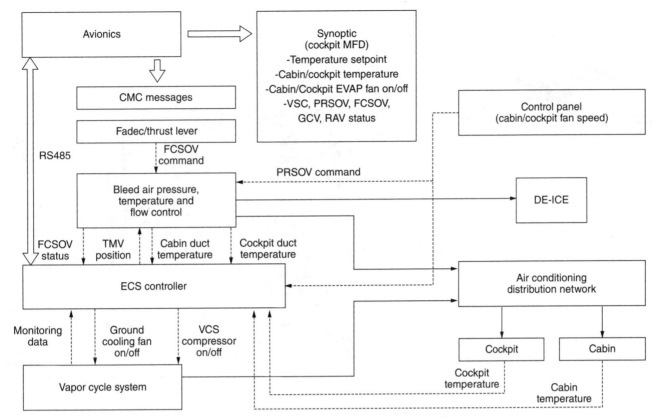

FIGURE 16-20 Environmental system block diagram.

the VCS to cool the aircraft. Cockpit and cabin temperature set-points on the synoptic page indicate failure (red "X") when the ECS temperature controller sends values out of the valid range. Avionics will flag these parameters as invalid if the ECS temperature controller sends temperature set-points lower than 14°C (57°F) or higher than 34°C (93°F).

The temperature of the regulated hot bleed air is modulated by the TCS through either the heat exchanger circuits or bypass lines depending on the heating or cooling requirements of each aircraft zone (cockpit and cabin). This modulation is made by the TMV of each circuit. The valve independently controls the temperature of the bleed air entering the cabin or cockpit. The TMV receives 28-V DC digital pulse signals from the ECS temperature controller to modulate the hot air flow either through the air-to-air heat exchanger or through a bypass, mixing the air as required to obtain the desired downstream temperature for cabin/cockpit heating. Figure 16-21 shows the ECS synoptic page on the multifunction display (MFD).

Air-Cycle Cooling

Modern large turbine-powered aircraft make use of air-cycle machines to adjust the temperature of the air directed into the passenger and crew compartments of these large aircraft. Although this portion of the discussion of air-cycle machines is directed to the ability to provide cabin-cooling air, it should be noted that the cabin can also be heated and pressurized by the use of an air-cycle machine. These operations are discussed in this chapter.

These large aircraft utilize air-cycle cooling because of its simplicity, freedom from troubles, and economy. In these systems, the refrigerant is air. Air-cycle cooling systems utilize the same principles of thermodynamics and the same laws of gases involved in vapor-cycle systems. One principal difference is that the air is not reduced to a liquid, as is the refrigerant in a vapor-cycle system.

The principle of cooling by means of a gas is rather simple. When a gas (air) is compressed, it becomes heated, and when the pressure is reduced, the gas becomes cooled. If a pressure cylinder is connected to an air compressor and compressed air is forced into the cylinder, one can observe that the cylinder becomes warm or even hot, depending upon the level of compression and the rate at which the air is compressed. If the cylinder filled with highly compressed air is then allowed to cool to ambient temperature, the pressure in the cylinder will be reduced to a certain degree as the air temperature is reduced. If a valve is then opened and the air is allowed to escape from the cylinder, the temperature of the escaping air will be much lower than the ambient temperature due to the air expanding as its pressure returns to the ambient value. This cold air can then be used as a cooling agent.

In an air-cycle system, the air is continuously compressed and then cooled by means of heat exchangers through which ram air is passed; then the pressure is reduced by passing the air through an expansion turbine. The air leaving the expansion turbine is at low pressure and low temperature. The cooled air is directed through ducting with control valves to regulate the amount of cooling air needed to produce the desired cabin temperature.

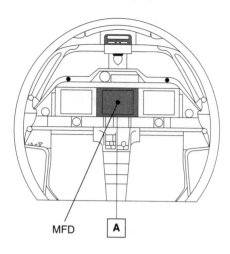

MFD A

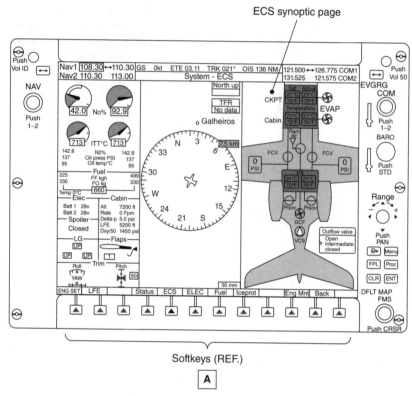

ECS synoptic page

Softkeys (REF.)

A

FIGURE 16-21 ECS synoptic page on MFD.

The turbine-compressor unit by which air is cooled is called an **air-cycle machine** (ACM). An illustration of how the ACM is connected into a "cooling pack" is shown in Fig. 16-22. Hot compressed air from the compressor of one of the turbine engines flows through the **primary heat exchanger**. The heat exchanger is exposed to ram air, which removes heat from the air. The cooled but still compressed air is then ducted to the compressor inlet of the ACM. The compressor further compresses the air and causes it to rise in temperature. This air is directed to the **secondary heat exchanger**, which, being exposed to ram air, removes heat from the compressed air. The compressed air is then directed to the **expansion turbine**. The expansion turbine absorbs energy from the air and utilizes this energy to drive the compressor. As the air exits the expansion turbine, it enters a large chamber, which allows the air to expand and causes a further reduction in the air temperature. Thus the air leaving the turbine is cooled by the loss of heat energy and by the expansion that takes place. The great reduction in temperature causes the moisture in the air to condense, and this moisture is removed by means of a **water separator**. The dried, cold air is then routed to ducting to be utilized as required to provide the desired temperature in the cabin. In the drawing of Fig. 16-23 note that a bypass duct with the cabin-temperature-control valve will bypass air around the cooling system when cooling is not required.

Electrically Driven Environmental Control System

Most modern transport aircraft use an air cycle cooling system that uses engine bleed air. The Boeing 787 is the first transport aircraft that uses electrically driven air cycle air-conditioning packs to provide the air conditioning and cabin pressurization function, with fresh air brought onboard via dedicated cabin air inlets. Instead of using bleed air from the engine compressor, air is delivered by electrically driven air compressors. There is no need to regulate down the supplied compressed air. Instead, the compressed air is produced by adjustable speed motor compressors at the required pressure without significant energy waste. The output of the cabin pressurization compressors flows through low-pressure air-conditioning packs. The environmental-control system air inflow can be adjusted in accordance with the number of airplane occupants to achieve the lowest energy waste while meeting the air-flow requirements. The electrically driven compressors are located together with the other air cycle cooling components in the lower aircraft fuselage.

CABIN-PRESSURIZATION SYSTEMS

The purpose of aircraft cabin pressurization is to maintain a comfortable environment for aircraft occupants while allowing the aircraft to operate efficiently at high altitudes.

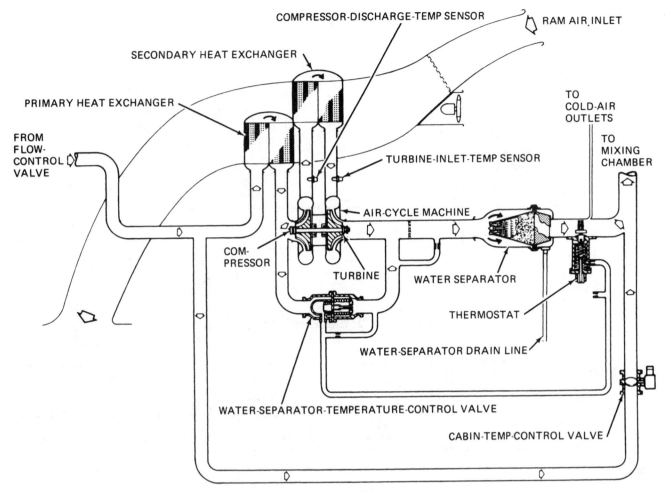

FIGURE 16-22 Air-cycle machine in a cooling system. (*AiResearch Mfg. Co.*)

At high altitudes the aircraft can fly above most of the weather conditions that contain turbulence and make flight uncomfortable for passengers and crew members. Additionally, the fuel efficiency of the aircraft is increased at high altitudes, resulting in less fuel being burned for any given cruising airspeed.

In order to make the cabin environment comfortable for the aircraft occupants, the cabin must normally be pressurized to maintain the cabin air pressure at the level reached at no higher than 8000 ft [2438 m]. This enables the crew and passengers to function without the use of supplemental oxygen and, with adjustments of the cabin air temperature, allows them to be in a "shirtsleeve" environment.

Along with allowing the aircraft and occupants to function in efficient environments, pressurization also permits the occupants to be insulated from the rapid changes in altitude associated with modern transport aircraft. For example, the aircraft normally climbs to cruising altitude as quickly as possible for maximum fuel efficiency, but the pressurized cabin gradually climbs to a selected lower-pressure altitude. A similar situation can take place during descents for landing, where the aircraft might descend very rapidly, but the pressure in the cabin increases slowly until both the cabin pressure and the ambient pressure are the same when the aircraft lands. Examples of the aircraft and

cabin altitudes and relative rates of change are shown in the flight profile in Fig. 16-23. By being able to control the cabin rate of climb and descent pressure independent of the aircraft climb and descent pressure, the aircraft occupants are spared any discomfort from rapid pressure changes.

Although the use of pressurization is normally associated with turbine-powered transport aircraft, pressurization may be found in aircraft ranging in size from light single-engine aircraft through light twins, and turboprop-powered aircraft. Although the basic controlling mechanisms for each of these types of aircraft are the same, the sources of pressure and the capabilities vary, depending on the purpose and complexity of the aircraft.

Aircraft Structures

In order for an aircraft to be pressurized, it must have enough structural strength in the pressurized section, called the "pressure vessel," to withstand the operational stresses. The normal limiting factor as to how high the aircraft can operate is often the maximum allowed cabin differential pressure. Cabin differential pressure is the difference in pressure between the ambient outside air pressure and the pressure inside the aircraft. The higher the allowed differential pressure, the stronger the airframe structure must be.

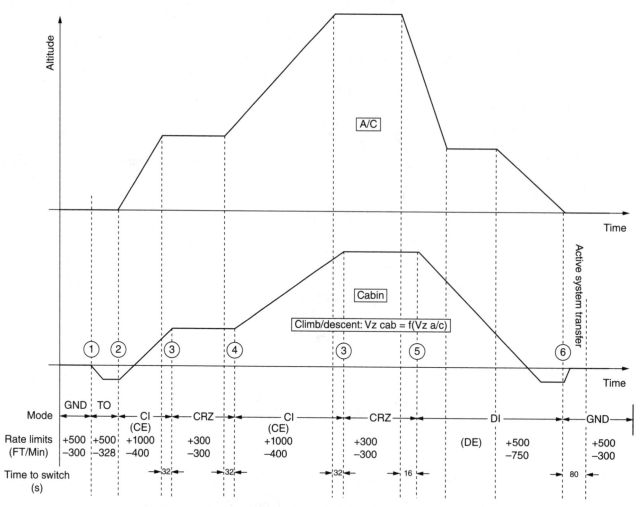

FIGURE 16-23 Comparison of aircraft and cabin altitudes during flight.

Light aircraft are generally designed to operate with a maximum cabin differential pressure of about 3 to 5 psi [20.7 to 34.5 kPa]. Turbine-powered transports are designed for a maximum differential pressure on the order of 9 psi [62.06 kPa].

Sources of Pressurization

The source of pressure for an aircraft varies, depending on the type of engine being used and the design requirements of the aircraft manufacturer. In all cases, the engine provides the power to pressurize the aircraft, but the means of pressurization varies.

Reciprocating engines can supply pressure from a supercharger, a turbocharger, or an engine-driven compressor. A **supercharger** is an engine-driven air pump, mechanically driven from the engine drive train, which compresses air for use by the engine in the combustion process. A portion of this supercharger air can be directed into the cabin pressurization system, provided that the supercharger is positioned in the induction system before fuel is introduced into the airflow.

A **turbocharger** is used in a similar manner as a supercharger except that the turbocharger is driven by exhaust gases from the engine, which drives an air compressor to supply an air charge to the engine.

Superchargers and turbochargers have disadvantages in that fumes and oil can be introduced into the pressurized air and foul the cabin air. Additionally, the use of either of these pressure sources may substantially reduce engine-power output, especially at higher altitudes where the pressure output of the pump is minimal.

Aircraft using turbine engines usually make use of **engine bleed air** to pressurize the cabin. Bleed air is pressurized air that is "bled" from the compressor section of the turbine engine. The loss of this compressor air causes some reduction in total engine-power output.

Independent cabin compressors are used in some aircraft to eliminate the problem associated with air contamination. These pumps are driven from the engine accessory section or by turbine engine bleed air. Two types of pumps used are the **Roots-type positive displacement pump**, shown in Fig. 16-24 and the **centrifugal cabin compressor**, shown in Fig. 16-25.

Pressurization System Components

The complexity of a particular pressurization system depends on the various functions that it is designed to serve. In addition to supplying pressure to maintain the desired cabin altitude,

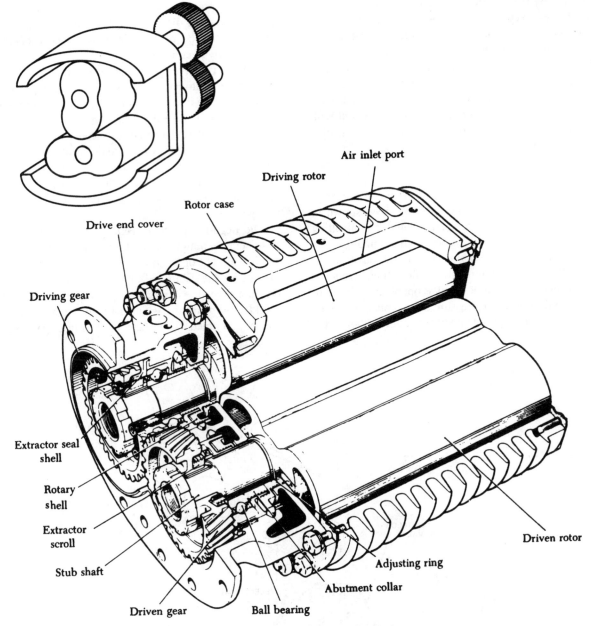

FIGURE 16-24   A Roots-type cabin supercharger.

Labels on figure: Air inlet port, Driving rotor, Rotor case, Drive end cover, Driving gear, Extractor seal shell, Rotary shell, Extractor scroll, Stub shaft, Driven gear, Ball bearing, Abutment collar, Adjusting ring, Driven rotor

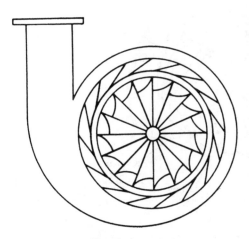

FIGURE 16-25   A centrifugal compressor.

a pressurization system may also incorporate components to heat and to cool the air prior to its entering the cabin. This is normally the case with transport aircraft. Smaller aircraft often incorporate vapor-cycle air conditioners and heaters in the system to provide the proper temperature adjustment for the aircraft cabin. Typical systems for different types of aircraft are presented in the following sections.

When the pressurized air exits the pressure source, it is usually at a high temperature and must be cooled to a usable temperature. The unit used to cool the air is a **heat exchanger**. The heat exchanger typically uses ambient air flowing over a radiator-like structure to remove heat from the pressurized air moving through the unit. In flight the ambient airflow occurs by ram air. When the aircraft is on the ground, fans or ejectors may be used to cause air to flow across the heat exchanger.

Cabin-Pressurization Systems   **573**

Since all of the air from the compressor must flow through the heat exchanger, the air may be cooled too much. To adjust the air temperature, a heat source may be used to bring the air back up to the desired value. This heat source may be an electric heater; heated air that has been used to cool equipment, such as from the avionics bay; bleed air from the engine; or air from an engine-exhaust heat muff or augmenter tube. The amount of heated air mixed with the cooled air may be automatically or manually controlled by a mixing valve.

After the air temperature is adjusted, the airflow is directed into the pressure vessel.

Several different valves are used to regulate the pressure in the pressure vessel. The **outflow valve**, shown in Fig. 16-26, is used as the primary means of controlling the cabin pressure. This valve controls the amount of air allowed to escape from the cabin. The outflow valve is controlled by the flight crew through the aircraft environmental control panel. The outflow valve opens and closes to maintain the desired cabin pressure and, in many systems, operates to maintain a preset maximum cabin altitude value

if the cabin pressure goes above that value. Outflow valves may be operated directly by pneumatic pressure, or their operation may be by electric motors whose operation is controlled by pneumatic pressure.

The **safety valve**, also called a **positive pressure-relief valve**, opens automatically and starts releasing cabin pressurization when its preset value is reached. This preset value is about 0.5 psi [3.4 kPa] higher than the maximum setting of the outflow valve. The safety valve prevents the cabin from being overpressurized, which could result in aircraft structural failure. In some aircraft the safety valve and the outflow valve are identical in design, with the only differences being the maximum pressure setting and the pneumatic connections for operation.

The **negative pressure-relief valve** prevents the cabin from being at a higher altitude (lower air pressure) than the ambient air. The operation of this valve is automatic. It opens to equalize the cabin and ambient pressure if the ambient pressure exceeds cabin pressure by more than about 0.3 psi [2.3 kPa]. Although this is called a "valve," the actual operation may be performed by door seals or other fuselage components.

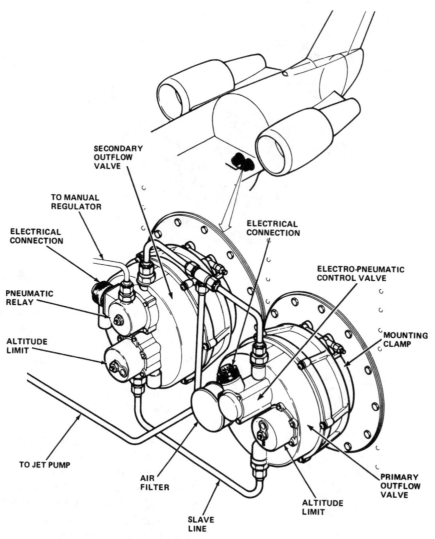

FIGURE 16-26 Outflow valves installed in a corporate jet. (*Canadair Inc.*)

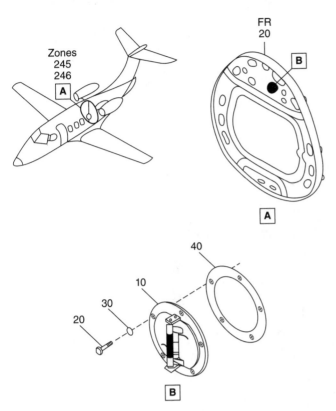

FIGURE 16-27 Negative relief valve.

Figure 16-27 shows the location of the negative pressure-relief valves on a jet transport.

A **dump valve** is used to release all cabin pressurization when the aircraft lands. This valve is commonly controlled by a landing-gear squat switch. When the landing-gear oleo is compressed, the squat switch causes the dump valve to open and equalizes the cabin and ambient atmospheric pressures. This prevents the cabin from being pressurized after landing. If the cabin were pressurized on the ground, it might not be possible to open the aircraft cabin doors.

Note that one pressurization control valve may serve more than one function in a specific aircraft design. For example, the outflow valve or the safety valve may be configured to act also as the dump valve, and the outflow valve may be designed to serve as a negative pressure-relief valve.

Pressurization-Control Operation

The control of the cabin pressurization is through various valves, as has already been mentioned. These valves are operated by differences in air pressure in order to achieve the desired level of pressurization or to prevent overstressing the aircraft structure. To understand the basic principles of the pressurization-control operation, look at a safety valve and then move on to an outflow valve and the controller. The components presented and described are intended to help the technician in understanding the basics of system operation. Before working on a specific system, the technician should be acquainted with that system's specific design.

The **safety valve** for a transport aircraft is shown in Fig. 16-28. This valve is designed to *release cabin*

pressurization above a preset value. This function is automatic and does not require the use of a separate controller. The small chamber (A) to the right of the differential diaphragm assembly is vented to cabin air pressure through a filter. The left side of this diaphragm assembly (B) is vented to atmospheric pressure. If cabin pressure exceeds the combined force of atmospheric pressure and the spring, the metering valve is lifted off its seat. When the round head of the metering valve is moved to the left, air from the reference chamber (C) on the left side of the outflow valve poppet assembly begins to flow out to atmospheric pressure through the hollow center of the outflow-valve pilot. The reduced pressure in the reference chamber allows the large spring to the right of the poppet assembly to move the assembly to the left. When the poppet is moved to the left, cabin air flows out around the base of the poppet valve, shown at the right of Fig. 16-28.

The **outflow valve** for a corporate jet aircraft is shown in Fig. 16-29. This valve is designed to serve as the **outflow valve**, **altitude limiter**, and **positive pressure-relief valve** for the aircraft. The outflow operation is controlled by the **reference air pressure** in the **reference chamber** (termed **actuator chamber** by the manufacturer) and is controlled through the **electropneumatic control valve** from the pilot's controller and the vacuum line. Air from the cabin air enters this chamber through the filter and flows out through the control valve to the controller. This keeps the reference chamber pressure below cabin pressure.

When the controller is set to pressurize the cabin, the control valve is closed or only slightly open so that the pressure in the reference chamber is great enough to hold the poppet valve on its seat. Since no air can escape, the cabin pressure increases. Once the cabin has reached the desired amount of pressurization, the controller commands the control valve to open and reduces the reference chamber pressure. With the reduced reference pressure, cabin air pressure in the chamber between the poppet valve and the baffle plate overcomes the reference pressure and spring pressure on the poppet valve. This opens the valve enough to release air from the cabin. The rate of this exhaust air is the same as that of the air flowing into the cabin and maintains the cabin at a certain pressure.

Although the system just discussed uses an electropneumatic control, many aircraft have the reference pressure established in the pilot's controller, and lines are used to provide the outflow valve with the reference pressure.

To prevent the cabin differential pressure from becoming too great, a positive pressure-relief control system is included in the valve design. This consists of the differential pressure–control diaphragm and spring and the metering pin. Cabin air pressure between the baffle plate and the poppet valve is applied to one side of the differential pressure diaphragm. If the force of this pressure exceeds that of the combined force of atmospheric pressure and the spring to the right of the diaphragm, the diaphragm moves to the right and allows the metering pin to move off its seat (as shown in Fig. 16-29). This allows air pressure in the reference chamber to escape through the passage now opened and causes the poppet valve to open. This lowers the cabin air pressure. The operation of this safety device is automatic.

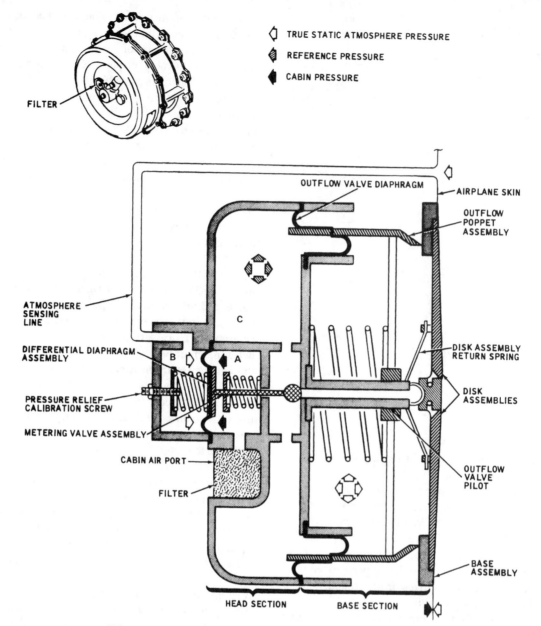

TRUE STATIC ATMOSPHERE PRESSURE

REFERENCE PRESSURE

CABIN PRESSURE

FILTER

OUTFLOW VALVE DIAPHRAGM

AIRPLANE SKIN

OUTFLOW POPPET ASSEMBLY

ATMOSPHERE SENSING LINE

DISK ASSEMBLY RETURN SPRING

DIFFERENTIAL DIAPHRAGM ASSEMBLY

DISK ASSEMBLIES

PRESSURE RELIEF CALIBRATION SCREW

METERING VALVE ASSEMBLY

CABIN AIR PORT

FILTER

OUTFLOW VALVE PILOT

BASE ASSEMBLY

HEAD SECTION   BASE SECTION

FIGURE 16-28 A safety valve. (*McDonnell Douglas Aircraft Co.*)

The cabin altitude is prevented from exceeding approximately 13,000 ft [3962.4 m] by the altitude-limit components of the valve. The altitude-limit chamber on the upper left of Fig. 16-30 is vented to cabin air pressure. As cabin air pressure decreases, the sealed bellows expands. If the cabin altitude exceeds the preset limit, the bellows expands enough to press on the small poppet just to the right of the bellows. This opens the reference chamber to cabin air pressure and causes the outflow valve to close. The valve remains closed as long as the cabin altitude is above the setting of the bellows. This operation can be overridden by the positive pressure-relief control system.

The **controller that operates the outflow valve** is located on the flight deck and is used by the pilot to **select the desired rate** of cabin **altitude change** and the **cabin pressure** altitude. The controller requires pressure inputs from the cabin air pressure and a venturi system and generates a reference pressure.

The vacuum pressure is usually a value generated by the Bernoulli effect as cabin air pressure escapes from the cabin to the outside air through a venturi. The vacuum line opens into the venturi chamber at the throat to create a low pressure. The value of the vacuum is determined by the difference in pressure between the cabin and the atmosphere.

The principle of operation of a controller can best be understood by studying the operation of a manual controller of the type shown in Fig. 16-30. When it is desired to decrease the cabin altitude, the manual control lever is moved down. This opens the cabin-pressure port and allows cabin pressure to be applied to the reference chamber in the outflow valve. This pressure in combination with the spring in the outflow valve holds the outflow valve closed, causing the cabin pressure to increase. The rate of airflow to the outflow valve is controlled by the manual-rate control-needle valve. If the pilot wishes to

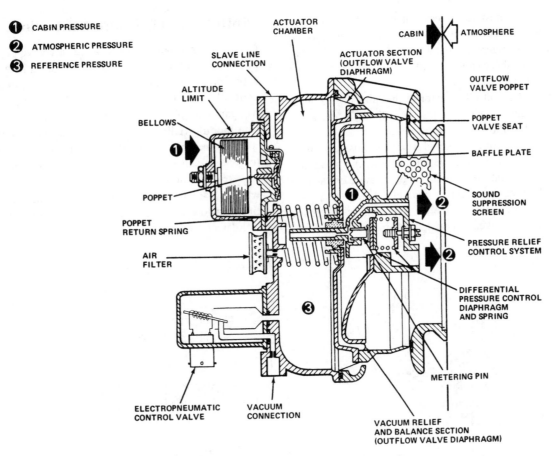

① CABIN PRESSURE
② ATMOSPHERIC PRESSURE
③ REFERENCE PRESSURE

FIGURE 16-29   An outflow valve. (*Canadair Inc.*)

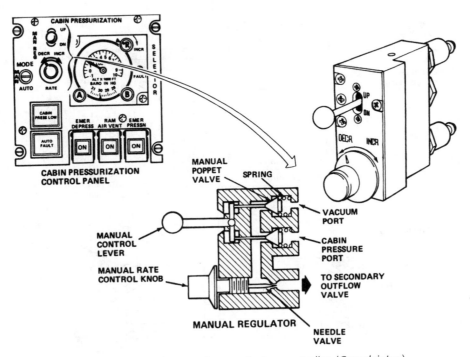

FIGURE 16-30   A manual pressurization controller. (*Canadair Inc.*)

decrease the cabin pressure (increase cabin altitude), the control lever is moved upward and the vacuum port is opened to the line leading to the reference chamber. Pressure in the chamber is decreased, and this causes the outflow valve poppet to open and release cabin pressure.

The automatic controller performs the same basic operation as the manual controller just described. In the automatic controller, the pilot sets a rate-control knob, if the manufacturer has not established a preset valve, and then he or she selects the desired cabin altitude. A spring and bellows arrangement then directs cabin pressure or vacuum to the outflow valve reference chamber. Once the preset altitude is reached, the controller maintains a reference chamber pressure to keep a constant cabin altitude.

Digital Cabin Pressure Controller

New corporate and transport aircraft have replaced the pneumatic controller used in older designs with digital cabin pressure controllers. These digital cabin pressure controllers (CPC) use inputs from the cabin pressure control panel, air data module (ADM), cabin pressure acquisition module (CPAM), flight management system (FMS), weight on wheels (WOW) system, and "doors closed" to generate an electrical signal to the outflow valves; see Fig. 16-31. The pressurization control is fully automatic with a manual backup control. Larger aircraft use multiple outflow valves that are electrically operated by multiple electrical motors.

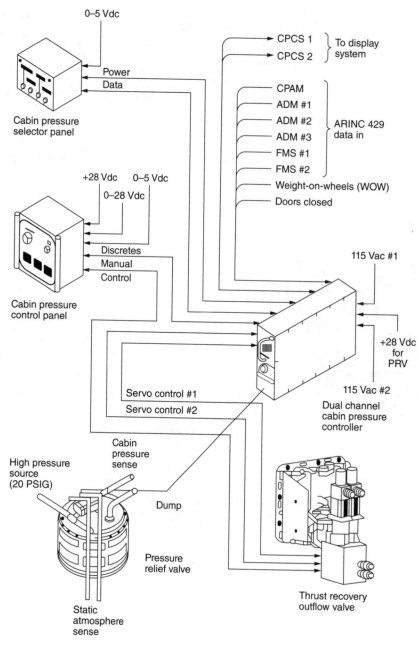

FIGURE 16-31 Digital pressurization control system.

CABIN ENVIRONMENTAL SYSTEM FOR A JET AIRLINER

A typical **air-conditioning system** for a transport aircraft is represented by the system employed for one model of the Boeing 777 airplane. This system provides conditioned, pressurized air to the crew cabin, passenger cabin, lower nose compartment, forward cargo compartment, air-conditioning distribution bay, and aft cargo compartment. The air supply is furnished by engine bleed air when in flight and from engine bleed air, a ground pneumatic supply cart, or a ground conditioned-air supply cart during ground operation. Part of the warm-air supply from the engines (or pneumatic cart, if on the ground) is passed through the **air-conditioning packs** to be cooled. The cold air is then mixed with the remainder of the warm air as required to obtain the temperature of air called for by the temperature control system. This conditioned air then passes into conditioned areas through the distribution system. After the air has passed through the cabins, it is exhausted through a number of outlets. The combined air from all outlets other than the pressurization outflow valves, however, is limited to a value less than that which enters the cabin from the air-conditioning system. The **outflow valves** are regulated to exhaust only that additional quantity of air required to maintain the desired pressure in the cabin.

Other air outlets include the galley vent, lavatory vents, equipment-cooling outlets, the ground conditioned-air condensate drain, water-separator drains, the pressurization controllers' cabin-to-ambient venturis, and the cargo-heat outflow valves.

The cargo-heat outflow valves normally remain open to provide warmth around the cargo compartments, but they may be closed by a switch on the third crew member's panel.

Bleed air from the engine compressors is used for delivering pressurized air to the cabin. The air will be conditioned in the air-conditioning pack and delivered to the aircraft cabin through risers and ducts. A number of valves will determine the output of the environmental control system (ECS). Figure 16-32 shows the engine air supply system. Compressed air is taken from the 8th and 15th stage engine compressor port and delivered to the following systems: engine starter, hydraulic reservoir, wing anti-ice (WAI), and the ECS. Figure 16-33 shows the location of various pneumatic components on the engine.

Air-Conditioning Packs

The ECS of the 777 aircraft uses two air-conditioning packs. These packs provide conditioned air to the cabin and also remove moisture. Air-conditioning components are located in the center fuselage fairing. The cooling devices used in the cooling packs consist of heat exchangers, air cycle machine (ACM), reheater/condensor unit, and a number of valves to control the airflow in the system. Figure 16-34 shows the basic layout of the pack system. High-pressure hot air from the engine compressor will flow through the two flow-control and shutoff valves to the primary heat exchanger. At high altitudes the upper flow control and shutoff valve is closed and the air will be forced to flow through the ozone converter (at high altitudes there is ozone in the atmosphere which needs to be removed) and the lower flow control and shutoff valve.

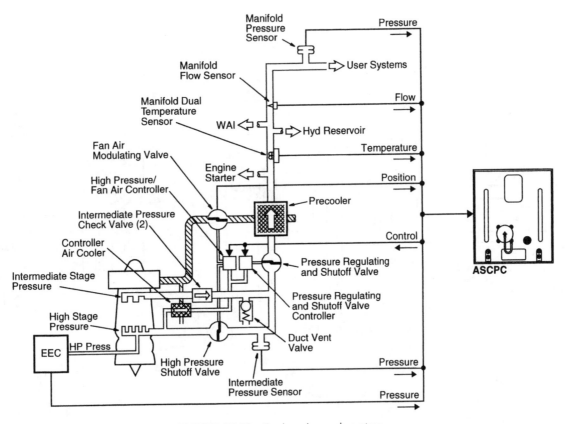

FIGURE 16-32 Engine air supply system.

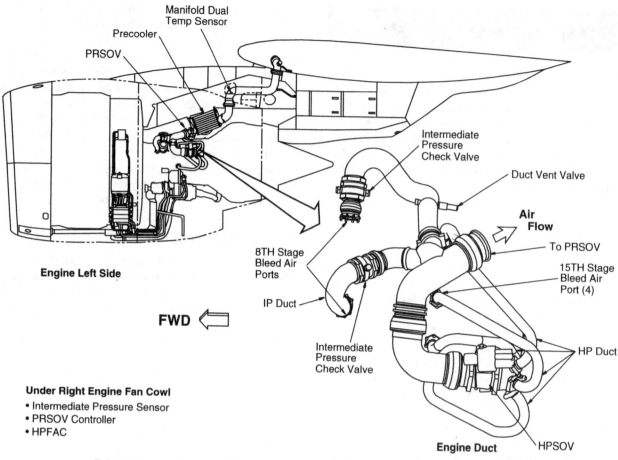

Engine Left Side

PRSOV
Precooler
Manifold Dual Temp Sensor

Intermediate Pressure Check Valve

Duct Vent Valve

Air Flow

To PRSOV

15TH Stage Bleed Air Port (4)

8TH Stage Bleed Air Ports

IP Duct

Intermediate Pressure Check Valve

HP Duct

FWD

Under Right Engine Fan Cowl
- Intermediate Pressure Sensor
- PRSOV Controller
- HPFAC

Engine Duct

HPSOV

FIGURE 16-33 Location of pneumatic components on transport aircraft powerplant.

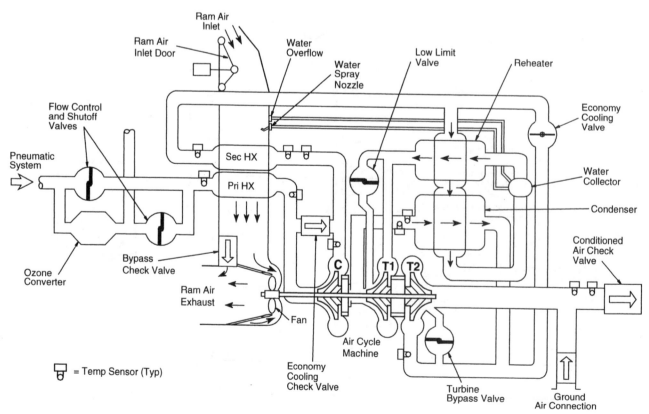

Ram Air Inlet

Ram Air Inlet Door

Water Overflow

Water Spray Nozzle

Low Limit Valve

Reheater

Economy Cooling Valve

Flow Control and Shutoff Valves

Water Collector

Pneumatic System

Sec HX

Pri HX

Condenser

Conditioned Air Check Valve

Ozone Converter

Bypass Check Valve

C

T1   T2

Ram Air Exhaust

Fan

Economy Cooling Check Valve

Air Cycle Machine

Turbine Bypass Valve

Ground Air Connection

🔲 = Temp Sensor (Typ)

FIGURE 16-34 Air-conditioning pack schematic.

At lower altitudes the lower control and shutoff valve is closed and the air bypasses the ozone converter. The air will flow in the air-to-air primary heat exchanger which is cooled by ram air during flight and a ram air exhaust fan on the ground. The ram air exhaust fan is driven by the ACM shaft. After the primary heat exchanger the cooled high-pressure air flows to the compressor of the ACM where the air will be compressed to a higher pressure and as a result the air temperature will increase. The ACM is a unit with one compressor and two turbine wheels. Air is increased in pressure and temperature in the compressor and expanded and cooled in the turbine. The energy that the turbines extract from the air drives the compressor and ram air fan. The air will flow from the compressor to the secondary heat exchanger where the air will be cooled. After the secondary heat exchanger the air will flow through the reheater/condensor unit where the air will be cooled and condensed; this is necessary because the water droplets are easier to remove from the air. In the water collector the water will be removed and the water will be sprayed in the ram air inlet for improved cooling. The air will flow through the reheater where it will increase in temperature to gain energy. The air will flow to the turbine wheel 1 (T1) where the air will expand rapidly and as a result the air temperature will lower substantially. If the output temperature falls below 34°F the low limit valve will open to add warm air to the very cold air to prevent freezing. After the air leaves T1 it will flow through the condenser and enter T2 where the air will expand and temperature will decrease.

If pack temperature is too low the turbine bypass valve will open to add warm air. After T2 the air flows to the mixing manifold. During pack start-up or if the ACM is defective, the economy cooling check valve will open to bypass the ACM compressor and air can flow through the check valve directly to the secondary heat exchanger. The economy cooling valve opens to let air go around the air cycle machine turbine 1 and the water separation parts of the pack. The valve opens when the pack functions in the economy cooling mode or the standby cooling mode. Figure 16-35 shows the location of the components of the left pack.

Conditioned Air Distribution

After the conditioned air leaves the air-conditioning packs a small portion of the pack air flows to the flight deck zone; the rest of the air flows to the mixing manifold where conditioned air from the packs will be mixed with air that is already in the cabin. This cabin air is filtered before it enters the mixing manifold. Four recirculation fans draw the cabin air into the mixing manifold. The final air temperature for the flight deck and cabin zones is adjusted by trim air valves which can add some warm air from the pneumatic bleed air system. Figure 16-36 shows the temperature control and distribution system. The distribution system also provides gasper air, ventilation of lavatories and galleys to remove odors, and equipment cooling. The distribution system includes ducts and fans that move conditioned

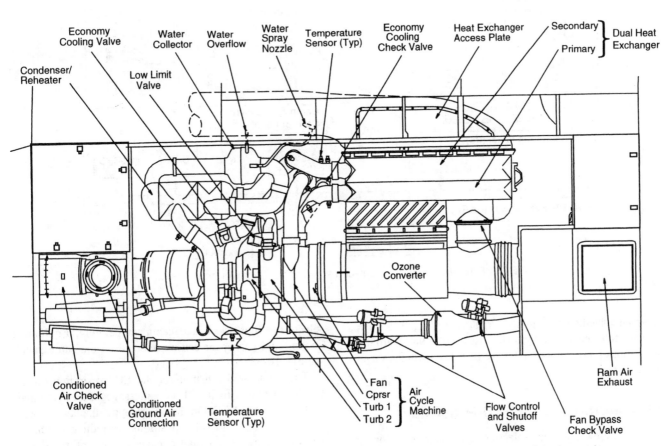

FIGURE 16-35 The location of air-conditioning components in an transport aircraft air-conditioning pack.

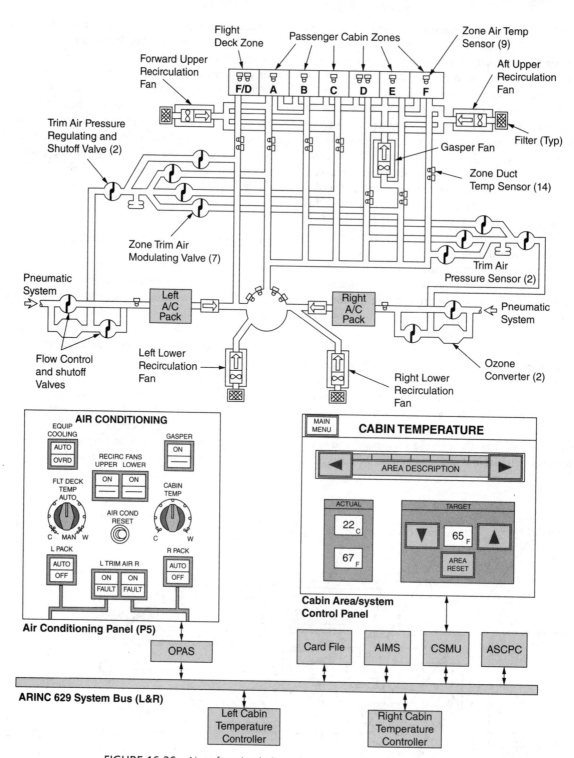

FIGURE 16-36 Aircraft recirculation schematic and temperature control.

and recirculated air through the airplane; see Fig. 16-37. The ECS is a fully automatic system that uses two digital controllers and various sensors imbedded in the ECS to control the valves in the system.

Air-Conditioning Indication (EICAS)

The flight crew can make selections on the air-conditioning panel. The system is fully automatic and the flight crew only

selects the flight deck and cabin temperature. Equipment cooling and gasper air can be selected on the same panel; see Fig. 16-38.

The fight crew can monitor the ECS operation on the EICAS display, which will show messages if there are malfunctions in the ECS. On the air synoptic display temperatures, pressures, and valve positions are indicated. On the air-conditioning maintenance page system information is displayed. Figure 16-39 shows the EICAS display.

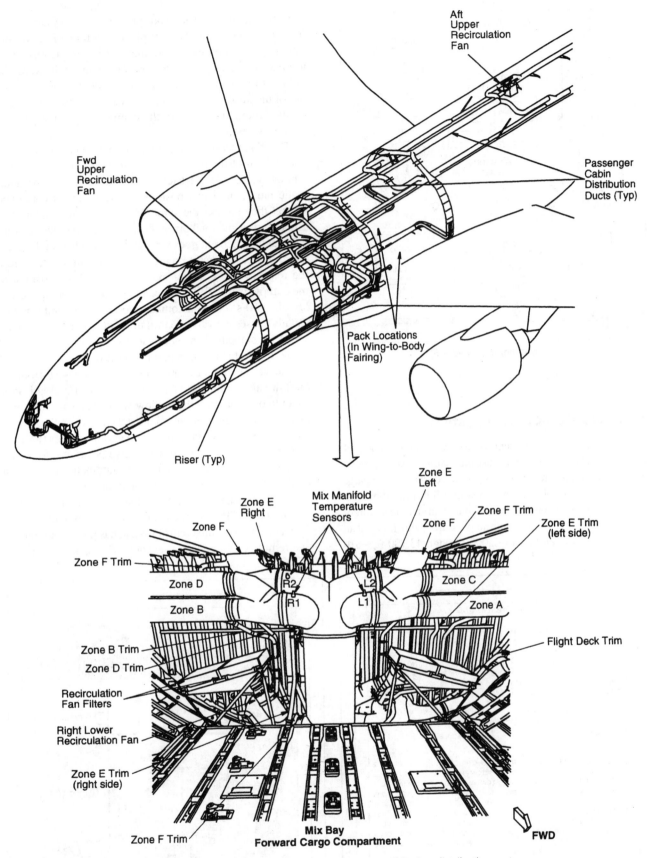

FIGURE 16-37 Component location of the air-conditioning distribution system.

Cabin Environmental System for a Jet Airliner   **583**

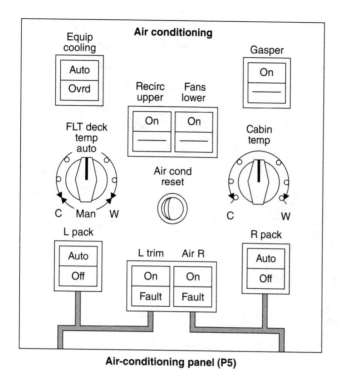

Air-conditioning panel (P5)

FIGURE 16-38   Air-conditioning panel.

Pressurization-Control System

The cabin pressurization system controls cabin altitude and protects the fuselage structure against internal and external overpressurization. The pressurization system consists of two parts: cabin pressure control and cabin pressure relief. The cabin pressure control system controls the amount of air that flows out of the airplane. This keeps the cabin air pressure within limits at all airplane-operating altitudes (−2,000 through 43,100 ft, −610

through 13,146 m). The cabin pressure relief system keeps the internal pressure of the airplane to a limit. The air-conditioning packs provide more than sufficient quantity of air to pressure the airplane. Outflow valves in the forward and aft areas of the airplane control the amount of air that goes out of the airplane. This keeps the cabin air pressure within limits. Figure 16-40 shows the components of the pressurization system.

The flight crew sets automatic or manual control for one or both of the outflow valves from the flight deck. The outflow valves are electric/mechanical valves. In the automatic mode, the usual position of the valves is controlled so that they are not the same. Valve position is set so that 20 percent of the total air outflow goes through the forward valve and 80 percent goes through the aft valve. In the automatic mode, the motor control logic uses information from the related ASCPC to set motor speed and the position of the outflow valve. The valve can open or close completely in less than 10 s for the automatic mode. ASCPC information usually causes the valve control unit (VCU) to change motor speed for smooth changes in cabin pressure. In the manual mode, the flight crew controls the position of one or both of the valves. The motor control logic uses the signal from the outflow valve manual switches on the bleed air/pressurization panel to open or close the outflow valve. The valve can open or close completely in approximately 27 s for the manual mode. Two rotary variable differential transformers (RVDTs) that are part of the gearbox send valve position information back to the VCU and ASCPCs in automatic and manual modes. The VCU continuously monitors for outflow valve faults. If the VCU finds a fault, the VCU supplies the fault data to the ASCPCs for indication and to keep in memory.

The cabin pressure control system has these components: one remote cabin pressure sensor, flight deck controls, and

SHOW PG MENU	**AIR CONDITIONING**							
	MASTER TEMP 75			SEATS	426			
	F/D	A	B	C	D	E	F	AFT	BULK
ZONE TEMP	70	75	75	75	72	72	72	45	70
TRGT TEMP	70	75	75	75	72	72	72	45	70
DUCT TEMP	72	77	77	77	70	70	70	--	--
TRIM VLV	0.15	0.35	0.35	0.35	0.00	0.10	0.10	--	--
CTRL CH	1	1	2	1	2	1	2	--	--

LEFT LOWER RECIR FAN	ON	FWD UPPER RECIR FAN	ON
RIGHT LOWER RECIR FAN	ON	AFT UPPER RECIR FAN	ON
MIX MANIFOLD TMEP	70	FLOW SCHEDULE	1

	L	R		L	R
PACK FLOW-VOLUME	2700	2700	PACK CTRL CH	1	2
PACK FLOW-MASS	200.0	200.0	PACK IN PRESS	55.0	55.0
PACK OUT TEMP	68	68	LOW LIM VLV POS	0.10	0.10
PRI HX IN TEMP	385	385	TURB BYP VLV	0.15	0.15
PRI HX OUT TEMP	350	350	RAM AIR INLET	0.35	0.35
CPRSR OUT TEMP	400	400	RAM AIR EXIT	0.35	0.35
SEC HX OUT TEMP	300	300	ECON COOL VLV	CLSD	CLSD
CONDENSER IN TEMP	59	59	LOWER FLOW CTRL VLV	OPEN	OPEN
STG 2 TURB IN TEMP	77	77	UPPER FLOW CTRL VLV	CLSD	CLSD
TRIM AIR PRESS	5.0	5.0			
A/C TEMP ZONE			DATE 23 JUN 90	UTC 18:54:04	

Air conditioning Maintenance Page

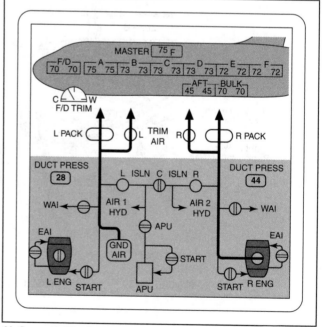

Air Synoptic Display

FIGURE 16-39   Air-conditioning system indications on EICAS.

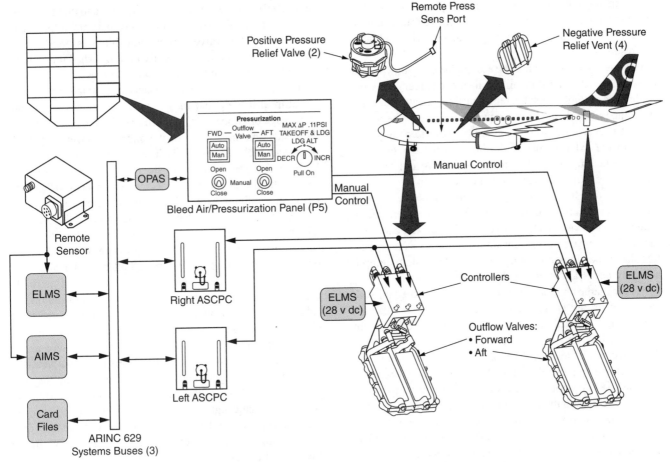

FIGURE 16-40 Cabin pressurization control system and components.

two outflow valves. The air supply cabin pressure controllers (ASCPCs) provide control, indications, and BITE for this system. The controllers also have an internal sensor that they use to monitor cabin pressure. The sensor is used for control and to give an output on the ARINC 629 buses; see Fig. 16-40. The remote cabin pressure sensor sends data to the aircraft information management system (AIMS), ASCPCs, and the electrical load management system (ELMS). AIMS uses the remote sensor data as an alternative to the ASCPC cabin pressure data. The ASCPCs use the remote cabin pressure sensor to crosscheck data from the internal cabin pressure sensor. ELMS sends the remote cabin pressure sensor data on the ARINC system buses. ELMS also uses the remote cabin pressure sensor data as an alternative to the ASCPC cabin pressure data. The ASCPCs usually get landing altitude from the flight management computing function (FMCF) of the AIMS. As an alternative to the FMCF, the flight crew can manually set landing altitude from the bleed air/pressurization panel.

Positive and Negative Relief Valves

Positive pressure relief valves and negative pressure relief vents in the forward part of the airplane keep the difference between cabin pressure and ambient pressure in limits. The positive pressure relief valve opens when cabin air pressure is higher than ambient pressure by a set value. This permits

air to go out of the airplane to keep differential pressure below a safe limit. When cabin pressure differential at the remote ambient pressure sense port increases to 8.95 psi, the relief valve opens. This permits air to go out of the airplane. If the remote pressure sense port does not function, the relief valve opens when cabin pressure differential at the integral ambient sense port increases to 9.42 psi. When cabin pressure differential goes below the limit, the positive pressure relief valve closes.

The negative pressure relief vent opens when cabin pressure is less than ambient pressure. This prevents negative cabin pressure. Too much negative cabin pressure could damage the fuselage structure. The negative pressure relief vent is a spring-loaded door. The door starts to open at a differential pressure of 0.2 psi. It is full open at a differential pressure of 0.5 psi.

SUMMARY OF PRESSURIZATION AND AIR-CONDITIONING SYSTEMS

The foregoing section contains a brief description of the principal functions and components of a typical air-conditioning system for a jet airliner. Numerous details have been omitted because space does not permit their description. The technician is reminded to consult the manufacturer's maintenance manual for service instructions and for information on the finer details.

As explained previously, air conditioning may include heating, ventilating, cooling, and pressurization. The methods and devices involved have been described but are reviewed briefly in this summary.

Heating for an aircraft may be accomplished by drawing air across a heated exhaust manifold (heat exchanger), drawing air through a combustion heater, or utilizing heated air from the compressor of one or more turbine engines. The use of exhaust heat from a piston engine involves the danger of a crack in the exhaust manifold, allowing exhaust fumes to enter the heating ducts. Frequent inspections and tests are required to ensure that leaks do not occur.

Combustion heaters may become inoperable because of fuel-system failure, ignition-system failure, or overheating. If such a heater overheats, an overheat sensor will cause the fuel to be shut off to prevent damage due to excess heat. Regular inspection and testing of the heater system components are necessary to ensure dependable operation.

Heating large commercial aircraft is accomplished through the use of heated air produced by the compressors of gas-turbine engines. The heated air is distributed to the cabin, flight compartment, cargo compartment, and other areas to provide the temperatures desired. When a particular temperature has been selected by a member of the flight crew, the temperature is maintained by automatic controls that sense the temperature and adjust valves in the ducting to increase or decrease hot or cold airflow and mix the air to provide the correct temperature. The valve controlling the cold-hot air mixture is called a **mixing valve**, and its function is to direct the correct proportions of cold and hot air to a **mixing chamber** or other mixing device to produce the temperatures desired. In some systems, **trim heat** is employed to raise the temperature in a particular area. A trim-heat manifold carries hot air, which can be released through outlets in any one of several areas. Thus it is possible to have a higher temperature in one area than in other areas. As explained previously, the heating, cooling, ventilation, and pressurization systems are all interconnected to provide the air-conditioning or cabin atmosphere system. So, it may be said that the air-conditioning system consists of four subsystems.

Cooling in an air-conditioning system is usually provided by means of either a vapor-cycle refrigeration system or an ACM, which utilizes air as a cooling medium. Cool air from the cooling unit or units is mixed with warm air to provide the desired temperature. The amount of cooling produced and utilized is also controlled by bypassing air around the cooling unit. The amount of bypassed air is controlled through a cabin temperature-control valve. Commercial passenger aircraft incorporate a gasper system, which enables each passenger to control cooled airflow to his or her location. An adjustable gasper valve is located overhead at each location or in the back of the seat ahead. The gasper system draws cold air directly from the cooling unit output before it is mixed with warmer air.

Pressurization and ventilation systems are closely related. The flow of air that provides pressurization also provides ventilation.

Pressurization of small aircraft designed for high-altitude operation is usually accomplished by bleeding air from the engine turbocharger(s) or super charger(s) and ducting it into the cabin. Manual and automatic controllers are employed to adjust the outflow valve for desired cabin pressure. Pressurization for large aircraft is usually accomplished through the use of bleed air from the gas-turbine engines.

The pressure in the cabin of an airplane is usually designated as **cabin altitude** rather than in terms of pounds per square inch gauge or kilopascals. The **differential pressure** between the inside and the outside of the aircraft is given as pounds per square inch or kilopascals. The differential pressure must be limited to the value established by engineers as safe for the structural strength of the fuselage, which leads to the slang term "an aluminum balloon." The automatic pressure controller incorporates a valve that limits the value of the differential pressure to a given level. If this pressure-differential valve should not operate, a differential pressure-relief valve will open at a slightly higher pressure and release cabin air.

The cabin altitude is selected by a member of the flight crew by means of the cabin altitude selector on the automatic cabin-pressure controller. The **isobaric**-control system in the controller maintains a constant cabin altitude until the altitude of the aircraft produces a differential pressure above that which is safe for the aircraft. At this time, the pressure-differential valve opens and reduces the pressure in the control chamber of the controller. This causes the controller to send a pneumatic signal to the outflow valve(s), requiring an increased opening and a greater outflow of air from the cabin.

In addition to a differential pressure-relief valve in the fuselage of a pressurized aircraft, a negative pressure-relief valve is installed. The purpose of this valve is to allow outside air to enter the fuselage at any time outside pressure is greater than the pressure in the airplane.

As shown in the drawing of Figs. 16-41 and 16-42, an air-conditioning system includes a large number of flow-control valves, regulating valves, and check valves. This drawing shows how the cooling packs are incorporated in the air-conditioning system. The control valves are operated manually or automatically, either from the flight station or from points in the aircraft where special conditions are required. The check valves ensure that the air can flow in only one direction. This prevents properly conditioned air from backflowing into inactive parts of the system or escaping overboard through ground service connections.

As explained previously, the cabin pressure is designated as cabin altitude. A particular cabin altitude is usually selected for the main part of the flight, and this cabin altitude is usually much greater (lower pressure) than the point of takeoff. It is, therefore, necessary for the cabin altitude to change gradually from the altitude at the point of takeoff to the cabin altitude selected for the flight. This change is made by the **cabin altitude rate selector**, which adjusts the rate system built into the automatic cabin pressure controller. The rate system is also called the **automatic cabin rate-of-climb control system.**

When an airplane is not in flight and air conditioning is required, the air is supplied by the auxiliary power unit (APU) on board the aircraft or by a ground service unit. Both the APU and the ground service unit are small gas-turbine

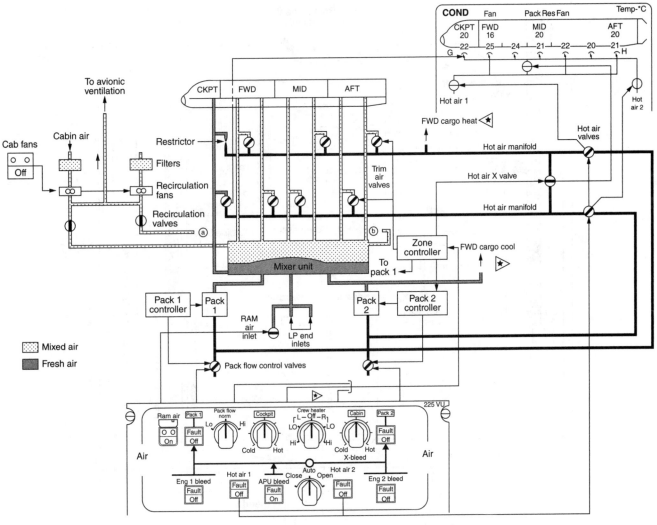

FIGURE 16-41 Environmental control system of transport aircraft.

engines designed to produce a high volume of compressed air. The units also drive alternators that supply necessary electrical power. The air supply from the APU or the ground service unit is also used to operate the pneumatic turbine starters for some aircraft engines. APUs are discussed in greater detail in Chap. 18.

In recent years, highly efficient screw-type compressors have been developed for ground air supplies. These are driven by diesel engines or electric motors and are mounted in mobile service units. Some airports have been equipped with fixed air supplies, where the air is compressed at a central location and then it is delivered to individual aircraft stations through air ducts.

Inspection, service, and maintenance for the APU are similar to that prescribed for other gas-turbine engines. Instructions are provided in the maintenance manual for the aircraft.

OXYGEN SYSTEMS

Oxygen systems are required on airplanes that fly for extended periods at altitudes substantially above 10 000 ft

[3048 m]. Although the normal human body can survive without a special supply of oxygen at altitudes of over 15 000 ft [4573 m], the mental and physical capacities of a human being are reduced when the usual supply of oxygen is not available in the air. It is particularly important that the pilot and crew of an airplane have an adequate supply of oxygen when operating an unpressurized airplane at altitudes in excess of 10 000 ft [3048 m].

A lack of oxygen causes a person to experience a condition called **hypoxia**. This condition results in "lightheadedness," headaches, dizziness, nausea, unconsciousness, or death, depending upon its duration and degree. When permanent physical damage results from lack of oxygen, the condition is defined as **anoxia**.

The importance of oxygen, especially when flying at higher altitudes, is not appreciated by many persons who fly, including pilots. It is generally known that the human body requires oxygen to sustain life, but the effects of a lack of sufficient oxygen on various functions of the body are not understood by a large percentage of the flying public.

Studies have shown that the effects of hypoxia become apparent at approximately 5000 ft [1500 m] altitude in

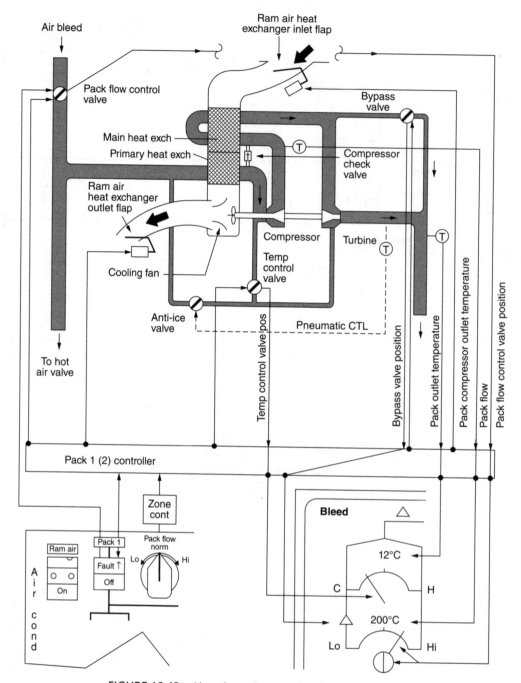

FIGURE 16-42 Air cycle cooling system of transport aircraft.

the form of reduced night vision. It is recommended, therefore, that a pilot flying above 5000 ft altitude at night use oxygen. As stated before, pilots flying above 10 000 ft [3048 m] altitude should use oxygen. Requirements for oxygen in aircraft are set forth in FAR Parts 23, 25, and 91.

Two principal factors affect the amount of oxygen that a person will absorb. These are (1) the amount of oxygen in the air the person is breathing and (2) the pressure of the air and oxygen mixture. Normal air contains approximately 21 percent oxygen, and this provides adequate oxygen for the human body at lower altitudes. At 34 000 ft [10 363 m]

altitude, a person must be breathing 100 percent oxygen to absorb the same amount of oxygen as when breathing air at sea level. It is, therefore, apparent that the percentage of oxygen in the air that a person is breathing must be increased in keeping with altitude if the person is to receive an adequate supply of oxygen for optimum functioning of physical and mental faculties and functions. To adjust for variations in cabin altitude, oxygen systems are often equipped with barometric regulators, which increase the flow of oxygen as cabin altitude increases. In a nonpressurized aircraft, the cabin altitude is the same as the aircraft altitude, and the oxygen flow is adjusted for aircraft altitude.

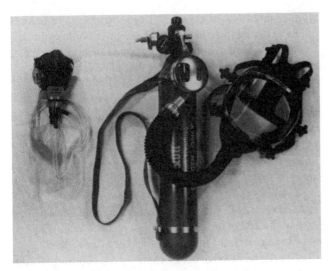

FIGURE 16-43 A portable oxygen system with two regulators and masks. (*Scott Aviation.*)

Types of Oxygen Systems

Oxygen systems, classified according to source of oxygen supply, may be described as **stored-gas**, **chemical** or **solid-state**, and **liquid oxygen** (LOX) systems. Systems for private and commercial aircraft are of the stored-gas or chemical type. LOX systems are limited to military aircraft and are not discussed in this text.

Oxygen systems may be **portable** or **fixed**. The fixed system is permanently installed in an airplane where a need for oxygen may exist at any time during flight at high altitudes. Commercial airplanes are always equipped with fixed systems, augmented by a few portable units for crew members, who must be mobile, and for emergency situations where only one or two persons may require oxygen for unusual physical reasons.

A typical portable unit or system is shown in Fig. 16-43. The simplest type of portable oxygen system includes a Department of Transportation– (DOT-) approved oxygen cylinder of either 11 ft³ [311.5 L] capacity or 22 ft³ [623 L] capacity, a regulator assembly, a pressure gauge, an ON-OFF valve, hose couplings, flow indicator, and one or two oronasal masks. This system is charged to 1800 psi [12 411 kPa] and is suitable for altitudes up to 28 000 ft [8536 m]. Portable oxygen systems are available with automatic flow-control regulators, which adjust oxygen flow in accordance with altitude.

Oxygen systems are also classified according to the type of regulator that controls the flow of oxygen. The mask employed must be compatible with the type of regulator. The majority of oxygen systems for both private and commercial aircraft are of the **continuous-flow** type. The regulator on the oxygen supply provides a continuous flow of oxygen to the mask. The mask valving provides for mixing of ambient air with the oxygen during the breathing process. As mentioned previously, some continuous-flow systems adjust flow rate in accordance with altitude.

Demand and **diluter-demand regulators** used with demand masks supply oxygen to the user during inhalation. When the individual using the equipment inhales, he or she causes a reduction of pressure in a chamber in the regulator. This reduction in pressure activates the oxygen valve and supplies oxygen to the mask. A flow indicator shows when oxygen flow is taking place. The diluter-demand regulator automatically adjusts the percentage of oxygen and air supplied to the mask in accordance with altitude. The demand masks cover most of the user's face and create an airtight seal. This is why a low pressure is created when the user inhales. These masks are used primarily by crew members because they use the oxygen more efficiently and have higher-altitude capabilities.

Pressure-demand regulators contain an aneroid mechanism, which automatically increases the flow of oxygen into the mask under positive pressure. This enables

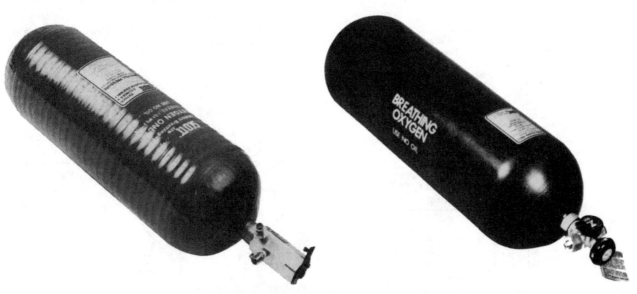

FIGURE 16-44 A composite oxygen bottle (left) and a steel bottle (right). (*Scott Aviation.*)

the user to absorb more oxygen under the conditions at very high altitudes. This type of equipment is normally used at altitudes above 40 000 ft [13 632 m]. The additional pressure is needed to enable the user to absorb oxygen at a greater rate than would be absorbed at ambient pressure. A **pressure-demand mask** must be worn with a pressure-demand regulator. By action of special pressure-compensating valves, the mask provides for a buildup of oxygen pressure from the regulator and creates the required input of oxygen into the lungs.

Oxygen Bottles

Oxygen cylinders, also called **oxygen bottles**, are the containers used to hold the aircraft gaseous oxygen supply. The cylinders may be designed to carry oxygen at a high or low pressure.

High-pressure cylinders are designed to contain oxygen at a pressure of approximately 1800 psi [12 411 kPa]. These cylinders can be identified by their green color and by the words "Aviators' Breathing Oxygen" on the side of the cylinder. Because of the high pressure in the cylinders, they must be very strong to withstand the operational stress without shattering. Three types of construction are used: a high-strength, heat-treated, steel-alloy cylinder; a wire-wrapped metal cylinder; and a Kevlar-wrapped aluminum cylinder. Figure 16-44 shows typical oxygen cylinders.

High-pressure cylinders are manufactured in several sizes and shapes. The cylinders are designed for a maximum operating pressure of about 2000 psi [13 740 kPa] but are normally serviced to a pressure of between 1800 and 1850 psi [12 411 and 12 755.8 kPa]. To obtain the proper pressure for the ambient air temperature, aircraft manufacturers often supply charts similar to that shown in Fig. 16-45. When using one of these charts, make sure that it is the correct one for the system you are servicing.

Low-pressure cylinders are made either of stainless steel with stainless steel bands seam-welded to the body of the cylinder or of low-alloy steel. These cylinders are painted yellow to distinguish them from the high-pressure cylinders. The low-pressure cylinders are designed to store oxygen at a maximum of 450 psi [3100.5 kPa], although they are not normally filled above 425 psi [2928 kPa].

There are several types of cylinder valves in use. The hand-wheel type has a wheel on the top of the valve and operates like a water faucet. The valve opens as the wheel is turned counterclockwise. If the cylinder does not incorporate a hand-wheel design with a slow-opening feature, a surge of pressure can be sent into the oxygen system and cause damage.

Another type of valve is of the self-opening design. When the valve is attached to the oxygen system, a check valve is moved off of its seat, allowing the cylinder to charge the system.

A third type of valve uses a cabin-operated push-pull control to operate a control lever on the top of the valve. This eliminates the necessity of always having the oxygen system charged but allows the pilot to activate the system whenever needed. This valve also can incorporate a pressure regulator so that there is no damage from a pressure surge in the system when the valve is opened.

Oxygen cylinders are often fitted with safety disks, which rupture if the pressure in the cylinder becomes too great. (The pressure can rise significantly due to high ambient air temperatures heating the cylinder.) The released oxygen is vented overboard through a discharge line. The discharge line should exit the aircraft through a discharge indicator, shown in Fig. 16-46, so that the discharge can be readily detected during preflight and maintenance inspections.

Regulators

Regulators for the pressure and flow of oxygen are incorporated in stored-gas systems because the oxygen is stored in high-pressure cylinders under pressures of 1800 psig [12 411 kPa] or more. The high pressure must be reduced to a value suitable for application directly to a mask or to a breathing regulator. This lower pressure is usually in the range of 40 to 75 psig [279 to 517 kPa], depending upon the system. One type of pressure regulator is illustrated in

Ambient Temperature, °F	Filling Pressure, psig
0	1650
10	1700
20	1725
30	1775
40	1825

Ambient Temperature, °F	Filling Pressure, psig
50	1875
60	1925
70	1975
80	2000
90	2050

FIGURE 16-45 Oxygen bottle–filling chart. (*Cessna Aircraft Co.*)

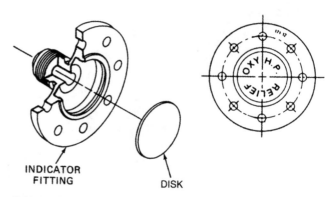

FIGURE 16-46 Oxygen overboard discharge indicator. (*Scott Aviation.*)

Fig. 16-47. This pressure regulator is similar in design to many other gas- or air-pressure regulators in that it utilizes a diaphragm balanced against a spring to control the flow of gas. This regulator consists of a **housing**, **diaphragm**, **regulator spring**, **link actuator assembly**, **relief valve**, and an **inlet valve**. With no inlet pressure on the regulator, spring tension on the diaphragm through the link actuator assembly forces the inlet valve to the open position. When oxygen is flowing, regulated pressure in the lower diaphragm chamber acts against the diaphragm, causing it to move upward and compress the regulator spring. The link actuator assembly then mechanically causes the regulator valve to move toward the closed position, thus reducing the flow of oxygen. When the pressure in the lower chamber of the diaphragm is equal to the regulator spring force, the diaphragm ceases to move and positions the inlet valve to maintain the proper oxygen flow.

Regulators for demand systems include both pressure regulators and demand or diluter-demand regulators. The demand regulator is sometimes built into the demand mask. Regulated oxygen pressure from the pressure regulator is applied to the demand regulator, which, in turn, delivers it to the user while inhaling through the demand mask. As explained previously, the diluter-demand regulator "dilutes" the pure oxygen with air in accordance with the cabin altitude.

A diluter-demand regulator is shown in Fig. 16-48. When the user inhales, a slight negative pressure is created in the chamber to the right of the demand diaphragm. This pressure reduction causes the diaphragm to move to the right and opens the demand valve. This causes a negative pressure to be applied to the chamber under the reducing valve diaphragm, moving the diaphragm to the left. When the diaphragm moves to the left, the pressure-reducing valve is lifted off of its seat, allowing oxygen to enter the regulator and flow toward the mask.

The mixing of air with oxygen is caused by the **aneroid** in the mixing chamber. The aneroid is a sealed metal bellows. At sea level the aneroid is compressed by atmospheric pressure so that the oxygen-metering port is closed and the air-metering port is open. As atmospheric pressure is decreased, the aneroid expands, opening the oxygen-metering port and reducing the air-metering port. At an

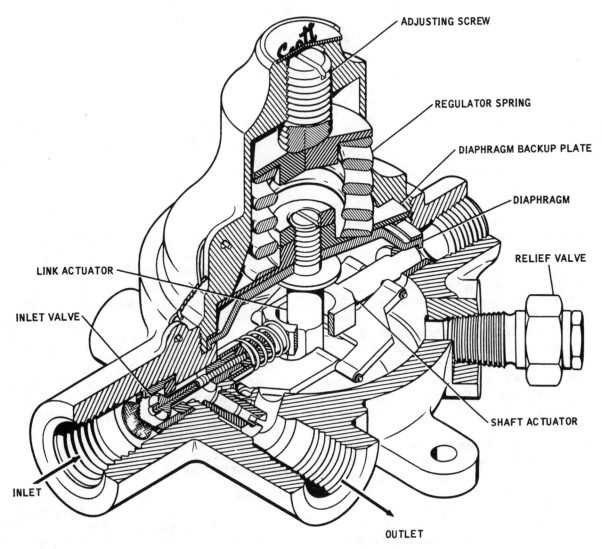

ADJUSTING SCREW

REGULATOR SPRING

DIAPHRAGM BACKUP PLATE

DIAPHRAGM

RELIEF VALVE

SHAFT ACTUATOR

LINK ACTUATOR

INLET VALVE

INLET

OUTLET

FIGURE 16-47   Crew oxygen-supply regulator for an airliner.

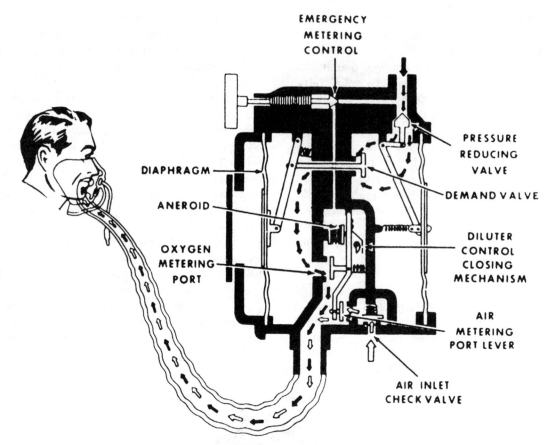

FIGURE 16-48 Diluter-demand oxygen regulator.

altitude of approximately 34 000 ft [10 363.2 m], the air-metering port is completely closed and the user is receiving only oxygen.

The diluter-control closing mechanism can be used to override the diluter-control operation of the system by mechanically closing off the air-metering port and fully opening the oxygen-metering port. Additionally, if the mechanism should malfunction, the user can open the emergency metering control, bypass the diluter mechanism, and be supplied with pure oxygen.

The pressure regulator for the continuous-flow passenger oxygen system installed in an airliner is shown in Fig. 16-49. This regulator is attached to the high-pressure oxygen-storage-cylinder hand shutoff valve. It is an altitude-compensating type, which varies supply-line pressure in accordance with cabin altitude. The regulator is actuated automatically by sensing a rise in cabin-pressure altitude or manually by controls located on the body of the regulator. A relief valve in the regulator will open to prevent outlet pressure from exceeding approximately 150 psig [1034 kPa].

At 10 500- to 12 000-ft [3200- to 3656-m] cabin-pressure altitude, the **automatic-opening aneroid** expands, thus causing the valve to open and supply pressure to the **pressure surge unit**. Increased pressure through the surge unit actuates the **door-release check valve**, unlatches the mask container doors, and pressurizes the dispensing manifolds. On descent, at a cabin pressure altitude of 6000 to 10 000 ft [1829 to 3048 m], the automatic-opening aneroid contracts, causing the valve to

close and shutting off the oxygen supply. It is then necessary to place the manual control in the ON position to supply supplemental oxygen to passengers as necessary. If the automatic-opening valve fails to function properly, the system is pressurized by placing the regulator manual control in the ON position and turning the manual oxygen-door release knob to the full-rotated position.

Oxygen entering the regulator passes through the **pressure reducer** and the **automatic-opening valve** and then enters the **altitude-compensating aneroid chamber**. As the aneroid expands and contracts owing to the changes in cabin pressure altitude, the outlet flow and pressure vary. When the cylinder is fully charged to approximately 1850 psig [12 756 kPa], the maximum outlet flow at the regulator is 430 L/min, and the maximum pressure is approximately 80 psig [552 kPa] at 35 000 ft [10 668 m].

First-aid oxygen can be made available during normal operation of the pressurized airplane if necessary. This is accomplished by a crew member placing the regulator manual control in the ON position.

Oxygen Masks

Oxygen masks vary considerably in size, shape, and design; however, each is designed for either a **demand** system or a continuous-flow (constant-flow) system. An oxygen mask for a demand system must fit the face closely, enclosing both the mouth and nose, and must form an airtight seal with the face.

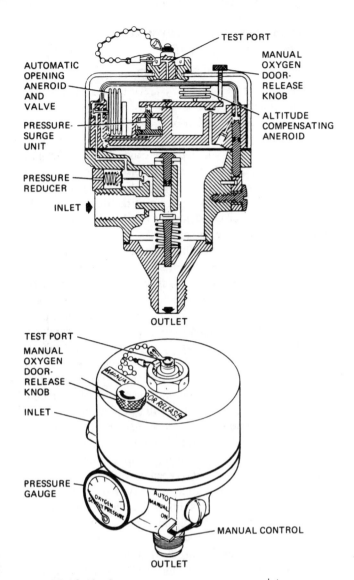

FIGURE 16-49　Passenger oxygen-pressure regulator.

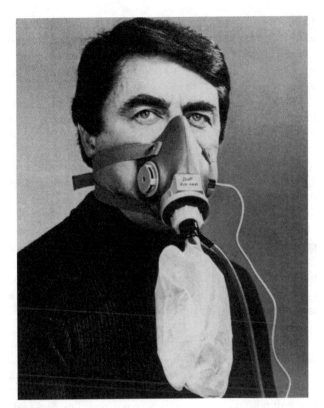

FIGURE 16-50　Two types of continuous-flow oxygen masks. (*Scott Aviation.*)

Inhalation by the user will then cause a low pressure in the demand regulator, which results in the opening of an oxygen valve and a flow of oxygen to the mask. When the user exhales, the flow of oxygen is cut off.

An oxygen mask for a constant-flow system is designed so that some ambient air is mixed with the oxygen. The complete mask usually includes an oronasal face piece, a reservoir bag, valves, a supply hose, and a coupling fitting. Some models include a **flow indicator** in the supply hose.

When the oxygen is turned on to a constant-flow mask, it fills the reservoir through a valve. When inhaling, the user draws oxygen directly from the reservoir bag. When the oxygen in the reservoir bag is depleted, the user breathes cabin air. When the user exhales, the reservoir bag refills with oxygen. The oxygen from the supply line flows continuously into the mask, sometimes filling the reservoir bag and at other times being breathed by the user. Exhaled oxygen and air are discharged from the mask into the cabin. Typical examples of continuous-flow oxygen masks are shown in Fig. 16-50. Masks of these types are usually provided with space for the installation of a microphone.

Passenger oxygen masks for airliners are of the constant-flow type and the face piece is oronasal in design; that is, it is designed to cover both the nose and the mouth. They are referred to as **phase-dilution masks** because of the characteristics of their operation.

When oxygen is turned on to the passenger mask, it enters the bottom of the reservoir bag and causes it to inflate. When inhaling, the user draws oxygen from the reservoir bag until it is deflated. At that time, the user begins to breathe cabin air plus a small amount of oxygen, which is flowing through the reservoir. Thus, there are two phases of oxygen consumption during inhalation. The first and largest part of the inhalation draws almost pure oxygen into the lungs. When the reservoir bag has deflated, the user continues to inhale but is breathing primarily cabin air. The first part of the inhalation provides a very rich oxygen mixture, which goes deep into the lungs. The last part of the inhalation, in which cabin air is being breathed, affects only the upper part of the lungs, the bronchial tubes, and the windpipe (trachea). Since these parts of the respiratory system do not contribute to the absorption of oxygen by the blood, the low oxygen content of the cabin air breathed during the last part of inhalation is of little consequence.

When the user of the mask exhales, the air is discharged through an exit valve in the mask to the cabin atmosphere. At this time, the reservoir bag refills with oxygen; however, the bag does not always fill completely, particularly if the user is breathing rapidly. If the flow rate of oxygen is only 1 L/min, which is normal for a cabin altitude of

15 000 ft [4573 m], the reservoir bag does not have time to fill between each inhalation by the passenger. This has caused concern among some passengers because they have thought they were not receiving an adequate flow of oxygen.

A passenger wearing an oxygen mask on a DC-10 airliner is shown in Fig. 16-51. This particular mask has a built-in flow indicator at the bottom of the reservoir bag. A small section of the bag has been partially sealed off so it will inflate immediately when even a small amount of oxygen is flowing. The indicator is colored a bright green so the passenger can see at once whether oxygen is flowing into the mask.

Oxygen masks on airliners are stowed in overhead compartments or in a compartment at the top of the seat back. If the cabin should depressurize, the compartments open automatically and present oxygen masks to the passengers. If the automatic system fails to work, a backup electrical system can be activated by a member of the crew to open the oxygen compartments.

Pressurized aircraft are normally equipped with diluter-demand oxygen systems for use by the flight-deck crew. The masks used by the crew are of an oronasal design and contain microphones and a strap or harness arrangement that will hold the mask securely in position. For some aircraft, which operate at very high altitudes, quick-donning masks are used. These masks can be put on in 5 s or less. Figure 16-52 shows the pneumatic harness type of diluter-demand masks.

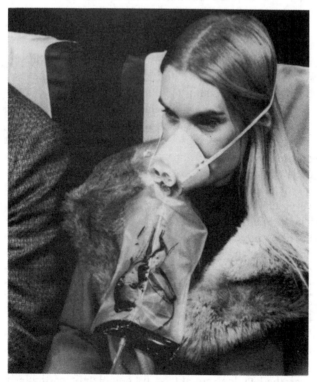

FIGURE 16-51 A passenger wearing a continuous-flow oxygen mask on an airliner. (*Scott Aviation.*)

Gaseous Oxygen Systems for Unpressurized Aircraft

Unpressurized aircraft that are capable of flying at altitudes requiring the use of oxygen by crew and passengers may be equipped with portable or fixed gaseous oxygen systems. The system includes a high-pressure oxygen tank, a pressure regulator, pressure gauge, manifold, and various types of outlets (fittings) to which tubing connected to masks may be attached. The regulators may be of the demand type or the constant-flow type. It is preferred that the regulators and masks for the pilot and copilot be of the demand type. In most cases, however, constant-flow regulators and masks are used for both crew and passengers.

A permanently installed (fixed), stored-gas oxygen system for a light twin airplane is shown in Fig. 16-53. This system consists of a high-pressure oxygen cylinder with a regulator, an altitude-compensating regulator, a filler valve, an overboard discharge indicator, a control cable and knob, a cylinder pressure gauge, outlets, oxygen masks, and required plumbing. The supply regulator attached to the oxygen cylinder reduces the high cylinder pressure to a lower, constant pressure. The altitude-compensating regulator reduces oxygen expenditure at lower altitudes, thus increasing oxygen supply duration. The pressure gauge shows actual cylinder pressure.

The control knob on the instrument panel is used to open a valve and allow controlled oxygen pressure to flow to the mask outlets. When a mask fitting is plugged into an outlet, a continuous flow of oxygen is available at the mask. The masks include flow indicators for visual verification of oxygen flow. The masks, hoses, and flow indicators are stored in plastic bags, where they are readily available to crew and passengers.

The oxygen outlets are installed in the overhead console and above the passenger seats. Each outlet contains a spring-loaded valve that prevents oxygen flow until the mask hose is engaged with the outlet.

The oxygen filler valve is usually located under an access panel on the outside of the fuselage and near the oxygen cylinder. The filler valve consists of the valve incorporating a filter and valve cap. A check valve is installed in the high-pressure line at the regulator to prevent the escape of oxygen from the cylinder at the filler line port. A typical service panel is shown in Fig. 16-54.

The overboard discharge indicator is located on the bottom or side of the aircraft near the oxygen bottle. A low-pressure [60 + 20 psi (413.7 + 137.9 kPa)] green disk is provided to prevent dust and contamination from entering the line. The indicator line is connected to the high-pressure rupture fitting of the regulator. A disk disappearance in the indicator indicates that oxygen cylinder overpressure existed and oxygen has been routed overboard.

Gaseous Oxygen Systems for Pressurized Aircraft

Oxygen systems for pressurized aircraft are primarily installed for emergency use in case of cabin pressurization failure or cabin decompression. The oxygen supply is sufficient to take

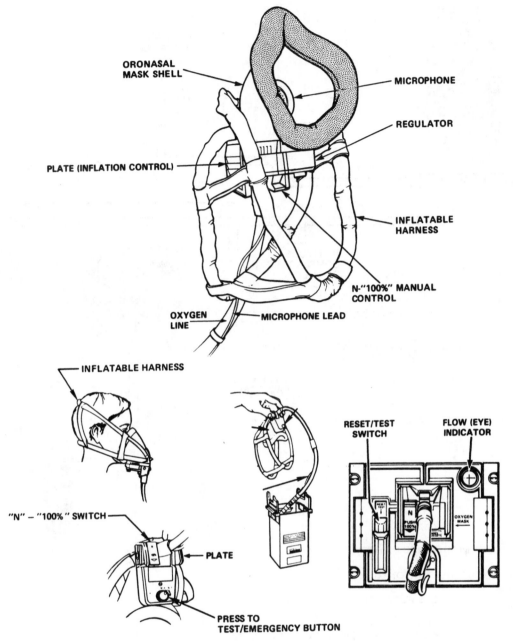

ORONASAL MASK SHELL

MICROPHONE

REGULATOR

PLATE (INFLATION CONTROL)

INFLATABLE HARNESS

N-"100%" MANUAL CONTROL

OXYGEN LINE

MICROPHONE LEAD

INFLATABLE HARNESS

RESET/TEST SWITCH

FLOW (EYE) INDICATOR

"N" – "100%" SWITCH

PLATE

PRESS TO TEST/EMERGENCY BUTTON

FIGURE 16-52   A quick-donning oxygen mask with a pneumatic harness. (*Canadair Inc.*)

care of all passenger and crew until the airplane is at low altitude, where oxygen is no longer necessary. Aircraft are equipped with a flight station oxygen system, a cabin (passengers) oxygen system, and a portable oxygen system for cabin crew.

Transport aircraft typically use a gaseous oxygen system in the flight compartment and either a gaseous oxygen system or a chemical oxygen system in the cabin. The following is a brief description of a gaseous oxygen system used in a commercial transport aircraft flight station; see Fig. 16-55. Oxygen for the flight crew system is supplied from a high-pressure gaseous oxygen cylinder located in the lower left-hand fuselage. The composite cylinder has a pressure gauge and an integrated pressure regulator/shutoff valve that supply oxygen to the individual mask-regulator compartment. The maximum cylinder pressure is 1800 psi. Figure 16-55

shows a schematic of the oxygen system. The oxygen system has two overpressure safety systems that vent oxygen overboard through a safety port when the pressure gets too high, and a supply solenoid valve that allows the crew to shut off the distribution system.

Oxygen line pressure downstream of regulator is indicated by the OXY LINE PRESS gauge on the aft overhead panel. The full-face oxygen mask with pneumatic harness, stowed in the mask-regulator compartment, contains a built-in microphone and a single oxygen-regulator control knob (Fig. 16-56).

The face seal includes integral purge valve assemblies. These purge valves automatically open when the control knob is in the EMER selection to allow oxygen to enter the face seal purging smoke and fumes. The mask-regulator

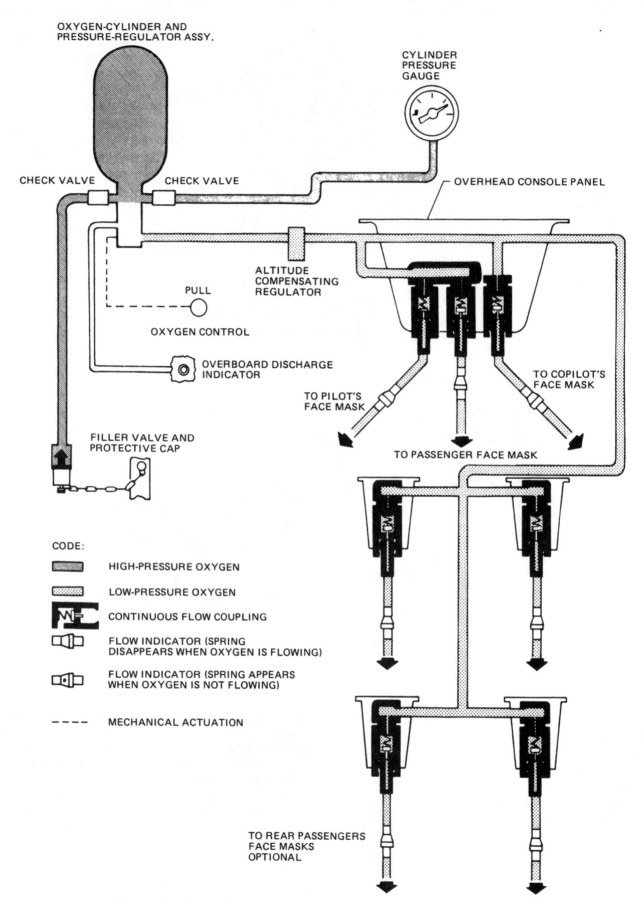

OXYGEN-CYLINDER AND
PRESSURE-REGULATOR ASSY.

CYLINDER
PRESSURE
GAUGE

CHECK VALVE   CHECK VALVE

OVERHEAD CONSOLE PANEL

ALTITUDE
COMPENSATING
REGULATOR

PULL

OXYGEN CONTROL

OVERBOARD DISCHARGE
INDICATOR

TO PILOT'S
FACE MASK

TO COPILOT'S
FACE MASK

TO PASSENGER FACE MASK

FILLER VALVE AND
PROTECTIVE CAP

CODE:

HIGH-PRESSURE OXYGEN

LOW-PRESSURE OXYGEN

CONTINUOUS FLOW COUPLING

FLOW INDICATOR (SPRING
DISAPPEARS WHEN OXYGEN IS FLOWING)

FLOW INDICATOR (SPRING APPEARS
WHEN OXYGEN IS NOT FLOWING)

— — — — MECHANICAL ACTUATION

TO REAR PASSENGERS
FACE MASKS
OPTIONAL

FIGURE 16-53   Fixed, stored-gas oxygen system for a light twin airplane. (*Cessna Aircraft Co.*)

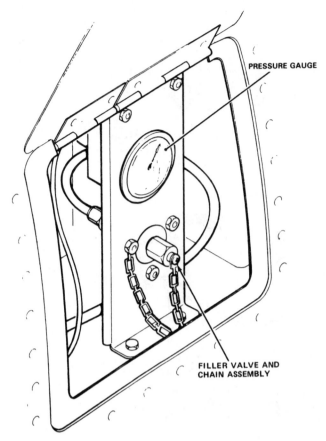

FIGURE 16-54 Oxygen ground servicing panel. (*Canadair Inc.*)

compartment has an in-place check feature, which allows the oxygen flow and oxygen mask microphone to be functionally tested without removing the mask from its compartment.

Operation of the Crew Oxygen System

The crew member squeezes the red grips to pull the mask out of its box which will inflate the mask harness. The mask is equipped with a mask-mounted regulator that supplies a mixture of air and oxygen or pure oxygen, or performs emergency pressure control. With the regulator set to NORMAL, the user breathes a mixture of cabin air and oxygen up to the cabin altitude at which the regulator supplies 100% oxygen. The user can select 100% oxygen, in which case the regulator supplies pure oxygen at all cabin altitudes. In case of smoke in the flight station the crew member can use the emergency overpressure rotating knob and receive pure oxygen at positive pressure. Figure 16-57 shows the donning operation.

Some transport aircraft also use a gaseous oxygen system for the cabin. The Airbus 330 uses a fixed gaseous system in the cabin to supply oxygen to the passengers in case of depressurization. The system stores its oxygen in interconnected cylinders (three, plus up to five additional) behind the right-hand sidewall lining in the forward cargo compartment. The oxygen flows to the mask containers in the cabin via two main supply lines and a network of pipes. The containers are above the passenger seats, in the lavatories, in each galley and at each cabin crew station. Each container has two, three, four, or five masks. An altimetric flow regulation device in each mask container controls the flow rate. The quantity calculation and control unit (QCCU) supplies the value of the average, temperature-compensated pressure for indication on the ECAM system page.

Normally the system is unpressurized. A pneumatic-controlled ventilation valve releases any residual pressure. Each container has an electrical latching mechanism that opens automatically to allow the masks to drop, if the cabin pressure altitude exceeds 14 000 ft. When this happens the system pressurizes, the masks drop, and (if installed) prerecorded instructions sound automatically over the passengers address system. The flight crew can operate the system manually. The generation of oxygen begins when the passenger pulls the mask toward the passenger seat. The mask receives pure oxygen under positive pressure at a rate governed by the cabin pressure altitude, a flow-regulating device in each container. The length of time that the oxygen supply will last after the cabin suffers decompression depends on the number of bottles installed, the number of masks in use, and cabin altitude. At a cabin pressure altitude of less than 10 000 ft, there will be no oxygen flow. Figure 16-58 shows the cabin oxygen system.

ECAM Information

The ECAM page provides information about the system and regulator pressure of the flight and cabin oxygen systems (see Fig. 16-59).

Chemical Oxygen Generators

Chemical oxygen generators, also termed **solid-state generators** because the chemicals involved are solid, have proved to be effective, lower in cost than stored gaseous oxygen, safe, and comparatively maintenance free. The chemical oxygen generator burns a mixture of sodium chlorate ($NaClO_3$) and iron (Fe) to produce pure oxygen suitable for human use. Sodium chlorate, when heated to 478°C [892°F], decomposes into ordinary salt and oxygen. The fuel to produce the heat for the decomposition of the sodium chlorate is iron. When sufficient heat is applied to the mixture to start the burning of the iron with a portion of the released oxygen, the burning continues until the chemical process is completed. The burned iron becomes iron oxide (FeO).

The construction of a chemical oxygen generator is shown in Fig. 16-60. The fuel unit is sometimes called a "candle." since it burns somewhat like a candle, starting at the ignited end and burning slowly from one end to the other. The term candle is disappearing from usage, however. Ignition of the generator is accomplished by means of an electrical squib (detonator) or a mechanically fired detonator, as is used to fire a rifle cartridge. The mixture of the material at the end of the generator where ignition takes place is "enriched" to make it start burning easily.

Note in the drawing of Fig. 16-60 that the generator is larger at the starting end than it is farther along. The purpose of this shape is to provide an increased flow of oxygen at the beginning of operation to ensure that the user will receive an adequate supply as soon as possible after ignition. Thereafter, the size of the generator is such

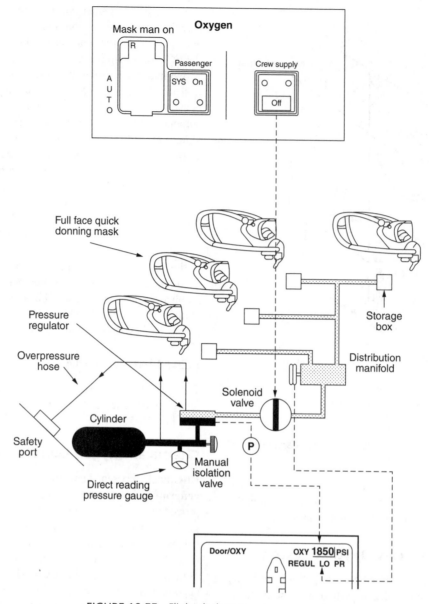

FIGURE 16-55 Flight deck gaseous oxygen system.

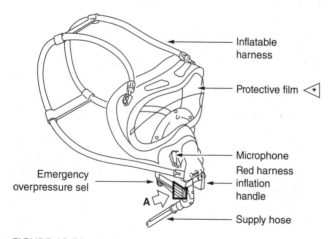

FIGURE 16-56 Oxygen mask.

that it provides oxygen sufficient for the number of people served from the generator. If more oxygen is produced than is required, the unneeded amount is vented through the relief valve to the cabin. The generator is surrounded by a filter and thermal insulation. The filter is to ensure that only pure oxygen and no particulate matter is delivered to the user's mask. The thermal insulation is to prevent the heat of the burning generator from damaging the surrounding area. The exterior surface of the generator may reach a temperature as high as 500°F [260°C], and so it is provided with a heat shield to prevent a user from touching it. A warning placard is mounted on the heat shield.

Generators come in several sizes and configurations. The generator shown in Fig. 16-61 is designed for use in portable units. The generators shown in Fig. 16-62 are designed for use in fixed installations of transport aircraft. These fixed generators come in sizes designed to accommodate two, three, or four people.

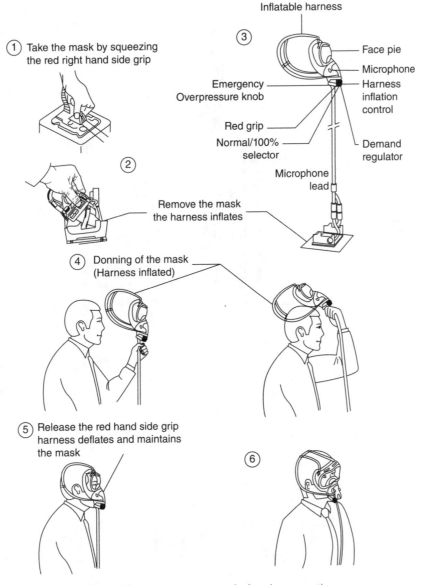

1. Take the mask by squeezing the red right hand side grip

2. Remove the mask the harness inflates

3. Inflatable harness
   Face pie
   Microphone
   Emergency Overpressure knob
   Harness inflation control
   Red grip
   Normal/100% selector
   Demand regulator
   Microphone lead

4. Donning of the mask (Harness inflated)

5. Release the red hand side grip harness deflates and maintains the mask

6.

FIGURE 16-57   Oxygen mask-donning operation.

Chemical Oxygen Systems

Many transport aircraft use a chemical oxygen system for the cabin. The advantages of the chemical system over the gaseous system include low maintenance, reduced fire hazard, and reliability. Maintenance activity is decreased because of the elimination of cylinders, valve, regulators, and distribution lines and their associated periodic maintenance tests and inspections. The fire hazard has been reduced by the elimination of possible ruptured oxygen lines adding to a fire and the very low pressure at which oxygen flows from the oxygen generators. Additionally, no oxygen is generated to contribute to combustion without the generator being activated. Reliability is improved by the elimination of the possibility of thermal discharge, the independent nature of each unit, and the indefinite life of the generator.

For instance, the Boeing 737 NG uses a chemical system to supply oxygen to the passenger, attendant stations, and lavatory service units. The passenger oxygen masks and chemical oxygen generators are located above the passenger seats in passenger service units (PSUs) as shown in Fig. 16-63. The passenger oxygen system is supplied by individual chemical oxygen generators located at each PSU. Four continuous flow masks are connected to each generator. A generator with two masks is located above each attendant station and in each lavatory. The system is activated automatically by a pressure switch at a cabin altitude of approximately 14 000 ft or when the passenger oxygen switch on the aft overhead panel is positioned to ON. When the system is activated, the PASS OXY ON light illuminates and OVERHEAD illuminates on the master caution system. When the high-altitude landing switch is placed in the ON position the cabin altitude warning horn will sound at 14 000 ft and automatic oxygen mask activation occurs at 14 650 ft. Activating the system causes the masks to drop from the stowage compartments. The oxygen generators are activated when any

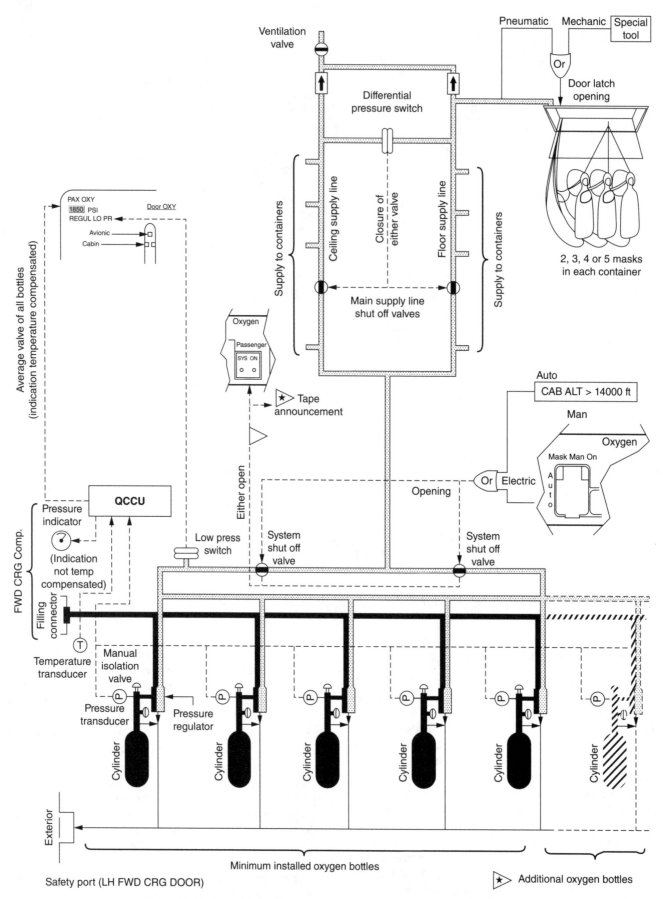

FIGURE 16-58 Cabin station gaseous oxygen system for transport aircraft.

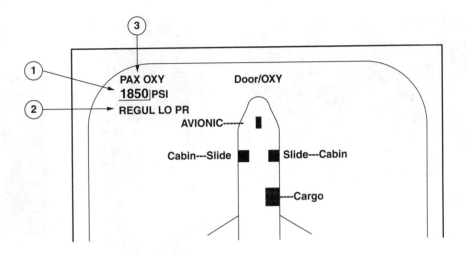

1 OXY high pressure indication

Green (steady) : When pressure is ≥ 1200 psi
Green (pulsing) : When pressure is ≥ 200 psi and < 1200 psi
Amber : When pressure is < 200 psi
An amber half frame appears when oxygen pressure is < 1600 psi

2 REGUL LO PR indication

Appears amber if:
– The oxygen pressure on the low-pressure circuit is low (64 psi), or if the low pressure switch fails.
– One of the system shutoff valves is not fully closed and the system has not been activated.
– One of the main supply line shutoff valves is not fully open and the system has not been activated.
– Both main supply line shutoff valves are closed when the system has been activated.

3 PAX OXY indication

Normally white
Becomes amber when:
– Pressure goes below 200 psi.
– Low oxygen pressure is detected on the low-pressure circuit.

FIGURE 16-59   ECAM oxygen page.

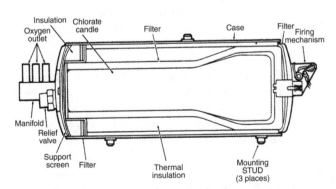

FIGURE 16-60   Construction of a chemical oxygen generator. (*Douglas Aircraft Co.*)

FIGURE 16-61   Chemical generator for a portable system. (*Scott Aviation.*)

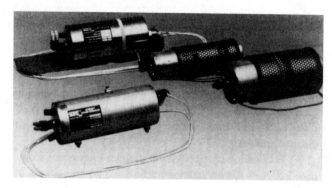

FIGURE 16-62 Various chemical generators used in airliners. (*Scott Aviation.*)

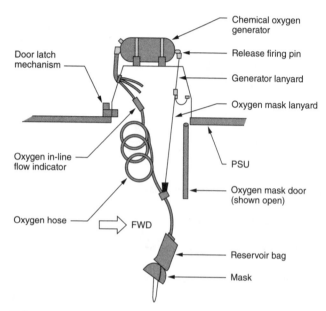

FIGURE 16-63 Chemical oxygen generator location.

Labels in Figure 16-63:
- Chemical oxygen generator
- Release firing pin
- Generator lanyard
- Oxygen mask lanyard
- Door latch mechanism
- PSU
- Oxygen in-line flow indicator
- Oxygen mask door (shown open)
- Oxygen hose
- FWD
- Reservoir bag
- Mask

FIGURE 16-64 Portable chemical oxygen generators.

mask in the unit is pulled down. Pulling one mask down causes all masks in that unit to come down and 100% oxygen flows to all masks. A green in-line flow indicator is visible in the transparent oxygen hose whenever oxygen is flowing to the mask. Oxygen flows for approximately 12 min and cannot be shut off. If the passenger oxygen is activated and a PSU oxygen mask compartment does not open, the masks may be dropped manually.

A heat-sensitive paint is applied to the chemical oxygen generator's corrosion-resistant steel case. This paint is typically white. When the generator is expended, the paint will change color, indicating that replacement is necessary.

Portable chemical oxygen generators are available for emergency use and for use on private and business-aircraft. Typical of these units are those shown in Fig. 16-64.

Service and Maintenance of Oxygen Systems

The service and maintenance of oxygen systems should be performed only by technicians who are qualified through training and experience. Oxygen is not an explosive and is not flammable in its pure state, but it supports combustion, sometimes violently, when in contact or mixed with other materials. It is, therefore, important that certain precautions be exercised when working with oxygen systems. The following are essential:

1. Smoking, open flames, or items that may cause sparks must not be permitted near aircraft when maintenance is being performed on the oxygen system.
2. All electrical power must be off and the airplane must be grounded.
3. Oxygen must not be permitted to come in contact with oils, greases, or solvents. Such contacts can cause spontaneous explosions.

Typical instructions for the maintenance and service of a fixed system in a light twin airplane are as follows:

1. Use extreme caution to ensure that every port in the oxygen system is kept thoroughly clean and free of water, oil, grease, and solvent contamination.
2. Cap all openings immediately upon removal of any component. Do not use masking or electrical tape or caps that will attract moisture.
3. Lines and fittings must be clean and dry. Manufacturers specify various methods that may be used to clean oxygen lines. One method is to use a vapor-degreasing solution of stabilized perchlorethylene conforming to MIL-T-7003 and then to blow the tubing clean and dry with a jet of nitrogen gas (BB-N411), Type 1, Class 1, Grade A, or technical argon (MIL-A-18455). Lines can also be cleaned with naphtha, Spec. TT-N95, after which they are blown clean and dry as just described or with dry, clean, filtered air. Air from a standard compressed-air supply usually contains minute quantities of entrained oil, so it is not suitable for blowing out oxygen lines.

Oxygen lines may be flushed with hot inhibited alkaline cleaner, washed with clean, hot water, and dried as previously explained. Lines should be capped as soon as they are cleaned and dried.

4. Fabrication of replacement oxygen lines is not recommended. New factory parts identified by part number should be used.

5. Thin Teflon fluorocarbon resin tape conforming to MIL-T-27730 may be used to aid in sealing tapered pipe threads. No compounds shall be used on aluminum-alloy flared fittings. No compound is used on the coupling sleeves or on the outside flares. Sealant tape is not applied to the first three threads of a coupling.

6. Technicians must be sure that their hands are free of dirt or grease before they handle oxygen tubing, fittings, or other components.

7. Tools used for the installation of oxygen lines and fittings must be sparkless and free of all dirt, grease, or oil.

8. In all cases, instructions provided for the system by the manufacturer must be followed.

High-pressure oxygen cylinders must be inspected and checked for conformity with regulations. Each cylinder must have a Department of Transportation (DOT) designation number. **Standard-weight cylinders** are marked DOT-3AA-1800 or ICC-3AA-1800. These cylinders have no total life limitations and may be used until they fail hydrostatic testing at five-thirds of their operating pressure. The hydrostatic test must be made every 5 years. **Lightweight cylinders** are designated DOT-3HT-1850 and must be hydrostatically tested every 3 years to five-thirds of their operating pressure. These cylinders must be retired from service after 24 years or 4380 filling cycles, whichever occurs first. Composite cylinders (aluminum shell wrapped with Kevlar) must conform to DOT-E-8162. These cylinders are rated for 1850 psi [12 746.5 kPa] and have a life of 15 years or 10 000 filling cycles, whichever occurs first. Each filling of an oxygen cylinder requires an entry in the aircraft logbook.

Oxygen cylinders are inspected for damage such as nicks, dents, corroded fittings, hydrostatic test date, DOT designation, and leakage. Cylinders that do not meet requirements must be completely disassembled and inspected in an approved facility. DOT numbers, serial numbers, and dates of hydrostatic testing are stamped on the shoulder or neck of each cylinder.

Whenever a component of a high-pressure oxygen system has been removed and replaced or whenever the system has been disassembled in any way, the system must be tested for leaks and purged. **Leak testing** is accomplished by applying an approved leak detector conforming to MIL-L-25567A, Type 1, to all fittings, as instructed. The leak-detector solution is completely removed after each test. Where leaks are found, fittings must be repaired or replaced.

Purging the oxygen system involves fully charging the system in accordance with service instructions and then releasing oxygen from the system. The airplane should be outdoors if possible; otherwise it should be isolated in a well-ventilated building with no smoking or open flame permitted in the area.

No grease or other lubricants should be near enough to come in contact with the oxygen. The doors and windows of the airplane must be open.

After the system is fully charged with oxygen, all oxygen masks are plugged into their outlets and the oxygen is allowed to flow for about 10 min. When the oxygen flowing from the masks is odorless, the purging is complete and the oxygen is shut off. The masks are removed from their outlets and the system is recharged. Recharging is usually accomplished through a filling valve and fitting mounted conveniently on the lower part of the airplane or accessible through the baggage area. The cap is removed from the fitting and the refill hose is connected. Oxygen is allowed to flow into the system until it is automatically or manually shut off when the required pressure is registered on the high-pressure oxygen gauge. **Only aviator's breathing oxygen conforming to MIL-O-27210 can be used.** Do *not* assume that because a cylinder is colored green, it contains breathing oxygen. Cylinders containing other gases are sometimes colored green.

The foregoing information applies in general to stored-gas oxygen systems. It must be emphasized that for any specific system, the appropriate manufacturer's instructions must be followed. This is particularly true for the large, complex systems installed in commercial transport airplanes.

Chemical oxygen systems require little or no maintenance and much less in the area of inspections than stored-gas systems. Replacement of expended generators and resetting firing mechanisms are the usual services required. Expended generators are identified by a heat-sensitive paint mark on the generator. This mark is white before the generator is used and black afterward. Procedures for replacement of generators are provided with the equipment. In fixed installations such as those on the DC-10 and the L-1011, manufacturer's service instructions should be followed.

REVIEW QUESTIONS

1. What are the sources of heat for heating systems?
2. Describe a simple *exhaust heating system.*
3. Explain the need for regular and careful inspections of exhaust heating systems.
4. Describe a *combustion heater.*
5. What fuel is used in a combustion heater?
6. Describe a typical inspection for a combustion heater system.
7. Explain the *vapor-cycle cooling system* principle.
8. What are the principal units required in a vapor-cycle cooling system?
9. What refrigerant is commonly employed in a vapor-cycle cooling system?
10. Describe the use of each vapor-cycle-system unit.
11. How would a technician determine that the refrigerant supply is low?
12. Why is a dryer essential in a vapor-cycle system?
13. What is meant by *subcooling*?
14. What controls the size of the orifice in a thermal expansion valve?

15. What is meant by the *high side* and the *low side* in a vapor-cycle system?

16. List the precautions that must be taken when servicing a vapor-cycle refrigeration system.

17. During inspection of a vapor-cycle system, what is the indication of leaking refrigerant?

18. If a vapor-cycle refrigeration system operates satisfactorily for a short time and then stops cooling, what is the likely cause?

19. What is the purpose of the *service manifold* and how is it connected into a refrigeration system?

20. What is the purpose of the *service valves* in a vapor-cycle system?

21. What precaution should be observed when connecting a fitting to a service valve prior to servicing?

22. Explain *purging* of a refrigeration system and the procedure for doing this.

23. What precaution should be taken with regard to ventilation when purging a vapor-cycle system?

24. What should be done when any unit of a refrigeration system is disconnected?

25. How is a refrigeration system *evacuated* (pumped down)?

26. How is a vapor-cycle system recharged?

27. Why is aircraft pressurization necessary?

28. What are the sources of aircraft-pressurization air pressure?

29. Describe the principle of *air-cycle cooling*.

30. Describe an *air-cycle machine* (ACM).

31. What is the function of the *heat exchangers* in an air-cycle cooling system?

32. What is the function of the *water separator* in an air-cycle cooling system?

33. Describe a *cooling pack*.

34. What is a *gasper system* and what is its air source?

35. What device limits the maximum cabin altitude in an airline-pressurization system?

36. Explain the need for limiting the pressure differential between the inside of the cabin and the outside.

37. What is the purpose of a *negative pressure-relief valve*?

38. Describe the operation of an *outflow valve*.

39. What are the functions of the *automatic* and *manual controllers* in the pressurization system?

40. What is the purpose of the *barometric correction selector*?

41. Explain the purpose of the *positive pressure-relief valve*.

42. Describe the operation and purpose of a *jet pump*.

43. How is air conditioning on an airliner accomplished when the airplane is on the ground and the engines are not running?

44. At what altitudes may a person begin to feel the effects of *hypoxia* in an unpressurized airplane?

45. What factors affect the amount of oxygen a person will absorb from the air he or she breathes?

46. Name three types of oxygen systems with respect to the method by which oxygen is stored on the aircraft.

47. Describe a typical *portable* oxygen system.

48. Explain the difference between a *continuous-flow* oxygen system and a *demand-diluter* system.

49. Describe an oxygen *pressure regulator*.

50. Explain the operation of a *diluter-demand regulator*.

51. Describe the automatic-opening provision of the pressure regulator for a typical airliner oxygen system.

52. How can an oxygen system bottle be distinguished from other gas bottles?

53. What advantages are associated with a chemical oxygen system when compared to a gaseous oxygen system?

54. Explain the difference between an oxygen mask for a demand system and a mask for a continuous-flow system.

55. Describe the operation of a continuous-flow oxygen mask.

56. Explain the production of oxygen in a chemical oxygen system.

57. What are the advantages of a chemical oxygen system?

58. What precaution must be observed after a chemical oxygen generator has been fired?

59. Why is the *chemical candle* for an oxygen generator made larger at the starting end than it is farther along?

60. Describe a fixed oxygen system for a light airplane.

61. What indicator is used to show the escape of oxygen overboard in the system for a light airplane?

62. What are the range markings of the oxygen-pressure gauge?

63. What are the two oxygen systems installed in some models of the B777 airliner?

64. What is the function of the *thermal-expansion safety discharge valve*?

65. How can a crew member determine that a safety valve has discharged oxygen from a cylinder?

66. Describe a *crew oxygen system*.

67. How many oxygen outlets are available to the passengers in an airliner?

68. Why are portable oxygen units required?

69. To what pressure is a stored oxygen system tank usually charged?

70. Give the precautions to be observed when working with oxygen systems.

71. List typical maintenance practices for an oxygen system.

72. Why is specially dried and filtered compressed air required for blowing out oxygen lines?

73. What sealing compound may be used on oxygen fittings?

74. How are oxygen cylinders marked for approval?

75. How often should oxygen cylinders be hydrostatically tested? To what pressure?

76. For what defects are oxygen cylinders inspected?

77. How is *leak testing* accomplished in an oxygen system? What material must be used?

78. Describe the *purging* of an oxygen system.

79. Explain the importance of capping all open oxygen lines and fittings.

80. How can the technician tell that the candle in a chemical oxygen generator has been expended?

Aircraft Instruments and Instrument Systems 17

INTRODUCTION

From the time that human beings began to fly, the need for instruments of various types has been apparent. It was quickly obvious that a pilot needed some indication of the airspeed of the airplane, even though many old-time pilots swore that they could tell the airspeed by the sound of the air flowing past the rigging wires. Instruments such as engine revolutions per minute (rpm), oil pressure, and oil temperature were found to be vital signs of engine-operating performance that are still used today to determine whether the engine is operating satisfactorily. If any of these instruments indicated an abnormal operating condition, the pilot would know that something was wrong and could land before an engine failure occurred.

With the increase in air traffic, today's pilots need to know exactly how high they are, how fast the airplane is traveling, in what direction the airplane is headed, how fast they are climbing or descending, the attitude of the airplane, how the engine is performing, and numerous other factors. These demands have led to the development of instruments, the prime requisites of which have been sensitivity and dependability plus lightweight construction. The aircraft instruments of today are masterpieces of the designer's art. Even with the extreme conditions of pressure, temperature, velocity, and altitude, modern aircraft instruments continue to function accurately and dependably.

Figure 17-1 illustrates an instrument panel for a light, single-engine airplane. Although all the instruments shown are not required by regulation, each contributes to the safety and convenience of the pilot and passengers. Instruments required for aircraft are listed in FAR Part 91, Sec. 91.33.

Instruments on an airplane may be classified according to function in three principal categories: **powerplant instruments, flight and navigational instruments**, and **systems instruments**. Powerplant instruments provide information concerning the operation of the engines and the powerplant systems. Flight and navigational instruments give the pilot information concerning flight speed, altitude, airplane attitude, heading, rate of ascent or descent, and other indications pertaining to the airplane and its flight path. In addition, some of the information provided to the pilot by the flight and navigational instruments is also automatically transmitted to air traffic–control centers to assist the controller in maintaining safe aircraft flight operations. The aircraft's instruments provide information concerning the hydraulic system and its pressures, the air-conditioning system, the electric system, and other special systems that may be installed in the aircraft.

There are two particularly important reasons why the principles of operation and the art of interpreting the readings of instruments used in the modern aircraft should be thoroughly understood by the technician. First, the safety of the aircraft, the crew, and the passengers depends on the proper installation and correct operation of the instruments. The powerplant instruments can forewarn the pilot or flight engineer of impending engine failure, and the flight instruments indicate any irregularity of flight attitude or direction. It is the duty of the aviation maintenance technician to check the instruments regularly.

The second reason why the technician must understand the operating principles of instruments is that they provide many indications of powerplant or system condition. Just as the physician diagnoses an illness by means of instruments, the aircraft technician can diagnose possible troubles in an aircraft powerplant or system by an understanding of instrument indications. Through a complete analysis of instrument readings that were recorded during flight, the technician can determine what units or systems are not functioning correctly. When a pilot returns from a flight and reports engine trouble, it is the duty of the technician to determine the cause of the trouble. The technician can eliminate several possible causes by determining such factors as the altitude at which the airplane has been flying, the temperature of the powerplant concerned,

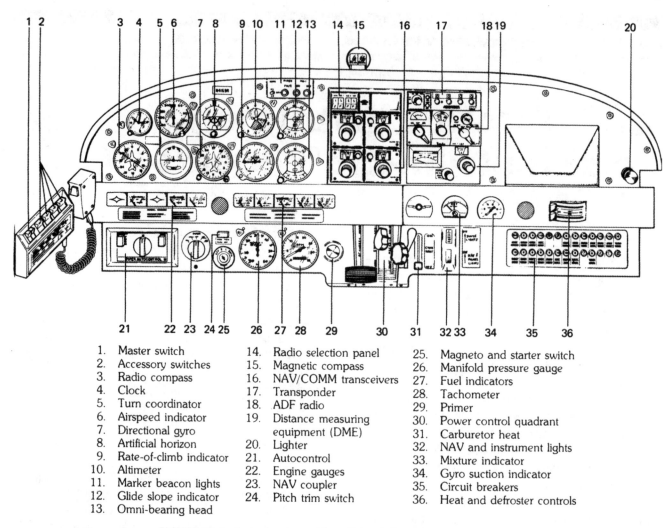

1. Master switch
2. Accessory switches
3. Radio compass
4. Clock
5. Turn coordinator
6. Airspeed indicator
7. Directional gyro
8. Artificial horizon
9. Rate-of-climb indicator
10. Altimeter
11. Marker beacon lights
12. Glide slope indicator
13. Omni-bearing head

14. Radio selection panel
15. Magnetic compass
16. NAV/COMM transceivers
17. Transponder
18. ADF radio
19. Distance measuring equipment (DME)
20. Lighter
21. Autocontrol
22. Engine gauges
23. NAV coupler
24. Pitch trim switch

25. Magneto and starter switch
26. Manifold pressure gauge
27. Fuel indicators
28. Tachometer
29. Primer
30. Power control quadrant
31. Carburetor heat
32. NAV and instrument lights
33. Mixture indicator
34. Gyro suction indicator
35. Circuit breakers
36. Heat and defroster controls

FIGURE 17-1 Instrument panel for a light airplane. (*Piper Aircraft Co.*)

the rpm and the manifold pressure of the powerplant, and the recorded fuel and oil pressure. From an understanding of the instruments and their function, the technician can often determine where the trouble lies. Instruments are useful for troubleshooting regardless of whether the airplane concerned is a small private airplane or a large jet airliner. It is, therefore, the responsibility of the technician to ensure, as far as possible, continuous and accurate operation of all instruments.

Instruments may also be classified according to the principles used in giving their indication. Instruments classified by principle of operation fall into four main categories: *pressure-type instruments, mechanical instruments, gyro instruments*, and *electrical* and *electronic instruments*. The basic principles of operation of each of these types of instruments are covered in this chapter. This chapter also describes the various instruments using these basic mechanisms for their operation, along with the installation and marking.

PRINCIPLES OF INSTRUMENT OPERATIONS

Aircraft instruments are required to monitor a wide variety of aircraft conditions. To do this, many different methods are used to collect information and display the information for the flight crew. It is important that technicians have a basic understanding of the principles of operation of the common aircraft instruments so that they may efficiently maintain the systems in proper operating condition.

This section deals with the primary operating principles of pressure instruments, gyro instruments, electrically powered resistance indicators, and temperature indicators. Mechanical principles are applied to instrumentation in both a primary and supportive role. When applied in a supportive role, mechanical principles are employed to convert other primary principles into a readable (calibrated) format. When mechanical principles are the primary operating principle, they are often used in a design unique to a specific application.

These principles of operation are applied to many different types of instrument systems in later sections of this chapter.

The principles of operation associated with mechanically operated instruments, where the principles are uniquely applied, are discussed individually in this chapter as the purpose, use, and operations of each mechanical instrument are reviewed.

Pressure Instruments

Pressure instruments are used to monitor many aircraft flight situations, such as altitude and airspeed, as well as the condition of systems such as hydraulic pressure and engine oil pressure. There are two basic types of mechanical pressure-measuring instruments: bourdon tubes and diaphragms/bellows.

Instruments that measure pressures in relatively high-pressure fluid systems are usually operated through a mechanism known as a **bourdon tube**. Among the indicators requiring this type of instrument are hydraulic pressure, engine oil pressure, oxygen pressure, and any other indicator of comparatively high pressures. The bourdon tube is constructed of metal and is oval or flattened in cross-sectional shape, with the tube itself formed into a crescent or part circle.

Figure 17-2 indicates the general construction of a bourdon tube. One of the ends of the tube is open, and the other end is closed. The open end is attached to a casting that is anchored to the case of the instrument, thus making the open end of the tube stationary. The closed end of the tube is free to move and is attached to a series of linkages such as levers and gears. When fluid under pressure enters the open end of the bourdon tube, it causes a pressure against the closed end, and this tends to straighten the tube. The principle involved here is well illustrated by a familiar party novelty—the rolled-up paper tube that uncoils when you blow into it. The pressure on the closed end of the paper causes it to uncoil, and the springs inside cause it to coil up again when the pressure is released. The bourdon tube is constructed of a metal such as spring-tempered brass, bronze, or beryllium copper. These metals have a strong spring effect that causes the bourdon tube to return to its original position when pressure is released.

In Fig. 17-3 a simplified diagram of a bourdon-tube-instrument mechanism is shown. When pressure enters the bourdon tube, the tube tends to straighten out; as it does so, it moves the mechanical linkage connected to the sector gear. The movement of the sector gear causes the spur

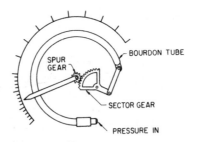

FIGURE 17-3 Bourdon-tube instrument mechanism.

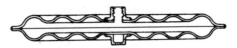

FIGURE 17-4 Cross section of an instrument diaphragm.

gear to rotate, and this, in turn, moves the indicating needle along the scale to give a reading of the pressure. The indicating needle is mounted on the hand staff, which is rotated by the spur gear. Instruments such as this one are ruggedly constructed, with little maintenance required.

Pressure gauges designed to provide readings of comparatively low pressures are usually of the **diaphragm** or **bellows** type. In some cases, both a diaphragm and a bellows are used in the same instrument. A drawing of the cross section of a typical instrument diaphragm is shown in Fig. 17-4. The diaphragm consists of two disks of thin metal corrugated concentrically and sealed together at the edges to form a cavity or capsule. The diaphragm in Fig. 17-4 is designed with an opening through one of the disks to admit the pressure to be measured. The opposite side is provided with a bridge that can bear against a rocking shaft lever, through which the movement is transmitted to the indicating needle.

In some instruments, such as a simple barometer, the diaphragm is sealed with dry air or an inert gas inside at sea-level atmospheric pressure. Changes in the pressure of the air outside the diaphragm cause it to expand or contract, thus producing a movement that is converted to a dial reading through the instrument mechanism. A bellows capsule is illustrated in Fig. 17-5. The bellows is made of thin metal with corrugated sides and formed into a cylindrical capsule, as shown. This unit operates in much the same manner as the diaphragm, but it provides a greater range of movement. Examples of the types of instruments that typically use a diaphragm or bellows mechanism include altimeters, airspeed indicators, and manifold pressure gauges.

The principle of operation for pressure instruments allows measurements to be made in relationship to three base pressures. Pressure instruments may be designed to indicate pressure as a value referenced to the ambient (surrounding atmospheric)

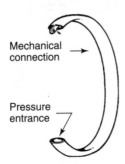

FIGURE 17-2 A bourdon tube.

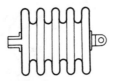

FIGURE 17-5 Bellows capsule.

air pressure [pounds per square inch gauge (psig)], to absolute zero [pounds per square inch absolute (psia)], or to some other reference value [pounds per square inch differential (psid)]. If a value on a pressure gauge is identified as indicating pressure in psig, then the pressure indicated is a gauge pressure—that is, the pressure above or below ambient pressure. Instruments such as hydraulic system pressure and vacuum system values are psig values. If the face of the gauge does not show that a pressure differential is indicated and if the gauge reads zero when the system is inactive, the instrument is indicating gauge pressure. For example, an empty oxygen cylinder gauge would read zero, indicating that there is no difference between the internal pressure of the cylinder and the ambient pressure.

Instruments such as manifold pressure gauges are referenced to absolute zero and indicate the actual pressure in the engine intake manifold. That is why the manifold pressure gauge at sea level indicates about 30 in Hg (inches of mercury) [762 mm Hg] when the engine is not running. Standard day conditions at sea level give an atmospheric pressure of 29.92 in Hg [760 mm Hg]. The absolute pressure thus indicated is labeled psia.

Other instruments such as the engine-pressure-ratio indicator for a turbine engine compare the ram air pressure at the inlet of the engine to the ram air pressure of the exhaust gases at the engine outlet to give a differential pressure indication (psid).

Gyroscopic Instruments

One of the most essential devices for the navigation of aircraft is the **gyroscope**. The gyroscope is *a device consisting of a wheel having much of its mass concentrated around the rim, mounted on a spinning axis.* The characteristic of a gyroscope that makes it valuable as a navigation device is its ability to remain rigid in space, thus providing a directional reference.

The principle of rigidity in space, illustrated in Fig. 17-6, is based on the fact that a spinning mass tends to stabilize with its axis fixed in space and will continue to spin in the

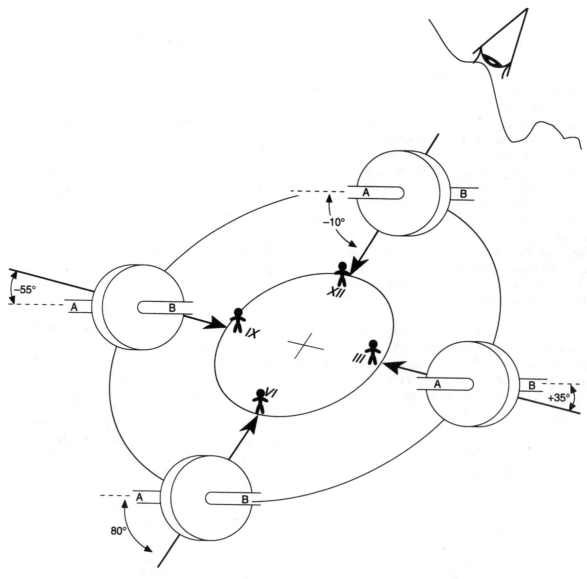

FIGURE 17-6 Rigidity in space.

same plane unless outside forces are applied to it. If a gyro free from outside influences were started running in a vertical plane (10° below the horizontal axis) at 12:00 noon standing on a sundial and then observed 3 h later, the axis of rotation would appear to have shifted 45°. However, from a fixed point in space the axis would appear to maintain its same position. The same is true at any time the axis of rotation is compared to the original starting position. From the sundial the axis of rotation appears to have changed, but when observed from space no rotation is observed. The foregoing action is the result of the rigidity of the gyro, and it is this characteristic that makes the device effective as a navigation reference.

The **rigidity** of a gyro is the force of the gyro that opposes any other force that tends to change its plane of rotation in space. The rigidity is increased as the mass at the rim of the gyro wheel is increased and/or as the speed of rotation is increased. A gyro built in the form of a heavy-rimmed wheel will have much more rigidity than a gyro shaped like a sphere or cylinder.

The gyroscopic instrument's case, like all other instrument cases, must be mounted to the aircraft. The internal mechanisms of the instrument must be secured to the instrument case. However, in order to take advantage of the gyroscope's rigidity in space, it must be free to rotate about one or both of two axes perpendicular to each other and to the spinning axis. In order to allow the necessary freedom of movement, a gyroscope is mounted in rings so that the mountings are free to rotate in any direction without disturbing the gyro; this is called a **free gyro**. The rings in which the gyro is mounted that permit it to move are called **gimbal rings**. Figure 17-7 shows gimbal rings and the progression of mountings that lead to a free gyro, along with the three axes of rotation. Figure 17-7a shows the spinning of an unmounted gyro.

Figure 17-7b shows the mounting of the gyro to a single ring. If the ring is secured to the aircraft's structure so that the axis of rotation (after the ring is mounted) is parallel to the lateral (pitch) axis of the aircraft, the aircraft may pitch up and down without the mounting mechanism affecting the gyro's ability to maintain rigidity in space. In the figure this axis of rotation is arbitrarily defined as X.

Figure 17-7c depicts the same ring illustrated in Fig. 17-7b but mounted to the aircraft at pivot points parallel to the longitudinal (roll) axis. This axis is designated Y. These pivot points now allow the aircraft to rotate around both the pitch and roll axes without affecting the gyro's ability to maintain rigidity in space.

In Fig. 17-7d the mounting ring in Fig. 17-7c is mounted to a second ring, maintaining the pivot point parallel to the Y axis. The additional mounting ring also has pivot points parallel to the vertical (yaw) axis, indicated as Z. The mounting of the second ring to the aircraft allows rotation of the aircraft around all three axes of rotation without affecting the gyro's rigidity in space. The spinning axis of the gyro is the X axis, which is horizontal in the illustration. The Y axis is on an axis parallel to the horizon but is perpendicular to the spinning axis. The Z axis is vertical and is therefore perpendicular to the two other axes.

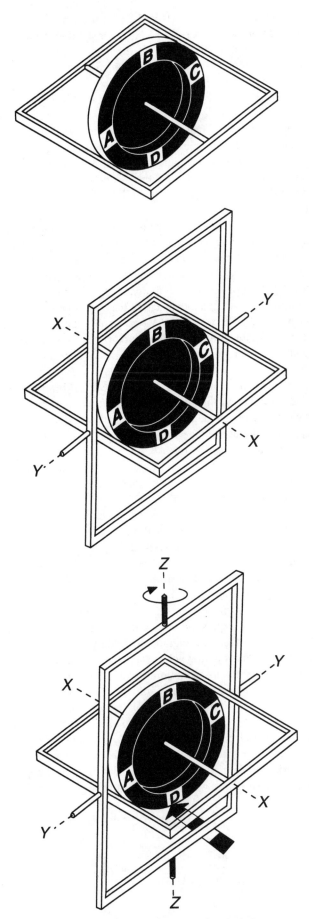

FIGURE 17-7 Buildup and operation of a gimbal.

Another important characteristic of the gyro is its tendency to *precess* if external forces are applied to it. **Precession** can be described briefly as the tendency of a gyro to react to an applied force 90° in the direction of rotation from the point the force is applied.

Precession may be visualized by thinking of two free gyros side by side, each gyro similar in construction to the gyro shown in Fig. 17-7. One gyro is not spinning, while the other is spinning at gyroscopic speed. Applying a force at point D to the nonspinning gyro would result in the gyro wheel moving around the Y axis. Looking at the gyro wheel from the front point B would move point B toward the viewer and point D away from the viewer.

However, if the gyro is revolving at gyroscopic speed in a clockwise direction, precessive forces make the gyro react differently to the same force. As just stated, precession results in the tendency of a gyro to react to an applied force 90° in the direction of rotation. In the preceding example, precession would result in a movement of the gyro at point A rather than point D. The wheel of the gyro would rotate around the Z axis. Again, looking at the gyro wheel from the front point C would move point C toward the viewer and point A away from the viewer.

This phenomenon is also illustrated in Fig. 17-8. In Fig. 17-8 the gyro is mounted on a shaft and is spinning in a clockwise direction, as shown by the arrows. If a force F is applied at the side of the wheel at approximately the 9 o'clock position, the precession force P will act at a point 90° in the direction of rotation, at approximately 12 o'clock, from the applied force. The precessive force will cause a shift in the axis of rotation, as shown by the arrows A and B.

It is important to note that the precession applies only when the gyro is in motion and reaches gyroscopic speed. Even though a detailed analysis of the cause of precession is beyond the scope of this discussion, a simple explanation of the forces causing precession is appropriate. The forces of centrifugal motion (see *Aircraft Basic Science*) play a significant role in precession. In centrifugal motion the force vector that represents the motion of the mass lies along the tangent line of the arc of motion. Adding a perpendicular force to the mass at any point results in the addition of two force vectors, the mass motion vector and the applied force vector. A combination of these forces and the limited mobility of a gyroscope combine to result in precession.

Gyroscope rotors are designed to be operated either pneumatically or electrically. Pneumatically operated gyro rotors are made from a heavy material such as brass. They are disk-shaped, with the majority of their mass located on the outer edge of the disk and with a thin web of material between the

FIGURE 17-9 Gyro rotor of an air-driven turn-and-bank indicator.

rim and the central shaft of the disk. The outer edge of the disk has notches, or "buckets," cut into the surface, as shown in Fig. 17-9, so that a flow of air can strike the buckets and cause the rotor to spin. The speed of a pneumatic rotor is on the order of 8000 rpm.

Electrically operated gyro rotors are actually the rotors of electric motors. The rotor is normally of a laminated soft-iron material using a squirrel-cage design inside a steel motor shell. The rotor can be operated by 12- or 24-V dc or 115-V ac, 400-Hz power sources. The speed of an electrically operated gyro can be in the area of 24 000 rpm. With this high rpm, the electric gyros can be smaller than the pneumatic gyros and still provide equal performance.

The central shaft of a gyro rotor is supported on fine bearing surfaces, with the materials and shape selected to keep friction to a minimum. Some bearing designs incorporate a spring mechanism to maintain the proper loading of the bearing against the shaft and compensate for wear and changes in temperature and the resulting variation in component size.

Electrically Powered Sensor Instruments

Many indicating systems rely on variations in electrical resistance to denote variations in fluid levels and temperatures. These types of instruments can be direct-indicating or can rely on Wheatstone bridge or ratiometer circuits to indicate changes in the condition of a system.

Resistance Circuits

Direct-indicating electrical **resistance systems** are often used in fuel-quantity transmitters. An example of such a system is shown in Fig. 17-10. Note that because the float level decreases as the fuel is used, the position of the wiper arm changes on the resistance wire. As the wiper moves up the resistance wire, the resistance in the circuit decreases and the

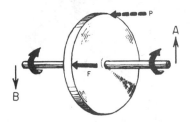

FIGURE 17-8 Principle of precession.

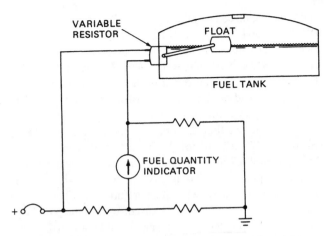

FIGURE 17-10 Float-type fuel-level indicator with variable resistor.

quantity gauge indicates a lower value. (Some systems are designed to have an increase in the circuit resistance as the quantity decreases.)

A similar type of position sensor can be used to monitor the position of aircraft components such as secondary flight controls and cowl flaps.

Wheatstone Bridge Circuit

A **Wheatstone bridge circuit** is a type often used to monitor operating temperatures that do not exceed 300°F [149°C]. This includes measurement of coolant temperatures, free-air temperatures, and carburetor-air temperatures. The use of a bridge circuit gives a more accurate reading than is possible with other direct-reading electrical-measurement circuits.

The circuit for a Wheatstone bridge is shown in Fig. 17-11. In this figure the bridge consists of three fixed resistors of 100 Ω each and one variable resistor whose resistance is 100 Ω at a fixed point such as 0°C, or 32°F. With the bridge circuit connected to a battery power source as shown, it can be seen that if all four resistances are equal, current flow through each side of the bridge will be equal and there will be no differential in voltage between points A and B. When the

variable resistor in the temperature-sensing bulb is exposed to heat, the resistance increases, which causes more current to flow through the top portion of the bridge than through the bottom portion. This causes a voltage differential between points A and B, and current then flows through the galvanometer indicator. Likewise, if the bulb is exposed to a low temperature, its resistance decreases, and current increases through the bottom portion of the circuit and through the indicator in the opposite direction. The instrument is calibrated to give a correct reading of the temperature sensed by the variable resistor in the sensing bulb. If any one segment of the bridge circuit should be broken, the indicator needle would move to one end of the scale, depending on which circuit was broken.

Ratiometers

Ratiometers are used where the variations in electrical voltage in the aircraft electrical system cause the bridge circuit to give inaccurate readings beyond acceptable limits. The ratiometer eliminates the voltage-induced error by using the aircraft electrical system to power two electromagnetic fields. Since both fields have the same power source, their comparative accuracy is not affected by variations in voltage.

The circuit for a ratiometer instrument is shown in Fig. 17-12. This circuit has two sections, each supplied with current from the same source. As shown in the drawing, a circular iron core is located between the poles of a magnet in such a manner that the gap between the poles and the cores varies in width. The magnetic flux density in the narrower parts of the gap is much greater than the flux density in the wider portions of the gap. Two coils are mounted opposite each other on the iron core, and both coils are fixed to a common shaft on which the indicating needle is mounted. The coils are wound to produce magnetic forces that oppose each other. When equal currents are flowing through the two coils, their opposing forces will balance each other, and they will be in the center position, as shown. If the current flowing in coil A is greater than that flowing in coil B, the force produced by coil A will be greater than the force of coil B, and coil A will move toward the wider portion of the gap,

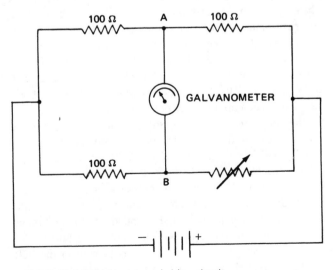

FIGURE 17-11 Wheatstone bridge circuit.

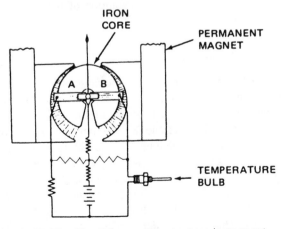

FIGURE 17-12 Circuit for a ratiometer-type instrument.

where a lower-flux density exists. This moves coil B into an area of higher-flux density at a distance sufficient to create a force that will balance the force of coil A. At this point the position of the indicating needle gives the appropriate temperature reading.

Like the sensing bulb for the Wheatstone bridge instrument, the bulb for the ratiometer instrument contains a coil of fine resistance wire, which increases in resistance as the temperature rises. The resistance wire is sealed in a metal tube and is connected to pin connectors for the electrical contacts.

FLIGHT INSTRUMENTS

Flight instruments are those instruments used by the pilot to determine the attitude, heading, altitude, and rate of changes in attitude, altitude, and heading of the aircraft. This portion of the chapter discusses the various flight instruments and their method of operation.

Altimeter

The **altimeter** is used to indicate the height of the aircraft above sea level. The instrument also provides information for determining the true airspeed of the aircraft, proper engine-power settings, proper clearance above the terrain, and proper flight altitude to avoid the flight path of other aircraft.

As altitude increases, the density of the air decreases, resulting in a decrease in pressure. Figure 17-13 shows the relationship between altitude and pressure on a standard day. A standard day at sea level is 29.92 in Hg [101.33 kPa] and 59°F. Note that the definition of a standard day includes not only pressure but also temperature. This is important because as temperature increases, so does pressure. All other things being equal, the pressure at sea level would be greater at 80°F than 59°F.

In general there are two types of altimeters that operate on the barometric principle: the nonsensitive, or simple, altimeter and the sensitive altimeter. The nonsensitive altimeter is rarely used in aircraft because of the very compressed altitude scale. This altimeter usually indicates a change of 10 000 ft [3048 m] for each revolution of the pointer, as shown in Fig. 17-14.

Sensitive Altimeter

The **sensitive altimeter**, so named because of its sensitivity, is the commonly used type of altimeter for aircraft and is illustrated in Fig. 17-15. To aid in understanding of how to read a sensitive altimeter, the face and pointers of the sensitive altimeter may be thought of in the same manner as the face and hands of a clock. One complete revolution of the longer pointer (or minute hand) registers 1000 ft [1304.8 m]. The face of the sensitive altimeter is typically indexed so this hand can easily show differences of 10 ft [3.05 m] in altitude. The shorter, wider pointer (or hour hand) rotates

one-tenth the distance of the longer pointer; hence it registers 10 000 ft [3048 m] for a complete revolution around the face of the instrument. Each numbered increment of the dial represents thousands of feet when read from this hand. The small hand and the center of the sensitive altimeter are arranged to read in increments of 10 000 or 40 000 ft [3048 or 12 192 m] for a complete revolution.

The sensitive altimeter in Fig. 17-15 indicates an altitude of about 1770 ft [540 m]. If the altimeter were showing 15 770 ft [4806.7 m], the smallest hand in the center would be slightly past the midpoint between the 1 and the 2 on the outer scale; the short, wide pointer (hour hand) would be pointing about three-fourths the distance from the 5 to the 6; and the longer pointer (minute hand) would point to the same position shown.

The pressure element of the sensitive altimeter consists of two or three diaphragm capsules in series, and there may be either two or three dial pointers. The pointers register hundreds, thousands, or tens of thousands, as explained in the preceding paragraphs.

Face and cutaway views of the interior of a Kollsman sensitive altimeter are shown in Fig. 17-15. This instrument utilizes three diaphragm capsules to impart movement to the rocking shaft as changes in altitude occur. The rocking shaft is linked to a sector gear, which drives the multiplying mechanism. Note that the rotation of the rocking shaft must be multiplied many times, since the large hand must rotate 50 times to indicate an altitude of 50 000 ft [15 240 m].

An altimeter must be adjusted for atmospheric pressure changes if it is to give true indications under all conditions. This compensation must be done by mechanical means by the pilot or another member of the flight crew with the **baronietric-pressure-setting** mechanism in the instrument. The setting device rotates the entire indicator drive mechanism within the case of the instrument and at the same time drives a barometric indicating dial that is visible through a window in the face of the instrument. When the adjusting knob is turned, the altimeter's dial face remains stationary, but the drive mechanism and pointers change position.

The reason for changing the barometric setting on an altimeter can be understood when the effect of barometric pressure on the altitude indication is considered. Assume that a certain airport is known to be at sea level with **standard** atmospheric pressure existing on a particular day. Before taking off in an airplane, the pilot sets the barometric scale to 29.92 in Hg [101.33 kPa] (standard sea-level atmospheric pressure). Then all altimeter pointers point to the zero mark on the altimeter scale. During the time the airplane is in the air, perhaps 2 h, the barometric pressure at the airport drops rapidly and is only 29.38 in Hg [99.49 kPa]. The pilot is ready to come in for a landing. If the pilot does not contact the field to obtain the corrected barometric pressure and leaves the altimeter at the setting made before taking off, the wheels of the airplane will hit the ground when the altimeter registers an altitude of approximately 500 ft [152 m].

The correct procedure, therefore, is to contact the field for the corrected existing barometric pressure, corrected to

TABLE OF U.S. STANDARD ATMOSPHERE					
ft	in of Hg	mm of Hg	psi	°C	°F
0	29.92	760.0	14.69	15.0	59.0
2,000	27.82	706.7	13.66	11.0	51.9
4,000	25.84	656.3	12.69	7.1	44.7
6,000	23.98	609.1	11.91	3.1	37.6
8,000	22.23	564.6	11.77	−0.8	30.5
10,000	20.58	522.7	10.10	−4.8	23.4
12,000	19.03	483.4	9.34	−8.8	16.2
14,000	17.58	446.5	8.63	−12.7	9.1
16,000	16.22	412.0	7.96	−16.7	1.9
18,000	14.95	379.7	7.34	−20.7	−5.1
20,000	13.76	349.5	6.76	−24.6	−12.3
22,000	12.65	321.3	6.21	−28.6	−19.4
24,000	11.61	294.9	5.70	−32.5	−26.5
26,000	10.64	270.3	5.22	−36.5	−33.6
28,000	9.74	237.4	4.78	−40.5	−40.7
30,000	8.90	226.1	4.37	−44.4	−47.8
32,000	8.12	206.3	3.99	−48.4	−54.9
34,000	7.40	188.0	3.63	−52.4	−62.0
36,000	6.73	171.0	3.30	−55.0	−69.7
38,000	6.12	155.5	3.00	−55.0	−69.7
40,000	5.56	141.2	2.73	−55.0	−69.7
42,000	5.05	128.3	2.48	−55.0	−69.7
44,000	4.59	116.6	2.25	−55.0	−69.7
46,000	4.17	105.9	2.05	−55.0	−69.7
48,000	3.79	96.3	1.86	−55.0	−69.7
50,000	3.44	87.4	1.70	−55.0	−69.7
55,000	2.71	68.8	1.33	Temperature Remains Constant	
60,000	2.14	54.4	1.05		
63,000	1.95	46.9	.907		
64,000	1.76	44.7	.86		
70,000	1.32	33.5	PSF 113.2		
74,000	1.09	27.7	77.3		
80,000	.82	20.9	58.1		
84,000	.68	17.3	47.9		
90,000	.51	13.0	35.9		
94,000	.43	10.9	29.7		
100,000	.33	8.0	22.3		

Source: The ARDC Model Atmosphere — 1959
in of Hg = inches of mercury
mm of Hg = millimeters of mercury
psi = pounds per square inch

°C = degrees Celsius
°F = degrees Fahrenheit
psf = pounds per square foot

FIGURE 17-13   Altitude-pressure relationship.

FIGURE 17-14 A nonsensitive altimeter. (*U.M.A., Inc.*)

sea level, before coming in for a landing. Upon receipt of the information in the foregoing case, the pilot sets the barometric scale at 29.38 in Hg [99.49 kPa] and can now expect the wheels to touch the ground when the altimeter indicates 0 ft altitude. When an altimeter is corrected for the existing altitude corrected to sea level, the altitude is said to be the "indicated altitude."

However, aircraft fly for extended periods of time at high altitudes and as a result may pass through a variety of weather conditions that could result in changes to the barometric pressure corrected to sea level. At high altitudes, a key factor in flight safety is aircraft separation, so it is important that all flight crews use the same pressure in the Kollsman window. To ensure that this is accomplished, all aircraft flying at or above 18 000 feet [5486 m] MSL are to adjust their Kollsman window to 29.92 in Hg [101.33 kPa]. Even if the setting is not technically correct, this procedure ensures that the error is consistent between aircraft and that the proper separation is maintained. The term *pressure altitude* is used to refer to altitudes in this situation. The altitude is referred to as a *flight level,* which is expressed in units of 100 ft [30.5 m]. Flight level 300 is a pressure altitude of 30 000 feet [9144 m].

Temperature compensation is automatically accomplished in the altimeter mechanism. Although the amount of error due to temperature change would normally be small because of the materials used in the construction of the mechanism, this small error is offset by a bimetallic strip that neutralizes any expansion or contraction of the parts.

In a standard instrument-panel configuration, the altimeter is mounted at the top right corner of the flight group. Static pressure is supplied to the altimeter from the static-pressure port through tubing. This tubing must be of sufficient strength to withstand the pressurization of the cabin in any pressurized airplane.

Compensated Altimeter

Because of the high speeds and altitudes at which many aircraft fly, the altimeters employed in such aircraft must be much more accurate than those used on aircraft designed to operate at more moderate altitudes. To provide an altimeter that will be sufficiently accurate for all conditions of operation, certain corrections and compensations must be made. First, an accurate altimeter must be corrected for installation or position error. This error applies to a particular airplane. Second, the instrument must have automatic compensation for temperature changes. Third, the changes in static pressure due to changes in Mach number must be compensated. An altimeter designed to make the necessary adjustments for accuracy is the mechanically compensated, pneumatically operated altimeter.

The mechanically compensated altimeter operates from the aircraft pilot–static pressure system and consists of a combination of altimeter and machmeter mechanisms, as shown in Fig. 17-16a. The indicated pressure altitude is sensed by an aneroid, and the Mach number is obtained from the combined motion of the **altitude-sensing aneroids** and a **differential-pressure-diaphragm capsule**. Aircraft-position-error information is applied to a two-dimensional cam and is algebraically added to the indicated pressure altitude by means of a mechanical differential. Thus, correction in feet of altitude is made to the altimeter indication for a given Mach number. This is required because of the change in static pressure at a given altitude for a change in Mach number. True altitude is computed by means of a mechanism actuated by the altitude- and Mach-sensing elements, as shown in Fig. 17-16.

As the aircraft changes altitude, the two aneroid capsules sense the change in static pressure. Deflection of these capsules is transmitted through the **bimetallic units** to the rocking shaft. Hence, as altitude changes, the rocking shaft is caused to rotate.

When the aircraft changes velocity, the ram air pressure on the pilot tube changes, and this pressure differential ΔP is sensed by the **differential capsule**. The deflection of the capsule is transmitted through a **link** to the link follower, which rotates a rocking shaft and a sector gear. This gear changes the position of the **altitude-correction cam**. This mechanism is shown in Fig. 17-16b. The correction applied by the mechanism is proportional to altitude error caused by changes in Mach number and is the input to the **altitude-correction differential**. The differential output is geared to the dials and pointers to read directly in corrected altitude.

Temperature compensation for the altimeter is provided through the operation of the bimetallic elements in the linkage between the aneroid capsules and the main rocking shaft. Calibration of the instrument is accomplished by making adjustments of the eccentric attachments between the bimetallic units and the rocking-shaft lever arms. The lever arms can also be rotated on the shaft to cause a change in the linkage starting angle.

The **barometric set** mechanism, shown in Fig. 17-16a, consists of an adjusting knob, gear train, cam, and rotating

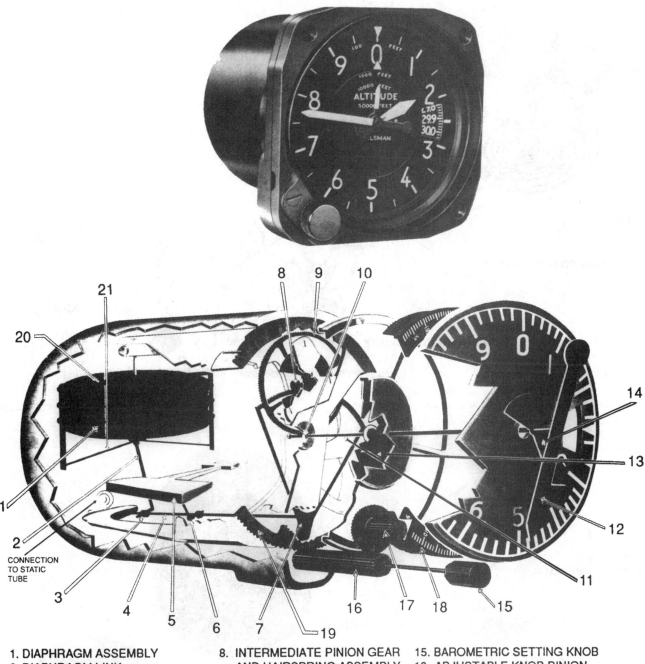

FIGURE 17-15 Face and cutaway views of a sensitive altimeter. (*Kollsman.*)

1. DIAPHRAGM ASSEMBLY
2. DIAPHRAGM LINK
3. DIAPHRAGM CALIBRATING ARM
4. ROCKING SHAFT
5. BALANCE
6. BALANCED CONNECTING LINK AND ARM
7. SECTOR GEAR

8. INTERMEDIATE PINION GEAR AND HAIRSPRING ASSEMBLY
9. INTERMEDIATE WHEEL
10. HAND-STAFF PINION
11. HAND STAFF
12. LARGER POINTER
13. REDUCING GEAR TRAIN FOR SMALL HAND
14. SMALL HAND

15. BAROMETRIC SETTING KNOB
16. ADJUSTABLE KNOB PINION
17. IDLER GEAR AND PINION
18. BAROMETRIC SETTING SCALE
19. MECHANISM BODY GEAR
20. THERMOMETAL COMPENSATOR BRACKET
21. PUSH RODS

CONNECTION TO STATIC TUBE

mechanism assembly. The range of ground-level pressure indications is 28.1 to 31.0 in Hg [713.7 to 787.4 mm Hg], with stops incorporated to limit the travel of the barometric counter to the range of desired operation.

Adjustment of the zero-setting system is accomplished by unlocking the knob shaft from the front of the instrument by a single screwdriver lock. This allows the pinion to disengage the counter gear train while still engaging the rotating frame assembly. The adjusting knob then rotates the pointers to the corrected setting.

The three-pointer dial display shown in Fig. 17-16c consists of two pointers and a disk with a pointer extension

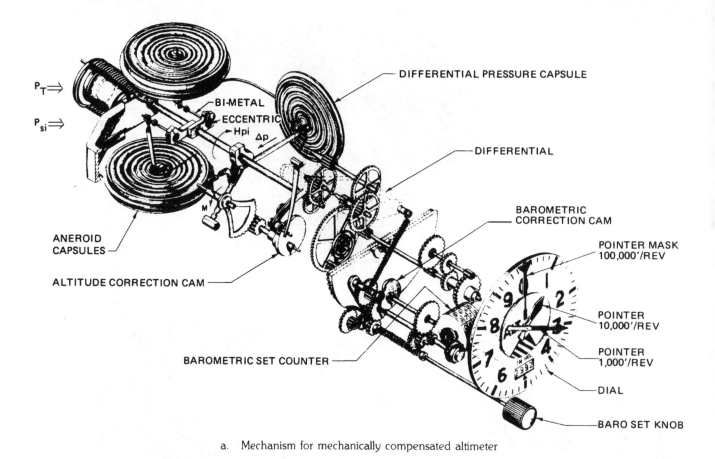

a. Mechanism for mechanically compensated altimeter

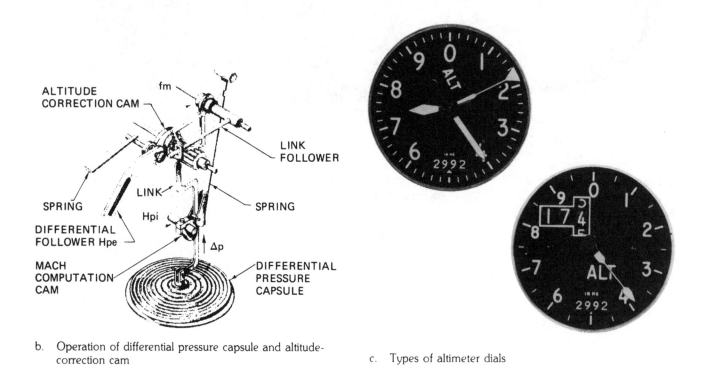

b. Operation of differential pressure capsule and altitude-correction cam

c. Types of altimeter dials

FIGURE 17-16   Bendix compensated altimeter.

indicating 1000, 10 000, and 100 000 ft [304.8, 3048, and 30 480 m] of altitude, respectively, for one revolution. An opening in the disk exposes a cross-hatched area, warning that the indication is below 15 000 ft.

If a counter-type readout is more desirable from the human engineering standpoint, it can be provided as an alternative to the three-pointer presentation. This dial is also shown in Fig. 17-16c and consists of a three-digit counter and a rotating pointer. The two large digits indicate thousands of feet altitude, whereas the smaller third digit and the pointer indicate hundreds of feet. The small digit rotates continuously with pointer rotation, whereas the two larger digits index with an intermittent motion. One complete revolution of the pointer indicates 1000 ft of altitude, thereby providing a sensitive indication of altitude change to the pilot. The dial presentation in the illustration indicates an altitude of 17 390 ft [5300.5 m].

Encoding Altimeter

Because of the need to determine accurately the position and altitude of aircraft in high-density traffic areas, the **encoding altimeter** has been developed to provide a signal for Air Traffic Control (ATC) that shows up on the ATC radar screen. The encoding altimeter includes electronic circuitry that operates in conjunction with the aircraft transponder. When the transponder is interrogated by ATC, the identification and altitude of the aircraft are shown on the radar screen.

Encoding altimeters are required on aircraft that fly in controlled airspace above 12 500 ft [3810 m] altitude, aircraft that are equipped with Category II and Category III landing systems, and aircraft operating in Group I and Group II Terminal Controlled Areas.

Encoding altimeters are available as **indicating instruments** and as **blind instruments**. Indicating instruments display the altitude indication on the face of the instrument and also contain the electronic circuitry necessary to provide the altitude signal for the transponder. Blind encoders may be located in any area of the aircraft where electrical power and a static pressure line are available. The blind encoder has no external display (hence the term "blind") and functions only as an electronic sensor of static air pressure and a signal transmitter to provide the transponder with altitude information.

Airspeed Indicator

An **airspeed indicator** is required on all certificated aircraft except free balloons. The purpose of the airspeed indicator is to show the speed of the aircraft through the air. The airspeed indicator is located in the upper left corner of a standard flight instrument grouping and is operated by a combination of pitot and static air pressure.

Some of the uses of the airspeed indicator are as follows:

1. It gives the pilot a definite indication of the attitude of the airplane with reference to the horizontal flight path. There will always be an increase in airspeed if the nose is down and a decrease in airspeed if the nose is high, provided that there is no change in throttle setting.

2. It assists the pilot in determining the best throttle setting for most efficient flying speeds.

3. Indications of airspeed are necessary in estimating or calculating ground speed (speed of the airplane with reference to positions on the ground).

4. Every airplane has certain maximum speeds recommended by the manufacturer, and without the airspeed indicator these limits of design might be exceeded without the pilot's being aware of it. This is particularly true with respect to flap and landing-gear extension speeds.

5. The airspeed indicator shows the correct takeoff and landing speeds, and it warns when the airplane is approaching stalling speed. It also indicates angle of attack, because higher speed generally means lower angle of attack.

The mechanism of an airspeed indicator is shown in Fig. 17-17. It consists of an airtight diaphragm enclosed in an airtight case, with linkages and gears designed to multiply the movement of the diaphragm and provide an indication on the dial of the instrument. The rear portion of the diaphragm is attached to the instrument case, and the forward portion is free to move. Unlike the diaphragm in an aneroid barometer, this one is provided with a connection for dynamic pressure from the pitot tube. Impact pressure from the pitot tube entering the diaphragm causes expansion that is transmitted through the multiplying mechanism to the indicating needle. Tubing connects the static side of the pitot head or other static ports to the instrument case, allowing static pressure to enter the case and surround the diaphragm. Thus the instrument actually measures **differential pressure** between

a. Pitot pressure connection	d. Rocking shaft
b. Diaphragm	e. Hairspring
c. Rocking-shaft arm	f. Sector gear

FIGURE 17-17 Mechanism of an airspeed indicator.

TABLE 17-1 Airspeed and Impact Pressure at Sea Level on a Standard Day

Airspeed, mph [km/h]	Ram Air Pressure			
	in Hg	in H$_2$O	psi	kPa
20 [32.2]	0.0146	0.1985	0.007	0.048
40 [64.4]	0.0582	0.7912	0.028	0.193
60 [96.6]	0.1309	1.7796	0.064	0.441
80 [128.8]	0.2330	3.1676	0.114	0.786
100 [161]	0.3646	4.9567	0.180	1.241
120 [193.2]	0.5263	7.1550	0.258	1.779
140 [225.4]	0.7178	9.7585	0.352	2.427
160 [257.6]	0.940	12.7793	0.462	3.185
180 [289.8]	1.193	16.2188	0.586	4.040
200 [322]	1.477	20.0798	0.725	4.999
250 [402.5]	2.331	31.6899	1.145	7.895
300 [483]	3.397	46.1822	1.673	11.535
350 [563.5]	4.67	63.4887	2.29	15.79
400 [644]	6.19	84.153	3.04	20.96
500 [805]	9.97	135.542	4.89	33.72
600 [966]	15.02	204.197	7.38	50.89

the inside of the diaphragm and the inside of the instrument case. Table 17-1 shows the relationship between airspeed and impact pressure on the pitot head under standard conditions at sea level.

In level flight, as the speed of the airplane increases, the impact pressure becomes greater, but the static pressure remains the same. This causes the diaphragm to expand more and more as the pressure increases inside. When the diaphragm expands, the rocking shaft picks up the motion and transmits it to the sector. The sector turns the pinion, which is fastened to the tapered shaft (hand-staff). The pointer is pressed onto this tapered shaft and moves around a calibrated dial. The hairspring keeps the linkage taut and causes it to follow the movement of the diaphragm as the airspeed increases or decreases. The dial on the face of the instrument is calibrated in **knots**, miles per hour, or both.

There are three types of airspeed of which the aviation maintenance technician should be aware: *indicated, calibrated,* and *true*. **Indicated** airspeed is the airspeed indicated on the instrument.

Calibrated airspeed is indicated airspeed corrected for instrument and sensor position error. This correction is a result of the location of the pitot tube and static ports and the changing airflow around these sensors with different aircraft flight attitudes, various flap positions, and door, window, and vent positions, as shown in Fig. 17-18. Each model aircraft and possibly each individual aircraft has a different set of calibrated airspeed values for specific indicated airspeed values.

Since the indicated and calibrated airspeeds are based on static and impact (pitot) air pressure, they indicate the true aircraft speed through the air only under standard day conditions at sea level. As the air density changes, the true airspeed of the aircraft changes. To find the true airspeed, the pilot must perform calculations with a flight computer or make use of a "true" speed airspeed indicator.

AIRSPEED CALIBRATION
NORMAL STATIC SOURCE

CONDITION:
Power required for level flight or maximum rated RPM dive.

FLAPS UP													
KIAS	40	50	60	70	80	90	100	110	120	130	140	150	160
KCAS	50	56	63	71	80	89	99	109	119	129	139	149	160

FLAPS 10°													
KIAS	40	50	60	70	80	90	100	110	---	---	---	---	---
KCAS	49	55	62	71	80	90	99	108	---	---	---	---	---

FLAPS 40°													
KIAS	40	50	60	70	80	85	---	---	---	---	---	---	---
KCAS	48	55	63	72	82	87	---	---	---	---	---	---	---

AIRSPEED CALIBRATION
ALTERNATE STATIC SOURCE

HEATER/VENTS AND WINDOWS CLOSED

FLAPS UP											
NORMAL KIAS	40	50	60	70	80	90	100	110	120	130	140
ALTERNATE KIAS	39	51	61	71	82	91	101	111	121	131	141

FLAPS 10°											
NORMAL KIAS	40	50	60	70	80	90	100	110	---	---	---
ALTERNATE KIAS	40	51	61	71	81	90	99	108	---	---	---

FLAPS 40°											
NORMAL KIAS	40	50	60	70	80	85	---	---	---	---	---
ALTERNATE KIAS	38	50	60	70	79	83	---	---	---	---	---

HEATER/VENTS OPEN AND WINDOWS CLOSED

FLAPS UP											
NORMAL KIAS	40	50	60	70	80	90	100	110	120	130	140
ALTERNATE KIAS	36	48	59	70	80	89	99	108	118	128	139

FLAPS 10°											
NORMAL KIAS	40	50	60	70	80	90	100	110	---	---	---
ALTERNATE KIAS	38	49	59	69	79	88	97	106	---	---	---

FLAPS 40°											
NORMAL KIAS	40	50	60	70	80	85	---	---	---	---	---
ALTERNATE KIAS	34	47	57	67	77	81	---	---	---	---	---

WINDOWS OPEN

FLAPS UP											
NORMAL KIAS	40	50	60	70	80	90	100	110	120	130	140
ALTERNATE KIAS	26	43	57	70	82	93	103	113	123	133	143

FLAPS 10°											
NORMAL KIAS	40	50	60	70	80	90	100	110	---	---	---
ALTERNATE KIAS	25	43	57	69	80	91	101	111	---	---	---

FLAPS 40°											
NORMAL KIAS	40	50	60	70	80	85	---	---	---	---	---
ALTERNATE KIAS	25	41	54	67	78	84	---	---	---	---	---

FIGURE 17-18 An airspeed calibration chart. (*Cessna Aircraft Co.*)

A true speed airspeed indicator is equipped with a rotating true airspeed scale next to the standard airspeed scale. The true airspeed scale is read through a window that is provided in the fixed scale. The movable scale can be positioned by the pilot to compensate for variations in temperature and pressure altitude by aligning the ambient temperature (indexed on the fixed scale) and the pressure altitude (located on the movable scale). The pilot can then read the "true" airspeed of the aircraft directly from the scale. A true speed airspeed indicator is shown in Fig. 17-19.

The dial or cover glass of an airspeed indicator is required to be marked with color arcs and radial lines to indicate various operating speeds.

FIGURE 17-19 A true speed airspeed indicator.

FIGURE 17-20 Typical airspeed indicator markings.

A *red radial line* is applied at the never-exceed speed (V_{ne}).

A *yellow arc* is applied through the caution range of speeds. This arc extends from the red radial line just mentioned to the green arc described in the next paragraph.

A *green arc* is placed to indicate the normal operating range of speeds, with the lower limit at the stalling speed with maximum weight and with landing gear and wing flaps retracted (V_{sl}) and the upper limit at the maximum structural cruising speed (V_{no}).

A *white arc* is used for the flap-operating range with the lower limit at the stalling speed at maximum weight with the landing gear and flaps fully extended (V_{so}) and the upper limit at the flaps' extended speed (V_{fe}).

For multiengine aircraft, a *blue radial line* is used for the one-engine-inoperative best-rate-of-climb speed (V_{yse}) and a *red radial line* indicates the one-engine-inoperative minimum control speed (V_{mc}).

These markings are normally placed on the face card of the instrument. Where the required markings are placed on the instrument cover glass, an indicator line must be painted or otherwise applied from the glass to the case so that any slippage of the glass is apparent. Figure 17-20 shows the markings on the face of an airspeed indicator.

Airspeed-Angle-of-Attack Indicator

The instrument whose dial is shown in Fig. 17-21 is called an airspeed-angle-of-attack indicator. Its function is to indicate airspeed, maximum allowable indicated airspeed, and angle of attack. **Indicated airspeed** (IAS) is shown on the dial by means of a pointer driven by a conventional pressure diaphragm. This diaphragm is labeled "Airspeed Diaphragm" in the schematic drawing of Fig. 17-22.

Angle-of-attack information is obtained by reading the indicated airspeed pointer against a segment rotating about the

FIGURE 17-21 Airspeed-angle-of-attack indicator. (*Kollsman.*)

periphery of the instrument's dial. This movable segment is positioned relative to the IAS pointer by the servo-mechanism, which responds to a signal from the **angle-of-attack sensor**. The IAS pointer therefore shows both indicated airspeed and angle of attack, the latter either in degrees or by markers showing best glide speed, and best cruise, approach, over-the-fence, and stall angles. The angle-of-attack sensor provides electric signals that control the servomechanism driving the angle-of-attack sector on the dial.

The maximum allowable pointer shows "maximum allowable" and/or "never exceed" indicated airspeed. It is

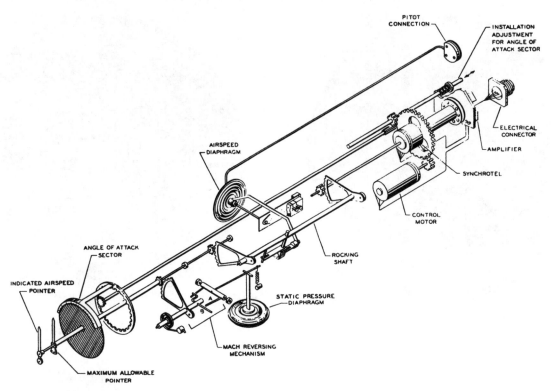

FIGURE 17-22 Schematic diagram of airspeed-angle-of-attack indicator. (*Kollsman.*)

actuated by a static pressure diaphragm and mechanism with a special calibration, so that the pointer reads maximum allowable indicated airspeed as a function only of absolute pressure. Maximum allowable indicated airspeed increases as altitude increases to about 25 000 ft [7620 m] and then decreases because it must not be allowed to approach Mach 1 (the speed of sound); that is, the airspeed is Mach-limited at altitudes above 25 000 ft [7620 m].

Machmeter

Because many of the modern jet aircraft fly near the speed of sound and some fly faster than the speed of sound, it is essential that such aircraft be equipped with an instrument that will compare the speed of the aircraft with the speed of sound. The flight characteristics of these high-speed aircraft change substantially as the speed of sound is approached or passed. These changes are associated with the occurrence of locally supersonic flow and shock waves. The result is that the pilot must substantially change the handling of the aircraft controls and retrim the airplane to meet the new conditions.

If an airplane is not designed to withstand the violent stresses that are sometimes imposed as a result of shock-wave formation during transonic flight, it may be severely damaged or may even disintegrate. In flying such an airplane, the pilot must know when the craft is approaching the speed of sound.

Since the speed of sound varies according to temperature and altitude, it is obvious that an airspeed indicator will not give the pilot an accurate indication of the airplane speed

relative to the speed of sound. It has, therefore, become necessary to use an instrument that gives the airplane speed in proportion to the speed of sound under the atmospheric conditions in which the airplane is flying. The instrument designed to accomplish this is the **Machmeter**, or Mach indicator, and it measures the ratio of true airspeed to the speed of sound. This ratio is called the **Mach number**, named for the Austrian physicist Ernst Mach (pronounced "mock"), who made the first studies of fast-moving projectiles.

The Machmeter, illustrated in Fig. 17-23a, shows the Mach number directly on its dial. For example, the number .7 is read as "Mach point seven," and it means that the airplane is flying at seven-tenths the speed of sound. The number 1.0 is read "Mach one" and means that the speed is equal to the speed of sound.

The Machmeter is similar in construction to an airspeed indicator; however, it includes an additional expanding diaphragm that modifies the magnifying ratio of the mechanism in proportion to altitude. As the altitude of the airplane increases, the altitude diaphragm expands and moves the floating rocking shaft. This movement changes the position of the rocking-shaft lever in relation to the sector lever, thus modifying the movement of the indicating needle to account for the effect of altitude. Figure 17-23b clearly shows the arrangement of the diaphragm with respect to the rocking shaft. Observe that expansion of the diaphragm will decrease the proportional movement of the indicating needle.

The ram air pressure is registered through a diaphragm that rotates the rocking shaft as it expands or contracts. Temperature correction is applied through a thermometal fork mounted on the altitude diaphragm.

(a)

(b)

FIGURE 17-23 Machmeter.

Vertical-Speed Indicator

Also referred to as a **rate-of-climb indicator**, a **vertical-speed indicator** is illustrated in Fig. 17-24. This instrument is valuable during instrument flight because it indicates the rate at which the airplane is climbing or descending. Level flight can be maintained by keeping the pointer of the instrument on zero, and any change in altitude is indicated on the dial in *feet per minute*. In this manner it assists the pilot in establishing a rate of climb that is within the prescribed limits of the engine. Likewise, when the pilot is coming in for a landing or descending to a lower altitude, the rate of descent can be controlled.

Like the airspeed indicator, the climb indicator is a differential-pressure instrument. It operates from the differential between atmospheric pressure and the pressure of a chamber that is vented to the atmosphere through a small, calibrated capillary restriction. Modern rate-of-climb indicators are "self-contained," with all the necessary mechanisms enclosed in the instrument case. However, some of the outside-cell or bottle types are still being used on older aircraft.

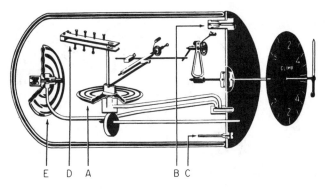

FIGURE 17-24 Vertical-speed indicator mechanism.

The rate-of-climb (vertical-speed) indicator illustrated in Fig. 17-24 shows the internal mechanism of the instrument. Changes in pressure due to changes in altitude are transmitted quickly through a large tube to the inside of a diaphragm A and slowly through an orifice assembly B and capillary C to the inside of the case. This creates a pressure differential that causes the diaphragm to expand or contract according to the rate of change of altitude. Adjustable restraining springs D are provided, which control the diaphragm deflection and permit accurate calibration.

The action of the diaphragm is transmitted through a lever-and-gear system to the pointer. As the plane assumes level flight, the pressures equalize and the pointer returns to zero. An overpressure diaphragm and valve E prevent excessive rates of climb or descent from damaging the mechanism.

It is characteristic of the rate-of-climb indicator to lag in its readings; that is, it does not respond instantaneously to changes in rate of climb. This lag occurs because it takes a few seconds for the differential pressures to develop after a change in rate of climb takes place. The more modern instruments used in recently manufactured aircraft have little or no lag.

Temperature changes have a tendency to cause variations in the flow of air through the capillary restrictors in the rate-of-climb instrument. Therefore, manufacturers have devised various methods of compensating the instrument for temperature. Likewise, altitude changes must be compensated for, owing to a tendency toward a greater lag when flying at higher altitudes.

Although the rate-of-climb indicator is one of the most sensitive differential-pressure instruments, it is not easily damaged during steep dives or violent maneuvers because of mechanical stops that have been incorporated in the instrument. Vertical-speed indicators are made with ranges up to 2000 ft/min [609.6 m/min] for aircraft that operate at slow rates of climb or descent and at more than 10 000 ft/min [3048 m/min] for higher-performance aircraft.

Variometers and **total-energy variometers** are sensitive vertical-speed indicators usually associated with gliders and sailplane operations. A variometer contains a sealed chamber with a balanced air vane inside, as shown in Fig. 17-25. One side of the chamber is connected to the static system of the aircraft. The other side is connected to an insulated

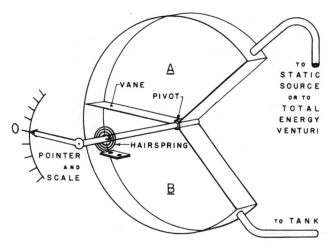

FIGURE 17-25 Diagram illustrating the operation of a variometer. (*Soaring Society of America.*)

air chamber, referred to as a "thermos bottle." When the aircraft increases in altitude, the air in the thermos bottle is at a higher pressure than the static air pressure, so the air flows out of the bottle. To get out of the bottle, the air must push the variometer vane out of the way. This works against the balance spring used to center the vane. The faster the aircraft is climbing, the greater the pressure differential and the higher the rate of flow out of the bottle. As the rate of flow varies, the amount of vane deflection varies. The vane drives an indicator needle on the instrument face and the rate of vertical speed is indicated in feet per minute or knots. If the aircraft descends, the static pressure increases. This causes the vane to deflect in the opposite direction and give an indication of a descent.

A total-energy variometer can be an electronic or mechanical device. It is designed to indicate the change in aircraft total energy (combination of potential and kinetic energy) by indicating a rate of climb or descent. This system is based on a variometer but includes connections to the pitot system. If the aircraft is at some fixed airspeed and altitude, it has some amount of total energy based on speed and distance above the ground. If the aircraft enters a rising current of air and maintains the same airspeed, the potential energy of the aircraft is increased by its increase in altitude. In this case the total energy system would indicate a climb. On the other hand, if the air were smooth with no vertical movement and the pilot pulled back on the control stick to use excess airspeed to increase altitude, the total energy of the aircraft would not change. The aircraft would only have exchanged kinetic energy for potential energy. The total-energy instrument would not indicate a climb.

Accelerometer

The **accelerometer** is frequently used on new airplanes during test flights to measure the acceleration loads on the aircraft structure. It serves as a basis for stress analysis because it gives an accurate indication of stresses imposed on the airplane during flight. Its function is to measure in gravity units the accelerations of gravity being exerted on the airplane. It is also used as a service instrument on many aerobatic airplanes to indicate any excessive stress that might have occurred during flight. If an airplane is designed to withstand loads of 6 g (six times gravity) and it is found after a particular flight that 8 g has been imposed on the airplane, it is necessary to subject the airplane to a very careful inspection to determine whether damage has been done as a result of the excessive loading. The most likely cause of excessive loads during flight is rough weather. Severe updrafts or downdrafts can cause very high loads on the aircraft, and the accelerometer gives an indication of the extent of the loads.

The accelerometer does not indicate any changes in velocity that take place in a line coinciding with the longitudinal axis of the aircraft. It indicates changes only along the aircraft vertical axis.

The accelerometer is usually graduated in gravity units up to plus 12 g, with −5 g as the lowest reading. Occasionally the instrument has a range from −1 to +8 g. Minus readings occur when the airplane is being nosed over into a dive or in level flight when a downward air current is encountered.

Vertical acceleration can be illustrated by carrying a heavy parcel in an elevator. When the elevator starts its downward motion, the parcel seems to lose weight. When the elevator assumes a steady rate of descent, the parcel again feels normal in weight. When the elevator slows down and just before it stops, the parcel seems to be excessively heavy. If these apparent changes in weight could actually be measured, the result would be the vertical acceleration of the parcel.

Some accelerometer instruments have two hands, but the more modern types are equipped with three hands. One hand measures the continuous acceleration, another measures the maximum acceleration reached at any time during the flight, and the third hand measures the minimum acceleration, or the largest minus reading, during the flight. The instrument is calibrated to read 1 g when the airplane is in normal flight or on the ground. This is the actual weight, or the vertical acceleration, of the airplane itself. This means that a load of 1 g is imposed upon the airplane as a result of its normal weight.

The principle of the accelerometer is illustrated in Fig. 17-26, where the mass weight is free to slide up and down on two mass shafts. As it moves up and down, it pulls

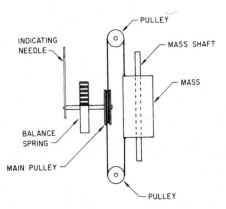

FIGURE 17-26 Diagram illustrating the principle of an accelerometer.

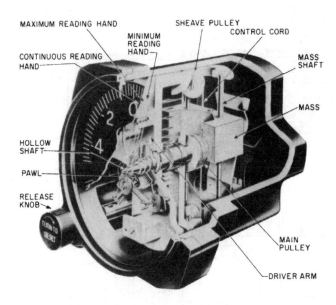

FIGURE 17-27 Cutaway view of an accelerometer.

on the cord that passes around three pulleys. The main pulley is attached to the shaft that carries the indicating needle. Thus, when the mass moves, it turns the main pulley and causes movement of the indicating needle. The mass is held in the 1-g position by the balance spring, which is of the flat-coiled type, like a clock spring. The inside of this spring is attached to the shaft on which the main pulley and indicating needle are mounted, and the outer end is attached to a stationary part of the case.

Figure 17-27 shows a cutaway view of an accelerometer. In this instrument the main shaft is enclosed in a hollow shaft that is somewhat shorter than the main shaft. Attached to this hollow shaft is the maximum-reading hand. Also fixed to the hollow shaft is a ratchet. As the main shaft rotates, a driver arm pushes a small tab that is a part of the ratchet, thus turning the hollow shaft along with the main shaft. This means that both the continuous-reading hand and the maximum-reading hand move simultaneously up to the point where the vertical acceleration on the weight is counterbalanced by the calibrating spring. The maximum-reading hand is stopped on the high point by means of the pawl, which engages one of the ratchet teeth. Obviously, the ratchet and hand will stay in this position until such time as a greater vertical acceleration is produced to move it farther around the scale. When the knob in front of the instrument is turned, the pawl is disengaged and the maximum-reading hand can return to its normal position.

The mechanism described in the foregoing section is only one of the many types of accelerometers, but in all cases the weight or pendulum weight works against a calibrated spring that determines the degree of travel of the mass weight. In inertial guidance systems for missiles and spacecraft and aircraft inertial reference systems, accelerometers are used to develop signals indicating what degree of acceleration is taking place. In a stable platform for an inertial guidance and reference system, there are usually three accelerometers: one for vertical acceleration, one for longitudinal acceleration, and one for lateral acceleration. These are arranged so that the mass can move up and down, fore and aft, and from side to side, respectively. The accelerometer signals are sent to the computer section of the system, where correction signals are developed.

Magnetic Compass

The **magnetic compass** is an independently operating instrument described here as mechanical because it requires no power from any aircraft source. The construction and face of a typical magnetic compass are shown in Fig. 17-28.

The magnetic compass is a comparatively simple device designed to indicate the direction in which an aircraft is headed. This indication is accomplished by one or more bar magnets mounted on a pivot with a circular **compass card**, which is usually marked in increments of 5°. The movable assembly is contained in a compass bowl filled with a light petroleum oil, which dampens oscillations and lubricates the pivot bearing.

The compass card assembly is mounted on a jeweled pivot bearing on the jewel post assembly. Immediately below the card can be seen the ends of the two parallel, permanent bar magnets that provide the magnetic direction indication. An expansion chamber is built into the compass to provide for the expansion and contraction of the liquid that results from changes in temperature and altitude. The flexible diaphragm moves as required to provide more or less space for the liquid. Between the aneroid back and the diaphragm is the air chamber, which permits the expansion and contraction. The liquid in the compass bowl (case) dampens the oscillations of the card, thus preventing the card from turning violently and vibrating. Compensating magnets are in the compensator assembly at the bottom of the case.

The compass provides an indication of *magnetic* direction rather than *true* direction. The bar magnet (or magnets) in the compass aligns itself (or themselves) with the magnetic lines of force in the earth's magnetic field. Since the magnetic poles of the earth are not located at the geographical poles, the magnetic compass reading is subject to **variation**, depending upon its location on the earth's surface. This variation must be taken into account by the pilot who is navigating the aircraft. For example, in the extreme western portion of the United States, the variation may be from 15 to 20° east of true north. If the variation is 15°E and the compass reading is 280°, the pilot must subtract 15° from 280° to find the true heading.

Magnetic compasses are subject to **deviation**, which is an inaccuracy caused by magnetic influences in the airplane. These are caused primarily by the proximity of magnetic materials such as iron and steel and by current flowing in nearby electrical circuits. Deviation is corrected as much as possible by the use of small **compensating magnets** mounted in the case. These magnets are adjustable to correct for north-south deviation and east-west deviation. The process of compass compensating is done by "swinging" the compass on a **compass rose**. The compass rose is a circle on which are marked magnetic directions in degrees, the 0°

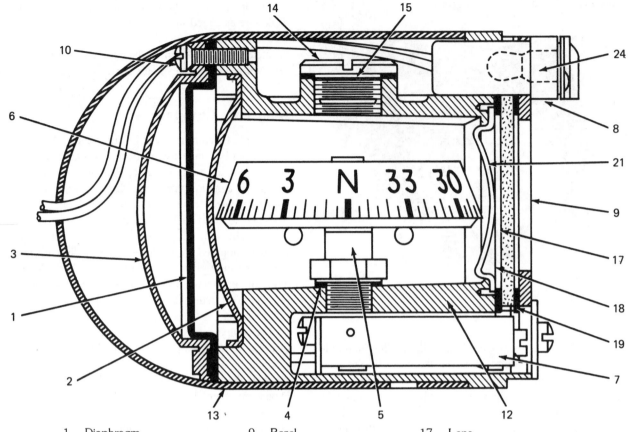

1. Diaphragm	9. Bezel	17. Lens
2. Diaphragm stop	10. Bezel and aneroid back screw	18. Bezel sealing gasket
3. Aneroid back		19. Bezel spacer gasket
4. Jewel post washer	11. Right coverplate screw	20. Compensator coverplate
5. Jewel post	12. Case	21. Lubber line wire
6. Compass card	13. Streamline housing	22. Light screw
7. Compensator assembly	14. Filler plug	23. Left coverplate screw
8. Light	15. Filler plug washer	24. Lamp
	16. Insert nut	

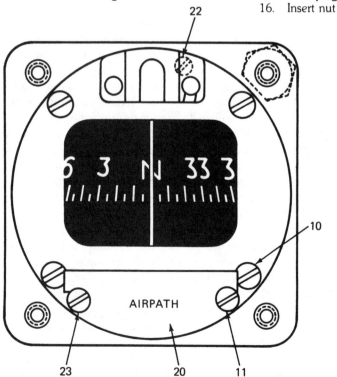

FIGURE 17-28 Construction of a magnetic compass. (*Airpath Instrument Co.*)

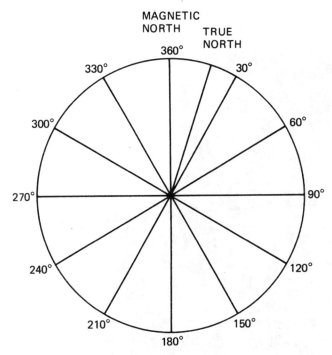

FIGURE 17-29   A typical compass rose.

FIGURE 17-30   Dial of a magnetic direction indicator.

mark showing the magnetic North direction. A typical compass rose is illustrated in Fig. 17-29.

Instructions for compensating a compass vary to some extent, depending upon the airplane involved and the equipment installed. Typical instructions are as follows.

Before attempting to compensate the compass, every effort should be made to place the aircraft in simulated level flight conditions. Check to see that the doors are closed, flaps are in the retracted position, engines are running, throttles are set at cruise position, and the aircraft is in level-flight attitude. All electrical switches, alternators, and radios should be on.

1. Set adjustment screws of the compensator on zero. The adjustment screws are in zero position when the dot of the screw is lined up with the dot of the frame.

2. Head the aircraft in a magnetic north direction. Adjust the N-S adjustment screw until the compass reads exactly north (0°). Use only a nonmagnetic screwdriver.

3. Head the aircraft in a magnetic east direction and adjust the E-W screw until the compass reads exactly east.

4. Head the aircraft in a magnetic south direction and note the resulting south error. Adjust the N-S screw until one-half the error has been removed.

5. Head the aircraft in a magnetic west direction and adjust the E-W screw to remove one-half the E-W error.

6. Head the aircraft in successive magnetic 30° headings and record the compass readings on the appropriate **deviation card.** Deviations must not exceed +10° on any heading.

The compass on an airplane should be checked (swung) whenever there is an installation or removal of any radio equipment or any other equipment that could have an effect on the compass. In addition, the compass and correction card

should be checked for deviation at least once a year at the annual inspection.

The compass is usually installed on the centerline of the fuselage. It is normally mounted on the top of the instrument panel or suspended from a mount on the windshield or near the top of the windshield. These mounting locations are chosen so that the compass is as far away as possible from items that might affect its operation, such as the radios and electrical circuits, but still in easy view of the pilot. Only rarely is the compass mounted in the instrument panel with the other instruments.

Vertical-Dial Compass

The vertical-dial direction indicator is actually a direct-reading magnetic compass; instead of the direction's being read from the swinging compass card, however, the reading is taken from a vertical dial, as shown in Fig. 17-30.

The reference index on the vertical dial can be set at any desired heading by turning the knob at the bottom of the dial, and it is necessary only to match the indicator needle with the reference pointer to hold a desired course. The design of the dial provides easy reading of direction and also quick and positive indication of deviation from a selected heading. The compass liquid in the instrument is contained in a separate chamber, and the dial, instead of floating in the liquid as in other types of magnetic compasses, is completely dry. This makes the indicator easier to read and eliminates fluid leakage around the dial.

An important feature of the vertical-dial direction indicator is its stability in comparison to the conventional compass. Both the period and the swing of this indicator are less than half that of the ordinary magnetic compass.

The construction of the vertical-dial direction indicator is shown in Fig. 17-31. The float assembly (20), which contains the directive magnet (27), is located in a fluid-filled bowl. This assembly rides on a steel pivot (23), which in turn rests on a stud containing a cup jewel (24). The jewel stud rides in a guide and is supported on a spring (26), which absorbs external vibration. A change in the position of the instrument in relation to the directional magnet transmits this motion to the indicator needle (8) through the follower magnets (18), the bevel gears (16), and the horizontal bevel-gear shaft (11) onto which the indicator needle is pressed.

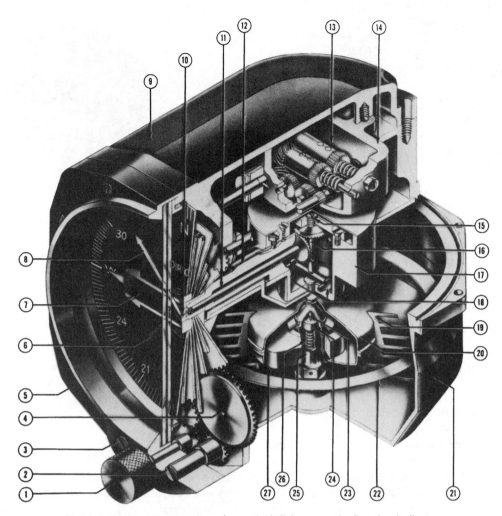

FIGURE 17-31 Construction of a vertical-dial, magnetic direction indicator.

The follower magnets and bevel gears are all contained in the housing (17). Both bevel-gear shafts are carefully balanced and ride on steel pivots in jeweled bearings.

A compensation system [polycompensator (13)] is mounted directly over the bevel-gear housing. The compensation-magnet gear assemblies are turned through gear trains to two slotted adjustment stem pinions (2 and 3). The reference lines (7) are set through a gear train by means of the adjustment knob (1).

As temperature varies, expansion of the compass liquid surrounding the float assembly takes place through a small hole in the baffle plate, and the level of the liquid in the expansion chamber above the baffle plate will rise and fall with changes in temperature. The normal level of the liquid is about even with the hole, which is used to fill the chamber.

The indicator is equipped with rim lighting, which utilizes a special diffusing lens at the lamp socket to distribute the light evenly over the entire dial. Three-volt lamps are used and are accessible from the front to facilitate replacement.

Flux Gate Compass System

The Gyro Flux Gate compass system is one type of remote-indicating earth-inductor compass system. Its advantage is that it is comparatively free of the disadvantages of the standard magnetic compass. It consists of a flux gate (flux valve) transmitter, master direction indicator, amplifier, junction box, caging switch, and one or more compass repeaters (see Fig. 17-32).

The **flux gate** is a special three-section transformer, which develops a signal whose characteristics are determined by the position of the unit with respect to the earth's magnetic field. The flux gate element consists of three highly permeable cores arranged in the form of an equilateral triangle with a primary and secondary winding on each core. The primary winding of the flux gate is energized by a single-phase 487.5-Hz, 1.5-V power supply from an oscillator. This current saturates the three sections of the core twice each cycle. During the time that the core is not saturated, the earth's magnetic flux can enter the core and affect the induction of a voltage in the secondary. The effect of the primary excitation is to *gate* the earth's magnetic flux into and out of the core. This cycle occurs twice during each cycle of the excitation current; hence, the voltage induced in the secondary windings of the flux gate has twice the frequency of the primary current, because the core is saturated twice each cycle and has no excitation flux twice each cycle. The resulting voltage in the secondary windings has a frequency of 975 Hz.

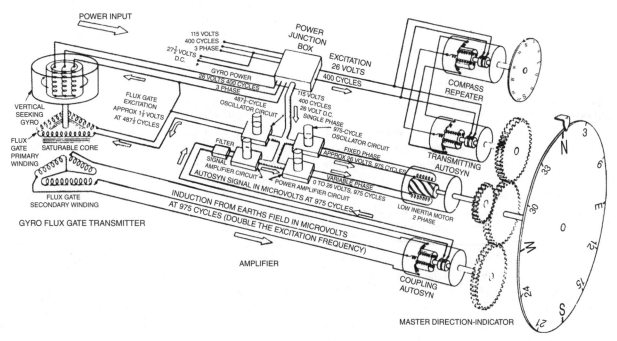

FIGURE 17-32   Diagram of the Gyro Flux Gate system. (*Bendix Corp.*)

The 975-Hz secondary signal is developed in the flux gate, as illustrated in Fig. 17-33. At point 1 in the time cycle, the excitation voltage is zero, the earth's magnetic flux in the core of the flux gate is maximum, and the induced signal in the secondary is zero. There is no induced secondary signal because there is no flux change in the core at this instant. Between points 1 and 2, excitation voltage in the primary increases from zero to maximum. This saturates the core and causes the earth flux in the core to change from maximum to minimum, thus inducing a voltage in the secondary, which rises to a maximum (A) in one direction and decreases to zero when there is no further change at point 2.

From point 2 to point 3 the earth flux increases to maximum; hence, the secondary voltage at B is in a direction opposite that developed at A. So, there are two cycles of induced voltage in the secondary for each one cycle of primary excitation voltage.

The ratios of the output voltages in the three secondary coils are determined by the position of the flux gate element with respect to the earth's magnetic field, with only one possible position for any combination of voltages.

In the **transmitter**, shown in Fig. 17-34, the flux gate is held in a horizontal position by means of a gyro. This is necessary to provide a uniform signal from the flux gate element. If the gyro should be *tumbled*, that is, moved out of the horizontal position, it can be erected by means of the **caging switch**.

The 975-Hz output of the secondary windings of the flux gate is connected by three leads to the stator of a **coupling Autosyn** in the **master direction indicator**. The currents in these leads set up a magnetic field in the stator of the coupling Autosyn similar in direction and strength to that appearing in the flux gate element. This field shifts its position whenever

FIGURE 17-33   Development of the signal in a flux gate. (*Bendix Corp.*)

FIGURE 17-34   Flux gate transmitter. (*Bendix Corp.*)

any change occurs in the stator voltages as a result of a change in the heading of the aircraft.

The rotor of the coupling Autosyn normally lies in a null (neutral) position; in this position no signal is induced in it by the stator field. When the stator field shifts as a result of a change in aircraft heading, the rotor will no longer be in the null position, and a signal will be induced in the rotor winding. This signal is fed to the amplifier and then to the variable phase of a low-inertia, two-phase motor in the master direction indicator. This causes a torque to be produced, and the motor turns in a direction determined by the phase of the current in the variable-phase winding. The motor shaft turns the indicator dial to provide a visual indication of the magnetic heading of the aircraft; at the same time, it drives the rotor of the coupling Autosyn to a new null position. This reduces the rotor signal to zero, and the low-inertia motor stops turning until another change in the heading of the aircraft causes a new signal to be induced in the rotor of the coupling Autosyn.

The **master direction indicator,** shown in Fig. 17-35, is the indicating unit for the Gyro Flux Gate system. It contains the coupling Autosyn, the low-inertia, two-phase motor, a transmitting Autosyn, and a course Autosyn synchro to supply a signal for the operation of the automatic pilot.

The **compass repeater** (Fig. 17-36) contains a single Autosyn, which receives its signal from the transmitting Autosyn in the master direction indicator. The dial of the repeater shows the four cardinal headings as well as intermediate headings marked every 5° and numbered every 30°.

When it is desired to provide direction information at more than one location, the compass repeater is used. The repeater is a single Autosyn receiver having an indicating dial attached to the rotor shaft. As stated before, the repeater receives signals from the Autosyn transmitter in the master direction indicator and responds by reproducing the heading information on its dial.

The **amplifier** serves to increase the power of the magnetic direction signal from the coupling Autosyn in the master direction indicator. It also furnishes a 487.5-Hz power supply for the excitation of the primary windings of the flux gate transmitter and a 975-Hz power supply for the excitation of the fixed phase of the motor of the master direction indicator.

The caging switch is used to control the cage-uncage cycle of the gyro in the flux gate transmitter. The gyro is said to be **caged** when it is held in a fixed position by mechanical means; it is **uncaged** when it is allowed complete freedom of motion.

The need for caging a gyro may be easily understood. The gyro in the flux gate transmitter must spin on an axis perpendicular to the earth's surface to keep the flux gate in the correct position with respect to the earth's flux. Since the gyro is free to move away from the vertical axis with respect to the airplane in which it is installed, there are many times when it is not in the correct position for operation. At these times it is necessary to erect the gyro by means of the caging switch.

To erect the gyro, the MOMENTARY-CONTACT-SWITCH button on the switch box is depressed and released. This starts the cage-uncage cycle, which then continues automatically until the cycle is complete. An indicator light on the switch box comes on when the gyro is caged.

FIGURE 17-35   Master direction indicator. (*Bendix Corp.*)

FIGURE 17-36   Compass repeater. (*Bendix Corp.*)

Directional Gyro

The **directional gyro** is a gyro-operated directional reference instrument designed to eliminate some of the problems associated with the magnetic compass. The gyro does not seek the north pole; however, it will continue to tell the pilot of an aircraft whether the aircraft is holding a particular heading. The directional gyro must be reset by the pilot to the magnetic compass indication from time to time to correct for precession or drift after a period of operation, usually after every 15 min.

The airborne directional gyro consists of a gyro rotor mounted in a set of gimbals so that the position of its spin axis can be maintained independent of the case of the instrument. The spin axis is normally horizontal. The axis of the inner gimbal is also horizontal and is perpendicular to the spin axis. The outer gimbal is pivoted on a vertical axis, so that all three axes are mutually perpendicular. It is the relative angle between the outer gimbal and a reference point on the case that is measured by the instrument and presented on the dial. Since the spin axis is horizontal and the case can be considered aligned with the directional axis

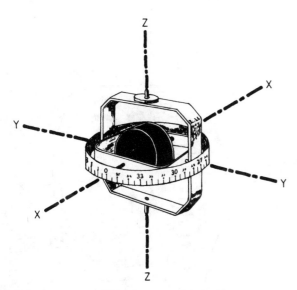

FIGURE 17-37   Simplified drawing of a directional gyro.

FIGURE 17-38   A modern heading indicator. (*Edo Corp.*)

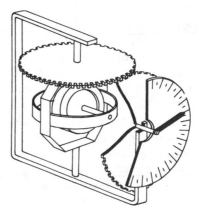

FIGURE 17-39   Internal components of a heading indicator.

of the aircraft, the angle presented on the dial is also the angle between the heading of the aircraft and the direction of the spin axis of the gyro. Any change in the heading of the aircraft is indicated on the dial, since the position of the gyro is rigid in space and is not affected by the motion of the aircraft.

The principal parts of a directional gyro are illustrated in Fig. 17-37. Observe the gyro with the X axis, the inner gimbal with the Y axis, and the outer gimbal with the Z axis. When the gyro is spinning, it will maintain its rigid position in space even though the case of the instrument is rotated about it in any direction. When the instrument is mounted in an aircraft and in operation, it is easy to understand that as the aircraft turns to the right or left, there will be a relative movement between the dial mounted on the outer gimbal and the instrument case. The amount of this movement will indicate to the pilot the number of degrees through which the airplane has turned.

The designs of the dial faces for directional gyros vary widely, but they all indicate a value in degrees from 0° to 360°. A knob is provided on the instrument that permits the reading to be changed by the pilot. The reading is usually made to coincide with the average reading of a magnetic compass. The directional-gyro instrument is used as a reference during flight instead of the magnetic compass, since the magnetic compass fluctuates and oscillates considerably during maneuvers. The directional gyro is a stable reference: hence the readings will be comparatively stable for normal flight maneuvers.

Heading Indicator

A **heading indicator** is a refined version of the directional gyro. Its principles of operation are the same as for the directional gyro, but the display presentation is a compass card, (see Fig. 17-38). This compass card display allows the pilot to visualize the heading of the aircraft in relation to heading changes that may be required. With this visualization,

the mental workload of the pilot is reduced and flying by reference to the instruments is made easier. Figure 17-39 shows a simplified cutaway of a heading indicator.

As mentioned previously, the gyros used in various instruments can be driven by air or by means of an electric current. If the gyro is electrically driven, the gyro itself is a part of the electric motor. Figure 17-40 is a photograph of the rotor mechanism in an air-driven directional gyro. To drive this gyro, the case of the instrument is connected to a vacuum or pressure source. The resulting airflow through the instrument is directed onto the buckets machined on the rim of the rotor. Within a short time after the airflow is started, the rotor is turning at a high rate of speed.

Gyro instruments must necessarily be limited in the degree of movement through which the gimbals can travel, except in the case of the directional gyro, where the Z axis can turn through a complete circle. If the degree of permitted movement is exceeded because of violent maneuvers of the airplane, the gyro rotor will be moved out of its normal

FIGURE 17-40   Rotor mechanism in a directional gyro.

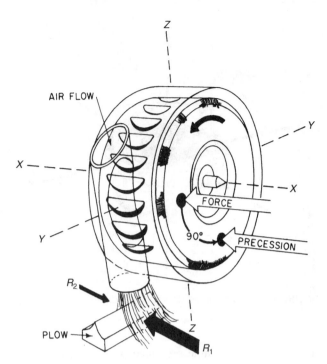

FIGURE 17-41   Utilizing air to level a gyro.

position of rigidity, and it is then said to be **tumbled**. In order to restore the gyro to its correct alignment, a **caging** mechanism is installed. When the gyro is caged, it is locked into its correct position mechanically. The caging mechanism must then be released to permit the gyro to function normally.

To be most effective as an indicator of azimuthal direction, a directional gyro should have its spin axis horizontal. In more complex equipment this would be accomplished by slaving the gyro gimbals to a gravity-detecting device. However, in order to keep the construction as simple as possible, most directional gyros are leveled so that the inner gimbal is perpendicular to the outer gimbal. This is usually satisfactory, since the instrument is installed in the aircraft with its outer gimbal approximately vertical. This means that in straight and level flight, the spin axis is about horizontal. Furthermore, gimbal error and X-axis frictional effects are at a minimum when the gimbals are perpendicular.

In the past, several devices have been used to obtain perpendicularity, or the level position of the gyro. When a gyro is air-driven, the jets of air that spin the rotor are used to keep it level. In one model, there are two parallel jets that are directed at the buckets on the gyro. If one of the spin axes rises, the buckets move out of the jet stream until one of the jets is directing its stream of air at the side of the gyro. This jet then applies force about the vertical axis, thus causing the gyro to precess back to the horizontal position. One of the faults of this system is that the jet also tends to apply a force about the horizontal axis. This causes the gyro to precess in azimuth and causes an erroneous heading indication.

In a later design for an air-driven gyro, the airstream flows over the top of the rotor and around in the direction of spin toward the lower portion of the vertical gimbal. On this gimbal there is a **plow** in the form of a wedge, which splits the airstream into two sections. This method is shown in Fig. 17-41.

If the gyroscope is level, the force on each face of the plow is equal; but if the gyro is titled, then the airstream strikes only one side of the plow. The resulting unbalanced force attempts to rotate the outer gimbal about its vertical axis, but instead the gyroscope is processed back to its level position.

Another air-driven gyro instrument employs **pendulous vanes**, which act as air valves to direct air for correction of position. As long as the gyro is level, the airstreams are balanced. If the gyro shifts position, the vanes move by gravity to a position that changes the airflow and applies a corrective force.

Electrically driven directional gyros are leveled by means of a torque motor located above the outer gimbal. This motor, by applying torque to the gimbal about its vertical axis, causes the gyro to be precessed to a level position. The motor is usually energized by a switch located between the inner and outer gimbals.

The directional gyro is usually mounted at the bottom center position of the flight instrument group.

Horizontal Situation Indicator

A **horizontal situation indicator,** or HSI, is an instrument that combines the information supplied by a heading indicator with radio navigation information. This provides more flight information for the pilot on one instrument and reduces the eye motion required when flying by reference to instruments. This type of instrument also reduces both the number of instruments that must be installed and the demand on panel space, since one instrument can replace two or more older displays.

The simplest HSI has an electrically or pneumatically driven gyro for the heading indicator portion of the display. Information from the lateral navigation radio is displayed through a deviation bar and a selected-course pointer, as

FIGURE 17-42 A horizontal situation indicator. (*Edo Corp.*)

shown in Fig. 17-42. Information from the glideslope radio receiver is displayed on the left side of the unit. "Flags," such as NAV, HDG, and GS, are displayed whenever that function is inoperative. For example, if the HDG flag is visible, the gyro is not operating properly. When the gyro is operating properly, this flag is retracted out of sight.

The more sophisticated HSI displays incorporate lateral navigation and glideslope information, distance-measuring-equipment information, and bearing pointers to indicate the location of other navigation ground stations in relation to the aircraft. An example of this type of display is given in Fig. 17-43.

Artificial Horizon

The **gyro horizon indicator**, also called the **artificial horizon**, **attitude indicator**, or **attitude gyro**, is designed to provide a visual reference horizon for an airplane that is flying "blind"—that is, with no visible reference outside the airplane. This instrument provides a horizontal reference similar to the natural horizon, thus making it possible for the pilot of an aircraft to see the position of the aircraft with respect to the horizon. A white bar across the face of the instrument represents the horizon, and a small figure of an airplane in the center of the dial represents the aircraft. The position of the airplane symbol relative to the horizon bar indicates the actual position of the aircraft with respect to the natural horizon. The face of an old-style artificial horizon is shown in Fig. 17-44. A more modern display is shown in Fig. 17-45.

A properly installed gimbaled gyroscope can measure motion about any axis except its spin axis. For example, the directional gyro has its spin axis maintained in a horizontal position, so that any change in the heading of the aircraft (motion about the Z axis) can be detected, but this

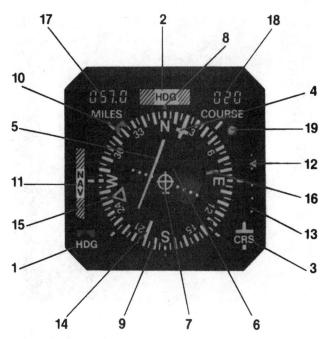

(*Aeronetics*)
1. Heading select knob.
2. Heading warning flag.
3. Course select knob.
4. Selected course pointer.
5. NAV lateral deviation bar.
6. To-From indicator.
7. Symbolic aircraft.
8. Lubber line.
9. Compass card.
10. Heading select marker.
11. NAV warning flag.
12. VERT deviation indicator.
13. VERT warning flag.
14. Reciprocal course pointer.
15. RMI bearing pointer.
16. RMI bearing reciprocal pointer.
17. Digital DME display.
18. Digital course display.
19. Synchronizing annunciator.

FIGURE 17-43 Components of a horizontal situation indicator. (*Aeronautics.*)

instrument cannot measure roll motion about the X axis. In order to detect pitch-and-roll motion of an aircraft, a gyro is used with a vertical spin axis and horizontal gimbal axis. The gyro horizon indicator is one of several instruments with such a vertical gyroscope. With the spin axis of the gyro maintained in a vertical direction, roll motion is detected by motion of the case about the outer gimbal (Z axis) and pitch motion is detected between the inner and outer gimbals (Y axis).

A gyro erection mechanism is incorporated into the design of an artificial horizon. The purpose of this mechanism is to correct for any force such as precession, which might cause the gyroscope to tilt out of its vertical plane. Two methods commonly used for this purpose are the **ball erector** and the **pendulous vane** mechanisms.

FIGURE 17-45 A modern artificial horizon. (*Edo Corp.*)

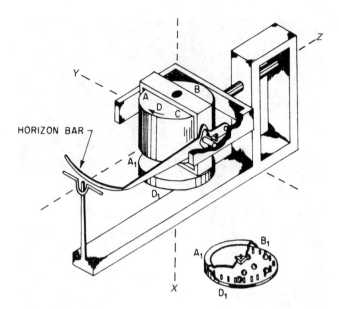

FIGURE 17-46 Gyro horizon indicator with ball erector.

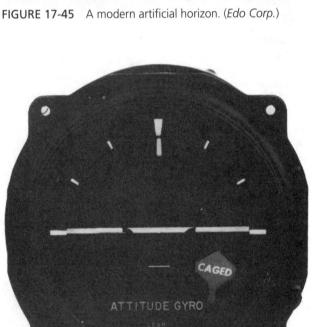

FIGURE 17-44 Face of a gyro horizon indicator.

The ball erector uses a concave plate attached to the bottom of the gyro support. When the gyro tilts out of the vertical, metal balls on the plate move to the low side of the plate. The movement of the balls causes a force to be exerted on the gyro axis, which causes the gyro to return to the vertical position. Figure 17-46 shows the location of the ball erector along with a simplified view of an artificial horizon.

A pendulous vane mechanism can be used with pneumatically operated gyroscopes. A pendulous cover plate partially blocks the escape of the operating airflow from the gyro. As the axis of the gyro tilts out of the vertical, the plates either cover more of the opening or less of the opening, depending on the direction of gyro tilt. The resulting difference in airflow causes the gyro to return to the vertical position. The operation of a pendulous vane mechanism is shown in Fig. 17-47.

The artificial horizon can be operated electrically or pneumatically. The instrument is usually mounted in the top center of the flight-instrument group. Sophisticated versions of the artificial horizon, such as the one shown in Fig. 17-48, may include more than just pitch and bank attitude information. Additional information may include localizer and glideslope deviation needles, radar altimeter information, steering bars, and an inclinometer. Instruments with all of those features are called **flight directors**.

Turn-and-Bank Indicators

The **turn-and-bank indicator**, sometimes called the "needle and ball," actually consists of two instruments. The "needle" part of the instrument is a gyro instrument that indicates to the pilot the rate at which the aircraft is turning. The "ball" part of the instrument is an inclinometer, which denotes the quality of the turn by indicating whether or not the turn is coordinated.

The turn-indicating section of the turn-and-bank indicator is actually a **rate gyro**, and it produces an indication in proportion to the **rate of turn**. The face of the instrument normally includes marks, or "doghouses," on each side of the centered needle position. When the needle is pointing at one of these marks, the aircraft is in a standard-rate turn of 3°. Some rate indicators are designed to indicate a turn at one-half of standard rate when the needle points at the "doghouse." These instruments can be identified by the words "4 MINUTE TURN" at the bottom of the instrument, meaning that a 360° change in direction will take 4 min rather than 2 min. These 4-min indicators are normally

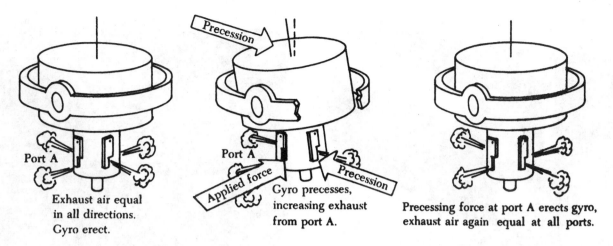

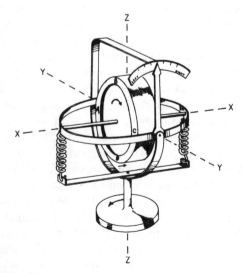 Port A

Exhaust air equal
in all directions.
Gyro erect.

Gyro precesses,
increasing exhaust
from port A.

Precessing force at port A erects gyro,
exhaust air again equal at all ports.

FIGURE 17-47 Erecting mechanism of a vacuum-driven attitude indicator.

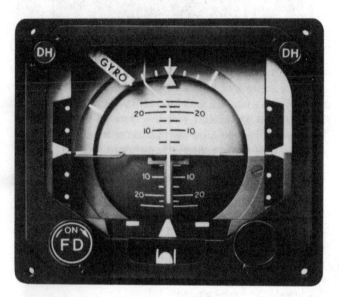

FIGURE 17-48 An artificial horizon with flight director capabilities. (*Edo Corp.*)

FIGURE 17-49 How a gyro provides a turn indication.

found in high-speed aircraft. If there are no doghouses on the instrument face, one needle-width deflection is the standard-rate position.

Figure 17-49 shows how a gyro serves to indicate a rate of turn. This figure does not show the actual arrangement of a turn-and-bank indicator gyro, but it does demonstrate the principle of operation. The gyro is set up to measure the rate at which its base is being turned about the Z axis. If the base is not turning, there is no force exerted on the gyro, since the springs attached to the X axis are adjusted so that under these conditions they are in balance.

Turning the gyro assembly to the left, as shown by the arrow on the base, a force will be applied to the left at A and to the right at C. Recalling the rule for precession, the point B on the gyro will move to the left and the point D will move to the right, thus precessing the gyro about the Y axis and causing the indicating needle to show a left turn. The rate of turn indicated will depend upon how rapidly the base of the unit is turned.

As the gyro precesses around the Y axis, the spring on the right end of the spin axle will be extended. The stretched spring will then apply a downward force on the end of the axle and, in effect, will apply a force at B and D. Thus force at B and D results in a precessional motion about the Z axis in the direction in which the base is turning. When the gyro is precessing about the Z axis at the same rate of speed as the base is being turned externally, there is no longer any force being exerted at points C and A. This means that the gyro will stop precessing about the Y axis and will hold its position. If, however, the rate at which the base is being turned is increased, once again there will be a force applied at C and D and the gyro will precess farther about the Y axis. Of course, the spring being stretched farther will cause a more rapid precession about the Z axis, and this will again match the external turning. Therefore, whenever the base is being turned, one of the springs is displaced enough to precess the gyro at a rate equal to the rate of turn. Whenever the base is stopped, the springs restore the gyro rotor to the neutral position.

Figure 17-50 shows a standard turn-and-bank indicator. This instrument is gyro-operated and functions on a principle

FIGURE 17-50   A turn-and-bank indicator.

FIGURE 17-51   Suction gauge.

similar to that described in the previous paragraphs. During a turn, the precession of the gyro causes the "needle" to swing to the right or left, depending upon the direction of the turn. The amount of deflection is proportional to the rate of turn. This enables the pilot to determine the rate of turn by the instrument.

Turn Coordinator

A **turn coordinator** is similar to a turn-and-bank indicator in that the pilot uses it to determine the rate of turn. The external difference between the two instruments is that the turn coordinator uses the outline of an airplane to indicate the rate of turn instead of a needle, as shown in Fig. 17-50. The turn coordinator has replaced the turn-and-bank indicator as the standard turn-rate-indicating instrument on many aircraft.

Internally, the gimbal that holds the gyroscope coordinated in the turn is tilted back at about a 30° angle rather than being vertical, as in the turn-and-bank instrument. This tilting allows the gyro to indicate both yaw about the vertical axis and roll about the longitudinal axis of the aircraft.

Both types of turn indicators can be powered electrically or pneumatically. The position of the turn-rate-indicating instrument in the standard flight instrument group is in the lower left corner, below the airspeed indicator and to the left of the directional gyro.

Suction Gauges

Suction or **vacuum gauges** are used in airplanes for the purpose of indicating a reduction of pressure. In other words, they indicate the amount of vacuum being created by the vacuum pumps or venturi tube and assist in the proper setting of the relief valve in the vacuum system. A typical suction gauge is illustrated in Fig. 17-51. The suction gauge gives pilots warning of impending failures of the system and warns them not to rely on the instruments that are operated by vacuum. It will be learned later that gyroscopes will operate for a short

period of time even after the source of vacuum has been shut off or partially cut off, but the indications may be erroneous. Some aircraft have an alternative source of vacuum available in the event of failure of the vacuum source being used. The suction gauge indicates any fluctuation or impending failure of the system, thus giving the pilot warning to switch over to the alternative vacuum source. For example, if the engine-driven vacuum pump on one engine of a multiengine aircraft is being used, the alternative pump may be located on the opposite engine. On the other hand, if the airplane has but one engine, the alternative source may be a **venturi tube**, discussed later in this chapter.

The suction gauge is vented to the atmosphere or to the air filter, if one is used in the system. The tube leads from the pressure-sensitive diaphragm in the instrument tees into the vacuum line as close to the instrument as possible. It is recommended by many manufacturers that it tee into the line leading directly to the gyro horizon in aircraft that include this instrument in the instrument panel. Theoretically, if sufficient vacuum is being created to operate the gyro horizon, all other instruments in the system will function satisfactorily. The reduction of pressure caused by the vacuum pump or venturi tube tends to collapse the diaphragm in the gauge. This movement is transferred through the rocking shaft, sector, and pinion to give a continuous indication on the dial. The range of the dial is usually from 0 to 10 and represents inches of mercury below the surrounding atmospheric pressure.

If the suction gauge is mounted on a shockproof panel and connected to another instrument on the same panel, it is not necessary to use rubber tubing for the connection. However, when the connecting tube is fastened to a rigid part of the aircraft structure, flexible tubing must be used. The amount of vacuum that should be registered on the suction gauge is dependent on the instruments that operate on that particular system and on the manufacturer's specifications. In instances where one instrument should have less vacuum than another but both operate from the same source of vacuum, a small restriction is placed in the line leading directly to the instrument requiring the lesser vacuum.

The relief valve is set to furnish the maximum amount of vacuum required for any one instrument. The suction gauge is set to show 4 in Hg [13.55 kPa] for artificial horizon and the direction gyro. The bank-and-turn instrument is set to 2 in Hg [6.77 kPa] by a suction gauge connected temporarily to a fitting screwed into the bank-and-turn needle valve that is in the back of the indicator.

Many aircraft are now equipped with electrically operated gyros, and it is not necessary for such aircraft to be equipped with vacuum systems.

FLIGHT INSTRUMENT SYSTEMS

Some flight instruments must be provided with electrical or pneumatic power to allow them to operate. Other instruments must be provided information from remote sensors so that they may provide the pilot with the proper indications. **Flight instrument systems** include instrument systems that use these power sources and sensor mechanisms.

Pitot-Static Systems

The purpose of a **pitot-static system** is to provide the pressures necessary to operate such instruments as the altimeter, airspeed indicator, and vertical-speed indicator. Static pressure is also required for such control units as air data transducers and automatic pilots.

Figure 17-52 illustrates a pitot head for a standard system on some older airplanes. The purpose of the head is to pick up indications of dynamic (ram) air pressure and static (ambient) air pressure to be transmitted through tubing to the instruments requiring these pressures for operation. The dynamic pressure caused by the movement of the airplane through the air is picked up through port A (in Fig. 17-52). This pressure is carried through the head and out tube D to the airspeed indicator and Mach number–sensing units. The baffle plate B helps keep water from entering the system. Water that goes by the baffle plate is stopped by the water trap C. Water is drained from the head through drain holes E and F.

Static pressure is picked up through the holes at G and H and is carried through the tube I to instruments such as the altimeter, vertical-speed indicator, and others requiring static pressure.

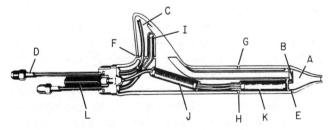

FIGURE 17-52 Typical pitot head.

The pitot head shown is provided with heaters to prevent the formation of ice. These heaters are shown at J and K, and the electrical connecting plug is indicated by L in the illustration. On many aircraft the dynamic pressure only is picked up by the pitot heads. The heads are mounted on the aircraft where they will provide the most accurate indications of dynamic pressure. The static pressure is obtained through one or more ports on the fuselage.

A pitot-static system as installed in a light, twin-engine airplane is illustrated in Fig. 17-53. In this system, the pitot head mounted under the wing supplies only ram air pressure. The static pressure is supplied through the perforated static buttons (ports) mounted flush with the fuselage skin toward the rear of the airplane.

An important feature of the pitot pressure system is that it have provision for removing all water before it can reach the instruments. Baffles in the pitot head and drains in the pressure line are used to remove the water.

Pitot-static systems are often provided with an alternative static pressure source in the cabin or cockpit. These are usually satisfactory except for pressurized aircraft. The pressure supplied to the static system must be that existing outside the airplane. If the airplane is pressurized, the pressure inside the airplane will be substantially greater than that existing outside, and the instruments will read at a lower value than normal if the static pressure is picked up inside the airplane. If there is a leak in the static line inside a pressurized airplane, pressure from the cabin will leak into the static line, and the instruments will register false readings. The altimeter will show an altitude below the actual altitude, the vertical-speed indicator will not give the correct rate of climb or descent, and the airspeed indicator will show the airplane speed below the actual speed.

Maintenance and testing practices for pitot-static systems are described later in this chapter.

Gyro Electrical Power Sources

All three of the basic gyro instruments are available in designs that rely on electrical power for operation. These instruments normally require one of four types of power sources: 14 or 28 V dc or 120 or 208 V ac.

Many of the electrically powered instruments have the word electric on the face, and all electrically operated instruments should have a flag visible when electrical power is not being supplied to the instrument, as shown in Fig. 17-54.

Gyro Pneumatic Power Sources

Many of the older and most of the lower-priced gyro instruments are operated by a flow of air moving over the buckets on the rotor disk. This flow of air can be caused by drawing air through the instrument with a venturi or vacuum pump or by forcing air through the instrument from a pressure source. Regardless of the type of pump used, the air inlet to the instruments should be filtered to keep contaminants out of the instruments and out of the pump.

A **venturi**, shown in operation in Fig. 17-55, generates a vacuum by moving air through a restricted passage. As the

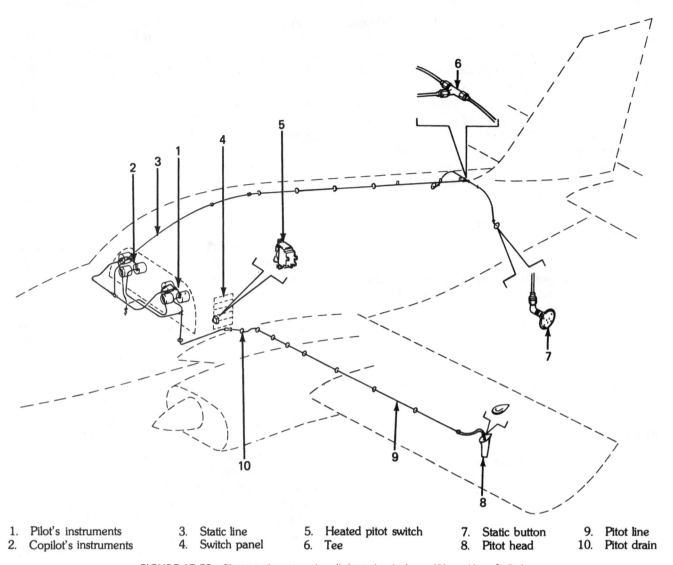

1.	Pilot's instruments	3.	Static line	5.	Heated pitot switch	7.	Static button	9.	Pitot line
2.	Copilot's instruments	4.	Switch panel	6.	Tee	8.	Pitot head	10.	Pitot drain

FIGURE 17-53 Pitot-static system in a light twin airplane. (*Piper Aircraft Co.*)

FIGURE 17-54 The "OFF" flag indicates that the instrument is not receiving electrical power. (*Aeronautics.*)

air accelerates through the passage, the sidewall pressure is reduced, and air can be drawn into the venturi core through a hole placed on the side of the tube at the restriction. If a hole in the side of the venturi throat is connected to a pneumatically operated gyro instrument, air is drawn through the instrument, spins the gyro, and causes the instrument to operate.

A venturi is the simplest type of vacuum "pump" and is normally found on aircraft not intended for instrument flight. Its primary advantages are simple installation, no operation expense, and reliability. The drawbacks include requiring 100 mph of airspeed to operate the instrument properly, no means of powering the instrument before flight, and the possibility of the instrument's being closed by ice.

Venturis come in two sizes designated by the amount of suction they can generate, a 2-in [5.08-cm] size and an 8-in [20.32-cm] size. The 2-in size is suitable for operating one turn-and-bank-turn coordinator. The 8-in venturi can handle an artificial horizon and a directional gyro. A suction gauge is normally connected to the line between the venturi and the instruments so that the pilot can monitor the suction on the system. The desired value for the artificial horizon and the directional

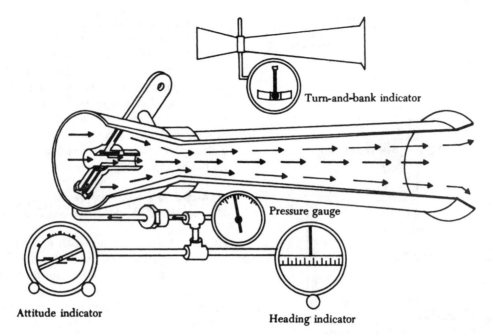

FIGURE 17-55 Venturi vacuum system.

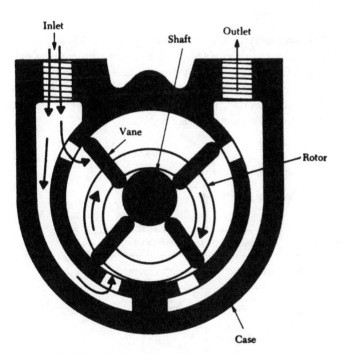

FIGURE 17-56 Cross section of a vacuum pump.

gyro is about 4 in Hg [10.16 cm Hg] of vacuum. If a turn-and-bank-turn coordinator is used with the large venturi, a pressure regulator is often required to drop the vacuum down to 2 in Hg for these instruments to operate properly.

Aircraft to be used for instrument flying and those with operational speeds much above 100 mph usually rely on engine-driven vacuum or pressure pumps to power the pneumatically operated instruments. A cutaway view of an engine-driven vacuum pump is shown in Fig. 17-56.

Vacuum pumps can be either the **wet-pump** or **dry-pump** design. A wet pump relies on engine oil to lubricate the

operating mechanism. The oil that is introduced into the air must be separated from the oil before the air is vented overboard or directed into other aircraft systems. This reclaimed oil is returned to the engine oil reservoir. The primary drawback associated with a wet pump is the oil contamination of pressure systems and oil appearing on the airframe as a result of inefficient oil separators.

A dry pump relies on the proper selection of construction materials to provide lubrication. The materials selected for the vanes are normally carbon-based, which means they provide low friction and proper sealing against the metal case walls. Dry pumps have become very popular because of the elimination of oil contamination of the system and the need for an oil separator. Figure 17-57 shows a schematic representation of a typical light aircraft vacuum system.

Aircraft intended for high-altitude flight often have pneumatically operated instruments powered by air pressure rather than a vacuum. This is because of the inability of vacuum pumps to draw the approximate 4 in [10.16 cm] of vacuum required when the air is already at a relatively low pressure. A pressure source may be as simple as connecting to the pressure side of the vacuum pump and connecting the pressure line to the inlet to the gyro instruments rather than to the outlet of the instrument, as is done with a vacuum line.

Air Data Computer

An **air data computer** is used on most turbine-powered airplanes to power indication and control systems needing information about the ambient air and aircraft speed. The inputs to the system are pitot pressure, static pressure, and ambient air temperature. This information is processed by a computer and signals are sent to various systems so that the aircraft performs properly.

Systems that typically use data from the air data computer on transport aircraft are flight instruments such as the

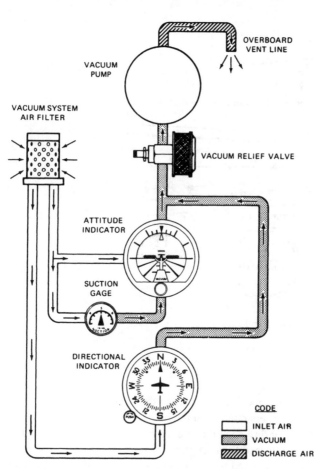

FIGURE 17-57 Schematic drawing of a light aircraft vacuum system. (*Cessna Aircraft Co.*)

true airspeed indicator, machmeter, and altimeter; navigation systems such as the inertial navigation systems, autopilots, and flight directors; and warning units such as mach limit warning and altitude alert. A schematic drawing for an air data computer for an airliner is shown in Fig. 17-58.

The type of system shown in Fig. 17-58 allows compensation for compressibility at high airspeeds and the friction heating of probes. The system also allows the information collected to be changed to electrical and electronic signals for use by various flight computer systems.

ENGINE INSTRUMENTS

Engine instruments are necessary in order that the flight crew can properly adjust and monitor engine system operations. These instruments are also useful to the technician when evaluating and troubleshooting engine systems. The principles of operation of most engine instruments are the same as have been discussed for many other aircraft instruments. These include the use of electrical temperature sensors and bourdon tubes or bellows to measure pressures.

Tachometers

Tachometers are used to measure engine rpm. When discussing reciprocating engines, the crankshaft rpm is measured and indicated with the unit's rpm indicator. For helicopters and free-turbine propellers, the rpm of the rotor or propeller is indicated in rpm. For turbine engines, the engine rpm is indicated in *percent of rated rpm*.

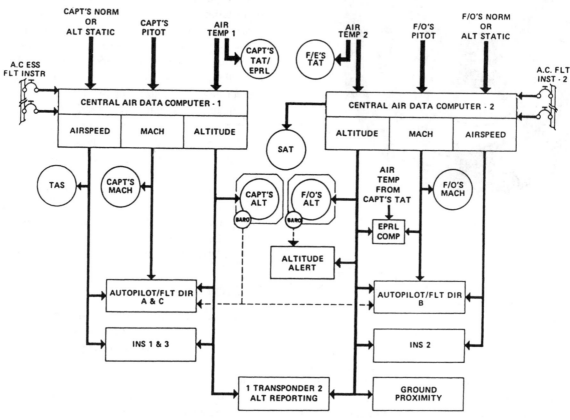

FIGURE 17-58 Schematic drawing of a typical system incorporating an air data computer. (*Boeing Commercial Aircraft Co.*)

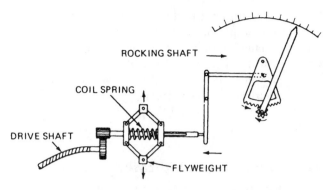

FIGURE 17-59   A mechanically operated tachometer.

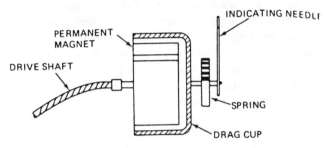

FIGURE 17-60   Illustration of magnetic tachometer.

Tachometers may be mechanically or electrically operated. Mechanically operated tachometers may be driven directly by the use of a flyweight mechanism, as Fig. 17-59 shows, or by a magnetic drag cup, as shown in Fig. 17-60. The magnetic tachometer is the most popular type in use in general aviation. This tachometer is operated by a flexible shaft that is rotated by a drive on the rear of the engine. As the shaft turns, it rotates a magnet inside of an aluminum drag cup in the instrument. The rotation of the magnet creates eddy currents in the drag cup and causes it to move in the same direction as the magnets. A light coil spring opposes this rotation. The faster the speed of rotation, the greater the deflection of the drag cup and the higher the indication of the needle attached to the drag cup.

Drag-cup tachometer cables should be routed so as to avoid kinks. If the tachometer oscillates, or "bounces," the cable should be inspected for binding, damage, or lack of lubrication.

Electric tachometers are used on large aircraft where the distance from the engine to the instrument panel makes mechanical tachometers impractical. A common type of this tachometer, shown in Fig. 17-61, uses a three-phase ac alternator to generate a signal. This signal drives a synchronous motor inside the instrument, which synchronizes its speed with that of the alternator. This motor turns a drag cup, and the drag cup positions the needle on the face of the instrument.

Oil-Pressure Indicators

Oil-pressure indicators can be mechanically operated or electrically powered. A mechanically operated gauge uses an oil-pressure line from the engine to the instrument to operate a bourdon tube and gear segment to position the indicator needle. The oil line should have a restrictor at the engine to prevent rapid oil loss if the line should break. Some aircraft use a light oil in the line between the gauge and the engine so that there will be no delay in oil pressure indication due to cold engine oil being in the line.

Electric oil pressure sensors use a pressure sensor on the engine, which varies in resistance as the pressure changes. As this pressure signal is generated, the pressure is indicated by one of the electrical indicating methods previously discussed.

Oil-Temperature Indicators

Oil-temperature indicators can be electrical or mechanical. To operate electrically, a resistance probe is placed in the oil line where the oil enters the engine. By one of the methods discussed at the beginning of this chapter, the oil temperature is derived by the change in probe resistance due to the temperature change.

Another method used to measure oil temperature is with a volatile liquid in a sealed sensor bulb and capillary tube. In this system a sensor bulb is placed in the oil line at the inlet to the engine.

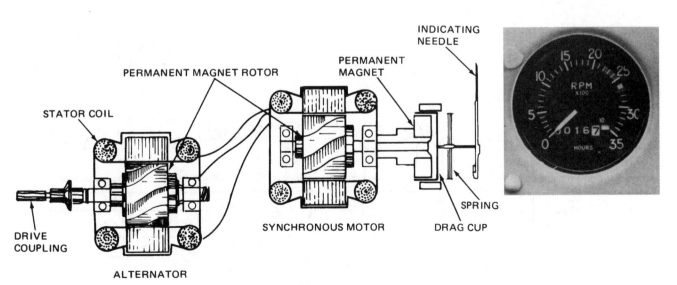

FIGURE 17-61   Electric tachometer system. (AC.)

The temperature of the oil causes the volatile liquid to vaporize and increase the pressure in the capillary tube. The instrument is permanently attached to the other end of the tube, and the change in gas pressure causes a bourdon tube to move and indicate the oil temperature on the face of the instrument.

Exhaust-Gas Temperature

Reciprocating-engine and turbine-engine exhaust-gas-temperature systems are used to monitor the performance of the engines and make flight and maintenance adjustments. These systems operate by placing thermocouples in the stream of exhaust gases exiting the engine. The thermocouples generate a current, which drives the indicator. The amount of current is usually very low and is amplified or adjusted in order to drive the indicator display.

For a reciprocating engine, one probe may be used for the engine or a probe may be mounted near the exhaust of each cylinder. If multiple probes are used, a selector switch on the instrument is used to allow selection of the cylinder being monitored. In some displays all cylinder temperatures are displayed at the same time in the form of a vertical light bar, with the length of each bar corresponding to the exhaust temperature of one cylinder.

In turbine-engine installations, the system uses several probes in parallel, as shown in Fig. 17-62. This ensures an average reading of the temperature, and the indication does not change substantially if one probe should fail. The exhaust gas temperature is a primary instrument for monitoring turbine engine operation.

Engine-Pressure Ratio

Engine-pressure ratio (EPR) is used to indicate the amount of thrust being generated by a turbine engine. EPR is determined by measuring the total pressure at the engine inlet and comparing it to the total pressure of the engine exhaust. These pressures are measured by pitot tube–like probes pointing into the airflow. The total pressure of the exhaust stream divided by the total pressure of the inlet stream gives the EPR. These pressures are normally transmitted electrically to the indicating system, where they are compared electronically and the proper value is displayed on the instrument.

Manifold-Pressure Indicator

The **manifold-pressure gauge** measures the sidewall pressure in a reciprocating-engine intake manifold downstream of the carburetor throttle. This pressure is a measure of the engine power output. Most smaller aircraft use direct-indicating instruments for the manifold-pressure gauge. The gauge uses a bellows, which expands and contracts with changes in the manifold pressure. This drives an indicator needle on the face of the instrument.

The manifold-pressure gauge is an important instrument for reciprocating engines when determining proper power settings. Without this gauge, the pilot of a turbocharged or supercharged engine would not be able to tell if he or she were exceeding the allowed maximum manifold pressure. Pilots flying aircraft with constant-speed propellers rely on this gauge in conjunction with the tachometer to establish the proper powerplant control settings.

Cylinder Head Temperature Indicators

A **cylinder head temperature gauge** is used to determine if the engine is operating at the proper temperature. This instrument system uses a thermocouple under a spark plug or mounted on the side of the cylinder head, as shown in Fig. 17-63. The thermocouple is attached to two wires,

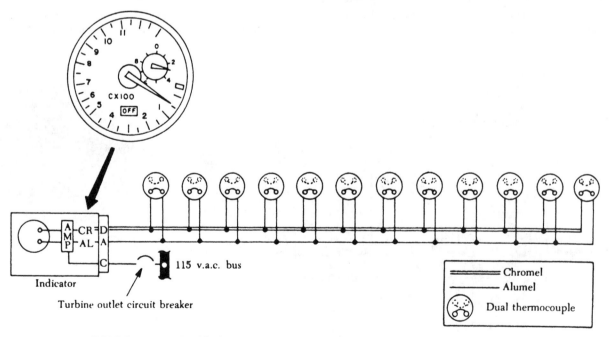

FIGURE 17-62 Simplified schematic drawing of a turbine-engine exhaust gas system.

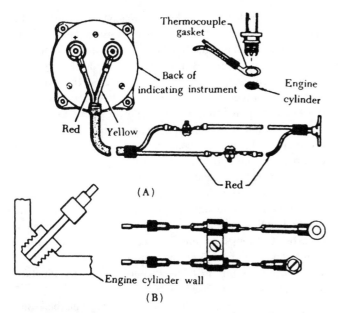

FIGURE 17-63 Gasket and bayonet sensors for a reciprocating-engine thermocouple-operated cylinder head temperature gauge.

a set of iron and constantan or Alumel and Chromel, which will generate a current based on the temperature of the thermocouple. In its basic form this instrument does not need a power source because the current generated is sufficient to power the indicating instrument. The wires are of a set length and must not be added to or shortened except with specified lengths of additional wire of the same type.

FUEL-QUANTITY INDICATORS

There are various fuel-quantity-indicating systems used in aircraft. They range from simple mechanical, direct-indicating systems used with light aircraft to large electronic systems used with transport aircraft. Each has its own specific characteristics, of which the technician should be aware.

Mechanical Quantity Indicators

Float-type and direct-reading gauges have been used on many light aircraft since the time that the first fuel gauge was installed. These gauges have many variations, depending upon the location of the fuel tanks.

The **inverted float gauge** is used on aircraft equipped with the gasoline tank located in the upper wing. One gauge of this type has a rod suspended from the float. The rod has a disk or small piece of metal on the lower end, and this disk is visible through a glass tube extending below the wing. The glass tube is usually graduated in gallons to show the amount of fuel remaining in the tank and is protected by a metal frame that surrounds the tube extending below the wing. A slot in the frame wide enough to ensure good visibility of the tube allows the operator of the aircraft to obtain readings from his or her location in the cockpit. In the event the fuel tank is located in a lower wing, the same type of float gauge can be used, with the tube extending above the wing.

The **rotating dial gauge** has a relatively simple mechanism. The indicator is a round drum-shaped device that has a revolving disk inside. The tank unit consists of a float on the end of a long arm, the other end of which transmits motion to the disk in the instrument by means of a semicircular gear. The semicircular gear is moved by the float rising and falling with the level of the fuel.

The **upright float gauge** has the float arm directly connected to the pointer by a link arm. Thus, the pointer is moved up and down with the float and float arm. The tank must be located below the cockpit if this gauge is used, thus placing the gauge in clear view of the pilot.

The **sight-glass gauge** is basically identical to the water-level gauge on a water tank. It can be used only when the fuel tank is approximately at the same level as the cockpit. The indicator is simply a glass tube that runs from the top to the bottom of the fuel tank and is connected to the tank at both top and bottom. This allows the fuel to assume the same level in the tube that exists in the tank. Graduations are painted or inscribed on the tube, or they can be painted on a metal scale adjacent to it. For safety, shutoff valves should be at each end of the tube to prevent fuel leakage if the glass is broken.

Magnetic direct-indicating gauges use a metal or cork float on an arm to operate a gear segment. The gear segment turns a rotating bar magnet, which is rotated by the float gear segment. The magnet is close to the mounting plate surface, which is used to hold the gauge in the fuel tank. When the gauge is installed, the outer surface of the mount has an indicator dial and pointer on the outside of the tank. When the float moves up or down and causes the bar magnet to rotate, the magnetic force passes through the plane of the mount and positions the indicator needle.

Resistance Quantity Indicators

A **resistance-type fuel-quantity indicator** uses a float on an arm attached to a tank mount. At the tank mount the float arm moves a wiper across a resistance wire winding. As the float moves up and down, the amount of resistance between the float arm and the indicator in the cabin varies.

The circuit, shown in Fig. 17-10, can be designed to indicate a full tank when the resistance is high or when the resistance is low. The circuit normally is grounded through the fuel-tank float arm. Considering the current to flow from the ground to the power source, the current flows from the grounded float arm through the wiper and the resistance winding. From there it flows through the instrument to the power source. The coil in the instrument causes the pointer to indicate the fuel quantity according to the current flow through the circuit.

Some adjustments are made to this system by bending the float arm as necessary to obtain the correct full and empty indications.

This type of system is used in fuel tanks that are regular in shape and proportionally deep when compared to their length and width. The system does not work well for fuel tanks that extend along the length of a wing or where the tank is irregular in shape, such as for large aircraft with tapered integral fuel tanks.

Capacitance Quantity Indicators

A **capacitance fuel-quantity-indicating system** is also referred to as an **electronic quantity indicating system**. This system measures the weight of fuel in the tanks rather than the volume. The system consists of one or more capacitance probes placed at designated locations inside of the tank. Each probe has a specific capacitance value. The probes are connected in parallel and placed in a capacitance bridge circuit. A simplified schematic drawing of this system is shown in Fig. 17-64.

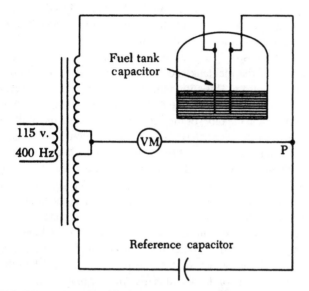

FIGURE 17-64 Simplified capacitance fuel-quantity-indicating bridge circuit.

Fuel has a different dielectric constant than air, so the capacitance value of the probe changes as the tank is filled or drained of fuel and the fuel is replaced with air. When the parallel probe circuit has a different capacitance compared to the reference capacitor, a difference in potential will exist across the voltmeter, causing the needle to deflect. Note that this circuit requires an ac electrical power source to operate.

In a complete system, the signal across the voltmeter position is fed to a signal conditioner. The signal conditioner or processor unit corrects any known irregularities in a specific system response and sends an indicator signal to the aircraft fuel-quantity indicators. Among these adjustments are provisions for the technician to establish the zero-fuel and full-fuel indications for the system. Until this adjustment is made, full tanks or zero fuel may not cause the proper quantity to be indicated on the instruments. Notice that the system shown in Fig. 17-65 uses a densitometer to adjust for different types of fuel used.

A capacitance system has an advantage over the resistance system in that it can be used in tanks of irregular shape and small dimensions. It also has the advantage of measuring the fuel quantity by weight rather than volume. This is important because engines consume fuel based on the fuel's weight rather than its volume. For large aircraft, this difference can be significant, because the volume of fuel expands or contracts with changes in temperature. If the ambient temperature is high and the aircraft is filled to volume capacity, it will have less fuel on board and will not be able to fly as far when compared to an aircraft filled to volume capacity on a cold day. The capacitance fuel system allows the crew to determine just how much weight is on board and thus determine the operational range of the aircraft.

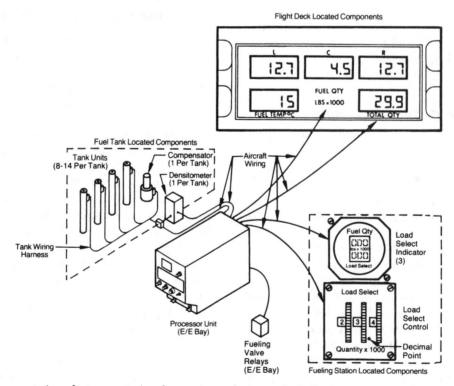

FIGURE 17-65 Representation of a transport-aircraft capacitance fuel-quantity-indicating system. (*Boeing Commercial Aircraft Co.*)

FUEL-SYSTEM-MONITORING INSTRUMENTS

In order to determine if the fuel system is operating properly, various fuel-pressure-, flow-, and temperature-indicating systems are used. The number and types of instruments on any particular aircraft will vary. A gravity-feed fuel system for a light aircraft will probably not have any of these indicators. A pressure-feed system will incorporate at least a fuel-flow or fuel-pressure indicator. Large turbine-powered aircraft designed to fly at high altitudes will use several different monitors of this type to ensure a proper fuel supply to the engine.

Fuel Pressure

Several different types of fuel-pressure gauges are in use for aircraft engines, each designed to meet the requirements of the particular engine fuel system with which it is associated. The technician should know the type of gauge before making judgments regarding its operation or indications.

Any fuel system utilizing an engine-driven or electric fuel pump must have a fuel-pressure gauge to ensure that the system is working properly. If a float-type carburetor is used with the engine involved, the fuel-pressure gauge will be a basic type, probably with a green range marking from 3 to 6 psi [20.69 to 41.37 kPa]. If the engine is equipped with a pressure-discharge carburetor, the fuel-pressure-gauge range marking will be placed in keeping with the fuel pressure specified for the carburetor. This will probably be between 15 and 20 psi [103.43 and 137.9 kPa]. A red limit line will be placed at each end of the pressure-range band to indicate that the engine must not be operated outside the specified range.

If an engine is equipped with either a direct-fuel-injection system or a continuous-flow fuel-injection system, the fuel pressure is a direct indication of power output. Since engine power is proportional to fuel consumption and since fuel flow through the nozzles is directly proportional to pressure, fuel pressure can be translated into engine power or fuel-flow rate or both. The fuel-pressure gauge, therefore, can be calibrated in terms of percent of power. A gauge of this type is shown in Fig. 17-66. Note that the instrument indicates a wide range of pressures at which the engine can operate. It is calibrated in pounds per square inch and also indicates percent of power and altitude limitations. The face of the instrument is color-coded blue to show the normal cruise range during which the engine can be operated in AUTO LEAN and green to show the range during which the AUTO RICH mixture setting should be used.

Fuel-pressure gauges are constructed similar to other pressure gauges used for relatively low pressures. The actuating mechanism is either a diaphragm or a pair of bellows. An illustration of a bellows is shown in Fig. 17-5. The advantage of the bellows is that it provides a greater range of movement than does a diaphragm. In a typical fuel gauge, the mechanism includes two bellows capsules joined end to end. One capsule is connected to the fuel-pressure line, and

FIGURE 17-66   A fuel-pressure gauge.

the other is vented to ambient pressure in the airplane. The fuel pressure causes the fuel bellows to expand and move toward the air capsule. The movement is transmitted to the indicating needle by means of conventional linkage.

Some fuel-pressure gauges on light aircraft are marked as if they were fuel-flow gauges. These gauges use fuel pressure at a fuel-injector distributor block to operate a gauge identical to a fuel-pressure gauge. However, the face is marked with a fuel-flow scale. These instruments can often be identified by small markings indicating minimum and maximum fuel pressures. For example, in Fig. 17-67, the minimum fuel pressure is marked as 1.5 psi and the maximum fuel pressure is 17.5 psi.

The reason that the technician should be able to identify these types of fuel-flow gauges is that if a fuel line should start to leak, this would cause an increase in system fuel flow. But the loss of pressure would show up as a decrease in the gauge indication on the fuel-flow scale. On the other hand, if a cylinder line is bent or blocked, the pressure in the system would increase, indicating a higher fuel flow, when the actual situation is a reduced fuel flow.

FIGURE 17-67   Pressure gauge used as a fuel-flow indicator.

Fuel Flowmeter

Fuel flowmeters are used with fuel systems to show the amount of fuel in gallons or pounds consumed per hour. These indicators are usually calibrated to read in pounds per hour. The face of a typical fuel-flow indicator is shown in Fig. 17-68.

A fuel-flow indicator system consists of a fuel-flow transmitter located in the fuel line leading from the tank to the engine and an indicator located on the instrument panel or on the flight engineer's panel. The transmitter signal may be developed by a single movable vane mounted in the fuel-flow path in such a manner that its movement will be proportional to fuel flow. As fuel flow increases, the vane must move to allow more fuel to pass, and this movement is linked to a synchro unit, which develops the electrical signal to be sent to the indicator. A schematic diagram of such a system is shown in Fig. 17-69. An in-line fuel-flow transmitter, such as the one shown in this figure, must have a bypass valve so that the full flow of the fuel required by the engine can pass even though the transmitter becomes obstructed.

The fuel-flow transmitter used in a modern jet-airliner fuel system to handle the large volume of fuel required for jet-engine operation is somewhat more complex than the vane type just described (see Fig. 17-70). In one such system the transmitter consists of an impeller and a turbine mounted in the main fuel line leading from the tank to the engine. The impeller is driven by a specially designed motor, which utilizes simulated three-phase four-cycle current to drive the motor at 240 rpm. The motor drives the impeller through a 4:1 reduction gear; hence the impeller rotates at 60 rpm. Electric power for the motor is provided by a special **fuel-flow power supply unit**, which consists of a permanent-magnet dc motor driving a commutator to chop the dc power into pulses that are fed to the three-phase impeller motor.

During operation, fuel flows through the passages in the impeller and then continues on through passages in the turbine. In passing through the impeller, the fuel is given a velocity at right angles to the direction of flow, because the impeller is rotated at a constant speed of 60 rpm by the impeller motor. This imparts an angular velocity to the fuel, which constitutes a change in momentum that is directly proportional to the mass of fuel flowing. The angular momentum of the fuel is removed while passing through the turbine, and the energy required for this change is converted to torque on the turbine. This torque is directly proportional to the **mass rate of flow**. Restraining springs on the turbine oppose the torque and produce a deflection of the turbine proportional to the flow rate. This deflection is transmitted to the fuel-flow indicator in the crew cabin.

Observe in Fig. 17-70 that the stators of both the transmitter and the indicator receive power from the same 115-V ac source. The angular position of the cross flux in the indicator follows the position of the transmitter magnet. The permanent-magnet rotor of the indicator lines up with the cross flux to provide the indication of fuel flow.

The measurement of the **mass flow rate** of fuel is particularly important for large turbine-engine-powered airplanes because energy consumed is proportional to fuel weight consumed rather than fuel volume consumed. The weight per unit volume and volume of fuel change considerably as temperature changes. For this reason, if the flowmeter shows fuel flow in gallons per hour, the flight crew could not determine accurately the power produced by the engines from a reading of fuel flow. On the other hand, the mass-flow fuel flowmeter provides a direct indication of energy or power produced by the engines.

MISCELLANEOUS INSTRUMENTS

There are a few instruments that do not fall into any specific grouping but are used to inform the pilot of data not related to specific systems. Among these are clocks and ambient air temperature gauges.

FIGURE 17-68   Face of a typical fuel-flow indicator.

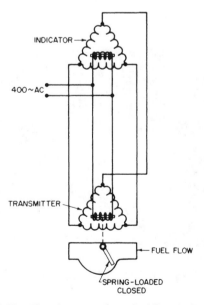

FIGURE 17-69   Electric system for a fuel-flow indicator.

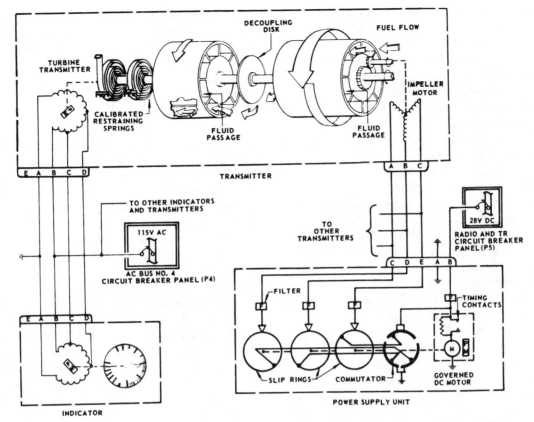

FIGURE 17-70 Fuel-flow-indicating system for a jet airliner. (*Boeing Commercial Aircraft Co.*)

Clocks

Clocks, or **time indicators**, are necessary items for instrument flying and long-distance navigation. When the pilot is flying on instruments, the clock becomes an integral part of the basic flight group. Its main function is to aid navigation. Some of the specific uses of the clock are as follows:

1. At all times pilots must be able to report their position and time accurately to air traffic–control facilities.
2. Operators familiar with their engines and rate of fuel consumption can estimate the fuel consumption by time if the fuel gauge becomes inoperative. Hence, they can compute remaining fuel at any time.
3. Ground-speed checks cannot be made without knowing the elapsed time between two points.
4. The clock is necessary for maintaining schedules, complying with times for radio communications, and making checks against aircraft instruments, such as the rate-of-climb and turn-and-bank indicators.
5. All types of navigation require accurate knowledge of elapsed time, especially when flight is over unfamiliar territory; hence a good clock must be available for long-distance flying.

All standard instruments panels are equipped with at least one 8-day clock. These clocks must be carefully shock-mounted to ensure long life and accurate performance. Although the clocks being built today are ruggedly constructed and temperature-compensated, vibration is one of the worst conditions to which a clock can be subjected. Variance on properly shock-mounted and properly calibrated clocks does not exceed 5 s per day.

The normal maintenance of clocks by the technician consists only of winding, setting, and checking them for accuracy unless continued trouble has been experienced because of faulty installation. When a clock is wound, no great force should be exerted after the winding stem offers any appreciable resistance to further winding. In a check for accuracy, the second hand is checked against another clock of known accuracy for a period of not less than 1 h.

Several different types and makes of clocks are being successfully used in airplanes. All are required to have the **hour**, **minute**, and **sweep-second hands**. Others are equipped with five hands. In addition to the three hands already mentioned, these clocks include a **stopwatch** hand and an **elapsed-time** hand.

Under no condition should a technician open an aircraft clock for inspection or adjustment. This type of work should be done only in an instrument overhaul shop, where conditions and equipment are suitable for the work.

Temperature Gauge

Temperature gauges in modern aircraft are often operated electrically by means of the output of the thermocouple, ratiometer, or Wheatstone bridge, as previously discussed. Other temperature gauges utilize the expansion and contraction of various materials to provide a temperature indication.

Nearly all solid materials expand when heated and contract when cooled. Since the results of such expansions are more important than the causes, it is not necessary to describe the molecular activities that take place when heat is applied to a solid. The principal point to remember is that some metals expand more than others when subjected to heat. If the amount of expansion that takes place in various solids is measured for a given increase in temperature, a specific value can be assigned to each material. This value is called the **coefficient of expansion**. The coefficient of linear expansion is defined as the **change in length of each unit of length for a rise of temperature of 1°C**.

Two metals commonly used in bimetallic thermometers are brass and iron. The coefficient of linear expansion of these two metals is, respectively, 0.0000105 and 0.0000067. Therefore, if a strip of brass and a strip of iron of equal length at room temperature were securely brazed or otherwise bonded together, they could be used as a thermometer. As the temperature rises, both metals expand; but the length of the brass strip increases faster than the length of the iron strip, which causes the bimetallic combination to curve. Conversely, if the temperature drops lower than room temperature, the curve is in the opposite direction.

When one end of the bimetallic strip is attached to the instrument case and the other end is attached to a pointer that is free to rotate around a calibrated dial, the temperature changes can be converted to actual temperature measurements in Celsius or Fahrenheit scales on the instrument. Some of these instruments have a range of −40 to +104°F [−40 to +40°C]. A diagram of a bimetallic thermometer is shown in Fig. 17-71.

In all types of bimetallic thermometers, there is some lag in registering rapid temperature changes because the thermal unit is enclosed. A few minutes must be allowed for the pointer to reach the correct position after a sudden temperature change takes place.

ELECTRONIC INSTRUMENTS

Electronic display flight and system instruments have started to appear in production aircraft on a regular basis. Although normally associated with airline and corporate aircraft, these video instrument displays are gradually working their way down into the smaller aircraft and will eventually make their way into light twins and the more expensive single-engine aircraft. Electronic display instruments are basically color television displays that are presented according to signals generated by a computer. These computers receive information from a gyro reference source, an air data computer, and navigation radios.

The computer is commonly referred to as a *symbol generator*. Its purpose is to accept information about the condition of the aircraft (attitude, outside-air temperature, pitot and static pressure, radio navigation signals, engine condition, etc.) and convert the information into a set of signals to drive a video display. The internal workings of a symbol generator are beyond the scope of this text. Suffice it to say that the symbol generator is a computer that analyzes information about the condition of the aircraft and causes the appropriate information to be shown on a video display for the flight crew to use in operating the aircraft.

The video display, referred to as the cathode-ray-tube display or multifunctional display, receives signals from the symbol generator and presents a combination of text and symbolic displays for the flight crew to use in operating the aircraft. Many of the displays are conventional, such as a representation of an artificial horizon, whereas other displays are simplified vertical or horizontal displays incorporated with other displays.

Through a keyboard or the selection of different flight situations (cruise, descent, landing), the flight crew can display or eliminate from the display various portions of the information available. For example, in cruising flight the pilot would normally have no need of instrument landing system information, such as the glideslope display, so this is normally eliminated from the display until selecting an approach mode. The video displays are usually divided into three basic types: electronic attitude director indicators, electronic horizontal situation indicators, and electronic system monitoring and alerting displays.

Typical display positions on the instrument panel are shown in Fig. 17-72.

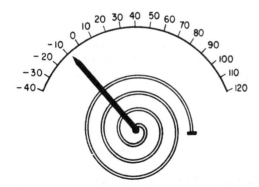

FIGURE 17-71 Bimetal thermometer.

FIGURE 17-72a Instrument panel for light aircraft.

FIGURE 17-72b Instrument panel for regional jet.

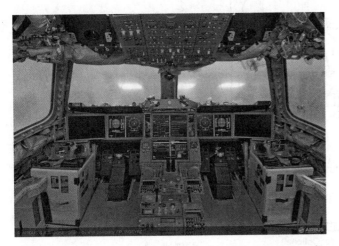

FIGURE 17-72c Instrument panel for large transport aircraft.

Reference Sources

Electronic flight instrument systems can be designed to accept information from any type of sensor system. Typically, pitot-static and navigation radios provide reference information. The gyro information can be provided from conventional mechanical gyro platforms, but a common method for the more modern transports is to use a laser gyro platform.

EADI

The **electronic attitude director indicator** (EADI) is a primary flight instrument used to supply attitude information. While each system manufacturer varies the type and method of information displayed on the EADI, the instrument typically contains an attitude indicator display, airspeed information, glideslope and localizer information, and annunciators indicating systems that have been selected for operation, such as autopilot mode and navigation mode. A typical jet transport display is shown in Fig. 17-73.

The EADI is normally located directly in front of the pilot and above the electronic horizontal situation indicator.

EHSI

The **electronic horizontal situation indicator** (EHSI) is used to display lateral guidance information along with the location of ground facilities and weather in relation to the aircraft, depending on the sophistication of the system being used. The EHSI has several modes of operation, which can be selected by the flight crew according to the amount of information that they require. Typical displays that can be selected by the flight crew are shown in Fig. 17-74. The information display may include a magnetic compass display, directional gyro display, relative bearings to ground navigation facilities,

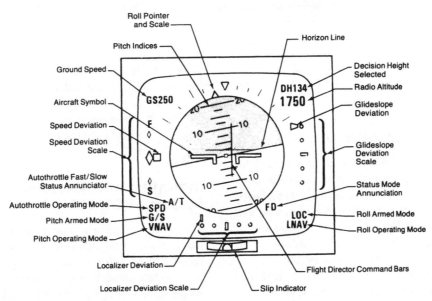

FIGURE 17-73 Electronic attitude director indicator. (*Boeing Commercial Aircraft Co.*)

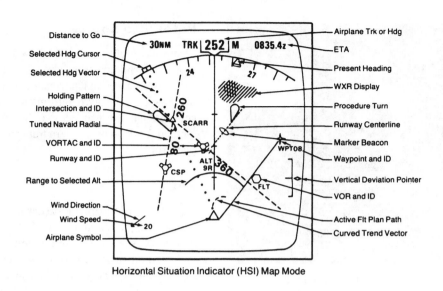

Horizontal Situation Indicator (HSI) Map Mode

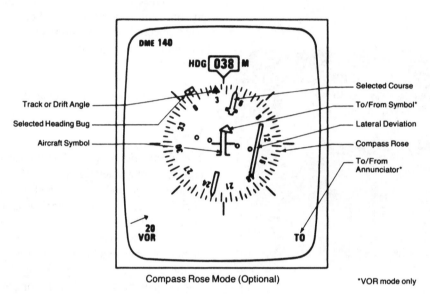

Compass Rose Mode (Optional)

*VOR mode only

Standard compass rose
DME and ILS information provided

FIGURE 17-74 Two of the several displays available on an electronic horizontal situation indicator. (*Boeing Commercial Aircraft Co.*)

distance-measuring equipment information, times, and wind information. The EHSI is normally located directly below the EADI.

Electronic Systems Monitor Displays

The various systems manufacturers have devised several different types of system-monitoring displays. Two of the common types of displays include the **electronic centralized aircraft monitor** (ECAM) and the **engine indicator and crew-alerting system** (EICAS). These systems are similar in the type of information displayed and are treated in this text as a generic display for aircraft system monitoring.

An ECAM-EICAS system provides both text and graphic displays of system condition, checklists for different operations, suggested sequences of actions during

system malfunctions, and the various flight-support information, such as route and approach data.

Different types of displays associated with ECAM-EICAS systems are illustrated in Fig. 17-75. These displays are normally located between the pilot positions, above the engine power levers.

Indicating and Recording System of CRJ 700

The indicating and recording systems consist of components that provide visual indications of system operation and that record aircraft information. Data from the aircraft systems and the full authority digital engine control (FADEC) on each engine is received and processed by two data concentrator units (DCUs) located in the avionics compartment.

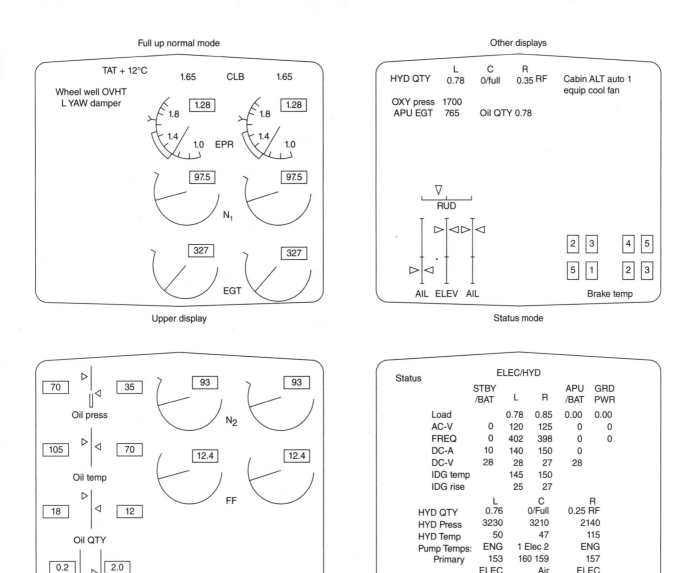

FIGURE 17-75 Typical displays on a Boeing 767 EICAS. (*Boeing Commercial Aircraft Co.*)

The DCUs provide information to the engine indication and crew alerting system (EICAS). Master warning and caution lights on the glareshield enhance the indication system. Audio signals are generated within the DCUs and are heard through the flight deck speakers. The DCUs also provide interface with the flight data recorder system (FDR), the lamp driver unit (LDU), and the maintenance data computer (MDC) via the integrated avionic processor (IAP). Figure 17-76 shows a schematic of the indicating and recording system.

Engine Indication and Crew Alerting System

The engine indicating and crew alerting system provides the crew with two electronic displays to monitor engines, control surfaces, and all major aircraft systems. The EICAS system also provides the crew with alerting system messages that are posted on the EICAS displays in the form of warning, caution, advisory, and status messages. All warning and caution messages will also illuminate the MASTER WARNING or MASTER CAUTION lights on the glareshield. Some crew alerts are also accompanied by aural tones and voice advisories. The EICAS system can also illuminate switch lights on specific system control panels to provide component/system status or to prompt corrective crew action.

The EICAS primary page displays the following information (Fig. 17-77):

- Engine compressor and turbine speeds (N1 and N2 rpm)
- Engine temperature (ITT)
- Fuel flow (FF)
- Oil pressure and temperature
- Engine vibration data

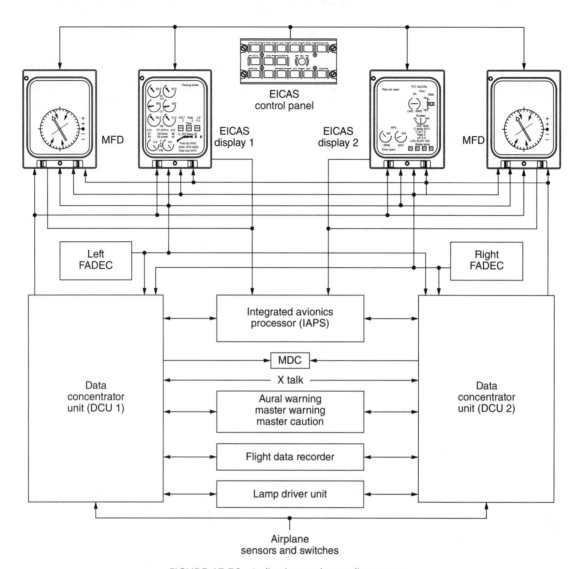

FIGURE 17-76 Indicating and recording system.

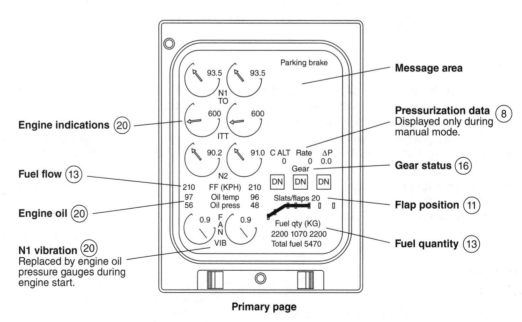

Primary page

FIGURE 17-77 EICAS primary page.

- Pressurization data
- Landing gear position
- Slat/flap position
- Fuel tank quantities and total fuel
- Crew alerting system (CAS) messages in the form of red warning and amber caution messages

The EICAS status page displays the following information (Fig. 17-78):

- Flight control trim indications
- Auxiliary power unit (APU) indications such as APU rpm, exhaust gas temperature EGT, and APU inlet door status
- Pressurization data such as cabin altitude, cabin rate of change, cabin pressure differential, and landing field elevation
- Oxygen system pressure
- Brake system temperature readouts
- Aircraft systems synoptic pages (via the EICAS control panel)
- MENU page (via the EICAS control panel) allows reset of the fuel used indicator and displays the engine oil quantity
- Crew alerting system (CAS) messages in the form of green advisory and white status messages

The EICAS control panel in Fig. 17-79 is used to select different modes of the EICAS system. The flight crew or maintenance crew can use the EICAS control panel to access the system synoptic pages. Aircraft system information is presented in the form of synoptic pages. Synoptic pages are simplified top-level schematic diagrams used for pilot and maintenance information. The synoptic pages are dynamic displays of the aircraft systems status and operation which includes all major components and parameter values. When a malfunction occurs, the affected component and/or parameter value will change color. System flow lines are green to indicate flow and white to indicate no flow. Status and malfunction messages are also included on the synoptic pages. The synoptic pages are selected by dedicated keys on the EICAS control panel (ECP) or by using the STEP key to sequence through the pages.

INSTALLATION AND MAINTENANCE OF INSTRUMENTS

Although the technician's maintenance activities do not normally involve the inspection or repair of an instrument's internal components, he or she is responsible for the proper instrument installation in the aircraft and for determining proper instrument operation.

Position of Instruments in the Panel

When installing instruments in the instrument panel, they are normally grouped according to basic purpose: flight instruments, powerplant instruments, and aircraft-system instruments. Be aware that many older aircraft had the instruments arranged in rather random patterns, with no special regard for the pilot's instrument-scanning requirements.

The modern flight instrument group is arranged as shown in Fig. 17-1. The grouping of the six instruments is centered around the attitude indicator (artificial horizon), which is the pilot's primary reference instrument for determining aircraft attitude. The instruments are grouped for maximum visibility and minimum eye movement so that the pilot's workload in reading the instruments is held to a minimum. It is becoming standard practice with many aircraft manufacturers to locate the instrument suction gauge within the flight instrument group so that the pilot can immediately be made aware of any loss of instrument suction for the pneumatically operated gyro instruments.

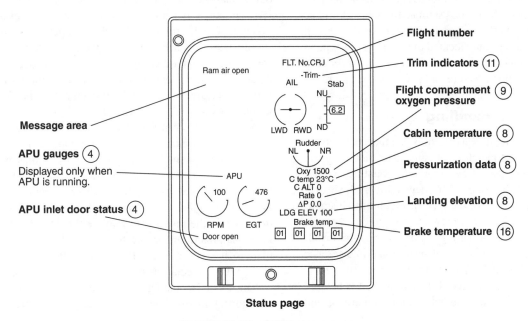

Status page

FIGURE 17-78 EICAS status page.

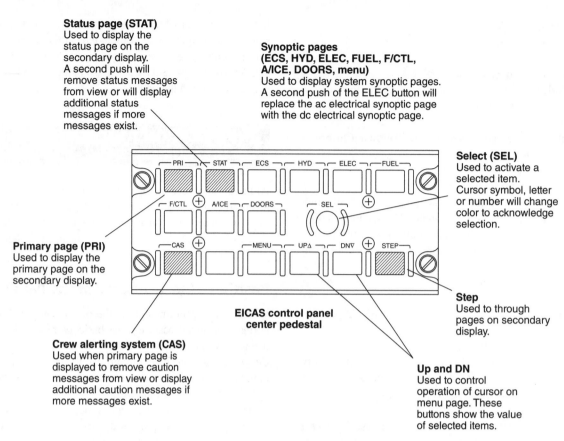

Status page (STAT)
Used to display the status page on the secondary display. A second push will remove status messages from view or will display additional status messages if more messages exist.

Synoptic pages (ECS, HYD, ELEC, FUEL, F/CTL, A/ICE, DOORS, menu)
Used to display system synoptic pages. A second push of the ELEC button will replace the ac electrical synoptic page with the dc electrical synoptic page.

Select (SEL)
Used to activate a selected item. Cursor symbol, letter or number will change color to acknowledge selection.

Primary page (PRI)
Used to display the primary page on the secondary display.

EICAS control panel center pedestal

Step
Used to through pages on secondary display.

Crew alerting system (CAS)
Used when primary page is displayed to remove caution messages from view or display additional caution messages if more messages exist.

Up and DN
Used to control operation of cursor on menu page. These buttons show the value of selected items.

FIGURE 17-79 EICAS control panel.

The grouping of powerplant instruments is not as standardized among the aircraft manufacturers as is the grouping of the flight instruments. However, all engine instruments that are directly affected by the operation of cockpit controls (i.e., tachometer, manifold pressure, and fuel flow) should be located as close to the controls as is practical so that the pilot will be able to monitor the instrument as he or she adjusts the control. Other powerplant and system instruments should be grouped according to the system being monitored—i.e., oil temperature and oil pressure located together—or by type of value being measured if only one instrument is used for each system—i.e., hydraulic pressure and pneumatic pressure.

Installation and Handling

The installation of instruments requires that they be mounted to a metal instrument panel or subpanel. The mounting can be achieved by the use of two or more mounting screws, circumferential clamps, or brackets, depending on the instrument design. Regardless of the type of mount, the installation should allow the pilot or crew member to clearly view the instrument from a normal flight position, and the installation should cause a minimum of operational interference with control systems and other instruments.

Because of the large variety of installations, it is beyond the scope of this text to deal with all installations, and

therefore only the more common types of installations are discussed here.

While some instruments require panel openings as small as 1 in [2.45 cm] in diameter and others require 4-in [10.16-cm] openings, the majority of instruments require panel openings $3\frac{1}{8}$ or $2\frac{1}{4}$ in. in diameter. Openings in the panel should be laid out so that the installation of the instrument will not interfere with any other instrument or any of the aircraft-control systems. The location of electrically operated instruments should take into account any effects on other equipment, such as the magnetic compass. Another consideration when determining the exact position of instruments is allowing sufficient clearance around the instrument so that the installation hardware will be accessible.

Mounting instruments by the use of screws usually involves positioning the instrument from behind the panel through the opening in the panel. Screw holes will have to be drilled at appropriate positions around the opening. Nonmagnetic screws, usually 440 size brass screws, are then used to attach the instrument to the panel. The mounting holes in the instrument may be threaded by the instrument manufacturer, or one may use a Tinnerman-type nut plate to accept the screw. If a separate fastener is required, brass nuts or special instrument mounting nuts can be used.

The use of a mounting bracket is common for some system-monitoring instruments such as engine oil pressure and oil temperature. These instruments are passed through the front of the panel until the flange of the instrument bears against the panel. The mounting bracket is then attached to the rear of the instrument and clamps the instrument to the panel.

Circumferential clamps are used in many modern instrument installations to allow ease of installation positioning and removal. The clamping mechanism is mounted to the instrument panel, and the instrument is then inserted into the clamp through the front of the instrument panel. The instrument can be rotated in the clamp until it is properly positioned, and the clamp is then tightened around the instrument by a screw on the front of the panel. Caution should be exercised when tightening the clamp, as excessive tightening may bind the instrument case against the instrument-operating mechanism. This may cause incorrect instrument indications or damage to the instrument.

Some aircraft manufacturers have mounted the instrument subpanel on sliding or hinged mounts so that the rear of the instrument panel is accessible without the technician having to work in the cramped and cluttered space behind the instrument panel. An example of this type of installation is shown in Fig. 17-80.

When handling instruments, treat them as delicate mechanisms. Avoid hitting or dropping the instrument, and avoid sudden movements of the instruments, since this may cause internal components to contact internal stops with some force. Never apply air pressure, suction, or electrical power to instruments unless directed to by the manufacturer's instructions or other competent references. When installing the instruments, never force them into position. If they do not fit properly in position, correct the fault with the mounting panel and then install the instrument.

When instruments are removed from an aircraft, all lines and openings should be covered with protective caps or plugs and all electrical connections should be properly protected. If the instrument is not going to be immediately reinstalled, it should be tagged with appropriate information as to time in service, when removed, and reason for removal. The instrument should then be placed in a plastic bag sealed against dirt and dust and stored in a safe place.

If the instrument is to be shipped, it should be placed in a well-padded container. The container should be marked with words such as "delicate instrument" or "fragile."

Marking Instruments

Many instruments require that marks be placed on the face of the instrument or on the cover glass to indicate operational ranges and limits. A technician is not normally permitted to remove the cover glass from an instrument and must, therefore, mark ranges and values on the exposed side of the cover glass of the instrument. Instrument-repair stations normally are involved in placing markings directly on the face of instruments.

Operational ranges and limitations can be obtained from the aircraft operations manual, maintenance manual, and Type Certificate Data Sheet/Aircraft Specification. Minimum and maximum operational limitations are indicated by a red radial line on the instrument. A green arc is used to indicate a normal operating range. A yellow arc on the instrument indicates a caution range. This caution range may require the use of a placard next to the instrument to describe the nature of the caution. This range may limit operation for specific periods of time or only under certain flight conditions.

The airspeed indicator has several types of markings that are not normally found on other instruments. These markings include a white arc to indicate the airspeed range through which full flaps can be used. A blue radial line is used on the airspeed indicators of multiengine aircraft to indicate the best single-engine rate-of-climb speed. Sailplanes often use a yellow triangle to indicate the minimum landing approach speed or best gliding speed.

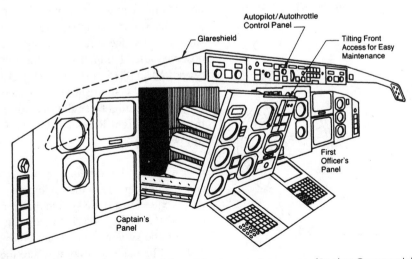

FIGURE 17-80   Some instrument panels slide out for easy maintenance. (*Boeing Commercial Aircraft Co.*)

A tachometer may incorporate a red arc to indicate a critical vibration range. This range should not be used for continuous operation.

When marking instructions by placing markings on the cover glass, a slippage mark should be placed on the glass and onto the case next to the glass. This mark will show if the glass has rotated from its original position, indicating that the markings no longer show the correct values.

Testing Instruments and Systems

Certain tests and adjustments can be made on some instruments by a qualified aviation maintenance technician in the field. In this case, a qualified aviation maintenance technician is one who is familiar with the particular instrument and the way the tests and adjustments should be made. The adjustments that can be made do not involve opening the case of the instrument. Whatever the instruments, the technician should consult the manufacturer's manual and FAA AC-43.13-1A & 2A.

Airspeed indicators are tested by applying controlled pressure to the pressure port of the instrument. A chart such as that shown in Table 17-1 is used as a guide. The pressure can be obtained from a properly regulated instrument test unit or by using a bulb-type hand pump. The pressure source and an accurate pressure gauge (or manometer) are connected to the pressure port of the airspeed indicator, as shown in Fig. 17-81. When a pressure of 0.1309 in Hg [3.32 mm Hg], 1.78 in H_2O [45.2 mm H_2O or 0.441 kPa] is applied to the instrument, the airspeed indication should be 60 mph [52.1 kn]. When the chart is followed, additional checks should be made at airspeed increments specified. This method of testing can be used on the airplane by applying the test pressure to the pitot pressure line.

An altimeter is tested by comparing the altitude reading on the instrument with the actual altitude at the point of testing when the local barometric pressure is adjusted into the instrument. For example, if the barometric setting of the altimeter is placed at 29.92 in Hg and the actual barometric pressure at the test location is 29.92 in Hg [76 cm Hg], then the altimeter should show an altitude of 0. If suction is applied to the suction port of the altimeter to a level of 28.86 in Hg [73.3 cm Hg], then the altimeter reading should be 1000 ft [305 m]. Additional tests should be made at specified intervals of altitude. *Positive pressure must not be applied to an altimeter* except in the small amount that may be specified by the manufacturer.

If it is found that the barometric setting and the altitude reading of an altimeter do not correlate, the instrument can be adjusted. A recommended procedure is as follows:

1. Rotate the barometric setting knob until the altimeter indicates the known altitude, exactly, of the test location.

2. Following the manufacturer's instructions, release the locking mechanism of the barometric setting device. Pull the barometric setting knob out until it is disengaged from the internal gearing. Be careful not to rotate the knob during this process.

3. Rotate the knob to the current correct barometric setting and then gently press the knob back until the internal gearing is meshed and engage the locking mechanism.

4. Check the operation of the barometric setting mechanism by rotating the knob. The movement should be smooth, and no grinding or sticking should be apparent.

If an adjustment of more than 40 ft [12.2 m] altitude is required to correlate the altimeter reading and the barometric scale, the instrument should be tested for scale error. This should be accomplished in an instrument shop. Adjustments to altimeters should be recorded in the aircraft's permanent maintenance record.

Pitot-static systems on aircraft should be tested on a regular schedule in accordance with the manufacturer's instructions. The pitot system may be tested by sealing the drain holes and connecting the system to a regulated pressure source with a manometer, pressure gauge, or reliable airspeed indicator. A pressure that would produce a reading of 150 kn is applied to the system by means of a bulb pump or instrument pressure source. This pressure is 1.1 in Hg [2.79 cm Hg] or 14.9 in H_2O [37.85 cm H_2O]. The pressure source is clamped off, and the pressure gauge or altimeter observed after 1 min. The decrease in pressure should not exceed 10 kn [0.15 in Hg or 2.04 in H_2O].

The static system should be tested in accordance with FAR 43 and flight regulations for aircraft flying under Instrument Flight Rules (IFR). The basic procedure is as follows:

1. Connect the test equipment (vacuum source and vacuum gauge) directly to the static ports, if practicable. Otherwise, connect to a static system drain or a tee connection and seal off the static ports. If the test equipment is connected to the static system at any point other than the static ports, it should be made at a point where the connection may be readily inspected for system integrity.

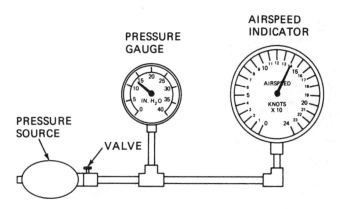

FIGURE 17-81   Test system for an airspeed indicator.

2. Apply a vacuum equivalent to 1000 ft [305 m] altitude [differential pressure of approximately 1.07 in Hg (3.62 kPa) or 14.5 in H_2O] and hold.

3. After 1 min, check to see that the leakage has not exceeded the equivalent of 100 ft [30.5 m] of altitude. This is approximately 0.105 in Hg [.36 kPa] or 1.45 in H_2O.

During the testing of pitot-static systems, care must be taken to see that positive pressure is not applied to the altimeter or negative pressure (vacuum) applied to the airspeed indicator. Positive pressure should not be applied to the static system, and negative pressure should not be applied to the pitot system when instruments are connected. Pressure in the pitot system must always be equal or greater than that in the static system.

When positive pressure is applied to the pitot system or negative pressure is applied to the static system, the rate of change of pressure applied should not exceed the design limits of the instruments connected to the system.

Upon completion of testing, the pitot and static systems must be restored to flight configuration. Static ports and drain holes should be unsealed, and drain plugs should be replaced. System inspection, instrument replacement, and required repairs are completed *before* the leak tests are made.

Inspection and Maintenance of Instruments and Systems

Instruments should be checked for proper operation, condition of the glass faces, condition and placement of range markings, condition of cases, cleanliness of case vent filters, security of mounting, and tightness of tube and electrical connections. The shock mounts by which panels or instruments are attached should be checked for looseness.

Gyro instruments should be checked for gyro erection time and unusual noise during operation. After operation, the run-down time provides an indication of condition. If the run-down time is shorter than normal, bearing wear or damage is indicated, or foreign matter has accumulated inside the instrument.

Instrument systems should be checked for condition of tubing, hose, hose clamps, and fittings. Bulkhead fittings should be checked for tightness in the bulkhead. When tubing fittings at bulkhead fittings are removed or replaced, the bulkhead fitting should be held with a wrench to prevent turning while the tubing fitting is being tightened.

Vacuum fittings should be checked carefully at all hose and fitting connections. If the hose is starting to deteriorate, particles can be drawn into the system and cause damage to the instruments and pumps in the system. When connecting fittings in a vacuum system, standard thread lubricant or thread sealers should not be employed. Particles of the material could be drawn into the system

and cause damage. Material such as silicone spray may be employed.

A vacuum system may be checked for operation by running the engine that drives the vacuum pump at a medium speed. The suction gauge should indicate the correct suction pressure (vacuum), about 5 ± 0.1 in Hg [16.93 kPa]. If the amount of suction is not correct, the engine should be stopped and the cause determined. The vacuum-regulator valve may have to be adjusted. The adjustment is made with a screw that is normally covered with a protective cap or secured with a locknut.

The routing of tubing and hoses for instrument systems should be checked for correct position. Tubing and hoses should be clear of any structure that could cause damage due to vibration. Tubing should be secured to the aircraft structure with approved clamps having rubber or plastic liners.

During complete inspections of instrument systems, it is usually advisable to blow air under pressure through the lines to remove water and dirt. *Before blowing air through a pitot or static system, all instruments connected to the system must be disconnected.* Air is then blown from the instrument end of the lines outward toward the static vents or pitot head. After the lines are blown clear, the instruments are reconnected and the system is pressure-tested for leakage.

During the operation of vacuum systems, the filters in the associated instrument cases or the air inlet lines become clogged with dust and dirt particles. When this occurs, the reading of the vacuum gauge increases. When the vacuum reaches the allowable limit, the filters in the instruments or in the inlet line must be removed and cleaned. In some cases, the filter element can be cleaned according to manufacturer's directions; in others, the filter element is replaced.

A typical vacuum system for a twin-engine airplane is shown in Fig. 17-82. A schematic drawing of the system is shown in Fig. 17-83. This system utilizes dry vacuum pumps, which require no lubrication. These are mounted on the accessory mount pads of each engine. The pump outlets are exhausted into the engine nacelles. The vacuum line plumbing is routed from the vacuum pumps through the nacelles and wings into the cabin and forward to the relief valves. The adjustable relief valves are provided to give the desired vacuum system pressure. From the relief valves, the lines are routed to the vacuum manifold located on the left side of the forward cabin bulkhead. The manifold has check valves included to prevent reverse flow in the event of failure of either vacuum pump. Hoses are routed from the manifold to the directional gyro, horizontal gyro, and suction gauge. Other hoses connect the gyros to the central vacuum air filter and suction gauge. The suction gauge indicates the amount of vacuum present in the system. The system is provided with in-operation indicator buttons for each pump. The vacuum air filter is provided to remove dust particles, vapor, and other contaminants such as tobacco smoke from the air. Experience has shown that the tars from tobacco smoke rapidly clog the filters and render them ineffective. Such filter elements should be discarded rather than cleaned.

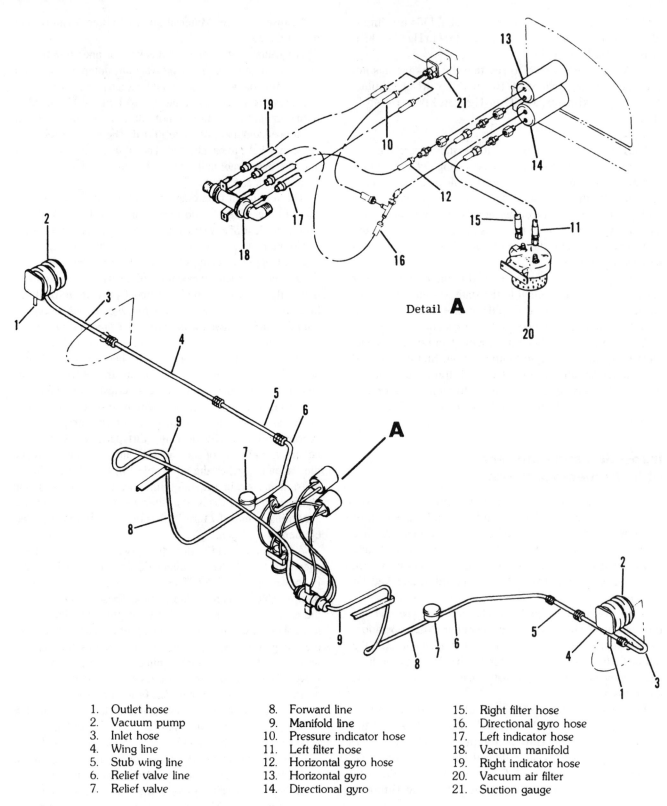

Detail **A**

1.	Outlet hose	8.	Forward line
2.	Vacuum pump	9.	Manifold line
3.	Inlet hose	10.	Pressure indicator hose
4.	Wing line	11.	Left filter hose
5.	Stub wing line	12.	Horizontal gyro hose
6.	Relief valve line	13.	Horizontal gyro
7.	Relief valve	14.	Directional gyro

15.	Right filter hose
16.	Directional gyro hose
17.	Left indicator hose
18.	Vacuum manifold
19.	Right indicator hose
20.	Vacuum air filter
21.	Suction gauge

FIGURE 17-82 Vacuum system for a twin-engine airplane. (*Cessna Aircraft Co.*)

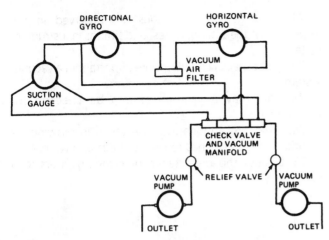

FIGURE 17-83 Schematic drawing of a vacuum system. (*Cessna Aircraft Co.*)

REVIEW QUESTIONS

1. In what ways are the knowledge and understanding of aircraft instruments and instrument systems important for the technician?

2. What type of mechanism is generally used for instruments that measure high pressures?

3. Describe the operation of a *bourdon tube.*

4. Explain the significance of *psia* and *psig.*

5. Describe a *diaphragm-type pressure instrument.*

6. Which pressure-sensing instruments are referenced to absolute zero?

7. List uses of an *airspeed indicator* in flight.

8. Explain the operation of an airspeed indicator.

9. What correction should be applied to the airspeed indication as altitude increases to obtain true airspeed?

10. Compare a *Machmeter* with an *airspeed indicator.*

11. Explain the operation of an *altimeter.*

12. Why is the barometric setting important in the use of an altimeter?

13. What is the purpose of an *encoding altimeter*?

14. What are the two basic types of altimeters and how do their displays differ?

15. How does a blind encoding altimeter differ from an encoding altimeter?

16. What is the difference between *indicated, calibrated,* and *true* airspeed?

17. Describe the operation of a *vertical-speed indicator.*

18. How does a *variometer* work?

19. Describe the purpose of a *total-energy variometer.*

20. What is the value of a *suction gauge* in an instrument system?

21. What methods may be used to power gyro instruments?

22. What is the purpose of a *venturi tube* in an instrument system?

23. Explain the function of a *pitot-static* system.

24. What is the purpose of a pitot system?

25. How is water removed from a pitot system?

26. Explain the alternative static system.

27. What would be the effect of a leaking or disconnected static line inside a pressurized airplane?

28. Describe the construction of a *magnetic compass.*

29. What is meant by *variation* in a magnetic compass?

30. What are causes of *deviation* in compass readings?

31. Describe the procedure for *swinging* a magnetic compass?

32. Explain the basic operational concept of a *flux gate.*

33. Explain the operation of a *bimetallic temperature gauge.*

34. What is the purpose of an *accelerometer*?

35. Describe the operation of an accelerometer.

36. How does a horizontal situation indicator differ from a heading indicator?

37. What are two types of *gyro erection mechanisms* and what is their purpose?

38. Explain the features of a spinning gyroscope that make it useful in some types of instruments.

39. Explain *precession.*

40. Explain the operation of a *directional gyro.*

41. Describe the operation of a *turn-and-bank indicator.*

42. How does the gyro indicate a rate of turn?

43. How does a turn-and-bank indicator differ from a turn coordinator?

44. Which instrument is used to indicate a proper bank?

45. Describe a *gyro-horizon indicator.*

46. Discuss the handling of instruments to avoid damage.

47. How should instruments be prepared for shipping or storage?

48. How is foreign material prevented from entering instruments?

49. Explain the importance of shock mountings.

50. What is the purpose of an *air data computer*?

51. What are three types of *tachometers* and how does each operate?

52. Explain the basic operation of an engine-pressure ratio gauge.

53. How does a *resistance-type fuel-quantity indicator* operate?

54. Explain the operation of a *vapor-pressure temperature gauge.*

55. What are the advantages of a *capacitance fuel-quantity-indicating system* when compared to a resistance-type system?

56. Explain the operation of one type of fuel flowmeter.

57. Describe three types of mechanical fuel-quantity gauges.

58. What is a *symbol generator*?

59. What are the two flight displays in an electronic flight instrument system?

60. Draw the conventional flight instrument group arrangement.

61. Describe the different types of instrument-mounting methods.

62. Describe the different types of instrument range and limit markings.

63. What type of hardware is used to attach instruments to panels?

64. Explain how an airspeed indicator may be tested.

65. How is an altimeter tested and adjusted?

66. Describe the testing of a pitot system.

67. Describe the testing of a static system.

68. What precautions must be observed in applying pressures to pitot and static systems?

69. During the inspection of instruments, what conditions are examined?

70. What checks should be made during the operational tests of gyro instruments?

71. What general conditions are observed in the inspection of the tubing, hoses, and fittings in instrument systems?

72. What special conditions are important in a vacuum system?

73. How may a vacuum system be checked for operation?

74. What precautions must be taken in blowing air through pitot-static lines to clear out dirt and water?

75. Discuss the importance of cleaning filters in a vacuum system.

Auxiliary Systems 18

INTRODUCTION

The majority of modern aircraft, primarily the large transport type, are equipped with certain systems that are not necessary for the actual operation and flight of the airplane but are needed for the comfort and convenience of the crew and passengers and may be required by Federal Aviation Regulations (FARs). Some of these systems are important for the safe operation of the aircraft under a variety of conditions, and some are designed to provide for emergencies.

Systems not essential to the actual operation of the aircraft are commonly called **auxiliary systems**. Among such systems are ice and rain protection, fire-warning and fire-extinguishing systems, water and waste systems, position and warning systems, and auxiliary power units (APUs).

FIRE PROTECTION SYSTEMS

Fire protection systems on aircraft usually consist of two separate operating systems with associated controls and indicators. One system is for fire or overheat **detection**, and the other is for fire suppression or extinguishing. In some cases, the systems can be interconnected so extinguishing takes place automatically when a fire is detected.

Requirements for Overheat and Fire Protection Systems

The requirements for fire protection in aircraft are set forth in FAR Part 25. These regulations specify the types of detecting and suppression devices and systems required in accordance with the classification of areas of the aircraft and the conditions existing in these areas.

Certain general features and operational capabilities for fire warning and protection systems must be met or exceeded if such systems are to be used in certificated aircraft. These are as follows:

1. The fire warning system must provide an immediate warning of fire or overheat by means of a red light and an audible signal in the cockpit or flight compartment.

2. The system must accurately indicate that a fire has been extinguished and indicate if a fire reignites.

3. The system must be durable and resistant to damage from all the environmental factors that may exist in the location where it is installed.

4. The system must include an accurate and effective method for testing to ensure the integrity of the system.

5. The system must be easily inspected, removed, and installed.

6. The system and its components must be designed so the possibility of false indications is unlikely.

7. The system must require a minimum of electrical power and must operate from the aircraft electrical system without inverters or other special equipment.

Types of Fire or Overheat Detectors

High temperatures caused by fires or other conditions can be detected by a variety of devices. Among these are thermal switches, thermocouples, and tubular detectors. Tubular detectors are commonly employed in large aircraft for fire and heat detection wherever fire protection must be provided.

A thermal-switch fire detection system is simply a circuit in which one or more thermal switches are connected in an electrical circuit with a warning light and an aural alarm unit to warn the pilot or flight crew that an overheat condition exists in a particular area. If more than one thermal switch is in the circuit, the switches will be connected in parallel, so the closing of any one switch will provide a warning. A thermal switch is shown in Fig. 18-1, and the circuit for a thermal-switch fire and overheat warning system is shown in Fig. 18-2. Note that a test circuit is included so the system may be tested for operation.

The thermal switch, called a spot detector, works by the expansion of the outer case of the unit. When the detector is exposed to heat, the case becomes longer and causes the two contacts inside the case to be drawn together. When the contacts meet, the electrical circuit is completed and the alarm activates.

The thermocouple detection system, also called a "rate-of-rise" detection system, utilizes one or more thermocouples connected in series to activate an alarm system when there is a sufficiently high rate of temperature increase at the sensor. The thermocouple is made of two dissimilar metals, such as Chromel and constantan, which are twisted together

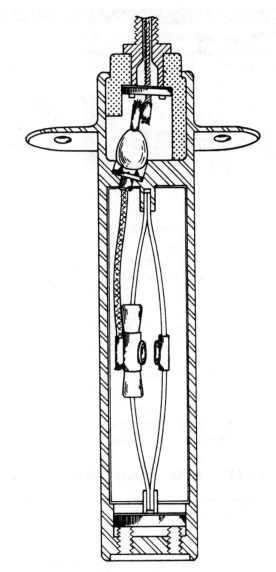

FIGURE 18-1 Fenwal spot detector.

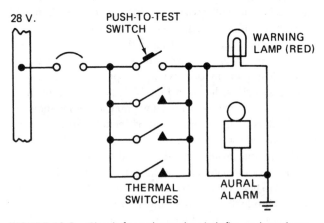

FIGURE 18-2 Circuit for a thermal-switch fire and overheat warning system.

and located inside an open frame, as shown in Fig. 18-3. The frame protects the sensing wires from damage while allowing a free flow of air over the wires. The exposed wires make up the hot junction. A cold junction is located behind insulating material in the sensor unit. Where there is a difference in

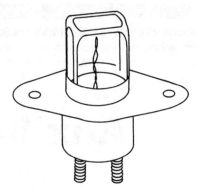

FIGURE 18-3 A thermocouple heat detector.

temperature between the hot junction and the cold junction, a current is created. When sufficient current is being generated (about 4 mA), a sensitive relay in a relay box closes, activating a slave relay and causing the alarm to activate. The basic circuit for this system is shown in Fig. 18-4. If the rate of temperature increase is slow enough so that the temperature of the cold junction increases along with the hot junction, the sensitive relay will not close and the alarm will not activate.

A test circuit is provided for the system through the use of a heater next to a thermocouple. When the heater is energized, the thermocouple will generate sufficient current to activate the system.

There are three types of tubular sensing devices, called "continuous-loop" systems, commonly employed in modern aircraft for detecting overheat or fire. These tubular sensors are manufactured in lengths from 18 in [0.46 m] to more than 15 ft [4 to 5 m]. The diameters of these sensing elements may be from less than 0.060 in [1.5 mm] to more than 0.090 in [2.3 mm].

Cross-sectional drawings to illustrate the operating principles of the three most commonly employed sensors are shown in Fig. 18-5. These are the sensors manufactured by the Fenwal Company, the Walter Kidde Company, and the Systron-Donner (Meggit safety systems) Company.

The Fenwal sensor consists of a small [0.089-in OD (2.3-mm)], lightweight, flexible Inconel tube with a pure nickel wire–center conductor. The space between the nickel conductor and the tubing wall is filled with a porous aluminum-oxide, ceramic insulating material. The voids and clearances between the tubing and the ceramic material are saturated with a eutectic salt mixture, which has a low melting point.

The nickel wire in the center of the tube is insulated from the tube wall by the ceramic and eutectic salt materials. The tube is hermetically sealed at both ends with insulating material, and threaded fittings are located at each end of the tube.

When heated sufficiently, current can flow between the center wire and the tube wall because the eutectic salt melts, and its resistance drops rapidly when the temperature reaches a given level. The elevated temperature will cause a response at any point along the entire length of the sensing element. The increased current flow between the nickel center wire and the tubing wall provides the signal, which is utilized in

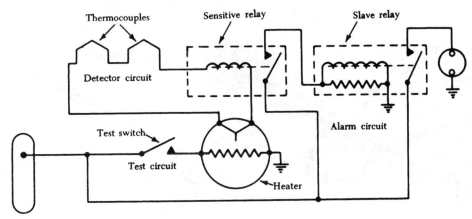

FIGURE 18-4 Thermocouple fire warning circuit.

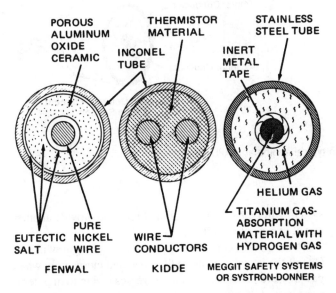

FIGURE 18-5 Drawings to illustrate the operation of overheat and fire sensors.

the electronic **control unit** to produce the output signal that actuates the alarm system.

When the fire is extinguished or the overheat condition is corrected, the eutectic salt in the sensing element increases in resistance and the system returns to the standby condition. If the fire should reignite, the sensor would again produce a signal for an alarm.

The sensing element of the Kidde system consists of an Inconel tube filled with a thermistor material. Two electrical conductors are embedded in this material, and one of the conductors is grounded to the outer shell of the connector at the end of the tube. Electrical connectors are provided at both ends of the tube.

In the Kidde sensing element the resistance of the thermistor material decreases rapidly when a high temperature is applied. This change in resistance is sensed by the electronic control circuit monitoring the system, and the control provides the warning signal to illuminate the fire warning light and activate the aural warning device. This sensor returns to a normal (standby) condition when the fire or overheat condition is corrected.

The **sensing element assembly** utilized in the Kidde fire and overheat warning system consists of a pair of sensing elements mounted on a preshaped, rigid support tube. The sensing elements are held in place by dual clamps riveted to the tube assembly approximately every 6 in [15 cm]. Asbestos grommets are provided to cushion the sensor tubing in the clamps. Four nuts per assembly are used to secure the sensing element end connectors to the bracket assemblies at either end of the support-tube assembly. The nuts are secured with safety wire.

The two sensing elements mounted on the support-tube assembly provide separate sensing circuits, and each is connected to its own electronic control circuit mounted on a separate circuit card. This arrangement provides for complete redundancy, thus ensuring operation of the system even though one side may fail.

A Kidde (left) and Fenwal (right) sensor assembly is shown in Fig. 18-6. These assemblies are preshaped to fit the contours of the area in which they are installed.

A drawing of the control unit for the Kidde system is shown in Fig. 18-7. Signals from the sensing units are processed in the control unit to provide visual and audible warnings in the flight compartment. The control unit is basically a transistorized electronic device consisting of two component board assemblies, a test switch, test jacks, wiring harness, and an electrical receptacle, all enclosed in a metal case.

There are two identical circuits in the control unit, one monitoring sensing element loop A and the other monitoring sensing element loop B. Each circuit is contained on its own component board assembly. Remember that in the Kidde system, two sensing elements are mounted on a preformed support-tube assembly. If one circuit should fail, the other circuit will provide the overheat or fire warning signal.

The sensing element produced by Meggit safety systems also known as Lindbergh or Systron-Donner, shown in Fig. 18-8, is pneumatic in operation. The pressure of the gas inside the element is increased by heat, and the increased pressure actuates a diaphragm switch inside the responder, which closes the circuit and provides the warning signal.

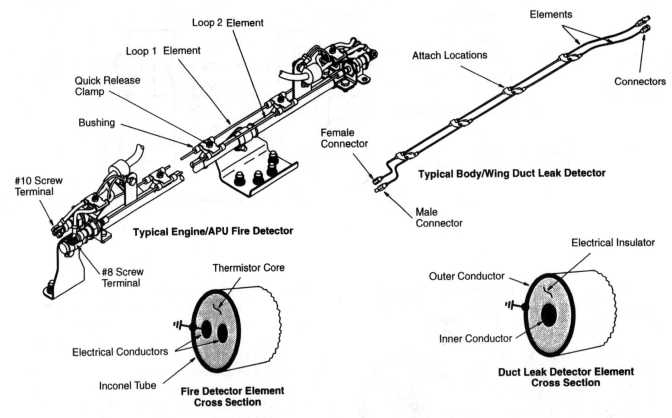

FIGURE 18-6 Fire and duct leak detectors used in typical engine and APU fire detection systems.

The sensing element consists of a stainless steel tube containing two separate gases plus a gas-absorption material in the form of a wire inside the tube. Under normal conditions, the tube is filled with helium gas under pressure. The titanium center wire, which is the gas-absorption material, contains hydrogen gas. The wire is wrapped in a helical fashion with an inert metal tape for stabilization and protection. Gaps between the turns of the tape allow for rapid release of the hydrogen gas from the wire when the temperature reaches the required level.

The sensor responds in accordance with the law of gases. If the volume of gas is held constant, its pressure will increase as temperature increases. The helium gas in the tube exerts a pressure proportional to the average temperature along the entire length of the tube. This is the **averaging**, or **overheat, function** of the sensor. If the average temperature exceeds a specified level, the helium gas pressure will be such that it closes the pneumatic switch in the responder and signals an overheat condition. If there is a very high temperature, as a fire would cause, anywhere along the sensing element, the center wire in the tube will release a large quantity of hydrogen gas. This will increase the total gas pressure in the tube to a level that will close the pneumatic switch. This is called the **discrete function** of the sensor. These operations are shown in Fig. 18-9.

When a fire is extinguished and the temperature begins to drop, the specially processed titanium wire in the tube will reabsorb the hydrogen gas and reduce the pressure in the tube. This will cause the pneumatic switch to open, and the system will be back to normal and ready to provide another signal in case of reignition.

The responder contains two identical diaphragm switches. One of the switches is normally open and closes only when gas pressure in the sensor tube increases owing to high temperature or fire in the area where the sensor is installed. The other switch is held closed by the normal helium pressure in the sensor tube. If the helium pressure should be lost, the switch opens the test circuit. When the test switch is closed by a member of the crew to check the circuit, the alarm will not respond and the loss of helium will be revealed.

The continuous-loop (Kidde and Fenwal) and continuous-length (Systron-Donner) types of fire detection mechanisms are considered superior to the spot and thermocouple systems where large areas must be covered, such as around a jet engine. The continuous-loop systems do have a disadvantage in that damage to the tubing wall, which will cause it to be closer to the center elements, may cause false fire signals to be generated.

The continuous-length sensor is not as sensitive to damage as are the continuous-loop sensors. The outer tubing of these units can be bent, dented, kinked, and otherwise distorted without their effectiveness being affected. The wire inside the tube prevents complete collapse, so there is always room for gas to flow. The only cause of failure is loss to the helium gas because the tube is worn or cut sufficiently to allow the gas to escape.

Each of the different sensor mechanisms can be selected for various operating temperatures. For example, the normal

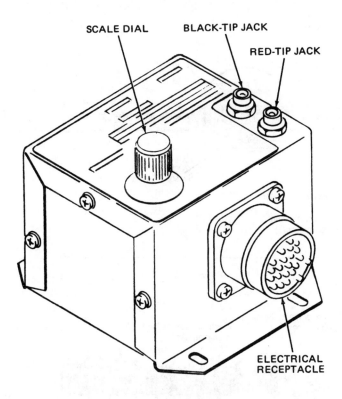

SCALE DIAL   BLACK-TIP JACK

RED-TIP JACK

ELECTRICAL
RECEPTACLE

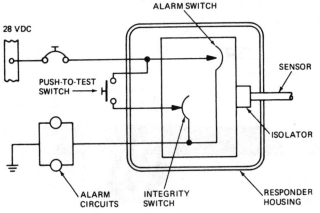

ALARM SWITCH

28 VDC

PUSH-TO-TEST
SWITCH

SENSOR

ISOLATOR

ALARM
CIRCUITS

INTEGRITY
SWITCH

RESPONDER
HOUSING

FIGURE 18-7   Control unit for Kidde system. (*Walter Kidde & Co.*)

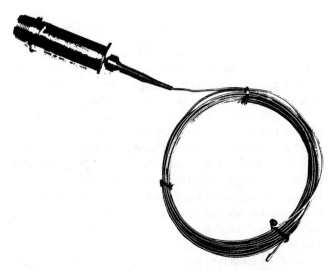

FIGURE 18-8   A Systron-Donner pneumatic fire detection unit. (*Systron-Donner Corp.*)

temperatures surrounding the turbine-engine combustion chamber will be much higher than the normal temperatures in the area of the engine inlet. A sensor selected for the inlet area of the engine should actuate the alarm system at a lower temperature than the sensor used near the engine combustion chamber. Different temperature-rating detectors may be included in a common alarm system such as the one shown in Fig. 18-10. The section from point 1 to point 2 uses a detector that will activate the alarm at 425°F [218°C], the area from 2 to 3 will activate the system at about 640°F [338°C], and the section between 3 and 4 will indicate a fire or overheat when the temperature exceeds 450°F [232°C]. For this reason, when installing or inspecting sensor units, always be sure that the proper heat rating is being used.

Installation and Routing of Sensing Units

The installation of overheat and fire warning sensing units must be done strictly in accordance with the instructions provided by the manufacturer. The routing of the sensors has been designed by engineers to provide the most effective performance and the detection of overheat or fire in the most likely areas. Routing must also take into consideration possible damage to the sensors. For example, in a baggage or cargo compartment, the sensors must be placed where there is no chance that cargo or luggage will strike them and cause damage.

Sensors must be supported with specially designed clamps in which the small tubing is held in rubber or soft plastic grommets. These prevent damage due to vibration and wear. The clamps with grommets must be spaced as specified in the appropriate instructions.

A typical method for installing the sensing element in an airplane is shown in Fig. 18-11. Electrical connector fittings are employed to connect the sensor into the circuit. The tubing of the sensor is held in place by means of mounting clips. The grommet placed around the tubing in the clip is essential to prevent damage due to chafing and vibration. The sensing element must be routed as described in the aircraft maintenance manual. A Systron-Donner system would be mounted in a similar manner except that connections are only made at the responder.

Typical installations for continuous systems are shown in Figs. 18-12 and 18-13.

Smoke- and Toxic-Gas-Detection Systems

Smoke-detection systems are usually installed to monitor the condition of the air in cargo and baggage compartments, where considerable smoke may be generated before the heat level reaches the point at which it sets off the overheat and fire warning system. Detectors (sensors) are placed in locations at which they are most likely to provide the earliest possible warning.

Toxic-gas (carbon monoxide) detectors are usually installed in cockpits and cabins, where the presence of the gas would affect the flight crew and passengers. Carbon monoxide (CO) is a clear gas and is not detectable by the type of smoke detectors that rely on a change in the transmission of light through a sample of air.

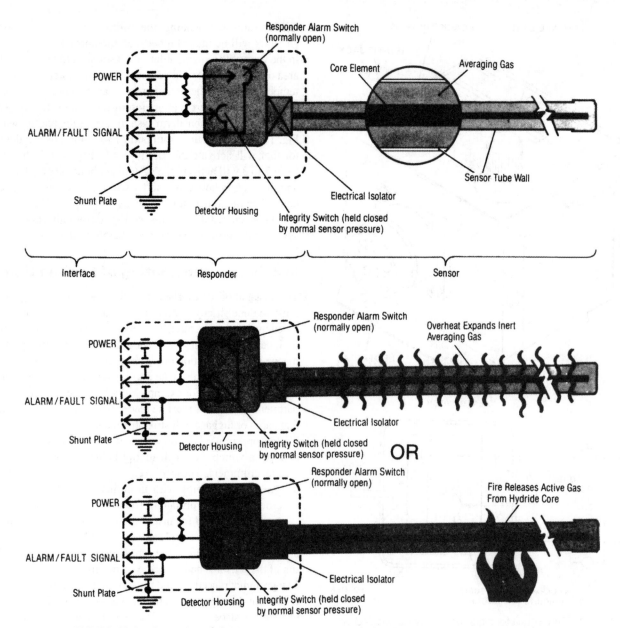

FIGURE 18-9 Operation of the Systron-Donner pneumatic fire detector. (*Systron-Donner Corp.*)

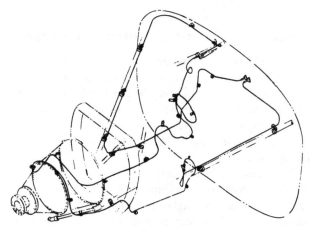

FIGURE 18-10 The routing of a continuous-loop fire detection assembly in an engine compartment. (*Cessna Aircraft Co.*)

Smoke and flame detectors operate according to several different principles. Among these are light detection, light refraction, ionization, and the change in resistance of a solid-state semiconductor material.

In a light-detection system, a photoelectric cell such as the one in Fig. 18-14 is placed in a location where it can "see" the surrounding area and produce a change of current flow when there is a change in the visible light or infrared radiation striking the cell. (Units are designed to detect either visible light or infrared radiation.) This change in current flow is used to activate an amplifier circuit, which produces the visual and aural alarm signals. This type of system is activated only by an open flame.

In a light-refraction detector, shown in Fig. 18-15, a light beam is passed through the detection chamber, and a photoelectric cell is placed in the chamber, where it is shielded

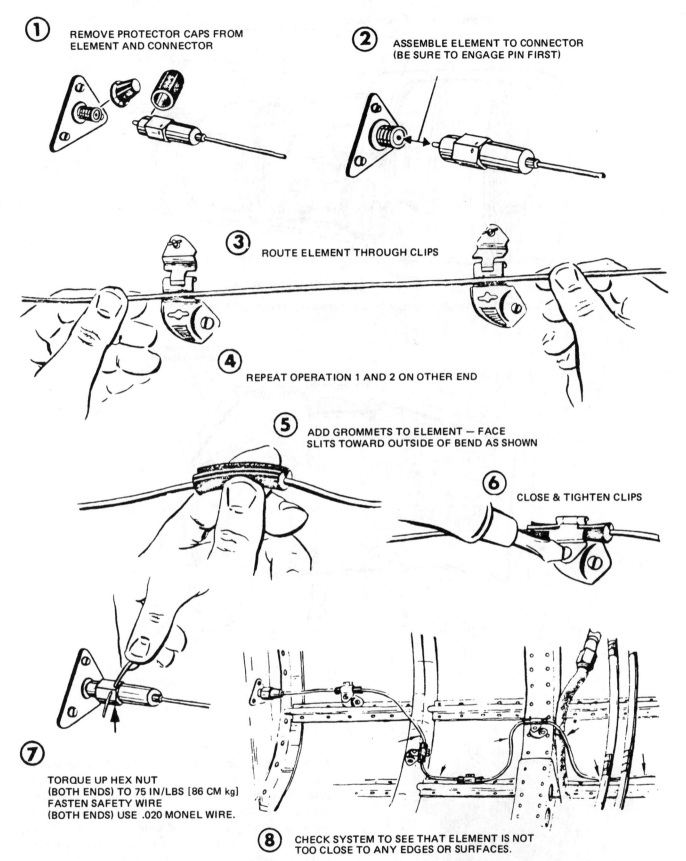

① REMOVE PROTECTOR CAPS FROM ELEMENT AND CONNECTOR

② ASSEMBLE ELEMENT TO CONNECTOR (BE SURE TO ENGAGE PIN FIRST)

③ ROUTE ELEMENT THROUGH CLIPS

④ REPEAT OPERATION 1 AND 2 ON OTHER END

⑤ ADD GROMMETS TO ELEMENT — FACE SLITS TOWARD OUTSIDE OF BEND AS SHOWN

⑥ CLOSE & TIGHTEN CLIPS

⑦ TORQUE UP HEX NUT (BOTH ENDS) TO 75 IN/LBS [86 CM kg] FASTEN SAFETY WIRE (BOTH ENDS) USE .020 MONEL WIRE.

⑧ CHECK SYSTEM TO SEE THAT ELEMENT IS NOT TOO CLOSE TO ANY EDGES OR SURFACES.

FIGURE 18-11   Installation procedure for a tubular sensing unit. (*Fenwal, Inc.*)

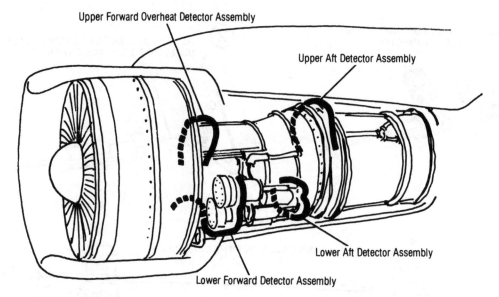

FIGURE 18-12   Pneumatic sensor locations around a jet engine. (*Systron-Donner Corp.*)

Upper Forward Overheat Detector Assembly

Upper Aft Detector Assembly

Lower Aft Detector Assembly

Lower Forward Detector Assembly

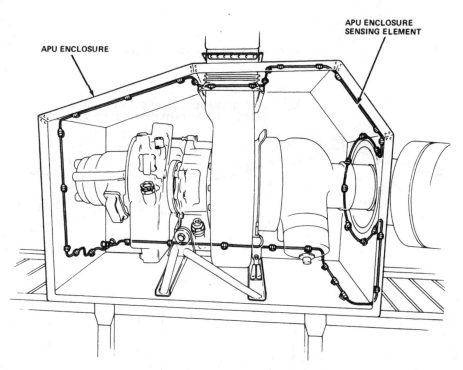

FIGURE 18-13   Continuous-loop sensor in an auxiliary power unit (APU) compartment. (*Canadair Inc.*)

APU ENCLOSURE

APU ENCLOSURE
SENSING ELEMENT

FIGURE 18-14   A flame detector. (*Pyrotector.*)

from the direct light of the light source. Air from the area being monitored for smoke and fire is passed through the detector unit by normal airflow or by means of a fan or other device. When smoke is introduced into the chamber, the light from the particles of smoke is reflected into the photoelectric cell. This changes the resistance (conductivity) of the photoelectric cell and changes the flow of current through the cell. This change in the flow of current activates the amplifier and produces the visual and aural warning signals. A schematic drawing of the chamber is shown in Fig. 18-16 and a typical system wiring diagram is shown in Fig. 18-17.

FIGURE 18-15 A smoke detector. (*Pyrotector.*)

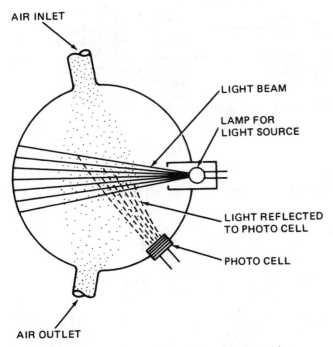

AIR INLET

LIGHT BEAM

LAMP FOR
LIGHT SOURCE

LIGHT REFLECTED
TO PHOTO CELL

PHOTO CELL

AIR OUTLET

FIGURE 18-16 Detector utilizing light refraction with a photoelectric cell.

In an ionization-type smoke detector, a small amount of radioactive material is used to bombard the oxygen and nitrogen molecules in the air within the detection chamber. The ionization that takes place permits a small current to flow through the chamber and in the external circuit. Air from the area being checked for smoke is passed through the chamber, as shown in Fig. 18-18. If smoke is in the air,

small particles of the smoke attach themselves to the oxygen and nitrogen ions and reduce the flow of current. When the current level is reduced by a predetermined amount, the alarm circuit will be triggered to produce the visual and aural alarm.

The solid-state type of smoke or toxic-gas detection utilizes two heated, solid-state detecting elements. Each element is approximately 1 mm [0.0394 in] in diameter and consists of a heating coil encased in a coating (substrate) of the solid-state (semiconductor) material. The composition of the sensors is such that ions of carbon monoxide or nitrous oxide will be absorbed into the solid-state coating of the sensing element and change its current-carrying ability.

The system operates by comparing the electrical output of the two sensors, one of which is exposed to outside air and the other to cabin air. The sensors are connected in separate legs of a modified bridge circuit. When both are conducting equally, there will be no output and no signal. If the sensor sampling cabin air absorbs toxic gases from the cabin air inlet, the bridge circuit will be unbalanced, and this will produce the current flow that activates the warning system. The warning is provided by an illuminated annunciator unit. An aural signal can also be produced.

A circuit to illustrate how electrical-type sensors are connected to develop a warning signal is shown in Fig. 18-19. The bridge part of the circuit is designed to be in balance when the two sensing elements are conducting the same current. If one of the sensors changes its resistance or conductivity because of exposure to light or smoke, the bridge is unbalanced, and a small current flow will occur through the sensitive relay. When this current reaches the predetermined alarm level, the relay will close and send current to the main signal relay. This relay closes the circuit to the alarm elements (light and aural warning unit) to provide the warning.

Fire-Extinguishing Agents

Fire-extinguishing agents are those chemicals that are injected into a compartment or area to extinguish a fire. These agents work by either displacing the oxygen or chemically combining with the oxygen to prevent combustion. Some additional extinguishing effect can occur by the low temperature at which the agents are discharged.

The commonly used agents are carbon dioxide (CO_2), Freon (a chlorinated hydrocarbon), and Halon 1301 (monobromotrifluoromethane—CF_3Br). Nitrogen (N_2) is also an extinguishing agent but is used primarily in current systems as a propellant for one of the other chemicals. Freon and Halon are in a liquid state when under sufficient pressure but become gaseous when released to atmospheric pressure. Liquid Freon and Halon must not be allowed to come into contact with the skin because they will cause frostbite due to the extremely low temperatures attained when the liquid evaporates.

Most modern aircraft extinguishing systems make use of Halon 1301 as the extinguishing agent, whereas some manufacturers specify Freon. The use of CO_2 is usually limited to older, reciprocating-engine-powered transports.

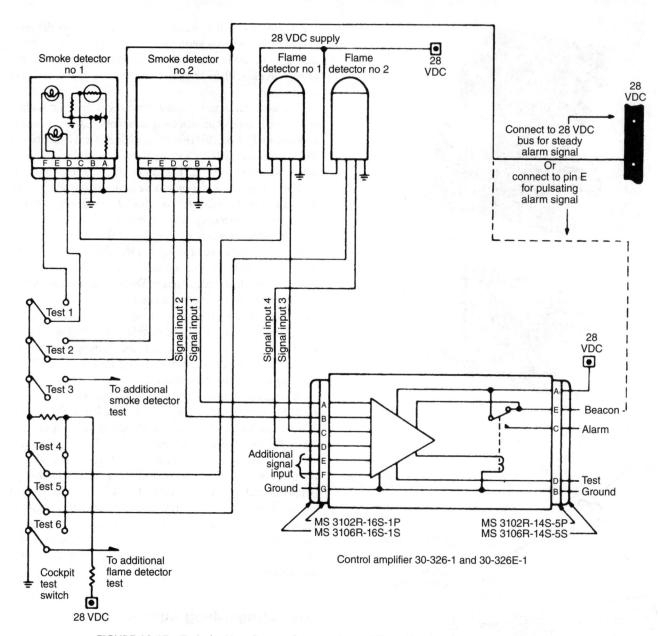

FIGURE 18-17 Typical wiring diagram for a smoke and flame fire detector system. (*Pyrotector.*)

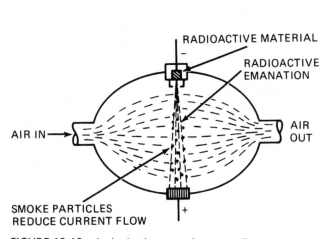

FIGURE 18-18 An ionization-type detector cell.

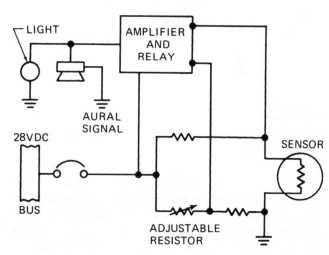

FIGURE 18-19 Circuit to illustrate the use of a resistance bridge to compare sensor resistance with that of a reference resistor.

Dry-chemical extinguishing agents are not used for aircraft fire-extinguishing systems, and their use for hand-held extinguishers has decreased greatly. Dry chemical agents are toxic and corrosive, whereas other agents, being gaseous, do not cause corrosive action to occur. Some of the gaseous agents may be considered toxic while they are present in an area in large quantities because of the displacement of free oxygen.

While Halon 1301 is still in service, the production of this ozone-depleting agent has been restricted. Although not required, consider replacing Halon 1301 extinguishers with Halon replacement extinguishers when discharged. Halon replacement agents found to be compliant to date include the halocarbons HCFC Blend B, HFC-227ea, and HFC-236fa.

Fire-Suppression Systems

Fire-suppression or fire-extinguishing systems usually consist of a fire-extinguishing agent stored in pressurized containers, tubing to carry the extinguishing agent to areas that require protection, control valves, indicators, control circuitry, and associated components. Systems vary considerably on different aircraft; however, the basic elements are similar.

Extinguishing-agent containers for modern aircraft are usually spherical or cylindrical in shape. Typical containers are shown in Fig. 18-20. The containers shown are pressure tested hydrostatically at 1500 psi [10 342 kPa]. When charged, the spherical container holds 60 lb [27.22 kg] of Halon, and the cylindrical container holds 105 lb [47.63 kg] of Halon. An additional 300-psi [2068-kPa] nitrogen charge is placed in the containers to expel the agent. A discharge head containing an explosive cartridge is installed on one or both of the container necks to discharge the container by rupturing the disk when the cartridge is activated. One type of cartridge is shown in Fig. 18-21. The squib is the explosive charge that drives the slug through the disk. A screen in the discharge head prevents particles of the disk from entering the deployment lines.

A thermal pressure relief fitting is located on each bottle. If the bottle pressure exceeds approximately 1400 psi [9653 kPa],

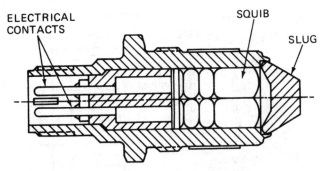

FIGURE 18-21 Explosive cartridge. (*Walter Kidde & Co.*)

the fitting will discharge the bottle contents either overboard or into the compartment containing the bottles.

Older aircraft still using CO_2 as the extinguishing agent use one or more gas cylinders equipped with quick-acting flood valves. When activated, the flood valves rapidly discharge the bottles into the distribution lines. The bottles have a thermal relief mechanism that automatically discharge the bottles when the internal pressure exceeds 2650 psi [18271.75 kPa].

Two methods are used to allow the pilot or technician to determine if the bottle has discharged thermally: a pressure gauge or a "blowout" disk. Some aircraft have a gauge for the fire bottles visible from the outside of the aircraft or through access panels. If the gauge does not indicate a charge, then the bottle has been discharged. Other aircraft have a discharge line running from the bottle to the outside of the aircraft or to an open area such as a wheel well. The discharge line opening is covered with a red disk. If the bottle has discharged, the disk will be blown out of the discharge fitting. The absence of the red disk indicates a thermal discharge.

A similar system is used to indicate if the system has been discharged by actuation of the extinguishing system. In this case a small line from the system distribution line leads to a disk-actuator line. This line is covered with a yellow disk. When the extinguishing system is activated, the yellow disk is pushed out by a plunger.

There are two classifications of fire-extinguishing systems, the *conventional system* and the *high-rate-of-discharge system* (HRD system).

The **conventional system** is normally found on older reciprocating-engine-powered aircraft and is based on the design concept of the systems first used on aircraft. Although this type of system is adequate, it is not as efficient as the HRD systems. The conventional system usually uses CO_2 as the extinguishing agent and makes use of a perforated ring and distributor-nozzle discharge arrangement. When the system is activated, the CO_2 bottles are opened and the gas flows through the lines to the selected engine. At the engine the gas flows out of the perforated ring and distributor nozzles to smother the fire.

The HRD system uses Freon or Halon 1301 and the spherical bottles actuated by explosive cartridges. The discharge tubes are configured to allow a rapid release of agent into the fire area and flood the compartment to eliminate the fire quickly. The advantage of this system over the conventional

FIGURE 18-20 Typical extinguishing-agent containers. (*Walter Kidde & Co.*)

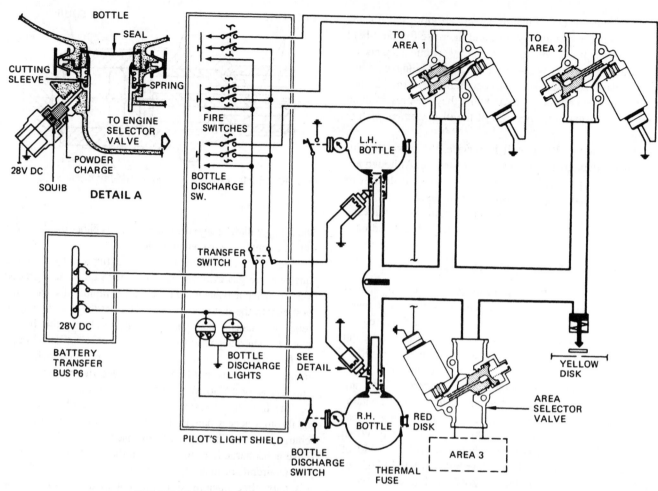

FIGURE 18-22 Schematic diagram of a fire-suppression system. (*Boeing Co.*)

system lies in the ability to flood a compartment much more quickly.

Both conventional and HRD systems may be designed to allow only one extinguishing agent to discharge into a fire area, or they may be designed to allow several discharges into an area in an attempt to extinguish the fire. In most aircraft only two discharges are possible for any one location. The exact configuration of a system is determined by the aircraft manufacturer.

A schematic diagram of a typical fire-suppression system for an airliner is shown in Fig. 18-22. The system consists of two steel containers charged with Freon and nitrogen gas to a pressure of 600 psi [4136 kPa], connecting deployment lines to the control valves and control circuitry. The system is activated when a member of the crew closes the fire switch to direct extinguishing agent to the area where a fire is indicated.

When the fire-extinguishing switch is closed, an explosive charge at the neck of the selected agent container is detonated and a cutter is driven through the sealing disk in the neck of the container. This instantly releases the extinguishing agent from the container and permits it to flow to the area selected. The pilot or other crew member will have selected the appropriate area by operating a switch on the fire control panel. This will direct the agent through the correct deployment line.

Figure 18-23 shows the extinguisher system used on a modern turboprop aircraft. Each engine system is independent, and only one discharge is available to extinguish the fire.

Figure 18-24 shows the system for a twin-engine helicopter. One or both bottles can be used to extinguish a fire in either engine area. By moving the engine toggle switch from NORM to ARM, fuel is shut off to the engine and the system is armed. When the springloaded-to-center toggle marked AGENT RLSE is moved to the right or the left, the appropriate bottle will discharge into the armed engine area.

Portable Fire Extinguishers

Portable or hand-held fire extinguishers must be available as specified for the aircraft. The fire extinguishers must be of approved types and must be appropriate for the kinds of fires that are likely to occur in the areas concerned. For example, if electrical fires are most likely, the fire extinguisher must be of the dry-chemical type, a dry-gas type (CO_2), or a Halon 1301 type. Extinguishing agents containing water must not be used, because water increases electrical conductivity and may cause more damage than good. Oil or fuel fires should be smothered with a foam-type agent or a dry-chemical agent.

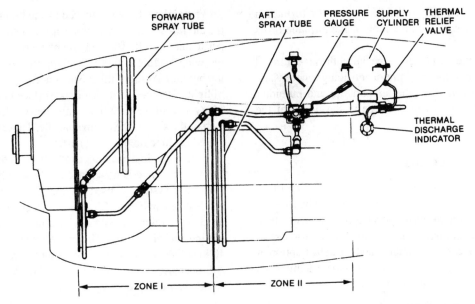

FIGURE 18-23 Fire-extinguishing system configuration for a twin-engine turboprop airplane. (*Flight Safety International.*)

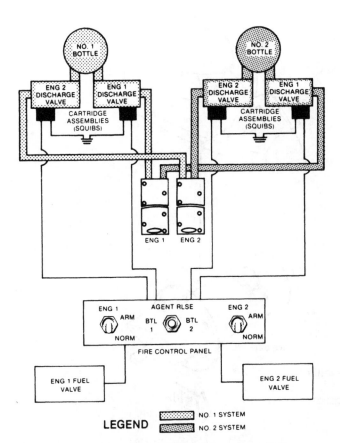

FIGURE 18-24 Fire-extinguishing system schematic for a twin-engine helicopter. (*Flight Safety International.*)

Each extinguisher for use in a personnel compartment must be designed to minimize the hazard of toxic gas concentrations. Readily accessible hand-held fire extinguishers must be available for use in each Class A or Class B cargo compartment. Class A and Class B cargo compartments are those accessible to members of the flight crew while the aircraft is in flight.

Inspection and Servicing of Fire Protection Systems

Inspections for fire warning and fire-suppression systems follow the general procedures for the inspections of other systems. Mechanical parts of systems are examined for damage, wear, security of mounting, and compliance with technical and regulatory requirements. Electrical control systems are inspected in accordance with approved practices for electrical systems and the special instructions given in the manufacturer's instructions. Certain elements must be electrically tested with voltmeters, ohmmeters, meggers, or otherwise, as specified in the maintenance instructions. Continuity of electrical circuits may be tested with ohmmeters or continuity test lights.

For additional information on the maintenance and service of fire protection systems, refer to the text *Aircraft Powerplants.*

For service of particular fire warning and fire-suppression systems, the technician must follow carefully the appropriate manufacturer's instructions. This is to ensure that the correct materials are employed and the proper procedures are followed.

ICE PROTECTION SYSTEMS

All aircraft that operate in weather conditions where ice is likely to form must be provided with ice protection. This protection may be in the form of anti-icing systems or deicing systems. An **anti-icing system** prevents the formation of ice on the airplane, and a **deicing system** removes ice that has already formed.

Among the parts of the airplane where ice prevention or removal is essential are the windshield, wing leading edges, tail airfoil leading edges, propellers, engine air inlets, pitot tubes, water drains, and any other part where the formation of ice can interfere with the operation of the airplane or its systems.

On piston-engine airplanes, especially those equipped with float-type carburetors, carburetor anti-icing is necessary, even in clear weather, when the temperature and humidity are conductive to the formation of ice in the throat of the carburetor.

Ice Detection Systems

Some method of ice detection is desirable for aircraft so that the ice-control systems are operated only when necessary. If ice-control systems were operated continuously, there would be a significant increase in operational expense due to increased wear and tear on equipment and the consumption of fluids and power unnecessarily. There are two basic methods used for ice detection: visual and electronic.

Visual detection is achieved by the flight crew monitoring the aircraft structures that first start to accumulate ice on their particular aircraft. This may involve no more than looking at the wing leading edge or checking the windshield wiper for ice buildup. At night this visual checking is aided by the use of lights designed to shine on the surface that accumulate ice. Many aircraft have ice lights mounted on the side of the fuselage or the side of the engine nacelle. These lights are usually aimed to shine on the wing surface.

Electronic instruments can be used to detect ice accumulation when there is no surface easily seen by the flight crew. One such system is used on the Canadair Challenger 601 and is shown in Fig. 18-25. The ice detector consists of a microprocessor circuit with an aerodynamic strut and probe extending into the slipstream. The probe vibrates at a frequency of 40 kHz. When ice starts to build on the probe, the frequency will decrease. When the frequency has decreased to a preset value, the microprocessor will turn on a red annunciator light

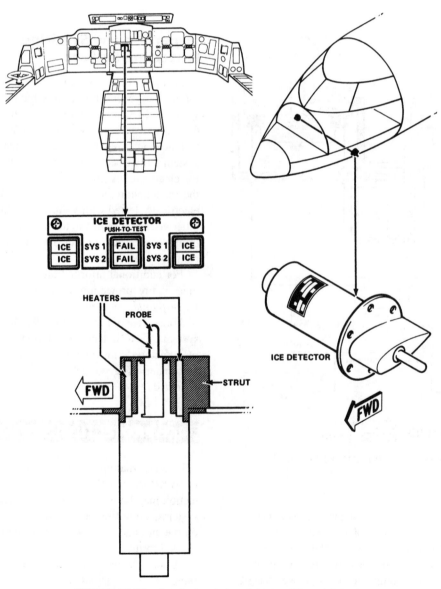

FIGURE 18-25   An electronic ice detector. (*Canadair Inc.*)

to advise the flight crew that the aircraft is in icing conditions. If the flight crew then turns on the ice-control systems, the red ICE light will go out and a white ICE light will illuminate. Once the probe detects ice, the microprocessor will energize a heating element in the probe to remove all ice so that the probe can recheck for icing conditions. As long as the probe continues to detect icing at each check, the ICE annunciator will remain on. When ice is no longer detected, the light will go out. The 601 has two of these detectors, one on each side of the forward fuselage section.

Pneumatic-Mechanical Deicing Systems

For many years, various airplanes have utilized mechanical deicing systems consisting of inflatable rubber "boots" formed to the leading edge of wings, struts, and stabilizers. The deicing boots are attached to the leading edge of the airfoils by means of cement and fasteners such as rivnuts, also called "bootnuts." Rivnuts are described in the section of this text covering structural repairs. Some aircraft use cement only for attaching the deicer boots to leading-edge surfaces.

The inflatable boots are usually constructed with several separate air passages or chambers, so that some can be inflated while alternate chambers are deflated. The inflation of the boot is accomplished by utilizing the output pressure from a vacuum pump for inflation and the inlet side of the pump for deflation. The control of the pressure and suction is accomplished by means of a distributor valve, which rotates and periodically changes the flow of air to or from the different section of the boots, or by flow-control valves. This results in alternate raising and lowering of sections of the boots, and this action cracks off any ice that has formed on the boots.

The pneumatic deicing system installed on one model of a twin-engine airplane provides a good example of the application of such a system. The system as installed in the airplane is illustrated in Fig. 18-26.

Deicer boots, also referred to as **deicers**, consist of fabric-reinforced rubber sheets containing built-in inflation tubes. The operation and design of these deicer boots are shown in Fig. 18-27. The boots used in this installation have spanwise inflation tubes. The deicers are attached by cement to the leading edges of the surfaces being protected. Either aluminum or flexible rubber air connections called **air-connection stems** are provided on the back side of each deicer. Each stem projects from the underside of the boot into the leading edge through a round hole provided in the metal skin. These provide for connection to the deicer pneumatic system.

Through the engine-driven vacuum pumps, the system normally applies vacuum to the deicer boots at all times except when the boots are being inflated. When it is desired to inflate the deicer boots, the pilot or copilot uses the deicer system control switch. This is a momentary ON–type switch that returns to OFF when released. When the control switch is turned ON, the **time module** (timer) energizes the pneumatic pressure control valve for 6 s. The boot solenoid valves are energized, and air pressure from the engine-driven pumps is supplied to the inflatable tubes in the boots. The inflation sequence is controlled by the time module (timer) and solenoid-operated valves located

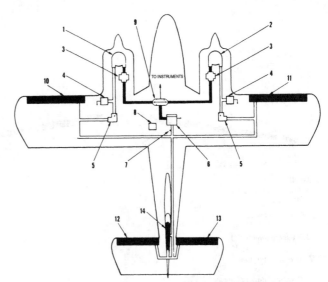

1. Vacuum pump, left engine
2. Vacuum pump, right engine
3. Vacuum regulator valve
4. Pressure-control valve
5. Flow check valve
6. Exhaust valve
7. Pressure switch
8. Time module
9. Vacuum manifold
10. Deicer boot, left wing
11. Deicer boot, right wing
12. Deicer boot, left stabilator
13. Deicer boot, right stabilator
14. Deicer boot, fin

FIGURE 18-26 Pneumatic-mechanical deicing system on a twin-engine airplane. (*Piper Aircraft Co.*)

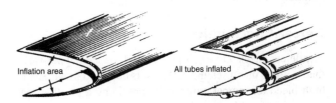

FIGURE 18-27 Construction and operation of deicer boots. (*Piper Aircraft Co.*)

near the deicer air inlets. Upon automatic de-energization of the control valves by the timer, the deicer solenoid valves permit the deicer pressurizing air to return to the solenoid valves and be exhausted overboard. Deicer pressure is normally about 18 psig [124 kPa]. After the boots have been inflated and the pressure released as described, vacuum is again applied to the boots.

The wing and horizontal stabilizer pneumatic deicing system for a small jet is discussed. Figure 18-28 shows the central part of the system that provides dried air and proper pressure for the wing and horizontal stabilizer deicing system.

This is accomplished by the water separator and pressure regulator/reliever. Two check valves avoid backflow when there is loss of bleed air from one engine. The low-pressure switch monitors the system pressure and indicates when inadequate (low) pressure is provided to the wing and horizontal stabilizer deicing system. The controller monitors the

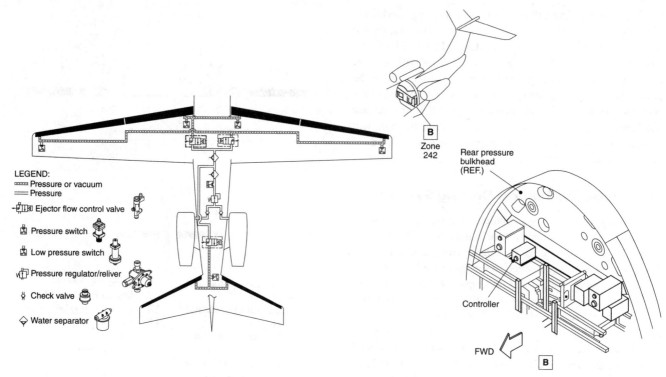

FIGURE 18-28 Deice system of small jet aircraft.

deice pressure switches and the low-pressure switch and indicates a fault to the flight crew in case of under-pressure, inflation, or deflation problems.

Deicer Check Valve

The deicer check valve (CV) is a spring-loaded flapper valve that, with its metal sealing surfaces, is designed to withstand high-bleed air temperatures while providing a positive seal against engine bleed air backflow with minimum leakage. The valve is also designed for minimum pressure drop and flow resistance in the forward flow condition. Check valves are used on duel-engine aircraft in which pressurized bleed air from both engines is supplied to a pneumatic deicing system through a common manifold. The check valves provide positive seal against engine bleed air backflow in the event of loss of bleed air from one engine. This ensures that the pressure to the deicing system is maintained in the event bleed air is only available from one engine.

Pressure Regulator/Reliever

The pressure regulator/reliever is designed to receive inlet air pressure from an engine bleed air source and to provide a constant outlet pressure for deicer inflation, over a wide range of flow rates and inlet pressures. Outlet pressure regulation is accomplished by means of a variable orifice which opens and closes to modulate the flow area. The orifice is formed by a poppet which moves relative to a fixed seat located in the regulator housing. The poppet is attached to a spring-loaded diaphragm which senses outlet pressure and closes the poppet in the event of high outlet pressure, and

opens the poppet in the event of low outlet pressure. The regulator is equipped with an over-pressure relief valve, or reliever, which serves as a fail-safe device for venting excess pressure overboard in the event the regulating poppet fails in the open position. The reliever consists of a self-resetting spring-loaded ball, which opens at a preset pressure. The reliever is installed to one of the two outlet ports of the regulator housing.

Water Separator

The water separator serves to eliminate the majority of liquid water from the air line by means of centrifugal separation in order to minimize the potential of valve freeze-up. The separator has a continuous drain port, which should be located downward (toward gravity). The separator drain port is equipped with a fitting to allow drain water to be routed to another location if desired. Its inlet pressure rating is 100 psig (maximum).

Low Pressure Switch

The switch is located upstream of the ejector flow control valves (EFCVs) and is used to monitor the system pressure to ensure a minimum pressure is being supplied to the EFCVs.

Controller

The controller provides the command signals to the EFCVs. The controller operates in a 1-min mode using the control switch. A single cycle mode can be initiated by momentarily toggling the control switch from the OFF position to the ON position and then back to OFF. This provides the pilot with immediate operation of the boot deicing system. Power input to the controller consists of 28-V dc. The controller will communicate status and fault messages to the data concentrator unit via ARINC 429. The fault messages will indicate a pressure fault or a stepping fault. A timed pressure fault is one or a combination of the following:

- Failure of a pressure switch to close within 4 s of activation of its associated ejector flow control valve, indicating failure of the deicer to inflate to operating pressure in proper time.
- Failure of a pressure switch to reopen within 4 s of deactivation of its associated ejector flow control valve, indicating failure of the deicer boot to deflate in the proper time.
- A deice pressure switch is not open when the controller is not cycling.
- The low-pressure switch is not open within 4 s after a valve is actuated.

Wing Deice System

The wing deicing system removes the formation of ice from the wing leading edges. Figure 18-29 shows the components of the wing deice system of a small jet aircraft.

The wing deicer boots cycle (inflate/deflate) in order to mechanically remove the formation of ice from the wing leading edges. The EFCV provides the boot inflation and deflation. Pressure switches monitor the boots pressure to ensure that they work properly.

Wing Ejector Flow Control Valve

The aircraft has three EFCVs. There are two EFCVs for the wing deicing system and one for the horizontal stabilizer deicing system. The EFCV controls the flow of air to and from the deicer boots. It is a two-position, solenoid-operated poppet valve that provides system pressure (energized position) or vacuum to the pneumatic deicers. When the solenoid valve is in the de-energized condition, the ejector section of the valve provides the vacuum necessary to maintain the deicing tubes in a deflated condition using a minimum amount of airflow. The vacuum generated is 5.0 at the inlet of 18 psig and 21°C [70°F].

Wing Deice Pressure Switch

The aircraft has five deice pressure switches. Four pressure switches for the wing deicing system and one pressure switch is dedicated to the horizontal stabilizer deicing system. The wing deice pressure switches are located at the inlets of both

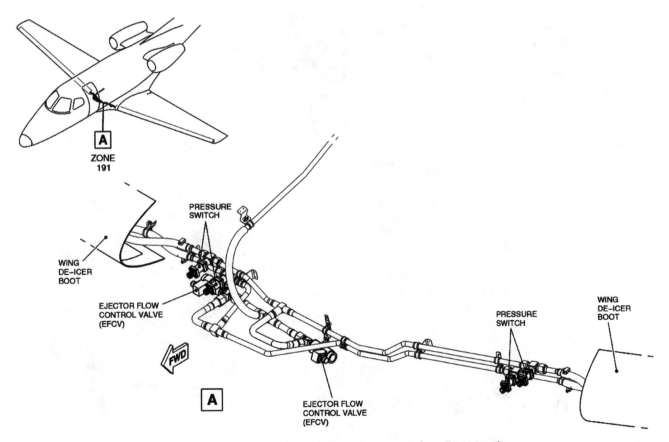

FIGURE 18-29 Wing deice system components of small jet aircraft.

the inboard and outboard LH (Left-Hand) and RH (Right-Hand) chambers of the wing deicer boots and downstream of the EFCVs. The pressure switches ensure a minimum pressure is being supplied to the deicers in a specific timing window.

Wing Boot Deicer

The wing deicer boots are silver polyurethane-surfaced pneumatic deicers consisting of a smooth rubber and fabric blanket containing span-wise deicing tubes. Each wing deicer boot is a single boot with separate inflatable chambers, one for the inboard wing section and one for the outboard wing section. Each wing deicer boot has an inboard and an outboard air connection for inflation. The LH and RH outboard chambers of the deicer boot will inflate simultaneously. The LH and RH inboard deicers will inflate simultaneously. The inflation pressure of the deicer boot is 20.0 ± 1.0 psig.

Horizontal Stabilizer Deicing System

The horizontal stabilizer deicing system removes the formation of ice from the horizontal stabilizer leading edges, as shown in Fig. 18-30.

The horizontal stabilizer deicer boots cycle (inflate/deflate) in order to mechanically remove the formation of ice from the horizontal stabilizer leading edges. The EFCV provides the boot inflation and deflation. A pressure switch monitors the boot pressure to make sure that the system works properly. The components of the horizontal stabilizer deicing system are similar to the components of the wing deicing system.

Inspection and Maintenance of Pneumatic-Mechanical Deicer Systems

The inspection of pneumatic-mechanical deicer systems requires an examination of the deicer boots for condition, adherence to the protected surface, and condition of the surface of the boots. If cuts, abrasion, or other damage is found, repairs should be made as specified by the manufacturer. Grease or oil found on the boots should be removed with an approved solvent, after which the boots should be scrubbed with soap and water and then rinsed with clean water.

Deicer boots are provided with a conductive neoprene coating to prevent the buildup of static charges on the boots. If this were not done, static charges would build up and then discharge to the metal surface of the airplane, causing radio interference. During inspection and maintenance, the technician should determine whether the conductive coating is intact and effective.

Removal and replacement of deicer boots depends upon the type of attachment involved. If the boots are attached with screws and rivnuts, removal of the screws permits

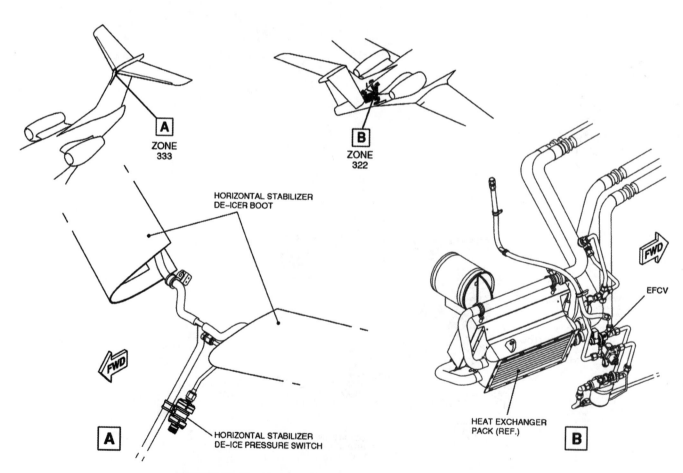

FIGURE 18-30 Horizontal stabilizer deicing system components.

removal of the boots. If the boots are cemented, the technician must consult the manufacturer's instructions regarding the type of solvent to be used for dissolving the cement and the procedure for applying the solvent.

Operational tests are performed as specified in appropriate instructions. Usually, the test can be made by running one or more engines and turning the system on in the normal manner. The inflation of the tubes in the boots can easily be observed. If any of the tubes should fail to inflate in sequence or at the proper time, the air supply to that tube should be checked for obstruction or damage.

Thermal Anti-Icing

Thermal anti-icing uses heated air flowing through passages in the leading edge of wings, stabilizers, and engine cowlings to prevent the formation of ice. The heat source for this operation normally comes from combustion heaters in reciprocating-engine-powered aircraft and from engine bleed air in turbine-powered aircraft. From the heat source the hot air is distributed along the leading edge of the item being anti-iced by the use of a perforated air duct called a *piccolo*, or *spray*, tube. When the air exits the piccolo tube, it is in contact with the leading-edge skin of the surface, as shown in Fig. 18-31. The skin is heated and ice is prevented from

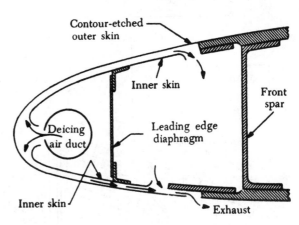

FIGURE 18-31   A typical heated leading edge.

forming. The air then flows out of the wing through openings in the end of the wing or on the bottom of the wing.

Figure 18-32 is a drawing of the layout of the Boeing 777 anti-ice system. The anti-ice system consists of ice detection, windshield rain removal, window heat, air data probe heat, engine anti-ice, wing anti-ice, and water and waste heat. Figure 18-33 shows the wing anti-ice schematic of a Bombardier 700 aircraft. The wing anti-ice system prevents ice formation on the wing-leading edge by heating the

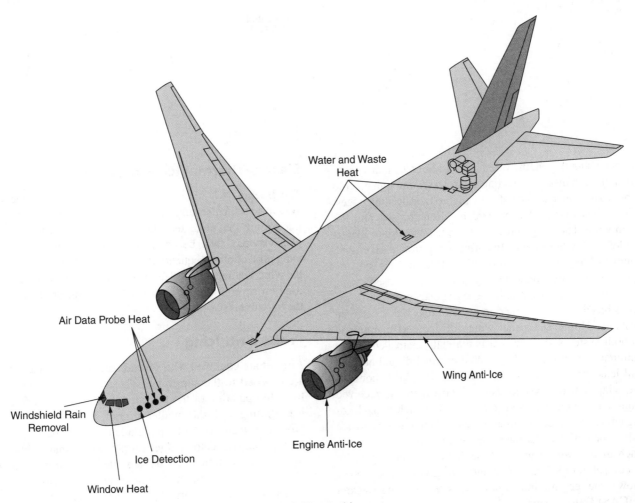

FIGURE 18-32   Anti-ice system.

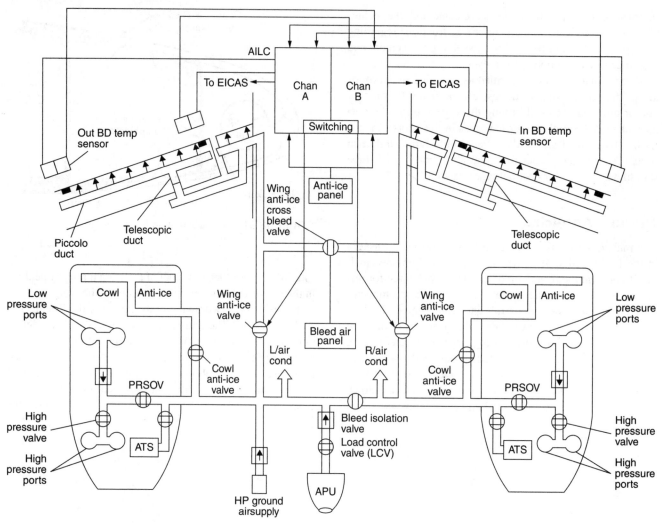

FIGURE 18-33 Wing anti-ice system.

surface using hot engine bleed air. The hot bleed air is supplied through insulated ducting and released through piccolo tubes to the inner surface of the wing and slat-leading edges. Air is then extracted overboard between the slat and the main wing. The wing anti-ice system is divided into identical left and right systems. In normal operation, each engine supplies bleed air to its respective wing anti-ice system. The systems are connected by a normally closed, wing anti-ice cross bleed valve. In the event one system fails, the cross bleed valve is opened to permit cross bleed between systems. This ensures that wing anti-icing is maintained to both systems. The system is manually activated and is automatically controlled by a dual channel digital anti-ice and leak detection controller (AILC). The AILC controls the wing anti-ice system using electrical inputs received from skin temperature sensors located at each wing-leading edge. The AILC modulates the respective wing anti-ice valve open or closed as necessary to prevent ice formation. Each of the two channels of the AILC has the capability to control both left and right anti-ice valves. Figure 18-34 shows the engine indication and crew-alerting system (EICAS) anti-ice page.

Electro-Thermal Ice Protection

The Boeing 787 utilizes an electro-thermal ice protection system in which several heating blankets are bonded to the interior of the protected slat-leading edges. The heating blankets may then be energized simultaneously for anti-icing protection or sequentially for deicing protection to heat the wing-leading edge. Electro-thermal ice protection is also used on several aircraft to provide anti-icing protection for the horizontal stabilizers.

Probe Anti-Icing

Aircraft are equipped with several sensors and devices that are exposed to the slipstream. These devices are required for safe operation of the aircraft and must be kept free of ice. Starting with the smaller aircraft, the items that need protection from ice buildup include the pitot mast and the stall warning indicator. Larger aircraft require that additional items such as static ports be kept clear, along with total-air-temperature probes and angle-of-attack indicators.

These probes are normally kept free of ice by the use of electric heaters. These heaters are controlled by switches at

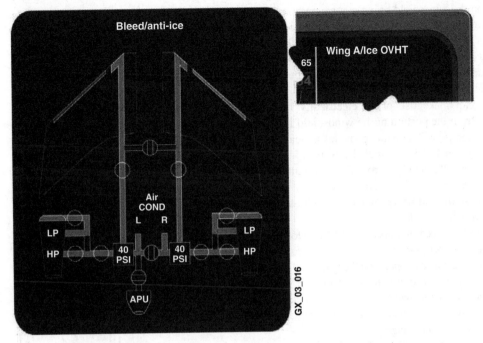

FIGURE 18-34 EICAS anti-ice page indicating overheat situation.

the pilot's station. The heater elements may be operated by either ac or dc electrical power.

When inspecting these devices, the technician should be aware that even the units on small aircraft are capable of generating sufficient heat to cause painful burns to the skin. Never touch one of these devices if its heater element is operating. Figure 18-35 shows the electrical heating system for the air data probes on a Boeing 777 aircraft.

Windshield-Ice Control

The control of ice buildup on windshields may be accomplished by one of two basic methods: by heating the windshield or by

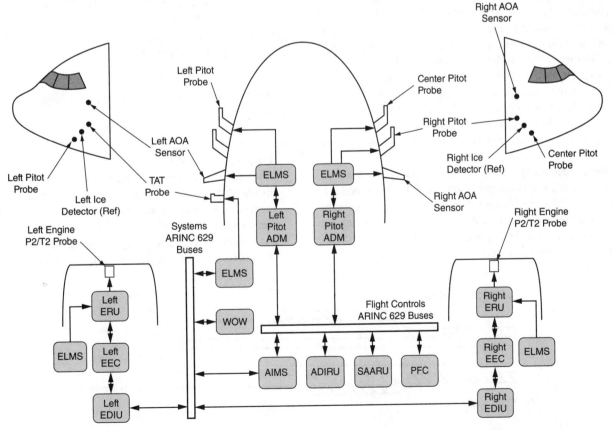

FIGURE 18-35 Air data probe heat system.

spraying a fluid on the windshield to remove ice and prevent the formation of any more ice. The heating of windshields is the more common method and may involve the use of a heated panel over the windshield surface, electric heater elements inside the windshield structure, or a flow of heated air between windshield surfaces.

Light aircraft can make use of an electrically heated panel to clear ice from the portion of the windshield immediately in front of the pilot. The panel provides a clear area of approximately 12 in [30.5 cm] high by 6 in [15.2 cm] wide, with the size of different units providing different clear areas. This system can be used on aircraft that are not available with internal windshield heater elements, such as most aircraft with plastic windshields.

The system must be placed into operation by the pilot prior to any large ice buildup, because the unit is designed for anti-icing rather than deicing. Once in operation the panel will maintain a surface temperature of approximately 100°F [37.8°C], with the temperature being regulated by a temperature-sensing circuit that cycles the heater elements on and off to maintain the designed operating temperature range.

The panel is mounted on a fixture in front of the windshield and can be removed when the aircraft is not going to be operated in icing conditions. The panel used on the Cessna Caravan is shown in Fig. 18-36. This unit includes a restraining strap (not shown), which prevents the panel from pivoting forward on its mount when reverse thrust is used during landing.

Aircraft with integral windshield heaters are normally equipped with glass or hardened acrylic outer layers reinforced internally with shatter-resistant materials such as acrylic and polycarbonate materials and a polyvinyl butyral interlayer, as shown in Fig. 18-37. A conductive coating is placed beneath the glass layer in the area of the windshield to be anti-iced, such as the area shown in Fig. 18-38. When in operation, current flows through the coating and generates heat due to the resistance of the coating. A portion of the windshield or complete panels can be kept ice-free by this method. This method of windshield-ice control is commonly used on turbine-powered corporate and airline-type aircraft.

Some older transport aircraft have windshields that consist of an inner and outer panel separated by an air passage. When windshield anti-icing is turned on, heated air from the cabin heating system is ducted through this passage, heating both panels of glass. The heated air exits the windshield area and flows into the flight deck area.

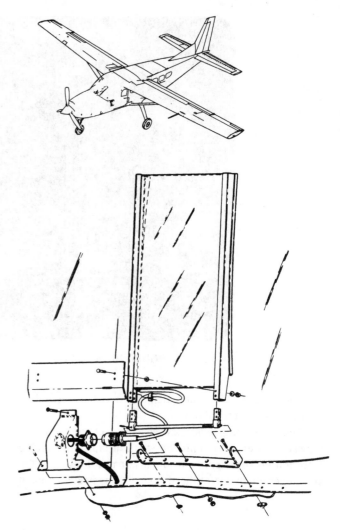

FIGURE 18-36 Heated panel mounted in front of a windshield. (*Cessna Aircraft Co.*)

The following items are important when inspecting heated windshields:

- Check for any delamination (separation of the layers) in the windshield panels.
- Look for any scratching of the windshield and determine its cause. Scratching is often caused by dirty windshield wipers moving across the panel. For this reason, always keep windshield wipers clean and in good condition.
- Any arcing that occurs during operating indicates that the conductive coating may be breaking down. This situation

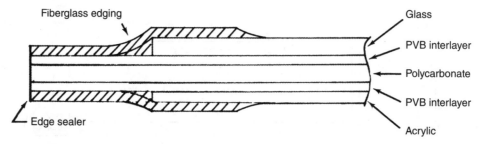

FIGURE 18-37 Cross section of a heated windshield. (*Bell Helicopter Textron.*)

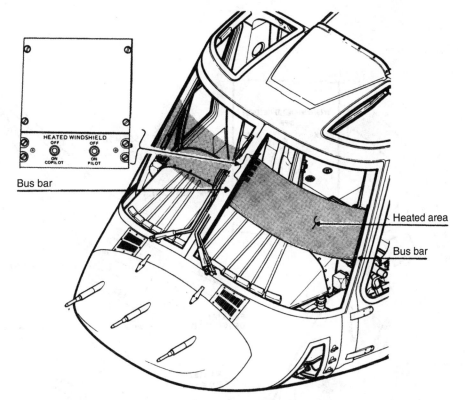

FIGURE 18-38 A heated windshield installation. (*Bell Helicopter Textron.*)

can lead to overheating of the panel where the arcing is occurring. Correct the problem immediately in accordance with the manufacturer's maintenance instructions.

- Discoloration of the windshield is not normally a problem unless the optical quality of the windshield is affected. Heated windshields are clear when looking directly through them, but they may have a gold, blue, or pink tint to them when viewed with light reflecting off them.

When evaluating any of these types of damage to a windshield, follow the recommendations of the aircraft manufacturer. The basic guidelines are that the damage should not decrease the strength of the windshield, change the optical quality of the windshield, or affect the normal operation of the system.

Alcohol is used on some aircraft to provide for windshield deicing. This system sprays isopropyl alcohol on the outside of the windshield. The alcohol can be used to remove ice, but the supply is not normally great enough to be used throughout the flight. This system is normally used to clear the windshield prior to landing. Operation of the system requires that a crew member turn on the fluid pump and adjust the rate of flow according to the amount of ice that has accumulated. Care should be taken when inspecting this system to ensure that all lines and fittings are in good condition. In-flight fires have resulted from the alcohol line rupturing and spraying onto electrical equipment.

RAIN-REMOVAL SYSTEMS

Rain-removal systems are designed to allow the pilot to have a clear view of the airport when taxiing and to allow him

or her to see the approach and departure paths and runway environment when taking off and landing during rain. The systems are not commonly used during flight at altitude.

Rain may be removed by the use of windshield wipers, chemical repellents in combination with windshield wipers, or by a flow of air.

Windshield-Wiper Systems

Windshield-wiper systems may be operated electrically, hydraulically, or pneumatically.

A typical electrically operated **windshield-wiper system** is illustrated in Fig. 18-39. This drawing shows the components of the wipers installed on an airliner. Each wiper on the airplane is operated by a separate system to ensure that clear vision through one of the windows will be maintained in the event of a system failure. The wiper blades clear a path 15 in [38.1 cm] wide through an arc of 40°.

Both wiper systems are electrically operated and controlled by a common gang switch located on the pilot's overhead panel. The switch provides a selection of four wiper-action speeds ranging from 190 to 275 strokes per minute and controls the stowing of the wiper blades in a PARK position when the system is not in use.

Each windshield-wiper system consists of a drive motor, a control switch, a resistor box, a flexible drive shaft, a torque converter, and a windshield-wiper assembly.

Speed control for the windshield wipers is accomplished by changing the voltage applied to the windshield-wiper motor by means of resistances arranged in the resistor box. The required resistance is connected into the motor circuit

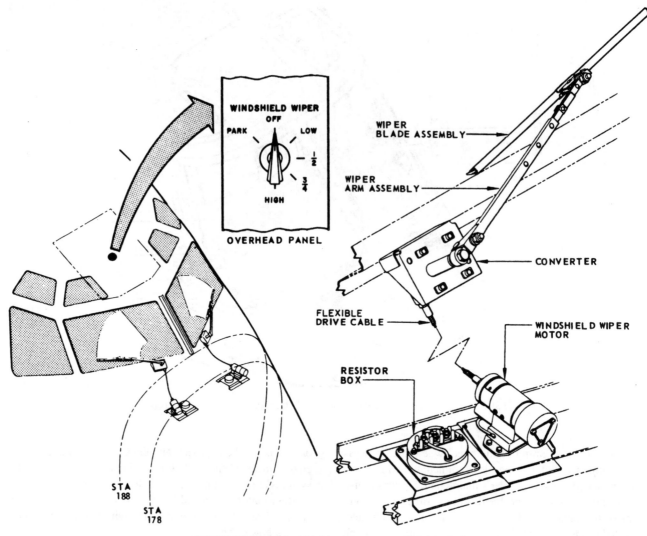

FIGURE 18-39 Windshield-wiper system. (*Boeing Co.*)

by turning the windshield-wiper switch to a selected speed. The rotary motion of the windshield-wiper motor is transmitted by the flexible shaft to the converter. The converter reduces the shaft speed and changes the rotary motion to an oscillating motion of the windshield-wiper arm. An electrical circuit for windshield wipers is shown schematically in Fig. 18-40.

Rain-Repellent Systems

To help maintain the clarity of vision through the windshield during rain conditions, a rain-repellent system is provided for the windshields of many modern aircraft. This system consists of pressurized fluid containers, a selector valve, solenoid-actuated valves, spray nozzles, push-button switches, a control switch, a time-delay relay, and necessary plumbing. A rain-repellent system and windshield wipers for an airliner are shown in Fig. 18-41. During rain conditions, the windshield wipers are turned on, and the repellent is sprayed on the windshield. The repellent is spread evenly by the wiper blades. The rain repellent should not be sprayed on the windshield unless the windshield is wet and the wipers

are operated, nor should the windshield wipers be operated on a dry windshield.

The effect of the rain repellent is to cause the water to form small globules, which are quickly blown away by the rush of air over the windshield in flight.

Flight deck windows of transport aircraft often use a permanent surface seal coating also called "hydrophobic coating" that is on the outside of the pilot/copilot windshield (see Fig. 15-41). The coating causes raindrops to bead up and roll off. This allows the flight crew to see through the windshield with very little distortion. The hydrophobic windshield coating reduces the need for wipers and gives the flight crew better visibility during heavy rain.

Pneumatic Rain-Removal System

Some turbine-powered aircraft use engine bleed air to prevent rain from striking the windshield. When the pilot turns on the rain-removal system, bleed air at a high temperature and pressure is directed to an outlet at the base of the windshield, as shown in Fig. 18-42. This flow of air over the windshield carries away the rain drops before they can strike

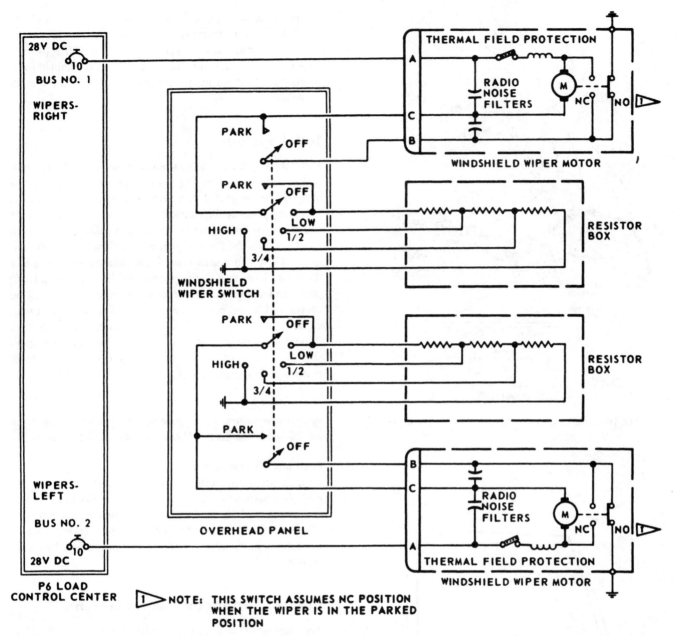

FIGURE 18-40 Electrical system for windshield wipers. (*Boeing Co.*)

the windshield. Any raindrops on the windshield when the system is turned on are also blown away.

WATER AND WASTE SYSTEMS

All modern airliners are required to incorporate water systems to supply the needs and comforts of the passengers and crew. Such systems include potable (drinkable) water for the galley and drinking fountains, water for the lavatories, and water for the toilet systems. Systems may include one or more tanks of water with connections to the various units that require a water supply. The passenger water system for one model of the Airbus 330 airplane is illustrated in Fig. 18-43.

Potable Water Supply

The water for drinking fountains or faucets is usually drawn from main pressurized water tanks, passed through filters to remove any impurities and solids, cooled by dry ice or other means of cooling, and delivered to the faucets and/or drinking fountains. Disposable drinking cups are supplied at each location in the forward and rear parts of the passenger cabin.

Lavatory Water

Water for the lavatories is also drawn from the main water tanks and passed directly through suitable plumbing and valves to the lavatories. Hot water for washing is provided by means of electric water heaters located beneath the lavatory bowls.

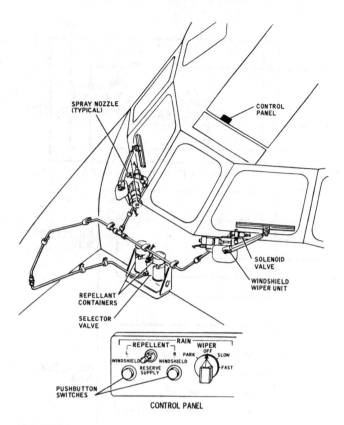

FIGURE 18-41　Rain-repellent system. (*Douglas Aircraft Co.*)

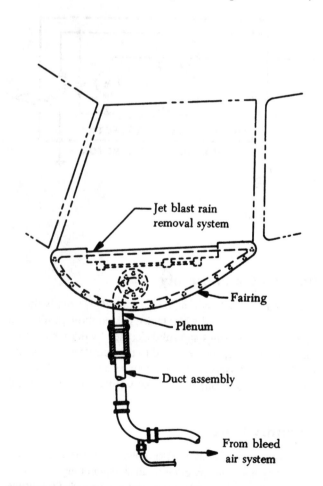

FIGURE 18-42　Pneumatic rain-repellent system.

A typical hot-water supply is contained in a 2-qt [1.89-L] tank, which includes the thermostatically controlled heating unit to maintain the water at a temperature of 110 to 120°F [43.3 to 48.9°C]. Drain water from the lavatories can be drained overboard through drain masts or can be drained into the toilet waste tanks.

Toilet System

Differential pressure forces waste from toilet bowls into two storage tanks (see Fig. 18-44). On the ground, and at altitudes below 16 000 ft, a vacuum generator produces the necessary pressure differential. Clear water from the potable water system flushes the toilets. A flush control unit in each toilet controls the flush sequence. The vacuum system controller (VSC) furnishes operational information (including the waste level in the storage tank) to the flight attendant's panel, and maintenance information and a test program to the centralized maintenance system. The waste tank has a total capacity of 700 L. Ground personnel services the waste tank through a single service panel under the fuselage. A manual shutoff valve on the lower right-hand side of the toilet bowl isolates an inoperative toilet.

POSITION AND WARNING SYSTEMS

Each aircraft system may incorporate a warning system to indicate when that particular system is not functioning properly. The purpose of this section is to discuss systems that do nothing but indicate the position of components and warn of unsafe conditions. These systems include control-surface indicator systems, takeoff warning systems, and stall warning systems.

Control-Surface Indicating Systems

The purpose of control-surface indicating systems is to allow the flight crew to determine if a control surface is in the correct position for some phase of flight and to determine if a flight control is moving properly.

The most common type of control-surface indicating system is that used for the elevator trim tab, and this will serve as an example for other mechanical indicating systems. This type of mechanism is found in most fixed-wing aircraft and is used by the pilot to indicate if the trim tab is properly positioned for some phase of flight, such as takeoff or landing, before that phase of flight is initiated. If the trim tab is out of position, the pilot may have to use large forces on the flight controls to achieve the desired action. If the trim tab is properly positioned, then it will provide most of the force necessary to hold a specific attitude and the pilot will need to make only small adjustments about the trimmed setting of the control surface.

These types of systems often use a mechanical indicator to show the control position. In some systems, a cable is attached to the control horn of the trim tab, and as the tab

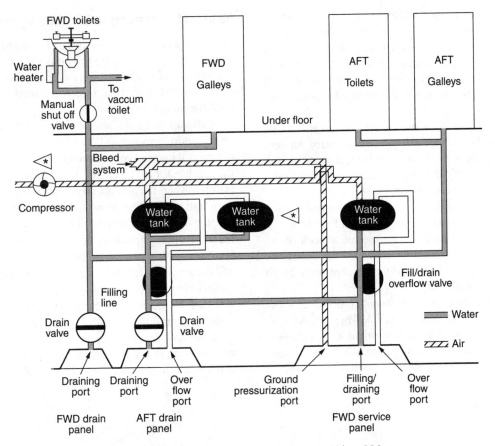

FIGURE 18-43 Passenger water system in Airbus 330.

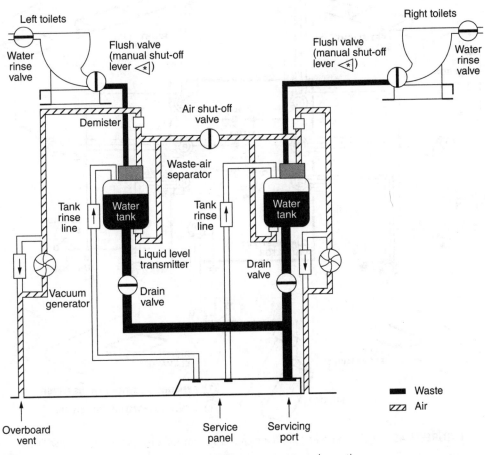

FIGURE 18-44 Toilet waste water schematic.

moves the cable pulls against a spring at the pilot's position. As the spring is stretched or released by the movement of the trim tab, a pointer located where the cable connects to the springs moves along a scale, indicating the trim-tab position. In other systems a spiral groove in the trim-tab control wheel causes a pointer to move, indicating the position of the trim tab.

Either of these systems or similar system designs can be used to monitor the position of any flight-control surface. When inspecting these mechanical indicating systems, the technician should verify that the indicator agrees with the position of the surface being monitored.

Large aircraft may make use of electric control-surface indicating systems, such as the one shown in Fig. 18-45. In this system, synchro transmitters are located at each of the control surfaces. The voltage inducted into the armatures by their exciter windings is used at the synchro receivers to position the indicator pointers.

Modern transport aircraft use glass cockpits and the electronic centralized aircraft monitor (ECAM) or EICAS system will display all control surface information on a monitor in the cockpit. Figure 18-46 shows the position of the spoilers, ailerons, elevators, elevator trim, and rudder.

Takeoff Warning Indicator Systems

A takeoff warning indicator system is used to advise the pilot that one or more items are not properly positioned for takeoff. When a takeoff warning system actuates, an intermittent horn is sounded until the incorrect situation is corrected or until the takeoff is aborted.

A simplified takeoff warning system schematic for a large jet transport is shown in Fig. 18-47. Note that if any one of the several items is not in the correct position for takeoff and the throttle for no. 3 engine is advanced beyond a certain position when the aircraft weight is on the landing gear, the horn will sound. The items that are checked by this system are that the elevator trim is in the takeoff range, the speedbrake (spoilers) handle is in the 0° position, the steerable fuselage landing gear is centered, the wing flaps are at 10°, and the leading-edge wing flaps are extended. The exact aircraft configuration monitored by a takeoff warning system will vary, but the intent of each system is to prevent a takeoff with the aircraft in an unsafe configuration.

In modern aircraft equipped with an EICAS system if the aircraft is in an unsafe takeoff configuration, configuration aural, warning messages, and MASTER WARNING lights

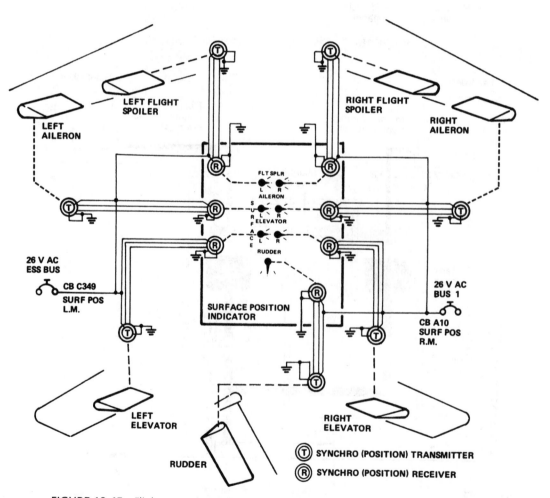

FIGURE 18-45 Flight-control-position indicating system for a jet aircraft. (*Canadair Inc.*)

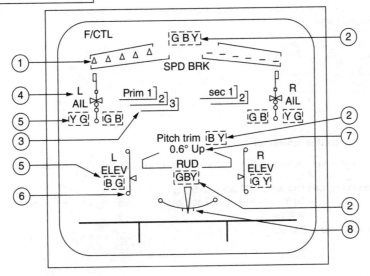

ECAM F/CTL page

① **Spoilers/speed brakes indication**

R	Δ	: Spoiler not retracted (green)
	–	: Spoiler retracted (green)
	Δ	: Spoiler fault deflected (amber)
	1 2 3 ...	: Spoiler fault retracted (amber)

Note: Same indications are displayed on WHEEL page.

② **Hydraulic system pressure indication**

Normally green. Becomes amber in case of hydraulic system low pressure (downstream the leak valves).

③ **PRIM/SEC indication**

– PRIM, SEC labels are always displayed in white.
– Computer number is normally green boxed grey. Number and box become amber in case of computer failure.

④ **Ailerons position indication**

White scale and green indexes. Index becomes amber when both (associated) servojacks are not available.

⑤ **Aileron/Elevator actuators indication**

G, B and Y are normally displayed in green.
Becomes amber in case of hydraulic system low pressure. It is partially boxed amber in case of electrical failure detected by the PRIM.

FIGURE 18-46 Flight control indications on ECAM page.

come on. Figure 18-48 shows the different failure messages that could be displayed on the EICAS monitor.

Stall Warning Indicators

A stall warning indicator is designed to indicate to the pilot when the aircraft is close to the stalling angle of attack. When the stall warning system is activated, the pilot will be given one or more of the following indications: A horn will sound, a light will illuminate, and/or a "stick-shaker" will start vibrating the control wheel column. The specific warning method varies with the aircraft design.

Although some light aircraft do not have stall warning indicator systems, most have a system using a sensor vane on the leading edge of the wing similar to that shown in Fig. 18-49. When the angle of attack of the wing causes the airflow to be

(6) Elevator position indication

White scale and green index. The index becomes amber when both associated actuators are not available.

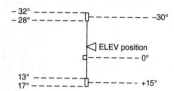

(7) Pitch trim position indication

PITCH TRIM label : Normally white. Becomes amber in case of THS electrical control loss.

Position indication : Varies from 2° down to 14° up.

Normally green. Becomes amber in case of B + Y system low pressure.

(8) Yaw control indications

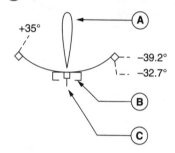

(A) Rudder position indication

It is normally in green.

The rudder symbol becomes amber, in case of blue + green + yellow hydraulic low pressure.

(B) Rudder travel limiter or PTLU indication

It is normally in green.

It becomes amber, when travel limiters 1 and 2 are faulty.

(C) Rudder trim position

It is normally in blue.

The position varies from −29.2 to +29.2°

It becomes amber, if rudder trim systems 1 and 2 are inoperative.

FIGURE 18-46 *(continued)*

from below the vane on the sensor, the vane moves up and completes a circuit to activate the stall warning indicator.

A pneumatic stall warning indication system is used by some light aircraft. A slot in the leading edge of the wing is connected to tubing routed to the cabin area. A horn reed is positioned in the end of the tubing. When the wing angle of attack increases to the point where a negative pressure exists on the slot, air flows through the reed and out the slot and creates an audible stall warning indication.

Modern transport aircraft use stall warning computer circuits to activate the stall warning system. Inputs to the computer include information about angle of attack, flap and slat position, and weight on landing gear. When the computer determines that a stall is imminent, a stick-shaker may start moving the control columns back and forth, an aural warning may be sounded, or an indicator may illuminate, depending on the specific system design.

AUXILIARY POWER UNITS

When a modern aircraft's engines are not operating, there are two available sources of power to operate its other systems. They are the aircraft's battery and the auxiliary power unit. Because of the limited capacity of the batteries, the amount of power they supply is insufficient to provide all but the very basic needs. Operation of an aircraft's air-conditioning system, for example, would require significantly more power than could be supplied by the batteries.

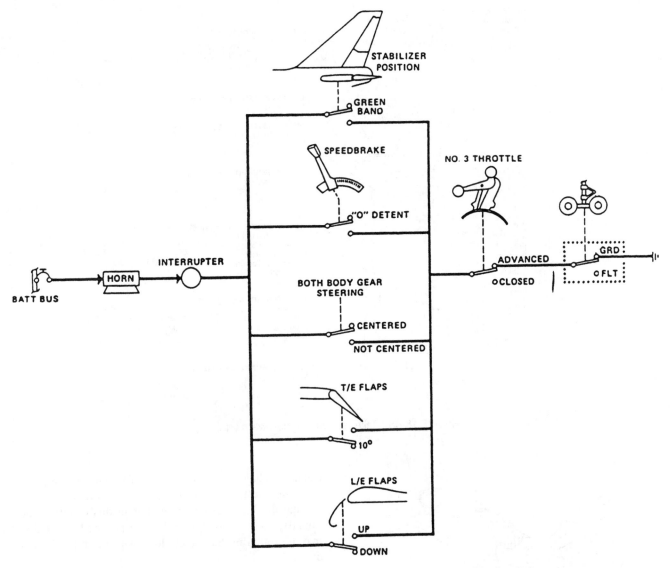

FIGURE 18-47 Takeoff warning system schematic for an airliner. (*Boeing Co.*)

To accommodate the needs of the aircraft on the ground for substantial amounts of energy while its engines are not operating, modern aircraft are equipped with auxiliary power units (APUs). The APUs are gas-turbine engines, using the aircraft's own fuel supply, which provide the power to run the attached generators. In addition, the APU is typically large enough to provide sufficient pneumatic power to start the aircraft's engines. The presence of an APU eliminates the need for ground power units (GPUs).

The Boeing 747 auxiliary power unit is manufactured by the AiResearch Division of Garret Corporation. This unit, shown in Fig. 18-50, is capable of producing approximately 660 lb/min airflow to the pneumatic systems and 90 kVA from the two attached generators to the aircraft's electrical system. The APU has a separate battery for starting and is protected by its own fire-protection system.

During operations, as with other gas-turbine engines, the exhaust is extremely high. In the case of the AiResearch

APU previously identified, the exhaust-gas temperature (EGT) ranges from 1652°F [900°C] during starting to 1265°F [685°C] during normal operations. Extreme care should be taken before operating the APU to ensure the exhaust-gas flume area is clear.

The principles for operating a typical APU parallel those of other gas-turbine engines. A more detailed discussion of gas-turbine operations and associated systems of fuel, lubrication, and ignition is found in *Aircraft Powerplants*. A description of the electrical generating systems and controls used with gas-turbine engines may be found in *Aircraft Electricity*.

APU Operation and Control

Operating controls and indicating systems are typically separated by the functions they serve. The gas turbine operations and the control switch for the air bleed valve are found on the same panel; see Fig. 18-51. The APU operations panel

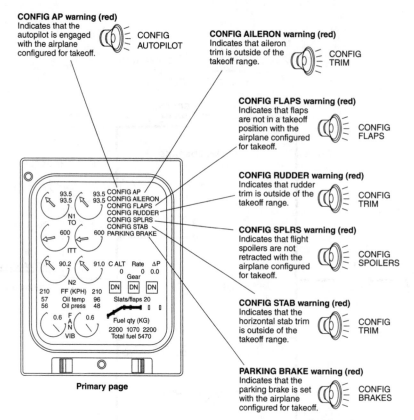

FIGURE 18-48 Takeoff configuration warnings.

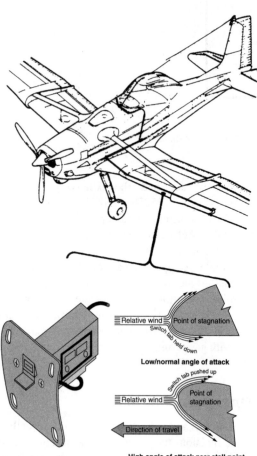

FIGURE 18-49 Stall warning indicator installation. (*Cessna Aircraft Co.*)

usually includes controls for the APU start and stop, fire-protection test and indicating, and the fire extinguisher manual discharge discussed earlier in this chapter. Additionally, controls for the fuel valve and APU inlet doors may be found on the panel along with the typical engine operation indicators.

The APU inlet door controls access to the gas-turbine compressor inlet and provides cooling air for the APU's accessories. This door may be powered by an electric motor or may use the suction caused by the operation of the gas turbine to open a spring loaded door.

Airflow and Bleed Air

The amount of airflow and its pressure are dependent upon the ambient temperature and the load on (power being drawn from) the APU. The greater the load and/or ambient temperature the lower the airflow and pressure.

The air-bleed valve (Fig. 18-52) should be closed during start-up and not operated until the APU is at approximately 95 percent power. When the APU bleed valve, also referred to as the "load-control valve," is opened by placing its switch in the cockpit in the OPEN position, airflow is supplied to the aircraft's pneumatic power-distribution system.

Accessory cooling is provided from inlet air through a shut-off valve, which is closed unless there is a specified amount of pressure being discharged by the APU's compressor.

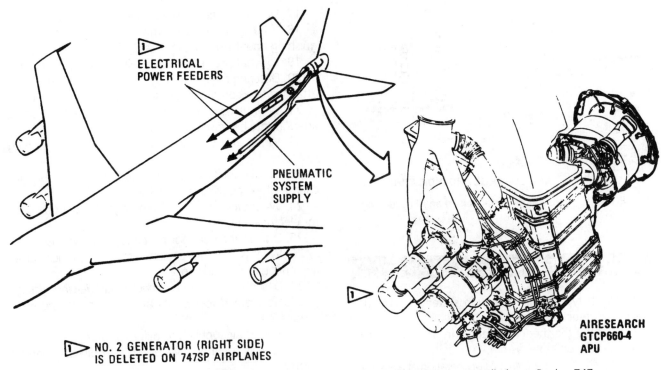

ELECTRICAL POWER FEEDERS

PNEUMATIC SYSTEM SUPPLY

AIRESEARCH GTCP660-4 APU

▷1 NO. 2 GENERATOR (RIGHT SIDE) IS DELETED ON 747SP AIRPLANES

FIGURE 18-50 Auxiliary Power Unit manufactured by AiResearch Division of Garret Corp., installed on a Boeing 747. (*Boeing Commercial Aircraft Co.*)

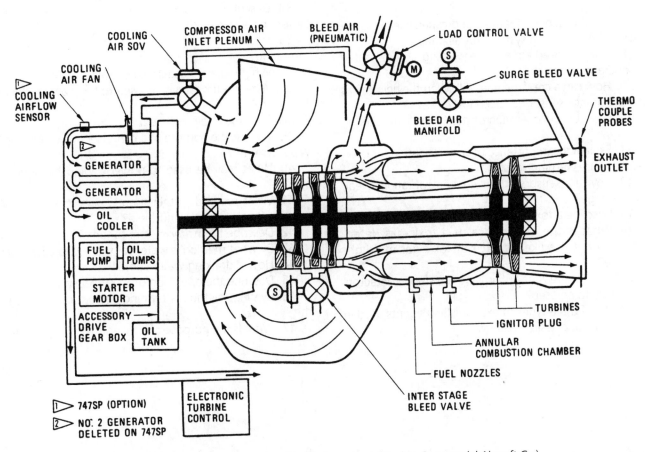

COOLING AIR SOV

COMPRESSOR AIR INLET PLENUM

BLEED AIR (PNEUMATIC)

LOAD CONTROL VALVE

SURGE BLEED VALVE

COOLING AIR FAN

▷1 COOLING AIRFLOW SENSOR

GENERATOR

GENERATOR

OIL COOLER

FUEL PUMP   OIL PUMPS

STARTER MOTOR

ACCESSORY DRIVE GEAR BOX   OIL TANK

BLEED AIR MANIFOLD

THERMO COUPLE PROBES

EXHAUST OUTLET

TURBINES

IGNITOR PLUG

ANNULAR COMBUSTION CHAMBER

FUEL NOZZLES

INTER STAGE BLEED VALVE

▷1 747SP (OPTION)
▷2 NO. 2 GENERATOR DELETED ON 747SP

ELECTRONIC TURBINE CONTROL

FIGURE 18-51 APU control and indicating panel. (*Boeing Commercial Aircraft Co.*)

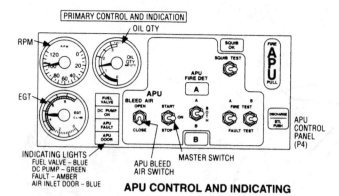

FIGURE 18-52 Cross-section schematic of an APU. (*Boeing Aircraft Co.*)

REVIEW QUESTIONS

1. Which type of fire-detection system is termed a *rate-of-rise* detection system?

2. Describe the theory of operation of a *spot detector* and a *thermocouple* system.

3. Describe the different types of operation of a Systron-Donner pneumatic fire-detection system.

4. Which fire-extinguishing agent is used on most modern aircraft?

5. What devices are used as overheat or fire detectors?

6. Describe three types of *tubular fire detectors*, or *sensors*.

7. List the general features and capabilities of overheat and fire warning systems for aircraft.

8. How are overheat and fire warning systems tested for operational capability?

9. How can a crew member tell when the helium gas in a pneumatic sensor has been lost?

10. Describe the operation of the *integrity monitoring circuit*.

11. Explain the importance of correct routing for overheat and fire warning sensors.

12. How are sensors protected from vibration, wear, and damage?

13. Describe a *toxic-gas detector.*

14. What is the importance of a toxic-gas detector and what is a likely source of toxic gas in a light airplane?

15. In what areas of an airplane are smoke detectors most likely to be used?

16. Name three types of extinguishing agents used in aircraft fire-suppression systems.

17. Describe extinguishing-agent containers.

18. What gas is usually employed to expel the extinguishing agent from a container?

19. To what pressure is a container usually charged?

20. Describe the method by which the discharge head attached to a container releases the extinguishing agent.

21. What is used to indicate that a fire-extinguishing agent bottle has discharged thermally?

22. What are two classifications of fire-extinguishing systems?

23. What indicators are required to show the discharge of extinguishing agent?

24. What type of extinguishing agent should be used for electrical fires?

25. Discuss the need for portable fire extinguishers.

26. What parts of an airplane are subject to ice collection during icing conditions?

27. Describe a *pneumatic-mechanical deicing system.*

28. How are *deicer boots* attached to leading-edge surfaces?

29. How is air supplied to deicer boots?

30. Why is a conductive coating required for deicer boots?

31. For what conditions are deicer boots inspected?

32. Discuss the removal and replacement of deicer boots.

33. How is an operational test of a deicer system performed?

34. Describe the operation of a *thermal anti-icing system* on a large airplane.

35. What is the usual source of heat for thermal anti-icing?

36. What units are usually electrically anti-iced?

37. What methods may be used to keep a windshield free of ice?

38. Describe a *rain-repellent system.*

39. What is the purpose of a *converter* in a windshield-wiper system?

40. What may be used to power a windshield-wiper system?

41. What methods, other than wipers, may be used to remove rain from a windshield?

42. Describe the *water system* for an airliner.

43. By what method is water for the lavatory heated?

44. How is drinking water cooled?

45. Describe the operation of the toilet system.

46. By what method is flushing liquid forced through the system?

47. What is the purpose of a takeoff warning system?

Troubleshooting Theory and Practice 19

INTRODUCTION

Whenever a malfunction or discrepancy is reported or comes to the attention of the technician, no matter how simple or how complex, the process of troubleshooting begins. The technician must determine the discrepancy's effect upon the airworthiness of the aircraft and the appropriate time and location for the corrective action to be taken.

THE TROUBLESHOOTING PROCESS

Troubleshooting is the process of identifying the **cause** of a malfunction or discrepancy, determining its **severity**, **eliminating the cause**, **replacing** or **repairing** discrepant components, systems, or structures, and, finally, **returning the aircraft to service**. The ultimate object of troubleshooting in aviation is to return aircraft to an airworthy condition offering a high probability that the malfunction or discrepancy will not recur. Although the technician cannot ensure that the malfunction or discrepancy will never recur, returning an aircraft to service with a high probability that the malfunction will be experienced again cannot be considered successful troubleshooting. In fact, if the discrepancy is an airworthiness item, returning the aircraft to service with a high probability of premature failure may be tantamount to returning the aircraft to service in an unairworthy condition.

Troubleshooting is more than just replacing a malfunctioning component or making a repair. Whether the thought process is evident or not, the first step in troubleshooting is to identify the **cause** of the discrepancy. The cause of the discrepancy can be as simple as "worn out through usage" or complex enough to involve more than one aircraft system.

For example, a landing-gear position-indicating switch may fail because of its frequent usage, or the failure may be premature, with the true cause of the failure being a voltage regulator or perhaps a leak in the hydraulic system allowing hydraulic fluid to drip on the switch. It is important that the technician identify the true cause of the discrepancy when determining the corrective action to be taken.

The second step in the troubleshooting process is to **evaluate** the reported discrepancy to determine if it has an adverse effect upon the aircraft's airworthiness. If the technician's evaluation determines that existence of a discrepancy does render the aircraft unairworthy, the technician must continue with the third step of the troubleshooting process. (It might be argued that the second step is truly the first step in the troubleshooting process. In the determination that an aircraft is unairworthy, this step may, indeed, be the first step. However, to determine that an aircraft is airworthy, a knowledge of the cause also is necessary.)

If the discrepancy adversely affects the airworthiness of the aircraft, **corrective action** must be taken before its next flight. The one exception is when the discrepancy is sufficient to render the aircraft unairworthy for normal flight, but under special FAA authorization (see *Aircraft Basic Science*) it may be flown (ferried) to a location where facilities are available for corrective action.

However, if the discrepancy does not affect the airworthiness of the aircraft, the technician is faced with another decision. This decision is the determination of whether or not the discrepancy should be repaired immediately or deferred to a later date. There are a number of factors that affect this decision-making process. The two major considerations are: *Is the appropriate equipment available? Is there sufficient time available to perform the required activities properly?*

The Use of Troubleshooting Charts

Many aircraft manufacturers include, as part of the maintenance and training manuals, troubleshooting charts designed to help the technician identify failed components. Figure 19-1 is an example of a troubleshooting chart supplied by Sikorsky Aircraft Corporation. It is important that the technician again note the distinction between failed component identification and identifying the cause of the discrepancy. These charts, when considered by themselves, typically fall short in the identification of the cause for the failure.

The location of this chapter as the final chapter in this book is not by accident. In order to take the necessary steps in the troubleshooting process, a thorough understanding of aircraft systems is required. The technician must use the information found in this text, personal experience and training, and information in the maintenance manuals, service bulletins, etc., to identify the cause of a discrepancy. Frequently, when manufacturers include a troubleshooting chart in the maintenance manual, a written description of the actions to be taken is included. It is important that the technician

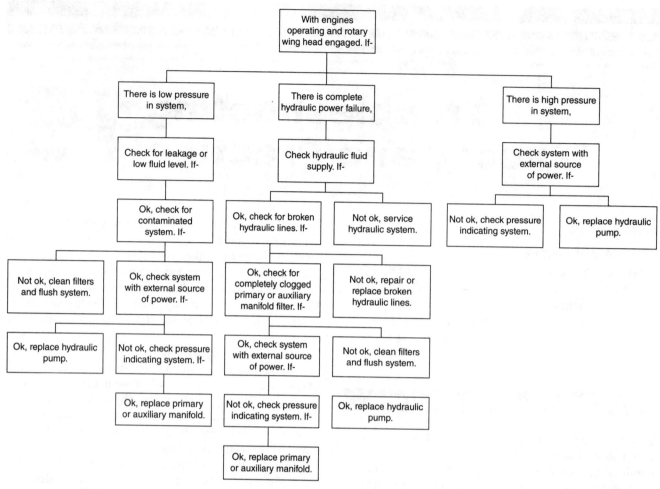

FIGURE 19-1 Troubleshooting chart. (*Sikorsky Aircraft Corp., Division of United Technologies Corp.*)

read the associated information and not refer solely to the graphic display.

For example, again referring to Fig. 19-1, assume that the following scenario exists: The system is experiencing low pressure, the fluid level is adequate, and no leakage is evident, but contamination is found in the system. According to the graphic troubleshooting chart, the technician is to clean filters and flush the system. If the technician follows only those instructions, the troubleshooting process has not been properly completed. The technician must analyze the contamination to determine its nature and source; the corrective action should include elimination of the source of the contamination.

Troubleshooting charts help identify common malfunctions. However, the technician may be confronted with situations that are not reflected by troubleshooting charts. In such instances, the technician must have a thorough knowledge of the system, its components, its operations, and its relationship to other systems.

Troubleshooting without a Chart

When attempting to troubleshoot a discrepancy without the aid of a troubleshooting chart, the technician's first step is to decide where to begin the analytical process. This decision may be based upon the technician's personal experience or made by heeding the advice of more experienced technicians. When experience does not suggest the point at which the troubleshooting should begin, the technician should consider the *divide-and-conquer* technique.

In the divide-and-conquer technique, the technician separates the system into two equal parts, either in regard to the number of components or in terms of a linear measurement, and then tests the operation of each half of the system to that point. Based upon the results of the test, the technician knows whether the discrepancy lies in the first or second half of the system being tested. The discrepant half of the system is then divided into two equal parts and the system is again tested. This process is repeated until the discrepancy is discovered. The use of the divide-and-conquer technique offers, in the long run, the quickest way to identify a discrepancy and its possible cause.

When coupled with experience, the divide-and-conquer technique is hard to beat. The technician may, after previous troubleshooting, note that one portion of a system rarely experiences failure. In such cases troubleshooting testing may begin at some other, more likely point.

Testing at any point in the system requires knowledge of the specific system involved. The technician must be able to determine the inputs and outputs of the system at the test point. Generally, it is better for the technician to determine what is to be expected at any point prior to making the test. Determining the acceptable test result criteria before testing lessens the tendency of the technician to accept invalid results. If the technician errs in establishing the test criteria, little harm is done if the system is operating properly up to that point; when the error is identified the technician will have a better understanding of a system's operations.

The selection of the appropriate corrective action is the next step in returning the aircraft to an airworthy condition. A discussion of corrective (repair) techniques is beyond the scope of this chapter. Previous chapters typically will contain the appropriate corrective techniques. However, the technician should remember that the repair is not complete until the cause and the symptom are corrected.

Figure 19-2 helps demonstrate the divide-and-conquer technique. Assume the pilot reports that a left-hand low-pressure warning light flickers. Also assume that all valves are in the position shown in the figure. The obvious things are checked first: fuel supply and warning light bulb, and both are functional. Note that the troubleshooting chart says nothing about flickering. From the system schematic, it is determined that there are four locations from which troubleshooting may be accomplished: the engine compartment, the cockpit, and either of the two fuel-tank areas.

Step 1A. Divide and conquer by selecting the cockpit as the middle location. Since the difficulty appears on the left-hand system, the right-hand emergency shutoff valve is closed and tagged (tagging of all switches and manual valves placed in nonoperational positions during the troubleshooting process is accomplished to ensure that they are returned to the operational positions before the aircraft is released for service). When the fuel-boost pumps (all four) are activated, the discrepancy is not duplicated.

Step 1B. When the left-hand emergency shutoff valve is also closed (something that is not likely to be done in flight) the low-pressure light illuminates but does not flicker. Chances are now good that the fault does not lie in the engine compartment. The emergency shutoff valve is opened.

Since no master warning of fuel bypass was reported, fuel bypass systems are assumed to be operational. The discrepancy may now be assumed to be in either of the fuel-tank systems.

Step 2A. Each system may be checked separately by use of the cross-feed valve. With the cross-feed valve open, the left-hand boost pumps are turned on (right-hand boost pumps are off). The discrepancy is not duplicated. From this the left-hand system and the right-tank check valve (CV) may be considered operational.

Step 2B. Turning off left-hand pumps and turning on the right-hand pumps results in the problem being duplicated. The problem lies in either the right-hand tank or the left-hand check valve.

Step 2C. Closing the left-hand manual shutoff valve, the discrepancy continues, so the left-hand check valve is probably operational.

Step 3. Closing the cross-feed valve and opening the right-hand emergency shutoff valve moves the discrepancy from the left-hand low-pressure light to the right-hand light. The discrepancy is definitely in the right-hand tank system. However, since the tank boost pumps do not indicate failure, the boost pumps are probably all right. The two remaining potential problems are the two check valves, either the fuel-ejector unit check valve or the check valve between the boost-pump check valves.

By dividing the system in half at the emergency shutoff valve, the portion of the system beyond the emergency shutoff valve was eliminated from consideration. By dividing the faulty portion of the system in half again, the right-hand tank system was identified as the culprit. A cross-check verified this fact. After dividing the system in half twice and doing a little analysis, the technician was thus able to limit the fault to one of two components from a potential of slightly fewer than 50 components.

The important aspect of this example is not the use of this particular schematic for this particular discrepancy, but rather to illustrate the advantages of using the divide-and-conquer technique as opposed to a hit-and-miss approach to troubleshooting.

Operating Data Helpful in Troubleshooting

In troubleshooting, it is most helpful to have a good overall knowledge of the particular airplane and aircraft type in which the trouble is encountered. Just as a physician, in diagnosing an illness of a patient, tries to obtain a knowledge of the patient's general health before prescribing a remedy for the particular ailment, so should technicians learn the various operating weaknesses (if any) of the airplanes they customarily repair. This information is obtained by personal observation, from conversations with the pilots, from the flight record forms, and from various technical publications.

Troubleshooting Intermittent Discrepancies

The experienced technician knows—and the novice technician will soon find out—that the most difficult discrepancies to locate and correct are intermittent discrepancies. One minute the discrepancy exists; the next minute everything seems to be working properly.

The first step in troubleshooting intermittent discrepancies is to gather as much information as possible regarding the nature, frequency, and duration of the discrepancy. Even the weather, especially humidity, precipitation, and temperature, before and during the discrepancy should be considered. To troubleshoot intermittent discrepancies without this information is guesswork, not troubleshooting.

Analyze this information for patterns. Intermittent discrepancies result most often as a result of heat, vibration, and weather. Look for patterns, such as during take-off and landings, changes in altitude, and during and after flight maneuvers.

If the discrepancy seems to relate to vibration, check components for security first. Operate as many moving parts

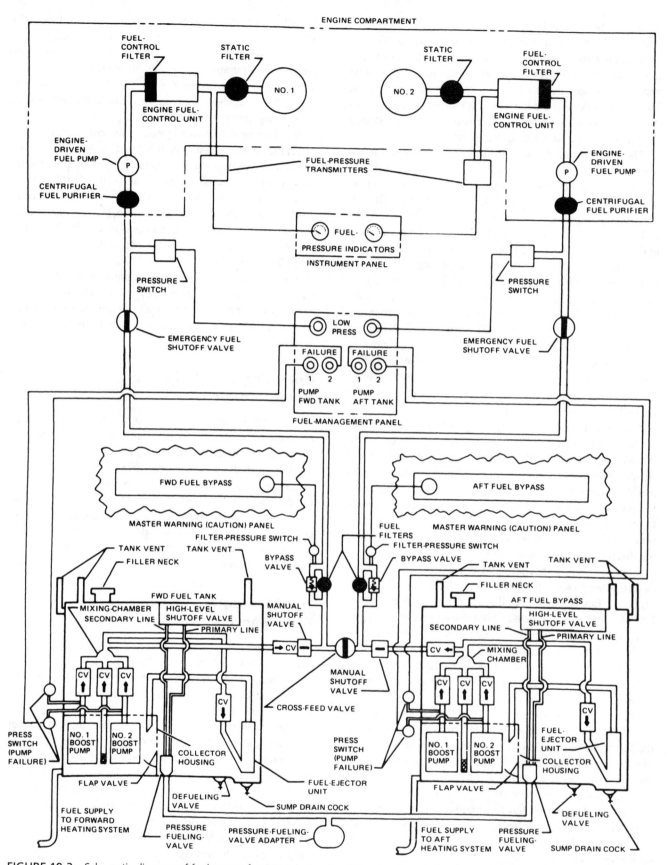

FIGURE 19-2 Schematic diagram of fuel system for S61-N helicopter. (*Sikorsky Aircraft Corp., Division of United Technologies Corp.*)

as possible, checking for security and wear. For example, a bolt that acts as the axis for a pulley may be worn in one spot, allowing it to rotate when there is vibration and no tension on the pulley. If, when the tension is applied to the pulley, the pulley mates with the bolt in the worn spot, the associated rigging is adversely affected. After the tension is released, the bolt may rotate randomly. If the next time tension is applied to the pulley, the pulley and bolt mate at a nonworn surface, the system might work properly.

Heat, another major cause of intermittent discrepancies, generally causes intermittent discrepancies due to thermal expansion. One component of heat-related intermittent discrepancies is the ambient temperature. A heat-sensitive discrepancy may exist when an engine, for a few minutes, ran hot on a hot day, but operated fine after the same amount of operation on a cool day. Information similar to this will help the technician know where to concentrate troubleshooting efforts. In this case, since the discrepancy occurs only when the powerplant has been running, the technician can concentrate troubleshooting in the areas where heat from the engine is concentrated. By noting the ambient temperature relationship in the discrepancy, the technician might start first in areas where ambient air provides engine compartment cooling.

In troubleshooting intermittent discrepancies, the technician should realize that the discrepancy is the effect of not one but at least two causes. One cause will be the failure of the system itself (or the thing that needs to be repaired or replaced); the other cause(s) will be intermittent, whether by design (such as vibration caused by landing) or chance (such as the way the sun hits an electrical component, providing just enough additional heat to cause the discrepancy to reoccur). The successful troubleshooter of intermittent discrepancies looks for both causes, not just one.

Validation of Troubleshooting Results

The final step in the troubleshooting process is the validation that the analytical steps of the troubleshooting process were properly interpreted. This often results in the necessity for some type of operational check. The procedures for such a check should be specified in the aircraft's maintenance manual. When operationally checking intermittent discrepancies, the check needs to include the suspected causes.

FORMAT OF TROUBLESHOOTING CHARTS

The generic use of troubleshooting charts was discussed earlier in this chapter. As part of that discussion, Fig. 19-1 was used to identify potential causes of a malfunction. Figure 19-1 is one of the three most popular formats used to display troubleshooting logic. These popular formats are the **flow**, **pick**, and **binary** formats.

The *flow format* used in Fig. 19-1 is a relatively simple format. Using this format the technician begins with an assumed condition, in this case "with engines operating and rotary wing head engaged," and then proceeds through a series of checks. Each check leads to an "OK" or "NOT OK" determination. The OK determinations lead the technician through the appropriate vertical logic path. The presence of a NOT OK determination changes the direction of the troubleshooting path to a horizontal path, leading to suggested corrective action.

Figure 19-3 is an example of a *pick-type* troubleshooting chart. This chart is composed of three columns. The first column indicates the nature of the discrepancy. The second column indicates the possible causes for the discrepancy. The third column indicates the corrective action typically required to correct the cause of the discrepancy. The vertical lines separate the causes and remedies into the appropriate trouble groupings.

There are two major differences between the flow and pick formats. The flow format has a tendency to follow the aircraft system's operational sequences. As a result, system schematics may be used to complement flow formats. Flow formats may also be constructed so that the most frequent cause for a discrepancy or malfunction is listed early in the troubleshooting sequence. Note that the pick chart in Fig. 19-3 does not identify the relative probabilities of the causes of problems.

The third format used for troubleshooting is the *binary logic chart*. Binary logic charts (as illustrated in Fig. 19-6) are used to troubleshoot systems that operate based upon a logic. This logic may be used when the status of two or more functions determines the operational condition of another function. Binary logic may be used in electrical, pneumatic, and hydraulic system applications. The major benefit derived from the use of a binary logic chart for troubleshooting is that component inputs and outputs are easily identified. This will become clearer later in this chapter. The major weakness of using a binary logic chart as a troubleshooting tool is the same as for the pick chart: The relative probability of a discrepancy's cause is not indicated.

Binary logic charts have six primary decision criteria called **gates**. A gate has channels through which inputs are received. The signal processed through the channel either exists or is absent. If a signal exists, then it is said to be input. The absence of a signal, although used by the gate in determining whether or not its decision criteria is met, is said not to have input. In mathematical terms, if the channel is producing input it has a value of 1. A channel not producing input has a value of 0, so it is said not to have input. Technicians familiar with Boolean algebra will recognize this logic.

There are two basic decision gates, one invert gate, three complementary gates, and an infinite number of gate variations. Four of these variations are included in Fig. 19-4 and are discussed in more detail later.

When the input meets the requirements of the decision criteria (gate), there is an output. If the gate's input requirements are not met, there is no output. Although there may be more than one input, there may be only one output.

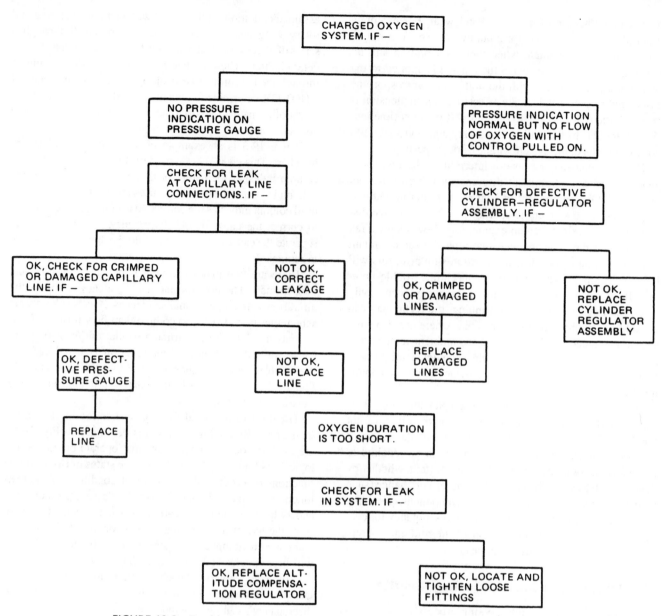

FIGURE 19-3   Troubleshooting chart for a stored-gas oxygen system in a light airplane.

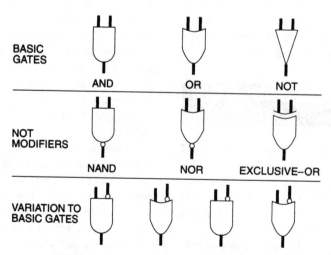

FIGURE 19-4   Symbols for binary logic gates.

There are three basic types of gates:

1. The AND gate has an output only when all inputs exist.
2. The OR gate has an output if any or all expected inputs exist.
3. The NOT gate simply inverts (or changes) the output.
4. NOT gates are also referred to as *invert gates*. If there is no input to a NOT gate, there is an output. If input exists, the NOT gate has no output.

The technician should note the circle at the bottom (output) of the NOT gate. A circle at an input or an output indicates an invert gate precedes (for inputs) or follows (for outputs) the basic gate decision criteria.

The three complementary gates are slight modifications to or combinations of the three basic gates:

1. The NAND gate has an invert (NOT) gate immediately prior to the AND gate. In other words, the NOT gate (indicated by the circle) output is the input for the AND gate.
2. The NOR gate places a NOT gate immediately prior to an OR gate.
3. The EXCLUSIVE-OR gate operates in the same manner as the OR gate, except that if both inputs meet the decision criteria, there is no output.

When reviewing the three complementary gates, the technician should note a pattern that repeats in the gate variations shown at the bottom of the figure—that is, there are two basic "decision" functions, AND and OR. A circle at an input places a NOT gate prior to the basic decision gate and as a result reverses the input (creates an input if one does not exist or deletes one if one exists).

For example, assume that a pneumatic system has the symbol variation at the lower right corner of Fig. 19-4. Because the system is a pneumatic system, the presence or absence of air pressure is the input and output. The basic gate is an AND gate, so if all inputs to the basic gate have air pressure, the output will be air pressure. If either input does not have pressure, there will be no output. In this variation note that the rightmost input has an invert gate. Possible scenarios for this gate are listed here, where P indicates the existence of pressure and NP indicates no pressure (input).

Scenario	INPUTS		OUTPUT
	Left	*Right*	
1	NP	NP	NP
2	NP	P	NP
3	P	NP	P
4	P	P	NP

Note that in scenario 4, even though both inputs to the basic gate might initially appear to be P, the right-hand invert gate changes the right-hand input to the AND decision

gate from pressure to no pressure. The opposite is true for scenario 3. The NP is changed to P before entering the decision gate.

The layout of all possible scenarios for a decision gate is called a **truth table**. A more complete discussion of logic gates, truth tables, and their use in electrical systems can be found in the text *Aircraft Electricity and Electronics*. The symbols used in binary logic charts for hydraulic and pneumatic systems follow the same pattern as those used in electricity and are shown in Fig. 19-4.

A pneumatic starting system for a three-engine gas-turbine aircraft is shown in Fig. 19-5. A review of the schematic shows that there are five pneumatic power sources that may be used to start any engine. These power sources are the auxiliary power unit (APU), a ground pneumatic unit (GPU) connected to either connection 1 or 2, and bleed air from either of the two remaining engines.

The binary logic chart shown in Fig. 19-6 depicts the binary logic corresponding to the pneumatic starting of engine 2. The broken lines in the figure would not normally be present in a binary logic chart found in an aircraft's maintenance manual. They have been included in this figure to segregate the power sources for the following discussion. The binary logic associated with the components used in an APU start is bordered by a simple dashed line. The binary logic for starting engine 2 with air supplied by engine 1 is bordered by a line consisting of a dash and a dot. The dotted lines segregate the binary logic for a start using engine 3 as the air source. The two GPU sources are between the APU and engine-start blocks. To help identify the various logic gates, logic decision levels have been segregated into levels. The levels are separated by lines with two dots between each dash.

Note that all five pneumatic power sources are routed through a series of EXCLUSIVE-OR gates. This is because only one source may be used at a time to start an engine pneumatically. If this were not so, the EXCLUSIVE-OR gates could be replaced with OR gates. Typically, this limitation would be accomplished as part of the associated electrical systems. However, multiple EXCLUSIVE-OR gates are shown in three levels in Fig. 19-6 to demonstrate the compounding effect of binary decision gates. In an aircraft's maintenance manual, decision levels 2 and 3 would most likely be combined and indicated by showing four inputs to a single gate.

Before the pneumatic starter will operate, two inputs must exist. The engine starter valve must be on and there must be a pneumatic air source. This fact is depicted in the binary logic chart in level 1 by the AND gate. The first logical step in troubleshooting the pneumatic starter system is to ensure that the starter valve is operational.

It is logical to assume that if trouble occurs when attempting to start the engine pneumatically and all five pneumatic systems are used in the attempt, the fault probably lies in the starter valve or the pneumatic starter motor systems. The following discussion assumes that the engine start valve is providing input (is operational and on) to the level 1 AND gate.

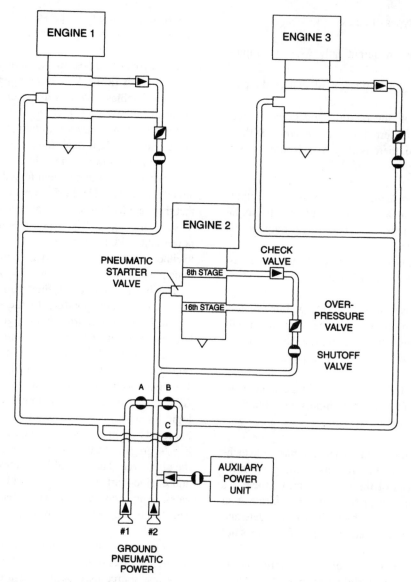

FIGURE 19-5 Pneumatic starting schematic for a three-engine gas-turbine aircraft.

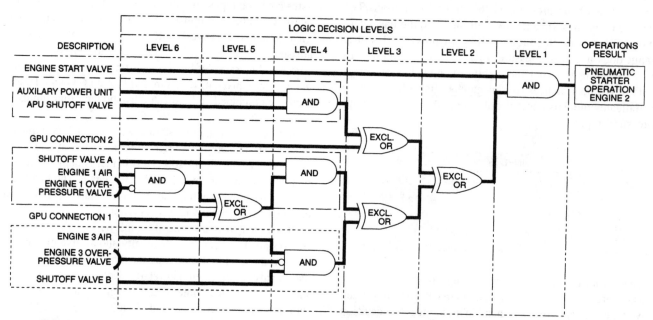

FIGURE 19-6 Binary logic chart for starting engine 2 using pneumatic start for a three-engine gas-turbine aircraft.

If the starter valve system is operational, the fault must lie within the binary logic of the associated air source. In the case of an APU start, either the APU shutoff valve is closed, the APU is not running, the APU is not supplying sufficient air to be considered a valid input, or there is a malfunction of the associated EXCLUSIVE-OR devices. Failure in the case of a GPU connector 2 start is most likely the fault of the associated EXCLUSIVE-OR devices.

In further discussions it is assumed that all logic decision gate devices are functional and operating properly.

Failure of the pneumatic starter system when attempting a start using engine 3 air might result from one of three sources: Engine 3 is not providing enough air to be interpreted by the logic gate as a valid input; the shutoff valve B is not open (thus not allowing an input); or the overpressure valve is open. If the technician is not careful in the interpretation of the logic chart, the last fault option might easily be misunderstood.

Note that the AND gate for an engine 3 start is a variation of the standard gate. There are three inputs to the logic gate, but the center input (from the overpressure valve) is an inverted input. If the overpressure valve is providing an output, then the AND gate does not receive an input because the invertor negates the input. On the other hand, if the overpressure valve is not generating an output, the invert gate will generate an output to the AND gate. If the overpressure valve generates an output when open, then the valve must be closed for the AND gate to receive an input. In the open position, the output of the overpressure valve will be reversed by the inverted gate and no input will be received by the AND gate, resulting in no output being generated by the AND gate.

For either a GPU 1 connection start or an engine 1 start, shutoff valve A must be open, providing an input to the AND gate. In addition, the level 5 EXCLUSIVE-OR gate must receive an input from either GPU connector 1 or the engine 1 air system, but not both. Input from GPU connector 1 goes directly to the gate. Engine 1 air input must pass through the level 5 AND gate. Note that the level 5 AND gate is a modified gate because there is an invert gate in the overpressure valve input to the gate.

Although the use of binary logic charts in troubleshooting may initially seem awkward to the technician, this is usually a result of the technician's lack of familiarity with the basic logic symbols. Once the technician becomes familiar with the symbols and is adept at interpreting basic logic charts, use of logic charts can simplify the troubleshooting process.

Transport aircraft are often equipped with a centralized maintenance system and the following discussion will describe the centralized fault display system (CFDS) used on Airbus aircraft. Boeing and other aircraft manufacturers have similar systems installed on their aircraft models.

Centralized Fault Display System

The purpose of the CFDS is to make the maintenance task easier by displaying fault messages in the cockpit and permitting the flight crew to perform some specific tests. The CFDS has two levels of maintenance: removal and replacement of equipment and troubleshooting.

The CFDS includes

- *BITE (Built-in test equipment)*. Modern transport aircraft and business jets use BITE for most of the avionics and electrically controlled systems. These BITE systems monitor themselves as well as the aircraft systems and indicate a system fault on the maintenance computer and ECAM/EICAS pages. Although the BITE system is a great resource for aircraft technicians, it does not mean that no system knowledge is required for troubleshooting. Just relying on BITE and changing parts is not a good solution. Often the BITE guides the technician in the right direction but the technician will still have to use a diagnostic manual and troubleshooting logic to solve the malfunction.

- A central computer, also called the centralized fault display interface unit (CFDIU).
- Two MCDUs (multi-purpose control and display units).
- One printer.

Figure 19-7 shows the architecture of the CFDS.

CFDS Operation

The CFDS operates in two main modes:

1. The NORMAL mode or REPORTING mode (in flight). In NORMAL mode, the CFDS records and displays the failure messages transmitted by each system BITE.

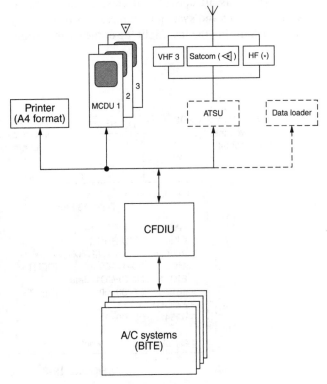

FIGURE 19-7 Architecture of CFDS.

2. The INTERACTIVE mode or MENU mode (on ground). In INTERACTIVE mode, the CFDS allows any BITE to be connected with the MCDI in order to display the maintenance data stored and formatted by the BITE or to initiate a test.

The CFDS identifies the faulty system and puts any failures or faults into one of three classes:

1. *Class 1.* Failures indicated to the flight crew by means of the ECAM, or other flight deck effect. These failures must be repaired or entered in the minimum equipment list (MEL) before the aircraft can depart.

2. *Class 2.* Faults indicated to maintenance personnel by the CFDS, and which trigger a MAINT status entry on the maintenance part of the ECAM maintenance page. The aircraft can operate with these faults but must be repaired within 10 days.

3. *Class 3.* Faults indicated to maintenance personnel by the CFDS, but which do not trigger a MAINT status. The aircraft operator may have these faults corrected at their convenience.

The main functions of the CFDS are as follows:

- Obtaining and storing messages transmitted by the connected systems BITEs, or by the flight warning computer
- Detailing the maintenance phases
- Presenting maintenance reports: last leg report, last leg ECAM report, previous leg report, avionics status, system report test, and post-flight report. Figure 19-8 shows a last leg report generated by the CFDS. The last leg report (on the ground) or the current leg report (in flight) lists all class 1 and class 2 faults and all system failures and system fault messages received by the CFDS during the last flight leg or the current flight

leg. The report can be printed out in the cockpit. One of the main advantages of the CFDS is that maintenance is already informed during the flight of possible maintenance issues and troubleshooting and ordering of replacement parts can be started before the aircraft arrives at its destination. Figure 19-9 shows a CFDS post-flight report which is used by maintenance to troubleshoot malfunctions.

REVIEW QUESTIONS

1. List the steps in the troubleshooting process.

2. Why is it important to identify the cause of a discrepancy?

3. If an aircraft's airworthiness is not affected by a discrepancy, what are the two major factors that determine when the discrepancy will be corrected?

4. What is the major weakness of troubleshooting charts?

5. What information must the technician use when troubleshooting a discrepancy?

6. Why should the technician pay attention to information on a troubleshooting chart other than the graphic logic of the chart?

7. When the technician must troubleshoot a discrepancy without the use of a chart or the information required is not on the troubleshooting chart, what process should be used?

8. Explain the benefit of using the *divide-and-conquer* troubleshooting technique.

9. What is the first step in troubleshooting *intermittent discrepancies*?

10. What must the technician look for when troubleshooting intermittent discrepancies?

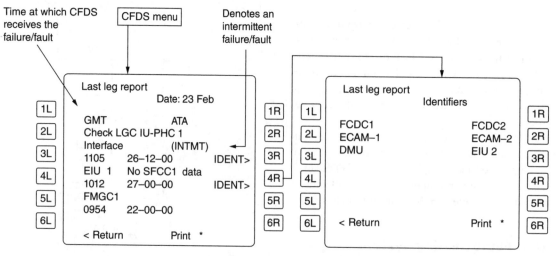

FIGURE 19-8 Last leg or current leg report.

11. What is the generic name given the symbol used to indicate a decision-making process in binary logic?

12. What is the function of a NOT gate?

13. Name the two primary decision gates.

14. What is the name of a table that lists all possible outputs of a binary logic gate?

15. How many outputs does a three-input AND gate produce?

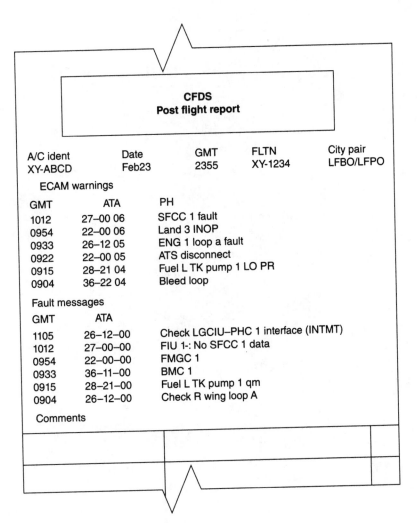

CFDS
Post flight report

A/C ident	Date	GMT	FLTN	City pair
XY-ABCD	Feb23	2355	XY-1234	LFBO/LFPO

ECAM warnings

GMT	ATA		PH
1012	27–00	06	SFCC 1 fault
0954	22–00	06	Land 3 INOP
0933	26–12	05	ENG 1 loop a fault
0922	22–00	05	ATS disconnect
0915	28–21	04	Fuel L TK pump 1 LO PR
0904	36–22	04	Bleed loop

Fault messages

GMT	ATA	
1105	26–12–00	Check LGCIU–PHC 1 interface (INTMT)
1012	27–00–00	FIU 1-: No SFCC 1 data
0954	22–00–00	FMGC 1
0933	36–11–00	BMC 1
0915	28–21–00	Fuel L TK pump 1 qm
0904	26–12–00	Check R wing loop A

Comments

FIGURE 19-9 Post-flight report.

Index

Note: Page numbers followed by *f* denote figures; page numbers followed by *t* denote tables.